3,000 Solved Problems in Physics

Alvin Halpern, Ph.D.
Brooklyn College

Schaum's Outline Series

New York   Chicago   San Francisco   Lisbon   London   Madrid
Mexico City   Milan   New Delhi   San Juan   Seoul
Singapore   Sydney   Toronto

CONTENTS

TO THE STUDENT

This book is intended for use by students of general physics, either in calculus- or noncalculus-based courses. Problems requiring real calculus (not merely calculus notation) are marked with a small superscript **c**.

The only way to master general physics is to gain ability and sophistication in problem-solving. This book is meant to make you a master of the art — and should do so if used properly. As a rule, a problem can be solved once you have learned the ideas behind it; sometimes these very ideas are brought into sharper focus by looking at sample problems and their solutions. If you have difficulty with a topic, you can select a few problems in that area, examine the solutions carefully, and then try to solve related problems before looking at the printed solutions.

There are numerous ways of posing a problem and, frequently, numerous ways of solving one. You should try to gain understanding of how to approach various classes of problems, rather than memorizing particular solutions. Understanding is better than memory for success in physics.

The problems in this book cover every important topic in a typical two- or three-semester general physics sequence. Ranging from the simple to the complex, they will provide you with plenty of practice and food for thought.

The *Chapter Skeletons with Exams,* beginning on the next page, was devised to help students with limited time gain maximum benefit from this book. It is hoped that the use of this feature is self-evident; still, the following remarks may help:

- The *Chapter Skeletons* divide the problems in this book into three categories: SCAN, HOMEWORK and EXAMS. (Turn to page ix to see an example.)

- To gain a quick overview of the basic ideas in a chapter, review the SCAN problems and study their printed solutions.

- HOMEWORK problems are for practicing your problem-solving skills; *cover the solution with an index card* as you read, and try to solve, the problem. Do *both sets* if your course is calculus based.

- No problem from SCAN or HOMEWORK is duplicated in EXAMS, and no two Exams overlap. Calculus-based students are urged also to take the Hard Exam. Exams run about 60 minutes, unless otherwise indicated.

- Still further problems constitute the two groups of Final Exams. Stay in your category(ies), and good luck.

CHAPTER SKELETONS WITH EXAMS

Chapter 1

S C A N		1.1, 1.2, 1.3, 1.4, 1.5, 1.6, 1.12, 1.13, 1.14, 1.15, 1.27, 1.60, 1.61, 1.83, 1.84, 1.85, 1.89, 1.90
H O M E W O R K	Everybody	1.8, 1.11, 1.16, 1.18, 1.23, 1.26, 1.31, 1.33, 1.35, 1.41, 1.42, 1.48, 1.50, 1.53, 1.65, 1.68, 1.74, 1.86, 1.92
	Calc.-based only	1.51, 1.53, 1.56, 1.57, 1.59, 1.75, 1.76, 1.77, 1.80, 1.81, 1.87, 1.93
E X A M S	Easy	1.10, 1.21, 1.39, 1.54
	Hard	1.40, 1.49, 1.67, 1.86
	Calc.-based only	1.58, 1.72, 1.88, 1.91

Chapter 2

S C A N		2.1, 2.2, 2.4, 2.12, 2.27, 2.28, 2.29 2.38
H O M E W O R K	Everybody	2.5, 2.6, 2.21, 2.25, 2.31, 2.32, 2.39
	Calc.-based only	2.14, 2.20, 2.23, 2.36, 2.40
E X A M S	Easy	2.3, 2.7, 2.30, 2.40
	Hard	2.15, 2.24, 2.33, 2.44
	Calc.-based only	2.16, 2.22, 2.37, 2.43

Chapter 3

S C A N		3.2, 3.4, 3.8, 3.19, 3.23, 3.29, 3.36, 3.37
H O M E W O R K	Everybody	3.5, 3.12, 3.16, 3.21, 3.28, 3.39, 3.41, 3.46, 3.49, 3.53
	Calc.-based only	3.7, 3.24, 3.38, 3.45, 3.55, 3.59, 3.60, 3.61
E X A M S	Easy	3.10, 3.26, 3.40, 3.51
	Hard	3.25, 3.30, 3.48, 3.52
	Calc.-based only	3.33, 3.54, 3.62

Chapter 4

S C A N		4.1, 4.2, 4.3, 4.4, 4.14, 4.16, 4.28, 4.37, 4.39, 4.66, 4.73, 4.93
H O M E W O R K	Everybody	4.5, 4.7, 4.15, 4.22, 4.38, 4.40, 4.45, 4.68, 4.69, 4.75
	Calc.-based only	4.31, 4.50, 4.52, 4.53, 4.58, 4.61, 4.72, 4.81, 4.97
E X A M S	Easy	4.10, 4.43, 4.77, 4.79
	Hard	4.17, 4.46, 4.80, 4.85, 4.86
	Calc.-based only	4.34, 4.54, 4.74, 4.95 (60 - 70 min.)

Chapter 5

S C A N		5.1, 5.5, 5.6, 5.10, 5.19, 5.34, 5.37, 5.52
H O M E W O R K	Everybody	5.2, 5.9, 5.14, 5.20, 5.38, 5.53
	Calc.-based only	5.31, 5.35, 5.40, 5.41, 5.44, 5.60
E X A M S	Easy	5.3, 5.21, 5.24, 5.51
	Hard	5.7, 5.8, 5.11, 5.22, 5.54 (60 - 70 min.)
	Calc.-based only	5.16, 5.18, 5.43, 5.59 (60 - 70 min.)

Chapter 6

S C A N		6.1, 6.7, 6.15, 6.21, 6.25, 6.27, 6.33
H O M E W O R K	Everybody	6.11, 6.18, 6.22, 6.29, 6.37, 6.42
	Calc.-based only	6.41, 6.45, 6.46, 6.50, 6.51, 6.55
E X A M S	Easy	6.2, 6.4, 6.23, 6.31
	Hard	6.9, 6.14, 6.28, 6.35
	Calc.-based only	6.16, 6.44, 6.47, 6.49

Chapter 7

S C A N		7.1, 7.4, 7.5, 7.8, 7.18, 7.29, 7.30, 7.32, 7.41, 7.43, 7.53, 7.61, 7.62, 7.67, 7.79, 7.83
H O M E W O R K	Everybody	7.9, 7.10, 7.14, 7.20, 7.33, 7.42, 7.48, 7.50, 7.51, 7.69, 7.75, 7.82, 7.84, 7.87, 7.92, 7.98
	Calc.-based only	7.17, 7.26, 7.35, 7.71, 7.76, 7.93, 7.105, 7.107, 7.110, 7.114, 7.117
E X A M S	Easy	7.19, 7.37, 7.65, 7.70, 7.88
	Hard	7.44, 7.45, 7.73, 7.77, 7.80
	Calc.-based only	7.74, 7.91, 7.106, 7.113

Chapter 8

S C A N		8.1, 8.2, 8.8, 8.19, 8.20, 8.21, 8.22, 8.24, 8.34
H O M E W O R K	Everybody	8.3, 8.5, 8.11, 8.13, 8.25, 8.28, 8.33
	Calc.-based only	8.10, 8.18, 8.27, 8.31
E X A M S	Easy	8.6, 8.7, 8.23, 8.36, 8.37
	Hard	8.9, 8.15, 8.29, 8.38
	Calc.-based only	8.12, 8.17, 8.30, 8.35

Chapter 9

S C A N		9.2, 9.4, 9.10, 9.16, 9.17, 9.19, 9.32, 9.54, 9.68, 9.87
H O M E W O R K	Everybody	9.5, 9.23, 9.24, 9.34, 9.56, 9.59, 9.70, 9.88, 9.95
	Calc.-based only	9.13, 9.27, 9.47, 9.50, 9.79, 9.97, 9.104
E X A M S	Easy	9.18, 9.55, 9.73, 9.98 (60 - 70 min.)
	Hard	9.39, 9.57, 9.69, 9.102 (60 - 70 min.)
	Calc.-based only	9.60, 9.85, 9.103, 9.106 (60 - 70 min.)

Chapter 10

S C A N		10.1, 10.7, 10.14, 10.32, 10.37, 10.44, 10.48 (also 9.107)
H O M E W O R K	Everybody	10.2, 10.8, 10.15, 10.33, 10.47, 10.58, 10.60 (also 9.113)
	Calc.-based only	10.10, 10.31, 10.64, 10.65, 10.66, 10.75, 10.79 (also 9.120)
E X A M S	Easy	10.17, 10.20, 10.39, 10.67
	Hard	10.22, 10.34, 10.38, 10.118
	Calc.-based only	10.24, 10.52, 10.57, 10.77 (60 - 70 min.)

Chapter 11

S C A N		11.10, 11.21, 11.26, 11.34, 11.52, 11.68, 11.72, 11.82
H O M E W O R K	Everybody	11.12, 11.22, 11.30, 11.35, 11.53, 11.69, 11.78, 11.83
	Calc.-based only	11.8, 11.40, 11.44, 11.46, 11.62, 11.75, 11.79, 11.84
E X A M S	Easy	11.23, 11.28, 11.59, 11.73
	Hard	11.37, 11.60, 11.81, 11.87
	Calc.-based only	11.45, 11.66, 11.76, 11.88

Chapter 12

S C A N		12.3, 12.6, 12.8, 12.12, 12.19, 12.28, 12.50, 12.56
H O M E W O R K	Everybody	12.2, 12.4, 12.11, 12.14, 12.20, 12.29, 12.52, 12.62
	Calc.-based only	12.15, 12.25, 12.34, 12.37, 12.49, 12.65, 12.69, 12.71
E X A M S	Easy	12.9, 12.17, 12.31, 12.58
	Hard	12.13, 12.23, 12.30, 12.59
	Calc.-based only	12.24, 12.43, 12.57, 12.72 (60 - 70 min.)

Chapter 13

S C A N		13.1, 13.2, 13.16, 13.17, 13.18, 13.19, 13.20, 13.21
H O M E W O R K	Everybody	13.5, 13.11, 13.22, 13.33, 13.43, 13.46
	Calc.-based only	13.34, 13.50
E X A M S	Easy	13.7, 13.25, 13.38, 13.45
	Hard	13.10, 13.32, 13.44, 13.47
	Calc.-based only	13.15, 13.36, 13.48, 13.51

Chapter 14

S C A N		14.1, 14.7, 14.11, 14.18, 14.27, 14.36, 14.44, 14.56
H O M E W O R K	Everybody	14.3, 14.9, 14.13, 14.21, 14.22, 14.37, 14.46, 14.58
	Calc.-based only	14.32, 14.34, 14.41, 14.47, 14.52, 14.59
E X A M S	Easy	14.17, 14.23, 14.24, 14.45, 14.54
	Hard	14.19, 14.20, 14.55, 14.57
	Calc.-based only	14.30, 14.35, 14.42, 14.53

Chapter 15

S C A N		15.1, 15.7, 15.14, 15.23, 15.39, 15.43, 15.49, 15.64
H O M E W O R K	Everybody	15.2, 15.8, 15.17, 15.24, 15.40, 15.44, 15.50, 15.63
	Calc.-based only	15.5, 15.26, 15.35, 15.36, 15.59
E X A M S	Easy	15.19, 15.41, 15.54, 15.65
	Hard	15.31, 15.32, 15.42, 15.58, 15.66
	Calc.-based only	15.27, 15.28, 15.62, 15.67 (60 - 70 min.)

Chapter 16

S C A N		16.1, 16.2, 16.3, 16.4, 16.5, 16.10, 16.15, 16.31, 16.40
H O M E W O R K	Everybody	16.7, 16.10, 16.12, 16.16, 16.18, 16.34, 16.41
	Calc.-based only	16.8, 16.27, 16.32, 16.33, 16.44, 16.45
E X A M S	Easy	16.6, 16.13, 16.24, 16.37
	Hard	16.18, 16.18, 16.20, 16.29, 16.43
	Calc.-based only	16.17, 16.22, 16.26, 16.51

Chapters 17 and 18

S C A N		17.1, 17.8, 17.14, 17.38, 18.2, 18.16, 18.24, 18.26, 18.39
H O M E W O R K	Everybody	17.2, 17.11, 17.20, 17.36, 18.5, 18.19, 18.27, 18.35, 18.46
	Calc.-based only	17.18, 17.28, 17.29, 18.12, 18.45
E X A M S	Easy	17.10, 17.37, 18.17, 18.33
	Hard	17.15, 17.40, 18.6, 18.36
	Calc.-based only	17.23, 17.44, 18.48, 18.51

Chapter 19

S C A N		19.1, 19.2, 19.7, 19.30, 19.31, 19.32, 19.38, 19.39, 19.40
H O M E W O R K	Everybody	19.3, 19.10, 19.15, 19.33, 19.34, 19.41
	Calc.-based only	19.17, 19.20, 19.22, 19.28
E X A M S	Easy	19.8, 19.13, 19.35, 19.44
	Hard	19.11, 19.14, 19.36, 19.48
	Calc.-based only	19.25, 19.29, 19.37, 19.50

Chapter 20

S C A N		20.1, 20.2, 20.7, 20.9, 20.20, 20.35, 20.37, 20.40, 20.45, 20.70
H O M E W O R K	Everybody	20.3, 20.8, 20.10, 20.22, 20.31, 20.32, 20.43, 20.48, 20.71, 20.73
	Calc.-based only	20.26, 20.38, 20.41, 20.55, 20.56, 20.62, 20.68, 20.82, 20.86
E X A M S	Easy	20.17, 20.25, 20.43, 20.74
	Hard	20.24, 20.27, 20.53, 20.77, 20.78
	Calc.-based only	20.44, 20.59, 20.69, 20.89 (60 - 70 min.)

Chapters 21 and 22

S C A N		21.1, 21.2, 21.3, 21.4, 21.6, 21.27, 21.42, 22.1, 22.2, 22.3, 22.4, 22.6, 22.14, 22.23
H O M E W O R K	Everybody	21.5, 21.6, 21.23, 21.30, 21.31, 21.46, 22.5, 22.8, 22.15, 22.24, 22.30
	Calc.-based only	21.14, 21.20, 21.26, 21.39, 21.40, 22.9, 22.18, 22.19
E X A M S	Easy	21.16, 21.29, 21.47, 22.11
	Hard	21.23, 21.34, 22.13, 22.25
	Calc.-based only	21.15, 21.45, 22.21, 22.28 (60 - 70 min.)

Final Exams for Chapters 1-22 (160 - 180 min.)

Easy A 1.47, 3.34, 5.23, 7.102, 10.18, 11.27, 13.39, 15.21, 15.22, 18.40, 20.15

Easy B 2.35, 4.90, 6.5, 8.32, 9.57, 12.51, 14.38, 16.23, 19.42, 21.35

Hard A 1.78, 3.44, 5.58, 7.99, 10.42, 12.61, 13.37, 16.21, 18.47, 20.28

Hard B 2.17, 4.91, 6.32, 7.111, 9.101, 11.54, 14.39, 15.53, 19.46, 22.7

Calc. A 1.79, 3.56, 6.20, 6.56, 7.116, 10.79, 12.70, 15.30, 17.40, 20.91

Calc. B 2.41, 4.78, 5.49, 6.58, 7.120, 9.61, 14.28, 14.29, 16.46, 19.49, 22.27

Chapters 23 and 24

S C A N		23.1, 23.4, 23.7, 23.8, 23.18, 23.24, 23.41, 23.51, 24.1, 24.11, 24.12, 24.16, 24.28
H O M E W O R K	Everybody	23.2, 23.5, 23.19, 23.21, 23.25, 23.42, 23.54, 24.2, 24.14, 24.18, 24.20, 24.33
	Calc.-based only	23.27, 23.29, 23.34, 23.37, 24.15, 24.19, 24.25, 24.26, 24.34, 24.39
E X A M S	Easy	23.6, 23.46, 24.13, 24.29
	Hard	23.20, 23.47, 24.24, 24.32
	Calc.-based only	23.36, 23.56. 24.21, 24.38

Chapter 25

S C A N		25.1, 25.2, 25.3, 25.5, 25.11, 25.16, 25.29, 25.32, 25.46, 25.59, 25.64
H O M E W O R K	Everybody	25.8, 25.12, 25.14, 25.18, 25.34, 25.36, 25.48, 25.60, 25.62, 25.68
	Calc.-based only	25.38, 25.39, 25.44, 25.53, 25.54, 25.57, 25.58, 25.61
E X A M S	Easy	25.15, 25.35, 25.45, 25.63
	Hard	25.20, 25.31, 25.50, 25.69
	Calc.-based only	25.23, 25.42, 25.65, 25.71, 25.72

Chapter 26

S C A N		26.1, 26.2, 26.4, 26.6, 26.8, 26.15, 26.33, 26.52, 26.53, 26.58, 26.64, 26.67, 26.75, 26.90, 26.91, 26.98
H O M E W O R K	Everybody	26.6, 26.7, 26.9, 26.16, 26.19, 26.20, 26.35, 26.59, 26.62, 26.65, 26.76, 26.80, 26.82, 26.93, 26.99
	Calc.-based only	26.3, 26.5, 26.21, 26.25, 26.26, 26.27, 26.39, 26.43, 26.47, 26.63, 26.66, 26.68, 26.85, 26.86, 26.106
E X A M S	Easy	26.17, 26.32, 26.60, 26.92, 26.95
	Hard	26.18, 26.36, 26.50, 26.101, 26.102
	Calc.-based only	26.24, 26.29, 26.37, 26.109 (60 - 70 min.)

Chapter 27

S C A N		27.1, 27.2, 27.4, 27.5, 27.11, 27.31, 27.38, 27.41, 27.49, 27.50, 27.55, 27.64, 27.65, 27.66, 27.67, 27.68, 27.69, 27.81, 27.82, 27.89, 27.92, 27.94, 27.95, 27.120
H O M E W O R K	Everybody	27.6, 27.13, 27.21, 27.25, 27.33, 27.43, 27.47, 27.51, 27.56, 27.70, 27.74, 27.83, 27.90, 27.99, 27.102, 27.117, 27.121
	Calc.-based only	27.3, 27.17, 27.18, 27.36, 27.87, 27.113, 27.124, 27.125, 27.126, 27.131, 27.141, 27.142
E X A M S	Easy	27.10, 27.42, 27.59, 27.104
	Hard	27.15, 27.46, 27.60, 27.105
	Calc.-based only	27.37, 27.80, 27.136, 27.145

Chapter 28

S C A N		28.1, 28.2, 28.3, 28.9, 28.30, 28.33, 28.43, 28.54, 28.62, 28.64, 28.87, 28.89, 28.90, 28.143
H O M E W O R K	Everybody	28.4, 28.6, 28.19, 28.34, 28.45, 28.65, 28.66, 28.68, 28.76, 28.92, 28.105, 28.106, 28.119, 28.134, 23.144, 28.146
	Calc.-based only	28.26, 28.49, 28.50, 28.60, 28.70, 28.72, 28.73, 28.82, 28.88, 28.112, 28.122, 28.131, 28.138
E X A M S	Easy	28.38, 28.47, 28.69, 28.94, 28.136
	Hard	28.24, 28.57, 28.77, 28.117
	Calc.-based only	28.21, 28.81, 28.118, 28.141 (60 - 70 min.)

Chapter 29

S C A N		29.1, 29.2, 29.3, 29.4, 29.20, 29.21, 29.24, 29.27, 29.34, 29.35, 29.36, 29.40, 29.43, 29.47
H O M E W O R K	Everybody	29.6, 29.9, 29.10, 29.16, 29.17, 29.30, 29.36, 29.37, 29.38, 29.44, 29.45, 29.46, 29.50
	Calc.-based only	29.5, 29.27, 29.28, 29.29, 29.32, 29.39, 29.48
E X A M S	Easy	29.7, 29.8, 29.18, 29.45, 29.55
	Hard	29.14, 29.25, 29.49, 29.55
	Calc.-based only	29.22, 29.23, 29.41, 29.51

Chapter 30

S C A N		30.1, 30.2, 30.11, 30.15, 30.21, 30.30, 30.31, 30.35, 30.66, 30.67, 30.74, 30.77 30.78, 30.80, 30.92, 30.100
H O M E W O R K	Everybody	30.4, 30.8, 30.12, 30.13, 30.16, 30.22, 30.39, 30.49, 30.50, 30.65, 30.76, 30.79, 30.83, 30.93, 30.102
	Calc.-based only	30.5, 30.14, 30.32, 30.37, 30.40, 30.44, 30.53, 30.64, 30.71, 30.110
E X A M S	Easy	30.19, 30.36, 30.82, 30.108
	Hard	30.17, 30.18, 30.38, 30.48, 30.86
	Calc.-based only	30.41, 30.54, 30.75, 30.105

Chapter 31

S C A N		31.1, 31.2, 31.10, 31.13, 31.26, 31.29, 31.30, 31.31, 31.38, 31.42, 31.59
H O M E W O R K	Everybody	31.4, 31.5, 31.14, 31.18, 31.27, 31.32, 31.33, 31.40, 31.60
	Calc.-based only	31.3, 31.15, 31.17, 31.25, 31.46, 31.48, 31.49, 31.53, 31.56
E X A M S	Easy	31.6, 31.20, 31.35, 31.40
	Hard	31.11, 31.21, 31.37, 31.44
	Calc.-based only	31.6, 31.28, 31.52, 31.54 (60 - 70 min.)

Chapter 32

S C A N		32.1, 32.2, 32.9, 32.10, 32.20, 32.34, 32.35, 32.37, 32.38, 32.39, 32.40, 32.42, 32.56, 32.57, 32.58, 32.77
H O M E W O R K	Everybody	32.4, 32.7, 32.13, 32.21, 32.24, 32.36 32.41, 32.43, 32.50, 32.54, 32.60, 32.61, 32.81
	Calc.-based only	32.6, 32.15, 32.17, 32.23, 32.28, 32.30, 32.85, 32.86, 32.87, 32.91
E X A M S	Easy	32.3, 32.12, 32.72, 32.74
	Hard	32.11, 32.25, 32.65, 32.66
	Calc.-based only	32.8, 32.31, 32.84, 32.94

Chapter 33

S C A N		33.8, 33.9, 33.17, 33.22, 33.30, 33.37, 33.44, 33.54, 33.55, 33.57, 33.69
H O M E W O R K	Everybody	33.10, 33.19, 33.33, 33.45, 33.46, 33.50, 33.58, 33.70
	Calc.-based only	33.1, 33.2, 33.4, 33.5, 33.6, 33.14, 33.28, 33.40, 33.41
E X A M S	Easy	33.11, 33.35, 33.36, 33.56, 33.61
	Hard	33.20, 33.38, 33.48, 33.71
	Calc.-based only	33.15, 33.42, 33.49, 33.53

Chapter 34

S C A N		34.1, 34.2, 34.10, 34.15, 34.24, 34.26, 34.33, 34.43, 34.45, 34.56, 34.58, 34.69, 34.71, 34.80, 34.81, 34.85, 34.88, 34.91
H O M E W O R K	Everybody	34.5, 34.13, 34.16, 34.21, 34.27, 34.37, 34.41, 34.44, 34.52, 34.59, 34.72, 34.82, 34.89, 34.90, 34.94
	Calc.-based only	34.6, 34.53, 34.54, 34.57, 34.64, 34.73, 34.77
E X A M S	Easy	34.11, 34.16, 34.63, 34.96
	Hard	34.14, 34.60, 34.83, 34.101
	Calc.-based only	34.55, 34.65, 34.74, 34.102

Chapter 35

S C A N		35.1, 35.2, 35.3, 35.4, 35.5, 35.15, 35.16, 35.22, 35.35, 35.46, 35.47, 35.48, 35.49, 35.54, 35.58, 35.67, 35.88, 35.96, 35.100, 35.106, 35.107, 35.112, 35.125, 35.130
H O M E W O R K	Everybody	35.6, 35.9, 35.12, 35.17, 35.18, 35.21, 35.33, 35.37, 35.50, 35.51, 35.52, 35.55, 35.57, 35.68, 35.74, 35.82, 35.87, 35.94, 35.97, 35.108, 35.113, 35.127, 35.133, 35.134
	Calc.-based only	35.10, 35.13, 35.43, 35.45, 35.64, 35.65, 35.78, 35.86, 35.121, 35.123
E X A M S	Easy	35.19, 35.39, 35.61, 35.111
	Hard	35.11, 35.23, 35.114, 35.132
	Calc.-based only	35.7, 35.34, 35.80, 35.129 (60 - 70 min.)

Chapter 36

S C A N		36.5, 36.6, 36.11, 36.28, 36.30, 36.33, 36.35, 36,50, 36.51, 36.56, 36.58, 36.66, 36.67
H O M E W O R K	Everybody	36.7, 36.15, 36.24, 36.31, 36.34, 36.37, 36.41, 36.52, 36.54, 36.59, 36.63, 36.65
	Calc.-based only	36.1, 36.20, 36.22, 36.26, 36.40, 36.42, 36.48, 36.62, 36.70, 36.74
E X A M S	Easy	36.8, 36.29, 36.36, 36.44, 36.53
	Hard	36.12, 36.25, 36.43, 36.53
	Calc.-based only	36.19, 36.27, 36.47, 36.72, 36.75

Chapter 37

S C A N		37.1, 37.3, 37.4, 37.13, 37.14, 37.21, 37.30, 37.47, 37.50, 37.51, 37.58
H O M E W O R K	Everybody	37.2, 37.5, 37.7, 37.15, 37.16, 37.18, 37.19, 37.24, 37.48, 37.53, 37.62, 37.64
	Calc.-based only	37.10, 37.11, 37.26, 37.32, 37.34, 37.36, 37.56, 37.65
E X A M S	Easy	37.12, 37.22, 37.28, 37.49, 37.52, 37.57
	Hard	37.6, 37.8, 37.31, 37.55
	Calc.-based only	37.27, 37.37, 37.59, 37.66 (60 - 70 min.)

Chapter 38

S C A N		38.1, 38.2, 38.18, 38.19, 38.23, 38.29, 38.37, 38.41, 38.43, 38.53, 38.66
H O M E W O R K	Everybody	38.3, 38.10, 38.17, 38.18, 38.24, 38.30, 38.38, 38.42, 38.44, 38.52, 38.56, 38.67
	Calc.-based only	38.9, 38.26, 38.33, 38.36, 38.47, 38.49, 38.58, 38.61, 38.62, 38.65, 38.71
E X A M S	Easy	38.8, 38.25, 38.34, 38.54
	Hard	38.22, 38.31, 38.57, 38.68
	Calc.-based only	38.14, 38.15, 38.32, 38.51, 38.55 (60 - 70 min.)

Chapter 39

S C A N		39.1, 39.4, 39.13, 39.14, 39.15, 39.30, 39.33, 39.38, 39.39, 39.42, 39.48, 39.54, 39.57, 39.64, 39.67, 39.73, 39.75, 39.76, 39.83
H O M E W O R K	Everybody	39.3, 39.5, 39.16, 39.18, 39.28, 39.34, 39.40, 39.43, 39.44, 39.50, 39.53, 39.55, 39.58, 39.61, 39.68, 39.81
	Calc.-based only	39.20, 39.23, 39.25, 39.27, 39.32, 39.36, 39.37, 39.46, 39.65, 39.72, 39.77, 39.78, 39.84
E X A M S	Easy	39.8, 39.41, 39.49, 39.62
	Hard	39.10, 39.21, 39.29, 39.60
	Calc.-based only	39.12, 39.17, 39.26, 39.52 (60 - 70 min.)

Final Exams for Chapters 23-39 (160 - 180 min.)

Easy A   23.52,   24,9,   26.57,   28.55,   30.20,   32.44,   32.45, 32.46,   34.92,   36.14,   38.13,   39.2

Easy B   24.10,   25.22,   27.44,   29.52,   31.9,   32.51,   33.72, 35.63,   37.9,   39.7

Hard A   23.23,   24.17,   26.38,   27.108,   28.48,   30.109,   32.63, 34.51,   36.23,   38.45

Hard B   24.30,   25.47,   27.39,   28.97,   29.53,   31.47,   33.47, 35.98,   37.29,   39.31

Calc. A   23.17,   26.88,   28.120,   30.72,   30.73,   32.19,   34.48, 38.60,   39.71

Calc. B   23.50,   25.40,   27.132,   28.125,   29.56,   31.55,   33.43, 35.103,   36.16,   37.33

1.1 PLANAR VECTORS, SCIENTIFIC NOTATION, AND UNITS

1.1 What is a scalar quantity?

▮ A scalar quantity has only magnitude; it is a pure number, positive or negative. Scalars, being simple numbers, are added, subtracted, etc., in the usual way. It may have a unit after it, e.g. mass = 3 kg.

1.2 What is a vector quantity?

▮ A vector quantity has both magnitude and direction. For example, a car moving south at 40 km/h has a *vector velocity* of 40 km/h southward.
   A vector quantity can be represented by an arrow drawn to scale. The length of the arrow is proportional to the magnitude of the vector quantity (40 km/h in the above example). The direction of the arrow represents the direction of the vector quantity.

1.3 What is the 'resultant' vector?

▮ The resultant of a number of similar vectors, force vectors, for example, is that single vector which would have the same effect as all the original vectors taken together.

1.4 Describe the graphical addition of vectors.

▮ The method for finding the resultant of several vectors consists in beginning at any convenient point and drawing (to scale) each vector arrow in turn. They may be taken in any order of succession. The tail end of each arrow is attached to the tip end of the preceding one.
   The resultant is represented by an arrow with its tail end at the starting point and its tip end at the tip of the last vector added.

1.5 Describe the parallelogram method of addition of two vectors.

▮ The resultant of two vectors acting at any angle may be represented by the diagonal of a parallelogram. The two vectors are drawn as the sides of the parallelogram and the resultant is its diagonal, as shown in Fig. 1-1. The direction of the resultant is away from the origin of the two vectors.

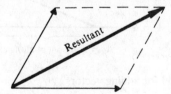

Fig. 1-1

1.6 How do you subtract vectors?

▮ To subtract a vector **B** from a vector **A**, reverse the direction of **B** and add it vectorially to vector **A**, that is, $\mathbf{A} - \mathbf{B} = \mathbf{A} + (-\mathbf{B})$.

1.7 Describe the trigonometric functions.

▮ For the right triangle shown in Fig. 1-2, by definition

$$\sin \theta = \frac{o}{h} \qquad \cos \theta = \frac{a}{h} \qquad \tan \theta = \frac{o}{a}$$

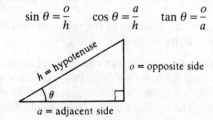

Fig. 1-2

1.8 Express each of the following in scientific notation: (*a*) 627.4, (*b*) 0.000365, (*c*) 20 001, (*d*) 1.0067, (*e*) 0.0067.

▎ (*a*) 6.274×10^2. (*b*) 3.65×10^{-4}. (*c*) 2.001×10^4. (*d*) 1.0067×10^0. (*e*) 6.7×10^{-3}.

1.9 Express each of the following as simple numbers $\times 10^0$: (*a*) 31.65×10^{-3} (*b*) 0.415×10^6 (*c*) $1/(2.05 \times 10^{-3})$ (*d*) $1/(43 \times 10^3)$.

▎ (*a*) 0.03165. (*b*) 415,000. (*c*) 488. (*d*) 0.0000233.

1.10 The diameter of the earth is about 1.27×10^7 m. Find its diameter in (*a*) millimeters, (*b*) megameters, (*c*) miles.

▎ (*a*) $(1.27 \times 10^7 \text{ m})(1000 \text{ mm}/1 \text{ m}) = \underline{1.27 \times 10^{10} \text{ mm}}$. (*b*) Multiply meters by $1 \text{ Mm}/10^6$ m to obtain $\underline{12.7 \text{ Mm}}$. (*c*) Then use $(1 \text{ km}/1000 \text{ m})(1 \text{ mi}/1.61 \text{ km})$; the diameter is $\underline{7.89 \times 10^3 \text{ mi}}$.

1.11 A 100-m race is run on a 200-m-circumference circular track. The runners run eastward at the start and bend south. What is the displacement of the endpoint of the race from the starting point?

▎ The runners move as shown in Fig. 1-3. The race is halfway around the track so the displacement is one diameter $= \underline{200/\pi = 63.7 \text{ m}}$ due south.

Fig. 1-3

1.12 What is a *component* of a vector?

▎ A component of a vector is its "shadow" (perpendicular drop) on an axis in a given direction. For example, the *p*-component of a displacement is the distance along the *p* axis corresponding to the given displacement. It is a scalar quantity, being positive or negative as it is positively or negatively directed along the axis in question. In Fig. 1-4, A_p is positive. (One sometimes defines a vector component as a *vector* pointing along the axis and having the size of the scalar component. If the scalar component is negative the vector component points in the negative direction along the axis.) It is customary, and useful, to resolve a vector into components along *mutually perpendicular* directions (*rectangular components*).

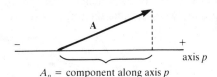

A_p = component along axis p **Fig. 1-4**

1.13 What is the component method for adding vectors?

▎ Each vector is resolved into its x, y, and z components, with negatively directed components taken as negative. The x component of the resultant, R_x, is the algebraic sum of all the x components. The y and z components of the resultant are found in a similar way.

1.14 Define the multiplication of a vector by a scalar.

▎ The quantity $b\mathbf{F}$ is a vector having magnitude $|b| F$ (the absolute value of b times the magnitude of $\mathbf{F}$); the direction of $b\mathbf{F}$ is that of $\mathbf{F}$ or $-\mathbf{F}$, depending on whether b is positive or negative.

1.15 Using the graphical method, find the resultant of the following two displacements: 2 m at 40° and 4 m at 127°, the angles being taken relative to the +*x* axis.

▎ Choose x, y axes as shown in Fig. 1-5 and lay out the displacements to scale tip to tail from the origin. Note that all angles are measured from the +*x* axis. The resultant vector, **R**, points from starting point to endpoint as shown. Measure its length on the scale diagram to find its magnitude, 4.6 m. Using a protractor, measure its angle θ to be 101°. The resultant displacement is therefore $\underline{4.6 \text{ m at } 101°}$.

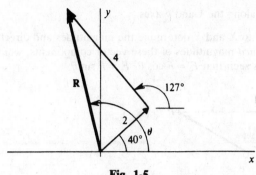

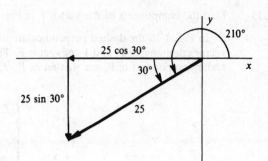

Fig. 1-5   Fig. 1-6

1.16 Find the x and y components of a 25-m displacement at an angle of 210°.

❚ The vector displacement and its components are shown in Fig. 1-6. The components are

$$x \text{ component} = -25 \cos 30° = \underline{-21.7 \text{ m}} \qquad y \text{ component} = -25 \sin 30° = \underline{-12.5 \text{ m}}$$

Note in particular that each component points in the negative coordinate direction and must therefore be taken as negative.

1.17 Solve Prob. 1.15 by use of rectangular components.

❚ Resolve each vector into rectangular components as shown in Fig. 1-7(a) and (b). (Place a cross-hatch symbol on the original vector to show that it can be replaced by the sum of its vector components.) The resultant has the scalar components

$$R_x = 1.53 - 2.40 = -0.87 \text{ m} \qquad R_y = 1.29 + 3.20 = 4.49 \text{ m}$$

Note that components pointing in the negative direction must be assigned a negative value.

The resultant is shown in Fig. 1-7(c); we see that

$$R = \sqrt{(0.87)^2 + (4.49)^2} = \underline{4.57 \text{ m}} \qquad \tan \phi = \frac{4.49}{0.87}$$

Hence, $\phi = \underline{79°}$, from which $\theta = 180° - \phi = \underline{101°}$.

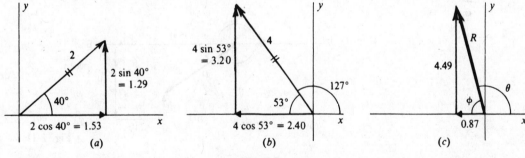

(a)   (b)   (c)

Fig. 1-7

1.18 Add the following two force vectors by use of the parallelogram method: 30 pounds at 30° and 20 pounds at 140°. (A *pound of force* is chosen such that a 1-kg object weighs 2.21 lb on earth. One pound is equivalent to a force of 4.45 N.)

❚ The force vectors are shown in Fig. 1-8. Construct a parallelogram using them as sides, as shown in Fig. 1-9. The resultant, **R**, is then shown as the diagonal. Measurement shows that **R** is $\underline{30 \text{ lb at } 72°}$.

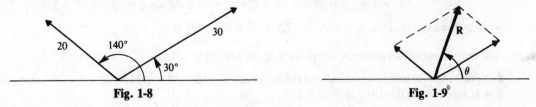

Fig. 1-8   Fig. 1-9

1.19 Find the components of the vector **F** in Fig. 1-10 along the x and y axes.

▌ In Fig. 1-10 the dashed perpendiculars from P to X and Y determine the magnitudes and directions of the vector components $\mathbf{F}_x$ and $\mathbf{F}_y$ of vector **F**. The *signed* magnitudes of these vector components, which are the scalar components of **F**, are written as F_x, F_y. It is seen that $F_x = F \cos \theta$, $F_y = F \sin \theta$.

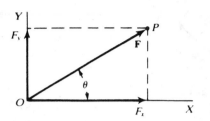

Fig. 1-10

1.20 (*a*) Let **F** have a magnitude of 300 N and make angle $\theta = 30°$ with the positive x direction. Find F_x and F_y.
(*b*) Suppose that $F = 300$ N and $\theta = 145°$ (**F** is here in the second quadrant). Find F_x and F_y.

▌ (*a*) $F_x = 300 \cos 30° = \underline{259.8 \text{ N}}$, $F_y = 300 \sin 30° = \underline{150 \text{ N}}$. (*b*) $F_x = 300 \cos 145° = (300)(-0.8192) = \underline{-245.75 \text{ N}}$ (in the negative direction of X), $F_y = 300 \sin 145° = (300)(+0.5736) = \underline{172.07 \text{ N}}$

1.21 A car goes 5.0 km east, 3.0 km south, 2.0 km west, and 1.0 km north. (*a*) Determine how far north and how far east it has been displaced. (*b*) Find the displacement vector both graphically and algebraically.

▌ (*a*) Recalling that vectors can be added in any order we can immediately add the 3.0-km south and 1.0-km north displacement vectors to get a net 2.0-km south displacement vector. Similarly the 5.0-km east and 2.0-km west vectors add to a 3-km east displacement vector. Because the east displacement contributes no component along the north-south line and the south displacement has no component along the east-west line, the car is -2.0 km north and 3.0 km east of its starting point. (*b*) Using the head-to-tail method, we easily can construct the resultant displacement **D** as shown in Fig. 1-11. Algebraically we note that

$$D = \sqrt{D_x^2 + D_y^2} = \sqrt{2^2 + 3^2} = \underline{3.6 \text{ km}} \qquad \tan \phi = -\tfrac{2}{3} \qquad \text{or} \qquad \tan \theta = \tfrac{2}{3} \qquad \theta = \underline{34°} \text{ south of east.}$$

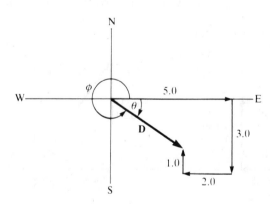

Fig. 1-11

1.22 Find the x and y components of a 400-N force at an angle of 125° to the x axis.

▌ Formal method (uses angle above positive x axis):

$$F_x = (400 \text{ N}) \cos 125° = -229 \text{ N} \qquad F_y = (400 \text{ N}) \sin 125° = 327 \text{ N}$$

Visual method (uses only acute angles above or below positive or negative x axis):

$$|F_x| = F \cos \phi = 400 \cos 55° = 229 \text{ N} \qquad |F_y| = F \sin \phi = 400 \sin 55° = 327 \text{ N}$$

By inspection of Fig. 1-12, $F_x = -|F_x| = \underline{-229 \text{ N}}$; $F_y = |F_y| = \underline{327 \text{ N}}$.

1.23 Add the following two coplanar forces: 30 N at 37° and 50 N at 180°.

▌ Split each into components and find the resultant: $R_x = 24 - 50 = -26$ N, $R_y = 18 + 0 = 18$ N. Then $R = \underline{31.6 \text{ N}}$ and $\tan \theta = 18/-26$, so $\theta = \underline{145°}$.

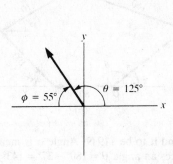

Fig. 1-12

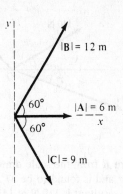

Fig. 1-13

1.24 For the vectors **A** and **B** shown in Fig. 1-13, find (*a*) **A** + **B**, (*b*) **A** − **B**, and (*c*) **B** − **A**.

▌ The components are $A_x = 6$ m, $A_y = 0$, $B_x = 12\cos 60° = 6$ m, and $B_y = 12\sin 60° = 10.4$ m.
(*a*) $(A + B)_x = 12$ m and $(A + B)_y = 10.4$ m, so that **A** + **B** = 15.9 m at 40.9°, (*b*) $(A − B)_x = 0$
and $(A − B)_y = 0 − 10.4$ so **A** − **B** = 10.4 m at −90°. (*c*) $(B − A)_x = 0$ and $(B − A)_y = 10.4 − 0$ so
B − **A** = 10.4 m at 90°.

1.25 For the vectors shown in Fig. 1-13, find (*a*) **A** + **B** + **C** and (*b*) **A** + **B** − **C**.

▌ The *x* and *y* components of **C** are 4.5 m and −7.8 m. (*a*) The *x* component is $A_x + B_x + C_x = 16.5$ and for
the *y* component we find 2.6, so the vector is 16.7 m at 9.0°. (*b*) $A_x + B_x − C_x = 7.5$ and the *y* component is
$0 + 10.4 − (−7.8) = 18.2$; changing this to a magnitude and angle, we find 19.7 m at 68°.

1.26 For the vectors shown in Fig. 1-13, find (*a*) **A** − 2**C**, (*b*) **B** − (**A** + **C**), and (*c*) −**A** − **B** − **C**.

▌ (*a*) The *x* component is $A_x − 2C_x = −3$ and the *y* component is $−2(−7.8) = 15.6$, giving 15.9 m at 101°.
(*b*) The *x* component $= 6 − (6 + 4.5) = −4.5$; the *y* component $= 10.4 − [0 + (−7.8)] = 18.2$; therefore
$(4.5^2 + 18.2^2)^{1/2} = 18.7$ m at 104°. (*c*) This is the negative of the vector of Prob. 1.25 (*a*), so that it = 16.7 m
at $9.0° + 180° = 189° = −171°$.

1.27 A displacement of 20 m is made in the *xy* plane at an angle of 70° (i.e., 70° counterclockwise from the +*x*
axis). Find its *x* and *y* components. Repeat if the angle is 120°; if the angle is 250°.

▌ In each case $s_x = s\cos\theta$ and $s_y = s\sin\theta$. The results are 6.8, 18.8 m; −10.0, 17.3 m; −6.8, −18.8 m.

1.28 It is found that an object will hang properly if an *x* force of 20 N and a *y* force of −30 N are applied to it.
Find the single force (magnitude and direction) which would do the same job.

▌ Adding components of the forces yields $R_x = 20$ N and $R_y = −30$ N. $R = (400 + 900)^{1/2} = 36$ N. Calling θ the
counterclockwise angle from the +*x* axis, $\tan\theta = −30/20$ and so $\theta = 303.7° = −56.3°$.

1.29 Find the magnitude and direction of the force which has an *x* component of −40 N and a *y* component of
−60 N.

▌ The resultant of these two forces is $R = (1600 + 3600)^{1/2} = 72$ N. The angle θ is $180° + \tan^{-1}(6/4) = 236.3°$.

1.30 Find the magnitude and direction of the sum of the following two coplanar displacement vectors: 20 m at 0°
and 10 m at 120°.

▌ Splitting each into components, $R_x = 20 − 5 = 15$ m and $R_y = 0 + 8.7 = 8.7$ m. Then $R = 17.3$ m with
$\tan\theta = 8.7/15$ giving $\theta = 30°$.

1.31 Four coplanar forces act on a body at point *O* as shown in Fig. 1-14(*a*). Find their resultant graphically.

▌ Starting from *O*, the four vectors are plotted in turn as shown in Fig. 1.14(*b*). Place the tail end of one
vector at the tip end of the preceding one. The arrow from *O* to the tip of the last vector represents the
resultant of the vectors.

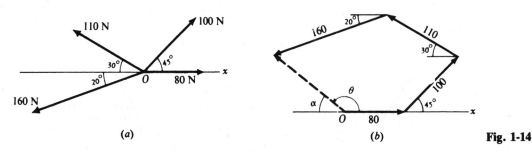

Fig. 1-14

Measure R from the scale drawing in Fig. 1.14(b) and find it to be 119 N. Angle α is measured by protractor and is found to be 37°. Hence the resultant makes an angle $\theta = 180° - 37° = 143°$ with the positive x axis. The resultant is <u>119 N</u> at <u>143°</u>.

1.32 Solve Prob. 1.31 by use of the rectangular component method.

▮ The vectors and their components are as follows.

magnitude, N	x component, N	y component, N
80	80	0
100	$100 \cos 45° = 71$	$100 \sin 45° = 71$
110	$-110 \cos 30° = -95$	$110 \sin 30° = 55$
160	$-160 \cos 20° = -150$	$-160 \sin 20° = -55$

Note the sign of each component. To find the resultant, we have

$$R_x = 80 + 71 - 95 - 150 = -94 \text{ N} \qquad R_y = 0 + 71 + 55 - 55 = 71 \text{ N}$$

The resultant is shown in Fig. 1-15; we see that $R = \sqrt{(94)^2 + (71)^2} = \underline{118 \text{ N}}$. Further, $\tan \alpha = 71/94$, from which $\alpha = 37°$. Therefore the resultant is 118 N at $180 - 37 = \underline{143°}$.

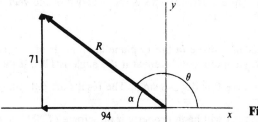

Fig. 1-15

1.33 Perform graphically the following vector additions and subtractions, where **A**, **B**, and **C** are the vectors shown in Fig. 1-16: (a) $\mathbf{A} + \mathbf{B}$. (b) $\mathbf{A} + \mathbf{B} + \mathbf{C}$. ($c$) $\mathbf{A} - \mathbf{B}$. (d) $\mathbf{A} + \mathbf{B} - \mathbf{C}$.

▮ See Fig. 1-16(a) through (d). In (c), $\mathbf{A} - \mathbf{B} = \mathbf{A} + (-\mathbf{B})$; that is, to subtract **B** from **A**, reverse the direction of **B** and add it vectorially to **A**. Similarly, in (d), $\mathbf{A} + \mathbf{B} - \mathbf{C} = \mathbf{A} + \mathbf{B} + (-\mathbf{C})$, where $-\mathbf{C}$ is equal in magnitude but opposite in direction to **C**.

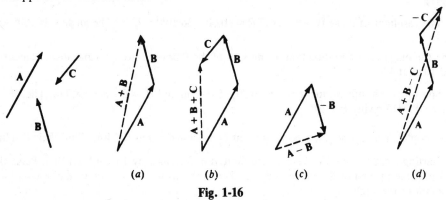

Fig. 1-16

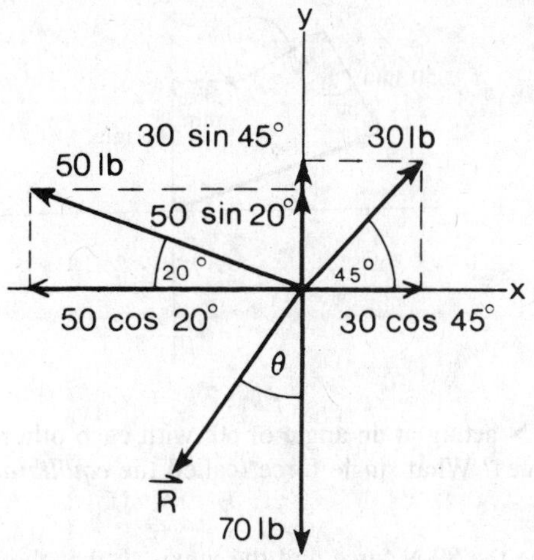

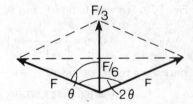

Fig. 1-17

1.34 Find the resultant **R** of the following forces all acting on the same point in the given directions: 30 lb to the northeast; 70 lb to the south; and 50 lb 20° north of west.

▌ Choose east as the positive x direction (Figure 1-17).

x components, lb	y components, lb
$30 \cos 45° = \quad 21.2$	$30 \sin 45° = \quad 21.2$
$-50 \cos 20° = \underline{-47.0}$	$50 \sin 20° = \quad 17.1$
Total $\qquad = -25.8$ lb	$-70 = \underline{-70}$
	Total $\qquad = -31.7$ lb

$$R = \sqrt{(-25.8)^2 + (-31.7)^2} = \sqrt{665.8 + 1004.9} = \underline{40.9 \text{ lb}} \qquad \tan \theta = \frac{25.8}{31.7} = 0.8139 \qquad \theta = \underline{39° \text{ west of south}}$$

1.35 Find the angle between two vector forces of equal magnitude, such that the resultant is one-third as much as either of the original forces.

▌ In the vector force diagram (Fig. 1-18), the diagonals of the rhombus bisect each other. Thus,

$$\cos \theta = \frac{F/6}{F} = \frac{1}{6} = 0.1667 \qquad \theta = 80.4° \qquad 2\theta = 160.8°$$

The angle between the two forces is $\underline{160.8°}$.

Fig. 1-18

1.36 Find the vector sum of the following four displacements on a map: 60 mm north; 30 mm west; 40 mm at 60° west of north; 50 mm at 30° west of south. Solve (a) graphically and (b) algebraically.

▌ (a) With ruler and protractor, construct the sum of vector displacements by the tail-to-head method as shown in Fig. 1-19. The resultant vector from tail of first to head of last is then also measured with ruler and protractor. *Ans.*: $\underline{97 \text{ mm}}$ at $\underline{67.7° \text{ W of N}}$. (b) Let **D** = resultant displacement.

$$D_x = -30 - 40 \sin 60° - 50 \sin 30° = -89.6 \text{ mm} \qquad D_y = 60 + 40 \cos 60° - 50 \cos 30° = +36.7 \text{ mm}$$

$$D = \sqrt{D_x^2 + D_y^2} = \underline{96.8 \text{ mm}} \qquad \tan \phi = \left| \frac{D_y}{D_x} \right| \Rightarrow \phi = \underline{22.3°} \qquad \text{above negative } x \text{ axis}$$

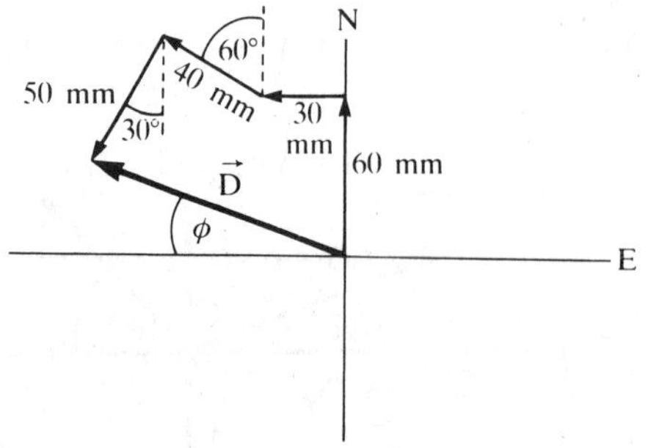

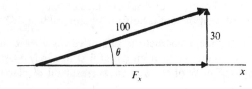

Fig. 1-19

1.37 Two forces, 80 N and 100 N acting at an angle of 60° with each other, pull on an object. What single force would replace the two forces? What single force (called the *equilibrant*) would balance the two forces? Solve algebraically.

▮ Choose the x axis along the 80-N force and the y axis so that the 100-N force 60° above the positive x axis has a positive y component. Then the single force **R** that replaces the two forces is the vector sum of these forces:

$$R_x = 80 + 100 \cos 60° = 130 \text{ N} \qquad R_y = 100 \sin 60° = 87 \text{ N}$$

$$R = \sqrt{R_x^2 + R_y^2} = \underline{156 \text{ N}} \qquad \theta = \tan^{-1}\left|\frac{R_y}{R_x}\right| = \underline{34°} \text{ above positive } x \text{ axis.}$$

The force that balances **R** is −**R** with a magnitude of 156 N but pointing in the opposite direction to **R**: 34° below negative x axis (or 214° above positive x axis).

1.38 Two forces act on a point object as follows: 100 N at 170° and 100 N at 50°. Find their resultant.

▮ $\mathbf{F}_1 = 100$ N at 170° above x axis; $\mathbf{F}_2 = 100$ N at 50° above x axis.

$$\mathbf{R} = \mathbf{F}_1 + \mathbf{F}_2 \qquad R_x = 100 \cos 170° + 100 \cos 50° = -34.2 \text{ N} \qquad R_y = 100 \sin 170° + 100 \sin 50° = 94.0 \text{ N}$$

$$R = \sqrt{R_x^2 + R_y^2} = \underline{100 \text{ N}} \qquad \theta = \tan^{-1}\frac{R_y}{R_x}$$

has two solutions: 290° and 110°. From a look at its components we see that **R** lies in the second quadrant, so the answer is $\underline{110°}$ (or $\underline{70°}$ above negative x axis). In a less formal approach we can find: $\phi = \tan^{-1}|R_y/R_x|$, which will give an acute angle solution, in this case of $\underline{70°}$, which always represents the angle with the positive or negative x axis, and either above or below that axis. Since we already know from the components which quadrant **R** lies in, we know the direction precisely. In our case the 70° is above the negative x axis.

1.39 A force of 100 N makes an angle of θ with the x axis and has a y component of 30 N. Find both the x component of the force and the angle θ.

▮ The data are sketched in Fig. 1-20. We wish to find F_x and θ. We know that

$$\sin \theta = \frac{o}{h} = \frac{30}{100} = 0.30$$

from which $\theta = \underline{17.5°}$. Then, since $a = h \cos \theta$, we have

$$F_x = 100 \cos 17.5° = \underline{95.4 \text{ N}}$$

Fig. 1-20

1.40 A boat can travel at a speed of 8 km/h in still water on a lake. In the flowing water of a stream, it can move at 8 km/h relative to the water in the stream. If the stream speed is 3 km/h, how fast can the boat move past a tree on the shore in traveling (*a*) upstream? (*b*) downstream?

▮ (a) If the water were standing still, the boat's speed past the tree would be 8 km/h. But the stream is carrying it in the opposite direction at 3 km/h. Therefore the boat's speed relative to the tree is $8 - 3 = \underline{5\,\text{km/h}}$. (b) In this case, the stream is carrying the boat in the same direction the boat is trying to move. Hence its speed past the tree is $8 + 3 = \underline{11\,\text{km/h}}$.

1.41 A plane is traveling eastward at an airspeed of 500 km/h. But a 90 km/h wind is blowing southward. What are the direction and speed of the plane relative to the ground?

▮ The plane's resultant velocity is the sum of two velocities, 500 km/h eastward and 90 km/h southward. These component velocities are shown in Fig. 1-21. The plane's resultant velocity is found by use of

$$R = \sqrt{(500)^2 + (90)^2} = \underline{508\,\text{km/h}}$$

The angle α is given by

$$\tan \alpha = \frac{90}{500} = \underline{0.180}$$

from which $\alpha = \underline{10.2°}$. The plane's velocity relative to the ground is 508 km/h at 10.2° south of east.

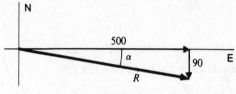
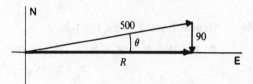

| **Fig. 1-21** | **Fig. 1-22** |

1.42 With the same airspeed as in Prob. 1.41, in what direction must the plane head in order to move due east relative to the earth?

▮ The sum of the plane's velocity through the air and the velocity of the wind must be the resultant eastward velocity of the plane relative to the earth. This is shown in the vector diagram of Fig. 1-22. It is seen that $\sin \theta = 90/500$, from which $\theta = 10.4°$. The plane should head $\underline{10.4°}$ north of east if it is to move eastward on the earth.

If we wish to find the plane's eastward speed, Fig. 1-22 tells us that $R = 500 \cos \theta = \underline{492\,\text{km/h}}$.

1.43 A child pulls a rope attached to a sled with a force of 60 N. The rope makes an angle of 40° to the ground. (a) Compute the effective value of the pull tending to move the sled along the ground. (b) Compute the force tending to lift the sled vertically.

▮ As shown in Fig. 1-23, the components of the 60 N force are 39 N and 46 N. (a) The pull along the ground is the horizontal component, $\underline{46\,\text{N}}$. (b) The lifting force is the vertical component, $\underline{39\,\text{N}}$.

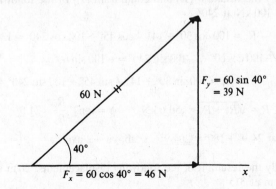

Fig. 1-23

1.44 Find the resultant of the coplanar force system shown in Fig. 1-24.

▮ $\mathbf{R} = \mathbf{F}_1 + \mathbf{F}_2 + \mathbf{F}_3$ $R_x = -40 + 80 \cos 30° + 0 = 29.3\,\text{lb}$ $R_y = 0 - 80 \sin 30° + 60 = 20\,\text{lb}$

$$R = \sqrt{R_x^2 + R_y^2} = \underline{35.4\,\text{lb}} \qquad \theta = \tan^{-1}\frac{R_y}{R_x} = \underline{34.3°} \qquad \text{above } +x \text{ axis}$$

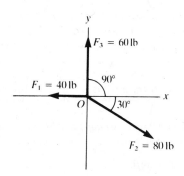

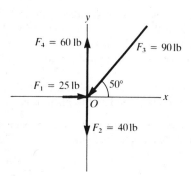

Fig. 1-24   Fig. 1-25

1.45   Repeat Prob. 1.44 for Fig. 1-25.

▮ $\mathbf{R} = \mathbf{F}_1 + \mathbf{F}_2 + \mathbf{F}_3 + \mathbf{F}_4$ $R_x = 25 + 0 - 90 \cos 50° + 0 = -32.8$ lb $R_y = 0 - 40 - 90 \sin 50° + 60 = -48.9$ lb

$$R = \sqrt{R_x^2 + R_y^2} = \underline{58.9 \text{ lb}} \qquad \phi = \tan^{-1}\left|\frac{R_y}{R_x}\right| = \underline{56.1°} \qquad \text{below } -x \text{ axis}$$

1.46   Repeat Prob. 1.44 for Fig 1-26.

▮ $\mathbf{R} = \mathbf{R}_1 + \mathbf{R}_2 + \mathbf{R}_3 + \mathbf{R}_4$ $R_x = -125 \cos 25° + 0 + 180 \cos 23° + 150 \cos 62° = 122.8$ lb

$$R_y = -125 \sin 25° - 130 - 180 \sin 23° + 150 \sin 62° = -120.7 \text{ lb}$$

$$R = \sqrt{R_x^2 + R_y^2} = \underline{172.2 \text{ lb}}, \qquad \phi = \tan^{-1}\left|\frac{R_y}{R_x}\right| = \underline{44.5°} \qquad \text{below } +x \text{ axis}$$

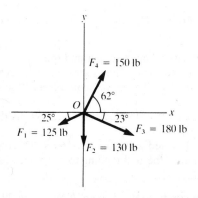

Fig. 1-26

1.47   Compute algebraically the resultant ($\mathbf{R}$) and equilibrant ($\mathbf{E}$) of the following coplanar forces: 100 kN at 30°, 141.4 kN at 45°, and 100 kN at 240°.

▮ $\mathbf{R} = \mathbf{F}_1 + \mathbf{F}_2 + \mathbf{F}_3$ $R_x = 100 \cos 30° + 141.4 \cos 45° + 100 \cos 240° = 136.6$ kN.

Note: $100 \cos 240° = -100 \cos 60°$ $100 \sin 240° = -100 \sin 60°$.

$$R_y = 100 \sin 30° + 141.4 \sin 45° + 100 \sin 240° = 63.4 \text{ kN}$$

$$R = \sqrt{R_x^2 + R_y^2} = \underline{150.6 \text{ kN}} \qquad \phi = \tan^{-1}\frac{R_y}{R_x} = \underline{24.9°} \qquad \text{above } +x \text{ axis.}$$

$\mathbf{E} = -\mathbf{R} = \underline{150.6 \text{ kN}}$ at $24.9° + 180° = \underline{204.9°}$ above $+x$ axis

1.48   Compute algebraically the resultant of the following displacements: 20 m at 30°, 40 m at 120°, 25 m at 180°, 42 m at 270°, and 12 m at 315°.

▮ $\mathbf{D} = \mathbf{d}_1 + \mathbf{d}_2 + \mathbf{d}_3 + \mathbf{d}_4 + \mathbf{d}_5 = $ resultant displacement $D_x = 20 \cos 30° + 40 \cos 120° - 25 + 0 + 12 \cos 315° = -19.3$ m

Note: 180° is along $-x$ axis; 270° is along $-y$ axis; $\cos 120° = -\cos 60°$; $\sin 120° = \sin 60°$; $\cos 315° = \cos 45°$; $\sin 315° = -\sin 45°$.

$$D_y = 20 \sin 30° + 40 \sin 120° + 0 - 42 + 12 \sin 315° = -5.8 \text{ m}$$

$$D_y = \sqrt{D_x^2 + D_y^2} = \underline{20.2\ \text{m}} \qquad \phi = \tan^{-1}\left|\frac{D_y}{D_x}\right| = \underline{16.7°} \qquad \text{below} -x \text{ axis}$$

or $\quad \theta = \tan^{-1}\dfrac{D_y}{D_x} = \underline{196.7°} \quad$ above $+x$ axis.

1.49 Refer to Fig. 1-27. In terms of vectors **A** and **B**, express the vectors **P**, **R**, **S**, and **Q**.

❚ We have here a parallelogram, so **R** = **B** and **P** = **A** + **R** = **A** + **B**. Clearly **S** = −**A** and **Q** is the sum of −**B** with **A** or **Q** = **A** − **B**.

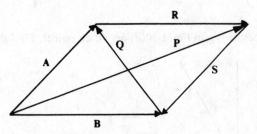

Fig. 1-27

1.50 Refer to Fig. 1-28. In terms of vectors **A** and **B**, express the vectors **E**, **D** − **C**, and **E** + **D** − **C**.

❚ Clearly −**E** = **A** + **B** or **E** = −(**A** + **B**) = −**A** − **B**. **D** − **C** = **D** + (−**C**) = **A**. Then **E** + **D** − **C** = **E** + **A** = −**B**.

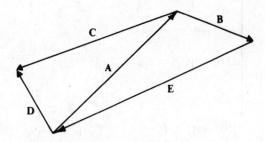

Fig. 1-28

1.51 A displacement **D** of 100 m from the origin at an angle of 37° above the x axis is the result of three successive displacements: $\mathbf{d}_1$, which is 100 m along the negative x axis; $\mathbf{d}_2$, which is 200 m at an angle of 150° above the x axis; and a displacement $\mathbf{d}_3$. Find $\mathbf{d}_3$.

❚ **D** = $\mathbf{d}_1 + \mathbf{d}_2 + \mathbf{d}_3$ $\quad D_x = d_{1x} + d_{2x} + d_{3x}$ or $100\cos 37° = -100 + 200\cos 150° + d_{3x}$ $\quad d_{3x} = 353$ m
(Note: $\cos 150° = -\cos 30°$) $\quad D_y = d_{1y} + d_{2y} + d_{3y}$ or $100\sin 37° = 0 + 200\sin 150° + d_{3y}$

$d_{3y} = -40$ m $\quad$ (Note: $\sin 150° = \sin 30°$). $\quad d_3 = \sqrt{d_{3x}^2 + d_{3y}^2} = \underline{355\ \text{m}} \quad \phi = \tan^{-1}\left|\dfrac{d_{3y}}{d_{3x}}\right| = \underline{6.5°} \quad$ below $+x$ axis

1.52 The resultant force due to the action of four forces is **R**, which is 100 N along the negative y axis. Three of the forces are 100 N, 60° above the x axis; 200 N, 140° above the x axis; 250 N, 320° above the x axis. Find the fourth force.

❚ **R** = $\mathbf{F}_1 + \mathbf{F}_2 + \mathbf{F}_3 + \mathbf{F}_4$ $\quad R_x = F_{1x} + F_{2x} + F_{3x} + F_{4x}$ $\quad$ or $\quad 0 = 100\cos 60° + 200\cos 140° + 250\cos 320° + F_{4x}$.
$F_{4x} = -88.3$ N $\quad R_y = F_{1y} + F_{2y} + F_{3y} + F_{4y}$ or $-100 = 100\sin 60° + 200\sin 140° + 250\sin 320° + F_{4y}$.

$F_{4y} = -154.5$ N $\quad F_4 = \sqrt{F_{4x}^2 + F_{4y}^2} = \underline{180\ \text{N}} \quad \phi = \tan^{-1}\left|\dfrac{F_{4y}}{F_{4x}}\right| = \underline{60°} \quad$ below $-x$ axis

1.53 A car whose weight is w is on a ramp which makes an angle θ to the horizontal. How large a perpendicular force must the ramp withstand if it is not to break under the car's weight?

❚ As shown in Fig. 1-29, the car's weight is a force **w** that pulls straight down on the car. We take components of **w** along the incline and perpendicular to it. The ramp must balance the force component $w \cos \theta$ if the car is not to crash through the ramp.

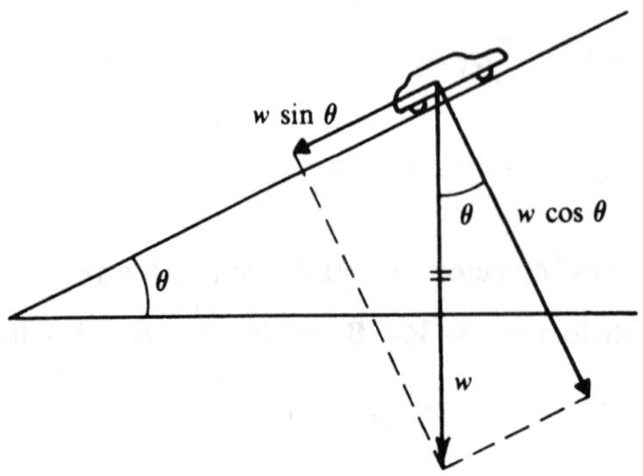

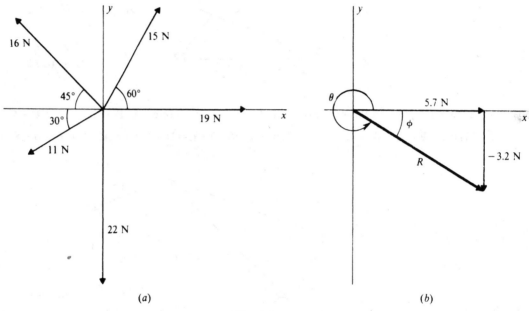

Fig. 1-29

1.54 The five coplanar forces shown in Fig. 1-30(a) act on an object. Find the resultant force due to them.

(a)

(b)

Fig. 1-30

▌ (1) Find the x- and y-components of each force.

magnitude, N	x component, N	y component, N
19	19.0	0
15	15 cos 60° = 7.5	15 sin 60° = 13.0
16	−16 cos 45° = −11.3	16 sin 45° = 11.3
11	−11 cos 30° = −9.5	−11 sin 30° = −5.5
22	0	−22.0

Note the signs to indicate + and − directions.
(2) The resultant **R** has components

$$R_x = \Sigma F_x = 19.0 + 7.5 - 11.3 - 9.5 + 0 = +5.7 \text{ N} \qquad R_y = \Sigma F_y = 0 + 13.0 + 11.3 - 5.5 - 22.0 = -3.2 \text{ N}$$

(3) Find the magnitude of the resultant from

$$R = \sqrt{R_x^2 + R_y^2} = 6.5 \text{ N}$$

(4) Sketch the resultant as shown in Fig. 1-29(b) and find its angle. We see that

$$\tan \phi = \frac{3.2}{5.7} = 0.56$$

from which $\phi = 29°$. Then we have $\theta = 360° - 29° = 331°$. The resultant is <u>6.5 N</u> at <u>331°</u> (or <u>−29°</u>).

1.55 Find algebraically the resultant (**R**) and equilibrant (**E**) of the following coplanar forces: 300 N at 0°, 400 N at 30°, and 400 N at 150°.

▌ $R = F_1 + F_2 + F_3$ $E = -R$ $R_x = 300 + 400 \cos 30° + 400 \cos 150° = 300 \text{ N}$
Note that $400 \cos 150° = -400 \cos 30°$ $400 \sin 150° = 400 \sin 30°$

$R_y = 0 + 400 \sin 30° + 400 \sin 150° = 400 \text{ N}$ $R = \sqrt{R_x^2 + R_y^2} = \underline{500 \text{ N}}$ $\phi = \tan^{-1}\left|\dfrac{R_y}{R_x}\right| = \underline{53°}$ above $+x$ axis.

Then $E = \underline{500 \text{ N}}$, $\phi_E = \underline{53°}$ below $-x$ axis.

1.2 THREE-DIMENSIONAL VECTORS; DOT AND CROSS PRODUCTS

1.56 Find the magnitude of the vector **A** in Fig. 1-31, whose tail lies at the origin and whose head lies at the point (7.0 m, 4.0 m, 5.0 m).

▌ First, note that the vector **A** and its vector component A_z are the hypotenuse and one side of a right triangle whose plane is perpendicular to the xy plane. Hence the pythagorean theorem gives $A^2 = B^2 + A_z^2$. But the vector **B** is itself the hypotenuse of the right triangle in the xy plane whose sides are the vector components A_x and A_y. Thus you can use the pythagorean theorem again to obtain $B^2 = A_x^2 + A_y^2$. Combining the two equations,

$$A^2 = A_x^2 + A_y^2 + A_z^2 \quad \text{or} \quad A = \sqrt{A_x^2 + A_y^2 + A_z^2}$$

which is the form taken by the pythagorean theorem in three dimensions.
For the data,

$$A = \sqrt{(7.0 \text{ m})^2 + (4.0 \text{ m})^2 + (5.0 \text{ m})^2} = \underline{9.5 \text{ m}}$$

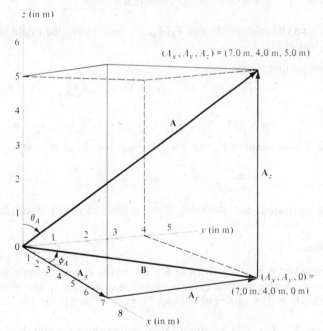

Fig. 1-31

1.57 Find the scalar components of the three-dimensional vector **F** in Fig. 1-32.

▌ In Fig. 1-32, F_x, F_y, F_z are the rectangular vector components of **F**; the scalar components F_x, F_y, F_z are given by

$$F_x = F \cos \theta_1 \qquad F_y = F \cos \theta_2 \qquad F_z = F \cos \theta_3$$

Or for convenience, writing $\cos \theta_1 = l$, $\cos \theta_2 = m$, $\cos \theta_3 = n$,

$$F_x = Fl \qquad F_y = Fm \qquad F_z = Fn$$

l, m, and n are referred to as the *direction cosines* of **F**. By the three-dimensional pythagorean theorem (Prob. 1.56), $l^2 + m^2 + n^2 = 1$.

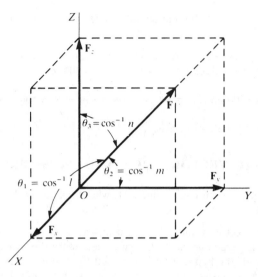

Fig. 1-32

1.58 In Fig. 1-32, let **F** represent a force of 200 N. Let $\theta_1 = 60°$, $\theta_2 = 40°$. Find F_x, F_y, and F_z.

■ $l = 0.5$ $m = 0.766$ $n = (1 - l^2 - m^2)^{1/2} = 0.404$

(assuming that F_z is positive; otherwise, $n = -0.404$), and the rectangular components of **F** are

$$F_x = (200)(0.5) = \underline{100\ N}   F_y = \underline{153.2\ N}   F_z = \underline{80.8\ N}$$

As a check, $(100^2 + 153.2^2 + 80.8^2)^{1/2} \approx 200$. Note that $\underline{\theta_3 = 66.17°}$.

1.59 Find the vector sum (**R**) of three vectors, $\mathbf{F}_1$, $\mathbf{F}_2$, $\mathbf{F}_3$, drawn from the origin of a three-dimensional coordinate system such as that of Fig. 1.32.

■ Following the component method,

$$R_x = F_{1x} + F_{2x} + F_{3x}   R_y = F_{1y} + F_{2y} + F_{3y}   R_z = F_{1z} + F_{2z} + F_{3z}$$

and

$$R = [(F_{1x} + F_{2x} + F_{3x})^2 + (F_{1y} + F_{2y} + F_{3y})^2 + (F_{1z} + F_{2z} + F_{3z})^2]^{1/2}$$

where F_{1x} is the X component of $\mathbf{F}_1$, etc. The direction cosines of **R** are given by

$$l = \frac{F_{1x} + F_{2x} + F_{3x}}{R}   m = \frac{F_{1y} + F_{2y} + F_{3y}}{R}   n = \frac{F_{1z} + F_{2z} + F_{3z}}{R}$$

The resultant (magnitude and direction) of any number of vectors drawn from O can be obtained in the same way.

1.60 Define a unit vector.

■ Any nonzero vector **F** may be written as $\mathbf{F} = F\mathbf{e}$, where F is the magnitude of **F** and where **e** is a *unit vector* (a vector whose magnitude is 1) in the direction of **F**. That is, the magnitude of **F** is indicated by F and its direction is that of **e**. If **F** carries units (e.g., N, m/s), F carries the same units; **e** is a dimensionless vector.

1.61 Express the vector **F** in Fig. 1.32 in terms of unit vectors along the coordinate axes.

■ In Fig. 1-32, let us introduce unit vectors **i**, **j**, **k** along X, Y, Z, respectively. Then the vector components of **F** can be written as

$$\mathbf{F}_x = F_x\mathbf{i}   \mathbf{F}_y = F_y\mathbf{j}   \mathbf{F}_z = F_z\mathbf{k}$$

according to Prob. 1.60. [If one of the scalar components is negative—say, $F_x = -|F_x| = -|\mathbf{F}_x|$—then we have $\mathbf{F}_x = -|\mathbf{F}_x|\,\mathbf{i} = |\mathbf{F}_x|\,(-\mathbf{i})$, which is still as prescribed by Prob. 1.60.] Since **F** is the resultant of its vector components, we obtain the very important expression $\mathbf{F} = F_x\mathbf{i} + F_y\mathbf{j} + F_z\mathbf{k}$. In this expression, $F_x = F \cos\theta_1 = Fl$, etc., as previously shown; and, as before, the magnitude and direction (direction cosines, that is) are obtained as

$$F = (F_x^2 + F_y^2 + F_z^2)^{1/2}   l = \frac{F_x}{F}   m = \frac{F_y}{F}   n = \frac{F_z}{F}$$

1.62 A force **F** has components $F_x = 100$ N, $F_y = 153.2$ N, $F_z = 80.8$ N. Express **F** in terms of unit vectors and find its magnitude and direction.

   ❚ The vector **F** can be written as $\mathbf{F} = 100\mathbf{i} + 153.2\mathbf{j} + 80.8\mathbf{k}$ with magnitude $F = (100^2 + 153.2^2 + 81^2)^{1/2} = 200$ N and direction:

$$l = \frac{100}{200} = 0.5 \qquad m = 0.766 \qquad n = 0.404$$

Strictly, we should have written

$$\mathbf{F} = (100\text{ N})\mathbf{i} + (153.2\text{ N})\mathbf{j} + (80.8\text{ N})\mathbf{k} \qquad \text{or} \qquad \mathbf{F} = 100\mathbf{i} + 153.2\mathbf{j} + 80.8\mathbf{k} \quad \text{N}$$

1.63 A force **A** is added to a second force which has x and y components 3 N and -5 N. The resultant of the two forces is in the $-x$ direction and has a magnitude of 4 N. Find the x and y components of **A**.

   ❚ Let $\mathbf{A} = A_x\mathbf{i} + A_y\mathbf{j}$. Then $A_x + 3 = -4$ and $A_y - 5 = 0$. So $A_x = \underline{-7\text{ N}}$ and $A_y = \underline{5\text{ N}}$.

1.64 Express **A**, **B**, and **C** of Fig. 1-13 in terms of the unit vectors **i**, **j**, and **k**.

   ❚ $\mathbf{A} = A_x\mathbf{i} + A_y\mathbf{j} + A_z\mathbf{k} = \underline{6\mathbf{i}\text{ m}}$, $\mathbf{B} = \underline{6\mathbf{i} + 10.4\mathbf{j}}\text{ m}$, and $\mathbf{C} = \underline{4.5\mathbf{i} - 7.8\mathbf{j}}\text{ m}$.

1.65 Find the components of a displacement which when added to a displacement of $7\mathbf{i} - 4\mathbf{j}$ m will give a resultant displacement of $5\mathbf{i} - 3\mathbf{j}$ m.

   ❚ We have $A_x + 7 = 5$ and $A_y - 4 = -3$, so $A_x = \underline{-2\text{ m}}$ and $A_y = \underline{1\text{ m}}$.

1.66 Find the magnitude and direction of the vector sum of the following three vectors: $2\mathbf{i} - 3\mathbf{j}$, $-9\mathbf{i} - 5\mathbf{j}$, $4\mathbf{i} + 8\mathbf{j}$.

   ❚ We have $(2 - 9 + 4)\mathbf{i} + (-3 - 5 + 8)\mathbf{j} = -3\mathbf{i}$. Thus, it is $\underline{3}$ units along the $\underline{-x\text{ direction}}$.

1.67 What must be the components of a vector which when added to the following two vectors gives rise to a vector $6\mathbf{j}$: $10\mathbf{i} - 7\mathbf{j}$ and $4\mathbf{i} + 2\mathbf{j}$?

   ❚ Call the vector **A**. Then $A_x + 10 + 4 = 0$ and $A_y - 7 + 2 = 6$. So $A_x = \underline{-14}$ and $A_y = \underline{11}$.

1.68 A certain room has a floor which is 5×6 m and the ceiling height is 3 m. Write an expression for the vector distance from one corner of the room to the corner diagonally opposite it. What is the magnitude of the distance?

   ❚ The x, y, z displacements in going from one corner to the other are 5, 6, and 3 m, respectively. Therefore, $\mathbf{D} = \underline{5\mathbf{i} + 6\mathbf{j} + 3\mathbf{k}\text{ m}}$. Also, $D^2 = D_x^2 + D_y^2 + D_z^2$, which gives $D = \underline{8.4\text{ m}}$.

1.69 Find the displacement vector from the point $(0, 3, -1)$ m to the point $(-2, 6, 4)$ m. Give your answer in **i**, **j**, **k** notation. Also give the magnitude of the displacement.

   ❚ The component displacements are $D_x = -2 - 0 = -2$, $D_y = 6 - 3 = 3$, and $D_z = 4 - (-1) = 5$; thus $\mathbf{D} = \underline{-2\mathbf{i} + 3\mathbf{j} + 5\mathbf{k}\text{ m}}$ and $D = \underline{6.2\text{ m}}$.

1.70 An object, originally at the point $(2, 5, 1)$ cm, is given a displacement $8\mathbf{i} - 2\mathbf{j} + \mathbf{k}$ cm. Find the coordinates of its new position.

   ❚ The new coordinates are $x = 2 + 8 = 10$; $y = 5 + (-2) = 3$; $z = 1 + 1 = 2$. The object is at $\underline{(10, 3, 2)\text{ cm}}$.

1.71 Find the resultant displacement caused by the following three displacements: $2\mathbf{i} - 3\mathbf{k}$, $5\mathbf{j} - 2\mathbf{k}$, and $-6\mathbf{i} + \mathbf{j} + 8\mathbf{k}$, all in millimeters. Give its magnitude as well as its **i**, **j**, **k** representation.

   ❚ $\mathbf{R} = (2 - 6)\mathbf{i} + (5 + 1)\mathbf{j} + (-3 - 2 + 8)\mathbf{k} = \underline{-4\mathbf{i} + 6\mathbf{j} + 3\mathbf{k}\text{ mm}}$, and so $R = 61^{1/2} = \underline{7.8\text{ mm}}$.

1.72 Give the **i**, **j**, **k** representation and magnitude of the force which must be added to the following two forces to give a force $7\mathbf{i} - 6\mathbf{j} - \mathbf{k}$: $2\mathbf{i} - 7\mathbf{k}$ and $3\mathbf{j} + 2\mathbf{k}$. All forces are in newtons.

   ❚ Call the force F. Then $F_x + 2 = 7$; $F_y + 3 = -6$; $F_z - 7 + 2 = -1$. Therefore $\mathbf{F} = \underline{5\mathbf{i} - 9\mathbf{j} + 4\mathbf{k}\text{ N}}$ and $F = \underline{11\text{ N}}$.

1.73 Vectors **A** and **B** are in the xy plane. If **A** is 70 N at 90° and **B** is 120 N at 210°, find (*a*) $\mathbf{A} - \mathbf{B}$, and (*b*) vector **C** such that $\mathbf{A} - \mathbf{B} + \mathbf{C} = 0$.

▮ A and B written in terms of unit vectors are $\mathbf{A} = 70\mathbf{j}$ and $\mathbf{B} = -120\cos 30°\mathbf{i} - 120\sin 30°\mathbf{j} = -104\mathbf{i} - 60\mathbf{j}$.
(a) $\mathbf{A} - \mathbf{B} = (0 + 104)\mathbf{i} + (70 + 60)\mathbf{j} = \underline{166\text{ N at }51°}$. (b) $\mathbf{C} = -(\mathbf{A} - \mathbf{B}) = -104\mathbf{i} - 130\mathbf{j} = \underline{166\text{ N at }231°}$.

1.74 If $\mathbf{A} = 2\mathbf{i} - 3\mathbf{j} + 5\mathbf{k}$ mm and $\mathbf{B} = -\mathbf{i} - 2\mathbf{j} + 7\mathbf{k}$ mm, find (in component form) (a) $\mathbf{A} - \mathbf{B}$. (b) $\mathbf{B} - \mathbf{A}$.
(c) Vector $\mathbf{C}$ such that $\mathbf{A} + \mathbf{B} + \mathbf{C} = 0$.

▮ (a) $\mathbf{A} - \mathbf{B} = [2 - (-1)]\mathbf{i} + [-3 - (-2)]\mathbf{j} + (5 - 7)\mathbf{k} = \underline{3\mathbf{i} - \mathbf{j} - 2\mathbf{k}\text{ mm}}$ (b) $\mathbf{B} - \mathbf{A} = -(\mathbf{A} - \mathbf{B}) = \underline{-3\mathbf{i} + \mathbf{j} + 2\mathbf{k}\text{ mm}}$ (c) $\mathbf{C} = -(\mathbf{A} + \mathbf{B}) = \underline{-\mathbf{i} + 5\mathbf{j} - 12\mathbf{k}\text{ mm}}$.

1.75 Vector $\mathbf{A} = 3\mathbf{i} + 5\mathbf{j} - 2\mathbf{k}$ and vector $\mathbf{B} = -3\mathbf{j} + 6\mathbf{k}$. Find a vector $\mathbf{C}$ such that $2\mathbf{A} + 7\mathbf{B} + 4\mathbf{C} = 0$.

▮ A vector is zero if and only if each of its components is zero. For example, $2A_x + 7B_x + 4C_x = 0$, so
$C_x = -1.5$. For C_y, solve $2(5) + 7(-3) + 4C_y = 0$ to obtain $C_y = 2.75$. Similarly, $C_z = -9.5$. Therefore
$\mathbf{C} = \underline{-1.5\mathbf{i} + 2.75\mathbf{j} - 9.5\mathbf{k}}$.

1.76 A certain vector is given by $3\mathbf{i} + 4\mathbf{j} + 7\mathbf{k}$. Find the angle it makes with the z axis.

▮ First find the projection of the vector in the xy plane; this is $(3^2 + 4^2)^{1/2} = 5$. The magnitude of the vector is
$(7^2 + 5^2)^{1/2} = 8.6$ and it makes an angle of $\tan^{-1}(5/7) = \underline{35.5°}$ to the z axis. Otherwise, from Prob. 1.61,

$$\cos \theta_3 = \frac{7}{\sqrt{3^2 + 4^2 + 7^2}} = 0.814 \qquad \theta_3 = \underline{35.5°}$$

1.77 What must be the relation between vectors $\mathbf{A}$ and $\mathbf{B}$ if the following condition is to be true:

$$\mathbf{A} - 2\mathbf{B} = -3(\mathbf{A} + \mathbf{B})$$

If vector $\mathbf{A} = 6\mathbf{i} - 2\mathbf{k}$ m, what is $\mathbf{B}$?

▮ One has $\mathbf{A} - 2\mathbf{B} = -3\mathbf{A} - 3\mathbf{B}$, so $\underline{4\mathbf{A} = -\mathbf{B}}$. Substituting $6\mathbf{i} - 2\mathbf{k}$ for $\mathbf{A}$ gives $\mathbf{B} = \underline{-24\mathbf{i} + 8\mathbf{k}\text{ m}}$.

1.78 What must be the relation between two vectors $\mathbf{A}$ and $\mathbf{B}$ if the magnitude of $\mathbf{A} + \mathbf{B}$ equals the magnitude of
$\mathbf{A} - \mathbf{B}$, that is,

$$|\mathbf{A} + \mathbf{B}| = |\mathbf{A} - \mathbf{B}|$$

▮ $\mathbf{A}$ and $\mathbf{B}$ must be mutually perpendicular. To see that, let $\mathbf{P}$ and $\mathbf{Q}$ be two vectors of the same magnitude.
As shown in Fig. 1-33, $\mathbf{P}$ and $\mathbf{Q}$ form a rhombus whose diagonals, $\mathbf{P} + \mathbf{Q}$ and $\mathbf{P} - \mathbf{Q}$, are necessarily
perpendicular. Now set $\mathbf{P} = \frac{1}{2}(\mathbf{A} + \mathbf{B})$, $\mathbf{Q} = \frac{1}{2}(\mathbf{A} - \mathbf{B})$.

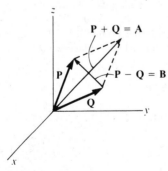

Fig. 1-33

1.79 The vector displacements of two points A and B from the origin are

$$s_A = 3\mathbf{i} - 2\mathbf{j} + 5\mathbf{k}\text{ cm} \qquad \text{and} \qquad s_B = -\mathbf{i} - 5\mathbf{j} + 2\mathbf{k}\text{ cm}$$

Find the magnitude and $\mathbf{i}$, $\mathbf{j}$, $\mathbf{k}$ representation of the vector from point A to point B.

▮ The vector in question has components $D_x = -1 - 3 = -4$; $D_y = -5 - (-2) = -3$; $D_z = 2 - 5 = -3$.
Thus $\mathbf{D} = \underline{-4\mathbf{i} - 3\mathbf{j} - 3\mathbf{k}\text{ cm}}$ and from $D^2 = 4^2 + 3^2 + 3^2$, we find $D = \underline{5.8\text{ cm}}$.

1.80 The rectangular components of an acceleration vector $\mathbf{a}$ are $a_x = 6$, $a_y = 4$, $a_z = 9$ m/s^2. Find the vector
expression for $\mathbf{a}$ and its direction cosines.

▮ In vector form, $\mathbf{a} = 6\mathbf{i} + 4\mathbf{j} + 9\mathbf{k}$ m/s². The magnitude of $\mathbf{a}$ is $a = (6^2 + 4^2 + 9^2)^{1/2} = \underline{11.53 \text{ m/s}^2}$, and the direction cosines of $\mathbf{a}$ are

$$l = \frac{6}{11.53} \qquad m = \frac{4}{11.53} \qquad n = \frac{9}{11.53}$$

1.81 Find a vector expression for a line segment.

▮ The straight line ab, Fig. 1-34, is determined by points P_1 and P_2. Regarding the line segment from P_1 to P_2 as a vector $\mathbf{s}$, we can write

$$\mathbf{s} = (x_2 - x_1)\mathbf{i} + (y_2 - y_1)\mathbf{j} + (z_2 - z_1)\mathbf{k}$$

with magnitude $s = [(x_2 - x_1)^2 + (y_2 - y_1)^2 + (z_2 - z_1)^2]^{1/2}$ and direction

$$l = \frac{x_2 - x_1}{s} \qquad m = \frac{y_2 - y_1}{s} \qquad n = \frac{z_2 - z_1}{s}$$

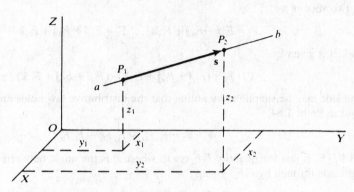

Fig. 1-34

An important special case of this is the so-called radius vector $\mathbf{r}$, the directed segment from the origin O to a point $P(x, y, z)$:

$$\mathbf{r} = x\mathbf{i} + y\mathbf{j} + z\mathbf{k}$$

with $r = (x^2 + y^2 + z^2)^{1/2}$ and

$$l = \frac{x}{r} \qquad m = \frac{y}{r} \qquad n = \frac{z}{r}$$

1.82 Consider the velocity vector $\mathbf{v} = 16\mathbf{i} + 30\mathbf{j} + 24\mathbf{k}$ m/s, with $v = (16^2 + 30^2 + 24^2)^{1/2} = 41.62$ m/s, direction given by $l = 16/41.62$, etc.
   Now let us multiply $\mathbf{v}$ by 10: $10\mathbf{v} = 160\mathbf{i} + 300\mathbf{j} + 240\mathbf{k} \equiv \mathbf{v}_1$. Find the magnitude and direction of $\mathbf{v}_1$.

▮ $$v_1 = [(160)^2 + (300)^2 + (240)^2]^{1/2} = (10)(41.62) = 10v$$

and the direction cosines of $\mathbf{v}_1$ are

$$l_1 = \frac{160}{(10)(41.62)} = \frac{16}{41.62} = l \qquad m_1 = m \qquad n_1 = n$$

which shows that $\mathbf{v}_1$ has the direction of $\mathbf{v}$.

1.83 Define the scalar or dot product of two vectors.

▮ The *dot product* of any two vectors, as $\mathbf{F}_1$ and $\mathbf{F}_2$, Fig. 1-35, is written as $\mathbf{F}_1 \cdot \mathbf{F}_2$ and is defined as the product of their magnitudes and the cosine of the included angle. That is, $\mathbf{F}_1 \cdot \mathbf{F}_2 = F_1 F_2 \cos \theta$, which is a scalar quantity. In Fig. 1-35, $F_1 = 75$, $F_2 = 100$, $\theta = 60°$. Thus, $\mathbf{F}_1 \cdot \mathbf{F}_2 = (75)(100)(0.5) = \underline{3750}$.

1.84 Find the dot products of the unit vectors along X, Y, Z.

▮ Since $\mathbf{i}, \mathbf{j}, \mathbf{k}$ are mutually perpendicular and of unit magnitude, the definition of the dot product gives

$$\mathbf{i} \cdot \mathbf{i} = \mathbf{j} \cdot \mathbf{j} = \mathbf{k} \cdot \mathbf{k} = 1 \qquad \mathbf{i} \cdot \mathbf{j} = \mathbf{i} \cdot \mathbf{k} = \mathbf{j} \cdot \mathbf{k} = 0$$

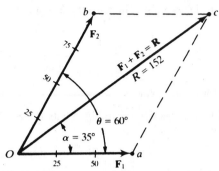

Fig. 1-35

1.85 Find the dot product of any two vectors in terms of rectangular components.

▌ Write any two vectors as

$$\mathbf{F}_1 = F_{1x}\mathbf{i} + F_{1y}\mathbf{j} + F_{1z}\mathbf{k} \qquad \mathbf{F}_2 = F_{2x}\mathbf{i} + F_{2y}\mathbf{j} + F_{2z}\mathbf{k}$$

Their dot product is given by

$$\mathbf{F}_1 \cdot \mathbf{F}_2 = (F_{1x}\mathbf{i} + F_{1y}\mathbf{j} + F_{1z}\mathbf{k}) \cdot (F_{2x}\mathbf{i} + F_{2y}\mathbf{j} + F_{2z}\mathbf{k})$$

The right-hand side may be simplified by noting that the distributive law holds and employing the values of $\mathbf{i} \cdot \mathbf{i}$, etc., found in Prob. 1.84.

$$\mathbf{F}_1 \cdot \mathbf{F}_2 = F_{1x}F_{2x} + F_{1y}F_{2y} + F_{1z}F_{2z}$$

To check that $\mathbf{F}_1 \cdot \mathbf{F}_2$ is just the quantity $F_1 F_2 \cos \theta$, where θ is the angle between $\mathbf{F}_1$ and $\mathbf{F}_2$, multiply and divide the right side through by $F_1 F_2$, giving

$$\mathbf{F}_1 \cdot \mathbf{F}_2 = F_1 F_2 \left(\frac{F_{1x}}{F_1} \frac{F_{2x}}{F_2} + \frac{F_{1y}}{F_1} \frac{F_{2y}}{F_2} + \frac{F_{1z}}{F_1} \frac{F_{2z}}{F_2} \right) = F_1 F_2 (l_1 l_2 + m_1 m_2 + n_1 n_2)$$

Now, the familiar addition formula in two dimensions,

$$\cos \theta = \cos (\theta_1 - \theta_2) = \cos \theta_1 \cos \theta_2 + \sin \theta_1 \sin \theta_2 = l_1 l_2 + m_1 m_2$$

extends to three dimensions as $\cos \theta = l_1 l_2 + m_1 m_2 + n_1 n_2$. Hence, the above becomes $F_1 F_2 \cos \theta$, and this method of multiplication is, as expected, in accord with the definition of the dot product.

1.86 Let $\mathbf{F}_1 = 10\mathbf{i} - 15\mathbf{j} - 20\mathbf{k}$, $\mathbf{F}_2 = 6\mathbf{i} + 8\mathbf{j} - 12\mathbf{k}$. Find their dot product and the angle between them.

▌ $$\mathbf{F}_1 \cdot \mathbf{F}_2 = (10)(6) + (-15)(8) + (-20)(-12) = \underline{180}$$

Now note that $F_1 = (10^2 + 15^2 + 20^2)^{1/2} = 26.93$, $F_2 = 15.62$. Hence, the angle θ between $\mathbf{F}_1$ and $\mathbf{F}_2$ is given by

$$\cos \theta = \frac{\mathbf{F}_1 \cdot \mathbf{F}_2}{F_1 F_2} = \frac{180}{(26.93)(15.62)} = 0.4279 \qquad \theta = \underline{64.66°}$$

Of course, the same value can be obtained from $\cos \theta = l_1 l_2 + m_1 m_2 + n_1 n_2$.

1.87 Find the projection of any vector along a straight line.

▌ The projection of vector $\mathbf{A} = (A_x, A_y, A_z)$ along the line determined by the radius vector $\mathbf{r} = (x, y, z)$ is $A_r = A \cos \theta$, where θ is the angle between $\mathbf{r}$ and $\mathbf{A}$. From the definition of the dot product, $\mathbf{A} \cdot \mathbf{r} = (Ar \cos \theta) = A_r r$. Hence, writing $\hat{\mathbf{r}} = (1/r)\mathbf{r} = (l, m, n)$ —a unit vector along $\mathbf{r}$—we have $A_r = \mathbf{A} \cdot \hat{\mathbf{r}} = A_x l + A_y m + A_z n$. This expression for A_r remains valid even when the line does not pass through the origin.

1.88 Find the projection of $\mathbf{A} = 10\mathbf{i} + 8\mathbf{j} - 6\mathbf{k}$ along $\mathbf{r} = 5\mathbf{i} + 6\mathbf{j} + 9\mathbf{k}$.

▌ Here $r = (5^2 + 6^2 + 9^2)^{1/2} = 11.92$ and Prob. 1.87 gives

$$A_r = A_x l + A_y m + A_z n = 10 \left(\frac{5}{11.92} \right) + 8 \left(\frac{6}{11.92} \right) - 6 \left(\frac{9}{11.92} \right) = \frac{44}{11.92} = \underline{3.69}$$

1.89 Define the cross product (or vector product) of two vectors.

▮ The *cross product* of two vectors, as $\mathbf{F}_1$ and $\mathbf{F}_2$, Fig. 1-36, written as $\mathbf{F} = \mathbf{F}_1 \times \mathbf{F}_2$, is defined as a vector $\mathbf{F}$ having a magnitude $F = F_1 F_2 \sin \theta$ and a direction which is the direction of advance of a right-hand screw when turned from $\mathbf{F}_1$ to $\mathbf{F}_2$ through angle θ, it being assumed that the axis of the screw is normal to the plane determined by $\mathbf{F}_1$ and $\mathbf{F}_2$ (the *right-hand screw rule*). Or, if the curled fingers of the right hand point from $\mathbf{F}_1$ to $\mathbf{F}_2$, the extended thumb points in the direction of $\mathbf{F}$ (*right-hand rule*).

Note that in accord with the right-hand screw rule, $\mathbf{F}_1 \times \mathbf{F}_2 = -(\mathbf{F}_2 \times \mathbf{F}_1)$.

1.90 Find the cross products of the coordinate unit vectors.

▮ Since $\mathbf{i}, \mathbf{j}, \mathbf{k}$ are mutually perpendicular and of unit magnitude, it follows from the definition of the cross product that

$$\mathbf{i} \times \mathbf{i} = \mathbf{j} \times \mathbf{j} = \mathbf{k} \times \mathbf{k} = 0$$

$$\mathbf{i} \times \mathbf{j} = \mathbf{k} \quad \mathbf{j} \times \mathbf{k} = \mathbf{i} \quad \mathbf{k} \times \mathbf{i} = \mathbf{j} \quad \mathbf{j} \times \mathbf{i} = -\mathbf{k} \quad \mathbf{k} \times \mathbf{j} = -\mathbf{i} \quad \mathbf{i} \times \mathbf{k} = -\mathbf{j}$$

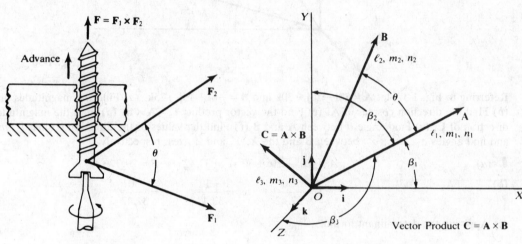

Fig. 1-36   **Fig. 1-37**

1.91 Given two vectors, as in Fig. 1-37,

$$\mathbf{A} = A_x \mathbf{i} + A_y \mathbf{j} + A_z \mathbf{k} \quad \mathbf{B} = B_x \mathbf{i} + B_y \mathbf{j} + B_z \mathbf{k}$$

find their cross product in rectangular coordinates.

▮
$$\mathbf{C} = \mathbf{A} \times \mathbf{B} = (A_x \mathbf{i} + A_y \mathbf{j} + A_z \mathbf{k}) \times (B_x \mathbf{i} + B_y \mathbf{j} + B_z \mathbf{k})$$

Applying the distributive law to the right-hand side and using the values of $\mathbf{i} \times \mathbf{i}$, etc., found in Prob. 1.90, we obtain

$$\mathbf{C} = \mathbf{A} \times \mathbf{B} = (A_y B_z - A_z B_y)\mathbf{i} + (A_z B_x - A_x B_z)\mathbf{j} + (A_x B_y - A_y B_x)\mathbf{k}$$

Equivalently, $\mathbf{A} \times \mathbf{B}$ may be expressed as a determinant:

$$\mathbf{C} = \mathbf{A} \times \mathbf{B} = \begin{vmatrix} \mathbf{i} & \mathbf{j} & \mathbf{k} \\ A_x & A_y & A_z \\ B_x & B_y & B_z \end{vmatrix}$$

as may be verified by expanding the determinant with respect to the first row. Note that the X, Y, Z components of $\mathbf{C}$ are

$$C_x = A_y B_z - A_z B_y \quad C_y = (A_z B_x - A_x B_z) \quad C_z = A_x B_y - A_y B_x$$

Hence the magnitude of $\mathbf{C}$ is $C = (C_x^2 + C_y^2 + C_z^2)^{1/2}$ and its direction cosines are

$$l = \frac{C_x}{C} \quad m = \frac{C_y}{C} \quad n = \frac{C_z}{C}$$

Vector $\mathbf{C}$ is, of course, normal to the plane of vectors $\mathbf{A}$ and $\mathbf{B}$.

1.92 Assuming that vectors **A** and **B**, Fig. 1-38, are in the XY plane, determine the magnitude and direction of $\mathbf{C} = \mathbf{A} \times \mathbf{B}$.

$$C = (200)(100) \sin (55° - 15°) = 20\,000 \sin 40° = 12\,855$$

and by the right-hand rule the direction of **C** is that of $+Z$. Vectorially we can write $\mathbf{C} = 12\,885\mathbf{k}$.

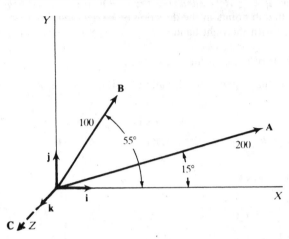

Fig. 1-38

1.93 Referring to Fig. 1-37, let $\mathbf{A} = 20\mathbf{i} - 10\mathbf{j} + 30\mathbf{k}$ and $\mathbf{B} = -6\mathbf{i} + 15\mathbf{j} - 25\mathbf{k}$. (*a*) Find the magnitudes of **A** and **B**. (*b*) Find the direction cosines of **A**. (*c*) Find the vector product $\mathbf{C} = \mathbf{A} \times \mathbf{B}$. (*d*) Find the magnitude and direction of **C**. (*e*) Find angle θ between **A** and **B** (*f*) Find the values of the direction cosines l_2, m_2, n_2 of **B**, and find angles α_{21}, α_{22}, α_{23} between **B** and the X, Y, and Z axes, respectively.

(*a*)
$$A = (20^2 + 10^2 + 30^2)^{1/2} = \underline{37.42} \qquad B = \underline{29.77}$$

(*b*)
$$l_1 = \frac{20}{37.42} \qquad m_1 = \frac{-10}{37.42} \qquad n_1 = \frac{30}{37.42}$$

(*c*) Applying the determinant formula,

$$\mathbf{C} = \begin{vmatrix} \mathbf{i} & \mathbf{j} & \mathbf{k} \\ 20 & -10 & 30 \\ -6 & 15 & -25 \end{vmatrix}$$

$$\mathbf{C} = \mathbf{i}[(-10)(-25) - (15)(30)] - \mathbf{j}[(20)(-25) - (30)(-6)] + \mathbf{k}[(20)(15) - (-10)(-6)]$$
$$= -200\mathbf{i} + 320\mathbf{j} + 240\mathbf{k} = 200(-\mathbf{i} + 1.6\mathbf{j} + 1.2\mathbf{k})$$

(*d*) The magnitude of **C** is
$$C = 200(1^2 + 1.6^2 + 1.2^2)^{1/2} = \underline{447.21}$$

The direction cosines are
$$l_3 = \frac{-200}{447.21} \qquad m_3 = \frac{320}{447.21} \qquad n_3 = \frac{240}{447.21}$$

Note that $\mathbf{C} = C(l_3\mathbf{i} + m_3\mathbf{j} + n_3\mathbf{k})$.

(*e*)
$$C = AB \sin \theta \qquad 447.21 = (37.42)(29.77) \sin \theta \qquad \sin \theta = 0.40145 \qquad \theta = \underline{23.67°}$$

(*f*)
$$\mathbf{B} = -6\mathbf{i} + 15\mathbf{j} - 25\mathbf{k} = B(l_2\mathbf{i} + m_2\mathbf{j} + n_2\mathbf{k})$$

Thus
$$Bl_2 = -6 \qquad Bm_2 = 15 \qquad Bn_2 = -25 \qquad B = (6^2 + 15^2 + 25^2)^{1/2} = \underline{29.766}$$
$$l_2 = -0.2016 \qquad m_2 = 0.5039 \qquad n_2 = -0.8399$$

Corresponding angles are
$$\alpha_{21} = 101.63° \qquad \alpha_{22} = 59.74° \qquad \alpha_{23} = 147.13°$$

CHAPTER 2
Equilibrium of Concurrent Forces

2.1 ROPES, KNOTS, AND FRICTIONLESS PULLEYS

2.1 The object in Fig. 2-1(a) weighs 50 N and is supported by a cord. Find the tension in the cord.

❚ Two forces act upon the object, the upward pull of the cord and the downward pull of gravity. Represent the pull of the cord by T, the tension in the cord. The pull of gravity, the weight of the object, is $w = 50$ N. These two forces are shown in the free-body diagram, Fig. 2-1(b).

The forces are already in component form and so we can write the first condition for equilibrium at once.

$$\sum F_x = 0 \quad \text{becomes} \quad 0 = 0$$
$$\sum F_y = 0 \quad \text{becomes} \quad T - 50\,\text{N} = 0$$

from which $T = \underline{50\,\text{N}}$.

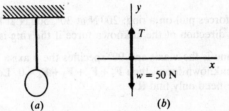

(a)   (b)   **Fig. 2-1**

2.2 As shown in Fig. 2-2(a), the tension in the horizontal cord is 30 N. Find the weight of the object.

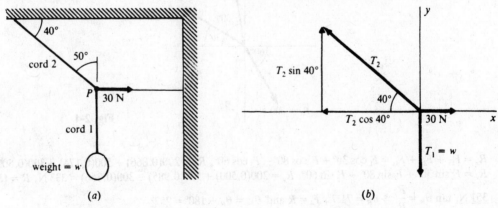

(a)   (b)

Fig. 2-2

❚ As seen in Prob. 2.1, the tension in cord 1 is equal to the weight of the object hanging from it. Therefore $T_1 = w$, and we wish to find T_1 or w.

Note that the unknown force, T_1, and the known force, 30 N, both pull on the knot at point P. It therefore makes sense to isolate the knot at P as our object. The free-body diagram showing the forces on the knot is drawn as Fig. 2-2(b). The force components are also found there.

Next write the first condition for equilibrium for the knot. From the free-body diagram,

$$\sum F_x = 0 \quad \text{becomes} \quad 30\,\text{N} - T_2 \cos 40° = 0$$
$$\sum F_y = 0 \quad \text{becomes} \quad T_2 \sin 40° - w = 0$$

Solving the first equation for T_2 gives $T_2 = 39.2$ N. Substituting this value in the second equation gives $w = \underline{25.2\,\text{N}}$ as the weight of the object.

2.3 For the system of Fig. 2-3(a), find the values of T_1 and T_2 if the weight is 600 N.

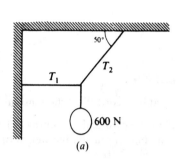 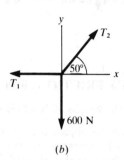

(a) (b) **Fig. 2-3**

▮ Consider the knot to be in equilibrium under the action of three forces, as shown in Fig. 2-3(b).

$$\sum F_x = 0 \quad \text{yields} \quad T_2 \cos 50° - T_1 = 0 \quad \text{or} \quad 0.643 T_2 = T_1$$

$$\sum F_y = 0 \quad \text{yields} \quad T_2 \sin 50° - 600 \text{ N} = 0 \quad \text{or} \quad 0.766 T_2 = 600 \text{ N}$$

This gives $T_2 = \underline{783 \text{ N}}$. Substituting into the $\sum F_x$ equation yields $T_1 = \underline{503 \text{ N}}$.

2.4 The following coplanar forces pull on a ring: 200 N at 30°, 500 N at 80°, 300 N at 240°, and an unknown force. Find the magnitude and direction of the unknown force if the ring is to be in equilibrium.

▮ Assume that the 0° line is the x axis and 90° specifies the y axis. The three known forces are then as shown in Fig. 2-4. If $\mathbf{F}_4$ is the unknown force, then $\mathbf{F}_1 + \mathbf{F}_2 + \mathbf{F}_3 + \mathbf{F}_4 = 0$. Let $\mathbf{R} = \mathbf{F}_1 + \mathbf{F}_2 + \mathbf{F}_3$, then $\mathbf{R} + \mathbf{F}_4 = 0 \Rightarrow \mathbf{F}_4 = -\mathbf{R}$. To find $\mathbf{F}_4$, we need only find $\mathbf{R}$.

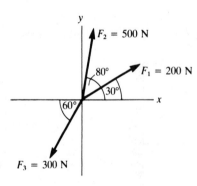

Fig. 2-4

$R_x = F_{1x} + F_{2x} + F_{3x} = F_1 \cos 30° + F_2 \cos 80° - F_3 \cos 60°$, $R_x = 200(0.866) + 500(0.174) - 300(0.500) = 110 \text{ N}$
$R_y = F_1 \sin 30° + F_2 \sin 80° - F_3 \sin 60°$, $R_y = 200(0.500) + 500(0.985) - 300(0.866) = 333 \text{ N}$, $R = (R_x^2 + R_y^2)^{1/2} = \underline{351 \text{ N}}$, $\tan \theta_R = \dfrac{R_y}{R_x} \Rightarrow \theta_R = 71.7°$, $F_4 = R$ and $\theta_{F4} = \theta_R + 180° = \underline{252°}$

2.5 In Fig. 2-5(a) the value of W is 180 N. Find the tensions in ropes A and B.

▮ Refer to Fig. 2-5(b). Summing forces in the x and y directions to zero yields $A = B \cos 53°$ and $B \sin 53° = W = 180$. From the latter, $B = \underline{225 \text{ N}}$, which when inserted into the former gives $A = \underline{135 \text{ N}}$.

2.6 If the identical ropes A and B in Fig. 2-5(a) can each support tensions no larger than 200 N, what is the maximum value that W can have? What is the tension in the other rope when W has this maximum value?

▮ From Prob. 2.5, B will experience the largest tension, 200 N in this case. Solve for forces along the vertical: $W = 200 \sin 53° = \underline{160 \text{ N}}$, and in the horizontal direction find $A = 200 \cos 53° = \underline{120 \text{ N}}$.

2.7 A rope extends between two poles. A 90-N boy hangs from it, as shown in Fig. 2-6(a). Find the tensions in the two parts of the rope.

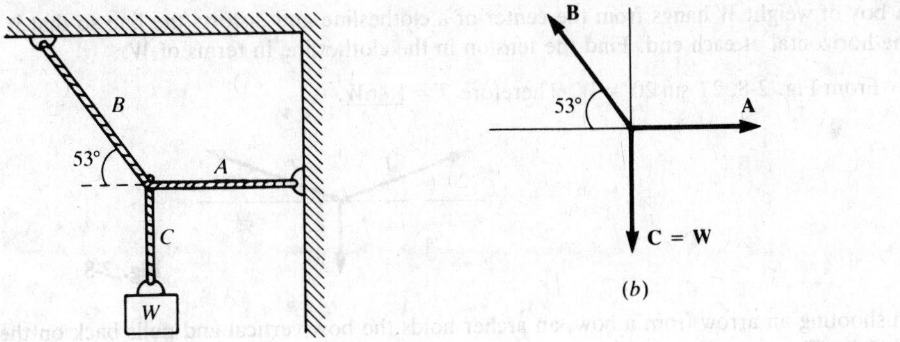

(a)

(b)

Fig. 2-5

(a)

(b)

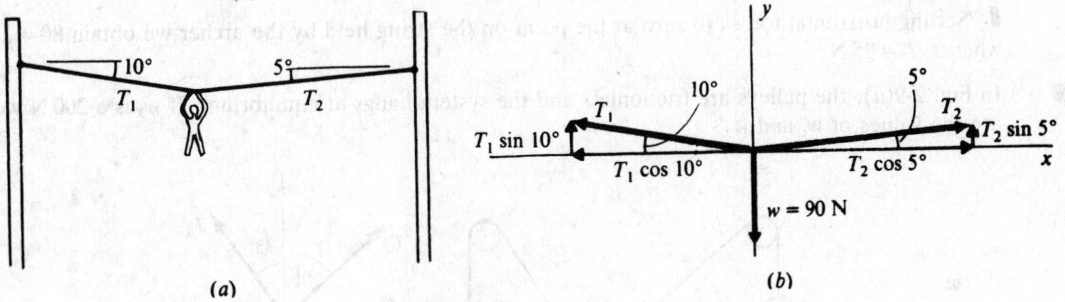

Fig. 2-6

❚ Label the two tensions T_1 and T_2, and isolate the rope at the boy's hands as the object. The free-body diagram for this object is shown in Fig. 2-6(b).

$$\sum F_x = 0 \quad \text{becomes} \quad T_2 \cos 5° - T_1 \cos 10° = 0$$

$$\sum F_y = 0 \quad \text{becomes} \quad T_2 \sin 5° + T_1 \sin 10° - 90 \text{ N} = 0$$

Evaluating the sines and cosines, these equations become

$$0.996T_2 - 0.985T_1 = 0 \quad \text{and} \quad 0.087T_2 + 0.174T_1 - 90 = 0$$

Solving the first for T_2 gives $T_2 = 0.989T_1$. Substituting this in the second equation gives

$$0.086T_1 + 0.174T_1 - 90 = 0$$

from which $T_1 = \underline{346 \text{ N}}$. Then, because $T_2 = 0.989T_1$, we have $T_2 = \underline{342 \text{ N}}$.

2.8 The tension in cord A in Fig. 2-7 is 30 N. Find the tension in B and the value of W.

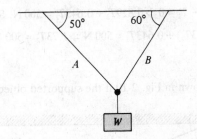

Fig. 2-7

❚ Draw a free-body diagram for the point on the rope where the cords meet; the equilibrium relations in the x and y directions are $T_A \cos 50° = T_B \cos 60°$ and $W = T_A \sin 50° + T_B \sin 60°$, where $T_A = 30$ N. Solving: $T_B = \underline{39 \text{ N}}$ and $W = \underline{56 \text{ N}}$.

2.9 In Fig. 2-7, how large are T_A and T_B if $W = 80$ N?

❚ The equilibrium equations have already been obtained in Prob. 2.8. The y-equilibrium equation when $W = 80$ N is $80 = T_A[\sin 50° + (\tan 60°)(\cos 50°)]$, where we have substituted for T_B from the horizontal equation. Thus the tensions are $T_A = \underline{43 \text{ N}}$ and $T_B = \underline{55 \text{ N}}$.

2.10 A boy of weight W hangs from the center of a clothesline and distorts the line so that it makes 20° angles with the horizontal at each end. Find the tension in the clothesline in terms of W.

▮ From Fig. 2-8, $2T \sin 20° = W$. Therefore $T = \underline{1.46W}$.

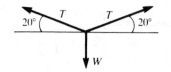

Fig. 2-8

2.11 In shooting an arrow from a bow, an archer holds the bow vertical and pulls back on the arrow with a force of 80 N. The two halves of the string make 25° angles with the vertical. What is the tension in the string?

▮ Setting horizontal forces to zero at the point on the string held by the archer we obtain $80 = 2T \sin 25°$, whence $T = \underline{95\,N}$.

2.12 In Fig. 2-9(a), the pulleys are frictionless and the system hangs at equilibrium. If w_3 is a 200 N weight, what are the values of w_1 and w_2?

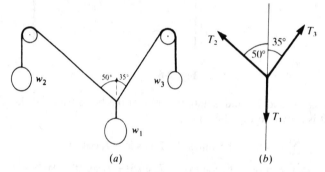

(a)   (b)   **Fig. 2-9**

▮ The knot above w_1 is in equilibrium under the action of three forces, as shown in Fig. 2-9(b). Since the pulleys are frictionless, $T_2 = w_2$; $T_3 = w_3$. Also $T_1 = w_1$. We are given $T_3 = w_3 = 200$ N. From $\Sigma F_x = 0$, $T_3 \sin 35° - T_2 \sin 50° = 0$. (Note the x-component equation involves sine functions because angles are with respect to y axis.) Then $200(0.574) = T_2(0.766) \Rightarrow T_2 = \underline{150\,N} = w_2$. From $\Sigma F_y = 0$, $T_3 \cos 35° + T_2 \cos 50° - T_1 = 0$ or $200(0.819) + 150(0.643) = T_1 \Rightarrow T_1 = \underline{260\,N} = w_1$.

2.13 Suppose w_1 in Fig. 2-9(a) weighs 500 N. Find the values of w_2 and w_3 if the system is to hang in equilibrium.

▮ Now $T_1 = 500$ N in Fig. 2-9(b). $\Sigma F_x = 0 \Rightarrow T_3 \sin 35° - T_2 \sin 50° = 0 \Rightarrow 0.574T_3 = 0.766T_2$. Or solving for T_3,

$$T_3 = 1.33T_2 \tag{1}$$

$\Sigma F_y = 0 \Rightarrow T_3 \cos 35° + T_2 \cos 50° - T_1 = 0 \Rightarrow 0.819T_3 + 0.643T_2 = 500$ N. Substituting $1.33T_2$ for T_3 we have

$$0.819(1.33T_2) + 0.643T_2 = 500\,N \Rightarrow 1.73T_2 = 500\,N \qquad T_2 = \underline{289\,N}$$

From (1), we get $T_3 = \underline{384\,N}$.

2.14 Find the tensions in the ropes shown in Fig. 2-10 if the supported object weighs 600 N.

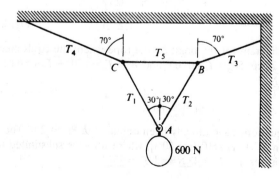

Fig. 2-10

I Let us select as our object the knot at A because we know one force acting on it. The weight pulls down on it with a force of 600 N and so the free-body diagram for the knot is as shown in Fig. 2-11(a). Applying the first condition for equilibrium to that diagram, we have

$$\sum F_x = 0 \quad \text{or} \quad T_2 \cos 60° - T_1 \cos 60° = 0 \quad \sum F_y = 0 \quad \text{or} \quad T_1 \sin 60° + T_2 \sin 60° - 600 = 0$$

The first equation yields $T_1 = T_2$. Substitution of T_1 for T_2 in the second equation gives $T_1 = \underline{346\,\text{N}}$, and this is also T_2.

Let us now isolate knot B as our object. Its free-body diagram is shown in Fig 2-11(b). We have already found that $T_2 = 346\,\text{N}$ and so the equilibrium equations are

$$\sum F_x = 0 \quad \text{or} \quad T_3 \cos 20° - T_5 - 346 \sin 30° = 0$$

$$\sum F_y = 0 \quad \text{or} \quad T_3 \sin 20° - 346 \cos 30° = 0$$

The last equation yields $T_3 = \underline{877\,\text{N}}$. Substituting this in the prior equation gives $T_5 = \underline{651\,\text{N}}$.

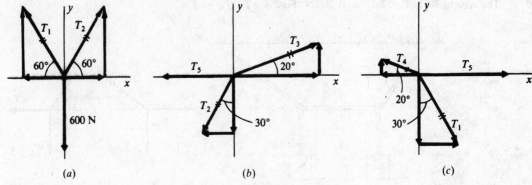

Fig. 2-11

We can now proceed to the knot at C and the free-body diagram of Fig 2-11(c). Recalling that $T_1 = 346\,\text{N}$,

$$\sum F_x = 0 \quad \text{becomes} \quad T_5 + 346 \sin 30° - T_4 \cos 20° = 0$$

$$\sum F_y = 0 \quad \text{becomes} \quad T_4 \sin 20° - 346 \cos 30° = 0$$

The latter equation yields $T_4 = 877\,\text{N}$.

[Note that from the symmetry of the system we could have deduced $T_1 = T_2$ and $T_4 = T_3$.]

2.15 If $w = 40\,\text{N}$ in the equilibrium situation shown in Fig. 2-12(a) find T_1 and T_2.

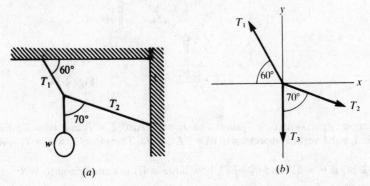

Fig. 2-12

I The knot is in equilibrium under the action of three forces, and the free-body diagram is as shown in Fig. 2-12(b). $T_3 = w = 40\,\text{N}$.

$$\sum F_x = 0 \Rightarrow T_2 \sin 70° - T_1 \cos 60° = 0 \quad \text{or} \quad (0.940)T_2 = (0.500)T_1, \quad T_1 = 1.88T_2$$

$$\sum F_y = 0 \Rightarrow T_1 \sin 60° - T_2 \cos 70° - T_3 = 0, \quad (0.866)T_1 - (0.342)T_2 = T_3 = 40\,\text{N}$$

Substituting for T_1,

$$(0.866)(1.88T_2) - (0.342)T_2 = 40\,\text{N} \qquad 1.29T_2 = 40\,\text{N} \qquad T_2 = \underline{31.0\,\text{N}} \qquad \text{and} \qquad T_1 = (1.88)(31.0) = \underline{58.3\,\text{N}}$$

2.16   Refer to Fig. 2-12(a). The cords are strong enough to withstand a maximum tension of 80 N. What is the largest value of w that they can support as shown?

▌ From Prob. 2.15 the equilibrium equations are, for any w_1,

$$T_1 = 1.88T_2 \tag{1}$$

$$0.866T_1 - 0.342T_2 = w \tag{2}$$

From Eq. (1) it is clear that $T_1 > T_2$ always. Therefore T_1 will reach the breaking point first. We thus set $T_1 = 80\,\text{N}$ to find the corresponding w. From (1),

$$1.88T_2 = 80\,\text{N} \Rightarrow T_2 = \underline{42.6}$$

From (2), $\qquad\qquad w = (0.866)(80\,\text{N}) - (0.342)(42.6\,\text{N}) = \underline{54.7\,\text{N}}$

2.17   The weight W_1 in Fig. 2-13(a) is 300 N. Find T_1, T_2, T_3, and W_2.

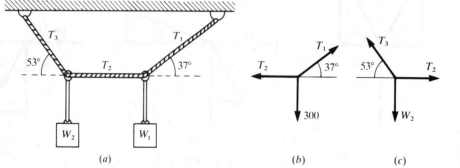

(a)   (b)   (c)   **Fig. 2-13**

▌ From Fig. 2-13(b): $T_1 \sin 37° = 300$ so $T_1 = \underline{500\,\text{N}}$. Also, $T_2 = T_1 \cos 37° = \underline{400\,\text{N}}$. From Fig. 2-13(c), $T_3 \cos 53° = T_2$, so $T_3 = \underline{670\,\text{N}}$. But $T_3 \sin 53° = W_2$, so $W_2 = \underline{530\,\text{N}}$. (Note answers are to two-place accuracy).

2.18   If $\theta_1 = \theta_2$ in Fig. 2-14, what can be said about T_1, T_2, T_3, W_1, and W_2 provided the pulley is frictionless?

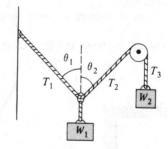

Fig. 2-14

▌ Where the three ropes join, $T_1 \sin \theta_1 = T_2 \sin \theta_2$, so $T_1 = T_2$. Also $T_2 = T_3$ and $W_2 = T_3$. Further, the equilibrium condition for the vertical direction is $W_1 = 2T_1 \cos \theta_1$. Therefore $T_1 = T_2 = T_3 = W_2 = W_1/(2 \cos \theta_1)$.

2.19   In reference to Fig. 2-14, if $\theta_1 = 53°$ and $\theta_2 = 37°$, how large is W_1 in comparison to W_2?

▌ Since $T_2 = T_3 = W_2$, the equilibrium equations are $W_2 \sin 37° = T_1 \sin 53°$ and $T_1 \cos 53° + W_2 \cos 37° = W_1$. Solving for T_1 in the first equation and placing it into the second yields $\underline{W_1 = 1.25W_2}$.

2.20   Suppose that $W_1 = W_2$ in Fig. 2-14 and that $\theta_1 = 53°$. Find θ_2.

▌ The equilibrium equations are $W_2 \sin \theta_2 = T_1 \sin 53°$ and $T_1 \cos 53° + W_2 \cos \theta_2 = W_2$. Eliminating T_1 in the second equation by use of the first yields

$$\sin \theta_2 \cos 53° + \cos \theta_2 \sin 53° = \sin 53° \qquad \text{or} \qquad \sin (\theta_2 + 53°) = \sin 53°$$

Thus, either $\theta_2 = 0$ [the two weights are on the same vertical line and $T_1 = 0$] or else $(\theta_2 + 53°) + 53° = 180° \Rightarrow \theta_2 = 74°$ [the "normal" answer].

2.21 Refer to Fig. 2-15(a). What is the tension in cord $\overline{AB}$?

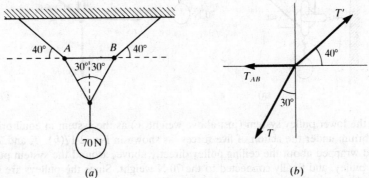

Fig. 2-15

(a)   (b)

▌ First find the tension in the cords below the cord in question by balancing vertical forces at the lower junction in Fig. 2-15(a): $2T \cos 30° = 70$, so $T = 40.4$ N. Equilibrium conditions for junction B [Fig. 2-15(b)] are $T' \cos 40° = T_{AB} + T \sin 30°$; $T' \sin 40° = T \cos 30°$. Substituting for T and eliminating T' from the two relations yields $T_{AB} = \underline{21.5\ N}$. From symmetry the same equations are found at junction A.

2.22 The weight w in Fig. 2-16(a) is 80 N and is in equilibrium. Find T_1, T_2, T_3, and T_4.

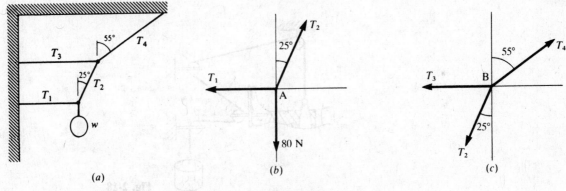

(a)   (b)   (c)

Fig. 2-16

▌ Labeling the lower knot A and the upper knot B, we have the free-body diagrams of Fig. 2-16(b) and (c). We address the equilibrium of knot A first because it involves the only known force.

$$\sum F_x = 0 \Rightarrow T_2 \sin 25° - T_1 = 0 \qquad T_1 = 0.423 T_2$$

$$\sum F_y = 0 \Rightarrow T_2 \cos 25° - 80\ N = 0 \qquad 0.906 T_2 = 80\ N \qquad T_2 = \underline{88.3\ N}$$

Then from before, $T_1 = \underline{37.4\ N}$.

Turning now to knot B, and remembering that we know T_2,

$$\sum F_x = 0 \Rightarrow T_4 \sin 55° - T_3 - T_2 \sin 25° = 0$$

or

$$0.819 T_4 - T_3 = 0.423 T_2 = 37.4\ N \qquad\qquad (1)$$

$$\sum F_y = 0 \Rightarrow T_4 \cos 55° - T_2 \cos 25° = 0 \Rightarrow 0.573 T_4 = 80\ N \Rightarrow T_4 = \underline{140\ N}$$

From (1), $T_3 = 0.819 T_4 - 37.4 = \underline{77\ N}$.

2.23 The pulleys shown in Fig. 2-17(a) have negligible weight and friction. What is the value of w if it remains supported as shown by the 70-N weight?

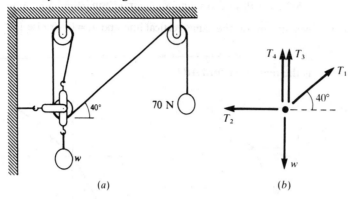

(a)   (b)   **Fig. 2-17**

▮ Consider the lower pulley system (just above weight w) as the system in equilibrium. Since it is weightless, it is in equilibrium under the action of five forces, as shown in Fig. 2-17(b). T_3 and T_4 and T_1 are due to the common cord wrapped about the ceiling pulley directly above, around the system pulley, and around the other ceiling pulley, and finally connected to the 70-N weight. Since the pulleys are frictionless, $T_1 = T_3 = T_4 = 70$ N. For the body in equilibrium,

$$\sum F_x = 0 \Rightarrow T_1 \cos 40° - T_2 = 0 \Rightarrow T_2 = (0.766)(70 \text{ N}) = \underline{53.6 \text{ N}}$$

$$\sum F_y = 0 \Rightarrow T_3 + T_4 + T_1 \sin 40° - w = 0 \Rightarrow w = 70 \text{ N} + 70 \text{ N} + (70 \text{ N})(0.642) = \underline{185 \text{ N}}$$

2.24 How large is the force that stretches the patient's leg in Fig. 2-18? How large an upward force does the device exert on foot and leg together? Assume frictionless and massless pulleys.

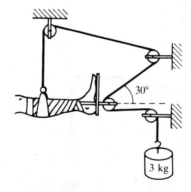

Fig. 2-18

▮ 3 kg weighs about 30 N. Since the pulleys are frictionless and with negligible mass, the tension T in the cord is the same everywhere. T holds up the weight, so $T = 30$ N. The forces on the leg and foot from the device are caused by the tensions in the cord. The horizontal or stretching force is $T + T \cos 30° = \underline{56 \text{ N}}$, while the upward force is $T + T \sin 30° = \underline{45 \text{ N}}$.

2.25 For the situation shown in Fig. 2-19, with what force must the 600-N man pull downward on the rope to support himself free from the floor? Assume the pulleys have negligible friction and weight.

▮ Call T the tension in the rope the man is holding; T is the same throughout the one piece of rope. The other vertical force on the man is the tension in the rope attached to the pulley above the man's head, which must be $2T$ for the pulley in equilibrium. The net vertical force is $3T$, which is balanced by his weight of 600 N. Therefore the man exerts a downward pull of $\underline{200 \text{ N}}$.

2.26 In the setup of Fig. 2-20, the mobile pulley and the fixed pulley, both frictionless, are associated with equal weights w. Find the angle θ.

▮ Since the tension in the cord is w, the condition for vertical equilibrium of the mobile pulley is $2w \sin \theta = w$, or $\sin \theta = \frac{1}{2}$, or $\theta = 30°$.

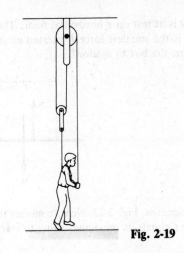

Fig. 2-19

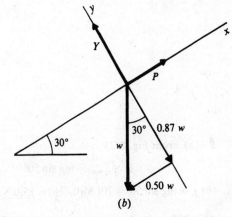

Fig. 2-20

2.2 FRICTION AND INCLINED PLANES

2.27 A 200-N wagon is to be pulled up a 30° incline at constant speed. How large a force parallel to the incline is needed if friction effects are negligible?

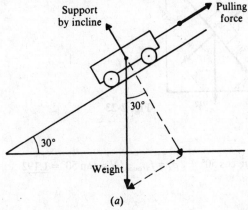

Fig. 2-21

▌ The situation is shown in Fig. 2-21(a). Because the wagon moves at constant speed along a straight line, its velocity vector is constant. Therefore the wagon is in translational equilibrium, and the first condition for equilibrium applies to it.

We isolate the wagon as the object. Three nonnegligible forces act on it: (1) the pull of gravity w (its weight), directed straight down; (2) the force P exerted on the wagon parallel to the incline to pull it up the incline; (3) the push Y of the incline that supports the wagon. These three forces are shown in the free-body diagram, Fig. 2-21(b).

For situations involving inclines, it is convenient to take the x axis parallel to the incline and the y axis perpendicular to it. After taking components along these axes, we can write the first condition for equilibrium.

$$\sum F_x = 0 \quad \text{becomes} \quad P - 0.50w = 0 \qquad \sum F_y = 0 \quad \text{becomes} \quad Y - 0.87w = 0$$

Solving the first equation and recalling that $w = 200$ N, we find that $P = 0.50w = \underline{100\,\text{N}}$. The required pulling force is 100 N.

2.28 A box weighing 100 N is at rest on a horizontal floor. The coefficient of static friction between the box and the floor is 0.4. What is the smallest force F exerted eastward and upward at an angle of 30° with the horizontal that can start the box in motion?

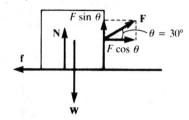

Fig. 2-22

▌ First draw a force diagram, Fig. 2-22. Next, consider the forces in the x direction and apply the conditions for equilibrium, noting f equals its maximum value to start motion.

$$\sum F_x = 0 \qquad F \cos \theta - f = 0 \qquad F \cos \theta = f \qquad 0.866F = f = \mu_s N = 0.4N$$

Now apply the conditions for equilibrium to the forces in y direction.

$$\sum F_y = 0 \qquad N + F \sin \theta - W = 0 \qquad N + 0.5F - 100 = 0 \qquad N = 100 - 0.5F$$

Substituting this equation for N in $0.866F = 0.4N$ above,

$$0.866F = 0.4(100 - 0.5F) \qquad 0.866F + 0.2F = 40 \qquad F = \underline{37.5\,\text{N}}$$

2.29 A block on an inclined plane just begins to slip if the inclination of the plane is 50°. (*a*) What is the coefficient of static friction? (*b*) If the block has a mass of 2 kg, what is the actual frictional force just before it begins to slip?

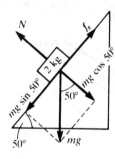

Fig. 2-23

▌ (*a*) From Fig. 2-23:

$$f_{s(\text{max})} = mg \sin 50° \qquad N = mg \cos 50° \qquad \mu_s = f_{s(\text{max})}/N = \tan 50° = \underline{1.192}$$

(*b*) $f_s = mg \sin 50° = 2(9.8)(0.766) = \underline{15.0\,\text{N}}$

2.30 A 50-N box is slid straight across the floor at constant speed by a force of 25 N, as shown in Fig. 2-24(*a*). How large a friction force impedes the motion of the box? How large is the normal force?

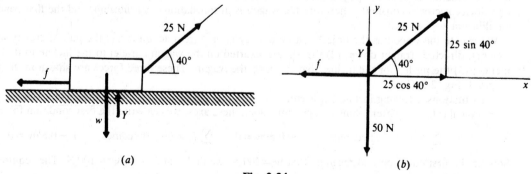

(*a*)   (*b*)

Fig. 2-24

▮ Note the forces acting on the box as shown in Fig. 2-24(a). The friction is f and the normal force, the supporting force exerted by the floor, is Y. The free-body diagram and components are shown in Fig. 2-24(b). Because the box is moving with constant velocity, it is in equilibrium. The first condition for equilibrium tells us that

$$\sum F_x = 0 \quad \text{or} \quad 25\cos 40° - f = 0$$

We can solve for f at once to find that $f = 19$ N. The friction force is <u>19 N</u>.
    To find Y we use the fact that

$$\sum F_y = 0 \quad \text{or} \quad Y + 25\sin 40° - 50 = 0$$

Solving gives the mormal force as <u>$Y = 34$ N</u>.

2.31 Each of the objects in Fig. 2-25 is in equilibrium. Find the normal force, Y, in each case.

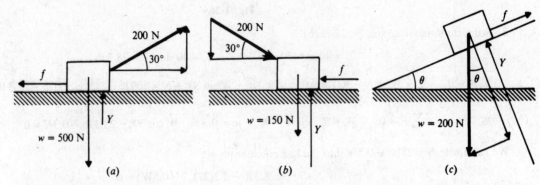

(a) (b) (c)

Fig. 2-25

▮ We apply $\sum F_y = 0$ in each case.

(a) $Y + 200\sin 30° - 500 = 0$ from which $Y = \underline{400\ N}$
(b) $Y - 200\sin 30° - 150 = 0$ from which $Y = \underline{250\ N}$
(c) $Y - 200\cos\theta = 0$ from which $Y = \underline{(200\cos\theta)\ N}$

2.32 For the situations of Prob. 2.31, find the coefficient of kinetic friction if the object is moving with constant speed.

▮ We have already found Y for each case in Prob. 2.31. To find f, the sliding friction force, we use $\sum F_x = 0$.

(a) $\qquad\qquad 200\cos 30° - f = 0 \quad$ and so $\quad f = 173$ N

Then, $\mu_k = f/Y = 173/400 = \underline{0.43}$.

(b) $\qquad\qquad 200\cos 30° - f = 0 \quad$ and so $\quad f = 173$ N

Then, $\mu_k = f/Y = 173/250 = \underline{0.69}$.

(c) $\qquad\qquad -200\sin\theta + f = 0 \quad$ and so $\quad f = (200\sin\theta)$ N

Then, $\mu_k = f/Y = (200\sin\theta)/(200\cos\theta) = \underline{\tan\theta}$.

2.33 Suppose that in Fig. 2-25(c) the block is at rest. The angle of the incline is slowly increased. At an angle $\theta = 42°$, the block begins to slide. What is the coefficient of static friction between the block and the incline? (The block and surface are not the same as in Probs. 2.31 and 2.32.)

▮ At the instant the block begins to slide, the friction force has its critical value. Therefore, $\mu_s = f/Y$ at that instant. Following the method of Probs. 2.31 and 2.32, we have

$$Y = w\cos\theta \quad \text{and} \quad f = w\sin\theta$$

Therefore, when sliding just starts,

$$\mu_s = \frac{f}{Y} = \frac{w\sin\theta}{w\cos\theta} = \tan\theta$$

But θ was found by experiment to be 42°. Therefore, $\mu_s = \tan 42° = \underline{0.90}$.

2.34 Two weights are hung over two frictionless pulleys, as shown in Fig. 2-26(a). What weight W will cause the 300-lb block to just start moving to the right?

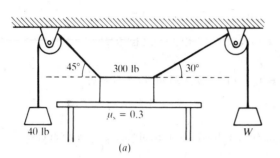
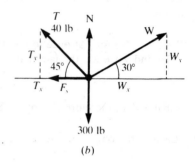

(a)   (b)

Fig. 2-26

▌ From the force diagram, Fig. 2-26(b),

$$T_x = 40 \cos 45° = 28.3 \text{ lb} \qquad T_y = 40 \sin 45° = 28.3 \text{ lb}$$

$$\sum F_y = 0 \qquad T_y + W_y + N - 300 \text{ lb} = 0 \qquad 28.3 \text{ lb} + W \sin 30° + N = 300 \text{ lb} \qquad N = 300 \text{ lb} - 28.3 \text{ lb} - 0.5W$$

$$\sum F_x = 0 \qquad W_x - T_x - \mu_s N = 0 \qquad \mu_s = 0.3 \qquad W \cos 30° - 28.3 \text{ lb} - 0.3N = 0$$

We substitute $N = 271.7 - 0.5W$ into the last equation to get

$$W \cos 30° - 28.3 \text{ lb} - 0.3(271.7 - 0.5W) = 0.$$

Solving for W gives $\underline{W = 108 \text{ lb}}$.

2.35 Assume that $W = 60$ lb, $\theta = 43°$, and $\mu_k = 0.3$ in Fig. 2-27(a). What push directed up the plane will move the block with constant speed (a) up the plane and (b) down the plane?

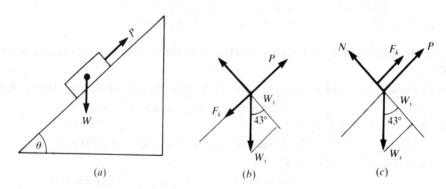

(a)   (b)   (c)   **Fig. 2-27**

▌ The weight components are:

$$W_x = (60 \text{ lb}) \sin 43° = \underline{40.92 \text{ lb}} \qquad W_y = (60 \text{ lb}) \cos 43° = \underline{43.88 \text{ lb}}$$

(a) Using Fig. 2-27(b),

$$\sum F_y = 0 \qquad N - W_y = 0 \qquad \underline{N = 43.88 \text{ lb}}$$

$$F_k = \mu_k N = 0.3(43.88) \qquad \underline{F_k = 13.16 \text{ lb}}$$

$$\sum F_x = 0 \qquad P - F_k - W_x = 0 \qquad P = 13.16 + 40.92 \qquad \underline{P = 54.1 \text{ lb}}$$

(b) The push is now just enough to keep motion down plane to constant speed. From Fig. 2-27(c),

$$\sum F_x = 0 \qquad P + F_k - W_x = 0 \qquad P = 40.92 \text{ lb} - 13.16 \text{ lb} \qquad \underline{P = 27.8 \text{ lb}}$$

2.36 What horizontal push P is required to just hold a 200-N block on a 60° inclined plane if $\mu_s = 0.4$?

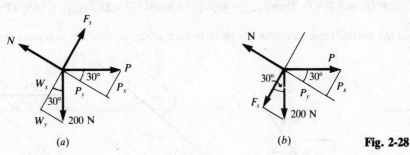

(a)   (b)   **Fig. 2-28**

❚ Figure 2-28(a) is the free-body diagram.

$$W_x = 200 \cos 30° = \underline{173.2 \text{ N}} \qquad W_y = 200 \sin 30° = \underline{100 \text{ N}} \qquad F_s = \mu_s N$$

$$\sum F_y = 0 \qquad N - P_y - W_y = 0 \qquad P_y = P \cos 30° \qquad N = P \cos 30° + W_y = 0.866P + 100 \text{ N}$$

$$\sum F_x = 0 \qquad P_x + \mu_s N - W_x = 0 \qquad \mu_s = 0.4$$

$$P \sin 30° + 0.4(0.866P + 100 \text{ N}) - 173.2 \text{ N} = 0 \qquad 0.5P + 0.346P + 40 \text{ N} - 173.2 \text{ N} = 0 \qquad \underline{P = 157 \text{ N}}$$

2.37 In Prob. 2.36, what horizontal push would just start the block moving up the plane?

❚ Now Fig. 2-28(b) applies.

$$\sum F_y = 0 \qquad N = 100 \text{ N} + 0.866P \qquad \sum F_x = 0 \qquad P \sin 30° - W_x - \mu_s N = 0$$

$$0.5P - 173.2 \text{ N} - (0.4)(0.866P + 100 \text{ N}) = 0 \qquad 0.5P - 173.2 \text{ N} - 0.346P - 40 \text{ lb} = 0 \qquad \underline{P = 1384 \text{ N}}$$

2.38 In Fig. 2-29(a), the system is in equilibrium. (a) What is the maximum value that w can have if the friction force on the 40-N block cannot exceed 12.0 N? (b) What is the coefficient of static friction between the block and tabletop?

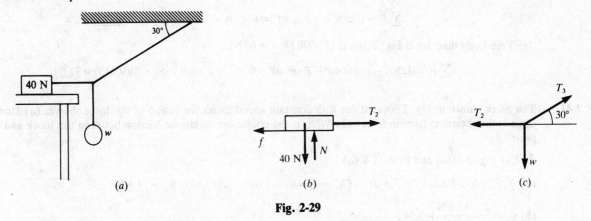

(a)   (b)   (c)

Fig. 2-29

❚ (a) The free-body diagrams for the block and the knot are shown in Fig. 2-29(b) and (c).

$$f_{s(\text{max})} = 12.0 \text{ N} \qquad \text{implies} \qquad T_{2(\text{max})} = 12.0 \text{ N}$$

For the knot:

$$T_2 = T_3 \cos 30° \qquad w = T_3 \sin 30°$$

Eliminating T_3, we get $w = T_2 \tan 30° = 0.577 T_2$. Thus $w_{(\text{max})} = 0.577 T_{2(\text{max})} = 6.92 \text{ N}$.
(b) $\mu_s = f_{s(\text{max})}/N = 12/40 = \underline{0.30}$.

2.39 The block of Fig. 2-29(a) is just on the verge of slipping. If $w = 8.0 \text{ N}$, what is the coefficient of static friction between the block and tabletop?

❚ At the verge of slipping, $f_s = f_{s(max)}$ and the corresponding hanging weight is $w = 8.0$ N. As in Prob. 2.38, $w = 0.577 T_2$ and $T_2 = f_s$. Thus $f_{s(max)} = w/0.577 = 8.0/0.577 = 13.9$ N; $\mu_s = 13.9/40 = \underline{0.347}$.

2.40 Find the normal force acting on the block in each of the equilibrium situations shown in Fig. 2-30.

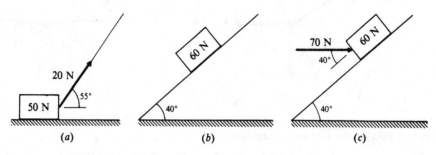

(a)   (b)   (c)   **Fig. 2-30**

❚ In each case friction is necessary for equilibrium. But friction does not enter in solving for N.

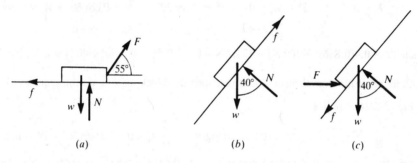

(a)   (b)   (c)   **Fig. 2-31**

(**a**) Free-body diagram is Fig. 2-31(a) ($F = 20$ N, $w = 50$ N).

$$\sum F_y = 0 \Rightarrow N - w + F \sin 55° = 0 \qquad N = 50\,\text{N} - (20\,\text{N})(0.819) = \underline{33.6\,\text{N}}$$

(**b**) Free-body diagram is Fig. 2-31(b) ($w = 60$ N).

$$\sum F_y = 0 \Rightarrow N - w \cos 40° = 0 \qquad N = (60\,\text{N})(0.766) = \underline{46.0\,\text{N}}$$

(**c**) Free-body diagram is Fig. 2-31(c): ($F = 70$ N, $w = 60$ N).

$$\sum F_y = 0 \Rightarrow N - w \cos 40° - F \sin 40° = 0 \qquad N = 60(0.766) + 70(0.643) = \underline{91.0\,\text{N}}.$$

2.41 The block shown in Fig. 2-30(a) slides with constant speed under the action of the force shown. (**a**) How large is the retarding friction force? (**b**) What is the coefficient of kinetic friction between the block and the floor?

❚ Use Fig. 2-31(a) and Prob. 2.40(a).

(**a**) $\sum F_x = 0 \Rightarrow F \cos 55° - f = 0 \qquad (N_x = w_x = 0) \qquad$ or $\qquad 20(0.573) = f = \underline{11.5\,\text{N}}$

(**b**) $\mu_k = \dfrac{f}{N} = \dfrac{11.5\,\text{N}}{34\,\text{N}} = \underline{0.34}$

2.42 The block shown in Fig. 2-30(b) slides at constant speed down the incline. (**a**) How large is the friction force that opposes its motion? (**b**) What is the coefficient of sliding (kinetic) friction between the block and plane?

❚ Use Fig. 2-31(b) and Prob. 2.40(b).

(**a**) $\sum F_x = 0 \Rightarrow f - w \sin 40° = 0 \qquad f = 60\,\text{N}(0.643) = \underline{38.6\,\text{N}}$

(**b**) $\mu_k = \dfrac{f}{N} = \dfrac{38.6\,\text{N}}{46\,\text{N}} = \underline{0.84}$

2.43 The block in Fig. 2-30(c) just begins to slide up the incline when the pushing force shown is increased to

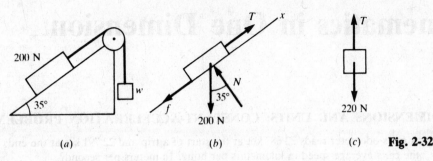

(a)   (b)   (c)   **Fig. 2-32**

70 N. **(a)** What is the critical static friction force on it? **(b)** What is the value of the coefficient of static friction?

❚ Use Fig. 2-31(c) and Prob. 2.40(c).

(a) $\sum F_x = 0 \Rightarrow F \cos 40° - w \sin 40° - f = 0$, $F = 70$ N and block just starts to move $\Rightarrow f = f_{s(max)} = 70(0.766) -$ 60(0.643) = $\underline{15.0 \text{ N}}$

(b) $\mu_s = \dfrac{f_{s(max)}}{N} = \dfrac{15.0 \text{ N}}{91 \text{ N}} = \underline{0.17}$

2.44   The system in Fig. 2-32(a) remains at rest when the hanging weight w is 220 N. What are the magnitude and direction of the friction force on the 200 N block? The pulley is frictionless.

❚ Since the pulley is frictionless, the tension is the same throughout the cord. The free-body diagrams for the two blocks are as in Fig. 2-32(b) and (c). In principle the frictional force f could be either down the incline or up the incline depending on the details of the problem. In this case we can quickly assert that it is down the incline with the following reasoning: $T = 220$ N. Opposing T along the incline is the component of the 200-N weight down the incline. This is surely less than the full 200 N and thus is insufficient to balance T. Therefore the help of friction down the incline is necessary. To obtain f we solve

$$\sum F_x = 0 \quad \text{(along incline)} \quad T - 200 \sin 35° - f = 0 \Rightarrow f = 220 - 200(0.574) = \underline{105 \text{ N}}$$

(Note that for this problem the normal force, N, does not enter the calculation.)

CHAPTER 3
Kinematics in One Dimension

3.1 DIMENSIONS AND UNITS; CONSTANT-ACCELERATION PROBLEMS

3.1 A car's odometer reads 22 687 km at the start of a trip and 22 791 km at the end. The trip took 4 h. What was the car's average speed in kilometers per hour? In meters per second?

▮ Average speed $= \dfrac{\text{distance traveled}}{\text{time taken}} = \dfrac{(22\,791 - 22\,687) \text{ km}}{4 \text{ h}} = \underline{26 \text{ km/h}}$

To convert; Average speed $= 26 \dfrac{\text{km} \times (1000 \text{ m/km})}{\text{h} \times (3600 \text{ s/h})} = \underline{7.2 \text{ m/s}}$

3.2 An auto travels at a rate of 25 km/h for 4 min, then at 50 km/h for 8 min, and finally at 20 km/h for 2 min. Find (a) the total distance covered in kilometers and (b) the average speed for the complete trip in meters per second.

▮ (a) Distance traveled $= d_1 + d_2 + d_3$, where

$$d_1 = \left(25 \dfrac{\text{km}}{\text{h}}\right)(4 \text{ min})\left(\dfrac{1 \text{ h}}{60 \text{ min}}\right) = 1\tfrac{2}{3} \text{ km} \qquad d_2 = \left(50 \dfrac{\text{km}}{\text{h}}\right)(8 \text{ min})\left(\dfrac{1 \text{ h}}{60 \text{ min}}\right) = 6\tfrac{2}{3} \text{ km}$$

$$d_3 = \left(20 \dfrac{\text{km}}{\text{h}}\right)(2 \text{ min})\left(\dfrac{1 \text{ h}}{60 \text{ min}}\right) = \tfrac{2}{3} \text{ km}$$

Thus $d_1 + d_2 + d_3 = \underline{9 \text{ km}}$.

(b) Average speed $= \dfrac{d_1 + d_2 + d_3}{\text{total time}} = (9 \text{ km} \times 1000 \text{ m/km})/(14 \text{ min} \times 60 \text{ s/min}) = \underline{10.7 \text{ m/s}}$

3.3 A runner travels 1.5 laps around a circular track in a time of 50 s. The diameter of the track is 40 m and its circumference is 126 m. Find (a) the average speed of the runner and (b) the magnitude of the runner's average velocity.

▮ (a) Average speed $= \dfrac{\text{distance}}{\text{time}} = \left(\dfrac{1.5 \text{ laps}}{50 \text{ s}}\right)(126 \text{ m/lap}) = \underline{3.78 \text{ m/s}}$.

(b) Average velocity is a vector. It is the displacement vector for the time lapse of interest divided by the time lapse—in this case, 50 s. Since $1\tfrac{1}{2}$ laps have taken place, the displacement points from the starting point on the track to a point on the track $\tfrac{1}{2}$ lap away, which is of course directly across a diameter of the track. The magnitude of the average velocity is therefore the magnitude of the displacement divided by the time lapse. Thus magnitude of average velocity $= 40 \text{ m}/50 \text{ s} = \underline{0.80 \text{ m/s}}$.

3.4 Use dimensional analysis to determine which of the following equations is certainly wrong:

$$\lambda = vt \qquad F = \dfrac{m}{a} \qquad F = \dfrac{mv}{t} \qquad h = \dfrac{v^2}{2g} \qquad v = (2gh)^{1/2}$$

where λ and h are lengths and $[F] = [MLT^{-2}]$. The other symbols have their usual meaning.

▮ $[vt] = [LT^{-1}][T] = [L]$, but $[\lambda] = [L]$, so equation $\lambda = vt$ can be correct. $[m/a] = [M]/[T^2 L^{-1}] = [MT^2 L^{-1}]$, but $[F] = [MLT^{-2}]$; hence $F = m/a$ is incorrect. $[mv/t] = [M][LT^{-1}][T^{-1}] = [MLT^{-2}]$ and $F = mv/t$ is dimensionally correct. $[v^2/2g] = [(L^2/T^2)/(L/T^2)] = [L]$; and since $[h] = [L]$, $h = v^2/2g$ is dimensionally correct. Since $[v] = [LT^{-1}]$, $[(2gh)^{1/2}] = [(L^{1/2}T^{-1})L^{1/2}] = [LT^{-1}]$, $v = (2gh)^{1/2}$ is also dimensionally correct. We note that pure numbers are dimensionless.

3.5 If s is distance and t is time, what must be the dimensions of C_1, C_2, C_3, and C_4 in each of the following equations?

$$s = C_1 t \qquad s = \tfrac{1}{2}C_2 t^2 \qquad s = C_3 \sin C_4 t$$

(*Hint:* The argument of any trigonometric function must be dimensionless.)

▮ The dimension of s is $[L]$, so all expressions on the right-hand sides of the equations must also have the dimensions of length. $[C_1] = [L/T]$, since $[C_1 t]$ are then $[(L/T)T] = L$. $[C_2] = [LT^{-2}]$, $[C_3] = [L]$ because the sine is dimensionless, and since the argument of a trigonometric function has no units, $[C_4] = [T^{-1}]$.

3.6 The speed v of a wave on a string depends on the tension F in the string and the mass per unit length m/ℓ of the string. If it is known that $[F] = [ML][T]^{-2}$, find the constants a and b in the following equation for the speed of a wave on a string: $v = (\text{constant})F^a(m/\ell)^b$.

▮ It is given that $[v] = [F]^a[m/\ell]^b$. After inserting into this expression $[v]$ and $[F]$, we have, recalling M^0 means no mass units involved, $[M^0L^1T^{-1}] = [MLT^{-2}]^a[ML^{-1}]^b = [M]^{a+b}[L]^{a-b}[T^{-2}]^a$. A fundamental dimension must appear to the same power on both sides of an equation; hence, $a + b = 0$, $a - b = 1$, and $-2a = -1$. From this $a = \frac{1}{2}$ and $b = -\frac{1}{2}$.

3.7 The frequency of vibration f of a mass m at the end of a spring that has a stiffness constant k is related to m and k by a relation of the form $f = (\text{constant})m^a k^b$. Use dimensional analysis to find a and b. It is known that $[f] = [T]^{-1}$ and $[k] = [M][T]^{-2}$.

▮ As in Prob. 3.6,

$$f \propto m^a k^b \Rightarrow [M^0T^{-1}] = [M^a][M^bT^{-2b}] = [M^{a+b}T^{-2b}]$$

so

$$a + b = 0 \quad \text{and} \quad -2b = -1 \quad \text{or} \quad b = -a = \tfrac{1}{2}$$

3.8 A body with initial velocity 8 m/s moves along a straight line with constant acceleration and travels 640 m in 40 s. For the 40 s interval, find (**a**) the average velocity, (**b**) the final velocity, and (**c**) the acceleration.

▮ Assume $x = 0$ at $t = 0$.

(**a**) Average velocity $= \dfrac{\text{displacement}}{\text{time elapsed}} = \dfrac{640\text{ m}}{40\text{ s}} = 16$ m/s

(**b, c**) $v_f = v_0 + at$. We know $v_0 = 8$ m/s and $t = 40$ s, but we don't know a. However,

$$x = v_0 t + \tfrac{1}{2}at^2 \quad \text{or} \quad 640\text{ m} = (8\text{ m/s})(40\text{ s}) + \tfrac{1}{2}a(40\text{ s})^2$$

and solving we have

$$a = \frac{2(640 - 320)\text{m}}{1600\text{ s}^2} = \underline{0.40\text{ m/s}^2}$$

Substituting into our velocity formula $v_f = 8$ m/s $+ (0.40$ m/s$^2)(40$ s$) = \underline{24\text{ m/s}}$.

3.9 A truck starts from rest and moves with a constant acceleration of 5 m/s^2. Find its speed and the distance traveled after 4 s has elapsed.

▮ $v_f = v_0 + at$, $v_0 = 0$, $a = 5$ m/s^2, $t = 4$ s

$$v_f = 0 + (5\text{ m/s}^2)(4\text{ s}) = \underline{20\text{ m/s}}$$

To get distance, which in this case is the same as the magnitude of the displacement, we have

$$x = v_0 t + \tfrac{1}{2}at^2 = 0 + \tfrac{1}{2}(5\text{ m/s}^2)(4\text{ s})^2 = \underline{40\text{ m}}$$

3.10 A box slides down an incline with uniform acceleration. It starts from rest and attains a speed of 2.7 m/s in 3 s. Find (**a**) the acceleration and (**b**) the distance moved in the first 6 s.

▮ Let x, v represent the displacement and velocity down the incline, respectively. Then, (**a**) $v_f = v_0 + at$ and we are given $v_0 = 0$, $v_f = 2.7$ m/s, $t = 3.0$ s; thus 2.7 m/s $= 0 + a(3.0$ s$)$, or $a = \underline{0.90\text{ m/s}^2}$. Now we are interested in time $t = 6.0$ s. So (**b**) $x = v_0 t + \tfrac{1}{2}at^2 = 0 + \tfrac{1}{2}(0.90\text{ m/s}^2)(6.0\text{ s})^2 = \underline{16.2\text{ m}}$.

3.11 A car starts from rest and coasts down a hill with a constant acceleration. If it goes 90 m in 8 s, find (**a**) the acceleration and (**b**) the velocity after 8 s.

▮ (**a**) $s = \dfrac{1}{2}at^2$ (since $v_0 = 0$) $\quad 90 = \dfrac{1}{2}a(8)^2 \quad a = \dfrac{180}{64} = \underline{2.8\text{ m/s}^2}$

(**b**) $v = at = 2.8 \times 8 = \underline{22.4\text{ m/s}}$

3.12 A car is accelerating uniformly as it passes two checkpoints that are 30 m apart. The time taken between checkpoints is 4.0 s and the car's speed at the first checkpoint is 5.0 m/s. Find the car's acceleration and its speed at the second checkpoint.

▮ Calling the first checkpoint the initial position and the second the final position, we have, $v_0 = 5.0$ m/s, $x = 30$ m, $t = 4.0$ s. To find a we use the displacement equation

$$x = v_0 t + \tfrac{1}{2}at^2 \quad \text{or} \quad 30 \text{ m} = (5.0 \text{ m/s})(4.0 \text{ s}) + \tfrac{1}{2}a(4.0 \text{ s})^2 \quad \text{or} \quad \underline{a = 1.25 \text{ m/s}^2}$$

Then $v_f = v_0 + at$ yields $v_f = 5.0$ m/s $+ (1.25$ m/s$^2)(4.0$ s$) = \underline{10 \text{ m/s}}$.

3.13 An auto's velocity increases uniformly from 6.0 m/s to 20 m/s while covering 70 m. Find the acceleration and the time taken.

▮ We are given $v_f = 20$ m/s and $v_0 = 6.0$ m/s and the associated displacement is 70 m. Since the elapsed time t is not given, we find a using $v_f^2 = v_0^2 + 2ax$, or $(20 \text{ m/s})^2 = (6.0 \text{ m/s})^2 + 2a(70 \text{ m})$. Thus $a = \underline{2.6 \text{ m/s}^2}$. Now t follows immediately from

$$v_f = v_0 + at \quad \text{or} \quad 20 \text{ m/s} = 6.0 \text{ m/s} + (2.6 \text{ m/s}^2)t \quad \text{and} \quad t = \underline{5.4 \text{ s}}$$

3.14 A plane starts from rest and accelerates along the ground before takeoff. It moves 600 m in 12 s. Find (a) the acceleration, (b) speed at the end of 12 s, (c) distance moved during the twelfth second.

▮ We are given $v_0 = 0$, $x = 600$ m, $t = 12$ s.
(a) $x = v_0 t + \tfrac{1}{2}at^2$, or 600 m $= 0 + \tfrac{1}{2}a(12$ s$)^2$. So $a = \underline{8.33 \text{ m/s}^2}$.
(b) $v_f = v_0 + at$, or $v_f = 0 + (8.33 \text{ m/s}^2)(12 \text{ s})$ and $v_f = \underline{100 \text{ m/s}}$.
(c) Remembering that the first second is between $t = 0$ and $t = 1$ s, we realize that the twelfth second is between $t = 11$ and $t = 12$ s. Since we already know $x(t = 12$ s$)$, we solve for $x(t = 11$ s$)$:

$$x = v_0 t + \tfrac{1}{2}at^2 \quad \text{or} \quad x = 0 + \tfrac{1}{2}(8.33 \text{ m/s}^2)(11 \text{ s})^2 = 504 \text{ m}$$

Our answer is then Δx(twelfth s) $= x(t = 12$ s$) - x(t = 11$ s$) = \underline{96 \text{ m}}$.

3.15 A train running at 30 m/s is slowed uniformly to a stop in 44 s. Find the acceleration and the stopping distance.

▮ Here $v_0 = 30$ m/s and $v_f = 0$ at $t = 44$ s. $v_f = v_0 + at$ yields $0 = 30$ m/s $+ a(44$ s$)$, or $a = \underline{-0.68 \text{ m/s}^2}$. $x = v_0 t + \tfrac{1}{2}at^2 = (30 \text{ m/s})(44 \text{ s}) + \tfrac{1}{2}(-0.68 \text{ m/s}^2)(44 \text{ s})^2 = \underline{662 \text{ m}}$.

3.16 An object moving at 13 m/s slows uniformly at the rate of 2.0 m/s each second for a time of 6 s. Determine (a) its final speed, (b) its average speed during the 6 s, and (c) the distance moved in the 6 s.

▮ Slowing "at the rate of 2.0 m/s each second" means $a = -2.0$ m/s^2. We are given $v_0 = 13$ m/s and $t = 6.0$.
(a) Then $v_f = v_0 + at = 13$ m/s $+ (-2.0 \text{ m/s}^2)(6.0 \text{ s}) = \underline{1.0 \text{ m/s}}$. Because instantaneous speed is the magnitude of instantaneous velocity, our answer is $\underline{1.0 \text{ m/s}}$.
(b, c) Average speed = distance/time. The distance will be the same as the magnitude of the displacement as long as the object does not backtrack, i.e., reverse direction. In our case the velocity at the end of 6 s is still positive, so there indeed has been no backtracking. For displacement, $x = v_0 t + \tfrac{1}{2}at^2 = (13 \text{ m/s})(6.0 \text{ s}) + \tfrac{1}{2}(-2.0 \text{ m/s}^2)(6.0 \text{ s})^2 = \underline{42 \text{ m}}$ [which incidentally solves part (c)]. Average speed = 42 m/6.0 s = $\underline{7.0 \text{ m/s}}$.

3.17 A rocket-propelled car starts from rest at $x = 0$ and moves in the $+$ direction of X with constant acceleration $a = 5$ m/s^2 for 8 s until the fuel is exhausted. It then continues with constant velocity. What distance does the car cover in 12 s?

▮ The distance from x_0 at the moment fuel is exhausted is $x_1 = (0)(8) + \tfrac{1}{2}(5)(8)^2 = \underline{160 \text{ m}}$, and at this point $v = (2ax_1)^{1/2} = 40$ m/s. Hence the distance covered in 12 s is $x_2 = x_1 + v(12 - 8) = 160 + (40)(4) = \underline{320 \text{ m}}$.

3.18 The particle shown in Fig. 3-1 moves along X with a constant acceleration of -4 m/s^2. As it passes the origin, moving in the $+$ direction of X, its velocity is 20 m/s. In this problem, time t is measured from the moment the particle is first at the origin. (a) At what distance x' and time t' does $v = 0$? (b) At what time is the particle at $x = 15$ m, and what is its velocity at this point? (c) What is the velocity of the particle at $x = +25$ m? at $x = -25$ m? Try finding the velocity of the particle at $x = 55$ m.

▌ (a) Applying $v = v_0 + at$, $0 = 20 + (-4)t'$, or $\underline{t' = 5\,s}$. Then $x' = v_0 t' + \frac{1}{2}at'^2 = (20)(5) + \frac{1}{2}(-4)(5)^2 =$ $\underline{50\,m}$. Or, from $v^2 = v_0^2 + 2ax$:

$$0 = (20)^2 + 2(-4)x' \qquad \text{or} \qquad \underline{x' = 50\,m}.$$

(b)
$$15 = 20t + \frac{1}{2}(-4)t^2 \qquad \text{or} \qquad 2t^2 - 20t + 15 = 0$$

Solving this quadratic,

$$t = \frac{20 \pm \sqrt{(20)^2 - 4(2)(15)}}{4} = \frac{1}{4}(20 \pm 16.7)$$

Thus $t_1 = 0.82\,s$, $t_2 = 9.17\,s$, where t_1 is the time from the origin to $x = 15\,m$ and t_2 is the time to go from O out beyond $x = 15\,m$ and return to that point. At $x = 15\,m$,

$$v_1 = 20 - 4(0.82) = +16.7\,m/s \qquad v_2 = 20 - 4(9.17) = -16.7\,m/s$$

Observe that the speeds are equal.
(c) At $x = 25\,m$, $v^2 = (20)^2 + 2(-4)(25)$, or $v = \pm 14.1\,m/s$; and at $x = -25\,m$, $v^2 = 20^2 + 2(-4)(-25)$, or $v = \underline{-24.5\,m/s}$. (Why has the root $v = +24.5\,m/s$ been discarded?)
    Assuming that $x = 55\,m$, $v^2 = 20^2 + 2(-4)(55)$, from which $v = \pm\sqrt{-40}$. The imaginary value of v indicates that x never reaches 55 m, as expected from the result of part (a).

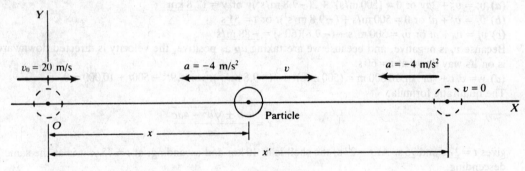

Fig. 3-1

3.19 A body falls freely from rest. Find (a) its acceleration, (b) the distance it falls in 3 s, (c) its speed after falling 70 m, (d) the time required to reach a speed of 25 m/s, (e) the time taken to fall 300 m.

▌ (a) Choose y downward as positive. Then $a = g = 9.8\,m/s^2$. (b) For $t = 3.0\,s$, $y = v_0 t + \frac{1}{2}at^2 =$ $0 + \frac{1}{2}(9.8\,m/s^2)(3.0\,s)^2 = \underline{44\,m}$. (c) Letting $y = 70\,m$, we have

$$v_f^2 = v_0^2 + 2ay = 0 + 2(9.8\,m/s^2)(70\,m) = 1372\,m^2/s^2 \qquad \text{or} \qquad v_f = \underline{37\,m/s}$$

(d) Letting v_f now equal 25 m/s, we have

$$v_f = v_0 + at \qquad \text{yields} \qquad 25\,m/s = 0 + (9.8\,m/s^2)t \qquad \text{or} \qquad t = \underline{2.55\,s}$$

(e) Now we let $y = 300\,m$ and we have

$$y = v_0 t + \frac{1}{2}at^2 \qquad \text{yields} \qquad 300\,m = 0 + \frac{1}{2}(9.8\,m/s^2)t^2 \qquad \text{or} \qquad t = \underline{7.8\,s}$$

3.20 A ball dropped from a bridge strikes the water in 5 s. Calculate (a) the speed with which it strikes and (b) the height of the bridge.

▌ Choose y downward as positive. Then $a = g = 9.8\,m/s^2$. We are given $v_0 = 0$, and $t = 5\,s$ to strike the water. Let $v = v_f$.
(a) $v = v_0 + at = 0 + (9.8\,m/s^2)(5\,s) = \underline{49\,m/s}$ (b) $y = v_0 t + \frac{1}{2}at^2 = 0 + \frac{1}{2}(9.8\,m/s^2)(5\,s)^2 = \underline{123\,m}$

3.21 A ball is thrown vertically downward from the edge of a high cliff with an initial velocity of 25 ft/s. (a) How fast is it moving after 1.5 s? (b) How far has it moved after 1.5 s?

▌ (a) $v = v_0 + at = 25 + 32(1.5)$ or $v = \underline{73\,ft/s}$ (where $a = g = 32\,ft/s^2$)
(b) For constant acceleration $v_{avg} = \frac{1}{2}(v + v_0) = \frac{1}{2}(73 + 25) = 49\,ft/s$

$$s = v_{avg}t = 49(1.5) = \underline{73.5\,ft} \qquad \text{or} \qquad s = v_0 t + \frac{1}{2}at^2 = 25(1.5) + \frac{1}{2}(32)(1.5)^2 = 37.5 + 36 = \underline{73.5\,ft}$$

3.22 A stone is thrown downward with initial speed 8 m/s from a height of 25 m. Find (*a*) the time it takes to reach the ground and (*b*) the speed with which it strikes.

▮ Choose downward positive. $a = g = 9.8$ m/s². We are given $v_0 = 8$ m/s.
(*a*) One might directly solve $y = v_0 t + \frac{1}{2}at^2$, or 25 m $= (8$ m/s$)t + \frac{1}{2}(9.8$ m/s²$)t^2$ for t; but it is easier first to find the final velocity:
(*b*) $v_f^2 = v_0^2 + 2ay = (8$ m/s$)^2 + 2(9.8$ m/s$)(25$ m$) = 554$ m²/s², or $v_f = \underline{23.5\text{ m/s}}$. Returning to (*a*), $v_f = v_0 + at$ yields 23.5 m/s $= 8$ m/s $+ (9.8$ m/s²$)t$, or $t = \underline{1.58\text{ s}}$.

3.23 A ball thrown vertically upward returns to its starting point in 4 s. Find its initial speed.

▮ Let us take *up* as positive. For the trip from beginning to end, $y = 0$, $a = -9.8$ m/s², $t = 4$ s. Note that the start and the endpoint for the trip are the same, so the displacement is zero. Use $y = v_0 t + \frac{1}{2}at^2$ to find $0 = v_0(4$ s$) + \frac{1}{2}(-9.8$ m/s²$)(4$ s$)^2$, from which $v_0 = \underline{19.6\text{ m/s}}$.

3.24 An antiaircraft shell is fired vertically upward with an initial velocity of 500 m/s. Neglecting friction, compute (*a*) the maximum height it can reach, (*b*) the time taken to reach that height, and (*c*) the instantaneous velocity at the end of 60 s. (*d*) When will its height be 10 km?

▮ Take *up* as positive. At the highest point, the velocity of the shell will be zero.
(*a*) $v_f^2 = v_0^2 + 2ay$ or $0 = (500$ m/s$)^2 + 2(-9.8$ m/s²$)y$ or $y = \underline{12.8\text{ km}}$
(*b*) $v_f = v_0 + at$ or $0 = 500$ m/s $+ (-9.8$ m/s²$)t$ or $t = \underline{51\text{ s}}$
(*c*) $v_f = v_0 + at$ or $v_f = 500$ m/s $+ (-9.8)(60$ s$) = \underline{-88\text{ m/s}}$
Because v_f is negative, and because we are taking *up* as positive, the velocity is directed downward. The shell is on its way down at $t = 60$ s.
(*d*) $y = v_0 t + \frac{1}{2}at^2$ or 10 000 m $= (500$ m/s$)t + \frac{1}{2}(-9.8$ m/s²$)t^2$ or $4.9t^2 - 500t + 10\,000 = 0$
The quadratic formula,

$$x = \frac{-b \pm \sqrt{b^2 - 4ac}}{2a}$$

gives $t = \underline{27\text{ s}}$ and $\underline{75\text{ s}}$. At $t = 27$ s, the shell is at 10 km and ascending; at $t = 75$ s, it is at the same height but descending.

3.25 A ballast bag is dropped from a balloon that is 300 m above the ground and rising at 13 m/s. For the bag, find (*a*) the maximum height reached, (*b*) its position and velocity 5 s after being released, and (*c*) the time before it hits the ground.

▮ The initial velocity of the bag when released is the same as that of the balloon, 13 m/s upward. Let us choose *up* as positive and take $y = 0$ at the point of release.
(*a*) At the highest point, $v_f = 0$. From $v_f^2 = v_0^2 + 2ay$, $0 = (13$ m/s$)^2 + 2(-9.8$ m/s²$)y$, or $y = \underline{8.6\text{ m}}$. The maximum height is $300 + 8.6 = \underline{308.6\text{ m}}$. (*b*) Take the endpoint to be its position at $t = 5$ s. Then, from $y = v_0 t + \frac{1}{2}at^2$, $y = (13$ m/s$)(5$ s$) + \frac{1}{2}(-9.8$ m/s²$)(5$ s$)^2 = -58$ m. So its height is $300 - 58 = \underline{242\text{ m}}$. Also, from $v_f = v_0 + at$, $v_f = 13$ m/s $+ (-9.8$ m/s²$)(5$ s$) = \underline{-36\text{ m/s}}$. It is moving downward at 36 m/s.
(*c*) Just before it hits the ground, the bag's displacement is -300 m. $y = v_0 t + \frac{1}{2}at^2$ becomes -300 m $= (13$ m/s$)t + \frac{1}{2}(-9.8$ m/s²$)t^2$, or $4.9t^2 - 13t - 300 = 0$. The quadratic formula gives $t = \underline{9.3\text{ s}}$ (and -6.6 s, which is not physically meaningful).

3.26 A stone is thrown vertically upward with velocity 40 m/s at the edge of a cliff having a height of 110 m. Neglecting air resistance, compute the time required to strike the ground at the base of the cliff. With what velocity does it strike?

▮ Choose upward as positive. $a = -g = -9.8$ m/s², $v_0 = 40$ m/s, $y_{\text{(base of cliff)}} = -110$ m. First obtain the final velocity: $v^2 = v_0^2 + 2ay = 40^2 + 2(-9.8)(-110) = 3756$ m²/s², whence $v = \underline{-61.3\text{ m/s}}$ (downward motion). Then, from $v = v_0 + at$,

$$-61.3 = 40 + (-9.8)t \qquad \text{or} \qquad t = \underline{10.3\text{ s}}$$

3.27 The hammer of a pile driver strikes the pile with a speed of 25 ft/s. From what height above the top of the pile did it fall? Neglect friction forces.

▮ Downward is positive. We assume it falls from rest, so $v_0 = 0$, $v = 25$ ft/s, $a = g = 32$ ft/s². We need not concern ourselves with the time of the fall t, since we can solve using

$$v^2 = v_0^2 + 2ay \quad \text{or} \quad (25 \text{ ft/s})^2 = 2(32 \text{ ft/s}^2)y$$

Then $y = \underline{9.8 \text{ ft}}$.

3.28 A baseball is thrown straight upward with a speed of 30 m/s. (*a*) How long will it rise? (*b*) How high will it rise? (*c*) How long after it leaves the hand will it return to the starting point? (*d*) When will its speed be 16 m/s?

▌ We choose upward as positive. $a = -g = -9.8 \text{ m/s}^2$, $v_0 = 30 \text{ m/s}$.
(*a*) At the highest point $v = 0$; so for time to reach highest point, $v = v_0 + at$ yields
$0 = 30 \text{ m/s} + (-9.8 \text{ m/s}^2)t$, or $t = \underline{3.06 \text{ s}}$. (*b*) $y = v_0 t + \frac{1}{2}at^2 = (30 \text{ m/s})(3.06 \text{ s}) + \frac{1}{2}(-9.8 \text{ m/s}^2)(3.06 \text{ s})^2 = \underline{46 \text{ m}}$,
[or $v^2 = v_0^2 + 2ay$ yields $0 = (30 \text{ m/s})^2 + 2(-9.8 \text{ m/s}^2)y$, and $y = 46 \text{ m}$]. (*c*) Without calculation: Since the time up equals the time down, we double the time up to get 6.12 s for the round trip. With calculation: For the final displacement, we have $y = 0$, so

$$y = v_0 t + \frac{1}{2}at^2 \quad \text{yields} \quad 0 = (30 \text{ m/s})t + \frac{1}{2}(-9.8 \text{ m/s}^2)t^2$$

This is a quadratic equation but easy to solve. One solution is $t = 0$, but this corresponds to $y = 0$ as it leaves the hand. The other solution is not 0, so we can divide out by t, leaving

$$0 = 30 \text{ m/s} - (4.9 \text{ m/s}^2)t \quad \text{or} \quad t = \underline{6.12 \text{ s}}$$

(*d*) Recalling that the speed is the magnitude of the velocity, we must consider the possible velocities: $v = \pm 16 \text{ m/s}$. $v = v_0 + at$ yields $\pm 16 \text{ m/s} = 30 \text{ m/s} + (-9.8 \text{ m/s})t$; and solving for t; $t_+ = \underline{1.43 \text{ s}}$, $t_- = \underline{4.7 \text{ s}}$.

3.29 The acceleration due to gravity at the surface of Mars is roughly 4 m/s². If an astronaut on Mars were to toss a wrench upward with a speed of 10 m/s, find (*a*) how long it would rise; (*b*) how high it would go; (*c*) its speed at $t = 3$ s; and (*d*) its displacement at $t = 3$ s.

▌ Let us choose up as positive. Then a equals -4 m/s^2, and v_0 equals $+10 \text{ m/s}$.
(*a*) By equation of motion $a = (v - v_0)/t$, we have

$$-4 \text{ m/s}^2 = \frac{0 - 10 \text{ m/s}}{t_{\text{rise}}} \qquad t_{\text{rise}} = \underline{2.5 \text{ s}}$$

(*b*) By equation of motion $v^2 = v_0^2 + 2as$, we have

$$0^2 = (10)^2 + 2(-4)s_{\text{max}} \quad \text{and} \quad s_{\text{max}} = \underline{12.5 \text{ m}}$$

(*c*) By equation of motion $a = (v - v_0)/t$, we have

$$-4 = \frac{v_3 - 10}{3} \quad \text{and} \quad v_3 = \underline{-2 \text{ m/s}} \quad \text{or} \quad 2 \text{ m/s downward}$$

(*d*) $s = v_0 t + \frac{1}{2}at^2$ and thus

$$s_3 = (10 \times 3) + \frac{1}{2}(-4)(9) = 30 - 18 = \underline{12 \text{ m}}$$

3.30 The acceleration due to gravity on the moon is 1.67 m/s². If a person can throw a stone 12.0 m straight upward on the earth, how high should the person be able to throw a stone on the moon? Assume that the throwing speeds are the same in the two cases.

▌ On earth we can write $v_E^2 = 2g_E(12)$ while on the moon $v_M^2 = 2g_M h_M$. The throwing velocities are the same, so the second expression can be divided by the first to give $h_M = 12(g_E/g_M) = 12(9.80/1.67) = \underline{70 \text{ m}}$.

3.31 A proton in a uniform electric field moves along a straight line with constant acceleration. Starting from rest it attains a velocity of 1000 km/s in a distance of 1 cm. (*a*) What is its acceleration? (*b*) What time is required to reach the given velocity?

▌ (*a*) $v_0 = 0$, $v = 10^6 \text{ m/s}$ in $x = 10^{-2} \text{ m}$ of displacement. Then $v^2 = v_0^2 + 2ax$ yields $(10^6 \text{ m/s})^2 = 0 + 2a(10^{-2} \text{ m})$, or $a = \underline{5.0 \times 10^{13} \text{ m/s}^2}$. (*b*) $v = v_0 + at$ yields $10^6 \text{ m/s} = 0 + (5.0 \times 10^{13} \text{ m/s}^2)t$, or $t = \underline{2.0 \times 10^{-8} \text{ s}}$.

3.32 A bottle dropped from a balloon reaches the ground in 20 s. Determine the height of the balloon if (*a*) it was at rest in the air and (*b*) it was ascending with a speed of 50 m/s when the bottle was dropped.

▮ Choose upward as positive for both (*a*) and (*b*). $a = -g = -9.8$ m/s^2.
(*a*) $v_0 = 0$. To find the height, let y be the displacement at time t (remember the $y = 0$ point is at the balloon)
and we are given $t = 20$ s. Then $y = v_0 t + \frac{1}{2}at^2 = 0 + \frac{1}{2}(-9.8$ m/s$^2)(20$ s$)^2 = -1960$ m. The height is $|y| = \underline{1960}$ m.
(*b*) Here the bottle initially has the velocity of the balloon so $v_0 = 50$ m/s. Now, $y = v_0 t + \frac{1}{2}at^2 =$
$(50$ m/s$)(20$ s$) + \frac{1}{2}(-9.8$ m/s$^2)(20$ s$)^2 = -960$ m. Again the height $= |y| = \underline{960}$ m.

3.33 A sandbag ballast is dropped from a balloon that is ascending with a velocity of 40 ft/s. If the sandbag reaches
the ground in 20 s, how high was the balloon when the bag was dropped? Neglect air resistance.

▮ Take downward as positive. Then $a = g = 32$ ft/s^2 and $v_0 = -40$ ft/s.
$$s = v_0 t + \tfrac{1}{2}at^2 = -40(20) + \tfrac{1}{2}(32)(20)^2 = -800 + 6400 = 5600 \text{ ft}$$
The balloon was $\underline{5600\text{ ft}}$ above the ground.

3.34 A stone is shot straight upward with a speed of 80 ft/s from a tower 224 ft high. Find the speed with which it
strikes the ground.

▮ Choose upward as positive. $a = g = -32$ ft/s^2, $v_0 = 80$ ft/s. The presumption is that the stone just misses
the edge of the tower on the way down and strikes the ground 224 ft below. We can avoid all reference to the
time in the problem by setting $y = -224$ ft in the equation
$$v^2 = v_0^2 + 2ay = (80 \text{ ft/s})^2 + 2(-32 \text{ ft/s}^2)(-224 \text{ ft}) = 20\,736 \text{ ft}^2/\text{s}^2$$
And $v = \pm 144$ ft/s, where the minus sign gives the physical solution. The speed is $|v| = \underline{144\text{ ft/s}}$.

3.35 A nut comes loose from a bolt on the bottom of an elevator as the elevator is moving up the shaft at 3.0 m/s.
The nut strikes the bottom of the shaft in 2 s. (*a*) How far from the bottom of the shaft was the elevator
when the nut fell off? (*b*) How far above the bottom was the nut 0.25 s after it fell off?

▮ Here the nut initially has the velocity of the elevator, so choosing upward as positive, $v_0 = 3.0$ m/s. Also
$a = -g = -9.8$ m/s^2.
(*a*) The time to hit bottom is 2.0 s, so $y = v_0 t + \frac{1}{2}at^2 = (3.0$ m/s$)(2.0$ s$) + \frac{1}{2}(-9.8$ m/s$^2)(2.0$ s$)^2 = -13.6$ m. The
bottom of the shaft is $\underline{13.6\text{ m}}$ below where the elevator was when the nut fell off.
(*b*) The new displacement y is for $t = 0.25$ s, so $y = v_0 t + \frac{1}{2}at^2 = (3.0$ m/s$)(0.25$ s$) +$
$\frac{1}{2}(-9.8$ m/s$^2)(0.25$ s$)^2 = \underline{0.44\text{ m}}$. Thus the nut is above its starting position. (This makes sense if we remember
the initial velocity.) The height above the bottom is thus $0.4 + 13.6 = \underline{14.0\text{ m}}$.

3.2 GRAPHICAL AND OTHER PROBLEMS

3.36 The graph of an object's motion (along a line) is shown in Fig. 3-2. Find the instantaneous velocity of the
object at points A and B. What is the object's average velocity? Its acceleration?

▮ Because the velocity is given by the slope, $\Delta x / \Delta t$, of the tangent line, we take a tangent to the curve at
point A. The tangent line is the curve itself in this case. For the triangle shown at A, we have
$$\frac{\Delta x}{\Delta t} = \frac{4 \text{ m}}{8 \text{ s}} = 0.50 \text{ m/s}$$
This is also the velocity at point B and at every other point on the straight-line graph. It follows that $\underline{a = 0}$
and $\bar{v}_x = v_x = \underline{0.50\text{ m/s}}$.

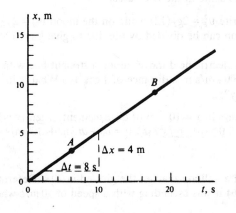

Fig. 3-2

3.37 Refer to Fig. 3-3. Find the instantaneous velocity at point F for the object whose motion the curve represents.

▌ The tangent at F is the dashed line GH. Taking triangle GHJ, we have

$$\Delta t = 24 - 4 = 20 \text{ s} \qquad \Delta x = 0 - 15 = -15 \text{ m}$$

Hence slope at F is

$$v_F = \frac{\Delta x}{\Delta t} = \frac{-15 \text{ m}}{20 \text{ s}} = -0.75 \text{ m/s}$$

The negative sign tells us that the object is moving in the $-x$ direction.

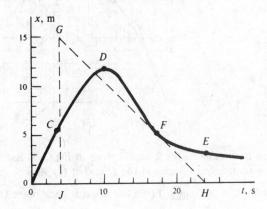

Fig. 3-3

3.38ᶜ Refer back to Fig. 3.3 for the motion of an object along the x axis. What is the instantaneous velocity of the object (**a**) at point D? (**b**) at point C? (**c**) at point E?

▌ (**a**) Point d is a maximum of the x-vs.-t curve. Therefore $v = dx/dt = 0$.
(**b**) Without the exact equation for x as function of t one cannot get a precise answer. The best we can do is to draw the tangent line at point c and get the slope in the same way as in Prob. 3.37. This yields the answer

$$v_C = \frac{dx}{dt}\bigg|_C \approx 1.3 \text{ m/s}$$

(**c**) We proceed as in part (**b**), but here the tangent line has a negative slope and the answer should be

$$v_E = \frac{dx}{dt}\bigg|_E \approx -0.13 \text{ m/s}$$

3.39 A girl walks along an east–west street, and a graph of her displacement from home is shown in Fig. 3-4. Find her average velocity for the whole time interval shown as well as her instantaneous velocity at points A, B, and C.

▌ The average velocity is zero, since the displacement vector is zero. The instantaneous velocities are just the slopes of the curve at each point. At A the velocity is $40/6 = 6.7$ m/min east. At B it is zero. At C it is $-65/5 = -13$ m/min east, or $+13$ m/min west.

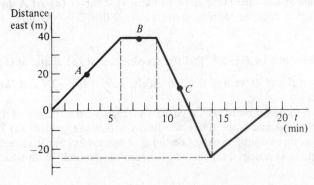

Fig. 3-4

3.40 Referring to Fig. 3-4, find **(a)** average velocity for the time interval $t = 7$ min to $t = 14$ min; the instantaneous velocity at **(b)** $t = 13.5$ min and **(c)** $t = 15$ min.

▌ **(a)** $\bar{v} = (-25 - 40)/(14 - 7) = -9.3$ m/min east; **(b)** the same as at point C, -13 m/min east; **(c)** the slope is $+25/(19 - 14) = \underline{5.0 \text{ m/min}}$ east. Note that negative velocity east means motion is west.

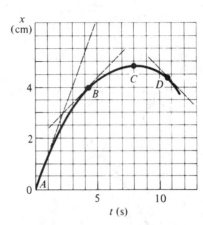

Fig. 3-5

3.41 The graph of a particle's motion along the x axis is given in Fig. 3-5. Estimate the **(a)** average velocity for the interval from A to C; instantaneous velocity at **(b)** D and at **(c)** A.

▌ **(a)** $\bar{v} = (4.8 - 0)/(8.0 - 0) = \underline{0.60 \text{ cm/s}}$. From the slope at each point **(b)** $v = \underline{-0.48}$ and **(c)** $\underline{1.3 \text{ cm/s}}$.

3.42 Figure 3-6 shows the velocity of a particle as it moves along the x axis. Find its acceleration at **(a)** A and at **(b)** C.

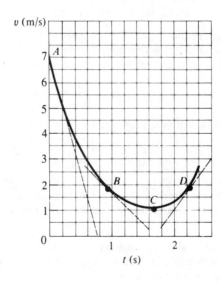

Fig. 3-6

▌ The acceleration at any time is the slope of the v-vs.-t curve. **(a)** At A the slope is, from where the tangent line through A cuts the coordinate axes, $a = -7.0/0.73 = \underline{-9.6 \text{ m/s}^2}$. **(b)** The slope and therefore a is zero.

3.43 For the motion described by Fig. 3-6, find the acceleration at **(a)** B and at **(b)** D.

▌ Taking slopes at B and D we find that the acceleration = **(a)** $\underline{-2.4}$ and **(b)** $\underline{3.2 \text{ m/s}^2}$.

3.44 A ball is thrown vertically upward with a velocity of 20 m/s from the top of a tower having a height of 50 m, Fig. 3-7. On its return it misses the tower and finally strikes the ground. **(a)** What time t_1 elapses from the instant the ball was thrown until it passes the edge of the tower? What velocity v_1 does it have at this time? **(b)** What total time t_2 is required for the ball to reach the ground? With what velocity v_2 does it strike the ground?

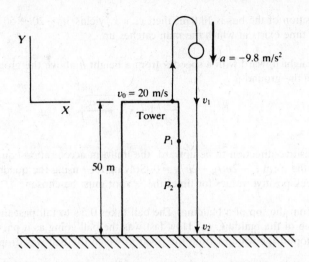

Fig. 3-7

▌ (a) For the coordinate system shown in Fig. 3-7, $y = v_0t + \frac{1}{2}at^2$. But at the edge of the roof $y = 0$, and thus $0 = 20t_1 + \frac{1}{2}(-9.8)t_1^2$, from which $t_1 = 0$, indicating the instant at which the ball is released, and also $t_1 = \underline{4.08\text{ s}}$, which is the time to go up and return to the edge. Then, from $v = v_0 + at$, $v_1 = 20 + (-9.8)(4.08) = \underline{-20\text{ m/s}}$, which is the negative of the initial velocity.

(b) $\qquad\qquad -50 = 20t_2 + \frac{1}{2}(-9.8)t_2^2 \qquad \text{or} \qquad t_2 = \underline{5.8\text{ s}} \qquad v_2 = 20 + (-9.8)(5.8) = \underline{-37\text{ m/s}}$

3.45 Refer to Prob. 3.44 and Fig. 3-7. (a) What is the maximum height above ground reached by the ball? (b) Points P_1 and P_2 are 15 and 30 m, respectively, below the top of the tower. What time interval is required for the ball to travel from P_1 to P_2? (c) It is desired that after passing the edge, the ball will reach the ground in 3 s. With what velocity must it be thrown upward from the roof?

▌ (a) Maximum height above ground: $h = y_{\max} + 50$. From $v_0^2 + 2ay_{\max} = 0$,

$$y_{\max} = \frac{-(20)^2}{-2(9.8)} = \underline{20.4\text{ m}}$$

Thus, $h = \underline{70.4\text{ m}}$.

(b) If t_1 and t_2 are the times to reach P_1 and P_2, respectively, $-15 = 20t_1 - 4.9t_1^2$ and $-30 = 20t_2 - 4.9t_2^2$. Solving, $t_1 = 4.723$ s, $t_2 = 5.248$ s, and the time from P_1 to P_2 is $t_2 - t_1 = \underline{0.525\text{ s}}$.

(c) If v_0 is the desired initial velocity, then $-v_0$ is the velocity upon passing the edge. Then, applying $y = v_0t + \frac{1}{2}at^2$ to the trip down the tower, we find $-50 = (-v_0)(3) - 4.9(3)^2$, or $v_0 = \underline{1.96\text{ m/s}}$.

3.46 A man runs at a speed of 4.0 m/s to overtake a standing bus. When he is 6.0 m behind the door (at $t = 0$), the bus moves forward and continues with a constant acceleration of 1.2 m/s². (a) How long does it take for the man to gain the door? (b) If in the beginning he is 10.0 m from the door, will he (running at the same speed) ever catch up? ever catch up?

▌ At $t = 0$ let the man's position be the origin, $x_{m0} = 0$. The bus door is then at $x_{b0} = 6.0$ m. The equations of motion for the man and the bus are

$$x_m = x_{m0} + v_{m0}t + \tfrac{1}{2}a_mt^2 \qquad x_b = x_{b0} + v_{b0}t + \tfrac{1}{2}a_bt^2$$

Now $\qquad\qquad v_{m0} = 4.0\text{ m/s} \qquad v_{b0} = 0 \qquad a_m = 0 \qquad a_b = 1.2\text{ m/s}^2$

Thus $\qquad\qquad x_m = 4.0t \qquad x_b = 6.0 + 0.6t^2$

When the man catches the bus, $x_m = x_b$, or $4.0t = 6.0 + 0.6t^2$.

This can be reexpressed as $3t^2 - 20t + 30 = 0$. Solving by the quadratic formula,

$$t = \frac{20 \pm \sqrt{400 - 360}}{6} = \frac{10 \pm \sqrt{10}}{3} = \underline{2.3\text{ s, }4.4\text{ s}}$$

Note that there are two positive time solutions. This can be understood as follows. The first time, $t_1 = 2.3$ s, corresponds to his first reaching the door. This is the real answer to the problem. However, the equations we have solved "don't know" he will stop running and board the bus; the equations have him continuing running at constant speed. He thus goes on past the bus; but since the bus is accelerating, it eventually builds up a larger velocity than the man and will catch up with him, $t_2 = 4.4$ s.

(b) If the initial position of the bus is 10.0 m, then $x_m = x_b$ yields $3t^2 - 20t + 50 = 0$, which has only complex roots. Thus no real time exists at which the man catches up.

3.47 A ball is thrown straight upward with a speed v from a height h above the ground. Show that the time taken for the ball to strike the ground is

$$\frac{v}{g}\left(1 + \sqrt{1 + \frac{2hg}{v^2}}\right)$$

▮ Assuming the positive direction to be upward, the uniform acceleration equation is $-h = vt - gt^2/2$; this can be rewritten in the form $t^2 - 2vt/g - 2h/g = 0$. Solving for t using the quadratic formula gives the desired answer. Since we seek positive values for time, the + root must be chosen.

3.48 A ball is dropped from the top of a building. The ball takes 0.5 s to fall past the 3-m length of a window some distance from the top of the building. (a) How fast was the ball going as it passed the top of the window? (b) How far is the top of the window from the point at which the ball was dropped?

▮ The velocities v_T at the top and v_B at the bottom of the window are related by the following equations: $\bar{v} = (v_T + v_B)/2 = 3/0.5 = 6$, so $v_T + v_B = 12$, and $v_B = v_T + g(0.5)$, so $v_B - v_T = 4.9$. Eliminating v_B between these two expressions yields $v_T = \underline{3.55 \text{ m/s}}$. The distance needed to reach this speed is

$$h = \frac{v_T^2}{2g} = \frac{(3.55)^2}{2(9.8)} = \underline{0.64 \text{ m}}$$

3.49 A truck is moving forward at a constant speed of 21 m/s. The driver sees a stationary car directly ahead at a distance of 110 m. After a "reaction time" of Δt, he applies the brakes, which gives the truck an acceleration of -3 m/s². (a) What is the maximum allowable Δt to avoid a collision, and what distance will the truck have moved before the brakes take hold? (b) Assuming a reaction time of 1.4 s, how far behind the car will the truck stop, and in how many seconds from the time the driver first saw the car?

▮ The total displacement of the truck from the instant the driver sees the car is $x = v_0 \Delta t + x_A$, where x_A is the displacement from the point of deceleration to the point of rest. We are given $v_0 = 21$ m/s and the deceleration is $a = -3$ m/s². x_A can be obtained from the equation $v_f^2 = v_0^2 + 2ax_A$, with $v_f = 0$. Thus $x_A = -v_0^2/2a$, or $x_A = -(21 \text{ m/s})^2/(-6 \text{ m/s}^2) = 73.5$ m.
(a) To find the maximum Δt, we note that $x_{max} = 110$ m and $x_{max} = (21 \text{ m/s}) \Delta t_{max} + x_A$, or $110 \text{ m} = (21 \text{ m/s}) \Delta t_{max} + 73.5$ m, and $\Delta t_{max} = \underline{1.74 \text{ s}}$. The distance moved before braking started is of course just $v_0 \Delta t = \underline{36.5 \text{ m}}$.
(b) If $\Delta t = 1.4$ s, then $x = (21 \text{ m/s})(1.4 \text{ s}) + 73.5 \text{ m} = 102.9$ m. The distance to the car is just $110 \text{ m} - 102.9 \text{ m} = \underline{7.1 \text{ m}}$. To find the time, we need to know the time t during which the truck accelerates. We have $v_f = v_0 + at$, with again $v_f = 0$ and $a = -3$ m/s². Then $0 = 21 \text{ m/s} - (3 \text{ m/s}^2)t$ and $t = 7$ s. The total time is $t + \Delta t = 7 + 1.4 = \underline{8.4 \text{ s}}$.

3.50 Just as a car starts to accelerate from rest with acceleration 1.4 m/s², a bus moving with constant speed of 12 m/s passes it in a parallel lane. (a) How long before the car overtakes the bus? (b) How fast will the car then be going? (c) How far will the car then have gone?

▮ The car starts with initial velocity zero and acceleration $a_c = 1.4$ m/s², while the bus has constant velocity $v_b = 12$ m/s. (b) Both travel the same distance x in time t, so set $a_c t^2/2 = v_b t$ to give $t = \underline{17 \text{ s}}$. (b) The final velocity of the car is $v = a_c t = \underline{24 \text{ m/s}}$. (c) The average velocity of the car (or the bus's fixed velocity) times 17 s yields $x = \underline{204 \text{ m}}$.

3.51 A monkey in a perch 20 m high in a tree drops a coconut directly above your head as you run with speed 1.5 m/s beneath the tree. (a) How far behind you does the coconut hit the ground? (b) If the monkey had really wanted to hit your toes, how much earlier should the coconut have been dropped?

▮ The time for the coconut to fall 20 m is given by $20 = gt^2/2$, or 2.02 s. Distance $x = (1.5 \text{ m/s})(2.02 \text{ s}) = \underline{3.03 \text{ m}}$. Since you are moving at a fixed speed the monkey should have dropped the coconut $\underline{2.02 \text{ s}}$ earlier.

3.52 Two balls are dropped to the ground from different heights. One ball is dropped 2 s after the other but they both strike the ground at the same time, 5 s after the first is dropped. (a) What is the difference in the heights at which they were dropped? (b) From what height was the first dropped?

■ The time to fall from the greater height h_1 is $t = 5$ s; the time from the lesser height h_2 is $t - 2 = 3$ s. Using $y = at^2/2$, we find **(a)** the height difference $h_1 - h_2 = g(5^2/2 - 3^2/2) = \underline{78\text{ m}}$ and **(b)** $h_1 = 9.8(25)/2 = \underline{123\text{ m}}$.

3.53 Two boys start running straight toward each other from two points that are 100 m apart. One runs with a speed of 5 m/s, while the other moves at 7 m/s. How close are they to the slower one's starting point when they reach each other?

■ The two boys meet at the same place and time. The time for the slower one to travel a distance x is $x/5$, while the other boy has $t = (100 - x)/7$. Equating the times yields $x = \underline{41.7\text{ m}}$.

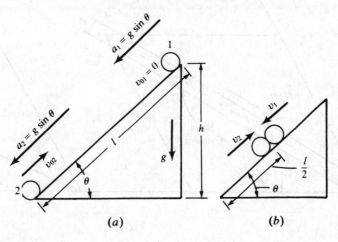

(a) (b) **Fig. 3-8**

3.54 Body 1, Fig. 3-8, is released from rest at the top of a smooth inclined plane, and at the same instant body 2 is projected upward from the foot of the plane with such velocity that they meet halfway up the plane. Determine **(a)** the velocity of projection and **(b)** the velocity of each body when they meet.

■ **(a)** In the common time t, body 1 travels the distance

$$\frac{l}{2} = (0)t + \frac{1}{2}(g \sin \theta)t^2$$

and body 2 travels the distance

$$\frac{l}{2} = v_{02}t + \frac{1}{2}(-g \sin \theta)t^2$$

Adding these two equations gives $l = v_{02}t$ or $t = l/v_{02}$. Substituting this value of t in the first equation and solving for v_{02}, we obtain $v_{02} = \sqrt{gl \sin \theta} = \sqrt{gh}$.

(b)
$$v_1^2 = 0^2 + 2(g \sin \theta)\frac{l}{2} \quad \text{or} \quad v_1 = \sqrt{gl \sin \theta} = \sqrt{gh}.$$

$$v_2^2 = v_{02}^2 + 2(-g \sin \theta)\frac{l}{2} = gl \sin \theta - gl \sin \theta = 0 \quad \text{or} \quad v_2 = 0$$

3.55 Two trains are headed toward each other on the same track with equal speeds of 20 m/s. When they are 2 km apart, they see each other and begin to decelerate. **(a)** If their decelerations are uniform and equal, how large must the acceleration be if the trains are to barely avoid collision? **(b)** If only one train slows with this acceleration, how far will it go before collision occurs?

■ **(a)** Each train stops in 1000 m, so using $v^2 - v_0^2 = 2ax$ yields, for $v_0 = 20$ m/s, an acceleration of $\underline{-0.2\text{ m/s}^2}$. **(b)** The decelerating train travels a distance $x = 20t - 0.1t^2$, while the other train travels $(2000 - x) = 20t$. Because they collide, times are the same. Eliminating t and solving for x by the quadratic formula yields $\underline{828\text{ m}}$.

3.56 A ball after having fallen from rest under the influence of gravity for 6 s crashes through a horizontal glass plate, thereby losing two-thirds of its velocity. If it then reaches the ground in 2 s, find the height of the plate above the ground.

■ From $v = v_0 + at$, the velocity just before striking the glass is $v_1 = 0 - 9.8(6) = -58.8$ m/s, and so the velocity after passing through glass is $(1/3)v_1 = -19.6$ m/s. Thus $-h = (-19.6)(2) - 4.9(2)^2$, or $\underline{h = 58.8\text{ m}}$.

3.57 An inclined plane, Fig. 3-9, makes an angle θ with the horizontal. A groove OA cut in the plane makes an angle α with OX. A short smooth cylinder is free to slide down the groove under the influence of gravity, starting from rest at the point (x_0, y_0). Find: **(a)** its downward acceleration along the groove, **(b)** the time to reach O, and **(c)** its velocity at O. Let $\theta = 30°$, $x_0 = 3$ m, $y_0 = 4$ m.

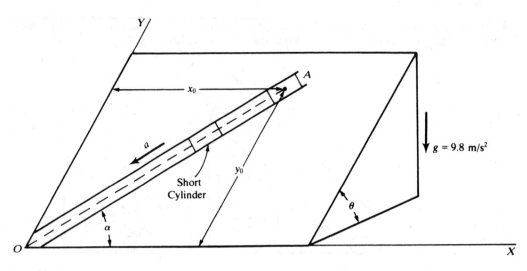

Fig. 3-9

▌ **(a)** The downward component of **g** parallel to OY is $g \sin \theta$; hence, the downward component along the groove is $a = g \sin \theta \sin \alpha$. Since

$$\sin \alpha = \frac{y_0}{(x_0^2 + y_0^2)^{1/2}} = 0.8 \quad \text{and} \quad \sin \theta = 0.5 \quad a = (9.8)(0.5)(0.8) = \underline{3.92 \text{ m/s}^2}$$

(b) $s = v_0 t + \frac{1}{2}at^2$, where $s = (x_0^2 + y_0^2)^{1/2} = 5$ m and $v_0 = 0$. Thus, $s = \frac{1}{2}(3.92)t^2$ or $t = \underline{1.597 \text{ s}}$.
(c) $v = 0 + (3.92)(1.597) = \underline{6.26 \text{ m/s}}$.

3.58 A bead, Fig. 3-10, is free to slide down a smooth wire tightly stretched between points P_1 and P_2 on a vertical circle of radius R. If the bead starts from rest at P_1, the highest point on the circle, find **(a)** its velocity v on arriving at P_2 and **(b)** the time to arrive at P_2 and show that this time is the same for any chord drawn from P_1.

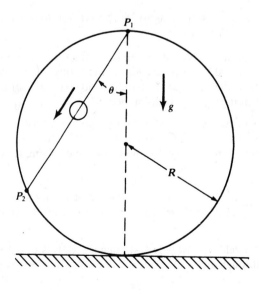

Fig. 3-10

(a) The acceleration of the bead down the wire is $g \cos \theta$ and the length of the wire is $2R \cos \theta$. Hence, $v^2 = 0^2 + 2(g \cos \theta)(2R \cos \theta)$, or $v = 2(\sqrt{gR}) \cos \theta$.

(b)
$$t = \frac{v}{a} = \frac{2\sqrt{gR} \cos \theta}{g \cos \theta} = 2\sqrt{\frac{R}{g}}$$

which is the same regardless of where P_2 is located on the circle.

3.59^c An object is forced to move along the X axis in such a way that its displacement is given by $x = 30 + 20t - 15t^2$ where x is in m and t is in s. **(a)** Find expressions for the velocity $\dot{x}$ and acceleration $\ddot{x}$. Is the acceleration constant? **(b)** What are the initial position and the initial velocity of the object? **(c)** At what time and distance from the origin is the velocity zero? **(d)** At what time and location is the velocity -50 m/s?

Remembering that all units are SI and that

$$\dot{x} \equiv \frac{dx}{dt} \qquad \ddot{x} \equiv \frac{d^2x}{dt^2}$$

(a) $v = \dot{x} = (20 - 30t)$ m/s; $a = \ddot{x} = -30$ m/s^2. Acceleration is constant.
(b) At $t = 0$, $x = 30$ m $= x_0$, $\dot{x} = 20$ m/s $\equiv v_0$.
(c) From the velocity equation, for $v = 0$, $0 = 20 - 30t$ and $t = \frac{2}{3}$ s. Substituting $t = \frac{2}{3}$ s into the displacement equation, $x = 30 + 20(\frac{2}{3}) - 15(\frac{2}{3})^2 = \underline{36.7 \text{ m}}$.
(d) Setting $v = -50$ m/s, $-50 = 20 - 30t$, and $t = 2\frac{1}{3}$ s. Then, $x = 30 + 20(2\frac{1}{3}) - 15(2\frac{1}{3})^2 = \underline{-5.0 \text{ m}}$.
(Note that a comparison of the displacement equation with $x = x_0 + v_0 t + \frac{1}{2}at^2$ would have made the use of calculus unnecessary.)

3.60^c A particle moving along the x axis has a velocity given by $v = 4t - 2.50t^2$ cm/s for t in seconds. Find its acceleration at **(a)** $t = 0.50$ s and **(b)** $t = 3.0$ s.

$a = dv/dt = 4 - 5.0t$, giving **(a)** $a = \underline{1.50 \text{ cm/s}^2}$ and **(b)** $\underline{-11.0 \text{ cm/s}^2}$.

3.61^c A ball is released from rest at the edge of a deep ravine. Assume that air resistance gives it an acceleration of $-b\dot{y}$, where y is measured positive downward. (This negative acceleration is proportional to its speed, $\dot{y}$; the positive constant b can be found by experiment.) The ball has a total acceleration of $-b\dot{y} + g$, and so

$$\ddot{y} = -b\dot{y} + g \tag{1}$$

is the differential equation of motion. **(a)** Show by differentiation and substitution that

$$y = k(e^{-bt} - 1) + (g/b)t \tag{2}$$

is a solution of (1) for an arbitrary value of the constant k and that (2) gives $y = 0$ for $t = 0$. **(b)** Since at $t = 0$, $\dot{y} = 0$, prove that $k = g/b^2$ and show that as $t \to \infty$, $\dot{y} \to g/b$; that is, the velocity reaches a limiting value such that the negative acceleration due to air resistance exactly offsets the positive acceleration of gravity and thus $\ddot{y} = 0$. **(c)** Assuming that $b = 0.1$ s^{-1}, find the distance fallen and the speed reached after 10 s. **(d)** Show that after 1 min the ball will have essentially reached its terminal velocity of 98 m/s.

(a) Differentiating (2) once, $\dot{y} = -bke^{-bt} + g/b$. Differentiating once more, $\ddot{y} = b^2ke^{-bt}$. Multiplying our expression for $\dot{y}$ by $-b$ and adding g yields b^2ke^{-bt}. Thus (1) is satisfied. Substituting $t = 0$ into (2) and recalling that $e^0 = 1$, we get $y = 0$.
(b) Since the ball is released from rest, $\dot{y} = 0$ at $t = 0$. Using our expression for $\dot{y}$ from **(a)**, we have $0 = -bk + g/b$, which yields $k = g/b^2$. As $t \to \infty$ the first term in $\dot{y}$ becomes infinitesimal and $\dot{y} \to g/b$. Thus $\ddot{y}$ must approach zero, as can be seen directly from the expression for $\ddot{y}$.
(c) If $b = 0.1$/s and using $g = 9.8$ m/s^2, we have $k = g/b^2 = 980$ m. Then at $t = 10$ s, $y = (980 \text{ m})(e^{-1} - 1) + (98 \text{ m/s})(10 \text{ s}) = \underline{360 \text{ m}}$. $v = \dot{y} = (-98 \text{ m/s})e^{-1} + (9.8 \text{ m/s}^2)/(0.1/\text{s}) = \underline{62 \text{ m/s}}$.
(d) At $t = 60$ s, $\dot{y} = (-98 \text{ m/s})e^{-6} + 98$ m/s. $e^{-6} \approx 0.0025$, so $\dot{y} \approx 98$ m/s.

3.62^c A ball is thrown vertically upward from the origin of axes (Y regarded + upward), with initial velocity $\dot{y}_0$. Assuming as in Prob. 3.61 an acceleration $-b\dot{y}$ due to air resistance, we write

$$\ddot{y} = -b\dot{y} - g \tag{1}$$

[Note that when $\dot{y}$ changes sign, so does $-b\dot{y}$; hence (1) is valid for the trip down as well as the trip up]. **(a)** Show that $y = k(e^{-bt} - 1) - (g/b)t$ is a solution of (1) for any value of k. **(b)** Prove that

$$k = -\frac{1}{b}\left(\dot{y}_0 + \frac{g}{b}\right)$$

(c) Assuming that $b = 0.1\,\text{s}^{-1}$ and $\dot{y}_0 = 50\,\text{m/s}$, find the height and speed at $t = 3\,\text{s}$. (d) How long does it take for the ball to attain its maximum height and what is the height?

▌ (a) $\dot{y} = -bke^{-bt} - g/b$, $\ddot{y} = b^2 ke^{-bt}$. Multiplying $\dot{y}$ by $-b$ and subtracting g yields $\ddot{y}$, which proves y is a solution of (1). (b) At $t = 0$, $\dot{y} = \dot{y}_0 = -bk - g/b$. Solving for k, $k = -(\dot{y}_0 + g/b)/b$.
(c) Again $b = 0.1\,\text{s}$, and we are told $\dot{y}_0 = 50\,\text{m/s}$. Then $k = -(50 + 98)/0.1 = -1480\,\text{m}$. At $t = 3\,\text{s}$, we have $y = -1480(e^{-0.3} - 1) - (98)(3) = \underline{89.6\,\text{m}}$. $\dot{y} = +148e^{-0.3} - 98 = \underline{11.6\,\text{m/s}}$.
(d) For maximum height, $\dot{y} = 0$, which yields $0 = 148e^{-0.1t} - 98$, or $e^{0.1t} = 1.51$. Thus $t = \underline{4.12\,\text{s}}$. Substituting $t = 4.12\,\text{s}$ into the y equation yields $y = -1480(e^{-0.412} - 1) - (98)(4.12) = \underline{96.0\,\text{m}}$.

3.63^c A mass at the end of a spring vibrates up and down according to the equation $y = 8 \sin 1.5t$ cm, where t is the time in seconds and the complete argument (angle) of the sine function, $1.5t$, is in radians. (a) What is the velocity of the mass at $t = 0.75\,\text{s}$? (b) At $t = 3.0\,\text{s}$? (c) What is the maximum velocity of the mass? (*Hint:* To express the angle in degrees, multiply it by $180/\pi$.)

▌ The velocity $v = dy/dt = (1.5)(8) \cos 1.5t$ cm/s; remember that $1.5t$ is expressed in radians. (a) $v = 12 \cos 1.13 = \underline{5.2\,\text{cm/s}}$. (b) $v = 12 \cos 4.5 = \underline{-2.5\,\text{cm/s}}$. (c) Maximum occurs where the cosine is ± 1, so $v = \underline{\pm 12.0\,\text{cm/s}}$.

CHAPTER 4

Newton's Laws of Motion

4.1 FORCE, MASS, AND ACCELERATION

4.1 A force acts on a 2-kg mass and gives it an acceleration of 3 m/s². What acceleration is produced by the same force when acting on a mass of (a) 1 kg? (b) 4 kg? (c) How large is the force?

▌ We first find the force F using $F = ma$ (one dimension). $F = (2\,\text{kg})(3\,\text{m/s}^2) = \underline{6\,\text{N}}$. Then noting $a = F/m$, we have for the different masses $m = 1\,\text{kg}$, $a = (6\,\text{N})/(1\,\text{kg}) = \underline{6\,\text{m/s}^2}$; $m = 4\,\text{kg}$, $a = \underline{1.5\,\text{m/s}^2}$. The answers are thus (a) $\underline{6\,\text{m/s}^2}$ (b) $\underline{1.5\,\text{m/s}^2}$ (c) $\underline{6\,\text{N}}$.

4.2 (Fill in the blanks.) The mass of a 300 g object is _____ (a). Its weight on earth is _____ (b). An object that weighs 20 N on earth has a mass on the moon equal to _____ (c). The mass of an object that weighs 5 lb on earth is _____ (d).

▌ (a) $\underline{300\,\text{g}}$ (b) $w = (300\,\text{g})(980\,\text{cm/s}^2) = 2.94 \times 10^5\,\text{dyn} = \underline{2.94\,\text{N}}$. (c) $m = w/g = (20\,\text{N})/(9.8\,\text{m/s}^2) = 2.04\,\text{kg}$; mass is the same anywhere. (d) $m = w/g = (5\,\text{lb})/(32.2\,\text{ft/s}^2) = \underline{0.155\,\text{slug}}$.

4.3 A resultant external force of 7.0 lb acts on an object that weighs 40 lb on earth. What is the object's acceleration (a) on earth? (b) on the moon?

▌ (a) We use $F = ma$, where we recall that F stands for the resultant of all forces acting on the mass m. To get m we note that $m = w/g = (40\,\text{lb})/(32.2\,\text{ft/s}^2) = 1.24\,\text{slug}$. Then $a = F/m = (7.0\,\text{lb})/(1.24\,\text{slug}) = \underline{5.64\,\text{ft/s}^2}$. (b) The acceleration on the moon is the same since the resultant force is still 7.0 lb and the mass is the same anywhere.

4.4 A horizontal cable pulls a 200-kg cart along a horizontal track. The tension in the cable is 500 N. Starting from rest, (a) how long will it take the cart to reach a speed of 8 m/s? (b) How far will it have gone?

▌ Assuming no friction, the tension in the cable is the only horizontal force. Then from $F_x = ma_x$ we get $a_x = (500\,\text{N})/(200\,\text{kg}) = 2.50\,\text{m/s}^2$. We now use kinematics to solve the problem.
(a) $v_x = v_{0x} + a_x t$; and since the cart starts from rest, $v_{0x} = 0$. Then $t = v_x/a_x = (8\,\text{m/s})/(2.50\,\text{m/s}^2) = \underline{3.2\,\text{s}}$.
(b) Letting the starting point be the origin we have $x = v_{0x}t + \frac{1}{2}at^2 = 0 + \frac{1}{2}(2.50\,\text{m/s}^2)(3.2\,\text{s})^2 = \underline{12.8\,\text{m}}$

4.5 A 900-kg car is going 20 m/s along a level road. How large a constant retarding force is required to stop it in a distance of 30 m?

▌ Here we start with the kinematical equation that allows us to find the acceleration a_x: $v_x^2 = v_{0x}^2 + 2a_x x$, where $v_x = 0$ when $x = 30\,\text{m}$ and $v_{0x} = 20\,\text{m/s}$. Solving we obtain $a_x = -6.67\,\text{m/s}^2$. Finally we solve for the retarding force using $F_x = ma_x = (900\,\text{kg})(-6.67\,\text{m/s}^2) = \underline{-6000\,\text{N}}$.

4.6 How much force does it take to give a 20 000-kg locomotive an acceleration of 1.5 m/s² on a level track with a coefficient of rolling friction of 0.03?

▌ Refer to Fig. 4-1. $F - \mu_k N = ma$, $N = mg$, $F = \mu_k mg + ma = 0.03(20\,000)(9.8) + 20\,000(1.5) = 5880 + 30\,000 = \underline{35.880\,\text{kN}}$

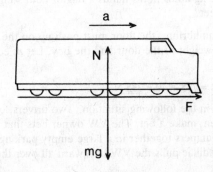

Fig. 4-1

4.7   A 12.0-g bullet is accelerated from rest to a speed of 700 m/s as it travels 20 cm in a gun barrel. Assuming the acceleration to be constant, how large was the accelerating force?

▐   From kinematics we have $v_x^2 = v_{0x}^2 + 2a_x x$. For our case $v_{0x} = 0$, and $v_x = 700$ m/s when $x = 0.20$ m. Solving we get $a_x = 1.23 \cdot 10^6$ m/s^2. We then obtain F_x by noting $m = 0.012$ kg and using $F_x = ma_x = \underline{14.8\ kN}$.

4.8   A 20-kg crate hangs at the end of a long rope. Find its acceleration when the tension in the rope is (a) 250 N, (b) 150 N, (c) zero, (d) 196 N.

▐   The crate is acted on by two vertical forces—the tension in the rope, T, upward and the weight of the crate, $w = mg$, downward. Noting that $w = (20\ \text{kg})(9.8\ \text{m/s}^2) = 196$ N and using $T - w = ma_y$, we get: (a) $a_y = \underline{2.7\ \text{m/s}^2}$; (b) $a_y = \underline{-2.3\ \text{m/s}^2}$; (c) $a_y = \underline{-9.8\ \text{m/s}^2}$; (d) $a_y = 0$. Note that negative acceleration is downward for our case.

4.9   A 40-kg trunk sliding across a floor slows down from 5.0 to 2.0 m/s in 6.0 s. Assuming that the force acting on the trunk is constant, find its magnitude and its direction relative to the velocity vector of the trunk.

▐   Letting the x axis be along the direction of motion, we have for the magnitude of the resultant force $F_x = ma_x$. To find a_x we use the kinematical relationship $v_x = v_{0x} + a_x t$, with $v_{0x} = 5.0$ m/s, $v_x = 2.0$ m/s, $t = 6.0$ s. Solving we get $a_x = -0.50$ m/s^2. Then $F_x = (40\ \text{kg})(-0.50\ \text{m/s}^2) = \underline{-20\ N}$. Noting that the resultant force in the y direction is zero since $a_y = 0$, we have our answer, **F** is 20 N in the direction opposite to the velocity.

4.10   A resultant force of 20 N gives a body of mass m an acceleration of 8.0 m/s^2, and a body of mass m' an acceleration of 24 m/s^2. What acceleration will this force cause the two masses to acquire if fastened together?

▐   From $F = ma$, with $F = 20$ N and $a = 8.0$ m/s^2, we get $m = 2.50$ kg. From $F = m'a'$ and $a' = 24.0$ m/s^2 we get $m' = 0.83$ kg. Combining the two masses yields $M = m + m' = 3.33$ kg and $F = MA$ yields $\underline{A = 6.0\ \text{m/s}^2}$.

4.11   An 1100-kg car travels on a straight highway with a speed of 30 m/s. The driver sees a red light ahead and applies her brakes, which exert a constant braking force of 4 kN. (a) What is the deceleration of the car? (b) In how many seconds will the car stop?

▐   (a) Use Newton's second law of motion. Take a retarding force to be negative.

$$F = ma \qquad -4 \times 10^3 = 1100a \qquad a = -3.636\ \text{m/s}^2$$

Thus, the deceleration is $\underline{3.636\ \text{m/s}^2}$.

(b) $a = \dfrac{v - v_0}{t}$ where v equals zero after t seconds, and v_0 equals 30 m/s.

$$-3.636 = \frac{0 - 30}{t} \qquad t = 30/3.636 = \underline{8.25\ s}$$

The car will come to a stop in $\underline{8.25\ s}$.

4.12   A force of 70 N gives an object of unknown mass an acceleration of 20 ft/s^2. What is the object's mass?

▐ $F = ma \qquad \dfrac{F}{a} = m \qquad m = \dfrac{70\ \text{N}}{(20\ \text{ft/s}^2)(0.305\ \text{m/ft})} = \dfrac{70\ \text{kg} \cdot \text{m/s}^2}{6.10\ \text{m/s}^2} = \underline{11.5\ \text{kg}}$

4.13   A boy having a mass of 75 kg holds in his hands a bag of flour weighing 40 N (Fig. 4-2). With what force does the floor push up on his feet?

▐   For the boy to be in equilibrium, the floor must push up on the boy's feet with a force F equal and opposite to the combined weight of the flour and the boy. Let m equal the mass of the boy and w the weight of the flour:

$$F = mg + w = 75(9.8) + 40 = 735 + 40 = \underline{775\ N}$$

4.14   Apply Newton's third law in the following situation: Two drivers, one owning a large Cadillac and the other owning a small Volkswagen, make a bet. The VW owner bets that his car can pull as hard as the Cadillac. They chain the two rear bumpers together in a large empty parking lot. Each driver gets into his car and applies full power. The Cadillac pulls the VW backward all over the lot. The driver of the VW later claims

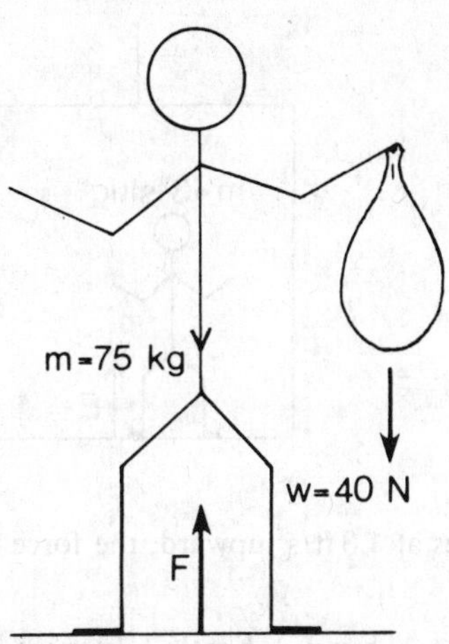

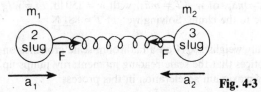

Fig. 4-2

that his car was pulling on the chain as hard as the Cadillac all the time. What does Newton's third law say in this case? Assume that the chain has negligible mass.

I Newton's third law says that the VW owner is right. Each car must pull on the chain with the same magnitude of force, one being the action and the other the reaction. The motion of the VW is a consequence of *all* the forces acting on it, not just the force of the chain. These other forces include, in particular, the frictional force between tires and road, which is quite different for the VW and the Cadillac.

4.15 A 2-slug mass pulls horizontally on a 3-slug mass by means of a lightly stretched spring (Fig. 4-3). If at one instant the 3-slug mass has an acceleration toward the 2-slug mass of 1.8 ft/s^2, find the net force on the 2-slug mass and its acceleration at that instant.

I The spring pulls on each mass with the same force F but in opposite directions. Using Newton's second law,

$$F = m_1 a_1 = m_2 a_2 \qquad F = m_2 a_2 = 3 \text{ slug} \times 1.8 \text{ ft/s}^2 = 5.4 \text{ slug} \cdot \text{ft/s}^2 \qquad F = \underline{5.4 \text{ lb}}$$

Using Newton's second law on the mass m_1,

$$F = m_1 a_1 = 5.4 \text{ lb} \qquad 2a_1 = 5.4 \qquad a_1 = \underline{2.7 \text{ ft/s}^2}$$

Fig. 4-3

4.16 A 96-lb boy is standing in an elevator. Find the force on the boy's feet when the elevator **(a)** stands still **(b)** moves downward at a constant velocity of 3 ft/s **(c)** accelerates downward with an acceleration of 4.0 ft/s^2, and **(d)** accelerates upward with an acceleration of 4.0 ft/s^2.

I **(a)** Find the mass of the boy in the proper unit, the slug:

$$m = \frac{\text{weight } (w)}{\text{acceleration of gravity } (g)} = \frac{96 \text{ lb}}{32 \text{ ft/s}^2} = 3 \text{ lb} \cdot \text{s}^2/\text{ft} = 3 \text{ slug}$$

When an elevator stands still, the force on the boy's feet, according to Newton's first law, is equal and opposite to his weight. The answer is therefore 96 lb upward.
(b) A constant velocity means zero acceleration. For zero acceleration the force of the elevator balances the earth's downward gravitational force, that is, the weight of the boy. Again by Newton's first law, the answer is 96 lb upward.
(c) When the acceleration of the elevator is downward, the boy's weight exceeds the upward force E of the elevator, yielding a net downward force F. From Fig. 4-4

$$F = ma = 3 \text{ slug} \times 4.0 \text{ ft/s}^2 = 12 \text{ lb s}^2/\text{ft} \times \text{ft/s}^2 \qquad F = 12 \text{ lb} \qquad E = w - F = 96 - 12 = \underline{84 \text{ lb}}$$

By Newton's second law, the floor of the elevator pushes upward on the boy's feet with a force of 84 lb.

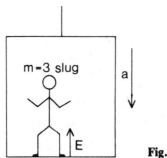

Fig. 4-4

(*d*) When the elevator accelerates at 4.0 ft/s² upward, the force E acting on the boy is greater than his weight w by an amount $F = 12$ lb:

$$E = w + F = 96 + 12 = \underline{108\ lb}\ \text{upward}$$

4.17 An elevator starts from rest with a constant upward acceleration. It moves 2.0 m in the first 0.60 s. A passenger in the elevator is holding a 3-kg package by a vertical string. What is the tension in the string during the accelerating process?

▌ To obtain the tension T in the string we apply the second law to the $m = 3.0$ kg package: $T - w = ma_y$, with $w = mg = \underline{29.4\ N}$. The acceleration a of the package is the same as that of the elevator; it is obtained from the displacement formula $y = v_{0y}t + \frac{1}{2}a_y t^2$, where $v_{0y} = 0$ and $y = 2.0$ m, when $t = 0.60$ s. Solving we get $a_y = \underline{11.1\ m/s^2}$. Substituting into our equation for T, we get $T = \underline{62.7\ N}$.

4.18 Just as her parachute opens, a 150-lb parachutist is falling at a speed of 160 ft/s. After 0.80 s has passed, the chute is fully open and her speed has dropped to 35 ft/s. Find the average retarding force exerted upon the chutist during this time.

▌ We choose downward as the positive y direction. The average acceleration of the chutist over the 0.80-s interval is

$$\bar{a}_y = \frac{(35 - 160)\text{ft/s}}{0.80\ \text{s}} = -156\ \text{ft/s}^2$$

Next we use: $\overline{\sum F_y} = m\bar{a}_y$, or $w - \bar{T} = m\bar{a}_y$, with $w = 150$ lb, $m = w/g = 4.69$ slug, and $\bar{T} =$ the average retarding force due to the chute. Solving we get $\bar{T} = \underline{881\ N}$.

4.19 A boy who normally weighs 300 N on a bathroom scale crouches on the scale and suddenly jumps upward. His companion notices that the scale reading momentarily jumps up to 400 N as the boy springs upward. Estimate the boy's maximum acceleration in this process.

▌ The maximum upward force exerted by the scale on the boy is 400 N. The net force on the boy is $(400 - 300)$ N and this equals ma. Using $m = 300/9.8$ yields $a = \underline{3.3\ m/s^2}$.

4.20 Shortly after leaping from an airplane a 91.8-kg man has an upward force of 225 N exerted on him by the air. Find the resultant force on the man.

▌ The resultant force on the man is the vector sum of two forces—the weight $w = mg = (91.8\ \text{kg})(9.8\ \text{m/s}^2) = 900$ N downward, and the 225-N force upward. Then the resultant force is $900 - 225 = \underline{675\ N\ downward}$.

4.21 To measure the mass of a box, we push it along a smooth surface, exerting a net horizontal force of 150 lb. The acceleration is observed to be 3.0 m/s². What is the mass of the box?

▌ Using $F_x = ma_x$ for the horizontal direction we get $150\ \text{lb} \times 4.45\ \text{N/lb} = m(3.0\ \text{m/s}^2)$ and $m = \underline{223\ kg}$.

4.22 A book sits on a horizontal top of a car as the car accelerates horizontally from rest. If the static coefficient of friction between car top and book is 0.45, what is the maximum acceleration the car can have if the book is not to slip?

▌ When the book of mass m is about to slide, the friction $f = \mu mg$. Friction is the only horizontal force acting, thus $f = ma$. Inserting $\mu = 0.45$ yields $a = \mu g = \underline{4.41\ m/s^2}$.

4.23 Prove the following for a car moving on a horizontal road: The magnitude of the car's acceleration cannot exceed μg, where μ is the coefficient of friction between tires and road. What is the similar expression for the acceleration of a car going up an incline whose angle is θ?

  ❚ Friction between tires and road supplies the force moving the car, so $f = ma$. On a horizontal road $F_N = mg$, so $f_{max} = \mu F_N = \mu mg$, therefore $a_{max} = \mu g$. On the incline, equations of motion parallel and perpendicular to the surface are $\mu F_N - mg \sin \theta = ma_{max}$ and $F_N - mg \cos \theta = 0$. Solve for a: $a_{max} = (\mu \cos \theta - \sin \theta)g$.

4.24 A 5-kg mass hangs at the end of a cord. Find the tension in the cord if the acceleration of the mass is (a) 1.5 m/s^2 up, (b) 1.5 m/s^2 down, and (c) 9.8 m/s^2 down.

  ❚ Choosing upward as positive we have $T - w = ma_y$, where T is the tension in the cord, $m = 5$ kg, and $w = mg = 49$ N is the weight of the mass.
(a) $T = \underline{56.5 \text{ N}}$ (b) $T = \underline{41.5 \text{ N}}$ (c) $T = \underline{0}$.

4.25 A 700-N man stands on a scale on the floor of an elevator. The scale records the force it exerts on whatever is on it. What is the scale reading if the elevator has an acceleration of (a) 1.8 m/s^2 up? (b) 1.8 m/s^2 down? (c) 9.8 m/s^2 down?

  ❚ Again choosing upward as positive and letting N represent the force of the scale on the man, we have $N - w = ma_y$. Noting that $w = 700$ N, and that $m = w/g = 71.4$ kg, we solve for N using the values of a_y given:
(a) $N = \underline{829 \text{ N}}$. (b) $N = \underline{571 \text{ N}}$. (c) $N = \underline{0}$.

4.26 Using the scale described in Prob. 4.25, a 65-kg astronaut weighs himself on the moon, where $g = 1.60 \text{ m/s}^2$. What does the scale read?

  ❚ Since g_{moon} is the acceleration of free fall on the moon, and w_{moon} is the force of gravity on the moon's surface, we have $w_{moon} = mg_{moon} = (65 \text{ kg})(1.60 \text{ m/s}^2) = \underline{104 \text{ N}}$. By Newton's first law scale reads $\underline{104 \text{ N}}$.

4.27 A rough rule of thumb states that the frictional force between dry concrete and a skidding car's tires is about equal to nine-tenths of the car's weight. If the skid marks left by a car in coming to rest are 20 m long, about how fast was the car going just before the brakes were applied?

  ❚ Since $|f| = \mu |F_N|$, and since $F_N = W$ on level ground, we have for $\mu = 0.9$, $f = -0.9W$. Find the car's acceleration from $F = ma$, which is $f = (W/g)a$ with $f = -0.9W$. Then $a = -0.9g$. Since x is 20 m, use $v^2 - v_0^2 = 2ax$ with $v = 0$, to give $v_0 = 6g^{1/2} = \underline{18.8 \text{ m/s}}$.

4.28 If the coefficient of friction between a car's wheels and a roadway is 0.70, what is the least distance in which the car can accelerate from rest to a speed of 15 m/s?

  ❚ Using $a = \mu g$ (see Probs. 4.23 and 4.27) in the kinematical formula $v^2 = v_0^2 + 2ax$,

$$x = \frac{v^2 - v_0^2}{2a} = \frac{15^2 - 0^2}{2(0.70)(9.8)} = \underline{16.4 \text{ m}}$$

4.29 A constant force accelerates an electron ($m = 9.1 \times 10^{-31}$ kg) from rest to a speed of 5×10^7 m/s in a distance of 0.80 cm. Determine this force. How many times larger than mg is it?

  ❚ First find the constant acceleration from $v^2 - v_0^2 = 2ax$; then $a = 25 \times 10^{14}/1.6 \times 10^{-2} = 1.56 \times 10^{17} \text{ m/s}^2$, so $F = ma = (9.1 \times 10^{-31})(1.56 \times 10^{17}) = \underline{1.43 \times 10^{-13} \text{ N}}$. Then $F/mg = a/g = 1.56 \times 10^{17}/9.8 = \underline{1.6 \times 10^{16}}$.

4.30 The 4.0-kg head of a sledge hammer is moving at 6.0 m/s when it strikes a spike, driving it into a log; the duration of the impact (or the time for the sledge hammer to stop after contact) is 0.0020 s. Find (a) the time average of the impact force, (b) the distance the spike penetrates the log.

  ❚ (a) The average force is a constant force that would effect the same result as the actual time-varying force over the time interval involved. For a constant force we can use $v_x = v_{0x} + a_x t$ to find the corresponding constant acceleration a_x. Here $v_{0x} = 6.0$ m/s, $v_x = 0$, and $t = 0.0020$ s. Solving we get $a_x = -3000 \text{ m/s}^2$. Then $F_x = ma_x$ yields $F_x = -12$ kN. Here F_x represents the force of the spike on the hammerhead. The reaction force on the spike is the impact force, which is $\underline{12 \text{ kN}}$.
(b) The spike moves the same distance as the hammerhead in the time interval in question. For the hammerhead we have $x = v_{0x}t + \frac{1}{2}a_x t^2 = (6.0 \text{ m/s})(0.0020 \text{ s}) + \frac{1}{2}(-3000 \text{ m/s}^2)(0.0020 \text{ s})^2 = 0.006$ m or $x = \underline{6.0 \text{ mm}}$.

4.31^c A body of mass m moves along Y such that at time t its position is $y(t) = at^{3/2} - bt + c$, where a, b, c are constants. (**a**) Calculate the acceleration of the body. (**b**) What is the force acting on it?

▌ (**a**) $a_y = d^2y/dt^2$. We first obtain $v_y = dy/dt = \frac{3}{2}at^{1/2} - b$ and then $d^2y/dt^2 = \frac{3}{4}at^{-1/2}$. To obtain the force we have (**b**) $F = ma_y = \frac{3}{4}mat^{-1/2}$

4.32^c Measurements on a 300-g object moving along the x axis show its position (in centimeters) to be given by $x = 0.20t - 5.0t^2 + 7.5t^3$, where t is the time in seconds. Find the net force that acted on the object during the time for which this expression applies.

▌ Because $F_{net} = ma$, we must find a; $a = dv/dt$ and $v = dx/dt$. Doing the differentiations, $v = 0.20 - 10.0t + 22.5t^2$, $a = -10.0 + 45t$; so $F_{net} = 300(-10.0 + 45t)$ dyn $= 0.00300(-10.0 + 45t)$ N

4.33^c A 50-g mass vibrating up and down at the end of a spring has its position given by $y = 0.150 \sin 3t$ m for t in seconds. Find the net force that acts on the mass to give it this motion.

▌ The acceleration is $d^2y/dt^2 = -1.35 \sin 3t$ m/s^2; so $F_{net} = 0.050(-1.35 \sin 3t) = \underline{-0.0675 \sin 3t}$ N. The minus sign indicates a restoring force.

4.34^c A body of mass m moves along X such that at time t its position is $x(t) = \alpha t^4 - \beta t^3 + \gamma t$, where α, β, γ are constants. (**a**) Calculate the acceleration of the body. (**b**) What is the force acting on it?

▌ (**a**) $$\dot{x} = 4\alpha t^3 - 3\beta t^2 + \gamma \quad \text{and} \quad \ddot{x} = 12\alpha t^2 - 6\beta t$$

(**b**) $$F_x = m\ddot{x} = 12m\alpha t^2 - 6m\beta t$$

4.2 FRICTION; INCLINED PLANES; VECTOR NOTATION

4.35 The breaking strength of a steel cable is 20 kN. If one pulls horizontally with this cable, what is the maximum horizontal acceleration which can be given to an 8-ton (metric) body resting on a rough horizontal surface if the coefficient of kinetic friction is 0.15?

▌ Let T be the cable force. Then $\sum F = ma$ becomes $T - \mu_k mg = ma$. For maximum acceleration $T = 2.0 \times 10^4$ N, so that 2.0×10^4 N $- 0.15(8000$ kg$)(9.8$ m/s$^2) = (8000$ kg$)a$. Solving we get $a = \underline{1.03 \text{ m/s}^2}$.

4.36 A 20-kg wagon is pulled along the level ground by a rope inclined at 30° above the horizontal. A friction force of 30 N opposes the motion. How large is the pulling force if the wagon is moving with (**a**) constant speed, and (**b**) an acceleration of 0.40 m/s^2?

▌ Let the pulling force of the rope be T. Using $\sum F_x = ma_x$ we have for our case $T \cos 30° - 30$ N $= ma_x$, where $m = 20$ kg. (**a**) For $a_x = 0$, $T = \underline{34.6 \text{ N}}$. (**b**) For $a_x = 0.40$ m/s^2, $T = \underline{43.9 \text{ N}}$

4.37 Suppose, as shown in Fig. 4-5, that a 70-kg box is pulled by a 400-N force at an angle of 30° to the horizontal. The coefficient of sliding friction is 0.50. Find the acceleration of the box.

▌ Because the box does not move vertically, $\sum F_y = ma_y = 0$. From Fig. 4-5, we see that this equation is $Y + 200$ N $- mg = 0$. But $mg = (70$ kg$)(9.8$ m/s$^2) = 686$ N. It follows that $Y = 486$ N.

We can find the friction force acting on the box by writing $f = \mu Y = (0.50)(486$ N$) = 243$ N. Now let us write $\sum F_x = ma_x$ for the box. It is $(346 - 243)$ N $= (70$ kg$)a_x$, from which $a_x = \underline{1.47 \text{ m/s}^2}$.

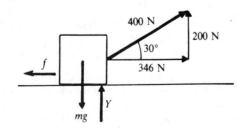

Fig. 4-5

4.38 As shown in Fig. 4-6, a force of 400 N pushes on a 25-kg box. Starting from rest, the box achieves a velocity of 2.0 m/s in a time of 4 s. Find the coefficient of sliding friction between box and floor.

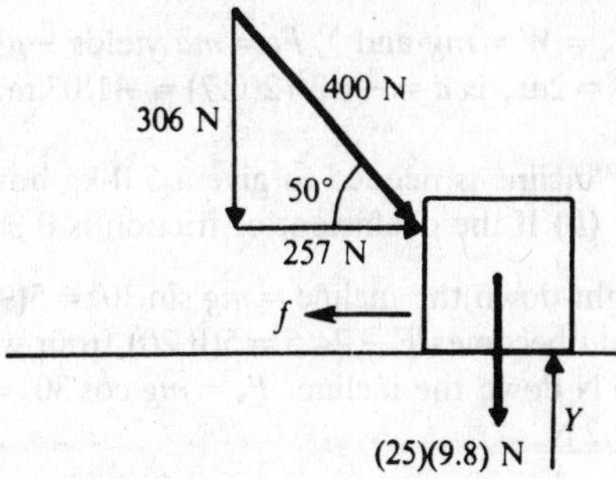

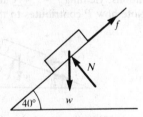

Fig. 4-6

$\blacksquare$ We must find f by use of $F = ma$. But first we must find a from a motion problem. We know that $v_0 = 0$, $v_f = 2 \text{ m/s}$, $t = 4 \text{ s}$. Using $v_f = v_0 + at$ gives

$$a = \frac{v_f - v_0}{t} = \frac{2 \text{ m/s}}{4 \text{ s}} = 0.50 \text{ m/s}^2$$

Now we can write $\Sigma F_x = ma_x$, where $a_x = a = 0.50 \text{ m/s}^2$. From Fig. 4-6, this equation is $257 \text{ N} - f = (25 \text{ kg})(0.50 \text{ m/s}^2)$, or $f = 245 \text{ N}$. We now wish to use $\mu = f/Y$. To find Y, we write $\Sigma F_y = ma_y = 0$, since no vertical motion occurs. From Fig. 4-6, $Y - 306 \text{ N} - (25)(9.8) \text{ N} = 0$, or $Y = 551 \text{ N}$. Then

$$\mu = \frac{f}{Y} = \frac{245}{551} = \underline{0.44}$$

4.39 A 12-kg box is released from the top of an incline that is 5.0 m long and makes an angle of 40° to the horizontal. A 60-N friction force impedes the motion of the box. (*a*) What will be the acceleration of the box and (*b*) how long will it take to reach the bottom of the incline?

$\blacksquare$ In Fig. 4-7 we show the three forces acting on the block: the frictional force $f = 60 \text{ N}$; the normal force N, which is perpendicular to the incline; and the weight of the block, $w = mg = (12 \text{ kg})(9.8 \text{ m/s}^2) = 118 \text{ N}$. We choose the x axis along the incline with downward as positive. Using $\Sigma F_x = ma_x$, we have $w \sin 40° - f = ma_x$, or $(118 \text{ N})(0.642) - (60 \text{ N}) = (12 \text{ kg})a_x$. Solving we have $a_x = \underline{1.31 \text{ m/s}^2}$. To find the time to reach the bottom of the incline, starting from rest, we use $x = v_{0x}t + \frac{1}{2}a_x t^2$, with $v_{0x} = 0$ and $x = 5.0 \text{ m}$. Solving we get $t = (7.63 \text{ s}^2)^{1/2} = \underline{2.76 \text{ s}}$.

4.40 For the situation outlined in Prob. 4.39, what is the coefficient of friction between box and incline?

$\blacksquare$ Again referring to Fig. 4-7 we have from $\Sigma F_y = 0$, $N - w \cos 40° = 0$, or $N = 90 \text{ N}$. Then, recalling that the coefficient of kinetic friction is given by $\mu_k = f/N$, we have $\mu_k = \underline{0.67}$.

Fig. 4-7

4.41 An inclined plane makes an angle of 30° with the horizontal. Find the constant force, applied parallel to the plane, required to cause a 15-kg box to slide (*a*) up the plane with acceleration 1.2 m/s² and (*b*) down the incline with acceleration 1.2 m/s². Neglect friction forces.

$\blacksquare$ Here we assume that the x axis is along the incline and positive upward. If P is the constant force referred to, then from $\Sigma F_x = ma_x$ we have $P - w \sin 30° = ma_x$, with $m = 15 \text{ kg}$ and $w = mg = 147 \text{ N}$. (*a*) For $a_x = 1.2 \text{ m/s}^2$, $P = \underline{91.5 \text{ N}}$. (*b*) $a_x = -1.2 \text{ m/s}^2$, $P = \underline{55.5 \text{ N}}$

4.42 A 400-g block originally moving at 120 cm/s coasts 70 cm along a tabletop before coming to rest. What is the coefficient of friction between block and table?

❚ For the block $\sum F_y = 0$ yields $F_N = W = mg$ and $\sum F_x = ma$ yields $-\mu F_N = ma$; hence, $\mu = -a/g$. The uniform acceleration, from $v^2 - v_0^2 = 2ax$, is $a = -1.2^2/2(0.7) = -1.03$ m/s^2; then $\mu = 1.03/9.8 = \underline{0.105}$.

4.43 How large a force parallel to a 30° incline is needed to give a 5.0-kg box an acceleration of 0.20 m/s^2 up the incline (**a**) if friction is negligible? (**b**) If the coefficient of friction is 0.30?

❚ (**a**) The component of the weight down the incline $= mg \sin 30° = 5(9.8)(0.5) = 24.5$ N, while the external force up the plane is F. So $F_{net} = ma$ becomes $F - 24.5 = 5(0.20)$ from which $F = \underline{25.5\text{ N}}$ (**b**) A friction force $= \mu F_N$ must be added to 24.5 N down the incline. $F_N = mg \cos 30° = 42.4$ N and $\mu F_N = 12.7$ N, so F is larger by this amount; thus $F = \underline{38.2\text{ N}}$.

4.44 An 8.0-kg box is released on a 30° incline and accelerates down the incline at 0.30 m/s^2. Find the frictional force impeding its motion. How large is the coefficient of friction in this situation?

❚ The component of the weight down the incline $= 8(9.8)(0.5) = 39.2$ N. Now $F = ma$ leads to $39.2 - f = 8(0.3)$, so that $f = 36.8$ N. The normal force equals the component of W perpendicular to the incline, $8(9.8)(0.867) = 67.9$ N. Therefore $\mu = 36.8/67.9 = \underline{0.54}$.

4.45 A horizontal force P is exerted on a 20-kg box in order to slide it up a 30° incline. The friction force retarding the motion is 80 N. How large must P be if the acceleration of the moving box is to be (**a**) zero and (**b**) 0.75 m/s^2?

❚ Here we choose the x axis along the incline with positive upward. All the forces on the block are shown in Fig. 4-8. From $\sum F_x = ma_x$ we have $P \cos 30° - w \sin 30° - f = ma_x$, where $m = 20$ kg, $w = mg = 196$ N, and $f = 80$ N. (**a**) For $a_x = 0$, $P = \underline{206\text{ N}}$. (**b**) For $a_x = 0.75$ m/s^2, $P = \underline{223\text{ N}}$

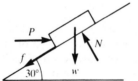

Fig. 4-8

4.46 A horizontal force of 200 N is required to cause a 15-kg block to slide up a 20° incline with an acceleration of 25 cm/s^2. Find (**a**) the friction force on the block and (**b**) the coefficient of friction.

❚ The situation is as shown in Fig. 4-9. We again choose the x axis along the incline, and the y axis perpendicular to the incline. Then from $\sum F_x = ma_x$ we get $P \cos 20° - f - w \sin 20° = (15\text{ kg})(0.25\text{ m/s}^2)$ and from equilibrium in the y direction we get $N - P \sin 20° - w \cos 20° = 0$. Noting that $P = 200$ N and $w = mg = 147$ N, we solve the equations, yielding $f = \underline{134\text{ N}}$ and $N = \underline{207\text{ N}}$, respectively. Finally $\mu_k = f/N = \underline{0.65}$ is the coefficient of friction. Note how P contributes to the normal force.

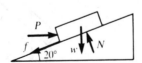

Fig. 4-9

4.47 What is the smallest force parallel to a 37° incline needed to keep a 100-N weight from sliding down the incline if the coefficients of static and kinetic friction are both 0.30?

❚ As usual, choose the x and y axes parallel and perpendicular to the incline. Let F be the unknown force, f the friction force, and N the normal force. If the weight is about to slide down the incline, then the frictional force is maximum and in the upward direction. Then

$$F - w \sin 37° + f_{max} = 0, \quad \text{where} \quad f_{max} = \mu_s N \quad \text{and} \quad N - w \cos 37° = 0.$$

Noting that $w = 100$ N and $\mu_s = 0.3$, we get $N = 80$ N and $f_{max} = 24$ N; and finally $F = \underline{36\text{ N}}$ is the minimum force needed along the incline.

4.48 Referring to Problem 4.47, indicate what parallel force is required to keep the weight moving up the incline at constant speed.

▮ Here we have kinetic friction down the incline, and $f = \mu_k N$. N is still 80 N (why?) and since $\mu_k = 0.30$, we have $f = 24$ N. Then $\Sigma F_x = 0$ yields $F - w \sin 37° - f = 0$; and solving, we get $F = \underline{84\,\text{N}}$.

4.49 For the same conditions as in Prob. 4.48 assume that the force F, up the incline, is 94 N. What is the acceleration of the object? If the object starts from rest, how far will it move in 10 s?

▮ As before, $f = \mu_k N = 24$ N and is down the incline opposing the motion. Then using $\Sigma F_x = ma_x$, we have $F - w \sin 37° - f = ma_x$. Noting $m = w/g = 10.2$ kg we solve, getting $a_x = \underline{0.98\,\text{m/s}^2}$ up the incline. Next we use the kinematic formula $x = v_{0x}t + \frac{1}{2}a_x t^2$, with $v_{0x} = 0$. Setting $t = 10$ s and using our a_x, we get $x = \underline{49\,\text{m}}$.

4.50 A 5-kg block rests on a 30° incline. The coefficient of static friction between the block and incline is 0.20. How large a horizontal force must push on the block if the block is to be on the verge of sliding up the incline?

▮ Let P be the horizontal force. We choose x and y axes ∥ and ⊥ to the incline. Since the block is on the verge of moving, it is in equilibrium: $\Sigma F_x = 0$ and $\Sigma F_y = 0$. Here we have maximum static frictional force down the incline. $P \cos 30° - w \sin 30° - f_{max} = 0$; $N - w \cos 30° - P \sin 30° = 0$, where N and f are the normal and friction forces, respectively. $w = mg = 49$ N. Substituting $f_{max} = \mu_s N$ in the first equation, and putting in all known quantities, we get

$$0.866P - 0.20N = 24.5\,\text{N} \qquad N - 0.50P = 42.4\,\text{N}$$

These two equations in two unknowns can be solved in a variety of ways. We multiply the first equation by 5 and add to the second equation to eliminate N. This yields $3.83P = 165$ N, or $P = \underline{43.1\,\text{N}}$.

4.51 Rework Prob. 4.50 for incipient motion down the incline.

▮ The only difference from Prob. 4.50 is that the maximum frictional force is up the incline. We therefore change the sign in front of f_{max} in the $\Sigma F_x = 0$ equation. We then get for our two equations:

$$0.866P + 0.20N = 24.5\,\text{N} \qquad N - 0.50P = 42.4\,\text{N}$$

We again multiply the first equation by 5, but now subtract the second equation to eliminate N. This yields $4.83P = 80.1$ N, or $P = \underline{16.6\,\text{N}}$. Note that in Probs. 4.50 and 4.51 the numerical valves of N are different, reflecting their dependence on P.

4.52 In Fig. 4-10, the 8-kg object is subject to the forces $F_1 = 30$ N and $F_2 = 40$ N. Find the acceleration of the object.

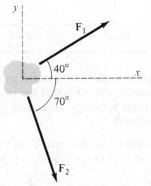

Fig. 4-10

▮ Newton's second law in component form is $F_{1x} + F_{2x} = ma_x$ and $F_{1y} + F_{2y} = ma_y$, or $30 \cos 40° + 40 \cos 70° = 8a_x$ and $30 \sin 40° - 40 \sin 70° = 8a_y$. Solving, $\mathbf{a} = \underline{4.6\mathbf{i} - 2.3\mathbf{j}\,\text{m/s}^2}$.

4.53 A 7-kg object is subjected to two forces, $\mathbf{F}_1 = 20\mathbf{i} + 30\mathbf{j}$ N and $\mathbf{F}_2 = 8\mathbf{i} - 50\mathbf{j}$ N. Find the acceleration of the object.

▮ $$\mathbf{F} = \mathbf{F}_1 + \mathbf{F}_2 = 28\mathbf{i} - 20\mathbf{j}\,\text{N}; \qquad \mathbf{a} = \frac{1}{m}\mathbf{F} = 4\mathbf{i} - (20/7)\mathbf{j}\,\text{m/s}^2.$$

4.54 The forces $\mathbf{F}_1$ and $\mathbf{F}_2$ shown in Fig. 4-10 give the 8-kg object the acceleration $\mathbf{a} = 3.0\mathbf{i}$ m/s². Find F_1 and F_2.

❙ Our motion equations take the form $F_{1x} + F_{2x} = 8(3)$, $F_{1y} + F_{2y} = 0$. From the latter, $F_1 \sin 40° = F_2 \sin 70°$; from the former, $F_1 \cos 40° = 24 - F_2 \cos 70°$. Dividing these relations we have $\tan 40° = F_2 \sin 70°/(24 - F_2 \cos 70°)$. Using these, $F_2 = \underline{16.4\,N}$ and $F_1 = (\sin 70°/\sin 40°)F_2 = \underline{24\,N}$.

4.55 Find the force needed to give a proton ($m = 1.67 \times 10^{-27}$ kg) an acceleration $2 \times 10^9 \mathbf{i} - 3 \times 10^9 \mathbf{j}$ m/s^2.

❙ $\mathbf{F} = m\mathbf{a} = 3.34 \times 10^{-18}\mathbf{i} - 5.01 \times 10^{-18}\mathbf{j}$ N

4.56 A 200-g object is subjected to a force $0.30\mathbf{i} - 0.40\mathbf{j}$ N. If the object starts from rest, what will be the velocity vector of the object after 6 s?

❙ $\mathbf{v} = \mathbf{v}_0 + t\mathbf{a} = 0 + \dfrac{t}{m}\mathbf{F} = \dfrac{6}{0.200}(0.30\mathbf{i} - 0.40\mathbf{j}) = \underline{9\mathbf{i} - 12\mathbf{j}}$ m/s

4.57 If the object of Prob. 4.56 started at the origin, what was its location at the end of the 6-s period?

❙ $\mathbf{s} = \mathbf{s}_0 + t\mathbf{v}_0 + \tfrac{1}{2}t^2\mathbf{a} = 0 + 0 + \dfrac{t^2}{2m}\mathbf{F} = \dfrac{36}{2(0.200)}(0.30\mathbf{i} - 0.40\mathbf{j}) = 27\mathbf{i} - 36\mathbf{j}$ m

Thus the object was found at the point $\underline{(27\,m, -36\,m)}$.

4.3 TWO-OBJECT AND OTHER PROBLEMS

4.58 In Fig. 4-11, find the acceleration of the cart that is required to prevent block B from falling. The coefficient of static friction between the block and the cart is μ_s.

❙ If the block is not to fall, the friction force, f, must balance the block's weight: $f = mg$. But the horizontal motion of the block is given by $N = ma$. Therefore,

$$\frac{f}{N} = \frac{g}{a} \qquad \text{or} \qquad a = \frac{g}{f/N}$$

Since the maximum value of f/N is μ_s, we must have $a \geq g/\mu_s$ if the block is not to fall.

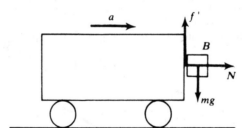

Fig. 4-11

4.59 A passenger on a large ship sailing in a quiet sea hangs a ball from the ceiling of her cabin by means of a long thread. Whenever the ship accelerates, she notes that the pendulum ball lags behind the point of suspension and so the pendulum no longer hangs vertically. How large is the ship's acceleration when the pendulum stands at an angle of 5° to the vertical?

❙ See Fig. 4-12. The ball is accelerated by the force $T \sin 5°$. Therefore $T \sin 5° = ma$. Vertically $\sum F = 0$, so $T \cos 5° = mg$. Solving for $a = g \tan 5°$ gives $a = 0.0875g = \underline{0.86\ \text{m/s}^2}$.

Fig. 4-12

4.60 A rectangular block of mass m sits on top of another similar block, which in turn sits on a flat table. The maximum possible frictional force of one block on the other is $2.0m$ N. What is the largest possible acceleration which can be given the lower block without the upper block sliding off? What is the coefficient of friction between the two blocks?

❙ $F_{\text{max}} = 2.0m = ma_{\text{max}}$, so $a_{\text{max}} = \underline{2.0\ \text{m/s}^2}$. Also $\mu = f/mg = 2.0m/9.8m = \underline{0.20}$.

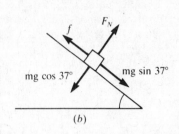

Fig. 4-13

(a) (b)

4.61 A block sits on an incline as shown in Fig. 4-13(a). (a) What must be the frictional force between block and incline if the block is not to slide along the incline when the incline is accelerating to the right at 3 m/s²? (b) What is the least value μ_s can have for this to happen?

▎ Resolve the forces and the 3-m/s² acceleration into components perpendicular and parallel to the plane [Fig. 4-13(b)]. Write $F = ma$ for each direction: $0.6mg - f = 3(0.8)m$ and $F_N - 0.8mg = 3(0.6)m$, which yield (a) $f = \underline{(3.48m)\,\text{N}}$, $F_N = (9.64m)\,\text{N}$

$$(b) \quad \mu_s = \frac{f}{F_N} = \frac{3.48m}{9.64m} = \underline{0.36}$$

4.62 In the absence of friction, would the block of Prob. 4.61 accelerate up or down the incline?

▎ *Down* [at 3.48 m/s² relative to incline, since total acceleration down incline is then 0.6 g by Prob. 4.61(a)].

4.63 The inclined plane shown in Fig. 4-14 has an acceleration **a** to the right. Show that the block will slide on the plane if $a > g \tan (\theta - \alpha)$, where $\mu_s \equiv \tan \theta$ is the coefficient of static friction for the contacting surfaces.

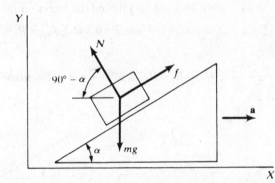

X   **Fig. 4-14**

▎ If the block is not to slide, it must have the same acceleration as the plane. Hence

$$f \cos \alpha - N \sin \alpha = ma \qquad f \sin \alpha + N \cos \alpha - mg = 0$$

From these,

$$f = m(a \cos \alpha + g \sin \alpha) \qquad N = m(g \cos \alpha - a \sin \alpha)$$

and

$$\frac{f}{N} = \frac{a \cos \alpha + g \sin \alpha}{g \cos \alpha - a \sin \alpha} = \frac{a + g \tan \alpha}{g - a \tan \alpha}$$

Now the maximum value of f/N in the absence of slipping is $\mu_s = \tan \theta$. Thus the acceleration a must satisfy

$$\frac{a + g \tan \alpha}{g - a \tan \alpha} \le \tan \theta \qquad \text{or} \qquad a \le g \frac{\tan \theta - \tan \alpha}{1 + \tan \theta \tan \alpha} = g \tan (\theta - \alpha)$$

If $a > g \tan (\theta - \alpha)$, the block will slide.

4.64 Objects A and B, each of mass m, are connected by a light inextensible cord. They are constrained to move on a frictionless ring in a vertical plane, as shown in Fig. 4-15. The objects are released from rest at the positions shown. Find the tension in the cord just after release.

▎ At the moment of release, A is constrained to move horizontally and B vertically, so that the two initial

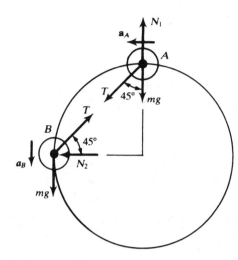

Fig. 4-15

accelerations are tangential as shown. Furthermore, the two accelerations have the same magnitude, a, since otherwise the cord would have to stretch. Thus, the horizontal force equation for A and the vertical force equation for B, at the indicated positions, are

$$T \sin 45° = ma \qquad mg - T \sin 45° = ma$$

Eliminating a,

$$T = \frac{mg}{2 \sin 45°} = \frac{mg}{\sqrt{2}}$$

4.65   If the system in Fig. 4-16(a) is given an acceleration, find the forces on the sphere, assuming no friction.

▮   From Fig. 4-16(b), $\Sigma F_{\text{ver}} = R_1 \cos 30° - w = ma_{\text{ver}} = 0$ and $\Sigma F_{\text{hor}} = R_2 - R_1 \sin 30° = ma$. Thus, the acting forces are

$$R_1 = \frac{w}{\cos 30°} = 1.15w \qquad R_2 = R_1 \sin 30° + \frac{w}{g}a = (1.15w)(0.5) + \frac{w}{g}a = w\left(0.58 + \frac{a}{g}\right)$$

and the weight, w.

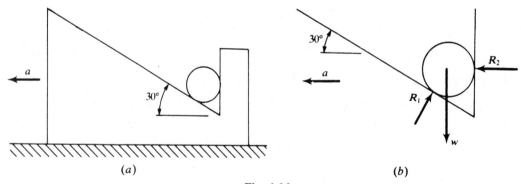

(a)   (b)

Fig. 4-16

4.66   In Fig. 4-17, mass A is 15 kg and mass B is 11 kg. If they are given an upward acceleration of 3 m/s² by pulling up on A, find the tensions T_1 and T_2.

▮   First apply Newton's second law to the system as a whole to find the force F_1 accelerating both masses upward.

$$F_1 = (m_A + m_B)a = (15 + 11)3 \qquad F_1 = 78 \text{ N}$$

Since F_1 is the resultant force, $F_1 = T_1 - m_A g - m_B g$, and the tension T_1 is the sum of the weights of A and B plus F_1.

$$T_1 = m_A g + m_B g + F_1 = 15(9.8) + 11(9.8) + 78 = 147 + 107.8 + 78 \qquad T_1 = \underline{332.8 \text{ N}}$$

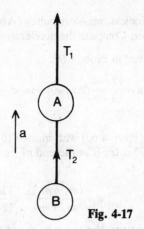

Fig. 4-17

Similarly for mass B only,

$$F_2 = m_B a = 11(3) = 33 \text{ N} \qquad T_2 = m_B g + F_2 = 11(9.8) + 33 = 107.8 + 33 = \underline{140.8 \text{ N}}$$

To check, for block A only,

$$T_1 = m_A g + m_A a + T_2 = 147 + 45 + 140.8 = \underline{332.8 \text{ N}}$$

4.67 Referring to Fig. 4-18, find the acceleration of the blocks and the tension in the connecting string if the applied force is F and the frictional forces on the blocks are negligible.

▮ Apply $F = ma$ to each block in turn to obtain $F - T = m_2 a$ and $T = m_1 a$. Solve for T and a to obtain $a = F/(m_1 + m_2)$ and $T = m_1 F/(m_1 + m_2)$.

4.68 In Fig. 4-18, if $F = 20$ N, $m_1 = m_2 = 3$ kg, and the acceleration is 0.50 m/s^2, what will be the tension in the connecting cord if the frictional forces on the two blocks are equal? How large is the frictional force on either block?

Fig. 4-18

▮ Write $F = ma$ for each block using f as the friction force on each block. Then we obtain $F - f - T = m_2 a$ and $T - f = m_1 a$. Use the given values and solve to find $\underline{T = 10 \text{ N}}$ and $\underline{f = 8.5 \text{ N}}$.

4.69 The device diagramed in Fig. 4-19 is called an *Atwood's machine*. In terms of m_1 and m_2 with $m_2 > m_1$, (*a*) how far will m_2 fall in time t after the system is released? (*b*) What is the tension in the light cord that connects the two masses? Assume the pulley to be frictionless and massless.

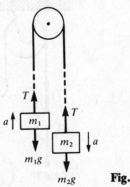

Fig. 4-19

▮ (*a*) Isolate the forces on each mass and write Newton's second law, choosing *up* as positive: $T - m_1 g = m_1 a$ and $T - m_2 g = -m_2 a$. Eliminating T gives $a = (m_2 - m_1)g/(m_1 + m_2)$. Now use $y = at^2/2$ to find the distance fallen in time t. (*b*) From the above equations, $T = 2m_1 m_2 g/(m_1 + m_2)$.

4.70 A cord passing over a frictionless, massless pulley (Atwood's machine) has a 4-kg block tied to one end and a 12-kg block tied to the other. Compute the acceleration and the tension in the cord.

▮ Using the formulas derived in Prob. 4.69,

$$a = \frac{12-4}{12+4}(9.8) = 4.9 \text{ m/s}^2 \qquad T = \frac{2(4)(12)}{4+12}(9.8) = 58.8 \text{ N}$$

4.71 For an Atwood's machine (Prob. 4.69) with masses 10 and 12 kg, find (**a**) the velocities at the end of 3 s and (**b**) the distances moved in 3 s. (**c**) If at the end of 3 s the string is cut, find the distances moved by the masses in the next 6 s.

▮ (**a**)
$$a = \frac{(m_2 - m_1)g}{m_1 + m_2} = \frac{12-10}{12+10}(9.8) = \underline{0.89 \text{ m/s}^2}$$

Since the acceleration is constant, the common speed at the end of 3 s is $v = v_0 + at = 0 + (0.89)(3) = \underline{2.67 \text{ m/s}}$. Mass 2 moves down and mass 1 moves up. (**b**) The distance moved by each mass in 3 s is

$$s = v_0 t + \tfrac{1}{2}at^2 = (0)(3) + \tfrac{1}{2}(0.89)(3)^2 = \underline{4 \text{ m}}.$$

(**c**) If the string is cut, the masses fall freely with initial velocities $v_{20} = -2.67$ m/s and $v_{10} = +2.67$ m/s, with *up* taken as positive. For mass 2, the displacement in 6 s is then

$$y_2 = v_{20}t - \tfrac{1}{2}gt^2 = (-2.67)(6) - \tfrac{1}{2}(9.8)(6)^2 = -192.4 \text{ m}$$

i.e., a downward distance of $\underline{192.4 \text{ m}}$. Mass 1 travels upward a distance

$$d' = \frac{v_{10}^2}{2g} = \frac{(2.67)^2}{2(9.8)} = 0.4 \text{ m}$$

before coming to a stop and then falling downward. The time of travel upward before coming to a stop for mass 1 is

$$t_{up} = \frac{v_{10}}{g} = \frac{2.67}{9.8} = 0.27 \text{ s}$$

It then travels downward 5.73 s for a distance

$$d'' = |\tfrac{1}{2}(-g)t^2| = \tfrac{1}{2}(9.8)(5.73)^2 = 160.9 \text{ m}.$$

The total distance traveled by mass 1 is then $d = d' + d'' = 0.4 + 160.9 = \underline{161.3 \text{ m}}$.

4.72 In Fig. 4-20, the weights of the objects are 200 and 300 N. The pulleys are essentially frictionless and massless. Pulley P_1 has a stationary axle but pulley P_2 is free to move up and down. Find the tensions T_1 and T_2, and the acceleration of each body.

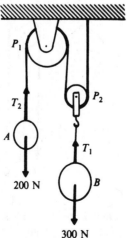

Fig. 4-20

▌ Mass B will rise and mass A will fall. You can see this by noting that the forces acting on pulley P_2 are $2T_2$ up and T_1 down. Therefore $T_1 = 2T_2$ (the inertialess object transmits the tension). Twice as large a force is pulling upward on B as on A.

Let a = downward acceleration of A. Then $\frac{1}{2}a$ = upward acceleration of B. [As the cord between P_1 and A lengthens by 1 unit, the segments on either side of P_2 each shorten by $\frac{1}{2}$ unit. Hence, $\frac{1}{2} = s_B/s_A = (\frac{1}{2}a_B t^2)/(\frac{1}{2}a_A t^2) = a_B/a_A$.] Write $\Sigma F_y = ma_y$ for each mass in turn, taking the direction of motion as positive in each case. We have

$$T_1 - 300\,\text{N} = m_B(\tfrac{1}{2}a) \qquad \text{and} \qquad 200\,\text{N} - T_2 = m_A a$$

But $m = w/g$ and so $m_A = (200/9.8)$ kg and $m_B = (300/9.8)$ kg. Further, $T_1 = 2T_2$. Substitution of these values in the two equations allows us to compute T_2 and then T_1 and a. The results are

$$T_1 = \underline{327\,\text{N}} \qquad T_2 = \underline{164\,\text{N}} \qquad a = \underline{1.78\,\text{m/s}^2}$$

4.73 An inclined plane making an angle of 25° with the horizontal has a pulley at its top. A 30-kg block on the plane is connected to a freely hanging 20-kg block by means of a cord passing over the pulley. Compute the distance the 20-kg block will fall in 2 s starting from rest. Neglect friction.

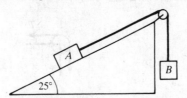

Fig. 4-21

▌ The situation is as shown in Fig. 4-21. We apply Newton's second law to each block separately. For block B we choose downward as positive, while for block A we choose our x axis along the incline with the positive sense upward. This choice allows us to use the same symbol, a, for the acceleration of each block. Then for block B, $w_b - T = m_b a$, where $m_b = 20$ kg and $w_b = 196$ N and T is the tension in the cord. Since the pulley is frictionless, the same tension T will exist on both sides of the pulley. Then for block A, $T - w_a \sin 25° = m_a a$, where $m_a = 30$ kg and $w_a = 294$ N. We can eliminate the tension T by adding the two equations, which yields $w_b - w_a \sin 25° = (m_a + m_b)a$. Substituting in the known values we solve, getting $a = \underline{1.44\,\text{m/s}^2}$. The equation for fall from rest is $y = v_{0y}t + \frac{1}{2}a_y t^2$, with $v_{0y} = 0$. Substituting in $a_y = 1.44$ m/s² and $t = 2$ s, we get $y = \underline{2.88\,\text{m}}$.

4.74 Repeat Prob. 4.73 if the coefficient of friction between block and plane is 0.20.

▌ The equation for block A is now $T - w_a \sin 25° - f = m_a a$. The block B equation is the same as before: $w_b - T = m_b a$. Adding the two equations we get this time $w_b - w_a \sin 25° - f = (m_a + m_b)a$. As soon as we obtain f we can solve for a. To obtain f we note that $f = \mu_k N$, where $\mu_k = 0.20$ is the coefficient of kinetic friction and N is the normal force exerted on the block by the incline. Noting $\Sigma F_y = 0$ for the direction perpendicular to the incline, we have $w_a \cos 25° - N = 0$, or $N = 266$ N. Then $f = 53$ N and solving for the acceleration $a = \underline{0.38\,\text{m/s}^2}$. Again using $y = v_{0y}t + \frac{1}{2}a_y t^2$ with $v_{0y} = 0$, $a_y = 0.38$ m/s², and $t = 2$ s, we get $y = \underline{0.76\,\text{m}}$.

4.75 In Fig. 4-22, the two boxes have identical masses, 40 kg. Both experience a sliding friction force with $\mu = 0.15$. Find the acceleration of the boxes and the tension in the tie cord.

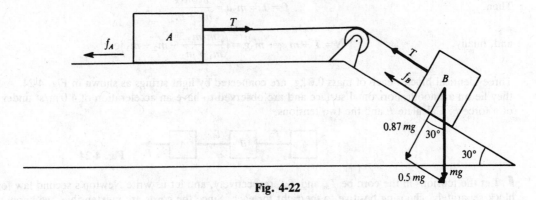

Fig. 4-22

❚ Using $f = \mu Y$, where Y = normal force, we find that the friction forces on the two boxes are

$$f_A = (0.15)(mg) \qquad f_B = (0.15)(0.87\,mg)$$

But $m = 40$ kg and so $f_A = 59$ N and $f_B = 51$ N.

Let us apply $\Sigma F_x = ma_x$ to each block in turn, taking the direction of motion as positive.

$$T - 59\,\text{N} = (40\,\text{kg})a \qquad \text{and} \qquad 0.5\,mg - T - 51\,\text{N} = (40\,\text{kg})a$$

Solving these two equations for a and T gives $a = \underline{1.08\,\text{m/s}^2}$ and $T = \underline{102\,\text{N}}$.

4.76 Two bodies, of masses m_1 and m_2, are released from the position shown in Fig. 4.23(a). If the mass of the smooth-topped table is m_3, find the reaction of the floor on the table while the two bodies are in motion. Assume that the table does not move.

❚ From Fig. 4-23(b), the force equations for the bodies are

Body 1: $\Sigma F_{\text{ver}} = w_1 - T = m_1 a$

Body 2: $\Sigma F_{\text{hor}} = T = m_2 a$

Table: $\Sigma F_{\text{ver}} = N - T - w_2 - w_3 = 0$, $\Sigma F_{\text{hor}} = T - f = 0$

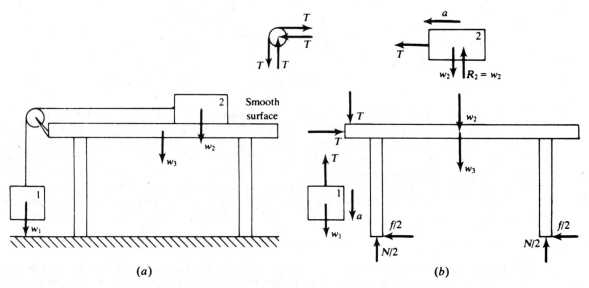

Fig. 4-23

where N and f are the vertical and horizontal (frictional) components of the force exerted by the floor on the table. [We assume in Fig. 4.23(b) that left and right legs share load equally. This does not affect our analysis.]

From the first two equations,

$$a = \frac{w_1}{m_1 + m_2} = \frac{m_1 g}{m_1 + m_2}$$

Then,

$$f = T = m_2 a = \frac{m_1 m_2 g}{m_1 + m_2}$$

and, finally,

$$N = T + m_2 g + m_3 g = \left(\frac{m_1 m_2}{m_1 + m_2} + m_2 + m_3\right)g$$

4.77 Three identical blocks, each of mass 0.6 kg, are connected by light strings as shown in Fig. 4-24. Assume that they lie on a smooth, horizontal surface and are observed to have an acceleration of 4.0 m/s² under the action of a force F. Calculate F and the two tensions.

$$\boxed{C} \;\overset{T_{bc}}{-}\; \boxed{B} \;\overset{T_{ab}}{-}\; \boxed{A} \overset{\longrightarrow}{F} \qquad \textbf{Fig. 4-24}$$

❚ Let the tensions in the cord be T_{ab} and T_{bc}, respectively, and let us write Newton's second law for each block separately, choosing positive to the right for each. Since the cords are inextensible, we know that they have the same acceleration, which we denote by a. Then for blocks A, B, C, respectively (letting $m = m_a = m_b = m_c$),

$$F - T_{ab} = ma \qquad T_{ab} - T_{bc} = ma \qquad T_{bc} = ma$$

(Note how the tensions appear with opposite signs in adjacent equations.) To solve these equations for a we add the equations, and the tensions cancel in pairs, leaving $F = 3ma = 3(0.6\,\text{kg})(4.0\,\text{m/s}^2) = \underline{7.2\,\text{N}}$. The tensions can now be obtained by substituting back into the individual equations:

$$T_{ab} = 7.2\,\text{N} - (0.6\,\text{kg})(4.0\,\text{m/s}^2) = \underline{4.8\,\text{N}} \qquad T_{bc} = (0.6\,\text{kg})(4.0\,\text{m/s}^2) = \underline{2.4\,\text{N}}$$

4.78 Three blocks with masses 6 kg, 9 kg, and 10 kg are connected as shown in Fig. 4-25. The coefficient of friction between the table and the 10-kg block is 0.2. Find (**a**) the acceleration of the system and (**b**) the tensions in the cord on the left and in the cord on the right.

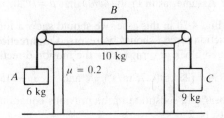

Fig. 4-25

I (**a**) Let the tension in the cord on the left be T_1 and on the right be T_2. The pulleys are assumed to be frictionless. We apply Newton's second law to each of the three blocks, choosing the positive sense of the axis for each block consistently. Thus we choose downward as positive for block C, to the right positive for block B, and upward as positive for block A. The frictional force on block B is to the left and can be obtained from $f = \mu_k N$, where $\mu_k = 0.20$ and the normal force N equals the weight $w_b = 98\,\text{N}$ from vertical equilibrium. Thus $f = 19.6\,\text{N}$. For our three equations we have

$$w_c - T_2 = m_c a$$

where a is the acceleration, $m_c = 9\,\text{kg}$, and $w_c = 88.2\,\text{N}$.

$$T_2 - T_1 - f = m_b a$$
$$T_1 - w_a = m_a a$$

where $m_a = 6\,\text{kg}$ and $w_a = 58.8\,\text{N}$.

As with earlier problems involving cords connecting blocks, the tensions in adjacent equations appear with opposite signs. Adding the three equations eliminates the tensions completely: $w_c - f - w_a = (m_a + m_b + m_c)a$. Note that this is equivalent to a one-dimensional problem involving a single block of mass $m_a + m_b + m_c$ acted on by a force w_c to the right and forces f and w_a to the left. Substituting the known masses, weights, and f gives $a = \underline{0.39\,\text{m/s}^2}$. (**b**) Substitute a back into the equations of motion for each block, most conveniently the first and third, to obtain $T_1 = \underline{61\,\text{N}}$, $T_2 = \underline{85\,\text{N}}$. The remaining equation can be used to check the results.

4.79 In Fig. 4-26, the coefficient of sliding friction between block A and the table is 0.20. Also, $m_A = 25\,\text{kg}$, $m_B = 15\,\text{kg}$. How far will block B drop in the first 3 s after the system is released?

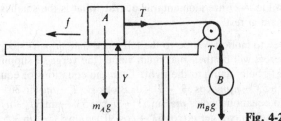

Fig. 4-26

▮ Since, for block A, there is no motion vertically, the normal force is $Y = m_A g = (25\ \text{kg})(9.8\ \text{m/s}^2) = 245\ \text{N}$. Then $f = \mu Y = (0.20)(245\ \text{N}) = 49\ \text{N}$.

We must first find the acceleration of the system and then we can describe its motion. Let us apply $F = ma$ to each block in turn. Taking the motion direction as positive, we have

$$T - f = m_A a \qquad \text{or} \qquad T - 49\ \text{N} = (25\ \text{kg})a$$

and

$$m_B g - T = m_B a \qquad \text{or} \qquad -T + (15)(9.8)\ \text{N} = (15\ \text{kg})a$$

We can eliminate T by adding the two equations. Then, solving for a, we find $a = 2.45\ \text{m/s}^2$.

Now we can work a motion problem with $a = 2.45\ \text{m/s}^2$, $v_0 = 0$, $t = 3\ \text{s}$.

$$y = v_0 t + \tfrac{1}{2}at^2 \qquad \text{gives} \qquad y = 0 + \tfrac{1}{2}(2.45\ \text{m/s}^2)(3\ \text{s})^2 = \underline{11.0\ \text{m}}$$

as the distance B falls in the first 3 s.

4.80 How large a horizontal force in addition to T must pull on block A in Fig. 4-26 to give it an acceleration of $0.75\ \text{m/s}^2$ *toward the left*? Assume, as in Prob. 4.79, that $\mu = 0.20$, $m_A = 25\ \text{kg}$, and $m_B = 15\ \text{kg}$.

▮ If we were to redraw Fig. 4-26 in this case, we should show a force P pulling toward the left on A. In addition, the retarding friction force f should be reversed in direction in the figure. As in Prob. 4.79, $f = 49\ \text{N}$.

We write $F = ma$ for each block in turn, taking the motion direction to be positive. We have

$$P - T - 49\ \text{N} = (25\ \text{kg})(0.75\ \text{m/s}^2) \qquad \text{and} \qquad T - (15)(9.8)\ \text{N} = (15\ \text{kg})(0.75\ \text{m/s}^2)$$

Solve the last equation for T and substitute in the previous equation. We can then solve for the single unknown, P, and find it to be $\underline{226\ \text{N}}$.

4.81 The two blocks shown in Fig. 4-27 have equal masses. The coefficients of static and dynamic friction are equal, 0.30 for both blocks. If the system is given an initial speed of 0.90 m/s to the left, how far will it move before coming to rest if the inclines are quite long?

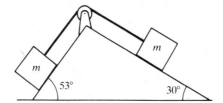

Fig. 4-27

▮ Again we assume that the pulley is frictionless and the tension in the cord is the same everywhere. We apply Newton's second law to each block to get the acceleration of the blocks. We choose our x axis for each block ‖ to the inclines and choose the positive sense to the left. Then we have $w \sin 53° - f_a - T = ma$ and $T - w \sin 30° - f_b = ma$, where m is the common mass and $w = mg$, the common weights of blocks A and B. The frictional forces on the two blocks are determined from the equilibrium conditions perpendicular to the inclines. Thus for block A, $f_a = \mu_k N_a$, with normal force $N_a = w \cos 53°$; and for block B, $f_b = \mu_k N_b$, with $N_b = w \cos 30°$. We eliminate the tension by adding our two equations to yield $w \sin 53° - f_a - f_b - w \sin 30° = 2ma$; or substituting for w, f_a, and f_b, $mg \sin 53° - \mu_k mg \cos 53° - \mu_k mg \cos 30° - mg \sin 30° = 2ma$. Dividing out by m and solving, we get $a = g(\sin 53° - \mu_k \cos 53° - \mu_k \cos 30° - \sin 30°)/2 = \underline{-0.694\ \text{m/s}^2}$. We now apply the kinematical equation $v_x^2 = v_{0x}^2 + 2a_x x$ to either block with $v_{0x} = 0.90\ \text{m/s}$, $a_x = -0.694\ \text{m/s}^2$, and $v_x = 0$ to get $x = \underline{0.583\ \text{m}}$.

4.82 If the blocks in Fig. 4-27 are momentarily at rest, what is the smallest coefficient of friction for which the blocks will remain at rest?

▮ The tendency to motion will be to the left since that slope is steeper. For minimum coefficient of friction the frictional forces will be their maximum value (the verge of slipping). Thus, $f_a = \mu_s mg \cos 53°$ and $f_b = \mu_s mg \cos 30°$, both acting to the right. Then the equations of equilibrium are
Block A: $mg \sin 53° - \mu_s mg \cos 53° - T = 0$. Block B: $T - mg \sin 30° - \mu_s mg \cos 30° = 0$
Adding the two equations yields $mg(\sin 53° - \mu_s \cos 53° - \sin 30° - \mu_s \cos 30°) = 0$. Dividing out mg and rearranging terms, we get $\mu_s(\cos 53° + \cos 30°) = \sin 53° - \sin 30°$, and $\mu_s = \underline{0.203}$.

4.83 A blimp is descending with an acceleration a. How much ballast must be jettisioned for the blimp to rise with the same acceleration a? There is a buoyant force acting upward on the blimp which is equal to the weight of the air displaced by the blimp; assume that the buoyant force is the same in both cases.

▌ From Fig. 4-28, the equations of motion are

Descending: $m_1g - F_b = m_1a$. Ascending: $F_b - m_2g = m_2a$

Adding gives $(m_1 - m_2)g = (m_1 + m_2)a$. But $m_1 - m_2 = m$, the mass of the discarded ballast. Therefore,

$$mg = [m_1 + (m_1 - m)]a \quad \text{or} \quad m = \left(\frac{2a}{g+a}\right)m_1$$

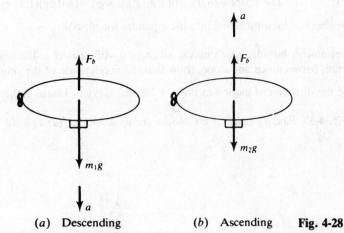

(a) Descending   (b) Ascending   **Fig. 4-28**

4.84ᶜ Show that the acceleration of the center of mass in Prob. 4.83 does not change when the ballast is ejected. Use this fact to confirm the value of m found in Prob. 4.83.

▌ Choose *up* as positive. Since $\sum \mathbf{F}$ on the system of blimp and ballast is the same before and after the ballast was thrown out, $\sum \mathbf{F} = m_1a_{cm}$, we must have $a_{cm} = -a$, before and after. Now, measured from some reference level, $y_{cm} = [m_2y_{blimp} + (m_1 - m_2)y_{ballast}]/m_1$. Then $\ddot{y}_{cm} = [m_2\ddot{y}_{blimp} + (m_1 - m_2)\ddot{y}_{ballast}]/m_1$. But $\ddot{y}_{blimp} = a$, $\ddot{y}_{ballast} = -g$, and $\ddot{y}_{cm} = a_{cm} = -a$; so $-a = [m_2a - (m_1 - m_2)g]/m_1$ or $(m_1 + m_2)a = (m_1 - m_2)g$ as before, yielding the same value for $m \equiv (m_1 - m_2)$.

4.85 Three blocks, of masses 2.0, 4.0, and 6.0 kg, arranged in the order lower, middle, and upper, respectively, are connected by strings on a frictionless inclined plane of 60°. A force of 120 N is applied upward along the incline to the uppermost block, causing an upward movement of the blocks. The connecting cords are light. What is the acceleration of the blocks?

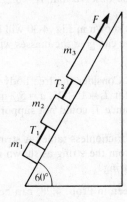

Fig. 4-29

▌ The situation is depicted in Fig. 4-29 with $F = 120$ N.

$$m_1 = 2.0 \text{ kg} \qquad m_2 = 4.0 \text{ kg} \qquad \text{and} \qquad m_3 = 6.0 \text{ kg}$$

Applying Newton's second law to each block, we have

$$F - T_2 - m_3g \sin 60° = m_3a \qquad T_2 - T_1 - m_2g \sin 60° = m_2a \qquad T_1 - m_1g \sin 60° = m_1a$$

Adding these equations, $F - (m_1 + m_2 + m_3)g \sin 60° = (m_1 + m_2 + m_3)a$; $120\,\text{N} - (12.0\,\text{kg})(9.8\,\text{m/s}^2)(0.866) = (12.0\,\text{kg})a$; $a = \underline{1.51\,\text{m/s}^2}$.

4.86 Refer to Prob. 4.85. What are the tensions between the upper and middle blocks, and the lower and middle blocks?

❚ Continuing from Prob. 4.85, substitute the value of a into the individual block equations, and solve for T_1 and T_2.

For block 1: $T_1 = (2.0\,\text{kg})(9.8\,\text{m/s}^2)(0.866) + (2.0\,\text{kg})(1.51\,\text{m/s}^2) = \underline{20.0\,\text{N}}$

For block 3: $T_2 = 120\,\text{N} - (6.0\,\text{kg})(9.8\,\text{m/s}^2)(0.866) - (6.0\,\text{kg})(1.51\,\text{m/s}^2) = \underline{60.0\,\text{N}}$

This can be checked by substituting into the equation for block 2.

4.87 A skier goes down a hillside, which makes an angle θ with respect to the horizontal. If μ_k is the coefficient of sliding friction between skis and slope, show that the acceleration of the skier is $a = g(\sin\theta - \mu_k \cos\theta)$.

❚ Reverse the direction of motion in Prob. 4.23, and apply to kinetic rather than static friction.

4.88 Refer to Fig. 4-30. Find T_1 and T_2 if the blocks are to accelerate (*a*) upward at 6.0 m/s² and (*b*) downward at 0.60 m/s².

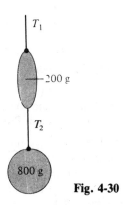

Fig. 4-30

❚ Treating the two masses and the connecting cord as an isolated object, $T_1 - 1.0g = 1.0a$; then isolating the 800-g mass, obtain $T_2 - 0.8g = 0.8a$; a is the same in both expressions. (*a*) $a = 6.0\,\text{m/s}^2$, so $T_1 = 6.0 + 9.8 = \underline{15.8\,\text{N}}$ and $T_2 = \underline{12.6\,\text{N}}$. (*b*) $a = -0.60\,\text{m/s}^2$, so $T_1 = 9.8 - 0.6 = \underline{9.2\,\text{N}}$ and $T_2 = \underline{7.4\,\text{N}}$. As a check on the answers, isolate the 200-g mass and observe that $T_1 - 0.2g - T_2 = 0.2a$.

4.89 The cords holding the two masses shown in Fig. 4-30 will break if the tension exceeds 15.0 N. What is the maximum upward acceleration one can give the masses without the cord breaking? Repeat if the strength is only 7.0 N.

❚ Note, $T_1 > T_2$ from Prob. 4.88. Consider the free body made up of both masses and the massless cord between them: $T_1 - 1.0\,\text{g} = 1.0a$; for $T_1 = 15.0\,\text{N}$, $a = \underline{5.2\,\text{m/s}^2}$; for $T_1 = 7.0\,\text{N}$, $a = \underline{-2.8\,\text{m/s}^2}$. (The system must be accelerating downward, since T_1 could not support the 9.8-N weight.)

4.90 A 6.0-kg block rests on a smooth frictionless table. A string attached to the block passes over a frictionless pulley, and a 3.0-kg mass hangs from the string as shown in Fig. 4-31. (*a*) What is the acceleration *a*? (*b*) What is the tension *T* in the string?

❚ (*a*) This type of problem, as seen in Prob. 4.78, can be treated as if it were in one dimension. Thus, Newton's second law takes the form

$$F = ma \qquad m_2g = (m_1 + m_2)a \qquad 3(9.8) = (6 + 3)a \qquad 29.4 = 9a \qquad a = \underline{3.27\,\text{m/s}^2}$$

(*b*) Applying Newton's second law to mass m_1 alone,

$$T = m_1a = 6(3.27) = \underline{19.6\,\text{N}}$$

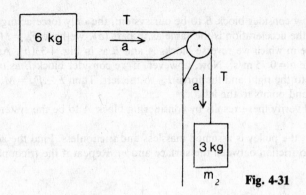

Fig. 4-31

To check this, apply Newton's second law to m_2 alone.

$$m_2g - T = m_2a \qquad 3(9.8) - T = 3(3.27) \qquad 29.4 - T = 9.8 \qquad T = \underline{19.6\,N}$$

As expected, the tension is the same.

4.91 A 6.0-kg block rests on a horizontal surface. Its coefficient of kinetic friction is 0.22. The block is connected by a string passing over a pulley to a 3.0-kg mass, as in Fig. 4-32. (a) What is the acceleration a? (b) What is the tension T in the string?

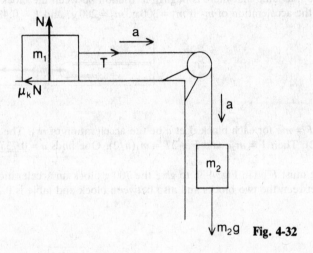

Fig. 4-32

❚ We treat the problem as if it were in one dimension.
(a) For the system as a whole,

$$\text{net } F = ma \qquad m = m_1 + m_2 = 6.0 + 3.0 = 9.0\,kg \qquad \text{net } F = m_2g - \mu_k N = ma \qquad N = m_1g$$
$$ma = m_2g - \mu_k m_1g \qquad 9.0a = 3.0(9.8) - 0.22(6.0)(9.8) = 1.68(9.8) \qquad a = \underline{1.83\,m/s^2}$$

(b) The force the string exerts on m_1 is

$$T = \mu_k N + m_1a = 0.22(6.0)(9.8) + 6.0(1.83) = 12.94 + 10.98 \qquad T = \underline{23.9\,N}$$

4.92 Suppose that blocks A and B have masses of 2 and 6 kg, respectively, and are in contact on a smooth horizontal surface. If a horizontal force of 6 N pushes them, calculate (a) the acceleration of the system and (b) the force that the 2-kg block exerts on the other block.

❚ (a) See Fig. 4-33(a). Considering the blocks to move as a unit, $M = m_a + m_b = 8\,kg$, $F = Ma = 6\,N$
$a = \underline{0.75\,m/s^2}$.

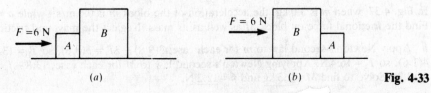

(a)   (b)   **Fig. 4-33**

(b) If we now consider block B to be our system, the only force acting on it is the force due to block A, F_{ab}. Then since the acceleration is the same as in part (a), we have $F_{ab} = M_b a = \underline{4.5\text{ N}}$. However, we must consider also the case in which we reverse blocks A and B as in Fig. 4-33(b). As before, considering the blocks as a unit we have $a = 0.75\text{ m/s}^2$. Now, however, if we consider block B as our system we have two forces acting, the force F to the right and the force F_{ab} to the left. Then $F - F_{ab} = M_b a$ and solving we get $F_{ab} = \underline{1.5\text{ N}}$ in magnitude and points to the left.

We could verify these results by considering block A to be the system for the two cases.

4.93 In Fig. 4-34, the pulley is assumed massless and frictionless. Find the acceleration of the mass m in terms of F if there is no friction between the surface and m. Repeat if the frictional force on m is f.

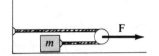

Fig. 4-34

▮ Note $T = F/2$. Newton's law for the block gives $T = ma$, hence $a = F/2m$. When friction is involved we have $F/2 - f = ma$, so $a = (F/2m) - (f/m)$.

4.94 In Fig. 4-35, assume that there is negligible friction between the blocks and table. Compute the tension in the cord and the acceleration of m_2 if $m_1 = 300$ g, $m_2 = 200$ g, and $F = 0.40$ N.

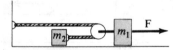

Fig. 4-35

▮ Write $F = ma$ for each block. Let a be the acceleration of m_2. The acceleration of m_1 is then $a/2$ (compare Prob. 4.72). Then $T = m_2 a$ and $F - 2T = m_1(a/2)$. One finds $a = \underline{0.73\text{ m/s}^2}$ and $T = \underline{0.145\text{ N}}$.

4.95 How large must F be in Fig. 4-36 to give the 700-g block an acceleration of 30 cm/s²? The coefficient of friction between the two blocks and also between block and table is 0.150.

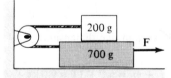

Fig. 4-36

▮ Converting to SI units, isolate each mass and note the forces that act on each. Vertically only the weights and normal forces are involved, $F_N = 0.2g$ on the upper block and $0.9g$ on the lower one. Friction forces are: $\mu(0.2g)$ between blocks and $\mu(0.9g)$ at the table. $F = ma$ for the blocks: $T - \mu(0.2g) = 0.2a$ and $F - T - \mu(0.2g) - \mu(0.9g) = 0.7a$; after eliminating tension T between these, $F = 0.9a + \mu(1.3g) = 0.9(0.30) + 0.150(1.3)(9.8) = \underline{2.18\text{ N}}$.

4.96 Assume in Fig. 4-36 that the coefficient of friction is the same at the top and bottom of the 700-g block. If $a = 70$ cm/s² when $F = 1.30$ N, how large is the coefficient of friction?

▮ Following Prob. 4.95, $F = 0.9a + \mu(1.3\,g)$; then we use $F = 1.30$ N and $a = 0.700$ m/s² to find $\mu = \underline{0.053}$.

4.97 In Fig. 4-37, when m is 3.0 kg, the acceleration of the block m is 0.6 m/s², while $a = 1.6$ m/s² if $m = 4.0$ kg. Find the frictional force on block M as well as its mass. Neglect the mass and friction of the pulleys.

▮ Apply Newton's second law to m for each case: $3(9.8) - 2T_1 = 3(0.6)$, so $T_1 = 13.8$ N; and $4(9.8) - 2T_2 = 4(1.6)$, so $T_2 = 16.4$ N. Applying Newton's second law to M for each case, $13.8 - f = M(1.2)$ and $16.4 - f = M(3.2)$. Solve to find $M = \underline{1.3\text{ kg}}$ and $f = \underline{12.2\text{ N}}$.

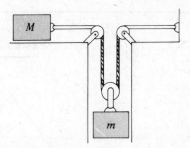

Fig. 4-37

4.98 In Fig. 4-38(a), block 1 is one-fourth the length of block 2 and weighs one-fourth as much. Assume that there is no friction between block 2 and the surface on which it moves and that the coefficient of sliding friction between blocks 1 and 2 is $\mu_k = 0.2$. After the system is released, find the distance block 2 has moved when only one-fourth of block 1 is still on block 2. Block 1 and block 3 have the same mass.

▮ From Fig. 4-38(b), the equations of motion are

$$\sum F_1 = T - \mu_k w_1 = ma_1 \qquad \sum F_2 = \mu_k w_1 = 4ma_2 \qquad \sum F_3 = w_1 - T = ma_1$$

Solve the first and third equations simultaneously to get $a_1 = (g/2)(1 - \mu_k)$; from the second equation, $a_2 = (g/4)\mu_k$. Then the displacements of blocks 1 and 2 are given by $x = \frac{1}{2}at^2$, i.e.,

$$x_1 = \frac{g}{4}(1 - \mu_k)t^2 \qquad x_2 = \frac{g}{8}\mu_k t^2$$

At the instant that one-fourth of block 1 remains on block 2, $x_2 + l = x_1 + (l/16)$, where l is the length of block 2. Therefore,

$$\frac{g}{8}\mu_k t^2 + l = \frac{g}{4}(1 - \mu_k)t^2 + \frac{l}{16} \qquad \text{or} \qquad t^2 = \frac{15l}{2g(2 - 3\mu_k)}$$

and

$$x_2 = \left(\frac{g}{8}\mu_k\right)\frac{15l}{2g(2 - 3\mu_k)} = \frac{15\mu_k}{16(2 - 3\mu_k)}l = \underline{\frac{l}{7.47}}$$

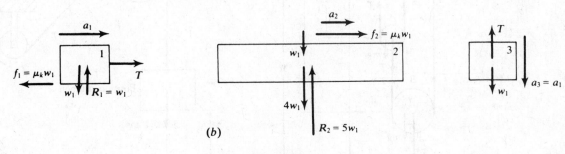

(a) Configuration at $t = 0$

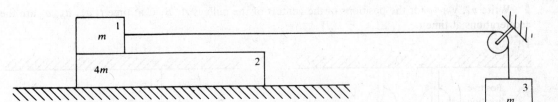

(b)

Fig. 4-38

4.99 A dinner plate rests on a tablecloth, with its center 0.3 m from the edge of the table. The tablecloth is suddenly yanked horizontally with a constant acceleration of 9.2 m/s² [Fig. 4-39(a)]. The coefficient of sliding friction between the tablecloth and the plate is $\mu_k = 0.75$. Find (a) the acceleration, (b) the velocity, and (c) the distance of the *plate* from the edge of the table, when the edge of the tablecloth passes under the center of the plate. Assume that the tablecloth just fits the tabletop.

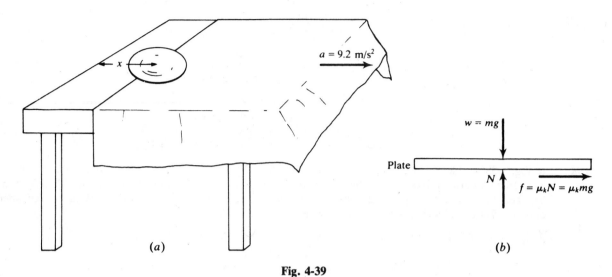

Fig. 4-39

▌ (a) from Fig. 4-39(b), the force equation for the plate is $\mu mg = ma_p$, or $a_p = \mu g = (0.75)(9.8) = \underline{7.35 \text{ m/s}^2}$. The plate slips, since a_p is less than 9.2 m/s^2. (b) At the time the edge of the tablecloth is at the center of the plate, the cloth and the plate are at the same distance from the edge of the table:

$$x_p = x_c \qquad 0.3 + \tfrac{1}{2}(7.4)t^2 = 0 + \tfrac{1}{2}(9.2)t^2$$

Solving, $t = 0.58 \text{ s}$ and $v_p = 0 + (7.35)(0.58) = \underline{4.26 \text{ m/s}}$.

(c) $$x_p = 0.3 + 0(0.58) + \tfrac{1}{2}(7.35)(0.58)^2 = \underline{1.54 \text{ m}}$$

4.100 In the pulley system shown in Fig. 4-40, the movable pulleys A, B, C are of mass 1 kg each. D and E are fixed pulleys. The strings are vertical and inextensible. Find the tension in the string and the accelerations of the frictionless pulleys.

▌ Write y_A, y_B, y_C for the positions of the centers of the pulleys A, B, C at time t; a_A, a_B, a_C are the accelerations at time t.

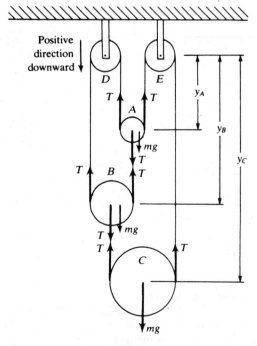

Fig. 4-40

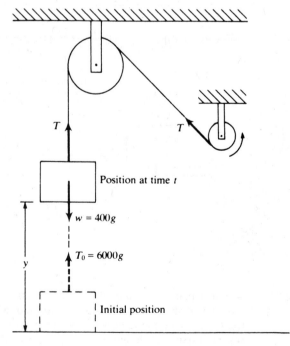

Fig. 4-41

Following the string from the end at the center of A to the end at the center of B, we get

$$(y_B - y_A) + y_B + 2y_A + y_C + (y_C - y_B) = \text{constant} \quad \text{or} \quad y_A + y_B + 2y_C = \text{constant}$$

Take the second time-derivative of this equation to get $a_A + a_B + 2a_C = 0$.
There is just one string and, thus, one tension T. The force equations are

$$T + mg - 2T = ma_A \qquad T + mg - 2T = ma_B \qquad mg - 2T = ma_C$$

Substituting $m = 1$ kg and solving the four equations for the four unknowns a_A, a_B, a_C, T, we obtain

$$a_A = a_B = -a_C = \frac{g}{3} = \underline{3.3 \text{ m/s}} \qquad T = \underline{6.5 \text{ N}}$$

4.101ᶜ A body of mass 400 kg is suspended at the lower end of a light vertical chain and is being pulled up vertically (see Fig. 4-41). Initially the body is at rest and the pull on the chain is $6000g$ N. The pull gets smaller uniformly at the rate of $360g$ N per each meter through which the body is raised. What is the velocity of the body when it has been raised 10 m?

▌ At time t, let y be the height (in meters) of the body above its initial position. The pull in the chain is then $T = (6000 - 360y)g$ and Newton's second law gives

$$T - 400g = 400\ddot{y} \qquad \text{or} \qquad (5600 - 360y)g = 400\ddot{y}$$

This equation may be changed into one for $\dot{y} = v$ (the velocity of the body) by use of the identity

$$2\ddot{y} = 2\frac{dv}{dt} = 2\frac{dv}{dy}\frac{dy}{dt} = 2v\frac{dv}{dy} = d(v^2)/dy$$

Thus

$$200\frac{d(v^2)}{dy} = (5600 - 360y)g \qquad \text{or} \qquad d(v^2) = g(28 - 1.8y)\,dy$$

Let V be the velocity at height 10 m. Then, on integrating

$$\int_0^{V^2} d(v^2) = g\int_0^{10}(28 - 1.8y)\,dy \qquad V^2 = g[28y - 0.9y^2]_0^{10} = g[28(10) - 0.9(100)] = 190g$$

$$V = +\sqrt{190g} = \underline{+43.2 \text{ m/s}}.$$

The choice of the $+$ sign for V (upward motion) should be checked. For $0 \le y \le 10$, the net force, $(5600 - 360y)g$, is positive, and so the acceleration is positive. Then, since the body started from rest, V must be positive.

CHAPTER 5
Motion in a Plane I

5.1 PROJECTILE MOTION

5.1 A marble with speed 20 cm/s rolls off the edge of a table 80 cm high. How long does it take to drop to the floor? How far, horizontally, from the table edge does the marble strike the floor?

▎ Choose downward as positive with origin at edge of table top.

$$v_{0x} = v_0 = 20 \text{ cm/s} \qquad v_{0y} = 0 \qquad a_y = +g = +980 \text{ cm/s}^2 \qquad a_x = 0$$

To find time of fall, $y = v_{0y}t + \frac{1}{2}gt^2$, or $80 \text{ cm} = 0 + (490 \text{ cm/s}^2)t^2$; $\underline{t = 0.40 \text{ s}}$. The horizontal distance is gotten from $x = v_{0x}t = (20 \text{ cm/s})(0.40 \text{ s}) = \underline{8.0 \text{ cm}}$.

5.2 How fast must a ball be rolled along a 70-cm-high table so that when it rolls off the edge it will strike the floor at this same distance (70 cm) from the point directly below the table edge?

▎ In the horizontal problem, $x = v_x t$ gives $v_x = 0.70/t$. In the vertical problem, choosing *down* as positive, $v_0 = 0$, $y = 0.7$ m, and $a = 9.8$ m/s². Use these values in $y = v_0 t + at^2/2$ to give $t = 0.378$ s. Then $v_x = \underline{1.85 \text{ m/s}}$.

5.3 A marble traveling at 100 cm/s rolls off the edge of a level table. If it hits the floor 30 cm away from the spot directly below the edge of the table, how high is the table?

▎ This is a projectile problem with $v_{0x} = 100$ cm/s. For the horizontal motion:

$$s_x = v_{0x}t \qquad 30 = 100t \qquad t = 0.30 \text{ s}$$

For the vertical motion:

$$s_y = v_{0y}t + \frac{1}{2}at^2 = 0 + \frac{1}{2}(980)(0.30)^2 = \underline{44.1 \text{ cm}} \qquad \text{(height of the table)}$$

5.4 In an ordinary television set, the electron beam consists of electrons shot horizontally at the television screen with a speed of about 5×10^7 m/s. How far does a typical electron fall as it moves the approximately 40 cm from the electron gun to the screen? For comparison, how far would a droplet of water shot horizontally at 2 m/s from a hose drop as it moves a horizontal distance of 40 cm?

▎ For the electron, the horizontal problem yields the time to hit the screen as $t = x/v_x = 0.40/(5 \times 10^7) = 8 \times 10^{-9}$ s. Then in the vertical problem, $y = v_0 t + at^2/2$, so $y = 0 + 4.9(64 \times 10^{-18}) = \underline{3.1 \times 10^{-16} \text{ m}}$. For a droplet, $t = 0.40/2 = 0.20$ s, and so $y = \underline{0.196 \text{ m}}$.

5.5 A body projected upward from the level ground at an angle of 50° with the horizontal has an initial speed of 40 m/s. How long will it be before it hits the ground?

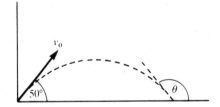

Fig. 5-1

▎ Choose *upward* as positive, and place the origin at the launch point (Fig. 5-1).

$$v_{0x} = v_0 \cos 50° = (40 \text{ m/s})(0.642) = 25.7 \text{ m/s} \qquad v_{0y} = v_0 \sin 50° = (40 \text{ m/s})(0.766) = 30.6 \text{ m/s}$$
$$a_y = -g = -9.8 \text{ m/s}^2 \qquad a_x = 0$$

To find the time in air, we have $y = v_{0y}t - \frac{1}{2}gt^2$ and since $y = 0$ at the end of flight, $0 = (30.6 \text{ m/s})t - (4.9 \text{ m/s}^2)t^2$, or $4.9t^2 = 30.6t$. The first solution $t = 0$ corresponds to the starting point, $y = 0$. The second solution is not zero and is obtained by dividing out by t. $4.9t = 30.6$ and $t = \underline{6.24 \text{ s}}$.

5.6 In Prob. 5.5, how far from the starting point will the body hit the ground, and at what angle with the horizontal?

▌ The horizontal distance traveled, or range R, is obtained from $R = v_{0x}t = (25.7 \text{ m/s})(6.24 \text{ s}) = \underline{160 \text{ m}}$. By symmetry it strikes at 50° (with negative x axis). This can also be seen by noting that $v_y = -v_{0y}$, $v_x = v_{0x}$, so $\tan\theta = -\tan\theta_0$ and therefore $180° - \theta = \theta_0$.

5.7 A body is projected downward at an angle of 30° with the horizontal from the top of a building 170 m high. Its initial speed is 40 m/s. How long will it take before striking the ground?

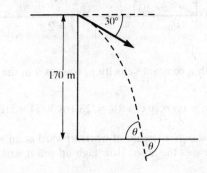

Fig. 5-2

▌ Choose *downward* as positive and origin at the top edge of building (Fig. 5-2).

$$v_{0x} = v_0 \cos 30° = (40 \text{ m/s})(0.866) = 34.6 \text{ m/s} \qquad v_{0y} = v_0 \sin 30° = (40 \text{ m/s})(0.500) = 20.0 \text{ m/s}$$
$$a_y = g = 9.8 \text{ m/s}^2 \qquad a_x = 0$$

We can solve for the time in different ways. Method 1:

$$y = v_{0y}t + \tfrac{1}{2}gt^2 \qquad \text{or} \qquad 170 \text{ m} = (20.0 \text{ m/s})t + (4.9 \text{ m/s}^2)t^2$$

We can solve the quadratic to yield

$$t = \frac{-20 \pm (400 + 3332)^{1/2}}{9.8} = \underline{4.2 \text{ s}}$$

(We keep only the positive solution; the negative time corresponds to a time before $t = 0$ when it would have been at ground level if it were a projectile launched so as to reach the starting position and velocity at $t = 0$.) Method 2: We avoid the quadratic. First find v_y just before impact:

$$v_y^2 = v_{0y}^2 + 2gy \qquad \text{or} \qquad v_y^2 = (20.0 \text{ m/s})^2 + 2(9.8 \text{ m/s}^2)(170 \text{ m}) \qquad v_y = \pm 61 \text{ m/s}$$

For our case $v_y = 61$ m/s. Next we find t:

$$v_y = v_{0y} + gt \qquad \text{or} \qquad 61 \text{ m/s} = 20.0 \text{ m/s} + (9.8 \text{ m/s}^2)t \qquad \text{or} \qquad t = \underline{4.2 \text{ s}}$$

5.8 In Prob. 5.7, find out how far from the foot of the building the body will strike and at what angle with the horizontal.

▌ $x = v_{0x}t = (34.6 \text{ m/s})(4.2 \text{ s}) = \underline{145 \text{ m}}$. We need the angle that the vector velocity makes with the x axis just before hitting the ground. We avoid having to directly deduce the quadrant this angle is in by solving for the acute angle θ made with the x axis (positive or negative; above or below).

$$\tan\theta = \left|\frac{v_y}{v_x}\right| = 1.76 \qquad \text{and} \qquad \underline{\theta = 60.4°}$$

Since v_y is negative and v_x is positive, this is clearly the angle below the positive x axis (see Fig. 5-2) and equals the angle we are looking for.

5.9 A body is projected from the ground at an angle of 30° with the horizontal at an initial speed of 128 ft/s. Ignoring air friction, determine (*a*) in how many seconds it will strike the ground, (*b*) how high it will go, and (*c*) what its range will be.

▌ (*a*) Find v_{0y} and then determine the time for a freely falling body projected upward at this velocity to return to the ground; that is, $s_y = 0$. From Fig. 5-3,

$$v_{0y} = v \sin\theta = 128 \sin 30° = 64 \text{ ft/s} \qquad s_y = v_{0y}t + \tfrac{1}{2}gt^2 \qquad 0 = 64t + \tfrac{1}{2}(-32)t^2 = (64 - 16t)t \qquad t = \underline{4 \text{ s}}$$

(*b*) Since time of ascent is equal to time of descent, the projectile reaches maximum height H at $t = 2$ s. Thus,

$$H = v_{0y}t + \tfrac{1}{2}gt^2 = 64(2) + \tfrac{1}{2}(-32)(2)^2 = \underline{64 \text{ ft}}$$

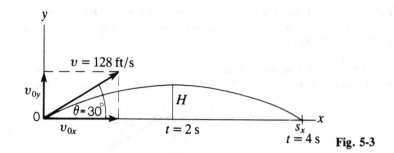

Fig. 5-3

(c) The projectile travels with a constant velocity $v_x = v \cos \theta$ in the x direction. It reaches the ground in 4 s. The range is

$$s_x = v_x t = (v \cos \theta)t = 128(\cos 30°)4 = 512(0.866) = \underline{443 \text{ ft}}$$

5.10   A hose lying on the ground shoots a stream of water upward at an angle of 40° to the horizontal. The speed of the water is 20 m/s as it leaves the hose. How high up will it strike a wall which is 8 m away?

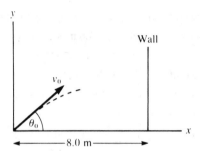

Fig. 5-4

■ Setting coordinates as shown in Fig. 5-4, with $v_0 = 20$ m/s and $\theta_0 = 40°$, we get

$$v_{0x} = v_0 \cos \theta_0 = (20 \text{ m/s}) \cos 40° = 15.3 \text{ m/s} \qquad v_{0y} = v_0 \sin \theta_0 = (20 \text{ m/s}) \sin 40° = 12.8 \text{ m/s}$$

$x = v_{0x}t$, and setting $x = 8$ m we find the time to hit the wall: $8 \text{ m} = (15.3 \text{ m/s})t$, yielding $t = 0.52$ s. To find the height at which it hits the wall, we use $y = v_{0y}t - \frac{1}{2}gt^2$, with $t = 0.52$ s. This yields $y = (12.8 \text{ m/s})(0.52 \text{ s}) - (4.9 \text{ m/s}^2)(0.52 \text{ s})^2 = \underline{5.33 \text{ m}}$.

5.11   A baseball batter hits a home run ball with a velocity of 132 ft/s at an angle of 26° above the horizontal. A fielder who has a reach of 7 ft above the ground is backed up against the bleacher wall, which is 386 ft from home plate. The ball was 3 ft above the ground when hit. How high above the fielder's glove does the ball pass?

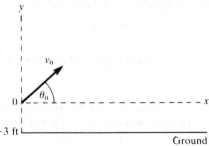

Fig. 5-5

■ The situation is depicted in Fig. 5-5, with the origin and x axis 3 ft above the ground. We must find the value of y on the ball's trajectory corresponding to $x = 386$ ft. Then we can subtract the height of the fielder's glove above the x axis, i.e., 7 ft − 3 ft = 4 ft. To find y we note that

$$v_{0x} = v_0 \cos \theta_0 = (132 \text{ ft/s}) \cos 26° = 119 \text{ ft/s} \qquad v_{0y} = v_0 \sin \theta_0 = (132 \text{ ft/s}) \sin 26° = 57.9 \text{ ft/s}$$

The time to reach $x = 386$ ft is given by $x = v_{0x}t$, or $386 \text{ ft} = (119 \text{ ft/s})t$, and $t = 3.24$ s.

Then $y = v_{0y}t - \frac{1}{2}gt^2 = (57.9 \text{ ft/s})(3.24 \text{ s}) - \frac{1}{2}(32 \text{ ft/s}^2)(3.24 \text{ s})^2 = 19.6 \text{ ft}$. Height above glove = 19.6 ft − 4 ft = 15.6 ft. [Note: The trajectory equation $y = (\tan \theta_0)x - gx^2/(2v_0^2 \cos^2 \theta_0)$ can also be used to find y.]

5.12   A ball is thrown upward at an angle of 30° to the horizontal and lands on the top edge of a building that is 20 m away. The top edge is 5 m above the throwing point. How fast was the ball thrown?

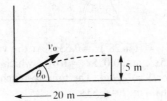

Fig. 5-6

▮   The situation is depicted in Fig. 5-6 with $\theta_0 = 30°$. We can use the trajectory equation $y = \tan \theta_0 x - gx^2/(2v_0^2 \cos^2 \theta_0)$, setting $x = 20 \text{ m}$ and $y = 5 \text{ m}$. Then $5 \text{ m} = (0.58)(20 \text{ m}) - (9.8 \text{ m/s}^2)(20 \text{ m})^2/(2v_0^2 \times 0.75)$, or $v_0 = \underline{20 \text{ m/s}}$. (Note: If you didn't remember the trajectory equation, you could still solve the problem by using the x-vs.-t and y-vs.-t equations.)

5.13   A projectile is fired with initial velocity $v_0 = 95 \text{ m/s}$ at an angle $\theta = 50°$. After 5 s it strikes the top of a hill. What is the elevation of the hill above the point of firing? At what horizontal distance from the gun does the projectile land?

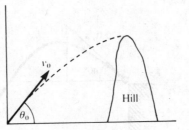

Fig. 5-7

▮   The situation is as shown in Fig. 5-7. $v_0 = 95 \text{ m/s}$; $\theta_0 = 50°$. At any time t, $y = v_{0y}t - \frac{1}{2}gt^2$, where $v_{0y} = v_0 \sin \theta_0 = 72.8 \text{ m/s}$. For $t = 5 \text{ s}$, we get $y = (72.8 \text{ m/s})(5.0 \text{ s}) - \frac{1}{2}(9.8 \text{ m/s}^2)(5.0 \text{ s})^2 = \underline{241 \text{ m}}$. The horizontal distance, x, is given by $x = v_{0x}t = v_0 \cos \theta_0 t = (61.1 \text{ m/s})t$. At $t = 5 \text{ s}$, we have $x = \underline{305 \text{ m}}$.

5.14   A ball is thrown with a speed of 20 m/s at an angle of 37° above the horizontal. It lands on the roof of a building at a point displaced 24 m horizontally from the throwing point. How high above the throwing point is the roof?

▮   The velocity components are $v_{0x} = 16 \text{ m/s}$ and $v_{0y} = 12 \text{ m/s}$. In the horizontal problem, $t = x/v_x = 24/16 = 1.5 \text{ s}$. Then in the vertical problem we have $y = v_{0y}t + at^2/2$, so $y = 12(1.5) - 4.9(2.25) = \underline{7.0 \text{ m}}$.

5.15   A hit baseball leaves the bat with a velocity of 110 ft/s at 45° above the horizontal. The ball hits the top of a screen at the 320-ft mark and bounces into the crowd for a home run. How high above the ground is the top of the screen? (Neglect air resistance.)

▮ $s_x = v_x t$ $v_x = 110 \cos 45° = $ a constant $320 = 110(0.707)t$ $t = 4.11 \text{ s}$

$s_y = v_{0y}t + \frac{1}{2}at^2 = (110 \sin 45°)(4.11) + \frac{1}{2}(-32)(4.11)^2 = (110)(0.707)(4.11) + \frac{1}{2}(-32)(4.11)^2$

$= 319.6 - 270.3$ $s_y = \underline{49 \text{ ft}}$ (height of screen)

5.16   A ball is thrown upward from a point on the side of a hill which slopes upward uniformly at an angle of 28°. Initial velocity of ball: $v_0 = 33 \text{ m/s}$, at an angle $\theta_0 = 65°$ (with respect to the horizontal). At what distance up the slope does the ball strike and in what time?

▮   The situation is depicted in Fig. 5-8. $v_0 = 33 \text{ m/s}$ and $\theta_0 = 65°$. The trajectory equation of the ball is

$$y_b = (\tan \theta_0)x - gx^2/2v_0^2 \cos^2 \theta_0) = 2.14x - 0.025x^2$$

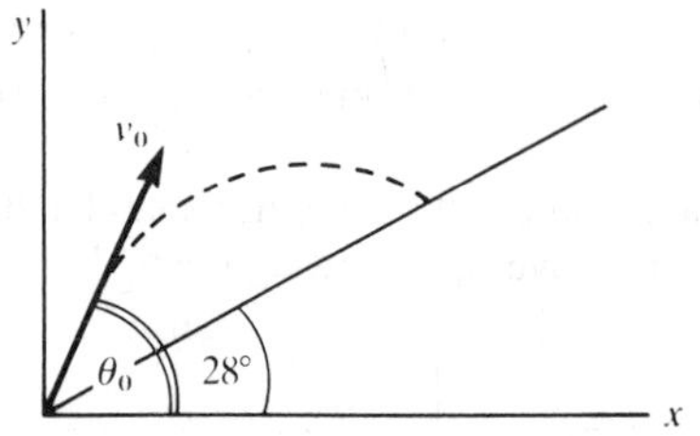

Fig. 5-8

The equation for the incline is $y_i = (\tan 28°)x = 0.53x$. At the value of x for which the ball hits the incline $y_b = y_i$, or $0.53x = 2.14x - 0.025x^2$, and $0.025x^2 = 1.61x$, which yields $x = 64.4$ m. The distance along the incline, S, obeys $x = S \cos 28°$ or $S = \underline{72.9 \text{ m}}$. The time to reach any x value is given by $x = v_{0x}t = (v_0 \cos \theta_0)t = 13.9t$. So for $x = 64.4$ m, $t = \underline{4.63 \text{ s}}$.

5.17 A projectile is to be shot at 50 m/s over level ground in such a way that it will land 200 m from the shooting point. At what angle should the projectile be shot?

▌ In the horizontal problem, $x = v_{0x}t$ gives $200 = (50 \cos \theta)t$, where θ is the angle we seek. In the vertical problem, $y = v_{0y}t + at^2/2$ gives $0 = 50 \sin \theta - 4.9t$. But $t = 200/(50 \cos \theta)$, and so $50 \sin \theta = 4.9(4/\cos \theta)$. This simplifies to $2 \sin \theta \cos \theta = 0.784$, and so $\sin 2\theta = 0.784$, from which $\theta = \underline{25.8°}$.

5.18 As shown in Fig. 5-9, a ball is thrown from the top of one building toward a tall building 50 ft away. The initial velocity of the ball is 20 ft/s at 40° above the horizontal. How far above or below its original level will the ball strike the opposite wall?

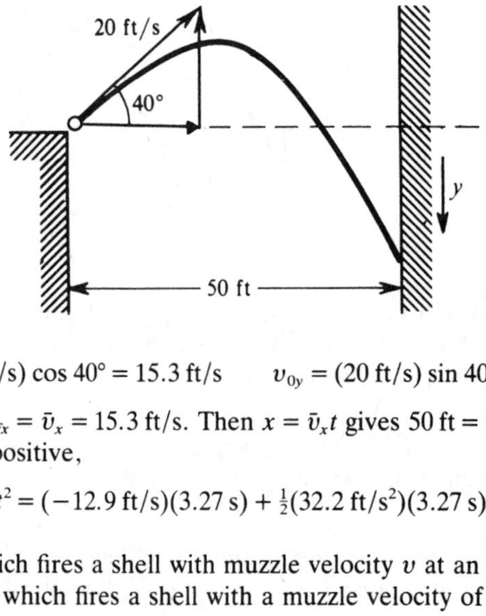

Fig. 5-9

▌ $$v_{0x} = (20 \text{ ft/s}) \cos 40° = 15.3 \text{ ft/s} \qquad v_{0y} = (20 \text{ ft/s}) \sin 40° = 12.9 \text{ ft/s}$$

In the horizontal motion, $v_{0x} = v_{fx} = \bar{v}_x = 15.3$ ft/s. Then $x = \bar{v}_x t$ gives $50 \text{ ft} = (15.3 \text{ ft/s})t$, or $t = 3.27$ s. In the vertical motion, taking *down* as positive,

$$y = v_{0y}t + \tfrac{1}{2}a_y t^2 = (-12.9 \text{ ft/s})(3.27 \text{ s}) + \tfrac{1}{2}(32.2 \text{ ft/s}^2)(3.27 \text{ s})^2 = \underline{130 \text{ ft below}}$$

5.19 **(a)** Find the range x of a gun which fires a shell with muzzle velocity v at an angle of elevation θ. **(b)** Find the angle of elevation θ of a gun which fires a shell with a muzzle velocity of 1.2 km/s at a target on the same level but 15 km distant. See Fig. 5-10.

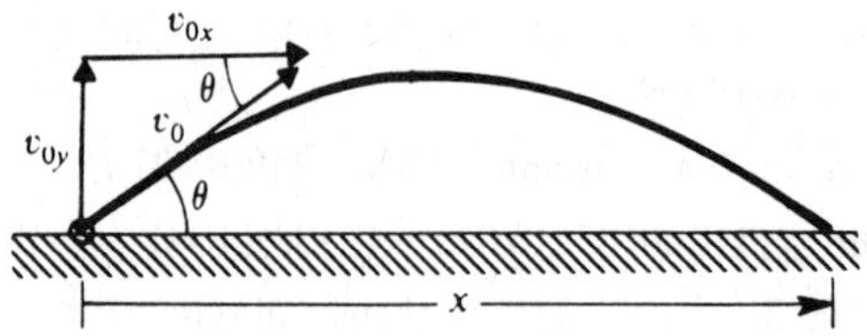

Fig. 5-10

▌ **(a)** Let t be the time it takes the shell to hit the target. Then, $x = v_{0x}t$ or $t = x/v_{0x}$. Consider the vertical motion alone, and take *up* as positive. When the shell strikes the target, vertical displacement $= 0 = v_{0y}t + \frac{1}{2}(-g)t^2$. Solving this equation gives $t = 2v_{0y}/g$. But $t = x/v_{0x}$, so

$$\frac{x}{v_{0x}} = \frac{2v_{0y}}{g} \qquad \text{or} \qquad x = \frac{2v_{0x}v_{0y}}{g} = \frac{2(v_0 \cos \theta)(v_0 \sin \theta)}{g} = \frac{v_0^2 \sin 2\theta}{g}$$

(b) From the range equation found in **(a)**,

$$\sin 2\theta = \frac{gx}{v_0^2} = \frac{(9.8 \times 10^{-3} \text{ km/s}^2)(15 \text{ km})}{(1.2 \text{ km/s})^2} = 0.102$$

whence $2\theta = 5.9°$ and $\theta = \underline{3.0°}$.

5.20 A rifle bullet has a muzzle velocity of 680 ft/s. **(a)** At what angle (ignoring air resistance) should the rifle be pointed to give the maximum range? **(b)** Evaluate the maximum range.

❚ **(a)** By Prob. 5.19(a), the range is a maximum when $\sin 2\theta = 1$, or $\theta = 45°$.

(b) $x_{max} = \dfrac{v_0^2}{g} = \dfrac{(680 \text{ ft/s})^2}{32 \text{ ft/s}^2} = \underline{14\,450 \text{ ft}}$.

5.21 A golf ball leaves the golf club at an angle of 60° above the horizontal with a velocity of 30 m/s. **(a)** How high does it go? **(b)** Assuming a level fairway, determine how far away it hits the ground.

❚ **(a)** With *up* positive, $v_{0y} = 30 \sin 60° = 15\sqrt{3}$ m/s and, at maximum height, v_y equals zero. Thus,

$$v_y^2 = v_{0y}^2 + 2ah \qquad 0 = 675 - 19.6h \qquad h = \frac{675}{19.6} = \underline{34.4 \text{ m}}$$

(b) By Prob. 5.19(a),

$$x = \frac{v_0^2 \sin 2\theta}{g} = \frac{900(\sqrt{3}/2)}{9.8} = \underline{79.5 \text{ m}}$$

5.22 Prove that a gun will shoot three times as high when its angle of elevation is 60° as when it is 30°, but will carry the same horizontal distance.

❚ We assume that v_0 is the same at both angles. Maximum height is given by the condition $v_y = 0$. Then, $v_y^2 = v_{0y}^2 - 2gy$ yields $y_{max} = v_{0y}^2/(2g)$. Noting that $v_{0y} = v_0 \sin \theta_0$, we have finally $y_{max} = (v_0^2 \sin^2 \theta)/2g$. Then $y_{max}(60°)/y_{max}(30°) = \sin^2 60°/\sin^2 30° = 3$. To show that the horizontal range is the same, we note that in the range formula [Prob. 5.19(a)] $\cos 60° \sin 60° = \cos 30° \sin 30°$ since 30° and 60° are complementary angles.

5.23 As shown in Fig. 5-11, a projectile is fired with a horizontal velocity of 330 m/s from the top of a cliff 80 m high. **(a)** How long will it take for the projectile to strike the level ground at the base of the cliff? **(b)** How far from the foot of the cliff will it strike? **(c)** With what velocity will it strike?

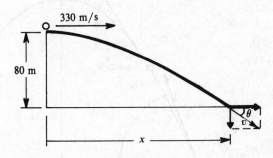

Fig. 5-11

❚ **(a)** The horizontal and vertical motions are independent of each other. Consider first the vertical motion. Taking *down* as positive we have $y = v_{0y}t + \frac{1}{2}a_yt^2$, or $80 \text{ m} = 0 + \frac{1}{2}(9.8 \text{ m/s}^2)t^2$, from which $t = \underline{4.04 \text{ s}}$. Note that the initial velocity had zero vertical component, and so $v_0 = 0$ in the vertical motion.
(b) Now consider the horizontal motion. For it, $a = 0$ and so $\bar{v}_x = v_{0x} = v_{fx} = 330$ m/s. Then, using the value of t found in **(a)**, $x = \bar{v}_x t = (330 \text{ m/s})(4.04 \text{ s}) = \underline{1330 \text{ m}}$.
(c) The final velocity has a horizontal component of 330 m/s. But its vertical velocity at $t = 4.04$ s is given by $v_{fy} = v_{0y} + a_yt$ as $v_{fy} = 0 + (9.8 \text{ m/s}^2)(4.04 \text{ s}) = 40$ m/s. The resultant of these two components is labeled v in Fig. 5-11; we have

$$v = \sqrt{(40 \text{ m/s})^2 + (330 \text{ m/s})^2} = \underline{332 \text{ m/s}}$$

Angle θ shown is given by $\tan \theta = 40/330$ to be $\underline{6.9°}$.

5.24 A stunt flier is moving at 15 m/s parallel to the flat ground 100 m below. How large must the horizontal distance x from plane to target be if a sack of flour released from the plane is to strike the target?

❚ Following the same procedure as in Prob. 5.23, use $y = v_{0y}t + \frac{1}{2}a_yt^2$ to get $100\,\text{m} = 0 + \frac{1}{2}(9.8\,\text{m/s}^2)t^2$, or $t = 4.5\,\text{s}$. Now use $x = v_x t = (15\,\text{m/s})(4.5\,\text{s}) = \underline{68\,\text{m}}$, since the sack's initial velocity is that of the plane.

5.25 A baseball is thrown with an initial velocity of 100 m/s at an angle of 30° above the horizontal. How far from the throwing point will the baseball attain its original level?

❚ By the range formula,

$$x = \frac{v_0^2 \sin 2\theta}{g} = \frac{10^4(\sqrt{3}/2)}{9.8} = \underline{890\,\text{m}}$$

5.26 A cart is moving horizontally along a straight line with constant speed 30 m/s. A projectile is to be fired from the moving cart in such a way that it will return to the cart after the cart has moved 80 m. At what speed (relative to the cart) and at what angle (to the horizontal) must the projectile be fired?

❚ To move horizontally with the cart, the projectile must be fired vertically with a flight time $= x/v_x = (80\,\text{m})/(30\,\text{m/s}) = 2.67\,\text{s}$. The initial velocity v_0 must satisfy $y = v_0t + at^2/2$, with $y = 0$ and $t = 2.67\,\text{s}$; thus $4.9t = v_0$, and so $v_0 = \underline{13.1\,\text{m/s}}$ at $\theta = \underline{90°}$.

5.27 In Fig. 5-12, α particles from a bit of radioactive material enter through slit S into the space between two large parallel metal plates, A and B, connected to a source of voltage. As a result of the uniform electric field between the plates, each particle has a constant acceleration $a = 4 \times 10^{13}\,\text{m/s}^2$ normal to and toward B. If $v_0 = 6 \times 10^6\,\text{m/s}$ and $\theta = 45°$, determine h and R.

❚ Here the electric force takes the place of gravity, but otherwise the analysis is the same. Choosing *upward* as positive, i.e., the direction from B to A, we have a regular trajectory problem with $v_0 = 6 \times 10^6\,\text{m/s}$; $\theta_0 = 45°$; $a_y = -4 \times 10^{13}\,\text{m/s}^2$; $a_x = 0$. Then $x = v_{0x}t = v_0 \cos\theta_0 t = (4.24 \times 10^6\,\text{m/s})t$; and similarly

$$y = v_{0y}t + \tfrac{1}{2}a_yt^2 = (4.24 \times 10^6\,\text{m/s})t - (2.0 \times 10^{13}\,\text{m/s}^2)t^2$$
$$v_y = v_{0y} + a_yt = (4.24 \times 10^6\,\text{m/s}) - (4.0 \times 10^{13}\,\text{m/s}^2)t$$

At the highest point, $y = h$, $v_y = 0$, and $t = 1.06 \times 10^{-7}\,\text{s}$. Thus $h = (4.24\times10^6)(1.06\times10^{-7}) - (2.0\times10^{13}) \times (1.06 \times 10^{-7})^2 = 0.225\,\text{m}$. The horizontal range R corresponds to x at the time the α particle returns to plate B. By symmetry, this is $2t = 2.12 \times 10^{-7}\,\text{s}$. Then $R = (4.24 \times 10^6\,\text{m/s})(2.12 \times 10^{-7}\,\text{s}) = \underline{0.90\,\text{m}}$.

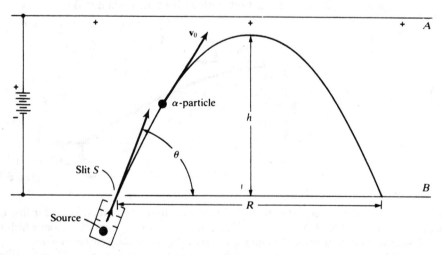

Fig. 5-12

5.28 A ball is thrown upward from the top of a 35-m tower, Fig. 5-13, with initial velocity $v_0 = 80$ m/s at an angle $\theta = 25°$. (*a*) Find the time to reach the ground and the distance R from P to the point of impact. (*b*) Find the magnitude and direction of the velocity at the moment of impact.

❚ (*a*) At the point of impact, $y = -35\,\text{m}$ and $x = R$. From $y = -35 = (80 \sin 25°)t - \frac{1}{2}(9.8)t^2$, $t = \underline{7.814\,\text{s}}$. Then $x = R = (80 \cos 25°)(7.814) = \underline{566.55\,\text{m}}$.
(*b*) At impact, $v_y = 80 \sin 25° - (9.8)(7.814) = -42.77\,\text{m/s}$ and $v_x = v_{0x} = 80 \cos 25° = 72.5\,\text{m/s}$. Thus $v = (42.77^2 + 72.5^2)^{1/2} = 84.18\,\text{m/s}$ and $\tan \beta = -42.77/72.5$, or $\beta = -30.54°$.

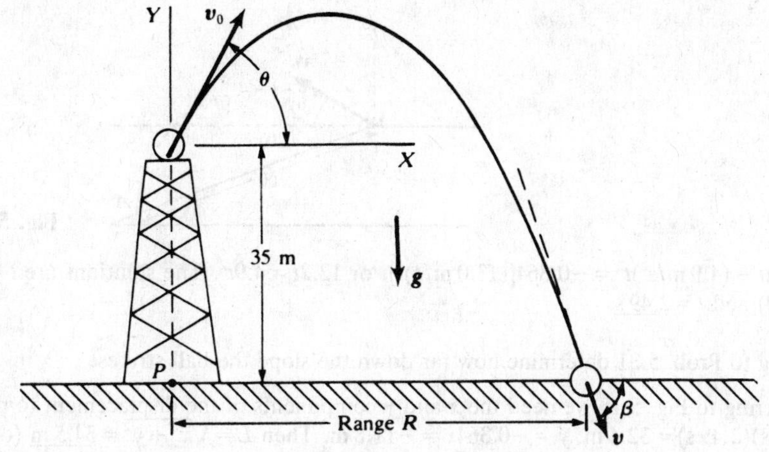

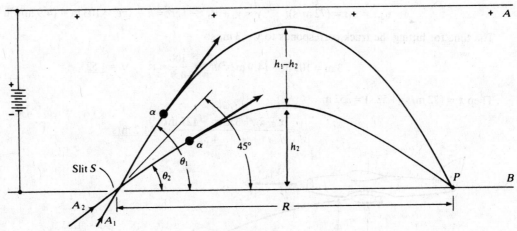

Fig. 5-13

5.29 The arrangement in Fig. 5-14 is the same as that in Fig. 5-12 except that α particles enter slit S from two sources, A_1 and A_2, at angles θ_1 and θ_2, respectively. v_0 and a are the same for both groups. Given that $v_0 = 6 \times 10^6$ m/s, $a = 4 \times 10^{13}$ m/s^2, $\theta_1 = 45° + 1°$, $\theta_2 = 45° - 1°$, show that all particles are "focused" at a single point P and find the value of R.

❚ Here we can use the horizontal range formula $R = 2v_0^2 \cos \theta_0 \sin \theta_0 / a$. Noting that for the first α particle $\theta_0 = \theta_1 = 46°$ and for the second α particle $\theta_0 = \theta_2 = 44°$, we see that the two are complementary angles. Then $\cos \theta_1 \sin \theta_1 = \sin \theta_2 \cos \theta_2$ and the ranges are the same. Indeed $R = 2(6 \times 10^6$ m/s$)^2(0.719)(0.695)/$ $(4 \times 10^{13}$ m/s$^2) = \underline{0.90 \text{ m}}$.

5.30 Again referring to Fig. 5-14, find $h_1 - h_2$.

❚ For the vertical heights we have $v_y^2 = v_{0y}^2 + 2a_y y$ with $v_y = 0$; $a_y = -|a| = 4 \times 10^{13}$ m/s^2. Then

$$h_1 = \frac{(v_0 \sin \theta_1)^2}{2 |a|} = 0.233 \text{ m} \qquad h_2 = \frac{(v_0 \sin \theta_2)^2}{2 |a|} = 0.217 \text{ m} \qquad h_1 - h_2 = 0.016 \text{ m} = 16 \text{ mm}$$

Fig. 5-14

5.31 A ball is thrown upward with initial velocity $v_0 = 15.0$ m/s at an angle of 30° with the horizontal. The thrower stands near the top of a long hill which slopes downward at an angle of 20°. When does the ball strike the slope?

❚ We choose the launch point at the origin (see Fig. 5-15). The equations of motion of the ball are

$$x = v_0 \cos 30°t = (13.0 \text{ m/s})t \qquad y = v_0 \sin 30° - \tfrac{1}{2}gt^2 = (7.5 \text{ m/s})t - (4.9 \text{ m/s}^2)t^2$$

The equation of the straight line incline is $y = -x \tan 20° = -0.364x$. We want the time at which the (x, y) values for the ball satisfy this equation. We thus substitute the time expressions for y and x:

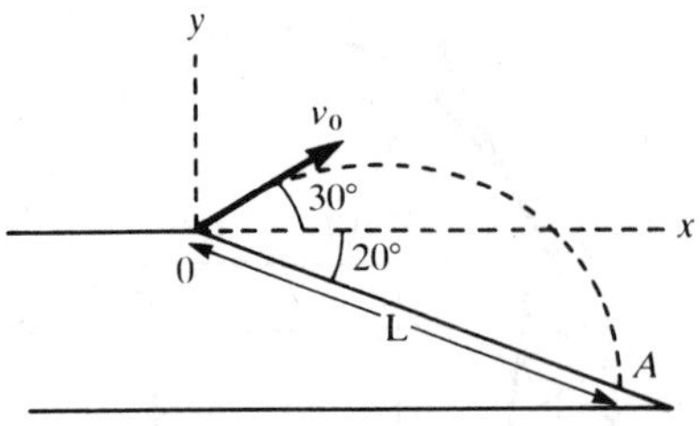

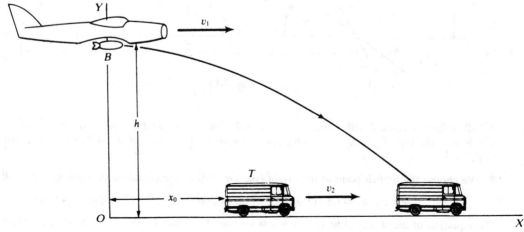

Fig. 5-15

$(7.5 \text{ m/s})t - (4.9 \text{ m/s}^2)t^2 = -0.364[(13.0 \text{ m/s})t]$, or $12.2t = 4.9t^2$. The solutions are $t = 0$ (corresponding to $x = y = 0$) and $t = \underline{2.49 \text{ s}}$.

5.32 Referring to Prob. 5.31 determine how far down the slope the ball strikes.

▮ Referring to Fig. 5-15 we need the x and y components of the displacement to position A: $x = (13.0 \text{ m/s})(2.49 \text{ s}) = 32.4 \text{ m}$; $y = -0.364x = -11.8 \text{ m}$. Then $L = \sqrt{x^2 + y^2} = \underline{34.5 \text{ m}}$ (or directly from x, $L = x/\cos 20° = \underline{34.5 \text{ m}}$).

5.33 Referring to Prob. 5.31 indicate with what velocity the ball hits.

▮ For the velocity of the ball just before impact, $\mathbf{v}_A$:

$$v_{Ax} = v_0 \cos 30° = 13.0 \text{ m/s} \qquad v_{Ay} = v_0 \sin 30° - gt = 7.5 \text{ m/s} - (9.8 \text{ m/s}^2)(2.49 \text{ s}) = -16.9 \text{ m/s}$$

Thus

$$v_A = \sqrt{v_{Ax}^2 + v_{Ay}^2} = \underline{21.3 \text{ m/s}} \qquad \theta_A = \tan^{-1}\left|\frac{v_{Ay}}{v_{Ax}}\right| = \underline{52.4°} \qquad \text{below positive } x \text{ axis}$$

5.34 A bomber, Fig. 5-16, is flying level at a speed $v_1 = 72 \text{ m/s}$ (about 161 mi/h), at an elevation of $h = 103 \text{ m}$. When directly over the origin bomb B is released and strikes the truck T, which is moving along a level road (the X axis) with constant speed. At the instant the bomb is released the truck is at a distance $x_0 = 125 \text{ m}$ from O. Find the value of v_2 and the time of flight of B. (Assume that the truck is 3 m high.)

▮ The equations for x and y motion of the bomb are

$$x = v_{0x}t = v_1 t = (72 \text{ m/s})t \qquad y = y_0 + v_{0y}t - \tfrac{1}{2}gt^2 = h - \tfrac{1}{2}gt^2 = 103 \text{ m} - (4.9 \text{ m/s}^2)t^2$$

The time for hitting the truck corresponds to $y = 3 \text{ m}$, so

$$3 \text{ m} = 103 \text{ m} - (4.9 \text{ m/s}^2)t^2 \qquad \frac{100}{4.9} = t^2 \qquad t = \underline{4.52 \text{ s}}$$

Then $x = (72 \text{ m/s})(4.52 \text{ s}) = 352 \text{ m}$.

$$v_2 = \frac{x - x_0}{t} = \frac{325 \text{ m} - 125 \text{ m}}{4.52 \text{ s}} = \underline{44.2 \text{ m/s}}$$

Fig. 5-16

5.35 A projectile, Fig. 5-17, is fired upward with velocity v_0 at an angle θ. (**a**) At what point $P(x, y)$ does it strike the roof of the building, and in what time? (**b**) Find the magnitude and direction of **v** at P. Let $\theta = 35°$, $v_0 = 40$ m/s, $\alpha = 30°$, and $h = 15$ m.

▐ First note that (using notation $\dot{x} \equiv v_x$, $\dot{y} \equiv v_y$)

$$\dot{x}_0 = v_0 \cos 35° = 32.7661 \text{ m/s} \qquad \dot{y}_0 = v_0 \sin 35° = 22.943 \text{ m/s}$$

and, from the equation of the roof,

$$y = h - x \tan \alpha = 15 - (0.57735)x \tag{1}$$

(**a**) Eliminating t from $y = \dot{y}_0 t - 4.9t^2$ by $x = \dot{x}_0 t$, we have

$$y = \left(\frac{\dot{y}_0}{\dot{x}_0}\right)x - \frac{4.9x^2}{\dot{x}_0^2} \tag{2}$$

for the path of the projectile. Equating y in (1) to y in (2) and inserting numerical values, $0.004564x^2 - 1.277558x + 15 = 0$, from which $x = \underline{12.28 \text{ m}}$. Then $y = h - (12.28) \tan \alpha = \underline{7.90 \text{ m}}$. The time to strike is given by $12.28 = 32.7661t$, or $t = \underline{0.375 \text{ s}}$.

(**b**) At P, $\qquad \dot{x} = \dot{x}_0 = 32.766$ m/s $\qquad \dot{y} = \dot{y}_0 - 9.8t = 22.943 - (9.8)(0.375) = 19.268$ m/s

Thus $v = (\dot{x}^2 + \dot{y}^2)^{1/2} = \underline{38.0 \text{ m/s}}$ and $\tan \beta = \dot{y}/\dot{x} = 0.588$, or $\beta = \underline{30.46°}$, where β is the angle that **v** makes with X at P.

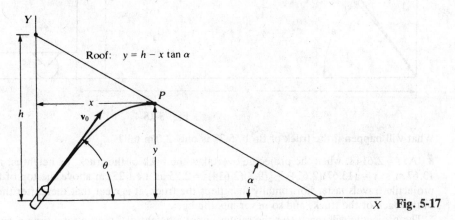

Fig. 5-17

5.36ᶜ In Prob. 5.35, angle θ can be adjusted. Find the value of θ for which the projectile strikes the roof in a minimum time.

▐ Again $\qquad y = \dot{y}_0 t - \frac{1}{2}gt^2 = (v_0 \sin \theta)t - \frac{1}{2}gt^2 \qquad$ and $\qquad y = h - x \tan \alpha$

Equating these two expressions for y and eliminating x by $x = \dot{x}_0 t = (v_0 \cos \theta)t$, we obtain the following equation for the time of striking: $\frac{1}{2}gt^2 - v_0(\cos \theta \tan \alpha + \sin \theta)t + h = 0$. Or, using the addition formula $\sin (\theta + \alpha) = \sin \theta \cos \alpha + \cos \theta \sin \alpha$,

$$\frac{1}{2}gt^2 - \left[\frac{v_0}{\cos \alpha} \sin (\theta + \alpha)\right]t + h = 0 \tag{1}$$

For a minimum t, we must have $dt/d\theta = 0$. Differentiating (1) with respect to θ and setting $dt/d\theta = 0$, we obtain

$$-\left[\frac{v_0}{\cos \alpha} \cos (\theta + \alpha)\right]t_{\min} = 0$$

which implies that (since $t_{\min} \neq 0$) $\cos (\theta + \alpha) = 0$, or $\theta = \underline{90° - \alpha}$.

This result means that the projectile should be aimed in the direction of minimum *distance*, just as though the acceleration of gravity did not exist. However, gravity cannot be ignored in this problem. If we seek to determine the value of $t_{\min}$ by substituting $\theta + \alpha = 90°$ into (1) and solving, we obtain

$$t_{\min} = \frac{v_0 - \sqrt{v_0^2 - 2gh \cos^2 \alpha}}{g \cos \alpha}$$

which is complex if $v_0 < \sqrt{2gh}\cos\alpha$. In other words, if $v_0 < \sqrt{2gh}\cos\alpha$, the projectile never reaches the roof, whatever the value of θ, and the concept of a minimum time becomes meaningless.

5.37 With reference to Fig. 5-18, the projectile is fired with an initial velocity $v_0 = 35$ m/s at an angle $\theta = 23°$. The truck is moving along X with a constant speed of 15 m/s. At the instant the projectile is fired, the back of the truck is at $x = 45$ m. Find the time for the projectile to strike the back of the truck, if the truck is very tall.

▎ In this case, the projectile hits the back of the truck at the moment of overtaking it, which is the moment at which the distance of the back of the truck, $x_1 = 45 + 15t$, equals the horizontal distance of the projectile, $x = (v_0\cos\theta)t = 32.22t$.

Thus
$$t = \frac{45}{32.22 - 15} = \underline{2.614\text{ s}}$$

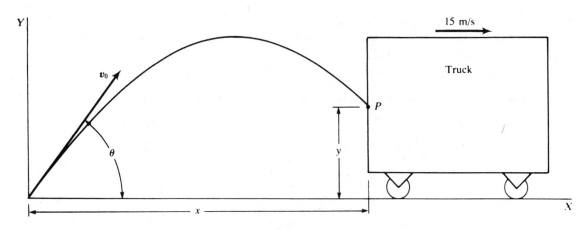

Fig. 5-18

5.38 What will happen if the truck of Prob. 5.37 is only 2.0 m tall?

▎ At $t = 2.614$ s, when the projectile overtakes the back of the truck, its height is, noting $v_0\sin\theta = 13.67$ m/s: $y = (13.67)(2.614) - \frac{1}{2}(9.8)(2.614)^2 = 2.25$ m, i.e., 25 cm above the top of the truck. Since the projectile travels faster horizontally than does the truck, it is clear that thereafter the projectile remains ahead of the back of the truck, and so never hits the back.

The projectile will reach (for the second time) a height of 2 m in a total time t_2 given by $2 = (13.67)t_2 - \frac{1}{2}(9.8)t_2^2$, or $t_2 = 2.635$ s, that is, $2.635 - 2.614 = 0.021$ s after overtaking the back of the truck. Thus the projectile hits the top of the truck of a distance of $(32.22 - 15)(0.021) = 0.36$ m $= \underline{36\text{ cm}}$ in front of the rear edge.

5.39 Referring to Prob. 5.37, find a value of v_0, all other conditions remaining the same, for which the projectile hits the truck at $y = 3$ m.

▎ The time taken to overtake the back of the truck is given by
$$45 + 15t = (v_0\cos\theta)t \qquad \text{or} \qquad t = \frac{45}{v_0\cos\theta - 15}$$

at which time $y = 3 = (v_0\sin\theta)t - \frac{1}{2}(9.8)t^2 = (v_0\sin\theta)\left(\frac{45}{v_0\cos\theta - 15}\right) - \frac{1}{2}(9.8)\left(\frac{45}{v_0\cos\theta - 15}\right)^2$

Inserting the numerical values of $\sin\theta$ and $\cos\theta$, we obtain the following quadratic equation for v_0: $v_0^2(4.55) - v_0(60.3) - 3532 = 0$. Solving, $v_0 = \underline{35.3\text{ m/s}}$.

5.40ᶜ The motion of a particle in the XY plane is given by $x = 25 + 6t^2$; $y = -50 - 20t + 8t^2$. Find the following initial values: $x_0, y_0, \dot{x}_0, \dot{y}_0, v_0$.

▎ At $t = 0$, $x = x_0 = 25$ m; $y = y_0 = -50$ m.
$$v_x \equiv \dot{x} = 12t \quad \text{and} \quad v_{0x} \equiv \dot{x}_0 = 0\text{ m/s} \quad v_y \equiv \dot{y} = -20 + 16t \quad \text{and} \quad v_{0y} \equiv \dot{y}_0 = -20\text{ m/s}$$
$$v_0 = (v_{0x}^2 + v_{0y}^2)^{1/2} = \underline{20\text{ m/s}}$$

5.41^c Find magnitude and direction of **a**, the acceleration of the particle in Prob. 5.40.

▌ $a_x \equiv \dot{v}_x = 12 \text{ m/s}^2$; $a_y \equiv \dot{v}_y = 16 \text{ m/s}^2$.

$$\tan \theta_a = \frac{a_y}{a_x} = \frac{16}{12} = \frac{4}{3}$$

implies that $\theta_a = \underline{53°}$ above the positive x axis. $a = (12^2 + 16^2)^{1/2} = \underline{20 \text{ m/s}^2}$.

5.42 Write an equation for the particle's path (find y as a function of x) in Prob. 5.40.

▌ We eliminate time between the two equations as follows: $t^2 = (x - 25)/6$, or $t = [(x - 25)/6]^{1/2}$. Then substituting into the y equation: $y = -50 - 20[(x - 25)/6]^{1/2} + 8(x - 25)/6$, or $6y = -500 + 8x - 120[(x - 25)/6]^{1/2}$.

5.43^c A particle moving in the XY plane has X and Y components of velocity given by

$$\dot{x} = b_1 + c_1 t \qquad \dot{y} = b_2 + c_2 t \tag{1}$$

where x and y are measured in meters and t in seconds. (**a**) What are the units and dimensions of the constants b_1 and b_2? of c_1 and c_2? (**b**) Integrate the above relations to obtain x and y as functions of time. (**c**) Denoting total acceleration as **a** and total velocity as **v**, find expressions for the magnitude and direction of **a** and of **v**. (**d**) Write **v** in terms of the unit vectors.

▌ (**a**) Inspection of (1) shows that b_1 and b_2 must represent velocities in meters per second; unit-dimensionally c_1 and c_2 must be $|\text{m/s}^2|$, thus accelerations.

(**b**)

$$x = x_0 + b_1 t + \tfrac{1}{2} c_1 t^2 \qquad y = y_0 + b_2 t + \tfrac{1}{2} c_2 t^2$$

where x_0, y_0 are the values of x and y at $t = 0$.
(**c**) Differentiating (1) with respect to t, $\ddot{x} = c_1$, $\ddot{y} = c_2$. Then

$$a = (\ddot{x}^2 + \ddot{y}^2)^{1/2} = (c_1^2 + c_2^2)^{1/2} \qquad \tan \alpha = \frac{\ddot{y}}{\ddot{x}} = \frac{c_2}{c_1}$$

where α is the angle **a** makes with X. Note that **a** is constant in magnitude and direction. For the velocity,

$$v = (\dot{x}^2 + \dot{y}^2)^{1/2} = [(b_1 + c_1 t)^2 + (b_2 + c_2 t)^2]^{1/2} \qquad \tan \beta = \frac{b_2 + c_2 t}{b_1 + c_1 t}$$

where β is the angle **v** makes with X. (**d**) $\mathbf{v} = (b_1 + c_1 t)\mathbf{i} + (b_2 + c_2 t)\mathbf{j}$.

5.44^c A particle moves in the XY plane along the path given by $y = 10 + 3x + 5x^2$. The X component of velocity, $\dot{x} = 4$ m/s, is constant, and at $t = 0$, $x = x_0 = 6$ m. (**a**) Write y and x as functions of t. (**b**) Find $\ddot{y}$ and $\ddot{x}$, the components of acceleration of the particle.

▌ (**a**) $\dot{y} = 3\dot{x} + 10x\dot{x}$; $\dot{x} = 4$ m/s at all times and $x_0 = 6$ m. Then

$$x = x_0 + \dot{x}t = 6 \text{ m} + (4 \text{ m/s})t.$$

$$y = 10 + (18 + 12t) + 5(6 + 4t)^2 = 208 \text{ m} + (252 \text{ m/s})t + (80 \text{ m/s}^2)t^2$$

(**b**)

$$\ddot{x} = 0 \qquad \ddot{y} = \underline{160 \text{ m/s}^2}$$

5.45 A ball, B_1, is fired upward from the origin of X, Y with initial velocity $v_1 = 100$ m/s at an angle $\theta_1 = 40°$. After $t_1 = 10$ s, as can easily be shown, the ball is at point $P(x_1, y_1)$, where $x_1 = 766.0$ m, $y_1 = 152.8$ m. Some time later, another ball, B_2, is fired upward, also from the origin, with velocity v_2 at angle $\theta_2 = 35°$. (**a**) Find a value of v_2 such that B_2 will pass through the point $P(x_1, y_1)$. (**b**) Find the time when B_2 must be fired in order that the two balls will collide at $P(x_1, y_1)$.

▌ (**a**) Let (x_1, y_1, t_1) refer to the coordinates and time of B_1 and (x_2, y_2, t_2) to those of B_2. Since B_2 is to pass through $P(x_1, y_1)$,

$$x_2 = (v_2 \cos 35°)t_2 = 766.0 \qquad y_2 = (v_2 \sin 35°)t_2 - 4.9 t_2^2 = 152.8$$

Eliminating t_2,

$$152.8 = (766.0) \tan 35° - (4.9)\left(\frac{766.0}{v_2 \cos 35°}\right)^2$$

from which $v_2 = \underline{105.69 \text{ m/s}}$.

(b) Inserting the value of v_2 in $x_2 = (v_2 \cos 35°)t_2 = 766.0$ m, we find $t_2 = 8.848$ s. Hence, with $v_2 = 105.69$ m/s and $\theta_2 = 35°$, B_2 passes through $P(x_1, y_1)$ 8.848 s after it is fired. But B_1 arrives at this point 10 s after starting Hence, if the two are to collide, the firing of B_2 must be delayed $10 - 8.848 = \underline{1.152\text{ s}}$.

5.46^c The motion of a particle in the XY plane is given by

$$x = 10 + 12t - 20t^2 \qquad y = 25 + 15t + 30t^2$$

Find values of x_0, $\dot{x}_0$; y_0, $\dot{y}_0$ and the magnitude and direction of $\mathbf{v}_0$.

▌ At $t = 0$, $x = x_0 = 10$; $y = y_0 = 25$. Differentiating we get $\dot{x} = 12 - 40t$; $\dot{y} = 15 + 60t$; $\dot{x}_0 = 12$; $\dot{y}_0 = 15$.

$$v_0 = (\dot{x}_0^2 + \dot{y}_0^2)^{1/2} = 19.2 \qquad \theta_0 = \tan^{-1}\frac{\dot{y}_0}{\dot{x}_0} = 51.3° \qquad \text{above positive } x \text{ axis}$$

5.47^c Referring to Prob. 5.46, find $\ddot{x}$, $\ddot{y}$, and $\mathbf{a}$, the vector acceleration.

▌ We differentiate the expressions for $\dot{x}$ and $\dot{y}$ to get $\ddot{x} = -40$, $\ddot{y} = 60$. Acceleration is thus constant and $a = \sqrt{40^2 + 60^2} = \underline{72}$. The direction of $\mathbf{a}$ is given by

$$\beta = \tan^{-1}\left|\frac{60}{40}\right| = \underline{56.3°} \qquad \text{above the negative } x \text{ axis}$$

5.48 Referring to Probs. 5.46 and 5.47, state whether or not the motion is along a straight line.

▌ No: Since the lines of $\mathbf{v}_0$ and $\mathbf{a}$ do not coincide the path must be parabolic, as in projectile motion. See Prob. 5.57.

5.49^c Can the motion of a particle be given by

$$x = 5 + 10t + 17t^2 + 4t^3 \qquad y = 8 + 9t + 20t^2 - 6t^3$$

if the particle is acted on by a constant force?

▌ $\dot{x} = 10 + 34t + 12t^2$; $\dot{y} = 9 + 40t - 18t^2$; $\ddot{x} = 34 + 24t$; $\ddot{y} = 40 - 36t$. But the acceleration is not constant, as it must be for a constant force. Hence, the motion is impossible.

5.2 RELATIVE MOTION

5.50 An elevator is moving upward at a constant speed of 4 m/s. A light bulb falls out of a socket in the ceiling of the elevator. A man in the building watching the cage sees the bulb rise for $(4/9.8)$ s and then fall for $(4/9.8)$ s; at $t = (4/9.8)$ s the bulb appears to be at rest to the man. Compute the velocity of the bulb at $t = (4/9.8)$ s from the view of an observer in the elevator.

▌ Taking *up* as positive $v_{\text{bulb/elev}} = v_{\text{bulb/bldg}} - v_{\text{elev/bldg}} = 0 - 4 = \underline{-4\text{ m/s}}$. Alternatively, in the inertial frame of the elevator,

$$v = v_0 + at = 0 + (-9.8)\left(\frac{4}{9.8}\right) = \underline{-4\text{ m/s}}$$

5.51 A ship is traveling due east at 10 km/h. What must be the speed of a second ship heading 30° east of north if it is always due north from the first ship?

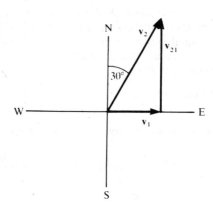

Fig. 5-19

▮ v_1 = velocity of first ship relative to the earth; v_2 = velocity of second ship relative to the earth. Let v_{21} = relative velocity of second ship to first ship. Then $v_2 = v_{21} + v_1$, (Fig. 5-19), where v_{21} is due north. Thus $v_2 \sin 30° = v_1 = 10$ km/h, and $v_2 = \underline{20 \text{ km/h}}$.

5.52 During a rainstorm, raindrops are observed to be striking the ground at an angle of 35° with the vertical. The wind speed is 4.5 m/s. Assuming that the horizontal velocity component of the raindrops is the same as the speed of the air, what is the vertical velocity component of the raindrops? What is their speed?

▮ Let v = velocity of raindrops relative to earth, v_w = wind velocity, $v_w = 4.5$ m/s, $v_x = v_w$. From Fig. 5-20,

$$v_y = \frac{v_w}{\tan 35°} = \underline{6.43 \text{ m/s}} \qquad v = \sqrt{v_x^2 + v_y^2} = \underline{7.84 \text{ m/s}}$$

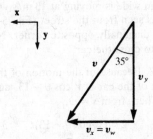

Fig. 5-20

5.53 A rowboat is pointing perpendicular to the bank of a river. The rower can propel the boat with a speed of 3.0 m/s *with respect to the water*. The river has a current of 4.0 m/s. (*a*) Construct a diagram in which the two velocities are represented as vectors. (*b*) Find the vector which represents the boat's velocity with respect to the shore. (*c*) At what angle is this vector inclined to the direction in which the boat is pointing? What is the boat's speed with respect to the launch point? (*d*) If the river is 100 m wide, determine how far downstream of the launch point the rowboat is when it reaches the opposite bank.

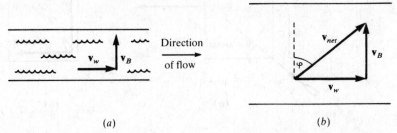

(a) (b) **Fig. 5-21**

▮ (*a*) See Fig. 5-21(*a*).
(*b, c*) The velocity with respect to the shore is given by $v_{net} = v_B + v_w$. Since v_B and v_w are perpendicular, we have $v_{net} = \sqrt{v_B^2 + v_w^2} = \sqrt{3^2 + 4^2} = \underline{5 \text{ m/s}}$. The angle φ shown in Fig. 5-21(*b*) is determined by $\tan \varphi = v_w/v_B$. For the speeds given, we find $\varphi = \underline{53.1°}$. The boat moves along a line directed 53.1° downstream from "straight across."
(*d*) Letting D = distance downstream, we have $D/100 \text{ m} = v_w/v_B = 4/3$, so that $D = \underline{133 \text{ m}}$.

5.54 A swimmer can swim at a speed of 0.70 m/s with respect to the water. She wants to cross a river which is 50 m wide and has a current of 0.50 m/s. (*a*) If she wishes to land on the other bank at a point directly across the river from her starting point, in what direction must she swim? How rapidly will she increase her distance from the near bank? How long will it take her to cross? (*b*) If, she instead decides to cross in the shortest possible time, in what direction must she swim? How rapidly will she increase her distance from the near bank? How long will it take her to cross? How far downstream will she be when she lands?

▮ (*a*) Let v_c be the velocity of the current, v_w be the velocity of the swimmer with respect to the water, and v_s be the velocity of the swimmer with respect to the shore. Then $v_s = v_w + v_c$. For a direct crossing, v_s must be perpendicular to v_c. Therefore $\sin \theta = v_c/v_w$, as shown in Fig. 5-22. We are given the values $v_c = 0.50$ m/s, $v_w = 0.70$ m/s, and $d = 50$ m; we find $\theta = 45.6°$ upstream from the direction "straight across." The swimmer will increase her distance from the near shore at the rate $v_s = v_w \cos \theta = 0.49$ m/s. She will cross the river in a time $t = d/v_s = 102$ s.

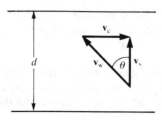

Fig. 5-22

(b) To maximize the component of her velocity perpendicular to the river bank, the swimmer should head straight across the stream. She will cross in a time $t = d/v_w = 71.4$ s, and she will land a distance $v_c t = (v_c/v_w)d = 35.7$ m downstream from her starting point.

5.55 An armored car 2 m long and 3 m wide is moving at 13 m/s when a bullet hits it in a direction making an angle arctan (3/4) with the car as seen from the street (Fig. 5-23a). The bullet enters one edge of the car at the corner and passes out at the diagonally opposite corner. Neglecting any interaction between bullet and car, find the time for the bullet to cross the car.

▮ Call the speed of the bullet V. Because of the motion of the car, the velocity of the bullet relative to the car in the direction of the length of the car is $V \cos \theta - 13$, and the velocity in the direction of the width of the car is $V \sin \theta$ (Fig. 5-23b). Then, from $s = vt$,

$$2 = (V \cos \theta - 13)t \qquad 3 = (V \sin \theta)t$$

Eliminate V to find

$$t = \frac{1}{13}\left(\frac{3}{\tan \theta} - 2\right) = \frac{2}{13} = \underline{0.15 \text{ s}}$$

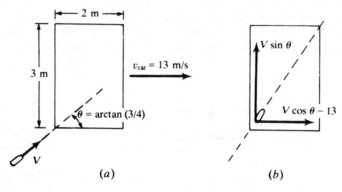

(a)   (b)   Fig. 5-23

5.56 Rain, pouring down at an angle α with the vertical, has a constant speed of 10 m/s. A woman runs against the rain with a speed of 8 m/s and sees that the rain makes an angle β with the vertical. Find the relation between α and β.

▮ From the vector diagram, Fig. 5-24,

$$\tan \beta = \frac{v_{woman} + v_{rain} \sin \alpha}{v_{rain} \cos \alpha} = \frac{8 + 10 \sin \alpha}{10 \cos \alpha}$$

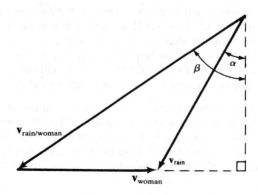

Fig. 5-24

5.57 Verify that the trajectory of Prob. 5.46 is a parabola by choosing coordinate axes parallel and perpendicular to the constant acceleration vector.

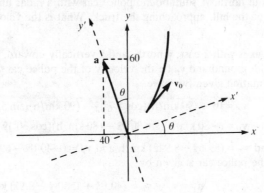

Fig. 5-25

▌ Figure 5-25 shows the new coordinate system; from analytic geometry we have the relations

$$x' = x \cos \theta + y \sin \theta = \frac{3}{\sqrt{13}} x + \frac{2}{\sqrt{13}} y \qquad y' = -x \sin \theta + y \cos \theta = -\frac{2}{\sqrt{13}} x + \frac{3}{\sqrt{13}} y$$

Hence the equations of motion in the primed coordinates are

$$\sqrt{13}\, x' = 3(10 + 12t - 20t^2) + 2(25 + 15t + 30t^2) = 80 + 66t \tag{1}$$

$$\sqrt{13}\, y' = -2(10 + 12t - 20t^2) + 3(25 + 15t + 30t^2) = 55 + 21t + 130t^2 \tag{2}$$

We can now solve (1) for t in terms of x' and substitute the result in (2), obtaining an equation

$$y' = ax'^2 + bx' + c \tag{3}$$

for definite constants a, b, c. Finally, completing the square on the right of (3), we transform it to

$$y' - \beta = K(x' - \alpha)^2 \tag{4}$$

which is recognized as the equation of a parabola, with vertex at $x' = \alpha$, $y' = \beta$ and with axis parallel to the y' axis.

5.58 Observer O drops a stone from the thirtieth floor of a skyscraper. Observer O', descending in an elevator at constant speed $V = 5.0$ m/s, passes the thirtieth floor just as the stone is released. At the time $t = 3.0$ s after the stone is dropped, find the position, the velocity, and the acceleration of the stone relative to O. Then find the position, the velocity, and the acceleration of the stone relative to O'.

▌ For O, the position of the stone is given by

$$x = x_0 + v_0 t + \frac{at^2}{2}$$

where $x = 0$ at the thirtieth floor with the downward direction as the positive x direction. Thus, at $t = 3.0$ s,

$$x = 0 + 0 + \frac{9.8 \text{ m/s}^2 \times (3.0 \text{ s})^2}{2} = +44 \text{ m}$$

Also, $v = v_0 + at$ gives $v = 0 + 9.8$ m/s$^2 \times 3.0$ s $= +29$ m/s.

The acceleration of a freely falling body, as seen by the observer O who is stationary with respect to the earth, is known to be the constant gravitational acceleration. (Indeed, this underlies the validity of the two calculations immediately above.) You thus have $a = +g = +9.8$ m/s^2.

O' measures position x', related to x via $x' = x - Vt$. Hence, after 3.0 s, $x' = 44$ m $- 5.0$ m/s $\times 3.0$ s $= +29$ m. That is, the stone is located 29 m below observer O' at the end of 3.0 s. The stone's velocity relative to O' is $v' = v - V$; hence, at $t = 3.0$ s,

$$v' = 29 \text{ m/s} - 5.0 \text{ m/s} = +24 \text{ m/s}$$

Since V is constant, $\mathbf{a}' = \mathbf{a}$, and $a' = +g = +9.8$ m/s^2. Observer O' sees the stone to have the same downward acceleration as that seen by O. (In general, accelerations are the same in all inertial frames.)

5.59 A truck is traveling due north and descending a 10 percent grade (angle of slope $= \tan^{-1} 0.10 = 5.7°$) at a constant speed of 90 km/h. At the base of the hill there is a gentle curve, and beyond that the road is level and heads 30° east of the north. A southbound police car with a radar unit is traveling at 80 km/h along the level road at the base of the hill, approaching the truck. What is the velocity vector of the truck with respect to the police car?

▌ We use coordinate axes with $\hat{\mathbf{x}}$ east, $\hat{\mathbf{y}}$ north, and $\hat{\mathbf{z}}$ vertically upward. We let $\mathbf{v}_T$ be the velocity of the truck with respect to the ground and $\mathbf{v}_P$ be the velocity of the police car with respect to the ground. According to the information given, we have

$$\mathbf{v}_T = 0\hat{\mathbf{x}} + (90 \text{ km/h})(\cos 5.7°)\hat{\mathbf{y}} - (90 \text{ km/h})(\sin 5.7°)\hat{\mathbf{z}} \quad \text{and}$$
$$\mathbf{v}_P = (-80 \text{ km/h})(\sin 30°)\hat{\mathbf{x}} - (80 \text{ km/h})(\cos 30°)\hat{\mathbf{y}} + 0\hat{\mathbf{z}}$$

Reducing these, we find $\mathbf{v}_T = (89.6\hat{\mathbf{y}} - 8.94\hat{\mathbf{z}})$ km/h and $\mathbf{v}_P = (-40.0\hat{\mathbf{x}} - 69.3\hat{\mathbf{y}})$ km/h. The velocity $\mathbf{u}_T$ of the truck with respect to the police car is given by

$$\mathbf{u}_T = \mathbf{v}_T - \mathbf{v}_P = (40.0\hat{\mathbf{x}} + 158.8\hat{\mathbf{y}} - 8.9\hat{\mathbf{z}}) \text{ km/h}$$

5.60ᶜ A bird, in level flight at constant acceleration $\mathbf{a}_0$ relative to the ground frame X, Y (Fig. 5-26), lets fall a worm from its beak. What is the path of the worm, as seen by the bird?

▌ In the bird's noninertial coordinate system X', Y' (Fig. 5-26), the equation of motion of the worm is

$$m\frac{d\mathbf{v}'}{dt} = m\mathbf{g} - m\mathbf{a}_0 \quad \text{or} \quad \frac{d\mathbf{v}'}{dt} = \mathbf{g} - \mathbf{a}_0 = \text{constant}$$

Thus the acceleration of the worm is constant, and its path is a straight line (supposing that it was dropped from rest). The slope of the line with respect to the horizontal is $\tan \theta = g/a_0$.

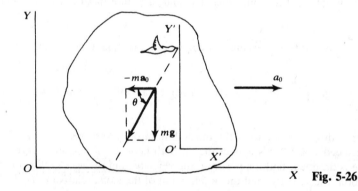

Fig. 5-26

5.61ᶜ Refer to Prob. 5.60 and Fig. 5-26. **(a)** Determine the path of the worm as seen from the ground. **(b)** Verify that the two descriptions of the path are equivalent.

▌ **(a)** In the ground frame X, Y, the worm has constant acceleration $\ddot{y} = -g$ and an initial velocity $\dot{x}_0 = v_0$, where v_0 is the speed of the bird at the instant the worm is released (call this time $t = 0$). Hence

$$x = x_0 + v_0 t \qquad y = y_0 - \tfrac{1}{2}gt^2 \tag{1}$$

and the path is a parabola.

(b) Let us suppose that at $t = 0$ the two coordinate frames coincide. At time t, O' will have advanced a distance $v_0 t + \tfrac{1}{2}a_0 t^2$ along the X axis, so that the coordinates (x, y) and (x', y') of the worm in the two systems are related by

$$x = x' + (v_0 t + \tfrac{1}{2}a_0 t^2) \qquad y = y' \tag{2}$$

The path in the X', Y' system is obtained by substituting the expressions (2) into (1):

$$x' + \cancel{v_0 t} + \tfrac{1}{2}a_0 t^2 = x_0 + \cancel{v_0 t} \qquad y' = y_0 - \tfrac{1}{2}gt^2$$

or

$$\frac{y' - y_0}{x' - x_0} = \frac{g}{a_0}$$

which is a straight line of slope g/a_0, as found in Prob. 5.60.

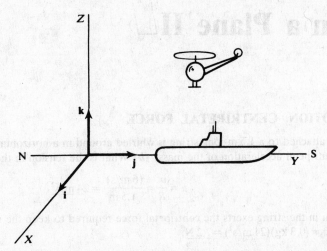

Fig. 5-27

5.62 A helicopter is trying to land on a submarine deck which is moving south at 17 m/s. A 12 m/s wind is blowing into the west. If to the submarine crew the helicopter is descending vertically at 5 m/s, what is its speed (a) relative to the water and (b) relative to the air. See Fig. 5-27.

(a) $\mathbf{v}_{hel/water} = \mathbf{v}_{sub/water} + \mathbf{v}_{hel/sub} = 17\mathbf{j} + (-5)\mathbf{k} = (17\mathbf{j} - 5\mathbf{k})$ m/s

(b) $\mathbf{v}_{hel/air} = \mathbf{v}_{hel/water} + \mathbf{v}_{water/air} = \mathbf{v}_{hel/water} - \mathbf{v}_{air/water} = (17\mathbf{j} - 5\mathbf{k}) - 12\mathbf{i} = (-12\mathbf{i} + 17\mathbf{j} - 5\mathbf{k})$ m/s

CHAPTER 6
Motion in a Plane II

6.1 CIRCULAR MOTION; CENTRIPETAL FORCE

6.1 A 0.3-kg mass attached to a 1.5 m-long string is whirled around in a horizontal circle at a speed of 6 m/s. **(a)** What is the centripetal acceleration of the mass? **(b)** What is the tension in the string? (Neglect gravity.)

(a)
$$a = \frac{v^2}{R} = \frac{(6 \text{ m/s})^2}{1.5 \text{ m}} = \underline{24 \text{ m/s}^2}$$

(b) The tension in the string exerts the centripetal force required to keep the mass in circular motion. This force is $T = ma = (0.3 \text{ kg})(24 \text{ m/s}^2) = \underline{7.2 \text{ N}}$.

6.2 A small ball is fastened to a string 24 cm long and suspended from a fixed point P to make a conical pendulum, as shown in Fig. 6-1. The ball describes a horizontal circle about a center vertically under point P, and the string makes an angle of 15° with the vertical. Find the speed of the ball.

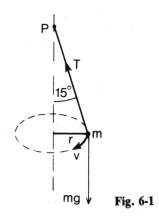

Fig. 6-1

$$T \cos 15° = mg \qquad T \sin 15° = \frac{mv^2}{r} \qquad \text{and hence} \qquad \tan 15° = \frac{v^2}{rg}$$

Since $r = 24 \sin 15° = 24(0.259) = 6.22$ cm,

$$\tan 15° = \frac{v^2}{6.22(980)} \qquad v = \underline{40.4 \text{ cm/s}}$$

6.3 In the Bohr model of the hydrogen atom an electron is pictured rotating in a circle (with a radius of 0.5×10^{-10} m) about the positive nucleus of the atom. The centripetal force is furnished by the electric attraction of the positive nucleus for the negative electron. How large is this force if the electron is moving with a speed of 2.3×10^6 m/s? (The mass of an electron is 9×10^{-31} kg.)

Force $= (9 \times 10^{-31} \text{ kg})(2.3 \times 10^6 \text{ m/s})^2/(5.0 \times 10^{-11} \text{ m}) = \underline{9.5 \times 10^{-8} \text{ N}}$

6.4 Find the maximum speed with which an automobile can round a curve of 80-m radius without slipping if the road is unbanked and the coefficient of friction between the road and the tires is 0.81.

First draw a diagram showing the forces (Fig. 6-2). If mg is the weight of the automobile, then the normal force is $N = mg$. The frictional force supplies the centripetal force F_c.

$$F_c = \mu_s N = 0.81 \, mg$$

Also,
$$F_c = \frac{mv^2}{r} \qquad 0.81 \, mg = \frac{mv^2}{80} \qquad v^2 = 0.81 \times 80 \times 9.8 = \underline{25.2 \text{ m/s}}$$

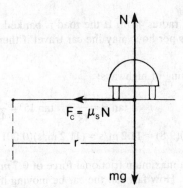

Fig. 6-2

6.5 What is the maximum velocity, in miles per hour, for an automobile rounding a level curve of 200-ft radius if μ_s between tires and roadbed is 1.0.

▌ $$\mu_s mg = \frac{mv^2}{r} \qquad \mu_s = \frac{v^2}{rg} \qquad v = \sqrt{\mu_s gr} = \sqrt{1.0 \times 32.2 \text{ ft/s}^2 \times 200 \text{ ft}} = 80 \text{ ft/s} = 80 \text{ ft/s} \times \frac{30 \text{ mi/h}}{44 \text{ ft/s}} = \underline{54.5 \text{ mi/h}}$$

6.6 A car is traveling 25 m/s (56 mi/h) around a level curve of radius 120 m. What is the minimum value of the coefficient of static friction between the tires and the road required to prevent the car from skidding?

▌ $$F = m\frac{v^2}{r} = F_f \leq \mu_s mg \qquad \mu_s \geq \frac{v^2}{gr} = \frac{(25 \text{ m/s})^2}{(9.8 \text{ m/s}^2)(120 \text{ m})} = \underline{0.53}$$

6.7 Traffic is expected to move around a curve of radius 200 m at 90 km/h. What should be the value of the banking angle if no dependence is to be placed on friction?

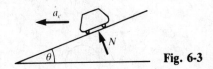

Fig. 6-3

▌ See Fig. 6-3. $$w = mg = N \cos \theta \qquad \text{and} \qquad F_c = \frac{mv^2}{r} = N \sin \theta$$

Dividing the second by the first equation, $\tan \theta = v^2/rg$. Substitute the data, changing kilometers per hour into meters per second:

$$\tan \theta = \frac{\left(\dfrac{90 \text{ km/h} \times 1000 \text{ m/km}}{3600 \text{ s/h}}\right)^2}{200 \text{ m} \times 9.8 \text{ m/s}^2} = 0.319 \qquad \text{so} \qquad \underline{\theta = 17.7°}$$

6.8 As indicated in Fig. 6-4, a plane flying at constant speed is banked at angle θ in order to fly in a horizontal circle of radius r. The aerodynamic lift force acts generally upward at right angles to the plane's wings and fuselage. This lift force corresponds to the tension provided by the string in a conical pendulum, or the normal force of a banked road. (*a*) Obtain the equation for the required banking angle θ in terms of v, r, and g. (*b*) What is the required angle for $v = 60$ m/s (216 km/h) and $r = 1.0$ km?

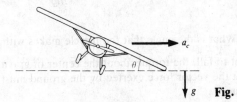

Fig. 6-4

▌ (*a*) As in Probs. 6.2 and 6.7, $\tan \theta = v^2/rg$.

(*b*) $$\theta = \tan^{-1} \frac{60^2}{(1.0 \times 10^3)(9.8)} = \underline{20.2°}$$

6.9 A car goes around a curve of radius 48 m. If the road is banked at an angle of 15° with the horizontal, at what maximum speed in kilometers per hour may the car travel if there is to be no tendency to skid even on very slippery pavement?

▌ Use the equation for banking of highways:

$$\tan \theta = \frac{v^2}{rg} \qquad \tan 15° = \frac{v^2}{48(9.8)}$$

$$v = 0.268(48)(9.8) = 11.2 \text{ m/s} = (11.2 \text{ m/s})(0.001 \text{ km/m})(3600 \text{ s/h}) = \underline{40.3 \text{ km/h}}$$

6.10 A certain car of mass m has a maximum frictional force of $0.7\,mg$ between it and pavement as it rounds a curve on a flat road ($\mu = 0.7$). How fast can the car be moving if it is to successfully negotiate a curve of 15-m radius?

▌ The centripetal force (mv^2/r) must be supplied by the frictional force. In the limiting case, $mv^2/r = f$ with $f = 0.7\,mg$. Thus, $v^2 = 0.7rg$ and $v = \underline{10 \text{ m/s}}$.

6.11 A crate sits on the floor of a boxcar. The coefficient of friction between the crate and the floor is 0.6. What is the maximum speed that the boxcar can go around a curve of radius 200 m without causing the crate to slide?

▌ As in other "unbanked-curve" problems (e.g., Prob. 6.5),

$$v_{max}^2 = \mu_s gr = (0.6)(9.8 \text{ m/s}^2)(200 \text{ m}) = 1176 \text{ m}^2/\text{s}^2 \qquad v_{max} = \sqrt{1176 \text{ m}^2/\text{s}^2} = \underline{34.3 \text{ m/s}}$$

6.12 A boy on a bicycle pedals around a circle of 22-m radius at a speed of 10 m/s. The combined mass of the boy and the bicycle is 80 kg. (*a*) What is the centripetal force exerted by the pavement on the bicycle? (*b*) What is the upward force exerted by the pavement on the bicycle? See Fig. 6-5.

▌ (*a*) $F_c = \dfrac{mv^2}{r} = \dfrac{80(10)^2}{22} = 364 \text{ N}$ \qquad (*b*) $N = mg = 80(9.8) = \underline{784 \text{ N}}$

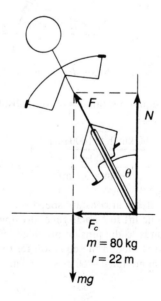

F

N

θ

F_c

$m = 80 \text{ kg}$

$r = 22 \text{ m}$

mg \qquad **Fig. 6-5**

6.13 Refer to Prob. 6.12. What is the angle that the bicycle makes with the vertical?

▌ For the bicycle not to fall, the torque about the center of gravity must be zero—see Chaps. 9 and 10—which means that the vector force exerted by the ground must have a line of action passing through the center of gravity. Thus

$$\tan \theta = \frac{F_c}{N} = \frac{364}{784} = 0.4643 \qquad \theta = \underline{25°}$$

6.14 A fly of mass 0.2 g sits 12 cm from the center of a phonograph record revolving at $33\frac{1}{3}$ rpm. (*a*) What is the magnitude of the centripetal force on the fly? **(b)** What is the minimum value of the coefficient of static friction between the fly and the record required to prevent the fly from sliding off?

(a)
$$v = \frac{2\pi r}{T} = 2\pi fr = 2\pi\left(\frac{33 \cdot 33 \text{ min}^{-1}}{60 \text{ s/min}}\right)(12 \times 10^{-2} \text{ m}) = 0.419 \text{ m/s}$$

$$F = ma = m\frac{v^2}{r} = \frac{(0.2 \times 10^{-3} \text{ kg})(0.419 \text{ m/s})^2}{0.12 \text{ m}} = 2.92 \times 10^{-4} \text{ N}$$

(b)
$$F_f = 2.92 \times 10^{-4} \text{ N} \leq \mu_s mg \qquad \mu_s \geq \frac{2.92 \times 10^{-4} \text{ N}}{(0.2 \times 10^{-3} \text{ kg})(9.8 \text{ m/s}^2)} = \underline{0.149}$$

6.15 Find **(a)** the speed and **(b)** the period of a spaceship orbiting around the moon. The moon's radius is 1.74×10^6 m, and the acceleration due to gravity on the moon is 1.63 m/s^2. (Assume that the spaceship is orbiting just above the moon's surface.)

(a)
$$\frac{v^2}{R_m} = \frac{GM_m}{R_m^2} \equiv g_m \qquad v = \sqrt{g_m R_m} = \sqrt{(1.63 \text{ m/s}^2)(1.74 \times 10^6 \text{ m})} = 1.68 \times 10^3 \text{ m/s} = \underline{1.68 \text{ km/s}}$$

(b) The circumference of the orbit is

$$d = 2\pi R_m = (6.28)(1.74 \times 10^6 \text{ m}) = 1.09 \times 10^4 \text{ km}$$

so the period is

$$t = \frac{d}{v} = \frac{1.09 \times 10^4 \text{ km}}{1.68 \text{ km/s}} = 6.5 \times 10^3 \text{ s} = \underline{108 \text{ min}}$$

6.16 At the equator, the effective value of g is smaller than at the poles. One reason for this is the centripetal acceleration due to the earth's rotation. The magnitude of the centripetal acceleration must be subtracted from the magnitude of the acceleration due purely to gravity in order to obtain the effective value of g. **(a)** Calculate the fractional diminution of g at the equator as a result of the earth's rotation. Express your result as a percentage. **(b)** How short would the earth's period of rotation have to be in order for objects at the equator to be "weightless" (that is, in order for the effective value of g to be zero)? **(c)** How would the period found in part **(b)** compare with that of a satellite skimming the surface of an airless earth?

(a) Using $R_e = 6.37 \times 10^6$ m and $T = 24$ h $= 86400$ s, we find $a = v^2/R_e = 4\pi^2 R_e/T^2 = 3.37 \times 10^{-2}$ m/s^2. Therefore $a/g = 3.44 \times 10^{-3}$. Since $g_{\text{eff}} = g - a$, the fractional diminution is $(g - g_{\text{eff}})/g = a/g = \underline{0.344 \text{ percent}}$. **(b)** In order that $g_{\text{eff}} = 0$, we need $a = g = 4\pi^2 R_e/T_1^2$. Solving for T_1, we find $T_1 = 2\pi\sqrt{R_e/g} = 5.06 \times 10^3$ s $= \underline{84.4 \text{ min}}$. **(c)** Since an orbiting satellite has $ma = mg$ and $g_{\text{eff}} = 0$, its period equals T_1.

6.17 A particle is to slide along the horizontal circular path on the inside of the funnel shown in Fig. 6-6. The surface of the funnel is frictionless. How fast must the particle be moving (in terms of r and θ) if it is to execute this motion?

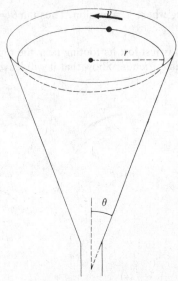

Fig. 6-6

❚ The funnel is equivalent to a road banked at an angle $90° - \theta$. Hence $v^2/rg = \tan(90° - \theta) = \cot\theta$, or $v = \sqrt{rg \cot\theta}$.

6.18 An automobile moves around a curve of radius 300 m at a constant speed of 60 m/s [Fig. 6-7(a)]. **(a)** Calculate the resultant change in velocity (magnitude and direction) when the car goes around the arc of 60°. **(b)** Compare the magnitude of the instantaneous acceleration of the car to the magnitude of the average acceleration over the 60° arc.

❚ **(a)** From Fig. 6-7(b), $\Delta v = \underline{60 \text{ ms}}$ and $\Delta\mathbf{v}$ makes a $\underline{120° \text{ angle}}$ with $\mathbf{v}_A$. **(b)** The instantaneous acceleration has magnitude

$$a = \frac{v^2}{r} = \frac{60^2}{300} = \underline{12 \text{ m/s}^2}$$

The time average acceleration is $\bar{\mathbf{a}} = \Delta\mathbf{v}/\Delta t$. Since

$$\Delta t = \frac{\Delta s}{v} = \frac{r\,\Delta\theta}{v} = \frac{300(\pi/3)}{60} = \frac{5\pi}{3} \text{ s}$$

we have

$$\bar{a} = \frac{\Delta v}{\Delta t} = \frac{60}{5\pi/3} = \underline{11.5 \text{ m/s}^2}$$

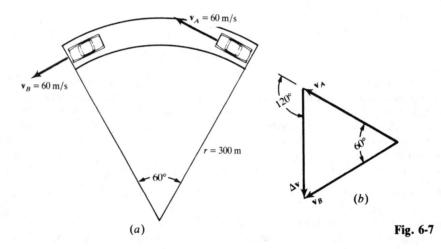

(a)

(b)

Fig. 6-7

6.19 While driving around a curve of 200 m radius, an engineer notes that a pendulum in the car hangs at an angle of 15° to the vertical. What should the speedometer read (in kilometers per hour)?

❚ $T \sin\theta = \dfrac{mv^2}{r}$; $T \cos\theta = mg$ where T = tension. Thus $\tan\theta = \dfrac{v^2}{rg}$ or $v = \sqrt{rg \tan\theta} = 23 \text{ m/s} = \underline{82.5 \text{ km/h}}$

6.20 The bug shown in Fig. 6-8(a) has just lost its footing near the top of the stationary bowling ball. It slides down the ball without appreciable friction. Show that it will leave the surface of the ball at the angle $\theta = \arccos\frac{2}{3} \approx 48°$.

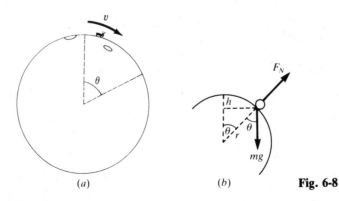

(a)

(b)

Fig. 6-8

I The centripetal force is given by

$$mv^2/r = mg \cos \theta - F_N \tag{1}$$

At angle θ, the decrease in potential energy, $mgh = mgr(1 - \cos \theta)$, must equal the increase in kinetic energy, $mv^2/2$; hence,

$$\frac{mv^2}{r} = 2mg(1 - \cos \theta) \tag{2}$$

Together, (1) and (2) give

$$3mg \cos \theta - F_N = 2mg \tag{3}$$

At the instant the bug loses contact with the ball, $F_N = 0$ and (3) yields $\cos \theta = \frac{2}{3}$.

6.21 A 180-lb pilot is executing a vertical loop of radius 2000 ft at 350 mi/h. With what force does the seat press upward against him at the bottom of the loop?

I
$$F - mg = \frac{mv^2}{r} \quad \text{or} \quad F = \frac{mv^2}{r} + mg$$

First we change miles per hour to feet per second:

$$350 \text{ mi/h} = 513 \text{ ft/s}$$

Substitute values:

$$F = \frac{180 \text{ lb} \times (513 \text{ ft/s})^2}{32.2 \text{ ft/s}^2 \times 2000 \text{ ft}} + 180 \text{ lb} = \underline{915 \text{ lb}}$$

6.22 How many g's must the pilot of the preceding problem withstand at the bottom of the loop?

I From the centripetal acceleration formula $a_c = v^2/r$. Substituting,

$$a_c = \frac{(513 \text{ ft/s})^2}{2000 \text{ ft}} = 132 \text{ ft/s}^2$$

Dividing this result by g (32.2 ft/s²), we obtain

$$a_c = \frac{132 \text{ ft/s}^2}{32.2 \text{ ft/s}^2} = \underline{4.1\,g}$$
$$\phantom{a_c = \frac{132 \text{ ft/s}^2}{}} 1\,g$$

6.23 The designer of a roller coaster wishes the riders to experience "weightlessness" as they round the top of one hill. How fast must the car be going if the radius of curvature at the hilltop is 20 m?

I To experience weightlessness, the gravitational force mg must exactly equal the required centripetal force mv^2/r. Equating the two and solving for v gives $\underline{14 \text{ m/s}}$.

6.24 A huge pendulum consists of a 200-kg ball at the end of a cable 15 m long. If the pendulum is drawn back to an angle of 37° and released, what maximum force must the cable withstand as the pendulum swings back and forth?

I The maximum tension will occur at the bottom when the cable must furnish a force $mg + mv^2/r$. To reach the bottom, the mass falls a distance $h = (15 - 15 \cos 37°) = 3.0$ m. Its speed there will be $v = (2gh)^{1/2} = (6g)^{1/2}$. Therefore the tension will be $T = 200g + 200(6g)/15 = \underline{2740 \text{ N}}$.

6.2 LAW OF UNIVERSAL GRAVITATION; SATELLITE MOTION

6.25 Two 16-lb shot spheres (as used in track meets) are held 2 ft apart. What is the force of attraction between them?

I In American engineering units, $16 \text{ lb} \Rightarrow 0.497$ slug and Newton's law of gravitation has the form $F = G[(m_1 m_2)/d^2]$, with $G = 3.44 \times 10^{-8}$ lb · ft²/slug². Thus,

$$F = \left(3.44 \times 10^{-8} \frac{\text{lb} \cdot \text{ft}^2}{\text{slug}^2}\right)\left(\frac{0.497 \text{ slug} \times 0.497 \text{ slug}}{(2 \text{ ft})^2}\right) = \underline{2.12 \times 10^{-9} \text{ lb}}$$

6.26 Calculate the force of attraction between two 90-kg spheres of metal spaced so that their centers are 40 cm apart.

▮ In SI units the gravitational constant has the value $G = 6.67 \times 10^{-11}$ N · m²/kg².

$$F = G\frac{m_1 m_2}{r^2} = (6.67 \times 10^{-11})\frac{90(90)}{(0.40)^2} = \underline{3.38 \times 10^{-6}\,\text{N}}$$

6.27 Compute the mass of the earth, assuming it to be a sphere of radius 6370 km.

▮ Let M be the mass of the earth, and m the mass of a certain object on the earth's surface. The weight of the object is equal to mg. It is also equal to the gravitational force $G(Mm)/r^2$, where r is the earth's radius. Hence, $mg = G[Mm/r^2]$, from which

$$M = \frac{gr^2}{G} = \frac{(9.8\,\text{m/s}^2)(6.37 \times 10^6\,\text{m})^2}{6.67 \times 10^{-11}\,\text{N} \cdot \text{m}^2/\text{kg}^2} = \underline{6.0 \times 10^{24}\,\text{kg}}$$

6.28 The average density of solids near the surface of the earth is $\rho = 4 \times 10^3$ kg/m³. On the (crude) assumption of a spherical planet of uniform density ρ, calculate the gravitational constant G.

▮ The mass of the spherical earth is given by $m_e = \rho V = \frac{4}{3}\pi\rho R_e^3$. Insert this value into $g = Gm_e/R_e^2$ and solve for G, obtaining:

$$G = \frac{3g}{4\pi\rho R_e} = \frac{3 \times 9.8\,\text{m/s}^2}{4\pi \times 4 \times 10^3\,\text{kg/m}^3 \times 6.4 \times 10^6\,\text{m}} = \underline{9 \times 10^{-11}\,\text{N} \cdot \text{m}^2/\text{kg}^2}.$$

This calculation most certainly overestimates G, since ρ (and hence m_e) are underestimates.

6.29 A mass $m_1 = 1$ kg weighs one-sixth as much on the surface of the moon as on the earth. Calculate the mass m_2 of the moon. The radius of the moon is 1.738×10^6 m.

▮ On the moon, m_1 weighs $\frac{1}{6}(9.8\,\text{N})$.

$$w_1 = G\frac{m_1 m_2}{r^2} \qquad \frac{1}{6}(9.8) = 6.67 \times 10^{-11}\frac{1 \times m_2}{(1.738 \times 10^6)^2}$$

$$m_2 = \frac{(9.8)(1.738 \times 10^6)^2}{(6)(6.67 \times 10^{-11})} = \underline{7.4 \times 10^{22}\,\text{kg}}$$

6.30 The earth's radius is about 6370 km. An object that has a mass of 20 kg is taken to a height of 160 km above the earth's surface. (**a**) What is the object's mass at this height? (**b**) How much does the object weigh (i.e., how large a gravitational force does it experience) at this height?

▮ (**a**) The mass is the same as that on the earth's surface. (**b**) As long as we are outside the earth's surface, the weight (force of gravity) varies inversely as the square of the distance from the center of the earth. Indeed $w = GmM/r^2$, where m, M are the masses of object and earth, respectively, and r is the distance to the center of the earth. Thus $w_2/w_1 = r_1^2/r_2^2$, since G, m, M are constant in this problem. For our case we set $r_1 = 6370$ km and $r_2 = 6530$ km and $w_1 = (20\,\text{kg})(9.8\,\text{m/s}^2) = 196$ N. This gives $w_2 = \underline{186.5\,\text{N}}$. Note that we could use the fact that $w = mg$ and that m is constant to find the acceleration of gravity at the two heights. That is, $g_2/g_1 = r_1^2/r_2^2$.

6.31 The radius of the earth is about 6370 km, while that of Mars is about 3440 km. If an object weighs 200 N on earth, what would it weigh, and what would be the acceleration due to gravity on Mars? Mars has a mass 0.11 that of earth.

▮ Newton's law of gravitation, $w = GmM/r^2$, gives $w_2/w_1 = (M_2/M_1)(r_1^2/r_2^2)$. Letting 1 refer to earth and 2 refer to Mars, we have $w_2 = 0.11(6370/3440)^2(200\,\text{N}) = \underline{75\,\text{N}}$. The acceleration is gotten from $w_2/w_1 = g_2/g_1$, or $g_2 = (75/200)(9.8\,\text{N}) = \underline{3.7\,\text{m/s}^2}$.

6.32 The moon orbits the earth in an approximately circular path of radius 3.8×10^8 m. It takes about 27 days to complete one orbit. What is the mass of the earth as obtained from these data?

▮ The gravitational attraction between the earth and moon provides the centripetal force; therefore, $mv^2/r = GMm/r^2$, where M is the earth's mass. Then $M = v^2 r/G = \omega^2 r^3/G$. Now $\omega = 1$ rev/27 days $= 2.7 \times 10^{-6}$ rad/s, $r = 3.8 \times 10^8$ m, and $G = 6.7 \times 10^{-11}$ in SI. Solving for M, it is $\underline{6.0 \times 10^{24}\,\text{kg}}$. (Compare Prob. 6.27.)

6.33 The sun's mass is about 3.2×10^5 times the earth's mass. The sun is about 400 times as far from the earth as the moon is. What is the ratio of the magnitude of the pull of the sun on the moon to that of the pull of the earth on the moon? It may be assumed that the sun–moon distance is constant and equal to the sun–earth distance.)

▌ Let m denote the moon's mass, M_s the sun's mass, M_e the earth's mass, r_{ms} the center-to-center distance from the sun to the moon, and r_{me} the center-to-center distance from the earth to the moon. We let F_{ms} denote the magnitude of the gravitational force exerted on the moon by the sun, and F_{me} denote the magnitude of the gravitational force exerted on the moon by the earth. Then $F_{ms} = GM_s m / r_{ms}^2$ and $F_{me} = GM_e m / r_{me}^2$, so that

$$\frac{F_{ms}}{F_{me}} = \frac{M_s r_{me}^2}{M_e r_{ms}^2}$$

Using the given numerical values, we find $F_{ms}/F_{me} = \underline{2}$.

6.34 Estimate the size of a rocky sphere with a density of 3.0 g/cm^3 from the surface of which you could just barely throw away a golf ball and have it never return. (Assume your best throw is 40 m/s.)

▌ The escape speed v_0 from a sphere of radius R and mass M is given by the energy-conservation equation

$$\frac{1}{2} m v_0^2 = \frac{GmM}{R}$$

Substitution of $M = \rho \frac{4}{3} \pi R^3$ and solution for R gives

$$R = v_0 \sqrt{\frac{3}{8 \pi G \rho}}$$

If $\rho = 3 \times 10^3 \text{ kg/m}^3$, then $R = 0.77 \times 10^3 \, v_0$, where R is in meters and v_0 is in meters per second. Estimating the highest speed at which a human can throw a golf ball as about 40 m/s, we find $R \approx 3 \times 10^4 \text{ m} = \underline{30 \text{ km}}$.

6.35 Newton, without knowledge of the numerical value of the gravitational constant G, was nevertheless able to calculate the ratio of the mass of the sun to the mass of any planet, provided the planet has a moon.
(a) Show that for a circular orbits

$$\frac{M_s}{M_p} = \left(\frac{R_p}{R_m}\right)^3 \left(\frac{T_m}{T_p}\right)^2$$

where M_s is the mass of the sun, M_p the mass of the planet, R_p the distance of the planet from the sun, R_m the distance of the moon from the planet, T_m the period of the moon around the planet, and T_p the period of the planet around the sun. (b) If the planet is the earth, $R_p = 1.50 \times 10^8$ km, $R_m = 3.85 \times 10^5$ km, $T_m = 27.3$ days, and $T_p = 365.2$ days. Calculate M_s/M_p.

▌ (a) Applying Newton's second law and the law of gravitation to each orbit, we find (expressing centripetal force in terms of period, T, using $v = 2\pi R/T$),

$$\frac{4\pi^2 M_p R_p}{T_p^2} = \frac{GM_s M_p}{R_p^2} \quad \text{and} \quad \frac{4\pi^2 m R_m}{T_m^2} = \frac{GM_p m}{R_m^2}$$

where m is the mass of the satellite. Solving the above equations for M_s/M_p, we obtain

$$\frac{M_s}{M_p} = \frac{(4\pi^2 R_p^3)/GT_p^2}{(4\pi^2 R_m^3)/GT_m^2} = \left(\frac{R_p}{R_m}\right)^3 \left(\frac{T_m}{T_p}\right)^2$$

as desired. (b) Inserting the given numerical values, we find

$$\frac{M_s}{M_e} = \left(\frac{1.50 \times 10^8}{3.85 \times 10^5}\right)^3 \left(\frac{27.3}{365.2}\right)^2 = \underline{3.30 \times 10^5}$$

6.36 (a) Find the orbital period of a satellite in a circular orbit of radius r about a spherical planet of mass M. (b) For a low-altitude orbit ($r \approx r_p$), show that for a given average planetary density $\langle \rho \rangle$ the orbital period is independent of the size of the planet.

▮ (a) Applying Newton's second law to the circular orbit, we have

$$\frac{mv^2}{r} \equiv \frac{4\pi^2 rm}{T^2} = \frac{GMm}{r^2}$$

where m is the satellite mass, v is its orbital speed, and T is its orbital period. Solving the above equation for the period, we obtain

$$T = \frac{2\pi r^{3/2}}{\sqrt{GM}}$$

or $T^2 \propto r^3$, which is Kepler's third law. (b) Since $M = 4\pi \langle \rho \rangle r_p^3 / 3$, the above equation for the period yields

$$T = \frac{2\pi r_p^{3/2}}{\sqrt{(4\pi G \langle \rho \rangle r_p^3)/3}} = \sqrt{\frac{3\pi}{G \langle \rho \rangle}}$$

which shows that the period of a low-altitude satellite is determined solely by the average density of the planet.

6.37 The rings of Saturn consist of myriad small particles, with each particle following its own circular orbit in Saturn's equatorial plane. The inner edge of the innermost ring is about 70 000 km from Saturn's center; the outer edge of the outermost ring is about 135 000 km from the center. Find the orbital period of the outermost particles as a multiple of the orbital period of the innermost particles.

▮ We denote the innermost and outermost orbital radii and periods by R_i, R_o, T_i, and T_o, respectively. Applying Kepler's third law [see Prob. 6.36(a)] to the ring system, we obtain

$$\frac{T_o}{T_i} = \left(\frac{R_o}{R_i}\right)^{3/2} = \left(\frac{135}{70}\right)^{3/2} = 2.68$$

That is, $T_o = \underline{2.68 T_i}$.

6.38 Refer to Prob. 6.37. Spectroscopic studies indicate that the outermost particles have a speed of 17 km/s. Find the mass of Saturn. Express your result in kilograms and as a multiple of the earth's mass.

▮ Applying Newton's second law and the law of gravitation to a particle of mass m at the outer edge of the rings, we have

$$\frac{mv_o^2}{R_o} = \frac{GM_s m}{R_o^2}$$

where M_s is Saturn's mass. Therefore

$$M_s = \frac{v_o^2 R_o}{G} = \frac{(17 \times 10^3 \text{ m/s})^2 (1.35 \times 10^8 \text{ m})}{6.67 \times 10^{-11} \text{ N} \cdot \text{m}^2/\text{kg}} = \underline{5.85 \times 10^{26} \text{ kg} = 97.7 M_e}$$

6.39 The acceleration due to gravity on the moon is only one-sixth that on earth. If the earth and moon are assumed to have the same average composition, what would you predict the moon's radius to be in terms of the earth's radius R_e?

▮ Since $g = GM/R^2$,

$$\frac{1}{6} = \frac{g_m}{g_e} = \frac{R_e^2}{R_m^2} \frac{M_m}{M_e} = \frac{R_e^2}{R_m^2} \frac{R_m^3}{R_e^3} = \frac{R_m}{R_e}$$

or $R_m = \frac{1}{6} R_e$. (Actually ρ of the moon is about three-fifths that of earth, so that $R_m = 0.27 R_e$.)

6.40 Two identical coins of mass 8 g are 50 cm apart on a tabletop. How many times larger is the weight of one coin than the gravitational attraction of the other coin for it?

▮ The force between the coins is Gm^2/d^2; when dividing the weight mg by this value, the ratio $= gd^2/Gm = 9.8(0.50)^2/(6.67 \times 10^{-11})(0.008) = \underline{4.6 \times 10^{12}}$.

6.41 There is a point along the line joining the center of the earth to the center of the moon at which the two gravitational forces cancel. Find this point's distance, x, from the earth's center. Use D for the earth–moon distance, and m_e and m_m as the masses of earth and moon, respectively.

■ $\qquad \dfrac{Gm_e}{x^2} = \dfrac{Gm_m}{(D-x)^2} \qquad x^2(m_e - m_m) - 2Dm_e x + m_e D^2 = 0 \qquad x = \dfrac{D[m_e - (m_e m_m)^{1/2}]}{m_e - m_m}$

6.42 Communication satellites are placed in orbit above the equator in such a way that they remain stationary above a given point on earth below. How high above the surface of the earth is such a *synchronous orbit*? ($R_e = 6400$ km, $M_e = 5.98 \times 10^{24}$ kg.)

■ The satellite must have the same angular velocity, $\omega = 1$ rev/day $= 7.27 \times 10^{-5}$ rad/s, about the earth's center as has the earth itself. As the gravitational force is the centripetal force that keeps the satellite in orbit, $GMm_s/(R_e + h)^2 = m_s \omega^2 (R_e + h)$. First solve for $(R_e + h)$; then find $h = \underline{35\,800 \text{ km}}$, or about $5.6R_e$.

6.43 Three identical point masses M lie in the xy plane at points $(0,0)$, $(0, 0.20 \text{ m})$, and $(0.20 \text{ m}, 0)$. Find the components of the gravitational force on the mass at the origin.

■ $\qquad F_1 = F_2 = \dfrac{GM^2}{0.04} = 1.67 \times 10^{-9} M^2 \qquad$ so the $\qquad \mathbf{F} = \underline{(1.67 \times 10^{-9} M^2)(\mathbf{i} + \mathbf{j}) \text{ N}}$

6.44 Figure 6-9 shows a uniform sphere of original total mass M in which a spherical hole of diameter R has been formed. Show by a superposition argument that it attracts the mass m with a force

$$F = \frac{GMm}{D^2}\left[1 - \frac{1}{8}\left(1 - \frac{R}{2D}\right)^{-2}\right]$$

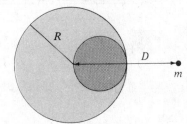

Fig. 6-9

■ The entire original sphere would exert a force

$$F' = \frac{GMm}{D^2}$$

The cut-out sphere, of radius $R/2$, would exert a force

$$F'' = \frac{G(M/8)m}{(D - R/2)^2}$$

By superposition $F + F'' = F'$, or $F = F' - F''$, and this leads to the desired result.

6.45 Consider an attractive force which is central but inversely proportional to the first power of the distance ($F \propto 1/r$). Prove that if a particle is in a circular orbit under such a force, its speed is independent of the orbital radius, but its period is proportional to the radius.

■ Denoting the proportionality constant by C, we have $F = -C/r$. Applying Newton's second law to a particle of mass m moving in a circular orbit with speed v, we obtain $mv^2/r = C/r$. Therefore $v = \sqrt{C/m}$, independent of the radius r. Then $T = 2\pi r/v = 2\pi r \sqrt{m/C}$, so the orbital period is proportional to r.

6.46^c A straight rod of length L extends from $x = a$ to $x = L + a$. Find the gravitational force it exerts on a point mass m at $x = 0$ if the mass per unit length of the rod is $\mu = A + Bx^2$.

■ From Fig. 6-10,

$$dF = \frac{Gm(\mu\, dx)}{x^2} \qquad \text{from which} \qquad F = Gm \int_a^{a+L} (A + Bx^2)\frac{dx}{x^2} = Gm\left[A\left(\frac{1}{a} - \frac{1}{a+L}\right) + BL\right]$$

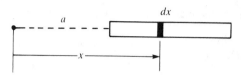

Fig. 6-10

6.47ᶜ   Repeat Prob. 6.46 if $\mu = Ax + Bx^2$.

▮
$$F = Gm \int_a^{a+L} (Ax + Bx^2)\frac{dx}{x^2} = Gm\left[A \ln\left(1 + \frac{L}{a}\right) + BL\right]$$

6.48   If the earth–moon distance is 3.8×10^5 km, compute the time (in days) it takes the moon to circle the earth. ($M_e = 5.98 \times 10^{24}$ kg.)

▮   Apply the formula of Prob. 6.36(a):
$$T = \frac{2\pi(3.8 \times 10^8)^{3/2}}{\sqrt{(6.67 \times 10^{-11})(5.98 \times 10^{24})}} \cdot \frac{1 \text{ day}}{24 \times 3600 \text{ s}} = \underline{27 \text{ days}}$$

6.49ᶜ   A standard 1-kg mass is suspended from each side of a sensitive beam balance, as in Fig. 6-11. The wire supporting the right mass goes through an opening in the floor so that it is 10.00 m below the left mass. **(a)** What is the fractional excess in the weight of the right mass over that of the left mass? (This is done most easily by using differentials.) **(b)** How many milligrams must be placed on the left mass to restore the balance?

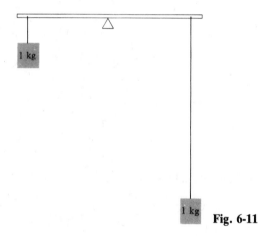

Fig. 6-11

▮   **(a)** The magnitude of the weight force W on a mass m located at distance R from the center of the earth is given by $W = (GM_e m)/R^2$. If two objects of equal mass are located at radial distances R_1 and $R_2 = R_1 + dR$, the difference in weight is
$$dW \equiv W_2 - W_1 = \left(\frac{dW}{dR}\right)_{R_1} dR = \frac{-2GM_e m}{R_1^3}\, dR$$

The fractional difference is
$$\frac{dW}{W_1} = \frac{-2\, dR}{R_1}$$

With $R_1 = 6.37 \times 10^6$ m and $dR = -10$ m, we find $dW/W_1 = (20 \text{ m})/(6.37 \times 10^6 \text{ m}) = \underline{3.14 \times 10^{-6}}$.
(b) To balance the excess weight force on the right-hand side in Fig. 6-11, we must increase the mass on the left by $dm = (dW/W_1)m = (3.14 \times 10^{-6})(1 \text{ kg}) = \underline{3.14 \text{ mg}}$.

6.3 GENERAL MOTION IN A PLANE

6.50ᶜ   Describe the kinematics and dynamics of motion along an arbitrary plane curve.

▮   In a general motion (Fig. 6-12), described in terms of a particle's distance $s = s(t)$ traveled along a curved

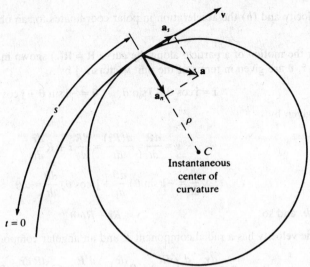

Fig. 6-12

Instantaneous
center of
curvature

$t = 0$

path, the particle has

$$\text{speed} \equiv v = \frac{ds}{dt} \qquad \text{tangential component of acceleration} \equiv a_s = \frac{d^2s}{dt^2}$$

$$\text{normal component of acceleration} \equiv a_n = \frac{v^2}{\rho} = \rho\omega^2$$

where ρ is the radius of curvature of the path and where $\omega = v/\rho$ is defined as the particle's angular speed of rotation about an axis through the instantaneous center of curvature. Newton's second law gives $F_n = ma_n$ and $F_s = ma_s$, where the resultant force acting on the particle has a normal component F_n and a tangential component F_s. Note that a positive F_n produces acceleration *toward* the center of curvature.

6.51^c The angular acceleration of the toppling pole shown in Fig. 6-13 is given by $\alpha = k \sin \theta$, where θ is the angle between the axis of the pole and the vertical, and k is a constant. The pole starts from rest at $\theta = 0$. Find **(a)** the tangential and **(b)** the centripetal acceleration of the upper end of the pole in terms of k, θ and l (the length of the pole).

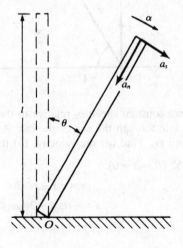

Fig. 6-13

▌ (a)
$$a_s = \frac{dv}{dt} = \frac{d}{dt}(l\omega) = l\alpha = lk \sin \theta$$

(b) From $d\omega/dt = \alpha$,
$$d\omega = k \sin \theta \, dt = k \sin \theta \frac{dt}{d\theta} d\theta = \frac{k}{\omega} \sin \theta \, d\theta$$

Then
$$\int_0^\omega \omega \, d\omega = k \int_0^\theta \sin \theta \, d\theta \qquad \text{or} \qquad \omega^2 = 2k(1 - \cos \theta)$$

and $a_n = l\omega^2 = 2kl(1 - \cos \theta)$.

6.52ᶜ Find (a) the velocity and (b) the acceleration in polar coordinates for an object moving in a curved path in a plane.

■ (a) Consider the motion of a particle along the curve $\mathbf{R} = \mathbf{R}(t)$ shown in Fig. 6-14. At a point of the curve, the unit vectors $\hat{\mathbf{r}}$, $\hat{\boldsymbol{\theta}}$ are given in terms of the unit vectors $\mathbf{i}$, $\mathbf{j}$ by

$$\hat{\mathbf{r}} = \mathbf{i}\cos\theta + \mathbf{j}\sin\theta \qquad \hat{\boldsymbol{\theta}} = -\mathbf{i}\sin\theta + \mathbf{j}\cos\theta$$

The velocity is given by

$$\mathbf{v} = \frac{d\mathbf{R}}{dt} = \frac{d(R\hat{\mathbf{r}})}{dt} = \frac{dR}{dt}\hat{\mathbf{r}} + R\frac{d\hat{\mathbf{r}}}{dt}$$

But

$$\frac{d\hat{\mathbf{r}}}{dt} = -\mathbf{i}(\sin\theta)\frac{d\theta}{dt} + \mathbf{j}(\cos\theta)\frac{d\theta}{dt} = \omega\hat{\boldsymbol{\theta}}$$

where $\omega = d\theta/dt$, and so

$$\mathbf{v} = \dot{R}\hat{\mathbf{r}} + R\omega\hat{\boldsymbol{\theta}}$$

It is seen that the velocity has a radial component $\dot{R}$ and an angular component $R\omega$.

(b)

$$\mathbf{a} = \frac{d\mathbf{v}}{dt} = \frac{d}{dt}\left(\frac{dR}{dt}\hat{\mathbf{r}} + R\frac{d\hat{\mathbf{r}}}{dt}\right) = \frac{d^2R}{dt^2}\hat{\mathbf{r}} + 2\frac{dR}{dt}\frac{d\hat{\mathbf{r}}}{dt} + R\frac{d^2\hat{\mathbf{r}}}{dt^2}$$

Substituting $d\hat{\mathbf{r}}/dt = \omega\hat{\boldsymbol{\theta}}$ and

$$\frac{d^2\hat{\mathbf{r}}}{dt^2} = \frac{d(\omega\hat{\boldsymbol{\theta}})}{dt} = \alpha\hat{\boldsymbol{\theta}} + \omega\frac{d\hat{\boldsymbol{\theta}}}{dt} = \alpha\hat{\boldsymbol{\theta}} + \omega[-\mathbf{i}(\cos\theta)\omega - \mathbf{j}(\sin\theta)\omega] = \alpha\hat{\boldsymbol{\theta}} - \omega^2\hat{\mathbf{r}}$$

we obtain

$$\mathbf{a} = (\ddot{R} - R\omega^2)\hat{\mathbf{r}} + (R\alpha + 2\dot{R}\omega)\hat{\boldsymbol{\theta}}$$

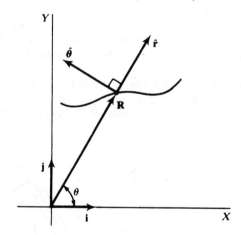

Fig. 6-14

6.53ᶜ A bead slides on a long bar with constant speed v_0 relative to the bar. From Fig. 6-15, $v_0 = \dot{r}$, where r is the distance of the bead from the axle through the end of the bar. At the same time, the bar rotates about the axle with constant angular speed ω_0. Find (a) the velocity, (b) the acceleration, and (c) the path of the bead.

■ Use the results of Prob. 6.52 ($\ddot{R} = \alpha = 0$).

(a)

$$\mathbf{v} = v_0\hat{\mathbf{r}} + r\omega_0\hat{\boldsymbol{\theta}}$$

(b)

$$\mathbf{a} = -r\omega_0^2\hat{\mathbf{r}} + 2v_0\omega_0\hat{\boldsymbol{\theta}}$$

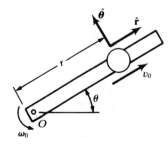

Fig. 6-15

(c) Integrating $\dot{r} = v_0$, $\dot{\theta} = \omega_0$, from time 0 to t, we get

$$r = r_0 + v_0 t \qquad \theta = \theta_0 + \omega_0 t,$$

where r_0, θ_0 are the initial values. Elimination of t gives the equation of the path:

$$r - r_0 = \frac{v_0}{\omega_0}(\theta - \theta_0)$$

which is a spiral.

6.54ᶜ A Coast Guard cutter in a fog at sea is notified by radio that an illegal trawler is at a particular position P, 12.5 km due west of the cutter. The trawler also hears the message and heads off immediately at 12.5 km/h. The captain of the cutter anticipates this speed but does not know the direction the trawler takes. He waits for 1 h and then begins to spiral around P at 48.5 km/h, with a component of velocity directed away from P equal to 12.5 km/h. What is the maximum time that it takes after the message is received to catch the trawler?

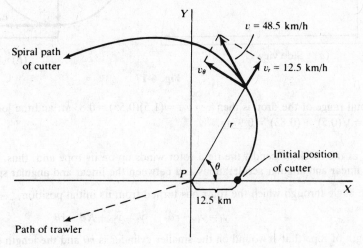

Fig. 6-16

▌ From Fig. 6.16, $\quad v_r = \dfrac{dr}{dt} = 12.5 \text{ km/h} \qquad v_\theta = r\dfrac{d\theta}{dt} = \sqrt{(48.5)^2 - (12.5)^2} = 46.85 \text{ km/h}$

From the first equation, $r = 12.5t$, where we have used the initial condition: $r = 12.5$ km at $t = 1$ h. Substituting for r in the second equation and integrating,

$$12.5t\frac{d\theta}{dt} = 46.85 \qquad \int_0^\theta d\theta = 3.75\int_1^t \frac{dt}{t} \qquad \theta = 3.75\ln t$$

The spiral path of the cutter must cross the radial path of the trawler at some moment, $t = \tau$, during the first revolution. At that moment, both ships will be at the same distance from P, so that the cutter will have indeed caught the trawler. Since $\theta \le 2\pi$ for $t = \tau$, $3.75\ln \tau \le 2\pi$, or $\tau \le e^{2\pi/3.75} = \underline{5.34 \text{ h}}$.

6.55 A wet open umbrella is held upright as shown in Fig. 6.17(a) and is twirled about the handle at a uniform rate of 21 rev in 44 s. If the rim of the umbrella is a circle 1 m in diameter, and the height of the rim above the floor is 1.5 m, find where the drops of water spun off the rim hit the floor.

▌ The angular speed of the umbrella is

$$\omega = \frac{21 \times 2\pi \text{ rad}}{44 \text{ s}} = 3 \text{ rad/s}$$

Then the tangential speed of the water drops on leaving the rim of the umbrella is $v_0 = r\omega = (0.5)(3) = 1.5$ m/s.

To calculate the time for a drop to reach the floor use $h = \frac{1}{2}gt^2$:

$$t = \sqrt{\frac{2h}{g}} = \sqrt{\frac{2(1.5)}{9.8}} = 0.553 \text{ s}$$

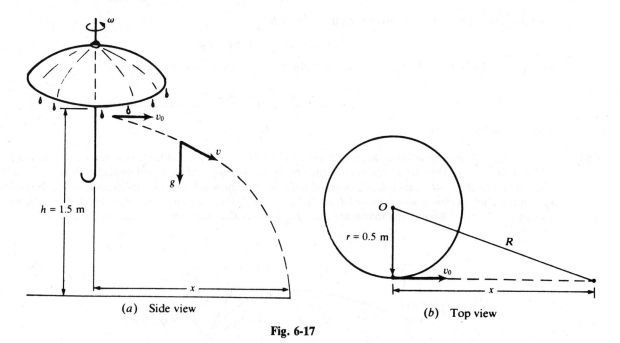

(a) Side view   (b) Top view

Fig. 6-17

The horizontal range of the drop is then $x = v_0 t = (1.5)(0.55) = 0.83$ m; and the locus of the drops is a circle of radius $R = \sqrt{(0.5)^2 + (0.83)^2} = \underline{0.97\ \text{m}}$.

6.56^c In Fig. 6.18, as the block descends, the rigid rotor winds up on its rope and, thus, ascends. Find the relations between the linear and angular accelerations and between the linear and angular speeds.

❚ If θ is the angle through which the rotor has turned from its initial position,

$$y_1 = y_{10} - r\theta \qquad y_2 = y_{20} + R\theta - r\theta$$

since the length of rope that is wound on the smaller cylinder is $r\theta$ and the length of rope that is unwound from the larger cylinder is $R\theta$. Take first and second time-derivatives of y_1 and y_2:

$$v_1 = \dot{y}_1 = -r\dot{\theta} = -r\omega \qquad v_2 = \dot{y}_2 = (R-r)\dot{\theta} = (R-r)\omega$$
$$a_1 = \ddot{y}_1 = -r\ddot{\theta} = -r\alpha \qquad a_2 = \ddot{y}_2 = (R-r)\ddot{\theta} = (R-r)\alpha$$

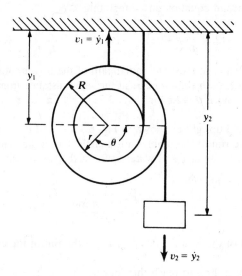

Fig. 6-18

6.57^c Figure 6-19 shows a ray of light that passes from air into water. The ray is bent upon passing into the water, according to Snell's law ($\sin \theta = n \sin \psi$). The angle θ increases at a constant rate of 10 rad/s, and $n = 1.3$. Find the angular speed ω and the angular acceleration α of the refracted ray for $\theta = 30°$.

▌ Take first and second time-derivatives of $\sin \theta = n \sin \psi$ to get $\omega = \dot{\psi}$ and $\alpha = \ddot{\psi}$; recall that $\ddot{\theta} = 0$.

$$\dot{\theta} \cos \theta = n \dot{\psi} \cos \psi \qquad \text{or} \qquad \dot{\psi} = \frac{\dot{\theta} \cos \theta}{n \cos \psi} = \frac{\dot{\theta} \cos \theta}{\sqrt{n^2 - \sin^2 \theta}}$$

$$-\dot{\theta}^2 \sin \theta = n \ddot{\psi} \cos \psi - n \dot{\psi}^2 \sin \psi \qquad \text{or} \qquad \ddot{\psi} = \frac{n \dot{\psi}^2 \sin \psi - \dot{\theta}^2 \sin \theta}{n \cos \psi} = \frac{(\dot{\psi}^2 - \dot{\theta}^2) \sin \theta}{\sqrt{n^2 - \sin^2 \theta}}$$

Substituting the data,

$$\dot{\psi} = \frac{10(\sqrt{3}/2)}{\sqrt{(1.3)^2 - (1/2)^2}} = 7.22 \text{ rad/s} \qquad \ddot{\psi} = \frac{[(7.22)^2 - (10)^2](1/2)}{\sqrt{(1.3)^2 - (1/2)^2}} = -20.0 \text{ rad/s}^2$$

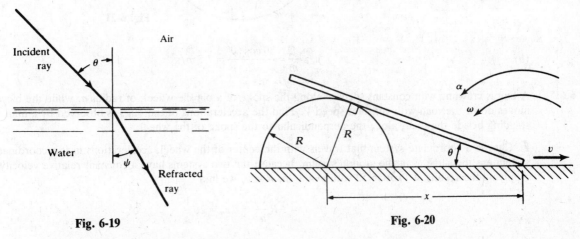

Fig. 6-19   Fig. 6-20

6.58ᶜ A rod leans against a stationary cylindrical body as shown in Fig. 6-20, and its right end slides to the right on the floor with a constant speed v. Find (**a**) the angular speed ω and (**b**) the angular acceleration α, in terms of v, x, and R.

▌ (**a**) From the geometry, $x = R/\sin \theta$. Also, $\omega = -\dot{\theta}$. Therefore,

$$v = \dot{x} = \frac{d}{dt}\left(\frac{R}{\sin \theta}\right) = \frac{-R\dot{\theta}\cos \theta}{\sin^2 \theta} = \frac{\omega R \cos \theta}{\sin^2 \theta} \qquad \omega = \frac{v \sin^2 \theta}{R \cos \theta} = \frac{Rv}{x\sqrt{x^2 - R^2}}$$

(**b**)

$$\alpha = \dot{\omega} = \frac{d}{dt}\left(\frac{Rv}{x\sqrt{x^2 - R^2}}\right) = \frac{-Rv^2(2x^2 - R^2)}{x^2(x^2 - R^2)^{3/2}}$$

6.59ᶜ A particle of mass m moves without friction along a semicubical parabolic curve, $y^2 = ax^3$, with constant speed v. Find the reaction force of the curve on the particle.

▌ The local radius of curvature of the curve is

$$\rho = \frac{[1 + (dy/dx)^2]^{3/2}}{d^2y/dx^2} = \frac{[1 + (9/4)ax]^{3/2}}{(3/4)a^{1/2}x^{-1/2}}$$

In this motion of the particle, the curve exerts a normal or centripetal force, causing the particle momentarily to move in an arc of a circle of radius ρ (see Fig. 6-12). Thus,

$$F = \frac{mv^2}{\rho} = \frac{3}{4}a^{1/2}x^{-1/2}\left(1 + \frac{9}{4}ax\right)^{-3/2}mv^2$$

6.60 A particle whose mass is 2 kg moves with a speed of 44 m/s on a curved path. The resultant force acting on the particle at a particular point of the curve is 30 N at 60° to the tangent to the curve, as shown in Fig. 6-21. At that point, find (**a**) the radius of curvature of the curve and (**b**) the tangential acceleration of the particle.

▌ (**a**)

$$\rho = \frac{mv^2}{F_n} = \frac{2(44)^2}{30 \sin 60°} = \underline{149 \text{ m}}$$

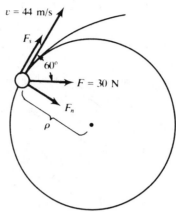

Fig. 6-21

(b)
$$a_s = \frac{F_s}{m} = \frac{30 \cos 60°}{2} = 7.5 \text{ m/s}^2$$

6.61^c   A bug is crawling with constant speed v along the spoke of a bicycle wheel, of radius a, while the bicycle moves down the road with constant speed V. Find the accelerations of the bug, as observed by a man standing beside the road, along the perpendicular to the spoke of the wheel.

▮ Choose a coordinate system that travels with the center of the wheel; accelerations in this coordinate system are the same as in the ground system, because the two systems have a constant relative velocity. Applying the results of Prob. 6.52 with $\omega = V/a$, $\dot{R} = v$, we find

$$a_r = -R\frac{V^2}{a^2} \qquad a_\theta = 2v\frac{V}{a}$$

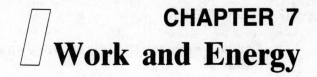

7.1 WORK DONE BY A FORCE

7.1 A force of 3 N acts through a distance of 12 m in the direction of the force. Find the work done.

▐ Force and displacement are in the same direction, so $W = Fs = (3\,\text{N})(12\,\text{m}) = \underline{36\,\text{J}}$.

7.2 A horizontal force of 25 N pulls a box along a table. How much work does it do in pulling the box 80 cm?

▐ Work is force times displacement through which the force acts. Here, force is in the same direction as the displacement, so $W = (25\,\text{N})(0.80\,\text{m}) = \underline{20\,\text{J}}$.

7.3 A child pushes a toy box 4.0 m along the floor by means of a force of 6 N directed downward at an angle of 37° to the horizontal. (*a*) How much work does the child do? (*b*) Would you expect more or less work to be done for the same displacement if the child pulled upward at the same angle to the horizontal?

▐ (*a*) Work $= Fs \cos\theta = 6(4)(0.80) = \underline{19.2\,\text{J}}$. (*b*) Less work; since the normal force on the block is less, the friction force will be less and the needed F will be smaller.

7.4 Figure 7-1 shows the top view of two horizontal forces pulling a box along the floor: (*a*) How much work does each force do as the box is displaced 70 cm along the broken line? (*b*) What is the total work done by the two forces in pulling the box this distance?

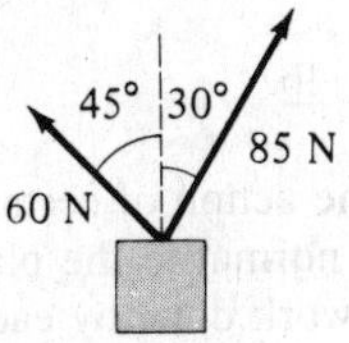

45° 30°
85 N
60 N

Fig. 7-1

▐ (*a*) In each case take the component of the force in the direction of the displacement:
$(85 \cos 30°\,\text{N})(0.70\,\text{m}) = 51.5\,\text{J}$, $(60 \cos 45°\,\text{N})(0.70\,\text{m}) = \underline{29.7\,\text{J}}$. (*b*) Work is a scalar, so add the work done by each force to give $\underline{81.2\,\text{J}}$.

7.5 A horizontal force F pulls a 20-kg carton across the floor at constant speed. If the coefficient of sliding friction between carton and floor is 0.60, how much work does F do in moving the carton 3.0 m?

▐ Because horizontal speed is constant, the carton is in horizontal equilibrium: $F = f = \mu F_N$. Normal force is the weight, $20(9.8) = 196\,\text{N}$. Therefore $W = Fx = 0.60(196)(3.0) = \underline{353\,\text{J}}$.

7.6 A box is dragged across a floor by a rope which makes an angle of 60° with the horizontal. The tension in the rope is 100 N while the box is dragged 15 m. How much work is done?

▐ Only the horizontal component of the tension, $T_x = 100 \cos 60°$, does work. Thus, $W = T_x x = (100 \cos 60°)(15) = \underline{750\,\text{J}}$.

7.7 An object is pulled along the ground by a 75-N force directed 28° above the horizontal. How much work does the force do in pulling the object 8 m?

▐ The work done is equal to the product of the displacement, 8 m, and the component of the force that is parallel to the displacement, $(75\,\text{N}) \cos 28°$.

$$\text{work} = [(75\,\text{N}) \cos 28°](8\,\text{m}) = \underline{530\,\text{J}}.$$

7.8 The coefficient of kinetic friction between a 20-kg box and the floor is 0.40. How much work does a pulling force do on the box in pulling it 8.0 m across the floor at constant speed? The pulling force is directed 37° above the horizontal.

▐ The work done by the force is $xF \cos 37°$, where $F \cos 37° = f = \mu F_N$. In this case $F_N = mg - F \sin 37°$, so that $F = \mu mg/(\cos 37° + \mu \sin 37°)$. For $\mu = 0.40$ and $m = 20$ kg, $F = 75.4$ N and $W = (75.4 \cos 37°)(8.0) = \underline{482\,J}$.

7.9 Repeat Prob. 7.8 if the force pushes rather than pulls on the box and is directed 37° below horizontal.

▮ $W = (F \cos 37°)(x) = F_x x$; thus $F \cos 37° = \mu F_N$, as in Prob. 7.8; but now $F_N = mg + F \sin 37°$; solve for F: $F = \mu mg/(\cos 37° - \mu \sin 37°) = 140$ N and $F_x = 112$ N. Thus $W = 112(8.0) = \underline{896\,J}$. [This larger value for the work accords with Prob. 7.3(b).]

7.10 How much work is done against gravity in lifting a 3-kg object through a distance of 40 cm?

▮ To lift a 3-kg object at constant speed, an upward force equal in magnitude to its weight, $mg = (3)(9.8)$ N, must be exerted on the object. The work done by this force is what we refer to as the work done against gravity: work against gravity $= mgh = [(3)(9.8)\,N](0.40\,m) = \underline{11.8\,J}$.

7.11 How much work is done against gravity in lifting a 20-lb object through a distance of 4.0 ft?

▮ As in Prob. 7.10, work against gravity $=$ (weight)(h) $= (20\,lb)(4\,ft) = \underline{80\,ft \cdot lb}$.

7.12 A 4-kg object is slowly lifted 1.5 m. (**a**) How much work is done against gravity? (**b**) Repeat if the object is lowered instead of lifted.

▮ (**a**) The lifting force is in the direction of the displacement and just balances the weight. $F = mg = 39.2$ N. $W = Fh = (39.2\,N)(1.5\,m) = \underline{58.8\,J}$. (**b**) If the object is lowered, F is opposite to the displacement. $W = -Fh = -(39.2\,N)(1.5\,m) = \underline{-58.8\,J}$.

7.13 A 400-lb load of bricks is to be lifted to the top of a scaffold 28 ft high. How much work must be done against gravity to lift it?

▮ $W = mgh = (400\,lb)(28\,ft) = \underline{11\,200\,ft \cdot lb}$.

7.14 A block moves up a 30° incline under the action of certain forces, three of which are shown in Fig. 7.2. F_1 is horizontal and of magnitude 40 N. F_2 is normal to the plane and of magnitude 20 N. F_3 is parallel to the plane and of magnitude 30 N. Determine the work done by each force as the block (and point of application of each force) moves 80 cm up the incline.

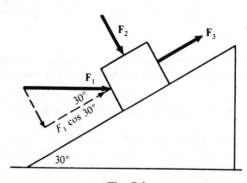

Fig. 7-2

▮ The component of F_1 along the direction of the displacement is $F_1 \cos 30° = (40\,N)(0.866) = 34.6$ N. Hence the work done by F_1 is $(34.6\,N)(0.80\,m) = \underline{28\,J}$. (Note that the distance must be expressed in meters.)
  Because it has no component in the direction of the displacement, F_2 does no work.
  The component of F_3 in the direction of the displacement is 30 N. Hence the work done by F_3 is $(30\,N)(0.80\,m) = \underline{24\,J}$.

7.15 Compute the useful work done by an engine as it lifts 40 L of tar 20 m. One cubic centimeter of tar has a mass of 1.07 g.

▮ Since each particle of tar is lifted the same distance, $h = 20$ m, we have $W = Mgh$, where M is the total mass of tar. Since $M = (40\,L)(10^3\,cm^3/L)(1.07 \times 10^{-3}\,kg/cm^3) = 42.8$ kg, $W = (42.8\,kg)(9.8\,m/s^2)(20\,m) = 8389\,J = \underline{8.389\,kJ}$.

7.16 A uniform rectangular marble slab is 3.4 m long and 2.0 m wide. It has a mass of 180 kg. If it is originally lying on the flat ground, how much work is needed to stand it on end?

❚ The work done by gravity is the work done *as if* all the mass were concentrated at the center of mass. The work necessary to lift the object can be thought of as the work done against gravity and is just $W = (mg)h$, where h is the height through which the center of mass is raised. $W = (180 \text{ kg})(9.8 \text{ m/s}^2)(1.7 \text{ m}) = \underline{3.0 \text{ kJ}}$.

7.17 In Fig. 7-3, evaluate the work done by the weight $m\mathbf{g}$ acting on a particle of mass m, as the particle is moved (by the application of other forces) from: **(a)** A to B; **(b)** B to A; **(c)** A to B to C; **(d)** A to C directly; **(e)** A to B to C to A.

❚ **(a)** For the path AB, $m\mathbf{g}$ is in the opposite direction to the direction of motion. Thus, $W_{AB} = -mgy$.

(b)
$$W_{BA} = -W_{AB} = mgy$$

(c)
$$W_{ABC} = W_{AB} + W_{BC} = -mgy + 0 = -mgy$$

(d) The component of force in the direction of motion is $-mg \cos \phi$ and $\overline{AC} = \Delta s = y/(\cos \phi)$.

$$W_{AC} = (-mg \cos \phi)\left(\frac{y}{\cos \phi}\right) = -mgy$$

(e)
$$W_{ABCA} = W_{AB} + W_{BC} + W_{CA} = -mgy + 0 + mgy = 0$$

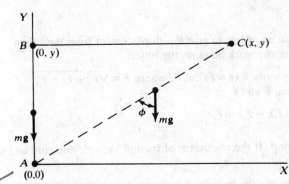

Fig. 7-3

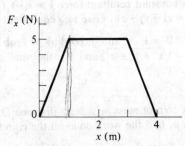

Fig. 7-4

7.18 The x-directed force that acts on an object is shown as a function of x in Fig. 7-4. Find the work done by the force in the interval **(a)** $0 \le x \le 1$ m, **(b)** $1 \le x \le 3$ m, **(c)** $0 \le x \le 4$ m.

❚ Work is the area under the F_x-vs-x curve. **(a)** The area is that of a right triangle of base 1 and height 5, so $W = \underline{2.5 \text{ J}}$. **(b)** The area of the rectangle is $2(5) = \underline{10 \text{ J}}$. **(c)** The total area under the curve in Fig. 7-4 is $W = \underline{15 \text{ J}}$.

7.19 The x-directed force that acts on an object is shown as a function of x in Fig. 7-5. Find the work done by the force in the interval **(a)** $0 \le x \le 3$ cm, **(b)** $3 \le x \le 5$ cm, **(c)** $0 \le x \le 6$ cm.

❚ **(a)** $W = (0.03 \text{ m})(5 \text{ N})/2 = \underline{0.075 \text{ J}}$; **(b)** $W = -0.02(3)/2 = \underline{-0.03 \text{ J}}$. **(c)** The work between 5 and 6 cm is $0.01(3) = 0.030 \text{ J}$; adding this to **(a)** and **(b)** gives a total work in the 6-cm interval of $\underline{0.075 \text{ J} = 75 \text{ mJ}}$.

7.20 If the x-directed force exerted on a cart by a boy varies with position as shown in Fig. 7-6, how much work does the boy do on the cart?

❚ The total work is the area under the curve. By counting entire squares and parts of squares we estimate about 34 squares under the curve, so work is about $(34 \text{ squares})(40 \text{ J/square}) = \underline{1360 \text{ J}}$.

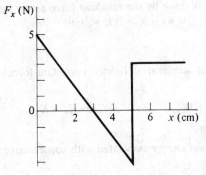

Fig. 7-5

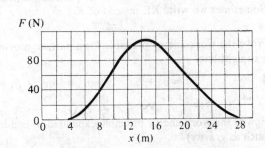

Fig. 7-6

7.21 How much work is done by a force of 40 N acting 37° above the horizontal in pulling a block 8 m along a horizontal surface?

$\blacksquare$ Work $= \mathbf{F} \cdot \mathbf{s} = Fs \cos \theta = (40\,\text{N})(8\,\text{m})(0.8) = \underline{256\,\text{J}}$.

7.22 In order to lift a 5.0-kg child through a vertical distance of 40 cm, how much work must be done?

$\blacksquare$ Work $= \mathbf{F} \cdot \mathbf{s} = (5 \times 9.8)(0.40) = \underline{19.6\,\text{J}}$.

7.23 If $\mathbf{A} = A_x\mathbf{i}$, $\mathbf{B} = B_x\mathbf{i} + B_y\mathbf{j} + B_z\mathbf{k}$, and $\mathbf{C} = C_x\mathbf{i} + C_y\mathbf{j} + C_z\mathbf{k}$, find (a) $\mathbf{A} \cdot \mathbf{B}$, and (b) $\mathbf{B} \cdot \mathbf{C}$.

$\blacksquare$ By Prob. 1.85, a dot product of two vectors can be evaluated by componentwise multiplication. Thus:
(a) $\mathbf{A} \cdot \mathbf{B} = A_x B_x + (0)(B_y) + (0)(B_z) = A_x B_x$. (b) $\mathbf{B} \cdot \mathbf{C} = B_x C_x + B_y C_y + B_z C_z$.

7.24 Find $\mathbf{A} \cdot \mathbf{B}$ if $\mathbf{A} = 3\mathbf{i} - 4\mathbf{j}$ and $\mathbf{B} = 6\mathbf{j} + 2\mathbf{k}$.

$\blacksquare$ $\mathbf{A} \cdot \mathbf{B} = (3)(0) + (-4)(6) + (0)(2) = \underline{-24}$. (If $\mathbf{A}$ is a force in newtons and $\mathbf{B}$ is a displacement in meters, $\mathbf{A} \cdot \mathbf{B}$ would represent work done in joules.)

7.25 Compute $\mathbf{C} \cdot \mathbf{D}$ if $\mathbf{C} = 3\mathbf{j} - 2\mathbf{k}$ and $\mathbf{D} = -8\mathbf{i} + 5\mathbf{k}$.

$\blacksquare$ $\mathbf{C} \cdot \mathbf{D} = (0)(-8) + (3)(0) + (-2)(5) = \underline{-10}$

7.26 A constant resultant force $\mathbf{F} = F_x\mathbf{i} + F_y\mathbf{j} + F_z\mathbf{k}$ acts on an object to give it a displacement from the origin $\mathbf{s} = x\mathbf{i} + y\mathbf{j} + z\mathbf{k}$. Give two equivalent expressions for the work done on the object.

$\blacksquare$ $W = \mathbf{F} \cdot \mathbf{s}$; and, according to Prob. 1.83, we may write $\mathbf{F} \cdot \mathbf{s} = Fs \cos \theta$, where $F = \sqrt{F_x^2 + F_y^2 + F_z^2}$, $s = \sqrt{x^2 + y^2 + z^2}$, and θ is the smaller angle between $\mathbf{F}$ and $\mathbf{s}$; or

$$\mathbf{F} \cdot \mathbf{s} = F_x x + F_y y + F_z z$$

7.27 A coin of mass m slides a distance D along a tabletop. If the coefficient of friction between the coin and table is μ, find the work done on the coin by friction.

$\blacksquare$ From the definition, $\Delta W = \mathbf{f} \cdot \Delta \mathbf{x}$. Because f opposes the motion, it is directed opposite to $\Delta \mathbf{x}$. The magnitude of $\mathbf{f}$ is $\mu F_N = \mu m g$, and thus $W = -\mu m g D$.

7.28 A 5.0-kg box is pulled across the floor at a constant speed of 20 cm/s by a horizontal force. If $\mu = 0.30$ between the box and floor, in each second how much work is done by (a) the pulling force? (b) the frictional force? (c) What is the total work per second done on the box?

$\blacksquare$ The speed is constant, thus Newton's first law applies and $\mathbf{F} = -\mathbf{f}$, where $f = \mu m g = 0.30(5.0)(9.8) =$ 14.7 N. (a) $\Delta W = \mathbf{F} \cdot \Delta \mathbf{x}$ yields $W = (14.7\,\text{N})(0.20\,\text{m}) = \underline{2.94\,\text{J}}$. (The displacement is the distance moved in 1 s, 0.20 m.) (b) $\Delta W = \mathbf{f} \cdot \Delta \mathbf{x} = -\mathbf{F} \cdot \Delta \mathbf{x}$, which gives $W = \underline{-2.94\,\text{J}}$. (c) The total work is the sum of parts (a) and (b), in this case zero.

7.2 WORK, KINETIC ENERGY, AND POTENTIAL ENERGY

7.29 What is the work-energy principle?

$\blacksquare$ The work-energy principle for a particle states that the work W done by the *resultant* force acting on the particle is equal to the change in the particle's kinetic energy: $W_{i \to f} = K_f - K_i = \Delta K = \frac{1}{2}mv_f^2 - \frac{1}{2}mv_i^2$. (Sometimes we write KE instead of K.)

7.30 Prove the work-energy principle for a particle moving at constant acceleration (under a constant force) along a straight line.

$\blacksquare$ $\qquad v_B^2 = v_A^2 + 2as \qquad \frac{1}{2}mv_B^2 = \frac{1}{2}mv_A^2 + Fs \qquad K_B = K_A + W_{AB} \qquad$ or $\qquad W_{AB} = K_B - K_A$

7.31 How is the work-energy principle expanded to include the *potential energy* associated with conservative forces (such as gravity)?

■ Let W_p represent the work done by those forces which are treated in terms of potential energy, U (or PE). Let W' represent the work due to all other forces. From the work-energy principle, $W'_{i\to f} + W_{p,i\to f} = K_f - K_i = \Delta K$. From the definition of U: $-W_{p,i\to f} \equiv U_f - U_i = \Delta U$. Then $W'_{i\to f} = (U_f - U_i) + (K_f - K_i) = \Delta U + \Delta K$ is the modified form of the work-energy principle when potential energy is included.

7.32 A car is moving at 100 km/h. If the mass of the car is 950 kg, what is its kinetic energy?

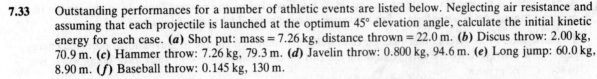

$$\text{KE} = \frac{1}{2}mv^2 = \frac{1}{2}(950 \text{ kg})\left(\frac{10^5 \text{ m}}{3600 \text{ s}}\right)^2 = 3.67 \times 10^5 \text{ J} = \underline{0.367 \text{ MJ}}$$

7.33 Outstanding performances for a number of athletic events are listed below. Neglecting air resistance and assuming that each projectile is launched at the optimum 45° elevation angle, calculate the initial kinetic energy for each case. (**a**) Shot put: mass = 7.26 kg, distance thrown = 22.0 m. (**b**) Discus throw: 2.00 kg, 70.9 m. (**c**) Hammer throw: 7.26 kg, 79.3 m. (**d**) Javelin throw: 0.800 kg, 94.6 m. (**e**) Long jump: 60.0 kg, 8.90 m. (**f**) Baseball throw: 0.145 kg, 130 m.

■ We recall that the maximum range for a projectile launched with speed v_0 from ground level is given by $R_{max} = v_0^2/g$. We use this expression, since no information is given on the launch elevations. The initial kinetic energy $K_0 \equiv \frac{1}{2}mv_0^2 = \frac{1}{2}mgR_{max}$. (**a**) $K_0 = \underline{783 \text{ J}}$, (**b**) $K_0 = \underline{695 \text{ J}}$, (**c**) $K_0 = 2.82 \text{ kJ}$, (**d**) $K_0 = \underline{371 \text{ J}}$, (**e**) $K_0 = \underline{2.62 \text{ kJ}}$, (**f**) $K_0 = \underline{92.4 \text{ J}}$.

7.34 A 150-g mass has a velocity $\mathbf{v} = (2\mathbf{i} + 6\mathbf{j})$ m/s at a certain instant. What is its kinetic energy?

$$K = \tfrac{1}{2}mv^2 = \tfrac{1}{2}m\mathbf{v}\cdot\mathbf{v} = \tfrac{1}{2}(0.150)(2^2 + 6^2) = \underline{3.0 \text{ J}}$$

7.35 The velocity of an 800-g object changes from $\mathbf{v}_0 = 3\mathbf{i} - 4\mathbf{j}$ to $\mathbf{v}_f = (-6\mathbf{j} + 2\mathbf{k})$ m/s. What is its change in kinetic energy?

■ Using $\mathbf{v}^2 = \mathbf{v}\cdot\mathbf{v}$, $v_f^2 = 40$ and $v_0^2 = 25$. Therefore, change in $K = 0.800(40 - 25)/2 = \underline{6.0 \text{ J}}$.

7.36 How large a force is required to accelerate a 1300-kg car from rest to a speed of 20 m/s in a distance of 80 m?

■ $W = \Delta K = \frac{1}{2}mv^2 - 0 = \frac{1}{2}(1300 \text{ kg})(20 \text{ m/s})^2 = 260 \text{ kJ}$. But $W = Fs = F(80 \text{ m})$; $F = \underline{3.25 \text{ kN}}$.

7.37 A crate of mass 50 kg slides down a 30° incline. The crate's acceleration is 2.0 m/s², and the incline is 10 m long. (**a**) What is the kinetic energy of the crate as it reaches the bottom of the incline? (**b**) How much work is spent in overcoming friction? (**c**) What is the magnitude of the frictional force that acts on the crate as it slides down the incline?

■ (**a**) We denote the crate's mass and acceleration by m and a, respectively. After sliding a distance s from rest, the crate has a kinetic energy K given by $K \equiv \frac{1}{2}mv^2 = \frac{1}{2}m(2as) = mas$. With $m = 50$ kg, $a = 2.0$ m/s², and $s = 10$ m, we obtain $K = \underline{1000 \text{ J}}$.
(**b**) The work W_g done on the crate by gravity is given by $W_g = mgh$, where h is the vertical distance descended by the crate. For an incline of angle θ, $h = s \sin\theta$, so we find $W_g = mgs \sin\theta$. With $\theta = 30°$ and the other given values, we find $W_g = 2450 \text{ J}$. The only other force which does work on the crate is friction. If we let W_{fr} denote the work done by friction, we have $W_g + W_{fr} = K$. Therefore $W_{fr} = K - W_g = 1000 - 2450 = -1450 \text{ J}$. The work spent in overcoming friction is $|-1450 \text{ J}| = \underline{1450 \text{ J}}$.
(**c**) The work W_{fr} done by friction is given by $W_{fr} = -F_{fr}s$, so that $F_{fr} = -W_{fr}/s = -(-1450)/(10) = \underline{145 \text{ N}}$.

7.38 Refer to Prob. 7.37. (**a**) What is the coefficient of kinetic friction between the crate and the incline? (**b**) At the base of the incline there is a horizontal surface with the same coefficient of kinetic friction. How far will the crate slide before coming to rest?

■ (**a**) Since the crate remains in contact with the incline, the normal force $N = mg \cos\theta$. Then the frictional force $F_{fr} = \mu_k N = \mu_k mg \cos\theta$. Solving for μ_k, we obtain $\mu_k = F_{fr}/(mg \cos\theta) = 145/[(50)(9.8)(0.866)] = \underline{0.342}$.
(**b**) On a horizontal surface, the frictional force will be $F'_{fr} = \mu_k mg$. The crate will slide a distance s' such that the work W_{fr} done by friction equals the negative of the kinetic energy K. That is, $-\mu_k mgs' = -K$, so that $s' = K/(\mu_k mg) = 1000/[(0.342)(50)(9.8)] = \underline{5.97 \text{ m}}$.

7.39 A 1200-kg car going 30 m/s applies its brakes and skids to rest. If the friction force between the sliding tires and the pavement is 6000 N, how far does the car skid before coming to rest?

■ $W = \Delta K = 0 - \frac{1}{2}mv^2 = -\frac{1}{2}(1200 \text{ kg})(30 \text{ m/s})^2 = -540 \text{ kJ}$. $W = -fx = -(6 \text{ kN})x$; $x = \underline{90 \text{ m}}$.

7.40 How much work is done in moving a body of mass 1.0 kg from an elevation of 2 m to an elevation of 20 m, (*a*) by the gravitational field of the earth? (*b*) by the external agent lifting the body?

▌ (*a*)
$$W = -\Delta U = -[(1.0)(9.8)(20) - (1.0)(9.8)(2)] = \underline{-176.4\,J}$$

The work is negative because the force opposes the motion.
(*b*) By Prob. 7.31, $W' = \Delta K + \Delta U = \Delta K + \underline{176.4\,J}$. Unlike the gravitational work, the external work depends on the change in speed of the body. If the body is unaccelerated ($\Delta K = 0$), then $W' = 176.4\,J$, the negative of the gravitational work.

7.41 A 200-kg cart is pushed slowly up an incline. How much work does the pushing force do in moving the object up along the incline to a platform 1.5 m above the starting point if friction is negligible?

▌ Work done by all forces other than gravity equals the combined change in gravitational potential energy and kinetic energy. Since the force F pushing the cart up the incline is the only such force doing work, we have

$$W_F = \Delta U + \Delta K = (mgh - 0) + (0) = (200\,kg)(9.8\,m/s^2)(1.5\,m) = \underline{2.94\,kJ}$$

7.42 Repeat Prob. 7.41 if the distance along the incline to the platform is 7 m and a friction force of 150 N opposes the motion.

▌ Now we must consider the work done by the frictional force, $f = 150\,N$, as well as that done by F (see Prob. 7.41). Thus we have $W_F + W_f = \Delta U + \Delta K = 2.94\,kJ$. Noting that $W_f = -(150\,N)(7\,m) = -1.05\,kJ$, we get $W_F = \underline{3.99\,kJ}$.

7.43 A ladder that is 3.0 m long and weighs 200 N has its center of gravity 120 cm from the foot. At its top end is a 50 N weight. Compute the work required to raise the ladder from a horizontal position on the ground to a vertical position.

▌ The work done (against gravity) consists of two parts—that needed to raise the center of gravity 1.20 m and that needed to raise the weight at the end through 3.0 m. Therefore work done = (200 N)(1.20 m) + (50 N)(3.0 m) = $\underline{390\,J}$.
  Another method. The center of gravity of the system is at a distance

$$\bar{x} = \frac{(200\,N)(1.20\,m) + (50\,N)(3.0\,m)}{250\,N} = \frac{390}{250}\,m$$

from the foot. To lift 250 N through 390/250 m requires 390 N · m = $\underline{390\,J}$ of work.

7.44 A 100 000-lb freight car is drawn 2500 ft up along a 1.2 percent grade at constant speed. Find the work done against gravity.

▌ Work done against gravity is only the increase in gravitational potential energy, $\Delta U = mgh$, where $h = 0.012(2500\,ft) = 30\,ft$. Thus

$$\Delta U = (100\,000\,lb)(30\,ft) = \underline{3.0 \times 10^6\,ft \cdot lb}$$

7.45 If a constant frictional force of 440 lb retards the motion in Prob. 7.44, how much work is done by the pulling force?

▌ Letting W_F be the work done by the drawbar and W_f the work done by friction, we have $W_F + W_f = \Delta U + \Delta K$ (see Prob. 7.31). Since the speed is constant, $\Delta K = 0$, and $W_f = -(440\,lb)(2500\,ft) = -1.1 \times 10^6\,ft \cdot lb$, then $W_F = \Delta U - W_f = \underline{4.1 \times 10^6\,ft \cdot lb}$.

7.46 A cylindrical tank 10 m high has an internal diameter of 4 m. How much work would be required to fill the tank with water if the water were pumped in (*a*) at the bottom and (*b*) over the top edge?

▌ It takes 1.23 MN of water to fill the tank, since $\rho g \pi r^2 h = (10^3\,kg/m^3)(9.8\,m/s^2)\pi(2\,m)^2(10\,m) = 1.23\,MN$. (*a*) In effect, the center of gravity of a 1.23-MN body of water is raised from height 0 to height 5 m: work required = (1.23 MN)(5 m) = $\underline{6.15\,MJ}$. (*b*) The CG must be raised from height 0 to height 10 m (after which it falls by itself to height 5 m): work required = $2 \times 6.15 = \underline{12.30\,MJ}$.

7.47 A box weighing 200 N is dragged up an incline 10 m long and 3 m high. The average force (parallel to the plane) is 120 N. (*a*) How much work is done? (*b*) What is the change in the potential energy of the box? in its kinetic energy? (*c*) What is the frictional force on the box?

∎ (a) The work done by the dragging force is $W_{i \to f} = \bar{F}s = (120)(10) = \underline{1200\,J}$. (b) The change in potential energy is $\Delta U = U_f - U_i = wh - 0 = (200)(3) = \underline{600\,J}$. (c) Because the box starts and finishes at rest, $\Delta K = 0$. The total work done by nonconservative forces is $W_{i \to f} + W$, where W is the work done on the box by friction. Then

$$\Delta K + \Delta U = W_{i \to f} + W \qquad 0 + 600 = 1200 + W \qquad W = \underline{-600\,J}$$

But $W = -fs$ (the friction force f opposes the motion of the box), and so

$$f = \frac{-600}{-10} = \underline{60\,N}$$

7.48 A rock weighing 20 N falls from a height of 16 m and sinks 0.6 m into the ground. From energy considerations, find the average force f between the rock and the ground as the rock sinks. See Fig. 7-7.

∎ Between A and C, nonconservative work $W' = -fh'$ is done on the rock.

$$\Delta K + \Delta U = W' \qquad 0 + [mg(-h') - mgh] = -fh' \qquad f = \frac{mg(h + h')}{h'} = \frac{20(16.6)}{0.6} = \underline{553\,N}$$

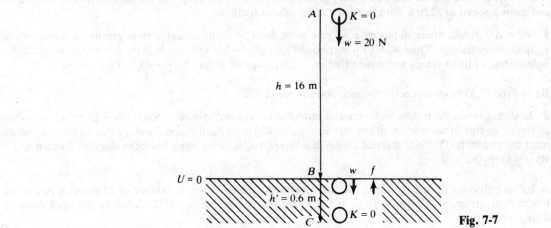

Fig. 7-7

7.49 A pistol fires a 3-g bullet with a speed of 400 m/s. The pistol barrel is 13 cm long. (a) How much energy is given to the bullet? (b) What average force acted on the bullet while it was moving down the barrel? (c) Was this force equal in magnitude to the force of the expanding gases on the bullet?

∎ (a) The kinetic energy of the bullet on leaving the barrel is $K_f = \frac{1}{2}mv^2 = \frac{1}{2}(0.003)(400)^2 = \underline{240\,J}$.
(b) The work done on the bullet is equal to the change in its kinetic energy. $W = \bar{F}x = K_f - K_i$, where $\bar{F}$ is the average (with respect to distance x) force exerted on the bullet. Thus,

$$\bar{F} = \frac{K_f - K_i}{x} = \frac{240 - 0}{0.13} = \underline{1846\,N}$$

since the bullet was at rest initially.
(c) No, since there are frictional forces in play as the bullet moves down the barrel.

7.50 A bullet having a speed of 153 m/s crashes through a plank of wood. After passing through the plank, its speed is 130 m/s. Another bullet, of the same mass and size but traveling at 92 m/s, is fired at the plank. What will be this second bullet's speed after tunneling through? Assume that the resistance of the plank is independent of the speed of the bullet.

∎ The plank does the same amount of work on the two bullets, and therefore decreases their kinetic energies equally.

$$\tfrac{1}{2}m(153)^2 - \tfrac{1}{2}m(130)^2 = \tfrac{1}{2}m(92)^2 - \tfrac{1}{2}mv^2 \qquad v^2 = 1955 \qquad v = \underline{44.2\,m/s}$$

7.51 A delivery boy wishes to launch a 2.0-kg package up an inclined plane with sufficient speed to reach the top of the incline. The plane is 3.0 m long and is inclined at 20°. The coefficient of kinetic friction between the package and the plane is 0.40. What minimum initial kinetic energy must the boy supply to the package?

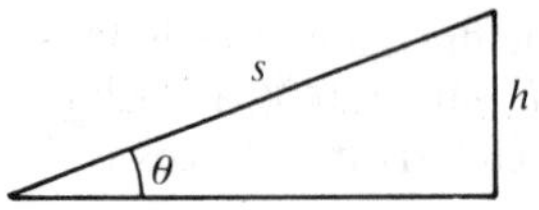

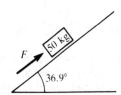

Fig. 7-8

■ The incline is shown in Fig. 7-8. If the package travels the entire length s of the incline, the frictional force will perform work $-\mu_k Ns$ on it. Furthermore, the gravitational potential energy of the package will increase by mgh. To reach the top, the package must have initial kinetic energy $\frac{1}{2}mv_0^2 \geq mgh + \mu_k Ns$ (see Prob. 7.31). Since $N = mg \cos \theta$ and $h = s \sin \theta$, we have $\frac{1}{2}mv_0^2 \geq mgs(\sin \theta + \mu_k \cos \theta)$. With $\theta = 20°$, $m = 2.0$ kg, $s = 3.0$ m, and $\mu_k = 0.40$, we find $(\frac{1}{2}mv_0^2)_{min} = (2.0)(9.8)(3.0)[(0.342) + (0.40)(0.940)] = \underline{42.2 \text{ J}}$.

7.52 A driver of a 1200-kg car notices that the car slows from 20 m/s to 15 m/s as it coasts a distance of 130 m along level ground. How large a force opposes the motion?

■ $W_F = -Fs = \Delta K$, where F is the force in question and $s = 130$ m. $\Delta K = (\frac{1}{2}mv_f^2 - \frac{1}{2}mv_i^2) = \frac{1}{2}(1200 \text{ kg})(225 \text{ m}^2/\text{s}^2 - 400 \text{ m}^2/\text{s}^2) = -105$ kJ. Thus $Fs = 105$ kJ and $F = \underline{0.81 \text{ kN}}$.

7.53 A 10-lb weight slides from rest down a rough inclined plane, 100 ft long and inclined 30° to the horizontal, and gains a speed of 52 ft/s. Find the work done against friction.

■ $W' = \Delta U + \Delta K$, where in this case W', the work done by all forces other than gravity, is simply W_f, the work done by friction. Then $W_f = (0 - mgL \sin 30°) + (\frac{1}{2}mv^2 - 0)$, with $mg = 10$ lb, $L = 100$ ft, $v = 52$ ft/s. Substituting in these values and noting that $m = 0.311$ slug, we obtain $W_f = \underline{-80 \text{ ft} \cdot \text{lb}}$.

7.54 Redo Prob. 7.53 from an energy balance point of view.

■ In sliding down the incline an amount of potential energy $mgL \sin 30° = 500$ ft · lb is given up. It reappears as kinetic energy at the bottom of the incline, $\frac{1}{2}mv^2 = 420$ ft · lb, and thermal energy due to friction, which must then be 80 ft · lb. Since thermal energy has increased, negative work has been done by friction and $W_f = \underline{-80 \text{ ft} \cdot \text{lb}}$.

7.55 A 2000-kg elevator rises from rest in the basement to the fourth floor, a distance of 25 m. As it passes the fourth floor, its speed is 3.0 m/s. There is a constant frictional force of 500 N. Calculate the work done by the lifting mechanism.

■ $W' = \Delta U + \Delta K$, where $W' = W_F + W_f$, the work done by the lift force and frictional force, respectively. Then, $W_F - fh = (Mgh - 0) + (\frac{1}{2}Mv^2 - 0)$, where $M = 2000$ kg, $h = 25$ m, and $f = 500$ N. Substituting in, we get $W_F - 12.5$ kJ $= 490$ kJ $+ 9$ kJ, or $W_F = \underline{511.5 \text{ kJ}}$.

7.56 Redo Prob. 7.56 from an energy-balance point of view.

■ The lifting mechanism supplies energy by doing positive work on the elevator, W_F. This energy appears in three forms: gravitational potential energy, kinetic energy, and thermal energy. The increase in potential energy is $Mgh = 490$ kJ; the increase in kinetic energy is $\frac{1}{2}Mv^2 = 9$ kJ; the increase in thermal energy (since friction does negative work) is $fh = 12.5$ kJ. Adding we get $W_F = \underline{511.5 \text{ kJ}}$.

7.57 The coefficient of sliding friction between a 900-kg car and pavement is 0.80. If the car is moving at 25 m/s along level pavement when it begins to skid to a stop, how far will it go before stopping?

■ $W_f = \Delta U + \Delta K$, where W_f is the work done by friction, and $\Delta U = 0$. Then $-\mu_k mgL = (0 - \frac{1}{2}mv^2)$, or $v^2 = (25 \text{ m/s})^2 = 2\mu_k gL = 2(0.8)(9.8 \text{ m/s}^2)L$. Solving, $L = \underline{40 \text{ m}}$.

7.58 In the setup of Fig. 7-9, the coefficient of friction between the mass and the plane is 0.2. (a) How much work is done by the force F in moving the mass up the plane a distance of 4 m? (b) If the force F is removed and the mass slides back down, what will its speed be when it reaches its original position?

Fig. 7-9

▮ (*a*) The weight $w = 50 \times 9.8 = 490$ N and

$$w \sin 36.9° = 294 \text{ N} \qquad w \cos 36.9° = 392 \text{ N}$$

The friction $f = \mu w \cos 36.9° = 0.2 \times 392 = 78.4$ N. Then the work $W_F = (294 + 78.4) \times 4 = \underline{1.49 \text{ kJ}}$, to just reach the top with no residual kinetic energy.

(*b*) $\qquad$ PE at top $= 294 \text{ N} \times 4 \text{ m} = 1176$ J $\qquad$ work against friction $= 78.4 \times 4 = 313.6$ J

$$(1176 - 313.6) \text{ J} = 862.4 \text{ J} = \tfrac{1}{2}mv^2 \qquad \tfrac{1}{2}(50)v^2 = 862.4 \qquad v^2 = 34.496 \text{ m}^2/\text{s}^2 \qquad v = \underline{5.87 \text{ m/s}}$$

7.59 A ski jumper glides down a 30° slope for 80 ft before taking off from a negligibly short horizontal ramp. If takeoff speed is 45 ft/s, what is the coefficient of kinetic friction on the slide?

▮ $\qquad$ PE $= wh = m \times 32 \times 80 \sin 30° = 1280m$ ft · lb $\qquad$ KE at takeoff $= \tfrac{1}{2}m(45)^2 = 1012.5m$ ft · lb

Thus $\qquad$ energy dissipated $= 267.5m$ ft · lb $\qquad f \times 80 = 267.5m \qquad f = \mu_k mg \cos 30°$

$$\mu_k(m \times 32 \times 0.866)80 = 267.5m \qquad \mu_k = \frac{267.5m}{80 \times 32 \times 0.866} = \underline{0.121}$$

7.60 A pump lifts water from a lake to a large tank 20 m above the lake. How much work does the pump do as it transfers 5 m³ of water to the tank? One cubic meter of water has a mass of 1000 kg.

▮ Work done by the pump, W, equals increase in potential energy of the water. $W = \Delta U = Mgh = (5000 \text{ kg})(9.8 \text{ m/s}^2)(20 \text{ m}) = \underline{0.98 \text{ MJ}}$. (Compare Prob. 7.46; here the height of the tank is neglected in comparison with 20 m.)

7.3 CONSERVATION OF MECHANICAL ENERGY

7.61 What is the law of conservation of mechanical energy?

▮ From Prob. 7.31 we have $W' = \Delta U + \Delta K$, where W' is the work done by all nonconservative forces and U is the potential-energy function for all the conservative forces. If $W' = 0$, we have

$$\Delta U + \Delta K = (U_f - U_i) + (K_f - K_i) = 0 \qquad \text{or} \qquad U_f + K_f = U_i + K_i \quad (law \ of \ conservation \ of \ mechanical \ energy)$$

Note that nonconservative forces may be acting (e.g., the normal force exerted by a smooth, curved wire on a bead sliding along it), but *so long as they do no work*, the conservation law holds.

7.62 Just before striking the ground, a 2.0-kg mass has 400 J of KE. If friction can be ignored, from what height was it dropped?

▮ $\qquad U_f + K_f = U_i + K_i \qquad 0 + K_f = mgh + 0 \qquad h = \dfrac{K_f}{mg} = \dfrac{400}{(2.0)(9.8)} = \underline{20.4 \text{ m}}$

7.63 A 40-g body starting from rest falls through a vertical distance of 25 cm to the ground. (*a*) What is the kinetic energy of the body just before it hits the ground? (*b*) What is the velocity of the body just before it hits the ground?

▮ (*a*) As the body falls its gravitational potential energy is converted to kinetic energy.

$$\text{PE} = mgh = (0.040 \text{ kg})(9.8 \text{ m/s}^2)(0.25 \text{ m}) = 0.098 \text{ J}$$

Therefore, the KE just before the body hits the ground is $\underline{0.098 \text{ J}}$.

(*b*) $\qquad$ KE $= \tfrac{1}{2}mv^2 \qquad 0.098 = \tfrac{1}{2}(0.04)v^2 \qquad v^2 = 4.9 \qquad v = \underline{2.21 \text{ m/s}}$

7.64 A boy throws a 0.15-kg stone from the top of a 20-m cliff with a speed of 15 m/s. Find its kinetic energy and speed when it lands in a river below.

▮ $\qquad$ (PE + KE) at top = KE at river $\qquad (0.15 \times 9.8 \times 20) + \tfrac{1}{2}(0.15)(15)^2 = \tfrac{1}{2}(0.15)v^2$

$$\text{KE at river} = 29.4 + 16.9 = \underline{46.3 \text{ J}} \qquad \tfrac{1}{2}(0.15)v^2 = 46.3 \qquad v = \underline{24.8 \text{ m/s}}$$

7.65 A 0.5-kg ball falls past a window that is 1.50 m in vertical extent. How much did the KE of the ball increase as it fell past the window? If its speed was 3.0 m/s at the top of the window, what was it at the bottom?

▌ $\Delta U + \Delta K = 0$; $\Delta U = -mgh = -7.35$ J and $\Delta K = \underline{7.35\ J}$; $\Delta K = (\frac{1}{2}mv_b^2 - \frac{1}{2}mv_t^2)$, with $v_t = 3$ m/s, so $\frac{1}{2}mv_b^2 = 7.35$ J $+ 2.25$ J $= 9.60$ J. Solving, we get $v_b = \underline{6.20\ m/s}$.

7.66 At sea level a nitrogen molecule in the air has an average translational KE of 6.2×10^{-21} J. Its mass is 4.7×10^{-26} kg. If the molecule could shoot straight up without striking other air molecules, how high would it rise? What is the molecule's initial speed?

▌ We first assume that the height is not so great that we have to use the exact law of gravity, $F = GMm/r^2$. Instead we assume that $F = mg$ and $U = mgh$ are valid. If we are wrong, it will be evident from the value of h. If the molecule shoots up unopposed, we have $\Delta U + \Delta K = 0$ and the loss of kinetic energy in reaching the highest point must equal the gain in potential energy. Thus 6.2×10^{-21} J $= mgh = (4.7 \times 10^{-26}$ kg$)(9.8$ m/s$^2)h$, and $h = \underline{13.5\ km}$. This is small compared with the radius of the earth (6400 km) and our assumption is reasonable. The initial speed of the molecule is obtained from $K = \frac{1}{2}mv^2 = 6.2 \times 10^{-21}$ J. $v = \underline{514\ m/s}$.

7.67 If the simple pendulum shown in Fig. 7-10 is released from point A, what will be the speed of the ball as it passes through point C?

▌ The tension in the cord does no work (refer to Prob. 7.61). When the ball reaches point C it has lost mgh_a in potential energy, where $h_a = 0.75$ m, and in its place gained $\frac{1}{2}mv_c^2$ in kinetic energy, with $mgh_a = \frac{1}{2}mv_c^2$. The m's cancel, leading to $v_c^2 = 2gh_a = 2(9.8$ m/s$^2)(0.75) = 14.7$ m^2/s^2, or $v_c = \underline{3.83\ m/s}$.

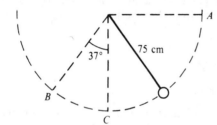

Fig. 7-10

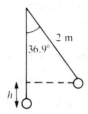

Fig. 7-11

7.68 Refer to Fig. 7-10. What is the speed of the ball at point B?

▌ With point C as the zero reference for potential energy, the conservation law gives $mgh_a + \frac{1}{2}mv_a^2 = mgh_b + \frac{1}{2}mv_b^2$, with $v_a = 0$, $h_a = 0.75$ m, and $h_b = (0.75$ m$)(1 - \cos 37°) = 0.15$ m. Our equation becomes $v_b^2 = 2gh_a - 2gh_b = 11.76$ m^2/s^2, and $v_b = \underline{3.43\ m/s}$.

7.69 A pendulum bob has a mass of 0.5 kg. It is suspended by a cord 2 m long which is pulled back through an angle of 36.9° (Fig. 7-11) and released. Find **(a)** its maximum potential energy relative to its lowest position, **(b)** its potential energy when the cord makes an angle of 10° with the vertical, **(c)** its maximum speed, and **(d)** its speed when the cord makes an angle of 10° with the vertical.

▌ $h = 2 - 2\cos 36.9° = 2 - 1.6 = 0.4$ m.
(a) $\text{PE}_{\text{max}} = (0.5 \times 9.8)0.4 = \underline{1.96\ J}$. **(b)** When $\theta = 10°$,

$$h_{10} = 2 - 2\cos 10° = 2(1 - 0.9848) = 0.0304\ \text{m} \qquad \text{PE}_{10°} = (0.5 \times 9.8) \times 0.0304 = \underline{0.149\ J}$$

(c) At the bottom $\qquad \frac{1}{2}(\frac{1}{2})v_{\text{max}}^2 = 1.96 \qquad v_{\text{max}} = 2 \times 1.4 = \underline{2.8\ m/s}$

(d) $\qquad\qquad (1.96 - 0.149)\ \text{J} = \frac{1}{2}(\frac{1}{2})v_{10}^2 \qquad v_{10}^2 = 7.24 \qquad v_{10} = \underline{2.69\ m/s}$

7.70 A 0.8-kg pendulum bob on a 2-m cord is pulled sideways until the cord makes an angle of 36.9° with the vertical (see Fig. 7-11). Find the work done on the bob and the speed of the bob as it passes through the equilibrium position after being released at rest.

▌ $h = 2 - 1.6 = 0.4$ m $\qquad$ work done by gravity $=$ loss in PE $\qquad$ loss in PE $= mgh = 0.8 \times 9.8 \times 0.4 = \underline{3.136\ J}$

$\qquad$ From the work-energy theorem $\qquad \frac{1}{2}mv^2 = mgh \qquad v^2 = 2gh = 2 \times 9.8 \times 0.4 = 7.84$ m^2/s$^2 \qquad v = \underline{2.8\ m/s}$

7.71 A toy car starts from rest at position 1 shown in Fig. 7-12(a) and rolls without friction along the loop *12324*. Find the smallest height h at which the car can start without falling off the track.

▌ When h has its critical value, the car will just lose contact with the track at position *3*. Then, at *3*, the

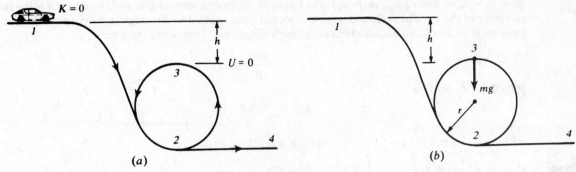

Fig. 7-12

normal force due to the loop vanishes and $mg = mv_3^2/r$, or $v_3^2 = gr$. By conservation of energy, the potential energy of the car at position 1 is the same as the kinetic energy of the car at position 3:

$$mgh = \frac{1}{2}mv_3^2 \qquad h = \frac{v_3^2}{2g} = \frac{r}{2}$$

7.72 Refer to Prob. 7.71. What speed does the car have at position 4?

▌ Applying conservation of energy between positions 1 and 4,

$$0 + mgh = \tfrac{1}{2}mv_4^2 + mg(-2r) \qquad v_4^2 = 2g(h + 2r) = 5gr \qquad v_4 = \sqrt{5gr}$$

7.73 In the track shown in Fig. 7-13, section AB is a quadrant of a circle of 1.0-m radius. A block is released at A and slides without friction until it reaches point B. (a) How fast is it moving at B, the bottom of the quadrant? (b) The horizontal part is not smooth. If the block comes to rest 3.0 m from B, what is the coefficient of kinetic friction?

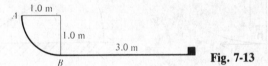

Fig. 7-13

▌ (a) Conservation of energy yields the equation $\tfrac{1}{2}mv_B^2 + mgh_B = mgh_A$, so that $v_B = \sqrt{2g(h_A - h_B)}$. With $h_A - h_B = 1.0$ m, we obtain $v_B = \sqrt{2(9.8)(1.0)} = \underline{4.43 \text{ m}}$.
(b) If the block slides distance d in coming to rest, the work-energy theorem gives $W_{\text{fr}} = -\mu_k mgd = -\tfrac{1}{2}mv_B^2$. Then $\mu_k = (\tfrac{1}{2}mv_B^2)/(mgd) = mg(h_A - h_B)/(mgd) = (h_A - h_B)/d$. With $d = 3.0$ m, $\mu_k = \tfrac{1}{3} = \underline{0.333}$.

7.74 Figure 7-14 shows the plan for a proposed roller coaster track. Each car will start from rest at point A and will roll with negligible friction. It is important that there be at least some small positive normal force (that is, a push) exerted by the track on the car at all points; otherwise the car would leave the track. What is the minimum safe value for the radius of curvature at point B?

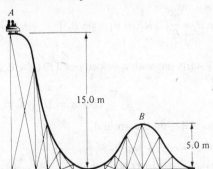

Fig. 7-14

▌ We use v_A and v_B to denote the speeds of the car at points A and B, and we use h_A and h_B to denote the elevations. If the car has mass m, the equation expressing energy conservation is $\tfrac{1}{2}mv_B^2 + mgh_B = \tfrac{1}{2}mv_A^2 + mgh_A$.

Since $v_A = 0$, we find $v_B^2 = 2g(h_A - h_B)$. The required centripetal force at B is mv_B^2/R, where R is the radius of curvature of the track at B. In order for the normal force exerted by the track to be positive, the centripetal force must be smaller in magnitude than the car's weight mg. That is, we must have

$$\frac{mv_B^2}{R} \leq mg$$

Solving for R, we find

$$R \geq R_{min} = \frac{v_B^2}{g} = 2(h_A - h_B)$$

With $h_A = 15.0$ m and $h_B = 5.0$ m, we obtain $R \geq 2(10) = \underline{20.0\ m}$.

7.75 Show that a pendulum bob, which has been pulled aside from its equilibrium position through an angle θ and then released, will pass through the equilibrium position with speed $v = \sqrt{2gl(1 - \cos\theta)}$, where l is the length of the pendulum.

▐ Using the conservation of energy law in the form

$$\text{loss of PE} = \text{gain of KE} \qquad mgh = \tfrac{1}{2}mv^2 \qquad gh = \tfrac{1}{2}v^2$$

From Fig. 7-15,

$$h = l - l\cos\theta \qquad g(l - l\cos\theta) = \tfrac{1}{2}v^2 \qquad 2gl(1 - \cos\theta) = v^2 \qquad v = \sqrt{2gl(1 - \cos\theta)}$$

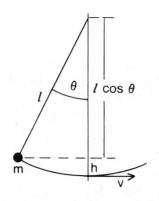

Fig. 7-15

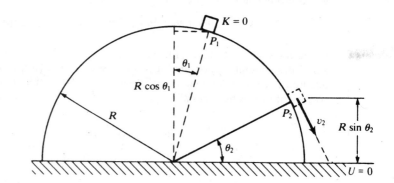

Fig. 7-16

7.76 A particle moves from rest at P_1 on the surface of a smooth circular cylinder of radius R (Fig. 7-16). At P_2 the particle leaves the cylinder. Find the equation relating θ_1 and θ_2 as shown.

▐ As the normal force does no work on the particle, its energy is conserved.

$$(K + U)_{P_1} = (K + U)_{P_2} \qquad 0 + mgR\cos\theta_1 = \tfrac{1}{2}mv_2^2 + mgR\sin\theta_2 \qquad v_2^2 = 2gR(\cos\theta_1 - \sin\theta_2)$$

At P_2, the normal force exerted by the surface vanishes, leaving the radial component of the particle's weight, $mg\sin\theta_2$, as the instantaneous centripetal force. Then, by Newton's second law,

$$mg\sin\theta_2 = \frac{mv_2^2}{R} \qquad mg\sin\theta_2 = \frac{m}{R}[2gR(\cos\theta_1 - \sin\theta_2)] \qquad \sin\theta_2 = 2\cos\theta_1 - 2\sin\theta_2 \qquad \sin\theta_2 = \tfrac{2}{3}\cos\theta_1$$

7.77 After being hit, a golf ball starts out with a velocity of 130 ft/s. If it reaches a maximum height of 180 ft, what is its velocity at that point?

▐ $$U_i + K_i = U_f + K_f$$

Taking U_i as zero when the ball is on the ground,

$$0 + \tfrac{1}{2}mv_0^2 = mgh + \tfrac{1}{2}mv^2 \qquad \tfrac{1}{2}v_0^2 = gh + \tfrac{1}{2}v^2 \qquad \tfrac{1}{2}(130)^2 = 32(180) + \tfrac{1}{2}v^2$$

$$130^2 - 64(180) = v^2 \qquad 16\,900 - 11\,520 = v^2 \qquad v^2 = 5380\ \text{ft}^2/\text{s}^2 \qquad v = \underline{73.3\ \text{ft/s}}$$

7.78 When a 300-g mass is hung from the end of a vertical spring, the spring's length is 40 cm. With 500 g hanging from it, its length is 50 cm. What is the spring constant of the spring?

⏸ Since the additional 0.20 kg stretched the spring by 0.10 m, $k = F/x = 0.20(9.8)/0.10 = \underline{19.6\,\text{N/m}}$.

7.79 A spring which stretches 10 cm under a load of 200 g requires how much work to stretch it 5 cm from its equilibrium position? How much work is required to stretch it the next 5 cm?

⏸ The spring constant $= k = F/x = 0.20(9.8)/0.10 = 19.6\,\text{N/m}$, the work is $kx^2/2 = 19.6(0.05)^2/2 = \underline{0.0245\,\text{J}}$ for the 5-cm case. The additional work is $[19.6(0.10)^2/2] - 0.0245 = \underline{0.0735\,\text{J}}$.

7.80 How much work by the pulling force is required to change the elongation of a spring from 10 to 20 cm if a load of 80 g elongates it 4.0 cm? All elongations are measured from its unstretched position.

⏸ Follow the same procedure as in Prob. 7.79. Then $k = (9.8 \times 0.08)/0.04 = 19.6\,\text{N/m}$. The respective works are 0.0980 J and 0.392 J. This gives a result of $\underline{0.294\,\text{J}}$.

7.81 A spring balance reads forces in newtons. The scale is 20 cm long and reads from 0 to 60 N. Find the potential energy of the spring **(a)** when it reads 40 N, **(b)** when it is stretched 20 cm, and **(c)** when a mass of 4 kg is suspended from the spring.

⏸ $\qquad\qquad F = kx \qquad 60\,\text{N} = k(0.2\,\text{m}) \qquad k = 300\,\text{N/m}$

(a) $\qquad\qquad 40 = 300x \qquad x = (4/30)\,\text{m} \qquad \text{PE} = \tfrac{1}{2}kx^2 = \tfrac{1}{2}(300)(4/30)^2 = \underline{2.67\,\text{J}}$

(b) $\qquad\qquad\qquad\qquad\qquad \text{PE} = \tfrac{1}{2}kx^2 = \tfrac{1}{2}(300)(0.2)^2 = \underline{6\,\text{J}}$

(c) $\qquad 4\,\text{kg weighs }39.2\,\text{N} \qquad 39.2 = kx = 300x \qquad \text{PE} = \tfrac{1}{2}(300)(39.2/300)^2 = \underline{2.56\,\text{J}}$

7.82 A block falls from a table 0.6 m high. It lands on an ideal, massless, vertical spring with a force constant of 2.4 kN/m. The spring is initially 25 cm high, but it is compressed to a minimum height of 10 cm before the block is stopped. Find the mass of the block.

⏸ $(\text{PE}_b + \text{PE}_s)_{\text{initial}} = (\text{PE}_b + \text{PE}_s)_{\text{final}}$, since KE $= 0$ at top and bottom. Then $m(9.8)(0.6) + 0 = m(9.8)(0.10) + \tfrac{1}{2}(2.4 \times 10^3)(0.15)^2$

Solving, $m = \underline{5.51\,\text{kg}}$.

7.83 The spring shown in Fig. 7-17 has a stiffness constant k and has been compressed a distance x_0. The mass M is free to leave the end of the spring and experiences negligible friction with the table. How fast will the mass be moving when it leaves the spring if the system is released?

⏸ All the compressional (potential) energy in the spring will be changed to K of the mass; $kx_0^2/2 = Mv^2/2$, which gives $v = x_0(k/M)^{1/2}$.

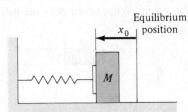

Fig. 7-17

7.84 The mass M shown in Fig. 7-17 is fastened to the end of the spring and moves without friction. When the spring is released from the position shown, the system oscillates between $\pm x_0$ from the equilibrium position. Find the speed of the mass when the elongation of the spring is **(a)** zero, **(b)** $2x_0/3$. The spring constant is k.

⏸ **(a)** From Prob. 7.83, $v = x_0(k/M)^{1/2}$. **(b)** $kx_0^2/2 = K$ of mass $+ U$ of spring, thus $kx_0^2/2 = Mv^2/2 + k(2x_0/3)^2/2$. This yields $v = x_0(5k/M)^{1/2}/3$.

7.85 A vertical spring with constant 200 N/m has a light platform on its top. When a 500-g mass is set on the platform, the spring compresses 0.0245 m. The mass is now pushed down 0.0755 m farther and released. How far above this latter position will the mass fly?

⏸ The spring constant $k = F/y = 200\,\text{N/m}$ as stated. If the mass goes higher than 0.10 m, it will fly loose from the spring. If it does, U_s at start $= U_g$ of mass at end, where zero U_g is at the lowest position. Then $200(0.10)^2/2 = 0.5(9.8)h$ gives $h = \underline{0.20\,\text{m}}$. Thus the mass leaves the spring and our assumption is justified.

7.86 Suppose a 300-g mass is dropped from a height of 40 cm onto the spring described in Prob. 7.85 and sticks to the platform. **(a)** How far will the spring compress? **(b)** How far will the spring be stretched as the mass and spring rebound?

▌ The mass will lose U_g as it moves from its top position. This loss in U_g appears as U_s. **(a)** Call y the spring compression distance, then $mg(0.4 + y) = 200y^2/2$ gives $y = \underline{0.124\text{ m}}$. **(b)** U_s stored in the spring in **(a)** is decreased by mgh, where h is the rebound distance. Hence $k(0.124)^2/2 = k(h - 0.124)^2/2 + mgh$. This yields $h = 0.219$ m. The spring will be stretched $h - y = \underline{0.095\text{ m}}$.

7.87 A body of mass 100 g is attached to a hanging spring whose force constant is 10 N/m. The body is lifted until the spring is in its unstretched state. The body is then released. Using the law of conservation of total mechanical energy, calculate the speed of the body when it strikes a table 15 cm below the release point.

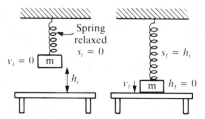

Fig. 7-18

▌ Figure 7-18 shows the system just as it is released and just as the body strikes the table. The potential energy of the system may be written as $U = mgh + \frac{1}{2}ks^2$; h is the height above the table and s is the amount of stretching of the spring. When the object is released, $h = h_i = 0.15$ m, $s = s_i = 0$, and the speed $v = v_i = 0$. When the object strikes the table, $h = h_f = 0$, $s = h_i = 0.15$ m, and $v = v_f$. The conservation of energy requires that

$$K_f + U_f = K_i + U_i \qquad \text{or} \qquad \tfrac{1}{2}mv_f^2 + \tfrac{1}{2}kh_i^2 = 0 + mgh_i$$

Solving for v_f, we find that $v_f = \sqrt{2gh_i - (kh_i^2/m)}$. With $k = 10$ N/m and $m = 0.100$ kg, we find that

$$v_f = \sqrt{2(9.80)(0.15) - [(10)(0.15)^2/(0.1)]} = \underline{0.831\text{ m/s}}.$$

7.88 A spring with negligible mass and a force constant of 600 N/m is kept straight by confining it within a smooth-walled guiding tube (Fig. 7-19). The tube is anchored in a horizontal position on a tabletop. The spring is compressed by 10.0 cm and held there by a latch pin inserted through the wall of the tube. A 200-g ball of the same diameter as the spring is placed in contact with the spring, as shown in Fig. 7-19. Then the latch pin is removed releasing the spring. What speed does the ball acquire?

Fig. 7-19

▌ We use x to denote the position of the free end of the spring, with $x = 0$ corresponding to the spring's relaxed position and $x > 0$ corresponding to stretching. Initially the spring is compressed ($x_0 = -0.100$ m) and the ball is at rest ($v_0 = 0$). The spring loses contact with the ball and the ball reaches its final speed v_f just as the spring would begin to pull rather than push. That is, $x_f = 0$. The conservation of energy requires that $\frac{1}{2}mv_f^2 + \frac{1}{2}kx_f^2 = \frac{1}{2}mv_0^2 + \frac{1}{2}kx_0^2$. With $v_0 = 0$ and $x_f = 0$, we find

$$v_f = |x_0|\sqrt{\frac{k}{m}}$$

With $k = 600$ N/m and $m = 0.200$ kg, we find $v_0 = (0.10)\sqrt{(600)/(0.200)} = \underline{5.48\text{ m/s}}$.

7.89 Refer to Prob. 7.88. If the same procedure is followed with the tube pointing vertically upward, what will be the speed of the ball as it leaves contact with the spring?

▌ Once again, the ball will lose contact with the spring just as the spring reaches its relaxed length. Measuring y positive upward from the relaxed position, the conservation of energy requires that

$\frac{1}{2}mv_f^2 + \frac{1}{2}ky_f^2 + mgy_f = \frac{1}{2}mv_0^2 + \frac{1}{2}ky_0^2 + mgy_0$. Using the values $v_0 = 0$ and $y_f = 0$, we obtain

$$v_f = \sqrt{\frac{ky_0^2}{m} + 2gy_0}$$

With $y_0 = -0.100$ m, we find

$$v_f = \sqrt{[(600)(-0.1)^2/(0.2)] + (2)(9.80)(-0.1)} = \underline{5.30 \text{ m/s}}$$

7.90ᶜ An object of mass m has a speed v as it passes through the origin on its way out along the $+x$ axis. It is subjected to a retarding force given by $F_x = -Ax$ $(A > 0)$. Find its x coordinate when it stops.

■ By the work-energy principle, work done by retarding force = change in KE:

$$\int_0^{x_f} (-Ax)\, dx = 0 - \frac{1}{2}mv^2 \qquad -\frac{1}{2}Ax_f^2 = -\frac{1}{2}mv^2 \qquad x_f = v\sqrt{\frac{m}{A}}$$

7.91 A 10-kg instrument packet is fired vertically to a height of 637 km above the earth's surface, which corresponds to a distance from the center of the earth of 1.1 earth radii. Find the weight w and potential energy (PE) of the instrument packet at its maximum altitude. (Take the earth's surface as the zero level of PE.)

■ At the earth's surface the weight is

$$mg = 10 \text{ kg} \times 9.8 \text{ m/s}^2 = G\frac{M_E(10 \text{ kg})}{R^2} \qquad \text{where } R \text{ is the radius of earth} \tag{1}$$

At $r = 1.1R$,
$$w = G\frac{M_E(10 \text{ kg})}{(1.1R)^2} \tag{2}$$

If we divide (2) by (1), we obtain $w/98 = 1/1.21$, or $w = \underline{81 \text{ N}}$.

$$\text{PE} = GM_E(10 \text{ kg})\left(\frac{1}{R} - \frac{1}{1.1R}\right) = GM_E(10 \text{ kg})\left(\frac{0.1}{1.1R}\right) = (6.67 \times 10^{-11} \text{ N} \cdot \text{m}^2/\text{kg}^2)(5.98 \times 10^{24} \text{ kg})(10 \text{ kg})$$

$$\times \frac{0.1}{1.1 \times 6.37 \times 10^6 \text{ m}} = \underline{57 \text{ MJ}}$$

(We note that relative to PE = 0 at $r = \infty$, the PE is $[-GM_E(10 \text{ kg})]/1.1R = -570$ MJ.)

7.92 Calculate the escape velocity of a body starting from the surface of earth. Ignore air friction. The mass of the earth is 5.98×10^{24} kg and its radius is 6370 km.

■ $\Delta\text{PE} = GMm(1/R - 1/r)$ where $R = 6.37 \times 10^6$ m. For escape, $r = \infty$, and $\Delta\text{PE} = GMm/R$. Since the final velocity is zero, energy conservation gives

$$\Delta\text{KE} + \Delta\text{PE} = 0 \qquad \left(0 - \frac{1}{2}mv_e^2\right) + \frac{GMm}{R} = 0 \qquad v_e = \sqrt{\frac{2GM}{R}} = \sqrt{\frac{2(6.67 \times 10^{-11})(5.98 \times 10^{24})}{6.37 \times 10^6}} = \underline{11.2 \text{ km/s}}$$

(This is approximately 25 000 mi/h.)

7.93 Show that if $h = r - R$ is a very small quantity compared with the radius of the earth R, the gravitational potential energy relative to that at the earth's surface, PE = $GMm(1/R - 1/r)$, reduces to PE = mgh, where g is the acceleration of gravity and h is vertical height.

■ $$\text{PE} = GMm\left(\frac{1}{R} - \frac{1}{R+h}\right) = \frac{GMmh}{R(R+h)} = m\frac{GM}{R^2}\frac{h}{(1+h/R)} \approx mgh \qquad \text{where} \qquad g = \frac{GM}{R^2}$$

7.94 The potential energy per kilogram for a mass m in the earth's gravitational field is shown in Fig. 7-20, where r is the distance of the object from the earth's center and $R_e = 6.4 \times 10^6$ m is the earth's radius. Zero potential energy is taken for $r \to \infty$. (a) If frictional effects of air are ignored, how much energy is required to free a 1-kg object from the earth (i.e., to move it from $r = R_e$ to $r \to \infty$)? (b) With what speed must an object be shot away from the earth if it is to escape?

■ From the graph, 62 MJ is needed to remove 1 kg to ∞; this is provided by K at the earth's surface, so $[(1 \text{ kg})v^2]/2 = 62 \times 10^6$ J. Solving, $v = \underline{11.1 \text{ km/s}}$, in agreement with Prob. 7.92.

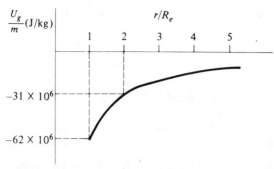

Fig. 7-20

7.95   Refer to Fig. 7-20. If a 1-kg object is released far from the earth, find its speed at a distance $r = 2R_e$ from the earth's center, where R_e is the earth's radius. Why is it allowable to ignore frictional effects of air?

▐   Between ∞ and $2R_e$ a 1-kg object loses 31 MJ in U_g, which is changed to K. Then, $[(1 \text{ kg})v^2]/2 = 31 \times 10^6$ and $v = \underline{7.87 \text{ km/s}}$. At $r = 2R_e$, we are essentially in vacuum and friction effects are negligible.

7.96   Consider a proton (positively charged) shot head on from a large distance at a heavy nucleus (also positively charged). To a first approximation, the heavy nucleus remains motionless as the proton approaches. Repulsion between the two charges slows the proton as it approaches. Eventually the proton stops at r_0 and reverses its motion. If the potential energy of the system varies with distance between particle centers (r) as shown in Fig. 7-21, (a) what energy must the proton be shot with if it is to reach r_0? (b) What will be the proton speed when it is first shot? (c) when it is at $2r_0$? (m of proton is 1.67×10^{-27} kg; $1 \text{ eV} = 1.6 \times 10^{-19}$ J.)

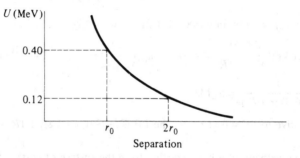

Fig. 7-21

▐   (a) Assuming no K given to the nucleus, at r_0 all the original K of the proton is changed into electric potential energy U; $0.40 \text{ MeV} = \underline{6.4 \times 10^{-14}}$ J would be required. (b) $mv^2/2 = 6.4 \times 10^{-14}$ J yields $v = \underline{8.75 \times 10^6 \text{ m/s}}$. (c) Its K value is $(0.40 - 0.12)$ MeV, thus $mv^2/2 = (0.28 \times 10^6)(1.6 \times 10^{-19})$, giving $v = \underline{7.3 \times 10^6 \text{ m/s}}$.

7.97   The potential energy curve for a proton in the vicinity of a large nucleus is shown in Fig. 7-21. If the proton is released from the position $r = r_0$, (a) what will be its speed at a large distance from the nucleus? (b) at $2r_0$? Assume the heavy nucleus remains stationary (see Prob. 7.96 for numerical data).

▐   (a) The U of the proton at r_0 will be changed to K as r nears infinity. Therefore $(0.40 \times 10^6)(1.6 \times 10^{-19}) = mv^2/2$, which gives $v = \underline{8.75 \times 10^6 \text{ m/s}}$. (b) Here, the energy lost in going from r_0 to $2r_0$ is changed to K; $mv^2/2 = (0.28 \times 10^6 \text{ eV})(1.6 \times 10^{-19} \text{ J/eV})$, from which $v = \underline{7.3 \times 10^6 \text{ m/s}}$.

7.4 ADDITIONAL PROBLEMS

7.98   A 5-lb block slides down a quarter-circular surface (Fig. 7-22). If the radius is 1.2 ft and the block has a speed of 6 ft/s at the bottom, what is the work done by the block against friction?

Fig. 7-22

▌ PE = 5 lb × 1.2 ft = 6 ft · lb at top. At bottom, KE = $\frac{1}{2}(5/32)(6)^2$ = 2.8125 ft · lb. Work done against friction = 6.000 − 2.8125 = 3.19 ft · lb.

7.99 A block of mass 3 kg starts from rest and slides down a surface which corresponds to a quarter circle of 1.6 m radius (Fig. 7-22). (*a*) If the curved surface is smooth, what is the speed at the bottom? (*b*) If the speed at the bottom is 4 m/s, what is the energy dissipated by friction in the descent? (*c*) After the block reaches the level region at 4 m/s, it slides to a stop in 3 m. Find the frictional force.

▌ (*a*) $\qquad\qquad$ PE at top = KE at bottom $\qquad (3 \times 9.8 \times 1.6) = \frac{1}{2}(3)v^2$

$$v^2 = 2 \times 9.8 \times 1.6 \qquad v = 5.6 \text{ m/s}$$

(*b*) $\qquad\qquad$ PE at top = KE at bottom + work against friction

$$(3 \times 9.8 \times 1.6) = \frac{1}{2}(3)(4)^2 + |W_f| \qquad \text{whence} \qquad |W_f| = 23.04 \text{ J}$$

(*c*) $\qquad\qquad \frac{1}{2}mv^2 = fs \qquad \frac{1}{2}(3)16 = f(3) \qquad f = 8 \text{ N}$

7.100 A 1200-kg car coasts from rest down a driveway that is inclined 20° to the horizontal and is 15 m long. How fast is the car going at the end of the driveway if friction is negligible?

▌ $\qquad \Delta K + \Delta U = (\frac{1}{2}mv^2 - 0) + (0 - mgh) = 0, \qquad$ where $\qquad h = (15 \text{ m}) \sin 20° = 5.13 \text{ m}$

$$v^2 = 2gh = 100 \text{ m}^2/\text{s}^2 \qquad \text{and} \qquad v = 10.0 \text{ m/s}$$

7.101 Repeat Prob. 7.100 assuming that a friction force of 3 kN opposes the motion.

▌ Now $W_f = \Delta K + \Delta U = (\frac{1}{2}mv^2 - 0) + (0 - mgh)$, where $W_f = -(3 \text{ kN})(15 \text{ m}) = -45 \text{ kJ}$ is the work done by friction. As in Prob. 7.100 $h = 5.13$ m. Then $-45 \text{ kJ} = \frac{1}{2}(1200 \text{ kg})v^2 - (1200 \text{ kg})(9.8 \text{ m/s}^2)(5.1 \text{ m})$, and $v = 5.06 \text{ m/s}$.

7.102 Figure 7-23 shows a bead sliding on a wire. How large must height h_1 be if the bead, starting at rest at A, is to have a speed of 200 cm/s at point B? Ignore friction.

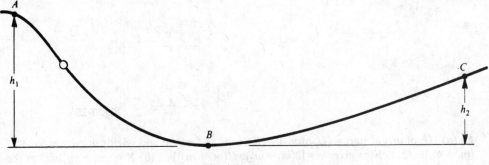

Fig. 7-23

▌ Since we ignore friction and since the normal force due to the wire does no work, we have conservation of mechanical energy, $\Delta U + \Delta K = 0$. Then $U_a + K_a = U_b + K_b$, or $mgh_1 + 0 = 0 + \frac{1}{2}mv_b^2$, or $v_b^2 = 2gh_1$. Substituting $v_b = 200$ cm/s and noting that $g = 980$ cm/s², we get $h_1 = 20.4$ cm.

7.103 In Fig. 7-23, $h_1 = 50$ cm, $h_2 = 30$ cm, and the length along the wire from A to C is 400 cm. A 3.0 g-bead released at A coasts to point C and stops. How large an average friction force opposed its motion?

▌ $W' = \Delta U + \Delta K$, where in this case $W' = W_f$, the work done by friction. Then $-fL = (mgh_2 - mgh_1) + (0 - 0)$, where f is the average frictional force, $L = 400$ cm is the length of the wire from A to C, and, as we have noted, the kinetic energy is zero at both A and C. Substituting in all the known quantities, we get $f = 147 \text{ dyn} = 1.47 \times 10^{-3} \text{ N}$.

7.104 An auto starting from rest reaches a kinetic energy K by accelerating without skidding along a horizontal road. (*a*) Neglect air resistance. Find the work done by the external forces which accelerate the car. (*b*) Is the result in (*a*) consistent with the conservation of energy?

▌ (a) The external forces that act on the auto are shown in Fig. 7-24. The net external force is $\mathbf{f}_1 + \mathbf{f}_2$ (the sum of the forces of static friction exerted on the tires by the road). This net force accelerates the auto, i.e., $\mathbf{f}_1 + \mathbf{f}_2 = M\mathbf{a}_{cm}$, where $\mathbf{a}_{cm}$ is the acceleration of the center of mass of the auto. Since the portions of the tires instantaneously in contact with the roadway are at rest relative to the road (no skidding), the forces $\mathbf{f}_1$ and $\mathbf{f}_2$ acting on these parts do no work. Thus, $W_{ext} = 0$.

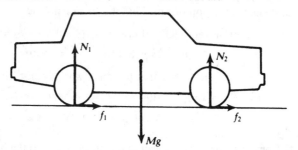

Fig. 7-24

(b) If no external work is done on the system, where does its kinetic energy K come from? We might speak of "internal work" W', such that $\Delta K = K - 0 = W'$. More likely, we would identify W' with the decrease in an "internal energy" Φ (the energy content of the gasoline) and write the conservation of energy as $\Delta K + \Delta \Phi = 0$. [It should be noted that although $\mathbf{f}_1$ and $\mathbf{f}_2$ do not actually perform work on the automobile, one can formally calculate the increase in kinetic energy of the automobile by multiplying $\mathbf{f}_1 + \mathbf{f}_2$ by the distance the auto moves through.]

7.105 If the masses in Fig. 7-25 are released from the position shown, (a) find an expression for the speed of either mass just before m_1 strikes the floor. Ignore the mass and friction of the pulley. (b) Repeat if m_1 has a downward velocity v_0 at the instant shown in the figure.

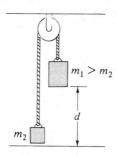

Fig. 7-25

▌ (a) (U_g of m_1) at start = (U_g of m_2) + (K of m_1 and m_2) at end, yields $m_1 g d = m_2 g d + (m_1 + m_2)v^2/2$. Solving gives $v = [2(m_1 - m_2)gd/(m_1 + m_2)]^{1/2}$. (b) $K = (m_1 + m_2)v_0^2/2$ at the start, so add it to the energy equation to yield the final velocity $v = [2(m_1 - m_2)gd/(m_1 + m_2) + v_0^2]^{1/2}$.

7.106 In Fig. 7-26, neither spring is distorted in the position shown. If now the mass is displaced 20 cm to point B and released, (a) what is the speed of the block as it passes through point A, and (b) how far does the block go to the left before stopping?

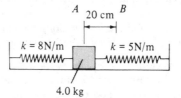

Fig. 7-26

▌ (a) (U_s in springs at B) = (K of mass at A) yields $[8(0.04)]/2 + [5(0.04)]/2 = 4.0v^2/2$, which gives $v = \underline{0.36 \text{ m/s}}$. (b) All the energy is stored in the springs when the mass is 20 cm to the left of A.

7.107 As shown in Fig. 7-27, a block of mass m is resting on a horizontal surface. The coefficients of static and kinetic friction between the block and the surface are μ_s and μ_k, respectively. The block is attached to a spring of negligible mass having spring constant k. Initially the block is at rest and the spring is relaxed. Then the block is struck sharply, so that it begins moving to the right with speed v_0. How far does the spring extend before the rightward motion is arrested?

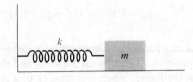

Fig. 7-27

▋ We denote the block's position by x (with $x > 0$ for a stretched spring). Initially $x = x_0 = 0$ and $v = v_0$. The initial total energy of the system is therefore $\frac{1}{2}mv_0^2$. When the rightward motion is arrested, $x_2 = x_M$ and $v = 0$, so that the total mechanical energy is $\frac{1}{2}kx_M^2$. During the rightward motion there is a decrease in the total mechanical energy due to the (negative) work done by the frictional force. Therefore we have

$$\frac{1}{2}kx_M^2 = \frac{1}{2}mv_0^2 - \mu_k mg\,|x_M - x_0| = \frac{1}{2}mv_0^2 - \mu_k mgx_M$$

Solving the quadratic equation for x_M and rejecting the negative root, we find

$$x_M = \sqrt{\left(\frac{\mu_k mg}{k}\right)^2 + \frac{mv_0^2}{k}} - \frac{\mu_k mg}{k}$$

7.108 Refer to Prob. 7.107. Find a criterion for determining whether the block begins to move back to the left or simply remains at the point of maximum extension. Apply it to the case $m = 10$ kg, $k = 100$ N/m, $\mu_s = 0.30$, $\mu_k = 0.15$, and $v_0 = 1.0$ m/s.

▋ The maximum static frictional force is $\mu_s mg$. If that maximum force equals or exceeds the leftward restoring force kx_M, the block remains at $x = x_M$. That is, if $kx_M \le \mu_s mg$, the block remains at $x = x_M$. If $kx_M > \mu_s mg$, the block starts to slide back. For $m = 10$ kg, $k = 100$ N/m, $\mu_s = 0.30$, $\mu_k = 0.15$, and $v_0 = 1.0$ m/s, we find

$$x_M = \sqrt{\left[\frac{(0.15)(10)(9.8)}{(100)}\right]^2 + \frac{(10)(1.0)^2}{(100)}} - \frac{(0.15)(10)(9.8)}{(100)} = 0.3487 - 0.1470 = \underline{0.202 \text{ m}}$$

The restoring force $kx_M = 20.2$ N, while the maximum available static frictional force $\mu_s mg = 29.4$ N, so the block remains at $x = x_M$.

7.109 As shown in Fig. 7-28, a smooth rod is mounted horizontally just above a tabletop. A 10-kg collar, which is able to slide on the rod with negligible friction, is fastened to a spring whose other end is attached to a pivot at O. The spring has negligible mass, a relaxed length of 10 cm, and a spring constant of 500 N/m. The collar is released from rest at point S. (*a*) What is its velocity as it passes point A? (*b*) Repeat for point B.

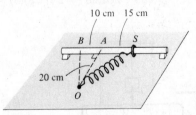

Fig. 7-28

▋ We denote the spring constant, the relaxed length, and the collar mass by k, l_0, and m, respectively. We let v_i and l_i represent the initial collar speed and spring length; v and l represent the instantaneous speed and spring length at some other time. The conservation of mechanical energy implies that $\frac{1}{2}mv^2 + \frac{1}{2}k(l - l_0)^2 = \frac{1}{2}mv_i^2 + \frac{1}{2}k(l_i - l_0)^2$. Since $v_i = 0$, we have

$$v = \sqrt{\frac{k}{m}}\,[(l_i - l_0)^2 - (l - l_0)^2]^{1/2}$$

Referring to Fig. 7-28, $l_i = \sqrt{(0.20)^2 + (0.15)^2} = 0.25$ m.

(a) As the collar passes point A, $l = l_A = 0.20$ m. With $l_0 = 0.10$ m, $k = 500$ N/m, and $m = 10$ kg, we obtain

$$v_A = \sqrt{\frac{500}{10}} \left[(0.25 - 0.10)^2 - (0.20 - 0.10)^2\right]^{1/2} = \sqrt{50} \sqrt{0.0225 - 0.0100} = \underline{0.791 \text{ m/s}}$$

(b) As the collar passes point B, $l = l_B = \sqrt{(0.20)^2 + (0.10)^2} = 0.2236$ m. Then we have

$$v_B = \sqrt{50} \left[(0.15)^2 - (0.1236)^2\right]^{1/2} = \underline{0.601 \text{ m/s}}$$

7.110 Figure 7-29 shows a pendulum of length l suspended at a distance $l - l_1$ vertically above a small peg C. Suppose that the bob is initially displaced by an angle β_0 and then released from rest. Find the speed v of the bob at the instant shown in the figure, when it is moving in a circular path of radius l_1 and has an angular displacement θ with respect to the vertical.

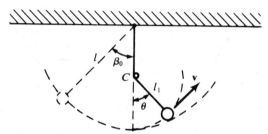

Fig. 7-29

▌ The tension that the string exerts on the bob carries out no work on the bob throughout its motion. The weight of the bob is the only force doing work. Take the lowest point in the path as the reference point of gravitational potential energy. The potential energy of the bob at the instant shown is $mgl_1(1 - \cos\theta)$. Since it starts out at rest and its initial potential energy is $mgl(1 - \cos\beta_0)$, the conservation of energy expression is

$$0 + mgl(1 - \cos\beta_0) = \tfrac{1}{2}mv^2 + mgl_1(1 - \cos\theta) \qquad v = \{2g[l(1 - \cos\beta_0) - l_1(1 - \cos\theta)]\}^{1/2}$$

7.111 A load W is suspended from a self-propelled crane by a cable of length d [Fig. 7-30(a)]. The crane and load are moving at a constant speed v_0. The crane is stopped by a bumper and the load on the cable swings out, as shown in Fig. 7-30(b). (a) What is the angle through which the load swings? (b) If the angle is 60° and $d = 5$ m, what was the initial speed of the crane?

▌ (a) The cable does no work on the load, so the load's energy is conserved.

$$K_i + U_i = K_f + U_f \qquad \frac{1}{2}\frac{W}{g}v_0^2 + 0 = 0 + W(d - d\cos\theta)$$

$$v_0^2 = 2gd(1 - \cos\theta) = 4gd\sin^2\frac{\theta}{2} \qquad \theta = 2\arcsin\left(\frac{v_0}{2\sqrt{gd}}\right)$$

(b)

$$v_0 = 2\sqrt{gd}\sin\frac{\theta}{2} = 2\sqrt{(9.8)(5)}\left(\frac{1}{2}\right) = \underline{7 \text{ m/s}}$$

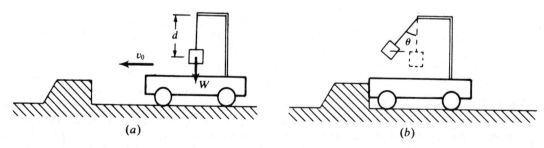

Fig. 7-30

7.112 An x-directed force $F_x = (21 - 3x)$ N displaces an object from $x = 0$ to $x = 7$ m. (a) Find the work done by the force. (b) Repeat for $x = 0$ to $x = 14$ m.

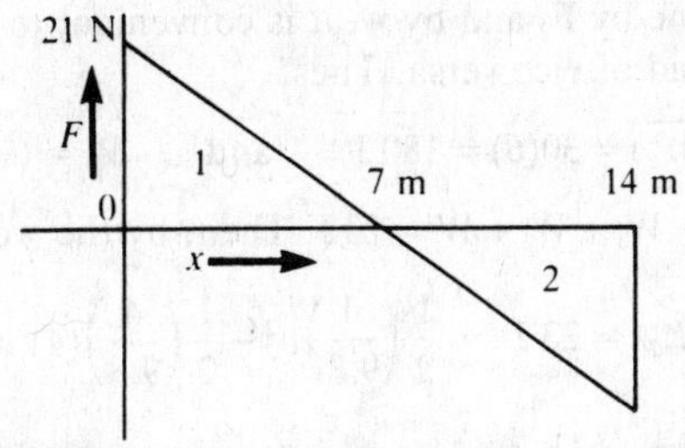

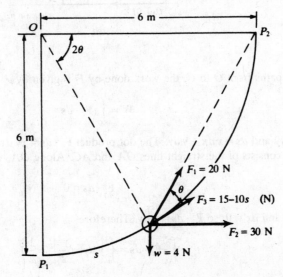

Fig. 7-31

▌ Use calculus or find the area under an F-vs-x curve, Fig. 7-31. **(a)** Work (area 1) = 73.5 J. **(b)** Force changes sign, so work (area 2) is negative. Total area under the curve and the work done are zero.

7.113^c A particle of mass m is subjected to an x-directed force given by $F_x = (3.0 + 0.50x)$ N. Find the work done by the force as the particle moves from $x = 0$ to $x = 4.0$ m.

▌
$$\text{Work} = \int \mathbf{F} \cdot d\mathbf{s} = \int_0^4 (3 + 0.5x)\, dx = [3x + 0.25x^2]_0^4 = \underline{16\ \text{J}}$$

(Or solve graphically, as in Prob. 7.112.)

7.114^c A smooth track in the form of a quarter-circle of radius 6 m lies in the vertical plane (see Fig. 7-32). A particle of weight 4 N moves from P_1 to P_2 under the action of forces $\mathbf{F}_1$, $\mathbf{F}_2$, and $\mathbf{F}_3$. Force $\mathbf{F}_1$ is always toward P_2 and is always 20 N in magnitude; force $\mathbf{F}_2$ always acts horizontally and is always 30 N in magnitude; force $\mathbf{F}_3$ always acts tangentially to the track and is of magnitude $(15 - 10s)$ N when s is in meters. If the particle has speed 4 m/s at P_1, what will its speed be at P_2?

▌ The work done by $\mathbf{F}_1$ is
$$W_1 = \int_{P_1}^{P_2} F_1 \cos\theta\, ds$$

From Fig. 7-32, $ds = (6\ \text{m})\, d(-2\theta) = -12\, d\theta$, and $F_1 = 20$. Hence,
$$W_1 = -240 \int_{\pi/4}^{0} \cos\theta\, d\theta = 240 \sin\frac{\pi}{4} = 120\sqrt{2}\ \text{J}$$

Note that $W_1 = (20\ \text{N})(6\sqrt{2}\ \text{m})$, just as if the chord P_1P_2 and not the circular arc were the path of integration. The reason for this is that $\mathbf{F}_1$ is a *conservative* force. For such a force, the work is the same along all paths joining two given points.

The work done by $\mathbf{F}_3$ is
$$W_3 = \int F_3\, ds = \int_0^{6(\pi/2)} (15 - 10s)\, ds = [15s - 5s^2]_0^{3\pi} = -302.8\ \text{J}$$

Fig. 7-32

To calculate the work done by $\mathbf{F}_2$ and by $\mathbf{w}$, it is convenient to take the projection of the path in the direction of the force, instead of vice versa. Thus,

$$W_2 = F_2(\overline{OP_2}) = 30(6) = 180 \text{ J} \qquad \text{and} \qquad W = (-w)(\overline{P_1O}) = (-4)(6) = -24 \text{ J}$$

The total work done is $W_1 + W_3 + W_2 + W = 23$ J. Then, by the work-energy principle,

$$K_{P2} - K_{P1} = 23 \text{ J} \qquad \frac{1}{2}\left(\frac{4}{9.8}\right)v_2^2 - \frac{1}{2}\left(\frac{4}{9.8}\right)(4)^2 = 23 \qquad v_2 = \underline{11.3 \text{ m/s}}$$

7.115^c In Prob. 7.114, does the particle move *solely* under the action of the four given forces? Explain.

▋ No; additional radial forces must operate to provide the necessary centripetal component. For instance, at P_1 the necessary centripetal force is

$$\frac{mv^2}{r} = \frac{(4/9.8)(4)^2}{6} = 1.09 \text{ N}$$

whereas $F_1 \cos 45° - w = 10.14$ N.

7.116^c Find the work done by a force given in newtons by $F_x = 5.0x - 4.0$ when this force acts on a particle that moves from $x = 1.0$ m to $x = 3.0$ m.

▋
$$W = \int F_x \, dx = \int_{1.0}^{3.0} (5.0x - 4.0) \, dx = [2.5x^2 - 4.0x]_{1.0}^{3.0} = \underline{12 \text{ J}}$$

7.117^c One of the forces acting on a certain particle depends on the particle's position in the xy plane. This force $\mathbf{F}_2$, expressed in newtons, is given by the expression $\mathbf{F}_2 = (xy\hat{\mathbf{x}} + xy\hat{\mathbf{y}})(1 \text{ N/m}^2)$, where x and y are expressed in meters. Calculate the work $\int_O^C \mathbf{F}_2 \cdot d\mathbf{s}$ done by this force when the particle moves from point O to point C in Fig. 7-33 along **(a)** the path OAC, which consists of two straight lines; **(b)** the path OBC, which consists of two straight lines; and **(c)** the straight line OC. **(d)** Is $\mathbf{F}_2$ a conservative force? Explain your answer.

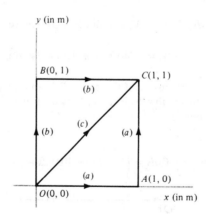

Fig. 7-33

▋ For each of the paths from O to C, the work done by $\mathbf{F}_2$ is given by

$$W = \int_O^C \mathbf{F}_2 \cdot d\mathbf{s}$$

where $\mathbf{F}_2 = xy\hat{\mathbf{x}} + xy\hat{\mathbf{y}}$ and $d\mathbf{s} = \hat{\mathbf{x}} \, dx + \hat{\mathbf{y}} \, dy$. The dot product $\mathbf{F}_2 \cdot d\mathbf{s} = xy \, dx + xy \, dy$.
(a) The path OAC consists of the straight lines OA and AC. Along OA, $y = 0$ and $dy = 0$, so

$$\int_{CA} \mathbf{F}_2 \cdot d\mathbf{s} = 0$$

Along AC, $x = 1$ m and $dx = 0$, so $\mathbf{F}_2 \cdot d\mathbf{s} = y \, dy$. Therefore

$$\int_{AC} \mathbf{F}_2 \cdot d\mathbf{s} = \int_0^1 y \, dy = \frac{y^2}{2}\bigg|_0^1 = \frac{1}{2} \text{ J}$$

The work done along the entire path OAC is given by

$$W_{OAC} = \int_{OA} \mathbf{F}_2 \cdot d\mathbf{s} + \int_{AC} \mathbf{F}_2 \cdot d\mathbf{s} = 0 + \frac{1}{2} J = \frac{1}{2} J$$

(b) The path OBC consists of the straight lines OB and BC. Along OB, $x = 0$ and $dx = 0$, so $\int_{OB} \mathbf{F}_2 \cdot d\mathbf{s} = 0$. Along BC, $y = 1$ m and $dy = 0$, so $\mathbf{F}_2 \cdot d\mathbf{s} = x \, dx$. Therefore

$$\int_{BC} \mathbf{F}_2 \cdot d\mathbf{s} = \int_0^1 x \, dx = \frac{x^2}{2} \Big|_0^1 = \frac{1}{2} J$$

The work done along the entire path OBC is given by

$$W_{OBC} = \int_{OB} \mathbf{F}_2 \cdot d\mathbf{s} + \int_{BC} \mathbf{F}_2 \cdot d\mathbf{s} = 0 + \frac{1}{2} J = \frac{1}{2} J$$

(c) Along the straight line OC, $y = x$, so $dy = dx$ and $\mathbf{F}_2 \cdot d\mathbf{s} = xy \, dx + xy \, dy = 2x^2 \, dx$. Then

$$W_{OC} = \int_{OC} \mathbf{F}_2 \cdot d\mathbf{s} = \int_0^1 2x^2 \, dx = \frac{2x^3}{3} \Big|_0^1 = \frac{2}{3} J$$

(d) The force $\mathbf{F}_2$ is *not* conservative because the work done by $\mathbf{F}_2$ between two points depends upon the particular path between the two points.

7.118^c Express the *gravitational potential* (GPE per unit mass) at a distance $r \geq R_e$ from the earth's center.

▮ By definition, the gravitational potential is the gravitational work done on a 1-kg body as it is taken at constant speed from the given location r to infinity (along an arbitrary path). Choosing a radial path,

$$GP = \int_r^\infty \left[-\frac{GM_e(1)}{s^2} \right] ds = -\frac{GM_e}{r}$$

7.119^c What is the formal relationship between a conservative force and its associated potential energy?

▮ A force $\mathbf{F}$ that can be derived from a function of position U by

$$F_s = -\frac{dU}{ds} \qquad (1)$$

is called a *conservative* force. Here, F_s is the component of the force in the direction of $d\mathbf{s}$, an arbitrary small displacement away from the point of observation. This follows from the fact that the work done by a conservative force is the negative of the change in the associated function U:

$$W_{AB} = \int_A^B F_s \, ds = -(U_B - U_A)$$

or, in differential form, $dU = -F_s \, ds$, leading to (1). The work W_{AB} is independent of the path joining the endpoints A and B; for a closed path, this work is zero. The work done by a *nonconservative* force (such as the force of kinetic friction) depends on the path taken between the endpoints.

7.120^c Calculate the forces $F(y)$ associated with the following one-dimensional potential energies: **(a)** $U = -\omega y$, **(b)** $U = ay^3 - by^2$, **(c)** $U = U_0 \sin \beta y$.

▮ **(a)**
$$F = -\frac{dU}{dy} = \omega$$

(b)
$$F = -\frac{dU}{dy} = -3ay^2 + 2by$$

(c)
$$F = -\frac{dU}{dy} = -\beta U_0 \cos \beta y$$

7.121^c The potential energy of a certain particle is given by $U = 20x^2 + 35z^3$. Find the vector force exerted on it.

▮ $F_x = -\partial U / \partial x = -40x$, $F_y = -\partial U / \partial y = 0$, and $F_z = -\partial U / \partial z = -105z^2$; therefore, $F = -40x\mathbf{i} - 105z^2\mathbf{k}$.

7.122ᶜ A particle in a certain conservative force field has a potential energy given by $U = 20xy/z$. Find the vector force exerted on it.

▊ $F_x = -20y/z$, $F_y = -20x/z$, and $F_z = 20xy/z^2$ after taking partial derivatives. $F = -(20y/z)\mathbf{i} - (20x/z)\mathbf{j} + (20xy/z^2)\mathbf{k}$.

7.123ᶜ A force $\mathbf{F} = x^2y^2\mathbf{i} + x^2y^2\mathbf{j}$ (N) acts on a particle which moves in the XY plane. (a) Determine if $\mathbf{F}$ is conservative and (b) find the work done by $\mathbf{F}$ as it moves the particle from A to C (Fig. 7-34) along each of the paths ABC, ADC, and AC.

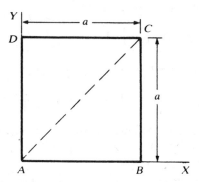

Fig. 7-34

▊ (a) If $\mathbf{F}$ is conservative, then

$$F_x = -\frac{\partial U}{\partial x} \qquad F_y = -\frac{\partial U}{\partial y} \qquad \text{and so} \qquad \frac{\partial F_x}{\partial y} = -\frac{\partial^2 U}{\partial y\,\partial x} = -\frac{\partial^2 U}{\partial x\,\partial y} = \frac{\partial F_y}{\partial x}$$

But, for the given force,

$$\frac{\partial F_x}{\partial y} = 2x^2y \qquad \frac{\partial F_y}{\partial x} = 2xy^2$$

Hence the given force is not conservative.
(b) The work done by $\mathbf{F}$ is given by

$$W = \int \mathbf{F} \cdot d\mathbf{s} = \int (x^2y^2\mathbf{i} + x^2y^2\mathbf{j}) \cdot (dx\,\mathbf{i} + dy\,\mathbf{j}) = \int x^2y^2\,dx + \int x^2y^2\,dy$$

Along AB, $y = 0$ and so $W_{AB} = 0$. Along BC, $dx = 0$ and

$$W_{BC} = \int_0^a a^2y^2\,dy = \frac{a^5}{3}$$

Thus

$$W_{ABC} = W_{AB} + W_{BC} = \frac{a^5}{3} \text{ (J)}$$

Along AD, $x = 0$ and so $W_{AD} = 0$. Along DC, $dy = 0$ and

$$W_{DC} = \int_0^a x^2a^2\,dx = \frac{a^5}{3}$$

Thus

$$W_{ADC} = W_{AD} + W_{DC} = \frac{a^5}{3} \text{ (J)}$$

Along AC, $x = y$ and $dx = dy$. Thus,

$$W_{AC} = 2\int_0^a x^4\,dx = \frac{2a^5}{5} \text{ (J)}$$

7.124ᶜ A block of mass m sits at rest on a frictionless table in a rail car that is moving with speed v_c along a straight horizontal track (Fig. 7-35). A person riding in the car pushes on the block with a net horizontal force F for a time t in the direction of the car's motion. (a) What is the final speed of the block according to a person in the car (i.e., the speed relative to the tabletop)? (b) According to a person standing on the ground outside the train (i.e., relative to the ground)? (c) How much did K of the block change according to the person in the car? (d) According to the person on the ground? (e) In terms of F, m, and t, how far did the force

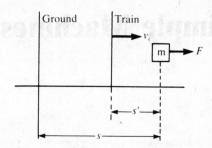

Fig. 7-35

displace the object according to the person in the car? (**f**) According to the person on the ground? (**g**) How much work does each say the force did? (**h**) Compare the work done to the K gain according to each person. (**i**) What can you conclude from this computation?

❚ Let primed coordinates refer to the train system.
(**a**) $a' = F/m$, so $v' = a't = Ft/m$. (**b**) Since velocities add, $v = v_c + v' = v_c + Ft/m$. (**c**) $\Delta K' = m(v')^2/2 = F^2t^2/2m$. (**d**) $\Delta K = m(v_c + v')^2/2 - mv_c^2/2 = F^2t^2/2m + v_cFt$. (**e**) s' is $a't^2/2 = Ft^2/2m$. (**f**) $s = s' + v_ct = Ft^2/2m + v_ct$. (**g**) $W' = \int F \, ds' = \int Fv' \, dt = \int F^2t \, dt/m = F^2t^2/2m$, while $W = \int F \, ds = \int Fv \, dt$; then from (**b**) above, $W = F^2t^2/2m + Fv_ct$. (**h**) Compare W and W' with ΔK and $\Delta K'$; they are respectively equal. (**i**) The work-energy theorem holds for moving observers. [Note that part (**g**) follows directly from $W' = Fs'$, $W = Fs$, and parts (**e**) and (**f**).]

CHAPTER 8
Power and Simple Machines

8.1 POWER

8.1 What is power?

▌ Power is the time rate of doing work:

$$P = \frac{dW}{dt} = \mathbf{F} \cdot \mathbf{v} = Fv \cos \theta$$

where $\mathbf{F}$ and $\mathbf{v}$ are the instantaneous force and velocity, respectively, and θ is the angle between $\mathbf{F}$ and $\mathbf{v}$. If the power does not vary with time, $P = W/t$.

   The units for power are joules per second. This combination of units is termed the *watt*. Therefore, $1\,\text{W} = 1\,\text{J/s} = 1\,\text{kg} \cdot \text{m}^2/\text{s}^3$. Another unit is horsepower; $1\,\text{hp} = 550\,\text{ft} \cdot \text{lb/s} = 746\,\text{W}$.

8.2 Calculate the average horsepower required to raise a 150-kg drum to a height of 20 m in a time of 1 min.

▌ $\bar{P} = W/t =$ average power, where W is work done in the time t. $W = (150\,\text{kg})(9.8\,\text{m/s}^2)(20\,\text{m}) = 29\,400\,\text{J}$ and $t = 60\,\text{s}$, so $\bar{P} = 490\,\text{W}$. Noting that $1\,\text{hp} = 746\,\text{W}$, we have $\bar{P} = (490\,\text{W})/(746\,\text{W/hp}) = \underline{0.657\,\text{hp}}$.

8.3 A certain tractor is said to be capable of pulling with a steady force of 14 kN while moving at a speed of 3.0 m/s. How much power in kilowatts and in horsepower is the tractor developing under these conditions?

▌ Power is force times velocity in the direction of the force.

$$P = (14\,\text{kN})(3.0\,\text{m/s}) = \underline{42.0\,\text{kW}} = (42.0\,\text{kW})\left(\frac{1\,\text{hp}}{0.746\,\text{kW}}\right) = \underline{56.3\,\text{hp}}$$

8.4 The 4 metric ton (4000-kg) hammer of a pile driver is lifted 1.0 m in 2.0 s. What power does the engine furnish to the hammer? Assume that there is no acceleration of the hammer while it is being lifted.

▌
$$P = \frac{W}{t} = \frac{Fs}{t} = \frac{mgs}{t} = \frac{(4.0 \times 10^3)(9.8)(1.0)}{2.0} = \underline{19.6\,\text{kW}}$$

8.5 While a boat is being towed at a speed of 20 m/s, the tension in the towline is 6 kN. What is the power supplied to the boat by the towline?

▌
$$P = Fv = (6 \times 10^3)(20) = \underline{120\,\text{kW}}.$$

8.6 At 8¢ per kilowatt hour, what is the cost of operating a 5.0-hp motor for 2.0 h?

▌
$$\text{cost} = (0.08)(5.0)(746 \times 10^{-3})(2.0) = \underline{\$0.60}.$$

8.7 Compute the power output of a machine that lifts a 500-kg crate through a height of 20 m in a time of 60 s.

▌
$$P = \frac{mgh}{t} = \frac{(500\,\text{kg})(9.8\,\text{m/s}^2)(20\,\text{m})}{60\,\text{s}} = \underline{1.63\,\text{kW}}$$

8.8 An engine expends 40 hp in propelling a car along a level track at 15 m/s. How large is the total retarding force acting on the car?

▌ The engine does positive work to just overcome the negative work of the retarding force f. Thus the forward thrust due to the engine, F, is equal in magnitude to f. From $P = Fv$ we have, converting to consistent units, $(40\,\text{hp})(746\,\text{W/hp}) = F(15\,\text{m/s})$, or $F = f = \underline{2.0\,\text{kN}}$.

8.9 A 1000-kg auto travels up a 3 percent grade at 20 m/s. Find the horsepower required, neglecting friction.

▌ Here the power must equal the time rate of increase of gravitational potential energy, since the kinetic energy remains the same. Then $P = mg(\Delta h/\Delta t)$. But $\Delta h/\Delta t = v \sin \theta$, where v is the velocity along the incline and θ is the angle of the incline. Thus $P = (1000 \text{ kg})(9.8 \text{ m/s}^2)(20 \text{ m/s})(0.03) = 5.88 \text{ kW}$. Noting that $0.746 \text{ kW} = 1 \text{ hp}$, we have $P = \underline{7.88 \text{ hp}}$.

8.10 A 30 000-kg airplane takes off at a speed of 50 m/s, and 5 min later it is at an elevation of 3 km and has a speed of 100 m/s. What average power is required during this 5 min if 40 percent of the power is used in overcoming dissipative forces?

▮ Energy supplied to plane in 300 s (5 min) is

$$(\text{PE} + \text{KE})_{\text{at altitude}} - \text{KE}_{\text{ground}} = (30\,000 \times 9.8 \times 3000) + \tfrac{1}{2}(30\,000)(100)^2 - \tfrac{1}{2}(30\,000)(50)^2$$

$$= 882 \times 10^6 + 150 \times 10^6 - 37.5 \times 10^6 = 994.5 \text{ MJ} \qquad 0.6\,P = \frac{994.5 \text{ MJ}}{300 \text{ s}} \qquad P = \underline{5.525 \text{ MW}}$$

8.11 An advertisement claims that a certain 1200-kg car can accelerate from rest to a speed of 25 m/s in a time of 8.0 s. What average power must the motor produce in order to cause this acceleration? Ignore friction losses.

▮ The work done in accelerating the car is given by

$$\text{work done} = \text{increase in KE} = \tfrac{1}{2}mv_f^2$$

The time taken for this work is 8 s. Therefore

$$\text{power} = \frac{\text{work}}{\text{time}} = \frac{\tfrac{1}{2}(1200 \text{ kg})(25 \text{ m/s})^2}{8 \text{ s}} = 46.9 \text{ kW} = (46.9 \text{ kW})\left(\frac{1 \text{ hp}}{0.746 \text{ kW}}\right) = \underline{63 \text{ hp}}$$

8.12 A motor having an efficiency of 90 percent operates a crane having an efficiency of 40 percent. With what constant speed does the crane lift an 880-lb bale if the power supplied to the motor is 5 kW?

▮ $\qquad$ overall efficiency $= (40\%)(90\%) = 36\%$ $\qquad$ useful power $= (0.36)(5 \text{ kW}) = 1.8 \text{ kW}$

$$\text{We have velocity} = \frac{\text{useful power}}{\text{force (weight)}} = \frac{(1.8 \text{ kW})(1 \text{ hp}/0.746 \text{ kW})(550 \text{ ft} \cdot \text{lb} \cdot \text{s}^{-1}/1 \text{ hp})}{880 \text{ lb}} = \underline{1.51 \text{ ft/s}}$$

8.13 In unloading grain from the hold of a ship, an elevator lifts the grain through a distance of 12 m. Grain is discharged at the top of the elevator at a rate of 2.0 kg each second and the discharge speed of each grain particle is 3.0 m/s. Find the minimum-horsepower motor that can elevate grain in this way.

▮ The work done by the motor each second is

$$\text{power} = mgh + \tfrac{1}{2}mv^2$$

where m is the mass of grain discharged (and lifted) each second. We are given that $m = 2.0 \text{ kg}$, $v = 3.0 \text{ m/s}$, and $h = 12 \text{ m}$. Substitution gives

$$\text{power} = 244 \text{ W} = 0.33 \text{ hp}$$

The motor must have an output of at least $\underline{0.33 \text{ hp}}$.

8.14 Water flows from a reservoir to a turbine 330 ft below. The efficiency of the turbine is 80 percent and it takes 100 ft³ of water each minute. Neglect friction in the pipe and the small KE of the exiting water and compute the horsepower output of the turbine. Water weighs 62.4 lb/ft³.

▮ The gravitational potential energy input to the turbine each minute is that of 100 ft³ of water at a height of 330 ft:

$$P_{\text{input}} = (62.4 \text{ lb/ft}^3)(100 \text{ ft}^3)(330 \text{ ft})/(60 \text{ s}) = (34\,320 \text{ ft} \cdot \text{lb} \cdot \text{s}^{-1})\left(\frac{1 \text{ hp}}{550 \text{ ft} \cdot \text{lb} \cdot \text{s}^{-1}}\right) = 62.4 \text{ hp}$$

Since the turbine efficiency is 80 percent, we have $P_{\text{output}} = 0.8 P_{\text{input}} = \underline{49.9 \text{ hp}}$.

8.15 Find the mass (in kilograms) of the block that a 40-hp engine can pull along a level road at 15 m/s if the friction coefficient between block and road is 0.15.

▮ Since the road is level and the speed is constant, all the energy expended by the engine goes into thermal energy. This, in turn, is just equal to $|W_f|$, where W_f is the work done by friction. $W_f = -\mu mgx$, where $\mu = 0.15$ and x is the horizontal distance moved through. In terms of power, the rate at which work is done, $P_{\text{engine}} = |P_f| = |-\mu mgv|$, or $(40 \text{ hp})(746 \text{ W/hp}) = 0.15(9.8 \text{ m/s}^2)(15 \text{ m/s})m$. Solving, we get $m = \underline{1353 \text{ kg}}$.

8.16 A 2400-lb automobile requires 50 hp to overcome retarding forces on a level road when traveling at a speed of 110 ft/s. Find the net retarding force on the automobile. If the power and retarding force remain the same, what is the speed of the car on a 5 percent grade, i.e., one which rises 5 ft in 100 ft of roadway?

▮ $P = fv$; $50 \times 550 = f(110)$, or $f = 250$ lb. On a 5 percent grade, retarding force is (including gravity) 250 lb + $(0.05 \times 2400) = 370$ lb. Now $(50 \times 550) = 370v$; $v = (74.3 \text{ ft/s})(60 \text{ mi/h})/(88 \text{ ft/s}) = \underline{50.7 \text{ mi/h}}$.

8.17 What power must a man expend on a 100-kg log that he is dragging down a hillside at a speed of 0.50 m/s? The hillside makes an angle of 20° with the horizontal and the coefficient of friction is 0.9.

▮ Here there is no change in kinetic energy and in terms of work-energy, gravity is helping the man: $W_{man} + W_f = \Delta U$, where $W_f = -\mu mgL \cos\theta$, $\Delta U = 0 - mgL \sin\theta$, L is length down along the incline, and θ is the angle of incline. In terms of power we have, using v for speed of the log, $P_{man} = \mu mgv \cos\theta - mgv \sin\theta = (100 \text{ kg})(9.8 \text{ m/s}^2)(0.5 \text{ m/s})(0.9 \cos 20° - \sin 20°)$, or $P_{man} = \underline{247 \text{ W}}$.

8.18ᶜ An engine pumps water continuously through a hose. If the speed with which the water passes through the hose nozzle is v, and if k is the mass per unit length of the water jet as it leaves the nozzle, what is the rate at which kinetic energy is being imparted to the water?

▮ During a small elapsed time, Δt, the mass of water ejected is $k(v \Delta t)$. The kinetic energy of this water is $\frac{1}{2}(kv \Delta t)v^2$. The rate at which kinetic energy is imparted is then

$$\lim_{\Delta t \to 0} \frac{\frac{1}{2}kv^3 \Delta t}{\Delta t} = \frac{1}{2}kv^3$$

8.2 SIMPLE MACHINES

8.19 What is a machine? What energy principle applies to a machine?

▮ A machine is any device by which the magnitude, direction, or method of application of a force is changed in order to achieve some advantage. Examples of simple machines are the lever, inclined plane, pulley, crank and axle, and jackscrew.
The principle of energy that applies to a continuously operating machine is work input = useful work output + friction losses.

8.20 Define the *mechanical advantage* of a machine.

▮ The *actual mechanical advantage* (AMA) of a machine is

$$\text{AMA} = \text{force ratio} = \frac{\text{force exerted by machine on load}}{\text{force used to operate machine}}$$

The *ideal mechanical advantage* (IMA) of a machine is

$$\text{IMA} = \text{distance ratio} = \frac{\text{distance moved by input force}}{\text{distance moved by load}}$$

Because friction is always present, the AMA is always less than the IMA.

8.21 What is the *efficiency* of a machine?

▮ $$\text{efficiency of a machine} = \frac{\text{work output}}{\text{work input}} = \frac{\text{power output}}{\text{power input}} = \frac{(\text{force} \times \text{distance})_{load}}{(\text{force} \times \text{distance})_{input}}$$

The efficiency is thus also equal to the ratio AMA/IMA (see Prob. 8.20).

8.22 A hoist lifts a 3000-kg load a height of 8 m in a time of 20 s. The power supplied to the engine is 18 hp. Compute (**a**) the work output, (**b**) the power output and the power input, (**c**) the efficiency of the engine and hoist system.

▮ (**a**) work output = (lifting force) × (height) = $[(3000 \times 9.8) \text{ N}](8 \text{ m}) = 235\,000 \text{ J} = \underline{235 \text{ kJ}}$.

(**b**) $$\text{power output} = \frac{\text{work output}}{\text{time taken}} = \frac{235 \text{ kJ}}{20 \text{ s}} = \underline{11.8 \text{ kW}}$$

$$\text{power input} = (18 \text{ hp})\left(\frac{0.746 \text{ kW}}{1 \text{ hp}}\right) = \underline{13.4 \text{ kW}}$$

(c) $\text{efficiency} = \dfrac{\text{power output}}{\text{power input}} = \dfrac{11.8\text{ kW}}{13.4\text{ kW}} = 0.88 = \underline{88\%}$ or

$\text{efficiency} = \dfrac{\text{work output}}{\text{work input}} = \dfrac{235\text{ kJ}}{(13.4\text{ kJ/s})(20\text{ s})} = 0.88 = \underline{88\%}$

8.23   What power in kilowatts is supplied to a 12-hp motor, having an efficiency of 90 percent, when it is delivering its full rated output?

❚   From the definition of efficiency (Prob. 8.21),

$$\text{power input} = \dfrac{\text{power output}}{\text{efficiency}} = \dfrac{(12\text{ hp})(0.746\text{ kW/hp})}{0.90} = \underline{9.95\text{ kW}}$$

8.24   A pulley system with an IMA of 4 requires a force of 15 N to lift a load of 50 N. Find the efficiency of the machine.

❚ $\text{AMA} = \dfrac{F_{\text{out}}}{F_{\text{in}}} = \dfrac{50}{15} = 3.33$ $\text{eff} = \dfrac{\text{AMA}}{\text{IMA}} = \dfrac{3.33}{4} = 0.83 = \underline{83\%}$

8.25   A man uses a pulley system to raise a 150-lb load to a height of 10 ft. If he exerts a force on the rope of 50 lb through a distance of 35 ft to accomplish the work, (a) how much work does he do? (b) what is the efficiency of the machine?

❚   (a) For the work done by the man,

$$W = FS \qquad 50(35) = \underline{1750\text{ ft}\cdot\text{lb}}$$

(b) $W_{\text{out}} = F_{\text{out}}d_{\text{out}} = 150(10) = 1500\text{ ft}\cdot\text{lb}$ $\text{efficiency} = \dfrac{W_{\text{out}}}{W_{\text{in}}} = \dfrac{1500}{1750} = 0.86 = \underline{86\%}$

8.26   In Prob. 8.25, if it takes the man 11 s to raise the load, what is his average power?

❚ $\text{average power} = \dfrac{W}{t} = \dfrac{1750}{11} = 159\text{ ft}\cdot\text{lb/s} = \dfrac{159}{550} = \underline{0.289\text{ hp}}$

8.27   A wooden wedge is pushed horizontally to raise a heavy object. All surfaces are frictionless. Apply the law of conservation of energy to determine the mechanical advantage in terms of the wedge angle θ.

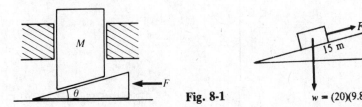

Fig. 8-1 $w = (20)(9.8)\text{ N}$ Fig. 8-2

❚   As indicated in Fig. 8-1, we assume that the object to be raised is held between fixed vertical guiding surfaces. If the wedge moves leftward a distance Δs under the application of the force **F**, the object is elevated by a distance $\Delta h = (\Delta s)\tan\theta$. We have $F\Delta s = Mg\,\Delta h$ from energy conservation. Therefore the mechanical advantage $Mg/F = \Delta s/\Delta h = \underline{\cot\theta}$.

8.28   The inclined plane shown in Fig. 8-2 is 15 m long and rises 3 m. (a) What force F parallel to the plane is required to slide a 20-kg box up the plane if friction is neglected? (b) What is the IMA of the plane? (c) Find the AMA and efficiency if a 64-N force is actually required.

❚   (a) Since there is no friction, the work done by the pushing force, $(F)(15\text{ m})$, must equal the increase of GPE, $(20\text{ kg})(9.8\text{ m/s}^2)(3\text{ m})$. Equating these two expressions and solving for F gives $F = \underline{39\text{ N}}$.

(b) $\text{IMA} = \dfrac{\text{distance moved by }F}{\text{distance }w\text{ is lifted}} = \dfrac{15\text{ m}}{3\text{ m}} = \underline{5.0}$

(c) $\text{AMA} = \text{force ratio} = \dfrac{w}{F} = \dfrac{196\text{ N}}{64\text{ N}} = \underline{3.1}$ $\text{efficiency} = \dfrac{\text{AMA}}{\text{IMA}} = \dfrac{3.1}{5.0} = 0.62 = \underline{62\%}$

Or as a check, $\text{efficiency} = \dfrac{\text{work output}}{\text{work input}} = \dfrac{(w)(3\text{ m})}{(F)(15\text{ m})} = 0.62 = \underline{62\%}$

8.29   Using the wheel and axle shown in Fig. 8-3, a 400-N load can be raised by a force of 50 N applied to the rim of the wheel. The radii of the wheel and axle are 85 cm and 6 cm, respectively. Determine the IMA, AMA, and efficiency of the machine.

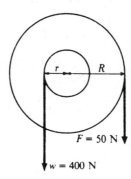

$F = 50$ N

$w = 400$ N   **Fig. 8-3**

▌ In one turn of the wheel-axle system, a length of cord equal to the circumference of the wheel or axle will be wound or unwound. Therefore

$$\text{IMA} = \frac{\text{distance moved by } F}{\text{distance moved by } w} = \frac{2\pi R}{2\pi r} = \frac{85\text{ cm}}{6\text{ cm}} = \underline{14.2} \qquad \text{AMA} = \text{force ratio} = \frac{400\text{ N}}{50\text{ N}} = \underline{8}$$

$$\text{efficiency} = \frac{\text{AMA}}{\text{IMA}} = \frac{8}{14.2} = 0.56 = \underline{56\%}$$

8.30   A windlass mechanism is shown in Fig. 8-4. (*a*) If friction is neglected, what is its mechanical advantage if the radius of the lower gear is N times the radius of the upper gear? (*b*) Evaluate your results for $N = 3.0$, $R = 40$ cm, and $r = 5.0$ cm.

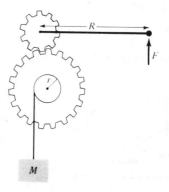

M **Fig. 8-4**

▌ (*a*) When the applied force F acts through a distance $\Delta s = 2\pi R$, the upper gear turns through 2π rad. Therefore the lower gear turns through $2\pi/N$ rad and the object of mass M is elevated by a distance $\Delta h = 2\pi r/N$. Since $Mg\,\Delta h = F\,\Delta s$, the mechanical advantage $Mg/F = \Delta s/\Delta h = 2\pi R \div (2\pi r/N) = \underline{NR/r}$. (*b*) With $N = 3.0$, $R = 40$ cm, and $r = 5.0$ cm, we find $NR/r = (3.0)(40)/(5.0) = \underline{24}$.

8.31   A differential pulley (chain hoist) is shown in Fig. 8-5. Two toothed pulleys of radii $r = 10$ cm and $R = 11$ cm are fastened together and turn on the same axle. A continuous chain passes over the smaller (10-cm) pulley, then around the movable pulley at the bottom, and finally around the 11-cm pulley. The operator exerts a downward force F on the chain to lift the load w. (*a*) Determine the IMA. (*b*) What is the efficiency of the machine if an applied force of 50 lb is required to lift a load of 700 lb?

▌ (*a*) Suppose that the force F moves down a distance sufficient to cause the upper rigid system of pulleys to turn one revolution. Then the smaller upper pulley unwinds a length of chain equal to its circumference, $2\pi r$, while the larger upper pulley winds a length $2\pi R$. As a result, the chain supporting the lower pulley is shortened by a length $2\pi R - 2\pi r$. The load w is lifted half this distance, $\frac{1}{2}(2\pi R - 2\pi r) = \pi(R - r)$ when the

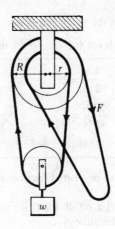

Fig. 8-5

input force moves a distance $2\pi R$. Therefore

$$\text{IMA} = \frac{\text{distance moved by } F}{\text{distance moved by } w} = \frac{2\pi R}{\pi(R-r)} = \frac{2R}{R-r} = \frac{22\text{ cm}}{1\text{ cm}} = \underline{22}$$

(b) From the data,

$$\text{AMA} = \frac{\text{load lifted}}{\text{input force}} = \frac{700\text{ N}}{50\text{ N}} = \underline{14}$$

Then

$$\text{efficiency} = \frac{\text{AMA}}{\text{IMA}} = \frac{14}{22} = 0.64 = \underline{64\%}$$

8.32 A motor drives a hoist which lifts a 3-ton load a distance of 40 ft in 50 s. The efficiency of the motor is 85 percent and that of the hoist is 45 percent. What power is supplied to the load? To the hoist? To the motor?

▌ Useful work $= 6000 \times 40 = 240\,000$ ft · lb   Power output $= \dfrac{240\,000\text{ ft} \cdot \text{lb}}{50\text{ s}} = 4800$ ft · lb/s $4800 = 0.45 P_{\text{hoist}}$

$P_{\text{hoist}} = 10\,667$ ft · lb/s $= 0.85 P_{\text{motor}}$ $P_{\text{motor}} = 12\,550$ ft · lb/s $= \underline{22.8\text{ hp}}$

8.33 For the three levers shown in Fig. 8-6, determine the vertical forces F_1, F_2, F_3 required to support the load $w = 90$ N. Neglect the weights of the levers. Also find the IMA, AMA, and efficiency for each system.

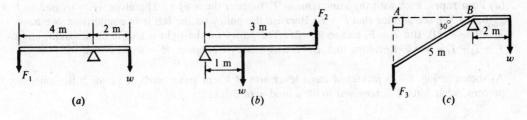

(a)   (b)   (c)

Fig. 8-6

▌ In each case, take torques about the fulcrum point as axis. If we assume that the lifting is occurring slowly at constant speed, then the systems are in equilibrium. The clockwise torques balance the counterclockwise torques.

clockwise torques = counterclockwise torques

(a) $(90\text{ N})(2\text{ m}) = (F_1)(4\text{ m})$ $F_1 = \underline{45\text{ N}}$
(b) $(90\text{ N})(1\text{ m}) = (F_2)(3\text{ m})$ $F_2 = \underline{30\text{ N}}$

(c) $(90\text{ N})(2\text{ m}) = (F_3)(5\text{ m})\dfrac{\sqrt{3}}{2}$ $F_3 = \underline{41.6\text{ N}}$

The IMA, AMA, and efficiency of the three levers are

	lever (a)	lever (b)	lever (c)
IMA	$\dfrac{4\,m}{2\,m} = 2$	$\dfrac{3\,m}{1\,m} = 3$	$\dfrac{4.33\,m}{2\,m} = 2.16$
AMA	$\dfrac{90\,N}{45\,N} = 2$	$\dfrac{90\,N}{30\,N} = 3$	$\dfrac{90\,N}{41.6\,N} = 2.16$
Eff.	1.00	1.00	1.00

The efficiencies are 1 because we have neglected friction at the fulcrums.

8.34 Determine the force F required to lift a 100-lb load, w, with each of the pulley systems shown in Fig. 8-7. Neglect friction and the weights of the pulleys.

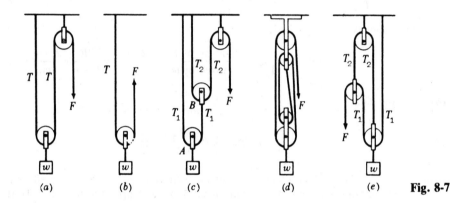

(a)   (b)   (c)   (d)   (e)   **Fig. 8-7**

▌ (a) Load w is supported by two ropes; each rope exerts an upward pull of $T = \frac{1}{2}w$. But, because the rope is continuous and the pulleys are frictionless, $T = F$. Then $F = T = \frac{1}{2}w = \frac{1}{2}(100 \text{ lb}) = \underline{50 \text{ lb}}$.
(b) Here, too, the load is supported by the tensions in two ropes, T and F, where $T = F$. Then $T + F = w$, or $F = \frac{1}{2}w = \underline{50 \text{ lb}}$. (c) Let T_1 and T_2 be the tensions around pulleys A and B, respectively. Pulley A is in equilibrium and so $T_1 + T_1 - w = 0$, or $T_1 = \frac{1}{2}w$.
Consider next pulley B. It, too, is in equilibrium and so $T_2 + T_2 - T_1 = 0$, or $T_2 = \frac{1}{2}T_1 = \frac{1}{4}w$. But $F = T_2$, and so $F = \frac{1}{4}w = \underline{25 \text{ lb}}$.
(d) Four ropes, each with the same tension T, support the load w. Therefore $4T_1 = w$, and so $F = T_1 = \frac{1}{4}w = \underline{25 \text{ lb}}$. (e) We see at once that $F = T_1$. Because the pulley on the left is in equilibrium, we have $T_2 - T_1 - F = 0$. But $T_1 = F$, and so $T_2 = 2F$. The pulley on the right is also in equilibrium, and so $T_1 + T_2 + T_1 - w = 0$. Recalling that $T_1 = F$ and that $T_2 = 2F$ gives $4F = w$, so $F = \underline{25 \text{ lb}}$.

8.35 As shown in Fig. 8-8, a jackscrew has a lever arm of 40 cm and a pitch of 5 mm. If the efficiency is 30 percent, what force F is required to lift a load w of 270 kg?

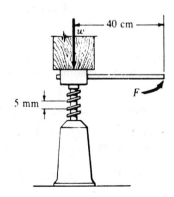

Fig. 8-8

❚ When the jack handle is moved around one complete circle, the input force moves a distance $2\pi r = 2\pi(0.40 \text{ m})$, while the load is lifted a distance of 0.005 m. The IMA is therefore

$$\text{IMA} = \text{distance ratio} = \frac{2\pi(0.40 \text{ m})}{0.005 \text{ m}} = \underline{500}.$$

Since efficiency = AMA/IMA, we have AMA = (efficiency)(IMA) = (0.30)(500) = $\underline{150}$. But AMA = (load lifted)/(input force), and so

$$F = \frac{\text{load lifted}}{\text{AMA}} = \frac{(270)(9.8) \text{ N}}{150} = \underline{17.6 \text{ N}}$$

8.36   A water bucket weighing 400 N is raised by a crank-and-axle arrangement. The axle has a radius of 75 mm, and the crank has a radius of 225 mm. If a force of 160 N is required on the crank, what is the actual mechanical advantage, the ideal mechanical advantage, and the efficiency?

❚

$$\text{AMA} = \frac{F_{\text{out}}}{F_{\text{in}}} = \frac{400}{160} = \underline{2.5} \qquad \text{IMA} = \frac{d_{\text{in}}}{d_{\text{out}}} = \frac{225 \times 2\pi \text{ mm/rev}}{75 \times 2\pi \text{ mm/rev}} = 3$$

$$\text{efficiency} = \frac{\text{AMA}}{\text{IMA}} = \frac{F_{\text{out}}d_{\text{out}}}{F_{\text{in}}d_{\text{in}}} = \frac{2.5}{3} = \underline{0.833}$$

8.37   An inclined plane (see Fig. 8-9) 12.5 m long is used as a simple machine to pull 4-kN boxes up to a height of 3.5 m. A force of 1.5 kN up to the plane is required to pull the boxes up. If it takes 50 s to bring a box to the top, what is the actual mechanical advantage of the inclined plane, the ideal mechanical advantage, the efficiency, and the power required?

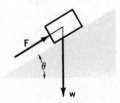

Fig. 8-9

❚

$$\text{AMA} = \frac{4000}{1500} = \underline{2.67} \qquad \text{IMA} = \frac{12.5}{3.5} = \underline{3.57} \qquad \text{efficiency} = \frac{\text{AMA}}{\text{IMA}} = \frac{8}{3} \times \frac{7}{25} = \frac{56}{75} = \underline{0.75}$$

$$P = \frac{W}{t} = \frac{1500 \times 12.5}{50} = \underline{375 \text{ W}}$$

8.38   An inclined plane 13 ft long (see Fig. 8-9) is used to slide a 390-lb box up to a loading platform 5 ft above ground level. A force of 200 lb is required. Find the efficiency of the inclined plane for this job. What force is required to overcome friction? What is the coefficient of friction between the plane and the box?

❚

$$\text{efficiency} = \frac{\text{work}_{\text{out}}}{\text{work}_{\text{in}}} = \frac{390 \times 5}{200 \times 13} = \underline{0.75}$$

$\frac{5}{13} \times 390 = 150$ lb is the component of weight down the plane. $200 = 150 + f$, so $f = 50$ lb for friction.

$$\mu = \frac{f}{N} = \frac{50}{390 \times 12/13} = \frac{5}{36} = \underline{0.139}$$

8.39   The upper end of an inclined ramp (Fig. 8-9) 7.5 m long is 2.1 m above the lower end. A force of 460 N is required to push a 900-N box up the incline. Find the coefficient of friction and the work done against friction. If the box is released at the top of the ramp, will it slide down? If so, what will its speed at the bottom be?

❚   Component of weight down plane = $900 \times 2.1/7.5 = 252$ N.

$$f = 460 - 252 = 208 \text{ N} \qquad \mu = \frac{208}{900 \times 7.2/7.5} = \underline{0.241}$$

Work against friction $= 208 \times 7.5 = \underline{1560\,J}$. Yes, because $252 > 208$.

$$44 = \frac{900}{9.8}a \qquad a = 0.479\,\text{m/s}^2 \qquad v^2 = v_0^2 + 2as \qquad v_0 = 0$$

so

$$v^2 = 2 \times 0.479 \times 7.5 = 7.187 \qquad v = \underline{2.68\,\text{m/s}}$$

8.40 The jackscrew in Fig. 8-10 has a pitch of 3 mm and a handle 600 mm long. A force of 60 N must be applied when a load of 25 kN is being lifted. Calculate the ideal mechanical advantage, the actual mechanical advantage, and the efficiency.

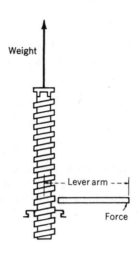

Weight

Lever arm

Force

Fig. 8-10

$$\text{IMA} = \frac{\text{distance force acts}}{\text{distance load moves}}$$

For one complete revolution,

$$\frac{d_{\text{in}}}{d_{\text{out}}} = \text{IMA} = \frac{2\pi \times 600}{3} = \underline{1257} \qquad \text{AMA} = \frac{25\,000}{60} = \underline{416.67} \qquad \text{efficiency} = \frac{\text{AMA}}{\text{IMA}} = \frac{416}{1257} = \underline{0.33}$$

8.41 A force of 1.2 kN is required on the bar of a capstan to obtain an output force of 10 kN for lifting an anchor. The axle of the capstan has a radius of 55 mm, and the input force is applied 0.88 m from the axis. Find the work done by the input force in lifting the anchor 15 m in a time of 2 min, the actual mechanical advantage, the ideal mechanical advantage, the efficiency of the capstan, and the power provided by the input force.

For one revolution, $d_{\text{out}} = 2\pi \times 0.055$ m and $d_{\text{in}} = 2\pi \times 0.88$ m. Thus for $d_{\text{out}} = 15$ m, we have $d_{\text{in}} = 240$ m.

The work done by the input force is

$$W = 1.2\,\text{kN} \times 240\,\text{m} = \underline{288\,\text{kJ}} \qquad \text{AMA} = \frac{10}{1.2} = \underline{8.33} \qquad \text{IMA} = \frac{0.88}{0.055} = \underline{16} \qquad \text{efficiency} = \frac{8.33}{16} = \underline{0.52}$$

$$\text{power} = \frac{W}{t} = \frac{288\,\text{kJ}}{120\,\text{s}} = \underline{2.4\,\text{kW}}$$

8.42 In the simple "machine" (for manipulating the gravitational force) shown in Fig. 8-11 a small car loops the loop. It is found experimentally that $h \geq 0.65R$ in order that the car can negotiate the loop. What is the efficiency of this machine when operating at minimal h?

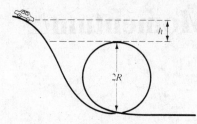

Fig. 8-11

I If v_t and N_t denote the speed and normal force at the top of the loop,

$$mg + N_t = \frac{mv_t^2}{R} = \frac{2}{R}\left(\frac{1}{2}mv_t^2\right)$$

But by energy conservation between the top of the track and the top of the loop, $mgh = \frac{1}{2}mv_t^2 + \Delta$, where $\Delta > 0$ is the frictional loss. Hence,

$$mg + N_t = \frac{2}{R}(mgh - \Delta) \tag{1}$$

At the critical value $h = 0.65R$, $N_t = 0$ and (1) gives $\Delta = 0.15mgR$. The input work for one cycle of the machine is the work required to return the car from the bottom of the track to the top, $mg(2R + 0.65R) = 2.65mgR$. Therefore,

$$\text{efficiency} = \frac{(\text{input work}) - (\text{losses})}{(\text{input work})} = \frac{2.65 - 0.15}{2.65} = \underline{0.94}$$

CHAPTER 9
Impulse and Momentum

9.1 IMPULSE–MOMENTUM

9.1 A mass m undergoes free fall. What is its linear momentum after it has fallen a distance h?

▋ The momentum changes from zero to mv due to the gravitational force acting for the time t that the mass falls a distance h. From $mv^2/2 = mgh$, solve for v and then $p = m(2gh)^{1/2}$.

9.2 Show that the linear momentum p and kinetic energy K of a mass m are related through $K = p^2/2m$.

▋ Since $p = mv$, multiply the numerator and denominator of $K = mv^2/2$ by m to get $K = (mv)^2/2m = p^2/2m$.

9.3 An earth satellite of mass M circles the earth with speed v. By how much does its momentum change (**a**) as it goes halfway around the earth? (**b**) as it goes all the way around? Ignore the earth's rotation.

▋ (**a**)
$$|Mv_f - Mv_i| = Mv - M(-v) = 2Mv$$

(**b**)
$$|Mv_i - Mv_i| = 0$$

9.4 While waiting in his car at a stoplight, an 80-kg man and his car are suddenly accelerated to a speed of 5 m/s as the result of a rear-end collision. Assuming the time taken to be 0.3 s, find (**a**) the impulse on the man and (**b**) the average force exerted on him by the back of the seat of his car.

▋ (**a**) Impulse $= \Delta(mv) = 80(5) = \underline{400\text{ N} \cdot \text{s}}$. (**b**) $\bar{F} = \Delta(mv)/\Delta t = 400/0.3 = \underline{1330\text{ N}}$.

9.5 A 3.0-kg block slides on a frictionless horizontal surface, first moving to the left at 50 m/s. It collides with a spring as it moves left, compresses the spring, and is brought to rest momentarily. The body continues to be accelerated to the right by the force of the compressed spring. Finally, the body moves to the right at 40 m/s. The block remains in contact with the spring for 0.020 s. What were the magnitude and direction of the impulse of the spring on the block? What was the spring's average force on the block?

▋ Impulse = change in momentum, which for our case is $(3.0\text{ kg})(40\text{ m/s}) - (3.0\text{ kg})(-50\text{ m/s}) = 270\text{ kg} \cdot \text{m/s}$ (i.e., to the right). Since the only horizontal force on the block is the spring force, the spring impulse is thus $\underline{270\text{ N} \cdot \text{s}}$ (to the right). Noting impulse = force × time and that $t = 0.020\text{ s}$, we have average force = $\underline{13.5\text{ kN to the right}}$.

9.6 A camper lets fall a heavy mallet of mass M from a height y upon the top of a tent stake of mass m and drives it into the ground a distance d. Find the resistance of the ground, assuming it to be constant and the stake and mallet to stay together on impact. See Fig. 9-1.

▋ The speed of the mallet on just striking the stake is $v = \sqrt{2gy}$. Momentum is conserved at the instant of collision, so that $Mv = (M + m)v'$, where v' is the speed of the stake plus mallet just after impact.

The resultant (upward) force on the stake plus mallet is $\sum F = f - (M + m)g$, where f is the resistive force of the ground. Then the work–energy principle gives

$$\Delta K = \left(\sum F\right)(-d) \qquad 0 - \frac{1}{2}(M + m)v'^2 = [f - (M + m)g](-d) \qquad f = (M + m)g + (M + m)\frac{v'^2}{2d}$$

Substituting $v' = Mv/(M + m)$ and $v^2 = 2gy$,

$$f = (M + m)g + (M + m)\frac{M^2v^2}{2(M + m)^2 d} = (M + m)g + \frac{M^2}{M + m}\frac{gy}{d}$$

9.7 Refer to Prob. 9.6. Find the time the stake is in motion.

▋ $\sum F = \Delta p/\Delta t$, where Δt is the time interval from just after (or just before, since momentum is conserved) the impact to cessation of motion of the stake plus mallet. Then,

$$\Delta t = \frac{\Delta p}{\sum F} = \frac{0 - [-(M + m)v']}{f - (M + m)g} = \frac{Mv}{\left(\dfrac{M^2}{M + m}\right)\dfrac{gy}{d}} = \frac{M + m}{M}d\sqrt{\frac{2}{gy}}$$

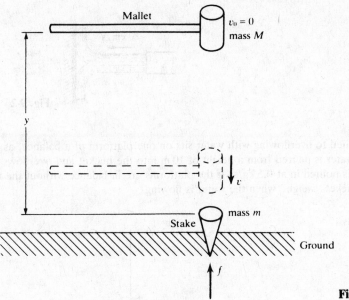

Fig. 9-1

9.8 Refer again to Prob. 9.6. Find the kinetic energy lost at impact.

▮ Just before impact the kinetic energy of the system was $\frac{1}{2}Mv^2$; just afterward it was

$$\frac{1}{2}(M+m)v'^2 = \frac{1}{2}\frac{M^2}{M+m}v^2$$

So the amount lost by the mallet was

$$\frac{1}{2}Mv^2 - \frac{1}{2}\frac{M^2}{M+m}v^2 = \frac{m}{M+m}\left(\frac{1}{2}Mv^2\right)$$

or the fraction lost was $m/(M+m)$.

9.9 A golfer hits a golf ball of mass 51 g and the ball leaves the club with a velocity of 80 m/s. Assuming that the ball and club are in contact for 0.006 s, find the ball's final momentum and the average force exerted by the club on the ball.

▮ Since impulse is equal to the change in momentum,

$$Ft = mv - mv_0 \qquad F(0.006) = 0.051(80) - 0 \qquad F = \frac{0.051(80)}{0.006} = \underline{680\,N}\text{ average force}$$

The final momentum is

$$mv = 0.051(80) = \underline{4.08\,kg \cdot m/s}$$

9.10 A 5.0-g bullet moving at 100 m/s strikes a log. Assume that the bullet undergoes uniform deceleration and stops in 6.0 cm. Find (a) the time taken for the bullet to stop, (b) the impulse on the log, and (c) the average force experienced by the log.

▮ (a) $t = s/\bar{v} = (0.06\,\text{m})/(50\,\text{m/s}) = \underline{1.2 \times 10^{-3}\,\text{s}}$. (b) Impulse $= \Delta(mv) = 100(0.005) = \underline{0.5\,N \cdot s}$.
(c) $\bar{F} = \Delta(mv)/t = 0.5/(1.2 \times 10^{-3}) = \underline{417\,N}$.

9.11 What force is exerted on a stationary flat plate held perpendicular to a jet of water as shown in Fig. 9-2? The horizontal speed of the water is 80 cm/s and 30 cm³ of the water hits the plate each second. Assume that the water moves parallel to the plate after striking it. One cubic centimeter of water has a mass of one gram.

▮ The plate exerts an impulse on the water and changes its horizontal momentum.

$$(\text{impulse})_x = \text{change in } x\text{-directed momentum} \qquad F_x t = (mv_x)_{\text{final}} - (mv_x)_{\text{initial}}$$

Let us take t to be 1 s so that m will be the mass that strikes in 1 s, namely, 30 g. Then the above equation becomes $F_x(1\,\text{s}) = (0.030\,\text{kg})(0\,\text{m/s}) - (0.030\,\text{kg})(0.80\,\text{m/s})$, from which $F_x = -0.024\,\text{N}$. This is the force of the plate on the water; by Newton's third law, the force on the plate is $\underline{+0.024\,\text{N}}$.

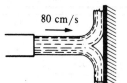

Fig. 9-2

9.12 A bucket filled to overflowing with water sits on one platform of a balance, as shown in Fig. 9-3. A constant stream of water is poured from a height of 10 m into the bucket and overflows on one side of the balance. The water is poured in at 0.5 kg/s. If the platforms are in balance without the flow of water, how much more does the bucket "weigh" when the water is flowing?

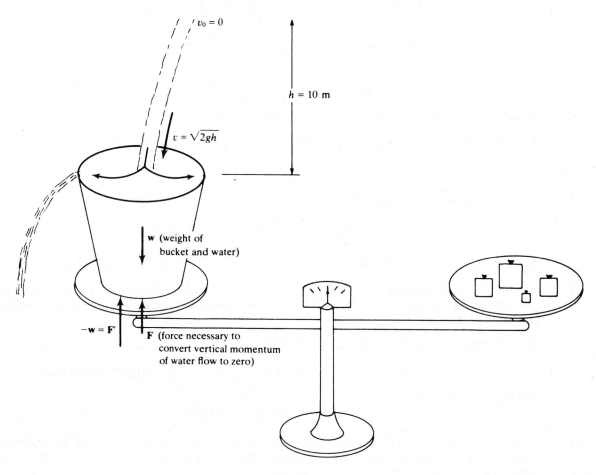

Fig. 9-3

▌ The balance supports the weight of the filled bucket and supplies the impulse to stop the downward flow of water. In time Δt the vertical momentum of a mass (0.5 kg/s) Δt of water, falling at speed $v = \sqrt{2gh}$, is changed to zero; thus,

$$I = F\,\Delta t = (0.5\,\Delta t)v \qquad \text{or} \qquad F = 0.5v = 0.5\sqrt{2(9.8)(10)} = \underline{7\,\text{N}}$$

The bucket is apparently 7 N heavier.

9.13ᶜ A uniform rope, of mass m per unit length, hangs vertically from a support so that the lower end just touches the tabletop shown in Fig. 9-4(a). If it is released, show that at the time a length y of the rope has fallen, the force on the table is equivalent to the weight of a length $3y$ of the rope.

▌ The descending part of the rope is in free fall; it has speed $v = \sqrt{2gy}$ at the instant all its points have

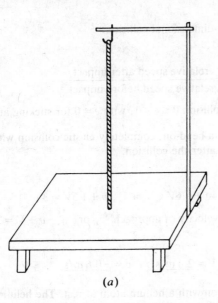
(a)

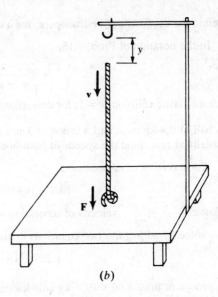
(b)

Fig. 9-4

descended a distance y. The length of the rope which lands on the table during an interval dt following this instant is $v\,dt$. The increment of momentum imparted to the table by this length in coming to rest is $m(v\,dt)v$. Thus, the rate at which momentum is transferred to the table is

$$\frac{dp}{dt} = mv^2 = (2my)g$$

and this is the force arising from stopping the downward fall of the rope. Since a length of rope y, of weight $(my)g$, already lies on the tabletop, the total force on the tabletop is $(2my)g + (my)g = (3my)g$, or the weight of a length $3y$ of rope.

9.14 An astronaut is doing maintenance work outside a space station. He is coasting along the station at a speed of 1.00 m/s. He wishes to change his direction of motion by 90° and to increase his speed to 2.00 m/s. His total mass is 100 kg, including his spacesuit and rocket belt, which provides a thrust of 50 N. (**a**) Find the magnitude and direction of the impulse needed to accomplish the desired change in motion. (**b**) What is the shortest time in which the astronaut can complete the change in motion? How must the rocket be pointed?

▌ (**a**) We let $\hat{x}$ be along the initial direction of motion and $\hat{y}$ be along the desired final direction. The initial momentum $m\mathbf{v}_i = (100\text{ kg})(1.00\text{ m/s})\hat{x} = (100\text{ kg}\cdot\text{m/s})\hat{x}$. The desired final momentum $m\mathbf{v}_f = (100\text{ kg})(2.00\text{ m/s})\hat{y} = (200\text{ kg}\cdot\text{m/s})\hat{y}$. The required impulse $I = m\mathbf{v}_f - m\mathbf{v}_i = (-100\text{ N}\cdot\text{s})\hat{x} + (200\text{ N}\cdot\text{s})\hat{y}$. The magnitude $I = 100\sqrt{5} = \underline{224\text{ N}\cdot\text{s}}$; the direction is at an angle of arccos $(-100/100\sqrt{5}) = \underline{116.6°}$ with respect to $\mathbf{v}_i$. (**b**) Since $\mathbf{I} = \sum \mathbf{F}\,\Delta t$ the shortest possible time occurs for $\mathbf{F}\parallel\mathbf{I}$ and is given by $T = I/F = (100\sqrt{5}\text{ kg}\cdot\text{m/s})/(50\text{ N}) = \underline{4.47\text{ s}}$. In order to accomplish the desired change in this minimum firing time, the rocket's exhaust must be pointed opposite to $\mathbf{I}$, or at an angle of $\underline{-63.4°}$ with respect to $\mathbf{v}_i$.

9.15 Suppose that the astronaut of Prob. 9.14 makes the change by decelerating to rest, turning the rocket exhaust by 90°, and then accelerating up to the desired final speed. How long would this take? How much rocket fuel is used, compared to the minimum?

▌ The deceleration to rest requires a time $t_1 = (mv_i/F)$; the subsequent acceleration to velocity $\mathbf{v}_f$ requires a time $t_2 = (mv_f/F)$. The total required time $T' = t_1 + t_2 = m(v_i + v_f)/F$. Using the given numerical values, $T' = (100)(1.00 + 2.00)/(50) = \underline{6.00\text{ s}}$. Since the firing time is $(6.00 - 4.47)/4.47 = 34$ percent longer than the minimum, the fuel consumption is $\underline{34\text{ percent more}}$ than the minimum.

9.2 ELASTIC COLLISIONS

9.16 Prove that relative velocity is reversed by a head-on elastic collision.

▌ If u_1 and u_2 are the initial velocities, and v_1 and v_2 are the final velocities of objects 1 and 2, then momentum conservation gives $m_1u_1 + m_2u_2 = m_1v_1 + m_2v_2$, or $m_1(u_1 - v_1) = m_2(v_2 - u_2)$. Energy conservation gives $\frac{1}{2}m_1u_1^2 + \frac{1}{2}m_2u_2^2 = \frac{1}{2}m_1v_1^2 + \frac{1}{2}m_2v_2^2$, or $m_1(u_1^2 - v_1^2) = m_2(v_2^2 - u_2^2)$, or $m_1(u_1 - v_1)(u_1 + v_1) = m_2(v_2 - u_2)(v_2 + u_2)$. By division of equations, $u_1 + v_1 = u_2 + v_2$, or $u_2 - u_1 = -(v_2 - v_1)$, the desired result.

9.17 Define the *coefficient of restitution, e,* for a one-dimensional collision.

▌ In the notation of Prob. 9.16,

$$e \equiv \frac{v_2 - v_1}{u_1 - u_2} = \frac{\text{relative speed after impact}}{\text{relative speed before impact}}$$

For an elastic collision, $e = 1$; for an inelastic collision $0 \le e < 1$, with $e = 0$ for sticking after collision.

9.18 A ball of 0.4-kg mass and a speed of 3 m/s has a head-on, completely elastic collision with a 0.6-kg mass initially at rest. Find the speeds of both bodies after the collision.

▌ Conservation of momentum:

$$(0.4 \times 3) + 0 = 0.4v + 0.6V \quad \text{or} \quad v + 1.5V = 3$$

Elastic: velocity of separation = −velocity of approach or $-v + V = 3$

We solve by adding the two equations to yield

$$2.5V = 6 \qquad V = \underline{2.4 \text{ m/s}} \qquad v = \underline{-0.6 \text{ m/s}}$$

9.19 A proton of mass 1.66×10^{-27} kg collides head-on with a helium atom at rest. The helium atom has a mass of 6.64×10^{-27} kg and recoils with a speed of 5×10^5 m/s. If the collision is elastic, what are the initial and final speeds of the proton and the fraction of its initial energy transferred to the helium atom?

▌ Mass of He = 4 (mass of proton). From conservation of momentum,

$$m_p v_i + 0 = m_p v_f + (4m_p)V_f \quad \text{and} \quad v_i = v_f + 4V_f$$

From elastic condition,

velocity of approach = −velocity of separation or $v_i = -v_f + V_f$

Adding the equations, we get

$$2v_i = 5V_f = 5(5 \times 10^5 \text{ m/s}) \quad \text{or} \quad v_i = \underline{1.25 \times 10^6 \text{ m/s}}$$

From momentum equation,

$$-v_f = 4V_f - v_i \quad \text{so} \quad v_f = 1.25 \times 10^6 - 2.00 \times 10^6 = \underline{-7.5 \times 10^5 \text{ m/s}}$$

$$\text{energy fraction} = \frac{\text{KE of He}}{\text{initial KE}} = \frac{\frac{1}{2}(4m_p)(5 \times 10^5)^2}{\frac{1}{2}m_p(1.25 \times 10^6)^2} = \underline{0.64}$$

9.20 A neutron of mass 1.67×10^{-27} kg collides head-on with a deuterium nucleus at rest. The deuterium has a mass of 3.34×10^{-27} kg. If the speed of the neutron is $v_1 = 8.8$ km/s before the collision and the collision is elastic, (a) what are speeds V_2 and v_2 of the deuterium nucleus and the neutron after the collision, and (b) what is the total kinetic energy?

▌ (a) Let the positive direction be that of the incident neutron. By conservation of momentum,

$$m_1 v_1 = m_1 v_2 + m_2 V_2 \quad \text{or} \quad 8.8 = v_2 + 2V_2$$

By reversal of relative velocity,

$$8.8 = V_2 - v_2$$

Adding the two equations, $17.6 = 3V_2$, or $V_2 = 5.87$ km/s; then, $v_2 = V_2 - 8.8 = -2.93$ km/s (in the negative direction). (b) KE $= \frac{1}{2}(m_1 v_1^2) = \frac{1}{2}(1.67 \times 10^{-27})(8800)^2 = \underline{6.47 \times 10^{-20} \text{ J}}$.

9.21 A fast neutron has an elastic head-on collision with a proton (hydrogen atom) at rest. Assuming both have the same mass, use the conservation of energy and the conservation of momentum laws to show algebraically what happens.

▌ With uppercase pertaining to the proton, apply the law of conservation of momentum:

$$mv_1 + 0 = mv_2 + mV_2 \qquad v_1 = v_2 + V_2 \qquad v_1^2 = v_2^2 + 2v_2V_2 + V_2^2$$

Next, apply the law of conservation of kinetic energy:

$$\tfrac{1}{2}mv_1^2 = \tfrac{1}{2}mv_2^2 + \tfrac{1}{2}mV_2^2 \qquad v_1^2 = v_2^2 + V_2^2 \qquad v_2^2 + V_2^2 = v_2^2 + 2v_2V_2 + V_2^2 \qquad 0 = 2v_2V_2$$

V_2 cannot be zero since the H atom (proton) must move when hit. Therefore, $v_2 = 0$. The neutron stands still after the collision, and the H atom recoils with a velocity $V_2 = v_1$; that is, the original velocities are exchanged.

9.22 Two perfectly elastic balls, one weighing 2 lb and the other 3 lb, are moving in opposite directions with speeds of 8 and 6 ft/s, respectively. Find their velocities after head-on impact.

▋ Conservation of momentum:

$$\left(\frac{2}{32} \times 8\right) - \left(\frac{3}{32}\right)6 = \frac{2}{32}v_f + \frac{3}{32}V_f \qquad \text{or} \qquad 2v_f + 3V_f = -2$$

Elasticity:

$$8 - (-6) = V_f - v_f \qquad \text{(velocity of approach = -velocity of separation)} \qquad \text{or} \qquad -v_f + V_f = 14$$

Solving the two equations yields

$$v_f = \underline{-8.8 \text{ ft/s}} \qquad \text{and} \qquad V_f = \underline{5.2 \text{ ft/s}}$$

9.23 Two identical balls collide head-on. The initial velocity of one is 0.75 m/s, while that of the other is -0.43 m/s. If the collision is perfectly elastic, what is the final velocity of each ball?

▋ Since the collision is head-on, all motion takes place along a straight line. Let us call the mass of each ball m. Momentum is conserved in a collision and so we can write

momentum before = momentum after $\qquad m(0.75 \text{ m/s}) + m(-0.43 \text{ m/s}) = mv_1 + mv_2$

where v_1 and v_2 are the final velocities. This equation simplifies to give

$$0.32 \text{ m/s} = v_1 + v_2 \qquad (1)$$

Because the collision is assumed to be perfectly elastic, KE is also conserved.

KE before = KE after $\qquad \frac{1}{2}m(0.75 \text{ m/s})^2 + \frac{1}{2}m(0.43 \text{ m/s})^2 = \frac{1}{2}mv_1^2 + \frac{1}{2}mv_2^2$

This equation can be simplified to give

$$0.747 = v_1^2 + v_2^2 \qquad (2)$$

We can solve for v_2 in (1) to give $v_2 = 0.32 - v_1$ and substitute this in (2). This yields

$$0.747 = (0.32 - v_1)^2 + v_1^2 \qquad \text{from which} \qquad 2v_1^2 - 0.64v_1 - 0.645 = 0$$

Using the quadratic formula we find that

$$v_1 = \frac{0.64 \pm \sqrt{(0.64)^2 + 5.16}}{4} = 0.16 \pm 0.59$$

from which $v_1 = 0.75$ m/s or -0.43 m/s. Substitution back into (1) gives $v_2 = -0.43$ m/s or 0.75 m/s. Two choices for answers are available:

$$(v_1 = 0.75 \text{ m/s}, v_2 = -0.43 \text{ m/s}) \qquad (v_1 = -0.43 \text{ m/s}, v_2 = 0.75 \text{ m/s})$$

We must discard the first choice because it implies that the balls continue on unchanged; that is to say, no collision occurred. The correct answer is therefore $v_1 = \underline{-0.43 \text{ m/s}}$ and $v_2 = \underline{0.75 \text{ m/s}}$, which tells us that in a perfectly elastic head-on collision between equal masses, the two bodies simply exchange velocities.

9.24 Solve Prob. 9.23 without the use of quadratics.

▋ By Prob. 9.17,

$$e = 1 = \frac{v_2 - v_1}{0.75 - (-0.43)} \qquad \text{which gives} \qquad v_2 - v_1 = 1.18 \text{ m/s} \qquad (3)$$

Equation (1) of Prob. 9.23 and (3) determine v_1 and v_2 uniquely: $v_2 = 0.75$ m/s, $v_1 = -0.43$ m/s.

9.25 In Fig. 9-5, the smaller ball has a mass of 0.3 kg and the larger one a mass of 0.5 kg. If the smaller one is pulled back and released so that it has a speed of 4 m/s just before collision, what are the velocities of both balls immediately after they have an elastic collision?

Fig. 9-5

❚ From momentum conservation,

$$(0.3 \times 4) + 0 = 0.3v_f + 0.5V_f \qquad \text{or} \qquad V_f + 0.6v_f = 2.4$$

From elastic condition   (velocity of separation = −velocity of approach) $V_f - v_f = 4$

Solving the pair of equations yields

$$v_f = -\underline{1\,\text{m/s}} \qquad \text{and} \qquad V_f = \underline{3\,\text{m/s}}$$

9.26   Two balls like those in Fig. 9-5 are pulled back and released together so that they collide elastically at the equilibrium position. The smaller ball has a mass of 0.15 kg and is moving to the right with a speed of 4 m/s, while the larger ball has a mass of 0.25 kg and is moving to the left at 3 m/s. Find the velocity of each after the head-on collision.

❚ From momentum conservation,

$$(0.15 \times 4) + (0.25 \times -3) = 0.15v_f + 0.25V_f \qquad \text{or} \qquad 5V_f + 3v_f = -3$$

From elasticity,   (velocity of separation = −velocity of approach) $V_f - v_f = 7$

Multiplying the second equation by 3 and adding to first equation yields

$$8V_f = 18 \qquad V_f = \underline{2.25\,\text{m/s}}$$

and from either equation $v_f = \underline{-4.75\,\text{m/s}}$.

9.27   Sphere A of mass m and velocity $\mathbf{v}_i$ has an oblique, perfectly elastic collision with an identical sphere B initially at rest. Show that the angle θ between the velocities of the spheres after collision is 90°.

❚ Vector conservation of momentum reads

$$m\mathbf{v}_i = m\mathbf{v}_f + m\mathbf{V}_f \qquad \text{or} \qquad \mathbf{v}_i = \mathbf{v}_f + \mathbf{V}_f$$

Hence

$$\mathbf{v}_i \cdot \mathbf{v}_i = (\mathbf{v}_f + \mathbf{V}_f) \cdot (\mathbf{v}_f + \mathbf{V}_f) = \mathbf{v}_f \cdot \mathbf{v}_f + 2\mathbf{v}_f \cdot \mathbf{V}_f + \mathbf{V}_f \cdot \mathbf{V}_f$$

or

$$v_i^2 = v_f^2 + 2v_f V_f \cos\theta + V_f^2 \tag{1}$$

But conservation of energy gives (after canceling the factor $m/2$)

$$v_i^2 = v_f^2 + V_f^2 \tag{2}$$

It follows that $\cos\theta = 0$, or $\theta = 90°$.

9.28   A body of mass m_1 traveling with velocity $\mathbf{v}_{1i}$ makes a head-on elastic collision with a stationary body of mass m_2. The velocities after collision are $\mathbf{v}_{1f}$ and $\mathbf{v}_{2f}$. Calculate $\mathbf{v}_{1f}$ and $\mathbf{v}_{2f}$ in terms of $\mathbf{v}_{1i}$.

❚ The conservation of momentum implies that $m_1\mathbf{v}_{1f} + m_2\mathbf{v}_{2f} = m_1\mathbf{v}_{1i}$. The reversal of relative velocity implies that $\mathbf{v}_{2f} - \mathbf{v}_{1f} = \mathbf{v}_{1i}$. Solving the two equations, we obtain

$$\mathbf{v}_{1f} = \left(\frac{m_1 - m_2}{m_1 + m_2}\right)\mathbf{v}_{1i} \qquad \text{and} \qquad \mathbf{v}_{2f} = \left(\frac{2m_1}{m_1 + m_2}\right)\mathbf{v}_{1i}$$

9.29   Given the situation of Prob. 9.28, **(a)** Calculate the ratio of the kinetic energy transferred to m_2 to the original kinetic energy. **(b)** For what value of m_2 is all the energy transferred to the stationary body? **(c)** If m_1

were a neutron and m_2 were a carbon atom whose mass is about 12 times the neutron's mass, what fraction of the neutron's energy would be transferred to the carbon atom in the head-on collision?

▌ (a) The initial kinetic energy of body 1 is $K_{1i} = \frac{1}{2}m_1 v_{1i}^2$. The final kinetic energy of body 2 is $K_{2f} = \frac{1}{2}m_2 v_{2f}^2$. The ratio K_{2f}/K_{1i} is given by (see Prob. 9.28)

$$\frac{K_{2f}}{K_{1i}} = \frac{\frac{1}{2}m_2[2m_1/(m_1 + m_2)]^2 v_{1i}^2}{\frac{1}{2}m_1 v_{1i}^2} = \frac{4m_1 m_2}{(m_1 + m_2)^2} = 1 - \left(\frac{m_1 - m_2}{m_1 + m_2}\right)^2$$

(b) By inspection of part (a), K_{2f}/K_{1i} assumes its maximum value (unity) when $m_2 = m_1$.
(c) For $m_2 = 12m_1$, the ratio $K_{2f}/K_{1i} = [(4m_1)(12m_1)]/(13m_1)^2 = \underline{0.284}$.

9.3 INELASTIC COLLISIONS AND BALLISTIC PENDULUMS

9.30   A 1-lb ball moving to the right at 12 ft/s collides head-on with a 2-lb ball moving at 12 ft/s in the opposite direction. If the two balls stick together after the collision, what is (a) their total momentum after the collision; (b) their final velocity, including the direction?

▌ (a) With *right* as positive,

$$\text{momentum after} = \text{momentum before} = (\tfrac{1}{32})(12) + (\tfrac{2}{32})(-12) = -0.375 \text{ slug} \cdot \text{ft/s}$$

(b) $(\tfrac{1}{32} + \tfrac{2}{32})V = -0.375$ or $V = -4.0 \text{ ft/s}$

9.31   An 8-g bullet is fired horizontally into a 9-kg block of wood and sticks in it. The block, which is free to move, has a velocity of 40 cm/s after impact. Find the initial velocity of the bullet.

▌ Consider the system (block + bullet). The velocity, and hence the momentum, of the block before impact is zero. The momentum conservation law tells us that

$$\text{momentum of system before impact} = \text{momentum of system after impact}$$

$$(\text{mass}) \times (\text{velocity of bullet}) + 0 = (\text{mass}) \times (\text{velocity of block} + \text{bullet})$$

$$(0.008 \text{ kg})v + 0 = (9.008 \text{ kg})(0.40 \text{ m/s})$$

where v is the velocity of the bullet. Solving gives $v = \underline{450 \text{ m/s}}$.

9.32   A 16-g mass is moving in the $+x$ direction at 30 cm/s while a 4-g mass is moving in the $-x$ direction at 50 cm/s. They collide head-on and stick together. Find their velocity after collision.

▌ Apply the law of conservation of momentum to the system consisting of the two masses.

$$\text{momentum before impact} = \text{momentum after impact}$$

$$(0.016 \text{ kg})(0.30 \text{ m/s}) + (0.004 \text{ kg})(-0.50 \text{ m/s}) = (0.020 \text{ kg})v$$

Note that the 4-g mass has negative momentum. Solving gives $v = \underline{0.14 \text{ m/s}}$.

9.33   A 2-kg rock is moving at a speed of 6 m/s. What constant force is needed to stop the rock in $7 \times 10^{-4} \text{s}$?

▌ impulse on rock = change in momentum of rock $Ft = mv_f - mv_0$ $F(7 \times 10^{-4}\text{ s}) = 0 - (2 \text{ kg})(6 \text{ m/s})$

from which $F = \underline{-1.71 \times 10^4 \text{ N}}$. The minus sign indicates that the force opposes the motion.

9.34   A 20-g bullet moving horizontally at 50 m/s strikes a 7-kg block resting on a table. The bullet embeds in the block after collision. Find (a) the speed of the block after collision and (b) the frictional force between the table and block if the block moves 1.5 m before stopping.

▌ (a) Apply momentum before = momentum after: $(0.020)(50) = 7.02v$, which gives $v = \underline{0.142 \text{ m/s}}$.
(b) Equate K right after the collision to work against friction: $[(7.02)(0.142)^2]/2 = f(1.5)$, which leads to $f = \underline{0.047 \text{ N}}$.

9.35   Suppose that a horizontal force of 0.70 N is required to pull a 5-kg block across a table at constant speed. What is the speed of a 20-g bullet, if the bullet embeds in the block and causes the block to slide 1.50 m before coming to rest?

▌ Conservation of momentum leads to $0.020v = 5.02V$. The velocity after collision is found from

$5.02V^2/2 = f(1.5)$, where $f = 0.70$ N (since $\sum F = 0$ for a constant-velocity object). Find V; then determine from the first expression that $v = \underline{162\ \text{m/s}}$.

9.36 A 2500-lb car moving at 55 mi/h in the positive direction collides head-on with a 3500-lb car moving at 25 mi/h in the negative direction. The two cars stick together after the collision. Find the momentum (magnitude and direction) of the wreckage immediately after the collision. Also, find the velocity of the cars immediately after collision.

▮ A velocity of 55 mi/h = 80.7 ft/s; −25 mi/h = −36.7 ft/s.

$$m_1 v_1 + m_2 v_2 = (m_1 + m_2)V \qquad (m_1 + m_2)V = \frac{2500}{32}(80.7) + \frac{3500}{32}(-36.7) = 6305 - 4014 = \underline{2291\ \text{slug} \cdot \text{ft/s}}$$

Hence
$$V = \frac{2291}{(2500/32) + (3500/32)} = \underline{12.2\ \text{ft/s}}$$

9.37 A railroad flatcar of mass M is coasting along a track at speed v when a large machine of mass m topples off a platform and falls straight down onto the car. How fast is the car moving after the machine comes to rest on it?

▮ The momentum conservation relation along the track becomes $Mv = (M + m)v_f$. Solving we obtain $v_f = Mv/(M + m)$.

9.38 A 1.0-kg steel ball 4.0 m above the floor is released, falls, strikes the floor, and rises to a maximum height of 2.5 m. Find the momentum transferred from the ball to the floor in the collision.

▮ Choose downward as positive. Let u and v be the speeds of the ball just before and after the collision. Then from conservation of mechanical energy from the time it is dropped to just before impact we have $\frac{1}{2}mu^2 = mgh_1$, or $u^2 = 2gh_1 = 78.4$ m^2/s^2, or $u = 8.85$ m/s. Repeating from the moment of rebound to the highest rise point we have $\frac{1}{2}mv^2 = mgh_2$, or $v^2 = 2gh_2 = 49$ m^2/s^2, or $v = 7.0$ m/s. Then, if P is the momentum imparted to the floor (i.e., to the earth), $mu = -mv + P$, or $P = m(u + v) = \underline{15.85\ \text{kg} \cdot \text{m/s}}$.

9.39 A golf ball is dropped on a sidewalk from a height of 2 m and rebounds to a height of 1.50 m. What was the impulse? Assuming that the ball was in contact with the concrete for 7 ms, find the average acceleration. What was the coefficient of restitution? The mass of the ball is 45.8 g.

▮ Choose *downward* as the positive direction. Let h_0 = original height, h_1 = rebound height. v_0 = velocity just before collision, v_1 = velocity just after collision. Then

$$v_0^2 = 2gh_0 = 2 \times 9.8 \times 2 = 39.2\ \text{m}^2/\text{s}^2 \qquad \text{and} \qquad v_0 = 6.26\ \text{m/s}$$
$$v_1^2 = 2gh_1 = 2 \times 9.8 \times 1.5 = 29.4\ \text{m}^2/\text{s}^2 \qquad \text{and} \qquad v_1 = -5.42\ \text{m/s}$$
$$\Delta v = v_1 - v_0 = -11.68\ \text{m/s} \qquad \text{impulse} = \Delta(mv) = (0.0458)(-11.68) = \underline{-0.535\ \text{N} \cdot \text{s}}$$
$$a_{\text{avg}} = \frac{\Delta v}{\Delta t} = -\frac{11.68}{0.007} = \underline{-1669\ \text{m/s}^2}$$

Coefficient of restitution is (see Prob. 9.17)

$$e = \left| \frac{\text{velocity of separation}}{\text{velocity of approach}} \right| = \left| \frac{v_1}{v_0} \right| = \frac{5.42}{6.26} = \underline{0.866}$$

[Note from above that we have obtained the formula $e = \sqrt{h_1/h_0}$ governing problems of this type.]

9.40 A mass A of 0.8 kg moving to the right with a speed of 5 m/s collides head-on with a mass B of 1.2 kg moving in the opposite direction with a speed of 4 m/s. After the collision A is moving to the left with a speed of 4 m/s. Find the velocity of B after collision and the coefficient of restitution.

▮ Momentum is conserved, so

$$(0.8 \times 5) + (1.2 \times -4) = (0.8 \times -4) + 1.2V_f \qquad \text{and} \qquad V_f = \underline{2\ \text{m/s}}$$

Coefficient of restitution is

$$e = \left| \frac{\text{velocity of separation}}{\text{velocity of approach}} \right| = \left| \frac{2 - (-4)}{5 - (-4)} \right| = \frac{6}{9} = \underline{0.667}$$

9.41 A hailstone strikes the roof of a parked car. The hailstone hits with a speed of 10 m/s and rebounds to a height of 0.20 m. What fraction of its initial kinetic energy is lost in the impact?

▮ Ignoring air resistance during the rebound of the hailstone, the kinetic energy $K_r = \frac{1}{2}mv_r^2$ at the beginning of the rebound is equal to mgh_r, where h_r is the rebound height. The kinetic energy at impact is $K_i = \frac{1}{2}mv_i^2$. The fraction that survives the impact is given by

$$\frac{K_r}{K_i} = \frac{mgh_r}{\frac{1}{2}mv_i^2} = \frac{2gh_r}{v_i^2} = \frac{2(9.8)(0.20)}{(10)^2} = 0.039$$

Therefore the fraction lost, $1 - (K_r/K_i)$, is 0.961.

9.42 The hailstone of Prob. 9.41 had radius $r = 5.0$ mm and density $\rho = 0.900$ g/cm³. Assume that during the collision the hailstone is decelerated to rest in the time it would take to travel its own diameter. What average force must the roof exert?

▮ The initial momentum has magnitude $mv_i = (4\pi\rho r^3/3)v_i$; we are told to assume a deceleration time $\tau = 2r/v_i$. The average decelerating force **F** must have magnitude

$$\frac{mv_i}{\tau} = \frac{mv_i^2}{2r} = \frac{2\pi\rho r^2 v_i^2}{3} = \frac{2(3.141)(900)(5.0 \times 10^{-3})^2(10)^2}{3} = 4.71 \text{ N}$$

9.43 With reference to Probs. 9.41 and 9.42, what pressure, in kilopascals, does the hailstone experience in the collision?

▮ It seems reasonable to assume that the force calculated in Prob. 9.42 is borne by the cross section of the hailstone.

$$\text{pressure} = \frac{F}{\pi r^2} = \frac{4.71 \text{ N}}{\pi(5.0 \times 10^{-3} \text{ m})^2} = 60.0 \text{ kN/m}^2 \equiv 60.0 \text{ kPa}$$

9.44 A tennis ball bounces down a flight of stairs, striking each step in turn and rebounding to the height of the step above. If the height of each step is d, what is the coefficient of restitution?

▮ By Prob. 9.39,

$$e = \sqrt{\frac{h_1}{h_0}} = \sqrt{\frac{d}{2d}} = \frac{1}{\sqrt{2}}$$

9.45 A ball strikes the floor vertically and rebounds with a coefficient of restitution of 0.8. If the mass of the ball is 0.011 slug and its velocity is 5 ft/s just before it hits, what is the momentum of the ball as it leaves the floor?

▮

$$e = \frac{v_2}{v_1} \qquad \text{where } e = \text{coefficient of restitution}$$

$$0.8 = \frac{v_2}{5} \qquad v_2 = 4 \text{ ft/s}$$

The final momentum is mv_2

$$mv_2 = 0.011(4) = 0.044 \text{ slug} \cdot \text{ft/s} \quad \text{(upward)}$$

9.46 A ball is dropped from a height h above a tile floor and rebounds to a height of $0.64h$. Find the coefficient of restitution between ball and floor.

▮ By Prob. 9.39, $e = \sqrt{0.64h/h} = 0.8$.

9.47 A ball is bouncing down a flight of stairs. The coefficient of restitution is e. The height of each step is d, and the ball descends one step at each bounce. After each bounce it rebounds to a height h above the next lower step. The height h is large enough compared with the width of a step that the impacts are effectively head on. Show that $h = d/(1 - e^2)$.

▮ The ball falls a distance h from its highest (rest) position and rebounds a distance $(h - d)$. Thus,

$$e = \sqrt{\frac{h - d}{h}} \qquad \text{or} \qquad e^2 = \frac{h - d}{h} \qquad \text{or} \qquad h = \frac{d}{1 - e^2}$$

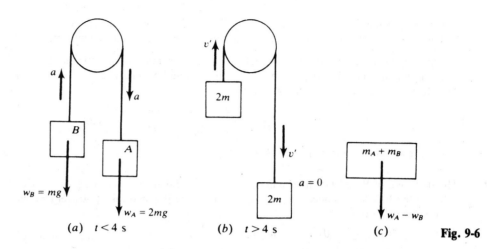

(a) $t < 4$ s   (b) $t > 4$ s   (c)   **Fig. 9-6**

9.48 Two bodies, of masses m and $2m$, are connected by a light inextensible cord passing over a smooth pulley and released, Fig. 9-6. At the end of 4 s a body of mass m is suddenly joined to the ascending body. Find (a) the resulting speed and (b) how much kinetic energy is lost by the descending body when the body of mass m is added.

❙ Because the only effect of the pulley is to change the direction of the tension in the cord, the system may be conveniently analyzed as a single body having mass $m_A + m_B$ acted on by the single force $w_A - w_B$ [Fig. 9-6(c)].

(a) For $t < 4$ s, $m_A = 2m_B = 2m$, and so the equation of motion is

$$mg = 3ma \qquad \text{or} \qquad a = \frac{g}{3}$$

The speed just before $t = 4$ s is then

$$v = 0 + at = \frac{4g}{3} \text{ (m/s)}$$

We assume that the addition of mass at $t = 4$ s is equivalent to a sticky collision between the system and a body of mass m which is *at rest*. Then, by conservation of momentum, the new speed is given by

$$3mv + 0 = 4mv' \qquad \text{or} \qquad v' = \tfrac{3}{4}v = g \text{ (m/s)}$$

(b) The loss in kinetic energy of A is $\frac{1}{2}(2m)v^2 - \frac{1}{2}(2m)v'^2 = \frac{7}{9}mg^2$ (J).

9.49 After falling from rest through a distance y, a body of mass m begins to raise a body of mass M ($M > m$) that is connected to it by means of a light inextensible string passing over a fixed smooth pulley. (a) Find the time it will take for the body of mass M to return to its original position. (b) Find the fraction of kinetic energy lost when the body of mass M is jerked into motion. See Fig. 9-7.

❙ As in Prob. 9.48 only *internal* forces are exerted as the mass of the system is increased, so that momentum is conserved for the instant.

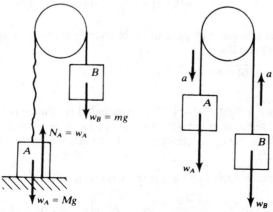

(a) Before the string tautens   (b) After the string tautens   **Fig. 9-7**

(a) The speed of body B just before the string becomes taut is $v = \sqrt{2gy}$, and its momentum, which is the momentum of the system, is mv. Immediately after the string becomes taut, the speed of the system (the common speed of the two bodies) is v'; by momentum conservation,

$$mv = (M + m)v' \quad \text{or} \quad v' = \frac{m}{M + m} v$$

Moreover, the acceleration of the system is given by

$$\sum F = mg - Mg = (M + m)a \quad \text{or} \quad a = -\frac{M - m}{M + m} g$$

where the positive direction is that used above for the momentum.

Applying the constant-acceleration formula $s = v_0 t + \frac{1}{2}at^2$, we find that the system returns to its original position when

$$0 = v't + \tfrac{1}{2}at^2 \quad \text{or} \quad t = -\frac{2v'}{a} = \frac{2m}{M - m} \sqrt{\frac{2y}{g}}$$

(b) The fractional loss of kinetic energy is

$$\frac{\frac{1}{2}mv^2 - \frac{1}{2}(M + m)v'^2}{\frac{1}{2}mv^2} = \frac{M}{M + m}.$$

9.50 A ball is dropped onto a fixed horizontal surface from height h_0. The coefficient of restitution is ϵ. Find the total distance D traveled by the ball before it comes to rest on the surface.

❚ Let h_i ($i = 1, 2, 3, \ldots$) be the height of rebound after the ith impact; then $\epsilon = \sqrt{h_i/h_{i-1}}$, or $h_i = \epsilon^2 h_{i-1}$, which gives $h_n = \epsilon^{2n} h_0$ ($n = 1, 2, 3, \ldots$).

$$D = h_0 + 2 \sum_{n=1}^{\infty} h_n = h_0 \left[1 + 2 \sum_{n=1}^{\infty} \epsilon^{2n} \right]$$

On the right is a geometric series with common ratio ϵ^2 and first term ϵ^2; thus,

$$D = h_0 \left[1 + 2 \frac{\epsilon^2}{1 - \epsilon^2} \right] = h_0 \frac{1 + \epsilon^2}{1 - \epsilon^2}$$

9.51 Find the duration τ of the process of Prob. 9.50.

❚ The initial descent requires time $t_0 = \sqrt{2h_0/g}$. For the descent after the nth impact, we have $t_n = \sqrt{2h_n/g} = \epsilon^n \sqrt{2h_0/g}$; the ascent also takes time t_n. Therefore,

$$\tau = t_0 + 2 \sum_{n=1}^{\infty} t_n = \sqrt{\frac{2h_0}{g}} \left(1 + 2 \sum_{n=1}^{\infty} \epsilon^n \right) = \sqrt{\frac{2h_0}{g}} \left(1 + 2 \frac{\epsilon}{1 - \epsilon} \right) = \sqrt{\frac{2h_0}{g}} \frac{1 + \epsilon}{1 - \epsilon}$$

9.52 What is the average speed of the ball of Probs. 9.50 and 9.51?

❚
$$\bar{v} = \frac{D}{\tau} = h_0 \frac{1 + \epsilon^2}{1 - \epsilon^2} \times \sqrt{\frac{g}{2h_0}} \frac{1 - \epsilon}{1 + \epsilon} = \sqrt{\frac{gh_0}{2}} \frac{1 + \epsilon^2}{(1 + \epsilon)^2} = \frac{v_0}{2} \frac{1 + \epsilon^2}{(1 + \epsilon)^2}$$

where $v_0 = \sqrt{2gh_0}$ is the speed just before the first impact.

9.53 A 1-kg ball moving at 12 m/s collides head-on with a 2-kg ball moving in the opposite direction at 24 m/s. Find the velocity of each after impact if (a) $e = \frac{2}{3}$, (b) the balls stick together, (c) the collision is perfectly elastic.

❚ In all three cases momentum is conserved and so we can write

momentum before = momentum after $(1 \text{ kg})(12 \text{ m/s}) + (2 \text{ kg})(-24 \text{ m/s}) = (1 \text{ kg})v_1 + (2 \text{ kg})v_2$

which becomes $-36 \text{ m/s} = v_1 + 2v_2$

(a) In this case $e = \frac{2}{3}$ and so

$$e = \frac{v_2 - v_1}{u_1 - u_2} \quad \text{becomes} \quad \frac{2}{3} = \frac{v_2 - v_1}{12 - (-24)}$$

from which $24\,\text{m/s} = v_2 - v_1$. Combining this with the momentum equation found above gives $v_2 = \underline{-4\,\text{m/s}}$ and $v_1 = \underline{-28\,\text{m/s}}$. (*b*) In this case $v_1 = v_2 = v$ and so the momentum equation becomes $3v = -36\,\text{m/s}$, or $v = \underline{-12\,\text{m/s}}$. (*c*) Here $e = 1$ and so

$$e = \frac{v_2 - v_1}{u_1 - u_2} \quad \text{becomes} \quad 1 = \frac{v_2 - v_1}{12 - (-24)}$$

from which $v_2 - v_1 = 36\,\text{m/s}$. Adding this to the momentum equation gives $v_2 = 0$. Using this value for v_2 then gives $v_1 = \underline{-36\,\text{m/s}}$.

9.54 As shown in Fig. 9-8, a 15-g bullet is fired horizontally into a 3-kg block of wood suspended by a long cord. The bullet sticks in the block. Compute the velocity of the bullet if the impact causes the block to swing 10 cm above its initial level.

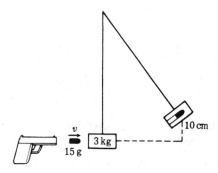

Fig. 9-8

▮ Consider first the collision of block and bullet. During the collision, momentum is conserved, so

$$\text{momentum just before} = \text{momentum just after} \quad (0.015\,\text{kg})v + 0 = (3.015\,\text{kg})V$$

where v is the initial speed of the bullet and V is the speed of block and bullet just after collision.
After the collision, mechanical energy is conserved:

$$\text{KE just after collision} = \text{final GPE} \quad \tfrac{1}{2}(3.015\,\text{kg})V^2 = (3.015\,\text{kg})(9.8\,\text{m/s}^2)(0.10\,\text{m})$$

From this we find $V = 1.40\,\text{m/s}$. Substituting this in the momentum equation gives $v = \underline{281\,\text{m/s}}$ for the speed of the bullet.

9.55 A 5-g bullet going 300 m/s strikes a 1.995-kg wooden block which is the bob of a ballistic pendulum. Find the speed at which block and bullet leave the equilibrium position and the height which the center of gravity of the bullet-block system reaches above the initial position of the center of gravity.

▮ $$\text{Momentum before collision} = \text{momentum after collision}$$

$$\frac{5}{1000}\,300 = (1.995 + 0.005)V \qquad V = \underline{0.75\,\text{m/s}}$$

$$\tfrac{1}{2}(M + m)V^2 = (M + m)gh \qquad \text{by conservation of energy}$$

$$\tfrac{1}{2}(0.75)^2 = 9.8h \qquad h = \underline{0.0287\,\text{m}}$$

9.56 A 4-g bullet is fired into a 2.996-kg block of wood which is the bob of a ballistic pendulum. If the bob leaves its equilibrium position with a speed of 0.5 m/s, what are the speed of the bullet and the height above the equilibrium position reached by the center of gravity of the block?

▮ $$\text{Momentum before collision} = \text{momentum after collision}$$

$$0.004v_B = (2.996 + 0.004)0.5 \qquad v_B = \underline{375\,\text{m/s}}$$

$$\tfrac{1}{2}(M + m)V^2 = (M + m)gh \qquad \tfrac{1}{2}(0.25) = 9.8h \qquad h = \underline{0.0128\,\text{m}}$$

9.57 A 5-g bullet traveling 250 m/s strikes and embeds itself in a 2.495-kg block held on a frictionless table by a spring with constant $k = 40$ N/m. Find the speed of block and bullet immediately after the collision and the distance the spring is compressed.

$\blacksquare$ $\qquad$ Momentum before collision = momentum after collision

$$0.005 \times 250 = (2.495 + 0.005)V \qquad V = \underline{0.5 \text{ m/s}}$$

Once the collision is completed, we have conservation of mechanical energy.

$$\text{KE at start = PE in spring at end} \qquad \tfrac{1}{2}(2.5)(0.5)^2 = \tfrac{1}{2}(40)x^2 \qquad x = \tfrac{1}{8} \text{ m} = \underline{125 \text{ mm}}$$

9.58 A 20-g bullet is fired horizontally with a speed of 600 m/s into a 7-kg block sitting on a tabletop; the bullet (b) lodges in the block (B). If the coefficient of kinetic friction between the block and the tabletop is 0.4, what is the distance the block will slide?

$\blacksquare$ By conservation of momentum, the momentum of the block-plus-bullet system just after the interaction is $p = m_b v_{0b}$; hence the kinetic energy of the system is

$$K = \frac{p^2}{2(m_B + m_b)} = \frac{m_b^2 v_{0b}^2}{2(m_B + m_b)}$$

The friction force does work $W_f = -fs = -\mu_k(m_B + m_b)gs$ in stopping the block. Hence

$$\Delta K = W_f \qquad 0 - \frac{m_b^2 v_{0b}^2}{2(m_B + m_b)} = -\mu_k(m_B + m_b)gs$$

$$s = \frac{1}{2\mu_k g}\left(\frac{m_b v_{0b}}{m_B + m_b}\right)^2 = \frac{1}{2(0.4)(9.8)}\left[\frac{(0.020)(600)}{7.020}\right]^2 = \underline{0.372 \text{ m}}$$

9.59 A 4-g bullet is fired horizontally with a speed of 300 m/s into a 0.8-kg block of wood at rest on a table. If the coefficient of friction between the block and the table is 0.3, how far will the block slide? What fraction of the bullet's energy is dissipated in the collision itself?

$\blacksquare$ From momentum conservation, $mv_B = (M + m)V$.

$$(0.004 \times 300) = (0.800 + 0.004)V \qquad \text{and} \qquad V = 1.493 \text{ m/s}$$

The frictional force is $f = \mu(M + m)g$. From the work-kinetic-energy relationship, $-fs = \Delta \text{KE}$, yielding

$$\tfrac{1}{2}(M + m)V^2 = fs = [\mu(M + m)g]s \qquad \tfrac{1}{2}(1.493)^2 = 0.3 \times 9.8s \qquad s = \underline{0.379 \text{ m}}$$

In the collision

$$\frac{K_f}{K_i} = \frac{(M + m)V^2}{mv_B^2} = \frac{m}{m + M}$$

Then

$$\text{fraction dissipated} = 1 - \frac{K_f}{K_i} = \frac{M}{m + M} = \frac{0.8}{0.804} = \underline{99.5\%}$$

9.60 A 0.3-kg block slides down a frictionless hemispherical bowl with a radius of 0.1 m. At the bottom it collides perfectly inelastically with a 0.4-kg mass initially at rest. What impulse is given the 0.4-kg mass? Find the maximum angle, α, which the radius vector to the two blocks will make with the vertical after the collision (Fig. 9-9).

$\blacksquare$ Let v_i = collision velocity; V = velocity just after the collision. From conservation of energy before collision,

$$v_i^2 = 2gh_i = 2 \times 9.8 \times 0.1 \qquad \text{and} \qquad v_i = 1.4 \text{ m/s}$$

Momentum is conserved during collision since no external horizontal forces are acting, and

$$(0.3 \times 1.4) = 0.7V \qquad V = 0.6 \text{ m/s}$$

Impulse given 0.4 kg = $\Delta(mv) = 0.4 \times (0.6 - 0) = \underline{0.24 \text{ N} \cdot \text{s}}$. From conservation of energy after collision,

$$h_f = \frac{V^2}{2g} = \frac{0.36}{2 \times 9.8} = 0.01837 \text{ m}$$

From Fig. 9-9, $h_f = 0.1(1 - \cos \alpha)$ and $\cos \alpha = 0.8163$; $\alpha = \underline{35.3°}$.

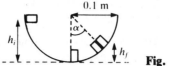

Fig. 9-9

9.61 A 3-g bullet with a speed of 300 m/s passes right through a 400-g block suspended on a long cord. The impulse gives the block a speed of 1.5 m/s. Find (*a*) the speed of the bullet after it has passed through the block, (*b*) the distance which the center of mass rises as the block swings upward after the bullet passes through, (*c*) the work done by the bullet in passing through the block, and (*d*) the mechanical energy converted into heat.

▮ (*a*) From momentum conservation,

$$\frac{3}{1000} \times 300 = \frac{400}{1000}1.5 + \frac{3}{1000}v_f \qquad \text{or} \qquad 3v_f = 900 - 600 = 300 \qquad v_f = \underline{100 \text{ m/s}}$$

(*b*) From conservation of energy for the block after the collision,

$$V_B^2 = 2gh_B \qquad \text{or} \qquad 1.5^2 = 2 \times 9.8 h_B \qquad \text{and} \qquad h_B = \underline{0.1148 \text{ m}}$$

(*c*) $$\text{Work} = \text{energy lost by bullet} = \frac{1}{2}\left(\frac{3}{1000}\right)(300)^2 - \frac{1}{2}\left(\frac{3}{1000}\right)(100)^2 = \underline{120 \text{ J}}$$

(*d*) Heat loss = (energy of bullet before collision) − (energy of bullet plus energy of block just after collision). Using part (*c*) result, heat loss = $[120 - \frac{1}{2}(0.4)(1.5)^2] = \underline{119.55 \text{ J}}$.

9.62 Refer to Fig. 9-10. The pendulum on the left is pulled aside to the position shown. It is then released and allowed to collide with the other pendulum, which is at rest. (*a*) What is the speed of the ball on the left just before collision? After collision, the two stick together. (*b*) How high, in terms of *h*, does the combination swing? Assume that the two balls have equal masses.

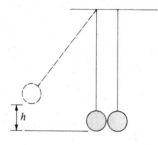

Fig. 9-10

▮ (*a*) $U_g = K$ leads to $v = \underline{(2gh)^{1/2}}$. (*b*) For the collision, write $\mathbf{p}_{\text{before}} = \mathbf{p}_{\text{after}}$, $mv = (2m)V$, so $V = \underline{v/2}$; the two-mass system will convert its $K = (2m)V^2/2$ to $U_g = (2m)gh'$. The new height $h' = V^2/2g = v^2/8g = \underline{h/4}$.

9.63 Suppose that the two pendulum balls in Fig. 9-10 have different masses; the ball on the left has a mass m_1. When the ball on the left is let swing from the height shown, it strikes and sticks to the ball on the right. The combination then swings to a height $h/3$. Find the mass m_2 of the ball on the right in terms of m_1.

▮ The velocity of m_1 before collision is $v = (2gh)^{1/2}$, while V of the combined masses is found from $\mathbf{p}_{\text{before}} = \mathbf{p}_{\text{after}}$, which gives a ratio of $v/V = (m_1 + m_2)/m_1$. The masses rise to $h/3$, thus $\Delta K = \Delta U_g$, $V = (2gh/3)^{1/2} = 3^{-1/2}v$ and $v/V = 3^{1/2}$. Equating velocity ratios yields $m_2 = (3^{1/2} - 1)m_1 = 0.732m_1$.

9.64 In Fig. 9-10, both masses are displaced to a height *h*, one to the left, as shown, and the other to the right. They are released simultaneously and undergo a perfectly elastic collision at the bottom. How high does each swing after collision? Both masses are identical.

▮ For *K* and **p** to be conserved, both masses rebound with equal but opposite velocities returning to their original heights.

9.65 As shown in Fig. 9-10, the mass on the left is pulled aside and released. Its speed at the bottom is v_0. It then collides with the mass on the right in a perfectly elastic collision. Find the velocities of the two masses just after collision if the mass on the left is 3 times as large as the mass on the right.

▮ Conservation of momentum yields $3v_0 = 3v_L + v_R$ and constant K gives $3v_0^2 = 3v_L^2 + v_R^2$. After common terms have been cancelled, solve simultaneously to find $v_L = \underline{v_0/2}$ and $v_R = \underline{3v_0/2}$ so both move to the right.

9.66 A 2.0-kg block rests over a small hole on a table. A 15.0-g bullet is shot through the hole into the block, where it lodges. How fast was the bullet going if the block rises 1.30 m above the table?

▮ $\sum \mathbf{p} = 0$ leads to $0.015v = 2.015V$. The velocity V of bullet and block is, using the energy-conservation equation, $2.015V^2/2 = 2.015(9.8)(1.30)$, which leads to $V = 5.05$ m/s. The bullet velocity v is then $\underline{678\ \text{m/s}}$.

9.67 Ball A in Fig. 9-11 is released from the point shown. It slides along the frictionless wire and collides with ball B. If the collision is perfectly elastic, find how high ball B will rise after the collision ($m_A = m_B/2$).

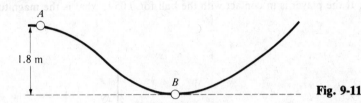

Fig. 9-11

▮ To find the speed of ball A before collision set $mgh_0 = mv_0^2/2$; $m(9.8)(1.8) = mv_0^2/2$, so that $v_0 = 5.94$ m/s. By conservation of momentum in the collision, $(m_B/2)(5.94) = (m_B/2)v_A + m_B v_B$. Because the collision is elastic, $(m_B/2)(5.94)^2/2 = (m_B/2)v_A^2/2 + m_B v_B^2/2$. Solving simultaneously gives $v_B = 3.96$ m/s. Use $m_B v_B^2/2 = m_B gh$ to find $h = v_B^2/2g = \underline{0.80\ \text{m}}$.

9.4 COLLISIONS IN TWO DIMENSIONS

9.68 A 1200-kg car is moving east at 30.0 m/s and collides with a 3600-kg truck moving at 20.0 m/s in a direction 60° north of east. The vehicles interlock and move off together. Find their common velocity.

▮ $m_1\mathbf{u}_1 + m_2\mathbf{u}_2 = (m_1 + m_2)\mathbf{v}$. Let x be eastward and y northward. For the x component, $(1200\ \text{kg})(30.0\ \text{m/s}) + (3600\ \text{kg})(20.0\ \text{m/s})(\cos 60°) = (4800\ \text{kg})v_x$, or $v_x = 15$ m/s. For the y component, $0 + (3600\ \text{kg})(20.0\ \text{m/s})(\sin 60°) = (4800\ \text{kg})v_y$, or $v_y = 13.0$ m/s. $v = (v_x^2 + v_y^2)^{1/2} = \underline{19.8\ \text{m/s}}$; and $\tan \theta = v_y/v_x$ yields $\theta = \underline{40.9°}$ north of east.

9.69 A 1.0-kg object, A, with a velocity of 4.0 m/s to the right, strikes a second object, B, of 3.0 kg, originally at rest. In the collision, A is deflected from its original direction through an angle of 50°; its speed after the collision is 2.0 m/s. Find the angle between B's velocity after the collision and the original direction of A, and find the speed of B after the collision.

▮ We choose our x axis to the right. Let $m_a = 1.0$ kg, $m_b = 3.0$ kg, $u_a = 4.0$ m/s, and $u_b = 0$. v_a and v_b are the magnitudes of the final velocities, and $v_a = 2.0$ m/s aimed 50° above the x axis. From conservation of momentum we have

x direction:

$$m_a u_a = m_a v_a \cos 50° + m_b v_b \cos \phi \qquad \text{where } \phi = \text{angle of } v_b \text{ below the } x \text{ axis}$$

Solving, we get $v_b \cos \phi = 0.907$ m/s.

y direction:
$$0 = m_a v_a \sin 50° - m_b v_b \sin \phi$$

Solving, we get $v_b \sin \phi = 0.511$ m/s. From these two equations for v_b and ϕ we get

$$v_b^2 = (0.907)^2 + (0.511)^2 \qquad \text{or} \qquad v_b = \underline{1.04\ \text{m/s}} \qquad \tan \phi = \frac{0.511}{0.907} \qquad \text{or} \qquad \phi = \underline{29.4°}$$

9.70 An object at rest explodes into three pieces of equal mass. One moves east at 20 m/s; a second moves southeast at 30 m/s. What is the velocity of the third piece?

▮ Let x be east and y be north. Then $m\mathbf{v}_1 + m\mathbf{v}_2 + m\mathbf{v}_3 = 0$, yields in x direction, 20 m/s + (30 m/s) $\cos 45° + v_3 \cos \theta = 0$; and in y direction, $0 - (30\ \text{m/s}) \sin 45° + v_3 \sin \theta = 0$, where θ is defined as angle of v_3 above the x axis. Solving, we get $v_3 \cos \theta = -41.2$ m/s; $v_3 \sin \theta = 21.3$ m/s. $\tan \theta = -0.516$ and $\theta = \underline{153°}$ or $\underline{27° \text{ north of west}}$. Solving for v_3 we get $v_3 = \underline{46.4\ \text{m/s}}$.

9.71 Three bodies form an isolated system. Their masses are m_1, $m_2 = 2m_1$, and $m_3 = 3m_1$. They have different directions of motion, but they all have the same initial speed v_0. One or more elastic collisions occur between pairs of the bodies which otherwise do not interact. Use energy considerations to find the maximum possible final speed of each of the three bodies.

▌ The initial total kinetic energy of the three particles is $\frac{1}{2}m_1v_0^2 + \frac{1}{2}(2m_1)v_0^2 + \frac{1}{2}(3m_1)v_0^2 = 3m_1v_0^2$. The maximum speed of any one of the particles is realized if the speeds of the other two particles are zero. Then $\frac{1}{2}m_i(v_{i,max})^2 = 3m_1v_0^2$, for $i = 1$, 2, and 3. Using this equation, we find $v_{1,max} = \sqrt{6}v_0 = \underline{2.45v_0}$, $v_{2,max} = \sqrt{3}v_0 = \underline{1.73v_0}$, and $v_{3,max} = \underline{1.41v_0}$.

9.72 A football of mass 0.42 kg is passed with a velocity of 25 m/s due south. A defending player lunges at the ball and deflects it so that the new velocity is 20 m/s, 36.9° west of south (Fig. 9-12). Find the magnitude of the impulse. If the player is in contact with the ball for 0.05 s, what is the magnitude of the average force he exerts?

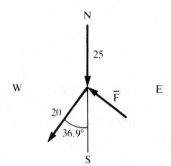

Fig. 9-12

▌ We choose south and west as positive. Final velocity has components 16 m/s south and 12 m/s west.

$$\text{Initial momentum} = 0.42 \times 25 = 10.5 \text{ kg} \cdot \text{m/s south}$$

$$\text{Final momentum} = (0.42 \times 16) \text{ south} \qquad \text{and} \qquad (0.42 \times 12) \text{ west}$$

$$[\Delta mv]_{south} = 0.42(16 - 25) = (-9 \times 0.42) \text{ kg} \cdot \text{m/s south} \qquad [\Delta mv]_{west} = 0.42 \times 12 = 5.04 \text{ kg} \cdot \text{m/s west}$$

$$|\Delta mv| = [(5.04)^2 + (3.78)^2]^{1/2} = 6.3 \text{ kg} \cdot \text{m/s} \qquad \text{impulse} = \Delta mv = \underline{6.3 \text{ N} \cdot \text{s}}$$

$$\text{impulse} = \bar{F}t = 0.05\bar{F} = 6.3 \qquad \bar{F} = 6.3 \text{ N} \cdot \text{s}/0.05 \text{ s} = \underline{126 \text{ N}}$$

9.73 A 7500-kg truck traveling at 5 m/s east collides with a 1500-kg car moving at 20 m/s in a direction 30° south of west. After collision, the two vehicles remain tangled together. With what speed and in what direction does the wreckage begin to move?

▌ The original momenta are shown in Fig. 9-13(a), while the final momentum, $M\mathbf{v}$, is shown in Fig. 9-13(b). Momentum must be conserved in both the north and east directions. Therefore

$$(\text{momentum before})_{east} = (\text{momentum after})_{east} \qquad (7500 \text{ kg})(5 \text{ m/s}) - (1500 \text{ kg})[(20 \text{ m/s}) \cos 30°] = Mv_{east}$$

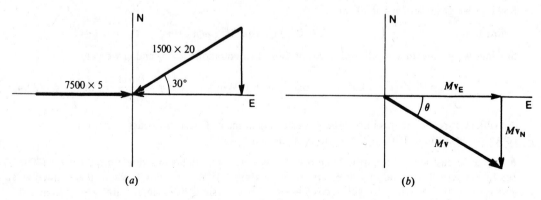

Fig. 9-13

where $M = 7500 + 1500 = 9000$ kg, and v_{east} is the eastward component of the velocity of the wreckage. Similarly,

(momentum before)$_{north}$ = (momentum after)$_{north}$ $(7500 \text{ kg})(0) - (1500 \text{ kg})[(20 \text{ m/s}) \sin 30°] = Mv_{north}$

The first equation gives $v_{east} = 1.28$ m/s and the second gives $v_{north} = -1.67$ m/s.
   The resultant velocity is

$$v = \sqrt{1.67^2 + 1.28^2} = \underline{2.1 \text{ m/s}}$$

The angle θ in Fig. 9-13(b) is given by

$$\theta = \arctan \frac{1.67}{1.28} = \underline{53°}$$

9.74 A 0.11-kg baseball is thrown with a speed of 17 m/s toward a batter. After the ball is struck by the bat, it has a speed of 34 m/s in the direction shown in Fig. 9-14(a). If the ball and bat are in contact for 0.025 s, find the magnitude of the average force exerted on the ball by the bat.

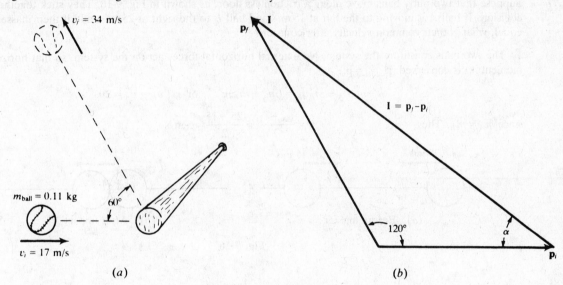

Fig. 9-14

▮ The impulse is $\mathbf{I} = \bar{\mathbf{F}} \Delta t$. The impulse-momentum relation is shown in Fig. 9-14(b). From the law of cosines,

$$I^2 = p_i^2 + p_f^2 - 2p_i p_f \cos 120° = [(0.11)(17)]^2 + [(0.11)(34)]^2 - 2[(0.11)(17)][(0.11)(34)](-0.5)$$

$$I = (0.11)(17)(\sqrt{7}) = 4.947 \text{ N} \cdot \text{s} \quad \text{and} \quad \bar{F} = \frac{I}{\Delta t} = \frac{4.947}{0.025} = \underline{197.90 \text{ N}}$$

9.75 A body of mass m collides with a frictionless surface. Its initial speed is v_i, and it strikes the surface at an angle θ_i. It bounces from the surface, but the collision is not elastic, so that after the impact the magnitude of the normal component of the velocity is only a fraction e of the original value $v_i \sin \theta$. (a) Find the impulse delivered by the surface to the body. (b) Find the angle θ_f at which the body leaves the surface.

▮ The situation is shown in Fig. 9-15. We let the plane of motion be the xy plane.

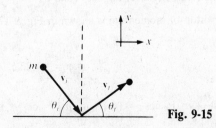

Fig. 9-15

(a) $|v_f \sin \theta_f| = e |v_i \sin \theta_i|$. Since the surface is frictionless, the momentum's x component is unchanged. The impulse $I_y = m[ev_i \sin \theta_i - (-v_i \sin \theta_i)] = mv_i[\sin \theta_i(1 + e)]$.

(b) The initial and final velocity components along x have the common value $v_i \cos \theta_i$. From Fig. 9-15,

$$\tan \theta_f = \frac{ev_i \sin \theta_i}{v_i \cos \theta_i} = e \tan \theta_i$$

so that $\underline{\theta_f = \tan^{-1}(e \tan \theta_i)}$.

9.76 Refer to Prob. 9.75. (a) Find the speed at which the body leaves the surface. (b) Express the ratio of final to initial kinetic energy in terms of e and θ_i.

▌ (a) Using the pythagorean theorem, the final speed v_f is given by

$$v_f = \sqrt{v_i^2 \cos^2 \theta_i + e^2 v_i^2 \sin^2 \theta_i} = v_i\sqrt{1 - \sin^2 \theta_i + e^2 \sin^2 \theta_i} = v_i\sqrt{1 - (1 - e^2) \sin^2 \theta_i}$$

(b) The kinetic energy ratio is given by $K_f/K_i = (v_f/v_i)^2$, so we have $(K_f/K_i) = 1 - (1 - e^2) \sin^2 \theta_i$.

9.77 Suppose that two putty balls move along a frictionless floor, as shown in Fig. 9-16. They stick together after collision. If ball A is moving to the left at 15 m/s and ball B to the right at 25 m/s, and if their masses are equal, what is their common velocity after collision?

▌ The two balls constitute the system. No external horizontal forces act on the system, so that horizontal momentum is conserved: $\mathbf{p}_{\text{initial}} = \mathbf{p}_{\text{final}}$.

$$-m_A v_A + m_B v_B = (m_A + m_B)v \qquad \text{or} \qquad v_B - v_A = 2v$$

since $m_A = m_B$. Thus,
$$v = \frac{v_B - v_A}{2} = \frac{25 - 15}{2} = 5 \text{ m/s}$$

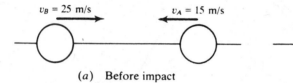

(a) Before impact (b) After impact

Fig. 9-16

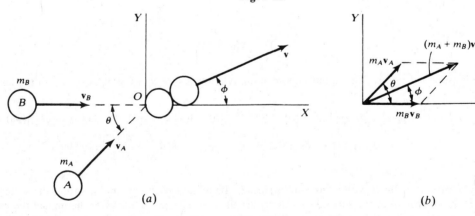

(a) (b)

Fig. 9-17

9.78 Suppose that the two putty balls of Prob. 9.77 collide obliquely, as shown in Fig. 9-17(a), and stick together after collision. Find their velocity after impact. Take $v_A = v_B = 45$ m/s and $\theta = 45°$.

▌ The vector conservation of momentum is shown in Fig. 9-17(b). Since $m_A = m_B$ and $v_A = v_B$, we have at once that

$$\phi = \frac{\theta}{2} = 22.5°$$

Then,
$$m_A v_A \cos \frac{\theta}{2} + m_B v_B \cos \frac{\theta}{2} = (m_A + m_B)v \qquad \text{or} \qquad v = v_A \cos \frac{\theta}{2} = 45(0.924) = \underline{41.58 \text{ m/s}}$$

9.79 The two balls shown in Fig. 9-18 collide and bounce off each other as shown. (*a*) What is the final velocity of the 500-g ball if the 800-g ball has a speed of 15 cm/s after collision? (*b*) Is the collision perfectly elastic?

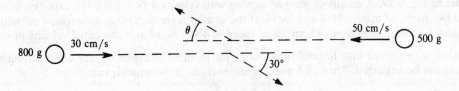

Fig. 9-18

▌ (*a*) From the law of conservation of momentum,

$$(\text{momentum before})_x = (\text{momentum after})_x$$

$$(0.80\text{ kg})(0.3\text{ m/s}) - (0.50\text{ kg})(0.5\text{ m/s}) = (0.8\text{ kg})[(0.15\text{ m/s})\cos 30°] + (0.5\text{ kg})v_x$$

from which $v_x = -0.228$ m/s. Also,

$$(\text{momentum before})_y = (\text{momentum after})_y \qquad 0 = -(0.8\text{ kg})[(0.15\text{ m/s})\sin 30°] + (0.5\text{ kg})v_y$$

from which $v_y = 0.120$ m/s. Then

$$v = \sqrt{v_x^2 + v_y^2} = \sqrt{0.228^2 + 0.120^2} = \underline{0.26\text{ m/s}}$$

Also, for the angle θ shown in Fig. 9-18,

$$\theta = \arctan\frac{0.120}{0.228} = \underline{28°}$$

(*b*)
$$\text{total KE before} = \tfrac{1}{2}(0.8)(0.3)^2 + \tfrac{1}{2}(0.5)(0.5)^2 = 0.0985\text{ J}$$
$$\text{total KE after} = \tfrac{1}{2}(0.8)(0.15)^2 + \tfrac{1}{2}(0.5)(0.26)^2 = 0.026\text{ J}$$

As KE is lost in the collision, the collision is not perfectly elastic.

9.80 Two roads at right angles to each other are carrying a 20-ton (metric) truck moving at 10 m/s and a 1-ton car moving at 20 m/s toward a collision at the intersection. After collision they stick together. Taking x and y coordinates along the original directions of motion of the truck and car, respectively, express the final velocity in terms of the unit vectors **i** and **j**.

▌ Momentum before = momentum after becomes $(20\,000)(10)\mathbf{i} + (1000)(20)\mathbf{j} = (21\,000)(v_x\mathbf{i} + v_y\mathbf{j})$. Solving, $\underline{\mathbf{v} = (9.52\mathbf{i} + 0.95\mathbf{j})\text{ m/s}}$.

9.81 A particle of mass m traveling with speed v_0 along the x axis suddenly shoots out one-third its mass with speed $2v_0$ parallel to the y axis. Express the velocity of the remainder of the particle in **i, j, k** notation.

▌ Momentum conservation becomes $mv_0\mathbf{i} = (2m/3)\mathbf{v} + (m/3)(2v_0)\mathbf{j}$, whence $\mathbf{v} = 1.5v_0\mathbf{i} - v_0\mathbf{j}$.

9.82 A particle of mass M moves along the x axis with speed v_0 and collides and sticks to a particle of mass m moving with speed v_0 along the y axis. Assuming **i** and **j** to be in the directions of motion, express the velocity of the combined particle after collision in **i, j, k** notation.

▌ Momentum before = momentum after, or $Mv_0\mathbf{i} + mv_0\mathbf{j} = (m + M)\mathbf{v}$. Solving, $\mathbf{v} = (Mv_0\mathbf{i} + mv_0\mathbf{j})/(m + M)$.

9.83 Three identical particles with velocities $v_0\mathbf{i}$, $-3v_0\mathbf{j}$, and $5v_0\mathbf{k}$ collide successively with each other in such a way that they form a single particle. Find the velocity of the resultant particle in **i, j, k** form.

▌ Initial momentum of system = final momentum of system: $v_0\mathbf{i} - 3v_0\mathbf{j} + 5v_0\mathbf{k} = 3\mathbf{v}$. This gives $\mathbf{v} = (v_0/3)(\mathbf{i} - 3\mathbf{j} + 5\mathbf{k})$.

9.84 A particle of mass m has a velocity $-v_0\mathbf{i}$, while a second particle of the same mass has a velocity $v_0\mathbf{j}$. After the particles collide, one particle is found to have a velocity $-\tfrac{1}{2}v_0\mathbf{i}$. (*a*) Find the velocity of the other. (*b*) Was the collision perfectly elastic?

❚ (a) Conservation of momentum gives $-mv_0\mathbf{i} + mv_0\mathbf{j} = -mv_0\mathbf{i}/2 + m\mathbf{v}$. Solving, $\mathbf{v} = -v_0\mathbf{i}/2 + v_0\mathbf{j}$.
(b) No. Kinetic energies are not the same before and after: $mv_0^2/2 + mv_0^2/2 \neq [m(v_0^2/4 + v_0^2)]/2 + [m(v_0^2/4)]/2$.

9.85 Refer to Fig. 9-19. A missile of mass M moving with velocity $\mathbf{v}$ ($v = 200$ m/s), explodes in midair, breaking into two parts, of masses $M/4$ and $3M/4$. If the smaller piece flies off at an angle of 60° with respect to the original direction of motion with an initial speed of 400 m/s, what is the initial velocity of the other piece?

❚ Over a very small time interval surrounding the moment of explosion, the effect of gravity (an external force) can be neglected. Then all forces are internal and momentum is conserved.

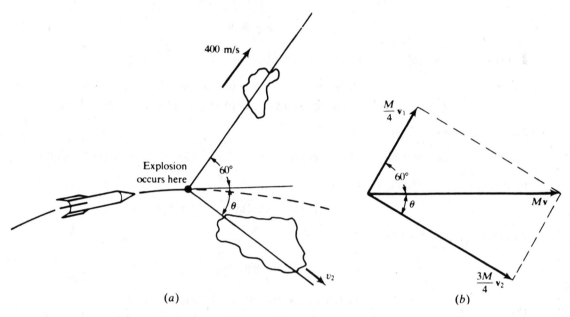

(a)   (b)

Fig. 9-19

The vector diagram for momentum conservation is shown in Fig. 9-19(b). We have

$$\frac{3M}{4}\mathbf{v}_2 = M\mathbf{v} - \frac{M}{4}\mathbf{v}_1 \qquad \text{or} \qquad \mathbf{v}_2 = \frac{4}{3}\mathbf{v} - \frac{1}{3}\mathbf{v}_1$$

Then

$$v_2^2 = \mathbf{v}_2 \cdot \mathbf{v}_2 = \frac{16}{9}(\mathbf{v}\cdot\mathbf{v}) - \frac{8}{9}(\mathbf{v}\cdot\mathbf{v}_1) + \frac{1}{9}(\mathbf{v}_1\cdot\mathbf{v}_1) = \frac{16}{9}(200)^2 - \frac{8}{9}(200)(400)(\cos 60°) + \frac{1}{9}(400)^2 = \frac{48\times10^4}{9}$$

$$\text{and} \qquad v_2 = \frac{400}{\sqrt{3}} = \underline{231 \text{ m/s.}} \qquad \text{Also} \qquad \mathbf{v}\cdot\mathbf{v}_2 = \frac{4}{3}\mathbf{v}\cdot\mathbf{v} - \frac{1}{3}\mathbf{v}\cdot\mathbf{v}_1 \qquad \text{gives}$$

$$(200)\left(\frac{400}{\sqrt{3}}\right)(\cos\theta) = \frac{4}{3}(200)^2 - \frac{1}{3}(200)(400)(\cos 60°) \qquad \cos\theta = \frac{\sqrt{3}}{2} \qquad \theta = \underline{30°}$$

9.5 RECOIL AND REACTION

9.86 An artillery piece, of mass m_1, fires a shell, of mass m_2. Find the ratio of the kinetic energies of the artillery piece and of the shell just after the firing. Assume the shell is fired horizontally.

❚ The system made up of the artillery piece and the shell is initially at rest, so that by conservation of momentum

$$\mathbf{p}_1 + \mathbf{p}_2 = 0 \qquad \text{or} \qquad \mathbf{p}_1 = -\mathbf{p}_2 \qquad \text{then} \qquad \frac{K_1}{K_2} = \frac{p_1^2/2m_1}{p_2^2/2m_2} = \frac{m_2}{m_1}$$

9.87 A 10-g bullet leaves a rifle barrel with a velocity of 600 m/s to the east. If the barrel is 1 m long and the gun has a mass of 4 kg, what is (a) the recoil velocity of the rifle; (b) the force exerted on the bullet; (c) the time during which the bullet was accelerated; and (d) the impulse supplied to the bullet, assuming a constant force on the bullet in the barrel?

❚ (a) By the conservation of momentum, the momentum before the trigger is pulled is equal to the momentum when the bullet leaves the gun. Let us take east as positive. Then, using SI units,

$$0 = m_B v_B + m_R v_R = (0.01 \text{ kg})(600 \text{ m/s}) + (4 \text{ kg})v_R \qquad v_R = \underline{-1.5 \text{ m/s}}$$

That is, 1.5 m/s west, as indicated by the minus sign.

(b) Using equation of motion,

$$v^2 = v_0^2 + 2as \qquad 600^2 = 0^2 + 2a(1)$$

since $s = 1$ m, the length of the barrel. Then,

$$a = 1.8 \times 10^5 \text{ m/s}^2 \qquad F = ma = (0.01)(1.8 \times 10^5) = \underline{1.8 \times 10^3 \text{ N}}$$

(c) Using the equation defining (constant) acceleration,

$$a = \frac{v - v_0}{t} \qquad 1.8 \times 10^5 \text{ m/s}^2 = \frac{600 \text{ m/s} - 0}{t} \qquad t = \underline{1/300 \text{ s}}$$

(d) $$\text{Impulse} = Ft = 1.8 \times 10^3 \times \tfrac{1}{300} = \underline{6 \text{ N} \cdot \text{s}}$$

Check:

$$Ft = mv - mv_0 = (0.01 \text{ kg})(600 \text{ m/s}) - 0 = 6 \text{ kg} \cdot \text{m/s} = 6 \text{ N} \cdot \text{s}$$

9.88 A machine gun fires six bullets per second into a target. The mass of each bullet is 3 g, and the speed 500 m/s. Find the average force required to hold the gun in position. What is the power delivered to the bullets?

❚ $$F_{\text{avg}} = \frac{\Delta(mv)}{\Delta t} = \frac{6(0.003)(500) \text{ kg} \cdot \text{m/s}}{1 \text{ s}} = \underline{9 \text{ N}}$$

To get power, we note that the energy given to each bullet is KE of 1 bullet $= \tfrac{1}{2}(0.003)(500)^2 = 375$ J. Then for power, $P = (375 \text{ J/bullet})(6 \text{ bullets/s}) = 2250 \text{ J/s} = \underline{2.25 \text{ kW}}$.

9.89 The nucleus of a certain atom has a mass of 3.8×10^{-25} kg and is at rest. The nucleus is radioactive and suddenly ejects from itself a particle of mass 6.6×10^{-27} kg and speed 1.5×10^7 m/s. Find the recoil speed of the nucleus left behind.

❚ The momentum of the system is conserved during the explosion.

$$\text{momentum before} = \text{momentum after} \qquad 0 = (3.73 \times 10^{-25} \text{ kg})(v) + (6.6 \times 10^{-27} \text{ kg})(1.5 \times 10^7 \text{ m/s})$$

where the mass of the remaining nucleus is 3.73×10^{-25} kg and its recoil velocity is v. Solving gives

$$v = -\frac{(6.6 \times 10^{-27})(1.5 \times 10^7)}{3.73 \times 10^{-25}} = \underline{-2.7 \times 10^5 \text{ m/s}}$$

9.90 Two blocks of masses 200 g and 500 g sit on a frictionless table with an essentially massless spring placed between them. They are pushed together until an energy of 3.0 J is stored in the spring. When released, the masses shoot off in opposite directions. What is the speed of (a) the center of mass and (b) the 500-g block?

❚ (a) Because the spring force is an internal force, the center of mass remains at rest. (b) The momentum before and after is zero; thus, $0.2(v_{0.2}) = 0.5(v_{0.5})$. Also, since the table is frictionless, the spring energy $U_s = 3.0 \text{ J} = [0.2(v_{0.2})^2]/2 + [0.5(v_{0.5})^2]/2$. Solving between the two expressions gives $v_{0.5} = \underline{1.85 \text{ m/s}}$.

9.91 A 500-lb gun fires a 2-lb projectile with a muzzle velocity of 1600 ft/s. Determine the initial recoil velocity v of the gun.

❚ Consider the system (gun + projectile). The momentum of the system before firing is zero. Therefore, since momentum is conserved,

$$\text{momentum before firing} = \text{momentum just after} \qquad 0 = \left(\frac{2}{32} \text{ slug}\right)(1600 \text{ ft/s}) + \left(\frac{500}{32} \text{ slug}\right)v$$

from which $v = \underline{-6.4 \text{ ft/s}}$.

9.92 For the situation in Prob. 9.91, if the recoil is against a constant resisting force of 400 lb, what are the time taken to bring the gun to rest and the distance it recoils?

I The loss in momentum during recoil is due to the impulse exerted on the gun by the 400-lb resisting force. Therefore, choosing the direction of recoil as positive,

$$\text{impulse} = mv_f - mv_0 \qquad (-400 \text{ lb})t = 0 - \left(\frac{500}{32} \text{ slug}\right)(6.4 \text{ ft/s})$$

from which $t = 0.25$ s.

Since the resisting force is constant, the gun's recoil is uniformly decelerated. We may therefore write $\bar{v} = \frac{1}{2}(0 + 6.4)$ ft/s $= 3.2$ ft/s. Then $x = \bar{v}t$ gives the recoil distance as $x = (3.2 \text{ ft/s})(0.25 \text{ s}) = \underline{0.80 \text{ ft}}$.

9.93 A 0.25-kg ball moving in the $+x$ direction at 13 m/s is hit by a bat. Its final velocity is 19 m/s in the $-x$ direction. The bat acts on the ball for 0.010 s. Find the average force F exerted on the ball by the bat.

I We have $v_0 = 13$ m/s and $v_f = -19$ m/s. The impulse equation then gives

$$Ft = mv_f - mv_0 \qquad F(0.01 \text{ s}) = (0.25 \text{ kg})(-19 \text{ m/s}) - (0.25 \text{ kg})(13 \text{ m/s})$$

from which $F = \underline{-800 \text{ N}}$.

9.94 A 500-g pistol lies at rest on an essentially frictionless table. It accidentally discharges and shoots a 10-g bullet parallel to the table. How far has the pistol moved by the time the bullet hits a wall 5 m away?

I Take the recoil direction as the positive x direction. Then, since the center of mass of the system remains at $x = 0$, $(500 \text{ g})x + (10 \text{ g})(-5 \text{ m}) = 0$, or $x = \underline{10 \text{ cm}}$.

9.95 While coasting along a street at a constant velocity of 0.50 m/s, a 20-kg girl in a 5-kg wagon sees a vicious dog in front of her. She has with her only a 3.0-kg bag of sugar which she is bringing from the grocery, and she throws it at the dog with a forward velocity of 4.0 m/s relative to her original motion. How fast is she moving after she throws the bag of sugar?

I Momentum conservation for the system of girl, wagon, and sugar is $(20 + 5.0 + 3.0)(0.50) = (20 + 5.0)v + 3.0(4.5)$; she is now moving at $v = \underline{0.020 \text{ m/s}}$.

9.96 A 60-kg man dives from the stern of a 90-kg boat with a horizontal component of velocity of 3.0 m/s north. Initially the boat was at rest. Find the magnitude and direction of the velocity acquired by the boat.

I Let 1 refer to the man and 2 to the boat. Before diving $u_1 = u_2 = 0$, then $m_1v_1 + m_2v_2 = 0$ and $(60 \text{ kg})(3.0 \text{ m/s}) + (90 \text{ kg})v_2 = 0$. $v_2 = -2.0$ m/s, or $\underline{2.0 \text{ m/s south}}$.

9.97 Suppose that a boy stands at one end of a boxcar sitting on a railroad track. Let the mass of the boy and the boxcar be M. He throws a ball of mass m with velocity $\mathbf{v}_0$ toward the other end, where it collides elastically with the wall and travels back down the length (L) of the car, striking the opposite side inelastically and coming to rest. If there is no friction in the wheels of the boxcar, describe the motion of the boxcar.

I All forces are internal (Fig. 9-20). Therefore, if $\mathbf{V}$ and $\mathbf{v}$ are the velocities of the boxcar plus boy and the

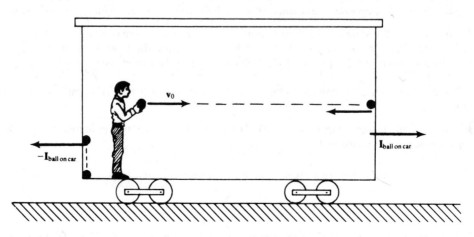

Fig. 9-20

ball, conservation of momentum gives

$$MV + mv = 0 \quad \text{or} \quad V = -\frac{m}{M}v \tag{1}$$

at all times.

Before the first collision, $v = v_0$, and so

$$V = -\frac{m}{M}v_0 \quad 0 < t < \frac{L}{\left(1 + \frac{m}{M}\right)v_0}$$

where we have on the right the time for the ball to reach the boxcar wall, traveling at speed

$$v_0 + V = v_0 + \frac{m}{M}v_0$$

with respect to the floor of the boxcar.

The effect of the first collision will be simply to reverse both velocity vectors (Eq. (1) and elasticity). Thus,

$$V = +\frac{m}{M}v_0 \quad \frac{L}{\left(1 + \frac{m}{M}\right)v_0} < t < \frac{2L}{\left(1 + \frac{m}{M}\right)v_0}$$

Finally, after the second collision, the ball and boxcar have a common velocity. Hence, $V = -(m/M)V$, or

$$V = 0 \quad t > \frac{2L}{\left(1 + \frac{m}{M}\right)v_0}$$

It is seen that the boxcar first moves to the left a distance

$$\left(\frac{m}{M}v_0\right)\frac{L}{\left(1 + \frac{m}{M}\right)v_0} = \left(\frac{m}{M + m}\right)L$$

and then moves an equal distance to the right, coming to rest at its starting point. This result—that there is no net displacement of the boxcar if the ball returns to its initial location within the boxcar—holds whether or not the first collision is elastic (because the center of mass of the system must remain at rest).

9.98 A 60-kg astronaut becomes separated in space from her spaceship. She is 15.0 m away from it and at rest relative to it. In an effort to get back, she throws a 500-g wrench with a speed of 8.0 m/s in a direction away from the ship. How long does it take her to get back to the ship?

▮ From $60V = 0.50(8.0)$, her velocity $V = (1/15)$ m/s toward the ship, 15.0 m away. The time is $t = 15.0/V = \underline{225\text{ s}}$.

9.99 The uranium-238 nucleus is unstable and decays into a thorium-234 nucleus and an α particle. The α particle is emitted with a speed of 1.4×10^6 m/s. What is the recoil speed of the thorium-234 nucleus, assuming that the uranium-238 atom is at rest at the time of decay? The thorium-234 and α particle masses are in the ratio 234 to 4.

▮ This is a recoil problem, as if the uranium were made up of a thorium nucleus and an alpha particle pressed together by a spring, and initially at rest. Then $0 = m_{th}v_{th} + m_\alpha v_\alpha$, or $v_{th}/v_\alpha = -m_\alpha/m_{th}$. Solving, we get $v_{th} = \underline{2.39 \times 10^4\text{ m/s}}$.

9.100 A 4-kg rifle fires a 6-g bullet with a speed of 500 m/s. What kinetic energy is acquired (a) by the bullet? (b) by the rifle? (c) Find the ratio of the distance the rifle moves backward while the bullet is in the barrel to the distance the bullet moves forward. See Fig. 9-21.

▮ (a) The kinetic energy of the bullet is

$$K_b = \tfrac{1}{2}m_b v_b^2 = \tfrac{1}{2}(0.006)(500)^2 = \underline{750\text{ J}}$$

(b) By conservation of momentum, since no external impulse forces act on the system (the bullet and rifle),

$$P_i = P_f \quad 0 = m_r v_r + m_b v_b \quad v_r = -\frac{m_b}{m_r}v_b$$

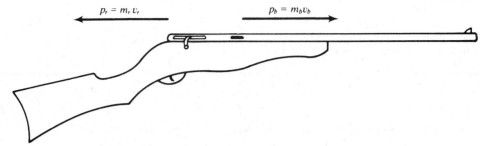

Fig. 9-21

The kinetic energy of the rifle then is

$$K_r = \frac{1}{2} m_r \frac{m_b^2}{m_r^2} v_b^2 = \frac{1}{2} \frac{(0.006)^2}{4} (500)^2 = \underline{1.125 \text{ J}}$$

(c) From **(b)** $\qquad 0 = m_r v_r + m_b v_b = m_r \dfrac{\Delta x_r}{\Delta t} + m_b \dfrac{\Delta x_b}{\Delta t} \qquad$ or $\qquad 0 = m_r\,\Delta x_r + m_b\,\Delta x_b$

which expresses the fact that the center of mass of the system remains at rest. Solving,

$$\frac{|\Delta x_r|}{|\Delta x_b|} = \frac{m_b}{m_r} = \frac{6}{4000} = \frac{1}{\underline{667}}$$

9.101 A missile launcher, of mass 4400 kg, fires horizontally a rocket of mass 110 kg and recoils up a smooth inclined plane, rising to a height of 4 m (see Fig. 9-22). Find the initial speed of the rocket.

❙ Since gravity was the only force acting on the launcher opposing its motion, the speed at which it starts up the inclined plane can be found from

$$0^2 = V^2 + 2(-g \sin \theta)s = V^2 - 2gh \qquad \text{or} \qquad V = \sqrt{2gh} = \sqrt{2(9.8)(4)} = \underline{8.85 \text{ m/s}}$$

All forces involved in the launching are internal forces. For the horizontal momentum: $\mathbf{p}_{\text{initial}} = \mathbf{p}_{\text{final}}$. Choosing the rocket's direction as positive,

$$0 = m_r v_0 - m_l V \qquad v_0 = \frac{m_l}{m_r} V = \left(\frac{4400}{110}\right)(8.85) = \underline{354 \text{ m/s}}$$

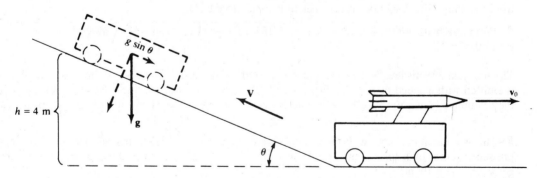

Fig. 9-22

9.102 A boy of mass m is standing on a toboggan of mass M which is moving with constant velocity $\mathbf{V}_i$ over the surface of a frozen lake. The boy then runs along the toboggan in the direction opposite to $\mathbf{V}_i$ and acquires a speed v *relative to the toboggan* as he jumps off the rear of the toboggan. What is the final speed of the toboggan relative to the ice?

❙ Let V_f be the final velocity of the toboggan. The velocity of the boy relative to the ice as he jumps is then $V_f - v$. From momentum conservation (in the ice-coordinate system), $(m + M)V_i = m(V_f - v) + MV_f = (m + M)V_f - mv$. Solving for V_f we get $V_f = V_i + mv/(M + m)$.

9.103 A rocket standing on its launch platform points straight upward. Its jet engines are activated and eject gas at a rate of 1500 kg/s. The molecules are expelled with a speed of 50 km/s. How much mass can the rocket initially have if it is slowly to rise because of the thrust of the engines?

▮ Because the motion of the rocket itself is negligible in comparison with the speed of the expelled gas, we can assume that the gas is accelerated from rest to a speed of 50 km/s. The impulse required to provide this acceleration to a mass m of gas is

$$Ft = mv_f - mv_0 = m(50\,000 \text{ m/s}) - 0 \quad \text{from which} \quad F = (50\,000 \text{ m/s})\frac{m}{t}$$

But we are told that the mass ejected per second (m/t) is 1500 kg/s and so the force exerted on the expelled gas is $F = (50\,000 \text{ m/s})(1500 \text{ kg/s}) = 7.5 \times 10^7 \text{ N}$.

An equal but opposite reaction force acts on the rocket and this is the upward thrust on the rocket. The engines can therefore support a weight of $7.5 \times 10^7 \text{ N}$ and so the maximum mass the rocket could have would be

$$M_{\text{rocket}} = \frac{\text{weight}}{g} = \frac{7.5 \times 10^7 \text{ N}}{9.8 \text{ m/s}^2} = 7.65 \times 10^6 \text{ kg} = \underline{7650 \text{ metric tons}}$$

9.104^c Consider a rocket in interstellar space, far from attracting bodies. If M is the instantaneous mass of the rocket and unused fuel, m is the mass of gas discharged by the rocket per second, and V_g is the speed of the discharged gas relative to the rocket, what is the speed-mass relation for the rocket?

▮ View the system (gas cloud + rocket) in an inertial frame, with the direction of the rocket chosen as positive. The total momentum of the system is constant. Now, in a small time interval dt, a mass $m\,dt$ of gas, traveling in the negative direction with speed $V_g - v$ (relative to our reference frame), joins the cloud. The cloud's change in momentum is therefore $m(v - V_g)\,dt$. By conservation, this change must exactly cancel the change in the rocket's momentum, $d(Mv)$; that is, $m(v - V_g)\,dt + d(Mv) = 0$. But $dM = -m\,dt$, and so

$$(V_g - v)\,dM + d(Mv) = 0 \quad \text{or} \quad V_g\,dM - v\,dM + M\,dv + v\,dM = 0 \quad \text{or} \quad V_g\,dM + M\,dv = 0$$

which is the desired differential relation between the velocity and mass of the rocket.

9.105^c Integrate the differential equation of Prob. 9.104, given the initial values v_0 and M_0.

▮ We have $dv = -V_g(dM/M)$; so, if V_g is assumed to be constant,

$$\int_{v_0}^{v} dv = -V_g \int_{M_0}^{M} \frac{dM}{M} \quad \text{or} \quad v = v_0 + V_g \ln \frac{M_0}{M}$$

9.106^c Refer to Probs. 9.104 and 9.105. A particular rocket is observed to lose mass according to the relation $m = \lambda M$ (λ = const). Obtain the equation of motion of the rocket.

▮ Since $dv = -V_g(dM/M)$ and $dM = -\lambda M\,dt$, we have

$$dv = \lambda V_g\,dt \quad \text{or} \quad \frac{dv}{dt} = \lambda V_g = \text{const}$$

which is linear motion at constant acceleration $a = \lambda V_g$. Hence, the displacement will be given by $s = s_0 + v_0 t + \frac{1}{2}\lambda V_g t^2$.

9.6 CENTER OF MASS (See also Chap. 10)

9.107 A system consisting of two masses connected by a massless rod lies along the x axis. A 0.4-kg mass is at $x = 2$ m while a 0.6-kg mass is at $x = 7$ m. Find the x coordinate of the center of mass.

▮
$$x_{\text{CM}} = \frac{m_1 x_1 + m_2 x_2}{m_1 + m_2} = \frac{(0.4 \times 2) + (0.6 \times 7)}{(0.4 + 0.6)} = \underline{5 \text{ m}}$$

9.108 Find the center of mass of the system of Fig. 9-23.

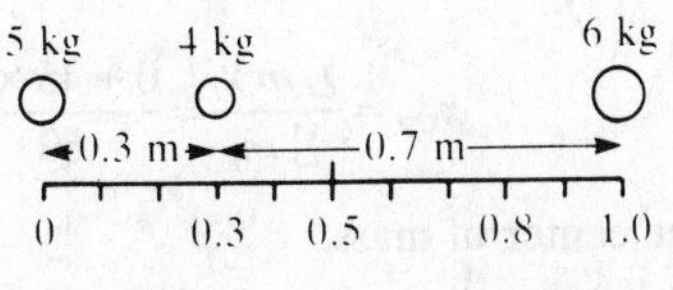

Fig. 9-23

�

$$x_{CM} = \frac{(5 \times 0) + (4 \times 0.3) + (6 \times 1.0)}{5 + 4 + 6} = \frac{7.2 \text{ kg} \cdot \text{m}}{15 \text{ kg}} = 0.48 \text{ m}$$

9.109 Find the distance from the center of the earth to the center of mass of the earth–moon system if the earth–moon separation is 3.8×10^5 km and the mass of the earth is 81.3 times the mass of the moon. To what fraction of the earth's radius of 6370 km does this distance correspond?

▮ Let the center of the earth be $x = 0$ and let masses be measured in units of the moon's mass.

$$x_{CM} = \frac{(81.3)(0) + (1)(3.8 \times 10^5)}{82.3} = \underline{4.617 \times 10^3 \text{ km}} \qquad \text{fraction} = \frac{4.617 \times 10^3}{6.37 \times 10^3} = \underline{0.725}$$

That is CM of earth–moon system is within the earth.

9.110 Assuming a dumbbell shape for the carbon monoxide (CO) molecule, find the distance of the center of mass of the molecule from the carbon atom in terms of the distance d between the carbon and the oxygen atom. Use 12 u for the atomic mass of carbon and 16 u for the atomic mass of oxygen.

▮ The position of the center of mass X_c is

$$X_c = \frac{m_1 x_1 + m_2 x_2}{m_1 + m_2} = \frac{12 \times 0 + 16d}{12 + 16} = \frac{16d}{28} = \underline{\frac{4}{7} d} \quad \text{from carbon atom}$$

9.111 Find the center of mass of the system shown in Fig. 9-24.

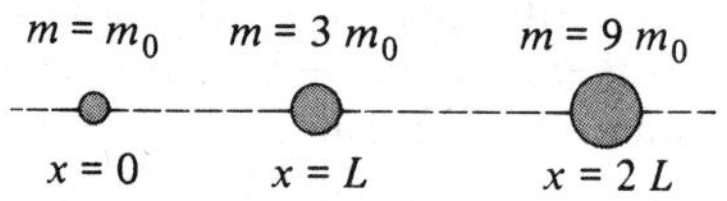

Fig. 9-24

$$x_{CM} = \frac{0 + 3m_0 L + 18m_0 L}{13m_0} = \underline{1.62L}$$

9.112 The separation between the centers of the hydrogen and chlorine atoms in the hydrogen chloride molecule (HCl) is about 0.13 nm. How far from the center of the hydrogen atom is the center of mass of the molecule? The atomic masses of H and Cl are 1 and 35 u, respectively.

$$x_{CM} = \frac{\sum m_i x_i}{\sum m_i} = \frac{0 + 35(0.13)}{36} = \underline{0.126 \text{ nm}}$$

9.113 Find the center of mass of the system shown in Fig. 9-25.

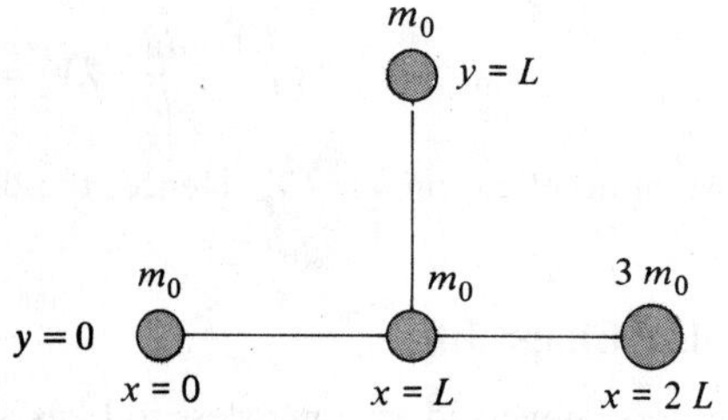

Fig. 9-25

$$x_{CM} = \frac{0 + m_0 L + m_0 L + 6m_0 L}{6m_0} = \underline{1.33L} \qquad \text{and} \qquad y_{CM} = \frac{0 + 0 + 0 + m_0 L}{6m_0} = \underline{0.167L}$$

9.114 A 45-kg boy is sitting on one end of a seesaw while a 15-kg girl sits on the other. They are 4 m apart. How far from the 15-kg girl is the center of mass? Where should the system be pivoted? Neglect the board's weight.

$$x_{CM} = \frac{\sum m_i x_i}{\sum m_i} = \frac{0 + 45 \times 4}{60} = \underline{3.0 \text{ m}}$$

The pivot is best placed at the center of mass.

9.115 A thin uniform rod of length L and mass M_1 has a uniform disk of radius a and mass M_2 fastened to the rod so that the disk and rod are coplanar (Fig. 9-26). The center of the disk is at the rod's end. Find the distance of the disk's center from the center of mass of the combination.

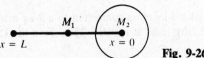

Fig. 9-26

▌ Take $x = 0$ at the attachment point. The disk's mass center coincides with its geometric center and similarly for the rod. Replacing them by two point masses we have

$$x_{CM} = \frac{\sum m_i x_i}{\sum m_i} = \frac{0 + (M_1 L)/2}{M_1 + M_2} = \frac{L M_1}{2(M_1 + M_2)}$$

9.116 A uniform wire is bent into the form of a rectangle with length L and width W. If two of its sides coincide with the $+x$ and $+y$ axes, where is its center of mass?

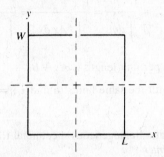

Fig. 9-27

▌ For any mass distribution, the x coordinate of the center of mass is defined by $\sum m_i(x_i - x_{CM}) = 0$. Now, if the mass distribution happens to be symmetric about the line $x = a$, we have $\sum m_i(x_i - a) = 0$; consequently, $x_{CM} = a$. For the problem at hand, this symmetry principle gives both $x_{CM} = L/2$ and $y_{CM} = W/2$. Thus, the CM is located at the geometric center of the rectangle.

9.117 Verify physically that the medians of an equilateral triangle intersect in a point (the center of the triangle).

▌ Imagine a length of uniform wire bent into the shape of the triangle. By the symmetry principle of Prob. 9.116, the CM of the wire must lie on each median. Thus, all three intersect in the CM.

9.118 The uniform solid sphere shown in Fig. 9-28 has a spherical hole in it. Find the position of its center of mass.

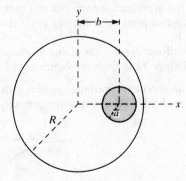

Fig. 9-28

▌ By symmetry, $y_{CM} = z_{CM} = 0$. To find x_{CM}, imagine the hole filled with matter so as to produce a uniform sphere of radius R and density ρ. The filled hole can then be represented by a point mass $\frac{4}{3}\pi a^3 \rho$ at $(b, 0, 0)$. The remainder of the sphere is equivalent to a point mass $\frac{4}{3}\pi(R^3 - a^3)\rho$ at $(x_{CM}, 0, 0)$. The CM of these two point masses must be the CM of the complete sphere, $(0, 0, 0)$. Thus,

$$\frac{4}{3}\pi(R^3 - a^3)\rho x_{CM} + \frac{4}{3}\pi a^3 \rho b = \frac{4}{3}\pi R^3 \rho(0) \qquad \text{or} \qquad x_{CM} = -\frac{a^3 b}{R^3 - a^3}$$

9.119 A rigid body consists of a 3-kg mass connected to a 2-kg mass by a massless rod. The 3-kg mass is located at $\mathbf{r}_1 = 2\mathbf{i} + 5\mathbf{j}$ m, and the 2-kg mass at $\mathbf{r}_2 = 4\mathbf{i} + 2\mathbf{j}$ m. Find the length of the rod and the coordinates of the center of mass.

▌ $$\mathbf{R}_{CM} = \frac{m_1 \mathbf{r}_1 + m_2 \mathbf{r}_2}{m_1 + m_2} = \frac{3}{5}(2\mathbf{i} + 5\mathbf{j}) + \frac{2}{5}(4\mathbf{i} + 2\mathbf{j}) = \frac{14}{5}\mathbf{i} + \frac{19}{5}\mathbf{j} \text{ m}$$

The length of the rod is just $\ell = |\mathbf{r}_1 - \mathbf{r}_2|$, or

$$\ell = \sqrt{[(2-4)^2 + (5-2)^2]} = \sqrt{13} = \underline{3.61 \text{ m}}$$

9.120ᶜ A straight rod of length L has one of its ends at the origin and the other at $x = L$. If the mass per unit length of the rod is given by Ax where A is a constant, where is its center of mass?

▌ $$x_{CM} = \frac{\int x\, dm}{\int dm} = \frac{\int_0^L x(Ax\, dx)}{\int_0^L Ax\, dx} = \frac{\int_0^L x^2\, dx}{\int_0^L x\, dx} = \frac{L^3/3}{L^2/2} = \frac{2}{3}L$$

9.121ᶜ Repeat Prob. 9.120 if the mass per unit length is $Ax + B$.

▌ Proceed as in Prob. 9.120 with the same limits to find $x_{CM} = [\int x(Ax + B)\, dx]/[\int (Ax + B)\, dx]$ $= [L(2AL + 3B)]/(3AL + 6B)$.

9.122 Prove that the center of mass of a homogeneous sheetmetal triangle lies at the intersection of the medians. (This point is called the *barycenter*.)

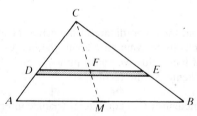

Fig. 9-29

▌ Refer to Fig. 9-29. The center of mass of any narrow strip of the triangle such as DE (which is parallel to side AB), lies at its midpoint F, and therefore lies on the median CM. From the definition of the center of mass, it can easily be shown that if the center of mass of each part of a composite system lies along a single line, then the center of mass of the entire system must lie along that same line. Since the triangle can be regarded as made up of strips such as DE, the center of mass of the entire triangle lies somewhere on the median CM. By repeating this argument, we see that the center of mass of the triangle lies somewhere on each of its medians. Therefore, it must lie at the one point common to all three medians—their intersection.

9.123ᶜ A hemispherical object of uniform density has radius R. Prove by integration that its center of mass lies along its axis of symmetry at a distance $3R/8$ from the center of the plane surface.

▌ As shown in Fig. 9-30, we choose a coordinate system with the origin at the center of the flat face and we let the yz plane be the plane of the face. We now imagine slicing the hemisphere into discs parallel to the yz

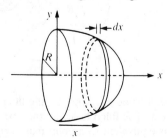

Fig. 9-30

plane. A disc of thickness dx, located a distance x from the plane face, has a radius $\sqrt{R^2 - x^2}$. Therefore the mass of the disk $dm = \pi\rho(R^2 - x^2)\,dx$, where ρ is the mass density of the hemisphere. The center of mass of the hemisphere has an x coordinate x_{CM} given by

$$x_{\text{CM}} = \frac{\int_0^R x \dfrac{dm}{dx}\,dx}{\int_0^R \dfrac{dm}{dx}\,dx} = \frac{\int_0^R \pi\rho x(R^2 - x^2)\,dx}{\int_0^R \pi\rho(R^2 - x^2)\,dx} = \frac{\left(\dfrac{x^2 R^2}{2} - \dfrac{x^4}{4}\right)\Big|_0^R}{\left(xR^2 - \dfrac{x^3}{3}\right)\Big|_0^R} = \frac{\dfrac{R^4}{4}}{\dfrac{2R^3}{3}} = \frac{3R}{8}$$

By Prob. 9.116, $y_{\text{CM}} = z_{\text{CM}} = 0$.

9.124^c Prove that the momentum of a system of N particles is the same as that of a single particle whose mass is that of the whole system, moving with the velocity of the center of mass. (It then follows that the velocity of the CM is constant if the system's momentum is conserved.)

❙ Writing

$$\sum_{i=1}^{N} m_i \equiv M$$

we have, by definition,

$$M\mathbf{R}_{\text{CM}} = \sum_{i=1}^{N} m_i \mathbf{r}_i$$

Differentiate this equation with respect to time:

$$M\dot{\mathbf{R}}_{\text{CM}} = \sum_{i=1}^{N} m_i \dot{\mathbf{r}}_i \qquad \text{or} \qquad M\mathbf{V}_{\text{CM}} = \sum_{i=1}^{N} \mathbf{p}_i \equiv \mathbf{P}$$

9.125 Consider the system composed of a 1-kg body and a 2-kg body initially at rest at a center-to-center distance of 1 m. All numerical values quoted in this exercise are to be considered exact, and the 2-kg body is to the right of the 1-kg body. (**a**) How far is the system's center of mass from the center of the 1-kg body? (**b**) Beginning at $t = 0$ s, a net rightward force of 2 N acts on the 2-kg body. What is its resultant acceleration? (**c**) How far does the 2-kg body move between $t = 0$ s and $t = 1$ s? (**d**) How far is the center of mass from the 1-kg body at $t = 1$ s? (**e**) How far did the center of mass move between $t = 0$ s and $t = 1$ s? (**f**) What is the acceleration of the center of mass, beginning at $t = 0$ s? (**g**) Suppose that all the mass in both objects were concentrated at the center of mass, and the net rightward force of 2 N acted on this concentrated mass. What would its acceleration be? (**h**) State the general theorem which is illustrated by the results of parts (**f**) and (**g**).

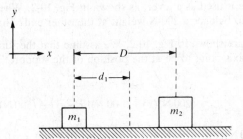

Fig. 9-31

❙ (**a**) The initial configuration is as shown in Fig. 9-31. The center of mass of the system is located between body 1 and body 2, at a distance d_1 from body 1, where $m_1 d_1 = m_2(D - d_1)$. With $D = 1$ m, $m_1 = 1$ kg, and $m_2 = 2$ kg, we find that $d_1 = \frac{2}{3}$ m. (**b**) Applying Newton's second law, we obtain $a_2 = F_2/m_2 = (2\ \text{N}) \div (2\ \text{kg}) = 1.00\ \text{m/s}^2$ rightward. (**c**) The constant-acceleration kinematic equations give $s_2 = \frac{1}{2}a_2 t^2 = 0.5(1.00)(1.00^2) = 0.50\ \text{m}$. (**d**) As before, we write $m_1 d_1' = m_2(D' - d_1')$, so that $d_1' = 2D'/3 = [2(1.5\ \text{m})]/3 = 1.00\ \text{m}$. (**e**) The rightward displacement is $1 - \frac{2}{3} = \frac{1}{3}$ m. (**f**) The x coordinate x_c of the center of mass is given by $x_c = (m_1 x_1 + m_2 x_2)/(m_1 + m_2) = (x_1 + 2x_2)/3$. Therefore $a_c = (a_1 + 2a_2)/3$. Since $a_1 = 0$, $a_c = 2a_2/3 = \frac{2}{3}\ \text{m/s}^2$ rightward. (**g**) In this case $a = F/(m_1 + m_2) = 2\ \text{N}/3\ \text{kg} = \frac{2}{3}\ \text{m/s}^2$ rightward. (**h**) The center of mass of any system moves with an acceleration $\mathbf{a}_c = \mathbf{F}/M$, where $\mathbf{F} \equiv \sum \mathbf{F}_{\text{ext}}$ is the resultant of all external forces acting on the system, and

$$M \equiv \sum_n m_n$$

is the total mass of the system.

CHAPTER 10
Statics of Rigid Bodies

10.1 EQUILIBRIUM OF RIGID BODIES

10.1 A uniform beam weighs 200 N and holds a 450-N weight as shown in Fig. 10-1. Find the magnitudes of the forces exerted on the beam by the two supports at its ends.

▌ The forces on the object being considered (the beam) are also shown in Fig. 10-1. Because the beam is uniform, its center of gravity is at its geometrical center. Thus the weight of the beam (200 N) is shown acting at the beam's center. (For a justification, see Prob. 10.80.) The forces F_1 and F_2 are exerted on the beam by the supports.

We have two equations to write for this equilibrium situation: $\sum F_y = 0$ and $\sum \tau = 0$. There are no x-directed forces on the beam. $\sum F_y = 0$ becomes $F_1 + F_2 - 200\,\text{N} - 450\,\text{N} = 0$. An axis for computing torques may be chosen arbitrarily; see Prob. 10.66. Choose it at A, because the unknown force F_1 will go through it and exert no torque. The torque equation is then $-(200\,\text{N})(L/2) - (450\,\text{N})(3L/4) + (F_2)(L) = 0$. Dividing through the equation by L and then solving for F_2 gives $F_2 = \underline{438\,\text{N}}$.

To find F_1 we substitute the value of F_2 in the first equation and obtain $F_1 = \underline{212\,\text{N}}$.

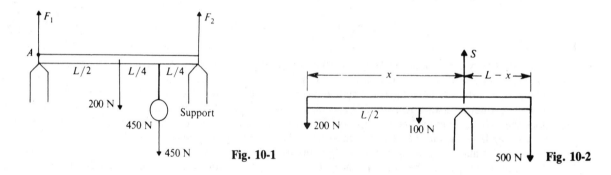

Fig. 10-1   **Fig. 10-2**

10.2 A uniform 100-N pipe is used as a lever, as shown in Fig. 10-2. Where must the fulcrum be placed if a 500-N weight at one end is to balance a 200-N weight at the other end? How much load must the support hold?

▌ The acting forces are shown in Fig. 10-2. We assume that the support point is at a distance x from one end. Let us take the axis point to be at the position of the support. Then the torque equation, $\sum \tau = 0$, becomes

$$(200\,\text{N})(x) + (100\,\text{N})\left(x - \frac{L}{2}\right) - (500\,\text{N})(L - x) = 0$$

This simplifies to $(800\,\text{N})(x) = (550\,\text{N})(L)$, and so we find that $x = 0.69L$. The support should be placed 0.69 of the way from the lighter-loaded end.

To find the load S held by the support, we use $\sum F_y = 0$, which gives $S - 200\,\text{N} - 100\,\text{N} - 500\,\text{N} = 0$, from which $S = \underline{800\,\text{N}}$.

10.3 Where must an 800-N weight be hung on a uniform, 100-N pole so that a boy at one end supports one-third as much as a man at the other end?

▌ In Fig. 10-3, we represent the force exerted by the boy as P and that by the man as $3P$. Take the axis point at the left end. Then the torque equation becomes $-(800\,\text{N})(x) - (100\,\text{N})(L/2) + (P)(L) = 0$.

A second equation we can write is

$$\sum F_y = 0 \qquad \text{or} \qquad 3P - 800\,\text{N} - 100\,\text{N} + P = 0$$

from which $P = 225\,\text{N}$. Substitution of this value in the torque equation gives $(800\,\text{N})(x) = (225\,\text{N})(L) - (100\,\text{N})(L/2)$ from which $x = \underline{0.22L}$. The load should be hung 0.22 of the way from the man to the boy.

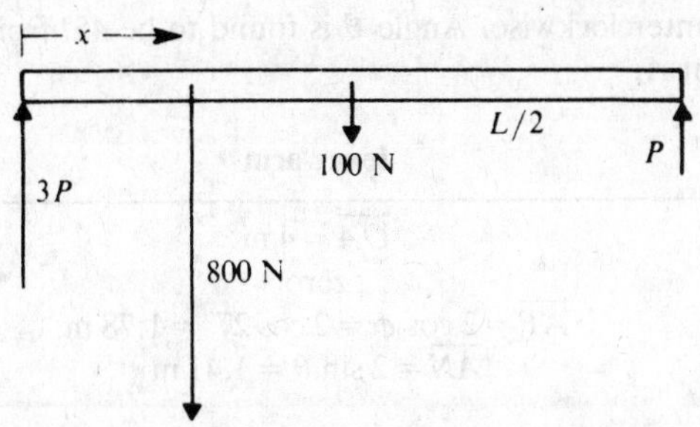

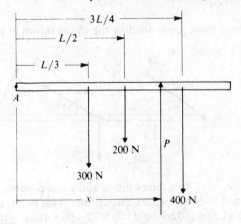

Fig. 10-3

10.4 A uniform 200-N board of length L has two weights hanging from it, 300 N at $L/3$ from one end and 400 N at $3L/4$ from the same end (Fig. 10-4). What single additional force acting on the board will produce equilibrium?

▍ For equilibrium, $\sum F_y = 0$ and so the equilibrant is of magnitude $P = 400\,\text{N} + 200\,\text{N} + 300\,\text{N} = 900\,\text{N}$. Because the board is to be in equilibrium, we are free to choose the axis anywhere. Choose it at end A. Then $\sum \tau = 0$ gives

$$(P)(x) - (400\,\text{N})(3L/4) - (200\,\text{N})(L/2) - (300\,\text{N})(L/3) = 0$$

Using $P = 900\,\text{N}$, we find that $x = 0.56L$. The required force is <u>900 N</u> upward at <u>0.56L</u> from the left end.

Fig. 10-4

10.5 In Fig. 10-5 are shown four forces acting on a sheet of wood; the forces are coplanar with the sheet. (*a*) Find the lever arm for each about point O as axis (i.e., the axis is perpendicular to the plane at O). (*b*) Repeat for point A as axis.

▍ (*a*) By definition, the lever arm is the length of the perpendicular dropped from the axis to the line of the force. For an axis at O:

force	lever arm	sense of rotation
F_1	zero	
F_2	$\overline{OA} = 4\,\text{m}$	ccw
F_3	zero	
F_4	$\overline{OM} = 2\sin\theta = 1.41$	ccw

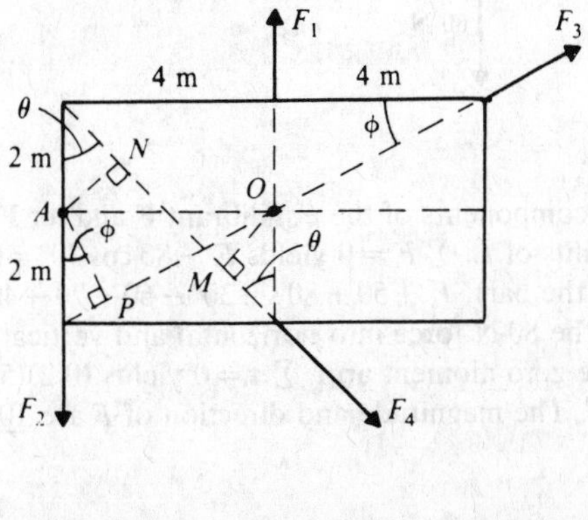

Fig. 10-5

where "ccw" means counterclockwise. Angle θ is found to be 45° from $\tan \theta = 4/4 = 1.00$.

(b) For an axis at point A:

force	lever arm	sense of rotation
F_1	$\overline{OA} = 4 \text{ m}$	ccw
F_2	zero	
F_3	$\overline{AP} = 2 \cos \phi = 2 \cos 27° = 1.78 \text{ m}$	ccw
F_4	$\overline{AN} = 2 \sin \theta = 1.41 \text{ m}$	cw

where "cw" stands for clockwise. Angle ϕ is determined from $\tan \phi = \frac{4}{8} = 0.50$.

10.6   Find the torques due to the forces shown in Fig. 10-5 about **(a)** point O as axis and **(b)** point A as axis.

▌ The lever arms were found in Prob. 10.5. Because ccw torques are positive, we have

force	(a) torque about O	(b) torque about A
F_1	0	$4F_1$
F_2	$4F_2$	0
F_3	0	$1.78F_3$
F_4	$1.41F_4$	$-1.41F_4$

10.7   Evaluate the torques about pivot A provided by the forces shown in Fig. 10-6, if $L = 3.0 \text{ m}$.

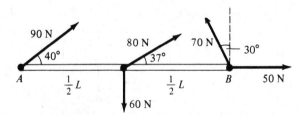

Fig. 10-6

▌ First break the 90-, 80-, and 70-N forces into x and y components. The line of action of the 90, 50, and x components of the 80- and 70-N forces pass through the pivot point; they cause no rotation. The torque about $A = (80 \sin 37°)(1.5) - 60(1.5) + (70 \cos 30°)(3.0) = 72 - 90 + 182 = \underline{+164 \text{ N} \cdot \text{m}}$. The + sign means rotation about A is ccw.

10.8   The uniform bar shown in Fig. 10-7 weighs 40 N and is subjected to the forces shown. Find the magnitude, location, and direction of the force needed to keep the bar in equilibrium.

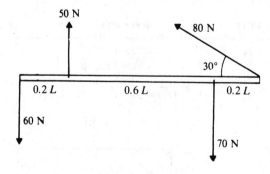

Fig. 10-7

▌ Let F_x and F_y be the components of the equilibrant **F** and let **F** act at a distance x from the left edge, with distances measured in units of L. $\sum F_x = 0$ yields $F_x - 80 \cos 30° = 0$, or $F_x = 69.3 \text{ N}$. $\sum F_y = 0$ yields (not forgetting the weight of the bar). $F_y + 50 + 80 \sin 30° - 60 - 70 - 40 = 0$, or $F_y = 80 \text{ N}$. Taking moments about the left edge, resolving the 80-N force into horizontal and vertical components, and noting that x components of forces on the bar have zero moment arm, $\sum \tau = 0$ yields $(0.2)(50) + (x)(80) + (80)(\sin 30°) - (0.8)(70) - (0.51)(40) = 0$, $x = 0.325$. The magnitude and direction of F are $\underline{106 \text{ N}}$ at $\underline{49°}$ above positive horizontal.

10.9 A force $\mathbf{F} = 6\mathbf{i} + 2\mathbf{j}$ N acts at the point $x = 0$, $y = 5$ m. Find its torque about the origin as pivot.

┃ The lever arm for the y component is zero while it is 5 m for the x component. Therefore, torque $= -(6\,\text{N})(5\,\text{m}) = \underline{-30\,\text{N} \cdot \text{m}}$.

10.10 A force $\mathbf{F} = -2\mathbf{i} + 3\mathbf{j}$ N acts at the point $x = 4$ m, $y = 5$ m. Find its torque about the origin as pivot.

┃ The torque about the origin is $F_y x - F_x y = (3\,\text{N})(4\,\text{m}) - (-2\,\text{N})(5\,\text{m}) = \underline{+22\,\text{N} \cdot \text{m}}$.

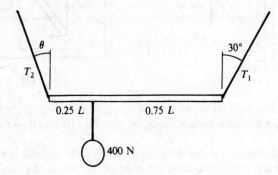

Fig. 10-8

10.11 The uniform 120-N board shown in Fig. 10-8 is supported by two ropes. A 400-N weight is suspended one-fourth of the way from the left end. Find T_1, T_2, and the angle θ made by the left rope.

┃ Remembering that the weight acts at the center of gravity, $\Sigma F_x = 0$ yields $T_1 \sin 30° - T_2 \sin \theta = 0$, or $T_2 \sin \theta = 0.50 T_1$. $\Sigma F_y = 0$ yields $T_1 \cos 30° + T_2 \cos \theta - 120 - 400 = 0$, or $T_2 \cos \theta + 0.866 T_1 = 520$ N. Taking moments about the left edge and resolving T_1 into x and y components, $\Sigma \tau = 0$ yields $L T_1 \cos 30° - (0.25L)(400) - (0.5L)(120) = 0$. Dividing out L and solving, we get $T_1 = \underline{185\,\text{N}}$. Substituting into our earlier equations, we get

$$T_2 \sin \theta = 92.5\,\text{N} \qquad \text{and} \qquad T_2 \cos \theta = 360\,\text{N}$$

Dividing the equations yields $\tan \theta = 0.257$, or $\theta = 14.4°$. Then $0.249 T_2 = 92.5$, and $T_2 = \underline{371\,\text{N}}$. One can always check moment problem results by taking moments about another point, such as the right end of the bar for this problem.

10.12 In reference to Fig. 10-9, the plank is uniform and weighs 500 N. How large must W be if T_1 and T_2 are to be equal?

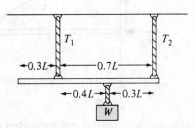

Fig. 10-9

┃ Taking the torques about the attachment point for W, one has $-T_1(0.4L) + T_2(0.3L) + 500(0.2L) = 0$, so that $T = 1000$ N, with $T = T_1 = T_2$. From $\Sigma F_y = 0$ one has $2T - W - 500 = 0$, so $W = \underline{1500\,\text{N}}$.

10.13 Under the assumptions of Prob. 10.12, find T_1 and T_2 if $W = 800$ N.

┃ As in Prob. 10.12, $-0.4 T_1 + 0.3 T_2 + 100 = 0$ and $T_1 + T_2 - 800 - 500 = 0$; solving simultaneously, $T_1 = \underline{700\,\text{N}}$ and $T_2 = \underline{600\,\text{N}}$.

10.14 A uniform beam, which weighs 400 N and is 5.0 m long, is hinged to a wall at its lower end by a frictionless hinge. A horizontal rope 3.0 m long is fastened between the upper end of the beam and the wall. Find the force T [see Fig. 10-10(b)] exerted on the beam by the rope, and also find the horizontal and vertical components of the force exerted on the beam by the hinge.

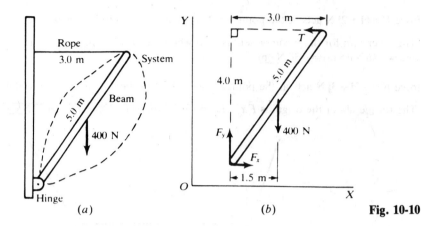

Fig. 10-10

❚ Let a frame of reference be set up with the X axis directed horizontally to the right and the Y axis directed vertically upward. Our system is the beam [see Fig. 10-10(a)] and we consider all forces on it [Fig. 10-10(b)].

Replace the force exerted on the beam by the hinge by its horizontal component F_x and its vertical component F_y. The weight acts at the center of mass of the beam. The rope pulls on the beam and therefore exerts a force T on the beam which acts to the left.

Consider torques about a horizontal axis perpendicular to the beam and passing through the hinge. The moment arms of F_x and F_y are zero and these forces have zero torque about this axis. The 400-N weight has a moment arm of 1.5 m. The moment arm of the force T is the distance along the wall from the hinge to the rope, $\sqrt{(5.0)^2 - (3.0)^2} = 4.0$ m. The torque condition is then

$$\sum \tau = T(4) - 400(1.5) = 0 \quad \text{or} \quad T = 150\,\text{N}$$

and the force conditions give

$$\sum F_x = F_x - T = 0 \quad \text{or} \quad F_x = T = 150\,\text{N} \qquad \sum F_y = F_y - 400 = 0 \quad \text{or} \quad F_y = 400\,\text{N}$$

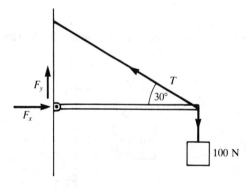

Fig. 10-11

10.15 A weightless strut is hinged at a wall with the far end supported by a cord (Fig. 10-11). A weight of 100 N hangs from the far end. Find the tension in the cord and the horizontal and vertical components of the force exerted on the strut at the hinge.

❚ First note that $F_y = 0$; this is a consequence of the fact that the only forces acting on the strut act at two points, the left and right end. Thus if we take the torque (moment) about an axis through the right end, the only force that can contribute is F_y. Then $\sum \tau = 0$ implies $F_y = 0$. $\sum F_x = 0$ yields $F_x - T \cos 30° = 0$, or $F_x = 0.866T$. $\sum F_y = 0$ yields $T \sin 30° - 100 = 0$, or $\underline{T = 200\,\text{N}}$. Then $\underline{F_x = 173\,\text{N}}$. Note that the vertical component of T just balances the 100-N weight so that the net force on the right end of the strut is also along the strut.

10.16 Rework Prob. 10.15 if the (uniform) strut weighs 60 N.

❚ $\sum F_x = 0$ yields $F_x - T \cos 30° = 0$, or $F_x = 0.866T$. $\sum F_y = 0$ yields $F_y + T \sin 30° - 60\,\text{N} - 100\,\text{N} = 0$, or $F_y + 0.50T = 160\,\text{N}$. Taking moments about an axis through the right end we have only F_y and the strut's weight with nonzero moment arms. $\sum \tau = 0$ yields $-LF_y + (L/2) 50\,\text{N} = 0$. Note that since the length L of the strut divides out of the equation, we do not need its value. Solving we get $\underline{F_y = 30\,\text{N}}$. From the force equations we obtain $\underline{T = 260\,\text{N}}$ and $\underline{F_x = 225\,\text{N}}$.

10.17   In Fig. 10-12(a) the uniform beam weighs 500 N. If the tie rope can support 1800 N, what is the maximum value the load w can have?

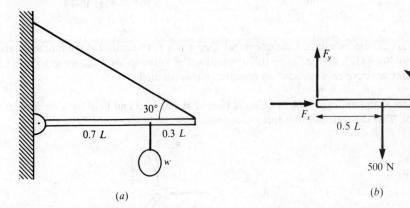

(a)   (b)   **Fig. 10-12**

▮   The forces on the beam are illustrated in Fig. 10-12(b). To find w_{max} we want a relationship between w and T. Taking moments about an axis through the left edge of the beam accomplishes this. We resolve T into vertical and horizontal components, realizing that T_x has no moment arm. $\sum \tau = 0$ yields $LT \sin 30° - (0.5L)(500) - 0.7Lw = 0$. Dividing out L and solving for w we get $w = 0.714T - 357\,\text{N}$. Setting $T = T_{max} = 1800\,\text{N}$ yields $\underline{w_{max} = 928\,\text{N}}$.

10.18   A light strut has its left end hinged to the wall and is supported at the other end by a cable as shown in Fig. 10-13. Find (a) the tension in the cable T, and (b) the direction of the force **F** exerted on the strut at the hinge.

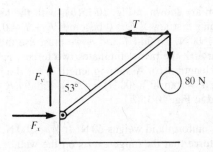

Fig. 10-13

▮   (a) To get T we take moments about an axis through the hinge point; assume that the strut has length L. $\sum \tau = 0$ yields $-(\sin 53°)(80) + (\cos 53°)(T) = 0$ or $\underline{T = 106\,\text{N}}$. (b) Taking moments about the right end, $\sum \tau = 0$ implies that either $F = 0$ or its moment arm is 0. Since F is clearly not 0, the moment arm must be 0, and **F** must point along the strut. (Again, this kind of argument fails if the strut has weight, since then there are forces acting at more than two points.)

10.19   The strut of Fig. 10-13 is now assumed to weigh 40 N and the center of gravity is one-third of the way along the strut from the left end. Find the tension T and the magnitude and direction of the force **F** acting on the strut at the hinge.

▮   (a) As in Prob. 10.18, find T by taking moments about an axis through the hinge. $\sum \tau = 0$ yields $-(\sin 53°)(80) - (\frac{1}{3})(\sin 53°)(40) + (\cos 53°)(T) = 0$, or $\underline{T = 124\,\text{N}}$. $\sum F_x = 0$ yields $F_x - T = 0$, or $\underline{F_x = T = 124\,\text{N}}$. $\sum F_y = 0$ yields $F_y - 40 - 80 = 0$, or $\underline{F_y = 120\,\text{N}}$. $F = (F_x^2 + F_y^2)^{1/2} = \underline{173\,\text{N}}$, and $\theta = \tan^{-1}(F_y/F_x) = \underline{44.1°}$ above positive x axis.

10.20   A man carries a bar 12 ft long which has two loads, one of 40 lb and the other of 60 lb, hung from its ends. The bar is uniform and weighs 30 lb. At which point should the man hold up the bar to keep it horizontal?

▮   The force diagram for the bar is given in Fig. 10-14. $\sum F_y = 0$ yields $F_{man} - 60 - 30 - 40 = 0$, or

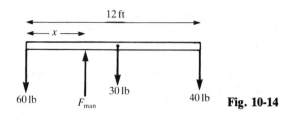

Fig. 10-14

$F_{man} = 130$ lb. There are no forces with x components so $\Sigma F_x = 0$ is not useful. Taking moments about the left end, $\Sigma \tau = 0$ yields $(6)(30) + (12)(40) - xF_{man} = 0$. Substituting the value in for F_{man}, we get $\underline{x = 5.1 \text{ ft}}$ from the 60-lb weight. (Note that we have chosen "cw" as positive in this problem.)

10.21 As shown in Fig. 10-15(a) the uniform 1600-N beam is hinged at one end and held by a tie rope at the other. Determine the tension T in the rope and the force components at the hinge.

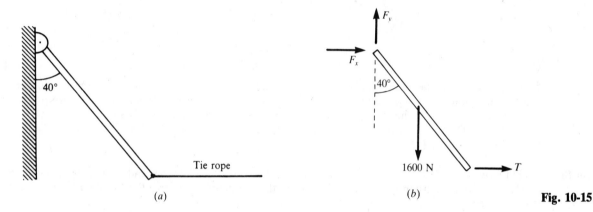

(a)   (b)   **Fig. 10-15**

▌ The forces on the beam are shown in Fig. 10-15(b), with the hinge force already separated into components. From Newton's first law $\Sigma F_x = 0$ becomes $F_x + T = 0$, and $F_x = -T$. $\Sigma F_y = 0$ becomes $F_y - 1600$ N $= 0$, and $\underline{F_y = 1600 \text{ N}}$. To find T and F_x we must use the second condition of equilibrium. Taking moments about an axis through the pivot eliminates two forces, F_x and F_y, from the torque equation since their moment arms are of length zero. Assuming length L for the beam, $\Sigma \tau = 0$ yields $(\cos 40°)(T) - (\frac{1}{2})(\sin 40°)(1600 \text{ N}) = 0$ so $T = (\tan 40°)(800 \text{ N}) = \underline{671 \text{ N}}$. Finally $\underline{F_x = -671 \text{ N}}$, indicating it is in the opposite direction to that presumed in Fig. 10-15(b).

10.22 In Fig. 10-16, the beam is uniform and weighs 60 N. If $W = 200$ N, find the tension in the tie rope and the x and y components of the force that the hinge exerts on the wall.

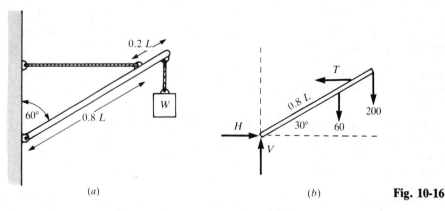

(a)   (b)   **Fig. 10-16**

▌ Taking torques about the hinge point, $-60(0.5L \cos 30°) - 200(L \cos 30°) + T(0.8L \sin 30°) = 0$, so $T = \underline{500 \text{ N}}$. Since $\Sigma F_x = 0$, $-T + H = 0$ and $H = \underline{500 \text{ N}}$. Since $\Sigma F_y = 0$, $V = 200 + 60 = \underline{260 \text{ N}}$.

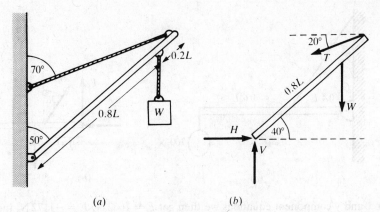

(a) (b) **Fig. 10-17**

10.23 Neglecting the weight of the beam in Fig. 10-17(a), find the tension in the tie rope and the force components at the hinge in terms of W. Repeat if the uniform beam weighs $\frac{1}{2}W$.

▮ Refer to Fig. 10-17(b). Taking torques about the hinge, $-W(0.8L\cos 40°) + (T\cos 20°)(L\sin 40°) - (T\sin 20°)(L\cos 40°) = 0$. This yields $T = \underline{1.80W}$. From $\sum F_x = 0$, $H - T\cos 20° = 0$, giving $H = \underline{1.69W}$. From $\sum F_y = 0$, $V - W - T\sin 20° = 0$, and so $V = \underline{1.62W}$. If the beam has weight, then an additional $W/2$ must be added acting vertically downward through the center of the beam. The torque about the hinge will be $-(W/2)(0.5L\cos 40°)$ due to the weight. With these additions to the equations above, we obtain $T = \underline{2.35W}$, from which $H = \underline{2.21W}$ and $V = \underline{2.30W}$.

10.24 In Fig. 10-17(a), if the tie rope can hold a maximum tension of 1000 N and if the beam is uniform and weighs 200 N, what is the maximum weight W which can be supported as shown?

▮ Solve in the same way as in Prob. 10.23, except with 200 N added at the center of gravity and $T = 1000$ N: $(-W)(0.8L\cos 40°) - (200)(0.5L\cos 40°) + (1000\cos 20°)(L\sin 40°) - (1000\sin 20°)(L\cos 40°) = 0$. This yields $W = \underline{430\text{ N}}$.

10.25 A display sign is supported by a guy wire making an angle of 37° with the horizontal and by a bracket at the upper end near the vertical wall. The sign weighs 150 N. See Fig. 10-18. Find the tension in the wire and the force exerted by the bracket on the sign.

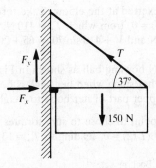

Fig. 10-18

▮ The forces acting on the sign are also displayed in Fig. 10-18, with the force due to the bracket broken into x and y components. Taking moments about an axis through the bracket and letting the width of the sign be L, we resolve T into x and y components. Then $\sum \tau = 0$ yields $T\sin 37° - (\frac{1}{2})(150) = 0$, or $T = \underline{125\text{ N}}$. $\sum F_x = 0$ yields $F_x - T\cos 37° = 0$, and $\underline{F_x = 100\text{ N}}$. $\sum F_y = 0$ yields $F_y + T\sin 37° - 150 = 0$, or $\underline{F_y = 75\text{ N}}$. Note that the vertical components of T and F_y are equal.

10.26 The uniform beam shown in Fig. 10-19(a) weighs 500 N and supports a 700-N load. Find the tension in the tie rope and the force of the hinge on the beam.

▮ The forces acting on the beam are shown in Fig. 10-19(b). $\sum F_x = 0$ yields $F_x - H\sin 35° = 0$, or $F_x = 0.574H$. $\sum F_y = 0$ yields $F_y + H\cos 35° - 500 - 700 = 0$, or $F_y + 0.819H = 1200$ N. Taking moments about an axis through the hinge yields an equation in which H is the only unknown. Breaking H into components and noting that the moment arm to H_x is zero and the one to $H_y = H\cos 35°$ is $0.4L$, we proceed. $\sum \tau = 0$ yields $0.4LH\cos 35° - (0.5L)(500\text{ N}) - L(700\text{ N}) = 0$. Dividing out L and solving for H, we get $\underline{H = 2896\text{ N}}$.

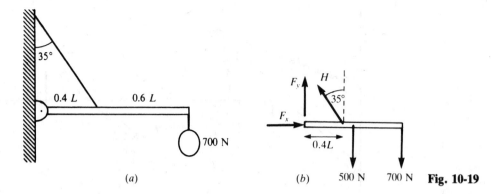

(a)

(b)   500 N   700 N   **Fig. 10-19**

From our x and y component equations we then get $F_x = \underline{1662\,N}$, $F_y = -\underline{1172\,N}$, indicating that the vertical force at the hinge pushes down. Using $F = (F_x^2 + F_y^2)^{1/2}$ and $\tan \theta = F_y/F_x$ we have for the magnitude and direction of **F**: $\underline{2036\,N}$ and $\underline{35°}$ below the positive horizontal.

10.27   A person holds a 20-N weight as shown in Fig. 10-20(a). Find the tension in the supporting muscle and the component forces at the elbow. Assume that the system can be approximated as shown in Fig. 10-20(b), where T_m is the tension in the muscle, the beam is the lower arm, and the lower arm weighs 65 N with center of gravity as shown.

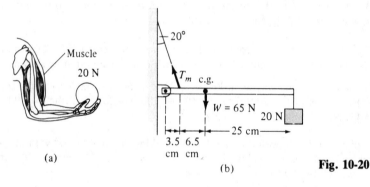

(a)

(b)   **Fig. 10-20**

▮ Let H and V be the forces exerted at the elbow. Take torques about the hinge and equate to zero: $+3.5T_m \sin 70° - 10(65) - 35(20) = 0$, from which $T_m = \underline{410\,N}$. Summing horizontal and vertical forces to zero we have $H = 410 \cos 70° = \underline{140\,N}$ and $V + 410 \sin 70° = 65 + 20$, which yields $V = \underline{300\,N}$ (down).

10.28   Consider a woman lifting a 60-N bowling ball as shown in Fig. 10-21(a). Find the tension in her back muscle and the compressional force in her spine when her back is horizontal. Approximate the situation as shown in Fig. 10-21(b) and assume the upper part of her body to weigh 250 N with center of gravity as indicated.

▮ Again the most convenient point to chose to sum torques is at the hinge. The torque equation is $+60L - (2L/3)(T_m \sin 12°) + 250(L/2) = 0$, leading to $T_m = \underline{1335\,N}$. Also $T \cos 12° - H = 0$, so $H = \underline{1305\,N}$.

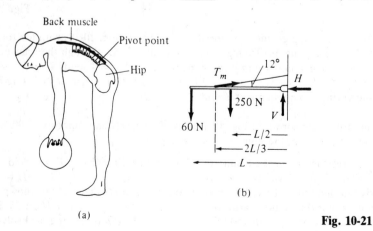

(a)

(b)

Fig. 10-21

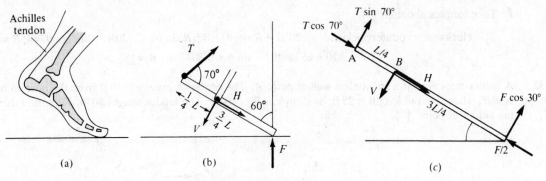

Fig. 10-22

10.29 When one stands on tiptoe and at equilibrium, the situation is much like that shown in Fig. 10-22(a). We can replace the actual situation by the model in Fig. 10-22(b). In terms of the push F of the floor on the toe, find (a) the tension in the Achilles tendon and (b) by the forces H and V at the ankle.

▌ Take the components of F and T parallel to and perpendicular to the beam (Fig. 10-22(c)). Torque about point $A = V(L/4) - LF \cos 30° = 0$, yielding $V = \underline{3.47F}$. Taking torques about B, $-(L/4)T \sin 70° + (3L/4)F \cos 30° = 0$ gives $T = \underline{2.77F}$. H can be found by setting forces parallel to the beam = 0; $H - F/2 + T \cos 70° = 0$, so $H = \underline{0.45F}$ and is directed toward point A.

10.30 (a) Consider the object of negligible weight shown in Fig. 10-23. If $T_1 = 40$ N, determine how large T_2, V, and H must be if the object is to remain in equilibrium? (b) Repeat if the force T_2 is pulling at an angle of 37° below the horizontal.

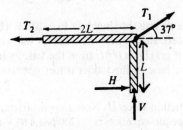

Fig. 10-23

▌ (a) Take torques about the upper-right-hand corner. Only H has a nonzero lever arm, so $HL = 0$, and therefore $H = 0$. Also $\Sigma F_x = 0$ gives $-T_2 + H + T_1 \cos 37° = 0$; $T_2 = 0.8T_1$, $\Sigma F_y = 0$ gives $V + T_1 \sin 37° = 0$, so $-V = 0.6T_1$. Since $T_1 = 40$ N we have $T_2 = \underline{32\,\text{N}}$, $V = \underline{-24\,\text{N}}$, $\underline{H = 0}$.
(b) The equations become $-HL - (0.6T_2)(2L) = 0$, $-0.8T_2 + H + 0.8T_1 = 0$, and $V + 0.6T_1 - 0.6T_2 = 0$. But $T_1 = 40$ N, and so solving these equations gives $T_2 = \underline{16}$, $V = \underline{-14.4}$, and $H = \underline{-19.2\,\text{N}}$.

10.31 A flat, uniform, cylindrical steel plate is 2.40 ft in diameter and weighs 45.0 lb. It is supported by a horizontal pivot at a point on the rim (labeled O in Fig. 10-24). A 20.0-lb weight is supported by a rope wrapped around the cylinder. Find the angle the radius OA makes with the vertical when the system is in equilibrium.

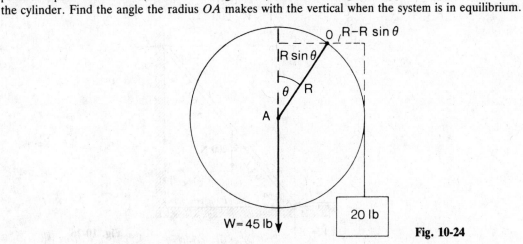

Fig. 10-24

▮ Take torques about O:

$$\text{clockwise} = \text{counterclockwise} \qquad 20(R - R\sin\theta) = 45(R\sin\theta) \qquad 20R - 20R\sin\theta = 45R\sin\theta$$

$$20 = 65\sin\theta \qquad \sin\theta = 0.3077 \qquad \theta = \underline{18°}$$

10.32 A ladder rests against a frictionless wall at point A. The center of gravity is 10 ft from the bottom of the ladder, whose overall length is 25 ft, as shown in Fig. 10-25. The ladder weighs 80 lb. Find the force acting on the ladder at point A.

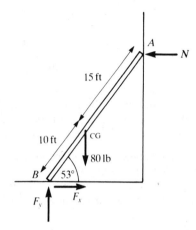

Fig. 10-25

▮ We show all forces acting on the ladder in the figure. Since we are interested only in N, the force due to the wall, we take moments about B, which gives an equation involving only the unknown, N. $\sum \tau = 0$ yields $(25\sin 53°)(N) - (10\cos 53°)(80) = 0$. Solving we have $N = \underline{24\text{ lb}}$.

10.33 A ladder leans against a smooth wall, as shown in Fig. 10-26. (By a "smooth" wall, we mean that the wall exerts on the ladder only a force that is perpendicular to the wall. There is no friction force.) The ladder weighs 200 N and its center of gravity is $0.4L$ from the base, where L is the ladder's length. (*a*) How large a friction force must exist at the base of the ladder if it is not to slip? (*b*) What is the necessary coefficient of static friction?

▮ (*a*) We wish to find the friction force H. Note that no friction force exists at the top of the ladder. Taking torques about point A, the torque equation is $-(200\text{ N})(\overline{AB}) + (P)(\overline{AC}) = 0$, or $-(200\text{ N})(0.4L\cos 50°) + (P)(L\cos 40°) = 0$. Solving gives $P = 67\text{ N}$. We can also write

$$\sum F_x = 0 \qquad \text{or} \qquad H - P = 0$$

$$\sum F_y = 0 \qquad \text{or} \qquad V - 200 = 0$$

and so $H = 67\text{ N}$ and $V = 200\text{ N}$. Since $V = Y$, the normal force,

(*b*)
$$\mu_s = \frac{f}{Y} = \frac{H}{V} = \frac{67}{200} = \underline{0.34}$$

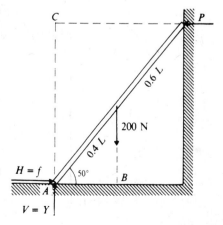

Fig. 10-26

10.34 A light ladder is supported on a rough floor and leans against a smooth wall, touching the wall at height h above the floor. A man climbs up the ladder until the base of the ladder is on the verge of slipping. The coefficient of static friction between the foot of the ladder and the floor is μ. What is the *horizontal* distance moved by the man?

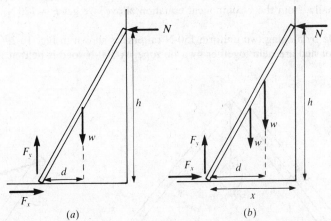

Fig. 10-27

▌ At the verge of slipping (see Fig. 10-27(a)), $F_x = F_{x,\max} = \mu F_y$. For equilibrium, $\sum F_y = 0$ yields $F_y - w = 0$, or $F_y = w$. $\sum F_x = 0$ yields $F_x - N = 0$, or $F_x = N$. Thus,

$$\mu = \frac{F_x}{F_y} = \frac{N}{w}$$

Taking moments about the bottom of the ladder, we have $hN - dw = 0$, or $N/w = d/h$. Thus $\mu = d/h$, $d = \mu h$.

10.35 Suppose that the ladder in Prob. 10.34 is uniform and has the same weight as the man, and the base of the ladder is a distance x from the wall, all else being the same. Find the horizontal distance the man moves to reach the point of just slipping. See Fig. 10-27(b).

▌ As in Prob. 10.34, $F_x = F_{x,\max} = \mu F_y$ and $\sum F_x = 0$ yields $F_x = N$. Now, however, $\sum F_y = 0$ yields $F_y = 2w$, so

$$\mu = \frac{F_x}{F_y} = \frac{N}{2w}$$

Taking the moments about the bottom of the ladder yields

$$hN - \frac{x}{2}w - dw = 0 \quad \text{or} \quad \frac{N}{w} = \frac{x/2 + d}{h} \qquad \mu = \frac{N}{2w} = \frac{x/2 + d}{2h} \quad \text{and} \quad d = 2\mu h - \frac{x}{2}$$

10.36 The foot of a ladder rests against a wall and its top is held by a tie rope, as shown in Fig. 10-28(a). The ladder weighs 100 N and its center of gravity is 0.4 of its length from the foot. A 150-N child hangs from a rung that

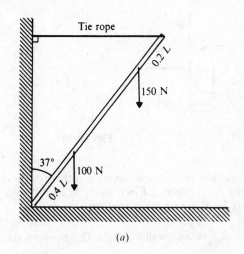

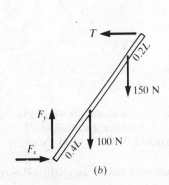

Fig. 10-28

is 0.2 of the length from the top. Determine the tension in the tie rope and the components of the force on the foot of the ladder.

❙ The forces acting on the ladder are shown in Fig. 10-28(b). $\sum F_x = 0$ yields $F_x - T = 0$, or $F_x = T$. $\sum F_y = 0$ yields $F_y - 100 - 150 = 0$, or $F_y = \underline{250\,\text{N}}$. We take moments about the bottom of the ladder. $\sum \tau = 0$ yields $L \cos 37° \, T - (0.4L \sin 37°)(100\,\text{N}) - (0.8L \sin 37°)(150\,\text{N}) = 0$. Dividing L out of all terms and solving, we get $T = \underline{121\,\text{N}}$. Finally, from the x-component equation above, we get $F_x = \underline{120\,\text{N}}$.

10.37 A truss is made by hinging two uniform 150-N rafters, as shown in Fig. 10-29(a). They rest on an essentially frictionless floor and are held together by a tie rope. A 500-N load is held at their apex. Find the tension in the tie rope.

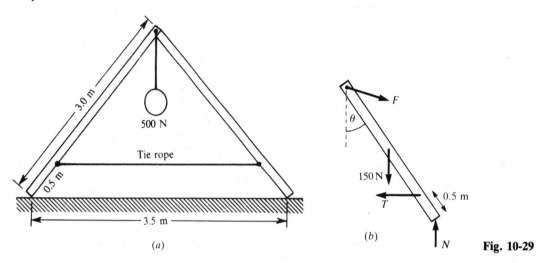

(a)

(b)

Fig. 10-29

❙ There are many approaches to this problem, but the one to take should yield the desired answer as quickly as possible. We are interested in only one unknown, the tension in the tie rope. Considering one of the rafters to be our system, we show all forces acting on it in Fig. 10-29(b). The force **F** includes the effect of the 500-N weight as well as the force due to the other rafter. If we take moments about an axis through the top of the rafter, we get $\sum \tau = 0$ yields $1.75N - (1.75/2)(150) - 2.5 \cos \theta T = 0$. Since $\sin \theta = (1.75/3.0)$, $\theta = 35.7°$, and $\cos \theta = 0.812$, therefore $2.03T = 1.75N - 131$. To find N we return to the complete system in Fig. 10-29(a) and appeal to symmetry. The normal force due to the floor on each rafter is the same. Furthermore, considering the entire system as one rigid body, $\sum F_y = 0$ yields $2N = (500 + 150 + 150)\,\text{N}$, and $N = \underline{400\,\text{N}}$. Substituting into our moment equation and solving we get $T = \underline{280\,\text{N}}$.

10.38 A sphere of radius 0.10 m and mass 10 kg rests in the corner formed by a 30° inclined plane and a smooth vertical wall. Calculate the forces that the two surfaces exert on the sphere.

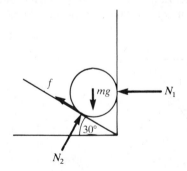

Fig. 10-30

❙ The possible forces are shown in Fig. 10-30. If we take moments about an axis through the center of the sphere, only f can have a torque and $\sum \tau = 0$ implies $f = 0$. Then $\sum F_y = 0$ yields $N_2 \cos 30° = mg = (10\,\text{kg})(9.8\,\text{m/s}^2)$; $\sum F_x = 0$ yields $N_2 \sin 30° - N_1 = 0$, or $N_1 = \underline{56.5\,\text{N}}$, $N_2 = \underline{113\,\text{N}}$.

10.39 A 13.0-m uniform ladder weighing 300 N rests against a smooth wall at height 12.0 m above the floor. The

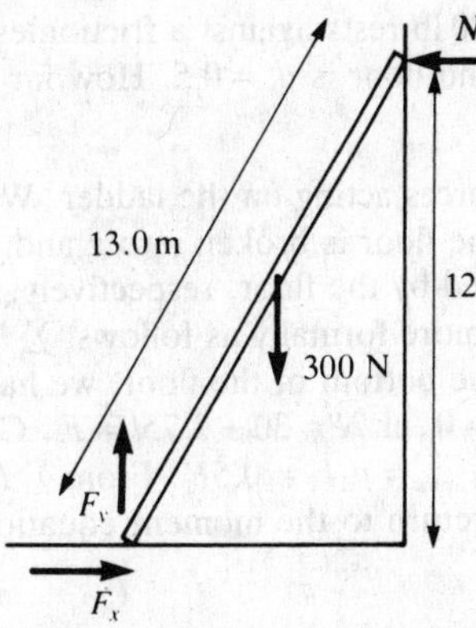

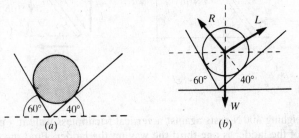

Fig. 10-31

floor is rough. Find the frictional force and the normal force exerted on the ladder by the floor and the normal force exerted on the ladder by the wall.

▌ The forces acting on the ladder are shown in Fig. 10-31. The force of the floor on the ladder is separated into x and y components which correspond to the frictional and normal forces of the floor, respectively. $\sum F_x = 0$ yields $F_x - N = 0$, or $F_x = N$. $\sum F_y = 0$ yields $F_y - 300 = 0$, or $F_y = 300\,\text{N}$. Taking moments about an axis through the point of contact with the floor, and noting the distance from this point to the wall is $\sqrt{13^2 - 12^2} = 5\,\text{m}$, $\sum \tau = 0$ yields $12.0N - (2.5)(300) = 0$, or $N = \underline{62.5\,\text{N}}$. Then $F_x = \underline{62.5\,\text{N}}$ as well.

10.40 The bowling ball shown in Fig. 10-32(a) weighs 70 N and rests against the walls of a *frictionless* groove as indicated. How large are the forces that the walls of the groove exert on the ball?

[Figure 10-32(a): a ball resting in a groove with walls at 60° and 40° to the horizontal]

[Figure 10-32(b): free-body diagram showing forces R and L at 60° and 40°, and weight W downward]

(a)   (b)   **Fig. 10-32**

▌ The force exerted by a frictionless surface must be normal to the surface. In addition to the normal forces due to the left and right hand surfaces, L and R, the 70-N weight acts straight down. All three forces act through the ball's center [Fig. 10-32(b)]. Then $\sum F_x = 0$ yields $L \sin 60° = R \sin 40°$ and $\sum F_y = 0$ gives $70 = R \cos 40° + L \cos 60°$. Solving, $R = \underline{62}$ and $L = \underline{46\,\text{N}}$.

10.41 A refrigerator with a mass of 100 kg measures 150 cm high by 75 cm wide by 75 cm deep. How much force applied horizontally at the top edge of the front will make it start tipping backward? Assume that the center of gravity is at the center of the refrigerator.

▌ Take torques about O in Fig. 10-33:

$$\text{clockwise} = \text{counterclockwise} \qquad mg(0.375) = 1.50F \qquad F = \frac{100(9.8)(0.375)}{1.50} = \underline{245\,\text{N}}$$

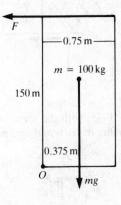

Fig. 10-33

10.42 A uniform 40-ft ladder weighing 80 lb rests against a frictionless wall as shown in Fig. 10-34(*a*). The coefficient of static friction between ladder and floor is $\mu_s = 0.5$. How far up the ladder can a 200-lb man go before the ladder slips?

▎ In Fig. 10-34(*b*) we show all forces acting on the ladder. We wish to find the distance *l* the man can climb up the ladder. The force due to the floor is broken into *x* and *y* components, which correspond to the frictional and normal forces exerted by the floor, respectively. While it is intuitive that F_x increases as the man moves up the ladder, we show it more formally as follows. $\Sigma F_x = 0$ yields $F_x - N = 0$, or $F_x = N$. Taking moments about an axis through the bottom of the floor, we have $\Sigma \tau = 0$, or $(40 \sin 53°)(N) - (20 \cos 53°)(80) - (l \cos 53°)(200) = 0$, or $N = 30 + 3.75l = F_x$. Clearly F_x increases with *l*. At maximum *l*, that is, just at the verge of slipping, $F_{x,\max} = \mu_s F_y = 0.5 F_y$. From $\Sigma F_y = 0$ we get $F_y - 80 - 200 = 0$, or $F_y = 280$ lb. Then $F_{x,\max} = 140$ lb $= N$, and we return to the moment equation to get $30 + 3.75l = 140$ lb, and $\underline{l = 29.3\ \text{ft}}$ up the ladder.

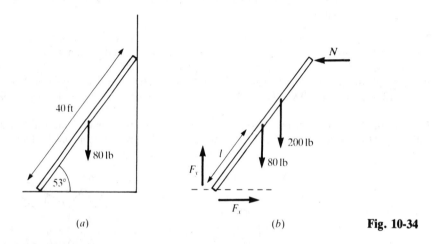

(*a*)   (*b*)   **Fig. 10-34**

10.43 A 21-m ladder weighing 400 N rests against a vertical frictionless wall at a point 16 m above the ground. The center of gravity of the ladder is one-third the way up the ladder. Find the coefficient of friction between the ladder and the horizontal ground at its base if an 80-kg man climbs halfway up before the ladder starts to slip.

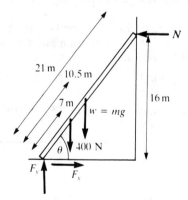

Fig. 10-35

▎ All the forces are shown in Fig. 10-35, with the force of the floor on the ladder resolved into horizontal and vertical components, corresponding to frictional and normal forces, respectively. F_x is assumed at its maximum value so

$$F_{x,\max} = \mu_s F_y \qquad \text{or} \qquad \mu_s = \frac{F_{x,\max}}{F_y}$$

Noting that the weight of the man is $w = mg = (80\,\text{kg})(9.8\,\text{m/s}^2) = 784\,\text{N}$, $\sum F_y = 0$ yields $F_y - 400\,\text{N} - 784\,\text{N} = 0$, or $F_y = \underline{1184\,\text{N}}$. $\sum F_x = 0$ yields $F_{x,\text{max}} - N = 0$, or $F_{x,\text{max}} = N$. Taking moments about the base of the ladder we have $(21\sin\theta)(N) - (10.5\cos\theta)(784) - (7\cos\theta)(400) = 0$. To solve for N we need θ. From the dimensions given, $\sin\theta = \frac{16}{21} = 0.762$ and then $\cos\theta = 0.648$. Solving, we have $N = \underline{447\,\text{N}} = F_{x,\text{max}}$, and, finally, $\mu_s = \frac{447}{1184} = \underline{0.38}$.

10.44 The hinges of a uniform door weighing 200 N are 2.50 m apart. One hinge is at a distance d from the top of the door, while the other is a distance d from the bottom. The door is 1.00 m wide. The weight of the door is supported by the lower hinge. Determine the forces exerted by the hinges on the door.

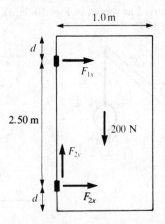

Fig. 10-36

▮ We naturally assume that the hinges are small in comparison with the lengths involved. Figure 10-36 shows all forces on the door. $F_{1y} = 0$ is in effect given. $\sum F_y = 0$ yields $F_{2y} = 200\,\text{N}$. $\sum F_x = 0$ yields $F_{1x} + F_{2x} = 0$, or $F_{1x} = -F_{2x}$. Taking moments about the lower hinge we have $\sum \tau = 0$ yields $2.50F_{1x} + (0.50)(200\,\text{N}) = 0$ and $F_{1x} = \underline{-40\,\text{N}}$. Then $F_{2x} = 40\,\text{N}$. Thus $\mathbf{F}_1$ pulls the door to the left. The magnitude and direction of $\mathbf{F}_2$ are 204 N, at $\theta = \tan^{-1}\frac{200}{40} = 79°$ above positive horizontal. Note that the distribution of weight between the hinges (F_{1y} and F_{2y}) must always be given in these kinds of problems: because F_{1y} and F_{2y} have the same line of action and have the same moment arm about any axis, you can never distinguish F_{1y} and F_{2y} without additional information; you can only calculate their sum.

10.45 In Fig. 10-37(a), the uniform 600-N beam is hinged at P. Find the tension in the tie rope and the components of the force exerted by the hinge on the beam.

▮ The forces acting on the beam are shown in Fig. 10-37(b). We represent the force exerted by the hinge by its components, H and V. We can work either with the tension T in the rope or with its components. In terms of T, the torque equation about P as axis is

$$(T)\left(\frac{3L}{4}\sin 40°\right) - (800\,\text{N})(L) - (600\,\text{N})\left(\frac{L}{2}\right) = 0$$

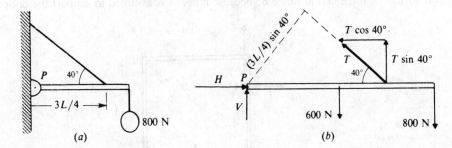

(a)　　　　　　　　　　(b)　　　　　　　　**Fig. 10-37**

We take the axis at P, because then H and V do not appear in the torque equation. Or we could write the torque equation in terms of the components of T. If we note that the component $T\cos 40°$ actually acts at the tie point and has zero lever arm, the torque equation in terms of the components of T is

$$(T\sin 40°)\left(\frac{3L}{4}\right) - (800\,\text{N})(L) - (600\,\text{N})\left(\frac{L}{2}\right) = 0$$

Note that the two equations are identical. Either one yields $T = \underline{2280\,\text{N}}$.
To find H and V we write

$$\sum F_x = 0 \qquad \text{or} \qquad -T\cos 40° + H = 0$$

$$\sum F_y = 0 \qquad \text{or} \qquad T\sin 40° + V - 600 - 800 = 0$$

Using the value found for T, these equations give $H = \underline{1750\,\text{N}}$ and $V = \underline{-66\,\text{N}}$.

10.46 A uniform 400-N boom is supported as shown in Fig. 10-38(a). Find the tension in the tie rope and the force exerted on the boom by the pin at P.

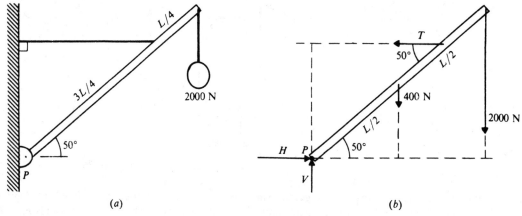

Fig. 10-38

▮ The forces acting on the boom are shown in Fig. 10-38(b). Taking the pin as axis, the torque equation is

$$(T)\left(\frac{3L}{4}\sin 50°\right) - (400\,\text{N})\left(\frac{L}{2}\cos 50°\right) - (2000\,\text{N})(L\cos 50°) = 0$$

from which $T = \underline{2460\,\text{N}}$. We now write $\sum F_x = 0$, or $H - T = 0$, and so $H = 2460\,\text{N}$. Also $\sum F_y = 0$, or $V - 2000\,\text{N} - 400\,\text{N} = 0$, and so $V = 2400\,\text{N}$. These are the components of the force at the pin. The magnitude of this force is $\sqrt{(2400)^2 + (2460)^2} = \underline{3440\,\text{N}}$. The tangent of the angle it makes with the horizontal is $\tan\theta = 2400/2460$, and so $\underline{\theta = 44°}$.

10.47 As shown in Fig. 10-39, the hinges A and B hold a uniform 400-N door in place. The upper hinge supports the entire weight of the door. Find the forces exerted on the door at the hinges. The width of the door is $h/2$, where h is the distance between the hinges.

▮ Only a horizontal force acts at hinge B, because hinge A is assumed to support the door's weight. Let us

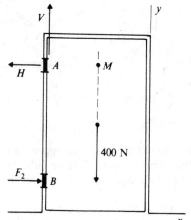

Fig. 10-39

take torques about A as axis. $\sum \tau = 0$ becomes $(F_2)(h) - (400 \text{ N})(h/4) = 0$, from which $F_2 = \underline{100 \text{ N}}$. We also have

$$\sum F_x = 0 \qquad \text{or} \qquad F_2 - H = 0$$

$$\sum F_y = 0 \qquad \text{or} \qquad V - 400 \text{ N} = 0$$

We find from these that $H = 100 \text{ N}$ and $V = 400 \text{ N}$.

To find the resultant force **R** on the hinge at A, we have $R = \sqrt{(400)^2 + (100)^2} = \underline{412 \text{ N}}$. The tangent of the angle that **R** makes with the negative x-direction is V/H and so the angle is arctan $4 = \underline{76°}$.

10.48 A *couple* consists of two forces having equal magnitudes, F, and opposite directions, with a distance d between the lines of action of the forces. Show that the torque exerted by a couple about any axis perpendicular to the plane of the forces has magnitude Fd and thus is independent of the position of the axis.

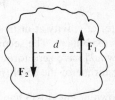

Fig. 10-40

▮ The two antiparallel forces are shown in Fig. 10-40. $F = F_1 = F_2$. Pick an axis through any point, say a distance x left of the line of action through $\mathbf{F}_2$. Take moments $\tau = F_1(x + d) - F_2 x = F(x + d) - Fx = Fd$, independent of x. Note that if $\mathbf{F}_1 + \mathbf{F}_2$ become $-\mathbf{F}_1$ and $-\mathbf{F}_2$, $\tau = F_2 x - F_1(x + d) = Fx - F(x + d) = -Fd$. Thus a couple can be positive or negative, indicating a net counterclockwise or clockwise rotational effect, but the magnitude is always Fd.

10.49 A baggage cart of weight w rests against a table of height h. Its center of gravity is a distance l from the wheel, as shown in Fig. 10-41(a). Neglecting friction at the wheel, find the upward force on the wheel and the horizontal and vertical forces at the table edge.

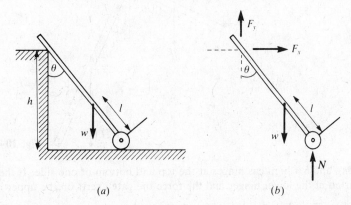

(a)   (b)   **Fig. 10-41**

▮ In Fig. 10-41(b) are shown the forces acting on the cart, with the force **F** due to the table resolved into components along the x and y axis. Since we ignore friction on the wheel, N is a vertical force. $\sum F_x = 0$ yields $F_x = 0$. $\sum F_y = 0$ yields $F_y + N - w = 0$, or $F_y = w - N$. Taking moments about an axis through the point of contact with the table and noting that the moment arm to **N** is $h \tan \theta$ and to **w** is $h \tan \theta - l \sin \theta$, we have from $\sum \tau = 0$: $(h \tan \theta)(N) - (h \tan \theta - l \sin \theta)(w) = 0$. Then $N = (1 - l \cos \theta / h)(w)$, and from before $F_y = w - N = (l \cos \theta / h)w$.

10.50 If the coefficient of static friction at the table edge in Prob. 10.49 is μ_s, what is the smallest angle, θ, for which the wheelbarrow won't slide?

▮ Since $F_x = 0$, **F** points along y. If we resolve **F** into components parallel (∥) and perpendicular (⊥) to the cart, $F_\parallel$ represents the frictional force of the table while $F_\perp$ is the normal force of the table. Indeed, $F_\parallel = F_y \cos\theta = (l\cos^2\theta/h)w$; $F_\perp = F_y \sin\theta$. $F_\parallel$ increases with decreasing θ, and the smallest θ before slipping corresponds to $F_{\parallel,\max}$. But $F_{\parallel,\max} = \mu_s F_\perp$ and $\mu_s = F_\parallel/F_\perp = \cot\theta$.

10.51 The uniform cylinder of radius a shown in Fig. 10-42 originally had a weight of 80 N. After an off-axis cylindrical hole was drilled through it as shown, it weighed 65 N. The axes of the two cylinders are parallel. Assuming the cylinder does not slip on the table determine what the tension T in the cord must be to keep it from moving.

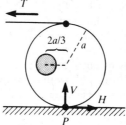

Fig. 10-42

I The 65-N body will have its center of gravity at distance d to the right of the contact point P. Taking torques about P, $T(2a) = (65\text{ N})d = (15\text{ N})(\frac{2}{3}a)$. The second equality expresses the fact that if the hole were refilled, the uniform solid cylinder would have zero torque about P, which implies that the torque of the refilled portion exactly balances the torque of the 65-N (hollowed) portion.

10.52 The right-angle rule (or square) shown in Fig. 10-43 hangs at rest from a peg as shown. It is made of uniform metal sheet. One arm is L cm long while the other is $2L$ cm long. Find the angle θ at which it will hang.

I If the rule is not too wide, we can approximate it as two thin rods of lengths L and $2L$ joined perpendicularly at A. Call the weight of each centimeter of arm γ. Then the forces acting on the rule are as indicated in Fig. 10-43, where P is the upward push of the peg.

Taking torques about point A, we have $(\gamma L)(\overline{MQ}) - (2\gamma L)(\overline{QN}) = 0$. But $\overline{MQ} = \frac{1}{2}L\cos\theta$ and $\overline{QN} = L\sin\theta$. Substituting and rearranging gives $\frac{1}{2}\cos\theta = 2\sin\theta$, and so $\tan\theta = 0.25$ and therefore $\underline{\theta = 14°}$.

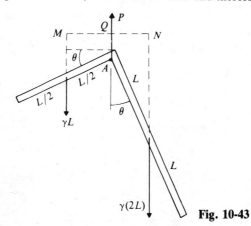

Fig. 10-43

10.53 A gate 9 ft long and 6 ft high has hinges at the top and bottom of one side. If the entire 90-lb weight of the gate is supported at the lower hinge, find the force the gate exerts on the upper hinge.

I First draw the force diagram, Fig. 10-44. Take torques about O, the lower hinge, assuming that the weight of the gate acts at its center of gravity at the center of the rectangle. The clockwise torque about O is $6F_1$.

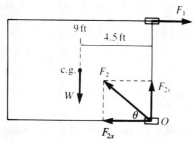

Fig. 10-44

The counterclockwise torque about O is $4.5W$. F_2 has no torque about O, since the lever arm is zero.

$$6F_1 = 4.5W = 4.5(90) \qquad F_1 = \underline{67.5\ lb}$$

10.54 Refer to Prob. 10.53. Find the total force the gate exerts on the lower hinge.

$$\sum F_y = 0 \qquad F_{2y} - W = 0 \qquad F_{2y} - 90 = 0 \qquad F_{2y} = 90\ lb$$

$$\sum F_x = 0 \qquad F_1 - F_{2x} = 0 \qquad 67.5 - F_{2x} = 0 \qquad F_{2x} = 67.5\ lb$$

$$F_2 = \sqrt{F_{2x}^2 + F_{2y}^2} = \sqrt{(67.5)^2 + (90)^2} = \sqrt{4556 + 8100} = \underline{112.5\ lb} \qquad \tan\theta = \frac{F_{2y}}{F_{2x}} = \frac{90}{67.5} = 1.333 \qquad \theta = \underline{53°}$$

10.55 Find T_1, T_2, and T_3 for the mechanism shown in Fig. 10-45(a). The boom is uniform and weighs 800 N.

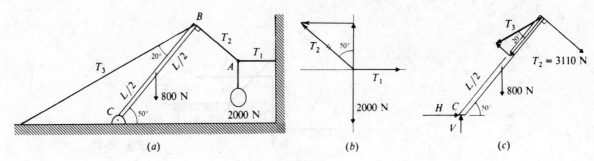

Fig. 10-45

Let us first apply the force condition to point A. The appropriate free-body diagram is shown in Fig. 10-45(b). $T_2 \cos 50° - 2000\ N = 0$ and $T_1 - T_2 \sin 50° = 0$. From the first of these we find $T_2 = \underline{3110\ N}$: then the second equation gives $T_1 = \underline{2380\ N}$.

Let us now isolate the boom and apply the equilibrium conditions to it. The appropriate free-body diagram is shown in Fig. 10-45(c). By resolving the vector $\mathbf{T}_3$ into the components shown we anticipate taking torques about point C. Then the component $T_3 \cos 20°$ exerts no torque about C. The torque equation is

$$(T_3 \sin 20°)(L) - (3110\ N)(L) - (800\ N)\left(\frac{L}{2}\cos 50°\right) = 0$$

Solving for T_3, we find it to be $\underline{9840\ N}$.

10.56 A light rod [Fig. 10-46(a)], which is suspended by a rope, itself suspends a 3114-N load at point B. The ends of the rod are in contact with smooth vertical walls. (a) If $l = 0.25$ m, find the tension in the rope CD and the reaction forces at A and E. (b) Determine the maximum value of l if the largest reaction force at E is 2224 N.

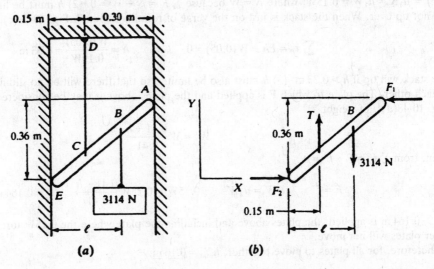

Fig. 10-46

▌ **(a)** The system is in static equilibrium. The reaction forces shown in Fig. 10-46(b) are in the X direction because the walls are assumed frictionless. The force condition in the Y direction gives $\sum F_y = T - 3114 = 0$, or $T = \underline{3144\,\text{N}}$.

Also, upon taking the sum of the torques about E (counterclockwise is positive), we get

$$\sum \tau_E = F_1(0.36) - 3114(0.25) + 3114(0.15) = 0 \quad \text{or} \quad F_1 = \underline{865\,\text{N}}$$

The force condition in the X direction then gives $\sum F_x = F_2 - F_1 = 0$, or $F_2 = \underline{865\,\text{N}}$. **(b)** With $F_2 = F_1 = 2224\,\text{N}$, the torque condition gives

$$\sum \tau_E = (2224)(0.36) - (3114)l + (3114)(0.15) = 0 \quad \text{or} \quad l = \underline{0.41\,\text{m}}.$$

10.57 Find the force F which when applied horizontally to the center of a wheel of radius R and weight W will pull it over a step of height $y < R$.

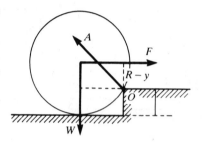

Fig. 10-47

▌ Figure 10-47 is the free-body diagram. Take moments of force around O, the edge of the step. The only clockwise moment about O is $F(R - y)$. The only counterclockwise moment about O is $W\sqrt{[R^2 - (R - y)^2]}$, where the square root is the lever arm of W. Set the clockwise moment equal to the counterclockwise moment and solve for F.

$$F(R - y) = W\sqrt{R^2 - (R - y)^2} = W\sqrt{2Ry - y^2} \qquad F = W\frac{\sqrt{y(2R - y)}}{R - y}$$

Note that F becomes infinite as $y \rightarrow R$.

10.58 A large set of plates is stacked on a table as shown in Fig. 10-48(a). The coefficients of friction between two adjacent plates, and between the bottom plate and the tabletop, are 0.25 and 0.15, respectively. The stack is to be moved to the left, without the plates tipping over or sliding with respect to each other, by applying a horizontal force **F**. Find the largest height h at which **F** may be applied.

▌ In order to move the stack as a rigid body the following must be true: (1) The force **F** must just overcome the friction force between the bottom plate and the tabletop. See Fig. 10-48(b). $\sum F_x = F - f = 0$, or $F = f = \mu_1 N = \mu_1 W = 0.15W$, where $N = W$ because $\sum F_y = N - W = 0$. (2) h must be limited so that the stack will not tip over. When the stack is just on the verge of rotating about P, Fig. 10-48(c),

$$\sum \tau_P = Fh - W(0.05) = 0 \quad \text{or} \quad h = \frac{W(0.05)}{0.15W} = 0.33\,\text{m}$$

The stack will tip if $h > 0.33\,\text{m}$. (3) h must also be limited so that there will be no sliding of plates relative to each other. The plate to which **F** is applied and the plates above it may be considered a free body [Fig. 10-48(d)], of weight

$$W' = W\frac{0.41 - h}{0.41}$$

Then, from $\sum F_x = F - f' = 0$,

$$F = f' \qquad \mu_1 W = \mu_2 W' \qquad 0.15W = 0.25W\frac{0.41 - h}{0.41} \qquad h = 0.164\,\text{m}$$

If $h > 0.164\,\text{m}$ is applied, the plates above and including the plate where the $0.15W$ force will slide and the lower plates will not move.

Therefore, for all plates to move together, $h_{\max} = \underline{0.16\,\text{m}}$.

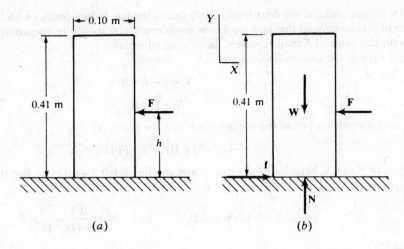

(a)   (b)

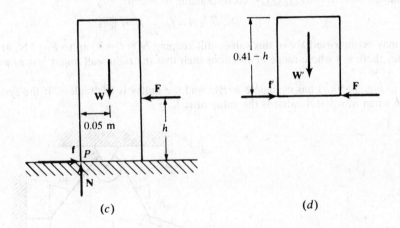

(c)   (d)   **Fig. 10-48**

10.59   A light T bar, 10 cm on each arm, rests between two vertical walls, as shown in Fig. 10-49. The left wall is smooth; the coefficients of static friction between the bar and floor, and between the bar and right wall, are 0.35 and 0.50, respectively. The bar is subject to a vertical load of 1 N, as shown. What is the smallest value of the vertical force F for which the bar will be in static equilibrium in the position shown?

❚   The solution of this problem is greatly simplified by an intuitive consideration of the situation when F is very small. The 1-N force then sets up a counterclockwise torque that because of the low frictional resistance

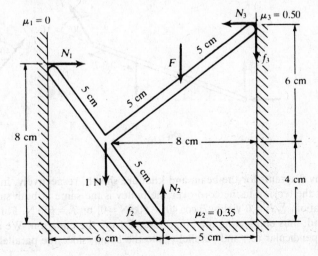

Fig. 10-49

offered by the left wall and the floor immediately causes the bar to lose contact with the right wall. Therefore, if a value of F can be found that puts the bar in equilibrium with its end just touching the right wall (i.e., $N_3 = f_3 = 0$), this value of F must represent the desired minimum.

With $N_3 = f_3 = 0$, the force conditions are

$$N_2 - 1 - F = 0 \tag{1}$$

$$N_1 - f_2 = 0 \tag{2}$$

and the torque condition (about the contact point with the floor) is

$$-N_1(0.08) + 1(0.03) - F(0.01) = 0 \tag{3}$$

Elimination of F and N_1 between these three equations gives $f_2 = 0.5 - 0.125N_2$. But the largest possible value of f_2 is $\mu_2 N_2 = 0.35N_2$. Hence,

$$0.5 - 0.125N_2 \le 0.35N_2 \qquad \text{or} \qquad N_2 \ge \frac{0.5}{0.475} = \frac{20}{19}\,\text{N}$$

and, from (1),

$$F = N_2 - 1 \ge \tfrac{1}{19}\,\text{N}$$

The minimum force is thus (1/19) N, corresponding to which

$$N_1 = \tfrac{7}{19}\,\text{N} = f_2 \qquad N_2 = \tfrac{20}{19}\,\text{N}$$

The force may be increased above this value, still keeping $N_3 = f_3 = 0$, up to $F = 3\,\text{N}$, at which point N_1 and f_2 vanish. Thus there is a whole range of solutions such that the right wall might just as well not be there.

10.60   The beam in Fig. 10-50(a) has negligible weight and the pulley is frictionless. If the system hangs in equilibrium when $w_1 = 500\,\text{N}$, what is the value of w_2?

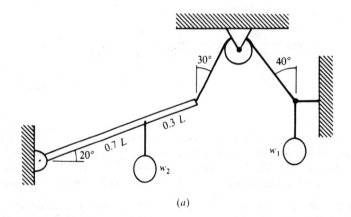

(a)

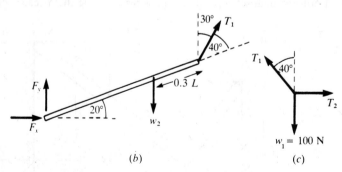

(b)   (c)   **Fig. 10-50**

▌ The free-body diagrams for the beam and knot are shown, respectively, in Figs. 10-50(b) and 10-50(c). We use the fact that the tension in the cord over the pulley is the same at both sides. For the knot we need only the vertical equation. $\Sigma F_y = 0$ yields $T_1 \cos 40° - 500\,\text{N} = 0$, or $T_1 = \underline{653\,\text{N}}$. For the beam we take moments about the left end, thus eliminating the hinge force $\mathbf{F}$ from the equation. We resolve $\mathbf{T}_1$ and $\mathbf{W}_2$ into components perpendicular ($\perp$) and parallel ($\parallel$) to the beam, since the parallel components have zero moment arm. Choosing the direction to the upper left as the positive ⊥ axis we get $T_{1\perp} = T_1 \sin 40° = 420$ N; $W_{2\perp} = -W_2 \cos 20°$ (just as for an inclined plane). $\Sigma \tau = 0$ yields $L420$ N $- 0.7LW_2 \cos 20° = 0$. Dividing out L and solving yields $W_2 = \underline{639\ N}$. (Note that it would be just as easy to take the moment of the full force, W_2, directly since the moment arm is $0.7L \cos 20°$. This clearly yields the same moment: $-0.7L \cos 20°\ W_2$.

10.61 Repeat Prob. 10.60, but now find w_1 if w_2 is 500 N. The beam now weighs 300 N and is uniform.

❚ Figure 10-50(b) must now include the weight of the beam. Taking moments about the left end, as before, we get an equation for T_1. $\Sigma \tau = 0$ yields $LT_1 \sin 40° - (0.7L \cos 20°)(500) - (0.5L \cos 20°)(300) = 0$, or $T_1 = \underline{731\ N}$. From the $\Sigma F_y = 0$ for the knot we get $T_1 \cos 40 - W_1 = 0$, or $W_1 = \underline{560\ N}$.

10.62 A 200-N uniform beam stands vertically. A light cable connects the midpoint of the beam to the floor. A weight $w = 100$ N is connected to the top of the beam by a horizontal rope which passes over a frictionless pulley. What are the horizontal and vertical forces exerted by the floor upon the beam?

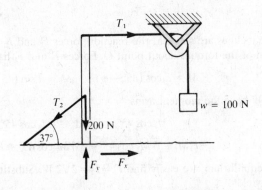

Fig. 10-51

❚ All the forces on the beam are included in Fig. 10-51. Since the pulley is frictionless, $T_1 = w = 100$ N. The force **F** due to the floor is resolved into horizontal and vertical components. $\Sigma F_x = 0$ yields $T_1 + F_x - T_2 \cos 37° = 0$, or $F_x - 0.80T_2 = -100$ N. $\Sigma F_y = 0$ yields $F_y - 200 - T_2 \sin 37° = 0$, or $F_y - 0.60T_2 = 200$ N. Taking moments about an axis through the midpoint of the beam, only the forces T_1 and F_x contribute. Letting the length of the beam be L, $\Sigma \tau = 0$ yields

$$\frac{L}{2}F_x - \frac{L}{2}T_1 = 0 \qquad \text{or} \qquad F_x = T_1 = \underline{100\ N}$$

We substitute F_x into our $\Sigma F_x = 0$ equation and get $T_2 = \underline{250\ N}$. Substituting T_2 into our $\Sigma F_y = 0$ equation we get $F_y = \underline{350\ N}$.

10.63 An object is subjected to the forces shown in Fig. 10-52. What single force applied at a point on the x axis will balance these forces? Where on the x axis should the force be applied?

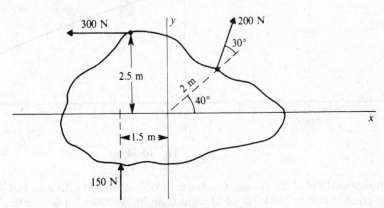

Fig. 10-52

❚ Let the unknown force have components F_x and F_y and assume it acts at point $(d, 0)$. $\Sigma F_x = 0$ yields $F_x + 200 \cos 70° - 300 = 0$, or $\underline{F_x = 232\ N}$. $\Sigma F_y = 0$ yields $F_y + 200 \sin 70° + 150 = 0$, or $\underline{F_y = -338\ N}$ (i.e., points downward). To find d we can use the torque equation about an axis through the origin. F_x does not contribute to this moment. The 200-N force can be resolved perpendicular ($\perp$) and parallel ($\parallel$) to its displacement from the origin (dashed line) and only the $\perp$ component contributes. Thus $\sum \tau = 0$ yields $(2.5)(300) + (2.0)(200)(\sin 30°) - (1.5)(150) - d(338) = 0$. Solving, we get $\underline{d = 2.14 \text{ m}}$.

10.64 Two spheres, of weights W and $3W$, are rigidly connected by a bar of negligible weight and are free to slide on the 45° inclines, as shown in Fig. 10-53(a). Find the angle ϕ that the bar makes with the horizontal when the system is in static equilibrium.

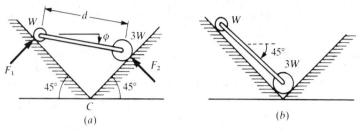

Fig. 10-53

❚ Assuming that the inclines are smooth, the reaction forces F_1 and F_2 indicated in Fig. 10.53(a) will be normal. Take the sum of the torques about point C. Forces F_1 and F_2 have respective moment arms

$$M_1 = d \cos (45° - \phi) \qquad M_2 = d \sin (45° - \phi)$$

and weights W and $3W$ have the moment arms

$$M_1 \cos 45° \qquad \text{and} \qquad M_2 \cos 45°$$

Thus,
$$-F_1 M_1 + W M_1 \cos 45° - 3 W M_2 \cos 45° + F_2 M_2 = 0 \tag{1}$$

But for translational equilibrium, we easily find $F_1 = F_2 = 2\sqrt{2}\, W$. Substituting these values in (1) and dividing through by $W M_1 / \sqrt{2}$ yields

$$-3 + \tan (45° - \phi) = 0 \tag{2}$$

from which $\underline{\phi = -26.6°}$. Since ϕ turns out to be negative, the heavier sphere actually lies higher than the lighter!

10.65 Refer to Fig. 10-53(b). This is obviously a configuration of static equilibrium. How is it to be reconciled to the result of Prob. 10.64?

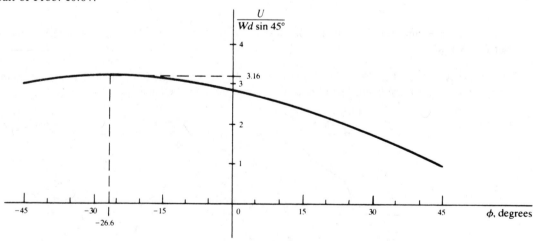

Fig. 10-54

❚ The gravitational PE of the system, $U = Wd \cos (45° - \phi) \sin 45° + 3Wd \sin (45° - \phi) \sin 45°$ relative to point C, is graphed in Fig. 10-54. [$U = 0$ at point C of Fig. 10-53(a).] It is seen that the solution determined in Prob. 10.64 corresponds to a *maximum* in the PE and thus also an *unstable* equilibrium (compare Prob. 10.72). The (boundary) minimum at $\phi = +45°$ corresponds to the stable equilibrium pictured in Fig. 10-53(b). This solution was not found in Prob. 10.64 because it requires a normal force on the heavier weight owing to the left incline.

10.66ᶜ Show that if the resultant external force on a rigid body is zero, the resultant torque has a fixed value, independent of the point about which the torque is computed.

▮ Figure 10-55 shows two arbitrary points, P and Q, which are fixed in space; they do not necessarily belong to the rigid body. The torque about P is given by

$$\tau_P = \int \mathbf{r}_P \times d\mathbf{F} = \int (\mathbf{D} + \mathbf{r}_Q) \times d\mathbf{F} = \mathbf{D} \times \int d\mathbf{F} + \int \mathbf{r}_Q \times d\mathbf{F} = 0 + \tau_Q = \tau_Q$$

since, by hypothesis, $\int d\mathbf{F} = 0$.

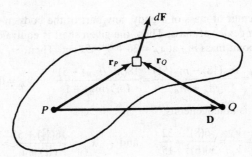

Fig. 10-55

10.2 CENTER OF MASS (CENTER OF GRAVITY)

10.67 Three particles, of masses 2 kg, 4 kg, and 6 kg, are located at the vertices of an equilateral triangle of side 0.5 m. Find the center of mass of this collection, giving its coordinates in terms of a system with its origin at the 2-kg particle and with the 4-kg particle located along the positive X axis.

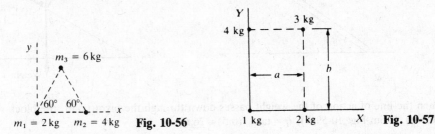

Fig. 10-56   **Fig. 10-57**

▮ The three masses are as shown in Fig. 10-56. To get the x component of the CM, $\bar{x}$, we have

$$\bar{x} = \frac{m_1 x_1 + m_2 x_2 + m_3 x_3}{m_1 + m_2 + m_3} \qquad \text{with} \qquad x_1 = 0 \qquad x_2 = 0.5 \text{ m} \qquad x_3 = 0.25 \text{ m}$$

Substituting we have $\bar{x} = \underline{0.29 \text{ m}}$. Similarly for the y component of the CM,

$$\bar{y} = \frac{m_1 y_1 + m_2 y_2 + m_3 y_3}{m_1 + m_2 + m_3} \qquad \text{with} \qquad y_1 = y_2 = 0 \qquad y_3 = 0.5 \sin 60° = 0.433 \text{ m}$$

Substituting, we have $\bar{y} = \underline{0.22 \text{ m}}$.

10.68 Four particles, of masses 1 kg, 2 kg, 3 kg, and 4 kg, are at the vertices of a rectangle of sides a and b (see Fig. 10-57). If $a = 1$ m and $b = 2$ m, find the location of the center of mass.

▮ Set up a cartesian coordinate system in the plane of the rectangle, with the origin at the 1-kg particle. The coordinates of the four particles are, in increasing order of their masses, $(0, 0)$, $(a, 0)$, (a, b), and $(0, b)$. The total mass M is $M = m_1 + m_2 + m_3 + m_4 = 10$ kg. Substituting into the center-of-mass equations, we find

$$x_{\text{cm}} = \frac{\sum m_i x_i}{M} = \frac{1}{10}(0 + 2a + 3a + 0) = \underline{0.5 \text{ m}} \qquad y_{\text{cm}} = \frac{\sum m_i y_i}{M} = \frac{1}{10}(0 + 0 + 3b + 4b) = \underline{1.4 \text{ m}}$$

10.69 Figure 10-58 is the side view of a small machine shaft made of homogeneous material. Find its center of mass.

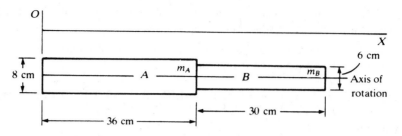

Fig. 10-58

■ In calculating the center of mass of a body, any part of the body may be treated as if all its mass were concentrated at its own center of mass. Thus, the given shaft is equivalent to a point mass m_A at $x_A = \frac{36}{2} = 18$ cm and a point mass m_B at $x_B = 36 + \frac{30}{2} = 51$ cm. Then,

$$x_{cm} = \frac{m_A x_A + m_B x_B}{m_A + m_B} = \frac{m_A(18) + m_B(51)}{m_A + m_B} = \frac{18(m_A/m_B) + 51}{(m_A/m_B) + 1} \qquad y_{cm} = 0 \qquad \text{(by symmetry about the } x \text{ axis)}$$

But, since the material is homogeneous,

$$\frac{m_A}{m_B} = \frac{V_A}{V_B} = \frac{36(8)^2}{30(6)^2} = \frac{32}{15} \qquad \text{and} \qquad x_{cm} = \frac{18(\frac{32}{15}) + 51}{(\frac{32}{15}) + 1} = \underline{28.53 \text{ cm}} \qquad y_{cm} = \underline{0}$$

10.70 The uniform rectangular block shown in Fig. 10-59(a) is twice as tall as it is wide. It is prevented from sliding by a small ridge. As θ is slowly increased, at what value of θ will the block topple over?

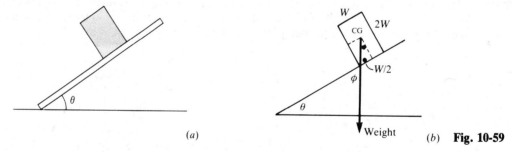

(a)   (b) Fig. 10-59

■ When the line of action of the weight passes down through the pivot point, the block will just be in equilibrium. From Fig. 10-59(b), $\theta = \tan^{-1}(0.5) = \underline{26.6°}$.

10.71 Suppose that the center of gravity of a car is a distance h above the roadway and the width of the car between wheel contact points with the road is d. If you try to tip the car over sideways, through how large an angle will you have to tilt it?

■ The critical tilt angle θ occurs when the weight of the car acting downward through the center of gravity is directly above the point where the wheels touch the ground. Using Fig. 10-59(b), $\tan \theta = (d/2)/h$, so $\theta = \tan^{-1}(d/2h)$.

10.72 A half-dollar is partly embedded in a large cork into which two forks have been stuck, as shown in Fig. 10-60. If the edge of the coin is placed on a needle, the system can oscillate on the point of the needle without falling over. Account for the stability of this system.

Fig. 10-60

■ Static equilibrium requires that the sum of the torques around the pivot be zero. Thus the center of mass of the system must be somewhere along a vertical line through the pivot point. If the center of mass lies *below* the pivot point, as it does in this system, then the system is tilted when the center of mass is raised, its potential energy increases, and the resulting torque restores the system to its equilibrium position. Such an

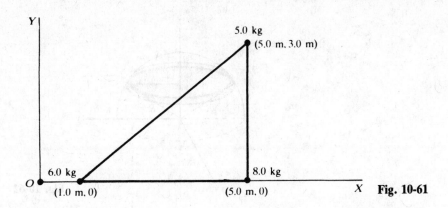

Fig. 10-61

equilibrium is *stable,* and the system can exhibit the oscillations described. (If the center of mass were *above* the pivot point, tilting the system would lower the potential energy. Such an equilibrium is *unstable.*)

10.73 Locate the center of mass of the three-particle system shown in Fig. 10-61.

▮ From Fig. 10-61, $m_1 = 6.0$ kg, $\mathbf{r}_1 = 1.0\mathbf{i}$; $m_2 = 8.0$ kg, $\mathbf{r}_2 = 5.0\mathbf{i}$; $m_3 = 5.0$ kg, $\mathbf{r}_3 = 5.0\mathbf{i} + 3.0\mathbf{j}$. Then, by definition,

$$(6.0 + 8.0 + 5.0)\mathbf{r}_{cm} = (6.0)(1.0\mathbf{i}) + (8.0)(5.0\mathbf{i}) + (5.0)(5.0\mathbf{i} + 3.0\mathbf{j}) \qquad 19.0\mathbf{r}_{cm} = 71.0\mathbf{i} + 15.0\mathbf{j}$$

$$\mathbf{r}_{cm} = \frac{71.0}{19.0}\mathbf{i} + \frac{15.0}{19.0}\mathbf{j} = \underline{3.7\mathbf{i} + 0.8\mathbf{j}}$$

or $x_{cm} = \underline{3.7\text{ m}}$, $y_{cm} = \underline{0.8\text{ m}}$.

10.74^c A thin bar of length L has a mass per unit length, λ, that increases linearly with distance from one end. If its total mass is M and its mass per unit length at the lighter end is λ_0, find the distance of the center of mass from the lighter end.

▮ Let the bar sit along the x axis with its lighter end at the origin. The mass/length is linear so $\lambda(x) = \lambda_0 + bx$. To find b we note that

$$M = \int_0^L \lambda(x)\,dx = \lambda_0 x + \frac{bx^2}{2}\Big|_0^L = \lambda_0 L + \frac{bL^2}{2} \qquad \text{or} \qquad b = \frac{2(M - \lambda_0 L)}{L^2}$$

To find the center of mass position, $\bar{x}$, we have

$$\bar{x} = \frac{\int_0^L x\lambda(x)\,dx}{M} = \frac{(\lambda_0 x^2/2 + bx^3/3)|_0^L}{M} = \frac{\lambda_0 L^2/2 + bL^3/3}{M}$$

Substituting the expression for b,

$$\bar{x} = \frac{\dfrac{\lambda_0 L^2}{2} + \dfrac{2(M - \lambda_0 L)L}{3}}{M} = \frac{2L}{3} - \frac{\lambda_0 L^2}{6M}$$

10.75^c Find the center of mass of a uniform solid hemisphere of radius R and mass M.

▮ Let the center of the sphere be the origin and let the flat of the hemisphere lie in the x, y plane, as shown in Fig. 10-62. By symmetry $\bar{x} = \bar{y} = 0$. Consider the hemisphere divided into a series of slices parallel to the x, y plane. Each slice is of infinitesimal thickness dz. The slice between z and $(z + dz)$ is a disk of radius $d = \sqrt{R^2 - z^2}$. If ρ is the constant density of the uniform sphere, then the mass of the slice is $dm = (\rho\pi d^2)\,dz = \rho\pi(R^2 - z^2)dz$ The $\bar{z}$ position of the center of mass is then obtained by getting the contributions of all the slices:

$$\bar{z} = \frac{\int_0^R z\,dm}{M} = \frac{\int_0^R \pi\rho(R^2 z - z^3)dz}{M} = \frac{\pi\rho(R^2 z^2/2 - z^4/4)|_0^R}{M} \qquad \bar{z} = \frac{\pi\rho(R^4/2 - R^4/4)}{M} = \frac{\rho\pi R^4/4}{M}$$

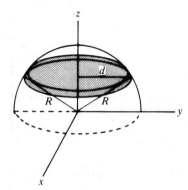

Fig. 10-62

Noting that $\rho(\frac{4}{3}\pi R^3) = 2M$,

$$\bar{z} = \frac{\rho\pi R^4/4}{\rho 2\pi R^3/3} = \frac{3}{8}R$$

10.76ᶜ Find the center of mass of a quadrant of a thin elliptical section made of material of mass per unit area σ. See Fig. 10-63.

▌ The quadrant is bounded by the ellipse

$$\frac{x^2}{a^2} + \frac{y^2}{b^2} = 1$$

and the coordinate axes; its area is $A = \pi ab/4$.

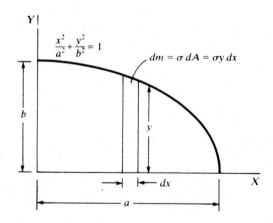

Fig. 10-63

$$x_{cm} = \frac{\displaystyle\int x\, dm}{\displaystyle\int dm} = \frac{\displaystyle\int_0^a \sigma xy\, dx}{\sigma A} = \frac{1}{A}\int_0^a xy\, dx$$

But, along the ellipse,

$$\frac{2x\, dx}{a^2} + \frac{2y\, dy}{b^2} = 0 \qquad \text{or} \qquad x\, dx = -\frac{a^2}{b^2}y\, dy$$

and so

$$x_{cm} = \frac{1}{A}\left(-\frac{a^2}{b^2}\right)\int_b^0 y^2\, dy = \frac{1}{A}\left(-\frac{a^2}{b^2}\right)\left(-\frac{b^3}{3}\right) = \underline{\frac{4a}{3\pi}}$$

By symmetry, $y_{cm} = \underline{4b/3\pi}$.

10.77ᶜ Locate the center of mass of a thin hemispherical shell of radius R (Fig. 10-64).

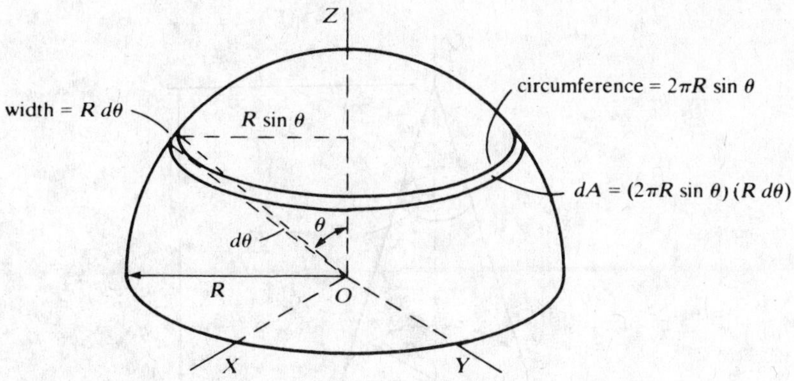

Fig. 10-64

❚ The coordinates of the center of mass are $x_{cm} = y_{cm} = 0$ and

$$z_{cm} = \frac{\int z\, dm}{\int dm} = \frac{\int z\sigma\, dA}{\sigma A}$$

where σ is the mass per unit area of the thin shell. Since $dA = 2\pi R^2 \sin\theta\, d\theta = A \sin\theta\, d\theta$ and $z = R\cos\theta$, we have

$$z_{cm} = \int_0^{\pi/2} R\cos\theta \sin\theta\, d\theta = -\frac{R}{2}[\cos^2\theta]_0^{\pi/2} = \frac{R}{2}$$

10.78^c Infer the result of Prob. 10.75 from that of Prob. 10.77.

❚ According to Prob. 10.77, a shell of radius r and thickness dr is equivalent to a point mass $dm = \frac{1}{2}(4\pi r^2\, dr)\rho$ located on the z axis at $z = r/2$. Hence, for the solid hemisphere of radius R,

$$M\bar{z} = \int z\, dm = \rho \int_0^R \pi r^3\, dr = \frac{\rho\pi R^4}{4}$$

as in Prob. 10.75. This is another application of the principle that the CM of a whole is the CM of the CMs of its parts.

10.79^c Find the center of mass of a right circular cone of height h, radius R, and constant density ρ.

❚ In Fig. 10-65, the base of the cone lies in the XY plane and the axis of symmetry is along Z. The coordinates of the center of mass are $x_{cm} = y_{cm} = 0$ and

$$z_{cm} = \frac{\int z\, dm}{\int dm} = \frac{\int z\rho\, dV}{\int \rho\, dV}$$

By similar triangles,

$$\frac{r}{h-z} = \frac{R}{h} \quad \text{or} \quad z = h - \frac{h}{R}r$$

so that $dV = -(\pi h/R)r^2\, dr$ and

$$z_{cm} = \frac{\int_R^0 \left(h - \frac{h}{R}r\right)r^2\, dr}{\int_R^0 r^2\, dr} = \frac{-h\dfrac{R^3}{3} + h\dfrac{R^3}{4}}{-\dfrac{R^3}{3}} = \frac{h}{4}$$

10.80^c Compute the torque about some fixed point O that a body of mass M experiences in a uniform gravitational field $\mathbf{g}$.

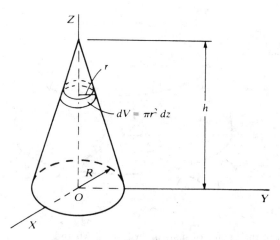

Fig. 10-65

▮ An element dm of the body has weight $dm\,\mathbf{g}$; so the differential torque is $d\boldsymbol{\tau} = \mathbf{r} \times dm\,\mathbf{g}$, where $\mathbf{r}$ is the radius vector from O to the element dm. Integrating over the entire body,

$$\boldsymbol{\tau} = \int \mathbf{r} \times dm\,\mathbf{g} = \left(\int \mathbf{r}\,dm \right) \times \mathbf{g} = (M\mathbf{r}_{cm}) \times \mathbf{g} = \mathbf{r}_{cm} \times M\mathbf{g}$$

Thus, the gravitational torque may be obtained *by considering the entire weight*, $M\mathbf{g}$, *of the body to act at its center of mass.*

11.1 ANGULAR MOTION AND TORQUE

11.1 Express the following angles in degrees, radians, and revolutions. (*a*) 20°; (*b*) 0.40 rad; (*c*) $\frac{1}{3}$ rev.

▌ (*a*) 20° = 20(2π/360) = <u>0.35 rad</u> = 0.35((1/2π) = <u>0.056 rev</u>; (*b*) 0.40 rad = 0.40(360/2π) = <u>23°</u> = 23($\frac{1}{360}$) = <u>0.064 rev</u>; (*c*) $\frac{1}{3}$ rev = $\frac{1}{3}$(360) = <u>120°</u> = 120(2π/360) = <u>2.09 rad</u>.

11.2 Express the following angular speeds in degrees per second, radians per second, and revolutions per second: (*a*) 0.020 rev/s; (*b*) 30°/s; (*c*) 1.40 rad/s.

▌ (*a*) 0.020 rev/s = 0.020(360) = <u>7.2°/s</u> = 0.020(2π) = <u>0.126 rad/s</u>; (*b*) 30°/s = 30(2π/360) = <u>0.52 rad/s</u> = 30($\frac{1}{360}$) = <u>0.083 rev/s</u>; (*c*) 1.40 rad/s = 1.40(360/2π) = <u>80°/s</u> = 1.40(1/2π) = <u>0.22 rev/s</u>.

11.3 The moon has a diameter of 3480 km and is 3.8×10^8 m from the earth. (*a*) How large an angle in radians does the diameter of the moon subtend to a person on earth? (*b*) If the diameter of the earth is 1.28×10^4 km, what is the angle subtended by the earth to a person on the moon?

▌ (*a*) $\theta = (3.48 \times 10^6 \text{ m})/(3.8 \times 10^8 \text{ m}) = \underline{0.0092 \text{ rad}}$; (*b*) $(1.28 \times 10^7 \text{ m})/(3.8 \times 10^8 \text{ m}) = \underline{0.034 \text{ rad}} \approx 2°$.

11.4 A tiny laser beam is directed from the earth to the moon. If the beam is to have a diameter of 2.50 m at the moon, how small must the divergence angle be for the beam?

▌ $\theta = (2.50 \text{ m})/(3.8 \times 10^8 \text{ m}) = \underline{6.6 \times 10^{-9} \text{ rad}}$.

11.5 A sphere is rotating about its axis in such a way that its angular velocity is given by $\omega = 3\mathbf{i} + 5\mathbf{j}$ rad/s. Find the angle between its axis and the *x* axis.

▌ $\omega = 3\mathbf{i} + 5\mathbf{j} = \omega_x\mathbf{i} + \omega_y\mathbf{j}$; since $\omega = (\omega_x^2 + \omega_y^2)^{1/2} = 34^{1/2} = 5.83$, we can find the angle from $\cos\theta = \omega_x/\omega = 3/5.83$, yielding $\theta = \underline{59°}$.

11.6 Two vectors **A** and **B** determine a parallelogram. Show that $|\mathbf{A} \times \mathbf{B}|$ is equal to the area of the parallelogram. Hence the area can be represented by a single vector $\mathbf{A} \times \mathbf{B}$. What is the spatial relationship between this vector and the plane of the parallelogram?

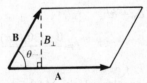

Fig. 11-1

▌ The area of the parallelogram with edges **A** and **B** is the product of its base and its altitude. The base is $|\mathbf{A}|$, while the altitude is the component of **B** perpendicular to **A**, $B_\perp = |\mathbf{B}| \sin\theta$ (Fig. 11-1). Then the area is $|\mathbf{A}| B_\perp = |\mathbf{A}| |\mathbf{B}| \sin\theta = |\mathbf{A} \times \mathbf{B}|$. The vector $\mathbf{A} \times \mathbf{B}$ has magnitude equal to the area of the parallelogram and is perpendicular to the plane of the parallelogram.

11.7 Find $\mathbf{A} \times \mathbf{B}$ if $\mathbf{A} = 3\mathbf{i} - 4\mathbf{j}$ N and $\mathbf{B} = 2\mathbf{j} + 6\mathbf{k}$ m. Repeat for $\mathbf{B} \times \mathbf{A}$.

▌ $(3\mathbf{i} - 4\mathbf{j}) \times (2\mathbf{j} + 6\mathbf{k}) = 6\mathbf{i} \times \mathbf{j} + 18\mathbf{i} \times \mathbf{k} - 8\mathbf{j} \times \mathbf{j} - 24\mathbf{j} \times \mathbf{k} = \underline{6\mathbf{k} - 18\mathbf{j} - 24\mathbf{i}}$ N·m. For $\mathbf{B} \times \mathbf{A}$, all products would be reversed and so all signs would be reversed.

11.8 Find the torque $\mathbf{r} \times \mathbf{F}$ if $\mathbf{r} = 2\mathbf{j} - 6\mathbf{k}$ m and $\mathbf{F} = 3\mathbf{i} + 4\mathbf{k}$ N. Repeat if $\mathbf{r} = 5\mathbf{i} + 2\mathbf{j} - 6\mathbf{k}$ m.

▌ $(2\mathbf{j} - 6\mathbf{k}) \times (3\mathbf{i} + 4\mathbf{k}) = 6\mathbf{j} \times \mathbf{i} + 8\mathbf{j} \times \mathbf{k} - 18\mathbf{k} \times \mathbf{i} - 24\mathbf{k} \times \mathbf{k} = \underline{8\mathbf{i} - 18\mathbf{j} - 6\mathbf{k}}$ N·m. $(5\mathbf{i} + 2\mathbf{j} - 6\mathbf{k}) \times (3\mathbf{i} + 4\mathbf{k}) = 0 - 20\mathbf{j} - 6\mathbf{k} + 8\mathbf{i} - 18\mathbf{j} + 0 = \underline{8\mathbf{i} - 38\mathbf{j} - 6\mathbf{k}}$ N·m.

11.9 A force of 200 N acts tangentially on the rim of a wheel 25 cm in radius. Find the torque. Repeat if the force makes an angle of 40° to a spoke of the wheel.

▮ Clearly the question refers to the torque about an axis through the center of the wheel. Then since the radius to the point of application of the force is the lever or moment arm, we have $\tau = (0.25\text{ m})(200\text{ N}) = \underline{50\text{ N}\cdot\text{m}}$. If the force makes an angle of 40° with a spoke, then the tangential component of the force is $F_t = (200\text{ N})(\sin 40°) = 128.6\text{ N}$. Then $\tau = (128.6\text{ N})(0.25\text{ m}) = \underline{32.1\text{ N}\cdot\text{m}}$, since the radial component of the force does not contribute to the moment.

11.2 ROTATIONAL KINEMATICS

11.10 A flywheel originally at rest is to reach an angular velocity of 36 rad/s in 6.0 s. **(a)** What constant angular acceleration must it have? **(b)** What total angle does it turn through in the 6.0 s?

▮ **(a)** Use the equation of angular motion, where ω_0 is zero:

$$\omega = \omega_0 + \alpha t \qquad 36 = \alpha(6.0) \qquad \alpha = \underline{6\text{ rad/s}^2}$$

(b) Use the equation of angular motion, where θ_0 and ω_0 are zero:

$$\theta = \theta_0 + \omega_0 t + \tfrac{1}{2}\alpha t^2 = \tfrac{1}{2}(6)(6^2) = \underline{108\text{ rad}}$$

or use

$$\theta = \theta_0 + \frac{\omega_0 + \omega}{2}t = \frac{0 + 36}{2}(6) = \underline{108\text{ rad}} \qquad \left(\bar{\omega} = \frac{\omega_0 + \omega}{2}\text{ for uniform angular acceleration}\right)$$

11.11 Derive $\phi = \phi_i + (\omega^2 - \omega_i^2)/2\alpha$ for constant α. (Sometimes ϕ instead of θ is used for the rotation angle and subscript i instead of subscript 0 is used to denote an initial value.)

▮ We start with $\omega = \omega_i + \alpha t$ and $\phi = \phi_i + \omega_i t + \tfrac{1}{2}\alpha t^2$. Solving the first equation for t, we obtain $t = (\omega - \omega_i)/\alpha$. Substituting this into the second equation, we find

$$\phi = \phi_i + \omega_i\frac{\omega - \omega_i}{\alpha} + \frac{1}{2}\alpha\frac{(\omega - \omega_i)^2}{\alpha^2} \qquad\text{or}$$

$$\phi = \phi_i + \frac{\omega - \omega_i}{\alpha}\left[\omega_i + \frac{1}{2}(\omega - \omega_i)\right] = \phi_i + \frac{(\omega - \omega_i)(\omega + \omega_i)}{2\alpha} = \phi_i + \frac{\omega^2 - \omega_i^2}{2\alpha}$$

11.12 A wheel turning with angular speed of 30 rev/s is brought to rest with a constant acceleration. It turns 60 rev before it stops. **(a)** What is its angular acceleration? **(b)** What time elapses before it stops?

▮ **(a)** The angular acceleration may be found from $\omega^2 = \omega_0^2 + 2\alpha(\theta - \theta_0)$:

$$\alpha = \frac{\omega^2 - \omega_0^2}{2(\theta - \theta_0)} = \frac{0^2 - [(30)(2\pi)]^2}{2(60)(2\pi)} = \underline{-47\text{ rad/s}^2}$$

(b) The time is found from $\omega = \omega_0 + \alpha t$:

$$t = \frac{0 - (30)(2\pi)}{-47} = \underline{4\text{ s}}$$

11.13 A spinning wheel initially has an angular velocity of 50 rad/s east; 20 s later its angular velocity is 50 rad/s west. If the angular acceleration is constant, what are **(a)** the magnitude and direction of the angular acceleration, **(b)** the angular displacement over 20 s, and **(c)** the angular speed at 30 s?

▮ **(a)** The direction of the angular acceleration is west, as shown in Fig. 11-2(a), since $\boldsymbol{\alpha}\,\Delta t = \boldsymbol{\omega}_f - \boldsymbol{\omega}_i = \boldsymbol{\omega}_f + (-\boldsymbol{\omega}_i)$ and both vectors on the right are to the west. The magnitude of the angular acceleration is

$$\alpha = \frac{\omega_f - \omega_i}{\Delta t} = \frac{50 - (-50)}{20} = \underline{5\text{ rad/s}^2}$$

(b) The angular displacement, from $\omega_f^2 = \omega_i^2 + 2\alpha(\theta - \theta_0)$, is

$$\theta - \theta_0 = \frac{\omega_f^2 - \omega_i^2}{2\alpha} = \frac{50^2 - 50^2}{2(5)} = \underline{0}.$$

This result also follows from the fact that the average angular velocity, $\tfrac{1}{2}(\omega_i + \omega_f)$, is zero. The only effect of the angular acceleration over the 20-s interval is to reverse the axis of rotation. **(c)** From Fig. 11-2(b), the angular speed at the end of 30 s is $\omega = \omega_0 + \alpha t = 50 + 5(30 - 20) = \underline{100\text{ rad/s}}$.

Once $\boldsymbol{\alpha}$ and $\boldsymbol{\omega}$ are parallel, the angular speed, but not the direction of the rotation axis, changes.

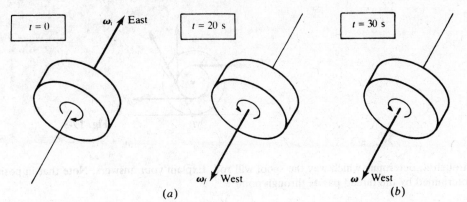

$t = 0$ ω_i ▲ East $t = 20$ s $t = 30$ s

ω_f ► West ω ► West

(a)   (b)   **Fig. 11-2**

11.14 A phonograph turntable rotating at 8.16 rad/s slows uniformly to a stop in 6 rev. Find the angular acceleration in rad/s^2.

▌ Let $\theta_0 = 0$. Then,

$$\omega^2 = \omega_0^2 + 2\alpha\theta \qquad 0 = 8.16^2 + 2\alpha(6 \times 2\pi) \qquad \alpha = \frac{-8.16^2}{24\pi} = \underline{-0.884 \text{ rad/s}^2}$$

The negative sign means a deceleration.

11.15 A belt runs on a wheel of 30-cm radius. During the time that the wheel coasts uniformly to rest from an initial speed of 2.0 rev/s, 25 m of belt length passes over the wheel. Find the deceleration of the wheel and the number of revolutions it turns while stopping.

▌ The respective angular quantities are $\theta = 25/0.3 = 83$ rad $= \underline{13.3 \text{ rev}}$; also $\omega_0 = 2.0$ rev/s and $\omega = 0$. Use $\omega^2 - \omega_0^2 = 2\alpha\theta$ to find $\alpha = \underline{-0.150 \text{ rev/s}^2}$, or $\alpha = \underline{-0.942 \text{ rad/s}^2}$.

11.16 In 7 s a car accelerates uniformly from rest to such a speed that its wheels are turning at a rate of 6.0 rev/s. What is the angular acceleration of a wheel? Through how many revolutions does the wheel turn?

▌ Given $\omega_0 = 0$, $\omega = 6.0$ rev/s, and $t = 7$ s. Use $\omega = \omega_0 + \alpha t$ to give $\alpha = \underline{0.86 \text{ rev/s}^2}$. Then $\theta = \bar{\omega}t = 3(7) = \underline{21 \text{ rev}}$. Equivalently, multiplying by 2π, $\alpha = \underline{5.38 \text{ rad/s}^2}$, $\theta = \underline{132 \text{ rad}}$.

11.17 A roulette wheel originally turning at 0.80 rev/s coasts to rest in 20 s. What was the acceleration of the wheel? Through how many revolutions did it turn in the process? (Assume uniform acceleration.)

▌ Given $\omega_0 = 0.80$ rev/s, $\omega = 0$, and $t = 20$ s. Use $\omega = \omega_0 + \alpha t$ to give $\alpha = \underline{-0.04 \text{ rev/s}^2}$. Also, $\theta = \bar{\omega}t = 0.40(20) = \underline{8 \text{ rev}}$.

11.18 Through how many revolutions must the 60-cm-diameter wheel of a car turn as the car travels 2.5 km?

▌ In this case, $s = 2500$ m and $r = 0.30$ m, so that $s = r\theta$ yields $\theta = 8330$ rad $= \underline{1330 \text{ rev}}$.

11.19 Two gear wheels which are meshed together have radii of 0.50 and 0.15 cm. Through how many revolutions does the smaller turn when the larger turns through 3 rev?

▌ The point of contact moves the same distance around either wheel; hence $(0.50 \text{ cm})(3 \text{ rev}) = (0.15 \text{ cm})\theta$, giving $\theta = \underline{10 \text{ rev}}$.

11.20 A car accelerates uniformly from rest to a speed of 15 m/s in a time of 20 s. Find the angular acceleration of one of its wheels and the number of revolutions turned by a wheel in the process. The radius of the car wheel is $\frac{1}{3}$ m.

▌ First find linear acceleration and distance from $v = v_0 + at$ and $s = \bar{v}t$ to be $a = 0.75$ m/s^2 and $s = 150$ m. Then transform to angular quantities through $\alpha = a/r = a/\frac{1}{3} = \underline{2.25 \text{ rad/s}^2}$ and $\theta = s/r = \underline{450 \text{ rad}} = \underline{72 \text{ rev}}$.

11.3 TORQUE AND ROTATION

11.21 A spool of thread rests on a level tabletop, as shown in Fig. 11-3. The thread is pulled gently so that there is no slippage at P, the point of contact between the spool and the tabletop. For each of the thread positions a

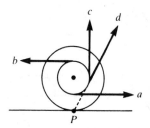

Fig. 11-3

through *d*, determine which way the spool will roll. Explain your answers. Note that in position *d* the line determined by the thread passes through point *P*.

❚ If the origin for calculating torques is taken at *P*, the torques due to gravity, friction, and the normal force are all zero.

(*a*) The torque about *P* is clockwise, so the spool rotates clockwise and rolls to the right, winding the string on the spool.

(*b*) The torque about *P* is counterclockwise, so the spool rotates counterclockwise and will roll to the left.

(*c*) The torque about *P* is counterclockwise, so the spool rotates counterclockwise and rolls to the left.

(*d*) The torque about *P* is zero. The spool will not rotate. Since we are told that it is pulled gently and therefore does not slip, it must be that the spool remains motionless.

11.22 A grindstone has a moment of inertia of 1.6×10^{-3} kg · m². When a constant torque is applied, the flywheel reaches an angular velocity of 1200 rev/min in 15 s. Assuming it started from rest, find (*a*) the angular acceleration; (*b*) the unbalanced torque applied; (*c*) the angle turned through in the 15 s; (*d*) the work *W* done on the flywheel by the torque.

❚ (*a*) At $t = 15$ s,

$$\omega = 1200 \text{ rev/min} \times 1 \text{ min}/60 \text{ s} \times 2\pi \text{ rad}/1 \text{ rev} = 40\pi \text{ rad/s} \qquad \alpha = \frac{\omega - \omega_0}{t} = \frac{(40\pi - 0) \text{ rad/s}}{15 \text{ s}} = \underline{8.38 \text{ rad/s}^2}$$

(*b*) $\tau = I\alpha = 1.6 \times 10^{-3} \text{ kg} \cdot \text{m}^2 \times 8.38 \text{ rad/s}^2 = \underline{0.0134 \text{ m} \cdot \text{N}}$

(*c*) $\theta = \omega_{avg}t = \dfrac{(40\pi + 0) \text{ rad/s}}{2} \times 15 \text{ s} = \underline{942 \text{ rad}}$

(*d*) $W = \tau\theta = 0.0134 \times 942 = \underline{12.6 \text{ J}}$

or, by the work-energy principle,

$$W = \text{KE} = \tfrac{1}{2}I\omega^2 = \tfrac{1}{2}(1.6 \times 10^{-3})(40\pi)^2 = \underline{12.6 \text{ J}}$$

11.23 A nearly massless rod is pivoted at one end so it can swing freely as a pendulum. Two masses, $2m$ and m, are attached to it at distances b and $3b$, respectively, from the pivot. The rod is held horizontal and then released. Find its angular acceleration at the instant it is released.

❚ Torque $= g(2mb + 3mb) = 5mgb$ and $I = 2mb^2 + m(9b^2) = 11mb^2$. Since $\tau = I\alpha$, we have $\alpha = \underline{5g/11b}$.

11.24 Three children are sitting on a seesaw in such a way that it balances. A 20- and a 30-kg boy are on opposite sides at a distance of 2.0 m from the pivot. If the third boy jumps off, thereby destroying the balance, what is the initial angular acceleration of the board? (Neglect the weight of the board.)

❚ Take torques about the pivot point; $30(9.8)(2) - 20(9.8)(2) = I\alpha$. But $I = 20(4) + 30(4) = 200$ kg · m². Substituting gives $\alpha = \frac{196}{200} = \underline{0.98 \text{ rad/s}^2}$.

11.25 A pendulum consists of a small mass m at the end of a string of length L. The pendulum is pulled aside to an angle θ with the vertical and released. At the instant of release, using the suspension point as axis, find (*a*) the torque on the pendulum and (*b*) its angular acceleration.

❚ Torque = (force)(lever arm) = $mgL \sin \theta$, since $\tau = mr^2\alpha$ with $r = L$ in this case; one has $\alpha = (g \sin \theta)/L$.

11.26 Determine the constant torque that must be applied to a 50-kg flywheel, of radius of gyration 40 cm, to give it an angular speed of 300 rev/min in 10 s.

▮ $\tau = I\alpha$. By definition of radius of gyration k, $I = Mk^2 = (50\,\text{kg})(0.40\,\text{m})^2 = 8\,\text{kg}\cdot\text{m}^2$. We need only α to obtain τ. But $\omega = \omega_0 + \alpha t$, with $\omega_0 = 0$ and $t = 10\,\text{s}$, and $\omega = 2\pi f = 2\pi(300\,\text{rev/min}) = 600\pi\,\text{rad/min} = 10\pi\,\text{rad/s}$. Then $10\pi\,\text{rad/s} = \alpha(10\,\text{s})$ and $\alpha = \pi\,\text{rad/s}^2 = 3.14\,\text{rad/s}^2$, yielding finally $\tau = (8\,\text{kg}\cdot\text{m}^2)(3.14\,\text{rad/s}^2) = \underline{25\,\text{N}\cdot\text{m}}$.

11.27 An 80-lb wheel of 2-ft radius of gyration is rotating at 360 rev/min. The retarding frictional torque is 4 lb · ft. Compute the time it will take the wheel to coast to rest.

▮ We start with $\omega = \omega_0 + \alpha t$, with $\omega = 0$, and $\omega_0 = 2\pi f$, or $\omega_0 = 2\pi(360\,\text{rev/min})/(60\,\text{s/min}) = 12\pi\,\text{rad/s}$. If we knew α, we could obtain t. From dynamics $\tau = I\alpha$ and we are given $\tau = -4\,\text{lb}\cdot\text{ft}$, and $I = Mk^2 = (80\,\text{lb}/32\,\text{ft/s}^2)(2\,\text{ft})^2 = 10\,\text{slug}\cdot\text{ft}^2$; so $\alpha = (-4\,\text{lb}\cdot\text{ft})/(10\,\text{slug}\cdot\text{ft}^2) = -0.4\,\text{rad/s}^2$. Finally $0 = 12\pi\,\text{rad/s} - (0.4\,\text{rad/s}^2)t$, and $t = \underline{94\,\text{s}}$.

11.28 A 500-g wheel that has a moment of inertia of $0.015\,\text{kg}\cdot\text{m}^2$ is initially turning at 30 rev/s. It coasts to rest after 163 rev. How large is the torque that slowed it?

▮ $\tau = I\alpha$, with $I = 0.015\,\text{kg}\cdot\text{m}^2$ given. To find α we note that $\omega^2 = \omega_0^2 + 2\alpha\theta$, with $\omega = 0$, $\omega_0 = 2\pi f = (2\pi\,\text{rad})(30\,\text{s}^{-1}) = 60\pi\,\text{rad/s}$, and $\theta = (2\pi\,\text{rad/rev})(163\,\text{rev}) = 326\pi\,\text{rad}$. Then $\alpha = -(60\pi\,\text{rad/s})^2/(652\pi\,\text{rad})$, or $\alpha = -17.3\,\text{rad/s}^2$. Then $\tau = -(0.015\,\text{kg}\cdot\text{m}^2)(17.3\,\text{m/s}^2) = \underline{-0.26\,\text{N}\cdot\text{m}}$.

11.29 A certain 8-kg wheel has a radius of gyration of 25 cm. What is its moment of inertia? How large a torque is required to give it an angular acceleration of 3 rad/s²?

▮ $I = Mk^2 = (8\,\text{kg})(0.25\,\text{m})^2 = \underline{0.50\,\text{kg}\cdot\text{m}^2}$. $\qquad \tau = I\alpha = (0.50\,\text{kg}\cdot\text{m}^2)(3\,\text{rad/s}^2) = \underline{1.5\,\text{N}\cdot\text{m}}$.

11.30 A constant horizontal force of 1.2 N is applied tangentially to the shaft of a solid disk rotating about a vertical axis (Fig. 11-4). The radius of the shaft is 3 cm, the radius of the disk is 8 cm, and its mass is 4 kg. Calculate the angular acceleration of the disk, ignoring the mass of the shaft.

▮ First compute the moment of inertia of the disk:

$$I = \tfrac{1}{2}mR^2 = \tfrac{1}{2}(4)(0.08^2) = 0.0128\,\text{kg}\cdot\text{m}^2$$

Now use Newton's second law for rotational motion:

$$\tau = I\alpha \qquad 1.2(0.03) = 0.0128\alpha \qquad \alpha = \underline{2.81\,\text{rad/s}^2}$$

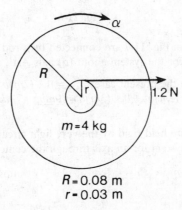

$R = 0.08\,\text{m}$
$r = 0.03\,\text{m}$ **Fig. 11-4**

11.31 A disk of 10-cm radius has a moment of inertia of $0.02\,\text{kg}\cdot\text{m}^2$. A force of 15 N is applied tangentially to the periphery of the disk to give it an angular acceleration, α. Find α in rad/s².

▮ $\tau = I\alpha$; $15(0.10) = 0.02\alpha$; $\alpha = \underline{75\,\text{rad/s}^2}$

11.4 MOMENT OF INERTIA

11.32 Two thin hoops of masses m_1 and m_2 have radii a_1 and a_2. They are mounted rigidly on a frame of negligible mass, as shown in Fig. 11-5. Find the system's moment of inertia about an axis through the center and perpendicular to the page.

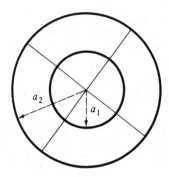

Fig. 11-5

▌ The definition of I is $\sum_1^N r_i^2 \, \Delta m_i$; for our hoops write this in two parts as

$$a_1^2 \sum_{i_1}^{N_1} \Delta m_{i_1} + a_2^2 \sum_{i_2}^{N_2} \Delta m_{i_2} = m_1 a_1^2 + m_2 a_2^2$$

11.33 Four coplanar, large, irregular masses are held by a rigid frame of negligible mass, as shown in Fig. 11-6. Taking an axis through P and perpendicular to the page, show that the system's moment of inertia is $I = I_1 + I_2 + I_3 + I_4$, where I_1 is the moment of inertia of object 1 alone about the axis and similarly for the others. What general rule could you prove in this way?

▌ $I = \sum r_i^2 \, \Delta m_i = \sum r_{i1}^2 \, \Delta m_{i1} + \sum r_{i2}^2 \, \Delta m_{i2} + \cdots$, where the first summation is over mass 1, the second over mass 2, and so on. The first summation is I_1, the second I_2, etc., so $I = I_1 + I_2 + I_3 + I_4$. *Rule:* Moments of inertia about an axis are added algebraically.

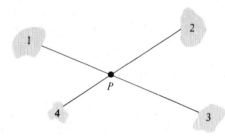

Fig. 11-6

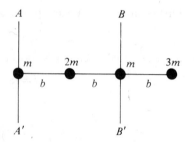

Fig. 11-7

11.34 The four point-masses shown in Fig. 11-7 are connected by a rod of negligible mass. Find the moment of inertia and radius of gyration for the system about **(a)** axis AA', **(b)** axis BB'.

▌ **(a)** $I = \sum m_i r_i^2 = k^2 \sum m_i$. In the present case, $I_{AA'} = 0 + (2m)b^2 + m(4b^2) + 3m(9b^2) = \underline{33mb^2} = k^2(7m)$ so $k = \underline{2.17b}$. **(b)** $I_{BB'} = m(4b^2) + (2m)b^2 + 0 + (3m)b^2 = \underline{9mb^2} = 7mk^2$, so $k = \underline{1.13b}$.

11.35 The four masses in Fig. 11-8 are held rigid by the very light circular frame shown. Find the moment of inertia and radius of gyration of the system for an axis through the center of the circle and perpendicular to the page.

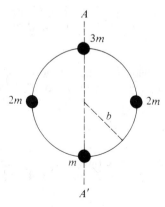

Fig. 11-8

How large a torque must be applied to the system to give it an angular acceleration α about this axis, provided it is free to turn? Repeat for the axis AA'.

▌ By definition $I = \sum m_i r_i^2 = \sum m_i b^2 = \underline{8mb^2}$. By comparison with $I = (\sum m_i)k^2$, one has $k = \underline{b}$. Since $\tau = I\alpha$, one finds $\tau = \underline{8mb^2\alpha}$. $I_{AA'} = (2m)(b^2) + (2m)(b^2) = \underline{4mb^2}$, which is $k_{AA'}^2(8m)$, so one finds that $k_{AA'} = \underline{b/2^{1/2}}$ and $\tau_{AA'} = \underline{4mb^2\alpha}$.

11.36   A nitrogen molecule can be thought of as two point masses (m of each $= 14\,u = 14 \times 1.67 \times 10^{-27}\,kg$) separated by a distance of $1.3 \times 10^{-10}\,m$. In air at room temperature the average rotational kinetic energy of such a molecule is about $4 \times 10^{-21}\,J$. Find the moment of inertia of such a molecule about its center of mass and its speed of rotation in revolutions per second.

▌ $I = \sum m_i r_i^2 = 2(2.34 \times 10^{-26})(0.65 \times 10^{-10})^2 = \underline{1.98 \times 10^{-46}\,kg \cdot m^2}$. Also, $I\omega^2/2 = 4 \times 10^{-21}$ gives $\omega = 6.4 \times 10^{12}\,rad/s = \underline{1.0 \times 10^{12}\,rev/s}$.

11.37   Three thin uniform rods each of mass M and length L lie along the x, y, z axes with one end of each at the origin. Find I about the z axis for the three-rod system.

▌ Only the rods along the x and y axes give rise to a moment of inertia about the z axis. From the parallel-axis theorem the moment of inertia of a rod about its end is $ML^2/12 + M(L/2)^2 = ML^2/3$. So, by Prob. 11.33, $\underline{I_z = 2ML^2/3}$.

11.38   Show that I, the moment of inertia of the system of Fig. 11-6 about an axis perpendicular to the page at P, is equal to $\sum (I_{cmi} + m_i r_i^2)$, where r_i is the distance from P to the ith center of mass.

▌ Call r_j the location with respect to P of the jth center of mass. The parallel-axis theorem gives $I_j = I_{cmj} + m_j r_j^2$, so

$$I = \sum_{j=1}^{4} (I_{cmj} + m_j r_j^2)$$

11.39   Knowing that $I = \frac{2}{5}Mr^2$ for a sphere with axis through its center, find I for an axis tangent to the sphere.

▌ Use the parallel-axis theorem: $I = I_{cm} + Mr^2 = \underline{7Mr^2/5}$.

11.40   A rod of length L is composed of a uniform length $\frac{1}{2}L$ of wood whose mass is m_w and a uniform length $\frac{1}{2}L$ of brass whose mass is m_b. (a) Find I for the rod about an axis perpendicular to the rod and through its center. (b) Repeat for a parallel axis through the wood end. (*Hint:* $I_{cm} = ma^2/12$ for a uniform rod of length a.)

▌ (a) First treat each half separately. For the wood, I about the end is, from the parallel-axis theorem, $[m_w(L/2)^2]/12 + m_w(L/4)^2 = (m_w L^2)/12$. For the brass it would be $(m_b L^2)/12$. Adding these we obtain in the first part $(m_w + m_b)(L^2/12)$. (b) The brass contribution to I about the wood end is $[m_b(L/2)^2]/12 + m_b(3L/4)^2 = (7m_b L^2)/12$ where we used the parallel-axis theorem again. To this we add $(m_w L^2)/12$, so $I = [L^2(m_w + 7m_b)]/12$.

11.41ᶜ   A flat hoop of mass M and radius R is shown in Fig. 11-9. The hoop lies in the xy plane and is centered at the origin O. In the figure, the z axis rises from O directly toward the viewer. Consider the mass element dm. It contributes an amount $dI_z = R^2\,dm$ to the moment of inertia about the z axis. The element dm also

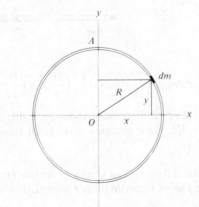

Fig. 11-9

contributes an amount $dI_x = y^2\,dm$ to the moment of inertia about the x axis; similarly, $dI_y = x^2\,dm$. **(a)** Find an equation relating dI_x, dI_y, and dI_z. **(b)** Find an equation relating I_x, I_y, and I_z. **(c)** Use symmetry considerations to obtain an equation relating I_x and I_y. **(d)** Determine I_x, I_y, and I_z for the hoop. Express your results in terms of M and R. **(e)** Determine the moment of inertia of the hoop about an axis that passes through point A and runs parallel to the x axis.

▌ **(a)** Referring to Fig. 11-9, since $x^2 + y^2 = R^2$, we have $x^2\,dm + y^2\,dm = R^2\,dm$, or $\underline{dI_y + dI_x = dI_z}$.
(b) The relationship obtained in **(a)** may be integrated over the ring to give $\underline{I_x + I_y = I_z}$. **(c)** Symmetry demands that the moment of inertia I_x about the x axis equal the moment of inertia I_y about the y axis. That is, $\underline{I_y = I_x}$.
(d) It is easily seen that $I_z = \int R^2\,dm = R^2 \int dm = MR^2$. Since $I_z = I_x + I_y = 2I_x$, we find $I_x = I_y = \underline{MR^2/2}$.
(e) Referring to Fig. 11-9, we apply the parallel-axis theorem to find that $I_{Ax} = I_x + MR^2 = \underline{3MR^2/2}$.

11.42^c Generalize the arguments involved in Prob. 11.41 to prove the following theorem: Any *flat* object in the xy plane has a moment of inertia I_z about the z axis which is given by $I_z = I_x + I_y$, where I_x and I_y are the object's moments of inertia about the x axis and the y axis, respectively.

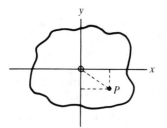

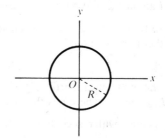

Fig. 11-10

Fig. 11-11

▌ Figure 11-10 indicates the general situation. The positive z axis rises directly toward the viewer. An infinitesimal mass dm located at the arbitrary point P contributes an amount $dI_x = y^2\,dm$ to the body's moment of inertia about the x axis, and an amount $dI_y = x^2\,dm$ to the moment of inertia about the y axis. It also contributes an amount $dI_z = R^2\,dm$ to the moment of inertia about the z axis. Since $R^2 = y^2 + x^2$, we have $dI_z = dI_x + dI_y$. Integrating this equation over the entire object, we obtain $\underline{I_z = I_x + I_y}$. This result is known as the *perpendicular-axis theorem*.

11.43 Employ the perpendicular-axis theorem of Prob. 11.42 to relate the moment of inertia of a thin circular disk (about any diameter) to its moment of inertia about its central axis.

▌ A thin uniform-density disk of radius R is confined to the xy plane, as shown in Fig. 11-11. The perpendicular-axis theorem implies that $I_{xO} + I_{yO} = I_{zO}$. But symmetry implies that $I_{xO} = I_{yO} = I_d$, where I_d denotes the moment of inertia about any diameter. Therefore, we have $2I_d = I_{zO}$. The moment of inertia of a solid cylinder or disk of radius R about the symmetry axis is $\frac{1}{2}MR^2$. Thus $I_{zO} = \frac{1}{2}MR^2$ and $I_d = \frac{1}{4}MR^2$.

11.44^c Using integration, compute the moment of inertia of a uniform thin rod of length L about an axis perpendicular to the rod at a point $L/4$ from one of its ends. Show how the same result could be obtained by use of the parallel-axis theorem.

Fig. 11-12

▌ Refer to Fig. 11-12.

$$I = \int r^2\,dm = \int_{-L/4}^{3L/4} r^2 \lambda\,dr = \frac{\lambda}{3}\left[\left(\frac{3L}{4}\right)^3 + \left(\frac{L}{4}\right)^3\right] = \frac{7(\lambda L)(L^2)}{48} = \frac{7ML^2}{48}$$

The thin rod has $I = (ML^2)/12$ about its center. About a parallel axis $L/4$ away, $I = (M)(L/4)^2 + (ML^2)/12 = (7ML^2)/48$. (The result $I_{CM} = ML^2/12$ is assumed here, but it could also be found by direct integration, as above, from $-L/2$ to $+L/2$.)

11.45^c A uniform hollow cylinder has a density ρ, a length L, an inner radius a, and an outer radius b. Show that its moment of inertia about the axis of the cylinder is $I = \frac{1}{2}\pi\rho L(b^4 - a^4) = \frac{1}{2}M(b^2 + a^2)$, where M is the mass of the cylinder.

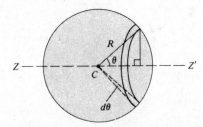

Fig. 11-13

▌ Use Fig. 11-13, noting that the mass M of the cylindrical shell is ρ times the volume of the shell $\rho\pi(b^2 - a^2)L$. The moment of inertia of a shell of thickness dr is $dI = (\rho 2\pi rL\, dr)r^2$; the term in the parentheses being the mass of the thin shell. Then

$$I = 2\pi\rho L \int_a^b r^3\, dr = \frac{2\pi\rho L(b^4 - a^4)}{4}$$

Substituting for ρ, write

$$I = \frac{\rho\pi L(b^2 - a^2)(b^2 + a^2)}{2} = \frac{M(b^2 + a^2)}{2}$$

11.46ᶜ Calculate the moment of inertia of a homogeneous thin-walled spherical shell of mass M and radius R, about a diameter.

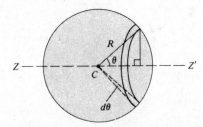

Fig. 11-14

▌ The total mass M is given by the integral of the surface density σ over the area of the shell: $M = \int \sigma\, dA$. Since the shell is homogeneous, the surface density σ is constant. Therefore $M = \sigma \int dA = \sigma A = 4\pi R^2\sigma$. This implies that $\sigma = M/4\pi R^2$.

Refer to Fig. 11-14. The hoop shown has a radius $R\sin\theta$ and therefore a circumference of $2\pi R\sin\theta$. The width of the hoop is $R\, d\theta$, so the area $dA = (2\pi R\sin\theta)(R\, d\theta)$, and the mass $dM = \sigma\, dA = (M/4\pi R^2)(2\pi R^2\sin\theta\, d\theta) = (M/2)\sin\theta\, d\theta$.

We use dI to denote the moment of inertia of the hoop about the axis ZZ'. The perpendicular distance from the axis to each part of the hoop is $R\sin\theta$. Therefore $dI = (R\sin\theta)^2\, dM = (MR^2/2)\sin^3\theta\, d\theta$.

The moment of inertia of the entire shell is the sum of the moments of inertia of the various hoops:

$$I = \int dI = \int_0^\pi \frac{MR^2}{2}\sin^3\theta\, d\theta = \frac{MR^2}{2}\int_0^\pi (1 - \cos^2\theta)\sin\theta\, d\theta = \frac{MR^2}{2}\left[-\cos\theta\big|_0^\pi + \int_0^\pi \cos^2\theta(-\sin\theta\, d\theta)\right]$$

$$= \frac{MR^2}{2}\left\{[-(-1) - (-1)] + \frac{\cos^3\theta}{3}\Big|_0^\pi\right\} = \frac{MR^2}{2}\left[2 + \frac{-1^3}{3} - \frac{1^3}{3}\right] = \frac{MR^2}{2}\frac{4}{3} = \underline{\frac{2MR^2}{3}}$$

11.47ᶜ Calculate the moment of inertia of a homogeneous sphere of mass M and radius R, about a diameter.

▌ By Prob. 11.46, a spherical subshell of radius r has moment of inertia

$$dI = \frac{2}{3}r^2 \times \text{mass} = \frac{2}{3}r^2\left(\frac{4\pi r^2\, dr}{\frac{4}{3}\pi R^3}\right)M = \frac{2M}{R^3}r^4\, dr$$

The moment of inertia of the sphere is the sum of the moments of inertia of the various shells:

$$I = \int dI = \frac{2M}{R^3}\int_0^R r^4\, dr = \frac{2M}{R^3}\frac{R^5}{5} = \underline{\frac{2MR^2}{5}}$$

11.48ᶜ The very thin uniform sheet shown in Fig. 11-15 has sides a and b. Its total mass is M. Show that its moment of inertia about the axis indicated is $Mb^2/12$.

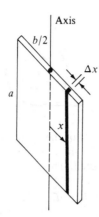

Axis

Fig. 11-15

▌ For a thin uniform slab the mass/area is $\sigma = M/ab$. The contribution of the mass element of thickness dx to the moment of inertia is $(\sigma a\,dx)x^2$, so that $I = \int_{-b/2}^{b/2} \sigma ax^2\,dx = \sigma ab^3/12$; but since $\sigma = M/ab$, $I = \underline{Mb^2/12}$.

11.49 A frictionless pivot at the center of the disk shown in Fig. 11-16 allows the system to turn freely. The rod is uniform and has a length L. Find the angular acceleration of the system when released from the position shown. The mass of the disk is m_1 and that of the rod is m_2.

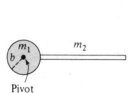

Pivot   **Fig. 11-16**

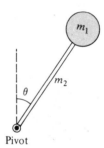

Pivot   **Fig. 11-17**

▌ I about the pivot $= I_{cm\ disk} + I_{cm\ rod} + m_2[b + (L/2)]^2 = m_1b^2/2 + m_2L^2/12 + m_2b^2 + m_2bL + m_2L^2/4$. The torque due to the weight of the rod $= m_2g[b + (L/2)]$. Then, after some algebra, $\alpha = \tau/I = [g(2b + L)]/[(m_1/m_2)b^2 + 2(b^2 + bL + L^2/3)]$.

11.50 A uniform rod of length L and mass m_2 is pivoted at one end, as shown in Fig. 11-17. To its other end is fastened a uniform disk of mass m_1 and radius b. Find the angular acceleration of the system just after it is released from the position shown.

▌ $I_{pivot} = I_{rod} + I_{CM\ disk} + m_1(L + b)^2 = m_2L^2/3 + m_1b^2/2 + m_1(L^2 + 2Lb + b^2)$. The lever arm for m_1 is $(L + b)\sin\theta$ and for m_2, $(L/2)\sin\theta$, so that $\tau = (m_2gL\sin\theta)/2 + m_1g(L + b)\sin\theta$. Solving for $\alpha = \tau/I$ we obtain $\alpha = g\sin\theta[m_2L + 2m_1(L + b)]/[2m_2L^2/3 + m_1(3b^2 + 4Lb + 2L^2)]$.

11.5 TRANSLATIONAL–ROTATIONAL RELATIONSHIPS

11.51 A watch has a second hand which is 2.0 cm long. (*a*) What is the frequency of revolution of the hand? (*b*) What is the speed of the tip of the second hand relative to the watch?

▌ (*a*) If T is the period in seconds,

$$f = \frac{1}{T} = \frac{1}{60} = 0.017\ \text{rev/s}$$

(*b*)
$$v = r\omega = r2\pi f = (2.0 \times 10^{-2})(2\pi)\left(\frac{1}{60}\right) = \underline{2.1 \times 10^{-3}\ \text{m/s}}$$

11.52 The earth rotates on its axis once every 24 h. Its radius is 6370 km. (*a*) What is the earth's angular velocity in radians per second? (*b*) If an airplane flies due west at the equator, at what speed (kilometers per hour) must it fly to just "keep up with the sun"?

▌ Earth turns 2π rad in 24 h. (a) $\omega = 2\pi$ rad/(24 h × 3600 s/h) = 7.272×10^{-5} rad/s. (b) The plane must fly at the same speed as the earth's "rim speed." Thus we use $v = \omega r$. Substituting, $v = (7.272 \times 10^{-5}$ rad/s)(6370 km)(3600 s/h) = 1667 km/h.

11.53 A 90-cm-radius roulette wheel initially turning at 3.0 rev/s slows uniformly and stops after 26 rev. (a) How long did it take to stop? (b) What was its angular deceleration? (c) What was the initial tangential speed of a point on its rim? (d) The initial radial acceleration of a point on the rim? (e) The magnitude of the resultant initial acceleration of a point on its rim?

▌ Given $r = 0.90$ m, $\omega_0 = 6\pi$ rad/s, and $\theta = 26$ rev = 52π rad. (a) $t = \theta/\bar{\omega} = 52\pi/3\pi = \underline{17.3}$ s. (b) $|\alpha| = \omega_0/t = 6\pi/17.3 = \underline{1.09}$ rad/s^2. (c) $v_t = \omega_0 r = 6\pi(0.90) = \underline{17.0}$ m/s. (d) $a_R = \omega_0^2 r = \underline{320}$ m/s^2. (e) The initial tangential acceleration is $\alpha r = \underline{0.98}$ m/s^2, so the total initial acceleration is $(a_T^2 + a_R^2)^{1/2} = \underline{320}$ m/s^2.

11.54 As shown in Fig. 11-18(a), a girl on a rotating platform holds a pendulum in her hand. The pendulum is at a radius of 6.0 m from the center of the platform. The rotational speed of the platform is 0.020 rev/s. It is found that the pendulum hangs at an angle θ to the vertical as shown. Find θ.

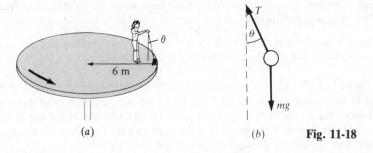

(a)   (b)   **Fig. 11-18**

▌ The external forces on the pendulum ball are shown in Fig. 11-18(b). In the equilibrium situation $\mathbf{F} = m\mathbf{a}$ can be written as $T \sin \theta = m\omega^2 r$ and $T \cos \theta - mg = 0$. Dividing the first expression by the second gives $\tan \theta = \omega^2 r/g$. Solving for θ, we get $\underline{0.55°}$.

11.55 A mass m is fastened to one end of a massless spring of natural length a and stiffness k. Holding the other end, a man whirls the apparatus in a horizontal circle at angular velocity ω. What is the radius of the circle?

▌ The tension in the spring, $k(r - a)$, must supply the centripetal force, $m\omega^2 r$, where $r - a$ is the elongation of the spring. Equating yields $r = ka/k(1 - m\omega^2/k)$. For a stiff spring, $r = a(1 + m\omega^2/k)$ provided that $m\omega^2/k \ll 1$. (Note: $1/(1 - x) \approx 1 + x$ for $|x| \ll 1$.)

11.56 Figure 11-19 shows a possible design for a space colony of the future. It consists of a 6-km-diameter cylinder of length 30 km floating in space. Its interior is provided with an earthlike environment. To simulate gravity, the cylinder spins on its axis as shown. What should be the rate of rotation of the cylinder (in revolutions per hour) so that a person standing on the landmass will press down on the ground with a force equal to his or her weight on earth?

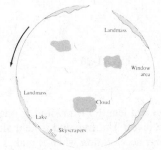

Fig. 11-19

▌ Using $m\omega^2 R = mg$, we find $\omega^2 = g/R = 9.8/3000 = 3.27 \times 10^{-3}$/s^2, from which $\omega = (0.0572/s)(3600/2\pi) = \underline{32.8}$ rev/h.

11.57 The rotation speed of the earth is 1 rev/day, or 1.16×10^{-5} rev/s, and the earth's radius is 6.37×10^6 m. If a man at the equator is standing on a spring scale, by what percent would his apparent weight increase if the earth were to stop rotating? A man at the north pole?

❚ We note $\sum F = ma = m\omega^2 r$ inward along the radial direction. Thus the scale supplies a force $(mg - m\omega^2 r)$ to support the man. When the earth stops rotating, the force will increase to mg. The ratio we need is $m\omega^2 r/mg = (1.16 \times 10^{-5} \times 2\pi)^2(6.37 \times 10^6)/9.8 = 3.5 \times 10^{-3}$, giving a change of 0.35 percent. At the north pole, the man will be on the axis and so $r = 0$ and the change is zero.

11.58 The red blood cells and other particles suspended in blood are too light in weight to settle out easily when the blood is left standing. How fast (in revolutions per second) must a sample of blood be rotating at a radius of 10 cm in a centrifuge if the centripetal force needed to hold one of the particles in a circular path is 10 000 times the weight of the particle, mg? Why do the particles separate from the solution in a centrifuge?

❚ Centripetal force $= m\omega^2 r$, which is given as $10^4 mg$. Solving, $\omega = 990$ rad/s $= 158$ rev/s. The required centripetal force is supplied primarily not by buoyant forces but by viscous forces. The cells therefore move slowly to larger radii and settle against the end of the centrifuge tube.

11.59 A 20-mg bug sits on the smooth edge of a 25-cm-radius phonograph record as the record is brought up to its normal rotational speed of 45 rev/min. How large must the coefficient of friction between the bug and record be if the bug is not to slip off?

❚ For just not slipping the friction force $f = \mu F_N = \mu mg$ supplies the centripetal force, so $\mu mg = m\omega^2 r$, which with $\omega = 45$ rev/min $= 4.71$ rad/s; then $r = 0.25$ m gives $\mu = 0.566$.

11.60 A cylinder of 5 cm radius is rolling along the floor with a constant speed of 80 cm/s. (**a**) What is the rotational speed of the cylinder about its axis? (**b**) What are the magnitude and direction of the acceleration of a point its surface? (**c**) At the instant a certain point on its surface is at the top of the cylinder, what is the velocity of the point? (**d**) Repeat if the point is at the contact with the floor. (**e**) Repeat if the point is midway between top and floor and at the forward surface of the cylinder.

❚ Refer to Fig. 11-20. The tangential velocity v_T with respect to the center equals the forward speed of the rolling cylinder. (**a**) $\omega = v_T/r = 0.80/0.05 = 16$ rad/s. (**b**) Since $\alpha = 0$, there is only radial acceleration, $\omega^2 r = 12.8$ m/s². To obtain instantaneous velocities at various points on the rim we add the forward speed of 0.80 m/s of the center to the velocity of a point v_T with respect to the center, giving (**c**) 1.6 m/s, (**d**) zero, and (**e**) 1.13 m/s at 45° below the horizontal.

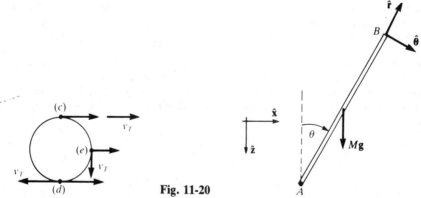

Fig. 11-20   Fig. 11-21

11.61 A slender uniform rod of mass m and length l is pivoted at one end so that it can rotate in a vertical plane. There is negligible friction at the pivot. The free end is held almost vertically above the pivot and then released. What is the rod's angular acceleration when it makes an angle θ with the vertical?

❚ Using the notation indicated in Fig. 11-21 and calculating torques about point A, with clockwise being positive, we have

$$I_A \alpha = \frac{Mgl}{2} \sin \theta$$

where $I_A = Ml^2/3$ as can easily be seen (Prob. 11.37). The angular acceleration is therefore given by

$$\alpha = \frac{3g}{2l}\sin\theta$$

11.62 Refer to Prob. 11.61. At the same angle θ, what is the magnitude of the translational acceleration of the free end of the rod?

┃ The tangential component of acceleration, a_t, is readily obtained from α, but the centripetal component of acceleration, a_c, requires a knowledge of ω. This can be obtained most readily from energy considerations. Recognizing that the force at the pivot does no work and using the conservation of energy, we find that

$$\frac{1}{2}I_A\omega^2 = \frac{Mgl}{2}(1-\cos\theta)$$

Solving for ω^2, we find

$$\omega^2 = \frac{3g}{l}(1-\cos\theta)$$

Using the unit vectors $\hat{\boldsymbol\theta}$ and $\hat{\mathbf r}$ shown in Fig. 11-21, we find that the linear acceleration a of a point on the rod a distance r from the pivot is given by

$$\mathbf{a} = a_t\hat{\boldsymbol\theta} + a_c\hat{\mathbf r}$$

where the tangential acceleration $a_t = r\alpha$ and the centripetal acceleration $a_c = -\omega^2 r$. Using the expressions already obtained for α and ω, and setting $r = l$, we find that the free end of the rod has an acceleration

$$\mathbf{a} = \left(\frac{3g\sin\theta}{2}\right)\hat{\boldsymbol\theta} + [-3g(1-\cos\theta)]\hat{\mathbf r}$$

The magnitude of this vector is

$$a = 3g\sqrt{\tfrac{5}{4} - 2\cos\theta + \tfrac{3}{4}\cos^2\theta}$$

11.63 In Fig. 11-22 the mass m is held by two strings and the system is rotating with angular velocity ω. Find the tensions in the two strings in terms of m, ω, r, and θ.

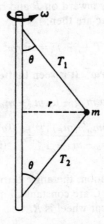

Fig. 11-22

┃ Examining the forces on the ball, we have in the vertical direction $T_1\cos\theta - T_2\cos\theta = mg$, while horizontally the centripetal force $m\omega^2 r = T_1\sin\theta + T_2\sin\theta$. Solving these two simultaneously, we find that the tensions are

$$\frac{m}{2\sin\theta}(\omega^2 r \pm g\tan\theta)$$

where the $+$ sign applies to T_1 and the $-$ sign to T_2.

11.64 In the turntable arrangement shown in Fig. 11-23, block A has a mass of 0.9 kg and block B has a mass of 1.7 kg, and the blocks are 13 cm from the axis of rotation. The coefficient of static friction between the blocks,

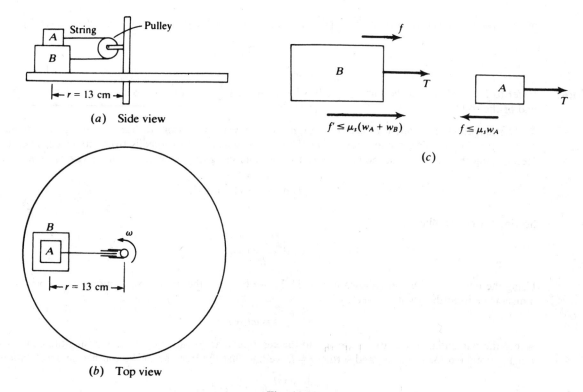

(a) Side view

(b) Top view

(c)

Fig. 11-23

and between the blocks and the turntable, is $\mu_s = 0.1$. Consider the friction and the mass of the pulley as negligible. Find the angular speed of rotation of the turntable for which the blocks just begin to slide.

▌ Everything depends on correctly predicting the direction of the frictional force between A and B. Since B is more massive than A, we extrapolate to the case where A is very light: B would tend to move radially outward, pulling A radially inward. The friction force f between the two surfaces would act to oppose their relative motion; it would act radially inward on B and radially outward on A, as shown in Fig. 11-23(c).

The force equations for no slipping are then:

$$\sum F_B = T + f + f' = m_B r \omega^2 \qquad \sum F_A = T - f = m_A r \omega^2$$

By subtraction, $2f + f' = (m_B - m_A)r\omega^2$. It is seen that ω can increase until both f and f' attain their maximum values. Thus

$$2\mu_s m_A g + \mu_s (m_A + m_B)g = (m_B - m_A)r\omega^2_{max}$$

or

$$\omega_{max} = \left[\frac{\mu_s g(3m_A + m_B)}{r(m_B - m_A)} \right]^{1/2} = \left[\frac{(0.1)(9.8)(2.7 + 1.7)}{(0.13)(1.7 - 0.9)} \right]^{1/2} = \underline{6.4 \text{ rad/s}}$$

11.65^c During the initial part of its acceleration, the angle through which a car's wheel turns as a function of time is given by $\theta = Bt + Ct^2$, where B and C are constants. Find the linear displacement of the car and its speed as a function of time. The radius of the car wheel is R.

▌ The linear distance is $R\theta = R(Bt + Ct^2)$. The car's speed is the tangential velocity of the rim with respect to the center of the wheel, $v = R\omega = R(d\theta/dt) = \underline{R(B + 2Ct)}$.

11.66^c The angle which a pendulum of length L makes with the vertical varies with time according to $\theta = \theta_0 \sin 2\pi ft$, where θ_0 is the maximum angle of swing and f is the frequency of the pendulum; both are constants. Find the tangential speed and acceleration of the pendulum ball as functions of time.

▌ The velocity and tangential acceleration of the ball are $L\omega$ and $L\alpha$. Since $\omega = d\theta/dt$, we have $v_T = Ld(\theta_0 \sin 2\pi ft)/dt = 2\pi fL\theta_0 \cos 2\pi ft$. The angular acceleration $\alpha = d\omega/dt$, so that $a_T = -(2\pi f)^2 L\theta_0 \sin 2\pi ft$, or $\underline{-(2\pi f)^2 L\theta}$.

11.67^c A smooth horizontal tube of length l rotates about a vertical axis as shown in Fig. 11-24(a). A particle placed

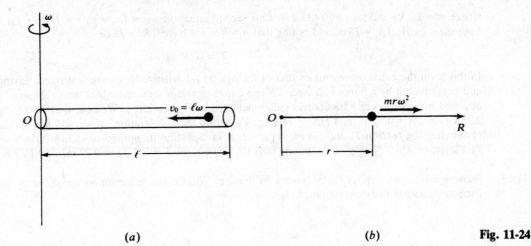

(a)   (b)   Fig. 11-24

at the extreme end of the tube is projected toward O with a velocity $l\omega$, while at the same time the tube rotates about the axis with constant angular speed ω. Determine the path of the particle.

▮ Since the tube is smooth, there is *no radial force* on the particle; the force, and hence the acceleration, is purely in the circumferential direction. This suggests viewing the motion in the noninertial frame that rotates with the tube, thereby "getting rid of" the circumferential force.

When the particle is at a distance r from O in the noninertial frame [Fig. 11-24(b)], the only force on it is the inertial force ("centrifugal force") $mr\omega^2$, directed as shown. Newton's second law becomes

$$m\ddot{r} = 0 + mr\omega^2 \quad \text{or} \quad \ddot{r} = \omega^2 r$$

Multiplying by $\dot{r}\, dt = dr$ and integrating,

$$\tfrac{1}{2}\int d(\dot{r}^2) = \omega^2 \int r\, dr \qquad \tfrac{1}{2}\dot{r}^2 = \tfrac{1}{2}\omega^2 r^2 + c$$

When $r = l$, $\dot{r} = -l\omega$, which gives $c = 0$ and $\dot{r} = -\omega r$, the minus sign being taken because r is decreasing. Finally,

$$\int \frac{dr}{r} = -\omega \int dt \qquad \ln r = -\omega t + c' \qquad r = c'' e^{-\omega t}$$

When $t = 0$, $r = l$, whence $c'' = l$ and $r = l e^{-\omega t}$.

11.6 PROBLEMS INVOLVING CORDS AROUND CYLINDERS, ROLLING OBJECTS, ETC.

11.68   A 25-kg wheel has a radius of 40 cm and turns freely on a horizontal axis. The radius of gyration of the wheel is 30 cm. A 1.2-kg mass hangs at the end of a cord that is wound around the rim of the wheel. This mass falls and causes the wheel to rotate. Find the acceleration of the falling mass and the tension in the cord.

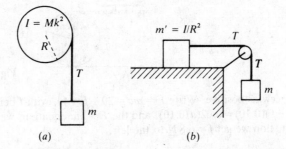

(a)   (b)   Fig. 11-25

▮ The situation is depicted in Fig. 11-25(a). We choose downward as positive for the block and for consistency, clockwise as positive for the wheel. We have two dynamical equations. For the block:

$$mg - T = ma \tag{1}$$

where $m = 1.2$ kg and thus $mg = 11.8$ N. Our second equation is $\tau = I\alpha$, with $\tau = TR$, or $TR = I\alpha$. Multiplying both sides by R, $TR^2 = IR\alpha$; and noting that $a = R\alpha$, we have $TR^2 = Ia$ or

$$T = (I/R^2)a \qquad (2)$$

In this form the equation resembles that of a block on a horizontal frictionless surface, having a mass (I/R^2) and being pulled by a horizontal force T. The entire problem then resembles the problem of two blocks attached by a cord over a frictionless pulley depicted in Fig. 11-25(b). We note that $I = (25\text{ kg})(0.30\text{ m})^2 = 2.25$ kg·m², $R = 0.40$ m, and $I/R^2 = 14.1$ kg. To obtain the acceleration, a, we add Eqs. (1) and (2), eliminating the tension T, and get $mg = (m + I/R^2)a$. Substituting numerical values, $11.8\text{ N} = (1.2\text{ kg} + 14.1\text{ kg})a$, or $a = \underline{0.77\text{ m/s}^2}$. Substituting into Eq. (2) we get $T = (14.1\text{ kg})(0.77\text{ m/s}^2) = \underline{10.9\text{ N}}$.

11.69 Starting from rest, a sphere rolls down a 30° incline. What is the minimum value of the coefficient of static friction if there is to be no slipping?

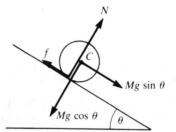

Fig. 11-26

▮ The problem is diagrammed in Fig. 11-26. We let α denote the angular acceleration (clockwise positive) and a denote the linear acceleration along the incline. If there is no slipping, then $v = \omega R$, where v and ω are the linear and angular velocities. Similarly, $a = \alpha R$, where a and α are the linear and angular accelerations. The translational acceleration is given by

$$a = g\sin\theta - \frac{f}{M},$$

where f is the frictional force. Neither the sphere's weight nor the normal force exerts any torque about C, the center of the (uniform) sphere. Therefore the angular acceleration is given by

$$\alpha = \frac{fR}{I_c} = \frac{5f}{2MR} \quad \text{and} \quad a = \frac{5f}{2M}$$

Solving the displayed equations for f, we find $f = \frac{2}{7}Mg\sin\theta$. The normal force $N = Mg\cos\theta$, so $f/N = \frac{2}{7}\tan\theta$. Since the maximum ratio (f_{max}/N) consistent with pure rolling is μ_s, we must have $\mu_s \geq \frac{2}{7}\tan\theta$. For $\theta = 30°$, $\tan\theta = 1/\sqrt{3}$, so we find $(\mu_s)_{min} = \underline{2/(7\sqrt{3}) = 0.165}$.

11.70 The rope shown in Fig. 11-27 is wound around a cylinder of mass 4.0 kg and $I = 0.020$ kg·m², about the cylinder axis. If the cylinder rolls without slipping, what is the linear acceleration of its center of mass? What is the frictional force? Use an axis along the cylinder axis for your computation.

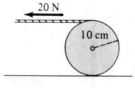

Fig. 11-27

▮ Choose left and ccw as positive. Write $F = ma = 20 + f = 4a$, with f being the friction force at the floor. From $\tau = I\alpha = (20 - f)(0.10) = 0.02(a/0.10)$, and the $F = ma$ equation, we find a to be $\underline{6.7\text{ m/s}^2}$. Substituting back into either equation we get $f = \underline{6.8\text{ N}}$ to the left.

11.71 Repeat Prob. 11.70 if the frictional force between table and cylinder is negligible. Choose any axis for your computation.

▮ In the case $f = 0$ in Prob. 11.70, the first equation gives $a = \underline{5.0\text{ m/s}^2}$. Note that $a = \alpha r$ is not applicable when slippage occurs.

11.72 A solid cylinder and a thin-walled pipe are simultaneously released from rest at the upper end of a ramp of inclination θ. Each object rolls without slipping. (**a**) Find the acceleration of the center of mass of the solid cylinder. (**b**) Find the acceleration of the center of mass of the pipe. (**c**) When the cylinder has rolled a distance s_c, how far has the pipe rolled?

▮ Figure 11-26 serves to indicate the general situation in each case. The acceleration a down the incline is determined by three equations. The first is Newton's second law:

$$Ma = Mg \sin \theta - f$$

where f is the frictional force. The second equation is the rotational form of Newton's second law:

$$I\alpha = fR$$

where I is the moment of inertia of the object about its symmetry axis. The third equation is the rolling constraint:

$$a = \alpha R$$

Solving the three displayed equations for a, we find

$$a = \frac{g \sin \theta}{(1 + I/MR^2)}$$

(**a**) For a solid cylinder $I = I_c = \frac{1}{2}MR^2$, so $a_c = (g \sin \theta)/(1 + \frac{1}{2}) = \underline{(2g/3) \sin \theta}$.
(**b**) For a thin-walled pipe, $I = I_p = MR^2$, so $a_p = (g \sin \theta)/(1 + 1) = \underline{(g/2) \sin \theta}$.
(**c**) The distances traveled are in the same ratio as the accelerations:

$$\frac{s_p}{s_c} = \frac{a_p}{a_c} = \frac{\frac{1}{2}}{\frac{2}{3}} = \frac{3}{4}$$

That is, $s_p = \underline{3s_c/4}$.

11.73 A heavy wheel of radius 20 cm is mounted on a horizontal axle. A rope wrapped around its rim is pulled straight downward with a constant force of 50 N. The rope moves a distance of 50 cm in 1.0 s. (**a**) What is the angular acceleration of the wheel? (**b**) What is the moment of inertia of the wheel? (**c**) The wheel is a homogeneous, solid disk. What is its mass?

▮ (**a**) Using the kinematic equation $s = \frac{1}{2}at^2$, we find $a = 2s/t^2 = 1 \text{ m/s}^2$ and $\alpha = a/R = \underline{5.00 \text{ rad/s}^2}$.
(**b**) The force F exerts a torque $FR = I\alpha$, so $I = FR/\alpha = [(50 \text{ N})(0.2 \text{ m})]/5.0 = \underline{2.00 \text{ kg} \cdot \text{m}^2}$.
(**c**) A uniform disk of radius R has a moment of inertia $I = \frac{1}{2}MR^2$ about the symmetry axis. Therefore $M = 2I/R^2 = [2(2.00)]/0.20^2 = \underline{100 \text{ kg}}$.

11.74 Refer to Prob. 11.73. (**a**) Suppose that an object whose weight is 50 N is attached to the rope, and the system is released from rest. What would the angular acceleration of the wheel be in that case? (**b**) Account for the difference between the results of part a of Prob. 11.73.

▮ (**a**) In this case the acceleration a' of the object is given by $Ma' = W - F'$, where F' is the tension in the rope. Furthermore, $F'R = I\alpha'$. The linear and angular accelerations are related as before: $a' = \alpha'R$. Solving for α', we find

$$\alpha' = \frac{W}{\left(\dfrac{WR}{g} + \dfrac{I}{R}\right)} = \frac{50}{\left[\dfrac{(50)(0.2)}{(9.80)} + \dfrac{2.00}{0.20}\right]} = \underline{4.54 \text{ rad/s}^2}$$

(**b**) The angular acceleration $\alpha' < \alpha$ because part of the weight force W is "expended" in accelerating the suspended mass. Thus $F' = W - Ma' < W$, so $\alpha' < \alpha$.

11.75 An upright hoop is projected onto a pavement with an initial horizontal speed v_0 but without spin, so that it slides. The resulting frictional force causes the hoop to lose translational speed and to acquire an angular speed. Eventually the hoop rolls without slipping. Prove that when the hoop ceases to slip, it has speed $v_0/2$.

▮ The situation is shown in Fig. 11-28. We take $t = 0$ to denote the instant at which the hoop is projected with velocity $v_x = v_0$. For translation, $f_x = -f = Ma_x$, and

$$v_x(t) = v_0 - \frac{ft}{M}$$

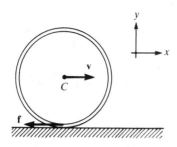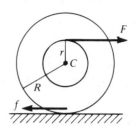

Fig. 11-28   **Fig. 11-29**

where M is the mass of the hoop. Taking clockwise rotation as positive, the rotational form of Newton's second law gives $\alpha = fR/I_c = f/MR$, since $I_c = MR^2$ for a hoop. Since $\omega = 0$ at $t = 0$, we have $\omega = ft/MR$. The hoop will stop sliding when $v_x = \omega R$, or when

$$v_x(t) = v_0 - \frac{ft}{M} = \frac{ft}{M}$$

Therefore the hoop will slide until $t = Mv_0/2f$, at which time its velocity will be $v_0 - (f/M)(Mv_0/2f) = \underline{v_0/2}$.

11.76   A spool of mass M is resting on a horizontal surface. The spool has moment of inertia MG_C^2 about its axis of symmetry. The spool is subjected to a rightward horizontal force of magnitude F, applied at a distance r above the axis. (Fig. 11-29) (a) Show that if there is to be no slippage between the spool and the supporting surface, a leftward frictional force $f = F(G_C^2 - rR)/(G_C^2 + R^2)$ must act on the spool. (b) Show that the required frictional force f has the value zero for a particular value r_0 of the distance r.

▌ (a) The translational form of Newton's second law implies that $Ma_C = F - f$, while the rotational form implies that $Fr + fR = MG_C^2\alpha$, where α is the angular acceleration. If there is no slippage, then $\alpha = a_C/R$. Solving these three equations for f, we find

$$f = \frac{F(G_C^2 - rR)}{(G_C^2 + R^2)}$$

as desired. (b) When $r \equiv r_0 = \underline{G_C^2/R}$, the frictional force vanishes.

11.77   Refer to Prob. 11.76: (a) Interpret the result of part a when r exceeds the value r_0 found in part b. (b) Show that the rightward translational acceleration of the center of the mass of the spool a_C exceeds F/M when $r > r_0$. Explain how this can happen.

▌ (a) When $r > r_0$, we find that $f < 0$; that is, the frictional force is directed toward the right, with magnitude equal to $F[(rR - G_C^2)/(G_C^2 + R^2)]$. (b) The translational acceleration is given by

$$a_C = \frac{F - f}{M} = \frac{F}{M}\left(1 + \frac{rR - G_C^2}{G_C^2 + R^2}\right)$$

which exceeds F/M when $r > r_0 = G_C^2/R$. This occurs because the frictional force can be directed toward the right. (A frictional force always opposes the *relative* motion of the two surfaces that are in contact. It does not necessarily oppose the translational motion of a spinning object.)

11.78   The moment of inertia of the wheel in Fig. 11-30 is $8.0\ \text{kg} \cdot \text{m}^2$. Its radius is 40 cm. Find the angular acceleration of the wheel caused by the 10.0-kg mass if the frictional force between the mass and the incline is 30 N.

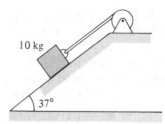

10 kg

37°

Fig. 11-30

▌ Along the incline $F = ma$ yields $ma = mg \sin 37° - T - 30$, while for the wheel $\tau = I\alpha$ becomes $rT = I\alpha$. Using $a = \alpha r$ we solve $F = ma$ to obtain $T = mg \sin 37° - 30 - m\alpha r$, which is placed into the torque equation yielding $\alpha = [r(mg \sin 37° - 30)]/(I + mr^2)$. Inserting values for m, r, g, and I, we get $\alpha = \underline{1.20\ \text{rad/s}^2}$.

11.79 A string is wound around an otherwise unsupported homogeneous, horizontal cylinder of mass M and radius R. As the string unwinds and the cylinder spins, the end of the string is continually pulled vertically upward with a force just sufficient to keep the cylinder from descending relative to the ground. **(a)** What is the tension in the vertical portion of the string? **(b)** What is the angular acceleration of the cylinder? **(c)** What is the upward acceleration of any given point along the vertical portion of the string?

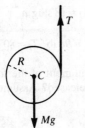

Fig. 11-31

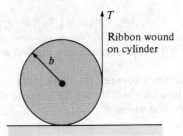

Ribbon wound on cylinder

Fig. 11-32

▌ **(a)** The situation is shown in Fig. 11-31. Since the center of mass of the spool is not accelerating, the tension T in the string must be equal to the weight: $\underline{T = Mg}$. **(b)** Taking torques about the central axis of the spool, we have $I_c\alpha = (-Mg)(0) + TR = \underline{MgR}$, where $I_c = \frac{1}{2}MR^2$ is the moment of inertia of the homogeneous spool. Therefore the angular acceleration is given by $\underline{\alpha = 2g/R}$. **(c)** The length of string unwound when the cylinder turns through angle $\theta(t)$ equals $R\theta(t)$. Therefore the upward linear acceleration a of any part of the string must be given by $a = R\alpha = \underline{2g}$.

11.80 For the cylinder shown in Fig. 11-32, $I_{CM} = \frac{1}{2}Mb^2$ and $T < Mg$. Describe the translational and rotational motion of the cylinder if **(a)** the cylinder doesn't slip on the floor and **(b)** there is no friction between floor and cylinder.

▌ **(a)** For the friction case, the rolling cylinder will spin faster as it accelerates to the left. The equations of motion are $(T - f)b = I\alpha$ and $f = Ma$. Since $a = \alpha b$ and I is given as $Mb^2/2$, the friction f can be eliminated, leaving $a = 2T/3M$ and $\alpha = 2T/3Mb$. The largest value T can have is mg; therefore the maximum translational acceleration is $2g/3$. **(b)** When no friction is present $Tb = I\alpha$ and $\alpha = 2T/Mb$; and since $f = Ma$, $a = 0$. If the cylinder starts from rest it will spin in place.

11.81 A wheel of rim radius 40 cm and mass 30 kg is mounted on a frictionless horizontal axle. When a 0.100-kg mass is suspended from a cord wound on the rim, the mass drops 2.0 m in the first 4.0 s after the mass is released. What is the radius of gyration of the wheel?

▌ We first find a of the 0.100-kg mass from kinematics: $v_0 = 0$, $y = 2$ m, $t = 4$ s, and we use $y = v_0t + at^2/2$ to obtain $a = 0.25$ m/s². Then $mg - T = ma$, or $0.10(9.8) - T = 0.10(0.25)$ to give $T = 0.955$ N. Next we use $Tr = \tau = I\alpha = I(a/r) \Rightarrow 0.955(0.40) = I(0.25/0.4)$ to give $I = 0.61$ kg · m². Then since $I \equiv Mk^2$, $k = (0.61/30)^{1/2} = \underline{0.143}$ m.

11.82 In Fig. 11-33 the pulley is free to rotate on a horizontal axis through its center. There is no slippage between the cord and the pulley. Assume $I = 8$ slug · ft². Find the acceleration of block m_2 and the tension in each side of the cord. (Such a setup is called an *Atwood's machine*.)

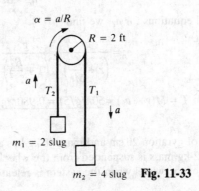

$\alpha = a/R$

$R = 2$ ft

a ↑ T_2 T_1

↓ a

$m_1 = 2$ slug

$m_2 = 4$ slug **Fig. 11-33**

❚ We choose downward as positive for m_2, upward as positive for m_1, and clockwise as positive for the pulley. Then

$$m_2g - T_1 = m_2a \qquad T_2 - m_1g = m_1a \qquad (T_1 - T_2)R = I\alpha = \frac{Ia}{R} \qquad \text{or} \qquad T_1 - T_2 = \left(\frac{I}{R^2}\right)a$$

Adding the three equations we have

$$(m_2 - m_1)g = \left(m_2 + m_1 + \frac{I}{R^2}\right)a \qquad \text{or} \qquad (2 \text{ slug})(32 \text{ ft/s}^2) = (8 \text{ slug})a$$

Solving, $a = \underline{8 \text{ ft/s}^2}$. Substituting into the first and second equations, we get $T_1 = \underline{96 \text{ lb}}$, $T_2 = \underline{80 \text{ lb}}$.

11.83   The frictional force between the block and table in Fig. 11-34 is 20 N. If the moment of inertia of the wheel is 4.0 kg · m², how long will it take the block to drop 60 cm after the system is released? Assume no slippage between rope and wheel.

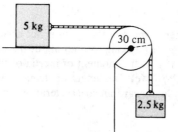

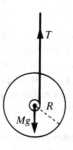

Fig. 11-34   **Fig. 11-35**

❚ Call the tension in the upper rope T_1 and in the lower T_2. Then

$$T_1 - 20 = 5a \qquad 2.5(9.8) - T_2 = 2.5a \qquad \text{and} \qquad (T_2 - T_1)(0.30) = I\alpha = I\frac{a}{r} = 4\left(\frac{a}{0.3}\right)$$

Solving the three equations for acceleration gives $a = 0.087 \text{ m/s}^2$. From the kinematic data, $v_0 = 0$, $a = 0.087 \text{ m/s}^2$, and $s = 0.60 \text{ m}$, and from $s = v_0t + at^2/2$, we find $t = \underline{3.71 \text{ s}}$.

11.84   Consider a Yo-Yo with outside radius R equal to 10 times its spool radius r (Fig. 11-35). The moment of inertia I_c of the Yo-Yo about its spool is given with good accuracy by $I_c = \frac{1}{2}MR^2$, where M is the total mass of the Yo-Yo. The upper end of the string is held motionless. (**a**) Compute the acceleration of the center of mass of the Yo-Yo. How does it compare with g? (**b**) Find the tension in the string as the Yo-Yo descends. How does it compare with Mg?

❚ (**a**) The situation is shown in Fig. 11-35. Taking counterclockwise rotations and downward translations as positive, we have the translational equation

$$Ma_c = Mg - T$$

the rotational equation

$$I_c\alpha = Tr$$

where $I_c = \frac{1}{2}MR^2$, and the constraint equation

$$a_c = \alpha r$$

Solving the displayed equations for a_c, we find

$$a_c = \frac{gr}{\left(r + \frac{I_c}{Mr}\right)} = \frac{g}{\left(1 + \frac{R^2}{2r^2}\right)} = \frac{g}{51} = \underline{0.192 \text{ m/s}^2}$$

(**b**) The string tension $T = M(g - a_c) = \underline{50Mg/51 = 0.980Mg}$.

11.85   A wheel with radius of gyration 20 cm and mass 40 kg has a rim radius of 30 cm and is mounted vertically on a horizontal axis. A 2-kg mass is suspended from the wheel by a rope wound around the rim. Find the angular acceleration of the wheel when the system is released.

❚ Write $F = ma$ for the mass and $\tau = I\alpha$ for the wheel: $2(9.8) - T = 2.0(0.3\alpha)$ and $T(0.30) = I\alpha$. We know $I = Mk^2 = 40(0.04) = 1.6 \text{ kg} \cdot \text{m}^2$. Solving for α gives $\underline{3.3 \text{ rad/s}^2}$.

11.86 The uniform wheel of moment of inertia I shown in Fig. 11-36 is pivoted on a horizontal axis through its center so that its plane is vertical. As shown, a small mass m is stuck on the rim of the wheel. Find the angular acceleration of the wheel when the mass is at point A. Repeat for points B and C. (Assume that I is not changed much by the presence of m.)

❚ The torque lever arm is b at A, zero at B, and $0.80b$ at C. The force in each case is mg. Hence, for A, $\tau = I\alpha$ gives $\alpha = \underline{mgb/I}$. For B one has $\alpha = \underline{0}$, while at C it is $\underline{0.80mgb/I}$.

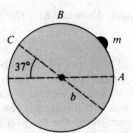

Fig. 11-36

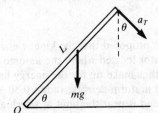

Fig. 11-37

11.87 A uniform thin rod (such as a meterstick) of length L stands vertically on one end on the floor. Its top is now given a tiny push so the rod begins to topple over (Fig. 11-37). Assume that the base of the rod does not slip, and find (**a**) the angular acceleration of the rod when it makes an angle θ with the floor, (**b**) the tangential acceleration of the upper tip of the rod.

❚ (**a**) Taking the torque about pivot point O we have $Mg[(L\cos\theta)/2] = I_0\alpha$. By the parallel-axis theorem $I_0 = I_{\text{CM}} + M(L/2)^2 = ML^2/3$. Solve for α to find $\alpha = (3g\cos\theta)/2L$. (**b**) $a_T = \alpha L = (3g\cos\theta)/2$.

11.88 Refer to Fig. 11-38. The moment of inertia of the pulley system is $I = 1.70 \text{ kg} \cdot \text{m}^2$, while $r_1 = 50$ cm and $r_2 = 20$ cm. Find the angular acceleration of the pulley system and the tensions T_1 and T_2.

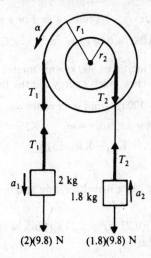

$(2)(9.8)$ N $(1.8)(9.8)$ N   **Fig. 11-38**

❚ Note at the beginning that $a = \alpha r$ gives $a_1 = (0.50 \text{ m})\alpha$ and $a_2 = (0.20 \text{ m})\alpha$. We shall write $F = ma$ for both masses and also $\tau = I\alpha$ for the wheel, taking the direction of motion to be the positive direction:

$$(2)(9.8) \text{ N} - T_1 = 2a_1 \qquad\qquad 19.6 \text{ N} - T_1 = 1.0\alpha$$
$$T_2 - (1.8)(9.8) \text{ N} = 1.8a_2 \quad \text{or} \quad T_2 - 17.6 \text{ N} = 0.36\alpha$$
$$(T_1)(r_1) - (T_2)(r_2) = I\alpha \qquad\quad 0.5T_1 - 0.2T_2 = 1.70\alpha$$

These three equations have three unknowns. Solve for T_1 in the first equation and substitute it in the third to obtain $9.8 - 0.5\alpha - 0.2T_2 = 1.70\alpha$. Solve this equation for T_2 and substitute in the second equation to obtain $-11\alpha + 49 - 17.6 = 0.36\alpha$, from which $\alpha = \underline{2.76 \text{ rad/s}^2}$.

We can now go back to the first equation to find $T_1 = \underline{16.8 \text{ N}}$, and to the second to find $T_2 = \underline{18.6 \text{ N}}$.

CHAPTER 12
Rotational Motion II: Kinetic Energy, Angular Impulse, Angular Momentum

12.1 ENERGY AND POWER

12.1 Find the rotational energy of the earth about the sun due to its orbit about the sun. Data: $M_e = 6 \times 10^{24}$ kg, orbit radius = 1.5×10^{11} m, time for rotation = 365 days = 3.2×10^7 s.

▌ I about the sun = $M_e r^2$ while $\omega = (3.2 \times 10^7)^{-1}$ rev/s = 1.96×10^{-7} rad/s. Then use $K_r = (I\omega^2)/2 = \underline{2.6 \times 10^{33} \text{ J}}$.

12.2 It has been proposed that the kinetic energy stored in a flywheel be used to propel an automobile. (A small electric motor located where the automobile is parked overnight could be used to spin up the flywheel each night and thus make up for the energy used during the day.) Calculate the energy content in a cylindrical flywheel of uniform density, radius 0.50 m, mass 200 kg, and angular speed 20 000 rotations per minute. (This angular speed is near the limit at which a steel flywheel would break apart.) With the flywheel installed, an automobile has a total mass of 1000 kg. When the car is traveling on a level road at a speed of 100 km/h, the total frictional force acting on it has a magnitude equal to 10 percent of the weight of the automobile. Calculate how far it could travel before using up all the energy stored in the flywheel.

▌ The spin energy K_s stored in the flywheel is $K_s = \frac{1}{2}I\omega_s^2$. If this is used to do work against a frictional force F, the stored spin energy will be consumed after the car has traveled a distance d such that $Fd = K_s$. Therefore $d = (\frac{1}{2}I\omega_s^2)/F$. In the present exercise, we are given $I = \frac{1}{2}MR^2 = \frac{1}{2}(200)(0.50^2) = 25$ kg $\cdot$ m^2. Also, $\omega_s = [2\pi(2 \times 10^4)]/60 = 2.094 \times 10^3$ rad/s. Therefore $\frac{1}{2}I\omega_s^2 = \underline{5.483 \times 10^7}$ J. Since $F = 0.1Mg = (0.1)(1000)(9.80) = 980$ N, we find $d = (5.483 \times 10^7)/(0.980 \times 10^3) = 5.595 \times 10^4$ m = $\underline{56.0 \text{ km}}$.

12.3 A thin ring of mass 2.7 kg and radius 8 cm rotates about an axis through its center and perpendicular to the plane of the ring at 1.5 rev/s. Calculate the kinetic energy of the ring.

▌ First convert to SI units: 1.5 rev/s = 3π rad/s = ω. For a thin ring,

$$I = mr^2 = 2.7(0.08)^2 = 0.0172 \text{ kg} \cdot \text{m}^2 \qquad \text{KE} = \frac{1}{2}I\omega^2 = \frac{1}{2}(0.0172)(3\pi)^2 = \underline{0.763 \text{ J}}$$

12.4 A flywheel having a moment of inertia of 900 kg $\cdot$ m^2 rotates at a speed of 120 rev/min. It is slowed down by a brake to a speed of 90 rev/min. How much energy did the flywheel lose in slowing down?

▌
$$\text{KE}_1 = \frac{1}{2}I\omega_1^2 \qquad 120 \text{ rev/min} = 4\pi \text{ rad/s} = \omega_1 \qquad \text{KE}_1 = \frac{1}{2}(900)(4\pi)^2 = 70.99 \text{ kJ}$$
$$90 \text{ rev/min} = 3\pi \text{ rad/s} = \omega_2 \qquad \text{KE}_2 = \frac{1}{2}I\omega_2^2 = \frac{1}{2}(900)(3\pi)^2 = 39.93 \text{ kJ}$$
$$\text{KE}_1 - \text{KE}_2 = \underline{31.0 \text{ kJ}} \qquad \text{loss of energy}$$

12.5 Compute the rotational KE of a 25-kg wheel rotating at 6 rev/s if the radius of gyration of the wheel is 0.22 m.

▌ $\text{KE} = \frac{1}{2}I\omega^2$. $I = Mk^2 = (25 \text{ kg})(0.22 \text{ m})^2 = 1.21$ kg $\cdot$ m^2. $\omega = 2\pi f = 12\pi$ rad/s. Thus KE = $\frac{1}{2}(1.21$ kg $\cdot$ m$^2)(12\pi$ rad/s$)^2 = \underline{860 \text{ J}}$.

12.6 When 100 J of work is done upon a flywheel, its angular speed increases from 60 rev/min to 180 rev/min. What is its moment of inertia?

▌ Work = $W = \Delta\text{KE}$, or $100 \text{ J} = (\frac{1}{2}I\omega_f^2 - \frac{1}{2}I\omega_i^2)$. $\omega_f = 2\pi f_f = 6\pi$ rad/s; $\omega_i = 2\pi f_i = 2\pi$ rad/s. Thus $100 \text{ J} = \frac{1}{2}I(36\pi^2 \text{ rad/s}^2 - 4\pi^2 \text{ rad/s}^2) = (16\pi^2 \text{ rad/s}^2)I$. Solving, $I = \underline{0.63 \text{ kg} \cdot \text{m}^2}$.

12.7 A 72-lb wheel, radius of gyration 9 in, is to be given a speed of 10 rev/s in 25 rev from rest. Find the constant unbalanced torque required. How much kinetic energy was gained?

▌
$$\tau = I\alpha \qquad I = Mk^2 = \frac{(72 \text{ lb})(0.75 \text{ ft})^2}{(32 \text{ ft/s}^2)} = 1.27 \text{ slug} \cdot \text{ft}^2$$

To find α we have $\omega^2 = \omega_0^2 + 2\alpha(\theta - \theta_0)$ with $\theta - \theta_0 = (25 \text{ rev})(2\pi \text{ rad/rev}) = 50\pi$ rad, and $\omega_0 = 0$, $\omega = 2\pi f = 20\pi$ rad/s. Then $\alpha = (20\pi)^2/100\pi = 12.5 \text{ rad/s}^2$. Finally $\tau = (1.27 \text{ slug} \cdot \text{ft}^2)(12.5 \text{ rad/s}^2) = \underline{15.9 \text{ N} \cdot \text{m}}$.

Starting from rest $\Delta\text{KE} = \frac{1}{2}I\omega^2 = \frac{1}{2}(1.27 \text{ slug} \cdot \text{ft}^2)(20\pi \text{ rad/s})^2 = \underline{2500 \text{ J}}$. (Or, from the work-energy theorem, $\tau(\theta - \theta_0) = \Delta\text{KE} = \frac{1}{2}I\omega^2$. Thus, $\frac{1}{2}I\omega^2 = (15.9 \text{ N} \cdot \text{m})(50\pi \text{ rad}) = 2500$ J.)

12.8 The rigid rod joining the three masses in Fig. 12-1 has negligible mass. It is pivoted at one end so that it can swing in a vertical plane. If it is released from the position shown, how fast will the bottom mass be moving when the rod is vertical?

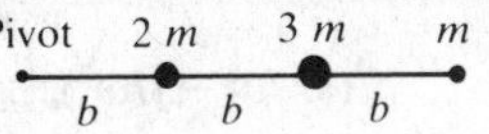

Fig. 12-1

▌ Let U_g = PE of gravity; K_r = rotational KE. The velocity of mass m is $3b\omega$. To find ω, equate the U_g lost by the three masses to K_r about pivot point so that $U_g = 2mbg + (3m)(2b)(g) + (m)(3b)(g) = K_r = (I\omega^2)/2$. About the pivot point I is $2mb^2 + (3m)(2b)^2 + (m)(3b)^2 = 23mb^2$, so that $\omega = (22g/23b)^{1/2} = (3.06/b^{1/2})$ $\underline{\text{rad/s}}$. From this, find $v = 3b\omega = \underline{9.2b^{1/2} \text{ m/s}}$.

12.9 When the system of Fig. 12-2 is released from rest, the 200-g mass slides down the incline against a frictional force of 0.50 N. If the moment of inertia of the wheel is $0.80 \text{ kg} \cdot \text{m}^2$, how fast will the block be moving after it has slid 100 cm along the incline?

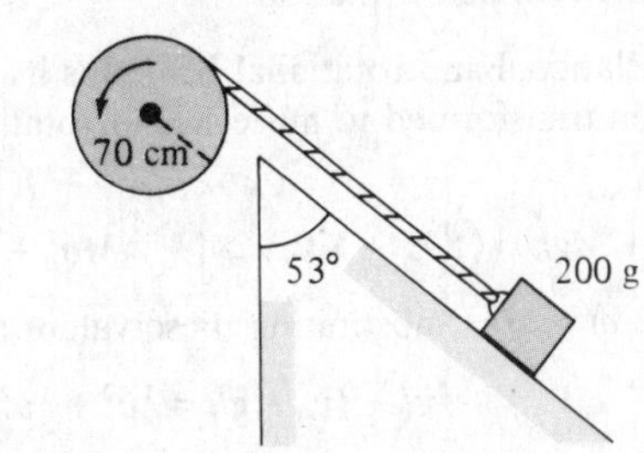

Fig. 12-2

▌ From an energy balance point of view, letting U_g = PE of gravity, K_r = rotational KE, K_t = KE of block, we have U_g lost by mass = friction loss + $K_r + K_t$: $0.2(9.8)(1.0 \cos 53°) = 0.50(1.0) + (0.80)(v/0.70)^2/2 + 0.20v^2/2$, so $v = \underline{0.86 \text{ m/s}}$.

12.10 Refer to Fig. 12-2. If the wheel at the instant shown there is rotating at a speed of 0.30 rev/s counterclockwise and the mass comes to rest at a vertical distance 80 cm above the position shown, how large is the moment of inertia of the wheel? Ignore losses due to friction.

▌ The $K_r + K_t$ lost by the wheel and mass = U_g gain of the mass. As an equation, $0.5I(0.60\pi)^2 + 0.5(0.20)(0.60\pi \times 0.70)^2 = (0.2 \times 9.8)(0.80)$, from which $I = \underline{0.78 \text{ kg} \cdot \text{m}^2}$.

12.11 A uniform 25-cm-long stick rotates freely about a horizontal axis through one of its ends (Fig. 12-3). It is released at an angle θ to the vertical. When it hangs straight down, the speed of the tip of the stick is 3.0 m/s. How large is θ?

▌ The center of mass has fallen through a distance of $(L/2)(1 + \cos\theta)$. So writing $U_g = K_r$ we have $mg(L/2)(1 + \cos\theta) = I\omega^2/2$, where $I = mL^2/3$ and $\omega = v_{\text{tip}}/L$. Solving we find $\cos\theta = (v^2/3gL) - 1$; then for $v = 3.0$ m/s and $L = 0.25$, $\cos\theta = 0.224$ m, $\theta = \underline{77°}$.

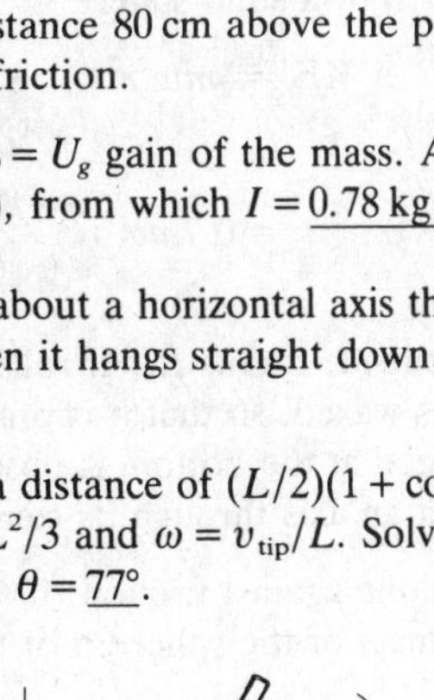

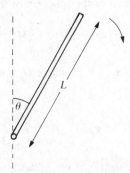

Fig. 12-3

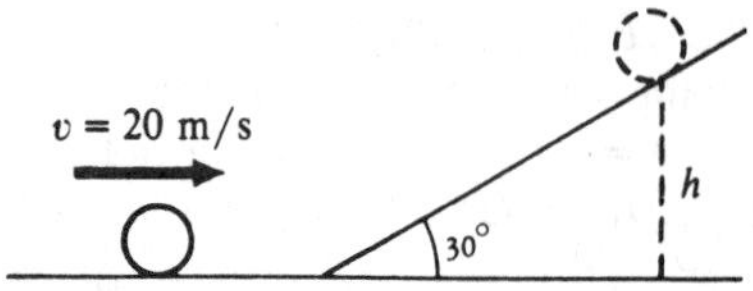

Fig. 12-4

12.12 As shown in Fig. 12-4, a uniform solid sphere rolls on a horizontal surface at 20 m/s. It then rolls up the incline shown. If friction losses are negligible, what will be the value of h where the ball stops?

▌ The rotational and translational KE of the ball at the bottom will be changed to GPE when the sphere stops. We therefore write

$$\left(\tfrac{1}{2}Mv^2 + \tfrac{1}{2}I\omega^2\right)_{\text{start}} = (Mgh)_{\text{end}}$$

But for a solid sphere, $I = \tfrac{2}{5}Mr^2$. Also, $\omega = v/r$. The equation becomes

$$\frac{1}{2}Mv^2 + \frac{1}{2}\left(\frac{2}{5}\right)(Mr^2)\left(\frac{v}{r}\right)^2 = Mgh \qquad \text{or} \qquad \frac{1}{2}v^2 + \frac{1}{5}v^2 = (9.8 \text{ m/s}^2)h$$

Using $v = 20$ m/s gives $h = \underline{28.6 \text{ m}}$. Note that the answer does not depend upon the mass of the ball or the angle of the incline.

12.13 As a solid disk rolls over the top of a hill on a track, its speed is 80 cm/s. If friction losses are negligible, how fast is the disk moving when it is 18 cm below the top?

▌ At the top, the disk has translational and rotational KE, plus its GPE relative to the point 18 cm lower. At the final point, the GPE has been transformed to more KE of rotation and translation. We therefore write, with $h = 18$ cm,

$$(\text{KE}_t + \text{KE}_r)_{\text{start}} + Mgh = (\text{KE}_t + \text{KE}_r)_{\text{end}} \qquad \tfrac{1}{2}Mv_0^2 + \tfrac{1}{2}I\omega_0^2 + Mgh = \tfrac{1}{2}Mv_f^2 + \tfrac{1}{2}I\omega_f^2$$

For a solid disk, $I = \tfrac{1}{2}Mr^2$. Also, $\omega = v/r$. Substituting these values and simplifying gives

$$\tfrac{1}{2}v_0^2 + \tfrac{1}{4}v_0^2 + gh = \tfrac{1}{2}v_f^2 + \tfrac{1}{4}v_f^2$$

But $v_0 = 0.80$ m/s and $h = 0.18$ m. Substitution gives $v_f = \underline{1.73 \text{ m/s}}$.

12.14 A solid sphere (mass m, radius r) rolls down an inclined plane. Show that its kinetic energy is $\tfrac{2}{7}$ rotational and $\tfrac{5}{7}$ translational.

▌ The translational kinetic energy of the rolling sphere is $\text{KE}_t = \tfrac{1}{2}mv_{\text{CM}}^2$. The rotational kinetic energy is $\text{KE}_r = \tfrac{1}{2}I_{\text{CM}}\omega^2$. For rolling, $v_{\text{CM}} = \omega r$. For a solid sphere

$$I_{\text{CM}} = \tfrac{2}{5}mr^2 \qquad \text{KE}_t = \tfrac{1}{2}m\omega^2 r^2 \qquad \text{KE}_r = (\tfrac{1}{2})(\tfrac{2}{5})m\omega^2 r^2 = \tfrac{1}{5}m\omega^2 r^2$$

Total kinetic energy is

$$\text{KE} = \text{KE}_t + \text{KE}_r = \tfrac{1}{2}m\omega^2 r^2 + \tfrac{1}{5}m\omega^2 r^2 = 0.7m\omega^2 r^2 \qquad \frac{\text{KE}_t}{\text{KE}} = \frac{\tfrac{1}{2}m\omega^2 r^2}{0.7m\omega^2 r^2} = \frac{0.5}{0.7} = \underline{\frac{5}{7}} \qquad \frac{\text{KE}_r}{\text{KE}} = 1 - \frac{5}{7} = \underline{\frac{2}{7}}$$

12.15 A thick-walled hollow sphere has outside radius R_o. It rolls down an incline without slipping, and its speed at the bottom is v_0. Now the incline is waxed, so that it is practically frictionless, and the sphere is observed to slide down (without rolling). Its speed at the bottom is observed to be $5v_0/4$. Determine the radius of gyration of the hollow sphere about an axis through its center.

▌ In neither descent is any work done against friction. If h is the height of the incline, the final kinetic energy $K_f = Mgh$, where M is the mass of the sphere. For the rolling descent, the kinetic energy is given by

$$K_f = \frac{1}{2}Mv_0^2 + \frac{1}{2}I_c\omega_0^2 = \frac{1}{2}Mv_0^2 + \frac{1}{2}MG_c^2\left(\frac{v_0^2}{R_o^2}\right)$$

$$= \frac{1}{2}Mv_0^2\left(1 + \frac{G_c^2}{R_o^2}\right) \qquad \text{where } G_c \text{ is the radius of gyration of interest}$$

For the sliding descent, we have

$$K_f = \frac{1}{2}M\left(\frac{5v_0}{4}\right)^2 = \frac{1}{2}Mv_0^2\left(\frac{25}{16}\right)$$

From these equations we find $G_c^2 = 9R_o^2/16$, or $G_c = \underline{3R_o/4}$.

12.16 For the sphere of Prob. 12.15, the central hollow has zero density and unknown radius R_i. For $R_i < r < R_o$, the density is uniform. Determine R_i/R_o. Compare the volume of the cavity to the total volume $4\pi R_o^3/3$.

❚ Let the density of the solid portion be ρ_0. Then the mass of the hollow sphere is given by

$$M = \frac{4\pi}{3}(\rho_0 R_o^3 - \rho_0 R_i^3)$$

and the moment of inertia is given by

$$I_c = \frac{2}{5}\left(\frac{4\pi}{3}\rho_0 R_o^3\right)R_o^2 - \frac{2}{5}\left(\frac{4\pi}{3}\rho_0 R_i^3\right)R_i^2$$

The radius of gyration G_c satisfies the equation

$$G_c^2 = \frac{I_c}{M} = \frac{2}{5}\left(\frac{R_o^5 - R_i^5}{R_o^3 - R_i^3}\right) = \frac{2}{5}R_o^2\left[\frac{1 - (R_i/R_o)^5}{1 - (R_i/R_o)^3}\right]$$

In Prob. 12.15 it was established that $G_c^2 = 9R_o^2/16$. Therefore we have

$$\frac{45}{32} = \frac{1 - (R_i/R_o)^5}{1 - (R_i/R_o)^3}$$

which leads to

$$\frac{13}{45} = \left(\frac{R_i}{R_o}\right)^3\left[1 - \frac{32}{45}\left(\frac{R_i}{R_o}\right)^2\right]$$

We solve by trial and error; a simple search algorithm rather quickly yields a solution good to three decimal places: $R_i/R_o = \underline{0.823}$. The corresponding volume ratio $V_i/V_o = (R_i/R_o)^3 = \underline{0.557}$.

12.17 A cord 3 m long is coiled around the axle of a wheel. The cord is pulled with a constant force of 40 N. When the cord leaves the axle, the wheel is rotating at 2 rev/s. Determine the moment of inertia of the wheel and axle. Neglect friction.

❚ Work $= W = \Delta\text{KE} = (\frac{1}{2}I\omega^2 - 0)$. $W = Fs = (40\text{ N})(3\text{ m}) = 120\text{ J}$. $\omega = 2\pi f = 4\pi$ rad/s. Thus $120\text{ J} = \frac{1}{2}I(4\pi \text{ rad/s})^2$. $I = \underline{1.52 \text{ kg} \cdot \text{m}^2}$.

12.18 An Atwood's machine consists of a pulley (radius $= b$) mounted on a horizontal axis, and two masses $m_1 > m_2$ hanging at ends of a cord that passes over the pulley. Find the moment of inertia I of the pulley in terms of m_1, m_2, b, and the time t taken for m_1 to fall a distance h after the system is released, assuming no slippage between cord and pulley.

❚ After time t the net potential energy lost by the masses is converted into kinetic energy of translation and rotation: $(m_1 - m_2)gh = (m_1 + m_2)(v^2/2) + (I\omega^2)/2$. Since $\omega = v/b$ and $v = 2\bar{v} = 2h/t$, ω and v can be eliminated. Solving for I gives $I = (m_1 - m_2)[(gb^2t^2)/2h)] - (m_1 + m_2)b^2$.

12.19 A cylinder of radius 20 cm is mounted on an axle coincident with its axis so as to be free to rotate. A cord is wound on it and a 50-g mass is hung from it. If, after being released, the mass drops 100 cm in 12 s, find the moment of inertia of the cylinder.

❚ Use energy conservation; first find the final speed of mass from $s = \bar{v}t = (v + v_0)(t/2)$ to obtain the final mass velocity $v = 0.167$ m/s. Then U_g lost by mass $= K_r + K_t$ or $mgh = (I\omega^2)/2 + (mv^2)/2$; $0.05(9.8)(1.0) = 0.5I(0.167/0.2)^2 + 0.5(0.05)(0.167)^2$. Then solving for I gives $\underline{1.4 \text{ kg} \cdot \text{m}^2}$.

12.20 A uniform disk of mass M and radius R is supported vertically by a pivot at its center. As shown in Fig. 12-5, a small dense object (also of mass M) is attached to the rim and raised to the highest point above the center. The (unstable) system is then released. What is the angular speed of the system when the attached object passes directly beneath the pivot?

❚ If we take the lowest point on the disk as the reference level for gravitational potential energy, the initial total energy is given by

$$E_i = K_i + U_i = U_i = MgR + Mg(2R) = 3MgR$$

When the object passes beneath the pivot, the potential energy is given by

$$U_f = MgR + (Mg)(0) = MgR$$

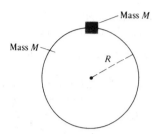

Fig. 12-5

The final kinetic energy of the system is given by

$$K_f = \tfrac{1}{2}I_d\omega_f^2 + \tfrac{1}{2}M(\omega_f R)^2 = \tfrac{3}{4}M\omega_f^2 R^2$$

since $I_d = \tfrac{1}{2}MR^2$ is the moment of inertia of the uniform disk. The system is conservative, so that $E_i = E_f = K_f + U_f$ and

$$K_f = \tfrac{3}{4}M\omega_f^2 R^2 = E_i - U_f = 3MgR - MgR$$

Solving for ω_f, we obtain $\quad \omega_f = \sqrt{8g/3R}$.

12.21 A wheel with $I = 20\ \text{kg} \cdot \text{m}^2$ is spinning at 3.0 rev/s on its axis. How large is the frictional torque if the wheel coasts 40 rev before stopping?

▮ The work done by a constant torque τ in turning through an angle θ is $\tau\theta$. Then K_r of wheel = work by stopping torque, so $(I\omega^2)/2 = \tau\theta$ and $\tau = [0.5(20)(6.0\pi)^2]/[40(2\pi)] = \underline{14.1\ \text{N} \cdot \text{m}}$.

12.22 The system shown in Fig. 12-6 is released from rest with the spring in the unstretched position. If friction is negligible, how far will the mass slide down the incline?

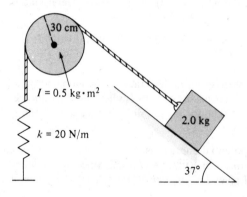

Fig. 12-6

▮ Since $K = 0$ initially and also when the mass stops sliding, the U_g lost by the mass equals the energy gained by the spring: $mgs \sin 37° = (ks^2)/2$, or $2.0(9.8)(0.60s) = 10s^2$, and $s = \underline{1.18\ \text{m}}$.

12.23 In the situation outlined in Prob. 12.22, what will be the speed of the mass when the mass has slid 1.0 m down the incline?

▮ In this case, both the wheel and mass have kinetic energy. Therefore, $mg(0.60s) = (ks^2)/2 + (I\omega^2)/2 + (mv^2)/2$. We have $\omega = v/0.30$ and so at $s = 1.0$ m, $v = \underline{0.68\ \text{m/s}}$.

12.24 For the situation outlined in Prob. 12.22, how far will the mass have slid when its speed is maximum? What will be its speed then?

▮ The energy equation in the general case is $mg(0.60s) = (ks^2)/2 + [I(v/0.3)^2]/2 + mv^2/2$. One can solve for v and find its maximum using calculus. However, without calculus, we know that the oscillation will be symmetric about the final rest position. The rest position is at s_0, where $ks_0 = mg \sin 37°$ (zero net force) or $s_0 = 0.588$ m. When oscillating, this is also the position of maximum kinetic energy. (Additional insight can be obtained by examining the system's parabolic potential energy curve: $U_p = (ks^2)/2 - mgs \sin 37°$. This has a minimum at $(\partial U_p)/\partial s = 0$, so $ks = mg \sin 37°$.) Use the energy equation given above to find v at $s = 0.588$ m to be $v = \underline{0.96\ \text{m/s}}$.

12.25 In Fig. 12-7 the large hoop (radius $3b$) is fastened rigidly to the table while the small hoop (mass m, radius b) rolls without slipping inside the large hoop. It is released from the position shown. What is the speed of the center of the hoop as it passes through its lowest point? Repeat for a spherical shell of the same radius rolling inside the large hoop.

▌ The change in U_g for both moving bodies is $2mgb = K_t + K_r$; for the hoop, $2mgb = [(mb^2)v^2/b^2]/2 + (mv^2)/2$, where $v = \omega b$, so $v = (2gb)^{1/2}$. For the spherical shell $I = (2mb^2)/3$, (see, e.g., Prob. 11.46) so $2mgb = [(2mb^2)/3][(v^2/b^2)/2] + (mv^2)/2 = (5mv^2)/6$, thus $v = [(12gb)/5]^{1/2}$.

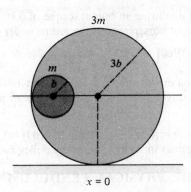

Fig. 12-7

12.26 Referring back to Fig. 12-7, we see two hoops. The inner one rolls without slipping inside the other, and that one, in turn, is free to move without friction on the table. The mass of the smaller one is m and the mass of the larger one is $3m$. They start from rest in the position shown. After release, the inner hoop rolls around with decreasing amplitude and finally comes to rest at the bottom. What is the x coordinate of the center of the outer hoop when the system comes to rest?

▌ In the initial state the center of mass of the two hoop system is $b/2$ to the left of the large hoop's center. This can be found from $m(2b - x) = 3mx$, giving $x = b/2$. No external horizontal forces act on the system so the horizontal coordinate of the CM does not change. In the final state the CMs of each hoop and the system lie on a vertical line. The outer hoop's center therefore moved $b/2$ to the left.

12.27 Suppose in Fig. 12-7 that the two hoops were rigidly fastened together. The system is released from the position shown, and there is negligible friction between the large hoop and the table. How fast relative to the table is the center of the large hoop moving when the centers of the two hoops lie on a vertical line?

▌ The center of mass is $b/2$ from the center of the large hoop, so $\Delta U_g = (4mgb)/2$. This is equal to $K_r + K_t$ for each hoop, $(mb^2\omega^2)/2 + [3m(3b)^2\omega^2]/2 + (mv^2)/2 + (3mV^2)/2$, where ω, the angular velocity of a hoop about its own CM, is the same for both hoops since they are fastened together; and v and V are the linear velocities of the centers of the small and large hoop, respectively. The horizontal velocity V of the center of the large hoop can be obtained by noting that the center of mass, $b/2$ below, has no horizontal velocity, and hence $V = \omega b/2$. Similarly, $v = 3\omega b/2$. Substituting for v and ω in terms of V and inserting into the energy equation gives rise to $V = (gb/31)^{1/2}$.

12.28 An electric motor runs at 900 rev/min and delivers 2 hp. How much torque does it deliver?

▌ Power $= P = \tau\omega$. We are given $P = 2$ hp $= (2)(746\text{ W}) = 1492$ W. $\omega = 2\pi f = [2\pi(900\text{ rev/min})]/(60\text{ s/min}) = 30\pi$ rad/s. Finally $\tau = (1492\text{ W})/(30\pi\text{ rad/s}) = \underline{15.8\text{ N}\cdot\text{m}}$.

12.29 The driving side of a belt has a tension of 1600 N and the slack side has 500-N tension. The belt turns a pulley 40 cm in radius at a rate of 300 rev/min. This pulley drives a dynamo having 90 percent efficiency. How much power is being delivered by the dynamo?

▌ Net torque $= (1600\text{ N} - 500\text{ N})(0.40\text{ m}) = 440\text{ N}\cdot\text{m}$. $\omega = 2\pi f = 10\pi$ rad/s. Net power delivered to dynamo $= \tau\omega = (440\text{ N}\cdot\text{m})(10\pi\text{ rev/s}) = 4400\pi$ W. Net power generated by dynamo $= 0.9(4400\pi\text{ W}) = \underline{12.4\text{ kW}}$.

12.30 It is proposed to use a uniform disk 50 cm in radius turning at 300 rev/s as an energy-storage device in a bus. How much mass must the disk have if it is to be capable, while coasting to rest, of furnishing the energy equivalent of a 100-hp motor operating for 10 min?

▌ $K_r = (I\omega^2)/2$ is to supply the needed work, with $I = \frac{1}{2}mR^2$. 100 hp for 10 min equals $(746\,\text{W/hp})(100\,\text{hp})(600\,\text{s}) = 4.48 \times 10^7$ J and $K_r = 0.5(0.5m \times 0.50^2)(300 \times 2\pi)^2 = 2.22 \times 10^5 m$; therefore, $m = \underline{202\,\text{kg}}$.

12.31 A small motor delivers 0.20 hp when its shaft is turning at 1400 rev/min. How large a torque is it capable of providing?

▌ The power delivered by the motor turning at angular velocity ω is $\tau\omega$. We have therefore $\tau = (0.20\,\text{hp})/(1400\,\text{rev/min}) = (149\,\text{W})/(147\,\text{rad/s}) = \underline{1.02\,\text{N} \cdot \text{m}}$.

12.32 A certain motor is capable of producing an output torque of 0.80 N · m. It operates at a speed of 1400 rev/min. What is the output horsepower of the motor under these conditions?

▌ Power $\tau\omega = 0.80\{[1400(2\pi)]/60\} = 117.3$ W and using 746 W = 1 hp we obtain $\underline{0.157\,\text{hp}}$.

12.33 A certain $\frac{1}{2}$-hp motor (based on output) normally operates at 1800 rev/min under load. How large a torque does the motor develop? If it drives a belt on a 5.0-cm-diameter pulley, what is the difference in tensions in the two portions of the belt?

▌ Obtain torque as power/ω; $\tau = [0.5(746)]/\{[1800(2\pi)]/60\} = \underline{1.98\,\text{N} \cdot \text{m}}$. This torque is applied to the pulley by the difference in tensions in the belt times the pulley radius, $\Delta T r$; thus, $\Delta T = 1.98/0.025 = \underline{79\,\text{N}}$.

12.2 ANGULAR IMPULSE; THE PHYSICAL PENDULUM

12.34 How far above its center should a billiard ball be struck in order to make it roll without any initial slippage? Denote the ball's radius by R, and assume that the impulse delivered by the cue is purely horizontal.

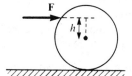

Fig. 12-8

▌ The situation is shown in Fig. 12-8. When the ball is struck, a horizontal force of arbitrarily large magnitude F (and arbitrarily short duration δt) is applied. Since any static frictional force supplied by the table is limited to μMg, the only sure way to avoid slippage is to create a situation in which *zero* frictional force is needed. Let v_0 be the speed and ω_0 be the angular speed just after the impulse $F\,\delta t$ has been delivered. Then we must have

$$F\,\delta t = Mv_0 \qquad \text{and} \qquad h(F\,\delta t) = I_c\omega_0$$

where $\omega_0 = v_0/R$. Since billiard balls are homogeneous, the moment of inertia about the center of the ball is given by $I_c = 2MR^2/5$. Combining these equations, we find $h(Mv_0) = [(2MR^2)/5](v_0/R)$, which leads to $\underline{h = 2R/5}$.

12.35 A uniform slender rod 1.00 m long is initially standing vertically on a smooth, horizontal surface (Fig. 12-9). It is struck a sharp horizontal blow at the top end, with the blow directed at right angles to the rod axis. As a result, the rod acquires an angular velocity of 3.00 rad/s. What is the translational velocity of the center of mass of the rod after the blow?

▌ Let the magnitude of the impulse delivered by the blow be represented by P. From the translational and rotational forms of Newton's second law, we find that $v_c = P/M$ and $I_c\omega_c = Pl/2$. Here v_c is the rightward velocity of the center of mass after the blow and ω_c is the (positive clockwise) angular velocity of the rod

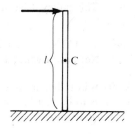

Fig. 12-9

after the blow. Using the fact that $I_c = (Ml^2)/12$, we find that $\omega_c l = 6P/M$. Therefore $v_c = (\omega_c l)/6$. Since $\omega_c = 3.00 \, \text{rad/s}$ and $l = 1.00 \, \text{m}$, we find $v_c = \underline{0.500 \, \text{m/s}}$.

12.36 Refer to Prob. 12.35. Which point on the rod is stationary just after the blow?

❚ Just after the blow is struck, a point on the rod a distance d from the struck end has a rightward velocity $v = v_c + \omega_c[(l/2) - d]$ which is valid even for $d > l/2$. We find that $v = 0$ for $d = (l/2) + (v_c/\omega_c)$. With $v_c = (\omega_c l)/6$, we find $d = (l/2) + (l/6) = 2l/3$. Since $l = 1 \, \text{m}$, we find that *the point located $\frac{2}{3}$ m from the struck end is initially stationary*. This point is called the "center of percussion" of the rod with respect to the top end. (The location of the center of percussion depends upon the site of the blow.)

12.37 A rigid body is free to rotate about a horizontal axis through A (into the paper) as shown in Fig. 12-10. Find an expression for the angular acceleration, α, in terms of: θ, the angle of rotation measured from the vertical; D, the distance from the pivot, A, to the center of mass; G_c, the radius of gyration about a horizontal axis through the center of mass; and g, the acceleration of gravity.

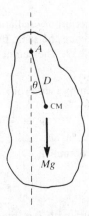

Fig. 12-10

❚ We choose counterclockwise as positive. The torque about A is $\tau_A = -MgD \sin \theta$, while the moment of inertia is, using the parallel-axis theorem, $I_A = MD^2 + I_{CM} = MD^2 + MG_c^2$. Then $\tau_A = I_A \alpha$ becomes $M(D^2 + G_c^2)\alpha = -MgD \sin \theta$, or

$$\alpha = -\left(\frac{gD}{D^2 + G_c^2}\right) \sin \theta$$

(Note that $G_A^2 = D^2 + G_c^2 \Rightarrow \alpha = -[(gD)/G_A^2] \sin \theta$, where G_A is the radius of gyration about A.)

12.38 For the situation of Prob. 12.37, describe qualitatively the motion if the rigid body is released from rest at some angle θ_A. Such a system is called a *compound*, or *physical*, *pendulum*. Assume that there are no frictional losses. If θ_A is small, what is the frequency of oscillation of the system?

❚ From conservation of mechanical energy we see that potential energy is converted to rotational kinetic energy as θ decreases to zero and the CM is lowered. The maximum KE occurs with the CM directly below A. The system will continue to rotate in the same direction until the CM rises to its original height, i.e., when $\theta = -\theta_A$. It will then swing back in the other direction returning to its original position θ_A. The motion will then repeat over and over again. If θ_A is small, then $\sin \theta \approx \theta$ for all points of the swing and from Prob. 12.37, $\alpha \approx -[(gD)/(D^2 + G_c^2)]\theta$. Since α is the time rate of change of the time rate of change of θ, and $\Omega^2 \equiv (gD)/(D^2 + G_c^2)$ is a constant, θ must be executing simple harmonic motion of frequency

$$v = \frac{\Omega}{2\pi} = \frac{1}{2\pi} \sqrt{\frac{gD}{D^2 + G_c^2}} = \frac{1}{2\pi} \sqrt{\frac{gD}{G_A^2}} \tag{1}$$

12.39 A uniform bar of length l and mass m is suspended from a very thin axle that passes through a hole near the top end A of the bar. How far from A should a blow be applied at right angles to the bar in order to start the bar rotating about A without breaking the axle?

❚ The bar is shown in Fig. 12-11. If we denote the magnitude of the impulse by P, then if the blow is placed so that there is no impulsive force on the axle, the center of mass C of the bar begins moving to the right with speed $v_c = P/m$. Using A as the axis for computing torques, the rotational form of Newton's second law gives

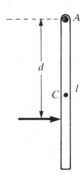

Fig. 12-11

$Pd = I_A \omega$, where $I_A = (ml^2)/3$ is the moment of inertia of the bar about the pivot. Since the distance from A to C is $l/2$, we also have $v_c = (\omega l)/2$. By combining the equations, we find $m[(\omega l)/2]d = [(ml^2)/3]\omega$, which implies that $d = \underline{2l/3}$. [Note the reciprocity with Prob. 12.36: a blow to the end (c.p.) causes rotation about the c.p. (end).]

12.40 Refer to Prob. 12.39. (*a*) What is the period of oscillation of the rod when it is suspended from A? (*b*) What is the length of the simple pendulum having the same period? The length you obtain here should be the same as the distance you obtained in Prob. 12.39. The center of percussion relative to A is also called the *center of oscillation* relative to A.

▌ (*a*) The oscillation frequency is given by (see Prob. 12.38)

$$v_A = \frac{1}{2\pi} \sqrt{\frac{(l/2)g}{G_A^2}}$$

where $G_A^2 = I_A/m = l^2/3$. Therefore we obtain

$$T_A = 2\pi \sqrt{\frac{2l}{3g}}$$

(*b*) The period of a simple pendulum of length L is given by $2\pi\sqrt{L/g}$. Therefore the length of a simple pendulum with the oscillation period T_A is $L = \underline{2l/3}$, which is equal to the distance found in Prob. 12.39.

12.41 A ring of mass M and radius R is hung from a knife edge, so that the ring can swing in its own plane as a physical pendulum. Find the period T_1 of small oscillations.

▌ The ring is shown in Fig. 12-12, with the knife edge at point A. We must find the period T_1 of small oscillations in the plane of the paper. Taking the origin of a coordinate system at O, the equilibrium position of the ring's center, with the positive z axis emerging toward the viewer, the moment of inertia $I_{zo} = MR^2$. By the parallel-axis theorem, the moment of inertia about the knife edge is given by $I_{zA} = I_{zo} + MR^2 = 2MR^2$; hence, $G_A^2 = 2R^2$.

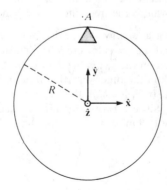

Fig. 12-12

Substituting this and $D = R$ in Eq. (1) of Prob. 12.38, we find

$$v_1 = \frac{1}{2\pi} \sqrt{\frac{g}{2R}} \qquad \text{and} \qquad T_1 = \frac{1}{v_1} = 2\pi \sqrt{\frac{2R}{g}}$$

12.42 Refer to Prob. 12.41. (*a*) Suppose that an identical ring is pivoted from an axis PP' lying in the ring plane

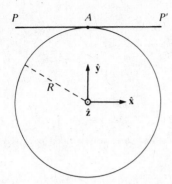

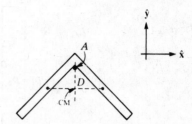

Fig. 12-13   **Fig. 12-14**

and tangent to the circumference (Fig. 12-13). This ring can execute oscillations in and out of the plane. Find the period T_2 of those small oscillations. **(b)** Which oscillation has the longer period? How much longer?

▌ **(a)** Now the ring is pivoted about the axis PP' as shown at right. By the parallel-axis theorem, $I_{PP'} = I_{xO} + MR^2$. As shown in Prob. 11.41, $I_{xO} = (MR^2)/2$, so $I_{PP'} = (3MR^2)/2$. With $G_A^2 = 3R^2/2$ and $D = R$, (1) of Prob. 12.38 gives $T_2 = 1/v_2 = 2\pi\sqrt{3R/2g}$. **(b)** $T_1/T_2 = \sqrt{\frac{4}{3}} = 1.1547$. The period for oscillations in the plane is 15.5 percent longer than the period for oscillations about the axis PP'.

12.43 A uniform right-angle iron is hung over a thin nail so that the iron pivots freely at the bend (Fig. 12-14). Each arm of the iron has mass m and length l. Find the period T of small oscillations (in the plane of the iron).

▌ The center of mass of the system is located on the angle bisector, at a distance $D = (l\sqrt{2}/4)$ from the pivot A. The moment of inertia I about an axis through A and perpendicular to the xy plane is given by $(ml^2)/3 + (ml^2)/3 = (2ml^2)/3$. Therefore the square of the gyration radius is given by $G^2 \equiv I/M = [(2ml^2)/3]/(2m) = l^2/3$. We can now apply (1) of Prob. 12.38:

$$T \equiv \frac{1}{v} = 2\pi\sqrt{\frac{G_A^2}{Dg}} = 2\pi\sqrt{\frac{l^2/3}{[(l\sqrt{2})/4]g}} = 2\pi\sqrt{\frac{2\sqrt{2}\,l}{3g}}$$

12.44 Figure 12-15 represents a three-dimensional object (not necessarily of uniform density) whose center of mass is at point C. The axis ZCZ' passing through point C has been chosen at random orientation. The body had gyration radius $G_C(ZZ')$ about the axis ZCZ': the notation $G_C(ZZ')$ makes explicit the fact that the gyration radius for an axis through C depends on the particular choice (ZZ') of axis.

Suppose that a physical pendulum is constructed by pivoting the body about the axis PP', which is parallel to ZCZ' at a distance D. Prove that the frequency v of small oscillations about equilibrium is given by

$$v = \frac{1}{2\pi}\sqrt{\frac{Dg}{G_C^2(ZZ') + D^2}}$$

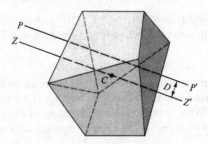

Fig. 12-15

▌ The desired result follows at once from (1) of Prob. 12.38. It is seen that the frequency of pendulum oscillations is the same for any choice of axis PP' on a cylinder of radius D centered on ZZ'.

12.45 Figure 12-16 depicts the *Kater pendulum*, used to measure the acceleration of gravity with high accuracy. It consists of a rigid rod on which a bob is mounted. The mass of the bob is sufficiently large so that the center of mass of the pendulum is fairly far from the middle of the rod. The bob is mounted on a slide; by moving the bob and then clamping it in position, the location of the center of mass of the pendulum can be adjusted. There are two very precisely honed knife-edges mounted on the rod. The pendulum can be swung from

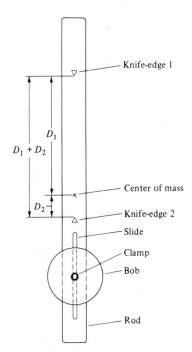

Fig. 12-16

knife-edge 1, and then reversed and swung from knife-edge 2. If G_0 is the gyration radius of the pendulum about an axis through its center of mass, what are the expressions for the periods of small oscillations about knife-edge 1 and knife-edge 2?

▌ From (1) of Prob. 12.38,

$$T_1 = \frac{1}{\nu_1} = 2\pi\sqrt{\frac{G_0^2 + D_1^2}{D_1 g}} \qquad T_2 = \frac{1}{\nu_2} = 2\pi\sqrt{\frac{G_0^2 + D_2^2}{D_2 g}}$$

12.46 Refer to Prob. 12.45; can $T_1 = T_2$ for $D_1 \neq D_2$?

▌ Yes. Consider T for an arbitrary D and fixed G_0, $T = 2\pi\sqrt{(G_0^2 + D^2)/(Dg)}$. If we plot T against D for D between 0 and ∞, we note that $T \rightarrow \infty$ as $D \rightarrow 0$ and $T \rightarrow \infty$ as $D \rightarrow \infty$. Thus T has a minimum value, T_{min}, for some $D = D_M$. Thus for any $T > T_{min}$, there are two D values, one greater and one less than D_M.

12.3 ANGULAR MOMENTUM

12.47 Find the rotational energy and angular momentum due to the daily rotation of the earth about its axis. Data: $M_e = 6 \times 10^{24}$ kg, $R_e = 6.4 \times 10^6$ m, $\omega = 1/86\,400$ rev/s. Assume the earth to be a uniform sphere.

▌ $K_r = I\omega^2/2 = [(2Mr^2)/5][(2\pi/86\,400)^2]/2 = \underline{2.6 \times 10^{29}}$ J. Angular momentum $= I\omega = [(2Mr^2)/5](2\pi/86\,400) = \underline{7.1 \times 10^{33}}$ kg $\cdot$ m^2/s. (The equivalent unit J $\cdot$ s is also used for angular momentum, especially in the atomic domain.)

12.48 Each of the wheels on a certain four-wheel vehicle has a mass of 30 kg and a radius of gyration of 30 cm. When the car is going forward and the wheels are turning at 5.0 rev/s, what is the rotational kinetic energy stored in the four wheels? What is the angular momentum of the vehicle about an axis parallel to the wheel axis and through the center of mass? Is the angular momentum vector directed toward the driver's right or left?

▌ $K_r = 4(I\omega^2/2) = 2(30)(0.30)^2(10\pi)^2 = \underline{5300}$ J. The angular momentum vector for each wheel is along the line described, so total $L = 4(I\omega) = 4(30)(0.09)(10\pi) = \underline{340}$ kg $\cdot$ m^2/s. Note that ω must be in radians per second. The total **L** will be in the same direction as the angular velocity, that is, to the driver's left.

12.49 In a common physics lecture demonstrations, a lecturer sits on a stool that can rotate freely about a vertical axis on low-friction bearings. The lecturer holds with extended arms two dumbbells, each of mass m, and kicks the floor so as to achieve an initial angular speed ω_1. The lecturer then pulls in the dumbbells, so that their distances from the rotation axis decrease from the initial value R_1 to the final value R_2. Determine the final angular speed ω_2, assuming that the moment of inertia about the rotation axis of the lecturer's body plus the stool does not change in the process. Then evaluate K_1 and K_2, the initial and final kinetic energies in the system. What is the source of the additional kinetic energy?

▌ We let I_0 be the moment of inertia of the lecturer plus the stool, but excluding the two dumbbells. We assume that I_0 does not change during the experiment. Since no torques can be exerted about the vertical axis, the vertical component of the total angular momentum of the system is conserved. Therefore $(I_0 + 2mR_1^2)\omega_1 = (I_0 + 2mR_2^2)\omega_2$. Solving for ω_2, we obtain

$$\omega_2 = \left(\frac{I_0 + 2mR_1^2}{I_0 + 2mR_2^2}\right)\omega_1$$

The initial and final kinetic energies are $K_1 = \frac{1}{2}(I_0 + 2mR_1^2)\omega_1^2$, and

$$K_2 = \frac{1}{2}(I_0 + 2mR_2^2)\omega_2^2 = \frac{1}{2}\frac{(I_0 + 2mR_1^2)^2}{I_0 + 2mR_2^2}\omega_1^2 = \left(\frac{I_0 + 2mR_1^2}{I_0 + 2mR_2^2}\right)K_1 > K_1$$

The additional kinetic energy $(K_2 - K_1)$ appears in the system by virtue of the work done by the lecturer to pull in the weights.

12.50 A man sits on a stool on a frictionless turntable holding a pair of dumbbells at a distance of 3 ft from the axis of rotation. He is given an angular velocity of 2 rad/s, after which he pulls the dumbbells in until they are 1 ft from the axis. The moment of inertia of the man plus stool plus turntable is 3 slug · ft² and may be considered constant. The dumbbells may be considered point masses of 0.5 slug each. (**a**) What is the initial angular momentum of the system? (**b**) What is the final angular velocity of the system?

▌ (**a**) $I_1\omega_1 = [3 + 2(0.5)(3^2)]2 = 24$ slug · ft²/s. (**b**) By conservation of angular momentum, $I_2\omega_2 = I_1\omega_1$, or

$$\omega_2 = \frac{I_1\omega_1}{I_2} = \frac{24 \text{ slug} \cdot \text{ft}^2/\text{s}}{[3 + 2(0.5)(1^2)] \text{ slug} \cdot \text{ft}^2} = \underline{6 \text{ rad/s}}$$

12.51 An ice skater spins with arms outstretched at 1.9 rev/s. Her moment of inertia at this time is 1.33 kg · m². She pulls in her arms to increase her rate of spin. If her moment of inertia is 0.48 kg · m² after she pulls in her arms, what is her new rate of rotation?

▌ $\qquad I_1\omega_1 = I_2\omega_2 \qquad 1.33(1.9) = 0.48\omega_2 \qquad \omega_2 = \underline{5.26 \text{ rev/s}}$

Note that the conservation equation allows the use of dissimilar units for I and ω.

12.52 It has been surmised that the sun was formed in the gravitational collapse of a dust cloud which filled the space now occupied by the solar system and beyond. If we assume that the original cloud was a uniform sphere of radius R_0 with an average angular velocity of ω_0, how fast should the sun be rotating now? For the present purposes, ignore the small mass of the planets and assume the sun to be a uniform sphere of radius R_s.

▌ Angular momentum conservation gives $I_0\omega_0 = I\omega$ with $I_0 = (2MR_0^2)/5$ and $I = (2MR_s^2)/5$. Then we find $\omega = \underline{(R_0/R_s)^2\omega_0}$.

12.53 Consider a satellite orbiting earth as shown in Fig. 12-17. Find the ratio of its speed at perihelion to that at aphelion.

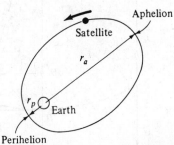

Fig. 12-17

▌ Since the line of action of the gravitational force on the satellite always passes through the earth, the angular momentum of the satellite about the earth must be conserved. (And the satellite's orbit must be confined to a plane perpendicular to **L**.) Thus, $mv_ar_a = mv_pr_p$ or $v_p/v_a = r_a/r_p$.

12.54 A merry-go-round in a park consists of an essentially uniform 200-kg solid disk rotating about a vertical axis. The radius of the disk is 6.0 m, and a 100-kg man is standing on its outer edge when it is rotating at a speed of 0.20 rev/s. How fast will the disk be rotating if the man walks 3.0 m in toward the center along a radius? What will happen if the man drops off the edge? Is it allowable to assume that the man acts like a point particle?

❚ Angular momentum conservation yields $I_0\omega_0 = I\omega$, with $I_0 = [200(6.0^2)]/2 + 100(6.0^2) = 7200$ kg · m^2 and $I = [200(36)]/2 + 100(9) = 4500$ kg · m^2. Therefore, $\omega = 0.20(7200/4500) = \underline{0.32 \text{ rev/s}}$. If the man falls off, rather than *pushes off* (viz. Prob. 12.56), the disk rotates as before and ω does not change. So long as the man's moment of inertia about a vertical axis passing through his center of mass is small compared with his mass times the distance of his center of mass to the vertical axis of the merry-go-round, squared, we can treat him as a point mass. This follows from the parallel-axis theorem.

12.55 Suppose that the merry-go-round of Prob. 12.54 has no one on it but is rotating at 0.20 rev/s. If now a 100-kg man quickly sits down on the edge of it, what will be its new speed?

❚ The disk exerts a torque on the man, setting him into rotation. Hence, by Newton's third law, the man exerts a countertorque on the disk, slowing it. In the collision of man and disk, angular momentum is conserved. Therefore, $I_d\omega_0 + 0 = I_d\omega + I_m\omega$. Then I_d is $(MR^2)/2$ and $I_m = 100(6.0)^2$. Using these values, the final $\omega = \omega_0/2 = \underline{0.10 \text{ rev/s}}$.

12.56 A 20-kg boy stands on and near the edge of a small merry-go-round with the system at rest. The system's total moment of inertia about the center is 120 km · m^2. The boy, at a radius of 2.0 m, jumps off the merry-go-round in a tangential direction with a speed of 1.5 m/s. How fast will the merry-go-round be rotating after the boy leaves it?

❚ At 2.0 m the boy has $I_c = 20(2.0)^2 = 80$ kg · m^2 leaving the $I_m = 120 - 80 = 40$ kg · m^2. When he jumps off tangentially in one direction his angular velocity $\omega_c = 1.5/2.0 = 0.75$ rad/s.
Because the angular momentum was zero initially, $I_c\omega_c + I_m\omega_m = 0$ yields $\omega_m = -(\frac{80}{40})(0.75) = \underline{-1.5 \text{ rad/s}}$. The minus sign means that the merry-go-round moves in a direction opposite to that of the boy.

12.57 Assume that the center of mass of a girl crouching in a light swing has been raised to 1.2 m (see Fig. 12-18). The girl weighs 400 N, and her center of mass is 3.7 m from the pivot of the swing while she is in the crouched position. The swing is released from rest, and at the bottom of the arc the girl stands up instantaneously, thus raising her center of mass 0.6 m (returning it to its original level). Find the height of her center of mass at the top of the arc.

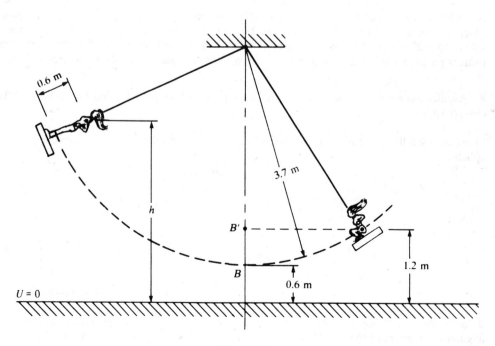

Fig. 12-18

▌ The torque due to gravity is the only external torque acting about the pivot of the swing. This torque is zero at B and remains zero for the instant during which the girl stands up. The angular momentum is conserved at B.

Considering the girl-swing system to be a rigid body just before and after she stands up, we have $I\omega = I'\omega'$. Letting $R = 3.7$ m and $R' = 3.1$ m be the distances from the pivot to the CM of the girl before and after, we have $I = mR^2 + I_{CM}$, $I' = mR'^2 + I'_{CM}$, where I_{CM} and I'_{CM} are the moments of inertia of the girl about her CM before and after. Since R, R' are large compared with the dimensions of the girl, we have $I_{CM} \ll mR^2$ and $I'_{CM} \ll mR'^2$. Then approximately $mR^2\omega = mR'^2\omega'$ or $mRv_B = mR'v'_B$, where v_B, v'_B are CM speeds. Thus $3.7mv_B = 3.1mv'_B$, or $v'_B = 1.2v_B$. From the conservation of energy, $\frac{1}{2}mv_B^2 = mg(1.2 - 0.6)$, or $v_B = 3.43$ m/s. Then $v'_B = 1.2v_B = 4.1$ m/s. Again we can use the conservation principle to write

$$\frac{1}{2}mv_B'^2 = mg(h - 1.2) \qquad \text{or} \qquad h = \frac{v_B'^2}{2g} + 1.2 = \underline{2.1\ \text{m}}$$

12.58 A boy stands on a freely rotating platform. With his arms extended, his rotation speed is 0.25 rev/s. But when he draws them in, his speed is 0.80 rev/s. Find the ratio of his moment of inertia in the first case to that in the second.

▌ Because there is no net torque on the system about the axis of rotation, the law of conservation of angular momentum tells us that

angular momentum before = angular momentum after $\qquad I_0\omega_0 = I_f\omega_f$

Or, since we desire I_0/I_f,

$$\frac{I_0}{I_f} = \frac{\omega_f}{\omega_0} = \frac{0.80\ \text{rev/s}}{0.25\ \text{rev/s}} = \underline{3.2}$$

12.59 A bead of mass m is constrained by a light inextensible cord to move in a circular path of radius R on a frictionless horizontal plane (see Fig. 12-19). The angular speed of the section of string from O to the bead is ω_0, initially. The pull T exerted on the string is increased until the distance from O to the bead is $R/4$. Find (a) the ratio of the final angular speed to the initial angular speed and (b) the ratio of the final tension in the string to the initial tension.

▌ (a) In this situation (a central force) there is no torque about O. Therefore, $L = $ constant, or

$$mR^2\omega_i = m\left(\frac{R}{4}\right)^2 \omega_f \qquad \text{or} \qquad \frac{\omega_f}{\omega_i} = \underline{16}$$

(b) Now, $T = mr\omega^2$, since T supplies the centripetal force. Thus

$$\frac{T_f}{T_i} = \frac{(R/4)\omega_f^2}{R\omega_i^2} = \frac{1}{4}(16)^2 = \underline{64}$$

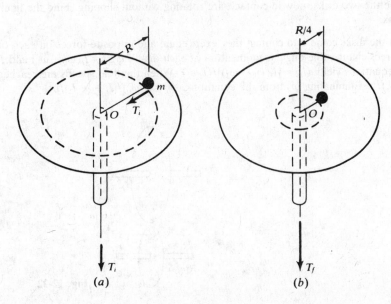

(a)   (b)   **Fig. 12-19**

12.60 The lower disk in Fig. 12-20 is coasting with angular speed ω_1. The combined moment of inertia of it and its axle is I_1. A second disk of moment of inertia I_2 is dropped onto the first and ends up rotating with it. Find the angular velocity of the combination if the original angular velocity of the upper disk was (a) zero, (b) ω_2 in the same direction as ω_1, and (c) ω_2 in a direction opposite to ω_1.

▌ Let ω_1 and ω_2 represent the magnitudes of the respective angular velocities. Conservation of angular momentum can be written as $I_1\omega_1 \pm I_2\omega_2 = (I_1 + I_2)\omega$. Therefore, (a) $\omega = (I_1\omega_1)/(I_1 + I_2)$; (b) $\omega = (I_1\omega_1 + I_2\omega_2)/(I_1 + I_2)$; (c) $\omega = (I_1\omega_1 - I_2\omega_2)/(I_1 + I_2)$. Note the analogy to a one-dimensional totally inelastic collision of two blocks.

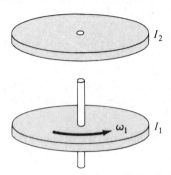

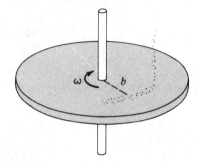

Fig. 12-20   **Fig. 12-21**

12.61 As shown in Fig. 12-21, sand drops onto a disk rotating freely about an axis. The moment of inertia of the disk about this axis is I, and its original rotational rate was ω_0. What is its rate of rotation after a mass M of sand has accumulated on the disk at radius b?

▌ No external torque is exerted on the falling sand or disk, so angular momentum is conserved. $I\omega_0 = (I + Mb^2)\omega$ and thus $\omega = \omega_0/(1 + Mb^2/I)$.

12.62 A large wooden wheel of radius R and moment of inertia I is mounted on an axle so as to rotate freely. A bullet of mass m and speed v is shot tangential to the wheel and strikes its edge, lodging in the rim. If the wheel were originally at rest, what would be its rotational rate just after collision.

▌ Just before collision, the angular momentum of the system about the axle is that of the bullet, $L_0 = R(mv)$. Just after collision, $L_f = (mR^2 + I)\omega$. Then $L_0 = L_f$ gives

$$\omega = \frac{mvR}{mR^2 + I}$$

12.63 The two uniform disks shown in Fig. 12-22 rotate separately on parallel axles. The upper disk has angular speed ω_0 and the lower disk is at rest. Now the two disks are moved together so that their rims touch. After a short time the two disks, now in contact, are rotating without slipping. Find the final rate of rotation of the upper disk.

▌ When the disks come into contact they exert equal and opposite forces on each other. The torque caused by the forces changes the angular momentum of each disk: $\bar{F}a\,\Delta t = I_1(\omega_0 - \omega_1)$ and $\bar{F}b\,\Delta t = I_2\omega_2$. The angular impulse equations yield $a/b = [I_1(\omega_0 - \omega_1)]/(I_2\omega_2)$. When the final speeds are reached $a\omega_1 = b\omega_2$, so $\omega_2 = (a\omega_1)/b$. Eliminating ω_2 from the equations, $\omega_1 = (I_1\omega_0)/[I_1 + (a^2I_2)/b^2]$.

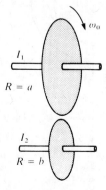

Fig. 12-22

12.64 A woman stands over the center of a horizontal platform that is rotating freely with speed 2.0 rev/s about a vertical axis through the center of the platform and straight up through the woman. She holds two 5-kg masses in her hands close to her body. The combined moment of inertia of platform, woman, and masses is 1.2 kg · m². The woman now extends her arms so as to hold the masses far from her body. In so doing, she increases the moment of inertia of the system by 2.0 kg · m². (*a*) What is the final rotational speed of the platform? (*b*) Was the kinetic energy of the system changed during the process? Explain.

▌ (*a*) Conservation of angular momentum dictates that $1.2(2.0) = (1.2 + 2.0)\omega$, so $\omega = \underline{0.75\ \text{rev/s}}$.
(*b*) Because $[1.2(2.0^2)]/2 > [3.2(0.75^2)]/2$, K is changed. The system did work in extending the 5-kg masses, thereby decreasing the kinetic energy.

12.65 A horizontal, homogeneous cylinder of mass M and radius R is pivoted about its axis of symmetry. As shown in Fig. 12-23(*a*), a string is wrapped several times around the cylinder and tied to a body of mass m resting on a support positioned so that the string has no slack. The body m is carefully lifted vertically a distance h, and the support is then removed, as shown in Fig. 12-23(*b*).
   (*a*) Just *before* the string becomes taut, evaluate the angular velocity ω_0 of the cylinder, the speed v_0 of the falling body m, and the kinetic energy K_0 of the system. (*b*) Evaluate the corresponding quantities, ω_1, v_1, and K_1, for the instant just *after* the string becomes taut.

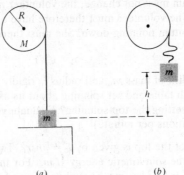

(a) (b) **Fig. 12-23**

▌ (*a*) Just *before* the string becomes taut, the cylinder is still at rest: nothing has caused it to begin turning. That is, $\omega_0 = 0$. Since the string has exerted no force on the body of mass m, it has fallen freely: $v_0 = \sqrt{2gh}$. The kinetic energy of the system is given by $K_0 = \frac{1}{2}mv_0^2 + \frac{1}{2}I_c\omega_0^2 = mgh$. (*b*) The string is assumed to be inextensible, so $v_1 = \omega_1 R$. Furthermore, the angular momentum of the system must be conserved, since the impulse due to the snap of the string is of very short duration and the weight mg contributes negligibly in that time. Thus we have $L_1 \equiv mv_1 R + \frac{1}{2}MR^2\omega_1 = L_0$, where $L_0 = mv_0 R = m\sqrt{2gh}\,R$. Solving for the angular speed, we find

$$\omega_1 = \frac{v_0}{R[1 + (M/2m)]} = \frac{\sqrt{2gh}}{R[1 + (M/2m)]}$$

The corresponding speed is given by

$$v_1 = \omega_1 R = \frac{v_0}{[1 + (M/2m)]} = \frac{\sqrt{2gh}}{[1 + (M/2m)]}$$

The final kinetic energy K_1 is given by

$$K_1 = \frac{1}{2}mv_1^2 + \frac{1}{2}I_c\omega_1^2 = \frac{1}{2}mv_1^2 + \frac{1}{2}\left(\frac{1}{2}MR^2\right)\left(\frac{v_1^2}{R^2}\right) = \frac{1}{2}\left(m + \frac{M}{2}\right)v_1^2 = \frac{1}{2}\left(\frac{mv_0^2}{1 + (M/2m)}\right) = \frac{K_0}{1 + (M/2m)}$$

12.66 Refer to Prob. 12.65. (*a*) Why is K_1 less than K_0? Where does the energy go? (*b*) If $M = m$, what fraction of the kinetic energy is lost when the string becomes taut?

▌ (*a*) The "lost kinetic energy" $K_0 - K_1$ is converted to heat energy and/or potential energy of deformation in the string and/or in the two massive objects. (The snap of the string leads to a dissipation of macroscopic kinetic energy that is analogous to the energy loss in completely inelastic two-body collisions.)
(*b*) If $M = m$, $K_1 = 2K_0/3$ so the fraction lost is $(K_0 - K_1)/K_0 = \frac{1}{3}$.

12.67   A student volunteer is sitting stationary on a piano stool with her feet off the floor. The stool can turn freely on its axle.

   (*a*) The volunteer is handed a nonrotating bicycle wheel which has handles on the axle. Holding the axle vertically with one hand, she grasps the rim of the wheel with the other and spins the wheel clockwise (as seen from above). What happens to the volunteer as she does this? (*b*) She now grasps the ends of the vertical axle and turns the wheel until the axle is horizontal. What happens? (*c*) Next she gives the rotating wheel to the instructor, who turns the axle until it is vertical with the wheel rotating clockwise, as seen from above. The instructor now hands the wheel back to the volunteer. What happens? (*d*) The volunteer grasps the ends of the axle and turns the axle until it is horizontal. What happens now? (*e*) She continues turning the axle until it is vertical but with the wheel rotating counterclockwise as viewed from above. What is the result?

   ▌ Since the axle of the piano stool is frictionless, there are no vertical torques exerted on the stool-volunteer system, so the vertical component of angular momentum is conserved. (There are horizontal torques; these result from forces that the floor exerts on the base of the stool.)

   (*a*) The initial angular momentum is zero, so the final angular momentum must also be zero. Therefore, the volunteer spins counterclockwise. (*b*) The angular momentum of the wheel is now horizontal. The volunteer's vertical component of angular momentum must now be zero, so she stops spinning. (*c*) When the wheel is handed back to the volunteer, the system of wheel and volunteer has a downward vertical angular momentum, all contributed by the wheel. The volunteer remains stationary. (*d*) Since the vertical component of the total angular momentum must not change, the volunteer must rotate clockwise. (*e*) The wheel's angular momentum is now upward. The volunteer must therefore have a downward vertical angular momentum to keep the total angular momentum pointing down. She must therefore spin clockwise. (Her spin rate is twice as fast as in part (*d*).)

12.68   A top consists of a uniform disk of mass m_0 and radius r_0 rigidly attached to an axial rod of negligible mass. The top is placed on a smooth table and set spinning about its axis of symmetry with angular speed ω_s. How much work must be done in setting the top spinning? Evaluate your result for $m_0 = 0.050$ kg, $r_0 = 2.0$ cm, and $\omega_s = 200\pi$ rad/s (or 6000 rotations per minute).

   ▌ The moment of inertia I_0 of the top is given by $I_0 = \frac{1}{2}m_0 r_0^2$. The work required to set the top spinning with angular speed ω_s is equal to the spin kinetic energy $\frac{1}{2}I_0\omega_s^2$. For the given numerical values, we find $I_0 = (0.50)(0.050)(2.0 \times 10^{-2})^2 = 10^{-5}$ kg·m². The work required is $(0.5)(10^{-5})(200\pi)^2 = \underline{1.97\ \text{J}}$.

12.69   Refer to Prob. 12.68. The center of the disk is a distance d from the top's point of contact with the table. The top is observed to precess steadily about the vertical axis with angular speed ω_p. Assuming that $\omega_p \ll \omega_s$, write ω_p in terms of r_0, d, ω_s, and g. Evaluate ω_p for $d = 3.0$ cm and $g = 9.80$ m/s², with the other quantities as given in Prob. 12.68. Is your result consistent with the assumption $\omega_p \ll \omega_s$?

   ▌ For $\omega_p \ll \omega_s$ the angular speed of steady precession is approximated by

$$\omega_p = \frac{gd}{G_0^2}\frac{1}{\omega_s}$$

where $G_0^2 = I_0/m_0 = \frac{1}{2}r_0^2$. With the numerical values given, we find

$$\omega_p = \frac{(9.80)(3.00 \times 10^{-2})}{(0.5)(2.00 \times 10^{-2})^2}\frac{1}{200\pi} = \underline{2.34\ \text{rad/s}}$$

The ratio of this angular speed to ω_s is $(\omega_p/\omega_s) = \underline{3.72 \times 10^{-3}}$. The assumption that $(\omega_p/\omega_s) \ll 1$ is fulfilled.

12.70   As shown in Fig. 12-24, a solid conical top of mass M, height h, and radius R is spinning about its symmetry axis OO' with spin angular speed ω_s. The axis OO' makes an angle α with the vertical.

   We note that for such a system the center of mass of the top is located along OO' at a distance $3h/4$ from the vertex O, and the moment of inertia I about the axis OO' is given by $I = \frac{3}{10}MR^2$.

   (*a*) Find the angular speed ω_p at which the top precesses about the vertical. (*b*) Consider a top for which $h = 10.0$ cm and $R = 3.0$ cm. The top is spinning at 5800 rotations per minute. Using $g = 9.80$ m/s², evaluate the precession angular speed ω_p.

   ▌ (*a*) Under the assumption that $\omega_p \ll \omega_s$, we can assume that

$$\omega_p \approx \frac{z_c g}{(I/M)\omega_s} = \frac{(3h/4)g}{(3R^2/10)\omega_s} = \frac{5hg}{2R^2\omega_s}$$

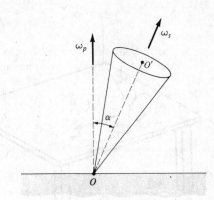

Fig. 12-24

independent of α. (b) Inserting $h = 10.0$ cm, $R = 3.0$ cm, and $\omega_s = 5800$ rotations per minute $= 607.4$ rad/s, we find

$$\omega_p = \frac{5(10.0)(980)}{2(3.0)^2(607.4)} = 4.48 \text{ rad/s}$$

Since $\omega_p/\omega_s = 7.38 \times 10^{-3}$, the condition $\omega_p \ll \omega_s$ is fulfilled.

12.71 A rocket, of mass 10^6 kg, has a speed of 500 m/s in the horizontal direction. If its altitude (y) is 10 km and its horizontal distance (x) from the chosen origin is 10 km, what is its angular momentum with respect to this origin?

❚ For this rocket, $\mathbf{r} = x\mathbf{i} + y\mathbf{j}$ and $\mathbf{v} = v\mathbf{i}$. Thus, $\mathbf{l} = \mathbf{r} \times m\mathbf{v} = m(x\mathbf{i} + y\mathbf{j}) \times v\mathbf{i} = -myv\mathbf{k} = -5 \times 10^{12}$ kg · m²/s.

12.72 The velocity of a particle of mass m is $\mathbf{v} = 5\mathbf{i} + 4\mathbf{j} + 6\mathbf{k}$ when at $\mathbf{r} = -2\mathbf{i} + 4\mathbf{j} + 6\mathbf{k}$. Find the angular momentum of the particle about the origin.

❚
$$\mathbf{l} = \mathbf{r} \times m\mathbf{v} = m \begin{vmatrix} \mathbf{i} & \mathbf{j} & \mathbf{k} \\ -2 & 4 & 6 \\ 5 & 4 & 6 \end{vmatrix} = \underline{m(42\mathbf{j} - 28\mathbf{k})}$$

12.73ᶜ Show that the angular momentum of a body may be expressed as the sum of two parts, one arising from the motion of the body's center of mass and the other from the motion of the body with respect to its center of mass.

❚ Relative to O, a fixed point in an inertial frame, the angular momentum is (see Fig. 12-25)

$$\mathbf{L}_O = \int \mathbf{r} \times \mathbf{v} \, dm = \int (\mathbf{r}_{cm} + \mathbf{r}') \times \mathbf{v} \, dm = \mathbf{r}_{cm} \times \int \mathbf{v} \, dm + \int \mathbf{r}' \times \mathbf{v} \, dm = \mathbf{r}_{cm} \times \mathbf{P} + \int \mathbf{r}' \times (\mathbf{v}' + \mathbf{v}_{cm}) \, dm$$

$$= \mathbf{r}_{cm} \times \mathbf{P} + \int \mathbf{r}' \times \mathbf{v}' \, dm + \left(\int \mathbf{r}' \, dm \right) \times \mathbf{v}_{cm} = \mathbf{r}_{cm} \times \mathbf{P} + \mathbf{L}_C$$

since, by definition of the center of mass, $\int \mathbf{r}' \, dm = 0$.

The first term, $\mathbf{r}_{cm} \times \mathbf{P}$, is the angular momentum of the center of mass, with respect to O; this is called the *orbital angular momentum* of the body (with respect to O). The second term is just the angular momentum of the body about its center of mass; this is called the *spin angular momentum* of the body. In short,

$$\mathbf{L} = \mathbf{L}_{orb} + \mathbf{L}_{spin}.$$

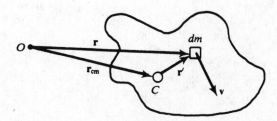

Fig. 12-25

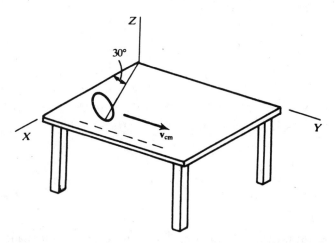

Fig. 12-26

12.74ᶜ A hoop of radius 0.10 m and mass 0.50 kg rolls across a table parallel to one edge with a speed of 0.50 m/s. Refer its motion to a rectangular coordinate system with the origin at the left rear corner of the table. At a certain time t, a line drawn from the origin to the point of contact of the hoop with the table has length 1 m and makes an angle of 30° with the X axis (Fig. 12-26). What is the angular momentum of the hoop with respect to the origin at this time t?

▌ Use the decomposition of Prob. 12.73. The position vector of the center of mass at the time t is

$$\mathbf{r}_{cm} = \mathbf{i}(\cos 30°) + \mathbf{j}(\sin 30°) + \mathbf{k}(0.10) = 0.866\mathbf{i} + 0.5\mathbf{j} + 0.10\mathbf{k}$$

and the total momentum of the hoop is

$$\mathbf{P} = m\mathbf{v}_{cm} = (0.50)(0.50\mathbf{j}) = 0.25\mathbf{j}$$

Thus

$$\mathbf{L}_{orb} = \mathbf{r}_{cm} \times \mathbf{P} = (0.866\mathbf{i} + 0.5\mathbf{j} + 0.10\mathbf{k}) \times 0.25\mathbf{j} = -0.025\mathbf{i} + 0.216\mathbf{k} \ \text{kg} \cdot \text{m}^2/\text{s}$$

To find the spin angular momentum, note that every element of mass of the hoop is at the same distance from the center of mass, $r' = 0.10$ m, and every element rotates about the center of mass with a velocity $\mathbf{v}'$ (of magnitude 0.50 m/s) perpendicular to $\mathbf{r}'$. Thus,

$$\mathbf{L}_{spin} = \int \mathbf{r}' \times \mathbf{v}' \, dm = \int r'v'(-\mathbf{i}) \, dm = -mr'v'\mathbf{i} = -0.025\mathbf{i} \ \text{kg} \cdot \text{m}^2/\text{s}$$

and

$$\mathbf{L} = \mathbf{L}_{orb} + \mathbf{L}_{spin} = \underline{-0.05\mathbf{i} + 0.216\mathbf{k} \ \text{kg} \cdot \text{m}^2/\text{s}}$$

12.75ᶜ Prove that if a particle moves under the central force field $\mathbf{F} = f(r)\mathbf{r}$, its trajectory is a plane curve (cf. Prob. 12.53).

▌ Newton's second law for the particle is $\mathbf{F} = m(d\mathbf{v}/dt)$, where $\mathbf{v} = d\mathbf{r}/dt$ is the velocity of the particle. Thus $\mathbf{r} \times \mathbf{F} = 0$ implies that

$$\mathbf{r} \times \frac{d\mathbf{v}}{dt} = 0$$

But then

$$\frac{d}{dt}(\mathbf{r} \times \mathbf{v}) = \left(\frac{d\mathbf{r}}{dt} \times \mathbf{v}\right) + \left(\mathbf{r} \times \frac{d\mathbf{v}}{dt}\right) = 0 + 0 = 0$$

or $\mathbf{r} \times \mathbf{v} = \mathbf{c}$, a constant vector. This last relation (conservation of angular momentum; $\mathbf{c} = \mathbf{l}/m$) shows us that $\mathbf{r}$ is always perpendicular to the fixed vector $\mathbf{c}$, which means that the motion is restricted to a plane through the origin.

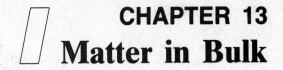

CHAPTER 13
Matter in Bulk

13.1 DENSITY AND SPECIFIC GRAVITY

13.1 Define *mass density*, *weight density*, and *specific gravity*. Discuss each with respect to water.

▮ The *mass density* (ρ) of a material is the mass per unit volume of the material.

$$\rho = \frac{\text{mass of body}}{\text{volume of body}} = \frac{m}{V}$$

The SI unit for mass density is kg/m³. Also used are g/cm³ (=g/mL) and, much less frequently, slug/ft³. 1000 kg/m³ = 1 g/cm³ = 1.94 slug/ft³. The density of water is close to 1000 kg/m³. The mass density is often referred to as simply "the density."

The *weight density* (D) of a material is the weight per unit volume of the material.

$$D = \frac{\text{weight of body}}{\text{volume of body}} = \frac{mg}{V} = \rho g$$

The units of D are N/m³, lb/ft³, or lb/in³. Water has a weight density close to 62.4 lb/ft³.

The *specific gravity* (sp gr) of a substance is the ratio of the density of the substance to the density of some standard substance. The standard is usually water (at 4° C) for liquids and solids, while for gases, it is usually air.

$$\text{sp gr} = \frac{\rho}{\rho_{\text{standard}}} = \frac{D}{D_{\text{standard}}}$$

Since sp gr is a dimensionless ratio, it has the same value for all units systems.

For liquids and solids the density in g/mL or g/cm³ is, numerically, approximately equal to the specific gravity, since $\rho_{\text{water}} \approx 1$ g/mL.

13.2 Find the density and specific gravity of ethyl alcohol if 63.3 g occupies 80.0 mL.

▮ $\rho = \text{density} = \dfrac{\text{mass}}{\text{volume}} = \dfrac{63.3 \text{ g}}{80.0 \text{ mL}} = 0.791 \text{ g/mL} = \underline{791 \text{ kg/m}^3}$

$\text{specific gravity} = \dfrac{\text{density of alcohol}}{\text{density of water}} = \dfrac{791 \text{ kg/m}^3}{1000 \text{ kg/m}^3} = \underline{0.791}$

13.3 Determine the volume of 200 g of carbon tetrachloride, for which sp gr = 1.60.

▮ Let ρ_c = density of carbon tetrachloride; ρ_ω = density of water.

$$\text{sp gr} = \frac{\rho_c}{\rho_\omega} = 1.60 \qquad \text{so} \qquad \rho_c = 1.60\rho_\omega = 1.60 \,(1 \text{ g/mL}) = 1.60 \text{ g/mL}.$$

$$\rho = \frac{m}{V} \qquad \text{so} \qquad V = \frac{m}{\rho} = \frac{2000 \text{ g}}{1.60 \text{ g/mL}} = \underline{125 \text{ mL}}.$$

13.4 The density of aluminum is 2.70 g/cm³. What volume does 2.0 kg occupy?

▮ $$\rho = \frac{m}{V} \qquad \text{or} \qquad V = \frac{m}{\rho} = \frac{2000 \text{ g}}{2.70 \text{ g/cm}^3} = \underline{741 \text{ cm}^3}$$

13.5 A drum holds 200 lb of water or 132 lb of gasoline. Determine for the gasoline (*a*) sp gr, (*b*) ρ in kg/m³, (*c*) D in lb/ft³.

▮ Weight of water = $D_w V = \rho_w g V$; weight of gasoline = $D_g V = \rho_g g V$.
(*a*) Dividing the corresponding quantities for gasoline and water,

$$\frac{132 \text{ lb}}{200 \text{ lb}} = 0.66 = \frac{\rho_g g V}{\rho_w g V} = \frac{\rho_g}{\rho_w}$$

Thus sp gr = <u>0.66</u>.
(**b**) $\rho_g = 0.66\rho_w = 0.66\ (1000\ \text{kg/m}^3) = \underline{660\ \text{kg/m}^3}$.
(**c**) $D_g = 0.66D_w$. From Prob. 13.1, $D_w = 62.4\ \text{lb/ft}^3$. Then $D_g = 0.66(62.4\ \text{lb/ft}^3) = \underline{41.2\ \text{lb/ft}^3}$.

13.6 Air has a density of 1.29 kg/m³ under standard conditions. What is the mass of air in a room with dimensions 10 m × 8 m × 3 m?

▌ $m = \rho V$; $V = (10\ \text{m})(8\ \text{m})(3\ \text{m}) = 240\ \text{m}^3$. Then $m = (1.29\ \text{kg/m}^3)(240\ \text{m}^3) = \underline{310\ \text{kg}}$.

13.7 What is the density of the material in the nucleus of the hydrogen atom? The nucleus can be considered to be a sphere of radius 1.20×10^{-15} m and its mass is 1.67×10^{-27} kg.

▌ $\rho = m/V$; $V = \frac{4}{3}\pi(1.20 \times 10^{-15}\ \text{m})^3 = 7.23 \times 10^{-45}\ \text{m}^3$. Thus

$$\rho = \frac{1.67 \times 10^{-27}\ \text{kg}}{7.23 \times 10^{-45}\ \text{m}^3} = \underline{2.31 \times 10^{17}\ \text{kg/m}^3}$$

13.8 A solid metal cylinder has $r = 0.60$ cm and $L = 8.0$ cm. Its mass is 71 g. Find the density of the metal.

▌ The cylindrical volume is $\pi r^2 L = \pi(0.60)^2(8.0) = 9.05\ \text{cm}^3$ and the density is $\rho = m/V = 71/9.05 = 7.85\ \text{g/cm}^3 = \underline{7850\ \text{kg/m}^3}$.

13.9 An aluminum sphere with $r = 0.25$ cm has a mass of 0.125 g. Is the sphere solid or does it have a bubble within it? If it is not solid, how much mass is missing from it? $\rho_{\text{Al}} = 2700\ \text{kg/m}^3$.

▌ The density of the sphere is $\rho = m/V = 0.125/[4\pi(0.25)^3/3] = 0.125\ \text{g}/0.0654\ \text{cm}^3 = 1.91\ \text{g/cm}^3$. The sphere is not solid because the density of aluminum is $\rho_{\text{Al}} = 2700\ \text{kg/m}^3 = 2.70\ \text{g/cm}^3$. If the sphere were solid, its mass would be $2.70(0.0654) = 0.177$ g. The missing mass is $0.177 - 0.125 = \underline{0.052\ \text{g}}$.

13.10 To determine the inner radius of a uniform capillary tube, the tube is filled with mercury. A column of mercury 2.375 cm long is found to "weigh" 0.242 g. What is the inner radius, r, of the tube? The density of mercury is 13 600 kg/m³.

▌ The weight given is actually the mass in grams. The radius r can be determined from $m = \rho V = \rho \pi r^2 h$, with $h = 2.375$ cm and, converting units, $\rho = 13.6\ \text{g/cm}^3$. Then $0.242\ \text{g} = (13.6\ \text{g/cm}^3)(3.14)(2.375\ \text{cm})r^2$, or $r^2 = 0.00239\ \text{cm}^2$, and $r = \underline{0.049\ \text{cm}}$.

13.11 Battery acid has sp gr = 1.285 and is 38% sulfuric acid by weight. What mass of sulfuric acid is contained in 1 L of battery acid?

▌ $\rho_{\text{battery acid}} = 1.285\ \rho_w = 1.285(1.00\ \text{kg/L}) = 1.285\ \text{kg/L}$. Of the 1.285 kg, $(0.38)(1.285) = 0.488$ kg is sulfuric acid.

13.12 A thin semitransparent film of gold ($\rho = 19\ 300\ \text{kg/m}^3$) has an area of 14.5 cm² and a mass of 1.93 mg. (**a**) What is the volume of 1.93 mg of gold? (**b**) What is the thickness of the film in angstroms, where $1\ \text{Å} = 10^{-10}$ m? (**c**) Gold atoms have a diameter of about 5 Å. How many atoms thick is the film?

▌ (**a**) $m = \rho V$, or $1.93 \times 10^{-3}\ \text{g} = (19.3\ \text{g/cm}^3)V$, and $V = 0.100 \times 10^{-3}\ \text{cm}^3 = \underline{0.100\ \text{mm}^3}$.
(**b**) $V = Ad$, with $A = 14.5\ \text{cm}^2$. Thus $d = 6.90 \times 10^{-6}$ cm. Noting that $1\ \text{Å} = 10^{-8}$ cm, we have $d = \underline{690\ \text{Å}}$.
(**c**) Dividing d by the diameter of a gold atom, we determine that the sample is <u>138 atoms</u> thick.

13.13 Under standard conditions, 22.4 m³ of helium gas contains 4.0 kg of helium in the form of 6.0×10^{26} atoms. What is the density of the gas, and what is the mass of each atom?

▌ The mass of each atom is $(4.0\ \text{kg})/(6.0 \times 10^{26}\ \text{atoms}) = \underline{6.7 \times 10^{-27}\ \text{kg}}$. The density $\rho = (4.0\ \text{kg})/(22.4\ \text{m}^3) = \underline{0.179\ \text{kg/m}^3}$.

13.14 The density within a neutron star is of the order of 1×10^{19} kg/m³. What would be the diameter of the earth if it were compressed to this density? ($M_e = 6 \times 10^{24}$ kg.)

▌ Find V from $V = (4\pi R^3)/3 = M_{\text{earth}}/\rho_{\text{star}} = (6 \times 10^{24})/(1 \times 10^{19}) = 6 \times 10^5\ \text{m}^3$. Solve for R to find the diameter $2R = \underline{105\ \text{m}}$.

13.15 In an unhealthy environment, there were 2.6×10^9 dust particles (sp gr = 3.0) per cubic meter of air.

Assuming the particles to be spheres of 2-μm diameter, calculate the mass of dust (**a**) in a 20 m × 15 m × 8 m room, (**b**) inhaled in each average breath of 400 cm³ volume. 1 μm = 10^{-6} m.

▮ (**a**) The mass of each dust particle is given by $m_p = \rho(\frac{4}{3}\pi r^3)$, where $\rho = 3.0\rho_w = 3.0(1000 \text{ kg/m}^3) = 3000$ kg/m³, and $r = 1.0 \times 10^{-6}$ m. Substituting we get $m_p = 12.6 \times 10^{-15}$ kg. The total volume of the room is $V = (20 \text{ m})(15 \text{ m})(8 \text{ m}) = 2400$ m³. Then the total particulate mass in the room is given by $M_p = (12.6 \times 10^{-15} \text{ kg/particle})(2.6 \times 10^9 \text{ particles/m}^3)(2400 \text{ m}^3) = 78 \times 10^{-3}$ kg = 78 g. (**b**) In one breath $V' = 4.00 \times 10^{-4}$ m³ and $M_p' = (12.6 \times 10^{-15})(2.6 \times 10^9)(4.00 \times 10^{-4}) = 131 \times 10^{-10}$ kg, or $M_p' = 13.1$ μg.

13.2 ELASTIC PROPERTIES

13.16 What are *stress* and *strain*?

▮ *Stress* is a measure of the strength of the agent that is causing a deformation. Precisely, if a force F is applied, possibly obliquely, to a surface of area A, then

$$\text{stress} = \frac{\text{force}}{\text{area of surface on which force acts}} = \frac{F}{A}$$

The units of stress are those of pressure; see Prob. 13.19.

Strain is the fractional deformation resulting from a stress. It is measured by the ratio of the change in some dimension of a body to the original dimension in which the change occurred.

$$\text{strain} = \frac{\text{change in dimension}}{\text{original dimension}}$$

Strain has no units because it is a ratio of like quantities. More exact definitions of strain for various situations are given in Probs. 13.18 to 13.20.

13.17 Define *elasticity*, *Hooke's law*, *elastic limit*, and *ultimate strength*.

▮ *Elasticity* is that property by virtue of which a body returns to its original size and shape when the forces that deformed it are removed.

Hooke's law can be stated in terms of stress and strain. If the system obeys Hooke's law, then stress $\propto$ strain. We then define a constant, called the *modulus of elasticity*, by the relation

$$\text{modulus of elasticity} = \frac{\text{stress}}{\text{strain}}$$

The modulus has the same units as stress. In the case of pure tension, the modulus is called *Young's modulus* (see Prob. 13.18).

The elastic limit is the smallest value of the stress required to produce a permanent distortion in a body. When a stress in excess of this limit is applied, the body will not return exactly to its original state after the stress is removed.

The ultimate strength is the smallest value of stress required to break the body.

13.18 What is *Young's modulus*?

▮ *Young's modulus* (or *tensile modulus*) describes the length elasticity of a material. Suppose a wire or rod of original length L and cross-sectional area A elongates an amount ΔL under a stretching force F applied to its end. Then

$$\text{tensile stress} = \frac{F}{A} \qquad \text{tensile strain} = \frac{\Delta L}{L}$$

and

$$\text{Young's modulus} = Y = \frac{\text{stress}}{\text{strain}} = \frac{F/A}{\Delta L/L} = \frac{FL}{A\Delta L}$$

The value of Y depends only on the material of the wire or rod and not on its dimensions.

13.19 What is the *bulk modulus*?

▮ *Bulk modulus* (B) describes the volume elasticity of a material. Suppose that a uniformly distributed compressive force acts on the surface of an object and is directed perpendicular to the surface at all points.

Then if F is the force acting on and perpendicular to an area A, we define

$$\text{pressure on } A = p = \frac{F}{A}$$

The SI unit for pressure is the *pascal* (Pa), where $1\,\text{Pa} = 1\,\text{N/m}^2$. Other units used for pressure are lb/in^2 and lb/ft^2.

Suppose that the pressure on an object of original volume V is increased by an amount Δp. The pressure increase causes a volume change ΔV, where ΔV will be negative. We then define

$$\text{volume stress} = \Delta p \qquad \text{volume strain} = -\frac{\Delta V}{V} \qquad \text{bulk modulus} = B = \frac{\text{stress}}{\text{strain}} = -\frac{\Delta p}{\Delta V/V} = -\frac{V\Delta p}{\Delta V}$$

The minus sign is used to cancel the negative numerical value of ΔV and thereby make B a positive number. The bulk modulus has the units of pressure.

The reciprocal of the bulk modulus is called the *compressibility, k,* of the substance. Bulk modulus is defined for liquids and gases as well as solids.

13.20   What is the *shear modulus*?

▮ *Shear modulus* (S) describes the shape elasticity of a material. Suppose, as shown in Fig. 13-1, that equal and opposite tangential forces F act on a rectangular block. These *shearing forces* distort the block as indicated, but its volume remains unchanged. We define

$$\text{shearing stress} = \frac{\text{tangential force acting}}{\text{area of surface being sheared}} = \frac{F}{A} \qquad \text{shearing strain} = \frac{\text{distance sheared}}{\text{distance between surfaces}} = \frac{\Delta L}{L}$$

Then

$$\text{shear modulus} = S = \frac{\text{stress}}{\text{strain}} = \frac{F/A}{\Delta L/L} = \frac{FL}{A\Delta L}$$

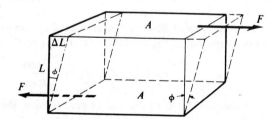

Fig. 13-1

Since ΔL is usually very small, the ratio $\Delta L/L$ is equal approximately to the shear angle ϕ in radians. In that case $S = F/(A\phi)$. More precisely, $\Delta L/L = \tan\phi$ and $S = F/(A\tan\phi)$. Like Y and B, S is measured in pascals.

13.21   An iron rod 4 m long and $0.5\,\text{cm}^2$ in cross section stretches 1 mm when a mass of 225 kg is hung from its lower end. Compute Young's modulus for the iron.

▮ $Y = (F/A)/(\Delta L/L)$, where $\Delta L/L = (1\times 10^{-3}\,\text{m})/(4\,\text{m}) = 0.25\times 10^{-3}$, $A = 0.5\times 10^{-4}\,\text{m}^2$, and $F = (225\,\text{kg})(9.8\,\text{m/s}^2) = 2205\,\text{N}$. Substituting, we get $Y = 1.76\times 10^{11}\,\text{N/m}^2 = 176\,000\,\text{MPa}$ (Because the pascal is a small unit—about 10^{-5} atmosphere—stresses and elastic moduli are usually given in kPa or MPa.)

13.22   If we consider the rod of Prob. 13.21 to be a spring, within its elastic limit, what is the effective force constant of this spring?

▮ $F = (YA\Delta L)/L \equiv k\Delta L$, where k is by its definition the force constant. Thus

$$k = \frac{YA}{L} = \frac{(1.76\times 10^{11}\,\text{N/m}^2)(0.5\times 10^{-4}\,\text{m}^2)}{4\,\text{m}} = 2.2\,\text{MN/m}$$

13.23   A coil spring is used to support a 1.8-kg mass. If the spring stretches 2 cm, what is the spring constant? What mass would be required to stretch the spring 5 cm?

❚ For a force F and displacement s,

$$k = \frac{F}{s} = \frac{mg}{s} = \frac{(1.8 \text{ kg})(9.8 \text{ m/s}^2)}{0.02} \qquad k = 882 \text{ N/m}$$

$$F = ks = (882 \text{ N/m})(0.05 \text{ m}) \qquad F = 44.1 \text{ N} \qquad m = \underline{4.50 \text{ kg}}$$

13.24 A 10-kg mass is supported by a spring whose constant is 12 N/cm. Compute the elongation of the spring.

❚

$$F = mg \qquad s = \frac{F}{k} = \frac{mg}{k} = \frac{(10 \text{ kg})(9.8 \text{ m/s}^2)}{12 \text{ N/cm}} \qquad s = \underline{8.17 \text{ cm}}$$

Note that the mixed units for k need not be converted since mg in newtons cancels the newtons in k.

13.25 A load of 100 lb is applied to the lower end of a steel rod 3 ft long and 0.20 in in diameter. How much will the rod stretch? $Y = 3.3 \times 10^7 \text{ lb/in}^2$ for steel.

❚

$$Y = \frac{F/A}{\Delta L/L} \quad \text{or} \quad \Delta L = \frac{LF}{AY} = \frac{(3 \text{ ft})(100 \text{ lb})}{\pi(0.1 \text{ in}^2)(3.3 \times 10^7 \text{ lb/in}^2)} = 2.9 \times 10^{-4} \text{ ft} \quad \text{or} \quad \Delta L = \underline{3.5 \times 10^{-3} \text{ in}}$$

Note that the units of L and A need not agree as long as the units of A and Y agree, since (distance^2) cancels between these two last quantities.

13.26 A wire whose cross section is 4 mm^2 is stretched by 0.1 mm by a certain weight. How far will a wire of the same material and length stretch if its cross-sectional area is 8 mm^2 and the same weight is attached?

❚ If the material and length are fixed as well as the stretching weight, then F, l, Y are fixed in the relation $\Delta l = (Fl)/(AY)$. Then

$$\Delta l \propto \frac{1}{A} \qquad A \, \Delta l = \text{const} \qquad \therefore \quad A_1 \, \Delta l_1 = A_2 \, \Delta l_2 \qquad \Delta l_2 = \frac{A_1}{A_2} \Delta l_1 = \frac{4 \text{ mm}^2}{8 \text{ mm}^2} 0.1 \text{ mm} \qquad \underline{\Delta l_2 = 0.05 \text{ mm}}$$

13.27 A steel wire is 4.0 m long and 2 mm in diameter. How much is it elongated by a suspended body of mass 20 kg? Young's modulus for steel is 196 000 MPa.

❚ Let ΔL be the elongation. Then, by Hooke's law,

$$\frac{F}{A} = Y \frac{\Delta L}{L}$$

where Y is Young's modulus. The elongation is

$$\Delta L = \frac{1}{Y} \frac{F}{A} L = \frac{mgL}{YA} = \frac{(20)(9.8)(4.0)}{(196 \times 10^9)\pi(0.001)^2} = 1.273 \times 10^{-3} \text{ m} = \underline{1.273 \text{ mm}}$$

13.28 A copper wire 2.0 m long and 2 mm in diameter is stretched 1 mm. What tension is needed? Young's modulus for copper is 117 600 MPa.

❚

$$\frac{F}{A} = Y \frac{\Delta L}{L} \qquad F = Y \frac{\Delta L}{L} A = (117.6 \times 10^9) \frac{0.001}{2} \pi(0.001)^2 = \underline{184.7 \text{ N}}$$

13.29 A wire is stretched 1 mm by a force of 1 kN. (*a*) How far would a wire of the same material and length but of four times that diameter be stretched? (*b*) How much work is done in stretching each wire?

❚ (*a*) The elongation is inversely proportional to the cross-sectional area, and so $\Delta L = (1)(\frac{1}{4})^2 = \frac{1}{16}$ mm.
(*b*) The work done in stretching the wire in the two cases is $W = \bar{F}x$. Since the force varies linearly with distance,

$$\bar{F} = \frac{F_i + F_f}{2} \qquad \text{so} \qquad W_1 = \frac{1000 + 0}{2}(0.001) = \underline{0.5 \text{ J}} \qquad W_2 = \frac{1}{16} W_1 = \underline{0.0313 \text{ J}}$$

13.30 Given a 2.0-m length of steel wire with 1.0-mm diameter, about how much will the wire stretch under a 5.0-kg load? $Y = 195 000$ MPa.

❚ Since tensile modulus is stress/strain, we have $\Delta L/L = (F/A)/Y = [5(9.8)]/[\pi(5 \times 10^{-4})^2(195 \times 10^9)] = 3.2 \times 10^{-4}$ and $\Delta L = \underline{6.4 \times 10^{-4} \text{ m}}$.

13.31 Approximately how large a force is required to stretch a 2.0-cm-diameter steel rod by 0.01 percent? $Y = 195\,000$ MPa.

▮ Solving Hooke's law for F, it is $F = AY(\Delta L/L) = \pi(0.01)^2(195 \times 10^9)(10^{-4}) = \underline{6100\,\text{N}}$.

13.32 A platform is suspended by four wires at its corners. The wires are 3 m long and have a diameter of 2.0 mm. Young's modulus for the material of the wires is $180\,000$ MPa. How far will the platform drop (due to elongation of the wires) if a 50-kg load is placed at the center of the platform?

▮ $\Delta L = (LF)/(AY)$, where $L = 3$ m, $A = \pi(1.0 \times 10^{-3}\,\text{m})^2 = 3.14 \times 10^{-6}\,\text{m}^2$, and since each wire supports one-quarter of the load, $F = [(50\,\text{kg})(9.8\,\text{m/s}^2)]/4 = 123$ N.

$$\Delta L = \frac{(3\,\text{m})(123\,\text{N})}{(3.14 \times 10^{-6}\,\text{m}^2)(1.8 \times 10^{11}\,\text{N/m}^2)} = 65 \times 10^{-5}\,\text{m} \qquad \text{or} \qquad \underline{0.65\,\text{mm}}$$

13.33 What is the minimum diameter of a brass rod if it is to support a 400-N load without exceeding the elastic limit? Assume that the stress for the elastic limit is 379 MPa.

▮ To find the minimum diameter, and hence minimum cross-sectional area, we assume that the force $F = 400$ N brings us to the elastic limit. Then from the stress, $F/A = 379 \times 10^6$ Pa, we get $A = (400\,\text{N})/(379 \times 10^6\,\text{Pa}) = 1.0554 \times 10^{-6}\,\text{m}^2$. Then

$$A = \frac{\pi D^2}{4} \qquad D^2 = \frac{4A}{\pi} = \frac{4(1.0554 \times 10^{-6}\,\text{m}^2)}{\pi} = 1.344 \times 10^{-6}$$

and

$$D = \sqrt{1.344 \times 10^{-6}\,\text{m}^2} = 1.16 \times 10^{-3}\,\text{m} = \underline{1.16\,\text{mm}}$$

13.34 A No. 18 copper wire has a diameter of 0.04 in and is originally 10 ft long. (**a**) What is the greatest load that can be supported by this wire without exceeding its elastic limit? (**b**) Compute the change in length of the wire under this load. (**c**) What is the maximum load that can be supported without breaking the wire? (**d**) What is the maximum elongation? (Assume that the elastic limit stress is $23\,000$ lb/in^2 and that the ultimate strength stress is $49\,000$ lb/in^2.)

▮ (**a**) $F/A = 23\,000$ lb/in^2, so $F_{\text{max}} = (23\,000\,\text{lb/in}^2)A$.

$$A = \frac{\pi D^2}{4} = \frac{\pi(0.04\,\text{in})^2}{4} = 4\pi \times 10^{-4}\,\text{in}^2 \qquad F_{\text{max}} = (23\,000)(4\pi \times 10^{-4}) = \underline{28.9\,\text{lb}}$$

(**b**) $\Delta l = (F/A)(l/Y) = (23\,000\,\text{lb/in})10\,\text{ft}/17 \times 10^6\,\text{lb/in} = \underline{0.0135\,\text{ft}}$
(**c**) $F'/A = (49\,000\,\text{lb/in}^2)$; $F' = (49\,000\,\text{lb/in}^2)(4\pi \times 10^{-4}\,\text{in}^2) = \underline{61.6\,\text{lb}}$
(**d**) $\Delta l' = (49\,000\,\text{lb/in}^2)(10\,\text{ft}/17 \times 10^6\,\text{lb/in}^2) = 0.0288\,\text{ft} = \underline{0.346\,\text{in}}$

13.35 A steel piano wire has an ultimate strength of about $35\,000$ lb/in^2. How large a load can a 0.5-in-diameter steel wire hold before breaking?

▮ $F = (35\,000\,\text{lb/in}^2)A \qquad A = \frac{\pi(0.05\,\text{in}^2)}{4} = 1.964 \times 10^{-3}\,\text{in}^2 \qquad F = (35\,000\,\text{lb/in}^2)(1.964 \times 10^{-3}\,\text{in}^2) = \underline{68.7\,\text{lb}}$

13.36^c A wire of original length L and cross-sectional area A is stretched, within the elastic limit, by a stress τ. Show that the density of stored elastic energy in the stretched wire is $\tau^2/2Y$.

▮ Let the total deformation be ΔL, so that $\tau = Y(\Delta L/L)$, by Hooke's law. Again by Hooke's law, the stretching force at deformation x is given by $F(x) = [(AY)/L]x$. Hence the stored elastic energy is

$$W = \int_0^{\Delta L} F(x)\,dx = \frac{AY}{L} \int_0^{(L\tau)/Y} x\,dx = \frac{\tau^2}{2Y}AL$$

But, neglecting a tiny change, the constant volume of the wire is $V = AL$; hence

$$\frac{W}{V} = \frac{\tau^2}{2Y}$$

Note that the energy density is independent of the wire dimensions.

13.37 Determine the fractional change in volume as the pressure of the atmosphere (0.1 MPa) around a metal block is reduced to zero by placing the block in vacuum. The bulk modulus for the metal is $125\,000$ MPa.

$$B = \frac{-\Delta p}{\Delta V / V} \quad \text{or} \quad \frac{\Delta V}{V} = -\frac{\Delta p}{B} = \frac{-(-0.1)}{125\,000} = \underline{8 \times 10^{-7}}$$

13.38 Compute the volume change of a solid copper cube, 40 mm on each edge, when subjected to a pressure of 20 MPa. The bulk modulus for copper is 125 000 MPa.

$$\Delta V = \frac{-V \Delta p}{B} = \frac{-(40\,\text{mm})^3 (20\,\text{MPa})}{125\,000\,\text{MPa}} = \underline{-10\,\text{mm}^3}$$

Note that the units of V and Δp need not agree because $\Delta p / B$ is dimensionless.

13.39 The pressure in an explosion chamber is 345 MPa. What would be the percent change in volume of a piece of copper subjected to this pressure? The bulk modulus for copper is 138 GPa ($= 138 \times 10^9$ Pa).

The bulk modulus is defined as $B = -\Delta p / (\Delta V / V)$, where the minus sign is inserted because ΔV is negative when Δp is positive.

$$100 \left| \frac{\Delta V}{V} \right| = 100 \frac{\Delta p}{B} = 100 \frac{345 \times 10^6}{138 \times 10^9} = \underline{0.25\%}$$

13.40 The compressibility of water is 5×10^{-10} m²/N. Find the decrease in volume of 100 mL of water when subjected to a pressure of 15 MPa.

We note that the compressibity, k, is simply the reciprocal of the bulk modulus. Then $\Delta V = -Vk\,\Delta p = -(100\,\text{mL})(5 \times 10^{-10}\,\text{m}^2/\text{N})(15 \times 10^6\,\text{N/m}^2) = -0.75\,\text{mL}$. The decrease is thus $\underline{0.75\,\text{mL}}$.

13.41 How large a pressure must be applied to water if it is to be compressed by 0.1 percent? What is the ratio of this pressure to atmospheric pressure, 101 kPa? The bulk modulus of water is 2100 MPa.

The volume strain $\Delta V / V_0 = 1.0 \times 10^{-3}$; set $B = p / (\Delta V / V_0)$ to find $p = (2.1 \times 10^9)(1.0 \times 10^{-3}) = 2100$ kPa. Dividing by the atmospheric pressure, the ratio is $\underline{21}$.

13.42 By what fraction will the volume of a steel bar increase as the air is evacuated from a chamber in which it rests? Standard atmospheric pressure = 0.101 MPa and B for steel is 160 000 MPa.

$$\Delta V / V_0 = -\Delta P / B = 0.101 / 160\,000 = \underline{6.3 \times 10^{-7}}$$

13.43 What increase in pressure is required to decrease the volume of 200 L of water by 0.004 percent? Find ΔV. ($B = 2100$ MPa)

$$\Delta V = 0.00004(200\,\text{L}) = 0.008\,\text{L} \quad \Delta p = B\left(-\frac{\Delta V}{V}\right) = (2100\,\text{MPa})\left(\frac{0.008\,\text{L}}{200\,\text{L}}\right) = 0.084\,\text{MPa} = \underline{84\,\text{kPa}}$$

13.44 Compute the compressibility of glycerin if a pressure of 290 lb/in² causes a volume of 64 in³ to decrease by 3×10^{-3} in³.

The compressibility is the reciprocal of the bulk modulus. Thus

$$k = -\frac{1}{\Delta p} \frac{\Delta V}{V_0} = \frac{3 \times 10^{-3}\,\text{in}^3}{(290\,\text{lb/in}^2)(64\,\text{in}^3)} = \underline{1.62 \times 10^{-7}\,\text{in}^2/\text{lb}}$$

13.45 Two parallel and opposite forces, each 4000 N, are applied tangentially to the upper and lower faces of a cubical metal block 25 cm on a side. Find the angle of shear and the displacement of the upper surface relative to the lower surface. The shear modulus for the metal is 80 GPa.

We use the approximate form $S = F/(A\phi)$ (Prob. 13.20), with $S = 8 \times 10^{10}$ N/m², $F = 4000$ N, and $A = (0.25\,\text{m})^2 = 6.25 \times 10^{-2}$ m². Solving for ϕ, we get

$$\phi = \frac{(4000\,\text{N})}{(6.25 \times 10^{-2}\,\text{m}^2)(8 \times 10^{10}\,\text{N/m}^2)} = \underline{8.0 \times 10^{-7}\,\text{rad}}$$

The displacement of the upper surface is given by $d = L\phi$, where L is an edge of the cube; $d = (8.0 \times 10^{-7})(25\,\text{cm}) = \underline{2.0 \times 10^{-5}\,\text{cm}}$.

13.46 The shear modulus for a metal is 50 000 MPa. Suppose that a shear force of 200 N is applied to the upper surface of a cube of this metal that is 3.0 cm on each edge. How far will the top surface be displaced?

▎ Shearing strain $= \Delta L/L = (F/A)/S = 200/[L^2(5 \times 10^{10})] = (4 \times 10^{-9})/L^2$; solve for ΔL with $L = 0.030$ m, $\Delta L = 1.33 \times 10^{-7}$ m $= \underline{0.133\ \mu m}$.

13.47 A block of gelatin is 60 mm by 60 mm by 20 mm when unstressed. A force of 0.245 N is applied tangentially to the upper surface, causing a 5-mm displacement relative to the lower surface (Fig. 13-2). Find (a) the shearing stress, (b) the shearing strain, and (c) the shear modulus.

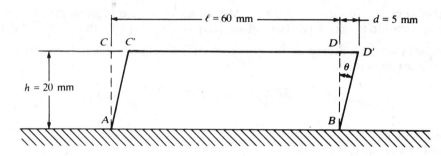

Fig. 13-2

▎ (a)
$$\text{stress} = \frac{F}{A} = \frac{0.245}{36 \times 10^{-4}} = \underline{68.1\ \text{Pa}}$$

(b)
$$\text{strain} = \tan\theta = \frac{d}{h} = \frac{5}{20} = \underline{0.25}$$

(c)
$$\text{shear modulus} = S = \frac{F/A}{\tan\theta} = \frac{68.4}{0.25} = \underline{272.4\ \text{N/m}^2}$$

13.48 Two sheets of aluminum on an aircraft wing are to be held together by aluminum rivets of cross-sectional area 0.25 in². The shearing stress on each rivet must not exceed one-tenth of the elastic limit for aluminum. How many rivets are needed if each rivet supports the same fraction of a total shearing force of 25 000 lb? Assume that the elastic limit stress is 19 000 lb/in².

▎ $F/A = \frac{1}{10}(19\,000\ \text{lb/in}^2) = 1900\ \text{lb/in}^2$ maximum stress allowed for each rivet. This means a shearing force of $F = (1900\ \text{lb/in}^2)(0.25\ \text{in}^2) = 475\ \text{lb/rivet}$. Number of rivets $= 25\,000\ \text{lb}/(475\ \text{lb/rivet}) = 52.7$, or $\underline{53\ \text{rivets}}$.

13.49 A 60-kg motor sits on four cylindrical rubber blocks. Each cylinder has a height of 3 cm and a cross-sectional area of 15 cm². The shear modulus for this rubber is 2 MPa. (a) If a sideways force of 300 N is applied to the motor, how far will it move sideways? (b) With what frequency will the motor vibrate back and forth sideways if disturbed?

▎ (a) We assume that the shear force is distributed evenly among the four cylinders. Then for a given cylinder $F = 75$ N.

$$\phi = \frac{F}{AS} = \frac{75\ \text{N}}{(15 \times 10^{-4}\ \text{m}^2)(2 \times 10^6\ \text{N/m}^2)} = 2.5 \times 10^{-2}\ \text{rad}$$

The displacement is then $d = L\phi = (3.0\ \text{cm})(2.5 \times 10^{-2}\ \text{rad}) = \underline{0.075\ \text{cm}}$. (b) Since the shear force on each cylinder is proportional to ϕ, it is also proportional to the horizontal displacement d. $F = AS\phi = [(AS)/L]d$. Since there are four cylinders, the total external horizontal force, F_T, is given by $F_T = [(4AS)/L]d$. This force just balances the elastic restoring force, or effective "spring" force, of the system $F_s = -[(4AS)/L]d$. If the shear force is removed, the system oscillates with an effective force constant $k = (4AS)/L = 4.0 \times 10^5$ N/m. Assuming that the masses of the cylinders are negligible, we have

$$f = \frac{1}{2\pi}\sqrt{\frac{k}{m}} = \frac{1}{2\pi}\sqrt{\frac{4.0 \times 10^5\ \text{N/m}}{60\ \text{kg}}} = \underline{13\ \text{Hz}}.$$

13.50 The twisting of a cylindrical shaft (Fig. 13-3) through an angle θ is an example of a shearing strain. An

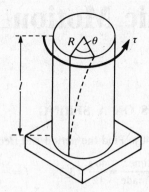

Fig. 13-3

analysis of the situation shows that the angle of twist (in radians) is given by

$$\theta = \frac{2\tau l}{\pi S R^4}$$

where τ = applied torque
l = length of cylinder
R = radius of cylinder
S = shear modulus for material

If a torque of 100 lb · ft is applied to the end of a cylindrical steel shaft 10 ft long and 2 in in diameter, through how many degrees will the shaft twist? Assume that $S = 12 \times 10^6$ lb/in^2.

▌ We are given $S = 12 \times 10^6$ lb/in^2, $R = 1$ in, $\tau = 100$ lb · ft = 1200 lb · in, $l = 10$ ft = 120 in. Using the given formula we find that

$$\theta = \frac{2\tau l}{\pi S R^4} = \frac{2(1200 \text{ lb} \cdot \text{in})(120 \text{ in})}{\pi (12 \times 10^6 \text{ lb/in}^2)(1 \text{ in})^4} = 7.64 \times 10^{-3} \text{ rad} = \underline{0.437°}$$

13.51 An engine delivers 140 hp at 800 rpm to an 8-ft solid-iron drive shaft 2 in in diameter. Find the angle of twist in the drive shaft. Assume that $S = 10 \times 10^6$ lb/in^2.

▌ First calculate the torque, then apply Prob. 13.50.

$$P = 140 \text{ hp}\left(550 \frac{\text{ft} \cdot \text{lb/s}}{\text{hp}}\right) = 7.7 \times 10^4 \text{ ft} \cdot \text{lb/s}; \qquad \omega = 2\pi f = 2\pi(800 \text{ rad/s})(1 \text{ min}/60 \text{ s}) = 83.78 \text{ rad/s},$$

$$P = \tau\omega, \qquad \text{so} \qquad \tau = \frac{P}{\omega} = \frac{7.7 \times 10^4 \text{ ft} \cdot \text{lb/s}}{83.78 \text{ rad/s}} = 919 \text{ lb} \cdot \text{ft} = 11\,000 \text{ lb-in.}$$

Finally

$$\theta = \frac{2\tau l}{\pi S R^4} = \frac{2(11\,000 \text{ lb} \cdot \text{in})(8 \text{ ft})(12 \text{ in/ft})}{\pi (10 \times 10^6 \text{ lb/in}^2)(1 \text{ in})^4} = 0.0674 \text{ rad} = \underline{3.86°}$$

CHAPTER 14
Simple Harmonic Motion

14.1 OSCILLATIONS OF A MASS ON A SPRING

14.1 A spring makes 12 vibrations in 40 s. Find the period and frequency of the vibration.

▌ $$T = \frac{\text{elapsed time}}{\text{vibrations made}} = \frac{40\,\text{s}}{12} = \underline{3.3\,\text{s}} \qquad f = \frac{\text{vibrations made}}{\text{elapsed time}} = \frac{12}{40\,\text{s}} = \underline{0.30\,\text{Hz}}$$

14.2 A 50-g mass hangs at the end of a Hookean spring. When 20 g more are added to the end of the spring, it stretches 7.0 cm more. (*a*) Find the spring constant. (*b*) If the 20 g are now removed, what will be the period of the motion?

▌ (*a*) Since the spring is linear,

$$k = \frac{\Delta F}{\Delta x} = \frac{(0.020\,\text{kg})(9.8\,\text{m/s}^2)}{0.07\,\text{m}} = \underline{2.8\,\text{N/m}}$$

(*b*) $$T = 2\pi\sqrt{\frac{m}{k}} = 2\pi\sqrt{\frac{0.050\,\text{kg}}{2.8\,\text{N/m}}} = \underline{0.84\,\text{s}}$$

14.3 A spring is stretched 4 cm when a mass of 50 g is hung on it. If a total of 150 g is hung on the spring and the mass is started in a vertical oscillation, what will the period of the oscillation be?

▌ First find the spring constant k:

$$k = \frac{(50\,\text{g})(980\,\text{cm/s}^2)}{4\,\text{cm}} = 12\,250\,\text{dyn/cm}$$

To find the period use

$$T = 2\pi\sqrt{\frac{m}{k}} = 2\pi\sqrt{\frac{150}{12\,250}} = 2\pi(0.1107) = \underline{0.695\,\text{s}}$$

14.4 A body of weight 27 N hangs on a long spring of such stiffness that an extra force of 9 N stretches the spring 0.05 m. If the body is pulled downward and released, what is its period?

▌ The spring constant is $k = 9/0.05 = 180\,\text{N/m}$, and so

$$T = 2\pi\sqrt{\frac{m}{k}} = 2\pi\sqrt{\frac{27/9.8}{180}} = \underline{0.78\,\text{s}}$$

14.5 A 3-lb weight hangs at the end of a spring which has $k = 25\,\text{lb/ft}$. If the weight is displaced slightly and released, with what frequency will it vibrate?

▌ Using $m = 3\,\text{lb}/(32.2\,\text{ft/s}^2) = \underline{0.093\,\text{slug}}$ gives

$$f = \frac{1}{2\pi}\sqrt{\frac{k}{m}} = \frac{1}{2\pi}\sqrt{\frac{25\,\text{lb/ft}}{0.093\,\text{slug}}} = \underline{2.61\,\text{Hz}}$$

14.6 A mass m suspended from a spring of constant k has a period T. If a mass M is added, the period becomes $3T$. Find M in terms of m.

▌ The period varies as the square root of the mass. Thus the mass must increase ninefold, making $M = \underline{8m}$.

14.7 A 0.5-kg body performs simple harmonic motion with a frequency of 2 Hz and an amplitude of 8 mm. Find the maximum velocity of the body, its maximum acceleration, and the maximum restoring force to which the body is subjected.

▌ We are given frequency v and amplitude R: $v = 2\,\text{Hz}$; $\omega = 2\pi v = 4\pi\,\text{rad/s}$; $R = 0.008\,\text{m}$. Then we have

$v_{max} = \omega R = 4\pi(0.008) = \underline{0.101 \text{ m/s}}$ and $a_{max} = \omega^2 R = 16\pi^2(0.008) = \underline{1.264 \text{ m/s}^2}$. From Newton's second law, $F_{max} = ma_{max} = 0.5 \text{ kg} \times 1.26 \text{ m/s}^2 = \underline{0.632 \text{ N}}$.

14.8 A mass of 250 g hangs on a spring and oscillates vertically with a period of 1.1 s. To double the period, what mass must be added to the 250 g? (Ignore the mass of the spring.)

▌ Reasoning as in Prob. 14.6, added mass $= (4 - 1)(250 \text{ g}) = \underline{750 \text{ g}}$.

14.9 A body describing SHM has a maximum acceleration of 8π m/s^2 and a maximum speed of 1.6 m/s. Find the period T and the amplitude R.

▌ We can express a_{max} and v_{max} in terms of T and R: $a_{max} = [(4\pi^2)/T^2]R = 8\pi$ m/s^2, $v_{max} = (2\pi/T)R = 1.6$ m/s. We solve by eliminating R:

$$\frac{a_{max}}{v_{max}} = \frac{2\pi}{T} = \frac{8\pi}{1.6} \quad \text{and} \quad T = \underline{0.4 \text{ s}}$$

Then

$$\frac{2\pi R}{T} = \frac{2\pi R}{0.4} = 1.6 \quad \text{and} \quad R = \frac{0.4 \times 1.6}{2\pi} = \underline{0.102 \text{ m}}$$

14.10 A vertically suspended spring of negligible mass and force constant k is stretched by an amount l when a body of mass m is hung on it. The body is pulled by hand an additional distance y (positive direction downward) and then released. Show that the motion of the body is governed by the equation $a = -ky/m$, so that the body executes harmonic motion about its equilibrium position, and show that the period of this motion is the same as that of a simple pendulum of length l.

▌ The initial static equilibrium is characterized by a balance between the spring force and the object's weight: $mg = kl$. We denote the instantaneous position by $y(t)$, with $y = 0$ corresponding to the equilibrium position and with $y > 0$ corresponding to additional stretching. Once the object has been released, Newton's second law is given by $ma = mg - k(l + y) = -ky$. Thus the combined effect of a hanging spring and gravity is the same as the spring alone measured from the equilibrium position. That is, $a = -ky/m$, as desired, and the body executes harmonic motion about its static equilibrium position; the period T of the motion is given by

$$T = 2\pi\sqrt{\frac{m}{k}} = 2\pi\sqrt{\frac{l}{g}} \quad \text{since } mg = kl$$

The last formula for the period is just that for a simple pendulum of length l.

14.11 In a certain engine, a piston undergoes vertical SHM with amplitude 7 cm. A washer rests on top of the piston. As the motor is slowly speeded up, at what frequency will the washer no longer stay in contact with the piston?

▌ The maximum downward acceleration of the washer will be that for free fall, g. If the piston accelerates downward faster than this, the washer will lose contact.

In SHM, the acceleration is given in terms of the displacement and the frequency by $a = -4\pi^2 f^2 x$. With the upward direction chosen as positive, the largest downward (most negative) acceleration occurs for $x = +x_0 = 0.07$ m, that is, $a_0 = 4\pi^2 f^2 (0.07 \text{ m})$. The washer will separate from the piston when a_0 first becomes equal to g. Therefore, the critical frequency is given by

$$4\pi^2 f_c^2 (0.07 \text{ m}) = 9.8 \text{ m/s}^2 \quad \text{or} \quad f_c = \frac{1}{2\pi}\sqrt{\frac{9.8}{0.07}} = \underline{1.88 \text{ Hz}}$$

14.12 A body of mass m is attached by a string to a suspended spring of spring constant k. Both the string and the spring have negligible mass, and the string is inextensible (it has a fixed length). The body is pulled down a distance A and then released. (*a*) Assuming that the string remains taut throughout the motion, find the maximum (downward) acceleration of the oscillating body. (*b*) The string will remain taut only as long as it remains under tension. Determine the largest amplitude A_{max} for which the string will remain taut throughout the motion. (*c*) Evaluate A_{max} for $m = 0.10$ kg and $k = 10$ N/m.

▌ (*a*) As long as the string remains taut, the body is subjected to a restoring force which varies linearly with vertical displacement from the position of static equilibrium. In this case the only effect of the string is to communicate the spring force to the body. The motion of the body is harmonic motion of amplitude A and radian frequency $\omega = \sqrt{k/m}$. The maximum magnitude of the acceleration (in either direction) is $\omega^2 A =$ kA/m. (b) If the tension in the string is $T(t)$, the downward acceleration $a(t) = g - \{[T(t)]/m\}$. Therefore the downward acceleration cannot exceed g if the string is to remain taut ($T > 0$). The corresponding maximum amplitude $A_{max} = mg/k$. Note that this amplitude is equal to the amount of stretching in the spring at equilibrium, which is consistent with our intuition. (c) With $m = 0.10$ kg and $k = 10$ N/m, we find $A_{max} = [(0.10)(9.80)]/10 = 9.80 \times 10^{-2}$ m = $\underline{98\text{ mm}}$.

14.13 A massive block resting on a tabletop is attached to an anchored horizontal spring. There is negligible friction between the massive block and the tabletop. The oscillation frequency is ν. A much less massive block is placed on top of the large block. The coefficient of static friction between the two blocks is μ_s. (a) What is the largest amplitude of oscillation of the large block which permits the small block to ride without slipping? (b) Evaluate your result for $\nu = 3.0$ Hz and $\mu_s = 0.60$.

▌ (a) If the amplitude of oscillation of the massive block is A, then its maximum acceleration is $\omega^2 A = 4\pi^2 \nu^2 A$. In order for the smaller block to follow this motion, the frictional force must be able to produce the above acceleration. The maximum acceleration that static friction can provide in this case is $\mu_s g$. Therefore we need $4\pi^2 \nu^2 A \leqslant \mu_s g$, so the maximum amplitude $A_{max} = (\mu_s g)/(4\pi^2 \nu^2)$. (b) With $\nu = 3.0$ Hz and $\mu_s = 0.60$, we find $A_{max} = [(0.60)(9.80)]/[4\pi^2(3.0)^2] = 1.65 \times 10^{-2}$ m = $\underline{16.5\text{ mm}}$.

14.14 A block of mass M_1 resting on a frictionless horizontal surface is connected to a spring of spring constant k that is anchored in a nearby wall. A block of mass $M_2 = \alpha M_1$ is placed on top of the first block. The coefficient of static friction between the two bodies is μ_s. (a) Assuming that the two bodies move as a unit, find the period of oscillation of the system. (b) What is the maximum oscillation amplitude A_{max} that permits the two bodies to move as a unit? (c) Evaluate your results of parts a and b for $k = 6.0$ N/m, $M_1 = 1.0$ kg, $\alpha = 0.50$, and $\mu_s = 0.40$.

▌ (a) Let x_1 and x_2 be the coordinates of objects 1 and 2, measured from the initial equilibrium (with the spring relaxed) and with $x_1 > 0$ corresponding to a stretched spring. If the two blocks move as a unit, then $x_1 = x_2 = x$, and Newton's second law reads $(M_1 + M_2)a_x = -kx$. The motion is a harmonic oscillation of period $T = 2\pi\sqrt{(M_1 + M_2)/k} = 2\pi\sqrt{[(1 + \alpha)M_1]/k}$. (b) The required maximum magnitude of the acceleration a_x of block 2, in order for it to move with block 1, is given by $\omega^2 A$. Since friction is the only horizontal force acting on block 2, the two blocks can move as a unit only if $M_2\omega^2 A \leqslant \mu_s M_2 g$. This implies a maximum amplitude $A_{max} = [\mu_s(M_1 + M_2)g]/k = [(1 + \alpha)M_1 g\mu_s]/k$. (c) With $k = 6.0$ N/m, $M_1 = 1.0$ kg, $\alpha = 0.50$, and $\mu_s = 0.40$, we find $T = 2\pi\sqrt{[(1.5)(1.0)]/(6.0)} = \pi = \underline{3.14\text{ s}}$. The maximum amplitude $A_{max} = [(1.5)(1.0)(9.8)(0.40)]/(6.0) = \underline{0.98\text{ m}}$.

14.15 A circle in the xy plane with center coincident with the origin of coordinates has a radius A. A particle moves around the circle with constant speed $A\omega$. Show that the particle's x coordinate is given by $x = A\cos(\omega t + \phi)$, where ϕ is the angular displacement of the particle from the x axis at $t = 0$. How is the angular frequency ω of the particle in the circle related to f, the frequency with which the x coordinate oscillates?

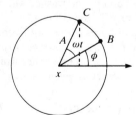

Fig. 14-1

▌ Refer to Fig. 14-1. At a time $t = 0$ the particle is at point B and it makes an angle ϕ with the horizontal axis. In a time t it moves an additional angular distance ωt to point C. The x coordinate of point C is $x = A\cos(\omega t + \phi)$. The time T for a complete cycle is distance/speed = $(2\pi A)/A\omega = 2\pi/\omega$. Since $f = 1/T$, $\omega = 2\pi f$; ω is called angular frequency for this reason.

14.16 In a newtonian experiment, a pail of water, at the end of a rope of length r, is whirled in a horizontal circle at constant speed v. A distant ground-level spotlight casts a shadow of the pail onto a vertical wall which is perpendicular to the spotlight beam. Show that the shadow executed SHM with angular frequency $\omega = v/r$.

▌ Figure 14-2 gives a top view of the situation. Relative to the indicated coordinate system, the

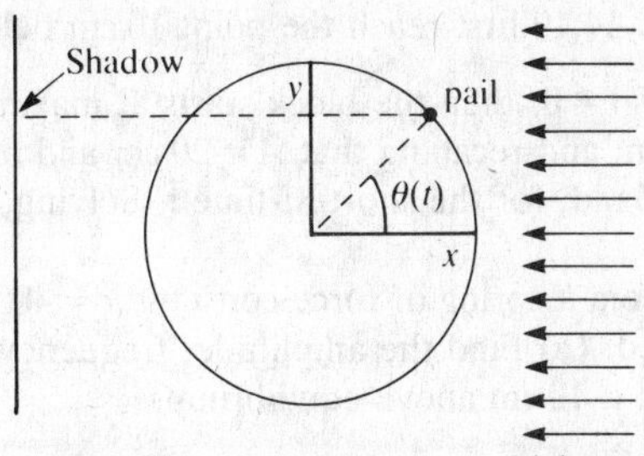

Fig. 14-2

instantaneous y coordinate of the pail gives $y(t)$ for the shadow. We let $\theta(t)$ denote the pail's angular position (in radians), measured counterclockwise from the positive x axis. Then $\omega = \pm v/r$, with the sign depending upon which way the pail is whirled. Letting $\theta(0) = \theta_0$, the angular position $\theta(t) = (\pm vt/r) + \theta_0$. But

$$y = r \sin \theta(t) = r \sin \left[\left(\pm \frac{vt}{r} + \theta_0 \right) \right] = \pm r \sin \left(\frac{v}{r} t \pm \theta_0 \right)$$

so, for either sense, we have SHM of amplitude $|\pm r| = r$ and angular frequency $\omega = v/r$.

14.17 A stone is swinging in a horizontal circle 0.8 m in diameter, at 30 rev/min. A distant light causes a shadow of the stone to be formed on a nearby wall. What is the amplitude of the motion of the shadow? What is the frequency? What is the period?

▌ The amplitude is the radius of the circle: $R = 0.8 \text{ m}/2 = \underline{0.4 \text{ m}}$. The frequency and period of the shadow are the same as that of the circular motion, so

$$30 \text{ rev/min} = 0.5 \text{ rev/s} = \pi \text{ rad/s} = \omega \qquad \text{and} \qquad v = \frac{\omega}{2\pi} = \frac{\pi}{2\pi} = \underline{0.5 \text{ Hz}}$$

Then

$$T = \frac{1}{v} \qquad \text{and} \qquad T = \frac{2\pi}{\omega} = \frac{2\pi}{\pi} = \underline{2 \text{ s}}$$

14.18 A ball moves in a circular path of 0.15-m diameter with a constant angular speed of 20 rev/min. Its shadow performs simple harmonic motion on the wall behind it. Find the acceleration and speed of the shadow (**a**) at a turning point of the motion, (**b**) at the equilibrium position, and (**c**) at a point 6 cm from the equilibrium position.

▌ Proceeding as in Prob. 14.17, $R = 0.075$ m; $\omega = (20/60) \times 2\pi = (2\pi/3)$ rad/s. Then from the general formulas for SHM,
(**a**) $\quad a = a_{max} = \omega^2 R = (4\pi^2/9)(0.075) = \underline{0.329 \text{ m/s}^2} \qquad v = 0.$

(**b**) $\quad a = 0 \qquad v = \omega R = \frac{2\pi}{3}(0.075) = \underline{0.157 \text{ m/s}}$

(**c**) $\quad |a| = \omega^2 y = \frac{4\pi^2}{9}(0.06) = \underline{0.263 \text{ m/s}^2}$

$$|v| = \omega \sqrt{(A^2 - y^2)} = \frac{2\pi}{3}\sqrt{[(0.075)^2 - (0.06)^2]} = \frac{2\pi}{3}(0.045) = \underline{0.0942 \text{ m/s}}$$

14.19 A block of mass 2 kg hangs from a spring of force constant $k = 800$ N/m. The block is pulled 20 cm from equilibrium and released. (**a**) What are the amplitude, angular frequency, and period of the motion. (**b**) What are the velocity and acceleration of the block when it is 12 cm from equilibrium?

▌ (**a**) Since the block is released from rest $A = \underline{20 \text{ cm}}$. The angular frequency is

$$\omega = \sqrt{\frac{k}{m}} = \sqrt{\frac{800 \text{ N/m}}{2 \text{ kg}}} = \underline{20 \text{ rad/s}}$$

(**b**) $a = -\omega^2 y$. For downward positive $y = 12$ cm when it is below equilibrium and $a = -4800 \text{ cm/s}^2$, that is, pointed upward. Above equilibrium $y = -12$ cm and $a = 4800 \text{ cm/s}^2$, that is, pointed downward. To obtain the velocity we use $v = \pm \omega \sqrt{A^2 - y^2}$. Then for each position, 12 cm above equilibrium or 12 cm below equilibrium, there are two velocities $v = \pm 320 \text{ cm/s}$ corresponding to passage on the way down or on the way up.

14.20 When does the block of Prob. 14.19 first reach the point 10 cm below the equilibrium point.

▌ If the moment of release is $t = 0$, then the block starts at maximum positive amplitude and we have $y = A \cos \omega t$. Setting $y = 10$ cm, and recalling that $A = 20$ cm and $\omega = 20/s$, we have $\cos(20t) = \frac{1}{2}$, or $20t = (60°)(2\pi \text{ rad}/360°) = \pi/3$ rad, for the shortest time t. Solving, $t = \underline{\pi/60 \text{ s}}$.

14.21 A block of mass 4 kg hangs from a spring of force constant $k = 400$ N/m. The block is pulled down 15 cm below equilibrium and released. (**a**) Find the amplitude, frequency, and period of the motion. (**b**) Find the kinetic energy when the block is 10 cm above equilibrium.

▌ (**a**) $A = 0.15$ m; $\qquad \omega = \sqrt{\dfrac{k}{m}} = 10$ rad/s; $\qquad f = \dfrac{\omega}{2\pi} = \dfrac{5}{\pi}$ Hz; $\qquad T = \dfrac{1}{f} = \dfrac{\pi}{5}$ s.

(**b**) $\frac{1}{2}mv^2 + \frac{1}{2}kx^2 = \frac{1}{2}kA^2$ from conservation of mechanical energy. $\frac{1}{2}mv^2 = \frac{1}{2}k(A^2 - x^2) = \underline{2.5 \text{ J}}$.

14.22 Refer to Prob. 14.21. How long does it take the block to go from 12 cm below equilibrium (on the way up) to 9 cm above equilibrium?

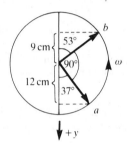

Fig. 14-3

▌ This problem can be solved by examining the reference circle, Fig. 14-3. The block's motion can be simulated by the shadow on the vertical diameter of a particle moving with uniform circular motion on the circle of radius $A = 15$ cm, with angular speed $\omega = 10$ rad/s. Let a be the point on the circle where the shadow is at $y = 12$ cm and b be the point where the shadow is at $y = -9$ cm. Since $12 = 15 \cos 37°$ and $9 = 15 \cos 53°$, the angles are as shown. In moving from $a \to b$ the particle sweeps through 90°. If t_{ab} represents the elapsed time, $\omega t_{ab} = (90°)(2\pi \text{ rad}/360°) = \pi/2$ rad, and $t_{ab} = \pi/2\omega = \frac{1}{4}(2\pi/\omega) = T/4 = \underline{\pi/20 \text{ s}}$. (Or, directly, 90° is one-quarter of a circle, so $t_{ab} = \frac{1}{4}T$.) t_{ab}, of course, also represents the time for the shadow to move from 12 cm below to 9 cm above the equilibrium position, and is the time of interest.

14.23 A body of mass 36 g moves with SHM of amplitude $A = 13$ cm and period $T = 12$ s. At time $t = 0$ the displacement, x, is $+13$ cm. (**a**) Find the velocity when $x = 5$ cm. (**b**) Find the force acting on the body when $t = 2$ s.

▌ (**a**) $v = \pm\omega\sqrt{A^2 - x^2}$, where $A = 13$ cm, $x = 5$ cm, and $\omega = 2\pi/T = \pi/6$ rad/s. Solving, $v = \pm 2\pi$ cm/s $= \underline{\pm 6.28 \text{ cm/s}}$. (**b**) Since $x = A$ at $t = 0$, we have $x = A \cos \omega t$. At $t = 2$ s, $x = (13 \text{ cm}) \cos(\pi/3 \text{ rad}) = (13 \text{ cm})(0.5) = 6.5$ cm. Since $F = ma$ and $a = -\omega^2 x$, we have $F = -(36 \text{ g})(\pi/6 \text{ rad/s})^2(6.5 \text{ cm}) = \underline{-64 \text{ dyn}}$.

14.24 Refer to Prob. 14.23. Find the shortest time of passage from $x = +6.5$ cm to $x = -6.5$ cm.

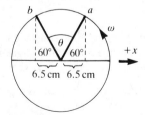

Fig. 14-4

▌ Following the reasoning of Prob. 14.22, we now consider the shadow on a horizontal diameter of a particle moving with $\omega = \pi/6$ rad/s on a circle of radius $A = 13$ cm (see Fig. 14-4). The time of interest, t_{ab}, is the time it takes to sweep through angle θ, which clearly equals 60°. Since this is precisely one-sixth of a complete circle, $t_{ab} = T/6 = \underline{2 \text{ s}}$. (Note, from Prob 14.23(b), that the time from $x = 13$ cm to $x = 6.5$ cm is 2s. The time from $x = 13$ cm to $x = 0$ is $T/4 = 3$ s. Then by subtraction the time from $x = 6.5$ cm $\to x = 0$ is 1 s. By symmetry, the time from $x = 6.5$ cm $\to x = -6.5$ cm is twice that, or $\underline{2 \text{ s}}$ in agreement with the reference circle result.)

14.25 A body of mass 100 g hangs from a long spiral spring. When pulled down 10 cm below its equilibrium position and released, it vibrates with a period of 2 s. (*a*) What is its velocity as it passes through the equilibrium position? (*b*) What is its acceleration when it is 5 cm above the equilibrium position?

▮ (*a*) At equilibrium $v = v_{max} = \omega A$, where $\omega = 2\pi/T = \pi$ rad/s, and $A = 10$ cm. Then $v_{max} = 10\pi$ cm/s.
(*b*) Acceleration $a = -\omega^2 x$. Assuming that downward is positive, $x = -5$ cm and $a = 5\pi^2$ cm/s². Note that the mass did not enter this problem.

14.26 When the body in Prob. 14.25 is moving upward, how long a time is required for it to move from a point 5 cm below its equilibrium position to a point 5 cm above it?

▮ If the system is released from rest at $t = 0$, the equation of motion is $y = A \cos \omega t = (10\text{ cm}) \cos \pi t$. We are interested in the time to go from $y_1 = +5$ cm to $y_2 = -5$ cm. Then

$$5 = 10 \cos \pi t_1 \quad \text{and} \quad \pi t_1 = 60°\left(\frac{2\pi \text{ rad}}{360°}\right) = \frac{\pi}{3}\text{ rad} \quad \text{or} \quad t_1 = \frac{1}{3}\text{ s}$$

$$-5 = 10 \cos \pi t_2 \quad \text{and} \quad \pi t_2 = 120°\left(\frac{2\pi \text{ rad}}{360°}\right) = \frac{2\pi}{3}\text{ rad} \quad \text{or} \quad t_2 = \frac{2}{3}\text{ s}$$

Then the time of interest $= t_2 - t_1 = \frac{1}{3}$ s. Equivalently we could note that the associated point on the reference circle (see, e.g., Prob 14.24) rotates through $120° - 60° = 60° = \frac{1}{6}$ circle. Therefore the time of interest $= \frac{1}{6}T = \frac{1}{3}$ s.

14.27 For the motion shown in Fig. 14-5, what are the amplitude, period, and frequency?

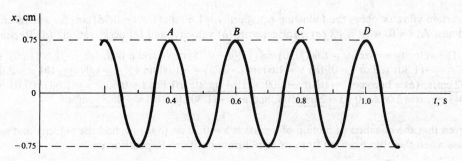

Fig. 14-5

▮ The amplitude is the maximum displacement from the equilibrium position and so it is 0.75 cm.
The period is the time for one complete cycle, the time from A to B, for example. Therefore the period is 0.20 s. The frequency is given by

$$f = \frac{1}{T} = \frac{1}{0.20\text{ s}} = 5\text{ Hz}$$

14.28 For the vibrational motion graphed in Fig. 14-6, find (*a*) amplitude, (*b*) frequency, (*c*) period. (*d*) Write a numerical equation of the form $y = y_0 \sin(2\pi ft + \theta_0)$ for it.

▮ (*a*) The amplitude is 4.0 cm. (*b*) There is 1 complete cycle in 1.2 s, so $f = 1/1.2 = 0.83$ Hz. (*c*) The period is 1.2 s. (*d*) The initial phase angle is $\pi/2$, because $y = y_0$ when $t = 0$; so $y = 4.0 \sin(5.2t + \pi/2)$ cm.

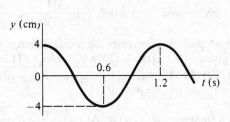

Fig. 14-6

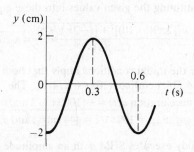

Fig. 14-7

14.29 Rework Prob. 14.28 for the motion of Fig. 14-7.

▮ (*a*) y_0 is the amplitude, 2.0 cm. (*b*) $f = 1/$time for one complete cycle $= 1/0.6 = 1.67$ Hz. (*c*) 0.6 s.
(*d*) $y = 2.0 \sin (10.5t - \pi/2)$ cm. Observe that $y(0) = -2.0$ cm, as required.

14.30ᶜ The equation of motion for a mass at the end of a particular spring is $y = 0.30 \cos 0.50t$ m. Find the displacement, velocity, and acceleration of the mass at (*a*) $t = 0$ and (*b*) $t = 3.0$ s.

▮ From $v = dy/dt = -0.15 \sin (0.50t)$ m/s, $a = dv/dt = -0.075 \cos (0.50t)$ m/s²; so (*a*) $y = 0.30$ m, $v = 0$, and $a = -0.075$ m/s². (*b*) For $t = 3.0$ s, $y = 0.021$ m, $v = -0.150$ m/s, and the acceleration $a = -0.0053$ m/s².

14.31ᶜ A particle attached to a spring undergoes SHM. The maximum acceleration of the particle is 18 m/s² and the maximum speed is 3 m/s. Find (*a*) the frequency of the particle's motion, and (*b*) the amplitude.

▮ (*a*) The equation of motion of the particle is $x = x_0 \cos (\omega t + \theta_0)$. Thus, using dot notation for time derivatives,

$$\dot{x} = -x_0\omega \sin (\omega t + \theta_0) \quad \text{and} \quad \ddot{x} = -x_0\omega^2 \cos (\omega t + \theta_0)$$

from which
$$\dot{x}_{max} = x_0\omega \quad \text{and} \quad \ddot{x}_{max} = x_0\omega^2$$

Thus, $\omega = \ddot{x}_{max}/\dot{x}_{max}$, and

$$f = \frac{\omega}{2\pi} = \frac{\ddot{x}_{max}}{2\pi\dot{x}_{max}} = \frac{18}{2\pi(3)} = 0.95 \text{ Hz}$$

(*b*)
$$x_0 = \frac{\dot{x}_{max}}{\omega} = \frac{3}{6} = 0.5 \text{ m}$$

14.32ᶜ A certain vibrator obeys the following equation: $y = 1.60 \sin (1.30t - 0.75)$ cm, for t in seconds and angles in radians. At $t = 0$, what is its (*a*) displacement, (*b*) velocity, and (*c*) acceleration? (*d*) Repeat for $t = 0.60$ s.

▮ The velocity $v = dx/dt = 1.60(1.30) \cos (1.30t - 0.75)$ cm/s and a is $dv/dt = -1.60(1.30)^2 \sin (1.30t - 0.75)$ cm/s² $= -(1.30)^2 y$. For $t = 0$: (*a*) $y = 1.60 \sin (-0.75) = -1.60 \sin 43° = -1.09$ cm. (*b*) $v = 2.08 \cos (-43°) = 1.52$ cm/s. (*c*) a becomes $-(1.30)^2(-1.09) = 1.84$ cm/s². (*d*) For $t = 0.60$ s: $y = 1.60 \sin 0.03 \approx 1.60(0.03) = 0.048$ cm, $v = 2.08 \cos 0.03 \approx 2.08$ cm/s, and a is $-(1.30^2)(0.048) = -0.081$ cm/s².

14.33ᶜ Given that the equation of motion of a mass is $x = 0.20 \sin (3.0t)$ m, find the velocity and acceleration of the mass when the object is 5 cm from its equilibrium position. Repeat for $x = 0$.

▮ Use $v = dx/dt = 0.60 \cos 3.0t$ and $a = dy/dt = -1.8 \sin 3.0t$. Now when $x = 0.05$, $\sin 3.0t = 0.05/0.2 = 0.25$ and so $3.0t = 14.5°$. Using this value in the expressions for v and a gives $v = 0.58$ m/s and $a = -0.45$ m/s². Repeating for $x = 0$, we have $v = 0.60$ m/s and $a = 0$.

14.34 An object is executing SHM with a frequency of 5.00 Hz. At $t = 0$ s its displacement is $x(0) = 10.0$ cm and its velocity is $v(0) = -314$ cm/s. (*a*) Obtain analytical expressions for the object's displacement $x(t)$, velocity $v(t)$, acceleration $a(t)$. (*b*) Find the maximum values of the object's displacement $x(t)$, velocity $v(t)$, acceleration $a(t)$.

▮ (*a*) Using the general expression $x(t) = A \cos (2\pi\nu t + \delta)$, we find $v(t) = -2\pi\nu A \sin (2\pi\nu t + \delta)$. Therefore $x(0) = A \cos \delta$ and $v(0) = -2\pi\nu A \sin \delta$. Solving for A and δ, we find

$$A = \frac{\sqrt{[2\pi\nu x(0)]^2 + [v(0)]^2}}{2\pi\nu} \quad \text{and} \quad \delta = \tan^{-1}\left[\frac{-v(0)}{2\pi\nu x(0)}\right]$$

Substituting the given values into these equations, we have

$$A = \frac{\sqrt{[(2\pi)(5)(10)]^2 + [(-2\pi)(50)]^2}}{(2\pi)(5)} = \sqrt{200} = 10\sqrt{2} = 14.1 \text{ cm} \quad \text{and} \quad \delta = \tan^{-1}\left[\frac{314}{(2\pi)(5)(10)}\right] = \frac{\pi}{4} \text{ rad}$$

since the initial conditions imply that both $\sin \delta$ and $\cos \delta$ are positive. Therefore the displacement is given by $x(t) = (14.1 \text{ cm})[\cos (10\pi t + \pi/4)]$. The velocity is given by $v(t) = -(100\pi\sqrt{2} \text{ cm/s})[\sin (10\pi t + \pi/4)]$. The acceleration $a(t) = -(1000\pi^2\sqrt{2} \text{ cm/s}^2)[\cos (10\pi t + \pi/4)]$. (*b*) The maximum values are $x_{max} = 10\sqrt{2} = 14.1$ cm, $v_{max} = 100\pi\sqrt{2} = 444$ cm/s, and $a_{max} = 1000\pi\sqrt{2} = 1.40 \times 10^4$ cm/s².

14.35 A body executes SHM with an amplitude of 2.00 cm and a frequency of 3.00 Hz. At $t = 0$ s, the displacement is $x(0) = 0$ cm and the velocity $v(0)$ is positive. (*a*) Obtain analytical expressions for the displacement $x(t)$, the velocity $v(t)$, the acceleration $a(t)$. (*b*) Evaluate the expressions found in part *a* for $t = 50$ ms.

(a) Using $x(t) = A \cos(2\pi v t + \delta)$, we have $v(t) = -2\pi v A \sin(2\pi v t + \delta)$. Since $x(0) = 0$, we have $\delta = \pi/2$ or $3\pi/2$. Because $v(0) > 0$, we must have $\delta = 3\pi/2$. Then, with the given values $A = 2.00$ cm and $v = 3.00$ Hz, we obtain $x(t) = (2.00 \text{ cm}) \cos(6.00\pi t + 3\pi/2)$, $v(t) = (-12.0\pi \text{ cm/s}) \sin(6.00\pi t + 3\pi/2)$, and $a(t) = (-72.0\pi^2 \text{ cm/s}^2) \cos(6.00\pi t + 3\pi/2)$. **(b)** At $t = 0.050$ s, $6.00\pi t + 3\pi/2 = 1.80\pi = 5.65$ rad. Therefore, we find $x = (2.00 \text{ cm}) \cos(1.80\pi) = \underline{1.62 \text{ cm}}$, $v = (-12.0\pi \text{ cm/s}) \sin(1.80\pi) = \underline{22.2 \text{ cm/s}}$, and $a = (-72.0\pi^2 \text{ cm/s}^2) \cos(1.80\pi) = \underline{-575 \text{ cm/s}^2}$.

14.36 A 0.2-kg mass suspended from a spring describes a simple harmonic motion with a period T of 3 s and amplitude R of 10 cm. At $t = 0$ the mass passes upward through the equilibrium position. **(a)** Find the force constant k of the spring. **(b)** Find the displacement, velocity, and acceleration of the mass when $t = 1$ s.

(a)
$$4\pi^2 v^2 = \frac{k}{m} \qquad k = 4\pi^2 v^2 m = \left(\frac{4\pi^2}{9}\right) 0.2 = \underline{0.88 \text{ N/m}}$$

(b) Let us choose upward as positive. $y = R \sin 2\pi v t$, so $y = 0$ and $v > 0$ at $t = 0$. Then at $t = 1$ s

$$y = R \sin 2\pi v t = 0.10 \sin\left[\left(\frac{2\pi}{3}\right)(1)\right] = 0.10 \sin 120° = \underline{0.0866 \text{ m}}$$

$$v = 2\pi v R \cos 2\pi v t = \left(\frac{2\pi}{3}\right)(0.10) \cos 120° = \underline{-0.105 \text{ m/s}}$$

$$a = -4\pi^2 v^2 R \sin 2\pi v t = -\left(\frac{4\pi^2}{9}\right)(0.10) \sin 120° = \underline{-0.38 \text{ m/s}^2}$$

14.37 Refer to Prob. 14.36. Verify that the sum of the potential and kinetic energies at $t = 1$ s is equal to $\frac{1}{2}kR^2$.

$$\tfrac{1}{2}kR^2 \overset{?}{=} \tfrac{1}{2}ky^2 + \tfrac{1}{2}mv^2$$
$$\tfrac{1}{2}(0.88)(0.1)^2 \overset{?}{=} \tfrac{1}{2}(0.88)(0.0866)^2 + \tfrac{1}{2}(0.2)(0.105)^2$$
$$4.4 \times 10^{-3} \text{ J} = (3.3 \times 10^{-3} + 1.1 \times 10^{-3}) \text{ J}$$

14.38 A 50-g mass vibrates in SHM at the end of a spring. The amplitude of the motion is 12 cm and the period is 1.70 s. Find: **(a)** the frequency, **(b)** the spring constant, **(c)** the maximum speed of the mass, **(d)** the maximum acceleration of the mass, **(e)** the speed when the displacement is 6 cm, **(f)** the acceleration when $x = 6$ cm.

(a)
$$f = \frac{1}{T} = \frac{1}{1.70 \text{ s}} = \underline{0.588 \text{ Hz}}$$

(b) Since $T = 2\pi\sqrt{m/k}$,
$$k = \frac{4\pi^2 m}{T^2} = \frac{4\pi^2(0.050 \text{ kg})}{(1.70 \text{ s})^2} = \underline{0.68 \text{ N/m}}$$

(c)
$$v_0 = x_0\sqrt{\frac{k}{m}} = (0.12 \text{ m})\sqrt{(0.68 \text{ N/m})/(0.050 \text{ kg})} = \underline{0.44 \text{ m/s}}$$

(d) From $a = -(k/m)x$ it is seen that a has maximum magnitude when x has maximum magnitude, that is, at the endpoints $x = \pm x_0$. Thus,
$$a_0 = \frac{k}{m}x_0 = \frac{0.68 \text{ N/m}}{0.050 \text{ kg}}(0.12 \text{ m}) = \underline{1.63 \text{ m/s}^2}$$

(e) From $|v| = \sqrt{(x_0^2 - x^2)(k/m)}$,
$$|v| = \sqrt{(0.12^2 - 0.06^2)\frac{0.68}{0.050}} = \underline{0.38 \text{ m/s}}.$$

(f)
$$a = -\frac{k}{m}x = -\frac{0.68 \text{ N/m}}{0.050 \text{ kg}}(0.06 \text{ m}) = \underline{-0.82 \text{ m/s}^2}$$

14.39 A spring-mass oscillator has a total energy E_0 and an amplitude x_0. **(a)** How large will K and U be for it when $x = \frac{1}{2}x_0$? **(b)** For what value of x will $K = U$?

The total energy is $E_0 = U + K$, or $kx_0^2/2 = kx^2/2 + K$. When $x = x_0/2$, $U = kx_0^2/8 = E_0/4$ and $K = E_0 - U = 3E_0/4$. Now, $U = K$ at $U = E_0/2 = kx_0^2/4 = kx^2/2$, therefore $\underline{x = x_0/2^{1/2}}$.

14.40 The energy of recoil of a rocket launcher of mass $m = 4536$ kg is absorbed in a recoil spring. At the end of the recoil, a damping dashpot engages so that the launcher returns to the firing position without any oscillation (critical damping). The launcher recoils 3 m with an initial speed of 10 m/s. Find the recoil spring constant (k) and the dashpot's coefficient for critical damping ($b = 2\sqrt{mk}$).

❚ To find the spring constant we may make use of the conservation of energy: $K + U_{\text{elastic}} = $ constant.

$$\frac{1}{2}mv_0^2 + 0 = 0 + \frac{1}{2}kx_{\text{max}}^2 \qquad k = \frac{mv_0^2}{x_{\text{max}}^2} = \frac{(4536)(10^2)}{3^2} = \underline{50.4 \text{ kN/m}}$$

The coefficient of critical damping is $b = 2\sqrt{mk} = 2\sqrt{(4536)(50\,400)} = \underline{30.24 \text{ kN} \cdot \text{s/m}}$.

14.41 Show that the combined spring energy and gravitational energy for a mass m hanging from a light spring of force constant k can be expressed as $\frac{1}{2}ky^2$, where y is the distance above or below the equilibrium position. (Compare Prob. 14.10.)

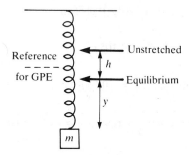

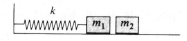

Fig. 14-8   **Fig. 14-9**

❚ We locate the reference level for gravitational potential energy midway between the unstretched and equilibrium positions (see Fig. 14-8); then the total potential energy is given by

$$U = \frac{1}{2}k(h+y)^2 - mg\left(\frac{h}{2} + y\right)$$

But the equilibrium position is defined by $mg = kh$; so

$$U = \frac{1}{2}k(h^2 + 2hy + y^2) - kh\left(\frac{h}{2} + y\right) = \frac{1}{2}ky^2$$

Since the choice of any other reference level for GPE changes U by a constant that can be discarded, our result is quite general.

14.42 The masses in Fig. 14-9 slide on an essentially frictionless table. m_1 but not m_2, is fastened to the spring. If now m_1 and m_2 are pushed to the left so that the spring is compressed a distance d, what will be the amplitude of the oscillation of m_1 after the spring system is released?

❚ Mass m_2 shoots off, carrying K away from the system. To find its speed, U_s of spring = maximum K_t of masses. Then $(kd^2)/2 = (m_1 + m_2)(v^2/2)$ giving $v^2 = (kd^2)/(m_1 + m_2)$. With m_1 alone on the spring, the energy balance equation becomes maximum $U_s = $ maximum K_t of m_1, so

$$\tfrac{1}{2}kA^2 = \frac{1}{2}m_1v^2 = \frac{1}{2}kd^2\frac{m_1}{m_1 + m_2} \qquad A = d\sqrt{\frac{m_1}{m_1 + m_2}}$$

Note that the amplitude is independent of the value of k.

14.2 SHM OF PENDULUMS AND OTHER SYSTEMS

14.43 Find the length (in meters) of a pendulum which has a period of 2.4 s.

❚ $T = 2\pi\sqrt{l/g}$. Thus $(2.4)^2 = (4\pi^2)(l/9.8)$, and $l = \underline{1.43 \text{ m}}$.

14.44 The period of a simple pendulum 35.90 cm long is $T = 1.200$ s. Find the value of g at this location.

❚
$$T = 2\pi\sqrt{\frac{l}{g}} \qquad T^2 = 4\pi^2\left(\frac{l}{g}\right) \qquad g = 4\pi^2\left(\frac{l}{T^2}\right) = \frac{4\pi^2(35.90)}{(1.200)^2} = \underline{984 \text{ cm/s}^2}$$

14.45 A certain pendulum clock keeps good time on the earth. If the same clock were placed on the moon, where objects weigh only one-sixth as much as on earth, how many seconds will the clock tick out in an actual time of 1 min?

▌ Because $2\pi f = (g/L)^{1/2}$, frequency on the moon is $1/(6^{1/2}) = 0.408$ times the frequency on the earth. The number of seconds ticked out will be $0.408(60) = \underline{24.5}$.

14.46 A simple pendulum 60 cm long has a 200-g ball. The pendulum is pulled aside 15° and released. If the timing clock is started just as the ball moves through its lowest position, write the equation of motion for the pendulum in terms of the pendulum angle in degrees.

▌ The amplitude is 15°, while $2\pi f = (g/L)^{1/2}$, which leads to $2\pi f = 4.04/s$. Since $\theta = 0$ when $t = 0$, this is a sine function. Therefore, $\theta = \underline{15 \sin 4.04t}$ degrees.

14.47 As shown in Fig. 14-10, a uniform bar is suspended in a horizontal position by a vertical wire attached to its center. When a torque of 5 N · m is applied to the bar as shown, the bar moves through an angle of 12°. If the bar is then released, it oscillates as a torsion pendulum with a period of 0.5 s. Determine its moment of inertia.

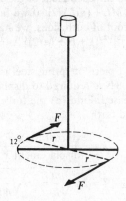

Fig. 14-10

▌ For a torsion pendulum, $T = 2\pi\sqrt{I/K}$. In our case the torsion constant

$$K = \frac{\text{external torque}}{\text{resulting angular displacement}} = \frac{5\,\text{N}\cdot\text{m}}{(12\,\text{deg})\left(\dfrac{2\pi\,\text{rad}}{360\,\text{deg}}\right)} = 23.9\,\text{N}\cdot\text{m/rad}$$

We then have, after squaring the equation for T and rearranging,

$$I = \left(\frac{T}{2\pi}\right)^2 K = \left(\frac{0.5\,\text{s}}{2\pi}\right)^2 (23.9\,\text{N}\cdot\text{m}) = \underline{0.151\,\text{kg}\cdot\text{m}^2}$$

14.48 Find the period of motion for a torsional pendulum consisting of a uniform 400-g sphere, of radius 8.0 cm, suspended from a torsion wire 20 ft long. The wire twists 90° when a torque of 0.40 N · m is applied to it.

▌ We know that $2\pi f = (k/I)^{1/2}$ and that $\tau = -k\theta$ which leads to $k = 0.40/0.5\pi = 0.25$. $I = 2mr^2/5 = 1.02 \times 10^{-3}\,\text{kg}\cdot\text{m}^2$. Then $T = 1/f = \underline{0.40\,\text{s}}$. (The length of a *torsional* pendulum is irrelevant.)

14.49 Find the frequency of oscillation when a meterstick is hung as a compound pendulum with the pivot at its 90-cm mark. Repeat for the pivot at the 50.10-cm mark.

▌ For a pendulum, $2\pi f = [(MgL)/I]^{1/2}$, where L is the distance from the axis of rotation to the center of mass, I is the moment of inertia about the axis of rotation, and M is the mass. For this case $L = 0.40$ m and $I = I_{cm} + M(0.40)^2$ from the parallel-axis theorem. Since $I_{cm} = [M(1.0)^2]/12$, one has that $I = 0.243M$. Then $f = \underline{0.64\,\text{Hz}}$. Similarly, for $L = 0.0010$ m, $I = 0.0833M$ and $f = \underline{0.055\,\text{Hz}}$.

14.50 As shown in Fig. 14-11, a uniform disk of radius R and mass M is attached to the end of a uniform rigid rod of length L and mass m. When the disk is suspended from a pivot as shown, what will be its period of motion?

▌ The equation of motion is $\tau = I\alpha$ where τ is the external torque, I is the moment of inertia, α is the

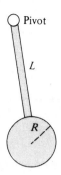

Pivot

L

R

Fig. 14-11

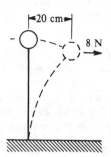

|←20 cm→|

8 N

Fig. 14-12

angular acceleration, and τ and I are about the pivot point. The contribution to τ is due to the rod, $-mg(L/2)\sin\theta$, and the disk, $-Mg(R+L)\sin\theta$ where θ is a small angular displacement from the vertical. Similarly $I = I_{rod} + I_{disk} = [(mL^2)/3] + [(MR^2)/2 + M(R+L)^2]$. For small θ, $\sin\theta \approx \theta$, so the equation of motion is $-g[MR + ML + (mL)/2]\theta = [(mL^2)/3 + (MR^2)/2 + MR^2 + 2MRL + ML^2]\alpha$. Solving for α and noting that $(2\pi f)^2 = (2\pi/T)^2$ is then the coefficient of the θ term, we have $T^2 = [(2\pi)^2/g][(mL^2)/3 + (3MR^2)/2 + 2MRL + ML^2]/[MR + ML + (mL)/2]$. If we factor L^2M from the numerator and LM from the denominator and take the square root of both sides, we have, with $a \equiv m/M$ and $b \equiv R/L$, the period $T = 2\pi(L/g)^{1/2}[a/3 + b^2/2 + (1+b)^2]^{1/2}/[1 + b + (a/2)]^{1/2}$.

14.51 As shown in Fig. 14-12, a long light piece of spring steel is clamped at its lower end and a 2-kg ball is fastened to its top end. A force of 8 N is required to displace the ball 20 cm to one side as shown. Assume the system to undergo SHM when released. Find (a) the force constant of the spring and (b) the period with which the ball will vibrate back and forth.

▌ (a)
$$k = \frac{\text{external force } F_{ext}}{\text{displacement } x} = \frac{8\,\text{N}}{0.20\,\text{m}} = \underline{40\,\text{N/m}}$$

(b)
$$T = 2\pi\sqrt{\frac{m}{k}} = 2\pi\sqrt{\frac{2\,\text{kg}}{40\,\text{N/m}}} = \underline{1.40\,\text{s}}$$

14.52 A rectangular block of wood is floating in a large pool of water. From Archimedes' principle we know that the water exerts an upward force on the bottom face of the block, of magnitude is $Ad\rho g$, where A is the area of that face, d is its depth beneath the surface of the water, ρ is the density of water, and g is gravitational acceleration. (a) The mass of the block is m. Find the value of d for which the block is in equilibrium. (b) Find an expression that describes the net force acting on the block for values of d that differ from the equilibrium value. Use it to show that the equilibrium is stable. (c) Show that if the block is depressed below its equilibrium depth (but not beneath the surface of the water) and then released, it will execute harmonic oscillations. (d) Determine the frequency of the oscillations.

▌ (a) We let d_0 denote the equilibrium depth. In equilibrium, the buoyant force $Ad_0\rho g$ must exactly balance the weight mg. Therefore $Ad_0\rho g = mg$, or $\underline{d_0 = m/\rho A}$. (b) Letting the downward direction be positive, the net force is given by $F = -\rho Agd + mg = -\rho Ag(d - d_0)$. From this equation, we see that if the block is displaced downward $(d > d_0)$, the net force is upward $(F < 0)$. If the block is displaced upward $(d < d_0)$, the net force is downward $(F > 0)$. Therefore the equilibrium is stable. (c) Assuming that we can neglect the fluid motion that accompanies the block's motion, we have $ma_y = -\rho Ag(d - d_0) = -\rho Agy$, where $y = d - d_0$. Then $a_y = -(\rho Ag/m)y$, which is the equation of harmonic oscillations. (d) Referring to the previous equation, we find that the angular frequency $\omega = \sqrt{\rho Ag/m}$, so the (ordinary) frequency $= (1/2\pi)(\sqrt{\rho Ag/m})$.

14.53 Nine kilograms of mercury are poured into a glass U tube, as shown in Fig. 14-13. The tube's inner diameter is 1.2 cm and the mercury oscillates freely up and down about its position of equilibrium $(x = 0)$. Compute (a) the effective spring constant for the oscillation, and (b) the period of oscillation. One m^3 of mercury has a mass $\mu = 13\,600$ kg. Ignore frictional and surface tension effects.

▌ (a) When the mercury is displaced x meters from its equilibrium position as shown, the restoring force is the weight of the unbalanced column of mercury with height $2x$. We have

$$\text{weight} = (\text{volume})(\text{weight of 1 m}^3 \text{ of mercury}) = [(\pi r^2)(2x)](\mu g)$$

Therefore Hooke's law, in the form $F = -(2\pi r^2 \mu g)x$, applies, from which we see that the effective spring

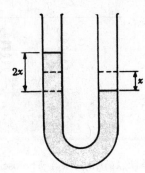

Fig. 14-13

constant for the system is

$$k = 2\pi r^2 \mu g = 2\pi (0.006 \text{ m})^2 (13.6 \times 10^3 \text{ kg/m}^3)(9.8 \text{ m/s}^2) = \underline{30 \text{ N/m}}$$

(b) The period of the vibration is

$$T = 2\pi \sqrt{\frac{m}{k}} = 2\pi \sqrt{\frac{9 \text{ kg}}{30 \text{ N/m}}} = \underline{3.4 \text{ s}}$$

14.54 A motor vehicle to carry astronauts on the surface of the moon has a spring suspension and has a natural up-and-down frequency of 0.40 Hz when fully loaded on the earth. Find its natural frequency on the moon, where it and everything in it will weigh only about one-sixth what it weighs on earth.

▌ The spring constant and the mass are independent of location, and so $\omega = 2\pi f = (k/m)^{1/2}$ is also independent of location. Because of the fact that the frequency of a spring system is independent of the initial constant force, the change in initial stretch due to the gravitational change is of no importance. $f = \underline{0.40 \text{ Hz}}$.

14.55 A pendulum has a period T for small oscillations. An obstacle is placed directly beneath the pivot, so that only the lowest one-quarter of the string can follow the pendulum bob when it swings to the left of its resting position (Fig. 14-14). The pendulum is released from rest at a certain point. How long will it take to return to that point? In answering this question, you may assume that the angle between the moving string and the vertical stays small throughout the motion.

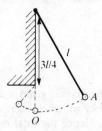

Fig. 14-14

▌ When the bob is located to the left of point O, the obstacle acts as the pivot point, and the pendulum is effectively of length $l' = l/4$. When the bob is located to the right of point O, the pendulum has its full length l. If the bob is released from point A, it will return after a total of one-half cycle of swinging as a pendulum of length l, and one-half cycle as a pendulum of length $l/4$. The elapsed time is therefore $\frac{1}{2}(2\pi\sqrt{l/g}) + \frac{1}{2}(2\pi\sqrt{l/4g})$. Since $2\pi\sqrt{l/g} = T$, the elapsed time is $T/2 + T/4 = \underline{3T/4}$.

14.56 Springs A and B have spring constants of 2000 N/m and 1000 N/m, respectively. Spring A is hung from a rigid horizontal beam and its other end is attached to an end of spring B. The pair of springs is then used to suspend a body of mass 50 kg from the lower end of spring B. What is the period of harmonic oscillation of the system?

▌ Since the springs are assumed to be massless, under the application of a given stretching force F, the compound spring stretches by distance $x = x_A + x_B$, where $x_A = F/k_A$ and $x_B = F/k_B$. Therefore

$$x = \frac{F}{k_A} + \frac{F}{k_B} = F\left(\frac{k_A + k_B}{k_A k_B}\right) = \frac{F}{k_{\text{eff}}} \quad \text{for an equivalent single spring}$$

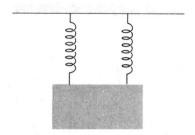

Fig. 14-15

The effective spring constant $k_{\text{eff}} = (k_A k_B)/(k_A + k_B)$. The oscillation period $T = 2\pi\sqrt{m/k_{\text{eff}}}$. With the given values, we find $k_{\text{eff}} = [(2000)(1000)]/(3000) = (2000/3)$ N/m. The period of oscillation of the 50-kg mass is therefore $2\pi\sqrt{[(50)(3)]/2000} = 2\pi\sqrt{0.075} = \underline{1.72\ \text{s}}$.

14.57 An object suspended from a spring exhibits oscillations of period T. Now the spring is cut in half and the two halves are used to support the same object, as shown in Fig. 14-15. Show that the new period of oscillation is $T/2$.

▌ Let the original spring constant be k. When the spring is cut in half, each piece is a spring whose constant $k' = 2k$. (Under the application of a given external force, each half of the spring stretches half as much as the entire spring would have stretched.) When the two pieces are used side by side to support an object, each piece exerts its restoring force on that object. Thus, the object is effectively supported by a spring of spring constant $k'' = 2k' = 4k$. Thus the suspended object will execute harmonic motion of period $T'' = 2\pi\sqrt{m/k''} = 2\pi\sqrt{m/4k} = (2\pi\sqrt{m/k})/2 = \underline{T/2}$.

14.58 Two identical springs each have $k = 20$ N/m. A 0.3-kg mass is connected to them as shown in Fig. 14-16(a) and (b). Find the period of motion for each system. Ignore friction forces.

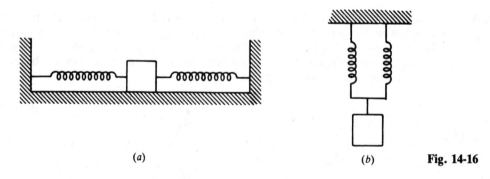

(a)　　　　　　　　　　　　　　(b)　　　**Fig. 14-16**

▌ (a) Consider what happens when the mass is given a displacement $x > 0$. One spring will be stretched x and the other will be compressed x. They will each exert a force of magnitude $(20\ \text{N/m})x$ on the mass in the direction opposite to the displacement. Hence the total restoring force is $F = -(20\ \text{N/m})x - (20\ \text{N/m})x = -(40\ \text{N/m})x$. Comparing with $F = -kx$ tells us that the system has a spring constant of $k = 40$ N/m. Hence

$$T = 2\pi\sqrt{\frac{m}{k}} = 2\pi\sqrt{\frac{0.3\ \text{kg}}{40\ \text{N/m}}} = \underline{0.54\ \text{s}}$$

(b) When the mass is displaced a distance y downward, each spring is stretched a distance y. The net restoring force on the mass is then $F = -(20\ \text{N/m})y - (20\ \text{N/m})y = -(40\ \text{N/m})y$. Comparison with $F = -ky$ shows k to be 40 N/m, the same as in (a). Hence the period in this case is also $\underline{0.54\ \text{s}}$.

14.59 The uniform stick in Fig. 14-17 has mass m and length L and is pivoted at its center. In the equilibrium position shown, the identical light springs have their natural length. (a) Show that the stick will undergo simple harmonic motion when turned through a small angle θ_0 from the position shown and released. (b) What is the frequency of the motion? (c) How fast will the tip of the stick be moving when the stick passes through the horizontal?

▌ If the stick is rotated through a small angle θ, each spring is stretched a distance $L\theta/2$. Each spring causes a torque $= (kL\theta)/2$ times $L/2$ with both torques in the same direction. The torque equation is

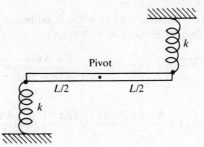

Fig. 14-17

$-2k\theta(L/2)(L/2) = I_{CM}\alpha$ where $I_{CM} = mL^2/12$; this yields $\alpha = -(6k/m)\theta$. This is the differential equation for SHM of frequency $f = (6k/m)^{1/2}/2\pi$. The solution for θ is $\theta_0 \cos 2\pi ft$, so maximum velocity $= (L/2)(2\pi f)\theta_0 = L\theta_0(1.5k/m)^{1/2}$.

14.60 From energy considerations, discuss the small motions of the system shown in Fig. 14-18. Find the natural frequency of the system, assuming no slippage of rope on disk.

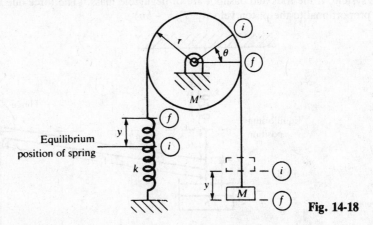

Fig. 14-18

▮ Let the (small) angle θ specify the configuration of the system. By Prob. 14.41, the potential energy of the system is $U = \frac{1}{2}ky^2 = \frac{1}{2}kr^2\theta^2$. Comparing this with $U = \frac{1}{2}K\theta^2$, we see that in effect the system executes torsional oscillations, with stiffness constant $K = kr^2$. Since the moment of inertia of the system about the axle is $I = mr^2$, where $m \equiv M + \frac{1}{2}M'$, we have at once for the natural frequency

$$\omega = \sqrt{\frac{K}{I}} = \sqrt{\frac{k}{m}}$$

14.61 Determine whether vertical SHM is possible for the system shown in Fig. 14-19(a). If so, find the natural frequency ω.

▮ The criterion for SHM is that the restoring force be proportional to the displacement *from the equilibrium position*. Because of the weight mg, the equilibrium configuration is as shown in Fig. 14-19(b), with the springs stretched to a distance $d' = \sqrt{d^2 + h^2}$.

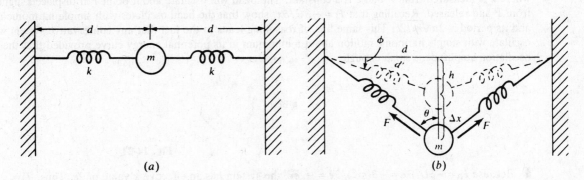

(a)   (b)

Fig. 14-19

If the mass m is displaced a distance Δx below the equilibrium position, where Δx is small compared with h and d', the force in each spring is $F = k(\sqrt{d^2 + (h + \Delta x)^2} - d')$, and so the restoring force on mass m is

$$F_{res} = -2F \cos \theta = -2k(\sqrt{d^2 + (h + \Delta x)^2} - d') \frac{h + \Delta x}{\sqrt{d^2 + (h + \Delta x)^2}} = -2kh\left(1 + \frac{\Delta x}{h}\right)\left[1 - \frac{1}{\sqrt{1 + \frac{2h(\Delta x) + (\Delta x)^2}{d'^2}}}\right]$$

$$\approx -2kh\left(1 + \frac{\Delta x}{h}\right)\left[\frac{1}{2}\frac{2h(\Delta x) + (\Delta x)^2}{d'^2}\right] \approx -2\frac{kh^2}{d'^2}(\Delta x) = -(2k \sin^2 \phi)\, \Delta x$$

where we have used the binomial expansion $(1 + u)^{-1/2} = 1 - \frac{1}{2}u + \cdots$, and then retained only terms of first order in small quantities.

Thus, SHM can occur, the effective spring constant being $k_{eff} = 2k \sin^2 \phi$, and

$$\omega = \sqrt{\frac{k_{eff}}{m}} = \sqrt{\frac{2k}{m}} \sin \phi$$

14.62^c The cranking apparatus shown in Fig. 14-20 moves with very small oscillations. Find the equation of motion of the system, if the rods and dashpot are of negligible mass. (The force due to the dashpot, F_D, is a damping force proportional to the piston velocity: $F_D = -bv$).

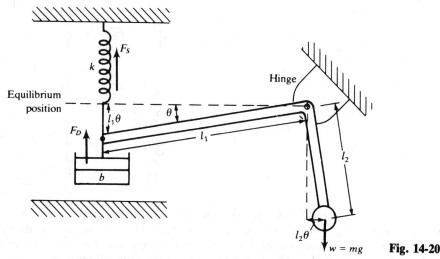

Fig. 14-20

▌ The torques of the three forces about the hinge are, for small θ,

$$\tau_S = -F_S l_1 = -kl_1^2 \theta \qquad \tau_D = -F_D l_1 = -bl_1^2 \dot{\theta} \qquad \tau_w = -mgl_2 \theta$$

The moment of inertia about the hinge is $I = ml_2^2$. Thus the equation of motion is (using dot notation for time derivatives)

$$\sum \tau = I\ddot{\theta} \qquad -(kl_1^2 + mgl_2)\theta - bl_1^2 \dot{\theta} = ml_2^2 \ddot{\theta}$$

Because of the dashpot term, this equation represents a damped angular harmonic motion.

14.63^c A bead of mass m slides on a frictionless wire as shown in Fig. 14-21. Because the shape of the wire near P can be approximated as a parabola, the potential energy of the bead is given by $U = cx^2$ in the vicinity of P, where x is measured from P and c is a constant. The bead will oscillate about point P if displaced slightly from P and released. Recalling that $F_x = -\partial U/\partial x$, show that the bead oscillates with simple harmonic motion and its period is $2\pi\sqrt{m/2c}$. This same line of reasoning leads to the fact that any conservative system will oscillate with simple harmonic motion about a minimum in its potential energy curve provided that the oscillation amplitude is small enough.

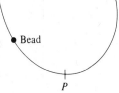

Fig. 14-21

▌ Because $F_x = -\partial U/\partial x = -\partial(cx^2)/\partial x = -2cx$, the system has an effective k value of $2c$. Thus, $T = 2\pi(m/2c)^{1/2}$.

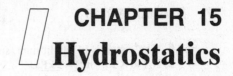

CHAPTER 15
Hydrostatics

15.1 PRESSURE AND DENSITY

15.1 A solid sphere has a radius of 1.5 cm and a mass of 0.038 kg. What is the specific gravity of the sphere?

I Use $1.5\,\text{cm} = 1.5 \times 10^{-2}\,\text{m}$ for the radius.

$$V = \frac{4}{3}\pi R^3 = \frac{4}{3}\pi(1.5 \times 10^{-2})^3 = 1.413 \times 10^{-5}\,\text{m}^3 \qquad d = \frac{m}{V} = \frac{0.038}{1.413 \times 10^{-5}} = 2690\,\text{kg/m}^3$$

$$\text{specific gravity} = \frac{2690\,\text{kg/m}^3}{1000\,\text{kg/m}^3} = \underline{2.69}$$

15.2 A certain pycnometer (a small flask used for density measurements) weighs 20.00 g when empty, 22.00 g when filled with water, and 21.76 g when filled with benzene. (**a**) Find the density of benzene. (**b**) For very accurate measurements of density, the weight of air in the empty flask must be taken into account. What mass of air fills the pycnometer?

I (**a**) Mass of water needed to fill $= 2 \times 10^{-3}$ kg. Since $p_w = 1000\,\text{kg/m}^3$, $V = 2 \times 10^{-6}\,\text{m}^3$. The mass of the same volume of benzene is 1.76×10^{-3} kg. Therefore $p = \text{mass/volume} = (1.76 \times 10^{-3})/(2 \times 10^{-6}) = \underline{880\,\text{kg/m}^3}$. (**b**) The density of air is $1.29\,\text{kg/m}^3$. Since the volume of the flask is $2 \times 10^{-6}\,\text{m}^3$, the mass of air $= pV = \underline{2.58 \times 10^{-6}\,\text{kg}}$.

15.3 The weight density of water is $62.5\,\text{lb/ft}^3$. A piece of cork has a specific gravity of 0.25 and weighs 4 lb in air. Find the volume in cubic feet of this piece of cork.

I

$$\text{Specific gravity} = \frac{\text{density of cork}}{\text{density of water}}$$

$$0.25 = \frac{\text{density of cork}}{62.5} \quad \text{and} \quad 15.6\,\text{lb/ft}^3 = \text{density of cork}$$

Next, $W = d_w V$, $4 = 15.6V$, and $V = \underline{0.26\,\text{ft}^3}$.

15.4 A liquid cannot withstand a shear stress. How does this imply that the surface of a liquid at rest must be level, that is, normal to the gravitational force?

I If the surface is not horizontal a fluid element at the surface will experience a component of gravitational force parallel to the surface. Since the adjacent liquid cannot exert a shear stress, this element will flow (as will all others) until the surface is horizontal.

15.5 A bowl of soup rests on a table in the dining car of a train. If the acceleration of the train is $g/4$ in the forward direction, what angle does the surface of the soup make with the horizontal?

I In the noninertial reference frame of the train (and the soup bowl), any mass m experiences a fictitious force of magnitude $ma = mg/4$ in a direction opposite to the horizontal forward acceleration of the train. The effective gravitational force is the resultant vector $m\mathbf{g}_{\text{eff}} = m\mathbf{g} - m\mathbf{a}$. If the soup is in equilibrium, its surface must be perpendicular to $\mathbf{g}_{\text{eff}}$ (Prob. 15.4). As shown in Fig. 15-1, this implies that the surface makes an angle $\theta = \tan^{-1}\frac{1}{4} = \underline{14.0°}$ with the horizontal, with the high side of the surface toward the rear of the train (opposite to the acceleration).

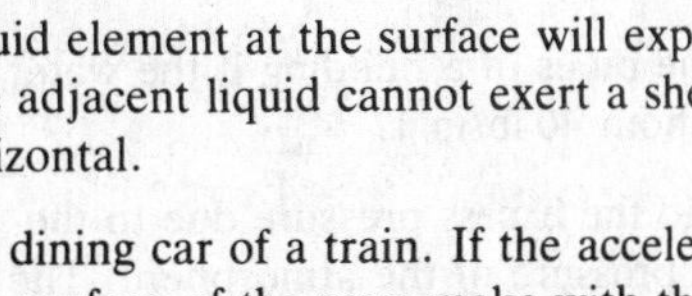

Fig. 15-1

15.6 Consider a gas confined to a container by means of a piston of area $40\,\text{cm}^2$. If a perpendicular force 20 N is exerted on the piston to keep the gas from expanding, find the gas pressure.

271

▮ The piston is in equilibrium, so the gas exerts a counterbalancing force of 20 N. Then $p = F/A =$ 20 N/$(40 \times 10^{-4} \, \text{m}^2) = \underline{5 \, \text{kPa}}$.

15.7 An 80-kg metal cylinder 2 m long and with each end of area 25 cm² stands vertically on one end. What pressure does the cylinder exert on the floor?

▮
$$p = \frac{\text{normal force}}{\text{area}} = \frac{(80 \, \text{kg})(9.8 \, \text{m/s}^2)}{25 \times 10^{-4} \, \text{m}^2} = 3.14 \times 10^5 \, \text{N/m}^2 = \underline{314 \, \text{kPa}}$$

15.8 Atmospheric pressure is about 100 kPa. How large a force does the air in a room exert on the inside of a window pane that is 40 cm × 80 cm?

▮ The atmosphere exerts a force normal to any surface placed in it. Consequently, the force on the window pane is perpendicular to the pane and is given by $F = pA = (100 \, \text{kN/m}^2)(0.40 \times 0.80 \, \text{m}^2) = \underline{32 \, \text{kN}}$.

15.9 At a height of 10 km (33 000 ft) above sea level, atmospheric pressure is about 210 mm of mercury. What is the resultant normal force on a 600 cm² window of an airplane flying at this height? Assume hydrostatic conditions and a pressure inside the plane of 760 mm of mercury. The density of mercury is 13 600 kg/m³.

▮ The resultant normal force (in the outward direction) $= N_{\text{outward}} - N_{\text{inward}}$, or $\rho_{\text{Hg}}gh_1A - \rho_{\text{Hg}}gh_2A =$ $(13\,600 \, \text{kg/m}^3)(9.8 \, \text{m/s}^2)(0.76 \, \text{m} - 0.21 \, \text{m})(0.06 \, \text{m}^2) = \underline{4398 \, \text{N}}$ (almost 1000 lb).

15.10 The pressure gauge shown in Fig. 15-2 has a spring for which $k = 60 \, \text{N/m}$, and the area of the piston is 0.50 cm². Its right end is connected to a closed container of gas at a gauge pressure of 30 kPa. How far will the spring be compressed if the region containing the spring is (a) in vacuum and (b) open to the atmosphere? Atmospheric pressure is 101 kPa.

▮ (a) The pressure to the right of the piston is (30 + 101) kPa, so the compressive force is $(131 \times 10^3 \, \text{Pa})(0.50 \times 10^{-4} \, \text{m}^2) = 6.55 \, \text{N}$, then dividing by $k = 60 \, \text{N/m}$, we have $\underline{10.9 \, \text{cm}}$. (b) The force is now due to the gauge pressure acting on the piston area, the spring will compress $3.0 \times 10^4(5 \times 10^{-5})/60 = \underline{2.5 \, \text{cm}}$.

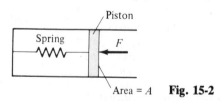

Area = A **Fig. 15-2**

15.11 The height of the mercury column in a barometer is 760 mm. Find the pressure of the atmosphere in pascals.

▮
$$p = \rho g h = (13.6 \times 10^3 \, \text{kg/m}^3)(9.80 \, \text{m/s}^2)(0.760 \, \text{m}) = \underline{1.013 \times 10^5 \, \text{Pa}}$$

15.12 How high will water rise in the pipes of a building if the water pressure gauge shows the pressure at the ground floor to be 270 kPa (about 40 lb/in²)?

▮ Water pressure gauges read the excess pressure due to the water, that is, the difference between the pressure in the water and the pressure of the atmosphere. The water pressure at the bottom of the highest column that can be supported is 270 kPa. Therefore, $p = h\rho g$ gives

$$h = \frac{p}{\rho g} = \frac{2.7 \times 10^5 \, \text{N/m}^2}{(1000 \, \text{kg/m}^3)(9.8 \, \text{m/s}^2)} = \underline{27.6 \, \text{m}}$$

15.13 Find the pressure at a depth of 10 m in water when the atmospheric pressure is that corresponding to a mercury column of height 760 mm. The densities of water and mercury are 10³ kg/m³ and 13.6 × 10³ kg/m³, respectively.

▮
$$p = p_{\text{atm}} + \rho g y = \rho_{\text{Hg}}gh_{\text{Hg}} + \rho g y = (13.6 \times 10^3)(9.8)(0.760) + (10^3)(9.8)(10) = 1.99 \times 10^5 \, \text{Pa} = \underline{199 \, \text{kPa}}$$

15.14 Atmospheric pressure is 101 kPa. How high a column of water could be supported by this pressure? Of mercury? (sp gr = 13.6.)

▮ In each case $\rho g h = 1.01 \times 10^5 \, \text{N/m}^2$. For water, $\rho = 1000 \, \text{kg/m}^3$ and $h = \underline{10.3 \, \text{m}}$; for mercury, $\rho = 13\,600 \, \text{kg/m}^3$, $h = \underline{0.76 \, \text{m}}$.

15.15 Find the ratio of a systolic blood pressure of 120 (in mmHg) to atmospheric pressure. Standard atmospheric pressure is 1.01×10^5 Pa.

▮ Pressure due to 0.120 m of mercury $= \rho g h = (13\,600)(9.8)(0.120) = 0.16 \times 10^5$ Pa. Ratio $= 0.16/1.01 = \underline{0.16}$.

15.16 If the blood vessels in a human being acted as simple pipes (which they do not), what would be the difference in blood pressure between the blood in a 1.80-m-tall man's feet and in his head when he is standing? Assume the specific gravity of blood to be 1.06.

▮ $\Delta p = \rho g h = 1060(9.8)(1.8) = \underline{18.7\ \text{kPa}}$.

15.17 What is the pressure due to the water 1 mi beneath the ocean's surface if we assume the mean density of seawater to be 1025 kg/m³? If its compressibility is the same as that of pure water (bulk modulus = 2000 MPa), by what percent has the density changed in going from the surface to this depth (1 mi = 1609 m)?

▮ $\Delta p = \rho g h = 1025(9.8)(1609) = 16.2$ MPa. By definition of the bulk modulus, $B = -\Delta p/(\Delta V/V)$. But, for a fixed mass m,

$$0 = \Delta m = \rho\,\Delta V + V\,\Delta\rho \qquad \text{whence} \qquad \frac{\Delta V}{V} = -\frac{\Delta\rho}{\rho}$$

Therefore $B = \Delta p/(\Delta\rho/\rho)$, or

$$\frac{\Delta\rho}{\rho} = \frac{\Delta p}{B} = \frac{16.2}{2000} = 0.0081 = \underline{0.81\%}$$

15.18 A cylindrical tank 3 ft in diameter and 4 ft high is filled with water whose weight density is 62.5 lb/ft³. Find the gauge pressure at the bottom of the tank.

▮ Use the equation for pressure at a depth y in a liquid. Ignore atmospheric pressure. $p_L = d_w y = 62.5\ \text{lb/ft}^3 \times 4\ \text{ft} = \underline{250\ \text{lb/ft}^2}$.

15.19 A tank contains a pool of mercury 0.30 m deep, covered with a layer of water that is 1.2 m deep. The density of water is 1.0×10^3 kg/m³ and that of mercury is 13.6×10^3 kg/m³. Find the pressure exerted by the double layer of liquids at the bottom of the tank. Ignore the pressure of the atmosphere.

▮ First find the pressure at the top of the mercury pool. For a point below the surface of the mercury this may be regarded as a source of external pressure p_{ext}. Thus

$$p_{\text{ext}} = \rho_{\text{water}}gh_{\text{water}} = (1.0 \times 10^3\ \text{kg/m}^3)(9.8\ \text{m/s}^2)(1.2\ \text{m}) = 12\ \text{kPa}$$

The pressure p_{int} exerted by the mercury column itself is found in the same manner:

$$p_{\text{int}} = \rho_{\text{merc}}gh_{\text{merc}} = (13.6 \times 10^3\ \text{kg/m}^3)(9.8\ \text{m/s}^2)(0.30\ \text{m}) = 40\ \text{kPa}$$

The total pressure at the bottom is thus $\underline{52\ \text{kPa}}$.

15.20 How high does a mercury barometer stand on a day when atmospheric pressure is 98.6 kPa?

▮
$$h = \frac{p_{\text{atm}}}{\rho_{\text{Hg}}g} = \frac{98.6 \times 10^3\ \text{N/m}^2}{(13.6 \times 10^3\ \text{kg/m}^3)(9.8\ \text{m/s}^2)} = 740\ \text{mm}$$

15.21 Two liquids which do not react chemically are placed in a bent tube, as shown in Fig. 15-3. Show that the heights of the liquids above their surface of separation are inversely proportional to their densities.

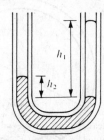

Fig. 15-3

▮ The pressure at the interface must be the same, calculated via either tube. Since both tubes are open to the atmosphere, we must have $\rho_1 gh_1 = \rho_2 gh_2$, or $h_1/h_2 = \rho_2/\rho_1$.

15.22 Assume that the two liquids in the U-shaped tube of Fig. 15-3 are water and oil. Compute the density of the oil if the water stands 19 cm above the interface and the oil stands 24 cm above the interface.

▎ We apply the result of Prob. 15.21:

$$\rho_{oil} = \left(\frac{h_w}{h_{oil}}\right)\rho_w = \frac{19\ cm}{24\ cm}\ (1000\ kg/m^3) = \underline{792\ kg/m^3}$$

15.23 A uniform glass tube is bent into a U shape such as that shown in Fig. 15-4. Water is poured into the tube until it stands 10 cm high in each tube. Benzene (sp gr = 0.879) is then added slowly to the tube on the left until the water rises 4 cm higher on the right. What length is the column of benzene when the situation is reached? (Water and benzene do not mix.)

▎ An 8-cm column of water balances the benzene column; so by Prob. 15.21,

$$h_b = \frac{h_w}{\rho_b/\rho_w} = \frac{8\ cm}{0.879} = \underline{9.1\ cm}$$

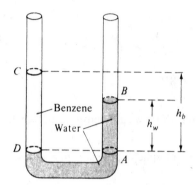

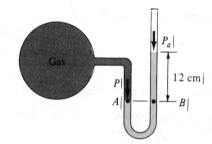

Fig. 15-4

Fig. 15-5

15.24 The manometer shown in Fig. 15-5 uses mercury as its fluid. If atmospheric pressure is 100 kPa, what is the pressure of the gas in the container shown on the left?

▎ At points A and B pressures must be equal since the fluid is not moving. Therefore $P = P_a + \rho gh = 100\ kPa + [(13.6)(9.8)(0.12)]\ kPa = \underline{116\ kPa}$.

15.25 A mercury barometer stands at 762 mm. A gas bubble, whose volume is 33 cm³ when it is at the bottom of a lake 45.7 m deep, rises to the surface. What is its volume at the surface of the lake?

▎ In terms of the weight density, ρg, of water,

$$p_{bottom} = \rho gy + p_{atm} = \rho gy + \rho\left(\frac{\rho_{Hg}}{\rho}\right)gh_{Hg} = \rho g[45.7 + (13.6)(0.762)] = 45.7\rho g + 10.4\rho g = 56.1\rho g$$

For the bubble, *Boyle's law* states that $pV = $ constant, assuming that the temperature stays fixed. Then,

$$V_{surface} = \frac{p_{bottom}}{p_{surface}}V_{bottom} = \frac{56.1\rho g}{10.4\rho g} \times 33 = \underline{178\ cm^3}$$

15.26 A small uniform tube is bent into a circle of radius r whose plane is vertical. Equal volumes of two fluids whose densities are ρ and σ ($\rho > \sigma$) fill half the circle (see Fig. 15-6). Find the angle that the radius passing through the interface makes with the vertical.

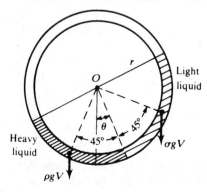

Fig. 15-6

❚ Of the external forces acting on the two fluid segments only the two weights, ρgV and σgV, have torques about the center O; the forces exerted by the container are purely radial. Thus, for equilibrium,

$$0 = \rho gVr \sin (45° - \theta) - \sigma gVr \sin (45° + \theta)$$

$$0 = \rho(\sin 45° \cos \theta - \cos 45° \sin \theta) - \sigma(\sin 45° \cos \theta + \cos 45° \sin \theta)$$

$$0 = \rho(1 - \tan \theta) - \sigma(1 + \tan \theta) \qquad \tan \theta = \frac{\rho - \sigma}{\rho + \sigma}$$

15.27 Rework Prob. 15.26 by requiring that the fluid pressures be equal at the interface.

❚ The pressure at the interface must be the same for the heavy and light liquids to have equilibrium. Referring to Fig. 15-6, we calculate the pressure at the interface due to the heavy liquid and set it equal to the pressure due to the light liquid. (The added contribution of the gas in the rest of the tube is the same for both liquids and can be ignored.) Then $\rho gh_1 = \sigma gh_2$, where h_1 and h_2 are the heights of the respective liquids above the interface. We have, from Fig. 15-6,

$$h_1 = r \cos \theta - r \cos (90° - \theta) = r(\cos \theta - \sin \theta) \qquad h_2 = r \cos \theta + r \sin \theta = r(\cos \theta + \sin \theta)$$

Substituting into the pressure equation and canceling g and r for both sides, we have $\rho(\cos \theta - \sin \theta) = \sigma(\cos \theta + \sin \theta)$, or dividing by $\cos \theta$, $\rho(1 - \tan \theta) = \sigma(1 + \tan \theta)$. Solving for $\tan \theta$, we get $\tan \theta = (\rho - \sigma)/(\rho + \sigma)$.

15.28ᶜ Water stands at a depth h behind the vertical face of a dam, Fig. 15-7(a). It exerts a resultant horizontal force on the dam, tending to slide it along its foundation, and a torque, tending to overturn the dam about the point O. Find **(a)** the horizonal force, **(b)** the torque about O, and **(c)** the height at which the resultant force would have to act to produce the same torque

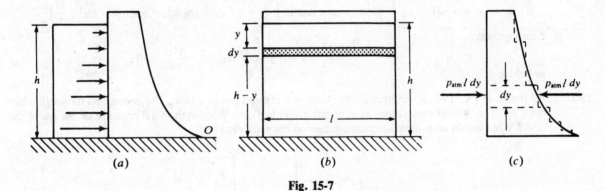

Fig. 15-7

❚ **(a)** Figure 15-7(b) is a view of the face of the dam from upstream. The pressure at depth y is $p = \rho gy$. We may neglect the atmospheric pressure, since it acts on the other side of the dam also. [The construction shown in Fig. 15-7(c) may be used to justify the neglect of atmospheric pressure.] The force against the shaded strip is $dF = p\,dA = \rho gyl\,dy$. The total force is

$$F = \rho gl \int_0^h y\,dy = \frac{\rho glh^2}{2}$$

(b) The torque of the force dF about an axis through O is, in magnitude, $d\tau = (h - y)\,dF = \rho gly(h - y)\,dy$. The total torque about O is

$$\tau = \rho gl \int_0^h y(h - y)\,dy = \frac{\rho glh^3}{6}$$

(c) If H is the height above O at which the total force F would have to act to produce this torque,

$$HF = \tau \qquad \text{or} \qquad H = \frac{\tau}{F} = \frac{\rho glh^3/6}{\rho glh^2/2} = \frac{h}{3}$$

15.29ᶜ A conical cup, $r = (b - z) \tan \alpha$, rests open-end-down on a smooth flat surface, as shown in Fig.15-8. The cup is to be filled to a height h with liquid of density ρ. What will be the lifting force on the cup?

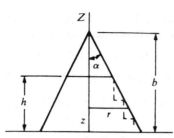

Fig. 15-8

■ Imagine that the inside surface of the cup consists of an infinite number of infinitesimal ring-shaped steps (Fig. 15-8). The pressure $p(z)$ acting on the vertical face of a step does not contribute to the lifting force, since it acts horizontally. Thus, the infinitesimal lifting force is only the pressure force on the horizontal face of the step:

$$dF_z = p(z)\,dA = \rho g(h-z)(2\pi r\,dr) = \rho g(h-z)[2\pi(b-z)\tan\alpha](-dz\tan\alpha)$$

Integrating to obtain the total lifting force, we get

$$F_z = -2\pi\rho g\tan^2\alpha\int_h^0 (h-z)(b-z)\,dz = \pi\rho g\left(bh^2 - \frac{h^3}{3}\right)\tan^2\alpha$$

15.30ᶜ   Redo Prob. 15.29 using the weight of the fluid and the force it exerts on the flat surface.

■ The *total* force exerted by the pressure of a static fluid on its container is equal to the fluid's weight. Hence, for this problem, $F_b - F_z = w$, where F_b is the downward force on the plane surface; F_z is the lifting force on the cup (by symmetry, the horizontal pressure forces on the cup cancel); and w is the weight of the liquid. Now,

$$F_b = p(0)A = \rho g h(\pi b^2\tan^2\alpha)$$

and

$$w = \rho g V = \rho g\int_0^h \pi r^2\,dz = \rho g\pi\tan^2\alpha\int_0^h (b-z)^2\,dz = (\rho g\pi\tan^2\alpha)\left(b^2 h - bh^2 + \frac{h^3}{3}\right)$$

Consequently,

$$F_z = F_b - w = (\rho g\pi\tan^2\alpha)\left(bh^2 - \frac{h^3}{3}\right)$$

as before.

15.31   As shown in Fig. 15-9, a weighted piston holds compressed gas in a tank. The piston and its weights have a mass of 20 kg. The cross-sectional area of the piston is 8 cm². What is the total pressure of the gas in the tank? What would an ordinary pressure gauge on the tank read?

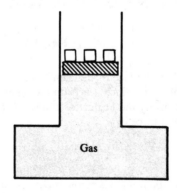

Gas

Fig. 15-9

■ The *total* pressure in the tank will be the pressure of the atmosphere (about 1.0×10^5 Pa) plus the pressure due to the piston and weights.

$$p = 1.0\times 10^5\,\text{N/m}^2 + \frac{(20)(9.8)\,\text{N}}{8\times 10^{-4}\,\text{m}^2} = 1.0\times 10^5\,\text{N/m}^2 + 2.45\times 10^5\,\text{N/m}^2 = 3.45\times 10^5\,\text{N/m}^2 = \underline{345\,\text{kPa}}$$

A pressure gauge on the tank would read the difference between the pressure inside and outside the tank: gauge reading $= 2.45\times 10^5\,\text{N/m}^2 = \underline{245\,\text{kPa}}$. It reads the pressure due to the piston and weights.

15.32   Refer to Prob. 15.31; assume that the area of the tank bottom is 20 cm². (*a*) Find the total force on the bottom of the tank and compare it with the weight of the piston plus the force of atmosphere on the piston. (The weight of the compressed air is negligible.) (*b*) How do you explain the result?

▌ (a) From Prob. 15.31, at the bottom of the tank, $p = 345$ kPa. Then $F = pA = (345 \times 10^3 \text{ N/m}^2)(0.002 \text{ m}^2) = \underline{690 \text{ N}}$. The weight of the piston is $(20 \text{ kg})(9.8 \text{ m/s}^2) = 196 \text{ N}$, to which we add the force of the atmosphere $= (1.0 \times 10^5 \text{ Pa})(0.0008 \text{ m}^2) = 80 \text{ N}$, for a total of $\underline{276 \text{ N}}$. (b) The force on the bottom of the tank is substantially more than that necessary to hold up the piston with the atmosphere pushing down. This can be understood by noting that there is additional downward force exerted on the gas by the horizontal portion of the tank adjacent to the vertical shaft. This force is just $(345 \text{ kPa})(0.0012 \text{ m}^3) = 414 \text{ N}$, which accounts for the missing force. (Compare Prob. 15.30.)

15.33 If a vessel and the liquid contained in it rotate uniformly about a vertical axis, show that the free surface of the liquid is a paraboloid (the surface formed by the rotation of a parabola about its axis).

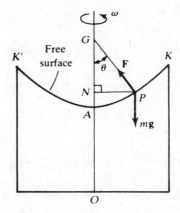

Fig. 15-10

▌ As shown in Fig. 15-10, let the surface of the liquid take the shape generated by the rotation of the curve APK about the axis of rotation OA. This surface is one of equal pressure, since it is in contact with the air, which exerts essentially the same pressure everywhere on the surface. The force F exerted on a surface element at P by the air and by the rest of the liquid is normal to the surface at P. (Otherwise the air and liquid would exert a shear, or tangential, force along the surface.) The only other force on the element is its weight, mg. The element moves in a circle of radius NP with angular speed ω. The equations of motion in the vertical and horizontal directions are then

$$\sum F_{\text{ver}} = F \cos \theta - mg = 0 \qquad \sum F_{\text{hor}} = F \sin \theta = m\omega^2 \overline{PN}$$

Combine these equations to get

$$\tan \theta = \frac{\omega^2 \overline{PN}}{g} \qquad \text{or} \qquad \overline{NG} = \frac{\overline{PN}}{\tan \theta} = \frac{g}{\omega^2} = \text{constant}$$

The subnormal of the curve AP is $\overline{NG}$ and is constant. Thus, the curve AP is a parabola, since a constant subnormal is a defining property of the parabola.

15.34ᶜ Relative to the coordinate axes shown in Fig. 15-11, show that the parabola of Prob. 15.33 has the equation $y = (\omega^2/2g)x^2$.

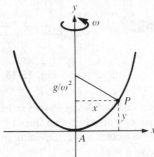

Fig. 15-11

▌ We know that the slope of the tangent to the curve at P is dy/dx; therefore the slope of the normal is $-dx/dy$. But, from Fig. 15-11, the slope of the normal is also given by $-(g/\omega^2)/x$. Hence

$$\frac{dx}{dy} = \frac{g/\omega^2}{x} \qquad \text{or} \qquad dy = \frac{\omega^2}{g} x \, dx \qquad \text{or} \qquad y = \frac{\omega^2}{2g} x^2$$

(The integration constant is zero because $y = 0$ when $x = 0$.)

15.35^c A student wishes to calculate the force with which the atmosphere presses one of the *Magdeburg hemispheres* against the other. He multiplies the atmospheric pressure, p_{atm}, by the surface area of the hemisphere, $2\pi R^2$. Why is this incorrect? What is the correct result?

❚ It is incorrect to simply multiply the pressure by the surface area, because the pressure force on each patch of the surface is directed at right angles to that patch. The various pressure forces must be added *vectorially*. In Fig. 15-12 we set up standard spherical coordinates, with the origin at the center of a hemisphere and with the polar axis (z axis) as the symmetry axis of the hemisphere. Writing the resultant force as

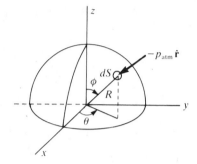

Fig. 15-12

$\mathbf{F} = F_x\hat{\mathbf{x}} + F_y\hat{\mathbf{y}} + F_z\hat{\mathbf{z}}$, we see that $F_x = F_y = 0$, by symmetry, while

$$F_z = -\iint p_{atm}\, dS \cos\phi = -p_{atm}R^2 \int_0^{2\pi} d\theta \int_0^{\pi/2} \sin\phi\cos\phi\, d\phi = -p_{atm}R^2\left(2\pi \times \frac{1}{2}\right) = -p_{atm}(\pi R^2)$$

The force is along the $-z$ axis, and its magnitude is the product of the pressure and the cross-sectional area of the hemisphere.

15.36 Obtain the result of Prob. 15.35 by physical reasoning (without calculus).

❚ Suppose that the (weightless) hemisphere of Fig. 15-12 had a flat base. Since the closed body would be in a state of equilibrium in the atmosphere, force on curved surface = −(force on flat surface) = $-p_{atm}(\pi R^2)$. Now, removing the flat base cannot change the force on the curved surface.

15.37 When at rest, a liquid stands at the same level in the tubes shown in Fig. 15-13(a). But, as indicated, a height differential occurs when the system is given an acceleration a toward the right. Show that $h = aL/g$.

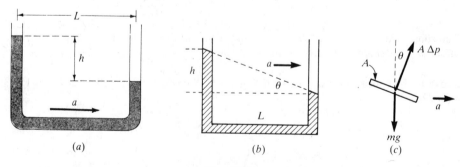

Fig. 15-13

❚ The water surface while accelerated will lie as shown in Fig. 15-13(b), with lines of constant pressure parallel to the surface. Take a disk of water between lines of constant pressure. From Fig. 15-13(c), Newton's equations, $A\,\Delta p \sin\theta = ma$ and $A\,\Delta p \cos\theta = mg$. Dividing one equation by the other, we get $\tan\theta = a/g$; but from Fig. 15-13(b), $\tan\theta = h/L$, so $h = aL/g$.

15.38 Redo Prob. 15.37 by applying $F = ma$ to the liquid in the horizontal portion of the tube.

❚ Let p_1 and p_2 be the pressures at the bottom of the left end and right end of the tube, respectively. Then $F = (p_1 - p_2)A = \rho ghA$, where A is the tube cross section. The mass of liquid in the horizontal portion is $m = \rho LA$; $\rho ghA = (\rho LA)a$, or $h = aL/g$, as before.

15.2 PASCAL'S AND ARCHIMEDES' PRINCIPLES; SURFACE TENSION

15.39 A hydraulic lift in a service station has a large piston 30 cm in diameter and a small piston 2 cm in diameter. *(a)* What force is required on the small piston to lift a load of 1500 kg? *(b)* What is the pressure increase due to the force in the confined liquid?

▌ *(a)* Pascal's principle says that the pressure change is uniformly transmitted throughout the oil, so $\Delta p = F_1/A_1 = F_2/A_2$, where F_1 and F_2 are the forces on the small and on the large pistons, respectively, and A_1 and A_2 are the respective areas. Thus,

$$\frac{F_1}{\pi(2^2/4)} = \frac{1500 \times 9.8}{\pi(30^2/4)}$$

Multiplying both sides by $\pi/4$ and solving for F_1, we obtain

$$F_1 = \frac{(1500)(9.8)(2^2)}{30^2} = \underline{65\,N}$$

(b)
$$\Delta p = \frac{F_1}{A_1} = \frac{65}{\pi(2^2/4)} = 21\,N/cm^2 = \underline{210\,kPa}$$

15.40 A hydraulic lift is to be used to lift a truck weighing 5000 lb. If the diameter of the large piston of the lift is 1 ft, what gauge pressure in lb/in^2 must be applied to the oil?

▌ The gauge pressure in the oil acts, by Pascal's principle, on the bottom of the large piston to produce the force that lifts the load.

$$\Delta p = \frac{F}{A} = \frac{5000}{\pi r^2} = \frac{5000}{\pi(0.5^2)} = 6370\,lb/ft^2 = \frac{6370}{144} = \underline{44.2\,lb/in^2}$$

15.41 In a hydraulic press the large piston has cross-sectional area $A_1 = 200\,cm^2$ and the small piston has cross-sectional area $A_2 = 5\,cm^2$. If a force of 250 N is applied to the small piston, what is the force F_1 on the large piston?

▌ By Pascal's principle,

$$\text{pressure under large piston} = \text{pressure under small piston} \qquad \frac{F_1}{A_1} = \frac{F_2}{A_2}$$

$$F_1 = \frac{A_1}{A_2} F_2 = \frac{200}{5}\,(250\,N) = \underline{10\,kN}$$

15.42 For the system shown in Fig. 15-14, the cylinder on the left, at L, has a mass of 600 kg and a cross-sectional area of 800 cm^2. The piston on the right, at S, has cross-sectional area 25 cm^2 and negligible weight. If the apparatus is filled with oil ($\rho = 0.78\,g/cm^3$), what is the force F required to hold the system in equilibrium as shown?

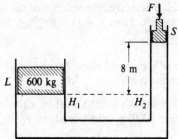

Fig. 15-14

▌ The pressures at points H_1 and H_2 are equal since they are at the same level in a single connected fluid. Therefore,

$$\text{pressure at } H_1 = \text{pressure at } H_2$$

$$\begin{pmatrix} \text{pressure due to} \\ \text{left piston} \end{pmatrix} = \begin{pmatrix} \text{pressure due to } F \\ \text{and right piston} \end{pmatrix} + (\text{pressure due to 8 m of oil})$$

$$\frac{(600)(9.8)\,N}{0.08\,m^2} = \frac{F}{25 \times 10^{-4}\,m^2} + (8\,m)(780\,kg/m^3)(9.8\,m/s^2)$$

Solving for F gives it to be $\underline{31\,N}$.

15.43 A block of wood weighing 71.2 N and of specific gravity 0.75 is tied by a string to the bottom of a tank of water in order to have the block totally immersed. What is the tension in the string?

▋ The block is in equilibrium under the action of three forces—the weight $w = 71.2$ N, the tension T, and the buoyant force B, with $B = w + T$. We can determine B since $B = \rho_L g V_B$, where ρ_L is the density of water and V_B is the volume of the totally immersed block. $w = \rho_B g V_B$, where ρ_B is the density of the block. Then $w/B = \rho_B/\rho_L = (\text{sp gr})_B = 0.75$, so $B = w/0.75 = 94.9$ N. Finally, from our equilibrium equation, 94.9 N $= 71.2$ N $+ T$, or $T = \underline{23.7\ \text{N}}$.

15.44 A metal ball weighs 0.096 N. When suspended in water it has an apparent weight of 0.071 N. Find the density of the metal.

▋ The desired density is given by $\rho = m/V$. But since the volume V of the ball is also the volume of displaced water, the buoyant force is given by $B = \rho_{\text{water}} g V$. Thus,

$$\rho = \frac{(mg)\rho_{\text{water}}}{B} = \frac{(0.096\ \text{N})(1 \times 10^3\ \text{kg/m}^3)}{(0.096 - 0.071)\ \text{N}} = 3840\ \text{kg/m}^3$$

15.45 A block of material has a density ρ_1 and floats three-fourths submerged in a liquid of unknown density. Show that the density ρ_2 of the unknown liquid is given by $\rho_2 = \frac{4}{3}\rho_1$.

▋ By Archimedes' principle, $\rho_1 V g = \rho_2 (3V/4)g$, which gives $\rho_2 = \frac{4}{3}\rho_1$.

15.46 The density of ice is 917 kg/m³, and the approximate density of the seawater in which an iceberg floats is 1025 kg/m³. What fraction of the iceberg is beneath the water surface?

▋ By Prob. 15.45, fraction submerged $= \rho_1/\rho_2 = \frac{917}{1025} = \underline{0.89}$.

15.47 A block of wood has a mass of 25 g. When a 5-g metal piece with a volume of 2 cm³ is attached to the bottom of the block, the wood barely floats in water. What is the volume V of the wood?

▋ The effective density of the system is $30/(V + 2)$ g/cm³; hence, by Probs. 15.45 and 15.46,

$$1 = \text{fraction submerged} = \frac{30/(V + 2)}{1}$$

Solving, $V = \underline{28\ \text{cm}^3}$.

15.48 A piece of wood weighs in mass units 10.0 g in air. When a heavy piece of metal is suspended below it, the metal being submerged in water, the "weight" of wood in air plus metal in water is 14.00 g. The "weight" when both wood and metal are submerged in water is 2.00 g. Find the volume and the density of the wood.

▋ Since in both cases the metal is submerged, the difference in weight is only the buoyant force on the wood, or $(12.00 \times 10^{-3})(9.8)$ N. Therefore, the volume of the wood can be found by equating the weight of displaced water, $1000(9.8)V$, to this. Then $V = \underline{12 \times 10^{-6}\ \text{m}^3}$. Its mass is 10×10^{-3} kg and so its density is $(10 \times 10^{-3})/(12 \times 10^{-6}) = \underline{830\ \text{kg/m}^3}$.

15.49 What is the minimum volume of a block of wood (density = 850 kg/m³) if it is to hold a 50-kg woman entirely above the water when she stands on it?

▋ The woman's weight plus the block's weight must be equal to the buoyant force on the just barely submerged block, $50g + 850Vg = 1000Vg$, which leads to $V = \underline{0.33\ \text{m}^3}$.

15.50 A man whose weight is 667 N and whose density is 980 kg/m³ can just float in water with his head above the surface with the help of a life jacket which is wholly immersed. Assuming that the volume of his head is $\frac{1}{15}$ of his whole volume and that the specific gravity of the life jacket is 0.25, find the volume of the life jacket.

▋ The man's volume is

$$V = \frac{w}{\rho g} = \frac{667}{(980)(9.8)} = 0.07\ \text{m}^3$$

Equating the buoyant force to the weight of the man plus the weight of the life jacket,

$$\rho_{\text{water}}g[\tfrac{14}{15}V + V_{1j}] = 667 + (0.25\rho_{\text{water}})gV_{1j}$$

Solving,

$$V_{lj} = \frac{(667/\rho_{\text{water}}) - (14/15)gV}{0.75g} = \frac{0.667 - (14/15)(9.8)(0.07)}{(0.75)(9.8)} = 0.004 \text{ m}^3 \quad \text{or} \quad \underline{4 \text{ L}}$$

15.51 An irregular piece of metal "weighs" 10.00 g in air and 8.00 g when submerged in water. (*a*) Find the volume of the metal and its density. (*b*) If the same piece of metal weighs 8.50 g when immersed in a particular oil, what is the density of the oil?

❚ Buoyant force = $(2 \times 10^{-3})g$ N, with $g = 9.8$ m/s^2. This equals the weight of the displaced water, $\rho_w Vg$. Equating and using $\rho_w = 1000$ kg/m^3 gives $V = 2 \times 10^{-6}$ m^3. The density = mass/$V = (10 \times 10^{-3})/(2 \times 10^{-6}) = \underline{5000 \text{ kg/m}^3}$. The buoyant force in oil is $(1.50 \times 10^{-3})g$ N. The weight of the displaced oil is $\rho_o(2 \times 10^{-6})g$, from which $\rho_o = \underline{750 \text{ kg/m}^3}$.

15.52 A beaker partly filled with water has a total mass of 20.00 g. If a piece of wood having a density of 0.800 g/cm^3 and volume 2.0 cm^3 is floated on the water in the beaker, how much will the beaker "weigh" (in grams)?

❚ The scale must support both the original 20 g and the (0.80 g/cm^3)(2 cm^3) = 1.60-g block, so the "weight" = 21.6 g.

15.53 A beaker when partly filled with water has total mass 20.00 g. If a piece of metal with density 3.00 g/cm^3 and volume 1.00 cm^3 is suspended by a thin string so that it is submerged in the water but does not rest on the bottom of the beaker, how much does the beaker then appear to weigh if it is resting on a scale?

❚ The tension in the thread is equal to the weight of the metal less the buoyant force. The buoyant force will be $(1 \times 10^{-6} \text{ m}^3)(1000 \text{ kg/m}^3)g = 10^{-3}g$ N, where $g = 9.8$ m/s^2. The weight of the metal is $3 \times 10^{-3}g$ N. Therefore, the thread exerts an upward force of $2 \times 10^{-3}g$ N. Hence the scale supports the total weight less the tension in the thread. Therefore, the apparent weight read by a scale will be $(23 - 2)$ g, or $\underline{0.206 \text{ N}}$.

Equivalently we can get the result by noting that if the water exerts an upward buoyant force of $10^{-3}g = 9.8 \times 10^{-3}$ N on the metal, by Newton's third law the metal exerts a like force downward on the water. Thus the scale balances the weight, 0.196 N, plus the downward force of the metal, 0.0098 N, for a total of $\underline{0.206 \text{ N}}$.

15.54 A solid cube of material is 0.75 cm on each edge. It floats in oil of density 800 kg/m^3 with one-third of the block out of the oil. (*a*) What is the buoyant force on the cube? (*b*) What is the density of the material of the cube?

❚ (*a*) The block is in equilibrium so $\rho_B V_B g = \text{BF} = \rho_0(2V_B/3)g$. Since $V_B = 4.22 \times 10^{-7}$ m^3 and $\rho_0 = 800$ kg/m^3, BF = $\underline{2.21 \times 10^{-3} \text{ N}}$. (*b*) Since BF = $\rho_B V_B g$, we find $\rho_B = 2\rho_0/3 = \underline{533 \text{ kg/m}^3}$.

15.55 A cubical copper block is 1.50 cm on each edge. (*a*) What is the buoyant force on it when it is submerged in oil for which $\rho = 820$ kg/m^3? (*b*) What is the tension in the string that is supporting the block when submerged? $\rho_{\text{Cu}} = 8920$ kg/m^3.

❚ (*a*) The buoyant force equals the weight of the displaced liquid, BF = $\rho_{\text{oil}} V_{\text{Cu}} g$, with the volume $V = 3.38 \times 10^{-6}$ m^3 and $\rho = 820$ kg/m^3. Thus BF = $\underline{0.027 \text{ N}}$. (*b*) The forces acting on the block are the tension T up, BF up, and the block weight $\rho_{\text{Cu}} V_{\text{Cu}} g = 8920(3.38 \times 10^{-6})(9.8) = 0.295$ N acting downward. Then $T = 0.295 - 0.027 = \underline{0.268 \text{ N}}$.

15.56 A balloon having a mass of 500 kg remains suspended motionless in the air. If the air density is 1.29 kg/m^3, what is the volume of the balloon in cubic meters?

❚ By equilibrium and Archimedes' principle, $500g = 1.29gV$, or $V = \underline{388 \text{ m}^3}$.

15.57 A cylindrical wooden buoy, of height 3 m and mass 80 kg, floats vertically in water. If its specific gravity is 0.80, how much will it be depressed when a body of mass 10 kg is placed on its upper surface?

❚ By Archimedes' principle, the submerged height, h, of the unloaded buoy is given by

$$\rho_{\text{water}}gAh = \rho_{\text{wood}}gA(3) \quad \text{or} \quad h = \frac{\rho_{\text{wood}}}{\rho_{\text{water}}}3 = (0.80)(3) = 2.40 \text{ m}$$

Under loading, the submerged height is directly proportional to the total weight or mass.

$$\frac{h + \Delta h}{h} = \frac{80 + 10}{80} \qquad \text{or} \qquad \Delta h = \frac{10}{80} h = \frac{10}{80} 2.40 = \underline{0.30\ \text{m}}$$

15.58 A tank contains water on top of mercury. A cube of iron, 60 mm along each edge, is sitting upright in equilibrium in the liquids. Find how much of it is in each liquid. The densities of iron and mercury are $7.7 \times 10^3\ \text{kg/m}^3$ and $13.6 \times 10^3\ \text{kg/m}^3$, respectively.

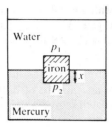

Fig. 15-15

▮ Let x equal the distance submerged in mercury. Then $(0.06 - x)$ equals the distance protruding into the water. The net vertical force due to the liquids is $p_2 A - p_1 A$, where p_2, p_1 are the pressures at the lower and upper face of the block (see Fig. 15-15) and A is the face area of the block. For equilibrium we have the weight of the iron $w_I = p_2 A - p_1 A = (p_2 - p_1)A$. The pressure difference is just $p_2 - p_1 = \rho_w g(0.06 - x) + \rho_{Hg} g x$. Then $(p_2 - p_1)A = \rho_w g(0.06 - x)A + \rho_{Hg} g x A$. [Note that the two terms on the right represent the weight of displaced water and mercury, respectively; and the expression on the left is the buoyant force. Thus Archimedes' principle holds even when two (or more) liquids are displaced. This is true for any shape object.] We can now solve for x from the equilibrium equation, noting that $w_I = \rho_I g V_I = \rho_w g(0.06 - x)A + \rho_{Hg} g x A$; or canceling the g, we have

$$(7.7 \times 10^3\ \text{kg/m}^3)(0.06\ \text{m})^3 = (1.0 \times 10^3\ \text{kg/m}^3)(0.06\ \text{m} - x)(0.06\ \text{m})^2 + (13.6 \times 10^3\ \text{kg/m}^3)x(0.06\ \text{m})^2$$

$$\text{and} \qquad 7.7(0.06) = (0.06 - x) + 13.6x \qquad 0.40 = 12.6x$$

$x = 0.032\ \text{m} = \underline{32\ \text{mm}}$ = depth submerged in mercury, and $\underline{28\ \text{mm}}$ protrudes into water.

15.59 A slender homogeneous rod of length $2l$ floats partly immersed in water, being supported by a string fastened to one of its ends, as pictured in Fig. 15-16. If the specific gravity of the rod is 0.75, what is the fraction of the length of the rod that extends out of the water?

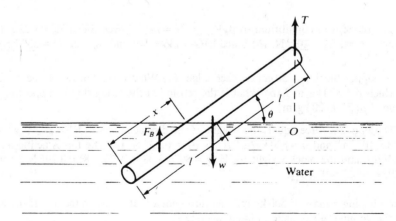

Fig. 15-16

▮ Since the buoyant force acts through the center of gravity of the displaced water, the condition for rotational equilibrium is, taking moments about a point O along the line of action of T,

$$0 = \sum \tau_O = wl \cos \theta - F_B\left(2l - \frac{x}{2}\right) \cos \theta = \rho_{rod} g A(2l)(l \cos \theta) - \rho_{water} g A x\left(2l - \frac{x}{2}\right) \cos \theta$$

$$= \left(\frac{1}{2}\rho_{water} g A \cos \theta\right)\left(x^2 - 4lx + 4\frac{\rho_{rod}}{\rho_{water}} l^2\right) \qquad \text{where } A = \text{cross sectional area}$$

From this, $x^2 - 4lx + 3.00l^2 = 0$, or $x = l, 3l$. Discarding the nonphysical root, we see that one-half the rod extends out of the water.

Strictly speaking, the above solution is only approximate since the water surface does not cut the rod perpendicularly. However, the error will be negligible if A is small.

15.60 The weight of a balloon and the gas it contains is 11.12 kN. If the balloon displaces $1132\ m^3$ of air and the weight of $1\ m^3$ of air is 12.3 N, what is the acceleration with which the balloon begins to rise?

❚ The equation of motion of the balloon is

$$\sum F = F_B - w = ma \qquad \text{or} \qquad a = \frac{F_B - w}{m} = \frac{(1132)(12.3) - (1.112 \times 10^4)}{(1.112 \times 10^4)/9.8} = \underline{2.47\ m/s^2}$$

Note that we have ignored viscous forces. This is justified if the velocity is negligible.

15.61 A small block of wood, of density $0.4 \times 10^3\ kg/m^3$, is submerged in water at a depth of 2.9 m. Find (a) the acceleration of the block toward the surface when the block is released, and (b) the time for the block to reach the surface. Ignore viscosity.

❚ (a) By Archimedes' principle, the net upward force on the block is $F = \rho_{water}gV - \rho_{wood}gV$, where V is the volume of the block. Then

$$a = \frac{F}{m} = \frac{\rho_{water}gV - \rho_{wood}gV}{\rho_{wood}V} = \left(\frac{\rho_{water}}{\rho_{wood}} - 1\right)g = \left(\frac{1}{0.4} - 1\right)(9.8) = \underline{14.7\ m/s^2}$$

(b)
$$s = \frac{1}{2}at^2 \qquad \text{or} \qquad t = \sqrt{\frac{2s}{a}} = \sqrt{\frac{2(2.9)}{14.7}} = \underline{0.63\ s}$$

15.62 A body of density ρ' is dropped from rest at a height h into a lake of density ρ, where $\rho > \rho'$. Neglect all dissipative effects and calculate (a) the speed of the body just before entering the lake, (b) the acceleration of the body while it is in the lake, and (c) the maximum depth to which the body sinks before returning to float on the surface.

❚ (a) The speed just before entering can be determined from conservation of mechanical energy in free fall, or directly from the kinematic equations of free fall, yielding $v = \sqrt{2gh}$. (b) The bouyant force of the lake, B, is greater than the weight of the body, w, since $\rho > \rho'$. Choosing upward as positive, we have

$$B - w = ma \qquad \text{where} \qquad B = \rho gV \qquad w = \rho'gV \qquad m = \rho'V \qquad \text{and} \qquad V = \text{volume of body}$$

Then, canceling V on both sides,

$$(\rho - \rho')g = \rho'a \qquad \text{or} \qquad a = g\frac{\rho - \rho'}{\rho'} = g\left(\frac{\rho}{\rho'} - 1\right), \qquad \text{upward}$$

(c) To find the maximum depth we have $v^2 = v_0^2 + 2ay$, with v_0 the velocity at $y = 0$ (from part a), $v = 0$ is the velocity at maximum depth, and y is the negative displacement from the surface to the maximum depth. Then

$$v_0^2 = -2ay \qquad \text{and} \qquad (-y) = \text{depth} = \frac{v_0^2}{2a} = \frac{2gh}{2g\left(\frac{\rho}{\rho'} - 1\right)} = \frac{h\rho'}{\rho - \rho'}.$$

15.63 A soap bubble has a radius of 5 cm. If the soap solution has a surface tension $T = 30 \times 10^{-3}\ N/m$, what is the gauge pressure within the bubble?

❚ Consider a hemisphere of the bubble (Fig. 15-17). The downward force of surface tension on each of the two bubble surfaces, inside and outside, is $2\pi rT$. For both surfaces the total force F is $F = 2(2\pi rT) = \Delta p\,A$, where A is the area of the flat circular face of the hemisphere. (See, e.g., Prob. 15.36.) Since $A = \pi r^2$,

$$4\pi rT = \Delta p(\pi r^2) \qquad \frac{4T}{r} = \Delta p \qquad \Delta p = \frac{4T}{r} = \frac{4(30 \times 10^{-3})}{0.05} = \underline{2.4\ Pa}$$

15.64 Find an expression for the height h that a liquid of density ρ will rise in a capillary tube of radius r if the surface tension of the liquid is γ and the meniscus makes an angle θ with the tube, as shown in Fig. 15-18.

❚ γ is a force per unit length and points in different directions around the tube, as shown. At equilibrium,

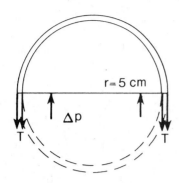

Fig. 15-17

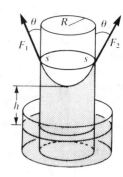

Fig. 15-18

the upward force is $2\pi r \gamma \cos \theta$ due to surface tension and is equal to the downward force mg due to gravity:

$$2\pi r \gamma \cos \theta = mg \qquad (1)$$

The mass of the column is $m = \rho \pi r^2 h$. Putting this expression for m into Eq. (1) we get $2\pi r \gamma \cos \theta = \rho \pi r^2 hg$, or

$$h = \frac{2\gamma \cos \theta}{\rho g r} \qquad (2)$$

If the force of adhesion is much greater than the force of cohesion, the contact angle θ is very small. In this case $\cos \theta$ is approximately 1, and Eq. (2) can be replaced by

$$h = \frac{2\gamma}{\rho g r} \qquad (3)$$

Equation (3) is valid for water in a glass capillary, where θ is 20° or less.

15.65 The surface tension of water is 0.07 N/m. Find the weight of water supported by surface tension in a capillary tube with a radius of 0.1 mm.

▌ Let T be the surface tension, and assume a contact angle of zero. $F = 2\pi r T = 2\pi(10^{-4})(0.07) = \underline{44\ \mu N}$.

15.66 How high does methyl alcohol rise in a glass tube 0.07 cm in diameter? The surface tension is 0.023 N/m and the density of alcohol is 0.8 g/cm³. Assume that the angle of contact is zero.

▌ This is a capillarity problem that can be solved using Eq. (3) of Prob. 15.64,

$$h = \frac{2(0.023)}{(800)(9.8)(0.035 \times 10^{-2})} = 0.017\ \text{m} \qquad \text{or} \qquad \underline{1.7\ \text{cm}}$$

15.67 A mercury barometer has a glass tube with an inside diameter equal to 4 mm. Since the contact angle for mercury with glass is 140°, capillarity depresses the column. How many millimeters of mercury must be added to the reading to correct for capillarity? Assume a surface tension $T = 0.545$ N/m and a density for mercury of 13.6 g/cm³.

▌ The height differential due to capillarity is given by Eq. (2) of Prob. 15.64: $h = (2T \cos \theta)/(rdg)$. Use the value of T given and the value of $d = 13\,600$ kg/m². Then

$$h = \frac{2(0.545) \cos 140°}{(13\,600)(9.8)(2 \times 10^{-3})} = \frac{(0.545)(-0.766)}{(13\,600)(9.8)(10^{-3})} = -0.0031\ \text{m}$$

Therefore, $\underline{3.1\ \text{mm}}$ must be added to the barometer reading.

CHAPTER 16
Hydrodynamics

16.1 EQUATION OF CONTINUITY, BERNOULLI'S EQUATION, TORRICELLI'S THEOREM

16.1 With regard to fluid flow, define *streamline, stream tube, steady flow, turbulent flow, incompressible flow,* and *irrotational flow*.

❚ A *streamline* is an imaginary line in a fluid, taken at an instant of time, such that the velocity vector at each point of the line is tangential to it. A *stream tube* is a tube whose surface is made up of streamlines, across which there is no transport of fluid (Fig. 16-1).

In *steady flow*, the fluid velocity at a given location is independent of time. (However, the velocity will in general vary from point to point.) Streamlines and stream tubes are fixed in steady flow, and individual particles flow along the streamlines and within the stream tubes. Steady flow is sometimes called *laminar flow*.

In *turbulent flow* the fluid velocity changes from moment to moment at any given location. Streamlines thus no longer characterize the paths of fluid particles. Turbulent flow is often characterized by constantly changing swirls or eddies of fluid.

A flow is *incompressible* if the fluid density ρ is constant; it is *irrotational* if there is no swirling or circular flow of fluid.

16.2 What is the equation of continuity?

❚ The conservation of mass requires that the net rate of flow of mass inward across any closed surface equal the rate of increase of the mass within the surface, assuming that there are no sources or sinks of matter within the surface. Applying this to a stream tube in steady flow (Fig. 16-1), we obtain the *continuity equation* in the form $\rho_1 A_1 v_1 = \rho_2 A_2 v_2$, where ρ is the density, assumed uniform over the cross section of area A, and v is the average velocity over the cross section (and normal to it).

If, besides being steady, the flow is incompressible, the continuity equation reduces to $A_1 v_1 = A_2 v_2$.

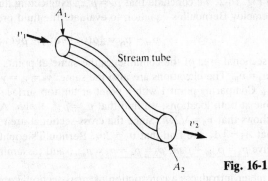

Stream tube

Fig. 16-1

16.3 What is Bernoulli's equation?

❚ For the steady flow of a nonviscous incompressible fluid, *Bernoulli's equation* relates the pressure p, the fluid speed v, and the height y at any two points *on the same streamline*, as follows:

$$p_1 + \tfrac{1}{2}\rho v_1^2 + \rho g y_1 = p_2 + \tfrac{1}{2}\rho v_2^2 + \rho g y_2$$

Note that each term in Bernoulli's equation has the dimensions of energy per unit volume. In fact, the equation simply states that the work done by pressure forces along a streamline is equal to the change in kinetic and potential energy (all per unit volume).

Bernoulli's equation can also be expressed in the form

$$\Delta p + \tfrac{1}{2}\rho \, \Delta v^2 + \rho g \, \Delta y = 0$$

where $\Delta p = p_2 - p_1$, $\Delta v^2 = v_2^2 - v_1^2$, $\Delta y = y_2 - y_1$.

16.4 Consider the flow of a fluid at speed v_0 through a cylindrical pipe of radius r. What would be the speed of this fluid at a point where, because of a constriction in the pipe, the fluid is confined to a cylindrical opening of radius $r/4$?

❚ If the radius decreases from r to $r/4$, the cross-sectional area decreases from πr^2 to $\pi r^2/16$. From the continuity equation, assuming an incompressible fluid, the velocity would have to *increase sixteen-fold* at the constriction.

16.5 Water is flowing smoothly through a closed-pipe system. At one point the speed of the water is 3.0 m/s, while at another point 1.0 m higher the speed is 4.0 m/s. If the pressure is 20 kPa at the lower point, what is the pressure at the upper point? What would the pressure at the upper point be if the water were to stop flowing and the pressure at the lower point were 18 kPa?

❚ Use Bernoulli's equation: $p_1 + \rho g h_1 + (\rho v_1^2)/2 = p_2 + \rho g h_2 + (\rho v_2^2)/2$. Using known values and $h_2 - h_1 = 1$ m, one obtains $p_2 = \underline{6.7\ \text{kPa}}$. For no flow, $v_1 = v_2 = 0$, and so $p_2 = p_1 + \rho g(h_1 - h_2) = \underline{8.2\ \text{kPa}}$.

16.6 Water flows out of a pipe at the rate of 3.0 cm³/s. Find the velocity of the water at a point in the pipe where its diameter is **(a)** 0.50 cm and **(b)** 0.80 cm.

❚ From the equation of continuity we have, since $\rho = $ constant, **(a)** $\pi(0.25)^2 v = 3.0$, so $v = \underline{15.3\ \text{cm/s}}$ and **(b)** $\pi(0.40)^2 v = 3.0$, from which $v = \underline{6.0\ \text{cm/s}}$.

16.7 In Fig. 16-2, the system is filled to height h with a liquid of density ρ. The atmospheric pressure is p_{atm}. Neglecting fluid friction, evaluate the pressure of the fluid at each of the points labeled 1, 2, 4, and 5. Compare the pressure at point 3 with that at point 5.

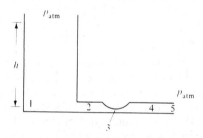

Fig. 16-2

❚ Referring to Fig. 16-2, we conclude that $p_5 = p_{atm}$. Neglecting fluid friction and assuming that the flow is steady, we can employ Bernoulli's equation to evaluate the fluid pressure at various other locations:

$$p_M - p_N = \tfrac{1}{2}\rho(v_N^2 - v_M^2) + \rho g(y_N - y_M)$$

Since the cross-sectional area of the container is the same at points 2, 4, and 5, the flow speeds must also be the same: $v_2 = v_4 = v_5$. The elevations are also the same: $y_2 = y_4 = y_5$. Bernoulli's equation therefore implies that $p_2 = p_4 = p_5$. Comparing point 1 with a point at the top surface of the liquid, and assuming that the flow speed is negligible at both locations, we find that $p_1 = p_{atm} + \rho g h$. Application of Bernoulli's equation to points 1 and 5 shows that $v_5 = \sqrt{2gh}$. Since the cross-sectional area A_3 is less than A_5, we must have $v_3 > v_5$. If we assume that $A_3 = \tfrac{1}{2}A_5$, we find that $v_3 = 2v_5$. Bernoulli's equation then implies that $p_3 = p_5 - 3\rho g h$. In summary, we have $\underline{p_1 = p_{atm} + \rho g h}$, $\underline{p_2 = p_4 = p_5 = p_{atm}}$, and (assuming $\underline{A_3 = \tfrac{1}{2}A_5}$), $\underline{p_3 = p_{atm} - 3\rho g h}$.

16.8 A Venturi flow meter introduces a constriction of cross-sectional area a_2 in a pipe of cross-sectional area a_1. The meter records the difference in pressure, $p_1 - p_2$, between the ordinary fluid pressure, p_1, and the pressure at the constriction, p_2. From this, infer the fluid speed in the unconstricted pipe.

❚ We assume that the height levels, y_1 and y_2, at the normal flow point and the constricted point are the same. $p_1 + \tfrac{1}{2}\rho v_1^2 = p_2 + \tfrac{1}{2}\rho v_2^2$, from Bernoulli's equation, and $v_1 a_1 = v_2 a_2$, from the continuity equation. Then

$$p_1 - p_2 = \tfrac{1}{2}\rho(v_2^2 - v_1^2) = \tfrac{1}{2}\rho\left[\left(\frac{a_1}{a_2}\right)^2 - 1\right]v_1^2 \quad \text{or} \quad v_1^2 = \frac{2(p_1 - p_2)}{\rho}\frac{a_2^2}{a_1^2 - a_2^2}$$

Finally
$$v_1 = a_2\sqrt{\frac{2(p_1 - p_2)}{\rho(a_1^2 - a_2^2)}}$$

16.9 Using Prob. 16.8, find the fluid volume per unit time passing any cross section of the pipe.

❚ The volume flow is the same anywhere along the pipe. At position 1, with velocity v_1 and area a_1,

$$\text{volume flow} = a_1 v_1 = a_1 a_2 \sqrt{\frac{2(p_1 - p_2)}{\rho(a_1^2 - a_2^2)}}$$

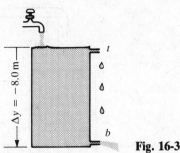

Fig. 16-3

16.10 The tank shown in Fig. 16-3 is kept filled with water to a depth of 8.0 m. Find the speed v_b with which the jet of water emerges from the small pipe just at the bottom of the tank.

▮ Apply Bernoulli's equation to the differences Δv^2, Δp, and Δy between the locations t and b in Fig. 16-3. Since the tank is large and the pipe is small, it may be assumed that the water has negligible speed until it is actually in the outlet pipe. Bernoulli's equation, $\Delta p + \frac{1}{2}\rho\,\Delta v^2 + \rho g\,\Delta y = 0$, can be solved by noting that $\Delta v^2 = v_b^2 - 0$ and $\Delta p = 0$ (both the jet and the top surface are at atmospheric pressure). Bernoulli's equation thus becomes $\frac{1}{2}\rho\,\Delta v^2 + \rho g\,\Delta y = 0$, or $v_b^2 = -2g\,\Delta y$, and so $v_b = \sqrt{2g(-\Delta y)}$.

The outlet speed is the free-fall speed—this is *Torricelli's theorem*.

Inserting the numerical values gives $v_b = \sqrt{2 \times 9.8 \text{ m/s}^2 \times 8.0 \text{ m}} = \underline{13 \text{ m/s}}$.

16.11 The upper pipe in Fig. 16-3 is located just under the surface of the water. Its free end is plugged except for a small hole through which water drips. Neglecting air resistance, find the speed v_t of a drop from the upper pipe just as it falls past the bottom of the tank.

▮ $v_t = 13 \text{ m/s}$, from free-fall kinematics or by Torricelli's theorem and the results of Prob. 16.10.

16.12 A water barrel stands on a table of height h. If a small hole is punched in the side of the barrel at its base, it is found that the resultant stream of water strikes the ground at a horizontal distance R from the barrel. What is the depth of water in the barrel?

▮ From Torricelli's theorem, $v = \sqrt{2gd}$, where v is the horizontal velocity of efflux from the barrel and d is the depth of interest. The efflux is horizontal, so the time to hit the ground is given by $h = \frac{1}{2}gt^2$, or $t = \sqrt{2h/g}$ and $R = vt = \sqrt{2gd}\sqrt{2h/g} = 2\sqrt{dh}$. Then $R^2 = 4dh$ and $d = R^2/(4h)$.

16.13 A hole of area 1 mm² opens in the pipe near the lower end of a large water-storage tank, and a stream of water shoots from it. If the top of the water in the tank is 20 m above the point of the leak, how much water escapes in 1 s?

▮ Using Torricelli's theorem for the escape speed, we have for the volume flow

$$vA = \sqrt{2gh}\,A = \sqrt{2(9.8 \times 10^3 \text{ mm/s}^2)(20 \times 10^3 \text{ mm})}\,(1 \text{ mm}^2) = 19\,800 \text{ mm}^3/\text{s} = \underline{19.8 \text{ mL/s}}$$

16.14 Figure 16-4 shows a crude type of perfume atomizer. When the bulb at A is compressed, air flows swiftly through the tiny tube BC, thereby causing a reduced pressure at the position of the vertical tube. Liquid then rises in the tube and enters BC and is sprayed out. If the pressure in the bulb is $P_a + P$, where P_a is atmospheric pressure, and v is the speed of the air in BC, find the approximate pressure in BC. How large would v need to be to cause the liquid to rise to BC? The density of air can be taken as 1.30 kg/m³.

▮ Use Bernoulli's equation. $P_a + P = P_{BC} + (\rho_a v^2)/2$ yields $P_{BC} = P_a + P - 0.65v^2$, where $\rho_a =$ density of air. But $P_{BC} = P_a - \rho gh$, and so equating these two values for P_{BC} yields $v = \{[(P + \rho gh)]/0.65\}^{1/2}$.

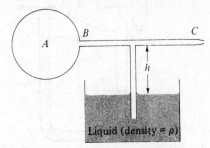

Fig. 16-4

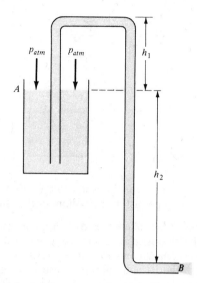

Fig. 16-5

16.15 A tube of uniform cross section is used to siphon water from a vessel, as in Fig. 16-5. Derive an expression for the speed with which the water leaves the tube at B.

▋ As long as water fills the tube, as shown, and points A and B are open to the atmosphere, we can apply Torricelli's theorem between A and B to obtain $v_B = \sqrt{2gh_2}$.

16.16 Refer to Prob. 16.15. If $h_2 = 3.0$ m, what is the greatest value of h_1 for which the siphon will work?

▋ We apply Bernoulli's theorem to determine quantities (labeled with a subscript 1) at the highest part of the tube in terms of those at level A.

$$p_A = p_1 + \tfrac{1}{2}\rho v_1^2 + \rho g h_1 \tag{1}$$

Since the tube has a constant cross section and the fluid is incompressible, the steady-state flow speed in the tube must equal the exit speed. Thus, by Prob. 16.15,

$$v_1 = v_B = \sqrt{2gh_2} \tag{2}$$

Using $p_A = p_{\text{atm}}$ and Eq. (2) in Eq. (1), we find that the pressure at the highest point in the tube is given by

$$p_1 = p_{\text{atm}} - \tfrac{1}{2}\rho(\sqrt{2gh_2})^2 - \rho g h_1 = p_{\text{atm}} - \rho g(h_1 + h_2) \tag{3}$$

The pressure value p_1 must be nonnegative or else the water could not reach the top of the siphon. (Liquids cannot withstand tension.) Setting $p_1 = 0$ in Eq. (3), we find the maximum value of h_1:

$$\max{(h_1)} = \frac{p_{\text{atm}}}{\rho g} - h_2 = \frac{(1.0 \times 10^5 \text{ N/m}^2)}{(1.0 \times 10^3 \text{ kg/m}^3)(9.8 \text{ m/s}^2)} - 3.0 \text{ m} = 10.2 \text{ m} - 3.0 \text{ m} = \underline{7.2 \text{ m}}$$

16.17 Air flows through the horizontal main tube of the *Venturi meter* of Fig. 16-6 from left to right. If the U tube of the meter contains mercury, find the mercury-level difference h between the two arms. Let the radii of the wide and narrow parts of the main tube be $r_1 = 1.0$ cm and $r_2 = 0.50$ cm, respectively, and let the speed of the

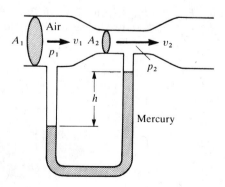

Fig. 16-6

air entering the meter be $v_1 = 15.0$ m/s. The density of air is $\rho_{air} = 1.3$ kg/m^3, and that of mercury is $\rho_{merc} = 13.6 \times 10^3$ kg/m^3.

▮ Since the air moves horizontally, $\Delta y = 0$, and Bernoulli's equation becomes

$$\Delta p = -\tfrac{1}{2}\rho_{air}\,\Delta v^2 \tag{1}$$

But, by the continuity equation,

$$v_2 = v_1 \frac{A_1}{A_2} = v_1 \frac{r_1^2}{r_2^2} \qquad \text{giving} \qquad \Delta v^2 = v_2^2 - v_1^2 = v_1^2\left(\frac{r_1^4}{r_2^4} - 1\right)$$

Inserting this value into Eq. (1) leads to the relation

$$\Delta p = -\frac{1}{2}\rho_{air}v_1^2\left(\frac{r_1^4}{r_2^4} - 1\right) \tag{2}$$

This difference in pressure between the two ends of the mercury-containing U tube produces a mercury-level difference h given by $|\Delta p| = \rho_{merc}gh$. Taking the absolute value of both sides of Eq. (2) and combining the result with this equation, you have

$$h = \frac{\rho_{air}v_1^2}{2\rho_{merc}g}\left(\frac{r_1^4}{r_2^4} - 1\right) \tag{3}$$

For the data,

$$h = \frac{1.3 \text{ kg/m}^3 \times (15.0 \text{ m/s})^2}{2 \times 13.6 \times 10^3 \text{ kg/m}^3 \times 9.80 \text{ m/s}^2}\left[\left(\frac{1.0 \text{ cm}}{0.50 \text{ cm}}\right)^4 - 1\right] = 0.016 \text{ m} = \underline{1.6 \text{ cm}}$$

16.18 A water-filled can sits on a table. The water squirts out of a small hole in the side of the can, located a distance y below the water surface. The height of the water in the can is h. At what distance R from the base of the can, directly below the hole, does the water strike the table top? Neglect air resistance.

▮ By Torricelli's theorem the water leaves the hole at speed $v = \sqrt{2gy}$. Each element of the water stream follows the trajectory of a particle launched horizontally at the same speed and elevation. The "fall time" t satisfies the equation $\tfrac{1}{2}gt^2 = h - y$, so that $t = \sqrt{2(h-y)/g}$. The horizontal distance traveled during the fall is therefore given by

$$R = vt = \sqrt{2gy}\,\sqrt{\frac{2(h-y)}{g}} = 2\sqrt{y(h-y)} \tag{1}$$

16.19 Refer to Prob. 16.18. (**a**) How far from the bottom of the can must a second small hole be located if the water coming out of this hole is to have the same range R? (**b**) How far from the surface of the water must a hole be located to give the maximum range?

▮ (**a**) Inspection of Eq. (1) of Prob. 16.18 shows that *two holes have the same range if one is as far above the bottom of the can as the other is below the water surface.*
(**b**) Because $R(y)$ is symmetric about $y = h/2$, the midheight furnishes either a maximum or a minimum range. But $R(0) = R(h) = 0$, so $R(h/2) = $ max.

16.20 A horizontal pipe system connects a pipe of cross-sectional area A_1 to a pipe of area A_2 and the other end of this latter pipe opens to the air. If atmospheric pressure is P_0 and if viscous effects can be ignored, how large a pressure is required in the first pipe to cause the water to flow at a speed v_2 out of the open end? What will be the speed of the water in the first pipe? How much water will flow out of the pipe in a time Δt? Express your answers in terms of P_0, v_2, A_1, A_2, and Δt.

▮ According to Bernoulli's equation, $P_1 + (\rho v_1^2)/2 = P_2 + (\rho v_2^2)/2$, from which $P_1 = P_2 + 500(v_2^2 - v_1^2)$. The continuity equation gives $v_1A_1 = v_2A_2$, and so $P_1 = P_0 + 500[1 - (A_2/A_1)^2]v_2^2$. A cylinder of length $v_2\,\Delta t$ and area A_2 flows out each second; therefore, $\Delta Q = v_2A_2\,\Delta t$ is the amount of water flowing out in time Δt.

16.21 A hose shoots water straight up for a distance of 2.5 m. The end opening on the hose has an area of 0.75 cm^2. What is the speed of the water as it leaves the hose? How much water comes out in 1 min?

▮ To find v equate KE$_{bottom}$ to PE$_{top}$, giving $v = (2gh)^{1/2} = 7.0$ m/s $= \underline{700 \text{ cm/s}}$. The flow rate is Av, so in 1 min we have $(0.75 \text{ cm}^2)(700 \text{ cm/s})(60 \text{ s}) = 3.15 \times 10^4 \text{ cm}^3 = \underline{31.5 \text{ L}}$.

16.22 Suppose that the gas in the explosion chamber of a rocket ship is kept at a density ρ_1 and a pressure p_1, and that it exudes from the chamber into empty space through an opening of area a at one end of the rocket ship. Find (a) the exhaust speed of the gas relative to the rocket, in terms of p_1 and ρ_1; (b) the thrust produced on the rocket ship.

▌ (a) The situation is somewhat contrived, but we assume that at the instant the gas exudes it maintains its density in the chamber, ρ_1. Then applying Bernoulli's equation between the interior of the chamber (where $v \approx 0$) and the point of egress ($p = 0$), and neglecting gravitational potential energy in empty space, we get

$$0 + \tfrac{1}{2}\rho_1 v^2 = p_1 + 0 \qquad \text{or} \qquad v = \sqrt{\frac{2p_1}{\rho_1}}$$

(b) The thrust = time rate of change of momentum. All particles exit with velocity v, and the total mass/time leaving is $\rho_1 a v$. Then the momentum/time = $(\rho_1 a v)v = \rho_1 a v^2 = 2p_1 a$.

16.23 In Fig. 16-7, the opening through which water leaves the vessel has a sharp edge, causing a neck or *vena contracta*, in the discharging stream. For sharp openings, the observed minimal cross-sectional area for which Torricelli's theorem holds is 0.62 times the actual opening area. Suppose that the vessel is filled to a height h. What is the flux through the opening?

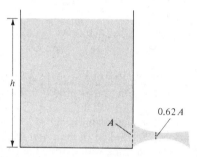

Fig. 16-7

▌ The mass flux is $\Phi = \rho A v_A = \rho A' v$, where v is the velocity of flow at the vena contracta, ρ is the density, and A' is the area at the vena contracta. Assuming that the flow is uniform over the cross section of the vena contracta, and applying Torricelli's theorem, $\Phi = 0.62\rho A\sqrt{2gh}$.

16.24 Water leaves a faucet with a downward velocity of 3.0 m/s. As the water falls below the faucet, it accelerates with acceleration g. The cross-sectional area of the water stream leaving the faucet is 1.0 cm². What is the cross-sectional area of the stream 0.50 m below the faucet?

▌ We denote the initial speed by v_0 and the initial cross-sectional area by A_0. After the freely falling stream has descended a distance h, its speed $v_1 = \sqrt{v_0^2 + 2gh}$. Under steady-flow conditions, the mass fluxes at the locations are equal: $\rho_0 v_0 A_0 = \rho_1 v_1 A_1$. Since the water is effectively incompressible, $\rho_0 = \rho_1$ and therefore $A_1 = (v_0/v_1)A_0$. Inserting numerical values, we find $v_1 = \sqrt{(3.0)^2 + 2(9.8)(0.50)} = 4.34$ m/s. Then $A_1 = (3.0/4.34)(1.0 \text{ cm}^2) = \underline{0.69 \text{ cm}^2}$.

16.25 A cylindrical tank 0.9 m in radius rests atop a platform 6 m high, as shown in Fig. 16-8. Initially the tank is filled with water ($\rho = 1 \times 10^3$ kg/m³) to a depth $h_0 = 3$ m. A plug whose area is 6.3 cm² is removed from an orifice in the side of the tank at the bottom. What is the speed of the stream as it strikes the ground?

▌ Applying Bernoulli's equation between the top of the tank and ground level,

$$p_\text{atm} + \tfrac{1}{2}\rho(0^2) + \rho g(h_0 + H) = p_\text{atm} + \tfrac{1}{2}\rho v_3^2 + \rho g(0)$$

whence $v_3 = \sqrt{2g(h_0 + H)} = \sqrt{2(9.8)(9)} = \underline{13.28 \text{ m/s}}$ (as though in free fall all the way). This result holds even if we lose laminar flow in the stream to the ground, since it follows from conservation of energy for each droplet of water.

16.26ᶜ Refer to Prob. 16.25. How long does it take to empty the tank entirely?

▌ By the continuity equation, $-a_1(dh/dt) = a_2 v_2$. Assuming that the top surface always moves slowly ($|dh/dt| \ll v_2$), Torricelli's theorem gives $v_2 = \sqrt{2gh}$; hence,

$$dt = -\frac{a_1}{a_2}\frac{dh}{\sqrt{2gh}} \qquad t = -\frac{a_1}{a_2\sqrt{2g}}\int_{h_0}^{0}\frac{dh}{\sqrt{h}} = \frac{a_1}{a_2}\sqrt{\frac{2h_0}{g}} = \frac{\pi(0.9)^2}{6.3 \times 10^{-4}}\sqrt{\frac{2(3)}{9.8}} = 3160 \text{ s} = \underline{52.7 \text{ min}}$$

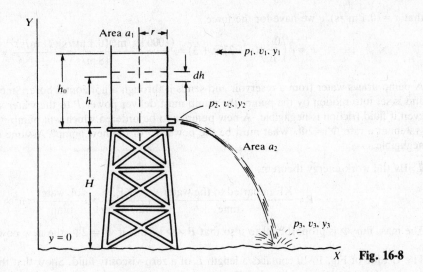

Fig. 16-8

16.27 The opening near the bottom of the vessel in Fig. 16-9 has an area a. A disk is held against the opening to keep the liquid, of denisty ρ, from running out.
 (a) With what net force does the liquid press on the disk? (b) The disk is moved away from the opening a short distance. The liquid squirts out, striking the disk inelastically. After striking the disk, the water drops vertically downward. Show that the force exerted by the water on the disk is twice the force in part **a**.

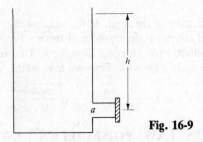

Fig. 16-9

▌ (a) The hydrostatic pressure on the inside surface of the disk is given by $p_i = p_{atm} + \rho g h$. The air pressure of the outside of the disk is $p_o = p_{atm}$. Since the disk has area a, the net outward force is $(p_i - p_o)a = \underline{\rho g h a}$.
(b) Once the disk is removed, the fluid quickly attains a (relatively) steady flow. Torricelli's theorem implies that the exit speed is $v = \sqrt{2gh}$. Since friction is being ignored, this is the exit speed across the entire opening. Therefore the mass flux $\Phi = \rho a v = \rho a \sqrt{2gh}$ and the flux of rightward momentum is $(\rho a v)v = 2\rho g h a$. If the water loses its entire rightward momentum as it strikes the disk, the disk must absorb momentum at the rate $2\rho g h a$. That is, it will experience a rightward force $\underline{2\rho g h a}$, which is twice the hydrostatic force found in part **a**. Note that the force due to the atmosphere cancels on the left and right of the disk.

16.28 A flat plate moves normally toward a discharging jet of water at the rate of 3 m/s. The jet discharges water at the rate of $0.1 \text{ m}^3/\text{s}$ and at a speed of 18 m/s. (a) Find the force on the plate due to the jet and (b) compare it with that if the plate were stationary.

▌ We do part **b** first. With no other information we assume the plate stops the forward motion, but there is no bounce back, i.e., the water splashes along the plate at right angles to the original motion. Then the force normal to the plate equals the time rate of change of momentum along the direction of the water jet, or $F = (\rho a v)v$, where the term in parentheses is the mass/time hitting the plate and a is the cross-sectional area of the jet. We are given $v = 18 \text{ m/s}$ and $av = 0.1 \text{ m}^3/\text{s}$. $\rho = 1000 \text{ kg/m}^3$. $F = (1000 \text{ kg/m}^3)(0.1 \text{ m}^3/\text{s})(18 \text{ m/s}) = \underline{1800 \text{ N}}$.
 In part **a** the plate is moving toward the stream at 3 m/s. Two things effect a change in the momentum change/time. First if the liquid again splashes at right angles to the plate it has picked up a velocity of 3 m/s opposite to the jet's direction. The total change in forward velocity is therefore not $v = 18 \text{ m/s}$ but $= 18 + 3 = 21 \text{ m/s}$. Second, the mass of water hitting the plate per second increases from av to $a(v + 3)$. Noting

that $a = (0.1 \text{ m}^3/\text{s})/v$ we have for the force

$$F = \rho\left(\frac{0.1}{v}\right)(v+3)(v+3) = \frac{(1000 \text{ kg/m}^3)(0.1 \text{ m}^3/\text{s})(21 \text{ m/s})^2}{18 \text{ m/s}} = \underline{2450 \text{ N}}$$

16.29 A pump draws water from a reservoir and sends it through a horizontal hose. Since the water starts at rest and is set into motion by the pump, the pump must deliver power P to the water when the flow rate is Φ, even if fluid friction is negligible. A new pump is to be ordered which will pump water through the same system at a rate $\Phi' = 2\Phi$. What must be the power P' of the new pump? Assume that friction is still negligible.

❚ By the work-energy theorem,

$$P = \frac{\text{KE imparted to the water}}{\text{time}} = \frac{\text{KE}}{\text{vol. water}} \times \frac{\text{vol. water}}{\text{time}} \propto v^2 \times v = v^3$$

The mass flux Φ is proportional to v so that $P \propto \Phi^3$. Thus if $\Phi' = 2\Phi$, the new power $P' = 8P$.

16.30 The U tube of Fig. 16-10 contains a length L of a zero-viscosity fluid. Show that the fluid column oscillates like a simple pendulum of length $L/2$.

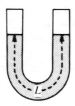

Fig. 16-10

❚ Let ρ denote the density of the fluid, M the mass of the fluid, and A the cross-sectional area of the tube. If one end of the fluid column is depressed a distance x below the equilibrium level, the other end must (assuming incompressibility) rise an equal distance x. The weight due to the height difference, $2x$, is a restoring force, $F = -\rho g A(2x) \equiv -kx$. Thus we have SHM, of frequency

$$f = \frac{1}{2\pi}\sqrt{\frac{k}{M}} = \frac{1}{2\pi}\sqrt{\frac{2\rho g A}{M}} = \frac{1}{2\pi}\sqrt{\frac{2\rho g A}{\rho A L}} = \frac{1}{2\pi}\sqrt{\frac{g}{L/2}}$$

16.2 VISCOSITY, STOKES' LAW, POISEUILLE'S LAW, TURBULENCE, REYNOLDS NUMBER

16.31 A number of tiny spheres made of steel with density ρ_s, and having various radii r_s, are released from rest just under the surface of a tank of water, whose density is ρ.

(a) Show that the "net gravitational force" acting on a sphere (the combined effect of weight and buoyancy) has magnitude $(4\pi/3)r_s^3(\rho_s - \rho)g$. (b) Assuming that the fluid flow around each descending sphere is laminar, find the terminal speed v of a sphere in terms of r_s, ρ_s, ρ, and the viscosity η of the water.

❚ (a)
$$(\text{weight}) - (\text{buoyant force}) = \rho_s g\left(\frac{4}{3}\pi r_s^3\right) - \rho g\left(\frac{4}{3}\pi r_s^3\right) = \frac{4\pi}{3}r_s^3(\rho_s - \rho)g.$$

(b) When the sphere is descending at terminal speed, the net force must vanish, so we can equate the downward, "net gravitational force" to the upward viscous drag, which by Stokes' law $= 6\pi\eta r_s v$. Thus

$$\frac{4\pi r_s^3}{3}(\rho_s - \rho)g = 6\pi\eta r_s v$$

Solving for the terminal speed, we obtain

$$v = \frac{2r_s^2}{9\eta}(\rho_s - \rho)g \tag{1}$$

16.32 Describe an idealized experiment for defining the coefficient of viscosity, η.

❚ Figure 16-11 shows two very large parallel plates A and B, separated by a distance d. The space between them is filled with fluid. A constant force F must be applied to plate B to keep it moving at a constant speed v_0 with respect to plate A. If a thin lamina of fluid at a uniform distance y from plate A moves with a speed

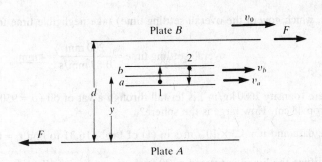

Fig. 16-11

$v(y) = v_0 y/d$, the flow is laminar. Then it is found that σ_s, the shear stress on the liquid, is proportional to v_0/d and the proportionality constant is the viscosity, η. Thus

$$\sigma_s = \eta v_0/d \qquad (1)$$

From (1) it is seen that the SI unit of viscosity is the *pascal · second* or *poiseuille*: $1\,Pl = 1\,Pa \cdot s = 1\,N \cdot s/m^2$.

16.33 A rotating-cylinder viscometer is employed to measure the coefficient of viscosity of castor oil at a temperature of 20 °C. The radius of the inner cylinder is $r_1 = 4.00$ cm, and the radius of the outer cylinder is $r_2 = 4.28$ cm. The inner cylinder is submerged in the oil to a depth $h = 10.2$ cm. When the outer cylinder is rotating at 20.0 rev/min, the torsion balance reads a torque $T = 3.24 \times 10^{-2}\,N \cdot m$. Find the viscosity of the castor oil.

▮ The cylindrical space between the outer and inner cylinders of the viscometer is not a bad approximation to the ideal flat-plate system of Fig. 16-11. Hence, by (1) of Prob. 16.32, $\eta = \sigma_s(d/v_0)$.

The wetted area of the inner cylinder is $a = 2\pi r_1 h$. Thus you have $\sigma_s = F/2\pi r_1 h$, where F is the drag force applied to the inner cylinder by the fluid, which is driven by the outer cylinder. This force results in a torque $T = r_1 F$, which can be read on the scale of the torsion balance. In terms of this torque, the shear stress is $\sigma_s = T/(2\pi r_1^2 h)$.

Since the inner cylinder is at rest, the relative speed v_0 of the two cylinders is given by $v_0 = \omega r_2$, where ω is the angular speed of the outer cylinder. And the distance between the cylinders is $d = r_2 - r_1$.

Using the above values of σ_s, v_0, and d, you obtain

$$\eta = \frac{T(r_2 - r_1)}{2\pi\omega r_1^2 r_2 h}$$

Inserting the numerical values gives

$$\eta = \frac{(3.24 \times 10^{-2}\,N \cdot m)(0.28 \times 10^{-2}\,m)}{2\pi\left(\dfrac{20.0\ \text{rev/min} \times 2\pi\ \text{rad/rev}}{60\ \text{s/min}}\right)(4.00 \times 10^{-2}\,m)^2(4.28 \times 10^{-2}\,m)(10.2 \times 10^{-2}\,m)} = \underline{0.99\ Pl}$$

16.34 How fast will an aluminum sphere of radius 1 mm fall through water at 20 °C once its terminal speed has been reached? Assume laminar flow. [sp gr (Al) = 2.7; $\eta_{\text{water}} = 8 \times 10^{-4}\,Pl$.]

▮ Substituting the data in (1) of Prob. 16.31, we find $v = 4.6$ m/s. (Actually, the assumption of laminar flow for this problem is not realistic; see Prob. 16.51.)

16.35 A typical riverborne silt particle has a radius of 20 μm and a density of 2×10^3 kg/m^3. The viscosity of water is approximated by 1.0 mPl. Find the terminal speed with which such a particle will settle to the bottom of a motionless volume of water. (Unless the speed of internal fluid motions is smaller than this settling speed, the silt particles will not settle to the bottom.)

▮ Using (1) of Prob. 16.31,

$$v_f = \frac{2(20 \times 10^{-6})^2[(2.0 - 1.0) \times 10^3](9.80)}{9(1.0 \times 10^{-3})} = 8.7 \times 10^{-4}\,m/s = \underline{0.87\ mm/s}$$

16.36 Refer to Prob. 16.35. Suppose that you filled a 1-L bottle of 50-cm^2 cross section with water from a muddy river, such as the lower Mississippi. After all internal motions of the water itself had stopped, about how long would it take for all the silt to settle to the bottom?

▮ Since the height of the bottle will be about 200 mm, and since all silt particles (in particular, those at the

294 ☐ CHAPTER 16

top of the bottle, which govern the overall settling time) take negligible time to achieve the terminal speed,

$$\text{overall settling time} = \frac{200 \text{ mm}}{0.87 \text{ mm/s}} \approx \underline{4 \text{ min}}$$

16.37 A tiny glass sphere (density 2600 kg/m³) is let fall through a vat of oil ($\rho = 950$ kg/m³, $\eta = 0.21$ Pl). In 100 s it is observed to drop 43 cm. How large is the sphere?

▮ Substitute the data and $v = 4.3 \times 10^{-3}$ m/s in (1) of Prob. 16.31 to find $r_s = \underline{0.50 \text{ mm}}$.

16.38 In a certain centrifuge the liquid is rotated at 20 rev/s at a radius of 10 cm from the axis of rotation. Tiny spherical particles of radius b and density 1020 kg/m³ in a dilute water solution (density 1000 kg/m³) are placed in the centrifuge. Find the terminal speed with which they settle out of the solution. Ignore the effect of gravity. $\eta_{\text{water}} = 8.0 \times 10^{-4}$ Pl.

▮ The particle needs a centripetal force $m\omega^2 r = [(4\pi b^3 \rho)/3](20 \times 2\pi)^2(0.10) = 6.75 \times 10^6 b^3$ N to keep it moving in a circle. A liquid (water-composed) particle also needs a centripetal force to keep it moving in a circle. Since only the surrounding liquid can supply the centripetal force, particles more dense than water will pass through to the outer edge of the centifuge. This is then very similar to particles falling through water in a gravitational field. The actual static force supplied by the water in the centrifuge is just the effective "buoyant force" and must equal the centripetal force on an equivalent volume of water (using the same reasoning as that for Archimedes' principle in a gravitational field). Hence, BF $= (1000/1020)6.75 \times 10^6 b^3 = 6.61 \times 10^6 b^3$. The remaining force is supplied by the viscous friction force ($6\pi\eta bv$). Hence, $6\pi\eta bv = 0.14 \times 10^6 b^3$, with $\eta = 8.0 \times 10^{-4}$ Pl, yields $v = \underline{9.3 \times 10^6 b^2 \text{ m/s}}$.

16.39 The viscous force on a liquid passing through a length L of pipe in laminar flow is given by $F_v = 4\pi\eta L v_m$, where η is the liquid viscosity and v_m is the maximum velocity of the liquid (i.e., along the central axis of the pipe). Find an expression for v_m in a horizontal segment of pipe in terms of p_1 and p_2, the pressure at the back and forward ends of the pipe, and in terms of η, L, and r, the radius of the pipe.

▮ In steady flow the viscous force is balanced by the force due to the pressure difference at the back and front ends. Thus $p_1 \pi r^2 - p_2 \pi r^2 = 4\pi\eta L v_m$. Solving for v_m we have $v_m = [(p_1 - p_2)r^2]/(4\eta L)$.

16.40 What is the pressure drop (in mmHg) in the blood as it passes through a capillary 1 mm long and 2 μm in radius if the speed of the blood through the center of the capillary is 0.66 mm/s? (The viscosity of whole blood is 4×10^{-3} Pl).

▮ By the result of Prob. 16.39,

$$p_1 - p_2 = \frac{(4)(4 \times 10^{-3} \text{ Pa} \cdot \text{s})(10^{-3} \text{ m})(6.6 \times 10^{-4} \text{ m/s})}{(2 \times 10^{-6} \text{ m})^2} = 2600 \text{ Pa} = (2600 \text{ Pa})\left(\frac{1 \text{ mmHg}}{133 \text{ Pa}}\right) = \underline{19.5 \text{ mmHg}}$$

16.41 Using the result of Prob. 16.39, find an expression for the volume flow rate of fluid through a segment of pipe. In laminar flow the average flow velocity over a cross section is $v_m/2$.

▮ The volume flow rate, H, is the product of the cross-sectional area of the pipe and the average flow velocity. Thus $H = (\pi r^2 v_m)/2$. Using the expression for v_m from Prob. 16.39, we have $H = [\pi r^4(p_1 - p_2)]/(8\eta L)$. This is known as *Poiseuille's law*.

16.42 How much power is delivered at the back end of the capillary of Prob. 16.40 in pushing the blood through? Assume $p_1 = 10.0$ kPa.

▮ $P \equiv \text{power} = \text{force} \times \text{velocity} = (p_1 \pi r^2)(v_m/2) = p_1 H$, by Prob. 16.41. Thus

$$P = \frac{(1.0 \times 10^4 \text{ N/m}^2)(3.14)(2 \times 10^{-6} \text{ m})^4(0.26 \times 10^4 \text{ N/m}^2)}{8(4 \times 10^{-3} \text{ Pl})(10^{-3} \text{ m})} = \underline{4.1 \times 10^{-11} \text{ W}}$$

16.43 Assuming all else remains the same, how would the flow rate, H, change if the radius of a pipe is doubled? How would the power necessary to push the fluid through change? How would the fluid velocity change?

▮ Referring to Poiseuille's law (Prob. 16.41), and assuming that p_1, p_2, η, and L remain constant, we see that doubling r will increase H sixteen-fold. For the power P we have from Prob. 16.42 $P = p_1 H$ and the power increases sixteen-fold as well. The velocity varies as r^2 (see Prob. 16.39) and therefore only quadruples.

16.44 Verify that the Reynolds number, $R = \rho v_0 d / \eta$, is dimensionless.

▌ $$\text{Dim } [R] = \frac{(\text{kg/m}^3)(\text{m/s})(\text{m})}{(\text{N/m}^2)(\text{s})} = \frac{\text{kg} \cdot \text{m}}{\text{N} \cdot \text{s}^2} = \frac{\text{N}}{\text{N}} = \text{dimensionless}$$

16.45 A submarine is 100 m long. The shape of its hull is roughly cylindrical, with a diameter of 15 m. When it is submerged, it cruises at a speed of about 40 knots, or 20 m/s. Is the flow of water around the hull laminar or turbulent? The viscosity of water is $\eta = 1.0 \times 10^{-3}$ Pa · s.

▌ Even though the water in question is seawater rather than pure water, the values of the viscosity η and density ρ for pure water are sufficiently good approximations for the purposes of calculating the Reynolds number. The characteristic linear dimension, d, will be between 15 m and 100 m; choose $d = 30$ m. Then

$$R = \frac{\rho v_0 d}{\eta} = \frac{(1 \times 10^3 \text{ kg/m}^3)(20 \text{ m/s})(30 \text{ m})}{1 \times 10^{-3} \text{ Pa} \cdot \text{s}} \approx 1 \times 10^9$$

It is seen from Table 16-1 that this value is far above the critical Reynolds numbers given for transition from laminar to turbulent flow, so the flow must be turbulent and the drag force is probably proportional to a power of v greater than the second.

TABLE 16-1 Some Critical Reynolds Numbers

R (approx)	phenomenon
10	Upper limit for strict conformance to Stokes' law for a sphere
1200	Onset of turbulent flow in a cylindrical pipe with an irregular inlet
3000	Onset of turbulent flow in a long cylindrical pipe
2×10^4	Onset of turbulent flow in a pipe with entrance of optimized shape
3×10^5	Upper limit for v^2-dependence of viscous force in turbulent flow

16.46 Calculate the speed at which the flow of water in a long cylindrical pipe of diameter 2 cm becomes turbulent. Assume that the temperature is 20 °C, so the viscosity $\eta = 1.0 \times 10^{-3}$ Pa · s.

▌ In terms of the Reynolds number, the flow speed $v = (\eta R)/(\rho d)$. From Table 16-1, $R_{\text{crit}} = 3000$ for flow in a long cylindrical pipe; hence,

$$v_{\text{crit}} = \frac{\eta R_{\text{crit}}}{\rho d} = \frac{(1.0 \times 10^{-3} \text{ Pa} \cdot \text{s})(3 \times 10^3)}{(1.0 \times 10^3 \text{ kg/m}^3)(2 \times 10^{-2} \text{ m})} = \underline{0.150 \text{ m/s}}$$

16.47 When water is flowing through the pipe of Prob. 16.46 at the critical speed, what is the approximate rate at which the pipe delivers water to a tank at its end?

▌ When a viscous fluid flows through a cylindrical pipe, the flow speed is a maximum on the central axis and decreases to zero at the inner surface of the pipe. It is appropriate to identify the flow speed found in Prob. 16.46 with the flow speed on the central axis. For simplicity, in the present problem we neglect the variation of flow speed with distance from the central axis. Then the volume per unit time carried by the pipe is given by

$$\text{volume flow} = \left(\frac{\pi d^2}{4}\right) v_{\text{crit}} = \left[\frac{\pi (2.0 \times 10^{-2})^2}{4}\right](0.150) = 4.71 \times 10^{-5} \text{ m}^3/\text{s} = \underline{2.83 \text{ L/min}}$$

16.48 Refer to Probs. 16.46 and 16.47. (*a*) What increase in pumping power would be required to double the output to the tank? (*b*) Compare your answer in (*a*) to that holding for laminar flow.

▌ (*a*) In turbulent flow the drag is proportional to the *square* of the flow speed, so that the power required to maintain the flow increases in proportion to the *cube* of the flow speed. That is, when $v \geq v_{\text{crit}}$, an eightfold increase in power input is needed in order to double the fluid volume delivered per unit time.
(*b*) In the laminar regime a fourfold increase in power input suffices to double the fluid volume delivered per unit time, since the drag is proportional to v (Prob. 16.39).

16.49 Evaluate the Reynolds number of the flow in the rotating-cylinder viscometer of Prob. 16.33. Assume that the density of castor oil at 20 °C is 0.96×10^3 kg/m^3. Compare your result with the critical Reynolds numbers given in Table 16-1.

▮ To obtain the Reynolds number we must identify a characteristic length d and a characteristic velocity v_0. We take $v_0 = \omega r_2$, the relative speed of the two cylinders, and $d = r_2 - r_1$, the distance between the two cylinders. Using the numerical values $\rho = 0.96 \times 10^3\,\text{kg/m}^3$, $\omega = 20.0\,\text{rev/min} = 2.09\,\text{rad/s}$, $r_1 = 4.00 \times 10^{-2}\,\text{m}$, and $r_2 = 4.28 \times 10^{-2}\,\text{m}$, we find $v_0 = 8.96 \times 10^{-2}\,\text{m/s}$ and $d = 2.8 \times 10^{-3}\,\text{m}$. Using the measured viscosity value $\eta = 0.99\,\text{Pa}\cdot\text{s}$, we obtain a value for the Reynolds number:

$$R \equiv \frac{\rho v_0 d}{\eta} = \frac{(0.96 \times 10^3)(8.96 \times 10^{-2})(2.8 \times 10^{-3})}{0.99} = \underline{0.24}$$

This value is smaller than any of the critical Reynolds numbers listed in Table 16-1.

16.50 Refer to Probs. 16.33 and 16.49. **(a)** Is the assumption of laminar flow in Prob. 16.33 a valid one? **(b)** How could the viscometer be used to check the assumption of laminar flow?

▮ **(a)** The Reynolds number found in Prob. 16.49 is subcritical, so the assumption of laminar flow appears to be valid. (Note that the flow must be laminar in order for the viscometer to correctly indicate the viscosity.) **(b)** One could operate the turntable at various angular speeds. In laminar flow, the torque on the inner cylinder is proportional to the speed of the outer cylinder. If the flow is laminar and the torsion fiber is within its range of linear response, the angular displacement of the pointer will be directly proportional to the angular speed of the turntable.

16.51 Refer to Prob. 16.31. **(a)** Find the Reynolds number corresponding to the terminal speed v. For what range of radii r_s is it correct to assume strictly laminar flow? (See Table 16-1.) **(b)** Obtain numerical results for v, R, and $r_{\max}$, given that $\rho_s = 7.9 \times 10^3\,\text{kg/m}^3$, $\eta = 1.0 \times 10^{-3}\,\text{Pl}$.

▮ **(a)** Using the sphere diameter $2r_s$ as the characteristic length, we find the Reynolds number:

$$R = \frac{2\rho v r_s}{\eta} = \frac{4r_s^3}{9\eta^2}(\rho_s - \rho)\rho g$$

The flow is laminar for $R < R_c$, or for

$$r_s < \left[\frac{9\eta^2 R_c}{4(\rho_s - \rho)\rho g}\right]^{1/3}$$

According to Table 16-1, $R_c = 10$ for a sphere, so we conclude that the flow will be laminar for sphere radii less than

$$r_{\max} = \left[\frac{45\eta^2}{2(\rho_s - \rho)\rho g}\right]^{1/3}$$

(b) With $g = 9.80\,\text{m/s}^2$, $\rho = 1.00 \times 10^3\,\text{kg/m}^3$, $\rho_s = 7.9 \times 10^3\,\text{kg/m}^3$, and $\eta = 1.0 \times 10^{-3}\,\text{Pa}\cdot\text{s}$, we find that the terminal speed is given by $v = \underline{(1.5 \times 10^7\,\text{m}^{-1}\cdot\text{s}^{-1})r_s^2}$, the Reynolds number is given by $R = \underline{(3 \times 10^{13}\,\text{m}^{-3})r_s^3}$, and the maximum radius for laminar flow is equal to $r_{\max} = \underline{69\,\mu\text{m}}$.

CHAPTER 17

Temperature and Thermal Expansion

Tables 17-1 and 17-2 give data for many of the problems in Chaps. 17 and 18. Note that as a unit of temperature *difference*, the degree celsius (°C) is identical to the kelvin (K). Thus, e.g., the specific heat of aluminum may be cited as 0.22 kcal/kg · K. Note also that tabular data may be converted to SI by use of the relation 1 cal = 4.184 J (exactly).

TABLE 17-1 Heat Constants of Solids

substance	melting point, °C	coefficient of linear expansion, °C^{-1}	specific heat, kcal/kg · °C	heat of fusion	
				kcal/kg	BTU/lb
Aluminum	660	0.0000255	0.22	76.8	140
Bismuth	271	0.000013	0.030	12.6	22.7
Brass		0.0000193	0.090		
Copper	1084	0.0000167	0.093	43	77
Glass		0.0000083	0.20		
Gold	1064	0.0000142	0.031		
Ice	0	0.000051	0.50	79.8	144
Iron	1535	0.000012	0.11	30	54
Lead	327	0.000029	0.031	5.4	9.7
Mercury	−38.9		0.033	2.8	5.4
Nickel	1452	0.000013	0.106	4.6	8.3
Platinum	1772	0.000009	0.032	27	48.6
Silver	962	0.000019	0.057	22	39
Steel		0.000012	0.11		
Tungsten	3410	0.0000045	0.032		
Zinc	420	0.000032	0.092	28.1	50.6

TABLE 17-2 Heat Constants of Liquids

substance	boiling point, °C	volume expansion, °C^{-1}	specific heat, kcal/kg · °C	heat of vaporization	
				kcal/kg	BTU/lb
Alcohol (ethyl)	78.1	0.0011	0.55	205	369
Ammonia	−34			294	529
Aniline	184		0.514	110	198
Benzene	80.3	0.00124	0.34	94.4	170
Chloroform	61	0.00126	0.232	58	106
Ether (ethyl)	34.5	0.00163	0.56	88.4	159
Gasoline	70–90	0.0012		71–81	128–146
Glycerin	290	0.00053	0.58		
Mercury	358	0.000182	0.0332	68	122
Turpentine	159	0.00094	0.42	70	126
Water	100	0.00030	1.00	540	970

17.1 TEMPERATURE SCALES; LINEAR EXPANSION

17.1 Change each of the following to the Celsius and Kelvin scales: 68 °F, 5 °F, 176 °F.

▮ $t_C = \frac{5}{9}(t_F - 32)$; $T_K = t_C + 273.15 \approx t_C + 273$. Then

$$t_F = 68 °F \Rightarrow t_C = \underline{20 °C} \quad T_K = \underline{293 K}$$
$$t_F = 5 °F \Rightarrow t_C = \underline{-15 °C} \quad T_K = \underline{258 K}$$
$$t_F = 176 °F \Rightarrow t_C = \underline{80 °C} \quad T_K = \underline{353 K}$$

17.2 Change each of the following to the Fahrenheit and Rankine scales: 30 °C, 5 °C, −20 °C.

▮ $t_F = \frac{9}{5}t_C + 32$; $T_R = t_F + 459.67 \approx t_F + 460$. Then

$$t_C = 30 °C \Rightarrow t_F = 86 °F \quad T_R = 546 °R$$
$$t_C = 5 °C \Rightarrow t_F = 41 °F \quad T_R = 501 °R$$
$$t_C = -20 °C \Rightarrow t_F = -4 °F \quad T_R = 456 °R$$

17.3 At what temperature do the Celsius and Fahrenheit readings have the same numerical value?

▮ $t_F = \frac{9}{5}t_C + 32$. Let $t_F = t_C = x$. Then

$$x = \frac{9}{5}x + 32 \quad \text{or} \quad \frac{4}{5}x = -32 \quad \text{and} \quad x = -40$$

Thus $t_F = \underline{-40 °F}$ and $t_C = \underline{-40 °C}$ are the same temperature.

17.4 Ethyl alcohol boils at 78.5 °C and freezes at −117 °C under a pressure of 1 atm. Convert these temperatures to the (*a*) Kelvin scale and (*b*) Fahrenheit scale.

▮ (*a*) We have Kelvin temperature = Celsius temperature + 273.15 = 78.5 + 273.15 = $\underline{351.7 K}$, where the answer is read as "351.7 kelvins." Similarly, for −117 °C, Kelvin temperature = −117 + 273.15 = $\underline{156 K}$.
(*b*) We use Fahrenheit temp = $\frac{9}{5}$(Celsius temp) + 32.

boiling point = $\frac{9}{5}(78.5) + 32 = 141 + 32 = \underline{173 °F}$ freezing point = $\frac{9}{5}(-117) + 32 = -211 + 32 = \underline{-179 °F}$

17.5 Hydrogen may be liquefied at −235 °C under a pressure of 20 atm. What is this temperature on the Fahrenheit scale?

▮
$$t_F = 1.8t_C + 32 = 1.8(-235) + 32 = \underline{-391°}$$

17.6 The temperature of a room is 77 °F. What would it be on the Celsius scale?

▮
$$t_F = 1.8t_C + 32 \quad 77 = 1.8t_C + 32 \quad t_C = 45/1.8 = \underline{25°}$$

17.7 The temperature of liquid hydrogen is 20 K. What is this temperature on the Fahrenheit scale?

▮
$$20 K = 20 - 273 = -253 °C \quad t_F = 1.8t_C + 32 = 1.8(-253) + 32 = \underline{-423 °F}$$

17.8 Compute the increase in length of 50 m of copper wire when its temperature changes from 12 to 32 °C.

▮ $\Delta L = \alpha L_0 \, \Delta T = (1.67 \times 10^{-5} °C^{-1})(50 \text{ m})(20 °C) = \underline{16.7 \text{ mm}}$. (Values of α can be found in Table 17-1.)

17.9 A rod 3 m long is found to have expanded 0.091 cm in length for a temperature rise of 60 °C. What is α for the material of the rod?

▮
$$\alpha = \frac{1}{L_0}\frac{\Delta L}{\Delta T} = \frac{0.091 \times 10^{-2} \text{ m}}{(3 \text{ m})(60 \text{ K})} = \underline{5.1 \times 10^{-6} \text{ K}^{-1}}$$

17.10 At 15 °C, a bare wheel has a diameter of 30.000 in and the inside diameter of a steel rim is 29.930 in. To what temperature must the rim be heated so as to slip over the wheel?

▮ $\Delta L = \alpha L_0 \, \Delta T$. For our case $L_0 = 29.930$ in and $\Delta L = 0.070$ in. Then $\Delta T = 0.070$ in/ $[(1.2 \times 10^{-5} °C^{-1})(29.930 \text{ in})] = \underline{195 °C}$. Finally $t_C = 15 °C + 195 °C = \underline{210 °C}$.

17.11 A brass rod is 0.70 m long at 40 °C. Find the length of this rod at 50 °C.

▌ $L = L_0(1 + \alpha \, \Delta T) = (0.70 \text{ m})\{1 + (19.3 \times 10^{-6}/°\text{C})(10 °\text{C})\}$. $L = \underline{0.70013 \text{ m}}$.

17.12 A nickel-steel rod at 21 °C is 0.624 06 m in length. Raising the temperature to 31 °C produces an elongation of 121.6 μm. Find the length at 0 °C and the coefficient of linear expansion.

▌ To find the length at 0 °C, we could first solve for α and calculate ΔL for $\Delta T = -21$ °C. Instead we note that $\Delta L/L \sim \Delta T$. Then, starting from the same length L, for two different temperature changes we have for corresponding length changes $\Delta L_1/\Delta L_2 = \Delta T_1/\Delta T_2$. Letting $\Delta T_1 = -21$ °C (to get to 0 °C) and $\Delta T_2 = 10°$ (to get to 31°) we have

$$\Delta L_1 = \left(\frac{-21}{10}\right)(121.6 \times 10^{-6} \text{ m}) = -255 \times 10^{-6} \text{ m}$$

Then $L(0 °\text{C}) = 0.62406 \text{ m} - 0.00026 \text{ m} = \underline{0.62380 \text{ m}}$. To obtain α we note that

$$\alpha = \frac{1}{L}\frac{\Delta L}{\Delta T} \qquad \alpha = \frac{121.6 \times 10^{-6}}{(0.624)(10 °\text{C})} = \underline{19.5 \times 10^{-6} °\text{C}^{-1}}$$

17.13 Suppose that the standard of length was a 1-m-long bar of iron. What would be the maximum temperature variation of the bar if its length were to be preserved to an accuracy of 1 part per million?

▌ We want $\Delta L/L = \pm 10^{-6} = \alpha \, \Delta T = 12 \times 10^{-6} \, \Delta T$, so $\Delta T = \underline{\pm 0.083 °\text{C}}$.

17.14 A cylinder of diameter 1.000 00 cm at 30 °C is to be slid into a hole in a steel plate. The hole has a diameter of 0.999 70 cm at 30 °C. To what temperature must the plate be heated?

▌ The plate will expand in the same way whether or not there is a hole in it. Hence the hole expands in the same way a circle of steel filling it would expand. We want the diameter of the hole to change by $\Delta L = (1.000 00 - 0.999 70)$ cm $= 0.000 30$ cm. Using $\Delta L = \alpha L_0 \, \Delta T$,

$$\Delta T = \frac{\Delta L}{\alpha L_0} = \frac{0.000 30 \text{ cm}}{(1.2 \times 10^{-5} °\text{C}^{-1})(0.999 70 \text{ cm})} = 25.0 °\text{C}$$

The temperature of the plate must be $30 + 25 = \underline{55 °\text{C}}$.

17.15 An iron ball has a diameter of 6 cm and is 0.010 mm too large to pass through a hole in a brass plate when the ball and plate are at a temperature of 30 °C. At what temperature, the same for ball and plate, will the ball just pass through the hole?

▌ We let I stand for the iron ball and B stand for the brass plate. $L_I = 6$ cm and $L_I - L_B = 0.001$ cm at $t = 30$ °C. Since the brass plate expands uniformly, the hole must expand in the same proportion. Then heating both the ball and the plate leads to increases in the diameters of the ball and the hole, with the hole increasing more, since $\alpha_B > \alpha_I$. We require $\Delta L_B - \Delta L_I = 0.001$ cm. $\Delta L_B = \alpha_B L_B \, \Delta t$; $\Delta L_I = \alpha_I L_I \, \Delta t$. We can approximate L_B in this formula by 6 cm $= L_I$. Then

$$\Delta L_B - \Delta L_I = (\alpha_B - \alpha_I)L_I \, \Delta t = 0.001 \text{ cm}$$

or $\quad [(1.9 \times 10^{-5} °\text{C}^{-1}) - (1.2 \times 10^{-5} °\text{C}^{-1})](6 \text{ cm}) \, \Delta t = 0.001$ cm

Solving, $\Delta t = 23.8$ °C, and finally $t = 30 °\text{C} + 23.8 °\text{C} = \underline{53.8 °\text{C}}$.

17.16 It is desired to put an iron rim on a wooden wheel. The diameter of the wheel is 1.1000 m and the inside diameter of the rim is 1.0980 m. If the rim is at 20 °C initially, to what temperature must it be heated to just fit onto the wheel?

▌ From Table 17-1, α for iron is found to be $1.2 \times 10^{-5} °\text{C}^{-1}$.

$$\Delta L = 1.1000 - 1.0980 = 0.0020 \text{ m} = \alpha L \, \Delta t \qquad 0.0020 = (1.2 \times 10^{-5})(1.098) \, \Delta t$$

$$\Delta t = 152 °\text{C} \qquad \Delta t + 20 = \underline{172 °\text{C}}$$

17.17 A steel rail 30 m long is firmly attached to the roadbed only at its ends. The sun raises the temperature of the rail by 50 °C, causing the rail to buckle. Assuming that the buckled rail consists of two straight parts meeting in the center, calculate how much the center of the rail rises.

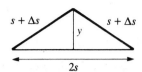

Fig. 17-1

❚ As indicated in Fig. 17-1, we let the initial length be $2s$ and the final total length be $2(s + \Delta s)$. The height of the center of the buckled rail is denoted by y. Assuming that the standard coefficient α of linear expansion can be used (in spite of the fact that the ends are anchored), we have $\Delta s = \alpha s \, \Delta T$. By the Pythagorean theorem,

$$y = \sqrt{(s + \Delta s)^2 - s^2} = \sqrt{2s \, \Delta s + (\Delta s)^2} = s\sqrt{2\alpha \, \Delta T + (\alpha \, \Delta T)^2}$$

With $s = 15.0 \text{ m}$, $\alpha = 12 \times 10^{-6} \text{ K}^{-1}$, and $\Delta T = 50 \text{ K}$, we obtain

$$y = (15.0 \text{ m})\sqrt{(12 \times 10^{-4}) + (6 \times 10^{-4})^2} = \underline{0.52 \text{ m}}$$

17.18 Consider two parallel bars [Fig. 17-2(a)] of different metals, having linear expansion coefficients α', α'' and fastened together so as to keep them at a fixed distance d apart. A change of temperature will cause their bending into two circular arcs intercepting an angle θ [Fig. 17-2(b)]. Find their mean radius of curvature, R.

❚ Assume as the temperature scale zero-point that temperature at which the bars are straight. If their common length at this temperature is L_0, their lengths at any other temperature T will be

$$L' = \theta R' = L_0(1 + \alpha'T) \qquad L'' = \theta R'' = L_0(1 + \alpha''T)$$

where R', R'' are the radii of curvature of the bars and θ is the angle subtended at the center of curvature by the joined bars. Also, $R' - R'' = d$.

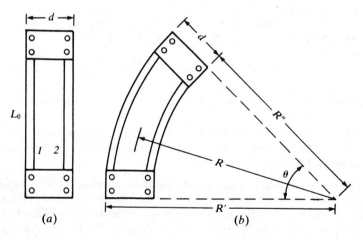

(a)   (b)   Fig. 17-2

Subtract the second of the above equations from the first to get

$$\theta(R' - R'') = (\alpha' - \alpha'')L_0 T \qquad \text{or} \qquad \theta = \frac{(\alpha' - \alpha'')L_0 T}{d}$$

Now add the two equations to obtain

$$\theta(R' + R'') = 2L_0 + (\alpha' + \alpha'')L_0 T$$

The mean radius of curvature is

$$R = \frac{R' + R''}{2} = \frac{2L_0 + (\alpha' + \alpha'')L_0 T}{2\theta} = \frac{2L_0 + (\alpha' + \alpha'')L_0 T}{2(\alpha' - \alpha'')L_0 T / d} = \frac{2 + (\alpha' + \alpha'')T}{2(\alpha' - \alpha'')T} d$$

But, $(\alpha' + \alpha'')T \ll 2$; so

$$R \approx \frac{2d}{2(\alpha' - \alpha'')T} = \frac{d}{(\alpha' - \alpha'')T}$$

17.19 (a) An aluminum measuring rod, which is correct at $5 \,°\text{C}$, measures a certain distance as 88.42 cm at $35 \,°\text{C}$.

Determine the error in measuring the distance due to the expansion of the rod. (*b*) If this aluminum rod measures a length of steel as 88.42 cm at 35 °C, what is the correct length of the steel at 35 °C?

▌ (*a*) The 88.42-cm mark on the aluminum rod is really at a greater distance from the zero position than indicated because of the increase in temperature, $\Delta t = 30$ °C. This increased length is $\Delta L = \alpha_{Al} L_{Al} \Delta t = (2.55 \times 10^{-5}\,°C^{-1})(88.42\,\text{cm})(30\,°C) = \underline{0.068\,\text{cm}}$.

(*b*) At 35 °C the measuring rod at the 88.42-cm mark is, from part *a*, actually 88.49 cm long. It thus measures an $\underline{88.49\,\text{cm}}$ length of steel at 35 °C.

17.20 A steel tape measure is calibrated at 70 °F. The width of a building lot is measured with the tape when the temperature is 15 °F, and a reading of 150 ft is obtained. How great an error is produced by the temperature difference? The coefficient of linear expansion of steel is $(6.7 \times 10^{-6})/°F$.

▌ $\Delta L = \alpha L \, \Delta t = (6.7 \times 10^{-6})(150)(15° - 70°) = (6.7 \times 10^{-6})(150)(-55°) \qquad \Delta L = -0.055\,\text{ft}$

The tape is 0.055 ft too short, and the reading obtained is too large.

17.21 A steel tape measures the length of a copper rod as 90.00 cm when both are at 10 °C, the calibration temperature for the tape. What would the tape read for the length of the rod when both are at 30 °C?

▌ At 30 °C, the copper rod will be of length $L_0(1 + \alpha_c \, \Delta T)$, while adjacent "centimeter" marks on the steel tape will be separated by a distance of $(1\,\text{cm})(1 + \alpha_s \, \Delta T)$. Therefore, the number of "centimeters" read on the tape will be

$$\frac{L_0(1 + \alpha_c \, \Delta T)}{(1\,\text{cm})(1 + \alpha_s \, \Delta T)} = \frac{(90\,\text{cm})[1 + (1.7 \times 10^{-5}\,°C^{-1})(20\,°C)]}{(1\,\text{cm})[1 + (1.2 \times 10^{-5}\,°C^{-1})(20\,°C)]} = 90\,\frac{1 + 3.4 \times 10^{-4}}{1 + 2.4 \times 10^{-4}}$$

Using the approximation $1/(1 + x) \approx 1 - x$, for x small compared with 1, we have

$$90\,\frac{1 + 3.4 \times 10^{-4}}{1 + 2.4 \times 10^{-4}} \approx 90(1 + 3.4 \times 10^{-4})(1 - 2.4 \times 10^{-4}) \approx 90(1 + 3.4 \times 10^{-4} - 2.4 \times 10^{-4}) = 90 + 0.009$$

The tape will read $\underline{90.01\,\text{cm}}$.

17.22 A steel tape is calibrated at 20 °C. On a cold day when the temperature is -15 °C, what will be the percent error in the tape?

▌ For a temperature change from 20 °C to -15 °C, we have $\Delta T = -35$ °C. Then,

$$\frac{\Delta L}{L_0} = \alpha \, \Delta T = (1.2 \times 10^{-5}\,°C^{-1})(-35\,°C) = -4.2 \times 10^{-4} = \underline{-0.042\%}$$

17.23 A steel wire of 2.0 mm² cross section is held straight (but under no tension) by attaching it firmly to two points a distance 1.50 m apart at 30 °C. If the temperature now decreases to -10 °C, and if the two tie points remain fixed, what will be the tension in the wire? For steel, $Y = 200\,000$ MPa.

▌ If free to do so, the wire would contract a distance ΔL as it cooled, where

$$\Delta L = \alpha L_0 \, \Delta T = (1.2 \times 10^{-5}\,°C^{-1})(1.5\,\text{m})(40\,°C) = 7.2 \times 10^{-4}\,\text{m}$$

But the ends are fixed. As a result, forces at the ends must, in effect, stretch the wire this same length, ΔL. Therefore, from $Y = (F/A)/(\Delta L/L)$, we have

$$\text{tension} = F = \frac{YA \, \Delta L}{L} = \frac{(2 \times 10^{11}\,\text{N/m}^2)(2 \times 10^{-6}\,\text{m}^2)(7.2 \times 10^{-4}\,\text{m})}{1.5\,\text{m}} = \underline{192\,\text{N}}$$

Strictly, we should have substituted $(1.5 - 7.2 \times 10^{-4})$ m for L in the expression for the tension. However, the error incurred in not doing so is negligible.

17.24 When a building is constructed at -10 °C, a steel beam (cross-sectional area 45 cm²) is put in place with its ends cemented in pillars. If the sealed ends cannot move, what will be the compressional force in the beam when the temperature is 25 °C? For steel, $Y = 200\,000$ MPa.

▌ As in Prob. 17.23,

$$\frac{\Delta L}{L_0} = \alpha \, \Delta T = (1.2 \times 10^{-5}\,°C^{-1})(35\,°C) = 4.2 \times 10^{-4}$$

Then
$$F = YA\frac{\Delta L}{L_0} = (2 \times 10^{11}\ \text{N/m}^2)(45 \times 10^{-4}\ \text{m}^2)(4.2 \times 10^{-4}) = \underline{380\ \text{kN}}$$

17.25 The two ends of a horizontal iron girder 8 m long and having a solid cross section of 150 cm² are anchored firmly in massive concrete posts when the temperature is 4 °C. How large a horizontal force will the girder exert on the post when the temperature is 30 °C? Use $Y = 200\,000$ MPa for iron.

❚ From the definition of Young's modulus we have $F/A = Y\,\Delta L/L$, so that $F = YA\alpha\,\Delta T$; substituting we have $(2 \times 10^{11}\ \text{Pa})(1.5 \times 10^{-2}\ \text{m}^2)(12 \times 10^{-6}\ {}^\circ\text{C}^{-1})(26) = \underline{940\ \text{kN}}$.

17.26 A flask of mercury is sealed off at 20 °C and is completely filled with the mercury. Find the pressure within the flask at 100 °C. Ignore the expansion of the glass. (Assume the bulk modulus for mercury to be $B = 2500$ MPa. The volume expansivity, γ, can be found in Table 17-2.)

❚ We first find $\Delta V/V = \gamma\,\Delta T = (182 \times 10^{-6})(80) = 1.46 \times 10^{-2}$; then pressure can be found from the expression for the bulk modulus $p = B\,|\Delta V/V|$ to be $(2500\ \text{MPa})(1.46 \times 10^{-2}) = \underline{36\ \text{MPa}}$.

17.27 A grandfather clock has a pendulum made of brass. The clock is adjusted to have a period of 1 s exactly at 20 °C. If operated at 30 °C, how much will the clock be in error 1 week after it is set? Will it be fast or slow?

❚ At 20 °C, $T_0 = 2\pi(L_0/g)^{1/2} = 1$ s. The brass elongates to $L = L_0[1 + (19.3 \times 10^{-6})(10\ ^\circ\text{C})]$ so the new period $T = 2\pi(L/g)^{1/2} = T_0(1 + 1.93 \times 10^{-4})^{1/2} = 1.0001 T_0$. At 30 °C the clock loses $(T - T_0)(1\ \text{week}) = (10^{-4}\ \text{s})(60 \times 60 \times 24 \times 7) = \underline{60\ \text{s}}$. (Note that an increase in period means that the clock ticks off fewer seconds and hence is slow.)

17.28 In the all-solid pendulum clock shown in Fig. 17-3, this pendulum is compensated for temperature change by using differential expansion so as to keep the center of oscillation of the pendulum at a fixed distance below the point of suspension. Find the dimensions of this pendulum if it is to tick off seconds.

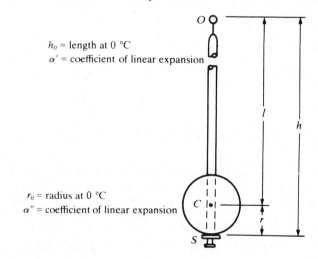

h_0 = length at 0 °C
α' = coefficient of linear expansion

r_0 = radius at 0 °C
α'' = coefficient of linear expansion

Fig. 17-3

❚ The light rod OS supports the heavy bob by means of the regulating screw S. Regarding the center of the bob C as approximately the center of oscillation of the pendulum (Prob. 0.00), the effective length of the pendulum is $l = h - r$. Now,

$$h = h_0(1 + \alpha'\,\Delta T) \qquad r = r_0(1 + \alpha''\,\Delta T)$$

where $\alpha'' \gg \alpha'$. Then $l = h_0 - r_0 + (h_0\alpha' - r_0\alpha'')\,\Delta T$, which will be independent of temperature if $h_0\alpha' - r_0\alpha'' = 0$. Also $l_0 = h_0 - r_0$, where l_0 is the length that must be maintained. Solve these last two equations for h_0 and r_0:

$$h_0 = \frac{\alpha''l_0}{\alpha'' - \alpha'} \qquad r_0 = \frac{\alpha'l_0}{\alpha'' - \alpha'}$$

If the clock is to tick off seconds, the period (2 s) of the pendulum is

$$P = 2\pi\sqrt{\frac{l_0}{g}} \qquad \text{or} \qquad 2 = 2\pi\sqrt{\frac{l_0}{g}} \qquad \text{or} \qquad l_0 = \frac{g}{\pi^2}$$

For the dimensions of the pendulum, we have:

$$h_0 = \frac{\alpha'' l_0}{\alpha'' - \alpha'} = \frac{\alpha''(g/\pi^2)}{\alpha'' - \alpha'} \quad \text{and} \quad r_0 = \frac{\alpha'(g/\pi^2)}{\alpha'' - \alpha'}$$

17.29^c The balance (timing) wheel of a mechanical wristwatch has a frequency of oscillation given by

$$\nu = \frac{1}{2\pi}\sqrt{\frac{k}{I}} = \frac{1}{2\pi G}\sqrt{\frac{k}{M}}$$

where I is its moment of inertia and G is its gyration radius. The wristwatch keeps accurate time at 25 °C. How many seconds would it gain a day at −25 °C if the balance wheel were made of aluminum? If it were made of platinum?

❚ Assuming that the torsion constant k does not vary with temperature, the change in frequency $d\nu$ corresponding to a change in the gyration radius of the balance wheel is given by

$$\frac{d\nu}{\nu} = \frac{1}{\nu}\frac{d\nu}{dG}\,dG = \frac{1}{\nu}\left(\frac{-1}{2\pi G^2}\sqrt{\frac{k}{M}}\right)dG = \frac{-dG}{G}$$

Since the balance wheel is homogeneous in composition, it changes in size but *not* in shape in response to temperature variations. Therefore its gyration radius G varies with temperature in the same manner that any of its linear dimensions do:

$$\frac{dG}{G} = \alpha\,dT$$

Here α is the coefficient of linear expansion. Assuming that $\alpha\,dT$ is much less than unity and that α does not vary rapidly with temperature, the above equations are sufficiently accurate. Then $d\nu/\nu = -\alpha\,dT$ and the number of seconds gained per day is $(86\,400\text{ s/day})(d\nu/\nu) = (8.64 \times 10^4\text{ s/day})(-\alpha\,dT)$. In the present problem, $dT = -50$ °C. If the balance wheel were made of aluminum, we would have $\alpha = 25.5 \times 10^{-6}/$°C, and therefore the watch would gain $(8.64 \times 10^4)(-25.5 \times 10^{-6})(-50) = \underline{110.2\text{ s/day}}$. If the balance wheel were made of platinum, for which $\alpha = 9.0 \times 10^{-6}/$°C, the watch would gain $(8.64 \times 10^4)(-9.0 \times 10^{-6})(-50) = \underline{38.9\text{ s/day}}$.

17.2 AREA AND VOLUME EXPANSION

17.30 A sheet of copper has an area of 500 cm^2 at 0 °C. Find the area of this sheet at 80 °C.

❚ The coefficient of area expansion is to a good approximation 2α. Thus

$$\Delta A = 2\alpha A\,\Delta t = 2(1.67 \times 10^{-5})(500)(80° - 0°) = 1.34\text{ cm}^2$$

The new area is $500 + 1.34 = \underline{501.3\text{ cm}^2}$.

17.31 A uniform solid brass sphere of radius b_0 and mass M is set spinning with angular speed ω_0 about a diameter. If its temperature is now increased from 20 to 80 °C without disturbing the sphere, what will be its new (a) angular speed and (b) rotational kinetic energy?

❚ (a) From conservation of angular momentum, $I_0\omega_0 = I\omega$, where $I_0 = (2Mb_0^2)/5$ and $I = (2Mb^2)/5$ with $b^2 = b_0^2(1 + 2\alpha\,\Delta T)$. Hence $\omega = (I_0\omega_0)/I = \omega_0/(1 + 2\alpha\,\Delta T)$ from which $\omega = \omega_0/[1 + 2(19.3 \times 10^{-6})(60)] = 0.997\,69\omega_0$. (b) The kinetic energy $= (I\omega^2)/2 = [0.997\,69(I_0\omega_0^2)]/2$.

17.32 If an anisotropic solid has coefficients of linear expansion α_x, α_y, and α_z for three mutually perpendicular directions in the solid, what is the coefficient of volume expansion for the solid?

❚ Consider a cube, with edges parallel to X, Y, Z, of dimension L_0 at $T = 0$. After a change in temperature $\Delta T = (T - 0)$, the dimensions change to

$$L_x = L_0(1 + \alpha_x T) \qquad L_y = L_0(1 + \alpha_y T) \qquad L_z = L_0(1 + \alpha_z T)$$

and the volume of the parallelopiped is

$$V = V_0(1 + \alpha_x T)(1 + \alpha_y T)(1 + \alpha_z T) \approx V_0[1 + (\alpha_x + \alpha_y + \alpha_z)T]$$

where $V_0 = L_0^3$. Therefore, the coefficient of volume expansion is given by $\beta \approx \alpha_x + \alpha_y + \alpha_z$. (Note: The symbols β and γ are often used for the coefficient of volume expansion.)

17.33 Calculate the increase in volume of 100 cm³ of mercury when its temperature changes from 10 to 35 °C.

▮ $\Delta V = \beta V_0 \Delta T$. The volume coefficient of expansion of mercury can be gotten from Table 17-2: $\beta = 1.8 \times 10^{-4} \, °C^{-1}$. Then $\Delta V = (1.8 \times 10^{-4} \, °C^{-1})(100 \, cm^3)(25 \, °C) = \underline{0.45 \, cm^3}$.

17.34 The coefficient of linear expansion of glass is found in Table 17-1. If a bottle holds 50.000 cm³ at 15 °C, what is its capacity at 25 °C?

▮ The glass expands uniformly in all dimensions, so the capacity expands in proportion to the glass. The volume coefficient of expansion is related to the linear coefficient by $\beta = 3\alpha$. Then $V = V_0(1 + 3\alpha \, \Delta T) = (50.000 \, cm^3)[1 + 3(8.3 \times 10^{-6} \, °C^{-1})(10 \, °C)]$. $V = \underline{50.012 \, cm^3}$.

17.35 Determine the change in volume of a block of iron 5 cm × 10 cm × 6 cm, when the temperature changes from 15 °C to 47 °C.

▮ $V_0 = 300 \, cm^3$; $\beta = 3\alpha$. Then

$$\Delta V = 3\alpha V_0 \, \Delta T = 3(1.2 \times 10^{-5} \, °C^{-1})(300 \, cm^3)(32 \, °C) = \underline{0.35 \, cm^3}$$

17.36 An open aluminum 300-mL container is full of glycerin at 20 °C. What volume of glycerin overflows when the container is heated to 110 °C? Use the expansion coefficients given in Tables 17-1 and 17-2.

▮ The volume of glycerin that overflows is equal to the difference in the volume expansion of glycerin ΔV_g and the volume expansion of aluminum ΔV_a.

$$\Delta V_g = \beta_g V \, \Delta t \qquad \Delta V_a = 3\alpha V \, \Delta t \qquad \Delta V_g - \Delta V_a = \beta_g V \, \Delta t - 3\alpha V \, \Delta t = (\beta_g - 3\alpha)V \, \Delta t$$

From Table 17-1, α equals $2.55 \times 10^{-5} \, °C^{-1}$, and from Table 17-2, β_g equals $5.3 \times 10^{-4} \, °C^{-1}$.

$$\Delta V_g - \Delta V_a = [(5.3 \times 10^{-4}) - 3(2.55 \times 10^{-5})](300)(110° - 20°) = (4.535 \times 10^{-4})(27\,000) = 12.2 \, mL$$

Thus, $\underline{12.2 \, mL}$ of glycerin overflows.

17.37 A glass vessel is filled with exactly 1 L of turpentine at 20 °C. What volume of the liquid will overflow if the temperature is raised to 86 °C?

▮ As the temperature rises, both the capacity of the glass vessel and the volume of the turpentine increase. The amount of overflow is the difference in these increases: $\Delta V_T - \Delta V_G = (\beta_T - \beta_G)V \, \Delta T$. Noting that $\beta_G = 3\alpha_G$, and using the data in Tables 17-1 and 17-2, we have

$$\Delta V_T - \Delta V_G = (9.4 \times 10^{-4} \, °C^{-1}) - (0.25 \times 10^{-4} \, °C^{-1})(1 \, L)(66 \, °C) = \underline{60 \, mL}$$

17.38 The density of gold is 19.30 g/cm³ at 20 °C. Compute the density of gold at 90 °C.

▮
$$d = \frac{M}{V} \qquad \frac{d_1}{d_2} = \frac{V_2}{V_1}$$

since mass is constant. The indices 1 and 2 refer to 90 °C and 20 °C, respectively.

$$\frac{d_1}{d_2} = \frac{V_2}{V_2(1 + 3\alpha \, \Delta T)} = \frac{1}{1 + 3\alpha \, \Delta T} \approx 1 - 3\alpha \, \Delta T \qquad (\text{since } 3\alpha \, \Delta T \ll 1)$$

$$d_1 = d_2(1 - 3\alpha \, \Delta T) = (19.3 \, g/cm^3)[1 - 3(14.2 \times 10^{-6} \, °C^{-1})(70 \, °C)] \qquad d_1 = \underline{19.24 \, g/cm^3}$$

17.39 The density of mercury at 0 °C is 13 600 kg/m³. Calculate the density of mercury at 50 °C.

▮ Let $\rho_0 =$ density of mercury at 0 °C, $\rho_1 =$ density of mercury at 50 °C, $V_0 =$ volume of m kg of mercury at 0 °C, $V_1 =$ volume of m kg of mercury at 50 °C. By conservation of mass, $m = \rho_0 V_0 = \rho_1 V_1$, from which

$$\rho_1 = \rho_0 \frac{V_0}{V_1} = \rho_0 \frac{V_0}{V_0 + \Delta V} = \rho_0 \frac{1}{1 + (\Delta V/V_0)}$$

But

$$\frac{\Delta V}{V_0} = \beta \, \Delta T = (1.82 \times 10^{-4} \, °C^{-1})(50 \, °C) = 0.0091$$

Substitution gives

$$\rho_1 = (13\,600 \, kg/m^3) \frac{1}{1 + 0.0091} = \underline{13\,477 \, kg/m^3}$$

17.40 Let Δl_1 be the length of the column in a mercury thermometer which corresponds to a rise in temperature Δt if the expansion of the glass is neglected. Let Δl_2 be the actual length which allows for the expansion of the glass. Calculate the numerical value of $(\Delta l_1 - \Delta l_2)/\Delta l_1$. Only the bulb is immersed in the object whose temperature is being measured.

▌ If the expansion of the bulb is neglected, the change Δl_1 in the length of the mercury column is given by

$$\Delta l_1 = \frac{\gamma_{Hg} V_{Hg} \Delta t}{A} \qquad (1)$$

where γ_{Hg} is the coefficient of volume expansion of mercury, V_{Hg} is the volume of mercury in the bulb, and A is the cross-sectional area of the capillary. However, the bulb's interior volume will increase by an amount $3\alpha_g V$, where α_g is the coefficient of linear expansion of glass and V is the volume of the bulb. Therefore the change in length of the mercury column is not Δl_1 but the smaller amount Δl_2 given by

$$\Delta l_2 = \frac{(\gamma_{Hg} V_{Hg} - 3\alpha_g V)\,\Delta t}{A} = (\gamma_{Hg} - 3\alpha_g)\frac{V\,\Delta t}{A} \qquad (2)$$

where we have used the fact that $V_{Hg} = V$. From Eqs. (1) and (2), we find

$$\frac{\Delta l_1 - \Delta l_2}{\Delta l_1} = \frac{\gamma_{Hg} - (\gamma_{Hg} - 3\alpha_g)}{\gamma_{Hg}} = \frac{3\alpha_g}{\gamma_{Hg}} \qquad (3)$$

Referring to Tables 17-1 and 17-2, we find $\alpha_g = (8.3 \times 10^{-6})\ \mathrm{K}^{-1}$ and $\gamma_{Hg} = (182 \times 10^{-6})\ \mathrm{K}^{-1}$, so that $(\Delta l_1 - \Delta l_2)/\Delta l_1 = \underline{0.14}$.

17.41 When the expansion of a liquid in a vessel is measured to obtain the coefficient γ, what is actually obtained directly is γ relative to the material of which the container is made. Figure 17-4 illustrates an apparatus from which the correct value of γ can be found without any knowledge of γ for the container. **(a)** Show that $\gamma = (h_t - h_0)/h_0 t$. **(b)** If $h_t - h_0 = 1.0\ \mathrm{cm}$, $h_0 = 100\ \mathrm{cm}$, and $t = 20\,°\mathrm{C}$, what is the value of γ for the liquid?

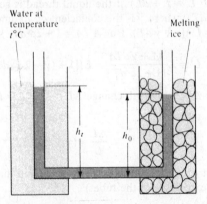

Water at temperature $t\,°\mathrm{C}$

Melting ice

h_t

h_0

Fig. 17-4

▌ **(a)** Hydrostatic equilibrium of the liquid in the horizontal section of the tube implies that $\rho_t g h_t = \rho_0 g h_0$, where ρ_t is the density of the liquid in the warm column and ρ_0 is the density of the liquid in the cold column. Using the definition of the coefficient of volume expansion γ, the volumes of a given mass M of liquid at temperatures t and $0\,°\mathrm{C}$ are related by $V_t = V_0(1 + \gamma t)$. Since $\rho_t V_t = \rho_0 V_0$, the densities are related by $\rho_0 = \rho_t(1 + \gamma t)$. But then, since $\rho_t h_t = \rho_0 h_0$, we have $h_t = h_0(1 + \gamma t)$. Solving this for γ, we find $\gamma = (h_t - h_0)/(h_0 t)$, as desired. **(b)** With $h_t - h_0 = 1.0\ \mathrm{cm}$, $h_0 = 100\ \mathrm{cm}$, and $t = 20\,°\mathrm{C}$, we obtain $\gamma = 1.0/[(100)(20)] = \underline{500 \times 10^{-6}/°\mathrm{C}}$.

17.42 The mercury in a thermometer has a volume of $210\ \mathrm{mm}^3$ at $0\,°\mathrm{C}$, at which temperature the diameter of the bore is $0.2\ \mathrm{mm}$. How far apart are the degree marks on the stem?

▌ We assume that the temperature rises from $0\,°\mathrm{C}$ to t, so $\Delta t = t$. Then $\Delta V_{Hg} = \beta_{Hg} V_{Hg} \Delta t$, $\Delta V_G = 3\alpha_G V_G \Delta t$, where here V_{Hg} is the volume occupied by the mercury at $0\,°\mathrm{C}$ and V_G is the volume of the glass container occupied by the mercury at $0\,°\mathrm{C}$. Thus $V_{Hg} = V_G = 210\ \mathrm{mm}^3$. The mercury rises in the tube of the thermometer because $\Delta V_{Hg} > \Delta V_G$. Indeed $hA_t = \Delta V_{Hg} - \Delta V_G$, where h is the height that the mercury rises and A_t is the cross section of the bore at temperature t. $A_t = \pi r^2(1 + 2\alpha_G \Delta t)$ with $r = 0.1\ \mathrm{mm}$, the radius at $0\,°\mathrm{C}$. Then

$$h = \frac{(\beta_{Hg} - 3\alpha_G)V_{Hg} t}{\pi r^2 (1 + 2\alpha_G t)} = \frac{[(182 - 25) \times 10^{-6}\,°\mathrm{C}^{-1}](210\ \mathrm{mm}^3)t}{0.0314[1 + (16.6 \times 10^{-6}\,°\mathrm{C}^{-1})t]}$$

We can clearly ignore the second term in brackets in the denominator since for a reasonable range of t it is $\ll 1$. Then

$$h = (1.05 \text{ mm} \cdot {}^\circ C^{-1})t \quad \text{or} \quad \frac{h}{t} = \underline{1.05 \text{ mm}/{}^\circ C} \quad \text{spacing}$$

17.43   A glass beaker holds exactly 1 L at $0\,{}^\circ C$. (*a*) What is its volume at $50\,{}^\circ C$? (*b*) If the beaker is filled with mercury at $0\,{}^\circ C$, what volume of mercury overflows when the temperature is raised to $50\,{}^\circ C$?

▌ (*a*) The volume of the beaker after the temperature change is

$$V_{\text{beaker}} = V_0(1 + 3\alpha_G\,\Delta T) = (1)[1 + 3(8.3 \times 10^{-6})(50)] = 1.001 \text{ L}.$$

(*b*) For the mercury expansion,

$$V_{\text{mercury}} = V_0(1 + \beta\,\Delta T) = (1)[1 + (1.82 \times 10^{-4})(50)] = 1.009 \text{ L}$$

The overflow is thus $1.009 - 1.001 = 0.008$ L, or $\underline{8\text{ mL}}$.

17.44^c   A thread of liquid in a uniform capillary tube is of length L, as measured by a ruler. The temperature of the tube and thread of liquid is raised by ΔT. Show that the increase in the length of the thread, again measured with a ruler, is $\Delta L = L(\gamma - 2\alpha)\,\Delta T$, where γ is the coefficient of volume expansion of the liquid and α is the coefficient of linear expansion of the tube material.

▌ Let us use differential notation in determining incremental information. Denoting the radius of the capillary by r and its cross-sectional area by A, the increase dr due to thermal expansion is given by $dr = r\alpha\,dT$, where α is coefficient of linear expansion of the tube material, and dT is the temperature increase. We assume that $dr/r \ll 1$. Since $A = \pi r^2$, we find that $dA/A = 2dr/r$, or that $dA \equiv A\beta\,dT = A(2\alpha)\,dT$, where β is used here as the symbol for the area coefficient of expansion. If the temperature is changed from T to $T' \equiv T + dT$, the cross-sectional area changes from A to $A' \equiv A + dA = A(1 + 2\alpha\,dT)$. The change in length (from L to $L' \equiv L + dL$) of the liquid thread is governed by the volume expansion of the liquid. Since $V' = V(1 + \gamma\,dT)$, where γ is the coefficient of volume expansion for the liquid, we have $L'A' = V' = V(1 + \gamma\,dT) = LA(1 + \gamma\,dT)$. But $A'/A = 1 + 2\alpha\,dT$, so

$$L' = L\frac{(1 + \gamma\,dT)}{(1 + 2\alpha\,dT)} = L\{[1 + (\gamma - 2\alpha)\,dT + O[(\alpha\,dT)^2]\}$$

For a temperature increase ΔT, the length change $\Delta L \equiv L' - L \approx L(\gamma - 2\alpha)\,\Delta T$, and the *relative* change in length is

$$\frac{\Delta L}{L} \approx (\gamma - 2\alpha)\,\Delta T$$

Here we have assumed that $\alpha\,\Delta T \ll 1$. (Note: The specification that a ruler is used for the measurement is to avoid the use of scale that expands with the tube.)

17.45   Show that $\Delta V = \gamma V\,\Delta T$ implies the following for the temperature variation of density: $\Delta\rho = -\gamma\rho\,\Delta T$. What is the meaning of the minus sign?

▌ $\Delta\rho = \rho - \rho_0 = m/V - m/V_0 = [m(V_0 - V)]/VV_0 = (-m\,\Delta V)/(VV_0)$, but $\Delta V/V_0 = \gamma\,\Delta T$, so $\Delta\rho = -(m/V)\gamma\,\Delta T = -\rho\gamma\,\Delta T$. The minus sign indicates that the density decreases with increasing temperature.

17.46   Show that the volume thermal expansion coefficient for an ideal gas at constant pressure is $1/T$.

▌ From $pV = nRT$, we have two equations, $pV_0 = nRT_0$ and $p(V_0 + \Delta V) = nR(T_0 + \Delta T)$. Subtract one from the other to obtain $p\,\Delta V = nR\,\Delta T$. But $(nR)/p = V_0/T_0$ and so this can be written $\Delta V/V_0 = (1/T_0)\,\Delta T$. Since $\Delta V/V_0 = \gamma\,\Delta T$, we see that $\gamma = 1/T_0$. Since T_0 is an arbitrary temperature, we can write for any T: $\gamma = 1/T$.

CHAPTER 18

Heat and Calorimetry

18.1 HEAT AND ENERGY; MECHANICAL EQUIVALENT OF HEAT

18.1 A quantity of lead shot is placed in a vertical cardboard tube 1.0 m long. When the tube is turned end for end 15 times, the rise in temperature of the lead shot is measured and found to be 1.0 °C. What value does this crude experiment give for the mechanical equivalent of heat? What is the main source of error?

▌ If the tube is turned fairly quickly, the lead shot falls through a distance $h = 1.0$ m each time the tube is turned. In $N = 15$ turns of the tube, the total energy converted from potential energy into kinetic energy and then into heat is given by $Nmgh$, where m is the mass of the lead shot. Assuming that all the heat remains in the lead shot, the temperature rise $\Delta t = 1.0$ °C of the shot is related to the mechanical energy release by $mc\,\Delta t = Nmgh$, where c is the specific heat capacity of lead. From Table 17-1, we have $c = 0.031$ kcal/kg · °C. Therefore we have

$$1 = \frac{Ngh}{c\,\Delta t} = \frac{(15)(9.80 \text{ m/s}^2)(1.0 \text{ m})}{(0.031 \text{ kcal/kg} \cdot \text{°C})(1 \text{°C})} = 4.7 \text{ kJ/kcal}$$

According to this experiment, 4.8 J of mechanical energy is equivalent to 1 cal of heat. The main source of error in the experiment is the fact that not all the heat generated is within the lead shot. During the inelastic impacts of the shot with the stoppers at the ends of the tube, some of the energy is deposited in the stoppers rather than in the shot. Furthermore, some of the heat generated within the lead shot is transferred to the slightly cooler surroundings. Both these effects tend to result in an *overestimate* for the mechanical equivalent of heat. An additional source of uncertainty is the difficulty of accurately measuring a slight temperature rise in a small amount of a solid.

18.2 In the Joule experiment, a mass of 20 kg falls 1.5 m at a constant velocity to stir the water in a calorimeter. If the calorimeter has a water equivalent of 2 g and contains 12 g of water, what is $\mathscr{J}$, the mechanical equivalent of heat, for a temperature rise of 5.0 °C?

▌ Expressing ΔPE in joules and Q in cal we have

$$\mathscr{J} = \frac{\Delta \text{PE}}{Q} = \frac{mgy}{m_w c\,\Delta t} = \frac{20(9.8)(1.5)}{(12+2)(1)(5.0)} = \underline{4.2 \text{ J/cal}}$$

18.3 Victoria Falls in Africa is 122 m in height. Calculate the rise in temperature of the water if all the potential energy lost in the fall is converted to heat.

▌ Consider mass m of water falling.

$$mgy = mc\,\Delta t \qquad gy = c\,\Delta t$$

We express both sides in joules by noting

$$c = 1 \text{ kcal/kg} \cdot \text{K} = 4184 \text{ J/kg} \cdot \text{K}$$

Then

$$9.8(122) = 4184\,\Delta t \qquad \text{and} \qquad \Delta t = \underline{0.29 \text{ K}}$$

18.4 A 2.2-g lead bullet is moving at 150 m/s when it strikes a bag of sand and is brought to rest. (*a*) If all the frictional work is transferred to thermal energy in the bullet, what is the rise in temperature of the bullet as it is brought to rest? (*b*) Repeat if the bullet lodges in a 50-g block of wood that is free to move.

▌ (*a*) $K = \frac{1}{2}mv^2 = mc\,\Delta T$ for the lead bullet. The left side is calculated in joules and converted to calories: $K = 0.0022(150)^2/2 = 24.8$ J $= (24.8/4.184)$ cal $= 5.91$ cal. Then 5.91 cal $= (2.2 \text{ g})(0.031 \text{ cal/g} \cdot \text{°C})\,\Delta T$ and $\Delta T = \underline{87 \text{°C}}$. (*b*) Here some of the initial kinetic energy is carried away by the bullet-block combination. From momentum conservation the velocity of the system after the collision is given by $(2.2 \text{ g})(150 \text{ m/s}) = (50 \text{ g} + 2.2 \text{ g})V$ and $V = 6.32$ m/s. $K_f = \frac{1}{2}(0.0522 \text{ kg})(6.32 \text{ m/s})^2 = 1.04$ J $= 0.25$ cal. ΔQ is now $5.91 - 0.25 = 5.66$ cal $= 2.2(0.031)\,\Delta T'$ and $\Delta T' = \underline{83 \text{°C}}$. (Note that the symbols Q and ΔQ are used to denote heat transfer. Sometimes the symbols H and ΔH are used instead.)

18.5 A lead bullet of mass m is fired at a tree trunk and emerges on the other side. The speed of the bullet is 500 m/s as it enters and 300 m/s as it emerges. Assuming that 40 percent of the loss of kinetic energy is stored as heat in the bullet, calculate the rise in temperature of the bullet. The specific heat of lead is found in Table 17-1.

▮ Express kinetic energy and heat energy in joules and the mass in kilograms.

$$0.40(\Delta KE) = Q \qquad 0.40(\tfrac{1}{2}mv_1^2 - \tfrac{1}{2}mv_2^2) = mc\,\Delta t \qquad 0.40(\tfrac{1}{2}v_1^2 - \tfrac{1}{2}v_2^2) = c\,\Delta t$$
$$0.40(\tfrac{1}{2} \times 500^2 - \tfrac{1}{2} \times 300^2) = (0.031)(4186)\,\Delta t \qquad \text{and} \qquad \Delta t = 247\,°C$$

The temperature rise of the bullet is $\underline{247\,°C}$.

18.6 A mass m of lead shot is placed at the bottom of a vertical cardboard cylinder that is 1.5 m long and closed at both ends. The cylinder is suddenly inverted so that the shot falls 1.5 m. By how much will the temperature of the shot increase if this process is repeated 100 times? Assume no heat loss.

▮ Potential energy is converted into heat. $U_g = (100m)(9.8)(1.5) = 1470m$ J, where m is in kilograms. This equals $(130\ \text{J/kg} \cdot °C)(m\ \text{kg})\,\Delta T$. Solving for ΔT one obtains $\underline{11.3\,°C}$. Note that the mass m drops out of the calculation. (Care must be taken that m appear with the same units on both sides of the equation in this type of problem.)

18.7 As in Joule's experiment, a weight attached to a string which passes over a pulley causes a paddle wheel to turn in a water container as the weight falls. The paddle is so constructed that the weight falls slowly, thereby acquiring negligible kinetic energy. Its decrease in potential energy is therefore converted entirely into frictional work or heat energy in the water. Suppose that the 200-g mass fell 2.0 m. How much would it heat the 100 g of water in the container if no heat were lost to the paddle or container?

▮ We calculate the mechanical energy and convert to calories. Energy lost by mass $= mgh = 0.20(9.8)(2.0) = 3.92$ J $= 0.94$ cal. Now $\Delta Q = cm\,\Delta T$, where c is in cal/g · °C and m is in grams, leading to $0.94 = 1.0(100)\,\Delta T$, so $\Delta T = \underline{0.0094\,°C}$.

18.8 An iron rocket fragment initially at $-100\,°C$ enters the atmosphere almost horizontally and quickly fuses completely. Assuming no heat losses by the fragment, calculate the minimum velocity it must have had when it entered.

▮ From Table 17-1, the specific heat of iron is 0.11 kcal/kg · K, and the heat of fusion is 30 kcal/kg. The kinetic energy of the fragment is all changed to heat. Its melting point is 1535 °C. We express m in kilograms everywhere.

$$\tfrac{1}{2}mv^2 = [m(30) + m(0.11)(1535° + 100°)](4184) \qquad \tfrac{1}{2}v^2 = (30 + 179.9)(4184)$$
$$v^2 = 1.76 \times 10^6 \qquad v = \underline{1.32\ \text{km/s}}$$

18.9 A 60-kg boy running at 5.0 m/s while playing basketball falls to the floor and skids along on his leg until he stops. How many calories of heat are generated between his leg and the floor? Assume that all this heat energy is confined to a volume of 2.0 cm^3 of his flesh. What will be the temperature change of the flesh? Assume $c = 1.0$ cal/g · °C and $\rho = 950$ kg/m^3 for flesh.

▮ The boy's kinetic energy is changed to heat energy. Set $Q = (mv^2)/2 = [60(25)]/2 = 750$ J $= 179$ cal. From $Q = c\rho V\,\Delta T$, 179 cal $= (1.0\ \text{cal/g} \cdot °C)(0.950\ \text{g/cm}^3)(2.0\ \text{cm}^3)\,\Delta T$, whence $\Delta T = \underline{94\,°C}$.

18.10 The Btu is defined as the heat needed to raise 1 lb of water 1 F°. (**a**) Show that specific heats are numerically the same in cal/g · °C and Btu/lb · F°. (**b**) What is the conversion factor between Btu and cal?

▮ (**a**) The calorie is defined as the heat needed to raise 1 g of water 1 °C. To say then, for example, that $c_{Al} = 0.22$ cal/g · °C is to imply that only 0.22 as much heat is required to raise a given mass (say, 1 lb) of Al through a given temperature interval (say, 1 F°) as is required to raise 1 lb of H$_2$O through 1 F°. But this latter amount is, by definition, 1.0 Btu. Hence,

$$c_{Al} = \frac{0.22(1.0\ \text{Btu})}{(1\ \text{lb})(1\ \text{F}°)} = 0.22\ \text{Btu/lb} \cdot \text{F}°$$

(**b**) If

$$\frac{1\ \text{cal}}{(1\ \text{g})(1\ °C)} = \frac{1\ \text{Btu}}{(1\ \text{lb})(1\ °F)} = \frac{1\ \text{Btu}}{(453.6\ \text{g})(\tfrac{5}{9}\,°C)}$$

then

$$1\ \text{Btu} = (453.6)(\tfrac{5}{9}) = \underline{252\ \text{cal}}$$

18.11 How many Btu are removed in cooling each of the following from 212 to 68 °F? (a) 1 lb water, (b) 2 lb leather ($c = 0.36$ cal/g · °C), (c) 3 lb asbestos ($c = 0.20$ cal/g · °C).

▌ In view of Prob. 18.10(a): (a) $\Delta Q = (1)(1.0)(144) = \underline{144\text{ Btu}}$. (b) $\Delta Q = (2)(0.36)(144) = \underline{104\text{ Btu}}$. (c) $\Delta Q = (3)(0.20)(144) = \underline{86\text{ Btu}}$.

18.12ᶜ The specific heat of many solids at very low temperatures varies with absolute temperature T according to the relation $c = AT^3$, where A is a constant. How much heat energy is needed to raise the temperature of a mass m of such a substance from $T = 0$ to $T = 20$ K?

▌ Since $\Delta Q = cm\,\Delta T$, one has $\Delta Q = AT^3 m\,\Delta T$. In the limit as ΔT goes to zero, this becomes $dQ = AmT^3\,dT$. Integrating on T from 0 to 20 gives $Q = \underline{40\,000Am}$.

18.13 Assume that the total heat of vaporization (per gram) of water can be used to supply the energy needed to tear 1 g of water molecules apart from each other. How much energy is needed per molecule for this purpose? Find the ratio of this energy to kT at the boiling point.

▌ In 1 g of water there are $N_A(0.001/M)$ molecules (where M is the molecular weight in kg/kmol) = $(6.02 \times 10^{26})(0.001/18) = 3.34 \times 10^{22}$ molecules. Using the heat of vaporization of water from Table 17-2 of 540 cal/g, we have for the energy per molecule = $540/(3.34 \times 10^{22}) = 1.612 \times 10^{-20}$ cal $= 6.74 \times 10^{-20}$ J. But $kT = (1.38 \times 10^{-23})(373) = 5.15 \times 10^{-21}$ J, so the ratio is $\underline{13.1}$.

18.14 Cool water at 9.0 °C enters a hot-water heater from which warm water at a temperature of 80 °C is drawn at an average rate of 300 g/min. How much average electric power does the heater consume in order to provide hot water at this rate? Assume that there is negligible heat loss to the surroundings.

▌ Each second, $\frac{300}{60} = 5.0$ g are heated. Thus $\Delta Q = cm\,\Delta T = (1.0\text{ cal/g} \cdot \text{°C})(5.0\text{ g/s})(71\text{ °C})(4.184\text{ J/cal}) = \underline{1.48\text{ kW}}$.

18.15 An electric heater supplies 1.8 kW of power in the form of heat to a tank of water. How long will it take to heat the 200 kg of water in the tank from 10 to 70 °C? Assume heat losses to the surroundings to be negligible.

▌ The heat added is $(1.8\text{ kJ/s})t$ and the heat absorbed is $cm\,\Delta T = (4.184\text{ kJ/kg} \cdot \text{K})(200\text{ kg})(60\text{ K}) = 5.0 \times 10^4$ kJ. Equating heats, $t = 2.78 \times 10^4$ s $= \underline{7.75\text{ h}}$.

18.16 Starting at 20 °C, how much heat is required to heat 0.3 kg of aluminum to its melting point and then to convert it all to liquid?

▌ $Q = mc\,\Delta t + mL$. The specific heat of aluminum c is 0.22 kcal/kg · °C, the heat of fusion L is 76.8 kcal/kg, and the melting point is 660 °C.

$$Q = 0.3(0.22)(660° - 20°) + 0.3(76.8) = 42.24 + 23.04 = \underline{65.3\text{ kcal}}$$

18.17 A certain 6-g bullet melts at 300 °C and has a specific heat capacity of 0.20 cal/g · °C and a heat of fusion of 15 cal/g. How much heat is needed to melt the bullet if it is originally at 0 °C?

▌ The bullet must first be heated to 300 °C and then melted. Heat needed = $(6\text{ g})(0.20\text{ cal/g} \cdot \text{°C})(300 - 0)\text{ °C} + (6\text{ g})(15\text{ cal/g}) = \underline{450\text{ cal}}$, or $\underline{1880\text{ J}}$.

18.18 Refer to Prob. 18.17. What is the slowest speed at which the bullet can travel if it is to just melt when suddenly stopped?

▌ 1880 J is to be supplied as $K = (mv^2)/2$. 1880 J $= \frac{1}{2}(0.006\text{ kg})v_{min}^2$, so that $v_{min} = \underline{790\text{ m/s}}$.

18.19 How many calories are required to change exactly 1 g of ice at -10 °C to steam at atmospheric pressure and 120 °C? [Assume the specific heat of steam at a constant pressure of 1 atm is 0.481 cal/(g · °C).]

▌ The heat requirement is evidently the sum of five contributions. The first stage is the warming of the ice from -10 °C to the melting point (0 °C). According to Table 17-1, the specific heat capacity of ice is 0.50 cal/(g · °C). Therefore the heat required in the first stage is given by $\Delta H_1 = mc_1\,\Delta t_1 = (1.00\text{ g})[0.50\text{ cal/(g} \cdot \text{°C)}](10\text{ °C}) = \underline{5.0\text{ cal}}$. The second stage is the melting of the ice at 0 °C and 1 atm of pressure. According to Table 17-1, the latent heat for the melting of ice is 79.8 cal/g, so $\Delta H_2 = mL_2 = (1.00\text{ g})(79.8\text{ cal/g}) = \underline{79.8\text{ cal}}$. The third stage is the heating of the water from 0 to 100 °C, the boiling point, so the heat required is given by $\Delta H_3 = mc_3 \Delta t_3 = (1.00\text{ g})(1.000\text{ cal/g} \cdot {}^\circ\text{C})(100\,{}^\circ\text{C}) = \underline{100\text{ cal}}$. The fourth stage is the boiling of the water at a temperature of $100\,{}^\circ\text{C}$ and at a constant pressure of 1.00 atm. According to Table 17-2, the latent heat for the boiling of water at 1.00 atm is 540 cal/g, so $\Delta H_4 = mL_4 = (1.00\text{ g})(540\text{ cal/g}) = \underline{540\text{ cal}}$. The fifth and final stage is the heating of the steam from 100 to $120\,{}^\circ\text{C}$ (at a constant pressure of 1.00 atm). Assuming that between 100 and $120\,{}^\circ\text{C}$ the specific heat capacity of steam is constant and has the value 0.481 cal/(g $\cdot$ ${}^\circ\text{C}$) given, we find $\Delta H_5 = mc_5 \Delta t_5 = (1\text{ g})[0.481\text{ cal/(g} \cdot {}^\circ\text{C})](20\,{}^\circ\text{C}) = \underline{9.62\text{ cal}}$. The total heat requirement $\Delta H = \Delta H_1 + \Delta H_2 + \Delta H_3 + \Delta H_4 + \Delta H_5 = (5.0 + 79.8 + 100 + 540 + 9.62) = \underline{734.4\text{ cal}}$.

18.20 In an experiment performed at atmospheric pressure, 20 g of ice at $-50\,{}^\circ\text{C}$ is changed to steam at $150\,{}^\circ\text{C}$. If the specific heats of steam and of ice are both 0.5 kcal/kg $\cdot$ ${}^\circ\text{C}$, find the approximate number of kilocalories required.

▎ Using heats of fusion and vaporization from Tables 17-1 and 17-2 we have

$$Q = 0.02(0.5)[0^\circ - (-50^\circ)] + 0.020(80) + 0.020(100^\circ - 0^\circ) + 0.020(540) + 0.02(0.5)(150^\circ - 100^\circ)$$
$$= 0.5 + 1.6 + 2 + 10.8 + 0.5 = \underline{15.4\text{ kcal}}$$

18.21 How much heat must be removed by a refrigerator from 2 lb of water at $70\,{}^\circ\text{F}$ to convert it to ice cubes at $10\,{}^\circ\text{F}$?

▎ $Q = mc_w \Delta t_w + mc_i \Delta t_i + mL$, where L is the heat of fusion. From Table 17-1 and Prob. 18.10(a), c_i for ice is 0.50 Btu/(lb)(F°) and L is 144 Btu/lb; m is the weight of water.

$$Q = 2(1)(70^\circ - 32^\circ) + 2(0.50)(32^\circ - 10^\circ) + 2(144) = 76 + 22 + 288 = \underline{386\text{ Btu}}$$

18.22 A normal diet might furnish 2000 nutritionist's calories (2000 kcal) to a 60-kg person in a day. If this energy were used to heat the person with no losses to the surroundings, how much would the person's temperature increase? $c = 0.83$ cal/g $\cdot$ ${}^\circ\text{C}$ for the average person.

▎ The temperature rise Δt is obtained from $Q = mc\,\Delta t$, or $2000\text{ kcal} = (60\text{ kg})(0.83\text{ kcal/kg} \cdot {}^\circ\text{C})\,\Delta t$. Solving we have $\Delta t = \underline{40.1\,{}^\circ\text{C}}$.

18.23 If all the heat produced were used, how many liters of natural gas at STP (the heat of combustion is 37.3 MJ/m³) is needed to heat 4.54 kg of water in a 0.45 kg copper cup from 297.59 to 373.15 K?

▎ Let $V =$ volume of natural gas. From Table 17-1, the specific heat of copper is 389 J/kg $\cdot$ K. Then, $(37.3 \times 10^6\text{ J/m}^3)V = [(4.54\text{ kg})(4184\text{ J/kg} \cdot \text{K}) + (0.45\text{ kg})(389\text{ J/kg} \cdot \text{K})](75.56\text{ K})$. Solving, we get $V = 0.039\text{ m}^3 = \underline{39\text{ L}}$.

18.24 When 5 g of a certain type of coal is burned, it raises the temperature of 1000 mL of water from 10 to $47\,{}^\circ\text{C}$. Calculate the heat energy produced per gram of coal. Neglect the small heat capacity of the coal.

▎ Let Q' be the heat per gram of coal. Then

$$(5\text{ g})Q' = mc\,\Delta t = (1000\text{ g})(1.00\text{ cal/g} \cdot {}^\circ\text{C})(37\,{}^\circ\text{C}) \qquad Q' = \underline{7400\text{ cal/g}}$$

18.25 Furnace oil has a heat of combustion of 19 000 Btu/lb. Assuming that 70 percent of the heat is useful, how many pounds of oil are required to heat 1000 lb of water from 50 to $190\,{}^\circ\text{F}$?

▎ Let w be the weight of oil needed. Then $0.70w(19\,000\text{ Btu/lb}) = (1000\text{ lb})(1.00\text{ Btu/lb F}°)(190\,{}^\circ\text{F} - 50\,{}^\circ\text{F})$. Solving, $w = \underline{10.5\text{ lb}}$.

18.2 CALORIMETRY, SPECIFIC HEATS, HEATS OF FUSION AND VAPORIZATION

18.26 A copper container of mass 0.30 kg contains 0.45 kg of water. Container and water are initially at room temperature, $20\,{}^\circ\text{C}$. A 1-kg block of metal is heated to $100\,{}^\circ\text{C}$ and placed in the water in the calorimeter. The final temperature of the system is $40\,{}^\circ\text{C}$. Find the specific heat of the metal.

▎ $\begin{pmatrix}\text{heat lost by} \\ \text{metal block}\end{pmatrix} = \begin{pmatrix}\text{heat gained} \\ \text{by water}\end{pmatrix} + \begin{pmatrix}\text{heat gained by} \\ \text{copper calorimeter}\end{pmatrix}$ $\qquad (mc\,\Delta T)_{\text{metal}} = (mc\,\Delta T)_{\text{water}} + (mc\,\Delta T)_{\text{copper}}$

Note that each term in parentheses is positive as defined. Thus the ΔT values are all chosen positive. Using

Table 17-1, we have

$$(1\,\text{kg})c(100\,°C - 40\,°C) = (0.45\,\text{kg})(1.00\,\text{kcal/kg}\cdot°C)(40\,°C - 20\,°C)$$
$$+ (0.30\,\text{kg})(0.093\,\text{kcal/kg}\cdot°C)(40\,°C - 20\,°C)$$

Solving, we get $c = \underline{0.159\,\text{kcal/kg}\cdot°C}$.

18.27 A 0.70-kg metal dish contains 1 kg of water at 20 °C. A 0.5-kg iron bar at 120 °C is dropped into the water, and the final temperature is 24.9 °C. What is the material of the dish?

▌ We use Table 17-1 (in units of kcal, kg, °C) to calculate c for the unknown metal. We assume no losses to the surroundings.

$$\text{heat lost by iron} = \text{heat gained by dish and water}$$
$$0.5(0.11)(120° - 24.9°) = 0.70c(24.9° - 20°) + 1(1)(24.9° - 20°) \qquad 5.23 = 3.43c + 4.9$$
$$\text{or} \qquad 0.33 = 3.43c \qquad \text{and} \qquad c = \underline{0.096\,\text{kcal/(kg}\cdot°C)}.$$

The dish is probably copper.

18.28 What will be the final temperature if 50 g of water at 0 °C is added to 250 g of water at 90 °C?

▌ heat gained = heat lost. (We assume no heat transfer to or from container.)

$$(50\,\text{g})(1.00\,\text{cal/g}\cdot°C)(t - 0\,°C) = (250\,\text{g})(1.00\,\text{cal/g}\cdot°C)(90\,°C - t)$$

where t is the final equilibrium temperature.

$$50t = 22\,500 - 250t \qquad \text{or} \qquad 300t = 22\,500 \qquad t = \underline{75\,°C}$$

18.29 A student heated a 10-g iron nail for some time in a Bunsen-burner flame and then plunged the nail into 100 g of water at 10 °C. The water temperature rose to 20 °C. What was the temperature of the flame?

▌ The nail acquired the temperature t of the flame. When the nail was plunged into the cold water, it lost an amount of heat equal to the heat gained by the water. With the specific heat capacity of iron as given in Table 17-1, we have

$$(10\,\text{g})(0.11\,\text{cal/g}\cdot°C)(t - 20\,°C) = (100\,\text{g})(1.00\,\text{cal/g}\cdot°C)(20\,°C - 10\,°C)$$

Therefore $t - 20\,°C = (1000/1.1) = 909\,°C$, so that the result for the flame temperature is $\underline{929\,°C}$.

18.30 If 250 g of Ni at 120 °C is dropped into 200 g of water at 10 °C contained by a calorimeter of 20-cal/°C heat capacity, what will be the final temperature of the mixture?

▌ $$\text{heat lost by Ni} = \text{heat gained by water} + \text{calorimeter}$$

With the aid of Table 17-1,

$$0.250(0.106)(120° - t) = [(0.200)(1.00) + 0.020](t - 10\,°C)$$
$$3.18 - 0.027t = 0.220t - 2.20 \qquad 0.247t = 5.38 \qquad t = \underline{22\,°C}$$

18.31 How much water at 0 °C is needed to cool 500 g of water at 80 °C down to 20 °C?

▌ $$\text{heat lost} = \text{heat gained} \qquad c(500)(80 - 20) = cm(20)$$

which yields $m = \underline{1500\,\text{g}}$.

18.32 How much oil at 200 °C must be added to 50 g of the same oil at 20 °C to heat it to 70 °C?

▌ Heat lost = heat gained gives $cm(130) = c(50)(50)$, so that $m = \underline{19.2\,\text{g}}$.

18.33 A 500-g piece of iron at 400 °C is dropped into 800 g of oil at 20 °C. If $c = 0.40\,\text{cal/g}\cdot°C$ for the oil, what will be the final temperature of the system. Assume no loss to the surroundings. Use Table 17-1 for data.

▌ Heat lost = heat gained is written as $0.11(500)(400 - t) = 0.40(800)(t - 20)$, from which $t = \underline{75.7\,°C}$.

18.34 If 200 g of alcohol at 20 °C is poured into a 400-g aluminum container at −60 °C, what will be the final temperature of the system if interaction with the surroundings can be neglected? Use Tables 17-1 and 17-2 for data.

❚ Heat lost = heat gained gives $0.55(200)(20 - t) = 0.22(400)[t - (-60)]$; $2200 - 110t = 88t + 5280$; $198t = -3080$; $t = \underline{-15.5\,°C}$.

18.35 Initially 48.0 g of ice at 0 °C is in an aluminum calorimeter can of mass 2.0 g, also at 0 °C. Then 75.0 g of water at 80 °C is poured into the can. What is the final temperature?

❚ Here we use ΔH as the algebraic expression for heat gained by a system. Thus it is positive for heat entering a system and negative when it leaves a system. The heat gained by the can and the ice is equal to the heat lost by the hot water. Assuming that the hot water is sufficient in quantity to melt all the ice, the heat gained by the can and the ice is given by

$$\Delta H_{\text{can \& ice}} = m_{\text{ice}}L_{\text{ice}} + (m_{\text{ice}}c_{\text{H}_2\text{O}} + m_{\text{can}}c_{\text{Al}})(t_f - t_c)$$

Here m_{ice} is the mass of ice, m_{can} is the mass of the can, t_c is the initial temperature of the can and ice, t_f is the final temperature, L_{ice} is the latent heat of the melting of ice, $c_{\text{H}_2\text{O}}$ is the specific heat capacity of water, and c_{Al} is the specific heat capacity of aluminum. The heat lost by the hot water is

$$-\Delta H_{\text{hot water}} = m_{\text{hot water}}c_{\text{H}_2\text{O}}(t_h - t_f)$$

where $m_{\text{hot water}}$ is the mass of the hot water and t_h is its initial temperature. Setting $\Delta H_{\text{can \& ice}} = -\Delta H_{\text{hot water}}$, we have

$$m_{\text{ice}}L_{\text{ice}} + (m_{\text{ice}}c_{\text{H}_2\text{O}} + m_{\text{can}}c_{\text{Al}})t_f - (m_{\text{ice}}c_{\text{H}_2\text{O}} + m_{\text{can}}c_{\text{Al}})t_c = m_{\text{hot water}}c_{\text{H}_2\text{O}}t_h - m_{\text{hot water}}c_{\text{H}_2\text{O}}t_f$$

Solving for t_f, we find

$$t_f = \frac{m_{\text{hot water}}c_{\text{H}_2\text{O}}t_h + (m_{\text{ice}}c_{\text{H}_2\text{O}} + m_{\text{can}}c_{\text{Al}})t_c - m_{\text{ice}}L_{\text{ice}}}{m_{\text{hot water}}c_{\text{H}_2\text{O}} + m_{\text{ice}}c_{\text{H}_2\text{O}} + m_{\text{can}}c_{\text{Al}}}$$

Inserting the given numerical values, and those from Table 17-1, $c_{\text{H}_2\text{O}} = 1.00\,\text{cal/g} \cdot °C$, $c_{\text{Al}} = 0.22\,\text{cal/g} \cdot °C$, and $L_{\text{ice}} = 79.8\,\text{cal/g}$, we obtain

$$t_f = \frac{6000 + (48.0 + 0.44)(0) - (3830)}{(75.0 + 48.0 + 0.44)} = \underline{17.6\,°C}$$

(Note: If a negative value had been obtained for t_f, it would have shown the incorrectness of our a priori assumption that the hot water was sufficient to melt all of the ice.)

18.36 In an experiment to determine the latent heat of fusion of ice, 200.0 g of water at 30.0 °C in an iron can of mass 200.0 g is cooled by the addition of ice to a temperature of 10.0 °C. The can was weighed at the end of the experiment and found to have increased in mass by 50.0 g.
  Calculate the latent heat of fusion of ice.

❚ We let m_0 be the initial mass of water, m_{Fe} be the mass of the iron can, and Δm be the mass of ice added. We let t_h be the initial temperature of the water and the can, $t_c = 0\,°C$ be the initial temperature of the ice, and t_f be the common final temperature. The law of mixtures (heat gained = heat lost) takes the form

$$L\,\Delta m + \Delta m c_{\text{H}_2\text{O}}t_f = (m_0 c_{\text{H}_2\text{O}} + m_{\text{Fe}}c_{\text{Fe}})(t_h - t_f)$$

where L is the latent heat of fusion (melting) of ice. Solving for L, we have

$$L = \frac{(m_0 c_{\text{H}_2\text{O}} + m_{\text{Fe}}c_{\text{Fe}})(t_h - t_f) - \Delta m c_{\text{H}_2\text{O}}t_f}{\Delta m}$$

Using the values from Table 17-1 $c_{\text{Fe}} = 0.11\,\text{cal/g} \cdot °C$, $c_{\text{H}_2\text{O}} = 1.00\,\text{cal/g} \cdot °C$, and the given numerical values, we obtain

$$L = \frac{(200.0 + 22.0)(20.0) - (50.0)(1.00)(10.0)}{(50.0)} = 78.8\,\text{cal/g} = \underline{78.8\,\text{kcal/kg}}$$

18.37 Refer to Prob. 18.37. What is the advantage of stopping the addition of ice when the water temperature is 10.0 °C? Room temperature is 20.0 °C.

❚ During the first part of the experiment, heat escapes from the warm water to the room. This source of error can be offset by heat gained from the room if the water temperature is lower than the room temperature during the last part of the experiment. By conducting an experiment in which the final temperature is as much *below* room temperature as the initial temperature is *above* room temperature (and in which the ice is added steadily, so that the water temperature drops at a uniform rate), it can reasonably be hoped that the competing errors will (almost exactly) cancel.

18.38 Water at 212 °F weighing 7 lb is poured on a large block of ice at 32 °F. How much ice melts?

▌ We assume that no heat is lost to the surroundings. The final temperature is 32 °F. Use $Q = mc\,\Delta t$; heat of fusion for water = 144 Btu/lb (from Table 17-1).

$$\text{heat lost} = \text{heat gained} \qquad 7(1)(212° - 32°) = m_i(144) \qquad m_i = \frac{1260}{144} = \underline{8.75\,\text{lb}}$$

18.39 How many kilograms of ice at 0 °C must be added to 0.6 kg of water at 100 °C in an insulated 0.1-kg copper container in order to cool the container and its contents to 30 °C? The specific heats of water and copper are 4.2 and 0.39 kJ/kg · K, respectively; the heat of fusion of ice is 335 kJ/kg.

▌ Heat gained = heat lost. m_i = mass of ice, m_w = mass of original water.

$$m_i L_i + m_i c_w(30\,°C - 0\,°C) = m_w c_w(100\,°C - 30\,°C) + m_c c_c(100\,°C - 30\,°C)$$
$$m_i(335) + m_i(4.2)(30) = (0.6)(4.2)(70) + (0.1)(0.39)(70) \qquad 461 m_i = 179 \qquad m_i = \underline{0.39\,\text{kg}}$$

18.40 How many grams of ice at 0 °C must one add to a 200-g cup of coffee at 90 °C to cool it to 60 °C? Assume heat transfer with the surroundings to be negligible.

▌ We assume that the specific heat of coffee is that of water, and use Table 17-1 to get L_i. Heat gained = heat lost: $m(80) + 1.0(m)(60) = 1.0(200)(30)$, from which $m = \underline{43\,\text{g}}$.

18.41 How much cola can be cooled from 30 to 10 °C by a 25-g ice cube at 0 °C?

▌ We proceed as in Prob. 18.40. Heat gained = heat lost: $25(80) + (1.0)(25)(10) = 1.0(m)(20)$; this gives $m = \underline{112.5\,\text{g}}$.

18.42 Find the heat of fusion, L, of ice from the following calorimetric data:

Mass of calorimeter (aluminum)	30 g
Mass of warm water	400 g
Temperature of warm water	38 °C
Mass of ice added at 0 °C	158 g
Final temperature	5 °C

Assume no heat gain or loss from the surroundings.

▌ Set heat lost equal to heat gained, using $Q = mc\,\Delta t$. From Table 17-1, we find that c for aluminum is 0.22 kcal/kg · °C.

$$0.030(0.22)(38° - 5°) + 0.400(1)(38° - 5°) = 0.158L + 0.158(1)(5° - 0°)$$
$$0.218 + 13.2 = 0.158L + 0.79 \qquad 12.63 = 0.158L \qquad L = \underline{79.9\,\text{kcal/kg}}$$

18.43 A calorimeter of mass 125 g contains 130 g of water at 20 °C. A 6.1-g mass of steam at 100 °C is introduced into the calorimeter and condensed into water. What is the final temperature of the water? Assume that no heat is lost to the surroundings and that the value of c for the calorimeter is 0.10 kcal/kg · °C.

▌ From Table 17-2 we find that the heat of vaporization of water is 540 kcal/kg. Let T be the final temperature. Use $Q = mc\,\Delta t$.

$$\text{heat lost} = \text{heat gained} \qquad 0.0061(540) + 0.0061(1)(100° - T) = 0.130(1)(T - 20°) + 0.125(0.10)(T - 20°)$$
$$3.294 + 0.61 - 0.0061T = 0.130T - 2.60 + 0.0125T - 0.250 \qquad 6.754 = 0.1486T \qquad T = \underline{45.5\,°C}$$

18.44 A blacksmith plunges a glowing 2.0-kg horseshoe at 1200 °C into 0.8 kg of water at 50 °C. How much steam will be produced? Necessary data are in Tables 17-1 and 17-2.

▌ It will be assumed that in the final state, the system will consist of 100 °C water (with a 100 °C horseshoe immersed in it), plus some quantity of steam at 100 °C. (It is easy to believe that the heat transfer from the hot horseshoe to its surroundings will occur in such a way that "portions" of the water will heat up more rapidly than others, and that the final state will involve a somewhat lower temperature than 100 °C for the liquid. Of course, there is also the matter of heat transfer between the water bath and its surroundings. It is not possible to accommodate these complications without additional information.) The cooling of 2.0 kg of iron from 1200 to 100 °C releases $m_{Fe}c_{Fe} \Delta t_{Fe} = (2.0 \text{ kg})[0.11 \text{ kcal}/(\text{kg} \cdot °C)](1100 °C) = 242 \text{ kcal}$. The heating of 0.8 kg of water from 50 to 100 °C requires $m_{H_2O}c_{H_2O} \Delta t_{H_2O} = (0.8 \text{ kg})[1.000 \text{ kcal}/(\text{kg} \cdot °C)](50 °C)] = 40 \text{ kcal}$. Therefore there are $242 - 40 = 202$ kcal of heat available to produce steam. The mass of steam produced equals 202 kcal, divided by the latent heat of boiling for water (which is 540 kcal/kg). The quotient is 0.37 kg.

18.45 Ocean water is used to condense spent steam at 260 °F from a nuclear power plant into water at 140 °F. If the ocean temperature is 60 °F and the cooling water leaves the condenser at 100 °F, how many pounds of ocean water are needed to condense each pound of spent steam? Take the specific heat of ocean water to be 1 Btu/(lb)(F°) and the specific heat of steam to be 0.5 Btu/(lb)(F°).

❚ We take the heat of vaporization of water from Table 17-2 in Btu. Let m equal the weight of ocean water needed. Using heat lost = heat gained,

$$1(0.5)(260° - 212°) + 1(970) + 1(1)(212° - 140°) = m(1)(100° - 60°)$$
$$24 + 970 + 72 = 40m \qquad m = 26.7 \text{ lb}$$

18.46 How much steam at 120 °C is needed to heat 800 g of aluminum at 20 to 70 °C? Use Tables 17-1 and 17-2 for data. Specific heat of steam can be taken to be 0.46 cal/g · °C.

❚ Heat lost = heat gained: $(0.46m)(20) + 540m + (1.0m)(30) = 0.22(800)(50)$, from which $m = 15.2 \text{ g}$.

18.47 How much perspiration must evaporate from a 5.0-kg baby to reduce its temperature by 2 °C? The heat of vaporization for water at body temperature is about 580 cal/g. For the body, $c = 0.83$ cal/g · °C.

❚ The heat ΔQ needed to reduce the temperature of the baby is $0.83(5000)(2) = 8300$ cal. This heat is due to evaporation, $m(580)$, which yields $m = 14.3 \text{ g}$.

18.48 What is the final temperature when 200 g of ice at -20 °C is dropped into 350 g of water at 40 °C contained in a calorimeter of 50-g water equivalent. Use Table 17-1 for data.

❚ Let us assume that the final temperature is above 0 °C, so the ice must all melt. Then

$$\text{heat lost} = \text{heat gained} \qquad (350 + 50)(1)(40° - T) = 200(0.5)[0° - (-20°)] + 200(80) + 200T$$
$$16\,000 - 400T = 2000 + 16\,000 + 200T \qquad 600T = -2000$$

The fact that T comes out negative indicates that all the ice in fact did *not* melt. The final temperature must be 0 °C.

18.49 Refer to Prob. 18.48. How many grams of ice melt?

❚ Since the final temperature is 0 °C, we redo the problem, writing $m \equiv$ grams of ice that melt. Then

$$\text{heat lost} = \text{heat gained} \qquad (350 + 50)(1)(40) = 200(0.5)[0 - (-20)] + m(80)$$
$$16\,000 = 2000 + 80m \qquad m = 175 \text{ g}$$

18.50 In an experiment, 50.0 g of ice at -40 °C is mixed with 11.0 g of steam at 120 °C (and 1 atm pressure). Neglecting any heat exchange with the surroundings, what is the final temperature? Use Tables 17-1 and 17-2 for data. Assume that the specific heat of steam = 0.481 cal/g · °C.

❚ In terms of enthalpy change (heat at constant pressure), $\Delta H_{ice} = -\Delta H_{steam}$. We assume that the final temperature is greater than 0 °C and less than 100 °C, so that water is entirely in the liquid state. The heat gained by the ice is given by

$$\Delta H_{ice} = m_{ice}c_{ice} \Delta t_{ice} + m_{ice}L_{ice} + m_{ice}c_{water} \Delta t_{melted ice} \tag{1}$$
$$\Delta H_{ice} = (50.0)(0.50)(40) + (50.0)(79.8) + (50.0)(1.00)t_f = 4990 \text{ cal} + (50.0 \text{ cal}/°C)t_f \tag{2}$$

where t_f is the final temperature. The heat *lost* by the steam is given by

$$-(\Delta H_{steam}) = m_{steam}c_{steam} |\Delta t|_{steam} + m_{steam}L_{steam} + m_{steam}c_{water} |\Delta t|_{condensed steam} \tag{3}$$
$$-(\Delta H_{steam}) = (11.0)(0.481)(20) + (11.0)(540) + (11.0)(1.00)(100 - t_f)$$
$$= 7145.8 \text{ cal} - (11.0 \text{ cal}/°C)t_f \tag{4}$$

Equating Eqs. (2) and (4) we find $t_f = 2152.42/61 = \underline{35.3\,°C}$. (Note: A final result colder than $0\,°C$ or hotter than $100\,°C$ would have indicated the incorrectness of our a priori assumption that the final state involved only liquid water.)

18.51 A calorimeter whose water equivalent is 5 lb contains 45 lb of water and 10 lb of ice at $32\,°F$. What will be the final temperature if 5 lb of steam at $212\,°F$ is admitted to the calorimeter and contents? Use data from Tables 17-1 and 17-2.

▌ We try to judge if all the ice melts and/or all the steam condenses. The heat to melt all the ice is $(10\,\text{lb})(144\,\text{Btu/lb}) = 1440\,\text{Btu}$. The heat extracted in condensing all the steam is $(5\,\text{lb})(970\,\text{Btu/lb}) = 4850\,\text{Btu}$. Thus all the ice clearly melts. To check if all the steam condenses we note that $(4850\,\text{Btu} - 1440\,\text{Btu}) = 3410\,\text{Btu}$ are available to raise the temperature of calorimeter, original water, and melted ice. The heat necessary to raise their temperature to $212\,°F$ is $(60\,\text{lb})[1\,\text{Btu}/(\text{lb})(\text{F}°)](180\,°F) = 10\,800\,\text{Btu}$. Thus condensing all the steam leaves the calorimeter and contents well below boiling. The steam not only condenses to water at $212\,°F$ but continues to give up heat until an equilibrium temperature t is reached. Heat in = heat out yields

$$(1440\,\text{Btu}) + (60\,\text{lb})[(1\,\text{Btu}/(\text{lb})(\text{F}°)](t - 32\,°F) = (4850\,\text{Btu}) + (5\,\text{lb})[1\,\text{Btu}/(\text{lb})(\text{F}°)](212\,°F - t)$$

$$-480 + 60t = 5910 - 5t$$

$$65t = 6390 \qquad t = \underline{98\,°F}$$

18.52 Determine the final result when 200 g of water and 20 g of ice at $0\,°C$ are in a calorimeter whose water equivalent is 30 g and into which is passed 100 g of steam at $100\,°C$. Use data in Tables 17-1 and 17-2.

▌ We proceed to check on the melting of ice and condensing of steam as in Prob. 18.51.

$$\text{heat to melt ice} = (20\,\text{g})(80\,\text{cal/g}) = 1600\,\text{cal}$$

$$\text{heat to condense steam} = (100\,\text{g})(540\,\text{cal/g}) = 54\,000\,\text{cal}$$

All the ice melts, and the temperature of calorimeter, original water, and melted ice rises. Heat to raise calorimeter and contents to $100\,°C$ is $(250\,\text{g})(1\,\text{cal/g}\cdot°C)(100\,°C) = 25\,000\,\text{cal}$. Since $(25\,000 + 1600)\,\text{cal} = 26\,600\,\text{cal} < 54\,000\,\text{cal}$, not all the steam condenses. The mass m that does condense is obtained from $m(540\,\text{cal/g}) = 26\,600\,\text{cal}$; $m = \underline{49\,\text{g}}$. Thus 49 g of steam condense and the system ends in equilibrium at $100\,°C$.

18.53 Determine the final result when 400 g of water and 100 g of ice at $0\,°C$ are in a calorimeter whose water equivalent is 50 g and into which passed 10 g of steam at $100\,°C$.

▌ We proceed as in Probs. 18.51 and 18.52.

$$\text{heat to melt ice} = (100\,\text{g})(80\,\text{cal/g}) = 8000\,\text{cal}$$

$$\text{heat extracted to condense steam} = (10\,\text{g})(540\,\text{cal/g}) = 5400\,\text{cal}$$

Thus all the steam condenses to water at $100\,°C$. The heat extracted in dropping temperatures of condensed steam to $0\,°C = (10\,\text{g})(1\,\text{cal/g}\cdot°C)(100\,°C) = 1000\,\text{cal}$. Since $(5400 + 1000)\,\text{cal} = 6400\,\text{cal} < 8000\,\text{cal}$ not all the ice melts in reaching equilibrium. To find the mass m of ice that melts, $m(80\,\text{cal/g}) = 6400\,\text{cal}$; $m = \underline{80\,\text{g}}$. Thus 80 g of ice melt and the system is in equilibrium at $0\,°C$.

CHAPTER 19
Heat Transfer

19.1 CONDUCTION

19.1 Obtain the equation of one-dimensional heat conduction.

▮ Consider the slab of material shown in Fig. 19-1. Its thickness is L and its cross-sectional area is A. The temperatures of its two faces are T_1 and T_2, with $T_1 > T_2$. We call the quantity $(T_2 - T_1)/L$ the *temperature gradient*. It is the rate of change of temperature with distance.

The quantity of heat, ΔQ, transmitted from face 1 through face 2 in time Δt is proportional to Δt, the face area A, and the magnitude of the temperature gradient. Therefore

$$\Delta Q = k(\Delta t)A \frac{T_1 - T_2}{L}$$

or

$$H = \frac{\Delta Q}{\Delta t} = -kA \frac{T_2 - T_1}{L} \tag{1}$$

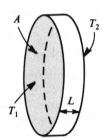

Fig. 19-1

where k is a proportionality constant which depends only on the nature of the substance. It is called the *coefficient of thermal conductivity*, or simply the *thermal conductivity*, of the substance. For a given substance, k is the amount of heat transmitted per unit time per unit perpendicular area per unit temperature gradient.

Typical units for k are cal/s · cm · °C and W/m · K, where 1 W = 1 J/s. Or, in the British system, the common units are Btu/h · ft · °F or Btu · in/ft² · h · °F.

In the language of calculus (1) may be expressed as

$$\frac{dQ}{dt} = -kA \frac{dT}{dx} \tag{2}$$

where the minus sign ensures flow from higher to lower temperature along the positive x axis.

19.2 What temperature gradient must exist in an aluminum rod in order to transmit 8 cal/s per square centimeter of cross section down the rod? k for aluminum is 0.50 cal/s · cm · °C.

▮
$$H = kA \left| \frac{\Delta T}{\Delta x} \right| \quad \text{or} \quad 8 \text{ cal/s} = (0.50 \text{ cal/s} \cdot \text{cm} \cdot °\text{C})(1 \text{ cm}^2) \left| \frac{\Delta T}{\Delta x} \right|$$

$$\left| \frac{\Delta T}{\Delta x} \right| = \text{magnitude of temperature gradient} = \underline{16 \, °\text{C/cm}}.$$

The actual sign of the gradient depends on whether the heat flow is in the positive or negative x direction, being negative or positive for the two cases, respectively.

19.3 A slab having a thickness of 4 cm and measuring 25 cm on a side has a 40°C temperature difference between its faces. How much heat flows through it per hour? The conductivity k is 0.0025 cal/s · cm · °C.

▮
$$Q = \frac{k \text{ (area)}(t_h - t_c)(\text{time})}{\text{thickness}} = \frac{0.0025(25 \times 25)(40)(3600)}{4} = \underline{56.25 \text{ kcal}}$$

19.4 A refrigerator door is 150 cm high, 80 cm wide, and 6 cm thick. If the coefficient of conductivity is 0.0005 cal/cm · s · °C, and the inner and outer surfaces are at 0 and 30 °C, respectively, what is the heat loss per minute through the door, in calories?

$$Q = \frac{kA(t_h - t_c)(\text{time})}{d} = \frac{0.0005(150 \times 80)(30° - 0°)(60)}{6} = \underline{1800 \text{ cal}}$$

19.5 A glass window 5 ft × 3 ft × 0.25 in has its inside surface at a temperature of 60 °F and its outside surface at 32 °F. The thermal conductivity of glass is 4.35 Btu · in/ft² · h · °F. How many Btu per hour are lost by conduction through this window?

▮ Use the heat conduction equation with BES units.

$$Q = \frac{kA(t_h - t_c)(\text{time})}{d} = \frac{4.35 \frac{\text{Btu} \cdot \text{in}}{\text{ft}^2 \cdot \text{h} \cdot °\text{F}} [(5 \times 3)\text{ft}^2](60 °\text{F} - 32 °\text{F})(1 \text{ h})}{0.25 \text{ in}} = \underline{7300 \text{ Btu}}$$

19.6 Masonite sheeting transmits 0.33 Btu/h through 1 ft² section when the temperature gradient is 1 °F per inch of thickness. How many Btu will be transmitted per day through a sheet having dimensions 3 ft × 6 ft × $\frac{3}{4}$ in if one face is at 38 °F and the opposite face is at 63 °F?

▮ Q is directly proportional to cross-sectional area, temperature gradient, and time; and inversely proportional to thickness. Hence

$$\text{heat transmitted per day} = (0.33 \text{ Btu})\left(\frac{18 \text{ ft}^2}{1 \text{ ft}^2}\right)\left(\frac{25 °\text{F}}{1 °\text{F}}\right)\left(\frac{24 \text{ h}}{1 \text{ h}}\right)\left(\frac{1 \text{ in}}{\frac{3}{4} \text{ in}}\right) = 4750 \text{ Btu}$$

19.7 What thickness of wood has the same insulating ability as 8 cm of brick? $k = 0.8$ W/m · K for brick and 0.1 W/m · K for wood.

▮ For the same ΔT and A, the $\Delta Q / \Delta t$ is the same in the two materials, so $k_w / L_w = k_b / L_b$, and the wood thickness is $L_w = (0.1/0.8)(8 \text{ cm}) = \underline{1 \text{ cm}}$.

19.8 How much water at 100 °C could be evaporated per hour by the heat transmitted through a 1 cm × 1 cm steel plate 0.2 cm thick, if the temperature difference between the plate faces is 100 °C? For steel, k is 0.11 cal/s · cm · °C.

▮ We first find the heat transferred across 1 cm² of the plate in 1 h.

$$Q = \frac{(0.11 \text{ cal/s} \cdot \text{cm} \cdot °\text{C})(1 \text{ cm}^2)(100 °\text{C})(3600 \text{ s})}{0.2 \text{ cm}} = 198 \text{ kcal}.$$

From Table 17-2 heat of vaporization of water is $L_v = 540$ kcal/kg. Then from $Q = mL_v$ we get $m = \underline{0.367 \text{ kg}}$.

19.9 To measure the thermal conductivity of a material, a cubical box (80 cm on edge and with walls 3 cm thick) is constructed from the material. When an electric heater at the center of the box delivers heat at the rate of 120 W, the steady temperature difference between inside and outside is 40 °C. What is the thermal conductivity of the material?

▮ The length of any inside face of the cube is $80 - 6 = 74$ cm. Then for $\Delta Q / \Delta t = 120$, $A = 6(0.74)^2$ and $\Delta T / \Delta x = 40/0.03$, we find $k = (\Delta Q / \Delta t)/A(\Delta T / \Delta x)$ to be $\underline{0.0274 \text{ W/m} \cdot \text{K}}$. This solution is only approximate because of the neglect of conduction where the walls join.

19.10 A boxlike cooler has 5.0-cm-thick walls made of plastic foam. Its total surface area is 1.5 m². About how much ice melts each hour inside the cooler to hold its temperature at 0 °C when the outside temperature is 30 °C? Take k for the plastic to be 0.040 W/m · K and $h_v = 80$ cal/g.

▮ The heat flow into the box is $\Delta Q / \Delta t = kA(\Delta T / \Delta x) = (0.040)(1.5)(30/0.05) = 36$ W; in an hour the total energy is 130 kJ. This equals $(m)(80 \text{ kcal/kg})(4.184 \text{ kJ/kcal})$, so $m = \underline{0.39 \text{ kg}}$ of ice will melt.

19.11 An ordinary refrigerator is thermally equivalent to a box of corkboard 90 mm thick and 5.6 m² in inner surface area. When the door is closed, the inside wall is kept, on the average, 22.2 °C below the temperature of the outside wall. If the motor of the refrigerator runs 15 percent of the time the door is closed, at what rate must heat be taken from the interior while the motor is running? The thermal conductivity of corkboard is $k = 0.05$ W/m · K.

∎ Consider a time interval Δt during which the door is closed. An an approximation, take the heat conduction to be steady over Δt. Then the rate of heat into the box is

$$\frac{Q}{\Delta t} = kA\left(\frac{\Delta T}{\Delta x}\right) = (0.05)(5.6)\left(\frac{22.2}{0.090}\right) = 69.1 \text{ W}$$

To remove this heat, the motor must, since it runs only for a time $(0.15)\,\Delta t$, cause heat to leave at the rate $69.1/0.15 = \underline{460 \text{ W}}$.

19.12 Consider a glass window of area 1 m^2 and thickness 0.50 cm. If a temperature difference of $20\,°\text{C}$ exists between one side and the other, how fast would heat flow through the window? Why is this result *not* applicable to a house window on a day when the temperature difference between inside and outside is $20\,°\text{C}$? Take $k = 0.80\text{ W/m}\cdot\text{K}$.

∎ The rate of heat flow through the glass is $(0.80\text{ W/m}\cdot\text{K})(1\text{ m}^2)(20\text{ K}/0.005\text{ m}) = \underline{3200 \text{ W}}$. This is not applicable since semistagnant air layers on each side also exist, making the actual temperature difference across the glass much less than 20 K.

19.13 Deep bore holes into the earth show that the temperature increases about $1\,°\text{C}$ for each 30 m of depth. If the earth's crust has a thermal conductivity of about $0.80\text{ W/}°\text{C}\cdot\text{m}$, how much heat flows out through the surface of the earth each second for each square meter of surface area?

∎ One has $\Delta Q/\Delta t = kA(\Delta T/\Delta x)$. Then $\Delta Q/\Delta t = (0.80\text{ W/m}\cdot\text{K})(1\text{ m}^2)(1\text{ K}/30\text{ m}) = \underline{27 \text{ mW/m}^2}$.

19.14 A certain double-pane window consists of two glass sheets, each $80\text{ cm} \times 80\text{ cm} \times 0.30$ cm, separated by a 0.30-cm stagnant air space. The indoor surface temperature is $20\,°\text{C}$, while the outdoor surface temperature is $0\,°\text{C}$. How much heat passes through the window each second? $k = 2 \times 10^{-3}\text{ cal/s}\cdot\text{cm}\cdot°\text{C}$ for glass and about $2 \times 10^{-4}\text{ cal/s}\cdot\text{cm}\cdot°\text{C}$ for air.

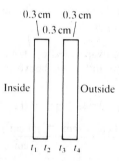

0.3 cm   0.3 cm
0.3 cm

Inside   Outside

t_1 t_2 t_3 t_4 **Fig. 19-2**

∎ A cross section of the window is shown in Fig. 19-2; the four surface temperatures are labeled, and $t_1 = 20\,°\text{C}$, $t_4 = 0\,°\text{C}$.

$$H_1 = \frac{(2 \times 10^{-3})(6400)(20 - t_2)}{0.30} = 42.7(20 - t_2) \qquad H_2 = \frac{(2 \times 10^{-4})(6400)(t_2 - t_3)}{0.30} = 4.27(t_2 - t_3)$$

$$H_3 = \frac{(2 \times 10^{-3})(6400)(t_3 - 0)}{0.30} = 42.7 t_3$$

Since the three rates of flow must be equal in a steady-state situation, $H_1 = H_2 = H_3$, we have

$$42.7(20 - t_2) = 4.27(t_2 - t_3) = 42.7 t_3 \qquad 10(20 - t_2) = (t_2 - t_3) = 10 t_3$$

yielding

$$t_2 = 11 t_3, \quad 20 - t_2 = t_3 \quad \Rightarrow \quad 20 = 12 t_3, \quad t_3 = \underline{1.67\,°\text{C}} \quad \text{and} \quad t_2 = 18.4\,°\text{C}$$

Finally

$$H = H_3 = (42.7)(1.67) = \underline{71.3 \text{ cal/s}}.$$

19.15 Two brass plates, each 0.50 cm thick, have a rubber spacer sheet between them which is 0.10 cm thick. The outer side of one brass plate is kept at $0\,°\text{C}$, while the outer side of the other is $100\,°\text{C}$. Find the temperatures of the two sides of the rubber spacer. (k_{Br} is 490 times k_{Ru}.)

■ Call T_1 and T_2 the temperatures at the two rubber-brass junctions. $\Delta Q/\Delta t$ and area A is the same in each section, so $k_{Br}(T_1 - 0°)/0.50 = k_{Ru}(T_2 - T_1)/0.10 = k_{Br}(100 - T_2)/0.50$. The first and third equations yield $T_1 + T_2 = 100$, while the first and second give $99T_1 - T_2 = 0$. From the latter two relations, $T_1 = \underline{1.0\,°C}$ and $T_2 = \underline{99.0\,°C}$.

19.16 Refer to Prob. 19.15. A 0.50-cm-thick sheet of brass has sealed to one face a 0.50-cm-thick rubber sheet. The other side of the brass sheet is connected to a bath maintained at 20 °C. The other rubber surface is attached to a circulating bath at 80 °C. Find the temperature at the rubber-brass junction.

■ Now $\Delta Q/\Delta t = kA(\Delta T/\Delta x)$ for both sheets; at equilibrium, the two $\Delta Q/\Delta t$'s are equal. Then $[k_{Br}(t - 20)]/0.50 = [k_{Ru}(80 - t)]/0.50$ giving $t = \underline{20.12\,°C}$ as the interface temperature.

19.17 Define *thermal resistance* and *R value* (or *R factor*) for heat conduction.

■ The conduction equation, $H \equiv \Delta Q/\Delta t = (kA\,\Delta T)/d$ can be reexpressed as $H = \Delta T/R$, where $R \equiv d/(kA)$ is called the thermal resistance of the slab. It has units of K/W in SI and units of °F · s/Btu in the English system.
  The *R value*, commonly used in the United States to grade insulating material, is the thermal resistance (*in English units*) of one square foot of insulating material: R value $\equiv d/k$, with k in Btu · in/(h · ft^2 · °F), and d in inches. If k and d are in SI units [W/(m · K) and m], R value $\equiv (5.68)(d/k)$. Clearly, thermal resistance = (R value)/A in English units; thermal resistance = (R value)/($5.68A$) in SI units. Units are usually omitted when giving an R-value.

19.18 A cubical box (70 cm on edge) is made from a foam plastic sheet that has an R factor of 4. About how much ice at 0 °C will melt inside the box per hour when the outside temperature is 25 °C? Heat of fusion $= h_f = 80$ kcal/kg.

■ The heat flow relation is $\Delta Q/\Delta t = A\,\Delta T/R_f$, with $A = 6(0.7)^2$, $\Delta T = 25$ K, and $R_f = 4/5.68 = 4(0.176)$. All these are in SI units so we calculate to find $\Delta Q/\Delta t = 104$ W, which in an hour gives $\Delta Q = 375$ kJ $= 89.6$ kcal. Now use $\Delta Q = mh_f$ to find $m = \underline{1.12\,kg}$ of ice.

19.19 A cubical box (60 cm on edge) contains ice at 0 °C. When the outside temperature is 20 °C, it is found that 250 g of ice melts each hour. About what is the R value for the material of the walls of the box?

■ The heat flow out of the box is $\Delta Q/\Delta t = mh_f/\Delta t = (250\text{ g})(80\text{ cal/g})/3600\text{ s} = 5.56$ cal/s $= 23.2$ W. Using $A = 6(0.6)^2$ m^2 and $\Delta T = 20$ °C, we have

$$R = 5.68\,\frac{A\,\Delta T}{\Delta Q/\Delta t} = \underline{10.5}$$

19.20 Show that the thermal resistance of a multiple layer is equal to the sum of the resistances of the individual laminas.

■ For two layers of equal cross section we have; $\Delta T_1 = H_1 R_{\text{therm},1}$ and $\Delta T_2 = H_2 R_{\text{therm},2}$. Noting that $H_1 = H_2 = H$ and ΔT (for both slabs) $= \Delta T_1 + \Delta T_2$, we have $\Delta T \equiv HR_{\text{therm}} = H(R_{\text{therm},1} + R_{\text{therm},2})$ or $R_{\text{therm}} = R_{\text{therm},1} + R_{\text{therm},2}$. This analysis can clearly be extended to multiple slabs.

19.21 Two sheets of insulation have R values of R_1 and R_2, respectively. Show that the R value of the combination of one sheet on top of the other is $R_1 + R_2$.

■ Since area A is the same for the two sheets, the R values add like the thermal resistances (see Prob. 19.20).

19.22 (a) Find the heat resistance of one square meter of window glass that is 0.5 cm thick. (b) Special insulated window pane consists of two 0.5-cm-thick sheets of glass separated by a 0.15-cm-thick layer of dry air. Calculate the resistance of one square meter of this pane. ($k_g = 0.60$ W/m · °C, $k_a = 0.025$ W/m · °C.)

■ (a) From the definition we have that the resistance R_g of a single sheet of glass is

$$R_g = \frac{d}{kA} = \frac{0.5 \times 10^{-2}\text{ m}}{(0.60\text{ W/m} \cdot °C)(1\text{ m}^2)} = \underline{0.0083\,°C/W}$$

(b) The resistance R_a of the layer of air between the two sheets of glass is

$$R_a = \frac{d}{kA} = \frac{0.15 \times 10^{-2}\text{ m}}{(0.025\text{ W/m} \cdot °C)(1\text{ m}^2)} = \underline{0.060\,°C/W}$$

Since the pane consists of two layers of glass and one layer of air, its total resistance R is

$$R = R_g + R_a + R_g = \underline{0.077\ °C/W}.$$

The resistance of the double-sheeted pane is more than 9 times greater than the resistance of the single-sheeted pane, and most of this increased resistance comes from the layer of air between the glass sheets.

19.23 The effective resistance of a barrier, such as a single pane of glass, is the resistance not only of the glass but of the stagnant air layers on either side. Assume 3 mm and 1.5 mm of stagnant air on the inside and outside surfaces of a pane 1 m^2 in area. Find the combined resistance of the stagnant air. $[k_{air} = 0.025\ W/(m \cdot °C)]$.

▌ $$R = R_{in} + R_{out} = \left(\frac{d}{kA}\right)_{in} + \left(\frac{d}{kA}\right)_{out} = \frac{0.003}{(0.025)(1)} + \frac{0.0015}{(0.025)(1)} = 0.18\ °C/W$$

19.24 The *effective resistance* R_{eff} of a barrier is the sum of the barrier resistance R and the surface resistance R_s: $R_{eff} = R + R_s$. What are the effective resistances of one square meter of the single-sheeted and double-sheeted window panes described in Prob. 19.22? (Take R_s from Prob. 19.23.)

▌ For the single pane $R_{eff} = (0.0083 + 0.18) = \underline{0.188\ °C/W}$. For the double pane $R_{eff} = (0.077 + 0.18) = \underline{0.257\ °C/W}$. It should be noted that under atmospheric conditions that deplete the stagnant air layers on either side of the window, the superiority of the double pane for insulation is even more apparent.

19.25 Consider the two insulating sheets with resistances R_1 and R_2 shown in Fig. 19-3. Show that

$$T' = \frac{R_2 T_1 + R_1 T_2}{R_1 + R_2},$$

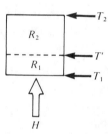

Fig. 19-3

▌ For the two sheets, $H_1 = (T_1 - T')/R_1$, $H_2 = (T' - T_2)/R_2$. Noting that $H_1 = H_2 = H$, we have $(T_1 - T')/R_1 = (T' - T_2)/R_2$. Cross multiplying we get $R_2(T_1 - T') = R_1(T' - T_2)$, or rearranging terms, $R_2 T_1 + R_1 T_2 = (R_1 + R_2)T'$, and $T' = (R_2 T_1 + R_1 T_2)/(R_1 + R_2)$.

19.26 A cup of tea cools from 150 to 145 °F in 1 min in a room at 72 °F. How long will the tea take to cool from 110 to 105 °F (in the same room)? (Hint: Use Newton's law of cooling.)

▌ Newton's law of cooling may be stated as

$$\frac{\text{change of temperature}}{\text{time}} \propto [(\text{initial temperature}) - (\text{ambient temperature})]$$

Thus, writing the proportionality constant as A,

$$\frac{150° - 145°}{1\ min} = A(150° - 72°) \qquad \frac{110° - 105°}{\tau} = A(110° - 72°)$$

Dividing the first equation by the second and solving, $\tau = 78/38 = \underline{2\ min}$ (approximate).

19.27 An 11-lb turkey takes 9 min to be heated from 35 to 40 °C in an oven maintained at 235 °C. How long does it take the same turkey to be heated from 55 to 60 °C in the same oven held at the same temperature?

▌ Use Newton's law of cooling (which also applied to heating), as in Prob. 19.26.

$$\frac{40° - 35°}{9\ min} = B(235° - 35°) \qquad \frac{60° - 55°}{\tau} = B(235° - 55°)$$

from which $\tau = (9\ min) \times (200/180) = \underline{10\ min}$.

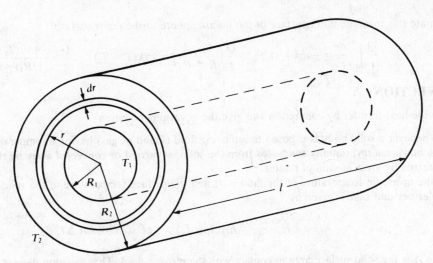

Fig. 19-4

19.28ᶜ Find the steady-state heat flow through the hollow circular cylinder shown in Fig. 19-4.

▮ Since the inner and outer cylindrical surfaces are at fixed temperatures T_1 and T_2, and ignoring effects at the two end surfaces, we must have symmetrical radial heat flow. Then for any cylindrical shell, say of radius r and thickness dr, we have $H = -k(2\pi rl)(dT/dr)$, where $(2\pi rl)$ is the effective area of the shell, and dT is the algebraic increase in temperature across the infinitesmal thickness dr. Since in steady state H is the same for all shells, we can separate the variables r and T into simple integrable form: $H(dr/r) = -2\pi kl \, dT$. Integrating the left side from R_1 to R_2, and the right side over the corresponding temperature range T_1 to T_2, we have

$$H\int_{R_1}^{R_2} \frac{dr}{r} = -2\pi kl \int_{T_1}^{T_2} dT \quad \text{or} \quad H \ln\left(\frac{R_2}{R_1}\right) = -2\pi kl(T_2 - T_1)$$

Finally

$$H = 2\pi kl \left[\frac{T_1 - T_2}{\ln(R_2/R_1)}\right]$$

19.29ᶜ Find the rate of heat flow through the hollow sphere shown in cross section Fig. 19-5.

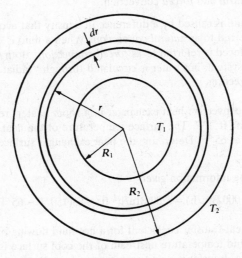

Fig. 19-5

▮ The steady-state heat flow (which is independent of the radius r) through a spherical shell of radius r and thickness dr is

$$\frac{Q}{t} = -k(4\pi r^2)\left(\frac{dT}{dr}\right) \quad \text{or} \quad \frac{Q}{t}\frac{dr}{r^2} = -4\pi k \, dT$$

Integrate this from the inner surface of the hollow sphere to the outer surface:

$$\frac{Q}{t}\int_{R_1}^{R_2}\frac{dr}{r^2}=-4\pi k\int_{T_1}^{T_2}dT \qquad \frac{Q}{t}\left(\frac{1}{R_1}-\frac{1}{R_2}\right)=4\pi k(T_1-T_2) \qquad \frac{Q}{t}=\frac{4\pi k(T_1-T_2)}{(1/R_1)-(1/R_2)}$$

19.2 CONVECTION

19.30 Describe heat transfer by convection and give the governing equation.

▌ Whenever a solid body is exposed to a moving fluid (liquid or gas) having a temperature lower than that of the body, energy, initially *conducted* from the solid, is carried or *convected* away by the fluid. Heat is thus transferred by gross motion of matter.

If the upstream temperature of the fluid is T_∞ and the surface temperature of the solid is T_s, the heat transfer per unit time is given by

$$H=\frac{\Delta Q}{\Delta t}=hA(T_s-T_\infty) \qquad \text{or} \qquad H=hA\,\Delta T, \tag{1}$$

where A is the solid surface area in contact with the moving fluid. This equation defines the *convective heat-transfer coefficient h*, which has units of Btu/h · ft^2 · °F or W/m^2 · K or kcal/m^2 · s · °C. Unlike thermal conductivity k, the convection coefficient h is not a property of the solid or fluid but depends upon many parameters of the system. It is known to vary with the geometry of the solid and its surface finish, the velocity of the fluid, the density of the fluid, and the thermal conductivity. Differences in the temperature and pressure of the fluid also affect the value for h.

The weak ΔT-dependence of h is indicated for certain geometries in Table 19-1.

TABLE 19-1 Convection Coefficients

geometry	h, kcal/m^2 · s · °C
Vertical plate	$(4.24\times10^{-4})\sqrt[4]{\Delta T}$
Horizontal plate	
Facing up	$(5.95\times10^{-4})\sqrt[4]{\Delta T}$
Facing down	$(3.14\times10^{-4})\sqrt[4]{\Delta T}$
Pipe of diameter D	
(in centimeters)	$(1.0\times10^{-3})\sqrt[4]{\Delta T/D}$

19.31 Distinguish between *natural* and *forced* convection.

▌ If the motion of a fluid is caused by a difference in density that accompanies a change in temperature, the current produced is referred to as *natural convection*. When a fluid is caused to move by the action of a pump or fan, the current produced is referred to as *forced convection*. Both types of convection are employed in a common home heating system: hot water is circulated through "radiators" by forced convection; the warmed air rises by natural convection.

19.32 Forced air flows over a convective heat exchanger in a room heater, resulting in a convective heat-transfer coefficient $h = 200$ Btu/h · ft^2 · °F. The surface temperature of the heat exchanger may be considered constant at 150 °F, and the air is at 65 °F. Determine the heat exchanger surface area required for 30 000 Btu/h of heating.

▌ $H = hA\,\Delta T$. From the information given

$$(30\,000\,\text{Btu/h}) = (200\,\text{Btu/h}\cdot\text{ft}^2\cdot\text{°F})(150\,\text{°F} - 65\,\text{°F})A \qquad A = \underline{1.765\,\text{ft}^2}$$

19.33 The forced convective heat-transfer coefficient for a hot fluid flowing over a cool surface is 40 Btu/h · ft · °F in a particular case. The fluid temperature upstream of the cool surface is 250 °F, and the surface is held at 50 °F. Determine the heat transfer per unit surface area from the fluid to the surface.

▌ Equation (1) of Prob. 19.30 applies to this problem, but the sign must be reversed to represent heat flux *from* the fluid *to* the surface. Thus

$$\frac{H}{A}=h(T_\infty-T_s)=\frac{40\,\text{Btu}}{\text{h}\cdot\text{ft}^2\cdot\text{°F}}[(250-50)\,\text{°F}]=8000\,\frac{\text{Btu}}{\text{h}\cdot\text{ft}^2}$$

19.34 A flat vertical wall 6 m² in area is maintained at a constant temperature of 116 °C, and the surrounding air on both sides is at 35 °C. How much heat is lost from the wall in 1 h by natural convection?

❚ We must first compute h for a vertical wall. From Table 19-1 we have

$$h = (4.24 \times 10^{-4})\sqrt[4]{116 - 35} = 1.27 \times 10^{-3} \text{ kcal/m}^2 \cdot \text{s} \cdot {}^\circ\text{C}$$

The quantity of heat transferred from each side can be found from

$$\Delta Q = hA\tau \, \Delta t = (1.27 \times 10^{-3} \text{ kcal/m}^2 \cdot \text{s} \cdot {}^\circ\text{C})(6 \text{ m}^2)(3600 \text{ s})(81 \, {}^\circ\text{C}) = 2.22 \times 10^3 \text{ kcal}$$

Since there are two sides, the total heat transferred is $\underline{4.44 \times 10^3 \text{ kcal}}$.

19.35 Rework Prob. 19.34 if the wall is horizontal.

❚ Only the temperature-independent coefficient in h is different (see Table 19-1); so, by proportion,

$$\Delta Q = \left(\frac{5.95}{4.24} + \frac{3.14}{4.24}\right)(2.22 \times 10^3 \text{ kcal}) = 4.76 \times 10^3 \text{ kcal}$$

19.36 A vertical steam pipe has an outside diameter of 8 cm and a height of 5 m. The outside temperature of the pipe is 94 °C, and the room temperature is 23 °C. How much heat is released to the air by convection in 1 h?

❚ $\Delta Q / \Delta t = hA \, \Delta T$. From Table 19-1, h for a pipe of outside diameter D is $h = (1.0 \times 10^{-3})\sqrt[4]{\Delta T / D}$ kcal/m² · s · °C. Using $\Delta T = 71 \, {}^\circ\text{C}$ and $D = 8$ cm, $h = 1.726 \times 10^{-3}$ kcal/m² · s · °C. Then $\Delta Q = (1.726 \times 10^{-3})(\pi \times 0.08 \times 5)(71)(3600) = \underline{554 \text{ kcal}}$.

19.37 The air in a room at 26 °C is separated from the outside air at −4 °C by a vertical glass window 3 mm thick and 10 m² in area. We must expect a small difference in temperature between the inner and outer surfaces of the glass. For the purposes of calculation, assume that the center of the glass is at the mean temperature (11 °C). (a) What is the steady-state rate of heat flow? (b) What are the inner and outer surface temperatures of the glass window? ($k = 2.5 \times 10^{-4}$ kcal/m · s · K for glass; use Table 19-1 for h.)

❚ (a) In the steady state, heat cannot pile up any place. So the transfer of heat per unit area per unit time must be the same by convection inside, by conduction through glass, and by convection outside:

$$\left(\frac{H}{A}\right)_{\text{inside}} = \left(\frac{H}{A}\right)_{\text{conduction}} = \left(\frac{H}{A}\right)_{\text{outside}}$$

Using Table 19-1, we have for the convective transfer from the room to the middle of the glass (a simplifying approximation)

$$\left(\frac{H}{A}\right)_{\text{inside}} = h \, \Delta T = (4.24 \times 10^{-4})(26^\circ - 11^\circ)^{5/4} = 1.25 \times 10^{-2} \text{ kcal/m}^2 \cdot \text{s}$$

Since this value is assumed universal, the steady-state flux is $H = (1.25 \times 10^{-2})(10) = \underline{0.125 \text{ kcal/s}}$.
(b) For the conduction through the glass,

$$0.125 \times 10^{-2} = \left(\frac{H}{A}\right)_{\text{conduction}} = \frac{k}{L}(T_{\text{in}} - T_{\text{out}}) = \frac{2.5 \times 10^{-4}}{3 \times 10^{-3}}(T_{\text{in}} - T_{\text{out}})$$

whence $T_{\text{in}} - T_{\text{out}} = 0.15 \, {}^\circ\text{C}$. But, by assumption, $\frac{1}{2}(T_{\text{in}} + T_{\text{out}}) = 11 \, {}^\circ\text{C}$. Thus, $\underline{T_{\text{in}} = 11.075 \, {}^\circ\text{C}}$, $\underline{T_{\text{out}} = 10.925 \, {}^\circ\text{C}}$.

19.3 RADIATION

19.38 Describe heat transfer by radiation and the mathematical law (Stefan's law) governing it.

❚ The rate P_e at which heat energy is emitted (as electromagnetic radiation) by an object with surface area A and absolute temperature T_e is

$$P_e = \epsilon \sigma A T_e^4 \quad \text{emission} \tag{1}$$

where $\sigma = 5.67 \times 10^{-8}$ W/m² · K⁴ is a universal constant called the *Stefan-Boltzmann constant*, and ϵ is a dimensionless parameter called the *emissivity*, which varies between 0 and 1 depending on the nature of the surface. The same object placed in an enclosure with walls at absolute temperature T_a will absorb radiation from the walls at the rate

$$P_a = \epsilon \sigma A T_a^4 \quad \text{absorption} \tag{2}$$

Thus if the object is hotter than the walls of the enclosure ($T_e > T_a$), there will be a net flow of energy from the object to the walls at the rate

$$P = P_e - P_a = \epsilon\sigma A(T_e^4 - T_a^4) \quad \text{net loss} \tag{3}$$

An object with the maximum emissivity of 1 is called a *blackbody* because it absorbs all the radiation incident on it. An object with an emissivity of zero is a perfect reflector, which absorbs none of the radiation incident on it. The emissivity ϵ is also called the *absorptivity*, for obvious reasons.

19.39 What is *Wien's displacement law?*

▌ The law states that the absolute temperature of a blackbody and the peak wavelength of its radiation are inversely proportional: $\lambda_m T = 2898 \, \mu\text{m} \cdot \text{K}$. By *peak wavelength* we mean that more energy is radiated in the wavelength interval $\lambda \pm \frac{1}{2}\Delta\lambda$ than in any other interval of width $\Delta\lambda$.

19.40 A spherical blackbody of 5 cm radius is maintained at a temperature of 327 °C. What is the power radiated?

▌ The surface area of a sphere is $4\pi r^2$. In this case, then, the area is $4\pi(25 \times 10^{-4}) = 0.01\pi \, \text{m}^2$. The power radiated is given by Stefan's law:

$$P = \sigma T^4 A = (5.67 \times 10^{-8} \, \text{W/m}^2 \cdot \text{K}^4)(600 \, \text{K})^4(0.01\pi \, \text{m}^2) = 231 \, \text{W}$$

19.41 Refer to Prob. 19.40. At what wavelength is the maximum energy radiated?

▌ By Wien's law, $\lambda_m(600 \, \text{K}) = 2898 \, \mu\text{m} \cdot \text{K}$ and $\lambda_m = \underline{4.82 \, \mu\text{m}}$.

19.42 A sphere of 3 cm radius acts like a blackbody. It is in equilibrium with its surroundings and absorbs 30 kW of power radiated to it from the surroundings. What is the temperature of the sphere?

▌ The power absorbed by a blackbody is $P_a = \sigma A T_a^4$, or $(30 \times 10^3 \, \text{W}) = (5.67 \times 10^{-8} \, \text{W/m}^2 \cdot \text{K}^4)4\pi(0.03 \, \text{m})^2 T_a^4$. $T_a^4 = 4.68 \times 10^{13}$; T_a = temperature of surroundings = 2600 K. Since the body is in equilibrium with its surroundings, it is at the same temperature, $\underline{2600 \, \text{K}}$.

19.43 An incandescent lamp filament has an area 50 mm² and operates at a temperature of 2127 °C. Assume that all the energy furnished to the bulb is radiated from it. If the filament acts like a blackbody, how much power must be furnished to the bulb when operating?

▌ All the energy furnished to the bulb is radiated from it. Nonetheless, the power radiated will be greater than that furnished by the filament because in the steady state, additional power being supplied by absorption from the surroundings is also being radiated. Since we are not given the temperature of the surroundings, we can reasonably assume room temperature $T_a \sim 300 \, \text{K}$. Clearly the effect of absorption is approximately $(\frac{1}{8})^4$ of the radiated energy, since $T_e = 2400 \, \text{K}$, and can be ignored. Then, power supplied to bulb is $P_e = \sigma A T_e^4 = (5.67 \times 10^{-8} \, \text{W/m}^2 \cdot \text{K}^4)(50 \times 10^{-6} \, \text{m}^2)(2400 \, \text{K})^4 = \underline{94 \, \text{W}}$.

19.44 A very small hole in an electric furnace used for treating metals acts nearly as a blackbody. If the hole has an area 100 mm², and it is desired to maintain the metal at 1100 °C, how much power travels through the hole?

▌ $$\text{power} = \sigma A T^4 = (5.67 \times 10^{-8})(10^{-4})(1373.15)^4 = \underline{20.2 \, \text{W}}$$

19.45 From measurements made on earth it is known that the sun has a surface area of $6.1 \times 10^{18} \, \text{m}^2$ and radiates energy at the rate of $3.9 \times 10^{26} \, \text{W}$. Assuming that the emissivity of the sun's surface is 1, calculate the temperature of the surface.

▌ $P_e = \epsilon\sigma A T_e^4$, with $\epsilon = 1$, $A = 6.1 \times 10^{18} \, \text{m}^2$, $P_e = 3.9 \times 10^{26} \, \text{W}$. $(3.9 \times 10^{26}) = (5.67 \times 10^{-8})(6.1 \times 10^{18})T_e^4$. Solving, $T_e = \underline{5800 \, \text{K}}$.

19.46 The surface of a household radiator has an emissivity of 0.55 and an area of 1.5 m². (*a*) At what rate is radiation emitted by the radiator when its temperature is 50 °C? (*b*) At what rate is the radiation absorbed by the radiator when the walls of the room are 22 °C? (*c*) What is the net rate of radiation from the radiator?

▌ (*a*) $P_e = \epsilon\sigma A T_e^4 = (0.55)(5.67 \times 10^{-8})(1.5)(323)^4 = 509 \, \text{W}$
(*b*) $P_a = \epsilon\sigma A T_a^4 = (0.55)(5.67 \times 10^{-8})(1.5)(295)^4 = 354 \, \text{W}$
(*c*) Net rate is $P_e - P_a = \underline{155 \, \text{W}}$. This result shows that radiation is not an important mechanism for heat transfer at these temperatues. The radiator, in spite of its name, transfers most of its heat to the room by convection (Prob. 19.31).

19.47 Use Stefan's law to calculate the total power radiated per square meter by a filament at 1727 °C having an absorption factor of 0.4.

▮ Stefan's law gives $R = \epsilon\sigma T^4 = 0.4(5.67 \times 10^{-8})(2000)^4 = \underline{0.36 \text{ MW/m}^2}$.

19.48 The sun's surface temperature is about 6000 K. The sun's radiation is maximum at a wavelength of 0.5 μm. A certain light bulb filament emits radiation with a maximum at 2 μm. If both the surface of the sun and of the filament have the same emissive characteristics, what is the temperature of the filament?

▮ Use Wien's displacement law, $\lambda_m T =$ constant:

$$2T_{\text{fil}} = (0.5)(6000) \quad \text{or} \quad T = \underline{1500 \text{ K}}$$

19.49 An aluminum sphere 5 cm in diameter is suspended by a fine thread inside an evacuated jar, so that it can lose heat only by radiation. The sphere's initial temperature is 100 °C, and the wall of the jar is always 22 °C. (a) What is the initial net rate of heat loss of the sphere? (b) At the rate in part a, how long will it take the sphere to go from 100 to 90 °C? (c) What is the net rate of heat loss at 90 °C? (d) At the rate in part c, how long will it take the sphere to go from 90 to 80 °C? (*Hint:* Use Table 17-1 and the facts that the density of aluminum is 2.7 g/cm^3 and its emissivity $\epsilon = 0.1$.)

▮ (a) The rate at which energy is lost to the sphere is given by $P_e - P_a = \epsilon\sigma A(T_e^4 - T_a^4)$, with $\epsilon = 0.10$, $\sigma = 5.67 \times 10^{-8}$ W/m$^2 \cdot$ K^4, $T_e = 373$ K, $T_a = 295$ K, $A = 4\pi r^2 = \pi d^2 = \pi(0.05 \text{ m})^2 = 0.0079$ m^2. Then $P_e - P_a = \underline{0.528 \text{ W}}$. (b) $(P_e - P_a)\,\Delta t = mc\,\Delta T$, where Δt is the time of interest, $\Delta T = 10$ K, $c = 0.92$ J/g $\cdot$ K is the specific heat of aluminum and the sphere's mass is $m = \rho_{\text{Al}}(\frac{4}{3}\pi r^3) = (2.7 \text{ g/cm}^3)(65.4 \text{ cm}^3) = 177$ g. Then

$$\Delta t = \frac{(177 \text{ g})(0.92 \text{ J/g} \cdot \text{K})(10 \text{ K})}{0.528 \text{ J/s}} = \underline{3084 \text{ s}} = \underline{51.4 \text{ min}}$$

(c) This is the same as part a, except that $T_e = 363$ K. Then $P_e - P_a = \underline{0.439 \text{ W}}$.
(d) This is the same as part b, except that we use the rate of part c. Then $\Delta t = \underline{3709 \text{ s}} = \underline{61.8 \text{ min}}$.

19.50 Show that when the difference $\Delta T = T_e - T_a$ between the temperature T_e of an object and the temperature T_a of the walls of its container is small compared with T_a, the net rate of heat transfer by radiation from the object to the walls can be written $P = 4\epsilon\sigma T_a^3\,\Delta T$ (Newton's law of cooling, Prob. 19.26 applied to radiation).

▮ Net rate of transfer of heat is given by $P_e - P_a = \epsilon\sigma A(T_e^4 - T_a^4)$. If $T_e = T_a + \Delta T$, we have

$$T_e^4 = T_a^4 = (T_a + \Delta T)^4 - T_a^4 = (T_a^4 + 4T_a^3\,\Delta T + 6T_a^2\,\Delta T^2 + 4T_a\,\Delta T^3 + \Delta T^4) - T_a^4$$

$$= 4T_a^3\,\Delta T\left[1 + \frac{6}{4}\frac{\Delta T}{T_a} + \left(\frac{\Delta T}{T_a}\right)^2 + \frac{1}{4}\left(\frac{\Delta T}{T_a}\right)^3\right].$$

Since $\Delta T \ll T_a$, the last three terms in the bracket are small compared with 1, and $T_e^4 - T_a^4 \approx 4T_a^3\,\Delta T$. Using this in our rate equation we get $P_e - P_a = 4\epsilon\sigma A T_a^3\,\Delta T$. This result is equivalent to the calculus result $\Delta y \approx (dy/dx)\,\Delta x$, where $x = T$ and $y = T^4$.

19.51 A blackbody is at a temperature of 527 °C. To radiate twice as much energy per second, its temperature must be increased to what value?

▮ Since $P \propto T^4$, the temperature must be increased to $2^{1/4}(800 \text{ K}) = \underline{951 \text{ K}}$.

19.52 What increase in radiated power results when the temperature of a blackbody is increased from 7 to 287 °C?

▮

$$\frac{P(560 \text{ K})}{P(280 \text{ K})} = \left(\frac{560}{280}\right)^4 = 16$$

CHAPTER 20
Gas Laws and Kinetic Theory

20.1 THE MOLE CONCEPT; THE IDEAL GAS LAW

20.1   What is a *mole*? What is *Avogadro's number*? What is the *atomic mass unit*?

▌ The SI unit for amount of substance is the mole (mol). By definition, 1 mol is the amount of substance that contains as many particles as there are atoms in exactly 0.012 kg ($= 12$ g) of the isotope carbon-12. Thus, the mass, in grams, of one mole of any substance numerically equals the atomic (or molecular, or formula) weight of that substance.

   The actual number of elementary units in a mole is called Avogadro's number; $N_A = 6.022 \times 10^{23}$ mol^{-1} = 6.022×10^{26} kmol^{-1}.

   The atomic mass unit (u) is defined to be $\frac{1}{12}$ the mass of a carbon-12 atom; thus,

$$1 \text{ u} = \frac{1}{12} \frac{0.012 \text{ kg/mol}}{6.022 \times 10^{23} \text{ mol}^{-1}} = 1.66 \times 10^{-27} \text{ kg}$$

It follows that the mass, in u, of any species of atom is numerically equal to the atomic weight of that species.

20.2   State the ideal gas law.

▌ The equation of state of a gas in thermal equilibrium (uniform temperature throughout the system) relates the pressure, the volume, and the temperature of the gas. At sufficiently low densities, all gases have the same equation of state, called the *ideal gas law*: $pV = NkT = nRT$, where N is the number of molecules in the gas, n is the number of moles of the gas, and T is the Kelvin temperature of the gas.

   The values of k, *Boltzmann's constant*, and R, the universal gas constant, are $k = 1.38 \times 10^{-23}$ J/K, $R = 8.314$ J/mol · K. The ratio of these constants is *Avogadro's number*,

$$N_A = \frac{R}{k} = 6.02 \times 10^{23} \text{ mol}^{-1}$$

which is the number of molecules in a mole.

   The ideal gas law for a pure substance is often expressed in the form $pV = (m/M)RT$, where m is the mass of the gas sample and M is its molecular weight (mass/mole). Dividing by V we get yet another form: $p = (\rho RT)/M$, where ρ is the mass density.

20.3   Find the mass (in kilograms) of an ammonia molecule NH_3.

▌ The molecular weight of NH_3 is 17.0. Hence,

$$m_{NH_3} = 17.0 \text{ u} = (17.0 \text{ u})(1.66 \times 10^{-27} \text{ kg/u}) = \underline{2.82 \times 10^{-26} \text{ kg}}$$

20.4   In the near future, when the world population becomes 6 billion (6×10^9), how many moles of humans will there be on earth?

▌ The number of moles of humans would be (6×10^9 people)/(6×10^{23} people/mol) = $\underline{1 \times 10^{-14} \text{ mol}}$.

20.5   What is the "atomic weight" of an electron? What is its mass in u? ($m_e = 9.1 \times 10^{-31}$ kg.)

▌ From the definition, $M = N_A m_e = 5.48 \times 10^{-4}$ kg/kmol. By Prob. 20.1, $m_e = \underline{5.48 \times 10^{-4} \text{ u}}$.

20.6   Consider a 60-kg young man to be a huge molecule. What is his mass in atomic mass units? What is his molecular weight?

▌ The mass of a kilomole of young men is $M = N_A(60 \text{ kg}) = \underline{3.6 \times 10^{28} \text{ kg/kmol}}$. Hence, the mass of one young man is $\underline{3.6 \times 10^{28} \text{ u}}$.

20.7   An ideal gas exerts a pressure of 1.52 MPa when its temperature is 298.15 K (25 °C) and its volume is 10^{-2} m^3 (10 L). (*a*) How many moles of gas are there? (*b*) What is the mass density if the gas is molecular hydrogen, H_2? (*c*) What is the mass density if the gas is oxygen, O_2?

∎ (a)
$$n = \frac{pV}{RT} = \frac{(1.52 \times 10^6)(10^{-2})}{(8.31)(298.15)} = \underline{6.135 \text{ mol}}$$

(b) The atomic mass of hydrogen is 1.008, so that 1 mol of hydrogen (H_2) contains 2.016 g, or 2.016×10^{-3} kg. The density of the hydrogen is then

$$\rho = \frac{nM}{V} = \frac{(6.13)(2.016 \times 10^{-3})}{10^{-2}} = \underline{1.24 \text{ kg/m}^3}$$

(c) The atomic mass of oxygen is 16, so that 1 mol of O_2 contains 32 g, or 32×10^{-3} kg. The density of the oxygen is then

$$\rho = \frac{nM}{V} = \frac{(6.13)(32 \times 10^{-3})}{10^{-2}} = \underline{19.6 \text{ kg/m}^3}$$

20.8 A compressor pumps 70 L of air into a 6-L tank with the temperature remaining unchanged. If all the air is originally at 1 atm, what is the final absolute pressure of the air in the tank?

∎ Use Boyle's law, since the temperature is constant.

$$p_0 V_0 = pV \qquad p_0 = 1 \text{ atm} \qquad V_0 = 70 + 6 = 76 \text{ L} \qquad V = 6 \text{ L}$$
$$1(76) = p(6) \qquad p = \underline{12.7 \text{ atm}} \text{ absolute pressure}$$

20.9 A 0.025-m^3 tank contains 0.084 kg of nitrogen gas (N_2) at a gauge pressure of 3.17 atm. Find the temperature of the gas in degrees Celsius. ($p_{atm} = 1.013 \times 10^5$ Pa).

∎ We note that the mass of 1 kmol of N_2 is 28 kg. From the ideal gas law, $pV = nRT$, with $n = 0.084/28$ kmol, $p = 3.17 + 1 = 4.17$ atm $= 4.17 \times 1.013 \times 10^5 \text{ N/m}^2$, $R = 8314 \text{ J/kmol} \cdot \text{K}$. Substituting,

$$4.17(1.013 \times 10^5)(0.025) = (0.084/28)(8314)T \qquad \text{and}$$
$$T = \frac{28(4.17)(1.013 \times 10^5)(0.025)}{0.084(8314)} = 423 \text{ K} = \underline{150 \text{ °C}}$$

20.10 A partially inflated balloon contains 500 m^3 of helium at 27 °C and 1-atm pressure. What is the volume of the helium at an altitude of 18 000 ft, where the pressure is 0.5 atm and the temperature is −3 °C?

∎ For a confined gas,

$$\frac{p_1 V_1}{T_1} = \frac{p_2 V_2}{T_2} \qquad \frac{1(500)}{300} = \frac{0.5 V_2}{270} \qquad V_2 = \frac{500(270)}{300(0.5)} = \underline{900 \text{ m}^3}$$

20.11 An air bubble released at the bottom of a pond expands to four times its original volume by the time it reaches the surface. If atmospheric pressure is 100 kPa, what is the absolute pressure at the bottom of the pond? Assume constant T.

∎
$$p_0 = p \frac{V}{V_0} = (100 \text{ kPa})(4) = \underline{400 \text{ kPa}}$$

20.12 A pressure gauge indicates the differences between atmospheric pressure and pressure inside the tank. The gauge on a 1.00-m^3 oxygen tank reads 30 atm. After some use of the oxygen, the gauge reads 25 atm. How many cubic meters of oxygen at normal atmospheric pressure were used? There is no temperature change during the time of consumption.

∎ Since the temperature is fixed we use Boyle's law to solve the problem. The total pressure in the tank has been reduced from 31 to 26 atm. At the latter pressure, the gas originally in the tank would occupy $\frac{31}{26}$ m^3. Since 1.00 m^3 remains in the tank, the amount of gas used was $\frac{5}{26}$ m^3 at 26 atm. At the same temperature and at atmospheric pressure, this would occupy a volume 26 times as large: $\underline{5.00 \text{ m}^3}$.

20.13 An air bubble of volume V_0 is released by a fish at a depth h in a lake. The bubble rises to the surface. Assume constant temperature and standard atmospheric pressure above the lake; what is the volume of the bubble just before touching the surface? The density of the water is ρ.

∎ Equating $p_0 V_0$ at depth h to pV at the surface leads to the equation $(p + \rho g h)V_0 = pV$, from which $V = (1 + \rho g h/p)V_0$.

20.14 If 2.1212 g of a monatomic gas occupies 1.49 L when the temperature is 0 °C and the pressure is 810.6 kPa, what is the gas?

■ $pV = (m/M)RT$, where m = mass of gas and M = molecular weight.

$$(810.6 \times 10^3 \, \text{Pa})(1.49 \times 10^{-3} \, \text{m}^3) = \frac{(2.1212 \times 10^{-3} \, \text{kg})(8.314 \, \text{J/mol} \cdot \text{K})(273.15 \, \text{K})}{M}$$

$$M = 3.99 \times 10^{-3} \, \text{kg/mol} = 3.99 \, \text{g/mol} \Rightarrow \underline{\text{gas is helium}}$$

20.15 Use the general gas law to compute the density of methane, CH_4, at 20 °C and 5-atm pressure. A kilomole of methane is 16.0 kg.

■ $pV = nRT$, with $T = 20 + 273 = 293$ K, $R = 8314$ J/kmol · K, $n = 1$ kmol, and $p = 5$ atm = $5(1.013 \times 10^5)$ N/m². Then

$$5(1.013 \times 10^5)V = 8314(293) \quad \text{and} \quad V = 4.8 \, \text{m}^3 \quad d = \left(\frac{m}{V}\right) = \frac{16.0}{4.8} = \underline{3.33 \, \text{kg/m}^3}$$

20.16 Solve Prob. 20.15 without finding the volume.

■ Use $pV = (m/M)RT$ and $\rho = m/V$ to give

$$\rho = \frac{pM}{RT} = \frac{(5 \times 1.013 \times 10^5 \, \text{N/m}^2)(16 \, \text{kg/kmol})}{(8314 \, \text{J/kmol} \cdot \text{K})(293 \, \text{K})} = \underline{3.33 \, \text{kg/m}^3}$$

20.17 A 100-ft³ volume of nitrogen at 27 °C and 15 lb/in² is compressed to fill a tank that is initially empty and has a volume of 5 ft³. If the final temperature of the nitrogen is 17 °C, what is the absolute pressure in the tank?

■ Use the Kelvin scale and the general gas law in the form

$$\frac{p_1 V_1}{T_1} = \frac{p_2 V_2}{T_2} \qquad \frac{15(100)}{273 + 27} = \frac{p_2(5)}{273 + 17} \qquad p_2 = \frac{1500(290)}{5(300)} = \underline{290 \, \text{lb/in}^2}$$

20.18 In the preparation of a sealed-off 20-mL tube at low temperatures, one drop (50 mg) of liquid nitrogen is accidentally sealed off in the tube. Find the nitrogen (N_2) pressure within the tube when the tube warms to 27 °C. Assume ideality. Express your answer in atmospheres (1 atm = 101.3 kPa).

■ Use $pV = (mRT)/M$. M is 28 for N_2. Use $T = 300$ K, $m = 5 \times 10^{-5}$ kg, $V = 2 \times 10^{-5}$ m³, and $R = 8.314$ kJ/kmol · K, giving $p = 223$ kPa = $\underline{2.20 \, \text{atm}}$.

20.19 A car tire is filled to a gauge pressure of 24 psi (lb/in²) when the temperature is 20 °C. After the car has been running at high speed, the tire temperature rises to 60 °C. Find the new gauge pressure within the tire if the tire's volume does not change.

■
$$p_1 = p_0 \frac{T_1}{T_0} = [(14.7 + 24) \, \text{psi}] \frac{333 \, \text{K}}{293 \, \text{K}} = 44.0 \, \text{psi} \qquad \text{or} \qquad \underline{29.3 \, \text{psi}} \text{ gauge pressure}$$

20.20 In a diesel engine, the cylinder compresses air from approximately standard pressure and temperature to about one-sixteenth the original volume and a pressure of about 50 atm. What is the temperature of the compressed air?

■ Use $(P_1 V_1)/T_1 = (P_0 V_0)/T_0$ to find $T_1 = (P_1/P_0)(V_1/V_0)T_0 = (50)(\frac{1}{16})(273) = \underline{853 \, \text{K}}$.

20.21 One way to cool a gas is to let it expand. Typically, a gas at 27 °C and a pressure of 40 atm might be expanded to atmospheric pressure and a volume 13 times larger. Find the new temperature of the gas.

■ As in Prob. 20.20, $T_1 = (P_1/P_0)(V_1/V_0)T_0 = (\frac{1}{40})(13)(300) = \underline{97.5 \, \text{K} \, (-176 \, °\text{C})}$.

20.22 A vertical right cylinder of height $h = 30.00$ cm and base area $A = 12.0$ cm² is sitting open under standard temperature and pressure. A 5.0-kg piston that fits tightly into the cylinder is now placed into the cylinder and allowed to fall to an equilibrium height within it. What then is the height of the piston and what is the pressure within the cylinder? Assume the final temperature to be 0 °C.

■ Pressure in the cylinder is increases by $\Delta p = (mg)/A = (5.0 \, \text{kg})(9.8 \, \text{m/s}^2)/(12.0 \times 10^{-4} \, \text{m}^2) = 41 \, \text{kPa}$, giving a new pressure of $p_{\text{atm}} + \Delta p = 142$ kPa. For constant T, $p_{\text{atm}}(30A) = 142\,hA$; solving, $h = [101(30.0)]/142 = \underline{21.4\,\text{cm}}$.

20.23 A cylinder whose inside diameter is 4.00 cm contains air compressed by a piston of mass $m = 13.0$ kg, which can slide freely in the cylinder (Fig. 20-1). The entire arrangement is immersed in a water bath whose temperature can be controlled. The system is initially in equilibrium at temperature $t_i = 20\,°C$. The initial height of the piston above the bottom of the cylinder is $h_i = 4.00$ cm.

The temperature of the water bath is gradually increased to a final temperature $t_f = 100\,°C$. Calculate the height h_f of the piston.

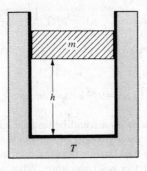

Fig. 20-1

▮ Since the pressure and the mass of the gas sample remain constant, Charles' law is applicable:

$$\frac{V_i}{T_i} = \frac{V_f}{T_f} \tag{1}$$

The cylindrical chamber has a constant cross-sectional area A, so that

$$\frac{V_f}{V_i} = \frac{Ah_f}{Ah_i} = \frac{h_f}{h_i} \tag{2}$$

Combining Eqs. (1) and (2), the final height is given by

$$h_f = h_i\left(\frac{V_f}{V_i}\right) = h_i\left(\frac{T_f}{T_i}\right)$$

With $t_i = 20\,°C$ and $t_f = 100\,°C$, we have $T_i = 293.15$ K and $T_f = 373.15$ K. Using $h_i = 4.00$ cm, we find that

$$h_f = (4.00\,\text{cm})\left(\frac{373.15\,\text{K}}{293.15\,\text{K}}\right) = \underline{5.09\,\text{cm}}$$

20.24 Refer to Prob. 20.23. Starting from the same initial conditions the temperature is again gradually raised, and weights are added to the piston to keep its height fixed at h_i. Calculate the mass that has been added when the temperature has reached $t_f = 100\,°C$.

▮ Since there is a fixed quantity of gas in the cylinder, the quantity $(pV)/T$ is constant; since V is held constant, the final pressure is given by

$$p_f = p_i\left(\frac{T_f}{T_i}\right) \tag{1}$$

In terms of the ambient exterior air pressure (which we take to be $p_{\text{atm}} = 1.013 \times 10^5$ Pa), the initial and final pressures in the cylinder are

$$p_i = p_{\text{atm}} + \frac{m_i g}{A} \tag{2}$$

and

$$p_f = p_{\text{atm}} + \frac{m_f g}{A} = p_{\text{atm}} + \frac{m_i g}{A} + \frac{\Delta m g}{A} \tag{3}$$

Solving Eqs. (1) to (3) for the necessary additional mass Δm, we find that

$$\Delta m = \frac{A}{g}(p_f - p_i) = \frac{A}{g}p_i\left(\frac{T_f}{T_i} - 1\right) = \frac{A}{g}\left(p_{\text{atm}} + \frac{m_i g}{A}\right)\left(\frac{T_f}{T_i} - 1\right) \tag{4}$$

The cross-sectional area $A = (\pi/4)(4.00 \text{ cm})^2 = 12.57 \text{ cm}^2 = 1.257 \times 10^{-3} \text{ m}^2$. Inserting $p_{\text{atm}} = 1.013 \times 10^5 \text{ Pa}$, $m_i = 13.0 \text{ kg}$, $g = 9.80 \text{ m/s}^2$, and the given temperatures, we obtain

$$\Delta m = \left(\frac{1.257 \times 10^{-3}}{9.80}\right)\left[1.013 \times 10^5 + \frac{(13.0)(9.80)}{1.257 \times 10^{-3}}\right]\left(\frac{373.15}{293.15} - 1\right) = \underline{7.09 \text{ kg}}$$

20.25 An ideal gas has been placed in a tank at 40 °C. The gauge pressure is initially 608 kPa. One-fourth of the gas is then released from the tank and thermal equilibrium is established. What will be the gauge pressure if the temperature is 315 °C? Take standard atmospheric pressure as 101 kPa.

I

$$pV = nRT \Rightarrow \frac{p_1 V_1}{n_1 T_1} = \frac{p_2 V_2}{n_2 T_2}$$

We have

$$T_1 = 273 + 40 = 313 \text{ K} \qquad p_1 = 101 \text{ kPa} + 608 \text{ kPa} = 709 \text{ kPa}$$
$$n_2 = \tfrac{3}{4}n_1 \qquad V_2 = V_1 \qquad T_2 = 273 + 315 = 588 \text{ K}$$

Then $P_2 = (709 \text{ kPa})(\tfrac{3}{4})(588 \text{ K})/(313 \text{ K}) = 999 \text{ kPa}$. The final gauge pressure is $999 - 101 = \underline{898 \text{ kPa}}$.

20.26 A mercury barometer whose scale is on a stand behind the glass tube, reads 740 mm. Because of the low reading, it is suspected that some air is present in the space above the mercury. The space is 60 mm long. The open end of the barometer is lowered farther into the mercury reservoir. When the barometer reading is 730 mm, the space above the mercury is 40 mm long. What is the true atmospheric pressure?

I We denote the true ambient pressure by p_a, the initial pressure of the trapped air by p_t, and the initial barometer reading by p_r. Hydrostatic equilibrium of the mercury at the surface of the reservoir implies that

$$p_a = p_t + p_r \qquad (1)$$

When the open end of the barometer is lowered farther into the reservoir, the pressure of the trapped air increases to p_t', while the barometer reading decreases to p_r'. The total pressure is unchanged:

$$p_a = p_t' + p_r' \qquad (2)$$

The volume of the trapped air decreases from V_t to $V_t' = 2V_t/3$, so Boyle's law implies that

$$p_t' = \frac{3p_t}{2} \qquad (3)$$

From Eqs. (1) to (3), we find that

$$p_t = 2(p_r - p_r') = 2(740 - 730) \text{ mmHg} = 20 \text{ mmHg}$$

Then Eq. (1) implies that $p_a = 740 + 20 = \underline{760 \text{ mmHg}}$.

20.27 The closed cylinder shown in Fig. 20-2 has a freely moving piston separating chambers 1 and 2. Chamber 1 contains 25 mg of N_2 gas, and chamber 2 contains 40 mg of helium gas. When equilibrium is established, what will be the ratio L_1/L_2? What is the ratio of the number of moles of N_2 to the number of moles of He? (Molecular weights of N_2 and He are 28 and 4.)

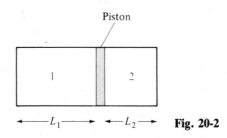

Piston

Fig. 20-2

I At equilibrium the pressure in each chamber is the same. The number of kilomoles in each part is $n_1 = (25 \times 10^{-6} \text{ kg})/(28 \text{ kg/kmol}) = 8.9 \times 10^{-7} \text{ kmol}$ and $n_2 = (40 \times 10^{-6})/4.0 = 1.0 \times 10^{-5}$. Writing the ideal gas law, with A as the cross-sectional area, $P = (n_1 RT)/(AL_1) = (n_2 RT)/(AL_2)$, from which $L_1/L_2 = n_1/n_2 = \underline{0.089}$.

20.28 Two gases occupy two containers, A and B. The gas in A, of volume 0.11 m^3, exerts a pressure of 1.38 MPa. The gas in B, of volume 0.16 m^3, exerts a pressure of 0.69 MPa. The two containers are united by a tube of negligible volume and the gases are allowed to intermingle. What is the final pressure in the container if the temperature remains constant?

❚ $$p_A V_A = n_A RT \qquad p_B V_B = n_B RT$$

After mingling and coming to equilibrium we have for the final pressure

$$p_f(V_A + V_B) = (n_A + n_B)RT = p_A V_A + p_B V_B$$

or $$p_f = \frac{p_A V_A + p_B V_B}{V_A + V_B} = \frac{(1.38 \text{ MPa})(0.11 \text{ m}^3) + (0.69 \text{ MPa})(0.16 \text{ m}^3)}{0.11 \text{ m}^3 + 0.16 \text{ m}^3} = \underline{0.97 \text{ MPa}}$$

20.29 Butane gas burns in air according to the following reaction: $2C_4H_{10} + 13O_2 \rightarrow 10H_2O + 8CO_2$. Suppose the initial and final temperatures are equal and high enough so that all reactants and products act as perfect gases. Two moles of butane are mixed with 13 mol of oxygen and then completely reacted. What will be the final pressure if the volume remains unchanged and the pressure before reaction is P_0?

❚ In the beginning there were 15 mol of gas, so $p_0 V_0 = 15RT_0$. In the end there are 18 mol of gas as the reaction equation shows; hence, $p_f V_f = 18RT_f$. Given that volume and temperature were unchanged, then $p_f = (18RT_f)/V_f = (18RT_0)/V_0 = (18)(p_0/15)$ where substitution for $(RT_0)/V_0$ was made from the first equation; thus, $p_f = \underline{1.20p_0}$.

20.30 Two samples x and y of the same ideal gas are in adjacent chambers, separated by a thermally insulating partition. The initial volumes, pressures, and temperatures of the samples are V_x, V_y, p_x, p_y, and T_x, T_y, respectively. The partition is removed, and the single chamber of combined gas is brought to a final temperature T_f. **(a)** Assume that the additional volume made available by the removal of the partition is negligible, so that the final volume is $V_x + V_y$. Find the final pressure p_f. Give your result in terms of V_x, V_y, p_x, p_y, T_x, T_y, and T_f. **(b)** Suppose that the removal of the partition makes available an additional volume V_p, so that the final volume is $V_x + V_y + V_p$. Modify the result of part (a) to allow for this.

❚ **(a)** The number of moles of sample x is given by $n_x = (p_x V_x)/(RT_x)$; similarly, $n_y = (p_y V_y)/(RT_y)$. After the partition is removed, the number of moles $n_f = n_x + n_y$. Applying the ideal gas law to the entire sample, and using $V_f = V_x + V_y$, we have

$$p_f = \frac{n_f RT_f}{V_f} = \frac{RT_f(n_x + n_y)}{V_x + V_y}$$

That is,

$$p_f = p_x \frac{T_f}{T_x} \frac{V_x}{V_x + V_y} + p_y \frac{T_f}{T_y} \frac{V_y}{V_x + V_y}$$

(b) With $V_f = V_x + V_y + V_p$, the result is

$$p_f = \frac{n_f RT_f}{V_f} = \frac{RT_f(n_x + n_y)}{V_x + V_y + V_p} = p_x \frac{T_f}{T_x} \frac{V_x}{V_x + V_y + V_p} + p_y \frac{T_f}{T_y} \frac{V_y}{V_x + V_y + V_p}$$

20.31 A glass bulb of volume 400 cm³ is connected to another of volume 200 cm³ by means of a tube of negligible volume. The bulbs contain dry air and are both at a common temperature and pressure of 20 °C and 1.000 atm. The larger bulb is immersed in steam at 100 °C; the smaller, in melting ice at 0 °C. Find the final common pressure.

❚ We let n_1 and n_2 denote the number of moles of gas in the large and small bulbs, in the final configuration. Denoting the final temperatures by T_1 and T_2 and the final pressure by p_f, the ideal gas law implies that

$$p_f V_1 = n_1 RT_1 \tag{1}$$

and

$$p_f V_2 = n_2 RT_2 \tag{2}$$

where $V_1 = 400 \text{ cm}^3$ and $V_2 = 200 \text{ cm}^3$. We also have

$$p_0 V_0 = nRT_0 \tag{3}$$

where p_0, $V_0 = V_1 + V_2$, and T_0 are the initial pressure, volume, and temperature. Using Eqs. (1) to (3) in the equation $n_1 + n_2 = n$, we find that

$$\frac{p_f V_1}{RT_1} + \frac{p_f V_2}{RT_2} = \frac{p_0 V_0}{RT_0}$$

Solving for p_f we obtain

$$p_f = \frac{p_0 V_0}{T_0\left(\dfrac{V_1}{T_1} + \dfrac{V_2}{T_2}\right)}$$

Inserting the numerical values $p_0 = 1.00$ atm, $T_0 = 293.15$ K ($t_0 = 20\,°C$), $T_1 = 373.15$ K ($t_1 = 100\,°C$), and $T_2 = 273.15$ K ($t_2 = 0\,°C$), we find

$$p_f = \frac{(1.00)(600)}{(293.15)\left[\left(\dfrac{400}{373.15}\right) + \left(\dfrac{200}{273.15}\right)\right]} = \underline{1.13\ \text{atm}}$$

20.32 A box of interior volume V_b has a heavy airtight hinged lid of mass M_l and area A_l. The box contains n_b kmol of a perfect gas at temperature T_0. The box is inside a chamber which also contains an additional n_c kmol of the gas at the same temperature. The gas in the chamber occupies a volume V_c. (**a**) Find the pressure p_b in the box in terms of n_b, V_b, and T_0. (**b**) Find the pressure p_c in the chamber in terms of n_c, V_c, and T_0. (**c**) Initially the hinged lid is closed. Show that this requires that $p_b - p_c \le M_l g / A_l$. (**d**) If the whole system is heated, at what temperature T_l will the gas pressure lift the hinged lid?

▐ (**a**) Applying the ideal gas law, we have $p_b = (n_b R T_0)/V_b$. (**b**) The ideal gas law implies that $p_c = (n_c R T_0)/V_c$. (**c**) The lid would lift if the net upward pressure force $(p_b - p_c)A_l$ exceeded the weight $M_l g$. The fact that the lid is closed therefore implies that $p_b - p_c \le (M_l g)/A_l$. (**d**) The lid will lift when $p_b - p_c = (M_l g)/A_l$. Using the results of parts a and b, this occurs for

$$\frac{n_b R T_l}{V_b} - \frac{n_c R T_l}{V_c} = \frac{M_l g}{A_l} \qquad \text{or when} \qquad T_l = \frac{M_l g}{R A_l [(n_b/V_b) - (n_c/V_c)]}$$

20.33 Refer to Prob. 20.32. Suppose that starting from T_0, the system is heated to a temperature $T' > T_l$, and then cooled back to temperature T_0. Assume that the hinged lid recloses as soon as the (changing) pressure in the box *fails* to exceed the (changing) chamber pressure by $M_l g / A_l$. Let n_b' denote the number of kilomoles remaining in the box after the heating and recooling. Show that

$$n_b' = \frac{(M_l g V_b V_c)/(A_l R T') + (n_b + n_c)V_b}{V_b + V_c}$$

▐ As the system is heated, gas escapes from the box to the surrounding chamber in amounts just sufficient to maintain an internal pressure excess of $(M_l g)/A_l$. Once the peak temperature T' is reached (and the cooling begins), the box remains closed. Therefore the final numbers of kilomoles (n_b' and n_c') in the box and in the surrounding chamber satisfy the equation

$$\left(\frac{n_b' R}{V_b} - \frac{n_c' R}{V_c}\right) T' = \frac{M_l g}{A_l}$$

Since the total number of kilomoles in the system is constant, we also have

$$n_c' = n_b + n_c - n_b'$$

Combining these two displayed equations, we obtain

$$n_b' V_c - (n_b + n_c - n_b') V_b = \frac{M_l g V_b V_c}{A_l R T'}$$

Solving for n_b', we find

$$n_b' = \frac{(M_l g V_b V_c)/(A_l R T') + (n_b + n_c)V_b}{V_b + V_c}$$

20.34 Refer to Prob. 20.33. Show that the expression for n_b' given approaches the limiting value $[(n_b + n_c)V_b]/(V_b + V_c)$ as $T' \to \infty$. Can you suggest a simple interpretation for this result?

▐ Examining the final expression in Prob. 20.33, we see that in the limit $T' \to \infty$, the first term in the numerator approaches zero, so that $n_b' \to [(n_b + n_c)V_b]/(V_b + V_c)$, as desired. That is,

$$\frac{n_b' N_A}{V_b} \to \frac{(n_b + n_c)N_A}{(V_b + V_c)}$$

where N_A is Avogadro's number. The expression on the left is the final number density of gas molecules in the box, while the expression on the right is the overall average number density for the whole system. These approach one another because the density difference corresponding to a given fixed pressure difference of $(M_l g)/A_l$ drops to essentially zero as $T' \to \infty$. Put another way the pressure difference becomes negligible compared with the actual pressures at $T' \to \infty$.

20.35 Justify *Dalton's law of partial pressures*.

❚ The ideal gas law can be expressed in terms of the total number of molecules, N: $pV = NkT$. Since this law presumes that we can ignore the detailed structure of the individual molecules and their interactions, N can represent a mixture of any number of nonreactive gases as long as the basic premises are valid. Suppose that we have such a mixture of gases distinguished by indices $1, 2, 3 \cdots$. Then $N = N_1 + N_2 + N_3 + \cdots$ is the sum of the number of molecules of each type. From above,

$$p = \frac{NkT}{V} = \frac{(N_1 + N_2 + N_3 + \cdots)kT}{V} = \frac{N_1 kT}{V} + \frac{N_2 kT}{V} + \frac{N_3 kT}{V} + \cdots$$

The terms on the right, however, represent the pressure due to each gas if it alone occupied the total volume at temperature T. Calling these the "partial pressures" $p_1, p_2, p_3, \ldots$, we have our result: $p = p_1 + p_2 + p_3 + \cdots$.

20.36 In a gaseous mixture at $20\,°C$ the partial pressures of the components are as follows: hydrogen, $200\,mmHg$; carbon dioxide, $150\,mmHg$; methane, $320\,mmHg$; ethylene, $105\,mmHg$. What are (a) the total pressure of the mixture and (b) the mass fraction of hydrogen? ($M_H = 2$, $M_{CO_2} = 44$, $M_{meth} = 16$, $M_{eth} = 30$ kg/kmol.)

❚ (a) According to Dalton's law, total pressure = sum of partial pressures = $200 + 150 + 320 + 105 = 775\,mmHg$. (b) From the gas law for a single species $m = M(pV/RT)$. The mass of hydrogen gas present is thus $m_H = M_H p_H(V/RT)$. The total mass of gas present, m_t, is the sum of similar terms:

$$m_t = (M_H p_H + M_{CO_2} p_{CO_2} + M_{meth} p_{meth} + M_{eth} p_{eth}) \frac{V}{RT}$$

The required fraction is

$$\frac{m_H}{m_t} = \frac{M_H p_H}{M_H p_H + M_{CO_2} p_{CO_2} + M_{meth} p_{meth} + M_{eth} p_{eth}} = \frac{(2)(200)}{(2)(200) + (44)(150) + (16)(320) + (30)(105)} = \underline{0.026}$$

20.2 KINETIC THEORY

20.37 What is meant by the *root-mean-square speed* of an ensemble of gas molecules? How is it related to the pressure of an ideal gas?

❚ Viewed in the center-of-mass reference frame, the molecules of a gas are in random motion, with a wide distribution of kinetic energies. The *root-mean-square speed* (or *thermal speed*) v_{rms} may be defined as the speed of a hypothetical molecule whose translational kinetic energy equals the average translational kinetic energy over the whole ensemble of gas molecules: $\frac{1}{2}mv_{rms}^2 \equiv \overline{\frac{1}{2}mv^2}$, or $v_{rms} = \sqrt{\overline{v^2}}$. For a dilute gas, obeying the ideal gas law, $\frac{1}{2}mv_{rms}^2 = \frac{3}{2}kT$. Thus, the absolute temperature of a macroscopic object is a measure of the average translational kinetic energy of its molecules, as determined in the object's center-of-mass frame. This last result can be combined with the ideal gas law, $pV = NkT$, to yield $pV = \frac{1}{3}Nmv_{rms}^2$. Noting that $Nm = M$, the total mass of the gas sample, we have $p = \frac{1}{3}\rho v_{rms}^2$.

20.38 What is the equipartition theorem?

❚ The equipartition theorem is a result derived from statistical mechanics ignoring quantum mechanical effects. It says that for a statistical ensemble of molecules in thermodynamic equilibrium, the average value of each "mode" of energy a molecule can have (averaged over all molecules) equals $\frac{1}{2}kT$, if the mode has quadratic dependence on a coordinate or velocity. A mode of energy refers to the energy associated with any coordinate or velocity. For example, translational kinetic energy has three modes: $\frac{1}{2}mv_x^2$, $\frac{1}{2}mv_y^2$, $\frac{1}{2}mv_z^2$. Thus average translational kinetic energy $\frac{1}{2}mv^2 = \frac{1}{2}mv_x^2 + \frac{1}{2}mv_y^2 + \frac{1}{2}mv_z^2 = \frac{3}{2}kT$. Another example is rotational kinetic energy of a diatomic molecule, where there are two axes of rotation and two angular velocity terms, and $\frac{1}{2}I\omega_\theta^2 + \frac{1}{2}I\omega_\phi^2 = 2(\frac{1}{2}kT) = kT$. Similarly, for a simple model of a solid with molecules which are free to move from their equilibrium positions with simple harmonic motion (in each dimension), we have six modes: $\frac{1}{2}kx^2$, $\frac{1}{2}ky^2$, $\frac{1}{2}kz^2$, $\frac{1}{2}mv_x^2$, $\frac{1}{2}mv_y^2$, $\frac{1}{2}mv_z^2$; and the total average energy is $6(\frac{1}{2}kT) = 3kT$.

20.39 Five molecules have speeds of 12, 16, 32, 40, and 48 m/s. Find **(a)** the average speed and **(b)** the root-mean-square speed, for these molecules. **(c)** Show that for any distribution of speeds, $v_{rms} \geq \bar{v}$.

▮ **(a)**
$$\bar{v} = \frac{12 + 16 + 32 + 40 + 48}{5} = \underline{29.6 \text{ m/s}}$$

(b)
$$v_{rms} = \left[\frac{12^2 + 16^2 + 32^2 + 40^2 + 48^2}{5} \right]^{1/2} = \underline{32.6 \text{ m/s}}$$

(c) Let v_i for $i = 1, \ldots, N$ represent the various speeds. Then

$$\bar{v} = \frac{1}{N} \sum_{i=1}^{N} v_i \quad \text{and} \quad v_{rms}^2 = \frac{1}{N} \sum_{i=1}^{N} v_i^2$$

We consider the expression
$$A = \frac{1}{N} \sum_{i=1}^{N} (v_i - \bar{v})^2$$

(which is the average of the squares of the deviations of the individual speeds from the average speed). This is inherently a nonnegative quantity. By multiplying out each term being summed, and noting $(v_i - \bar{v})^2 = v_i^2 - 2\bar{v}v_i + \bar{v}^2$ we have

$$A = \frac{1}{N} \sum_{i=1}^{N} v_i^2 - \frac{2\bar{v}}{N} \sum_{i=1}^{N} v_i + \frac{1}{N}(N\bar{v}^2)$$

or
$$A = v_{rms}^2 - 2\bar{v}^2 + \bar{v}^2 = v_{rms}^2 - \bar{v}^2 \Rightarrow v_{rms}^2 \geq \bar{v}^2 \quad \text{and} \quad v_{rms} \geq \bar{v}$$

(The inequality will hold unless all the v_i are equal.)

20.40 Calculate the root-mean-square speed of hydrogen molecules (H_2) at 373.15 K (100 °C).

▮ The mass of an H_2 molecule may be calculated from the molecular weight as

$$m = \frac{M}{N_0} = \frac{2.016 \times 10^{-3} \text{ kg/mol}}{6.02 \times 10^{23} \text{ mol}^{-1}} = 3.35 \times 10^{-27} \text{ kg}$$

Then
$$\frac{1}{2} m v_{rms}^2 = \frac{3}{2} kT \qquad v_{rms} = \sqrt{\frac{3kT}{m}} = \sqrt{\frac{3(1.38 \times 10^{-23})(373.15)}{3.35 \times 10^{-27}}} = \underline{2.15 \text{ km/s}}$$

20.41 Molecular speeds may be measured with the device shown in Fig. 20-3. In one experiment it is found that molecules will pass through the velocity selector with the disks separated by 0.50 m and with an angular displacement of 180° between the two slits, when the disks turn at the rate of 600 rev/s. What are the possible speeds of the molecules?

▮ A molecule that has passed through the first slit will pass through the second if the second disk turns through π, 3π, 5π, etc., while the molecule is in transit between the two disks. Thus

$$t = \frac{\theta}{\omega} = \frac{d}{v} \quad \text{or} \quad v = \frac{\omega d}{\theta} = \frac{(600 \times 2\pi)(0.50)}{(2n+1)\pi} = \frac{600}{2n+1} \text{ m/s}$$

where $n = 0, 1, 2, \ldots$.

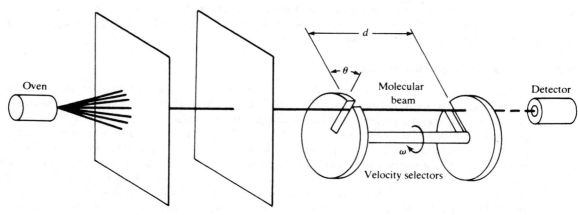

Fig. 20-3

20.42 The moving system of a D'Arsonval galvanometer consists of a coil of wire and a mirror suspended by a fine fiber and capable of rotation about a vertical axis. Random collisions of air molecules with the suspended system produce torques which are not equal and opposite at all instants. The result is that the angular position is continuously fluctuating and the system exhibits an unsteady "zero" (an example of Brownian motion). See Fig. 20-4. If the root-mean-square angular displacement of the system is $\theta_{rms} = 2 \times 10^{-4}$ rad and the torque constant for a fine quartz fiber is $K = 1 \times 10^{-13}$ N · m/rad, find the temperature of the air.

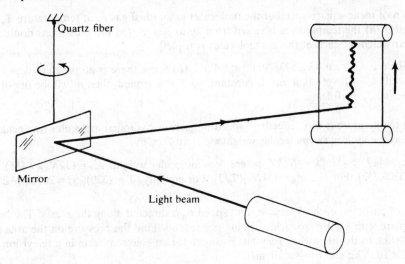

Quartz fiber

Mirror

Light beam

Fig. 20-4

▌ Let us assume that we have a large number of such fiber-mirror systems in equilibrium with a body of air of temperature T. The energy modes for these systems are rotational kinetic and torsional potential energy.

The equipartition theorem applies to both types of disordered energy, kinetic and potential. In its random oscillations the mirror-wire system has rotational kinetic energy, $\frac{1}{2}I\omega^2$, and elastic potential energy of the torsion of the wire, $\frac{1}{2}K\theta^2$. By the equipartition theorem, the mean value of each energy, averaged over the entire ensemble of fiber-mirror systems, is equal to $\frac{1}{2}kT$. Thus if we knew θ_{rms} for the ensemble, $\frac{1}{2}K\theta_{rms}^2 = \frac{1}{2}kT$. Since we have only one fiber-mirror system, we simulate the average over a large number of them at one instant of time, by the average for one of them over many instants of time. Since we have a steady-state system and the motions involved are due to random collisions with air molecules, the averages should be the same. Then, from the information given,

$$T = \frac{K\theta_{rms}^2}{k} = \frac{(1 \times 10^{-13})(2 \times 10^{-4})^2}{1.38 \times 10^{-23}} = 290 \text{ K} = \underline{17\,^\circ\text{C}}$$

20.43 What is the effective root-mean-square velocity of the molecules in a sample of oxygen at 200 °C?

▌ The mass of an O_2 is $m = 32 \text{ u} = 53 \times 10^{-27}$ kg.

$$\frac{1}{2}mv_{eff}^2 = \frac{3}{2}kT \qquad \frac{1}{2}(53 \times 10^{-27})v_{eff}^2 = \frac{3}{2}(1.38 \times 10^{-23})473$$

$$v_{eff} = \frac{3(1.38 \times 10^{-23})473}{5.3 \times 10^{-26}} = \underline{608 \text{ m/s}}$$

20.44 The pressure of a gas in a 100-mL container is 200 kPa and the average translational kinetic energy of each gas particle is 6.0×10^{-21} J. Find the number of gas particles in the container. How many moles are in the container?

▌ Solve $p = (2n_u/3)(\overline{mv^2/2})$ for the number density: $n_u = N/V$. $n_u = [3(2.0 \times 10^5)]/[2(6.0 \times 10^{-21})] = (5.0 \times 10^{25})/\text{m}^3$. This times $100 \times 10^{-6}\text{m}^3 = N = 5.0 \times 10^{21}$; number of moles $= N/N_A = \underline{8.3 \times 10^{-3} \text{ mol}}$.

20.45 A 2.00-mL-volume container contains 50 mg of gas at a pressure of 100 kPa. The mass of each gas particle is 8.0×10^{-26} kg. Find the average translational kinetic energy of each particle.

▌ To solve for K_{avg} for each particle, first find the particle number density $n_u = N/V$. The number of particles N is $(50 \times 10^{-6} \text{ kg})/(8.0 \times 10^{-26} \text{ kg/particle}) = 6.3 \times 10^{20}$. Then $n_u = N/(2.0 \times 10^{-6} \text{ m}^3) = (3.2 \times 10^{26})/\text{m}^3$. Gas pressure $p = (2n_u/3)K_{avg} = 1.0 \times 10^5$ Pa. Then, $K_{avg} = \underline{4.8 \times 10^{-22} \text{ J}}$.

20.46 Standard conditions (STP) are 1.01×10^5 Pa and 273 K. Under these conditions 22.4 m³ of an ideal gas contains 6.02×10^{26} molecules. Find the root-mean-square speed of a molecule in chlorine gas (Cl_2) under these conditions. Assume ideality. (Molecular wt of $Cl_2 = 71$.)

▮ The mass of the Cl_2 molecule is $71/(6.02 \times 10^{26}) = 1.18 \times 10^{-25}$ kg. We use $(m_0 \langle v^2 \rangle)/2 = \frac{3}{2}kT$, where $\langle \ \rangle$ means 'average of.' Then $\langle v^2 \rangle = [3(1.38 \times 10^{-23})(273)]/(1.18 \times 10^{-25}) = 9.6 \times 10^4$; thus $v_{rms} = \underline{309 \text{ m/s}}$.

20.47 Call the root-mean-square speed of the molecules in an ideal gas v_0 at temperature T_0 and pressure p_0. Find the speed if (**a**) the temperature is raised from 20 to 300 °C; (**b**) the pressure is doubled and $T = T_0$; (**c**) the molecular weight of each of the gas molecules is tripled.

▮ (**a**) Since $v \propto T^{1/2}$, $v = v_0(573/293)^{1/2} = 1.40 v_0$. (**b**) Since there is no pressure dependence at constant T, the speed is still v_0. (**c**) Note that mv^2 is constant, so if m is tripled, then v^2 will be one-third as large. The speed would be $(v_0/3)^{1/2} = \underline{0.58 v_0}$.

20.48 At what temperature is the "effective" speed of gaseous hydrogen molecules (molecular weight = 2) equal to that of oxygen molecules (molecular weight = 32) at 47 °C?

▮ $\frac{1}{2}mv_{eff}^2 = \frac{3}{2}kT \Rightarrow \frac{1}{2}Mv_{eff}^2 = \frac{3}{2}N_A kT$, where M = molecular weight. $v_{eff}^2 = (3kN_A)(T/M)$. For O_2, $v_{eff}^2 = (3kN_A)(320\text{K}/32)$. For H_2, $v_{eff}^2 = (3kN_A)(T/2)$. For equality, $T = (320)(\frac{2}{32}) = 20$ K $= \underline{-253 \text{ °C}}$.

20.49 A beam of particles, each of mass m_0 and speed v, is directed along the x axis. The beam strikes an area 1 mm square with 1×10^{15} particles striking per second. Find the pressure on the area due to the beam if the particles stick to the area when they hit. Evaluate for an electron beam in a television tube where $m_0 = 9.1 \times 10^{-31}$ kg and $v = 8 \times 10^7$ m/s.

▮ Each electron exerts an impulse of $m_0 v$ when it strikes and sticks to the surface. This times the number striking in unit time divided by the area is pressure $p = (10^{15}/10^{-6})m_0 v = 10^{21}m_0 v$. Inserting the momentum $(9.1 \times 10^{-31})(8 \times 10^7)$, we find $p = \underline{0.073 \text{ Pa}}$.

20.50 In the apparatus illustrated in Fig. 20-5, the length l of the rotating drum is 20.0 cm and the angular displacement ϕ between the entrance and exit of a molecule from the groove is 5.0°. Calculate the angular speed of the drum which will select molecules with speeds of 300 m/s. How many rotations per minute is this?

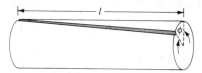

Fig. 20-5

▮ A molecule traveling with speed v from left to right parallel to the axis of the drum requires a time $t = l/v$ to travel the entire length of the drum. If a beam of molecules with various speeds is incident at the left end of the groove, the only molecules that can emerge are those for which the travel time t *also* equals ϕ/ω, where ω is the angular speed of the drum. Therefore, in order to select molecules with a particular speed v, the drum's rotation rate must be $\omega = (v\phi)/l$. With $l = 20.0$ cm $= 0.200$ m, $\phi = 5.0° = \pi/36$ rad, and $v = 300$ m/s, we find

$$\omega = \frac{(300 \text{ m/s})(\pi/36 \text{ rad})}{(0.200 \text{ m})} = \underline{130.9 \text{ rad/s}} = \underline{1250 \text{ rev/min}}$$

20.51 The temperature of outer space has an average value of about 3 K. Find the root-mean-square speed of a proton (a hydrogen nucleus) in space. ($m_p = 1.67 \times 10^{-27}$ kg.)

▮ Use the fact that $(m_0 \langle v^2 \rangle)/2 = (3kT)/2$. Then $\langle v^2 \rangle = [3(1.38 \times 10^{-23})(3)]/(1.67 \times 10^{-27}) = 7.44 \times 10^4$; thus $v_{rms} = \underline{272 \text{ m/s}}$.

20.52 The temperature of the corona of the sun, the luminous gaseous layer we see, is about 6000 K. Find the root-mean-square speed of a proton, one of the main constituents of the corona, at this temperature. ($m_p = 1.67 \times 10^{-27}$ kg.)

▮ Use the same approach as in Prob. 20.51 but with $T = 6000$ K. Then $v_{rms} = \underline{12.2 \text{ km/s}}$.

20.53 At what temperature is v_{rms} of H_2 molecules equal to the escape speed from the earth's surface

$(v_E = \sqrt{2gR_E})$? What is the corresponding temperature for escape of hydrogen from the moon's surface $(g_M = 1.6\,\text{m/s}^2,\; R_E = 6367\,\text{km},\; R_M = 1750\,\text{km})$?

❙ For the earth,

$$v_E = \sqrt{2gR_E} = \sqrt{2 \times 9.8 \times 6367 \times 10^3} = 11.2\,\text{km/s}$$

$$\frac{1}{2}mv_E^2 = \frac{3}{2}kT_E \quad\text{or}\quad \frac{1}{2}Mv_E^2 = \frac{3}{2}N_A kT_E = \frac{3}{2}RT_E \quad M = 2\,\text{kg/kmol}$$

$$T_E = \frac{Mv_E^2}{3R} = \frac{(2\,\text{kg/kmol})(11.2 \times 10^3\,\text{m/s})^2}{3(8314\,\text{J/kmol}\cdot\text{K})} = \underline{10\,059\,\text{K}}$$

For the moon, $v_M = \sqrt{2 \times 1.6 \times 1750 \times 10^3} = 2.37\,\text{km/s}$. Using $T_M/T_E = v_M^2/v_E^2$, we get $T_M = (2.37/11.2)^2(10\,059\,\text{K}) = \underline{450\,\text{K}}$.

20.54 A nuclear fusion reaction will occur in a gas of deuterium nuclei when the nuclei have an average kinetic energy of at least 0.72 MeV. What is the temperature required for nuclear fusion to occur with deuterium? $(1\,\text{eV} = 1.6 \times 10^{-19}\,\text{J}.)$

❙ $\frac{1}{2}mv^2 = \frac{3}{2}kT \quad (0.72 \times 10^6\,\text{eV})(1.6 \times 10^{-19}\,\text{J/eV}) = 1.5(1.381 \times 10^{-23}\,\text{J/K})T \quad T = \underline{5.57 \times 10^9\,\text{K}}$

20.55ᶜ The statistical energy-distribution underlying the ideal gas law is the *Maxwell–Boltzmann distribution*. It gives the number of molecules with kinetic energies between E and $E + dE$ as

$$N(E)\,dE = \frac{2N}{\sqrt{\pi}\,(kT)^{3/2}}\sqrt{E}\,e^{-E/kT}\,dE$$

Find the average kinetic energy over the collection of N molecules.

❙ $$\bar{E} = \frac{1}{N}\int_0^\infty EN(E)\,dE = \frac{2}{\sqrt{\pi}\,(kT)^{3/2}}\int_0^\infty E^{3/2}e^{-E/kT}\,dE = \frac{2kT}{\sqrt{\pi}}\int_0^\infty x^{3/2}e^{-x}\,dx$$

where we have made the change of variable $E = kTx$. The definite integral can be evaluated by standard techniques or looked up in definite integral tables, and equals $\frac{3}{4}\sqrt{\pi}$. Then $\bar{E} = \frac{3}{2}kT$.

20.56ᶜ Refer to Prob. 20.55 and reexpress the Maxwell–Boltzmann distribution as a function of velocity.

❙ We note that the Maxwell–Boltzmann distribution can be reexpressed in terms of speed v, using $E = \frac{1}{2}mv^2$. Then, noting $dE = mv\,dv$, we have for $N'(v)$, the number of particles between v and $v + dv$:

$$N'(v)\,dv = \sqrt{\frac{2}{\pi}}N\left(\frac{m}{kT}\right)^{3/2}v^2 e^{-mv^2/2kT}\,dv$$

Dividing by N we get the probability $P(v)\,dv$ of finding a particle between v and $v + dv$, with

$$P(v) = \sqrt{\frac{2}{\pi}}\left(\frac{m}{kT}\right)^{3/2}v^2 e^{-mv^2/2kT}$$

20.57 Using the graph of Fig. 20-6 (the Maxwell velocity distribution for mercury vapor), what will be the most probable speed of N_2 molecules at 460 K?

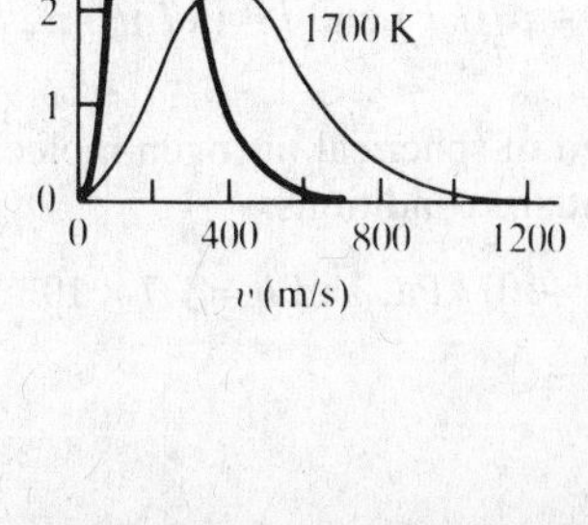

Fig. 20-6

❚ For Hg, the most probable speed is about 200 m/s. The probability density function $P(v)$ (see Prob. 20.56) can be written as a function of $(mv^2)/(2kT) \equiv x$ to give $P(v) = 4[m/(2\pi kT)]^{1/2} x e^{-x}$. Maximum probability occurs at $x_{max} = [m_{Hg}(200)^2]/[2k(460)]$, where we used the data for mercury. Since, at fixed T, x_{max} is the same for any molecule, at the maximum for N_2 we have $(m_N v^2)/[2k(460)] = [(m_{Hg})200^2]/[2k(460)]$. Using this gives $v^2 = (200)^2(m_{Hg}/m_N)$ from which we find $v = \underline{540 \text{ m/s}}$. [The exact value of x_{max} is 1.]

20.58 Using the graph of Fig. 20-6, determine what will be the most probable speed of Hg molecules at 920 K.

❚ As in Prob. 20.57 we note that the functional form of $P(v)$ depends only on $x = (mv^2)/(2kT)$; so once x_{max} is found, it is the same for any T. Proceeding as in Prob. 20.57, one has $(m_0 v^2)/[2k(920)] = [m_0(200^2)]/[2k(460)]$. Solving for v gives $v = \underline{280 \text{ m/s}}$.

20.59ᶜ Refer to Prob. 20.56. Show that the maximum value of $P(v)$ occurs when $v = \sqrt{(2kT)/m}$. Use the calculus criterion for a maximum.

❚ Set $dP/dv = 0$; $2v \exp[(-mv^2)/(2kT)] - (mv^3/kT)\exp[(-mv^2)/(2kT)] = 0$, which leads to $v_{max} = [(2kT)/m]^{1/2}$.

20.60ᶜ Show that the Maxwell velocity distribution given in Prob. 20.56 is normalized [that is, $\int P(v)\,dv = 1$, and hence $P(v)$ is a true probability density function].

❚
$$\int_0^\infty P(v)\,dv = \frac{4}{\pi^{1/2}} A^{3/2} \int_0^\infty v^2 e^{-Av^2}\,dv$$

where $A \equiv m/(2kT)$. But the integral on the right has the value $\pi^{1/2}/4A^{3/2}$, so $\int_0^\infty P(v)\,dv = 1$, as desired.

20.61 One of the earliest estimates of Avogadro's number was made by Perrin in 1908. He dispersed tiny uniform particles in a clear liquid and noted that when equilibrium had been reached, the number n of particles per unit volume decreased with height in the liquid. Show that the relation $n = n_0 e^{-(N_A v/RT)(\rho - \rho')gy}$ applies, where n_0 is n at $y = 0$, ρ and ρ' are the densities of the particle and liquid, respectively, and v is the volume of a particle. How would you graph your data of n versus y to obtain N_A?

❚ The energy mode of interest is the effective gravitational energy, taking buoyancy into account, since this is the only mode with height dependence. The particle density ratio from the Maxwell–Boltzmann distribution is $n/n_0 = \exp[-(E - E_0)/(kT)]$. The potential energy at $y = 0$ is taken to be zero. The energy at any height y would be due to the positive work done against gravity, $m_0 gy = \rho vgy$ and the negative work against the buoyant force $-\rho' vgh$, so that $E - E_0 = (\rho vgy - \rho' vgy) - 0$.

Recalling that $k = R/N_A$, we can substitute into the Maxwell–Boltzmann distribution to obtain the desired result; $n/n_0 = \exp\{[-N_A vgy(\rho - \rho')]/(RT)\}$. Taking reciprocals and then the natural logarithm of both sides yields $\ln(n_0/n) = \{[N_A vg(\rho - \rho')]/(RT)\}y$. Plotting $\ln(n_0/n)$ versus y will yield a straight line from which the slope is used to find N_A.

20.62 What is the *mean free path* of a gas molecule?

❚ The average distance traveled by a gas molecule between collisions. For an ideal gas of spherical molecules with radius b,

$$\text{mean free path} = \frac{1}{4\pi\sqrt{2}\,b^2(N/V)} \tag{1}$$

where N/V is the number of molecules per unit volume.

20.63 Show that the mean free path of molecules in an ideal gas can be written in terms of the gas pressure as

$$l = \frac{kT}{4\pi\sqrt{2}\,b^2 p} \tag{1}$$

❚ From Prob. 20.62 we have $l = 1/(4\pi 2^{1/2} n_u b^2)$ where $n_u = N/V$. Now $pV = nRT$ with $n/V = n_u/N_A$. But $R/N_A = k$, and so $n_u = (nN_A)/V = p/(kT)$ and $l = (kT)/(4\pi 2^{1/2} b^2 p)$.

20.64 Assuming the air to be composed of spherical nitrogen molecules with radius 1.7×10^{-10} m, find the mean free path of air molecules under standard conditions.

❚ Substitution of $T = 273$ K, $p = 101$ kPa, and $b = 1.7 \times 10^{-10}$ m in (1) of Prob. 20.63 yields $l = \underline{73 \text{ nm}}$.

20.65 Show that the molecular collision frequency for a given type of gas in a container of constant volume varies as the square root of the absolute temperature.

❚ From Prob. 20.62, the mean free path l is independent of temperature when the volume is held fixed. To solve the problem we must first find the time t between collisions. Using v_{rms} to get an ensemble average velocity, and the definition of mean free path, we have $v_{rms}t = l$, or $t = l/v_{rms}$. But $v_{rms} = \sqrt{(3kT)/m}$, so $t \propto 1/\sqrt{T}$. Since t represents elapsed time/collision, the collision frequency $f = (1/t) \propto \sqrt{T}$. Note that had we used any other sensible velocity to represent the ensemble average such as $\bar{v}$, or v_p = most probable velocity, the conclusion would still be the same, since these vary as $\sqrt{T}$ also.

20.66 A satellite sent into space samples the density of matter within the solar system and gets a value of 2.5 hydrogen atoms per cubic centimeter. What is the mean free path of the hydrogen atoms? Take the diameter of a hydrogen atom as $d = 0.24$ nm.

❚
$$l = \frac{1}{(N/V)\pi\sqrt{2}\,d^2} = \frac{1}{(2.5 \times 10^6)\pi\sqrt{2}\,(2.4 \times 10^{-10})^2} = \underline{1.56 \times 10^{12}\,\text{m}}$$

(about ten times the distance of the earth from the sun).

20.67 Ten small planes are flying at 150 km/h in total darkness in an air space that is $20 \times 20 \times 1.5$ km^3 in volume. You are in one of the planes, flying at random within this space with no way of knowing where the other planes are. On the average, about how long a time will elapse between near collisions with your plane? Assume for this rough computation that a plane can be approximated by a sphere with 6-m radius.

❚ The time $t = l/v$, where l is the mean free path given in Prob. 20.62, and v = plane velocity. The density of planes as seen by you is $9/[20(20)(1.5)] = 0.015$ km^{-3}. The collision time then will be about $1/[(4\pi 2^{1/2})(36 \times 10^{-6})(0.015)(150)] = \underline{700\,\text{h}}$.

20.68 A chamber contains monatomic helium at STP. (a) Determine the number density, n', of the atoms. (b) Relate the number density n' to the atomic mean free path l and *collision cross section* σ.

❚ (a) From the ideal gas law,

$$n' = \frac{p}{kT} = \frac{1.01 \times 10^5\,\text{N/m}^2}{(1.38 \times 10^{-23}\,\text{N}\cdot\text{m/K})(273\,\text{K})} = 2.68 \times 10^{25}\,\text{m}^{-3}$$

(b) As far as collisions are concerned, a moving atom can be treated as presenting a cross-sectional area σ. As the atom travels through one mean free path subsequent to a collision, it sweeps out a cylinder of volume σl before encountering another atom. This yields $1/\sigma l$ as the local density of atoms; hence,

$$\frac{1}{\sigma l} \approx n'$$

20.69 (a) For helium atoms at ordinary temperatures, the cross section σ is approximately 10^8 barns (10^{-20} m^2). Evaluate the mean free path for the sample of helium gas described in Prob. 20.68. (b) Suppose that the helium chamber is 1.0 m in diameter. At the given temperature of 273 K, how much would the pressure have to be reduced in order for the mean free path to be equal to the chamber diameter?

❚ (a) By Prob. 20.68,

$$l \approx \frac{1}{n'\sigma} = \frac{1}{(2.68 \times 10^{25}\,\text{m}^{-3})(1 \times 10^{-20}\,\text{m}^2)} = \underline{3.72\,\mu\text{m}}$$

(This is consistent with the result for air molecules, found in Prob. 20.64, which are much larger than helium atoms.)

(b) At constant temperature, the mean free path is inversely proportional to the pressure (see Prob. 20.63), so $lp = \text{constant} = 3.72 \times 10^{-6}$ m · atm. The pressure p' corresponding to a mean free path $l' = 1.00$ m is given by

$$p' = \frac{3.72 \times 10^{-6}\,\text{m}\cdot\text{atm}}{1.00\,\text{m}} = \underline{3.72 \times 10^{-6}\,\text{atm}}$$

20.3 ATMOSPHERIC PROPERTIES; SPECIFIC HEATS OF SOLIDS

20.70 A thermometer in a 10 m × 8 m × 4 m room reads 22 °C and a humidistat reads the relative humidity (R.H.) to be 35%. What mass of water vapor is in the room? Saturated air at 22 °C contains 19.33 g H_2O/m^3.

$$\%\text{R.H.} = \frac{\text{mass of water/m}^3}{\text{mass of water/m}^3 \text{ of saturated air}} \times 100 \qquad 35 = \frac{\text{mass/m}^3}{0.01933 \text{ kg/m}^3} \times 100$$

from which mass/m^3 = 6.77×10^{-3} kg/m^3. But the room in question has a volume of $10 \times 8 \times 4 = 320$ m^3. Therefore, the water in it is $(320 \text{ m}^3)(6.77 \times 10^{-3} \text{ kg/m}^3) = \underline{2.17 \text{ kg}}$.

20.71 On a certain day when the temperature is 28 °C, moisture forms on the outside of a glass of cold drink if the glass is at a temperature of 16 °C or lower. What is the relative humidity (R.H.) on that day? Saturated air at 28 °C contains 26.93 g/m^3 of water, while at 16 °C, it contains 13.50 g/m^3.

Dew forms at a temperature of 16° C or lower, so the dew point is 16° C. The air is saturated at that temperature and therefore contains 13.50 g/m^3. Then, ignoring the small drop in that density in the expanded air at 28° C:

$$\text{R.H.} = \frac{\text{mass present/m}^3}{\text{mass/m}^3 \text{ in saturated air}} = \frac{13.50}{26.93} = 0.50 = \underline{50\%}$$

20.72 Outside air at 5 °C and 20% relative humidity is introduced into a heating and air conditioning plant where it is heated to 20 °C and the relative humidity is increased to a comfortable 50%. How many grams of water must be evaporated into a cubic meter of outside air to accomplish this? Saturated air at 5 °C contains 6.8 g/m^3 of water, and at 20 °C it contains 17.3 g/m^3.

mass/m^3 of water vapor in air at 5 °C = 0.20×6.8 g/m^3 = 1.36 g/m^3

comfortable mass/m^3 at 20 °C = 0.50×17.3 g/m^3 = 8.65 g/m^3

one m^3 of air at 5 °C expands to (293/278) m^3 = 1.054 m^3 at 20 °C

mass of water vapor in 1.054 m^3 at 20 °C = 1.054 m^3 × 8.65 g/m^3 = 9.12 g

mass of water vapor to be added to each m^3 of air at 5 °C = $(9.12 - 1.36)$ g = $\underline{7.76 \text{ g}}$

20.73 In an air-conditioning plant, air at 15 °C and 90% relative humidity is cooled to 10 °C to remove some of the moisture and then heated to 20 °C and passed into an auditorium. What percentage of water vapor is removed from the air, and what is the final relative humidity?

Use data from Table 20-1. At 15°C the saturated vapor pressure is 1.77. The actual vapor pressure is then $0.90 \times 1.77 = 1.59$ kPa. At 10°C the saturated vapor pressure is 1.23 kPa. At a given absolute temperature T, the vapor pressure is proportional to the density of water vapor (from the ideal gas law). At T varies the vapor pressure stays fixed unless vapor condenses out or is added. Then

$$\text{fraction removed} = \frac{1.59 - 1.23}{1.59} = 0.23 = \underline{23\%}$$

At 20 °C the saturated vapor pressure is 2.31 kPa.

$$\text{final relative humidity} = \frac{1.23}{2.31} = \underline{53\%}$$

TABLE 20-1

temperature, °C	10	11	12	13	14	15	16	17	18	19	20	21	22
saturated vapor pressure of water, kPa	1.23	1.34	1.45	1.55	1.66	1.77	1.88	1.99	2.09	2.20	2.31	2.50	2.69
temperature, °C	23	24	25	26	27	28	29	30	31	32	33	34	35
saturated vapor pressure of water, kPa	2.88	3.07	3.26	3.45	3.63	3.82	4.01	4.20	4.51	4.83	5.14	5.45	5.76

20.74 A volume of 30 m³ of air with 85% humidity at 20 °C is passed through drying equipment that removes all the moisture. How many kilograms of water are removed?

❚ Use the ideal gas law and Table 20-1 to calculate the density of the water vapor at 20 °C:

$$\rho = \frac{Mp}{RT} = \frac{(18\text{ kg/kmol})(0.85 \times 2.31\text{ kN/m}^2)}{(8.314\text{ kN}\cdot\text{m/kmol}\cdot\text{K})(293\text{ K})} = 0.0145\text{ kg/m}^3$$

Hence, $(30\text{ m}^3)(0.0145\text{ kg/m}^3) = \underline{0.435\text{ kg water removed}}$.

20.75 If the air in a room has a dew point of 11 °C, what is its relative humidity at 21 °C?

❚ By the arguments of Probs. 20.71 and 20.73,

$$\text{R.H.} = \frac{\text{svp at }11\,^\circ\text{C}}{\text{svp at }21\,^\circ\text{C}} = \frac{1.34}{2.50} = 54\%$$

20.76 If a sample of air at 68 °F and 55% relative humidity is slowly cooled, condensation will occur at what temperature?

❚ We again use the reasoning of Prob. 20.73. From Table 20-1, the saturated vapor pressure at 68° F = 20° C is 2.31 kPa. (0.55)(2.31) = 1.27 kPa, which is the saturated vapor pressure at (appox.) 10°C = $\underline{50° F}$.

20.77 The relative humidity of a room is 75% at 23 °C. If the temperature falls to 19 °C, what will the relative humidity be?

❚ We again use the reasoning of Prob. 20.73. From Table 20-1 the saturated vapor pressure of water vapor at 23°C is 2.88 kPa, giving an actual pressure of (0.75)(2.88) = 2.16 kPa. From the table, the saturated vapor pressure at 19°C is 2.20 kPa. The new value is therefore relative humidity = 2.16/2.20 = $\underline{98\%}$.

20.78 What is the dew point if the air has a relative humidity of 60% at 25 °C?

❚ From Table 20-1 the saturated vapor pressure of water vapor at 25°C is 3.26 kPa; hence p = (0.60)(3.26) = 1.96 kPa. The dew point is the temperature at which the corresponding vapor pressure is the saturated vapor pressure. From the table, then, dew point ≈ $\underline{17°\text{C}}$.

20.79 If the normal lapse rate (decrease of temperature with altitude) prevails, and the temperature at ground level is 21 °C, what will the air temperature be at height of 1100 m?

❚ The normal decrease of temperature with altitude is about 1 °C per 110 m. (1100/110)(1 °C) = 10 °C decrease. 21 °C − 10 °C = $\underline{11\,^\circ\text{C}}$. (Note that 294 K → 284 K is a small *percentage* decrease on the Kelvin scale.)

20.80 (a) Pilots of light planes must be careful to calculate the loads on warm days. Why must pilots leaving from or landing at high elevation (for example, Denver, Colorado, or Mexico City) be particularly careful?
(b) Compare the density of the air at 0 °C to the density at 30 °C. Assume identical pressures.

❚ (a) In relatively sparse air, there will be less lift on the wings of the airplane. Other factors being equal, take-off runs will be longer, climbing rates will be smaller, and the descent during landing approaches will be more rapid in the less dense air at high elevations and/or on warm days. (b) At constant pressure, density and temperature are inversely proportional $\rho_2/\rho_1 = T_1/T_2$. With $T_2 = 273.15$ K ($t_2 = 0$ °C) and $T_1 = 303.15$ K ($t_1 = 30$ °C), we find $\rho_2/\rho_1 = 303.15/273.15 = 1.11$. That is, at 0 °C the density is 11 percent higher than it is at 30 °C.

20.81 Compare the density of the air at Logan Airport in Boston (elevation 0 m) at 0 °C to the density of air at Stapleton Field in Denver (elevation 1600 m) at 30 °C. At constant temperature, the atmospheric pressure p obeys approximately the equation $p = p_0 e^{-z/8150}$ if the elevation z is expressed in meters, and p_0 is the atmospheric pressure at 0 m.

❚ We begin by determining the pressure at Stapleton Field with the help of the hydrostatic isothermal profile. Using the subscripts l and s to refer to Logan and Stapleton, we have

$$p_s = p_l e^{-(z_s - z_l)/8150}$$

We now determine the density which corresponds to p_s and the given temperature T_s. If the temperatures were equal ($T_s = T_l$), the density ratio ρ_s/ρ_l would equal the pressure ratio p_s/p_l (by Boyle's law). Since $T_s \neq T_l$, we must use the more general relationship ($p/\rho T$) = constant (since the gas composition is assumed to be the same at the two locations). That is,

$$\frac{\rho_s}{\rho_l} = \frac{T_l}{T_s}\frac{p_s}{p_l} = \frac{T_l}{T_s}e^{-(z_s-z_l)/8150}$$

With $T_l = 0\,°C = 273.15$ K, $T_s = 30\,°C = 303.15$ K, and $z_s - z_l = 1600$ m, we obtain

$$\frac{\rho_s}{\rho_l} = \left(\frac{273.15}{303.15}\right)e^{-(1600/8150)} = (0.9010)e^{-0.1963} = \underline{0.74}$$

Under the given conditions and assumptions, the density is 26 percent lower at Stapleton Field than at Logan Airport. Equivalently, the density at Logan Airport is 35 percent higher than that at Stapleton Field.

20.82 Derive the law of atmospheres from the Maxwell–Boltzmann statistical law which states that the number of particles (in an equilibrium ensemble) with energy E is proportional to $e^{-E/kT}$. Assume constant temperature over the heights being considered. (See Prob. 20.79.)

▮ Since the only term in the energy of air molecules that depends on the vertical height, z, is the gravitational potential energy, mgz (measured from ground zero), where m is the average mass of a molecule of air, the particle density $n_u(z)$ obeys (Maxwell–Boltzmann): $n_u(z) \propto e^{-[E(z)/kT]} = e^{-mgz/kT}$, and $[n_u(z)]/[n_u(o)] = e^{-mgz/kT}/1$. Since the density of air, ρ, is proportional to n_u, we have $\rho(z)/\rho(o) = e^{-mgz/kT}$. Finally, at constant temperature, $p(z) \propto \rho(z)$; so $p(z) = p(o)e^{-mgz/kT}$, which is the law of atmospheres.

20.83 Find the uniform temperature at which the ratio of the densities of mercury vapor at the top and bottom of a 2.0-m-high tank would be $1/e$. (Assume an ideal gas could be obtained.) (Molecular weight of mercury = 201.)

▮ From the law of atmospheres, $\rho_1/\rho_2 = \exp[(-mgh)/(kT)]$. Hence $1/2.718 = \exp\{[-m(9.8)(2)]/(kT)\}$ where $m = 201/N_A = 3.34 \times 10^{-25}$ kg. Invert and take natural logarithms of each side to find $1 = 19.6m/kT$. Solving for T yields $T = \underline{0.47\text{ K}}$.

20.84 A gas of dust particles fills a 2.0-m-high tank. At equilibrium (27 °C), the density of particles at the top of the tank is $1/e$ the density at the bottom. Find the mass of a typical particle and find how many times more massive it is than a nitrogen molecule.

▮ Proceeding as in Prob. 20.83, $(mgh)/(kT) = 1$. This gives $m = (kT)/(gh) = 2.1 \times 10^{-22}$ kg. The mass of a nitrogen molecule is $28/N_A = 4.65 \times 10^{-26}$ kg and so the mass of the dust particle is $\underline{4500}$ times greater.

20.85ᶜ Infer the law of atmospheres from the ideal gas law.

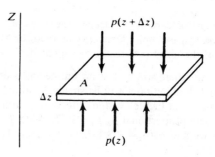

Fig. 20-7

▮ Figure 20-7 shows a thin slab or air at altitude z. For equilibrium of the slab,

$$[p(z) - p(z + \Delta z)]A = \rho g A\,\Delta z \qquad \text{or} \qquad \frac{dp}{dz} = -\rho g$$

From the ideal gas law,

$$p = \frac{\rho}{m}kT \qquad \text{or} \qquad \rho = \frac{m}{kT}p$$

where m is the average molecular mass for air. Thus

$$\frac{dp}{dz} = -\frac{mg}{kT} p$$

Ignoring the variation of T and g with altitude,

$$\int_{p_0}^{p} \frac{dp}{p} = -\frac{mg}{kT} \int_0^z dz \quad \text{and} \quad p = p_0 e^{-(mg/kT)z}.$$

20.86 Using the Dulong–Petit law, estimate the high-temperature specific heat capacity (in J/kg · K) for uranium metal ($M = 238$.)

❚ From the Dulong–Petit law the molar heat capacity at constant volume $C_V = 3R$. Thus, $c_V = C_V/M = [3(8314)]/238 = \underline{105 \text{ J/kg} \cdot \text{K}}$.

20.87 The Dulong–Petit law was used early in this century to determine the molecular weights of crystalline solids. A certain pure metal has a specific heat of 230 J/kg · K at high temperatures. What is the molecular weight of the metal?

❚ Again use $C_V = 3R$ and $M = C_V/c_V = 3(8314 \text{ J/kmol} \cdot \text{K})/(230 \text{ J/kg} \cdot \text{K}) = \underline{108 \text{ kg/kmol}}$.

20.88ᶜ Show how the equipartition theorem (Prob. 20.38) leads to the law of Dulong and Petit.

❚ At high temperatures, it may be supposed that essentially all the internal energy of a metal is due to the vibrations of the atoms about their equilibrium positions in the crystalline lattice. If we picture each atom as connected to its neighbors by springs, then the atom will have kinetic and potential energies along three mutually perpendicular directions—six modes in all. By the equipartition theorem, its total energy will be $6(\frac{1}{2}kT) = 3kT$, giving a molar energy of $E = N_A(3kT) = 3RT$ and a molar heat capacity of $C_V = dE/dT = 3R$.

20.89ᶜ The Dulong–Petit law holds true surprisingly well for solids if the temperature is high enough. To predict the behavior of the heat capacity at lower temperatures, a quantum-mechanical model must be used. One such model, originally used by Einstein, assumes that all atoms in a solid vibrate at the same frequency v. The total energy of a solid of N atoms is then the same as the energy of $3N$ one-dimensional oscillators. The correct quantum-mechanical expression for the average energy of this collection of oscillators is $\langle E \rangle = 3Nhv[\frac{1}{2} + 1/(e^{\beta hv} - 1)]$, where $\beta = 1/kT$, and $h = 6.63 \times 10^{-34}$ J · s. Show that this model gives the molar heat capacity as

$$C = 3R\left(\frac{\Theta}{T}\right)^2 \frac{e^{\Theta/T}}{(e^{\Theta/T} - 1)^2}$$

where $\Theta \equiv hv/k$.

❚ Writing A for Avogadro's number, we have, for one mole of material,

$$\langle E \rangle = 3Ahv\left[\frac{1}{2} + \frac{1}{e^{hv/kT} - 1}\right] \tag{1}$$

so the molar heat capacity is given by

$$C = \frac{d\langle E \rangle}{dT} = 3Ahv \frac{(-1)}{(e^{hv/kT} - 1)^2}\left(\frac{-hv}{kT^2}\right)e^{hv/kT} = 3(Ak)\left(\frac{hv}{kT}\right)^2 \frac{e^{hv/kT}}{(e^{hv/kT} - 1)^2} \tag{2}$$

Since $Ak = R$, the gas constant, Eq. (2) becomes

$$C = 3R\left(\frac{\Theta}{T}\right)^2 \frac{e^{\Theta/T}}{(e^{\Theta/T} - 1)^2} \tag{3}$$

where we have put $\Theta \equiv hv/k$.

20.90 Refer to Prob. 20.89, Determine the behavior of C in the limit $T \gg \Theta$, and in the limit $T \ll \Theta$.

❚ For $T \gg \Theta$, we have $\Theta/T \ll 1$, so Eq. (3) yields

$$C = 3R(\Theta/T)^2 \frac{[1 + (\Theta/T) + \cdots]}{\{[1 + (\Theta/T) + \cdots] - 1\}^2} \to 3R \frac{(\Theta/T)^2[1 + \Theta/T]}{(\Theta/T)^2} \to \underline{3R}$$

which is the Dulong–Petit result.

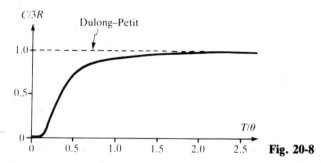

Fig. 20-8

For $T \ll \Theta$, we have $\Theta/T \gg 1$, so that

$$C \to 3R\left(\frac{\Theta}{T}\right)^2 \frac{e^{\Theta/T}}{(e^{\Theta/T})^2} = 3R\left(\frac{\Theta}{T}\right)^2 e^{-\Theta/T} \quad (\to 0 \text{ rapidly as } T \to 0)$$

which is much smaller than the Dulong–Petit result.

Figure 20-8 is the graph of (3), Prob. 20.89.

20.91ᶜ At low temperatures, the specific heat capacity of metals can be expressed as $c = \kappa_1 T + \kappa_3 T^3$, where T is in kelvins. For Cu, $\kappa_3 = 2.48 \times 10^{-7}$ cal/g · K⁴), $\kappa_1 = 2.75 \times 10^{-6}$ cal/g · K². How much heat energy is required to raise the temperature of a 15-g block of Cu from 5 to 30 K?

▐ The heat energy required is given by

$$\Delta H = m \int_{T_i}^{T_f} c(T)\, dT = m \int_{T_i}^{T_f} (\kappa_1 T + \kappa_3 T^3)\, dT = m\left[\frac{\kappa_1}{2}(T_f^2 - T_i^2) + \frac{\kappa_3}{4}(T_f^4 - T_i^4)\right]$$

Inserting the given numerical values, we find that

$$\Delta H = 15\left\{\left(\frac{2.75 \times 10^{-6}}{2}\right)(30^2 - 5^2) + \left(\frac{2.48 \times 10^{-7}}{4}\right)(30^4 - 5^4)\right\} = 0.0180 + 0.7527 = \underline{0.771 \text{ cal}}$$

CHAPTER 21

The First Law of Thermodynamics

21.1 BASIC THERMODYNAMIC CONCEPTS

TABLE 21-1 Specific Heats at Constant Volume

gas (at 15 °C)	c_v, kcal/kg $\cdot$ K	c_v, kJ/kg $\cdot$ K	*M, kg/kmol
He	0.75	3.100	4
Ar	0.075	0.310	40
O_2	0.155	0.650	32
N_2	0.177	0.740	28
H_2	2.40	10.000	2
CO_2	0.153	0.640	44
H_2O (200 °C)	0.359	1.500	18
CH_4	0.405	1.690	16

* $c_p \approx c_v + R/M$ ($R = 8.314$ kJ/kmol $\cdot$ K). The relation is exact for an ideal gas.
To obtain molar heat capacities (C_v, C_p) multiply specific heats (c_v, c_p) by M.

21.1 Describe the convention for work associated with a thermodynamic system.

▮ In physics it is usual to define W as the work done *by* a system on the environment. Thus W is positive when a chemical system expands against the environment and hence the system transfers energy to the surroundings; W is negative when the system contracts and the system absorbs energy from the surroundings. Sometimes, especially in chemistry, the opposite convention is adopted.

21.2 What is meant by the *internal energy* of a system?

▮ The internal energy (U) of a system is the total energy content of the system. It is the sum of the kinetic, potential, chemical, electric, nuclear, and all other forms of energy possessed by the atoms and molecules of the system. A form of energy may be categorized as organized or disorganized. *Organized energy* is associated with concerted behavior of the particles composing the system, e.g., macroscopic motion of the system. Also, chemical potential energy, where a definite amount of energy is to be released for each molecule formed, represents organized energy. *Disorganized energy* is associated with the random interactions ("collisions") of the particles. Thus, the temperature of an ideal gas measures its disorganized kinetic energy. Also called *thermal energy*, this category is of central interest in thermodynamics.

21.3 Give the first law of thermodynamics.

▮ The first law states the conservation of energy: If an amount of heat energy ΔQ flows into a system, then this energy must appear as increased internal energy ΔU for the system and/or work ΔW done by the system on its surroundings. As an equation, the first law is $\Delta Q = \Delta U + \Delta W$.

21.4 What is the relation between the specific heats (or heat capacities) at constant pressure and constant volume?

▮ When a gas is heated at *constant volume*, the entire heat energy supplied goes to increase the internal energy of the gas molecules. But when a gas is heated at *constant pressure,* the heat supplied not only increases the internal energy of the molecules but also does mechanical work in expanding the gas against the opposing constant pressure. Hence the specific heat of a gas at constant pressure, c_p, is greater than its specific heat at constant volume, c_v. It can be shown that for an ideal gas of molecular weight M, $c_p - c_v = R/M$ (ideal gas), where R is the universal gas constant.
    The ratio of specific heats at constant pressure and constant volume is important for various applications, especially those involving adiabatic processes, and is often given its own symbol γ ($\gamma = c_p/c_v$). As discussed above, this ratio is greater than unity for a gas. The kinetic theory of gases indicates that for monatomic gases (such as He, Ne, Ar), $\gamma = 1.67$. For diatomic gases (such as O_2, N_2), $\gamma = 1.40$ at ordinary temperatures.

21.5 Why is the heat capacity at constant volume considered to be more important than that at constant pressure?

∎ While it is often easier to perform experiments at constant pressure, especially with liquids and solids, the heat capacity at constant volume is more important because it is intimately related to the internal energy of a chemical system. This can be seen from the first law where $\Delta W = p \, \Delta V = 0$ if V is constant. Then $\Delta Q = nC_v \, \Delta T = \Delta U$. Letting $u = U/n =$ the molar internal energy, we have $C_v = \Delta u / \Delta T$. Thus C_v gives direct information on the internal energy.

21.6 Define *isothermal*, *isobaric*, *isovolumic*, and *adiabatic* processes.

∎ An isothermal process is one in which the system changes in such a way that the temperature remains constant throughout. It only makes sense to talk about an isothermal process for quasistatic processes in which there is a meaningful system temperature at all times, and the system thus progresses in orderly fashion from one state to another.

An isobaric process is one in which the pressure on the system remains unchanged throughout the process. It applies primarily to quasistatic processes where a definite system pressure (or pressure distribution) exists.

An isovolumic process, sometimes called isometric or isochoric, is one in which the volume of the system remains the same. It is meaningful in both quasistatic and more violent processes.

An adiabatic process is one in which no heat transfer takes place into or out of the system. It is meaningful in both quasistatic and more violent processes.

Almost all problems in this chapter refer to quasistatic processes.

21.7 A cubical room $2.5 \times 10 \times 7 \, m^3$ is filled with nitrogen gas at 100 kPa and 27 °C. What is the mass of gas in the room? How many calories are needed to increase the room temperature 1 °C if all the heat goes into the gas? Assume (*a*) constant volume, (*b*) constant pressure.

∎ From $pV = (m/M)RT$ we find that $m = 196$ kg. (*a*) Use Table 21-1 and $\Delta Q = c_v m \, \Delta T = 0.177(196)(1) =$ <u>34.8 kcal</u>. (*b*) Use $\Delta Q = c_p m \, \Delta T$, with $c_p = c_v + R/M = 0.740 + 8.314/28 = 1.037 \, kJ/kg \cdot K$; then $\Delta Q = 204 \, kJ$, or <u>48.7 kcal</u>.

21.8 If 20 g of CO_2 gas is confined to a cylinder by a piston, how much heat energy (in calories) is needed to heat this gas 5 °C (*a*) at constant volume, (*b*) at constant pressure? Assume $\gamma = 1.3$.

∎ (*a*) $\Delta Q = c_v m \, \Delta T = 0.153(20)(5) = $ <u>15.3 cal</u>. (*b*) Since $c_p/c_v = \gamma = 1.30$ for CO_2, one has $c_p = 1.30c_v = 0.199 \, cal/g \cdot °C$. Thus $\Delta Q = 0.199(20)(5) = $ <u>19.9 cal</u>.

21.9 Assume air to be composed of 78% nitrogen and 22% oxygen by weight. What should be the value of c_v for air?

∎ One gram of air should contain 0.78 g of N_2 and 0.22 g of O_2. The heat capacity should then be $c = 0.78(0.177) + 0.22(0.155) = $ <u>0.172 cal/g · °C</u>.

21.10 One-half mole of helium gas is confined to a container at standard pressure and temperature (1.01×10^5 Pa, 273 K). How much heat energy is needed to double the pressure of the gas?

∎ To double the pressure, the temperature must be doubled, since $pV = nRT$. Hence $\Delta T = 273$ K. From Table 21-1, the molar heat capacity of He is $C_V = Mc_V = (4 \, g/mol)(3.100 \, J/g \cdot K) = 12.4 \, J/mol \cdot K$. Therefore, $\Delta Q = nC_V \, \Delta T = \frac{1}{2}(12.4)(273) = $ <u>1693 J</u>.

21.11 What would you expect c_V and c_P to be for gaseous hydrogen bromide, HBr? Give your answer in cal/g · °C. Br has a molecular weight of 80.

∎ We note that there is a $\frac{1}{2}R \approx 1 \, cal/mol \cdot K$ contribution to C_V, the molar heat capacity, from each mode of energy. HBr being diatomic has $C_V = 5 \, cal/mol \cdot K$ and $C_p = 7 \, cal/mol \cdot K$. The specific heat capacities are $c_V = C_V/M$ and $c_p = C_p/M$, since $M = (1 + 80) \, g/mol$, $c_V = $ <u>0.062 cal/g · °C</u>, and $c_p = $ <u>0.086 cal/g · °C</u>.

21.12 From Table 21-1 we can see that the specific heats of hydrogen at constant pressure and at constant volume are, respectively, 3.4 kcal/kg · C and 2.4 kcal/kg · °C. If 100 g of hydrogen is heated from 10 to 20 °C at constant pressure, find the external work done.

∎ The work done is $W = m(c_p - c_V) \, \Delta t = 0.100(3.4 - 2.4)(20° - 10°) = $ <u>1 kcal</u>.

21.13 Solid sodium metal has a density of $970 \, kg/m^3$. (*a*) How many sodium atoms are there per cubic meter?

(b) Each sodium atom contributes one electron to a "free electron gas" within the metal. If we assume it to be an ideal gas, what is the contribution (in cal/g · °C) of the electron gas to the specific heat of sodium metal? The molecular weight of Na is 23.

❚ The number of sodium atoms in $1\,m^3 = (N_A m)/M = [(6 \times 10^{26})(970)]/23 = 2.5 \times 10^{28}$. The "free electron" has three translational degrees of freedom, so $C_V = 3$ and $C_p = 5$ cal/mol · K. Dividing by $M = 23$ for Na gives $c_V = \underline{0.13}$ and $c_p = \underline{0.22}$ due to the "electron gas" whereas the experimental value of c_p is 0.293 cal/g · °C.

21.14ᶜ A monatomic ideal gas is compressed from an initial volume V_i to a final volume V_f. During the compression, there is a transfer of heat which maintains the temperature of the gas at its initial value so that $T_f = T_i$. The sample contains n kmol of the gas. **(a)** Find the initial and final pressures p_i and p_f, in terms of n, V_i, T_i, and V_f. **(b)** What is the pressure $p(V)$ of the gas during this isothermal compression? **(c)** Find the work ΔW done by the gas during the compression. Express your result in terms of n, T_i, and the compression ratio V_i/V_f. **(d)** Find the heat ΔH added to the gas during the compression.

❚ **(a)** According to the ideal gas law, $p_i = (nRT_i)/V_i$ and $p_f = (nRT_i)/V_f$. **(b)** The ideal gas law gives the pressure as $p = (nRT_i)/V$. **(c)** The work done by the gas is given by

$$\Delta W = \int_{V_i}^{V_f} p(V)\, dV = nRT_i \int_{V_i}^{V_f} \frac{dV}{V} = \underline{nRT_i \ln\left(\frac{V_f}{V_i}\right)}$$

Observe that $\Delta W < 0$. **(d)** The energy of an ideal gas is a function solely of its temperature, so $\Delta E = 0$ during the isothermal compression. Therefore $\Delta H = \Delta W = \underline{nRT_i \ln(V_f/V_i)}$.

21.15ᶜ A cylinder fitted with a piston contains 0.10 mol of air at room temperature (20 °C). The piston is pushed so slowly that the air within the cylinder remains essentially in thermal equilibrium with the surroundings. Find the work done by the air within the cylinder if the final volume is one-half the initial volume.

❚ We have $pV = nRT$, with T constant.

$$W = \int_{V_i}^{V_f} p\, dV = nRT \int_{V_i}^{V_f} \frac{dV}{V} = nRT \ln\left(\frac{V_f}{V_i}\right)$$

$$W = (0.1\ \text{mol})(8.314\ \text{J/mol} \cdot \text{K})(293\ \text{K}) \ln \tfrac{1}{2} = \underline{-169\ \text{J}}$$

21.16 One mole of helium gas, initially at STP ($p_1 = 1$ atm $= 101.3$ kPa, $T_1 = 0\,°C = 273.15$ K), undergoes an isovolumetric process in which its pressure falls to half its initial value. **(a)** What is the work done by the gas? **(b)** What is the final temperature of the gas? **(c)** The helium gas then expands isobarically to twice its volume; what is the work done by the gas?

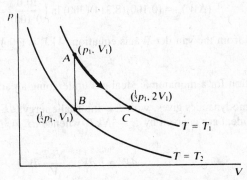

Fig. 21-1

❚ **(a)** Refer to Fig. 21-1. $W_{AB} = 0$, since $dW = p\, dV = 0$. **(b)** By the ideal gas law at constant volume,

$$\frac{T_2}{T_1} = \frac{p_2}{p_1} = \frac{1}{2} \qquad \text{or} \qquad T_2 = \frac{T_1}{2} = 136.58\ \text{K}$$

(c) The constant-pressure process returns the gas to the original temperature, $T_1 = 273.16$ K, since pressure is constant, and V doubles.

$$W_{BC} = \int_{V_1}^{2V_1} \frac{1}{2} p_1\, dV = \frac{1}{2} p_1 V_1 = \frac{1}{2} RT_1 = \frac{1}{2}(8.31)(273.15) = \underline{1135\ \text{J}}$$

21.17^e Refer to Prob. 21.16. Suppose that the gas undergoes a process from the initial state in (a) to the final state in (c) by an isothermal expansion; what is the work done by the gas?

▮ Along the isotherm $T = T_1$, $p = RT_1/V$; hence

$$W_{AC} = \int_{V_1}^{2V_1} p\,dV = RT_1 \int_{V_1}^{2V_1} \frac{dV}{V} = RT_1 \ln 2 = (8.31)(273.15)(0.693) = \underline{1573\text{ J}}$$

Note that $W_{AC} \neq W_{AB} + W_{BC}$; the work between states A and C is path-dependent.

21.18^c Derive an expression for the work done by a system undergoing isothermal compression (or expansion) from volume V_1 to V_2 for a gas which obeys the van der Waals equation of state, $(p + an^2/V^2)(V - bn) = nRT$. (Note that the ideal gas is included as the special case $a = b = 0$.)

▮ Solving the van der Waals equation of state for the pressure, we obtain

$$p = \frac{nRT}{V - nb} - \frac{an^2}{V^2} \tag{1}$$

We use Eq. (1) to calculate the work done by the gas as it changes volume from V_1 to V_2 isothermally:

$$(\Delta W)_{\text{vdw}} = \int_{V_1}^{V_2} p\,dV = nRT \int_{V_1}^{V_2} \frac{dV}{V - nb} - an^2 \int_{V_1}^{V_2} \frac{dV}{V^2} = nRT[\ln (V - nb)|_{V_1}^{V_2}] + an^2\left[\left(\frac{1}{V}\right)\Big|_{V_1}^{V_2}\right]$$

$$= nRT \ln\left(\frac{V_2 - nb}{V_1 - nb}\right) + an^2\left(\frac{V_1 - V_2}{V_1 V_2}\right) \tag{2}$$

21.19 Refer to Prob. 21.18. The numerical values of a and b in SI units for hydrogen gas (H_2) are approximately 24.8 kPa $\cdot$ m^6/kmol^2 and 0.0266 m^3/kmol, respectively. How large is the correction introduced by using the van der Waals equation of state when calculating the work (on the gas) required to compress 0.100 kmol of H_2 from 10.0 to 0.100 m^3 at a temperature of 300 K?

▮ From (2) of Prob. 21.18, with the sign reversed,

$$(\Delta W)_{\text{vdw}} = (0.100)(8.314)(300) \ln\left[\frac{10.0 - (0.100)(0.0266)}{0.100 - (0.100)(0.0266)}\right] - (24.8)(0.100^2)\left[\frac{10.0 - 0.100}{(10.0)(0.100)}\right] = 1153\text{ kJ}$$

The ideal gas result is (set $a = b = 0$)

$$(\Delta W)_{\text{ig}} = (0.100)(8.314)(300) \ln\left(\frac{10.0}{0.100}\right) = 1149\text{ kJ}$$

The correction obtained from the van der Waals equation, 4 kJ, is a 0.4 percent increase over the ideal-gas result.

21.20^c Determine the p-V relation for a monatomic ideal gas undergoing an adiabatic process.

▮ The first law of thermodynamics gives, with E = internal energy, $dE = -dW$ for the adiabatic process. The energy of a monatomic ideal gas is given by $E = \frac{3}{2}NkT$, whence $dE = \frac{3}{2}Nk\,dT$. Moreover, using the ideal gas law,

$$dW = p\,dV = \frac{NkT}{V}\,dV$$

Thus,

$$\frac{3}{2}Nk\,dT = -\frac{NkT}{V}\,dV \quad \text{or} \quad \frac{dT}{T} + \frac{2}{3}\frac{dV}{V} = 0$$

Integrating,

$$\ln T + \frac{2}{3}\ln V = \text{constant} \quad \text{or} \quad TV^{2/3} = \text{constant}$$

Finally, substituting for T from the ideal gas law, $\underline{pV^{5/3} = \text{constant}}$.

21.21 (a) Compute c_v for the monatomic gas argon, given $c_p = 0.125$ cal/g $\cdot$ °C and $\gamma = 1.67$. (b) Compute c_p for the diatomic gas nitric oxide (NO), given $c_v = 0.166$ cal/g $\cdot$ °C and $\gamma = 1.40$.

▌(a)

$$c_v = \frac{c_p}{\gamma} = \frac{0.125}{1.67} = \underline{0.0749 \text{ cal/g} \cdot {}^\circ C}$$

(b)

$$c_p = \gamma c_v = (1.40)(0.166) = \underline{0.232 \text{ cal/g} \cdot {}^\circ C}$$

21.22 If 150 L of nitrogen gas ($\gamma = 1.4$) expands adiabatically starting at a pressure of 1 atm and the final volume is 250 L, what is the final pressure?

▌ Use the adiabatic gas equation:

$$p_1 V_1^\gamma = p_2 V_2^\gamma \quad 1(150^{1.4}) = p_2(250^{1.4}) \quad p_2 = (0.6^{1.4}) = \underline{0.49 \text{ atm}}$$

21.23 As a sample of gas is allowed to expand quasistatically and adiabatically, its pressure drops from 120 to 100 kPa, and its temperature drops from 300 to 280 K. Is the gas monatomic or diatomic?

▌ The pressure-temperature relationship during an adiabatic process is described by $p_f V_f^\gamma = p_i V_i^\gamma$, or $p_f^{1/\gamma} V_f = p_i^{1/\gamma} V_i$. Using $V = (nRT)/p$, we get

$$p_f^{1/\gamma - 1} T_f = p_i^{1/\gamma - 1} T_i \tag{1}$$

Equation (1) implies that

$$\left(\frac{p_i}{p_f}\right)^{1/\gamma - 1} = \frac{T_f}{T_i} \tag{2}$$

or that

$$\frac{1}{\gamma} - 1 = \frac{\ln(T_f/T_i)}{\ln(p_i/p_f)} = \frac{\ln(280/300)}{\ln(120/100)} = -0.378 \tag{3}$$

Solving (3) for γ, we find that $\underline{\gamma = 1.61}$. Since this is closer to the monatomic value of 1.67 than to the diatomic value of 1.40, we conclude that the gas is *monatomic*.

21.24^c Show that if an ideal gas is compressed isothermally its compressibility is $1/p$, whereas if it is compressed adiabatically its compressibility is $1/\gamma p$.

▌ By definition, $\kappa = 1/B = -(1/V)(dV/dp)$ (B = bulk modulus). During an isothermal compression, the quantity pV is a constant (since $pV = nRT$). Therefore

$$0 = \left[\frac{d}{dp}(pV)\right]_{iso} = 1V + p\left(\frac{dV}{dp}\right)_{iso}$$

which immediately yields

$$\kappa_{iso} \equiv \frac{-1}{V}\left(\frac{dV}{dp}\right)_{iso} = \frac{1}{p}$$

During an adiabatic compression, the quantity pV^γ is a constant, so

$$0 = \left[\frac{d}{dp}(pV^\gamma)\right]_{ad} = 1V^\gamma + p\gamma V^{\gamma-1}\left(\frac{dV}{dp}\right)_{ad}$$

Solving for κ_{ad}, we find

$$\kappa_{ad} \equiv \frac{-1}{V}\left(\frac{dV}{dp}\right)_{ad} = \frac{1}{\gamma p}$$

21.25^c Find the work done by an ideal gas in expanding adiabatically from a state (p_1, V_1) to a state (p_2, V_2).

▌ For the adiabatic expansion,

$$pV^\gamma = \text{constant} = C \quad \text{or} \quad p = CV^{-\gamma}$$

Then

$$W = \int_{V_1}^{V_2} p \, dV = C\int_{V_1}^{V_2} V^{-\gamma} dV = \frac{CV_1^{1-\gamma} - CV_2^{1-\gamma}}{\gamma - 1}$$

But $C = p_1 V_1^\gamma = p_2 V_2^\gamma$, and so

$$W = \frac{p_1 V_1 - p_2 V_2}{\gamma - 1}$$

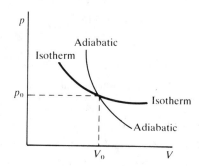

Fig. 21-2

21.26ᶜ In a p-V diagram (Fig. 21-2) an adiabatic and an isothermal curve for an ideal gas intersect. Show that the absolute value of the slope of the adiabatic is γ times that of the isotherm. Hence the adiabatic curve is steeper because the specific heat ratio γ is greater than 1.

▮ Denote the intersection point by (p_0, V_0). Then the isothermal curve is given by $pV = p_0V_0$, or

$$p_{iso} = p_0V_0V^{-1} \tag{1}$$

while the adiabatic curve is described by $pV^\gamma = p_0V_0^\gamma$, or

$$p_{ad} = p_0V_0^\gamma V^{-\gamma} \tag{2}$$

The slope of the isothermal curve at (p_0, V_0) is found by differentiating Eq. (1) and evaluating the derivative at $V = V_0$:

$$\left(\frac{dp}{dV}\right)_{iso} = -p_0V_0V^{-2}\bigg|_{V_0} = \frac{-p_0}{V_0} \tag{3}$$

Using Eq. (2), the slope of the adiabatic curve at (p_0, V_0) is

$$\left(\frac{dp}{dV}\right)_{ad} = -\gamma p_0V_0^\gamma V^{-\gamma-1}\bigg|_{V_0} = \frac{-\gamma p_0}{V_0} \tag{4}$$

Equations (3) and (4) show that

$$\left|\left(\frac{dp}{dV}\right)_{ad}\right| = \gamma \left|\left(\frac{dp}{dV}\right)_{iso}\right|$$

for the curves intersecting at (p_0, V_0).

21.2 THE FIRST LAW OF THERMODYNAMICS, INTERNAL ENERGY, p-V DIAGRAMS, CYCLICAL SYSTEMS

21.27 What is the change in internal energy of 0.100 mol of nitrogen gas as it is heated from 10 to 30 °C at (a) constant volume and (b) constant pressure?

▮ Internal energy for an ideal gas is linearly related to the temperature of the gas with $\Delta U = nC_V \Delta T$ so the answer to parts (a) and (b) is the same; $\Delta U = (0.100 \text{ mol})(4.96 \text{ cal/mol} \cdot \text{K})(20 \text{ K}) = 9.92 \text{ cal} = \underline{41.5 \text{ J}}$.

21.28 When 50 L of air at STP is isothermally compressed to 10 L, how much heat must flow from the gas? ($p_{atm} = 100$ kPa.)

▮ Since the process is isothermal, and we assume an ideal gas, $\Delta U = 0$. Then $\Delta Q = \Delta W$. But $\Delta W = nRT \ln (V_2/V_1) = P_1V_1 \ln (V_2/V_1)$. The substitution makes use of the ideal gas law, $P_1V_1 = nRT_1$, and recognizes that T is constant. In this case, $\Delta Q = (1 \times 10^5)(5 \times 10^{-2}) \ln 10/50 = -8050$ J; i.e., $\underline{8.05 \text{ kJ}}$ flows out.

21.29 An ideal gas in a cylinder is compressed adiabatically to one-third its original volume. During the process, 45 J of work is done on the gas by the compressing agent. (a) By how much did the internal energy of the gas change in the process? (b) How much heat flowed into the gas?

▮ In this case, $\Delta Q = 0$, so $\Delta U = -\Delta W = -(-45 \text{ J}) = \underline{45 \text{ J}}$; (b) the heat flow in the adiabatic process is zero.

21.30 In each of the following situations, find the change in internal energy of the system. (a) A system absorbs 500 cal of heat and at the same time does 400 J of work. (b) A system absorbs 300 cal and at the same time 420 J of work is done on it. (c) Twelve hundred calories is removed from a gas held at constant volume.

▮ *(a)* $\qquad \Delta U = \Delta Q - \Delta W = (500 \text{ cal})(4.184 \text{ J/cal}) - 400 \text{ J} = \underline{1700 \text{ J}}$

(b) $\qquad \Delta U = \Delta Q - \Delta W = (300 \text{ cal})(4.184 \text{ J/cal}) - (-420 \text{ J}) = \underline{1680 \text{ J}}$

(c) $\qquad \Delta U = \Delta Q - \Delta W = (-1200 \text{ cal})(4.184 \text{ J/cal}) - 0 = \underline{-5000 \text{ J}}$

Note that ΔQ is positive when heat is added to the system and ΔW is positive when the system does work. In the reverse cases, ΔQ and ΔW must be taken negative.

21.31 Rederive the result of Prob. 21.25 using the first law of thermodynamics and the facts that $C_p - C_V = R$ and $C_p/C_V = \gamma$ for an ideal gas, where C denotes molar heat capacity.

▮ Let the heat flow be denoted by ΔH, and the internal energy change by ΔE. Since the process is adiabatic, $\Delta H = 0$ and therefore $\Delta W = -\Delta E$. But $\Delta E = nC_V(T_B - T_A)$, and furthermore $R = C_p - C_V = (\gamma - 1)C_V$. Hence we can express the energy change as

$$\Delta E = \frac{nR}{(\gamma - 1)}(T_B - T_A)$$

But the ideal gas law yields $nRT = pV$, so we obtain

$$\Delta W = -\Delta E = \frac{-p_B V_B + p_A V_A}{\gamma - 1} = \frac{p_A V_A - p_B V_B}{\gamma - 1}.$$

in agreement with the result of Prob. 21.25.

21.32 Find ΔW and ΔU for a 6-cm cube of iron as it is heated from 20 to 300 °C. For iron, $c = 0.11 \text{ cal/g} \cdot °\text{C}$ and the volume coefficient of thermal expansion is $\beta = 3.6 \times 10^{-5} \, °\text{C}^{-1}$. The mass of the cube is 1700 g.

▮ $\qquad \Delta Q = cm\,\Delta T = (0.11 \text{ cal/g} \cdot °\text{C})(1700 \text{ g})(280 \, °\text{C}) = 52\,000 \text{ cal}$

The volume of the cube is $V = (6 \text{ cm})^3 = 216 \text{ cm}^3$. Using $(\Delta V)/V = \beta\,\Delta T$, we have

$$\Delta V = V\beta\,\Delta T = (216 \times 10^{-6} \text{ m}^3)(3.6 \times 10^{-5} \, °\text{C}^{-1})(280 \, °\text{C}) = 2.18 \times 10^{-6} \text{ m}^3$$

Then, assuming atmospheric pressure to be 1×10^5 Pa,

$$\Delta W = p\,\Delta V = (1 \times 10^5 \text{ N/m}^2)(2.18 \times 10^{-6} \text{ m}^3) = 0.22 \text{ J}$$

But the first law tells us that

$$\Delta U = \Delta Q - \Delta W = (52\,000 \text{ cal})(4.184 \text{ J/cal}) - 0.22 \text{ J} = 218\,000 \text{ J} - 0.22 \text{ J} \approx 218\,000 \text{ J}$$

Note how very small the work of expansion against the atmosphere is in comparison to ΔU and ΔQ. Often ΔW can be neglected when dealing with liquids and solids.

21.33 A cubic meter of helium originally at 0 °C and 1-atm pressure is cooled at constant pressure until the volume is 0.75 m³. How much heat was removed?

▮ Use the general gas law to find the final temperature. Then use the first law of thermodynamics.

$$\frac{p_1 V_1}{T_1} = \frac{p_2 V_2}{T_2} \qquad p_1 = p_2$$

Then

$$\frac{1}{273} = \frac{0.75}{T_2} \qquad T_2 = 205 \text{ K}$$

$\Delta Q = \Delta U + \Delta W$; and for an ideal gas, $\Delta U = mc_v\,\Delta T$. Thus, $\Delta Q = mc_v\,\Delta T + p\,\Delta V = nC_V\,\Delta T + p\,\Delta V$. Noting that at STP, 1 kmol occupies 22.4 m³,

$$\Delta Q = \frac{1}{22.4}(3)(205 - 273) + \frac{(1.013 \times 10^5)(0.75 - 1)}{4184} = -9.11 - 6.05 = \underline{-15.2 \text{ kcal}}$$

The minus sign means that heat is removed.

21.34 The volume of 1 kg of water at 100 °C is about 1×10^{-3} m³. The volume of the vapor formed when it boils at this temperature and at standard atmospheric pressure is 1.671 m³. *(a)* How much work is done in pushing back the atmosphere? *(b)* How much is the increase in the internal energy when the liquid changes to vapor?

▮ (a) The work done by the water is given by

$$\Delta W = p_{atm} \Delta V = (1.013 \times 10^5 \text{ N/m}^2)[(1.671 - 0.001) \text{ m}^3] = \underline{169 \text{ kJ}}.$$

(b) According to Table 17-2, the latent heat of vaporization of water is $L = 540$ kcal/kg. This is the heat per kilogram that must be added to vaporize 100-°C water at a constant pressure of 1 atm. The change in the internal energy of 1 kg of water when it is boiled at 1 atm is therefore given by

$$\Delta E = \Delta H - \Delta W = Lm - \Delta W = (540 \text{ kcal/kg})(1.00 \text{ kg})(4.184 \text{ kJ/kcal}) - (169 \text{ kJ}) = \underline{2090 \text{ kJ}}$$

21.35 A tank contains a fluid that is stirred by a paddle wheel. The power input to the paddle wheel is 2.24 kW. Heat is transferred from the tank at the rate of 0.586 kW. Considering the tank and the fluid as the system, determine the change in the internal energy of the system per hour.

▮ $$\Delta E = Q - W = -0.586 - (-2.24) = 1.654 \text{ kW} = 1.654 \text{ kJ/s} = 5954 \text{ kJ/h}$$

or about $\underline{6 \text{ MJ/h}}$.

21.36 A spring having a spring constant 5 N/m is compressed 0.04 m, clamped in this configuration, and dropped into a container of acid in which the spring dissolves. How much potential energy is stored in the spring, and what happens to it when the spring dissolves?

▮ The spring in being compressed acquired elastic potential energy of amount $U_{elastic} = \frac{1}{2}kx^2 = \frac{1}{2}(5)(0.04)^2 = \underline{0.004 \text{ J}}$. When the spring is dissolved, this ordered potential energy is converted into disordered potential and kinetic energy of the system. Overall, energy is conserved.

21.37 One pound of fuel, having a heat of combustion of 10 000 Btu/lb, was burned in an engine that raised 6000 lb of water 110 ft. What percentage of the heat was transformed into useful work?

▮ $$\text{efficiency} = \frac{\text{work done by engine}}{\text{work equivalent of heat supplied}} = \frac{(6000 \text{ lb})(110 \text{ ft})}{(10\,000 \text{ Btu})(778 \text{ ft} \cdot \text{lb/Btu})} = 0.085 = \underline{8.5\%}$$

21.38 A sample containing 1.00 kmol of the nearly ideal gas helium is put through the cycle of operations shown in Fig. 21-3. BC is an isothermal, and $p_A = 1.00$ atm, $V_A = 22.4$ m³, $p_B = 2.00$ atm. What are T_A, T_B, and V_C?

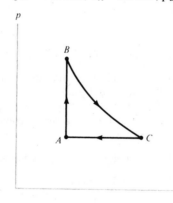

Fig. 21-3

▮ Applying the perfect gas law, we have

$$T_A = \frac{p_A V_A}{nR} = \frac{(1.013 \times 10^5 \text{ Pa})(22.4 \text{ m}^3)}{(1.00 \text{ kmol})(8.314 \times 10^3 \text{ J/kmol} \cdot \text{K})} = \underline{273 \text{ K}}$$

Because the process AB shown in Fig. 21-3 is isometric, $T_B = (p_B/p_A)T_A$. With $p_B = 2.00$ atm $= 2p_A$, we find that $T_B = \underline{546 \text{ K}}$. The process CA is isobaric, so Charles' law applies. Thus $V_C/T_C = V_A/T_A$, so that $V_C = (T_C/T_A)V_A$. But BC is an isothermal process, so that $T_C = T_B = 2T_A$. Therefore $V_C = 2V_A = \underline{44.8 \text{ m}^3}$.

21.39ᶜ Refer to Prob. 21.38 and calculate the work output during the cycle.

▮ The work done by the gas along AB is zero, since $dV = 0$. The work done on the gas along the isotherm BC is

$$\Delta W_{BC} = \int_{V_B}^{V_C} p \, dV = \int_{V_B}^{V_C} \frac{RT_B}{V} dV = RT_B \ln\left(\frac{V_C}{V_B}\right) = RT_B \ln 2 = (8.314 \text{ kJ/kmol} \cdot \text{K})(546 \text{ K})(0.6931) = 3150 \text{ kJ}$$

The work done along CA is

$$\Delta W_{CA} = \int_{V_C}^{V_A} p \, dV = p_A \int_{V_C}^{V_A} dV = p_A(V_A - V_C) = -p_A V_A = -(101.3 \text{ kPa})(22.4 \text{ m}^3) = -2270 \text{ kJ}$$

The total work done by the gas in one cycle is

$$\Delta W = \Delta W_{AB} + \Delta W_{BC} + \Delta W_{CA} = 0 + (3150) + (-2270) = \underline{880 \text{ kJ}}$$

21.40 Refer to Prob. 21.39 and show that the net work is equal to the net heat absorbed by the gas. Assume ideal monatomic gas conditions.

❚ Along AB, the gas is heated at constant volume, so $\Delta H_{AB} = nC_V(T_B - T_A)$. Since the gas is monatomic, $C_V = \frac{3}{2}R$. Therefore $\Delta H_{AB} = \frac{3}{2}nR(T_B - T_A) = (1.5)(1.00)(8.314)(273) = 3400 \text{ kJ}$. Along BC, $\Delta T = 0$, so the internal energy of the ideal gas is unchanged: $\Delta E_{BC} = 0$. The first law of thermodynamics then gives $\Delta H_{BC} = +\Delta W_{BC} = 3150 \text{ kJ}$. Along CA, the pressure is constant, so $\Delta H_{CA} = nC_p(T_A - T_C) = \frac{5}{2}nR(T_A - T_C) = (2.5)(1.00)(8.314)(-273) = -5670 \text{ kJ}$. The net heat input over the cycle is $\Delta H = \Delta H_{AB} + \Delta H_{BC} + \Delta H_{CA} = 3400 + 3150 - 5670 = \underline{880 \text{ kJ}}$, which equals the work output, as was to be shown.

21.41 Suppose that the cycle shown in Fig. 21-3 were reversed and the system went from A to C to B and back to A. What can you say about the work done by the system, and the heat entering the system, along each leg?

❚ Since the system is presumed to be describable at every instant by its thermodynamic variables: p, V, T, the process is quasistatic and reversible. This means that if heat entered along an original leg, the same amount leaves for the reversed leg, and similarly for work.

21.42 The p-V diagram in Fig. 21-4(a) represents a reversible cycle of operations performed by an ideal gas in which MN is an isothermal and NK an adiabatic. Fill in Fig. 21-4(b) for this cycle, using + to indicate an increase in the quantity listed, − to indicate a decrease, and 0 to indicate no change.

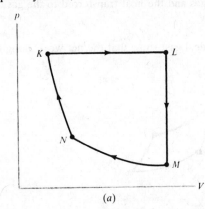

Path	ΔH	ΔW	ΔE	ΔT
KL				
LM				
MN				
NK				

(a)   (b)   **Fig. 21-4**

❚ Path KL: The process is an isobaric expansion (p remains constant while V increases). Clearly the gas does work on its surroundings, so $\Delta W > 0$. In order for the pressure to remain constant while the gas density decreases, the temperature must increase: $\Delta T > 0$. For an ideal gas, the energy is a function of only the temperature, and E increases with T, so $\Delta E > 0$. Because $\Delta H = \Delta E + \Delta W$, we must have $\Delta H > 0$.

Path LM: The process is isometric cooling (V remains constant while p—and with it T—decreases). Because the volume does not vary anywhere along the path, we have $\Delta V \equiv 0$ and therefore $\Delta W = 0$. As we already noted, $\Delta T < 0$, since that is the only way that p can decrease at constant volume. Since E is a monotonically increasing function of T, we have $\Delta E < 0$. Then $\Delta E + \Delta W = \Delta H < 0$.

Path MN: The process is isothermal compression, so $\Delta T = 0$. Therefore the ideal gas has $\Delta E = 0$. There is a monotonic volume decrease ($\Delta V < 0$), so the work $= \Delta W < 0$. Therefore $\Delta E + \Delta W = \Delta H < 0$.

Path NK: The process is adiabatic compression, so that $\Delta H = 0$. There is a monotonic volume decrease ($\Delta V < 0$), so $\Delta W < 0$. Therefore $\Delta H - \Delta W \equiv \Delta E > 0$. The temperature of the ideal gas must increase correspondingly: $\Delta T > 0$.

These results are summarized in Fig. 21-5.

21.43 n kmol of a monatomic ideal gas is taken quasistatically from state A to state C along the straight-line path shown in Fig. 21-6. For this process, calculate the work ΔW done by the gas, the increase ΔE of its internal energy, and the heat ΔH added to the gas. Express all answers in terms of p_A and V_A.

Path	ΔH	ΔW	ΔE	ΔT
KL	+	+	+	+
LM	−	0	−	−
MN	−	−	0	0
NK	0	−	+	+

Fig. 21-5

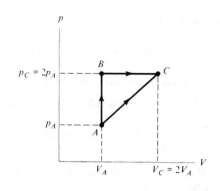

Fig. 21-6

▌ The work ΔW_{AC} done by the gas along the straight-line path AC is equal to the area $\int_A^C p\, dV$ under the path. That integral is the area $p_A(V_C - V_A)$ of a rectangle in Fig. 21-6, *plus* the area $\frac{1}{2}(p_C - p_A)(V_C - V_A)$ of a triangle. Since $p_C = 2p_A$ and $V_C = 2V_A$, we have $\Delta W_{AC} = p_A(V_A) + \frac{1}{2}(p_A)(V_A)$. Therefore $\underline{\Delta W_{AC} = \frac{3}{2}p_A V_A}$. For a monatomic ideal gas, $\Delta E_{AC} = \frac{3}{2}nR(T_C - T_A)$. Since $pV = nRT$, we find $\Delta E_{AC} = \frac{3}{2}(p_C V_C - p_A V_A) = \frac{3}{2}(4p_A V_A - p_A V_A) = \frac{9}{2}p_A V_A$. Finally, using the first law of thermodynamics, $\Delta H_{AC} = \Delta E_{AC} + \Delta W_{AC} = \underline{6p_A V_A}$.

21.44 (a) Repeat Prob. 21.43 if the gas is taken quasistatically from A to C along the path ABC. (b) Explain the similarities and differences between the results of (a) and Prob. 21.43.

▌ (a) The work ΔW_{ABC} done by the gas along the path ABC is the sum of two contributions, ΔW_{AB} and ΔW_{BC}. But $dV \equiv 0$ along AB, so $\Delta W_{AB} = 0$. Therefore $\Delta W_{ABC} = \Delta W_{BC} = p_C(V_C - V_A) = \underline{2p_A V_A}$. The energy change is independent of the path from A to C: $\Delta E_{ABC} = \Delta E_{AC} = \frac{3}{2}nR(T_C - T_A) = \frac{9}{2}p_A V_A$. The heat added to the gas is given by $\Delta H_{ABC} = \Delta E_{ABC} + \Delta W_{ABC} = \frac{13}{2}p_A V_A$. (b) The internal energy change is the same in the two processes because E is a state variable. However, both ΔW and ΔH vary from one process to the other because both the work done by the gas and the heat transferred to the gas depend upon the process. <u>Neither W nor H is a state variable.</u>

21.45 Refer to Fig. 21-7 for the Carnot cycle of an ideal gas; show that the net work along the adiabats NK and LM is zero.

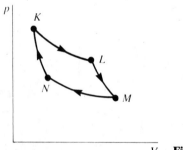

V **Fig. 21-7**

▌ A Carnot cycle consists of a closed path made up of two adiabatic and two isothermal legs as shown. Along the adiabat NK, the work done by the gas is (see Prob. 21.25)

$$\Delta W_{NK} = \frac{1}{\gamma - 1}(p_N V_N - p_K V_K) \tag{1}$$

Along the adiabat LM, the work done by the gas is

$$\Delta W_{LM} = \frac{1}{\gamma - 1}(p_L V_L - p_M V_M) \tag{2}$$

The net work done on the system during the two adiabatic parts of the Carnot cycle is given by

$$\Delta W = \Delta W_{NK} + \Delta W_{LM} = \frac{1}{\gamma - 1}[(p_L V_L - p_K V_K) + (p_N V_N - p_M V_M)] \tag{3}$$

Since KL and MN are isotherms, Boyle's law implies that $p_K V_K = p_L V_L$ and that $p_M V_M = p_N V_N$. Referring to Eq. (3), we conclude that $\Delta W = 0$.

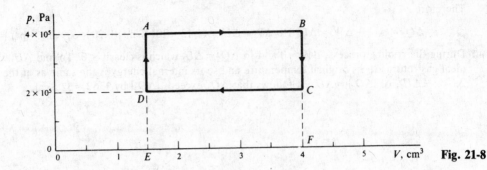

Fig. 21-8

21.46 The p-V diagram shown in Fig. 21-8 applies to a gas undergoing a cyclic change in a piston-cylinder arrangement. What is the work done by the gas in **(a)** portion AB of the cycle? **(b)** portion BC? **(c)** portion CD? **(d)** portion DA?

▮ In expansion, the work done is equal to the area under the pertinent portion of the p-V curve. In contraction, the work is numerically equal to the area but is negative.

(a) $$\text{work} = \text{area } ABFEA = [(4-1.5)\times10^{-6}\,\text{m}^3](4\times10^5\,\text{N/m}^2) = \underline{1.00\,\text{J}}$$

(b) $$\text{work} = \text{area under } BC = \underline{0}$$

In portion BC, the volume does not change; therefore $p\,\Delta V = 0$.
(c) This is contraction and so the work is negative.

$$\text{work} = -(\text{area } CDEFC) = -(2.5\times10^{-6}\,\text{m}^3)(2\times10^5\,\text{N/m}^2) = \underline{-0.50\,\text{J}}$$

(d) $$\text{work} = \underline{0}$$

21.47 For the thermodynamic cycle shown in Fig. 21-8, find **(a)** net output work of the gas during the cycle, **(b)** net heat flow into the gas per cycle.

▮ **(a)** Method 1: From Prob. 21.46, the net work done is $1.00\,\text{J} - 0.50\,\text{J} = \underline{0.50\,\text{J}}$.
Method 2: The net work done is equal to the area enclosed by the p-V diagram.

$$\text{work} = \text{area } ABCDA = (2\times10^5\,\text{N/m}^2)(2.5\times10^{-6}\,\text{m}^3) = \underline{0.50\,\text{J}}$$

(b) Suppose that the cycle starts at point A. The gas returns to this point at the end of the cycle, so there is no difference in the gas at its start and endpoint. For one complete cycle, ΔU is therefore zero. We have then, if the first law is applied to a complete cycle,

$$\Delta Q = \Delta U + \Delta W = 0 + 0.50\,\text{J} = \underline{0.50\,\text{J}}$$

21.48 A cylinder of ideal gas is closed by an 8-kg movable piston (area 60 cm²) as shown in Fig. 21-9. Atmospheric pressure is 100 kPa. When the gas is heated from 30 to 100 °C, the piston rises 20 cm. The piston is then fastened in place and the gas is cooled back to 30 °C. Calling ΔQ_1 the heat added to the gas in the heating process and $|\Delta Q_2|$ the heat lost during cooling, find the difference between ΔQ_1 and $|\Delta Q_2|$.

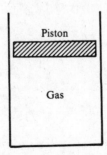

Piston

Gas

Fig. 21-9

▮ During the heating process, the internal energy changed by ΔU_1 and work ΔW_1 was done. The gas pressure was

$$p = \frac{8(9.8)\,\text{N}}{60\times10^{-4}\,\text{m}^2} + 1.00\times10^5\,\text{N/m}^2 = 1.13\times10^5\,\text{N/m}^2.$$

Therefore

$$\Delta Q_1 = \Delta U_1 + \Delta W_1 = \Delta U_1 + p\,\Delta V = \Delta U_1 + (1.13 \times 10^5\,\text{N/m}^2)(0.20 \times 60 \times 10^{-4}\,\text{m}^3) = \Delta U_1 + 136\,\text{J}$$

During the cooling process, $\Delta W = 0$ and so $\Delta Q_2 = \Delta U_2$ which is clearly <0. To find ΔU_2 we note that the ideal gas returns to its original temperature and so its internal energy is the same as at the start. Therefore, $\Delta U_2 = -\Delta U_1$, or $|\Delta Q_2| = \Delta U_1$. It follows that ΔQ_1 exceeds $|\Delta Q_2|$ by $136\,\text{J} = \underline{32.5\,\text{cal}}$.

CHAPTER 22
The Second Law of Thermodynamics

22.1 HEAT ENGINES, KELVIN–PLANCK AND CLAUSIUS STATEMENTS OF THE SECOND LAW

22.1 What is meant by a *reversible process* in thermodynamics?

▌ In an idealized *reversible process* a system goes from an initial equilibrium state to a final equilibrium state through a continuous sequence of equilibrium states. This means that at every instant during the process the system is in thermal and mechanical equilibrium with its surroundings. The direction of such a process can be reversed (hence, "reversible") at any instant by an infinitesimal change in external conditions.

An actual process approaches reversibility to the degree that it is quasistatic (i.e., extremely slow) and that dissipative effects (e.g., friction) are absent.

22.2 What is a *heat engine* and what is the *efficiency* of a heat engine?

▌ A *heat engine* is a device or system that converts heat into work. Heat engines operate by absorbing heat from a reservoir at a high temperature, performing work, and giving off heat to a reservoir at a lower temperature. The *efficiency* η of a cyclic heat engine is

$$\eta = \frac{W}{Q_{\text{hot}}} = 1 - \frac{Q_{\text{cold}}}{Q_{\text{hot}}}$$

where Q_{hot}, Q_{cold}, and W (see Fig. 22-1) represent, respectively, the heat absorbed per cycle from the higher-temperature reservoir, the heat rejected per cycle to the lower-temperature reservoir, and the work carried out per cycle. The last form for η follows from $W = Q_{\text{hot}} - Q_{\text{cold}}$, since $\Delta U = 0$ over a cycle.

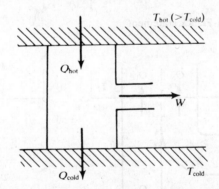

Fig. 22-1

The greatest possible thermal efficiency of an engine operating between two heat reservoirs is that of a *Carnot engine*, one that operates in the Carnot cycle (Prob. 21.45). This maximal efficiency is

$$\eta^* = 1 - \frac{T_{\text{cold}}}{T_{\text{hot}}}$$

22.3 What is a *refrigerator*, and what is its *coefficient of performance*?

▌ A *refrigerator* (or *heat pump*) is a heat engine operated backward; it takes heat from a low-temperature reservoir, is supplied work, and rejects heat to a high-temperature reservoir. When a refrigerator in one cycle takes *out* heat Q_{cold} from a cold reservoir, the work *input* W_i to the refrigerator satisfies $W_i = Q_{\text{hot}} - Q_{\text{cold}}$, where Q_{hot} represents the heat *ejected* to the hot reservoir. Then the coefficient of performance $\kappa = Q_{\text{cold}}/W_i = Q_{\text{cold}}/(Q_{\text{hot}} - Q_{\text{cold}})$. For a given Q_{cold}, the coefficient of performance is largest for a refrigerator operating between two reservoirs in a Carnot cycle;

$$\kappa^* = \left(\frac{T_{\text{hot}}}{T_{\text{cold}}} - 1\right)^{-1}$$

22.4 What are the two classical statements of the second law of thermodynamics?

▌ They are the Kelvin–Planck (engine) and Clausius (refrigerator) statements.
   Kelvin–Planck: It is impossible to construct an engine that operates in a cycle and produces no effect other than the extraction of heat from a reservoir and the performance of an equal amount of work.
   Clausius: It is impossible to construct an engine that operating in a cycle will produce no effect other than the transfer of heat from a cooler to a hotter body.

22.5   Show that the Clausius and Kelvin–Planck statements of the second law of thermodynamics are equivalent.

▌ Statements A and B are equivalent if a violation of A implies a violation of B, and conversely.
   Shown in Fig. 22-2(a) is an engine that would violate the Kelvin–Planck statement by extracting heat Q_2 from the reservoir at temperature T_2 and doing work $W = Q_2$. This engine could be used to drive an ordinary refrigerator between reservoirs at $T_1 < T_2$ and T_2. As shown in Fig. 22-2(b), the composite engine would transfer a net amount of heat Q_1 from T_1 to T_2 without any external work being supplied. It would therefore violate the Clausius statement.
   The Clausius statement would be violated by the refrigerator shown in Fig. 22-3(a). Using this refrigerator in conjunction with an ordinary heat engine, as in Fig. 22-3(b), one could convert an amount of heat $Q_2 - Q_1$ completely into work. This would violate the Kelvin–Planck statement.

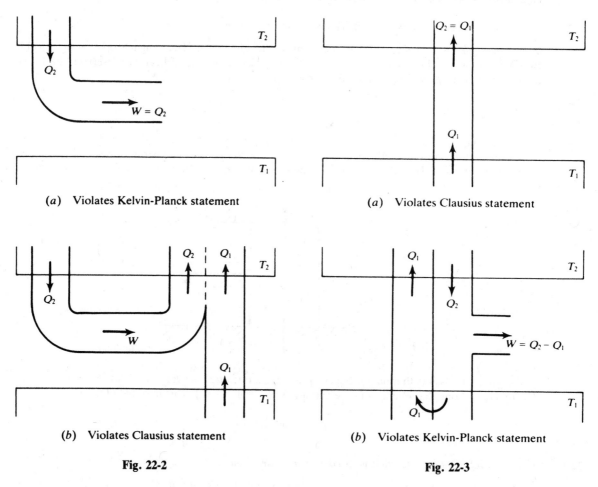

(a) Violates Kelvin-Planck statement	(a) Violates Clausius statement
(b) Violates Clausius statement	(b) Violates Kelvin-Planck statement
Fig. 22-2	**Fig. 22-3**

22.6   Prove that (as asserted in Prob. 22.2) any heat engine which operates cyclically between a hot reservoir and a cold reservoir has a thermal efficiency less than or equal to the Carnot efficiency.

▌ We use the reasoning of Prob. 22.5, remembering that a Carnot engine is reversible. In Fig. 22-4 assume that engine A, operating between the two reservoirs at temperatures T_2 and T_1 is more efficient than the Carnot engine C. Then if we construct the Carnot cycle, operating in reverse, to extract heat Q_1 from the cold reservoir, we must have $Q_2' < Q_2$. This follows from the fact that $\eta_A = 1 - (Q_1/Q_2) > \eta_C = 1 - (Q_1/Q_2')$ by hypothesis. Then W, the work done by system A is greater than W', the work done *on* system C, since $W = Q_2 - Q_1$ and $W' = Q_2' - Q$. Indeed $W - W' = Q_2 - Q_2'$. Thus, considering that the two systems form a

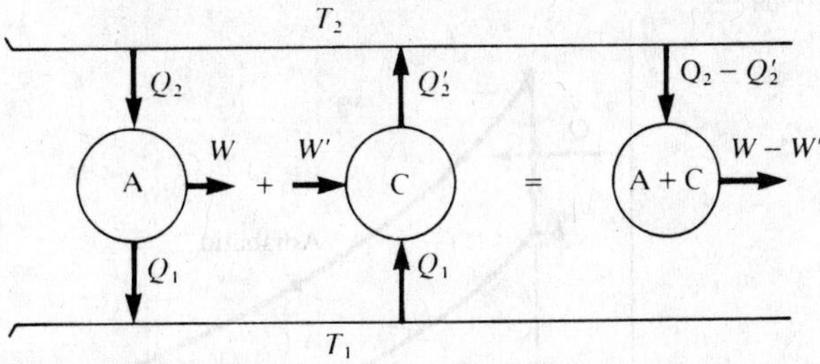

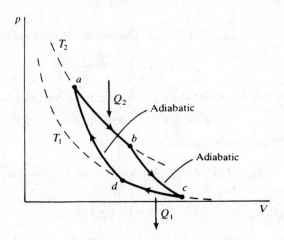

Fig. 22-4

single composite engine, as shown, we have violated the second law. Our hypothesis $\eta_A > \eta_C$ must therefore be false and $\eta_A \leq \eta_C \equiv \eta^*$.

22.7 Show that for the Carnot cycle, Fig. 22-5, $Q_2/Q_1 = T_2/T_1$. Assume 1 mol of ideal gas as the working substance. (Note that Q_1 and Q_2 are here defined to be inherently positive.)

Fig. 22-5

▌ During the stage $a \rightarrow b$, while the gas expands isothermally, its internal energy does not change, and the heat Q_2 absorbed from the T_2 reservoir is the same as the work carried out on it. Proceeding as in Prob. 21.15, we find that $Q_2 = RT_2 \ln (V_b/V_a)$. Similarly, in the stage $c \rightarrow d$, the heat rejected, Q_1, is $Q_1 = RT_1 \ln (V_c/V_d)$.

Along the paths $b \rightarrow c$ and $d \rightarrow a$ the processes are adiabatic, so

$$T_2 V_b^{\gamma-1} = T_1 V_c^{\gamma-1} \qquad T_1 V_d^{\gamma-1} = T_2 V_a^{\gamma-1}$$

Upon division of these equations, $V_b/V_a = V_c/V_d$. Thus

$$\frac{Q_1}{Q_2} = \frac{RT_1 \ln (V_c/V_d)}{RT_2 \ln (V_b/V_a)} = \frac{T_1}{T_2} \qquad \text{as required}$$

22.8 Prove that all Carnot engines operating between the same two thermal reservoirs have the same efficiency, $\eta^* = 1 - (T_1/T_2)$.

▌ With reference to Prob. 22.6, assume that both A and C are Carnot engines that use different working substances. Clearly $\eta_A \leq \eta_C$. Now let the two engines change place. By the same reasoning, $\eta_C \leq \eta_A$. Thus $\eta_A = \eta_C$ and our result is proved.

Note that from Prob. 22.7 we have for a Carnot engine using an ideal gas as a working substance $Q_1/Q_2 = T_1/T_2$. Therefore $\eta^* = 1 - (T_1/T_2)$ for an ideal gas. Since all Carnot engines are equally efficient, we have $\eta^* = 1 - (T_1/T_2) = 1 - (T_{\text{cold}}/T_{\text{hot}})$ for any Carnot cycle, as stated in Prob. 22.2.

22.9 For a gasoline engine undergoing the *Otto cycle* shown in Fig. 22-6, determine the thermal efficiency. Assume 1 mol of air as the working substance.

▌ We assume that the cycle is reversible and that C_v is constant for the working substance (air). Then, along the line $b \rightarrow c$, the heat into the system is $Q_1 = C_V \Delta T = C_v(T_c - T_b)$. Similarly, for process $d \rightarrow a$, the heat

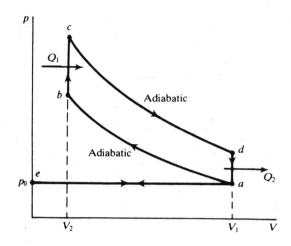

Fig. 22-6

out of the system is $Q_2 = -C_V \Delta T = C_V(T_d - T_a)$. The thermal efficiency is therefore

$$\eta = 1 - \frac{Q_2}{Q_1} = 1 - \frac{T_d - T_a}{T_c - T_b}$$

Now, assuming an ideal gas, the two adiabatic processes may be described by the equations

$$T_d V_1^{\gamma-1} = T_c V_2^{\gamma-1} \quad \text{and} \quad T_a V_1^{\gamma-1} = T_b V_2^{\gamma-1}$$

(see Prob. 21.23). These give, upon subtraction,

$$(T_d - T_a)V_1^{\gamma-1} = (T_c - T_b)V_2^{\gamma-1} \quad \text{or} \quad \frac{T_d - T_a}{T_c - T_b} = \left(\frac{V_2}{V_1}\right)^{\gamma-1}$$

Thus

$$\eta = 1 - \left(\frac{V_2}{V_1}\right)^{\gamma-1}$$

The Otto cycle is an example of a reversible cycle that differs from the Carnot cycle.

22.10 A Carnot engine operates between 317 and 67 °C. What is its efficiency?

❚ $$\text{efficiency} = \frac{T_h - T_c}{T_h} = \frac{590 - 340}{590} = \underline{42\%}$$

22.11 A Carnot-type engine is designed to operate between 480 and 300 K. Assuming that the engine actually produces 1.2 kJ of mechanical energy per kilocalorie of heat absorbed, compare the actual efficiency with the theoretical maximum efficiency.

❚ $$\text{maximum efficiency} = \frac{T_h - T_c}{T_h} = \frac{480 - 300}{480} = 37.5\% \qquad \text{actual efficiency} = \frac{\text{energy output}}{\text{energy input}} = \frac{1.2}{1 \times 4.184} = \underline{28.7\%}$$

The actual efficiency is about three-fourths of the maximum.

22.12 What is the maximum amount of work that a Carnot engine can perform per kilocalorie of heat input if it absorbs heat at 427 °C and exhausts heat at 177 °C?

❚ $$\text{efficiency} = \frac{Q_h - Q_c}{Q_h} = \frac{T_h - T_c}{T_h} \qquad Q_h = (1 \text{ kcal})(4.184 \text{ kJ/kcal}) = 4.184 \text{ kJ}$$

$$\text{efficiency} = \frac{Q_h - Q_c}{4.184 \text{ kJ}} = \frac{700 \text{ K} - 450 \text{ K}}{700 \text{ K}} \qquad \text{Then} \qquad W = Q_h - Q_c = \underline{1.49 \text{ kJ}}$$

22.13 An ideal Carnot engine takes heat from a source at 317 °C, does some external work, and delivers the remaining energy to a heat sink at 117 °C. If 500 kcal of heat is taken from the source, how much work is done? How much heat is delivered to the sink?

▮ $$\text{efficiency} = \frac{Q_h - Q_c}{Q_h} = \frac{T_h - T_c}{T_h} \qquad \frac{500 - Q_c}{500} = \frac{590 - 390}{590}$$

$$500 - Q_c = 169 \text{ kcal} \qquad \text{and} \qquad Q_c = 331 \text{ kcal delivered to the sink}$$

$$W = Q_h - Q_c = 169 \text{ kcal} = 169(4.184) = \underline{710 \text{ kJ}}.$$

22.14 A steam engine operating between a boiler temperature of 220 °C and a condenser temperature of 35 °C delivers 8 hp. If its efficiency is 30 percent of that for a Carnot engine operating between these temperature limits, how many calories are absorbed each second by the boiler? How many calories are exhausted to the condenser each second?

▮ $$\text{actual efficiency} = (0.30)(\text{Carnot efficiency}) = (0.30)\left(1 - \frac{308}{493}\right) = 0.113$$

But from the relation

$$\text{efficiency} = \frac{\text{output work}}{\text{input heat}} \qquad \text{input heat/s} = \frac{\text{output work/s}}{\text{efficiency}} = \frac{(8 \text{ hp})(746 \text{ W/hp})\left(\dfrac{1 \text{ cal/s}}{4.184 \text{ W}}\right)}{0.113} = 12.7 \text{ kcal/s}$$

To find the energy rejected to the condenser, we use the law of conservation of energy:

$$\text{input energy} = \text{output work} + \text{rejected energy}$$

Thus,

$$\text{rejected energy/s} = (\text{input energy/s}) - (\text{output work/s}) = (\text{input energy/s})[1 - (\text{efficiency})]$$

$$= (12.7 \text{ kcal/s})(1 - 0.113) = \underline{11.3 \text{ kcal/s}}$$

22.15 How many kilograms of water at 0 °C can a freezer with a coefficient of performance 5 make into ice cubes at 0 °C with a work input of 3.6 MJ (one kilowatt-hour)? Use Table 17-1 for data.

▮ By Prob. 22.3, the coefficient of performance is Q_c/W.

$$5 = \frac{mL}{3.6 \times 10^6 \text{ J}} = \frac{m(80 \text{ kcal/kg})(4184 \text{ J/kcal})}{3.6 \times 10^6 \text{ J}}$$

Solving, $m = \underline{54 \text{ kg}}$.

22.16 A refrigerator removes heat from a freezing chamber at −5 °F and discharges it at 95 °F. What is its maximum coefficient of performance?

▮ For a Carnot refrigerator, $Q_c/W = T_c/(T_h - T_c)$. Use absolute temperatures (Rankine). Then

$$\frac{Q_c}{W} = \frac{-5 + 460}{(95 + 460) - (-5 + 460)} = \frac{455}{100} = \underline{4.55 \text{ coefficient of performance}}$$

22.17 A freezer has a coefficient of performance of 5. If the temperature inside the freezer is −20 °C, what is the temperature at which it rejects heat? Assume an ideal system.

▮ $$\frac{Q_c}{W} = \frac{T_c}{T_h - T_c} \qquad 5 = \frac{253}{T_h - 253}$$

Solving, we get $T_h = 304 \text{ K} = \underline{31 \text{ °C}}$.

22.2 ENTROPY

22.18ᶜ Give a mathematical definition of entropy and discuss its relation to the second law of thermodynamics.

▮ Any thermodynamic system has a state function S, called the *entropy*. By this is meant that S—like p, V, and U—is always the same for the system when it is in a given equilibrium state. Entropy may be defined as follows. Let a system at absolute temperature T undergo an infinitesimal *reversible* process in which it absorbs heat ΔQ. Then the change in entropy of the system is given by

$$\Delta S = \frac{\Delta Q}{T} \qquad \text{or} \qquad dS = \frac{dQ}{T} \qquad \text{for infinitesimals}$$

Note that dQ is not the differential of a true function. Entropy will have the units J/K.

The *Clausius equation*, $dS = dQ/T$, holds only for reversible processes. However, since S is a state function, the entropy change accompanying an irreversible process can be calculated by integrating dQ/T along the path of an arbitrary *reversible* process connecting the initial and final states.

The importance of the entropy function is exhibited in the following form of the *second law of thermodynamics*: In any process, the total entropy of the system and its surroundings increases or (in a reversible process) does not change. The second law applies to the system alone if the system is isolated; that is, if it in no way interacts with its surroundings.

22.19 Define entropy in terms of order/disorder, and discuss briefly.

❚ The second law of thermodynamics indicates that entropy is a measure of irreversibility. Irreversibility is associated, on the molecular level, with the increase of disorder. Molecular systems tend, as time passes, to become chaotic, and it is extremely unlikely that a more organized state, once left, will ever be regained. Another, fully equivalent, definition of entropy can be given from a detailed molecular analysis of the system. If a system can achieve the same state (i.e., the same values of p, V, T, and U) in Ω different ways (different arrangements of the molecules, for example), then the entropy of the state is $S = k \ln \Omega$, where ln is the logarithm to base e and k is Boltzmann's constant, 1.38×10^{-23} J/K.

A state that can occur in only one way (one arrangement of its molecules, for example) is a state of high order. But a state that can occur in many ways is a more disordered state. To associate a number with disorder, the disorder of a state is taken proportional to Ω, the number of ways the state can occur. Because $S = k \ln \Omega$, the entropy is a measure of disorder.

Spontaneous processes in systems that contain many molecules always occur in a direction from

$$\begin{pmatrix} \text{state that can exist} \\ \text{in only a few ways} \end{pmatrix} \rightarrow \begin{pmatrix} \text{state that can exist} \\ \text{in many ways} \end{pmatrix}$$

Hence systems when left to themselves retain their original state of order or else increase their disorder.

22.20 For a heat engine, over one cycle, $\Delta S = \Delta E = 0$, since the engine returns to its original state. The first law of thermodynamics then gives for the work done by the engine per cycle.

$$W = Q_{\text{hot}}\left(1 - \frac{T_{\text{cold}}}{T_{\text{hot}}}\right) - T_{\text{cold}}\,\Delta S_{\text{total}} = W^* - T_{\text{cold}}\,\Delta S_{\text{total}} \tag{1}$$

where W^* is the work that would be done by a Carnot engine operating between the same two temperatures, and ΔS_{total} is the entropy change of the universe (in this case, the entropy change of the hot and cold reservoirs) during one cycle. Derive (1) and show its significance for the second law.

❚ For any engine operating between the temperature reservoirs, we have $W = Q_{\text{hot}} - Q_{\text{cold}}$, since $\Delta U = 0$ over a cycle. (Here Q_{cold} is defined as positive when heat leaves the engine, as is usual in engine and refrigerator problems.) In addition the changes in entropy of the hot and cold reservoirs are related to the heat transfers at constant temperature by $Q_{\text{hot}} = -T_{\text{hot}}\,\Delta S_{\text{hot}}$; $Q_{\text{cold}} = T_{\text{cold}}\,\Delta S_{\text{cold}}$. Then, noting $\Delta S_{\text{total}} = \Delta S_{\text{hot}} + \Delta S_{\text{cold}}$, we have $T_{\text{cold}}\,\Delta S_{\text{total}} = Q_{\text{cold}} - (T_{\text{cold}}Q_{\text{hot}})/T_{\text{hot}}$. Thus $W = Q_{\text{hot}} - Q_{\text{cold}} = Q_{\text{hot}}[1 - (T_{\text{cold}}/T_{\text{hot}})] - T_{\text{cold}}\,\Delta S_{\text{total}}$. Since the efficiency of a Carnot cycle is $W^*/Q_{\text{hot}} = 1 - (T_{\text{cold}}/T_{\text{hot}})$, we have the remainder of our result: $W = W^* - T_{\text{cold}}\,\Delta S_{\text{total}}$. We thus see that for a given Q_{hot}, $W \leq W^* \Leftrightarrow \Delta S \geq 0$. Or, in terms of efficiency, $\eta \leq \eta^* \Leftrightarrow \Delta S \geq 0$.

22.21 When 100 coins are tossed, there is one way that all can come up heads. There are 100 ways that only one tail is up. There are about 1×10^{29} ways that 50 heads can come up. One hundred coins are placed in a box with only one head up. They are shaken and then there are 50 heads up. What was the change in entropy of the coins caused by the shaking?

❚ From Prob. 22.19,

$$\Delta S = k(\ln \Omega_f - \ln \Omega_i) = (1.38 \times 10^{-23}\,\text{J/K})[\ln (1 \times 10^{29}) - \ln 100]$$

$$= (1.38 \times 10^{-23}\,\text{J/K})(27 \ln 10) = \underline{8.6 \times 10^{-22}\,\text{J/K}}$$

using $\ln 10 = 2.303$.

22.22^c The number Ω of states accessible to N atoms of a monatomic ideal gas with a volume V, when the energy of the gas is between E and $E + dE$, can be shown to be $\Omega = A(N)V^N E^{3N/2}$, where the factor $A(N)$ depends only on N. (a) Find the entropy S as a function of V and E. (b) Using this entropy function and the definition of the Kelvin temperature, $1/T = (\partial S/\partial E)_V$, show that $E = \frac{3}{2}NkT$.

▎ **(a)** The entropy is given by

$$S = k \ln \Omega = k \ln A(N) + Nk \ln V + \tfrac{3}{2} Nk \ln E$$

(b) The Kelvin temperature is given by

$$\frac{1}{T} = \left(\frac{\partial S}{\partial E}\right)_V = \frac{3}{2} Nk \frac{d(\ln E)}{dE} = \frac{3}{2} \frac{Nk}{E} \qquad \text{whence} \qquad E = \frac{3}{2} NkT$$

22.23 An ideal gas is confined to a cylinder by a piston. The piston is slowly pushed in so that the gas temperature remains at 20 °C. During the compression, 730 J of work is done on the gas. Find the entropy change of the gas.

▎ The first law tells us that $\Delta Q = \Delta U + \Delta W$. Because the process was isothermal, the internal energy of the ideal gas did not change. Therefore, $\Delta U = 0$ and $\Delta Q = \Delta W = -730$ J. (Because the gas was compressed, the gas did negative work, hence the minus sign.) Now we can write

$$\Delta S = \frac{\Delta Q}{T} = \frac{-730 \text{ J}}{293 \text{ K}} = \underline{-2.49 \text{ J/K}}$$

Note that the entropy change is negative. Disorder of the gas decreased as it was pushed into a smaller volume.

22.24 Why isn't the result of Prob. 22.23 a violation of the entropy statement of the second law, $\Delta S \geq 0$?

▎ The second law statement refers to the change in entropy *of the universe*. Clearly the heat that left the cylinder entered some other system in the environment, causing an increase in entropy in that system. The second law says that this increase was at least as great as the decrease in entropy of the gas in the cylinder.

22.25 A quantity of heat ΔH is transferred from a large heat reservoir at temperature T_1 to another large heat reservoir at temperature T_2, with $T_1 > T_2$ required for spontaneous transfer. The heat reservoirs have such large capacities that there is no observable change in their temperatures. Show that the entropy of the entire system has increased.

▎ Since the reservoir at constant temperature T_1 gives up heat ΔH, its entropy change is

$$\Delta S_1 = \frac{\Delta H_1}{T_1} = -\frac{\Delta H}{T_1}$$

The reservoir at constant temperature T_2 receives heat ΔH, so its entropy change is

$$\Delta S_2 = \frac{\Delta H_2}{T_2} = \frac{\Delta H}{T_2}$$

The overall entropy change $\Delta S = \Delta S_1 + \Delta S_2$ is therefore given by

$$\Delta S = -\frac{\Delta H}{T_1} + \frac{\Delta H}{T_2} = \Delta H \frac{T_1 - T_2}{T_1 T_2}$$

Since $T_1 > T_2$, ΔS is positive.

22.26 What is the change in the entropy of 2.00 kg of H_2O molecules when transformed at a constant pressure of 1 atm from water at 100 °C to steam at the same temperature? (Get data from Table 17-2.)

▎ Since heat is absorbed at constant temperature $\Delta S = \Delta Q/T = (mh_v)/T =$ (2.00 kg)(540 kcal/kg)(4.184 kJ/kcal)/(373 K), or $\Delta S = \underline{12.12 \text{ kJ/K}}$.

22.27ᶜ n kmol of an ideal gas expands from volume V_A to volume $V_B > V_A$ with initial and final temperatures the same. Find the change in entropy of the gas.

▎ In view of the discussion in Prob. 22.18, we can assume a reversible isothermal process. Since the process is isothermal, the internal energy of the ideal gas is constant: $dE = 0$. Hence $dH = dW = p\, dV$ along the isotherm. Then the entropy change is given by

$$\Delta S \equiv S_B - S_A = \int_A^B \frac{dH}{T} = \int_{V_A}^{V_B} \frac{p\, dV}{T_0} = \int_{V_A}^{V_B} \frac{nRT_0/V}{T_0}\, dV = nR \int_{V_A}^{V_B} \frac{dV}{V} = nR \ln \frac{V_B}{V_A}$$

22.28ᶜ A copper can, of negligible heat capacity, contains 1.000 kg of water just above the freezing point. A similar can contains 1.000 kg of water jsut below the boiling point. The two cans are brought into thermal contact. Find the change in entropy of the system.

▮ For the contents of either can, $dH = cm\,dT$, where, for water, $cm = 4184$ J/K. Since the heat capacities of the two masses are equal, the final temperature will be the average of the initial temperatures:

$$T_{1f} = T_{2f} = \frac{T_{1i} + T_{2i}}{2} = \frac{273\text{ K} + 373\text{ K}}{2} = 323\text{ K}$$

Then

$$\Delta S = \Delta S_1 + \Delta S_2 = \int_{T_{1i}}^{T_{1f}} cm\,\frac{dT_1}{T_1} + \int_{T_{2i}}^{T_{2f}} cm\,\frac{dT_2}{T_2}$$

$$= (4184\text{ J/K})[\ln{(323\text{ K}/273\text{ K})} + \ln{(323\text{ K}/373\text{ K})}] = \underline{+100\text{ J/K}}$$

22.29 Problem 22.22 shows that when an ideal gas of fixed volume is heated, its entropy increases (logarithmically). Explain from considerations of order/disorder.

▮ A higher temperature means a higher root-mean-square velocity, and hence generally higher velocities of the individual atoms. Since more velocity options are available to individual molecules, there are a larger number of microscopic arrangements for the same macroscopic state and hence greater disorder.

22.30 In the cycle shown in Fig. 22-7, AB and CD are isotherms. The working substance may be considered an ideal gas. For each of the four parts of the cycle, are the quantities ΔH, ΔW, ΔE, ΔT, and ΔS positive or negative according to the sign convention generally used in physics?

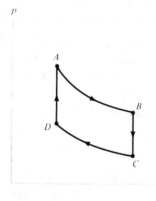

Fig. 22-7

▮ Path AB. The gas expands and does positive work, $\Delta W > 0$. Since AB is an isotherm, $\Delta T = 0$. The working substance is an ideal gas, so $\Delta E = 0$. Therefore $\Delta E + \Delta W = \Delta H$ is positive and $\Delta S = \Delta H/T$ is also positive.
    Path BC. The volume of the gas does not change, so $\Delta W = 0$. Since the pressure decreases at constant volume, ΔT is negative. Thus ΔE is negative and so is $\Delta E + \Delta W = \Delta H$. The entropy change $\Delta S = \int_b^c dH/T$ is also negative.
    Path CD. The gas contracts, so $\Delta W = \int_c^d p\,dV < 0$. Since CD is an isotherm traced in the direction *opposite* to that for path AB, the other results are $\Delta T = \Delta E = 0$, $\Delta H < 0$, and $\Delta S < 0$.
    Path DA. Since process DA is isometric heating while process BC is isometric cooling, all the signs should be reversed: $\Delta W = 0$, $\Delta T > 0$, $\Delta E > 0$, $\Delta H > 0$, and $\Delta S > 0$.

22.31 When the working substance of Prob. 22.30 returns to A after one complete cycle, are the five quantities positive or negative compared with their original value?

▮ Since E, T, and S are all state variables, we conclude that ΔE, ΔT, and ΔS all vanish over the complete cycle. Referring to the $p - V$ diagram, we can see that the positive work done *by* the gas along AB exceeds the positive work done *on* the gas along CD. Therefore, the net work done by the gas is positive: $\Delta W > 0$. Since $\Delta H = \Delta E + \Delta W$, we conclude that ΔH is positive.

22.32ᶜ A large heat reservoir at temperature T_f in contact with 1.0 kg of water warms it from T_i to T_f. Show that there is an increase in the entropy of the entire system.

▮ The heat ΔH_w added to the water is given by

$$\Delta H_w = m_w c_w \, \Delta T = (1.0 \text{ kg})(1.00 \text{ kcal/kg} \cdot \text{K})(T_f - T_i) = T_f - T_i$$

in units of kcal. The heat added to the reservoir is the negative of this: $\Delta H_r = -\Delta H_w$. Therefore the entropy increase of the reservoir is given by

$$\Delta S_r = \frac{\Delta H_r}{T_f} = -\frac{T_f - T_i}{T_f}$$

The entropy increase of the water must be found by integrating the differential expression $dS_w = dH_w/T$:

$$\Delta S_w = \int dS_w = \int \frac{dH_w}{T} = \int_{T_i}^{T_f} \frac{dT}{T}$$

Now, since $1/T$ is strictly decreasing over (T_i, T_f),

$$\Delta S_w > \frac{1}{T_f} \int_{T_i}^{T_f} dT = \frac{T_f - T_i}{T_f} = -\Delta S_r \qquad \text{or} \qquad \Delta S_{\text{net}} = \Delta S_w + \Delta S_r > 0$$

22.33ᶜ Find an expression for the entropy of a mole of an ideal gas having constant molar heat capacities C_p and C_v.

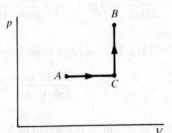

Fig. 22-8

▮ We have $\Delta S = \Delta Q/T$. To find $S_B - S_A$ for any two state points on a p-V diagram we can add up the ΔS contributions along any path on the p-V diagram. Since ΔQ for constant V is known and is $\Delta Q = C_v \, \Delta T$, while ΔQ for constant p is also known and is $\Delta Q = C_p \, \Delta T$, we choose the path ACB, as shown on Fig. 22-8. Then

$$S_C - S_A = \int_A^C \frac{dQ}{T} = C_p \int_{T_A}^{T_C} \frac{dT}{T} = C_p \ln \frac{T_C}{T_A} = C_p \ln \frac{V_C}{V_A}$$

this last since $pV = RT$ and p is constant. Similarly

$$S_B - S_C = \int_C^B \frac{dQ}{T} = C_v \int_C^B \frac{dT}{T} = C_v \ln \frac{T_B}{T_C} = C_v \ln \frac{p_B}{p_C}$$

this last since V is constant. Adding, we get (noting $p_C = p_A$ and $V_C = V_B$)

$$S_B - S_A = C_p \ln \frac{V_C}{V_A} + C_v \ln \frac{p_B}{p_C} = C_p \ln \frac{V_B}{V_A} + C_v \ln \frac{p_B}{p_A}$$

$$S_B - S_A = C_v \left(\frac{C_p}{C_v} \ln \frac{V_B}{V_A} + \ln \frac{p_B}{p_A} \right) = C_v \ln \frac{p_B V_B^\gamma}{p_A V_A^\gamma}$$

Holding A fixed and letting B be an arbitrary point on the p-V diagram, we get our result: $S = C_v \ln(pV^\gamma) + \text{constant}$. [The constant is $S_A - C_v \ln(p_A V_A^\gamma)$.]

22.34 From Prob. 22.33 infer that during a reversible adiabatic process the pressure and volume of an ideal gas are related through $pV^\gamma = \text{constant}$.

▮ From Prob. 22.33 $S = C_v \ln(pV^\gamma) + \text{constant}$. For an adiabatic process $\Delta S = 0$ and S is constant. Therefore $\ln(pV^\gamma) = \text{constant}$ if C_v is constant. Assuming that this latter is true, we get $pV^\gamma = \text{constant}$ for an adiabatic process.

CHAPTER 23
Wave Motion

23.1 CHARACTERISTIC PROPERTIES

23.1 The sound of a lightning flash is heard 6.0 s after the flash. Assume that the light travels much more swiftly than the sound. How far away was the lightning? (Assume that the speed of sound was 330 m/s.)

▮ It took sound 6.0 s to travel from the flash to the observer. Use $x = vt = (330 \text{ m/s})(6.0 \text{ s}) = \underline{1980 \text{ m}}$.

23.2 The average person can hear sound waves ranging in frequency from about 20 to 20 000 Hz. Determine the wavelengths at these limits, taking the speed of sound to be 340 m/s.

▮ $\lambda f = v$, so

$$\lambda_1 = \frac{340 \text{ m/s}}{20 \text{ Hz}} = \underline{17 \text{ m}} \qquad \lambda_2 = \frac{340 \text{ m/s}}{20\,000 \text{ Hz}} = 0.017 \text{ m} = \underline{1.7 \text{ cm}}$$

23.3 A radio station broadcasts at 760 kHz. The speed of radio waves is 3×10^8 m/s. What is the wavelength?

▮

$$\lambda = \frac{v}{f} = \frac{3 \times 10^8 \text{ m/s}}{760\,000 \text{ Hz}} = \underline{395 \text{ m}}$$

23.4 The velocity of sound in seawater is 1530 m/s. If a sound of frequency 1800 Hz is produced in seawater, what is its wavelength?

▮

$$v = v\lambda \qquad 1530 = 1800\lambda \qquad \lambda = \underline{0.85 \text{ m}}$$

23.5 Sound waves of wavelength λ travel from a medium in which their velocity is v into a medium in which their velocity is $4v$. What is the wavelength of the sound in the second medium?

▮ The frequency of the disturbance, ν, stays the same.

$$\nu = \frac{v}{\lambda} = \frac{4v}{\lambda_2} \qquad \text{whence } \lambda_2 = 4\lambda$$

23.6 (*a*) An ultrasonic transducer used in sonar produces a frequency of 40 kHz. If the velocity of the sound wave in seawater is 5050 ft/s, what is the wavelength? (*b*) The transducer is made to emit a short burst of sound and is then turned off. The receiver is turned on. The pulse is reflected from a lurking submarine and received 5.0 s after it was first emitted. How far away is the submarine?

▮ (*a*) $\qquad v = v\lambda \qquad 5050 = 40\,000\lambda \qquad \lambda = \underline{0.126 \text{ ft}}$

(*b*) If the submarine is at a distance d, the sonar wave must travel a total distance $2d$ from the transducer to the submarine and, after reflection, back to the receiver.

$$v = \frac{2d}{t} \qquad 5050 = \frac{2d}{5.0} \qquad d = \underline{12\,600 \text{ ft}}$$

23.7 The velocities of propagation of disturbances or waves in various media are determined by characteristic properties of those media. Express the velocity of propagation in terms of such properties for (*a*) transverse waves on a stretched string or wire; (*b*) longitudinal waves in liquids, solids, and gases.

▮ (*a*) $\qquad v = \sqrt{\dfrac{\text{tension in string}}{\text{mass per unit length of string}}}$

(b)

$$\text{In liquids: } v = \sqrt{\frac{\text{bulk modulus}}{\text{density of liquid}}} = \sqrt{\frac{B}{\rho}}$$

$$\text{In solid rods: } v = \sqrt{\frac{\text{Young's modulus}}{\text{density of solid}}} = \sqrt{\frac{Y}{\rho}}$$

$$\text{In gases: } v = \sqrt{\frac{\gamma \times (\text{pressure of gas})}{\text{density of gas}}} = \sqrt{\frac{\gamma p}{\rho}}$$

where γ, the ratio of specific heats c_p/c_v, is about 1.67 for monatomic gases such as helium and neon. It is about 1.40 for diatomic gases such as N_2, O_2, and H_2.

23.8 When driven by a 120-Hz vibrator, a string has transverse waves of 31-cm wavelength traveling along it. (a) What is the speed of the waves on the string? (b) If the tension in the string is 1.20 N, what is the mass of 50 cm of the string?

▎ (a) $v = \lambda f = (0.31\text{ m})(120\text{ Hz}) = \underline{37\text{ m/s}}$. (b) $v = \sqrt{S/\mu}$, where S is the tension. Then $\mu = S/v^2 = (1.20\text{ N})/(37\text{ m/s})^2 = 8.76 \times 10^{-4}$ kg/m. $M = \mu L = (0.50\text{ m})(8.76 \times 10^{-4}\text{ kg/m}) = 4.38 \times 10^{-4}\text{ kg} = \underline{0.44\text{ g}}$.

23.9 A copper wire 2.4 mm in diameter is 3 m long and is used to suspend a 2 kg mass from a beam. If a transverse disturbance is sent along the wire by striking it lightly with a pencil, how fast will the disturbance travel? The density of copper is 8920 kg/m³.

▎ $v = \sqrt{S/\mu}$ for a transverse wave in the wire. $S = (2\text{kg})(9.8\text{ m/s}^2) = 19.6$ N. $\mu = \rho A$, where A is the cross-sectional area of the wire. $\mu = (8920\text{ kg/m}^3)(3.14)(1.2 \times 10^{-3}\text{ m})^2 = 0.0403$ kg/m. Thus, $v = \sqrt{19.6/0.0403} = \underline{22\text{ m/s}}$.

23.10 An increase in pressure of 100 kPa causes a certain volume of water to decrease by 5×10^{-3} percent of its original volume. (a) What is the bulk modulus of water? (b) What is the speed of sound (compressional waves) in water?

▎ (a) $B = V |\Delta p/\Delta V| = \dfrac{100 \times 10^3\text{ Pa}}{5 \times 10^{-5}} = \underline{2000\text{ MPa}}$

(b) $v = \sqrt{\dfrac{B}{\rho}} = \sqrt{\dfrac{2.0 \times 10^9\text{ N/m}^2}{1000\text{ kg/m}^3}} = \underline{1400\text{ m/s}}$

23.11 If the bulk modulus of water is 2100 MPa, what is the speed of sound in water?

▎ $$v = \left(\frac{B}{\rho}\right)^{1/2} = \left[\frac{2.1 \times 10^9}{10^3}\right]^{1/2} = \underline{1450\text{ m/s}}$$

23.12 From the fact that the speed of sound in steel is about 5900 m/s, compute the bulk modulus of steel. The density of steel is 7900 kg/m³.

▎ For sound waves in a solid $v = (B/\rho)^{1/2}$. Therefore, $5900 = (B/7900)^{1/2}$, which gives $B = \underline{2.75 \times 10^{11}\text{ Pa}}$.

23.13 The speed of a wave on a string is given by $v = \sqrt{T/\mu}$. Show that the right-hand side of this equation has the units of speed. Repeat for $v = \sqrt{B/\rho}$.

▎ In the first case $[(T/\mu)^{1/2}] = [MLT^{-2}]^{1/2}[M^{-1}L]^{1/2} = [LT^{-1}] = [v]$. In the second case $[B/\rho]^{1/2} = [(F/A)/\rho]^{1/2} = [(MLT^{-2})(L^{-2})]^{1/2}[M^{-1}L^3]^{1/2} = [L/T] = [v]$.

23.14 A steel cable 3.0 cm in diameter is kept under a tension of 10 kN. The density of steel is 7.8 g/cm³. With what speed would transverse waves propagate along the cable?

▎ The cable has a linear mass density $\mu = \rho[(\pi d^2)/4]$, where ρ and d are the cable's mass density and diameter, respectively. With $\rho = 7.8 \times 10^3$ kg/m³ and $d = 0.03$ m, we find that $\mu = 5.51$ kg/m. Using the given tension $F = 1.0 \times 10^4$ N, we obtain the wave speed $v = \sqrt{F/\mu} = \underline{42.6\text{ m/s}}$.

23.15 Transverse waves pass along a stretched wire at 1000 ft/s. If the tension on the wire is quadrupled, what will the velocity be (in ft/s)?

▎ $v \propto T^{1/2}$; thus the velocity doubles and $v = \underline{2000\text{ ft/s}}$.

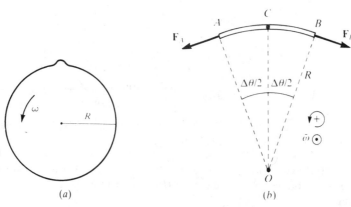

Fig. 23-1

(a)   (b)

23.16 A loop of rope is whirled at a high angular velocity, ω, so that it becomes a taut circle of radius R. A kink develops in the whirling rope; see Fig. 23-1(a). (a) Show that the tension in the rope is $F = \mu\omega^2 R^2$, where μ is the linear density of the rope. (b) Under what conditions does the kink remain stationary relative to an observer on the ground?

▌ (a) As shown in Fig. 23-1(b), we let ACB be a small (unkinked) section of the rope, subtending an angle $\Delta\theta$ at O, the center of the loop. We choose C at the midpoint of the arc. The centripetal force is provided by the sum of the two tension forces: $\mathbf{F}_{net} = \mathbf{F}_A + \mathbf{F}_B$. In order that the force be directed toward O, we must have $|\mathbf{F}_A| = |\mathbf{F}_B| = F$. The net inward radial force on the section ACB is given by $|\mathbf{F}_{net}| = 2F \sin(\Delta\theta/2)$. In the limit $\Delta\theta \to 0$, $|\mathbf{F}_{net}| \to 2F(\Delta\theta/2) = F\,\Delta\theta$. Equating this to the magnitude of the required centripetal force, we have $(\mu R\,\Delta\theta)\omega^2 R = F\,\Delta\theta$, since $\mu R\,\Delta\theta$ is the mass of the section. Solving this equation for F, we obtain $F = \mu\omega^2 R^2$, as desired. (b) The kink travels at speed $v = \sqrt{F/\mu} = R\omega$ relative to the rope. Hence, if the kink moves clockwise with respect to the rope, it will be stationary with respect to the ground.

23.17ᶜ A flexible steel cable of total length L and mass per unit length μ hangs vertically from a support at one end. (a) Show that the speed of a transverse wave down the cable is $v = \sqrt{g(L-x)}$, where x is measured from the support. (b) How long will it take for a wave to travel down the cable?

▌ (a) Since the tension, T, varies with length, so does v, the propagation velocity. For a wave down a string $v = (T/\mu)^{1/2}$. Tension $= g\mu(L-x)$ (the weight below station x), so $v = [g(L-x)]^{1/2}$. (b) The time dt for the wave to move distance dx is dx/v. Total time, is therefore

$$t = \int_0^L \frac{dx}{[g(L-x)]^{1/2}} = 2\left(\frac{L}{g}\right)^{1/2}$$

23.18 Describe mathematically a sinusoidal wave traveling in the positive x direction. Let y represent the displacement of a particle from its equilibrium position at x. (The wave may be either transverse or longitudinal.)

▌ The general expression for our wave is $y = A\cos(kx - \omega t + \phi)$. Here A is the amplitude of the disturbance, k is the wave number, ω is the "angular" frequency, and ϕ is a constant that depends on the position of the wave at $t = 0$. For traveling waves the value of ϕ is often uninteresting; and it is common to set $\phi = 0$, giving $y = A\cos(kx - \omega t) = A\cos(\omega t - kx)$; or $\phi = 3\pi/2$, giving $y = A\sin(kx - \omega t)$; or $\phi = \pi/2$, giving $y = A\sin(\omega t - kx)$. Since $\omega = 2\pi f$ (where f is the frequency) or $\omega = (2\pi/T)$ (T = period) and $k = (2\pi)/\lambda$ (λ = wavelength), we can reexpress our wave as

$$y = A\cos\left(\frac{2\pi x}{\lambda} - \frac{2\pi t}{T} + \phi\right) \qquad \text{etc.}$$

Also the velocity of propagation of the wave, c, is given by

$$c = \frac{\omega}{k} = \lambda f \quad \text{and} \quad y = A\cos\left[\frac{2\pi}{\lambda}(x - ct) + \phi\right] = A\cos\left[2\pi f\left(\frac{x}{c} - t\right) + \phi\right] \qquad \text{etc.}$$

(Note that while all the notation used is standard, the symbol ν is sometimes used for frequency, v for propagation velocity, and y_0 for amplitude.)

23.19 The wave shown in Fig. 23-2 is being sent out by a 60-Hz vibrator. Find the following for the wave: (a) amplitude, (b) frequency, (c) wavelength, (d) speed, (e) period.

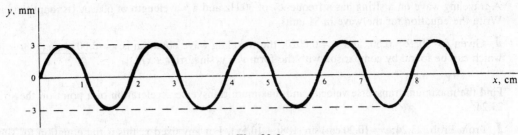

Fig. 23-2

▮ (a) Amplitude, $A = \underline{3\,mm} = \underline{0.3\,cm}$. (b) f is the same as the vibrator frequency, so $f = \underline{60\,Hz}$.
(c) λ = distance of one repeat of the wave (see Fig. 23-2). $\lambda = \underline{2\,cm}$. (d) Speed, $v = \lambda f = (2.0\,cm)(60\,Hz) = \underline{120\,cm/s}$. (e) Period, $T = 1/f = \underline{0.0167\,s}$.

23.20 For the wave $y = 5 \sin 30\pi[t - (x/240)]$, where x and y are in centimeters and t is in seconds, find the
(a) displacement when $t = 0$ and $x = 2\,cm$; (b) wavelength; (c) velocity of the wave; and
(d) frequency of the wave.

▮ (a)
$$y = 5 \sin 30\pi\left(0 - \frac{2}{240}\right) = 5\sin\left[\frac{-\pi}{4}\right] = 5(-0.707) = \underline{-3.535\,cm}$$

(b) Compare the given equation with (see Prob. 23.18) $y = R \sin 2\pi v(t - x/v)$. Then

$$30\pi = 2\pi v \qquad v = \underline{15\,Hz} \qquad \text{and} \qquad v = \underline{240\,cm/s}$$

Also, from $v = v\lambda$ we get $240 = 15\lambda$, and $\lambda = \underline{16\,cm}$. (c) $v = \underline{240\,cm/s}$ (d) $v = \underline{15\,Hz}$

23.21 A wave along a string has the following equation (x in meters and t in seconds):

$$y = 0.02 \sin (30t - 4.0x) \qquad m$$

Find its amplitude, frequency, speed, and wavelength.

▮ By comparison with the standard form $y = y_0 \sin [2\pi f(t - x/v)]$, we find $y_0 = \underline{0.02\,m}$, $f = \underline{4.78\,Hz}$, $v = \underline{7.5\,m/s}$; and since $\lambda = v/f$, $\lambda = \underline{1.57\,m}$.

23.22 The excess pressure in a traveling sound wave is given by the equation $p = 1.5 \sin \{[(2\pi)/\lambda](x - 330t)\}$, where x and λ are in meters, t is in seconds, and p is in pascals. (a) What is the velocity of the wave? (b) If $\lambda = 2\,m$, what is the frequency of the wave? (c) What is the maximum pressure (pressure amplitude)? (d) What is the pressure at $x = \frac{1}{6}\,m$ and $t = 0$?

▮ The standard equation of a wave is $y = y_0 \sin [2\pi v(t - x/v)]$. Rewrite the given equation in this form for comparison. $p = -1.5 \sin [2\pi(330/\lambda)(t - x/330)]$. (The minus sign is only a phase difference, see Prob. 23.18.)
(a) $v = \underline{330\,m/s}$. (b) $v = v\lambda$, $330 = v(2)$, $v = \underline{165\,Hz}$. (c) $p_0 = \underline{1.5\,Pa}$ (amplitude),

(d)
$$p = -1.5 \sin\left[2\pi \frac{330}{2}\left(0 - \frac{\frac{1}{6}}{330}\right)\right] = 1.5 \sin\frac{\pi}{6} = \underline{0.75\,Pa}.$$

23.23 For the wave shown in Fig. 23-3, find its amplitude, frequency, and wavelength if its speed is 300 m/s. Write the equation for this wave as it travels out along the $+x$ axis if its position at $t = 0$ is as shown.

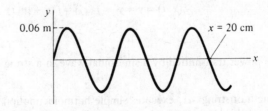

Fig. 23-3

▮ The amplitude = $\underline{0.06\,m}$ and $\frac{5}{2}\lambda = 20\,cm$, so $\lambda = \underline{0.080\,m}$; then $f = v/\lambda = 300/0.080 = \underline{3750\,Hz}$. Now $(2\pi)/\lambda = 78.5$ and $2\pi f = 23\,600$, so $y = \underline{0.06 \sin (78.5x - 23\,600t)}$ m.

23.24 A traveling wave on a string has a frequency of 30 Hz and a wavelength of 60 cm. Its amplitude is 2 mm. Write the equation for the wave in SI units.

▮ Given $f = 30$, $\lambda = 0.60$, and $y_0 = 0.0020$, the traveling wave equation is $y = \underline{0.0020 \sin (188t - 10.5x)}$ m, which can be found by comparison with the form $y = y_0 \sin [2\pi(ft - x/\lambda)]$.

23.25 Find the maximum transverse velocity and maximum transverse acceleration of a point on the string of Prob. 23.24.

▮ From Prob. 23.24, $y = (0.20 \text{ cm}) \sin (188t - 10.5x)$. For any fixed x, this is the equation for simple harmonic motion with amplitude $y_0 = 0.20$ cm and angular frequency $\omega = 188$ rad/s. Then $v_{y.\max} = \omega y_0 = \underline{37.6 \text{ cm/s}}$; $a_{y,\max} = \omega^2 y_0 = \underline{7070 \text{ cm/s}^2}$.

23.26ᶜ A traveling wave on a string obeys the following equation: $y = 0.27 \sin (12x - 500t)$ mm for x in meters and t in seconds. Find the equation for the transverse (a) velocity and (b) acceleration of the particle of string at $x = 20$ cm. (c) What is the displacement of this point when t is exactly 4 s?

▮ The transverse velocity of a particle of the string is $v = \partial y/\partial t = -0.27(500) \cos (12x - 500t)$ mm/s $= \underline{-0.135 \cos (2.4 - 500t)}$ m/s. (b) Transverse acceleration $a = \partial v/\partial t = -0.135(500) \sin (2.4 - 500t) = \underline{-67.5 \sin (2.4 - 500t)}$ m/s². (c) $y = 0.27 \sin [12(0.20) - 500(4.0)] = 0.27 \sin (-1997.6)$. The argument in radians divided by 2π yields -317.93 rev. But $-(0.93 \text{ rev})(360°/\text{rev}) = -334°$, so $y = 0.27 \sin 26° = \underline{0.118 \text{ mm}}$.

23.27ᶜ A sinusoidal wave traveling in the positive direction on a stretched string has amplitude 2.0 cm, wavelength 1.0 m, and wave velocity 5.0 m/s. At $x = 0$ and $t = 0$, it is given that $y = 0$ and $\partial y/\partial t < 0$. Find the wave function $y = f(x, t)$.

▮ We start with a general form for a rightward-moving wave:

$$y = A \cos \left[2\pi \left(\frac{x}{\lambda} - \frac{t}{T} \right) + \delta \right]$$

The amplitude given is $A = 2.0$ cm $= 0.020$ m, while the wavelength is given as $\lambda = 1.0$ m. The wave speed is 5.0 m/s, so $\lambda/T = 5.0$ m/s and $T = 0.20$ s. Therefore,

$$y = (0.020 \text{ m}) \cos [2\pi(x - 5.0t) + \delta]$$

We are told that for $x = 0$ and $t = 0$, $y = 0$ and $\partial y/\partial t < 0$. That is, $y = 0.020 \cos \delta = 0$ and $\partial y/\partial t = 0.2\pi \sin \delta < 0$. From these conditions, we may conclude that $\delta = (-\pi/2) + 2n\pi$, where n is an integer. Therefore

$$y = (0.020 \text{ m}) \cos \left[2\pi(x - 5t) - \frac{\pi}{2} \right] = \underline{(0.020 \text{ m}) \sin 2\pi(x - 5t)}$$

23.28 Derive directly the wave function $f(x, t) = f(x + |v| t)$ for a wave pulse traveling at speed $|v|$ in the negative x direction. Use an argument with an observer O' moving in the negative x direction.

▮ An observer O who watches a wave pulse traveling in the negative x direction with speed $|v|$ sees a wave function $y = f_l(x, t)$, where the subscript l indicates a leftward-moving wave. A second observer O', who is traveling to the left at speed $|v|$ with respect to O, sees a pulse whose location does not change with time. Therefore observer O' sees a wave function $y' = f(x')$, where x' is the coordinate of a point that is fixed with respect to a set of axes with origin at O'. Since O' travels to the left with respect to O at speed $|v|$, $x = x' - |v| t$, or $x' = x + |v| t$. Since the two observers are in relative motion along the x direction, the wave displacements that they measure are equal. That is, $y = y'$ and therefore

$$f_l(x, t) = y = y' = f(x') = f(x + |v| t)$$

as desired.

23.29 Prove that the average power transmitted by a sinusoidal wave in a string is $P_{\text{avg}} = \frac{1}{2}\omega^2 A^2 \mu c$, where μ is the mass per unit length.

▮ An infinitesimal length of string, dx, executes simple harmonic motion of amplitude A and angular frequency ω. The mass of this element is $\mu \, dx$, and from our study of simple harmonic motion its energy is equal to its maximum kinetic energy: $\frac{1}{2}(\mu \, dx)\omega^2 A^2$. The energy transmitted across a given imaginary boundary of string in a time t must supply all the elements of mass, $\mu \, dx$, beyond the boundary that are activated in that time. This corresponds to a length $L = ct$. Adding up the total energy in a length $L = ct$, we have $E = \frac{1}{2}\mu L\omega^2 A^2 = \frac{1}{2}\mu ct\omega^2 A^2$. To get the average power transmitted we divide by t:

$$P_{avg} = \frac{E}{t} = \frac{1}{2}\omega^2 A^2 \mu c$$

23.30 A certain 120-Hz wave on a string has an amplitude of 0.160 mm. How much energy exists in an 80-g length of the string? Assume that one wavelength of the string has a mass far smaller than 80 g.

▎ By Prob. 23.29, $E = \frac{1}{2}m\omega^2 A^2 = \frac{1}{2}(0.080 \text{ kg})(2\pi \times 120 \text{ s}^{-1})^2(16 \times 10^{-5} \text{ m})^2 = \underline{0.58 \text{ mJ}}$.

23.31 (a) If the frequency of the wave on a string shown in Fig. 23-3 is 150 Hz, what is its speed? (b) Assume that the string has a mass of 0.20 g/m. How much energy is sent down the string each second?

▎ (a) As in Prob. 23.23, $\lambda = 0.08$ m and so $v = \lambda f = \underline{12 \text{ m/s}}$. (b) $P = 2\pi^2 f^2 y_0^2 \mu v = 2\pi^2(150)^2(0.060)^2(0.00020)(12) = \underline{3.84 \text{ W}}$.

23.32^c Two wires having different densities are joined at $x = 0$ (see Fig. 23-4). An incident wave $y_i = A_i \sin(\omega t - k_1 x)$, traveling to the right in the wire $x \le 0$ is partly reflected and partly transmitted at $x = 0$. Find the reflected and transmitted amplitudes in terms of the incident amplitude.

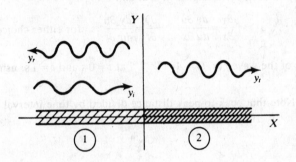

Fig. 23-4

▎ The reflected and transmitted waves will have the forms

$$y_r = A_r \sin(\omega t + k_1 x) \qquad y_t = A_t \sin(\omega t - k_2 x)$$

where A_r may possibly be negative (corresponding to a 180° phase change upon reflection). The boundary conditions at $x = 0$ are that the displacement y and the slope $\partial y/\partial x$ be continuous. Thus:

$$y_i\,|_{x=0} + y_r\,|_{x=0} = y_t\,|_{x=0} \qquad A_i + A_r = A_t \tag{1}$$

and

$$\left.\frac{\partial y_i}{\partial x}\right|_{x=0} + \left.\frac{\partial y_r}{\partial x}\right|_{x=0} = \left.\frac{\partial y_t}{\partial x}\right|_{x=0} \qquad -k_1 A_i + k_1 A_r = -k_2 A_t \tag{2}$$

Solving (1) and (2) simultaneously,

$$A_r = \frac{k_1 - k_2}{k_1 + k_2} A_i \qquad A_t = \frac{2k_1}{k_1 + k_2} A_i$$

It is seen that A_r is indeed negative when $k_2 > k_1$, which will be the case if wire 2 is denser than wire 1.

23.33^c Suppose that a transverse sinusoidal wave is traveling along a string. Prove that at any time t the slope $\partial y/\partial x$ of the string at any point x is equal to the negative of the instantaneous transverse velocity $\partial y/\partial t$ of the string at x divided by the wave velocity v. That is, show that $\partial y/\partial x = -(\partial y/\partial t)/v$.

▎ The general transverse sinusoidal traveling wave is given by

$$y = A \cos[k(x - vt) + \delta] \tag{1}$$

Here $k = (2\pi)/\lambda$ is the wave number, v is the vave velocity, and $k\,|v|$ is the angular frequency. The slope at point x at time t is

$$\frac{\partial y}{\partial x} = -kA \sin[k(x - vt) + \delta] \tag{2}$$

while the instantaneous transverse velocity is

$$\frac{\partial y}{\partial t} = kvA \sin \left[k(x - vt) + \delta \right] \tag{3}$$

Comparing Eqs. (2) and (3), we see that

$$\frac{\partial y}{\partial x} = -\frac{1}{v}\frac{\partial y}{\partial t} \tag{4}$$

as desired.

23.34ᶜ Give the general mathematical form of a wave traveling at constant speed, without dissipation, along the x axis, and show that it satisfies the standard wave equation.

▌ A wave $y(x, t)$ traveling along X with an unchanging form and speed v is given by $y = f(x \pm vt)$, where the minus and plus signs refer to wave propagation in the positive and negative X directions, respectively.
  These functions satisfy the one-dimensional *wave equation*

$$\frac{\partial^2 y}{\partial x^2} = \frac{1}{v^2}\frac{\partial^2 y}{\partial t^2}$$

as can immediately be seen by setting $u = x \pm vt$ and using

$$\frac{\partial y}{\partial x} = \frac{dy}{du}\frac{\partial u}{\partial x} \qquad \frac{\partial y}{\partial t} = \frac{dy}{du}\frac{\partial u}{\partial t} \qquad \text{for either choice of } u$$

23.35 Sketch the profile of the wave $f(x, t) = Ae^{-B(x-vt)^2}$ at $t = 0$ s and $t = 1$ s, using $A = 1.0$ m, $B = 1.0$ m^{-2}, and $v = +2.0$ m/s.

▌ See Fig. 23-5. Note that peak-to-peak distance divided by time interval equals v.

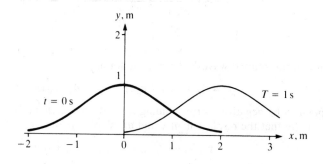

Fig. 23-5

23.36ᶜ Verify by partial differentiation that the wave function of Prob. 23.35 satisfies the one-dimensional wave equation.

▌ We begin with the proposed solution

$$f(x, t) = Ae^{-B(x-vt)^2} \tag{1}$$

Differentiating with respect to x, we find that

$$\frac{\partial f}{\partial x} = -2AB(x - vt)e^{-B(x-vt)^2} \tag{2}$$

and

$$\frac{\partial^2 f}{\partial x^2} = [-2AB + 4AB^2(x - vt)^2]e^{-B(x-vt)^2} \tag{3}$$

Differentiating with respect to t, we find that

$$\frac{\partial f}{\partial t} = 2ABv(x - vt)e^{-B(x-vt)^2} \tag{4}$$

The second derivative with respect to time is given by

$$\frac{\partial^2 f}{\partial t^2} = v^2 [-2AB + 4AB^2(x - vt)^2] e^{-B(x-vt)^2}$$

(5)

Comparing Eqs. (3) and (5), we see that

$$\frac{\partial^2 f}{\partial x^2} = \frac{1}{v^2} \frac{\partial^2 f}{\partial t^2}$$

(6)

which is the wave equation.

23.2 STANDING WAVES AND RESONANCE

23.37 Find the resultant disturbance in a long wire under tension when two transverse waves, of equal amplitudes and frequencies and traveling in opposite directions, pass through each other. Discuss your result.

▌ Let $y_1 = A \sin(\omega t - kx)$ and $y_2 = A \sin(\omega t + kx)$. (We lose no generality by ignoring a possible constant phase difference, ϕ, between the waves, since the waves are continually passing each other. The phase ϕ could only indicate a spatial shift in the origin for x.) Superimposing the two waves we have $y_T = y_1 + y_2$, or

$$y_T = A[\sin(wt - kx) + \sin(\omega t + kx)] = A[\sin \omega t \cos kx - \cos \omega t \sin kx + \sin \omega t \cos kx + \cos \omega t \sin kx]$$

$$= 2A \sin \omega t \cos kx$$

The disturbance y_T is called a *standing wave* because the wave form does not travel along the x axis, indeed there are points (when $\cos kx = 0$) where the string has zero amplitude for any time t. Since these occur for $kx = 0, \pm\pi, \pm2\pi, \ldots$, adjacent zeros are a distance Δx apart given by $k \Delta x = \pi$, or $\Delta x = \pi/k = \lambda/2$. These stationary points are called *nodes*. For any x the particle of the string executes simple harmonic motion with amplitude: $(2A \cos kx)$. Maximum amplitudes occur midway between the nodes, at points called *antinodes*. Between any two nodes all the points move in phase.

23.38 Show that two superimposed waves of the same frequency and amplitude traveling in the same direction cannot give rise to a standing wave.

▌ Using the expression for the sum of two sines, we have for two traveling waves $y_0 \sin(\omega t - kx) + y_0 \sin(kx - \omega t + \phi) = 2y_0 \cos(\phi/2) \sin(\omega t - kx + \phi/2)$, which is a traveling wave of amplitude $2|y_0 \cos(\phi/2)|$.

23.39 A wave given by $y_1 = A \sin[\omega t - (\omega x)/v]$ is sent down a string. Upon reflection it becomes $y_2 = -\frac{1}{2}A \sin[\omega t + (\omega x)/v]$. Show that the resultant of these two waves on the string can be written as a combination of a standing wave and a traveling wave.

▌ Let $y = y_1 + y_2$. We find, after dividing y_1 into two equal parts, $y = (A/2) \sin[\omega t - (\omega x)/v] + (A/2) \sin[\omega t - (\omega x)/v] - (A/2) \sin[\omega t + (\omega x)/v]$ or $y = -A \sin(\omega x/v) \cos(\omega t) + (A/2) \sin[\omega t - (\omega x)/v]$. The first term is the standing wave and the second is the traveling wave.

23.40 Sources separated by 20 m vibrate according to the equations

$$y_1' = 0.06 \sin \pi t \quad \text{(m)} \qquad y_2' = 0.02 \sin \pi t \quad \text{(m)}$$

They send out waves along a rod of speed 3 m/s. What is the equation of motion of a particle 12 m from the first source and 8 m from the second?

▌ Refer to Fig. 23-6. Let source *1* send out waves in the $+X$ direction, so that

$$y_1 = A_1 \sin 2\pi f_1 \left(t - \frac{x_1}{v}\right)$$

and let source *2* send out waves in the $-X$ direction, so that

$$y_2 = A_2 \sin 2\pi f_2 \left(t + \frac{x_2}{v}\right)$$

Here, $v = 3$ m/s; x_1, x_2 are measured from the respective sources, so equating y_1 at $x_1 = 0$ to y_1', and y_2 at $x_2 = 0$ to y_2', we obtain

$$A_1 = 0.06 \text{ m} \qquad f_1 = f_2 = \tfrac{1}{2}\text{Hz} \qquad A_2 = 0.02 \text{ m}$$

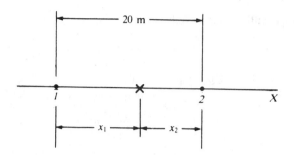

Fig. 23-6

The resultant disturbance at the point $x_1 = 12$ m, $x_2 = -8$ m is then

$$y = y_1 + y_2 = 0.06 \sin \pi\left(t - \frac{12}{3}\right) + 0.02 \sin \pi\left(t - \frac{8}{3}\right) = 0.06 \sin \pi t + 0.02 \sin\left(\pi t - \frac{2\pi}{3}\right)$$

$$= 0.06 \sin \pi t + 0.02\left(\sin \pi t \cos \frac{2\pi}{3} - \cos \pi t \sin \frac{2\pi}{3}\right)$$

$$= 0.06 \sin \pi t + 0.02\left[(\sin \pi t)\left(-\frac{1}{2}\right) - (\cos \pi t)\left(\frac{\sqrt{3}}{2}\right)\right]$$

$$= \underline{0.05 \sin \pi t - 0.0173 \cos \pi t}$$

23.41 Describe the nature of, and conditions for, *resonant standing waves.*

❚ When a wave is repeatedly reflected back and forth on a string, or in other similar situations, a standing wave (Prob. 23.37) of large amplitude is often produced. Such a large-amplitude wave, caused by multiple reinforcement, is called a resonant standing wave. A resonant standing wave results only if the wavelength has certain special values compared with the path length for the wave. In particular, for a string of length L, pinned at both ends, a whole number of half-wavelengths must equal L. It is common to use the shortened expression "standing waves" when referring to resonant standing waves.

A standing wave on a string causes the string to sweep out a pattern such as the one shown in Fig. 23-7. The string vibrates between the limits indicated. Points A, B, C, D are called *displacement nodes*. A displacement node is a point on a standing wave where the amplitude of vibration is zero. Points F, G, H, I, J are *displacement antinodes*. A displacement antinode is a point on a standing wave where the amplitude of vibration is greatest. The distance between two adjacent nodes (or antinodes) is a half wavelength. The region between two nodes is called a *segment.*

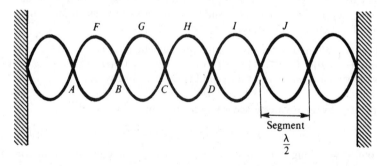

Fig. 23-7

23.42 A wire 0.5 m long and with a mass per unit length of 0.0001 kg/m vibrates under a tension of 4 N. Find the fundamental frequency.

❚

$$v = \sqrt{\frac{T_s}{\mu}} = \sqrt{\frac{4}{0.0001}} = 200 \text{ m/s}$$

A standing wave is of the fundamental frequency if one half-wavelength occupies the length of the wire; i.e., if $\lambda = 1.0$ m. Thus

$$v = \frac{v}{\lambda} = \frac{200 \text{ m/s}}{1.0 \text{ m}} = \underline{200 \text{ Hz}}$$

23.43 One end of a rubber tube that is 5.0 m long and has a mass-to-length ratio of 0.30 kg/m is fastened to a fixed support; at the other end, a tension of 100 N is applied. If a transverse blow is struck at one end of the tube, how long does it take to reach the other end?

$$\blacksquare \qquad v = \sqrt{\frac{T_s}{\mu}} = \sqrt{\frac{100}{0.30}} = 18.3 \text{ m/s} \qquad s = vt \qquad \text{so} \qquad 5.0 = 18.3t \qquad \text{and} \qquad t = \underline{0.27 \text{ s}}$$

23.44 Refer to Prob. 23.43. What frequency of vibration must be applied to the tube to produce a standing wave with four segments in the length of the tube? (This frequency is called the *fourth harmonic frequency*.)

    ■ We have $4(\lambda/2) = 5.0$ m, or $\lambda = 2.5$ m. Then $v = v/\lambda = 18.3/2.5 = \underline{7.3 \text{ Hz}}$.

23.45 Standing waves are produced in a rubber tube 12 m long. If the tube vibrates in five segments and the velocity of the wave is 20 m/s, what is (*a*) the wavelength of the waves, (*b*) the frequency of the waves?

    ■ (*a*)
$$5\left(\frac{\lambda}{2}\right) = 12 \text{ m} \qquad \text{or} \qquad \lambda = \underline{4.8 \text{ m}}$$

    (*b*)
$$v = \frac{20 \text{ m/s}}{4.8 \text{ m}} = \underline{4.17 \text{ Hz}}$$

23.46 A string has a length of 0.4 m and a mass of 0.16 g. If the tension in the string is 70 N, what are the three lowest frequencies it produces when plucked?

$$\blacksquare \qquad v = \sqrt{\frac{T_s}{\mu}} = \sqrt{\frac{70}{(0.00016/0.4)}} = \sqrt{175\,000} = 418 \text{ m/s}.$$

The fundamental frequency v_1 corresponds to a wavelength of $2(0.4) = 0.8$ m; thus,

$$v_1 = \frac{418 \text{ m/s}}{0.8 \text{ m}} = \underline{523 \text{ Hz}}$$

The second and third harmonic frequencies are then $2v_1 = \underline{1046 \text{ Hz}}$ and $3v_1 = \underline{1569 \text{ Hz}}$.

23.47 The third overtone produced by a vibrating string 2 m long is 1200 Hz. What are the frequencies of the lower overtones and of the fundamental? What is the velocity of propagation?

    ■ The third overtone is the fourth harmonic v_4. $v_4 = 1200$ Hz implies that $v_1 = 1200/4 = \underline{300 \text{ Hz}}$. Then $v_2 = 2v_1 = \underline{600 \text{ Hz}}$, and $v_3 = 3v_1 = \underline{900 \text{ Hz}}$. $v = \lambda_4 v_4 = (1 \text{ m})(1200 \text{ Hz}) = \underline{1200 \text{ m/s}}$ (or $v = \lambda_1 v_1 = (4 \text{ m})(300 \text{ Hz}) = 1200 \text{ m/s}$).

23.48 A 160-cm-long string has two adjacent resonances at frequencies of 85 and 102 Hz. (*a*) What is the fundamental frequency of the string? (*b*) What is the length of a segment at the 85-Hz resonance? (*c*) What is the speed of the waves on the string?

    ■ (*a*) The fundamental is $f_1 = v/(2L)$, the nth harmonic frequency is $nf_1 = 85$ Hz while the next harmonic frequency is $(n + 1)f_1 = 102$ Hz. From these we obtain $f_1 = \underline{17 \text{ Hz}}$ and $n = 5$. (*b*) For $n = 5$, there are five segments; the length of each is $160/5 = \underline{32 \text{ cm}}$. (*c*) From $f_1 = 17 = v/[2(1.6)]$, $v = \underline{54.4 \text{ m/s}}$.

23.49 A vertically suspended 200-cm length of string is given a tension equal to the weight of an 800-g mass. The string is found to resonate in three segments to a frequency of 480 Hz. What is the mass per unit length of the string?

    ■ In the third harmonic, $(3\lambda)/2 = 2.00$ m; so that $v = f\lambda = 480(1.33) = 640$ m/s. We then use the relation $v^2 = T/\mu$ to find $\mu = (mg)/v^2 = [0.800(9.80)]/640^2 = \underline{1.91 \times 10^{-5} \text{ kg/m}}$.

23.50 The equation for a particular standing wave on a string is $y = 0.15 \, (\sin 5x \cos 300t)$ m. Find the (*a*) amplitude of vibration at the antinode, (*b*) distance between nodes, (*c*) wavelength, (*d*) frequency (*e*) speed of the wave.

    ■ (*a*) By comparison with $y_0 = A \sin [(2\pi x)/\lambda] \cos (2\pi ft)$ we have $A = \underline{0.15 \text{ m}}$. (*b*) When the argument of the sine $= 0, \pi, 2\pi, \ldots$, we have nodes. Since $x = 0$ is a node we have for the next node $5x = \pi$ and $x = \pi/5 = \underline{0.628 \text{ m}}$. (*c*) The wavelength is twice this, so $\lambda = (2\pi)/5 = \underline{1.26 \text{ m}}$. (*d*) Since $2\pi f = 300$, $f = 150/\pi = \underline{47.7 \text{ Hz}}$ and (*e*) $v = \lambda f = [(2\pi)/5](150/\pi) = \underline{60 \text{ m/s}}$.

23.51 An organ pipe 1 ft long is open at both ends. If the velocity of sound is 1100 ft/s, what are the frequencies of the fundamental and of the first two overtones?

▌ As can be seen from Fig. 23-8, $n(\lambda/2) = L$ holds for an open organ pipe as well as a string. Thus, the fundamental frequency is $\nu_1 = v/2L = (1100 \text{ ft/s})/2 \text{ ft} = \underline{550 \text{ Hz}}$. The first overtone is the second harmonic, and $\nu_2 = 2\nu_1 = 2(550) = \underline{1100 \text{ Hz}}$. Similarly $\nu_3 = 3\nu_1 = 3(550) = \underline{1650 \text{ Hz}}$ (second overtone).

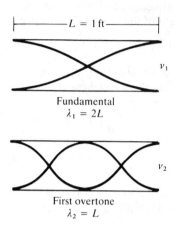

$\vdash L = 1 \text{ ft} \dashv$

ν_1

Fundamental
$\lambda_1 = 2L$

ν_2

First overtone
$\lambda_2 = L$

Fig. 23-8

23.52 A closed organ pipe 2.5 ft long is sounded. If the velocity of the sound is 1100 ft/s, what are the fundamental frequency and the first two overtones?

▌ For a "closed" pipe (that is, open at one end) $(2n-1)(\lambda/4) = L$, as can be seen from Fig. 23-9. Hence $\nu_{2n-1} = (2n-1)v/4L = (2n-1)\nu_1$, so only the odd harmonics appear. For our case, $\lambda_1 = 4L = 4(2.5) = 10$ ft. $v = \nu_1\lambda_1$ or $1100 = \nu_1(10)$ and $\nu_1 = \underline{110 \text{ Hz}}$. The first overtone is the third harmonic, so $\lambda_3 = 4L/3$ and $\nu_3 = v/\lambda_3 = 3v/4L = 3\nu_1 = 3(110) = \underline{330 \text{ Hz}}$. Similarly, $\nu_5 = 5\nu_1 = \underline{550 \text{ Hz}}$.

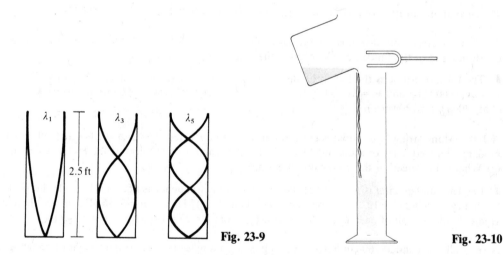

λ_1 λ_3 λ_5

2.5 ft

Fig. 23-9

Fig. 23-10

23.53 A sounding tuning fork whose frequency is 256 Hz is held over an empty measuring cylinder. See Fig. 23-10. The sound is faint, but if just the right amount of water is poured into the cylinder, it becomes loud. If the optimal amount of water produces an air column of length 0.31 m, what is the speed of sound in air to a first approximation?

▌ The loudest sound will be heard at resonance, when the frequency of vibration of the air column in the cylinder is the same as that of the tuning fork. Since the air column is open at one end and closed at the other, we conclude that the wavelength of the vibration is four times the length of the column: $\lambda = 4L = (4)(0.31 \text{ m}) = 1.24$ m. Here we have assumed that the observed resonant oscillation of the air column is its *fundamental* oscillation. Since the frequency $\nu = 256$ Hz, the sound speed $v = \nu\lambda = \underline{317 \text{ m/s}}$. This is an underestimate since the displacement antinode (or the pressure node) which is located a distance of one-quarter wavelength from the water surface actually lies outside the top of the cylinder, somewhat more than 0.31 m from the water.

23.54 Sound waves of frequency 320 Hz are sent into the top of a vertical tube containing water at a level that can be adjusted. If standing waves are produced at two successive water levels—20 cm and 73 cm—what is the speed of the sound waves in the air of the tube?

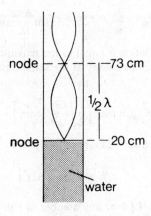

Fig. 23-11

❚ As Fig. 23-11 shows, the distance between water levels is the distance between successive nodes, or half of a wavelength.

$$\frac{\lambda}{2} = 73 - 20 = 53 \text{ cm} \quad \text{and} \quad \lambda = 106 \text{ cm} = \underline{1.06 \text{ m}} \quad v = v\lambda = 320 \times 1.06 = \underline{339 \text{ m/s}}$$

Note how this technique avoids the open-end effect discussed in Prob. 23.53, where the antinode is somewhat beyond the opening.

23.55 A 40-cm-long brass rod is dropped one end first onto a hard floor but is caught before it topples over. With an oscilloscope it is determined that the impact produces a 3-kHz tone. What is the speed of sound in brass?

❚ The brass rod is "open" at both ends, so the longitudinal wave will have $L = n(\lambda/2) = \underline{n[v/(2f)]}$. For the fundamental, $0.40 = (v/6000)$, from which $v = \underline{2400 \text{ m/s}}$.

23.56 A metal bar clamped at its center resonates in its fundamental to longitudinal waves of frequency 4 kHz. What will be its fundamental resonance frequency and its first two overtone frequencies when the clamp is moved to one end?

❚ With clamp in the center $L = \lambda/2$ for the fundamental, so $f = v/(2L) = 4$ kHz. Clamp on the end means $L = \lambda/4, 3\lambda/4, 5\lambda/4, \ldots, (2n-1)(\lambda/4)$ for $n = 1, 2, \ldots$ The frequencies $f_n = (2n-1)[v/(4L)] = (2n-1)(2 \text{ kHz})$, so $f_1 = \underline{2}$, $f_2 = \underline{6}$, and $f_3 = \underline{10 \text{ kHz}}$.

23.57 Write the equation for the fundamental standing sound wave in a tube that is closed at both ends if the tube is 80 cm long and the speed of the wave is 330 m/s. Represent the amplitude of the wave at an antinode by s_0. Repeat for the next two higher resonance frequencies.

❚ Here we have displacement nodes at both ends, as for a string. From $L = (n\lambda)/2$ we have $f_n = n[v/(2L)]$ so $\lambda_n = (2L)/n = (1.6/n)$m and $f_n = n(330/1.6) = (206n)$Hz. The standing-wave equation for nodes at each end is $s = s_0 \sin[(2\pi x)/\lambda_n] \cos(2\pi f_n t)$. The coefficients of x and t are $2\pi/\lambda_n = 3.93n$ and $2\pi f_n = 1295n$ leading to $s = s_0 \sin(3.93nx) \cos(1295nt)$, with $n = 1$ for the fundamental and $n = 2$ and 3 for the next two resonances.

23.58 A rather stiff wire is bent into a circular loop of diameter D. It is clamped by knife edges at two points opposite each other. A transverse wave is sent around the loop by means of a small vibrator which acts close to one clamp. Find the resonance frequencies of the loop in terms of the wave speed v and diameter D.

❚ Supports for the loop cause nodes at two points; for half the loop, $(\pi D)/2 = (n\lambda)/2$. Use $\lambda = v/f$, to find $f = n[v/(\pi D)]$.

23.59ᶜ A uniform string (length L, linear density μ, and tension F) is vibrating with amplitude A_n in its nth mode. Show that its total energy of oscillation is given by $E = \pi^2 v_n^2 A_n^2 \mu L$.

▌ The displacement of the string is given by

$$y(x, t) = A_n \sin \frac{n\pi x}{L} \cos (2\pi v_n t + \delta)$$

so the transverse velocity is

$$\frac{\partial y}{\partial t} = -2\pi v_n A_n \sin \frac{n\pi x}{L} \sin (2\pi v_n t + \delta)$$

The string's total energy of oscillation is equal to the maximum kinetic energy. (Note that *all* points on the string achieve their maximum kinetic energy at the same time, when $y = 0$ for all x.) Since $dm = \mu \, dx$, we have

$$E = K_{max} = \max \left[\frac{1}{2} \int \left(\frac{\partial y}{\partial t}\right)^2 dm\right] = \max \left[\frac{1}{2} \int_0^L \mu \left(\frac{\partial y}{\partial t}\right)^2 dx\right]$$

The maximum value occurs when $\sin^2 (2\pi v_n t + \delta) = 1$, so we find that

$$E = \frac{\mu}{2} (2\pi v_n A_n)^2 \int_0^L \sin^2 \frac{n\pi x}{L} \, dx$$

The average value of $\sin^2 [(n\pi x)/L]$ over any number of half cycles is given by $\frac{1}{2}$, so the integral has the value $L/2$. Therefore $E = 2\pi^2 \mu v_n^2 A_n^2 (L/2) = \underline{\pi^2 v_n^2 A_n^2 \mu L}$, as desired. [Compare with the result for a traveling wave, Prob. 23.29.]

23.60 A taut square membrane 85 cm on a side is fastened to a rigid frame on its edges. When tapped lightly at its center, it gives off a tone of about 200 Hz. Assuming it to be resonating in the mode shown in Fig. 23-12, what is the wave speed in the membrane?

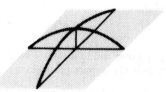

Fig. 23-12

▌ The resonant frequencies of a rectangular drumhead are given by the formula

$$f_{mn} = \frac{v}{2} \sqrt{\left(\frac{m}{L_x}\right)^2 + \left(\frac{n}{L_y}\right)^2}$$

where m and n are the respective numbers of half-wavelengths that fit into the dimensions L_x and L_y. Figure 23-12 shows the mode $m = n = 1$; hence,

$$200 = \frac{v}{2} \sqrt{\frac{2}{(0.85)^2}}$$

giving $\underline{v = 240 \text{ m/s}}$.

24.1 SOUND VELOCITY; BEATS; DOPPLER SHIFT

24.1 Helium is a monatomic gas that has a density of 0.179 kg/m^3 at a pressure of 76 cm of mercury and a temperature of 0 °C. Find the speed of compressional waves (sound) in helium at this temperature and pressure.

▋ $v = \sqrt{B/\rho}$, where B is the adiabatic bulk modulus. For an ideal gas, $B = \gamma p$ (see Prob. 21.24), and $\gamma = 1.67$ for a monatomic gas. Then

$$v = \sqrt{\frac{1.67(1.013 \times 10^5) \text{ N/m}^2}{0.179 \text{ kg/m}^3}} = \underline{972 \text{ m/s}}$$

24.2 Using the fact that hydrogen gas consists of diatomic molecules with $M = 2 \text{ kg/kmol}$, find the speed of sound in hydrogen at 27 °C.

▋ For ideal gases $v = [(\gamma p)/\rho]^{1/2} = [(\gamma RT)/M]^{1/2}$, so

$$v = [(1.40)(8314 \text{ J/kmol} \cdot \text{K})(300 \text{ K})/(2 \text{ kg/kmol})]^{1/2} = \underline{1321 \text{ m/s}}$$

24.3 From the fact that the molecular weight of oxygen molecules is 32 kg/kmol, find the speed of sound in oxygen at 0 °C.

▋ As in Prob. 24.2, $v = [(1.40)(8314)(273)/32]^{1/2} = \underline{315 \text{ m/s}}$

24.4 The velocity of sound in a container of hydrogen at −73 °C is approximately 4000 ft/s. What would the velocity be (in ft/s) if the temperature of the hydrogen were raised to 127 °C without a change in volume?

▋ Since $v = [(\gamma RT)/M]^{1/2}$ we have, assuming constant γ, $v = v_0\sqrt{T/T_0} = (4000 \text{ ft/s})\sqrt{(400 \text{ K})/(200 \text{ K})} = \underline{5657 \text{ ft/s}}$.

24.5 What is the speed of sound in air when the temperature is 35 °C? The speed of sound in air at 0 °C is 331 m/s.

▋ $v \propto \sqrt{T} = \sqrt{273 + t}$; so if v_0 = speed at $t = 0$ °C, $v = v_0(1 + t/273)^{1/2} = 331(1 + 35/273)^{1/2} = \underline{351.6 \text{ m/s}}$.

24.6 By how much must the temperature of air near 0 °C be changed to cause the speed of sound in it to change by 1 percent?

▋

$$\frac{(273 + t)^{1/2} - (273^{1/2})}{273^{1/2}} = 0.01$$

Noting $[1 + t/273]^{1/2} \approx 1 + t/546$, we get $(t/546)(100) = 1.0$, which leads to $t = \underline{5.5 \text{ °C}}$.

24.7 A certain gas mixture is composed of two diatomic gases (molecular weights M_1 and M_2). The ratio of the masses of the two gases in a given volume is $m_2/m_1 = r$. Show that the speed of sound in the gas mixture is as follows if the gases are ideal:

$$v = \sqrt{\frac{1.40RT}{M_1 M_2} \frac{M_2 + rM_1}{1 + r}}$$

▋ $v = [(\gamma p)/\rho]^{1/2}$, but $\rho = m/V = (m_1 + m_2)/V$, and (Dalton's law) $p = p_1 + p_2 = (n_1 + n_2)[(RT)/V)]$, with $n_1 = m_1/M_1$ and $n_2 = m_2/M_2$, so that $p/\rho = [(m_1/M_1 + m_2/M_2)RT]/(m_1 + m_2) = [(1/M_1 + r/M_2)RT]/(1 + r)$, with $r = m_2/m_1$. Substitute p/ρ and $\gamma = 1.40$ into the expression for v to find the required result.

24.8 Suppose (with Newton) that the compressional vibration in a gas were isothermal in character rather than adiabatic. Find the expression equivalent to $(\gamma p/\rho)^{1/2}$ for the speed of sound in that case.

▋ For pV a constant, $\Delta(pV) = p \Delta V + V \Delta p = 0$, or $\Delta V/\Delta p = -V/p$. But $\Delta V/\Delta p = -V/B$ from the definition of B. Therefore, $B = p$ and $v = (B/\rho)^{1/2} = (p/\rho)^{1/2}$. Thus $v_{ad}/v_{iso} = \gamma^{1/2}$.

24.9 A 1000-Hz sound wave in air strikes the surface of a lake and penetrates into the water. What are the frequency and wavelength of the wave in water? Assume that the speed of sound in water is 1500 m/s.

▌ The number of complete waves passing any point in air and in water in unit time is the same, so $f = 1000$ Hz for both media. Therefore,

$$\lambda_w = \frac{v_w}{f} = \frac{1500 \text{ m/s}}{1000 \text{ s}^{-1}} = \underline{1.5 \text{ m}}$$

24.10 An underwater sonar source operating at a frequency of 60 kHz directs its beam toward the surface. What is the wavelength of the beam in the air above? What frequency sound due to the sonar source does a bird flying above the water hear? Assume that v in air = 330 m/s.

▌ As indicated in Prob. 24.9, f remains constant—in this case at 60 kHz. The wavelength = v/f = $330/(6.0 \times 10^4) = \underline{5.5 \text{ mm}}$.

24.11 Define *pitch, loudness, quality, decibel, reverberation time, interference, beats, Doppler effect, supersonic velocity, shock wave,* and *Mach number.*

▌ *Pitch* is a sound characteristic that depends on the frequency of the fundamental. Higher pitch means higher frequency.

Loudness refers to the strength of the auditory sensation produced by a sound. It depends on the intensity and frequency of the sound.

The *quality* (or *timbre*) of a sound depends on the number and intensities of the overtones.

Decibel (dB) is the unit of intensity-level, *n*, of sound. One dB is ten times the log of an intensity ratio of 1.26:1. The equation for *n* is $n = 10 \log (I/I_0)$, where $I_0 = 10^{-12}$ W/m^2.

The *reverberation time* of a room is the time required for the sound level to fall 60 dB after the sound source is shut off; that is, for the intensity to decrease by a factor of a million.

Interference is the superposition of sound waves to produce either constructive or destructive addition.

Beats are fluctuations in sound intensity that occur when there is interference between two sound waves of equal intensity and slightly different frequency.

The *Doppler effect* is the apparent change in frequency as the source of a sound and an observer of the sound move relative to each other.

A *supersonic velocity* is one that is greater than the local velocity of sound.

A *shock wave* is the wave motion accompanying objects traveling at supersonic speeds.

Mach number is the ratio of the velocity of an object, or of a shock front, to the local velocity of sound.

24.12 Two closed organ pipes sounded simultaneously give five beats per second between the fundamentals. If the shorter pipe is 1.1 m long, find the length L of the longer pipe. Assume that v in air = 340 m/s.

▌ Each pipe is occupied by a quarter-wave (refer to Prob. 23.52); hence,

$$5 \text{ Hz} = v_1 - v_2 = \frac{340 \text{ m/s}}{4.4 \text{ m}} - \frac{340 \text{ m/s}}{4L}$$

Solving, $L = \underline{1.18 \text{ m}}$.

24.13 Two open organ pipes, one 2.5 ft and one 2.4 ft in length, are sounded simultaneously. How many beats per second will be produced between the fundamental tones if the velocity of the sound is 1100 ft/s?

▌ Each pipe is occupied by a half-wave (see Prob. 23.51), so

$$v_1 - v_2 = \frac{1100 \text{ ft/s}}{4.8 \text{ ft}} - \frac{1100 \text{ ft/s}}{5.0 \text{ ft}} = \underline{9 \text{ Hz}}$$

24.14 Four beats per second are heard when two tuning forks are sounded simultaneously. After attaching a small piece of tape to one prong of the second tuning fork, the two tuning forks are sounded again and two beats per second are heard. If the first fork has a frequency of 180 Hz, what must the original frequency of the second fork have been?

▌ The frequency of the second fork must be higher than that of the first fork or adding the tape would have *increased* the number of beats. Therefore, $v_2 - 180 = 4$ or $v_2 = \underline{184 \text{ Hz}}$.

24.15^c Some of the low keys of the piano have two strings. On a particular key one of the strings is tuned correctly to 100 Hz. When the two strings are sounded together, one beat per second is heard. By what percent must a piano tuner change the tension of the untuned string to make it match perfectly? (The beating is between the fundamental tones.)

▌ The fundamental frequency is $v = (\sqrt{F/\mu})/(2L)$. We assume that the two strings are identical in length, composition, and diameter, so the difference Δv is due to a tension difference ΔF. From the above equation, we have

$$\frac{dv}{dF} = \frac{1}{2L}\frac{1}{2}\sqrt{\frac{1}{F\mu}} = \frac{v}{2F}$$

so that for $|\Delta v| \ll v$, we have $\Delta v/v = \frac{1}{2}(\Delta F/F)$. In the present case, $v = 100$ Hz and $|\Delta v| = 1$ Hz. Therefore $|\Delta F|/F = 2|\Delta v|/v = 2(1/100) = \underline{2 \text{ percent}}$. (If the untuned string is "flat," its tension should be increased; if the string is "sharp," its tension should be lowered.)

24.16 A train is moving toward an observer with a speed of 100 ft/s (68 mi/h). The whistle of the locomotive has a frequency of 400 Hz, and the speed of the sound is 1100 ft/s. Find the frequency heard by the observer.

▌ Use the Doppler-effect equation:

$$\frac{v_L}{v + v_L} = \frac{v_s}{v - v_s}$$

where v_L = velocity of listener relative to medium toward source, and v_s = velocity of source relative to medium toward listener, v = speed of sound in medium. Since the listener is not moving, v_L is zero.

$$\frac{v_L}{1100} = \frac{400}{1100 - 100} \qquad v_L = \frac{400(1100)}{1000} = \underline{440 \text{ Hz}}$$

24.17 A car is traveling at a speed of 90 ft/s (61 mi/h) along a road paralleling a railroad track. Practically straight in front of the car is a locomotive waiting on a siding. If the speed of the sound is 1080 ft/s and the driver of the car hears a frequency of 400 Hz when the locomotive whistle is blown, what is the actual frequency emitted by the whistle?

▌ $\dfrac{v_L}{v + v_L} = \dfrac{v_s}{v - v_s}$ $\quad v_s = 0$ $\quad$ Then $\quad \dfrac{400}{1080 + 90} = \dfrac{v_s}{1080}$ $\quad$ and $\quad v_s = \dfrac{400(1080)}{1170} = \underline{369 \text{ Hz}}$

24.18 A hawk is flying directly away from a birdwatcher and directly toward a distant cliff at a speed of 15 m/s. The hawk produces a shrill cry whose frequency is 800 Hz. (*a*) What is the frequency in the sound that the birdwatcher hears directly from the bird? (*b*) What is the frequency that the birdwatcher hears in the echo that is reflected from the cliff?

▌ (*a*) We let v represent the emitted frequency, $|v|$ represent the sound speed, and $|v_s|$ represent the source speed. Since the bird is flying directly away from the observer, the received frequency v' is given by

$$v' = v\frac{|v|}{|v| + |v_s|}$$

Adopting $|v| = 340$ m/s, and using the given values $|v_s| = 15$ m/s and $v = 800$ Hz, we find

$$v' = \frac{(800)(340)}{340 + 15} = \underline{766 \text{ Hz}}$$

(*b*) Whatever frequency is incident upon the cliff is reflected without change. Therefore, the observer will receive the same frequency in the echo that another observer on the cliff would hear directly. The frequency v'' in the echo is therefore given by

$$v'' = v\frac{|v|}{|v| - |v_s|} = \frac{(800)(340)}{(340 - 15)} = \underline{837 \text{ Hz}}$$

24.19 A train sounds its whistle as it approaches and leaves a railroad crossing. An observer at the crossing measures a frequency of 219 Hz as the train approaches and a frequency of 184 Hz as the train leaves. The speed of sound is 340 m/s. Find the speed of the train and the frequency of its whistle.

▌ Only the source is moving with respect to the medium. Letting $|v_T|$ denote the speed of the train and v_o

denote the emitted frequency, the observer measures a frequency

$$v_a = v_0 \frac{|v|}{|v| - |v_T|} = 219 \text{ Hz} \qquad (1)$$

as the train approaches, and a frequency

$$v_r = v_o \frac{|v|}{|v| + |v_T|} = 184 \text{ Hz} \qquad (2)$$

as the train recedes from him. Dividing Eq. (1) by Eq. (2), we find that

$$\frac{v_a}{v_r} = \frac{|v| + |v_T|}{|v| - |v_T|} \qquad (3)$$

Solving this equation for the speed of the train we have

$$|v_T| = |v| \frac{v_a - v_r}{v_a + v_r} = 29.5 \text{ m/s} \qquad (4)$$

Using Eqs. (4) and (1), we find that

$$v_o = v_a \left(1 - \frac{|v_T|}{|v|}\right) = \frac{2v_a v_r}{v_a + v_r} = 200 \text{ Hz}$$

24.20 As shown in Fig. 24-1, an observer P is standing between two parallel train tracks when two trains approach from opposite directions. Locomotive A has a speed $|v_A| = 15$ m/s. It toots its whistle, which has a frequency $v_0 = 200$ Hz. Locomotive B has a speed $|v_B| = 30$ m/s. The speed of sound in the air is 340 m/s, and no breeze is blowing.

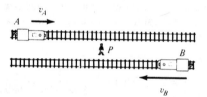

Fig. 24-1

(a) Find the wavelength λ_1 and frequency v_1 of the sound waves observer P receives from locomotive A.
(b) What frequency v_2 is heard by the engineer on locomotive B?

▮ (a) This is a case in which the source is moving but the medium and the observer are stationary. Since the source is approaching the observer with speed $|v_A|$, the received wavelength λ_1 is given by

$$\lambda_1 = \frac{|v| - |v_A|}{v_o}$$

where $|v|$ is the sound speed. The received frequency v_1 is given by

$$v_1 = v_o \frac{|v|}{|v| - |v_A|}$$

With $|v_A| = 15$ m/s and $v_o = 200$ Hz, we obtain $\lambda_1 = \underline{1.625 \text{ m}}$ and $v_1 = \underline{209 \text{ Hz}}$. (b) Here we have both a moving source and a moving observer. The general Doppler-effect formula is applicable. We take the rightward direction in Fig. 24-1 to be positive. The velocity v_m of the medium is zero. The observer's velocity $= -|v_B|$ since train B is moving to the left. The source velocity $v_s = |v_A|$ since train A is moving to the right. Then the frequency v_2 received by the engineer on train B is

$$v_2 = v_o \frac{|v| + |v_B|}{|v| - |v_A|} = (200) \frac{340 + 30}{340 - 15} = \underline{228 \text{ Hz}}$$

24.21 Suppose that in Prob. 24.20 some of the sound waves reaching locomotive B are reflected back toward observer P and locomotive A. (a) Find the wavelength λ_3 and the frequency v_3 of the reflected sound waves that observer P hears. (b) What frequency v_4 does the engineer on locomotive A hear in the reflected waves?

▮ (a) By reflecting the sound waves, locomotive B acts as a source of waves of frequency v_2 (as measured at the source). When these waves arrive at the observer P, the received wavelength λ_3 and frequency ν_3 are given by

$$\lambda_3 = \frac{|v| - |v_B|}{\nu_2} = \frac{340 - 30}{228} = \underline{1.36\,\text{m}} \quad \text{and} \quad \nu_3 = \nu_2 \frac{|v|}{|v| - |v_B|} = \frac{(228)(340)}{340 - 30} = \underline{250\,\text{Hz}}$$

(b) Since the source and the observer are moving directly toward each other, the received frequency is given by

$$\nu_4 = \nu_2 \frac{|v| + |v_A|}{|v| - |v_B|} = \frac{(228)(340 + 15)}{340 - 30} = \underline{261\,\text{Hz}}$$

24.22 A wave source of frequency ν and an observer are located a fixed distance apart. Both the source and the observer are stationary. However, the propagation medium (through which the waves travel at speed v) is moving at a uniform velocity $\mathbf{v}_m$ in an arbitrary direction. Find the frequency ν' received by the observer. Explain your result physically.

▌ The fundamental fact underlying the Doppler effect (in newtonian mechanics) is that the received frequency differs from the emitted frequency if and only if the time required for the wave to travel from source to observer is different for different wave fronts. In the present circumstance, with a uniform steady motion of the medium past the source and observer, *the transit time from source to observer is the same for all wave fronts*. This directly implies that $\nu' = \nu$.

24.23 A boy is walking away from a wall at a speed of 1.0 m/s in a direction at right angles to the wall. As he walks, he blows a whistle steadily. An observer toward whom the boy is walking hears 4.0 beats per second. If the speed of sound is 340 m/s, what is the frequency of the whistle?

▌ The wave approaching the observer directly from the boy undergoes a Doppler shift from the emitted frequency ν to a received frequency ν_1 given by

$$\nu_1 = \nu \frac{|v|}{|v| - |v_s|}$$

Here $|v|$ is the sound speed and $|v_s|$ is the speed at which the boy is approaching the observer. The observer also receives sound waves that have been reflected by the wall. Since the boy is receding from the wall, the waves strike the wall with the frequency ν_2 given by

$$\nu_2 = \nu \left(\frac{|v|}{|v| + |v_s|} \right)$$

Since the waves are reflected without any change of frequency, the observer will perceive beats at a frequency

$$\Delta\nu = \nu_1 - \nu_2 = \nu\,|v| \left(\frac{1}{|v| - |v_s|} - \frac{1}{|v| + |v_s|} \right) = \nu\,|v| \left(\frac{2\,|v_s|}{|v|^2 - |v_s|^2} \right)$$

Solving this equation for ν, we find that

$$\nu = \frac{\Delta\nu(|v| + |v_s|)(|v| - |v_s|)}{2\,|v|\,|v_s|}$$

With the given values $\Delta\nu = 4.0$ Hz, $|v| = 340$ m/s, and $|v_s| = 1.0$ m/s, we obtain

$$\nu = \frac{(4.0)(341)(339)}{2(340)(1.0)} = \underline{680\,\text{Hz}}$$

24.24 Two automobiles heading in opposite directions approach each other, the first at 88 ft/s and the second at 66 ft/s. The driver of the first automobile sounds a horn having a frequency of 400 Hz. (a) What frequency does the driver of the second car hear? (Take the velocity of the sound to be $v = 1100$ ft/s.) (b) After the two automobiles have passed each other, what frequency does the driver of the second car hear?

▌ (a) Use the Doppler-effect equation with both v_s and v_L positive.

$$\frac{\nu_L}{v + v_L} = \frac{\nu_s}{v - v_s} \qquad \frac{\nu_L}{1100 + 66} = \frac{400}{1100 - 88} \qquad \nu_L = \frac{400}{1012}\,1166 = \underline{461\,\text{Hz}}$$

(b) v_s and v_L are both negative.

$$\frac{v_L}{1100 - 66} = \frac{400}{1100 - (-88)} \qquad v_L = \frac{400}{1188} 1034 = \underline{348\ \text{Hz}}$$

24.2 POWER, INTENSITY, REVERBERATION TIME, SHOCK WAVES

24.25^c The equation of a sound wave traveling through water is $s = (3 \times 10^{-4}) \sin (2000t - 1.38x)$ m. Find the equations for the speed and acceleration of the water molecules as a result of the sound wave. Show that a force obeying Hooke's law is needed to cause this motion.

❚ The speed $= ds/dt = 0.60 \cos (2000t - 1.38x)$ m/s and acceleration $= d^2s/dt^2 = -1200 \sin (2000t - 1.38x)$ m/s^2. This is Hooke's law; $d^2s/dt^2 = -(\text{constant}) \cdot s$.

24.26 The intensity of a sound wave is the power per unit area perpendicular to the propagation direction, transmitted by the wave. Show that the intensity of sound transmitted in a given direction at a frequency f is given by $I = 2\pi^2 f^2 \rho v s_0^2$, where ρ is the density of the medium, v is the velocity of propagation, and s_0 is the displacement amplitude of the sound wave.

❚ Since the medium obeys Hooke's law (Prob. 24.25), the reasoning of Prob. 23.29 may be applied to a cross-sectional area a perpendicular to the propagation direction: $P_{\text{avg}}(\text{through } a) = \frac{1}{2}(2\pi f)^2 s_0^2(\rho a)v$, where $\rho a = \mu$, the mass/unit length for a tubular region of area a along the propagation direction. Then $I = P_{\text{avg}}/a = 2\pi^2 f^2 \rho v s_0^2$, as expected.

24.27 A certain loudspeaker has a circular opening with a diameter of 15 cm. Assume that the sound it emits is uniform and outward through this entire opening. If the sound intensity at the opening is $100\ \mu\text{W/m}^2$, how much power is being radiated as sound by the loudspeaker?

❚

$$P = IA = (100\ \mu\text{W/m}^2) \frac{\pi(0.15\ \text{m})^2}{4} = \underline{1.77\ \mu\text{W}}$$

24.28 A loud symphonic passage produces a sound level (intensity level) of 70 dB. A person speaking normally produces a sound level of 40 dB. How many times as many watts per square meter are there in the symphonic sound as there are in the person's speech?

❚ By Prob. 24.11, intensity level $n = 10 \log (I/I_0)$, or $I = I_0 10^{n/10}$, where I is intensity. Hence

$$\frac{I_1}{I_2} = 10^{(n_1 - n_2)/10} = 10^{(70-40)/10} = 10^3 = \underline{1000}$$

24.29 A sound intensity of about 1.2 W/m^2 can produce pain. To how many decibels is this equivalent?

❚ $n = 10 \log I/I_0 \qquad I_0 = 10^{-12}\ \text{W/m}^2 \qquad n = 10 \log \frac{1.2}{10^{-12}} = 10 \log (1.2 \times 10^{12}) = 10(12.079) = \underline{120.8\ \text{dB}}$

24.30 **(a)** What is the intensity of a 60-dB sound? **(b)** If the sound level is 60 dB close to a speaker that has an area of 120 cm^2, what is the acoustic power output of the speaker?

❚ **(a)** $60 = 10 \log (I/10^{-12})$ yields $I = \underline{1\ \mu\text{W/m}^2}$. **(b)** $P = (1 \times 10^{-6}\ \text{W/m}^2)(120 \times 10^{-4}\ \text{m}^2) = \underline{12\ \text{nW}}$.

24.31 About how many times more intense will the normal ear perceive a sound of 10^{-6} W/m^2 than one of 10^{-9} W/m^2?

❚ 10^{-6} W/m^2 corresponds to 60 dB, while 10^{-9} W/m^2 corresponds to 30 dB. Thus the former will sound about twice as loud, or intense, as the latter.

24.32 Assume that the average sound level in a certain room due to one person speaking is 40 dB. What will be the sound level when 20 people are speaking? Although it is not correct to do so, assume each of the 20 people speaks at the same level as did the single person.

❚ Let $I = $ intensity of one person. Then

$$n_{20} = 10 \log (20I/I_0) = 10 \log 20 + 10 \log (I/I_0) = 10(1 + \log 2) + 40 \approx \underline{53\ \text{dB}}$$

24.33 A rock band gives rise to an average sound level of 105 dB at a distance of 20 m from the center of the band. As an approximation, assume that the band radiates sound equally into a hemisphere. What is the sound power output of the band?

▌ The intensity at 20 m is found from 105 dB = $10 \log (I/10^{-12})$ to be $I = 0.0316$ W/m^2. Since area of the hemisphere is $2\pi R^2 = 2\pi(20)^2 = 2510$ m^2, $P = IA = \underline{79.4\ \text{W}}$.

24.34 If it were possible to generate a sinusoidal 3000-Hz sound wave in air that has a displacement amplitude of 0.200 mm, what would be the sound level of the wave? (Assume $v = 330$ m/s and $\rho_{\text{air}} = 1.29$ kg/m^3.)

▌ From Prob. 24.26 we have $I = 2\pi^2 f^2 \rho v s_0^2 = 2\pi^2(3000)^2(1.29)(330)(2.0 \times 10^{-4})^2 = (3.03 \times 10^3)$ W/m^2. This yields a large intensity level; $10 \log (I/I_0) = 10 \log (3 \times 10^{15}) = 10 (\log 3 + 15) \approx \underline{155\ \text{dB}}$.

24.35 What is the amplitude of motion for the air in the path of a 60-dB, 800-Hz sound wave? Assume that $\rho = 1.29$ kg/m^3 and $v = 330$ m/s.

▌ From Prob. 24.26, $I = 2\pi^2 \rho v s_0^2 f^2$. The intensity is found from 60 dB = $10 \log (I/I_0)$, which leads to 10^{-6} W/m^2. Then $s_0 = [10^{-6}/(2\pi^2)(1.29)(330)(800)^2]^{1/2} = 13.6 \times 10^{-9}$ m $= \underline{13.6\ \text{nm}}$.

24.36 Find the reverberation time of a room 10 m wide by 20 m long by 3 m high. The ceiling is acoustic, the walls are plaster, the floor is concrete, and there are 36 people in the room. [Sound-absorption coefficients are: acoustic ceiling—0.60, plaster—0.03, concrete—0.02. Take the absorbing power per person to be 0.5].

▌ First find the total absorbing power A:

$$A = 200(0.6) + 200(0.02) + (30 + 30 + 60 + 60)(0.03) + 36(0.5)(1) = 120 + 4 + 5.4 + 18 = 147.4$$

The volume V of the room is $V = (3)(10)(20) = 600$ m^3. Using the Sabine equation, $t_R = 0.16(V/A) = 0.16(600/147.4) = \underline{0.65\ \text{s}}$.

24.37 In Fig. 24-2, S_1 and S_2 are identical sound sources that send out their wave crests simultaneously (the sources are in phase). For what values of $L_1 - L_2$ will a loud sound be heard at point P?

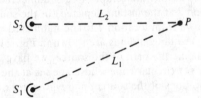

Fig. 24-2

▌ If $L_1 = L_2$, then the waves from the two sources will take equal times to reach P. Crests from one will arrive there at the same times as crests from the other. The waves will therefore be in phase at P and loudness will result.

If $L_1 = L_2 + \lambda$, then the wave from S_1 will be one wavelength behind the one from S_2 when they reach P. But because the wave repeats each wavelength, a crest from S_1 will still reach P at the same time that a crest from S_2 does. Once again the waves are in phase at P and loudness will exist there.

In general, loudness will be heard at P when $L_1 - L_2 = \pm n\lambda$, where n is an integer.

24.38 S_1 and S_2 are identical sources of sound of a single frequency. They are equidistant from O. When S_1 alone is sounding, the amplitude reaching O is A. (a) If S_1 and S_2 are both turned on and are operated "in phase," what is now the amplitude of the sound at O? (b) How does the energy flow at O with both sources sounding compare with the flow when only S_1 is sounding? (c) What is the increase in acoustic intensity in decibels when S_2 is turned on?

▌ (a) If the sources are operated in phase, then since the sources are equidistant from O, the waves also arrive in phase. The resultant amplitude A' is therefore twice the amplitude due to either source alone: $A' = \underline{2A}$. (b) We let $\langle S \rangle$ represent the magnitude of the average energy flux when only source 1 is operating, and we let $\langle S' \rangle$ represent the magnitude of the average energy flux when both sources are operating. If both waves approach O from the same direction, then

$$\frac{\langle S' \rangle}{\langle S \rangle} = \left(\frac{A'}{A}\right)^2 = \underline{4}$$

(This is the maximum possible value. If the waves approach O from different directions, the ratio $\langle S'\rangle/\langle S\rangle < 4$.) (c) When only source 1 is sounding, the acoustic intensity is $\alpha = 10\log(\langle S\rangle/S_0)$. With both sources sounding, the acoustic intensity $\alpha' = 10\log(\langle S'\rangle/S_0)$. The increase is given by

$$\alpha' - \alpha = 10\left(\log\frac{\langle S'\rangle}{S_0} - \log\frac{\langle S\rangle}{S_0}\right) = 10\log\left(\frac{\langle S'\rangle}{S_0}\frac{S_0}{\langle S\rangle}\right) = 10\log 4 \approx \underline{6\text{ dB}}$$

24.39 Consider a case where the speed $|v_s|$ of a point source through the medium is greater than the speed $|v|$ of sound waves with respect to the medium. Show that all the waves emanating from the source are tangent to a cone (the *Mach cone*) whose apex moves along with the source. This conical envelope of sound waves is the *shock wave*.

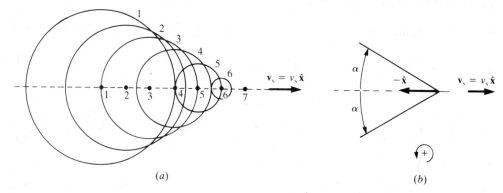

Fig. 24-3

▌ Figure 24-3(a) shows a set of six wave fronts in a stationary medium. The waves have been emitted by a source which is traveling toward the right with a speed $|v_s|$ equal to twice the wave speed $|v|$. The wave fronts are numbered and the positions of the source at the times of emission of the waves are indicated.

We observe that (by construction) the center of each numbered circular wave is located at a distance from the current position (location 7 at the instant depicted) of the source that is $(|v_s|/|v|)$ times the radius of the wave. The tangent lines drawn from location 7 to the numbered wave fronts all lie at the *Mach angle* α with $-\hat{x}$, where $\sin\alpha = |v|/|v_s|$. The tangent lines are shown in Fig. 24-3(b). Considering that the situation is as shown in every plane containing the path of the source, we have established that at each instant the wave fronts are all tangent to a right circular cone whose apex is at the current location of the source, whose axis passes through the prior locations of the source, and whose apex angle 2α is determined by the equation $\alpha = \sin^{-1}(|v|/|v_s|)$.

CHAPTER 25

Coulomb's Law and Electric Fields

25.1 COULOMB'S LAW OF ELECTROSTATIC FORCE

25.1 Describe Coulomb's law and the SI units involved.

▌ Suppose that two point charges, q and q', are a distance r apart in vacuum. If q and q' have the same sign, the two charges repel each other; if they have opposite signs, then they attract each other. The force experienced by one charge due to the other is given by *Coulomb's law*, $F = k(qq'/r^2)$, where k is a positive constant. Note that a positive F tends to increase r. In the SI, the unit of charge is the *coulomb* (C). To avoid very small numbers in practical work the *microcoulomb* ($1\,\mu C = 10^{-6}$ C) and the *nanocoulomb* ($1\,nC = 10^{-9}$ C) are frequently employed. In another commonly used system, Coulomb's law is written as $F = (qq')/r^2$, with F in dyn, r in cm, and q in esu. The conversion to coulombs is given by $1\,C = 3 \times 10^9$ esu.

The fundamental smallest charge found in nature is denoted e. Its value is $1.602\,19 \times 10^{-19}$ C. All other charges are integer multiples of e. The electron has a charge $-e$, while the proton charge is $+e$.

In SI units, the *Coulomb constant* k has the following value for charges in vacuum:

$$k = 8.988 \times 10^9 \, \text{N} \cdot \text{m}^2/\text{C}^2 \approx 9 \times 10^9 \, \text{N} \cdot \text{m}^2/\text{C}^2$$

Often k is replaced by $1/(4\pi\epsilon_0)$, where $\epsilon_0 = 8.85 \times 10^{-12}$ C²/N · m² is called the *permittivity of free space*. In terms of it, Coulomb's law becomes, for vacuum,

$$F = \frac{1}{4\pi\epsilon_0} \frac{qq'}{r^2}$$

In vector form, $\mathbf{F}_2$, the force on charge q_2 due to charge q_1, is given by

$$\mathbf{F}_2 = \frac{kq_1 q_2 \hat{\mathbf{r}}}{r^2} = \frac{kq_1 q_2 \mathbf{r}}{r^3}$$

where $\mathbf{r}$ is the vector displacement from q_1 to q_2 and $\hat{\mathbf{r}} = \mathbf{r}/r$ is the unit vector in the direction of $\mathbf{r}$.

25.2 What effect does an isotropic homogeneous medium have on Coulomb's law for charges embedded in it?

▌ When the surrounding medium is not a vacuum, forces caused by induced charges in the material reduce the force between point charges. If the material has a *dielectric constant* K, then ϵ_0 in Coulomb's law must be replaced by $K\epsilon_0 = \epsilon$, where ϵ is called the *permittivity* of the material. Then

$$F = \frac{1}{4\pi\epsilon} \frac{qq'}{r^2}$$

with $\epsilon = K\epsilon_0$. For vacuum, $K = 1$; for air, $K = 1.0006$, and is thus often taken to be 1.

25.3 What is meant by the *electric field* (sometimes called *electric intensity*) at a point in space?

▌ The electric field $\mathbf{E}$ at a point is the vector force experienced by a unit positive test charge placed at that point. The units of $\mathbf{E}$ are N/C or (see Chap. 26) V/m.

To find the electric field due to a point charge q, we make use of Coulomb's law. If a point charge q' is placed at a distance r from the charge q, it will experience a force

$$\mathbf{F} = \frac{1}{4\pi\epsilon_0} \frac{qq'\hat{\mathbf{r}}}{r^2} \quad \text{and} \quad \frac{\mathbf{F}}{q'} = \frac{1}{4\pi\epsilon_0} \frac{q\hat{\mathbf{r}}}{r^2} \equiv \mathbf{E}$$

This is the electric field at a distance r from a point charge q.

25.4 How many electrons are contained in -1 C of charge? What is the total mass of these electrons?

▌ From Prob. 25.1 the electron has charge $-e$, where $e = 1.6 \times 10^{-19}$ C. Therefore in -1.0 C of charge, there are $n = 1.0/(1.6 \times 10^{-19}) = \underline{6.2 \times 10^{18} \text{ electrons}}$. The mass of this many electrons, M, is just $M = nm_e = (6.2 \times 10^{18})(9.11 \times 10^{-31} \text{ kg}) = \underline{5.6 \times 10^{-12} \text{ kg}}$.

25.5 If two equal charges, each of 1 C, were separated in air by a distance of 1 km, what would be the force between them?

▌ $F = (kq_1q_2)/r^2$, where $k = 9.0 \times 10^9$ N · m²/C², $q_1 = q_2 = 1$ C, and $r = 1$ km. Then $F = [(9.0 \times 10^9)(1)(1)]/1000^2 = \underline{9.0\text{ kN}}$ (about 2000 lb). Note that even at a distance of 1 km, the repulsive force is substantial. This is indicative of the fact that the coulomb is a very large unit of charge.

25.6 Determine the force between two free electrons spaced 1 Å (0.1 nm) apart (a typical atomic dimension).

▌ F is repulsive and $F = (kq_1q_2)/r^2$, with $q_1 = q_2 = -1.6 \times 10^{-19}$ C, and $r = 10^{-10}$ m. Then $F = [(9.0 \times 10^9)(1.6 \times 10^{-19})^2]/(1.0 \times 10^{-10})^2 = 2.3 \times 10^{-8}$ N = $\underline{23\text{ nN}}$.

25.7 A copper sphere of mass 2.0 g contains about 2×10^{22} atoms. The charge on the nucleus of each atom is 29e. What fraction of the electrons must be removed from the sphere to give it a charge of $+2\,\mu$C?

▌ The total number of electrons is $29(2 \times 10^{22}) = 5.8 \times 10^{23}$. Electrons removed = $(2 \times 10^{-6}$ C)/$(1.6 \times 10^{-19}$ C/electron) = 1.25×10^{13}, so the ratio is $\underline{2.16 \times 10^{-11}}$.

25.8 What is the force of repulsion between two argon nuclei when separated by 1 nm (10^{-9} m)? The charge on an argon nucleus is +18e.

▌ $F = (kq^2)/r^2$ with $q = 18 \times 1.6 \times 10^{-19}$ C, and $r = 1.0 \times 10^{-9}$ m (=10 Å). Then $F = [(9 \times 10^9)(28.8 \times 10^{-19})^2]/(1.0 \times 10^{-18}) = \underline{75\text{ nN}}$.

25.9 The uranium nucleus contains a charge 92 times that of the proton. If a proton is shot at the nucleus, how large a repulsive force does the proton experience due to the nucleus when it is 1×10^{-11} m from the nucleus center? The nuclei of atoms are of order 10^{-14} m in diameter, so the nucleus can be considered a point charge.

▌ $F = [(9 \times 10^9)(92 \times 1.6 \times 10^{-19})(1.6 \times 10^{-19})]/(1 \times 10^{-11})^2 = \underline{2.1 \times 10^{-4}\text{ N}}$.

25.10 Two equally charged pith balls are 3 cm apart in air and repel each other with a force of 4×10^{-5} N. Compute the charge on each ball.

▌ $F = (kq^2)/r^2$, with $F = 4 \times 10^{-5}$ N and $r = 0.03$ m. Then $4 \times 10^{-5} = [(9 \times 10^9)q^2]/0.03^2 = 1.0 \times 10^{13}\,q^2$, or $q^2 = 4 \times 10^{-18}$ C², and finally $q = \underline{\pm 2\text{ nC}}$.

25.11 Two point charges Q_1 and Q_2 are 3 m apart, and their combined charge is 20 μC. (a) If one repels the other with a force of 0.075 N, what are the two charges? (b) If one attracts the other with a force of 0.525 N, what are the magnitudes of the charges?

▌ (a) $Q_1 + Q_2 = 20\mu$C. Since force is repulsive, $0.075 = (9 \times 10^9)[(Q_1Q_2)/3^2]$, which yields $Q_1Q_2 = 75 \times 10^{-12}$ C² = 75 μC². Substituting for Q_2, $Q_1(20 - Q_1) = 75$ or $Q_1^2 - 20Q_1 + 75 = 0$, so the Q's are $\underline{5\text{ and }15\,\mu\text{C}}$. (b) Force is attractive, so one charge is negative. The force equation is $-0.525 = (9 \times 10^9)[(Q_1Q_2)_2/9]$, or $Q_1Q_2 = -525\,\mu$C²; again substitute $Q_2 = 20 - Q_1$, giving $Q_1^2 - 20Q_1 - 525 = 0$ from which the Q's are $\underline{35\text{ and }-15\,\mu\text{C}}$ or $\underline{-35\text{ and }15\,\mu\text{C}}$.

25.12 A test charge $Q = +2\,\mu$C is placed halfway between a charge $Q_1 = +6\,\mu$C and a charge $Q_2 = +4\,\mu$C, which are 10 cm apart. Find the force on the test charge and its direction.

▌ Use Coulomb's law to find $\mathbf{F}_1$ and $\mathbf{F}_2$ and take the vector sum.

$$F_1 = k\frac{Q_1Q}{r^2} = (9 \times 10^9)\frac{(6 \times 10^{-6})(2 \times 10^{-6})}{0.05^2} = 43.2\text{ N} \qquad \text{away from } Q_1$$

$$F_2 = (9 \times 10^9)\frac{(4 \times 10^{-6})(2 \times 10^{-6})}{0.05^2} = 28.8\text{ N} \qquad \text{toward } Q_1$$

$$F = F_1 - F_2 = \underline{14.4\text{ N}} \qquad \text{away from } Q_1$$

25.13 Three $+20$-μC charges are placed along a straight line, successive charges being 2 m apart as shown in Fig. 25-1. Calculate the force on the charge on the right end.

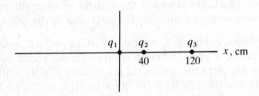

Fig. 25-1

$$F = F_1 + F_2 \qquad F_1 = \frac{kQ_1Q_3}{r^2} = \frac{(9 \times 10^9)(20 \times 10^{-6})^2}{4^2} = 0.225 \text{ N}$$

$$F_2 = \frac{kQ_2Q_3}{r^2} = \frac{(9 \times 10^9)(20 \times 10^{-6})^2}{2^2} = 0.9 \text{ N} \qquad F = 0.225 + 0.9 = \underline{1.125 \text{ N}} \qquad \text{to the right}$$

25.14 Three point charges are placed at the following points on the x axis: $+2 \, \mu C$ at $x = 0$, $-3 \, \mu C$ at $x = 40$ cm, $-5 \, \mu C$ at $x = 120$ cm. Find the force on the -3-μC charge.

▌ Figure 25-2 is a diagram of the setup with $q_1 = 2 \, \mu C$, $q_2 = -3 \, \mu C$, and $q_3 = -5 \, \mu C$. The force on q_2 is the vector sum of two contributions, the attractive force due to q_1 (toward q_1), and the repulsive force due to q_3 (also toward q_1). The sum of these two forces, taken algebraically, since they are along the same line, is

$$F = F_1 + F_2 = \frac{-k \, |q_1 q_2|}{(0.40)^2} + \frac{-k \, |q_2 q_3|}{(0.80)^2}$$

$$= -9 \times 10^9 \left[\frac{(2 \times 10^{-6} \text{ C})(3 \times 10^{-6} \text{ C})}{(0.40 \text{ m})^2} + \frac{(5 \times 10^{-6} \text{ C})(3 \times 10^{-6} \text{ C})}{(0.80 \text{ m})^2} \right] = \underline{-0.55 \text{ N}} \qquad \text{or} \qquad \underline{0.55 \text{ N}} \qquad \text{to left}$$

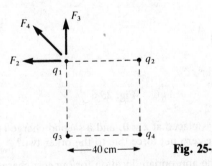

Fig. 25-2

25.15 In Prob. 25.14, what is the force on the -5-μC change?

▌ The force on q_3 is an attractive force to the left due to q_1 and a repulsive force to the right due to q_2.

$$F = F_1 + F_2 = k \left[\frac{-|q_1 q_3|}{(1.20 \text{ m})^2} + \frac{|q_2 q_3|}{(0.80 \text{ m})^2} \right]$$

$$= 9 \times 10^9 \left[\frac{-(2 \times 10^{-6})(5 \times 10^{-6})}{1.20^2} + \frac{(3 \times 10^{-6})(5 \times 10^{-6})}{0.80^2} \right] = \underline{0.15 \text{ N}} \qquad \text{or} \qquad \underline{0.15 \text{ N}} \qquad \text{to the right}$$

25.16 Four equal point changes, $+3 \, \mu C$, are placed at the four corners of a square that is 40 cm on a side. Find the force on any one of the charges.

▌ The situation is depicted in Fig. 25-3. We consider the forces acting on q_1, depicted in the diagram by F_2, F_3, F_4, where the labels identify the respective charges exerting the force. By symmetry $F_2 = F_3 = [(9 \times 10^9)(3 \times 10^{-6})^2]/0.40^2$, or $F_2 = F_3 = 0.51$ N. Since the directions of these forces are along the edges, as shown, their vector sum will lie along the diagonal from q_4 to q_1 and have magnitude $F_2 \cos 45° + F_3 \cos 45° = [2(0.51)]/\sqrt{2} = 0.72$ N. The remaining force F_4 is also along this diagonal, and $F_4 = [(9 \times 10^9)(3 \times 10^{-6})^2]/(0.40 \times \sqrt{2})^2 = 0.25$ N. The resultant of all three forces thus points along the diagonal and away from the square, and has magnitude $\underline{0.97 \text{ N}}$.

25.17 Four equal-magnitude point charges $(3\,\mu C)$ are placed at the corners of a square that is 40 cm on a side. Two, diagonally opposite each other, are positive and the other two are negative. Find the force on either negative charge.

▌ The new situation (Fig. 25-4) is identical to that of Prob. 25.16, except that now $q_1 = q_4 = -3\,\mu C$. Again we calculate the resultant force on q_1. F_2 and F_3 have the same magnitudes as in Prob. 25.16, but the directions are opposite, as shown. The vector sum of F_2 and F_3 is now 0.72 N pointed inward along the diagonal. F_4 is again repulsive and as before is 0.25 N. The resultant of all three forces is now $0.72\,\text{N} - 0.25\,\text{N} = \underline{0.47\,\text{N}}$ pointed inward along the diagonal (toward q_4).

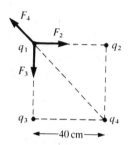
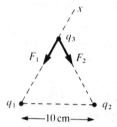

Fig. 25-4   Fig. 25-5

25.18 Charges of $+2$, $+3$, and $-8\,\mu C$ are placed at the vertices of an equilateral triangle of side 10 cm. Calculate the magnitude of the force acting on the -8-μC charge due to the other two charges.

▌ The situation is as shown in Fig. 25-5, with $q_1 = 2\,\mu C$, $q_2 = 3\,\mu C$, and $q_3 = -8\,\mu C$. For definiteness we let the side from q_1 to q_3 be our x axis. The net force on q_3 is the vector sum of F_1 and F_2, the forces due to q_1 and q_2. As shown these are attractive forces. In magnitude, $F_1 = [(9 \times 10^9)(2 \times 10^{-6}\,\text{C})(8 \times 10^{-6}\,\text{C})]/(0.10\,\text{m})^2$, or $F_1 = 14.4\,\text{N}$. Similarly $F_2 = [(9 \times 10^9)(3 \times 10^{-6})(8 \times 10^{-6})]/0.10^2 = 21.6\,\text{N}$. Let $\mathbf{F} = \mathbf{F}_1 + \mathbf{F}_2$; then $F_x = F_{1x} + F_{2x} = -14.4\,\text{N} - 21.6\,\text{N}\cos 60° = -25.2\,\text{N}$. $F_y = F_{1y} + F_{2y} = 0 + 21.6\,\text{N}\sin 60° = 18.7\,\text{N}$. $F = \sqrt{F_x^2 + F_y^2} = \sqrt{(25.2)^2 + (18.7)^2} = \underline{31.4\,\text{N}}$.

25.19 Refer to Fig. 25-6. Find the force on the 4-μC charge due to the other two charges.

▌ We label the forces by the magnitude of the charge exerting the force. Note that in Fig. 25-6 the hypotenuse $= 5/(\sin 30°) = 10\,\text{cm}$ and the base is 8.66 cm. Then $\mathbf{F}_3 = \{-(9 \times 10^9)[(3 \times 4 \times 10^{-12})/0.087^2]\}\mathbf{i} = (-14.3)\mathbf{i}$ and $\mathbf{F}_2 = (9 \times 10^9)[(2 \times 4 \times 10^{-12})/0.10^2] = 7.2$ at $-30°$. $\mathbf{F} = \mathbf{F}_3 + \mathbf{F}_2 = (-14.3 + 7.2\cos 30°)\mathbf{i} - (7.2\sin 30°)\mathbf{j} = -8.1\mathbf{i} - 3.6\mathbf{j}$ N, or $\underline{8.9\,\text{N at }204°}$.

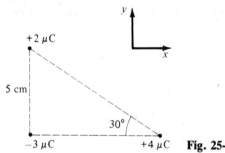
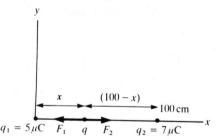

Fig. 25-6   Fig. 25-7

25.20 One charge $(+5\,\mu C)$ is placed at $x = 0$, and a second charge $(+7\,\mu C)$ at $x = 100\,\text{cm}$. Where can a third be placed and experience zero net force due to the other two?

▌ We assume that the appropriate location for our test charge is at position x as shown in Fig. 25-7. If the test charge q is negative, q_1 and q_2 will attract q in opposite directions; if it is positive, q_1 and q_2 will repel q in opposite directions. In either case the condition for q to experience no net force is that $F_1 = F_2$, or $(kqq_1)/x^2 = (kqq_2)/(1.0 - x)^2$, where we convert distance to meters. Dividing out k and q, we have $q_1/x^2 = q_2/(1.0 - x)^2$. (Note that our result, as might be expected, does not depend on q.) Substituting in numbers and cross-multiplying, we get $5(1 - x)^2 = 7x^2$, or $2x^2 + 10x - 5 = 0$. Solving, $x = (-10 \pm \sqrt{100 + 40})/4 = \{0.46\,\text{m}, -5.46\,\text{m}\}$. Since x for our case must lie between the two charges, our answer is $\underline{46\,\text{cm}}$. (At $x = -5.46\,\text{m}$ the two forces are equal in magnitude *and* direction.)

25.21   Discuss the nature of the equilibrium of Prob. 25.20.

▌ If q is positive, a small displacement from $x = 46$ cm along the x axis results in a net force on q tending to return it to the equilibrium position. But a small displacement parallel to the y axis sets up a force that accelerates q away from the equilibrium position. If q is negative, the opposite is true: now it is the x displacement that fails to provide a restoring force. In either case, then, the equilibrium is *unstable*. (This illustrates a general theorem to the effect that *stable equilibrium is impossible in an electrostatic field*.)

25.22   Two identical tiny metal balls carry changes of $+3$ nC and -12 nC. They are 3 cm apart. (**a**) Compute the force of attraction. (**b**) The balls are now touched together and then separated to 3 cm. Describe the forces on them now.

▌ (**a**) At first the force between them is *attractive* with magnitude $F = [(9 \times 10^9)(3 \times 10^{-9})(12 \times 10^{-9})]/0.03^2 = 3.6 \times 10^{-4}$N. (**b**) Since the two balls are metallic and identical, when they are touched the charges rearrange themselves to a new equilibrium distribution which must have a like charge on each ball. Since the total available charge is -9 nC, each ball has -4.5 nC of charge. When they are moved 3 cm apart, the new force is *repulsive* and has magnitude $F = [(9 \times 10^9)(4.5 \times 10^{-9})(4.5 \times 10^{-9})]/0.03^2 = \underline{2.0 \times 10^{-4}\, \text{N}}$.

25.23   The two balls shown in Fig. 25-8 have identical masses of 0.20 g each. When suspended from 50-cm-long strings, they make an angle of 37° to the vertical. If the charges on each are the same, how large is each charge?

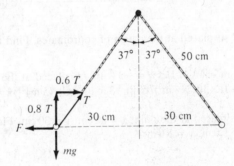

Fig. 25-8

▌ Since the system is at rest, we can apply the conditions for equilibrium to the ball on the left. Note that three forces act on the ball: its weight mg; the tension T in the string; and F, the repulsive force due to the charge on the other ball. We have the usual conditions for equilibrium: $\sum F_x = 0$, from which $F - 0.6T = 0$ and $\sum F_y = 0$, which gives $0.8T - (0.2 \times 10^{-3}\,\text{kg})(9.8\,\text{m/s}^2) = 0$, or $T = 2.45 \times 10^{-3}$ N. Using this to find F, we obtain $F = 1.47 \times 10^{-3}$ N. This is the force obeying Coulomb's law.

Substituting in Coulomb's law, we have

$$1.47 \times 10^{-3} = (9 \times 10^9)\frac{q^2}{(0.60)^2}$$

where all units are SI units. Solving for q, we find that $q \approx 2.4 \times 10^{-7}\,\text{C} = \underline{0.24\,\mu C}$.

25.24   In the Bohr model, taking the radius of the hydrogen atom as $r = 5.29 \times 10^{-11}$ m, find the strength of the force on the electron, its centripetal acceleration, and its orbital speed.

▌ From Coulomb's law,

$$F = \frac{1}{4\pi\epsilon_0}\frac{|q|\,|q|}{r^2} = (8.99 \times 10^9\,\text{N}\cdot\text{m}^2/\text{C}^2)\frac{(1.60 \times 10^{-19}\,\text{C})^2}{(5.29 \times 10^{-11}\,\text{m})^2} = \underline{82.3\,\text{nN}}$$

$$a = \frac{F}{m} = \frac{82.3 \times 10^{-9}\,\text{N}}{9.11 \times 10^{-31}\,\text{kg}} = \underline{9.03 \times 10^{22}\,\text{m/s}^2}$$

The orbital speed is given from the expression $a = v^2/r$ for centripetal acceleration; thus, it is

$$v = (ar)^{1/2} = (9.03 \times 10^{22}\,\text{m/s}^2 \times 5.29 \times 10^{-11}\,\text{m})^{1/2} = \underline{2.19 \times 10^6\,\text{m/s}}$$

Since this speed is less than 1 percent of the speed of light, the nonrelativistic form of Newton's second law, $a = F/m$, was appropriate.

25.25 Two positive point charges a distance b apart have sum Q. For what values of the charges is the Coulomb force between them a maximum?

▌ Call the two charges q_1 and q_2; then $F \propto q_1 q_2$. But $0 \le (q_1 - q_2)^2 = q_1^2 + q_2^2 - 2q_1 q_2$. Add $4q_1 q_2$ to both sides to obtain $4q_1 q_2 \le (q_1 + q_2)^2$, or $q_1 q_2 \le (Q/2)^2$. This shows that the force is greatest when $q_1 = q_2 = Q/2$.

25.2 THE ELECTRIC FIELD, CONTINUOUS CHARGE DISTRIBUTIONS, MOTION OF CHARGED PARTICLES IN AN ELECTRIC FIELD

25.26 A charge, $+6 \, \mu C$, experiences a force of 2 mN in the $+x$ direction at a certain point in space. (a) What was the electric field there before the charge was placed there? (b) Describe the force a $-2\text{-}\mu C$ charge would experience if it were used in place of the $+6 \, \mu C$.

▌ We, of course, assume that the charges giving rise to the electric field are stationary and not affected by the charge brought into the field. (a) Since $\mathbf{F} = q\mathbf{E}$, we must have $\mathbf{E}$ in the $+x$ direction with magnitude $E = F/q = (2 \times 10^{-3} \, \text{N})/(6 \times 10^{-6} \, \text{C}) = \underline{333 \, \text{N/C}}$. (b) $\mathbf{F}' = q'\mathbf{E}$ and since q' is negative, $\mathbf{F}'$ is along the $\underline{-x \, \text{direction}}$. In magnitude, $F' = (2 \times 10^{-6} \, \text{C})(333 \, \text{N/C}) = \underline{0.67 \, \text{mN}}$.

25.27 Find the electric field at a distance of 0.1 m from a charge of 2 nC.

▌ In magnitude,

$$E = \frac{kq}{r^2} = \frac{(9 \times 10^9)(2 \times 10^{-9})}{(0.1)^2} = \underline{1800 \, \text{N/C}}$$

25.28 A point charge, $-30 \, \mu C$, is placed at the origin of coordinates. Find the electric field at the point $x = 5$ m on the x axis.

▌ $\mathbf{E} = (kq\hat{\mathbf{r}})/r^2$ for a point charge. Let $q = -3 \times 10^{-5}$ C placed at the origin. At a point $x = 5$ m along the x axis, $\hat{\mathbf{r}} = \hat{\mathbf{x}}$. Then $\mathbf{E} = \{[9.0 \times 10^9 \, \text{N} \cdot \text{m}^2)/\text{C}^2][(-3 \times 10^{-5} \, \text{C})/(5 \, \text{m})^2]\}\hat{\mathbf{x}} = \underline{(10.8 \, \text{kN/C})(-\hat{\mathbf{x}})}$.

25-29 A 5.0-μC point charge is placed at the point $x = 20$ cm, $y = 30$ cm. Find the magnitude of $\mathbf{E}$ due to it (a) at the origin and (b) at $x = 1.0$ m, $y = 1.0$ m.

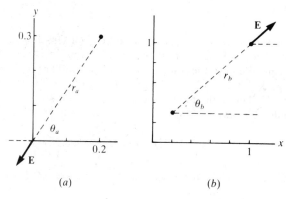

(a)   (b)   **Fig. 25-9**

▌ (a) See Fig. 25-9(a). $r_a = 0.36$ and $\theta_a = \tan^{-1} 1.5 = 56.3°$, so $\mathbf{E} = [(9 \times 10^9)(5 \times 10^{-6})]/r_a^2$ at $(180 + 56.3)°$ or $\underline{0.346 \, \text{MN/C at } 236°}$. (b) See Fig. 25-9(b). For $r_b = 1.06$ m and $\theta_b = \tan^{-1}(0.70/0.80)$, find $\mathbf{E} = \underline{40 \, \text{kN/C at } 41.2°}$.

25.30 A point charge, $-3 \, \mu C$, is placed at the point $(0, 2)$ m, while a charge $+2 \, \mu C$ is placed at $(2, 0)$ m. (a) Find the electric field at the origin. (b) What would be the magnitude of the force experienced by a proton at the origin?

▌ (a) $\mathbf{E}_3$ due to the $-3 \, \mu C = \{[(9 \times 10^9)(3 \times 10^{-6})]/4\}\mathbf{j} = 6.75\mathbf{j}$ kN/C and $\mathbf{E}_2$ due to the $2 \, \mu C = \{[-(9 \times 10^9)(2 \times 10^{-6})]/4\}\mathbf{i} = -4.5\mathbf{i}$ kN/C. Hence $\mathbf{E} = -4.5\mathbf{i} + 6.75\mathbf{j} = \underline{8.1 \, \text{kN/C at } \theta = 124°}$. (b) $F = eE = (1.6 \times 10^{-19})(8.1 \times 10^3) = \underline{1.30 \times 10^{-15} \, \text{N at } 124°}$.

25.31 Find the electric field at point P in Fig. 25-10(a) due to the charges shown.

▌ The fields due to these point charges obey the superposition principle. The fields due to each charge are

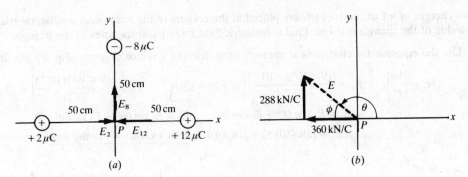

Fig. 25-10

shown in the figure. Using $E = (kq)/r^2$ for each charge we get

$$E_2 = 72 \text{ kN/C} \qquad E_8 = 288 \text{ kN/C} \qquad E_{12} = 432 \text{ kN/C}$$

We therefore find at P that $E_x = -360$ kN/C and $E_y = 288$ kN/C (see Fig. 25-10(b)). From this, $\underline{\mathbf{E} = 461 \text{ kN/C at } \theta = 141°}$.

25.32 Four equal-magnitude (4-μC) charges are placed at the four corners of a square that is 20 cm on each side (Fig. 25-11). Find the electric-field intensity at the center of the square, if the charges are all positive.

▌ $\mathbf{E} = \mathbf{E}_1 + \mathbf{E}_2 + \mathbf{E}_3 + \mathbf{E}_4$, where $\mathbf{E}_i = (kq\hat{\mathbf{r}}_i)/r_i^2$, $i = 1, 2, 3, 4$. ($r_i = 10\sqrt{2}$ cm, $i = 1, 2, 3, 4$.) With $q_1 = q_2 = q_3 = q_4 = 4$ μC, we have, at the center, $\mathbf{E}_1 = -\mathbf{E}_4$, $\mathbf{E}_2 = -\mathbf{E}_3$. Therefore $\underline{\mathbf{E} = \mathbf{0}}$.

25.33 Rework Prob. 25.32 if the charges alternate in sign around the perimeter of the square.

▌ $q_1 = q_4 = 4$ μC; $q_2 = q_3 = -4$ μC. Again, at the center, $\mathbf{E}_1 = -\mathbf{E}_4$, $\mathbf{E}_2 = -\mathbf{E}_3$, so $\underline{\mathbf{E} = \mathbf{0}}$.

25.34 Rework Prob. 23.32 for the sign sequence plus, plus, minus, minus.

▌ $q_1 = q_2 = 4$ μC; $q_3 = q_4 = -4$ μC. Now $\mathbf{E}_1 = \mathbf{E}_4$ since both have the same magnitude and point in the direction of $\hat{\mathbf{r}}_1$. Similarly $\mathbf{E}_2 = \mathbf{E}_3$, pointing along $\hat{\mathbf{r}}_2$. For the magnitudes, $E_i = (k\,|q_i|)/r^2$, we have $E_1 = E_2 = E_3 = E_4 = [(9 \times 10^9)(4 \times 10^{-6})]/(0.10 \times \sqrt{2})^2 = 1.80$ MN/C. Then $\mathbf{E} = (3.60 \text{ MN/C})\hat{\mathbf{r}}_1 + (3.60 \text{ MN/C})\hat{\mathbf{r}}_2$. From the diagram it is clear that $\mathbf{E}$ points *downward* from the *positive to negative side* with magnitude: $(3.60) \cos 45° + (3.60) \cos 45° = \underline{5.1 \text{ MN/C}}$.

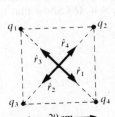

Fig. 25-11

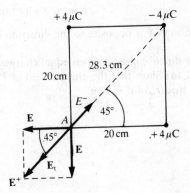

Fig. 25-12

25.35 Three charges are placed at three corners of a square as shown in Fig. 25-12. Find the electric-field strength at point A (magnitude and direction).

▌ Draw a vector diagram showing the fields produced at A by each of the three charges. Then find the vector sum E_t.

$$E = \frac{kQ}{r^2} = \frac{(9 \times 10^9)(4 \times 10^{-6})}{(0.20)^2} = 0.9 \text{ MN/C} \qquad E^+ = \sqrt{(9 \times 10^5)^2 + (9 \times 10^5)^2} = \sqrt{162 \times 10^{10}} = 1.27 \text{ MN/C}$$

$$E^- = \frac{(9 \times 10^9)(-4 \times 10^{-6})}{(0.283)^2} = -0.45 \text{ MN/C} \qquad E_t = E^+ + E^- = \underline{0.82 \text{ MN/C away from the negative charge.}}$$

25.36 Two charges of $+1\,\mu C$ and $-1\,\mu C$ are placed at the corners of the base of an equilateral triangle. The length of a side of the triangle is $0.7\,m$. Find the electric-field intensity at the apex of the triangle.

▮ Use the equation for electric-field intensity after drawing a vector diagram (Fig. 25.13). In magnitude:

$$E = \left|\frac{kq}{r^2}\right| \qquad E_1 = \left|\frac{(9 \times 10^9)(-10^{-6})}{(0.7)^2}\right| = +18.4\,kN/C \qquad E_2 = \left|\frac{(9 \times 10^9)(10^{-6})}{(0.7)^2}\right| = \underline{18.4\,kN/C}.$$

$$E_x = E_1 \cos 60° + E_2 \cos 60° \qquad E_y = E_1 \sin 60° - E_2 \sin 60° = 0$$

$$E_x = (18.4)(0.5) + (18.4)(0.5) = \underline{18.4\,kN/C} \qquad \text{to the right}$$

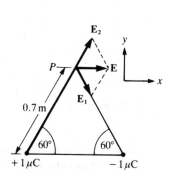

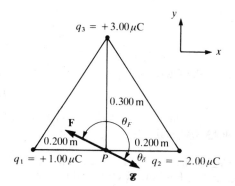

Fig. 25-13   **Fig. 25-14**

25.37 (a) Three point charges are fixed rigidly at the vertices of an isosceles triangle, as shown in Fig. 25-14. Find the electric field $\mathscr{E}$ at the midpoint P of the base of the triangle. (b) A point charge $q = -4.00\,\mu C$ is moved to P. What electric force F acts on this charge?

▮ (a) In vector form

$$\mathscr{E} = \frac{1}{4\pi\epsilon_0}\left(\frac{q_1}{r_1^2}\hat{r}_1 + \frac{q_2}{r_2^2}\hat{r}_2 + \frac{q_3}{r_3^2}\hat{r}_3\right)$$

where, from Fig. 25-14, $r_1 = (0.200\,m)\hat{x}$, $r_2 = (0.200\,m)(-\hat{x})$, and $r_3 = (0.300\,m)(-\hat{y})$. Then

$$\mathscr{E} = (8.99 \times 10^9\,N \cdot m^2/C^2)\left[\frac{+1.00 \times 10^{-6}\,C}{(0.200\,m)^2}(+\hat{x}) + \frac{-2.00 \times 10^{-6}\,C}{(0.200\,m)^2}(-\hat{x}) + \frac{+3.00 \times 10^{-6}\,C}{(0.300\,m)^2}(-\hat{y})\right]$$

$$= (8.99 \times 10^9)(75\hat{x} - 33.3\hat{y})\,\mu N/C = \underline{0.74\,MN/C\,\text{at}\,\theta_{\mathscr{E}} = -23.9°}$$

(b) $$F = |q|\,\mathscr{E} = (4.00 \times 10^{-6}\,C)(0.74 \times 10^6\,N/C) = \underline{2.96\,N}$$

The direction of F is opposite to the direction of $\mathscr{E}$, since q has a negative value.

25.38 An *electric dipole* consists of two equal charges q of opposite sign separated by a small distance a as shown in Fig. 25-15. (a) Show that the electric field at P is $[(1/4\pi\epsilon_0)(qa/r_p^3)]\mathbf{i}$ if $r_p \gg a$. (b) Show that the field at M is $-[(1/2\pi\epsilon_0)(qa/r_m^3)]\mathbf{i}$ if $r_m \gg a$.

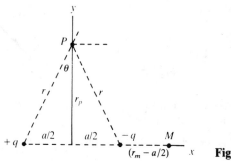

Fig. 25-15

▮ (a) At P, vertical components cancel; only an x component exists; $E_p = [(2q)/(4\pi\epsilon_0 r^2)]\sin\theta$. Substituting for $\sin\theta$ and r^2, $E_p = [2q(a/2)]/\{4\pi\epsilon_0[r_p^2 + (a/2)^2]^{3/2}\}$. If $r_p \gg a/2$, then $a/2$ can be neglected in the denominator and we have the desired result. (b) At point M the $-q$ charge is closer, which causes E_M to be in the $-x$ direction. $E_M = [q/(4\pi\epsilon_0)][1/(r_m - a/2)^2 - 1/(r_m + a/2)^2] = [q/(4\pi\epsilon_0 r_m^2)]\{1/[(1 - a/2r_m)^2 - 1/(1 + a/2r_m)^2]\}$.

For $r_p \gg a/2$, $(1 - a/2r_m)^{-2} \approx 1 + a/r_m$ and $(1 + a/2r_m)^{-2} \approx 1 - a/r_m$. These inserted into the E_M equation give the answer.

25.39ᶜ Find the electric field at the center of a uniformly charged semicircular arc, i.e., at point P in Fig. 25-16.

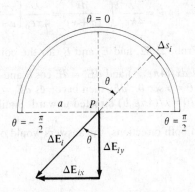

Fig. 25-16

❚ We assume the charge per unit length of arc to be λ and the radius of the arc to be a. Splitting the arc into small segments Δs_i, we have that the essentially point charge on each is $\lambda \, \Delta s_i$. The electric field due to the Δs_i shown is, from Coulomb's law,

$$\Delta E_i = \frac{1}{4\pi\epsilon_0} \frac{\lambda \, \Delta s_i}{a^2}$$

Each little portion of the arc will give a $\Delta \mathbf{E}_i$ in a different direction. We must therefore take components in order to find the total field at point P.

Before doing that, though, we note that E_x will be zero at point P. This is the result of the fact that the ΔE_{ix} shown in Fig. 25-16 will be canceled by the contribution from a symmetrically placed Δs on the left half of the arc. As a result, we need only compute E_y at point P in order to find the total E. We have

$$-\Delta E_{iy} = \frac{\lambda}{4\pi\epsilon_0 a^2} \Delta s_i \cos \theta$$

The negative sign arises because $\Delta \mathbf{E}_{iy}$ shown in Fig. 25-16 is in the $-y$ direction.

If we now sum over the whole arc and replace the sum by an integral in the usual way, we find that

$$-E_y = \frac{\lambda}{4\pi\epsilon_0 a^2} \int \cos \theta \, ds$$

The integrand involves two variables, θ and s. We can express s in terms of θ by recalling that an angle $d\theta$ subtends an arc length $ds = a \, d\theta$ along a circle of radius a. Therefore,

$$-E_y = \frac{\lambda}{4\pi\epsilon_0 a} \int_{-\pi/2}^{\pi/2} \cos \theta \, d\theta = \frac{\lambda}{4\pi\epsilon_0 a} \sin \theta \Big]_{-\pi/2}^{\pi/2} = \frac{\lambda}{2\pi\epsilon_0 a}$$

25.40ᶜ Repeat Prob. 25.39 if the charge distribution λ along the arc is nonuniform and is given by $\lambda = \lambda_0 \sin \theta$, with θ measured as shown in Fig. 25.16.

❚ Now $E_y = 0$ at P, since the charge distribution is antisymmetric about the y axis. For the incremental x component, we have

$$-\Delta E_{ix} = \frac{\lambda}{4\pi\epsilon_0 a^2} \Delta s_i \sin \theta$$

and so

$$-E_x = \frac{\lambda_0}{4\pi\epsilon_0 a} \int_{-\pi/2}^{\pi/2} \sin^2 \theta \, d\theta = \frac{\lambda_0}{4\pi\epsilon_0 a} \left[\frac{\theta}{2} - \frac{\sin 2\theta}{4} \right]_{-\pi/2}^{\pi/2} = \frac{\lambda_0}{8\epsilon_0 a}$$

25.41 A long thin rod is bent into a circle of radius b. It is uniformly charged along its length. Find the electric field at the center of the circle.

❚ Zero, by symmetry.

25.42ᶜ A rod lies along the x axis with one end at the origin and the other at $x \to \infty$. It carries a uniform charge λ C/m. Starting from Coulomb's law, find the electric field at the point $x = -a$ on the x axis.

❚ Because of the location of field point, $\mathbf{E} = E(-\mathbf{i})$ with

$$E = \frac{\lambda}{4\pi\epsilon_0} \int_0^\infty \frac{dx}{(a+x)^2} = \frac{\lambda}{4\pi\epsilon_0} \left[\frac{-1}{a+x}\right]_0^\infty = \frac{\lambda}{4\pi\epsilon_0 a}$$

25.43ᶜ For the situation given in Prob. 25.42, find E_x and E_y for the point on the y axis where $y = b$.

❚ From Fig. 25-17, $dE = \lambda\, dx/(4\pi\epsilon_0 r^2)$ and $dE_y = dE \cos\theta$ and $dE_x = dE \sin\theta$. Also note $x = b \tan\theta$, so $dx = b \sec^2\theta\, d\theta$; $r = b/\cos\theta = b \sec\theta$.. dE_y then becomes $dE_y = \{\lambda b \sec^2\theta\, d\theta/(4\pi\epsilon_0 b^2 \sec^2\theta)\} \cos\theta$. Then $E_y = [\lambda/(4\pi\epsilon_0 b)]\int_0^{\pi/2} \cos\theta\, d\theta = \lambda/(4\pi\epsilon_0 b)$ directed upward. Similarly $E_x = [\lambda/(4\pi\epsilon_0 b)]\int_0^{\pi/2} \sin\theta\, d\theta = \lambda/(4\pi\epsilon_0 b)$ to the left.

If the rod were infinite in both directions, the answers would obviously become $E_x = 0$, $E_y = \lambda/(2\pi\epsilon_0 b)$.

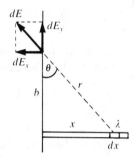
Fig. 25-17

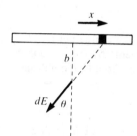
Fig. 25-18

25.44ᶜ A rod of length L carries a uniform charge λ per unit length. Find the electric field at a distance b from the rod on its perpendicular bisector. Show that in the limit of $b/L \ll 1$ this expression reduces to the value one obtains from a very long rod.

❚ As is clear from Fig. 25-18, $E_x = 0$ because of symmetry. To find E_y, $dE_y = [1/(4\pi\epsilon_0)]\lambda \cos\theta\, dx/(x^2 + b^2) = [\lambda b/(4\pi\epsilon_0)]\, dx/(x^2 + b^2)^{3/2}$. Integrate from $-L/2 \le x \le L/2$. We then have

$$\frac{\lambda b}{4\pi\epsilon_0} \int_{-L/2}^{L/2} \frac{dx}{(x^2 + b^2)^{3/2}}$$

As in Prob. 25.43 we use $x = b \tan\theta$, $dx = b \sec^2\theta\, d\theta$. Also $b/\cos\theta = (x^2 + b^2)^{1/2}$, so

$$E_y = \frac{\lambda b}{4\pi\epsilon_0} \int_{-\theta_0}^{\theta_0} \frac{b \sec^2\theta\, d\theta}{b^3 \sec^3\theta} = \frac{\lambda}{4\pi\epsilon_0 b} \int_{-\theta_0}^{\theta_0} \cos\theta\, d\theta = \frac{\lambda}{2\pi\epsilon_0 b} \sin\theta_0$$

where $b \tan\theta_0 = L/2$, or $\sin\theta_0 = [1 + 4(b/L)^2]^{-1/2}$. As $b/L \to 0$, $\sin\theta_0 \to 1$ and $E_y \to \lambda/(2\pi\epsilon_0 b)$, in agreement with the final result of Prob. 25.43.

25.45 Determine the acceleration of a proton ($q = +e$, $m = 1.67 \times 10^{-27}$ kg) in an electric field of intensity 500 N/C. How many times is this acceleration greater than that due to gravity?

❚ $F = qE = (1.6 \times 10^{-19}$ C$)(500$ N/C$) = 8.0 \times 10^{-17}$ N. $a = F/m = (8.0 \times 10^{-17}$ N$)/(1.67 \times 10^{-27}$ kg$) = \underline{4.8 \times 10^{10} \text{ m/s}^2}$. $a/g = (4.8 \times 10^{10})/9.8 = \underline{4.9 \times 10^9}$.

25.46 A tiny 0.60-g ball carries a charge of magnitude 8 μC. It is suspended by a thread in a downward electric field of intensity 300 N/C. What is the tension in the thread if the charge on the ball is (**a**) positive, (**b**) negative?

❚ The ball is presumably in equilibrium under the action of three forces: the tension in the thread, T; the pull of gravity on the ball, $w = mg = 5.9 \times 10^{-3}$ N; and the coulomb force, F_c, which is downward if the charge, q, on the ball is positive, as shown in Fig. 25-19, and upward if q is negative. (**a**) $F_c = qE = (8 \times 10^{-6}$ C$)(300$ N/C$) = 2.4 \times 10^{-3}$ N downward. Then $T - w - F_c = 0$, or $T = 5.9 \times 10^{-3}$ N $+ 2.4 \times 10^{-3}$ N $= \underline{8.3 \times 10^{-3} \text{ N}}$. (**b**) $F_c = |(-8 \times 10^{-6}$ C$)| (300$ N/C$) = 2.4 \times 10^{-3}$ N in magnitude, but points upward. Then $T + F_c - w = 0$, or $T = w - F_c = 5.9 \times 10^{-3}$ N $- 2.4 \times 10^{-3}$ N, $T = \underline{3.5 \times 10^{-3} \text{ N}}$.

25.47 The tiny ball at the end of the thread shown in Fig. 25-20 has a mass of 0.60 g and is in a horizontal electric

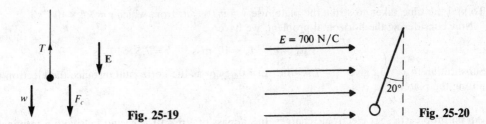

Fig. 25-19　　　　　　　　　　　　**Fig. 25-20**

field of intensity 700 N/C. It is in equilibrium in the position shown. What are the magnitude and sign of the charge on the ball?

▌ Let $w = mg = (0.60 \times 10^{-3}\,\text{kg})(9.8\,\text{m/s}^2) = 5.9\,\text{mN}$ be the weight of the ball. Let F_c be the coulomb force due to the charge q in the given electric field. Clearly F_c is to the left, so q is negative. From equilibrium we have $T \cos 20° = w = 5.9\,\text{mN}$, and $T \sin 20° = F_c$. Solving, $T = 6.3\,\text{mN}$ and so $F_c = 2.1\,\text{mN}$. Finally $F_c = |q|\,E$, or $|q| = (2.1 \times 10^{-3}\,\text{N})/(700\,\text{N/C}) = 3.1\,\mu\text{C}$. Thus $q = \underline{-3.1\,\mu\text{C}}$.

25.48　An electron ($q = -e$, $m = 9.1 \times 10^{-31}$ kg) is projected out along the $+x$ axis with an initial speed of 3×10^6 m/s. It goes 45 cm and stops due to a uniform electric field in the region. Find the magnitude and direction of the field.

▌ The force, and hence acceleration, are constant, and must be in the negative x direction. This is necessary because if there were a y component of force, the velocity component in that direction would continually increase and never become zero for $t > 0$. Furthermore the component in the x direction must act to slow down the initial velocity of the ball. From 1-D kinematics, $v_x^2 = v_{0x}^2 + 2a_x x$. For our case $v_{0x} = 3 \times 10^6$ m/s, $v_x = 0$ and $x = 0.45$ m. Then $a_x = -(3 \times 10^6)^2/(2 \times 0.45) = -1.0 \times 10^{13}$ m/s². The force causing this acceleration is $F_x = ma_x = (9.1 \times 10^{-31}\,\text{kg})(-1.0 \times 10^{13}\,\text{m/s}^2)$, or $F_x = -9.1 \times 10^{-18}$ N. F_x is due to the electrostatic force on the electron caused by the electric field, so $F_x = qE_x$. $E_x = (-9.1 \times 10^{-18}\,\text{N})/(-1.6 \times 10^{-19}\,\text{C}) = 57$ N/C. Finally since $F_y = 0$, we have the electric field $= 57$ N/C in the $+x$ direction. (Note: the force of gravity on the electron is completely negligible in comparison with the electric force and has therefore been ignored.)

25.49　A particle of mass m and charge $-e$ is projected with horizontal speed v into an electric field of intensity E directed downward. Find (a) the horizontal and vertical components of its acceleration, a_x and a_y; (b) its horizontal and vertical displacements, x and y, after time t; (c) the equation of its trajectory.

▌ (a) The electric field is downward so there is no horizontal force. Thus $a_x = 0$. Since $F_y = qE_y$ with q negative, it is clear that F_y is upward and $a_y = F_y/m = (eE)/m$, where E is the magnitude of the electric field, and $e = 1.6 \times 10^{-19}$ C. (b) $x = v_{0x}t = vt$. (We assume that $x = y = 0$ at $t = 0$.) $y = v_{0y}t + \frac{1}{2}a_yt^2 = [(eE)/2m]t^2$. (c) Eliminating t between the two equations of part b we get $y = \frac{1}{2}a_y(x/v)^2 = [(eE)/(2mv^2)]x^2$—a parabola as expected.

25.50　An electron is shot at 10^6 m/s between two parallel charged plates, as shown in Fig. 25-21. If E between the plates is 1 kN/C, where will the electron strike the upper plate? Assume vacuum conditions.

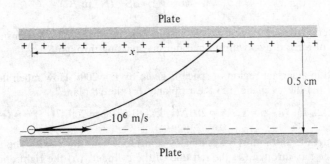

Fig. 25-21

▌ This is clearly a projectile problem, and the path followed by the particle will be as shown.
Taking upward as positive and using $e = 1.6 \times 10^{-19}$ C, $m = 9.1 \times 10^{-31}$ kg, and $E = 10^3$ N/C, we have for the vertical problem that

$$v_{0y} = 0 \qquad a = \frac{eE}{m} = 1.76 \times 10^{14}\,\text{m/s}^2 \qquad y = 0.5 \times 10^{-2}\,\text{m}$$

To find the time taken to strike the plate, use $y = v_0 t + \frac{1}{2}at^2$ from which $t = 7.5 \times 10^{-9}$ s.
Now considering the horizontal problem, we have

$$v_{0x} = v_x = \bar{v}_x = 10^6\,\text{m/s} \qquad t = 7.5 \times 10^{-9}\,\text{s}$$

Substitution in $x = \bar{v}_x t$ gives $x = 7.5 \times 10^{-3}\,\text{m} = \underline{0.75\,\text{cm}}$ as the horizontal distance the electron travels before hitting the plate.

25.51 An oil drop carries six electronic charges, has a mass of 1.6×10^{-12} g, and falls with a terminal velocity in air. What magnitude of vertical electric field is required to make the drop move upward with the same speed as it was formerly moving downward?

⫾ When the drop falls at its terminal velocity, the air friction force equals $-mg$. To make it move upward with the same speed, one has to overcome gravity as well as air friction in the downward direction, so it takes a force of 2 mg upward. Each electron carries 1.60×10^{-19} C of negative charge.

$$F = Eq = 2mg \qquad E(6 \times 1.60 \times 10^{-19}) = 2(1.6 \times 10^{-15})(9.8) \qquad E = \frac{2 \times 10^{-15}}{6 \times 10^{-19}}\,9.8 = 32.7\,\text{kN/C}$$

25.3 ELECTRIC FLUX AND GAUSS'S LAW

25.52 Use graph paper to trace a representative set of electric field lines and equipotential curves for two charges whose values and locations are $q_1 = +1$ C at $x = 0$ and $z = 0$; $q_2 = -4$ C at $x = 0$ and $z = 5$ (in cm).

⫾ Since both the electric-field lines and the equipotential curves will be symmetrical about the z axis passing through the two charges, you need perform the calculations for only the part of the xz plane lying on one side of that axis. Typical results are displayed in Fig. 25-22.

25.53 What is meant by *electric flux* and how is it related to "lines of force"?

⫾ In Fig. 25-23, Q represents the algebraic sum of charges (positive and negative) distributed throughout the region of free space shown. The dashed line indicates a surface S of arbitrary shape completely enclosing the charges Q. $\mathbf{E}$ is the electric field due to Q at some point P of S, and $d\mathbf{S} = \mathbf{n}\,dS$ is the vector element of area at P. The *electric flux* through the elementary area dS is defined as $d\psi = E_\perp\,dS$, where $E_\perp$ is the component of E perpendicular to the surface element dS. In vector form we have

$$d\psi = \mathbf{E} \cdot d\mathbf{S} \quad (\text{N} \cdot \text{m}^2/\text{C}). \tag{1}$$

In terms of lines of force (indicated by curved arrows in Fig. 25-23), which are lines to which $\mathbf{E}$ is tangent at each point, $d\psi$ can be interpreted as the number of lines cutting dS, provided that the lines are supposed drawn with normal density E; that is, the number of lines per unit perpendicular area is numerically equal to E at each location in space. (It can be shown that such lines can in principle be drawn continuously, starting on positive charges and ending on negative charges.)
The total flux through S in the outward direction is given by integration of (1) as

$$\psi = \int_S \mathbf{E} \cdot d\mathbf{S} \quad (\text{N} \cdot \text{m}^2/\text{C}) \tag{2}$$

In noncalculus terms the total flux through S is the sum of all the $E_\perp\,d\mathbf{S}$ terms around the surface, or the total number of lines of force drawn with normal density, passing through S.

25.54 The electric field in a certain region of space is given by $\mathbf{E} = 200\mathbf{i}$. How much flux passes through an area A if it is a portion of (*a*) the xy plane, (*b*) the xz plane, (*c*) the yz plane?

⫾ Refer to Fig. 25-24. (*a*) $\phi = \mathbf{E} \cdot \mathbf{A} = 200A(\mathbf{i} \cdot \mathbf{k}) = \underline{0}$ (*b*) $\phi = 200A(\mathbf{i} \cdot \mathbf{j}) = \underline{0}$ (*c*) $\phi = 200A(\mathbf{i} \cdot \mathbf{i}) = \underline{200A}$

25.55 A cylinder of length L and radius b has its axis coincident with the x axis. The electric field in this region is $\mathbf{E} = 200\mathbf{i}$. Find the flux through (*a*) the left end of the cylinder, (*b*) the right end, (*c*) the cylindrical wall, (*d*) the closed surface area of the cylinder.

⫾ Refer to Fig. 25-25. (*a*) $\mathbf{E} \cdot \mathbf{A} = (200\mathbf{i}) \cdot (-\pi b^2\mathbf{i}) = \underline{-200\pi b^2}$. (*b*) $(200\mathbf{i}) \cdot (\pi b^2\mathbf{i}) = \underline{200\pi b^2}$. (*c*) Zero since $d\mathbf{A}$ is perpendicular to $\mathbf{i}$. (*d*) Sum of *a*, *b*, and *c* is zero.

25.56 A cubical box sits in a uniform electric field as shown in Fig. 25-26. Find the net flux coming out of the box.

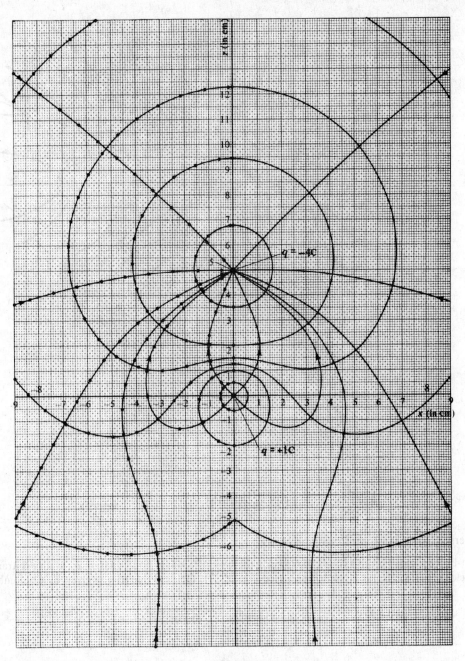

Fig. 25-22

▌ Because the field lines simply skim the four sides of the box, the flux through them is zero. The flux through the top is $\Phi_{top} = \mathbf{E} \cdot \mathbf{A}_t = EA_t$ since $\cos \theta = \cos 0° = 1$ in this case. For the bottom, $\Phi_{bottom} = \mathbf{E} \cdot \mathbf{A}_b = EA_b \cos 180° = -EA_b$. The negative sign arises because the flux is into the bottom area, not out of it.

Because $A_t = A_b$, the net flux from the box is $\Phi_{total} = EA_t - EA_b = \underline{0}$.

25.57 State Gauss' law if electric flux is defined as in Prob. 25.53.

▌ Gauss' law says if S is a closed surface and Q is the total charge enclosed by S, then the total net outward flux through S equals Q/ϵ_0. Thus

$$\Psi = \oint_S \mathbf{E} \cdot d\mathbf{S} = \oint_S \mathbf{E} \cdot \hat{\mathbf{n}} \, dS = \frac{Q}{\epsilon_0} \tag{1}$$

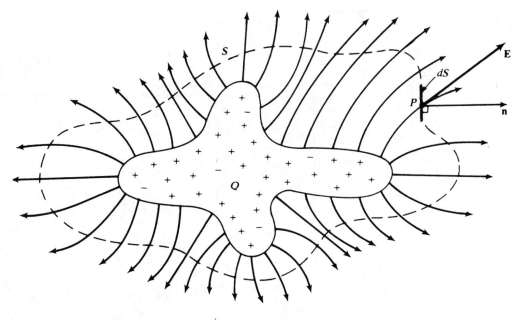

Fig. 25-23

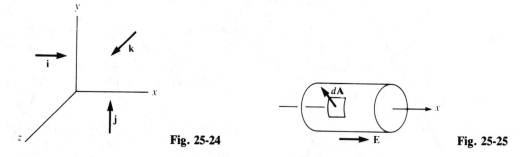

Fig. 25-24   Fig. 25-25

25.58° Derive Gauss' law from Coulomb's law.

❚ By superposition, it suffices to consider the case of a single point charge, q, at an arbitrary location inside a closed surface S. At a point P of S (see Fig. 25-27), the field due to q is given by Coulomb's law as

$$\mathbf{E} = \frac{q}{4\pi\epsilon_0 r^2}\,\mathbf{e}$$

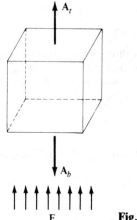

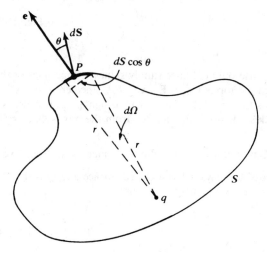

Fig. 25-26   Fig. 25-27

The flux through S is then

$$d\psi = \mathbf{E} \cdot d\mathbf{S} = \frac{q}{4\pi\epsilon_0 r^2} \mathbf{e} \cdot d\mathbf{S} = \frac{q}{4\pi\epsilon_0} \frac{dS \cos\theta}{r^2} = \frac{q}{4\pi\epsilon_0} d\Omega$$

where $d\Omega$ is the infinitesimal solid angle subtended by dS at the location of q. At that location, the total solid angle subtended by the entire closed surface S is 4π steradians. Thus,

$$\psi = \int_S d\psi = \frac{q}{4\pi\epsilon_0} \int_S d\Omega = \frac{q}{4\pi\epsilon_0}(4\pi) = \frac{q}{\epsilon_0}$$

which is Gauss' law.

Conversely, Coulomb's law can be derived from Gauss law (Prob. 25.59); hence the two laws are equivalent.

25.59   Apply Gauss' law to an appropriate surface and so obtain Coulomb's law.

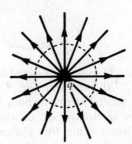

Fig. 25-28

▌ We consider a point charge, q_1 isolated in space, as shown in Fig. 25-28. Field lines have been drawn to show the spherical symmetry of the electric field. We assume a gaussian spherial surface of radius r centered at the charge q. Since the field strength E_r is equal in magnitude everywhere on the surface and is radially outward in direction, the total flux, ψ, is just $E_r A$, where A is the area of the spherical surface. Then from Gauss' law, $\psi = E_r A = 4\pi r^2 E_r = q/\epsilon_0$. Dividing by A we get $E_r = [1/(4\pi\epsilon_0)](q/r^2)$, which together with the radial direction of E_r is Coulomb's law.

Note that the flux, ψ, is not dependent on the radius of the spherical surface, r. This is of course a consequence of the $1/r^2$ nature of Coulomb's law and the fact that A increases as r^2. Furthermore, for *any* shape surface enclosing q, ψ would still be the same since the total number of lines of force (see Prob. 25.53) passing through every surface enclosing the charge is the same. These observations are at the heart of Gauss' law.

25.60   Use Gauss' law to verify that a charge Q, uniformly distributed over the surface of a sphere, is equivalent externally to a point charge Q at the center of the sphere.

▌ In Fig. 25-29, the sphere, of radius a, is surrounded by a concentric spherical gaussian surface, of radius

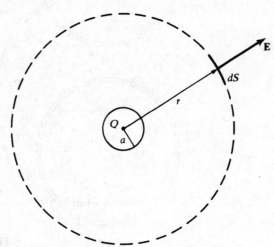

Fig. 25-29

$r > a$. By symmetry, $\mathbf{E}$ has constant magnitude E on the gaussian surface and is everywhere normal to the surface. Hence

$$\frac{Q}{\epsilon_0} = \psi = E(4\pi r^2) \qquad \text{or} \qquad E = \frac{Q}{4\pi\epsilon_0 r^2}$$

which is the same field as would be produced by a point charge Q located at the center of the sphere.

25.61 A solid metal body of arbitrary shape, as shown in Fig. 25-30, is given a certain charge. Use Gauss' law to show that this charge must distribute itself in such a way that when the distribution process ceases, all the charge is on the surface of the metal body.

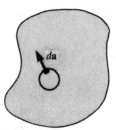

Fig. 25-30

❚ First you imagine a gaussian surface to be constructed around *any* region inside the metal body. Such a closed surface is indicated in Fig. 25-30, along with one of its typical surface elements $d\mathbf{a}$. When there is no longer a general motion of charge through the metal body because the charge has attained its equilibrium distribution, there can no longer be an electric field anywhere in the interior of the body. This is so because if there were such an electric field, then it would drive charges through the conductor, in violation of the statement that the charges are at rest. Hence, everywhere on the gaussian surface, $\mathcal{E} = 0$. And so

$$\int_{\text{closed surface}} \mathcal{E} \cdot d\mathbf{a} = \frac{q}{\epsilon_0} = 0$$

or $q = 0$, where q is the net charge contained within the gaussian surface. But this result is obtained no matter what region inside the metal body is enclosed by the gaussian surface. Therefore there is no net charge anywhere inside the metal body.

25.62 A spherical metal shell such as the one shown in Fig. 25-31(a) carries a uniform charge Q. Find the electric field at the points A and B indicated.

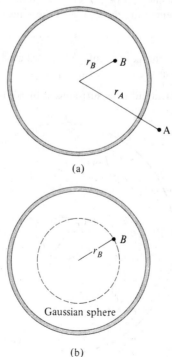

(a)

(b)

Fig. 25-31

I By Prob. 25.60,

$$E_A = \frac{Q}{4\pi\epsilon_0 r_A^2}$$

To find the field at B, we must take a gaussian surface through that point. In order to make use of the spherical symmetry, we take a gaussian sphere as shown in Fig. 25-31(b). There is no charge *inside* the sphere; and, on the sphere, $\mathbf{E}$ is radial, with $|\mathbf{E}| = E_B$. Thus

$$\psi = E_B(4\pi r_B^2) = 0 \qquad \text{or} \qquad E_B = 0$$

25.63 In Fig. 25-32, a small sphere carrying charge $+Q$ is located at the center of a spherical cavity in a large uncharged metal sphere. Find by Gauss' law the field E at points P_1 in the space between small sphere and cavity wall, at points P_2 in the metal of the large sphere, and at points P_3 outside the large sphere.

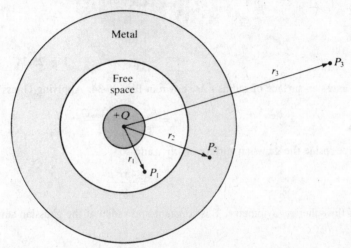

Fig. 25-32

I For a spherical gaussian surface of radius r_1 enclosing $+Q$,

$$\psi = \frac{Q}{\epsilon_0} = 4\pi r_1^2 E_1 \qquad \text{or} \qquad E_1 = \frac{Q}{4\pi\epsilon_0 r_1^2}$$

In the same way, we draw spherical gaussian surfaces at radius r_2 and again at r_3. Since the field at r_2 is inside the conducting material we must have $E_2 = 0$. Then the total charge enclosed is $= 0$, and a uniform negative charge $-Q$ must be induced on the inner surface of the hollow sphere. At radius r_3, again by symmetry and the fact that the large hollow sphere is uncharged, we again get $\psi = Q/\epsilon_0 = 4\pi r_3^2 E_3$, or $E_3 = Q/(4\pi\epsilon_0 r_3^2)$. Note that a uniform positive charge $+Q$ is induced on the outer surface of the hollow sphere. This follows from conservation of charge as well as from the last result for E_3. (See Prob. 25.61 to understand why no static charge resides in the interior of a conductor in electrical equilibrium.)

25.64 The plates of a capacitor, Fig. 25-33, carry charges $+Q$ and $-Q$ as shown. Each plate has area $A = 600\ \text{cm}^2$.

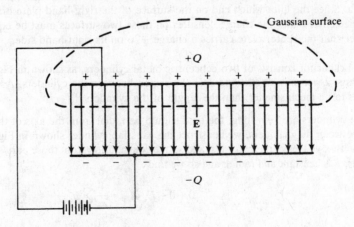

Fig. 25-33

Between the plates, the field is constant at $E = 300\,\text{kV/m}$ and the field is assumed zero outside the plates (i.e., no fringing). Evaluate Q.

▮ Employing the gaussian surface shown in Fig. 25-33 and assuming $\mathbf{E}$ normal to the inner surface of the plate, we have $Q/\epsilon_0 = \int_S \mathbf{E} \cdot d\mathbf{S} = EA$ (or no flux through outside regions and flux between plates $= EA$) or $Q = \epsilon_0 EA = (8.854 \times 10^{-12})(300 \times 10^3)(600 \times 10^{-4}) = 1.59 \times 10^{-7}\,\text{C} = \underline{0.159\,\mu\text{C}}$.

25.65 A sphere of radius a is made of insulating material and has charge distributed uniformly throughout its volume; the charge density is ρ. Find the field due to the charge for $r \leq a$.

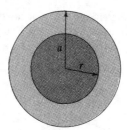

Fig. 25-34

▮ Choose a gaussian surface of radius r as shown in Fig. 25-34. Applying Gauss' law to it,

$$\psi = \frac{q_{\text{inside}}}{\epsilon_0} = \frac{\rho V_{\text{inside}}}{\epsilon_0}$$

But the volume inside the gaussian surface is $\frac{4}{3}\pi r^3$, and so

$$\psi = \frac{4\pi r^3 \rho}{3\epsilon_0}$$

Because of the spherical symmetry, $\mathbf{E}$ is constant and radial at the gaussian surface. The above equation becomes

$$\psi = EA = \frac{4\pi r^3 \rho}{3\epsilon_0}$$

But A is simply the area of the gaussian sphere, $4\pi r^2$. Substituting and solving gives

$$E = \frac{r\rho}{3\epsilon_0}$$

The field is due entirely to the charge inside the radius r.

25.66 The metal plate on the left in Fig. 25-35(a) carries a surface charge of $+\sigma$ per unit area. The metal plate on the right has a surface charge of -2σ per unit area. Find the charge densities on the two surfaces of the metal plate in the center. The center plate is assumed to be connected to the earth, so it need not be neutral. Assume the plates to be very large.

▮ We can solve this problem by drawing flux lines as shown in Fig. 25-35(b). The lines coming from the plate on the left must end on the left side of the center plate. Therefore the charge density there must be $-\sigma$ per unit area. Since the lines which end on the surface of the right-hand plate originate on the right-hand surface of the center plate, the charge densities on these two surfaces must be equal and opposite. As a result, we see that the center plate carries a charge $+2\sigma$ on its right-hand side.

25.67 A cylindrical capacitor consists of two concentric metal cylinders, as shown in Fig. 25-36. If the inner cylinder carries a charge λ per unit length, find E in the region between the cylinders. Assume the distance between the cylinders to be much smaller than the length of the cylinders.

▮ Since the cylinders are very long, the field should be radial from the axis of the cylinders. To find the field at point P between the cylinders, we construct the gaussian cylinder shown in Fig. 25-36. Note that $a < r < b$. As before, no flux will leave the two ends, and so the contribution from those parts of the gaussian surface will be zero. Further, E is constant and radial, and so we find

$$0 + 0 + E \cdot A_{\text{lat}} = \frac{1}{\epsilon_0} \sum q_i$$

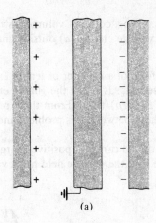

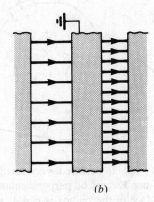

Fig. 25-35

The charge inside the gaussian cylinder is just λL, and the lateral area is $2\pi rL$. Hence

$$E(2\pi rL) = \frac{1}{\epsilon_0}\lambda L \quad \text{or} \quad E = \frac{\lambda}{2\pi r\epsilon_0}$$

25.68 A thin, long, straight wire carries a charge λ_1 per unit length. The wire lies along the axis of a long metal cylinder which carries a net charge λ_2 per unit length. The inner radius of the cylinder is b, and its outer radius is c. Find the electric field in the following three regions: $r < b$, $b < r < c$, $r > c$. How much charge per unit length exists on the inner surface of the cylinder? The outer surface?

▌ Use gaussian cylinders of length L. No flux passes through the ends. In each case $2\pi rLE = $ (charge enclosed)$/\epsilon_0$. For $r < b$, the enclosed charge is $\lambda_1 L$ and $E = \lambda_1/(2\pi\epsilon_0 r)$. For $b < r < c$, we are in the metal, so $E = 0$. At $r > c$, $(\lambda_1 + \lambda_2)L$ is the enclosed charge; $E = (\lambda_1 + \lambda_2)/(2\pi\epsilon_0 r)$. The charge on the inner surface is $-\lambda_1$, since λ_1 must be canceled by it (from Gauss' law case $b < r < c$). This leaves $\lambda_2 - (-\lambda_1) = \lambda_2 + \lambda_1$ on the outer surface.

25.69 Two long straight parallel wires carry charges λ_1 and λ_2 per unit length. The separation between their axes is b. Find the magnitude of the force exerted on unit length of one due to the charge on the other.

▌ The field at 2 due to 1 is $E = \lambda_1/(2\pi\epsilon_0 b)$ (see Prob. 25.68), exerting a force of $E\lambda_2$ on a unit length of 2; thus $F/L = (\lambda_1\lambda_2)/(2\pi\epsilon_0 b)$.

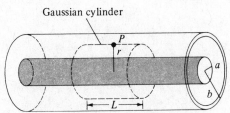

Fig. 25-36

25.70 A long straight cylinder of radius b has a constant volume distribution of charge ρ coulombs per unit volume. Find the electric field due to this volume charge (***a***) outside and (***b***) inside the cylinder. (***c***) Where is the field strongest? (***d***) Weakest?

▮ Use a gaussian surface in the form of a cylinder of length L. All the flux goes through the sides (not ends), so $2\pi r L E = $ (charge enclosed)$/\epsilon_0$. If $r > b$, the charge enclosed is $\pi b^2 L \rho$, so $E = (b^2 \rho)/(2\epsilon_0 r)$. Inside, the charge enclosed is $\pi r^2 L \rho$, so $E = (r\rho)/(2\epsilon_0)$. From these results, E is largest at the surface and zero on the axis. Note the structural similarly between this problem and Prob. 25.65.

25.71 A metal can suspended by a silk thread carries a positive charge (not uniformly distributed). At a certain point very near the surface of the can, the electric field has a value $E = 600\,\text{kV/m}$. Evaluate the surface charge density σ near that point.

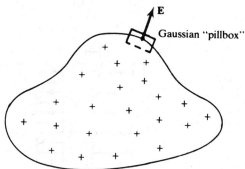

Fig. 25-37

▮ Apply Gauss' law, choosing as gaussian surface a very small, thin "pillbox" sunk halfway into the conducting surface (Fig. 25-37). Since **E** must be perpendicular to the surface of a metal in electric equilibrium, the flux through one face of the pillbox is $E\,\Delta A$, and the flux through the other two faces is zero. Hence,

$$E\,\Delta A = \frac{\sigma\,\Delta A}{\epsilon_0} \quad \text{or} \quad \sigma = \epsilon_0 E = (8.854 \times 10^{-12})(600 \times 10^3) = \underline{5.313\,\mu\text{C/m}^2}$$

25.72 The *electric strength* of air is about $3.0 \times 10^6\,\text{N/C}$. By this we mean that if the electric field exceeds this value, sparking will occur. What is the largest charge a 0.50-cm-radius metal sphere can hold if sparking is not to occur in the air surrounding it?

▮ Outside the sphere all the charge on the surface acts as if located at the center; thus we find Q from $E = [1/(4\pi\epsilon_0)](Q/r^2)$. With $r = 5.0 \times 10^{-3}\,\text{m}$ and $E = 3.0 \times 10^6\,\text{N/C}$, obtain $Q = \underline{8.34\,\text{nC}}$.

CHAPTER 26

Electric Potential and Capacitance

26.1 POTENTIAL DUE TO POINT CHARGES OR CHARGE DISTRIBUTIONS

26.1 What is the *electric potential energy* of a charge q in an electric field due to a group of fixed charges?

▐ Potential energy at a point P = minus the work done *by* the electric forces (or plus the work done *against* these forces) in taking an object (in this case the charge q) from a given reference point to point P. PE has meaning only for conservative forces. The Coulomb force is conservative, having the same basic form as the gravitational force, except that it can be either repulsive or atttractive. If the electric field is due to a single charge, q_1, and the reference point for potential energy is taken as $r = \infty$, we have $\mathbf{F} = q\mathbf{E} = [(kqq_1)/r_1^2]\hat{\mathbf{r}}_1$ ($\mathbf{r}_1$ points from q_1 to q), and in analogy to gravity, $PE = U = (kqq_1)/r_1$. (If q and q_1 have opposite signs, $\mathbf{F}$ is attractive and U is negative, as for gravity.) Since the work done by a vector sum of forces equals the sum of the works done by the individual forces, we have, for the field $\mathbf{E}$ due to a series of point charges, $q_1, q_2, \ldots, q_n$,

$$\mathbf{F} = q\mathbf{E} = \sum_{i=1}^{n} \frac{kqq_i}{r_i^2}\hat{\mathbf{r}}_i \quad \text{and} \quad PE = U = \sum_{i=1}^{n} \frac{kqq_i}{r_i}$$

26.2 (a) What is meant by the *electric potential* (often called *absolute electric potential*) at a point in space? (b) How is this related to the electric field?

▐ (a) The electric potential is the electric potential energy (Prob. 26.1), with respect to infinity, of a unit positive charge. The unit of electric potential is the *volt*, where $1\,\text{V} = 1\,\text{J/C}$. (b) In the notation of Prob. 26.1, write $V = U/q$ for the electric potential. Then, by definition of electric potential energy, a small change Δs (in a specified direction s) in the position of a unit positive charge gives rise to the change

$$\Delta V = -E_s\,\Delta s \tag{1}$$

in the electric potential. In particular, for displacements parallel to the coordinate axes, (1) gives

$$E_x = -\frac{\Delta V}{\Delta x} \qquad E_y = -\frac{\Delta V}{\Delta y} \qquad E_z = -\frac{\Delta V}{\Delta z} \tag{2}$$

Conversely, (1) follows from (2). Note that (1) implies the following relation among units: $1\,\text{V/m} = 1\,\text{N/C}$.

26.3^c Define electric potential energy and electric potential using the precise language of calculus.

▐ The electric force $\mathbf{F}$ exerted on a test charge q' by some stationary distribution of charges is a conservative force. Therefore, the test charge possesses electric potential energy U. With the reference point taken at infinity, the absolute potential energy of the test charge at the location $B(x, y, z)$ is given by

$$U(x, y, z) = -\int_{\infty}^{(x,y,z)} \mathbf{F} \cdot d\mathbf{s} = -\int_{\infty}^{(x,y,z)} F_x\,dx + F_y\,dy + F_z\,dz \quad \text{(J)}$$

with the integral taken along any path.

Conversely, if the electric potential energy is known as a function of position, then the component of the electric force along an arbitrary direction $d\mathbf{s}$ can be calculated as

$$F_s = -\frac{dU}{ds}$$

In particular, the components of $\mathbf{F}$ along the X, Y, and Z axes are given by

$$F_x = -\frac{\partial U}{\partial x} \qquad F_y = -\frac{\partial U}{\partial y} \qquad F_z = -\frac{\partial U}{\partial z}$$

so that, in vector form,

$$\mathbf{F} = -\left(\frac{\partial U}{\partial x}\mathbf{i} + \frac{\partial U}{\partial y}\mathbf{j} + \frac{\partial U}{\partial z}\mathbf{k}\right) \quad \text{(N)}$$

If the only force acting on a point charge is the electric force **F**, conservation of energy takes the form $\Delta K + \Delta U = 0$, where, as usual, K is the kinetic energy of the point charge.

The electric potential energy per unit test charge is called the *electric potential*, ϕ, or the *voltage*, V. Since $\phi = U/q'$ and $\mathbf{E} = \mathbf{F}/q'$, ϕ and $\mathbf{E}$ enjoy the same relation as U and $\mathbf{F}$:

$$\phi(x, y, z) = -\int_{\infty}^{(x,y,z)} \mathbf{E} \cdot d\mathbf{s} = -\int_{\infty}^{(x,y,z)} E_x \, dx + E_y \, dy + E_z \, dz \quad (\text{V}) \qquad \mathbf{E} = -\left(\frac{\partial \phi}{\partial x}\mathbf{i} + \frac{\partial \phi}{\partial y}\mathbf{j} + \frac{\partial \phi}{\partial z}\mathbf{k}\right) \quad (\text{V/m})$$

26.4   What is meant by the *potential difference* between two points in space?

▌ The potential difference from one point, A, to another point, B, is the work done *against* electric forces in carrying a unit positive test charge from A to B. Thus the potential difference from A to B is $V_B - V_A \equiv V$. Its units are those of potential, joules/coulomb = volts, and it is often called voltage or "voltage change."

Because work is a scalar quantity, so too is potential difference. Like work, potential difference may be positive or negative.

The work W done *against* electric forces in transporting a charge q from one point, A, to a second point, B, is $W = q(V_B - V_A) = qV$.

26.5ᶜ   Derive an expression for the potential due to a point charge.

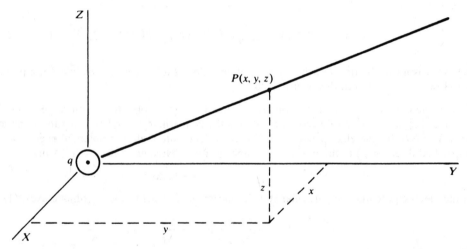

Fig. 26-1

▌ If the source of the field is a point charge q located at the origin (Fig. 26-1), then, the field at $P'(x', y', z')$ is given by

$$\mathbf{E}' = \frac{q}{4\pi\epsilon_0 r'^3}\mathbf{r}'$$

We choose as the path of integration the line from ∞ to $P(x, y, z)$ that, extended, meets q. Along this path,

$$ds = -dr' \qquad \text{and} \qquad E_s = -E' = -\frac{q}{4\pi\epsilon_0 r'^2}$$

Consequently,

$$\phi = -\int_{\infty}^{P} E_s \, ds = -\int_{\infty}^{P} \frac{q \, dr'}{4\pi\epsilon_0 r'^2} = -\frac{q}{4\pi\epsilon_0}\int_{\infty}^{r} \frac{dr'}{r'^2} = \frac{q}{4\pi\epsilon_0 r}$$

26.6   What is an *equipotential surface*?

▌ A surface (or, in two dimensions, a curve) on which the potential is constant is called an *equipotential surface (curve)*. Thus, in cartesian coordinates, the equipotential surfaces are given by the equation $\phi(x, y, z) = c$, with one surface for each value of the constant c. In terms of the equipotentials, the relation between ϕ and **E** may be geometrically expressed as follows: *the field lines and the equipotentials are everywhere perpendicular*. Indeed, for an infinitesimal displacement $d\mathbf{s}$ lying in an equipotential surface, $d\phi = 0 = \mathbf{E} \cdot d\mathbf{s}$; and so **E** must be perpendicular to $d\mathbf{s}$.

For conductors in electrostatic equilibrium (no moving charges) the surfaces are equipotential surfaces. (Indeed, since the electric field is zero inside the conductor, the entire conductor is at constant potential.)

26.7 Define the *electronvolt*.

▌ The work done in carrying a charge $+e$ (coulombs) through a potential rise of 1 V is defined to be 1 *electronvolt* (eV). Therefore,

$$1 \, eV = (1.602 \times 10^{-19} \, C)(1 \, V) = 1.602 \times 10^{-19} \, J$$

Equivalently,

$$\text{work or energy (in eV)} = \frac{\text{work (in J)}}{e}$$

26.8 If in Fig. 26-1, $q = 40 \, \mu C$, $x = 0.5 \, m$, $y = 0.8 \, m$, $z = 0.6 \, m$, what is the potential at point P?

▌ Here $r = (0.5^2 + 0.8^2 + 0.6^2)^{1/2} = 1.118 \, m$. Hence,

$$\phi = k\frac{q}{r} = (9 \times 10^9)\frac{40 \times 10^{-6}}{1.118} = 3.22 \times 10^5 \, V = \underline{322 \, kV}$$

26.9 In Prob. 26.8, a charge $q_1 = 9 \, \mu C$ is placed at P. Compute its electric potential energy.

▌ $$U = q_1 \phi = (9 \times 10^{-6})(3.22 \times 10^5) = \underline{2.9 \, J}$$

26.10 A point charge, $q_1 = +2 \, \mu C$, is placed at the origin of coordinates. A second, $q_2 = -3 \, \mu C$, is placed on the x axis at $x = 100 \, cm$. At what point (or points) on the x axis will the absolute potential be zero?

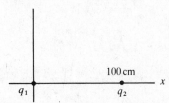

100 cm

q_1 q_2 **Fig. 26-2**

▌ The two charges are as shown in Fig. 26-2. At a point x along the x axis the potential is

$$V(x) = k\left(\frac{q_1}{|x|} + \frac{q_2}{|x - 1|}\right)$$

Setting $V = 0$ yields

$$\frac{2}{|x|} - \frac{3}{|x - 1|} = 0 \quad \text{or} \quad 2|x - 1| = 3|x|$$

We consider three cases $x > 1$, $0 < x < 1$, and $x < 0$.
For $x > 1$: $2(x - 1) = 3x \Rightarrow x = -2$ (contradiction and no solution)
For $0 < x < 1$: $2(1 - x) = 3x \Rightarrow x = 0.4$ or $x = \underline{40 \, cm}$
For $x < 0$: $2(1 - x) = -3x \Rightarrow x = -2$ or $x = \underline{-200 \, cm}$

26.11 Determine the absolute potential in air at a distance of 3 cm from a point charge of 50 nC.

▌ The dielectric constant of air can be taken as 1.0. Then, $V = [1/(4\pi\epsilon_0)](q/r) = (9 \times 10^9)(5 \times 10^{-8})/(0.03) = \underline{15 \, kV}$.

26.12 Show that the absolute potential at point P in Fig. 26-3 is zero.

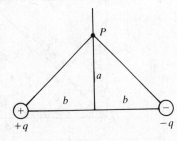

P

a

b b

$+q$ $-q$ **Fig. 26-3**

∎
$$\phi_P = \frac{1}{4\pi\epsilon_0}\frac{+q}{\sqrt{a^2+b^2}} + \frac{1}{4\pi\epsilon_0}\frac{-q}{\sqrt{a^2+b^2}} = 0$$

[Note how much simpler it is directly to add potentials (scalars) than it is to superpose electric fields (vectors) and then to integrate.]

26.13 What is the absolute potential of a 25-cm-radius metal sphere that carries a charge of 65 nC?

∎ The metal sphere is an equipotential surface and, by symmetry, the charge distributes itself uniformly on the surface. The potential is thus that of a point charge at the center:
$$V = [(9.0 \times 10^9)(65 \times 10^{-9})]/0.25 = \underline{+2340 \text{ V}}.$$

26.14 A hollow sphere has a radius of 10 cm and bears a charge of 5 nC. If the sphere is isolated, what is the potential a distance of 50 cm from the center of the sphere?

∎ As in Prob. 26.13, $V = (kQ)/r = [(9 \times 10^9)(5 \times 10^{-9})]/0.50 = \underline{90 \text{ V}}$.

26.15 What is the potential at the point midway between charges of +2 and +5 μC which are 6 m apart?

∎ From Fig. 26-4,
$$V = \frac{kq_1}{r_1} + \frac{kq_2}{r_2} = \frac{(9 \times 10^9)(2 \times 10^{-6})}{3} + \frac{(9 \times 10^9)(5 \times 10^{-6})}{3} = 21 \text{ kV}$$

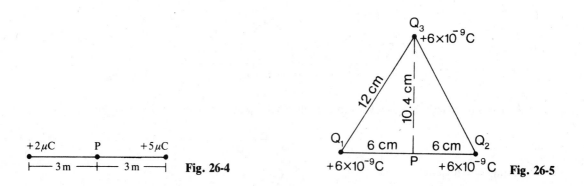

Fig. 26-4 Fig. 26-5

26.16 Three equal charges of +6 nC are located at the corners of an equilateral triangle whose sides are 12 cm long (Fig. 26-5). Find the potential at the center of the base of the triangle.

∎
$$V_P = (9 \times 10^9)\left(\frac{6 \times 10^{-9}}{0.06} + \frac{6 \times 10^{-9}}{0.06} + \frac{6 \times 10^{-9}}{0.104}\right) = \underline{2320 \text{ V}}$$

26.17 Three charges are placed at three corners of a square (Fig. 26-6). Find the potential at point A.

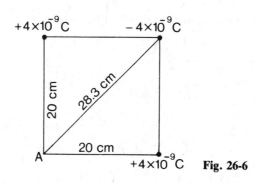

Fig. 26-6

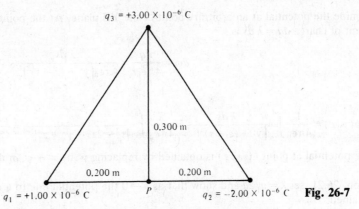

Fig. 26-7

$$V_A = (9 \times 10^9)\left(\frac{4 \times 10^{-9}}{0.20} + \frac{-4 \times 10^{-9}}{0.283} + \frac{4 \times 10^{-9}}{0.20}\right) = \underline{233\ V}$$

26.18 Figure 26-7 shows a triangular array of three point charges. Find the electric potential V of these source charges at the midpoint P of the base of the triangle.

$$V = (8.99 \times 10^9\ \text{N} \cdot \text{m}^2/\text{C}^2)\left(\frac{+1.00 \times 10^{-6}\ \text{C}}{0.200\ \text{m}} + \frac{-2.00 \times 10^{-6}\ \text{C}}{0.200\ \text{m}} + \frac{+3.00 \times 10^{-6}\ \text{C}}{0.300\ \text{m}}\right) = \underline{45.0\ \text{kV}}$$

26.19 A metal sphere 30 cm in radius is positively charged with $2\ \mu\text{C}$. Find the potential at the center of the sphere, on the sphere, and at 1 m from the center of the sphere.

▌ The potential of the sphere or of any point inside is the same (see Prob. 26.6).

$$V = \frac{kQ}{R} = \frac{(9 \times 10^9)(2 \times 10^{-6})}{0.30} = 60\ \text{kV} \qquad \text{on the surface or at the center}$$

For an external point, all charge may be considered concentrated at the center of the sphere.

$$V = \frac{kQ}{r} = \frac{(9 \times 10^9)(2 \times 10^{-6})}{1} = \underline{18\ \text{kV}} \qquad \text{at 1 m from the center}$$

26.20 Two metal spheres (radii a and b) are very far apart but are connected by a thin wire. Their combined charge is Q. (**a**) What is the charge on each? (**b**) What is their absolute potential?

▌ (**a**) Because a wire connects them, both spheres must be at the same potential. From Prob. 26.13, $V_a = (kQ_a)/a = V_b = (kQ_b)/b$. Also $Q = Q_a + Q_b$; so solving between these relations, $Q_a = (aQ)/(a + b)$ and $Q_b = (bQ)/(a + b)$. (**b**) Substitute into either V_a or V_b to find $V = (kQ)/(a + b)$.

26.21^c Find the potential due to a thin uniformly charged rod of length $2a$.

▌ Take the rod along the Z axis, as in Fig. 26-8. By the rotational symmetry of the problem, it is enough to

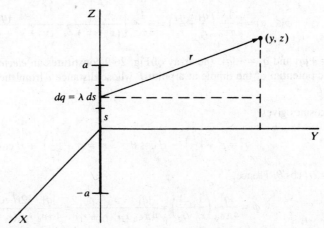

Fig. 26-8

determine the potential at an arbitrary point of the YZ plane. At the point (y, z), the potential due to the element of charge $dq = \lambda \, ds$ is

$$d\phi = \frac{dq}{4\pi\epsilon_0 r} = \frac{\lambda \, ds}{4\pi\epsilon_0 [y^2 + (z-s)^2]^{1/2}}$$

Hence,

$$\phi(y, z) = \frac{1}{4\pi\epsilon_0} \int_{-a}^{a} \frac{\lambda \, ds}{[y^2 + (z-s)^2]^{1/2}} = \frac{\lambda}{4\pi\epsilon_0} \int_{-a}^{a} \frac{ds}{[s^2 - 2zs + y^2 + z^2]^{1/2}} = \frac{\lambda}{4\pi\epsilon_0} \ln \frac{[y^2 + (z-a)^2]^{1/2} - (z-a)}{[y^2 + (z+a)^2]^{1/2} - (z+a)}$$

The potential at point (x, y, z) is obtained by replacing y^2 by $x^2 + y^2$ in the above expression.

26.22ᶜ In Prob. 26.21, set $\lambda = q/2a$ and show that as $a \to 0$ the potential due to a point charge q located at the origin is obtained.

▮
$$\phi(y, z) = \frac{q}{4\pi\epsilon_0} \left\{ \frac{1}{2a} \int_{-a}^{a} \frac{1}{[y^2 + (z-s)^2]^{1/2}} \, ds \right\}$$

The expression in brackets is just the mean value of $[y^2 + (z-s)^2]^{-1/2}$ over the interval $-a < s < a$; hence, as $a \to 0$, $\{ \ \} \to [y^2 + (z-0)^2]^{-1/2} = 1/r$, and so $\phi(y, z) \to q/(4\pi\epsilon_0 r)$. [The result may also be obtained from the logarithmic expression for ϕ.]

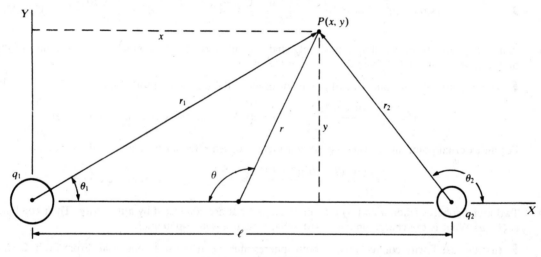

Fig. 26-9

26.23 Refer to Fig. 26-9 and write an expression for the potential $\phi(x, y)$ at a general point $P(x, y)$ in the XY plane.

▮ Since $r_1 = (x^2 + y^2)^{1/2}$ and $r_2 = [(l-x)^2 + y^2]^{1/2}$,

$$\phi(x, y) = \frac{1}{4\pi\epsilon_0} \left(\frac{q_1}{r_1} + \frac{q_2}{r_2} \right) = \frac{1}{4\pi\epsilon_0} \left[\frac{q_1}{(x^2 + y^2)^{1/2}} + \frac{q_2}{(x^2 + y^2 + l^2 - 2xl)^{1/2}} \right]$$

26.24 In the case $q_1 = +|q|$ and $q_2 = -|q|$, the array of Fig. 26-9 constitutes an *electric dipole* (of *moment* $\mu \equiv |q| \, l$). Find the electric potential of the dipole at a point P whose distance r from the center of the dipole is large compared with l.

▮ The law of cosines gives

$$r_1^2 = r^2 + \left(\frac{l}{2}\right)^2 - rl \cos \theta \qquad r_2^2 = r^2 + \left(\frac{l}{2}\right)^2 + rl \cos \theta$$

so that $r_2^2 - r_1^2 = 2rl \cos \theta$. Hence,

$$\phi = \frac{|q|}{4\pi\epsilon_0} \left(\frac{1}{r_1} - \frac{1}{r_2} \right) = \frac{|q|}{4\pi\epsilon_0} \frac{r_2^2 - r_1^2}{r_2 r_1 (r_2 + r_1)} = \frac{|q|}{4\pi\epsilon_0} \frac{2rl \cos \theta}{r_2 r_1 (r_2 + r_1)}$$

But, when $r \gg l$, we have $r_2 r_1 \approx r^2$ and $r_2 + r_1 \approx 2r$, giving

$$\phi \approx \frac{|q|}{4\pi\epsilon_0} \frac{2rl \cos\theta}{2r^3} = \frac{\mu \cos\theta}{4\pi\epsilon_0 r^2}$$

This approximate formula becomes exact when $l \to 0$ and $|q| \to \infty$ such that μ remains constant and the two charges form a *point dipole*.

26.2 THE POTENTIAL FUNCTION AND THE ASSOCIATED ELECTRIC FIELD

26.25ᶜ A circular disk carries a surface charge σ (C/m²). Show that the electric field at any point on the axis of the disk depends only on σ and the angle α subtended there by the disk.

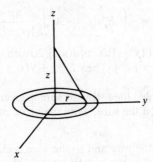

Fig. 26-10

❚ We consider the charged disk of radius R to lie in the xy plane centered at the origin (Fig. 26-10). Rather than calculate the field along the z axis directly, we first calculate the potential, $V(z)$. We consider the disk to be made up of concentric rings, each ring contributing to the potential:

$$dV = \frac{\sigma}{4\pi\epsilon_0} \frac{(2\pi r \, dr)}{(r^2 + z^2)^{1/2}}$$

We then integrate from $r = 0$ to $r = R$:

$$V(z) = \frac{\sigma}{2\epsilon_0} \int_0^R \frac{r \, dr}{(r^2 + z^2)^{1/2}} = \frac{\sigma}{2\epsilon_0} (r^2 + z^2)^{1/2} \Big|_0^R = \frac{\sigma}{2\epsilon_0} [(R^2 + z^2)^{1/2} - |z|] \tag{1}$$

Noting that $|z|/(R^2 + z^2)^{1/2} = \cos(\alpha/2)$ we have, factoring $|z|$ out of the brackets,

$$V(z) = \frac{\sigma |z|}{2\epsilon_0} \left[\frac{1}{\cos(\alpha/2)} - 1 \right]$$

Noting that by symmetry the electric field along the z axis is parallel to that axis, we have, using form (1),

$$\mathbf{E}(z) = E_z(z)\mathbf{k}, \quad \text{with}$$

$$E_z(z) = \frac{-dV(z)}{dz} \equiv -V'(z) = \begin{cases} \dfrac{\sigma}{2\epsilon_0}\left[1 - \dfrac{z}{(R^2 + z^2)^{1/2}}\right] & z > 0 \\[3mm] -\dfrac{\sigma}{2\epsilon_0}\left[1 + \dfrac{z}{(R^2 + z^2)^{1/2}}\right] & z < 0 \end{cases}$$

$$= \begin{cases} \dfrac{\sigma}{2\epsilon_0}\left(1 - \cos\dfrac{\alpha}{2}\right) & z > 0 \\[3mm] -\dfrac{\sigma}{2\epsilon_0}\left(1 - \cos\dfrac{\alpha}{2}\right) & z < 0 \end{cases}$$

26.26 Show that if the disk of Prob. 26.25 has a fixed charge, q, the potential on the z axis reduces to that of a point charge at the origin in the limit $R/|z| \to 0$ (i.e., far out on the z axis).

❚ By (1) of Prob. 26.25,

$$V(z) = \frac{\sigma}{2\epsilon_0}[(R^2 + z^2)^{1/2} - |z|]$$

We note that $\sigma = q/(\pi R^2)$ and that $(R^2 + z^2)^{1/2} = |z|(1 + R^2/z^2)^{1/2} \approx |z|[1 + \frac{1}{2}(R^2/z^2)]$, for $R/|z| \ll 1$. Then,

$$V(z) \approx \frac{q |z|}{2\pi\epsilon_0 R^2}\left(\frac{1}{2}\frac{R^2}{z^2}\right) = \frac{1}{4\pi\epsilon_0}\frac{q}{|z|}$$

(In general, any finite distribution of charge "looks like" a single point charge from a sufficient distance.)

26.27^c If the potential in the region of space near the point $(-2, 4, 6\,\mathrm{m})$ is $V = 80x^2 + 60y^2$ V, what are the three components of the electric field at that point?

▎ We have $E_x = -\partial V/\partial x = -160x = \underline{320\ \mathrm{V/m}}$, and $E_y = -\partial V/\partial y$ is $\underline{-480\ \mathrm{V/m}}$. Similarly, $E_z = \underline{0}$.

26.28^c The electric field **E** set up by a certain distribution of charges is two-dimensional, with X component

$$E_x = A[(x - l)(x^2 + y^2 + l^2 - 2xl)^{-3/2} - x(x^2 + y^2)^{-3/2}]$$

where A and l are constants. Determine E_y.

▎ We know that $E_y = -\partial\phi/\partial y$. To obtain ϕ we note that

$$\phi(x, y) = -\int_\infty^x E_x\,dx = -A\left[\int_\infty^x \frac{(x - l)\,dx}{[(x - l)^2 + y^2]^{3/2}} - \int_\infty^x \frac{x\,dx}{(x^2 + y^2)^{3/2}}\right]$$

Noting that $x\,dx = [d(x^2)]/2$ and $(x - l)\,dx = [d(u^2)]/2$, with $u \equiv x - l$, we get $\phi(x, y) = A[(x^2 + y^2 + l^2 - 2xl)^{-1/2} - (x^2 + y^2)^{-1/2}]$. Then $E_y = A[y(x^2 + y^2 + l^2 - 2xl)^{-3/2} - y(x^2 + y^2)^{-3/2}]$.

26.29^c The electric field outside a charged long straight wire is given by $E = -5000/r$ V/m and is radially inward. What is the sign of the charge on the wire? Find the value of $V_B - V_A$ if $r_B = 60$ cm and $r_A = 30$ cm. Which point is at the higher potential?

▎ The field is directed toward the wire and so the wire is charged negatively. Going from A to B is opposite to the direction of the field, so B is at a higher potential than A. Thus

$$V_B - V_A = -\int_{0.3}^{0.6} \frac{-5000}{r}\,dr = 5000\ln 2 = \underline{3470\ \mathrm{V}}$$

26.30 A 30-cm-diameter metal sphere hangs from a thread in a very large room, so its surroundings are essentially at infinity. If the electric field at its surface is to be equal to the breakdown strength of air, 3 MV/m, what must be the absolute potential of the sphere?

▎ E just outside the surface is 3×10^6 V/m; this is due to the charge Q on the uniformly charged sphere. Now $E = (kQ)/r^2$, while V just at the surface is $V = (kQ)/r$. Taking ratios, $V = Er = (3 \times 10^6)(0.15) = 4.5 \times 10^5$ V. No field exists in the metal, so $V = \underline{450\ \mathrm{kV}}$ throughout the volume of the metal.

26.31^c The potential at a point P a distance y above the bottom negative plate of a parallel-plate combination is $V(y) - V(0) = ky$. In this expression $V(0)$ is the potential at the lower plate, and $V(y)$ is the potential at the height y above it. Find E between the parallel plates.

▎ The situation can be simplified if we take $V(0) = 0$. This is allowable since the zero potential is arbitrary. We then have $V(y) = ky$. Making use of results of Prob. 26.3 we have

$$E_x = -\frac{\partial}{\partial x}(ky) = 0 \qquad E_y = -\frac{\partial}{\partial y}(ky) = -k \qquad E_z = -\frac{\partial}{\partial z}(ky) = 0$$

The field therefore points downward (in the $-y$ direction) between the plates and has a constant value k, as expected.

26.32 Suppose the parallel metal plates shown in Fig. 26-11 are spaced 0.50 cm apart and are connected to a 90-V battery. Find the electric field between them and the surface charge density on the plates.

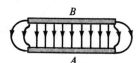

Fig. 26-11

▎ We know that $V_B - V_A = 90$ V and that E between the plates is constant. Thus

$$V_B - V_A = Ed \qquad \text{or} \qquad 90\ \mathrm{V} = E(5 \times 10^{-3}\ \mathrm{m})$$

Solving gives $E = \underline{18\ \mathrm{kV/m}}$. By Prob. 25.71, the charge density on the positive plate is $\sigma = \epsilon_0 E = (8.85 \times 10^{-12}\ \mathrm{C^2/N \cdot m^2})(18\,000\ \mathrm{N/C}) = \underline{159\ \mathrm{nC/m^2}}$; similarly, the value of σ at the negative plate is $\underline{-159\ \mathrm{nC/m^2}}$.

26.33 A potential difference of 150 V is applied to two parallel metal plates. If an electric field of 5000 V/m is produced between the plates, how far apart are the plates?

▌ E is constant so

$$E = \frac{V}{d} \quad \text{or} \quad 5000 = \frac{150}{d} \quad \text{and} \quad d = 0.03 \text{ m} = 3 \text{ cm}$$

26.34 The charge on an electron is 1.6×10^{-19} C in magnitude. An oil drop has a weight of 3.2×10^{-13} N. With an electric field of 5×10^5 V/m between the plates of Millikan's oil-drop apparatus, this drop is observed to be essentially balanced. What is the charge on the drop in electronic charge units?

▌ $E = 5 \times 10^5$ V/m $= 5 \times 10^5$ N/C. For equilibrium, $F = Eq = mg$

$$5 \times 10^5 q = 3.2 \times 10^{-13} \quad q = 6.4 \times 10^{-19} \text{ C} \quad \text{on the oil drop}$$

$$\frac{q}{e} = \frac{6.4 \times 10^{-19}}{1.6 \times 10^{-19}} = \underline{4 \text{ electrons}}$$

26.35 In the Millikan experiment, an oil drop carries four electronic charges and has a mass of 1.8×10^{-12} g. It is held almost at rest between two horizontal charged plates 1.8 cm apart. What voltage must there be between the two charged plates?

▌ $E = V/d$. Set the upward electric force equal to the downward force of gravity.

$$F = Eq = mg \quad \frac{Vq}{d} = mg \quad \frac{V(4 \times 1.60 \times 10^{-19})}{0.018} = (1.8 \times 10^{-15})(9.8) \quad \text{and} \quad V = \underline{496 \text{ V}}$$

26.36 Two large parallel metal plates (3.00 mm apart) are charged to a potential difference of 12 V. (*a*) What is the field between them? (*b*) They are now disconnected from the battery and pulled apart to 5.00 mm. What is the new electric field between them and what is now the potential difference?

▌ (*a*) From $V = Ed$ we have $E = 12.0/3.00 \times 10^{-3} = \underline{4000 \text{ V/m}}$. (*b*) Since the battery was removed, the charge remains the same, as does E ($E = \sigma/\epsilon_0$). Potential $= 4000(5.00 \times 10^{-3}) = \underline{20 \text{ V}}$.

26.37 Two very large flat metal plates are parallel and separated by a distance D. The side of the left plate that faces the right plate has a surface charge $+\sigma$. (*a*) What is the field between the plates? (*b*) The potential difference? (*c*) Another metal plate, uncharged, is placed between these two without altering the charge on the original plates. Its thickness is $d < D$. What is the field in the gap between it and the left plate? (*d*) In the other gap? (*e*) What is now the potential difference between the two outer plates?

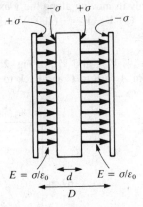

Fig. 26-12

▌ The situation for (*c*), (*d*), and (*e*) is depicted in Fig. 26-12. (*a*) Use Gauss' law; $E = \sigma/\epsilon_0$. (*b*) $E = $ constant so $V = ED = (\sigma D)/\epsilon_0$. (*c*) $E = \sigma/\epsilon_0$ since the same charge exists. (*d*) $+\sigma$ is induced on the right face of the inserted plate, so E in the gap is again σ/ϵ_0. (*e*) The potential between outer plates is $[\sigma(D - d)]/\epsilon_0$; the decrease from (*b*) occurs because the inserted plate is at constant potential (equivalently, the field vanishes in the plate).

26.38 A pair of horizontal metal plates are separated by a vertical distance of 10 cm, and the voltage different between them is 28 V. A small ball of 0.60-g mass hangs by a thread from the upper plate. What is the tension in the thread if the ball carries a charge of 20 μC? Two answers are possible. Find both.

▌ We use the condition for equilibrium: sum of forces equals zero. Tension T on the mass is up, weight mg

is down, and qE can be up or down. We have $mg = (6.0 \times 10^{-4})(9.8) = 5.88$ mN, $qE = (qV)/d = [(20\,\mu C)(28\,V)]/(0.10\,m) = 5.60$ mN, so that $T = mg \pm qE = \underline{11.5\,MN}$ or $\underline{0.28\,MN}$.

26.39 The electric field in a certain region is given by $\mathbf{E} = 5\mathbf{i} - 3\mathbf{j}$ kV/m. Find the difference in potential $V_B - V_A$ if A is at the coordinate origin and point B is (*a*) $(0, 0, 5)$ m, (*b*) $(4, 0, 3)$ m.

▮ Because the field is conservative, we can use straight-line paths parallel to the coordinate axes (see Fig. 26-13). (*a*) No work is done in carrying a unit positive charge, since there is no field component in the z direction. There is zero potential difference between the origin and $(0, 0, 5)$. (*b*) Following the path from $(0, 0, 0)$ to $(4, 0, 0)$ to $(4, 0, 3)$, the work done against the field in carrying a unit positive charge (the potential difference) is $-E_x(4 - 0) - E_z(3 - 0) = \underline{-20\,kV}$.

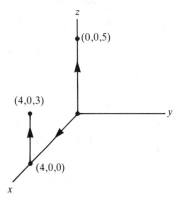

Fig. 26-13

26.40 For the field given in Prob. 26.39, find the potential difference $V_B - V_A$ between the two points A $= (0, 5, -1)$ and $B = (-3, 2, 2)$ m. Repeat for the points $A = (0, 5, -1)$ and $B = (3, 2, 2)$ m.

▮ Take as the path the broken line from $(0, 5, -1)$ to $(-3, 5, -1)$ to $(-3, 2, -1)$ to $(-3, 2, 2)$. We have $V_B - V_A = -(5000)(-3 - 0) - (-3000)(2 - 5) - (0)[2 - (-1)] = \underline{6\,kV}$. In the second case the x calculation is changed to $-(5000)(3 - 0)$, so the answer becomes $\underline{-24\,kV}$.

26.41 In a certain region of space, the electric field is directed in the $+y$ direction and has a magnitude of 4000 V/m. What is the potential difference from the coordinate origin to the following points? (*a*) $x = 0$, $y = 20$ cm, $z = 0$; (*b*) $x = 0$, $y = -30$ cm, $z = 0$; (*c*) $x = 0$, $y = 0$, $z = 15$ cm.

▮ In each case work is done only in moving along the y axis: $V - V_0 = -E_y(y)$. (*a*) $V - V_0 = -(4000\,V/m)(0.20\,m) = -800\,V$. (*b*) $V - V_0 = -(4000\,V/m)(-0.30\,m) = 1200\,V$. (*c*) $V - V_0 = -(4000\,V/m)(0) = 0\,V$.

26.42 Use $\Delta V = -E_s\,\Delta s$ to compute V_{AB}, V_{BC}, and V_{CA} in Fig. 26-14. Using these results, show that the work required to carry a charge q from A to B to C and back to A is zero.

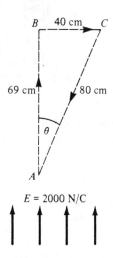

$E = 2000$ N/C

Fig. 26-14

▮ $V_{AB} = V_B - V_A = -E\,\Delta y = -2000(0.69) = -1380$ V; $V_{BC} = V_C - V_B = 0$ since $E_s = 0$ (E is vertical and Δs is horizontal); $V_{CA} = -(-2000\cos\theta)(0.08) = +2000(0.69) = 1380$ V where we have used $Es < 0$ along the direction from C to A. Net work is $q(V_{AB} + V_{CA}) = q(-1380 + 1380) = 0$.

26.43^c Give an informal proof of the following statement: *The electric potential cannot assume a minimum (or a maximum) value inside a charge-free region.*

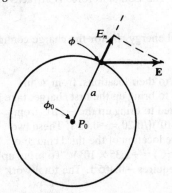

Fig. 26-15

▮ Suppose, on the contrary that the potential took on a local minimum value, ϕ_0, at the interior point P_0. Then (Fig. 26-15) we could enclose P_0 in a gaussian sphere of so small a radius, a, that (i) the sphere lies entirely within the charge-free region, and (ii) $\phi \ge \phi_0$ at every point of the spherical surface. Gauss' law, applied to this sphere, would give

$$0 = \int_S \mathbf{E}\cdot d\mathbf{S} = \int_S E_n\,dS \tag{1}$$

where E_n is the normal (radial) component of the field at the surface of the sphere. But, by definition of the electric potential,

$$E_n = -\frac{d\phi}{dn} \tag{2}$$

that is, the derivative of the potential in the radial direction. Now, by decreasing a if necessary, we can preserve conditions (i) and (ii) above and at the same time allow the derivative in (2) to be approximated by a difference-quotient to any required degree of accuracy. Thus,

$$E_n = -\frac{d\phi}{dn} \approx -\frac{\phi - \phi_0}{a} \tag{3}$$

and (1) becomes

$$0 = \int_S (\phi - \phi_0)\,dS \tag{4}$$

But (4) is impossible: $\phi - \phi_0$ is nonnegative at each point of S and so its integral over S must be positive. This contradiction establishes the desired result.

The implication is very significant: No charge placed in an electrostatic field can be in stable equilibrium, since that requires being at a minimum of potential energy. Note that *unstable* equilibrium (see, e.g., Prob. 25.21) does not demand a potential energy maximum, but only a saddle point.

26.3 ENERGETICS; PROBLEMS WITH MOVING CHARGES

26.44 It requires 50 μJ of work to carry a 2-μC charge from point R to S. What is the potential difference between the points? Which point is at the higher potential?

▮ $V_S - V_R = W/q = (5 \times 10^{-5})/(2 \times 10^{-6}) = \underline{25\ \text{V}}$. Point S is at the higher potential since work is needed to carry a positive charge from R to S.

26.45 The electron in a hydrogen atom is most probably at a distance $r = 5.29 \times 10^{-11}$ m from the proton, which is the nucleus of the atom. Evaluate the electric potential energy U of the atom.

▮ The proton is so much more massive than the electron that we identify U with the work needed to bring the electron from ∞ to point r in the field of the stationary proton. The potential established by the proton at distance r is

$$V = \frac{1}{4\pi\epsilon_0}\frac{e}{r} = (8.99 \times 10^9 \, \text{N} \cdot \text{m}^2/\text{C}^2)\frac{1.60 \times 10^{-19} \, \text{C}}{5.29 \times 10^{-11} \, \text{m}} = 27.2 \, \text{V}$$

Hence, $U = -eV = \underline{-27.2 \, \text{eV}}$. The Coulomb force is attractive, giving the atom a negative potential energy relative to infinity.

26.46 Compute the electric potential energy, U, for the charge configuration shown in Fig. 26-16.

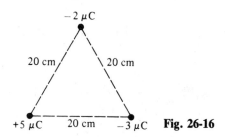

■ We bring the three charges to their locations, from ∞, in succession. No work is needed to bring up the first charge; take it to be the 5 μC. The work needed to bring up the $-2 \, \mu$C from infinity is $[(-2 \, \mu\text{C})(5 \, \mu\text{C})(9 \times 10^9)]/0.20 = -0.45 \, \text{J}$. These two charges set up a potential at the location of the third charge of $V = [(9 \times 10^9)/0.20](5 \, \mu\text{C} - 2 \, \mu\text{C}) = +1.35 \times 10^5 \, \text{V}$. To bring up the $-3 \, \mu$C charge therefore requires $-0.405 \, \text{J}$. The total work needed is $\underline{-0.855 \, \text{J}}$.

Fig. 26-16

Note that this is just the sum of the three pairwise potential energies, each calculated as if it were an isolated pair. It is *not*, however, the sum $q_1 V_1 + q_2 V_2 + q_3 V_3$, where V_1, V_2, V_3 are the potentials in each case due to the other two charges. This latter sum double-counts the energy.

26.47 Problem 26.46 suggests the formula

$$U = \tfrac{1}{2}(q_1 V_1 + q_2 V_2 + \cdots + q_n V_n)$$

for the potential energy of an arbitrary configuration of $n \geq 2$ point charges. Obtain this formula.

■ Let us introduce the notation $V_i(j)$ for the potential at the location of q_i due to the source charge q_j; from superposition,

$$V_i = V_i(1) + V_i(2) + \cdots + V_i(i-1) + V_i(i+1) + \cdots + V_i(n) \qquad (1)$$

for $i = 1, 2, \ldots, n$. First suppose the charges to be individually brought in from infinity in the order $1, 2, \ldots, n$. The work required is

$$U = 0 + q_2 V_2(1) + q_3[V_3(1) + V_3(2)] + \cdots + q_n[V_n(1) + V_n(2) + \cdots + V_n(n-1)] \qquad (2)$$

Now bring in the charges in the order $n, n-1, \ldots, 1$,

$$U = 0 + q_{n-1}V_{n-1}(n) + q_{n-2}[V_{n-2}(n) + V_{n-2}(n-1)] + \cdots + q_1[V_1(n) + V_1(n-1) + \cdots + V_1(2)] \qquad (3)$$

When (2) and (3) are added and divided by 2 and (1) is used to simplify the coefficient of q_i, the desired formula results.

For a continuous distribution of charge over a region $\mathscr{R}$, the calculus generalization of our formula is obviously

$$U = \frac{1}{2}\int_{\mathscr{R}} V \, dq = \frac{1}{2}\int_{\mathscr{R}} V\rho \, dv \qquad (4)$$

where ρ is the charge density and $dv = dx \, dy \, dz$ the element of volume.

26.48 Find the electric potential energy of an isolated metal sphere of radius R with total charge Q.

■ Recalling that the spherical surface is an equipotential, with $V = Q/(4\pi\epsilon_0 R)$, we have from (4) of Prob. 26.47,

$$U = \frac{1}{2}V\int_{\text{surface}} dq = \frac{1}{2}VQ = \frac{Q^2}{8\pi\epsilon_0 R}$$

More directly, we can suppose the charge brought in from infinity in tiny increments dq. If charge q has already been assembled, the work needed to bring in the next dq is $dU = V(q) \, dq = (q/4\pi\epsilon_0 R) \, dq$. Integrating from $q = 0$ to $q = Q$ gives $U = Q^2/8\pi\epsilon_0 R$, as before.

26.49 In Fig. 26-17, a sphere carrying a uniformly distrubuted charge $Q = 40 \, \mu$C is located at the origin; $r_1 = 0.5 \, \text{m}$ and $r_2 = 1.2 \, \text{m}$. **(a)** Find the potential at P_1 and at P_2 due to Q. What is the potential at P_1 with respect to P_2?

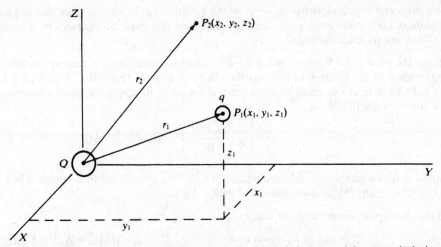

Fig. 26-17

(b) A small sphere carrying charge $q = 8\,\mu C$ is placed at P_1. What is its potential energy (relative to ∞)?
(c) Suppose that the small sphere moves freely from P_1 through P_2. In the trip from P_1 to P_2, what is its change in kinetic energy?

▌ (a)
$$\phi_1 = (9 \times 10^9)\left(\frac{40 \times 10^{-6}}{0.5}\right) = \underline{720\,kV} \qquad \phi_2 = (9 \times 10^9)\left(\frac{40 \times 10^{-6}}{1.2}\right) = \underline{300\,kV}$$

The potential at P_1 with respect to P_2 is the absolute potential at P_1 minus the absolute potential at P_2. We denote this relative potential as ϕ_{12} or V_{12}. Then

$$\phi_{12} = \phi_1 - \phi_2 = 720 - 300 = 420\,kV$$

(b)
$$U_1 = q\phi_1 = (8 \times 10^{-6})(720 \times 10^3) = \underline{5.76\,J}$$

(c)
$$K_2 - K_1 = U_1 - U_2 = q(\phi_1 - \phi_2) = q\phi_{12} = (8 \times 10^{-6})(420 \times 10^3) = \underline{3.36\,J}$$

26.50 The difference in potential, $V_{AB} \equiv V_A - V_B$, between wires A and B, Fig. 26-18, as measured by a voltmeter is 6000 V. A small sphere having mass 0.150 kg and carrying a charge $q = +500\,\mu C$ is released from rest at a point very near wire A and allowed to move to wire B. (a) Does V_{AB} depend on s, the distance between the wires? (b) What work W_E is done by electric forces on the sphere? (c) With what speed u will the sphere arrive at B? (d) What is the average field E_{avg} between A and B?

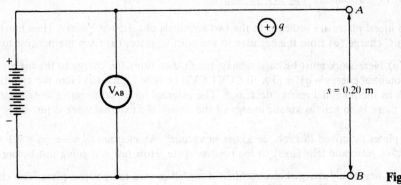

Fig. 26-18

▌ (a) No.

(b)
$$W_E = U_A - U_B = qV_{AB} = (500 \times 10^{-6})(6000) = 3.00\,J$$

(c)
$$W_T = W_E + W_G = \tfrac{1}{2}mu^2 \qquad W_G = mgh = (0.150)(9.8)(0.2) = 0.29\,J$$

Thus

$$\frac{1}{2}mu^2 = 3.29\,J \qquad \text{and} \qquad u = \sqrt{\frac{6.58}{0.15}} = \underline{6.6\,m/s}$$

(d)
$$E_{avg} = \frac{V_{AB}}{s} = \frac{6000}{0.20} = \underline{30\,kV/m}$$

26.51 An electron gun shoots electrons ($q = -e$, $m = 9.1 \times 10^{-31}$ kg) at a metal plate that is 4 mm away in vacuum. The plate is 5.0 V lower in potential than the gun. How fast must the electrons be moving as they leave the gun if they are to reach the plate?

▮ Since the plate is 5.0 V lower, and $q = -e$ is negative, the electron is moving "uphill." The potential energy gained by the electron in reaching the plate is thus $eV = (1.6 \times 10^{-19}\,\text{C})(5\,\text{V}) = 8.0 \times 10^{-19}$ J. Thus there must be at least this much kinetic energy at the start, to supply the needed potential energy gain. $KE = \frac{1}{2}mv^2 \geq 8.0 \times 10^{-19}$ J, or

$$v \geq \left[\frac{16.0 \times 10^{-19}}{9.1 \times 10^{-31}}\right]^{1/2} \text{m/s} = \underline{1330 \text{ km/s}}.$$

26.52 Suppose that two protons are released when 2.0×10^{-14} m apart. Find their speeds when they are 5.0×10^{-14} m apart. (Mass of proton $= 1.67 \times 10^{-27}$ kg.)

▮ From the law of conservation of energy,

$$\text{loss in } U_e = \text{gain in } K \qquad \text{or} \qquad (U_e)_{\text{start}} - (U_e)_{\text{end}} = K_{\text{end}} - K_{\text{start}}$$

Because K at the start is zero, and because the protons have identical kinetic energies,

$$\frac{e^2}{4\pi\epsilon_0}\left(\frac{1}{2.0 \times 10^{-14}\,\text{m}} - \frac{1}{5.0 \times 10^{-14}\,\text{m}}\right) = 2(\tfrac{1}{2}mv^2) - 0$$

Substituting known values for e and m yields $v = \underline{2034 \text{ km/s}}$.

26.53 The electron beam in a television tube consists of electrons accelerated from rest through a potential difference of about 20 kV. How large an energy do the electrons have? What is their speed? Ignore relativistic effects for this approximate calculation. ($m_e = 1.9 \times 10^{-31}$ kg.)

▮ The energy $= eV = \underline{20 \text{ keV}} = (1.6 \times 10^{-19})(20\,000) = 3.2 \times 10^{-15}$ J. To get the speed we have 3.2×10^{-15} J $= \frac{1}{2}(9.1 \times 10^{-31}\,\text{kg})v^2$. Solving, $v = \underline{8.4 \times 10^7 \text{ m/s}}$. $v = 28$ percent of speed of light, so our nonrelativistic approximation is crude but still reasonable.

26.54 A proton ($q = e$, $m = 1.67 \times 10^{-27}$ kg) is accelerated from rest through a potential difference of 1 MV. What is its final speed?

▮ The proton moves "downhill" through a 1.0×10^6 V potential difference. The loss in PE $= qV = (1.6 \times 10^{-19}\,\text{C})(1.0 \times 10^6\,\text{V}) = 1.6 \times 10^{-13}$ J, must equal the gain in KE $= \frac{1}{2}mv^2$, or $\frac{1}{2}(1.67 \times 10^{-27}\,\text{kg})v^2 = 1.6 \times 10^{-13}$ J. Solving, $\underline{v = 14 \times 10^6 \text{ m/s}}$.

26.55 Two metal plates are attached to the two terminals of a 1.50-V battery. How much work is required to carry a $+5$-μC charge **(a)** from the negative to the positive plate, **(b)** from the positive to the negative plate?

▮ **(a)** Here work must be done against the field to bring the charge to the higher voltage plate. $W =$ increase in potential energy $= qV = (5 \times 10^{-6}\,\text{C})(1.5\,\text{V})$, or $W = \underline{7.5 \,\mu\text{J}}$. **(b)** Here the electric forces are doing positive work as the potential energy decreases. The external forces do negative work to absorb this energy. Assuming that there is no gain in kinetic energy of the charge the external work is just $W = \underline{-7.5\,\mu\text{J}}$.

26.56 The plates described in Prob. 26.55 are in vacuum. An electron ($q = -e$, $m = 9.1 \times 10^{-31}$ kg) is released at the negative plate and falls freely to the positive plate. How fast is it going just before it strikes the plate?

▮ The negatively charged electron falls downhill toward the positive plate. The electric potential energy lost equals the gain in kinetic energy, so

$$\tfrac{1}{2}mv^2 = (1.6 \times 10^{-19}\,\text{C})(1.5\,\text{V}) = 2.4 \times 10^{-19}\,\text{J} \qquad v = \left[\frac{4.8 \times 10^{-19}}{9.1 \times 10^{-31}}\right]^{1/2} \text{m/s} = \underline{730 \text{ km/s}}$$

26.57 A lead pellet (mass, 2 g) fired from an air rifle has a speed of 150 ft/s. Through what difference in potential would this pellet have to fall to acquire the same speed, assuming it carries a charge of 1 μC?

▮ Kinetic energy $= \frac{1}{2}mv^2 = \frac{1}{2}(2 \times 10^{-3}\,\text{kg})(150\,\text{ft/s})^2(0.305\,\text{m/ft})^2 = 2.09$ J. This must equal qV. Thus $(1.0 \times 10^{-6}\,\text{C})V = 2.09$ J, or $V = \underline{2.09 \text{ MV}}$.

26.58 A proton ($q = e$) is released from a point P which is 10^{-14} m from a heavy nucleus which has a charge of $80e$.

(This would be the nucleus of a mercury atom.) How large will the kinetic energy of the proton be when it gets far away from the nucleus? What will be its speed?

❚ The proton will be repelled and caused to move radially toward infinity by the positive nucleus. In effect, it falls through the potential difference between point P and infinity. This potential difference is just the absolute potential at point P,

$$V = \frac{1}{4\pi\epsilon_0} \frac{80 \times 1.6 \times 10^{-19}}{10^{-14}} = 12 \times 10^6 \text{ V}$$

In falling through this potential difference, the proton will thus acquire a kinetic energy of $K = \underline{12 \text{ MeV}}$. In writing this, we assume the nucleus to remain nearly at rest (cf. Prob. 26.45). We then have

$$\tfrac{1}{2}mv^2 = (12 \times 10^6 \text{ eV})(1.602 \times 10^{-19} \text{ J/eV})$$

Using the proton mass 1.67×10^{-27} kg and solving for the proton's speed, we find that $v = \underline{4.8 \times 10^7 \text{ m/s}}$.

26.59 The nucleus of the radium atom has a charge of $+88e$ and a radius of about 0.007 pm. With what speed must a proton be shot at the atom if it is to reach a radius of 0.01 pm? The inner radius of the electron cloud, r_e, is about 50 pm.

❚ The potential due to the electrons alone is approximately constant and equal to $(-88e)/r_e$ within the inner radius. The potential due to the nucleus is $(+88e)/r$ with $r \ll r_e$.
   Therefore the approximate potential difference between infinity and $r = 0.01$ pm is simply the absolute potential at this radius from a point charge with $q = 88e$:

$$V = (9 \times 10^9) \frac{88(1.60 \times 10^{-19})}{10^{-14}} = 12.7 \text{ MV}$$

For the proton,

$$\text{kinetic energy at infinity} = \text{potential energy at } r \qquad \tfrac{1}{2}mv = qV$$

Using $m = 1.67 \times 10^{-27}$ kg, $q = 1.60 \times 10^{-19}$ C, and $V = 12.7 \times 10^6$ V in this equation gives $v = \underline{4.9 \times 10^7 \text{ m/s}}$.

26.60 The potential difference between the two plates in Fig. 26-19 is 100 V. If the system is in vacuum, what will be the speed of a proton released from plate B just before it hits plate A?

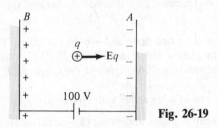

Fig. 26-19

❚ The mass and charge of a proton are 1.67×10^{-27} kg and 1.60×10^{-19} C, respectively. When the proton is moved from plate B to plate A, it loses a potential energy $q(V_B - V_A)$, where $V_B - V_A$ is 100 V in this case. This appears as kinetic energy of the proton at plate A. The law of conservation of energy therefore tells us that

$$\text{loss in potential energy} = \text{gain in kinetic energy} \qquad \text{or} \qquad q(V_B - V_A) = \tfrac{1}{2}mv^2$$

Placing in the values and solving for v, we find $\underline{140 \text{ km/s}}$.

26.61 The potential difference applied to the filament of a bulb of a 5-cell flashlight is 7.5 V. How much work is done by the flashlight cells (i.e., how much chemical energy is lost) in transferring 60 C of charge through the filament?

❚ $W = Vq = 7.5(60) = \underline{450 \text{ J}}$

26.62 A proton is accelerated from rest in a Van de Graaff accelerator by a potential difference of 0.9 MV. What is the kinetic energy of the proton after acceleration?

❚ The kinetic energy of the proton is equal to the work done on it by the potential difference through which it moved. $W = Vq = (9 \times 10^5)(1.60 \times 10^{-19}) = \underline{1.44 \times 10^{-13} \text{ J}}$.

26.63 In the Bohr model of the hydrogen atom, the electron was pictured to rotate in a circle of radius 0.053 nm about the nucleus. (*a*) How fast should the electron be moving in this orbit? (*b*) How much energy is needed to tear the electron loose from the nucleus of the atom?

▮ In order for the electron to travel in the circular orbit, a centripetal force must be furnished to it. This is provided by the coulomb attraction between it and the nucleus. We therefore have (after assuming that the nucleus remains stationary—a good assumption, since it is 1840 times more massive than the electron)

$$\frac{mv^2}{r} = \frac{1}{4\pi\epsilon_0}\frac{ee}{r^2} \tag{1}$$

In this expression m is the mass of the electron (9.1×10^{-31} kg), and e is the magnitude of the charge on the electron as well as on the nucleus, 1.6×10^{-19} C. After placing in the values and solving for v, we find that $v \approx 2.2 \times 10^6$ m/s. The energy needed to tear the electron loose (the *ionization energy* Φ) is given by energy conservation as

$$\Phi = \Delta PE + \Delta KE = [0 - (-e)V] + \left[0 - \frac{1}{2}mv^2\right] = e\frac{e}{4\pi\epsilon_0 r} - \frac{1}{2}\left(\frac{1}{4\pi\epsilon_0}\frac{ee}{r}\right) = \frac{e^2}{8\pi\epsilon_0 r} = \underline{2.15 \times 10^{-18}\,\text{J}}$$

26.4 CAPACITANCE AND FIELD ENERGY

26.64 What is a *capacitor* and how does one measure *capacitance*?

▮ A *capacitor, or condenser,* consists of two conductors with equal and opposite charges separated by an insulator or dielectric. The *capacitance* of a capacitor is defined by

$$\text{capacitance } C = \frac{\text{magnitude of charge } q \text{ on either conductor}}{\text{magnitude of potential difference } V \text{ between conductors}}$$

For q in coulombs and V in volts, C will be in *farads* (F). Convenient submultiples of the farad are:

$$1\,\mu\text{F} = 1 \text{ microfarad} = 10^{-6}\,\text{F} \qquad 1\,\text{pF} = 1 \text{ picofarad} = 10^{-12}\,\text{F}$$

26.65 Find an expression for the capacitance of a parallel-plate capacitor made up of two parallel conducting plates, each of area A, and separated by a distance d. Assume that d is much smaller than the dimensions of the plates.

▮ Assume a charge q on one plate and $-q$ on the other. The electric field between the plates is constant and perpendicular to the plates. Then, using Prob. 25.71, $V = Ed = (\sigma/\epsilon_0)d = [(d/(\epsilon_0 A)]q$, and so $C = q/V = (\epsilon_0 A)/d$. In this derivation we have ignored deviations in E at the edges of the plates.

26.66 Find the energy stored in a capacitor with charge q.

▮ We start with neutral plates and bring charge across in increments dq'. The energy U is then the total work done in bringing q units of charge across:

$$U = \int_0^q V(q')\,dq' = \frac{1}{C}\int_0^q q'\,dq' = \frac{q^2}{2C}$$

26.67 A plane-parallel capacitor has circular plates of radius $r = 10.0$ cm, separated by a distance $d = 1.00$ mm. How much charge is stored on each plate when their electric potential difference has the value $V = 100$ V? Discuss the accuracy of the calculation.

▮ From Prob. 26.65

$$C = \frac{\epsilon_0 A}{d} = \frac{(8.85 \times 10^{-12}\,\text{C}^2/\text{N} \cdot \text{m}^2)(3.14 \times 10^{-2}\,\text{m}^2)}{1.00 \times 10^{-3}\,\text{m}} = 2.8 \times 10^{-10}\,\text{F} = \underline{280\,\text{pF}} \tag{1}$$

It is not appropriate to quote the capacitance to more than two significant figures because (1) ignores the effects of the edges of the capacitor plates. Figure 26-20 shows that these effects occur in a region whose radial extent Δr is comparable to the separation d of the plates. Hence the ratio $\Delta r/r \approx d/r = 10^{-3}\,\text{m}/10^{-1}\,\text{m} = 1$ percent gives a measure of the accuracy to be expected from the equation.

The magnitude $|q|$ of the charge stored on either plate of the capacitor is $|q| = CV = (2.8 \times 10^{-10}\,\text{F})(1.00 \times 10^2\,\text{V}) = \underline{28\,\text{nC}}$.

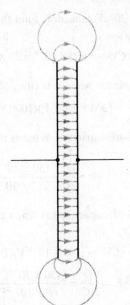

Fig. 26-20

26.68 A spherical capacitor is to be constructed by using a metal sphere of radius b as one plate and a concentric spherical metal shell as the other plate. The inner radius of the shell is $a > b$; show that the capacitance of the device is

$$C = \frac{4\pi\epsilon_0 ab}{a - b}$$

▌ We use the spherical symmetry and associated properties of E. Assume a charge Q on the inner plate. Then $E = Q/(4\pi\epsilon_0 r^2)$ and $\Delta V = [Q/(4\pi\epsilon_0)](1/b - 1/a)$. By definition C is $Q/\Delta V = (4\pi\epsilon_0)/(1/b - 1/a) = (4\pi\epsilon_0 ab)/(a - b)$.

26.69 Refer to Prob. 26.68. Show that if the separation of the spheres is very small in comparison with their radii, the capacitance is given by the parallel-plate relation $C = (\epsilon_0 A)/d$.

▌ Write $4\pi a^2 \approx 4\pi ab$ for the area of either sphere. Then $C = (\epsilon_0 A)/(a - b)$. But $a - b = d$, and so the desired result is obtained.

26.70 How much charge is stored in a capacitor consisting of two concentric spheres of radii 30 and 31 cm if the potential difference is 500 V? Assume $K = 1$ for air.

▌ For concentric spheres the capacitance is

$$C = \frac{Kab}{(9 \times 10^9)(a - b)} = \frac{(0.30)(0.31)}{(9 \times 10^9)(0.31 - 0.30)} = 1.033 \times 10^{-9}\,\text{F} = \underline{1.033\,\text{nF}}$$

$$Q = CV = (1.033 \times 10^{-9})(500) = \underline{517\,\text{nC}}$$

26.71 A metal sphere mounted on an insulating rod carries a charge of 6 nC when its potential is 200 V higher than its surroundings. What is the capacitance of the capacitor formed by the sphere and its surroundings?

▌

$$C = \frac{q}{V} = \frac{6 \times 10^{-9}\,\text{C}}{200\,\text{V}} = \underline{30\,\text{pF}}$$

26.72 A capacitor is charged with 9.6 nC and has a 120-V potential difference between its terminals. Compute its capacitance and the energy stored in it.

▌ $C = Q/V = (9.6 \times 10^{-9}\,\text{C})/(120\,\text{V}) = \underline{8.0 \times 10^{-11}\,\text{F}} = \underline{80\,\text{pF}}$. For the energy we can use any of three equivalent forms. $E = \frac{1}{2}CV^2 = \frac{1}{2}QV = \frac{1}{2}Q^2/C$. Using the middle form, $E = \frac{1}{2}(9.6 \times 10^{-9}\,\text{C})(120\,V) = 5.76 \times 10^{-7}\,\text{J} = \underline{576\,\text{nJ}}$.

26.73 A charge of 600 μC is placed on a 20-μF capacitor. Find the potential difference between the terminals of the capacitor.

$$Q = CV \qquad 600 \times 10^{-6} = 20 \times 10^{-6} V \qquad V = \underline{30\ V}.$$

26.74 What is the charge on a 300 pF capacitor when it is charged to a voltage of 1 kV?

$$q = CV = (300 \times 10^{-12}\ \text{F})(1000\ \text{V}) = 3 \times 10^{-7}\ \text{C} = \underline{0.3\ \mu C}.$$

26.75 A charge of 50 μC is placed on a 2-μF capacitor. What is the stored energy?

$$W = \frac{Q^2}{2C} = \frac{(50 \times 10^{-6})^2}{2(2 \times 10^{-6})} = 625\ \mu J$$

26.76 Compute the energy stored in a 60-pF capacitor (*a*) when charged to a potential difference of 2 kV, (*b*) when the charge on each plate is 30 nC.

(*a*)
$$E = \tfrac{1}{2}CV^2 = \tfrac{1}{2}(60 \times 10^{-12}\text{F})(2000\ V)^2 = \underline{1.2 \times 10^{-4}\ J}$$

(*b*)
$$E = \frac{1}{2}\frac{Q^2}{C} = \frac{\tfrac{1}{2}(30 \times 10^{-9}C)^2}{60 \times 10^{-12}\text{F}} = \underline{7.5 \times 10^{-6}\ J}$$

26.77 Find the energy stored in a 5-μF capacitor charged to a potential difference of 500 V.

$$W = \tfrac{1}{2}CV^2 = \tfrac{1}{2}(5 \times 10^{-6})(500^2) = \underline{0.625\ J}$$

26.78 If a 4-μF capacitor has a potential difference of 1000 V, what is its stored energy?

$$W = \tfrac{1}{2}CV^2 = \tfrac{1}{2}(4 \times 10^{-6})(1000^2) = \underline{2\ J}.$$

26.79 A 1.2 μF capacitor is charged to 3 KV. Compute the energy stored in the capacitor.

$$\tfrac{1}{2}CV^2 = \tfrac{1}{2}(1.2 \times 10^{-6}\ \text{F})(3000\ V)^2 = \underline{5.4\ J}$$

26.80 A capacitor of arbitrary shape with air ($K = 1$) between its plates has capacitance C. What is the capacitance when wax of dielectric constant K is between the plates?

▌ Let Q be the charge placed on one conductor and $-Q$ on the other. This will not change by placing dielectric between the conductors. What will change is the electric field in the region between the conductors. Because of induced charges on the interface between dielectric and conductor, the effective charge giving rise to the electric field is reduced by the dielectric constant factor, K. Since the field is reduced everywhere by the same factor, the potential difference will be reduced by the same factor. Thus if V is the potential difference without dielectric and V' with dielectric, $V' = V/K$. Then $C' = Q/V' = KQ/V = KC$.

26.81 A capacitor with air between its plates has a capacitance of 8 μF. Determine its capacitance when a dielectric with dielectric constant 6.0 is between its plates.

$$C \text{ with dielectric} = K(C \text{ with air}) = (6.0)(8\ \mu\text{F}) = \underline{48\ \mu F}$$

26.82 A certain parallel-plate capacitor consists of two plates, each with area 200 cm^2, separated by a 0.4-cm air gap. (*a*) Compute its capacitance. (*b*) If the capacitor is connected across a 500-V source, what are the charge on it, the energy stored in it, and the value of E between the plates? (*c*) If a liquid with $K = 2.60$ is poured between the plates so as to fill the air gap, how much additional charge will flow onto the capacitor from the 500-V source?

▌ (*a*) For a parallel-plate capacitor, $C = (\epsilon_0 A)/d = [(8.85 \times 10^{-12})(0.02)]/(0.004) = \underline{44\ pF}$.

(*b*)
$$q = CV = (4.4 \times 10^{-11}\ \text{F})(500\ \text{V}) = \underline{22\ nC} \qquad \text{energy} = \frac{1}{2}qV = \frac{1}{2}(2.2 \times 10^{-8}\ \text{C})(500\ \text{V}) = \underline{5.5\ \mu J}.$$

$$E = \frac{V}{d} = \frac{500\ \text{V}}{4 \times 10^{-3}\ \text{m}} = \underline{125\ kV/m}.$$

(*c*) The capacitor will now have a capacitance 2.60 times larger than before. Therefore. $q' = (2.60)(22) = 57$ nC, and so $q' - q = \underline{35\ nC}$ must flow onto it.

26.83 Two parallel conducting plates of area 100 cm² and 5 mm apart are given equal and opposite charges of 0.20 μC. The region between the plates is filled with a dielectric of $K = 5$. Compute (*a*) the capacitance of the system and (*b*) the voltage difference between the plates.

▌ (*a*)
$$C = \frac{KA}{k4\pi d} = \frac{5(0.01)}{(9 \times 10^9)(4\pi)(0.005)} = 88 \times 10^{-12}\,\text{F} = \underline{88\,\text{pF}}.$$

(*b*)
$$Q = CV \qquad 0.2 \times 10^{-6} = 88 \times 10^{-12}\,V \qquad V = \underline{2270\,\text{V}}.$$

26.84 A 5-μF capacitor with air between the metal plates is connected to a 30-V battery. The battery is then removed, leaving the capacitor charged. (*a*) Calculate the charge on the capacitor. (*b*) The air between the plates is replaced by oil with $K = 2.1$. Find the new value of the capacitance and the new potential difference between the plates.

▌ (*a*) $Q = CV = 5 \times 10^{-6}(30) = \underline{150\,\mu\text{C}}$. (*b*) The charge on the plates remains the same when the oil replaces the air. The capacitance increases by a factor K.

$$C' = KC = 10.5\,\mu\text{F} \qquad V' = \frac{V}{K} = \underline{14.3\,\text{V}}$$

26.85 Consider a parallel-plate capacitor with plate area A and charge Q. (*a*) Find the force on one plate because of the charge on the other. (*b*) Compute the work done in separating the plates from essentially zero separation to a separation d. (*c*) Compare this work with the energy stored in the capacitor as given in Prob. 26.66. (Assume that Q remains unchanged.)

▌ (*a*) Since each plate contributes equally to the field between the plates the field due to *one* plate is $\sigma/2\epsilon_0$ and this exerts a force $(\sigma/2\epsilon_0)\sigma A$ on the other, so $F = (\sigma^2 A)/(2\epsilon_0)$. (*b*) The work done is given by $Fd = [(\sigma^2 A)/(2\epsilon_0)]d$. (*c*) Note that $Q^2 = (\sigma A)^2$ and the capacitance is given by $(\epsilon_0 A)/d$; thus $Fd = Q^2/(2C)$, the usual expression for the stored energy.

26.86 Show that the electrical energy of a parallel-plate capacitor may be thought of as residing in the electric field, with an energy density of $\rho_e = (\epsilon_0 E^2)/2$ at any field point.

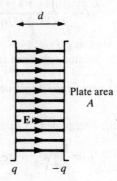

Plate area
A

q $-q$ **Fig. 26-21**

▌ The situation is depicted in Fig. 26-21. By Prob. 26.66,

$$U = \frac{1}{2}qV \qquad V = Ed \qquad E = \frac{\sigma}{\epsilon_0} = \frac{q}{\epsilon_0 A} \qquad \text{Thus} \qquad U = \frac{1}{2}(\epsilon_0 A E)(Ed) = \frac{\epsilon_0 E^2 \tau}{2}$$

where $\tau = Ad$ is the volume between the plates. Then $\rho_e = U/\tau = (\epsilon_0 E^2)/2$.

While this derivation is for the simple case of the constant field between two parallel plates, the result is true for the electrostatic energy of an arbitrary charge distribution.

26.87 Calculate the electric field, the electric-field energy density, and the energy stored in the plane-parallel capacitor of Prob. 26.67.

▌
$$E = \frac{V}{d} = \frac{100\,\text{V}}{1.00 \times 10^{-3}\,\text{m}} = \underline{100\,\text{kV/m}} \qquad \rho_e = \frac{\epsilon_0 E^2}{2} = \frac{(8.85 \times 10^{-12}\,\text{C}^2/\text{N} \cdot \text{m}^2)(10^5\,\text{N/C})^2}{2} = \underline{0.044\,\text{J/m}^3}$$

$$U = \rho_e \pi r^2 d = (4.4 \times 10^{-2}\,\text{J/m}^3)\pi(0.1\,\text{m})^2(0.001\,\text{m}) = \underline{1.4\,\mu\text{J}}$$

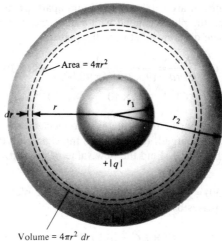

Fig. 26-22

Volume = $4\pi r^2 \, dr$

26.88^c Obtain expressions for the electric-field energy density ρ_e and energy content U for the spherical capacitor shown in Fig. 26-22, when the inner and outer spheres hold charges $+|q|$ and $-|q|$, respectively.

❚ By Gauss' law,

$$E = \frac{|q|}{4\pi\epsilon_0 r^2} \quad (r_1 \leq r \leq r_2) \quad \text{giving} \quad \rho_e = \frac{\epsilon_0 E^2}{2} = \frac{q^2}{32\pi^2\epsilon_0 r^4}$$

Using the volume element indicated in Fig. 26-22,

$$U = \iiint \rho_e \, dv = \int_{r_1}^{r_2} \rho_e (4\pi r^2 \, dr) = \frac{q^2}{8\pi\epsilon_0} \int_{r_1}^{r_2} \frac{dr}{r^2} = \frac{q^2}{8\pi\epsilon_0} \left(\frac{1}{r_1} - \frac{1}{r_2} \right)$$

As a check, we write $U = q^2/(2C)$, or

$$C = \frac{1}{2} \frac{q^2}{U} = \frac{1}{2} \frac{8\pi\epsilon_0}{(1/r_1) - (1/r_2)} = \frac{4\pi\epsilon_0 r_1 r_2}{r_2 - r_1}$$

in agreement with Prob. 26.68. Also, as $r_2 \to \infty$, $U \to q^2/(8\pi\epsilon_0 r_1)$ (in agreement with Probl 26.48) and $C \to 4\pi\epsilon_0 r_1$, the capacitance of an isolated sphere of radius r_1.

26.89 The unit of electrostatic energy density is the J/m^3, which is the same as the Pa, the unit of stress or pressure. Is this merely accidental?

❚ No: in fact, by Prob. 26.85, the energy density in a parallel-plate capacitor is

$$\rho_e = \frac{Fd}{Ad} = \frac{F}{A} = \text{pressure on either plate}$$

26.5 CAPACITORS IN COMBINATION

26.90 Show that capacitances add linearly when connected in parallel and reciprocally when connected in series.

❚ Figure 26-23(a) shows three capacitors connected in parallel. We wish to find the capacitance of a single

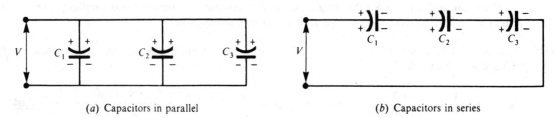

(a) Capacitors in parallel

(b) Capacitors in series

Fig. 26-23

capacitor that will behave equivalently to the combination: C_{eq}. If a voltage V is placed across the terminals we have $q_1 = C_1V$; $q_2 = C_2V$; $q_3 = C_3V$. The total charge stored is thus $q = q_1 + q_2 + q_3$. Hence $q = C_{eq}V = C_1V + C_2V + C_3V$, or dividing out V, $\underline{C_{eq} = C_1 + C_2 + C_3}$.

For the capacitors in series, as depicted in Fig. 26-23(b) we must have $q_1 = q_2 = q_3$. The voltages are $V_1 = q_1/C_1$, $V_2 = q_2/C_2$, and $V_3 = q_3/C_3$. The voltage across the equivalent capacitor is $V = V_1 + V_2 + V_3 = q/C_{eq}$, with $q = q_1 = q_2 = q_3$. Thus $q/C_{eq} = q/C_1 + q/C_2 + q/C_3$, and dividing out by q, $1/C_{eq} = 1/C_1 + 1/C_2 + 1/C_3$.

26.91 The parallel capacitor combination shown in Fig. 26-24 is connected across a 120-V source. Determine the equivalent capacitance C_{eq} and the charge on each capacitor.

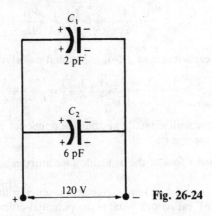

Fig. 26-24

❙ For a parallel combination,

$$C_{eq} = C_1 + C_2 = 6\,\text{pF} + 2\,\text{pF} = \underline{8\,\text{pF}}$$

Each capacitor has a 120-V potential difference impressed on it. Therefore,

$$q_1 = C_1V_1 = (2\,\text{pF})(120\,\text{V}) = 240\,\text{pC} \qquad q_2 = C_2V_2 = (6\,\text{pF})(120\,\text{V}) = 720\,\text{pC}$$

The charge on the combination is $q_1 + q_2 = \underline{960\,\text{pC}}$. Or we could write

$$q = C_{eq}V = (8\,\text{pF})(120\,\text{V}) = \underline{960\,\text{pC}}.$$

26.92 Determine the capacitance of a parallel combination of one 12-μF and two 6-μF capacitors.

❙
$$C = C_1 + C_2 + C_3 = 12 + 6 + 6 = \underline{24\,\mu\text{F}}$$

26.93 The series combination of two capacitors shown in Fig. 26-25 is connected across 1000 V. Compute

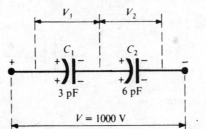

Fig. 26-25

(a) the equivalent capacitance C_{eq} of the combination, (b) the magnitudes of the charges on the capacitors, (c) the potential differences across the capacitors, (d) the energy stored in the capacitors.

❙ (a) $\qquad \dfrac{1}{C_{eq}} = \dfrac{1}{C_1} + \dfrac{1}{C_2} = \dfrac{1}{3\,\text{pF}} + \dfrac{1}{6\,\text{pF}} = \dfrac{1}{2\,\text{pF}}$

from which $\underline{C = 2\,\text{pF}}$.

(b) In a series combination, each capacitor carries the same charge, which is the charge on the combination. Thus, using the result of a,

$$q_1 = q_2 = q = C_{eq}V = (2 \times 10^{-12}\,\text{F})(1000\,\text{V}) = \underline{2\,\text{nC}}$$

(c) $V_1 = \dfrac{q_1}{C_1} = \dfrac{2 \times 10^{-9}\,\text{C}}{3 \times 10^{-12}\,\text{F}} = \underline{667\,\text{V}}$ $V_2 = \dfrac{q_2}{C_2} = \dfrac{2 \times 10^{-9}\,\text{C}}{6 \times 10^{-12}\,\text{F}} = \underline{333\,\text{V}}$

(d)   energy in $C_1 = \frac{1}{2}q_1 V_1 = \frac{1}{2}(2 \times 10^{-9}\,\text{C})(667\,\text{V}) = \underline{6.7 \times 10^{-7}\,\text{J}}$

energy in $C_2 = \frac{1}{2}q_2 V_2 = \frac{1}{2}(2 \times 10^{-9}\,\text{C})(333\,\text{V}) = \underline{3.3 \times 10^{-7}\,\text{J}}$

energy in combination $= (6.7 + 3.3) \times 10^{-7}\,\text{J} = \underline{10 \times 10^{-7}\,\text{J}}$

The last result is also directly given by $\frac{1}{2}qV$ or $\frac{1}{2}C_{eq}V^2$.

26.94 If a 6-μF capacitor and a 12-μF capacitor are connected in series, what is the capacitance of the combination?

∎ $\dfrac{1}{C} = \dfrac{1}{C_1} + \dfrac{1}{C_2} = \dfrac{1}{6} + \dfrac{1}{12} = \dfrac{3}{12}$ Thus $C = \underline{4\,\mu\text{F}}$

26.95 Find the equivalent capacitance of a 1-μ, a 2-μ, and a 6-μF capacitor connected in series.

∎ $\dfrac{1}{C} = \dfrac{1}{C_1} + \dfrac{1}{C_2} + \dfrac{1}{C_3} = \dfrac{1}{1} + \dfrac{1}{2} + \dfrac{1}{6} = \dfrac{6}{6} + \dfrac{3}{6} + \dfrac{1}{6} = \dfrac{10}{6}$ and $C = \underline{0.6\,\mu\text{F}}$

26.96 If you need a capacitor with $C = 0.25\,\mu$F, but the only ones in the storeroom have $C = 1.00\,\mu$F, must you delay finishing your experiment?

∎ No. You can connect four of the available capacitors in series.

26.97 Three capacitors (2, 3, and 4 μF) are connected in series with a 6-V battery. When the current stops, what is the charge on the 3-μF capacitor? What is the potential difference between the two ends of the 4-μF capacitor?

∎ The equivalent capacitance of the three in series $= \frac{12}{13} = 0.92\,\mu$F. Capacitors in series each carry the same charge, which is the same as the charge on the equivalent capacitor: $Q_{eq} = C_{eq}V = 0.92(6) = \underline{5.5\,\mu\text{C}}$. The V across 4 μf is $V = Q_{eq}/C = 5.5/4 = \underline{1.38\,\text{V}}$.

26.98 Three capacitors are connected as shown in Fig. 26-26. If a 12-V potential difference is applied to the terminals, what will the total capacitance be?

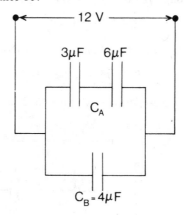

Fig. 26-26

∎ For the two capacitors in series,

$$\dfrac{1}{C_A} = \dfrac{1}{C_1} + \dfrac{1}{C_2} = \dfrac{1}{3} + \dfrac{1}{6} = \dfrac{2}{6} + \dfrac{1}{6} = \dfrac{3}{6}$$

and thus the capacitance of the upper branch is $C_A = 2\,\mu$F. For the two parallel branches

$$C = C_A + C_B = 2 + 4 = \underline{6\,\mu\text{F}} \quad \text{(capacitance of the system)}$$

26.99 Refer to Prob. 26.98 and find the charge on each capacitor.

∎ Use the general formula $Q = CV$ successively for C and for C_B.

$$Q = CV = 6 \times 10^{-6}(12) = \underline{72\,\mu\text{C}} \quad \text{(charge on the system)}$$

$$Q_B = C_B V = (4 \times 10^{-6})(12) = \underline{48\,\mu\text{C}} \quad \text{(charge on the 4-}\mu\text{F capacitor)}$$

The total charge on the system is the sum of the charges on the upper branch and on the lower branch; thus, $Q_A = Q - Q_B = \underline{24\ \mu C}$. Both the 3- and the 6-$\mu$F capacitors carry this same charge since they are in series.

26.100 Refer to Probs. 26.98 and 26.99 and find the voltage on each capacitor.

▌ The 4-μF capacitor has a potential difference of $\underline{12\ V}$, the applied voltage. Use $Q = CV$ for the 3-μF capacitor.

$$Q_A = (3 \times 10^{-6})V_3 \quad \text{or} \quad 2.4 \times 10^{-5} = (3 \times 10^{-6})V_3 \quad \text{so} \quad V_3 = \underline{8\ V}$$

Use $Q = CV$ for the 6-μF capacitor.

$$Q_A = (6 \times 10^{-6})V_6 \quad \text{or} \quad 2.4 \times 10^{-5} = (6 \times 10^{-6})V_6 \quad \text{and} \quad V_6 = \underline{4\ V}$$

Their sum must of course equal the terminal voltage, and indeed $8\ V + 4\ V = 12\ V$.

26.101 In the circuit of Fig. 26-27, find the total capacitance.

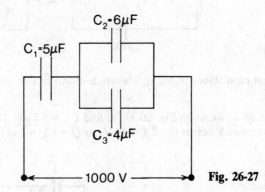

Fig. 26-27

▌ First find the capacitance C_P of the parallel section.

$$C_P = C_2 + C_3 = 6 + 4 = 10\ \mu F$$

C_P is in series with C_1, so

$$\frac{1}{C} = \frac{1}{C_1} + \frac{1}{C_P} = \frac{1}{5} + \frac{1}{10} = \frac{3}{10} \quad \text{and} \quad C = \underline{3.33\ \mu F}$$

26.102 Refer to Prob. 26.101 and find the potential difference across each capacitor.

▌ Find the total charge Q on the system first.

$$Q = CV = 3.33 \times 10^{-6}(1000) = 3.33 \times 10^{-3}\ C$$

Then

$$V_1 = \frac{Q}{C_1} = \frac{3.33 \times 10^{-3}}{5 \times 10^{-6}} = 0.667 \times 10^3 = \underline{667\ V} \qquad V_2 = V_3 = V - V_1 = 1000 - 667 = \underline{333\ V}.$$

As a check we can use the equivalent of the two parallel capacitors, C_p.

$$V_2 = V_3 = \frac{Q}{C_P} = \frac{3.33 \times 10^{-3}}{10 \times 10^{-6}} = \underline{333\ V}.$$

26.103 Find the equivalent capacitance of the combination shown in Fig. 26-28.

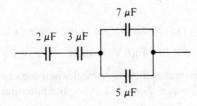

Fig. 26-28

▮ The 7 and 5 μF in parallel give an equivalent 12 μf which in turn is in series with the 2 and 3 μF; therefore, $1/C = \frac{1}{2} + \frac{1}{3} + \frac{1}{12} = \frac{11}{12}$, yielding $C = \underline{1.09\ \mu F}$.

26.104 Two capacitors in parallel, 2 and 4 μF, are connected, as a unit, in series with a 3-μF capacitor. The combination is connected across a 12-V battery. Find the equivalent capacitance of the combination and the potential difference across the 2-μF capacitor.

▮ Reduce the system as shown in Fig. 26-29, to find $C_{eq} = \underline{2.0\ \mu F}$. The charge on the equivalent capacitor $Q = CV = 2(12) = 24\ \mu C$. This is also the charge on the equivalent 6-μF capacitor. So V across it (and the 2 μF) is $Q/C = 24/6 = \underline{4.0\ V}$.

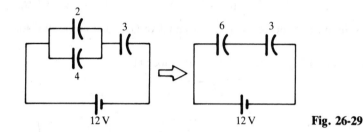

Fig. 26-29

26.105 Find the equivalent capacitance of the combination shown in Fig. 26-30. Also find the charge on the 4-μF capacitor.

▮ Reduce the circuit as shown in Fig. 26-31, to find $C_{eq} = \underline{5.4\ \mu F}$. The charge on the 4 μF is the same as on the 2.4 μF. The value of V across the 2.4 is 10 V, so $Q = CV = 2.4(10) = \underline{24\ \mu C}$.

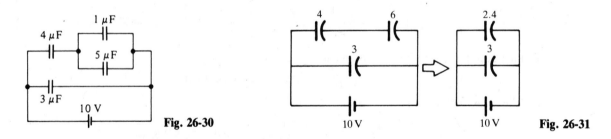

Fig. 26-30

Fig. 26-31

26.106 Two capacitors, 3 μF and 4 μF, are individually charged across a 6-V battery. After being disconnected from the battery, they are connected together with the negative plate of one attached to the positive plate of the other. What is the final charge on each capacitor?

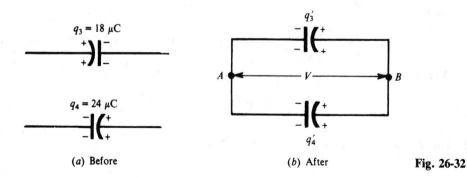

(a) Before   (b) After   Fig. 26-32

▮ The situation is shown in Fig. 26-32. Before being connected, their charges are

$$q_3 = CV = (3 \times 10^{-6}\ F)(6\ V) = 18\ \mu C \qquad q_4 = CV = (4 \times 10^{-6}\ F)(6\ V) = 24\ \mu C$$

As seen in the figure, the charges will partly cancel when the capacitors are connected together. Their final charges are given by $q_3' + q_4' = q_4 - q_3 = 6\ \mu C$. Also, the potential across each is now the same, so that

$V = q/C$ gives

$$\frac{q_3'}{3 \times 10^{-6}\,\text{F}} = \frac{q_4'}{4 \times 10^{-6}\,\text{F}} \qquad \text{or} \qquad q_3' = 0.75 q_4'$$

Substitution of this in the previous equation gives

$$0.75 q_4' + q_4' = 6\,\mu\text{C} \qquad \text{or} \qquad q_4' = \underline{3.43\,\mu\text{C}}$$

Then $q_3' = 0.75 q_4' = \underline{2.57\,\mu\text{C}}$.

26.107 Two capacitors, $C_1 = 3\,\mu\text{F}$ and $C_2 = 6\,\mu\text{F}$, are connected in series and charged by connecting a battery of voltage $V = 10\,\text{V}$ in series with them. They are then disconnected from the battery, and the loose wires are connected together. What is the final charge on each?

❚ The capacitors are charged in series so they originally have equal charges. When the loose wires are reconnected, they neutralize each other giving zero final charge.

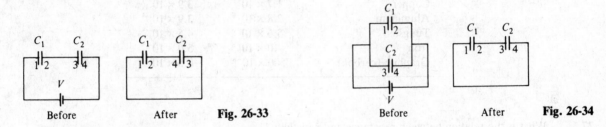

Before   After   **Fig. 26-33**   Before   After   **Fig. 26-34**

26.108 Repeat Prob. 26.107 if after being disconnected from the battery, the capacitors are disconnected from each other. They are now reconnected as shown in Fig. 26-33. What is the final charge on each?

❚ The original charge on each is $Q = C_{\text{eq}} V = 20\,\mu\text{C}$. After being connected as shown, $Q_1 + Q_2 = 2Q = 40\,\mu\text{C}$. Also $V_1 = V_2$ and so $Q_1/C_1 = Q_2/C_2$. Solving these two equations simultaneously gives for the two charges $\underline{26.7\,\mu\text{C}}$ on the $6\,\mu\text{F}$ and $\underline{13.3\,\mu\text{C}}$ on the $3\,\mu\text{F}$.

26.109 If two capacitors $C_1 = 4\,\mu\text{F}$ and $C_2 = 6\,\mu\text{F}$ are originally connected to a battery $V = 12\,\text{V}$, as shown in Fig. 26-34, and then disconnected and reconnected as shown, what is the final charge on each capacitor?

❚ Originally, $Q_1 = 48\,\mu\text{C}$, $Q_2 = 72\,\mu\text{C}$; hence, $Q_1' + Q_2' = 72 - 48 = 24\,\mu\text{C}$. Also, $V_1' = V_2'$ gives $Q_1'/C_1 = Q_2'/C_2$. Solving simultaneously gives $\underline{9.6}$ and $\underline{14.4\,\mu\text{C}}$.

CHAPTER 27
Simple Electric Circuits

27.1 OHM'S LAW, CURRENT, RESISTANCE

TABLE 27-1 Resistivities (ρ) at 20 °C and Temperature Coefficients (α).

material	$\rho, \Omega \cdot m$	$\alpha, °C^{-1}$
Silver	1.6×10^{-8}	3.8×10^{-3}
Copper	1.7×10^{-8}	3.9×10^{-3}
Aluminum	2.8×10^{-8}	3.9×10^{-3}
Tungsten	5.6×10^{-8}	4.5×10^{-3}
Iron	10×10^{-8}	5.0×10^{-3}
Graphite (carbon)	3500×10^{-8}	-0.5×10^{-3}

27.1 What is the relation between resistance and resistivity?

▋ The resistance R of a wire of length L and cross-sectional area A is

$$R = \rho \left(\frac{L}{A} \right),$$

where ρ is a constant called the *resistivity* and is a characteristic of the material from which the wire is made. For L in m, A in m², and R in Ω, the units of ρ are $\Omega \cdot m$.

27.2 How does the resistance of a conductor vary with temperature?

▋ If a wire has a resistance R_0 at a temperature T_0, then its resistance R at a temperature T is $R = R_0 + \alpha R_0 (T - T_0)$, where α is the *temperature coefficient of resistance* of the material of the wire. Usually α varies with temperature and so a linear relation is applicable only over a small temperature range. The units of α are K^{-1} or $°C^{-1}$.

A similar relation applies to the variation of resistivity with temperature. If ρ_0 and ρ are the resistivities at T_0 and T respectively, then $\rho = \rho_0 + \alpha \rho_0 (T - T_0)$.

Table 27-1 lists the resistivities of a number of conductors for $T_0 = 20\,°C$, as well as temperature coefficients of resistance.

27-3 How are current and current density related?

▋ The rate of flow of electric charge across a given area (within a conductor) is defined as the *electric current I* through that area. Thus,

$$I = \frac{dq}{dt} \quad \text{(A)}$$

The *electric current density* **J** at a point (within a conductor) is a vector whose direction is the direction of flow of charge at that point and whose magnitude is the current through a *unit area perpendicular to the flow direction* at that point. Thus, the current through an element of area dS, arbitrarily oriented with respect to the flow direction, is given by (see Fig. 27-1) $dI = \mathbf{J} \cdot d\mathbf{S} = J\, dA$, where $dA = dS \cos \theta$ is the projection of dS perpendicular to the flow direction. The total current through a surface S (e.g., a cross section of the conductor) is then

$$I = \int_S \mathbf{J} \cdot d\mathbf{S} = \int_S J\, dA$$

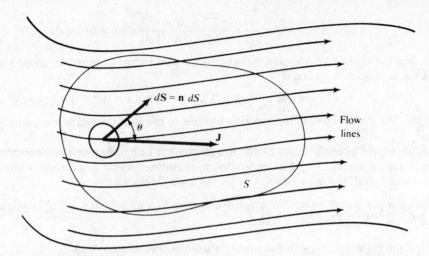

Fig. 27-1

27.4 Starting from the standard form of Ohm's law, $V = IR$, find the relation between $\mathbf{J}$, the current density, and $\mathbf{E}$, the electric field in a current-carrying conductor.

▮ We consider a conductor of uniform cross-sectional area A and length L. The resistance $R = \rho(L/A)$, where ρ is the resistivity. The current can be expressed as $I = JA$, and the potential drop across the resistor is related to the average electric field by $V = EL$. Then $V = IR$ becomes $EL = JA[\rho(L/A)]$, or $E = \rho J$. Often one specifies the conductivity, σ, instead of the resistivity, where $\sigma \equiv 1/\rho$. Then $J = \sigma E$. This result can be generalized to an arbitrary conductor in the vector form: $\mathbf{J} = \sigma\mathbf{E}$, which holds at each point in the conductor.

27.5 How many electrons per second pass through a section of wire carrying a current of 0.7 A?

▮ $I = 0.7$ A means 0.7 C/s. Dividing by $e = 1.6 \times 10^{-19}$ C, the magnitude of charge on a single electron, we get number of electrons per second $= 0.7/(1.6 \times 10^{-19}) = \underline{4.4 \times 10^{18}}$.

27.6 A current of 7.5 A is maintained in a wire for 45 s. In this time (a) how much charge and (b) how many electrons flow through the wire?

▮ (a) $q = It = (7.5\text{ A})(45\text{ s}) = \underline{337.5\text{ C}}$ (b) The number of electrons N is given by

$$N = \frac{q}{e} = \frac{337.5\text{ C}}{1.6 \times 10^{-19}\text{ C}} = \underline{2.1 \times 10^{21}}$$

where $e = 1.6 \times 10^{-19}$ C is the charge of an electron.

27.7 If 0.6 mol of electrons flow through a wire in 45 min, what are (a) the total charge that passes through the wire, and (b) the magnitude of the current?

▮ (a) The number N of electrons in 0.6 mol is

$$N = (0.6\text{ mol})(6.02 \times 10^{23}\text{ electrons/mol}) = 3.6 \times 10^{23}\text{ electrons}$$
$$q = Ne = (3.6 \times 10^{23})(1.6 \times 10^{-19}\text{ C}) = \underline{5.78 \times 10^{4}\text{ C}}$$

(b) $t = (45\text{ min})(60\text{ s/min}) = 2.7 \times 10^{3}\text{ s}$

$$I = \frac{q}{t} = \frac{5.78 \times 10^{4}\text{ C}}{2.7 \times 10^{3}\text{ s}} = \underline{21.4\text{A}}$$

27.8 An electron gun in a TV set shoots out a beam of electrons. The beam current is $10\,\mu\text{A}$. How many electrons strike the TV screen each second? How much charge strikes the screen in a minute?

▮ Let $n_e =$ number of electrons per second. $n_e = I/e = (1.0 \times 10^{-5}\text{ C/s})/(1.6 \times 10^{-19}\text{ C}) = 6.3 \times 10^{13}$ electrons per second. The charge Q striking the screen obeys $|Q| = IT = (10\,\mu\text{C/s})(60\text{ s}) = 600\,\mu\text{C}$. Since the charges are electrons, the actual charge is $Q = -600\,\mu\text{C}$.

27.9 In the Bohr model, the electron of a hydrogen atom moves in a circular orbit of radius 5.3×10^{-11} m with a speed of 2.2×10^{6} m/s. Determine its frequency f and the current I in the orbit.

❚
$$f = \frac{v}{2\pi r} = \frac{2.2 \times 10^6 \text{ m/s}}{2\pi(5.3 \times 10^{-11} \text{ m})} = 6.6 \times 10^{15} \text{ rev/s}$$

Each time the electron goes around the orbit, it carries a charge e aorund the loop. The charge passing a point on the loop each second is

$$\text{current} = I = ef = (1.6 \times 10^{-19} \text{ C})(6.6 \times 10^{15} \text{ s}^{-1}) = \underline{1.06 \text{ mA}}$$

Note that the current flows in the opposite direction to the electron, which is negatively charged.

27.10 A typical copper wire might have 2×10^{21} free electrons in 1 cm of its length. Suppose that the drift speed of the electrons along the wire is 0.05 cm/s. How many electrons would pass through a given cross section of the wire each second? How large a current would be flowing in the wire?

❚ Number per second = (number/length)(velocity) = $(2 \times 10^{21})(0.05) = \underline{1 \times 10^{20} \text{ electrons per second}}$.
$I = Q/t = (1 \times 10^{20})(1.6 \times 10^{-19}) = \underline{16 \text{ A}}$.

27.11 What is the current through an 8-Ω toaster when it is operating on 120 V?

❚ This is an application of Ohm's law: $V = IR$, $120 V = I(8\Omega)$, and $I = \underline{15 \text{ A}}$.

27.12 What potential difference is required to pass 3 A through 28 Ω?

❚
$$V = IR = (3 \text{ A})(28 \text{ }\Omega) = \underline{84 \text{ V}}$$

27.13 Determine the potential difference between the ends of a wire of resistance 5 Ω if 720 C passes through it per minute.

❚ First we determine the current, $I = Q/t$, or $I = 720 \text{ C}/60 \text{ s} = 12$ A. Then use Ohm's law, $V = IR$, or $V = (12 \text{ A})(5 \text{ }\Omega) = \underline{60 \text{ V}}$.

27.14 A copper bus bar carrying 1200 A has a potential drop of 1.2 mV along 24 cm of its length. What is the resistance per m of the bar?

❚ From Ohm's law, applied to 24 cm of the bar, $V_{24} = IR_{24}$, or $(1.2 \times 10^{-3} \text{ V}) = (1200 \text{ A})R$, and $R_{24} = 1.0 \text{ }\mu\Omega$. By proportion, $R_{100} = (100/24)R_{24} = \underline{4.2 \text{ }\mu\Omega}$.

27.15 A current of 3.0 A flows down a straight metal rod that has a 0.20-cm diameter. The rod is 1.5 m long, and the potential difference between its ends is 40 V. Find (**a**) current density and (**b**) field in the rod, and (**c**) resistivity of the material of the rod.

❚ (**a**) $J = I/A = 3/(\pi \times 10^{-6}) = \underline{9.55 \times 10^5 \text{ A/m}^2}$; (**b**) $E = V/d = 40/1.5 = \underline{27 \text{ V/m}}$; and (**c**) since $J = E/\rho$, $\rho = 2.8 \times 10^{-5} \text{ }\Omega \cdot m = \underline{28 \text{ }\mu\Omega \cdot m}$.

27.16 A 0.20-mm-diameter copper wire is sealed end to end to a 5.00-mm-diameter iron rod, and a current is sent lengthwise through them. If the current in the copper is 8.0 A, what are (**a**) the current and current density in the iron, and (**b**) the current density in the copper?

❚ (**a**) Since charge must be conserved, $I_{Cu} = I_{Fe} = 8.0$ A, and $J_{Fe} = I/A = 8.0/[\pi(5.0 \times 10^{-3})^2/4] = \underline{407 \text{ kA/m}^2}$. (**b**) By inverse proportion, $J_{Cu} = (5.00/0.20)^2 J_{Fe} = \underline{255 \text{ MA/m}^2}$.

27.17 A copper wire of 3.0-mm² cross-sectional area carries a current of 5.0 A. Find the magnitude of the drift velocity for the electrons in the wire.

❚ We have

$$J = \frac{I}{A} = \frac{5.0 \text{ A}}{3.0 \times 10^{-6} \text{ m}^2} = 1.67 \times 10^6 \text{ A/m}^2$$

The drift velocity is given by

$$v = \frac{J}{ne} = \frac{1.67 \times 10^6 \text{ A/m}^2}{n(1.60 \times 10^{-19} \text{ C})} = \frac{1}{n}(1.04 \times 10^{25} \text{ m}^{-2} \cdot \text{s}^{-1})$$

where n is the number of charge carriers per unit volume. To find n we must find the number of copper atoms per unit volume. Assuming one free electron per atom, $M = 63.5$ kg/kmol, and $\rho = 8920$ kg/m³, we have

$$n = \frac{(6.02 \times 10^{26} \text{ atoms/kmol})(8920 \text{ kg/m}^3)}{63.5 \text{ kg/kmol}} = 8.5 \times 10^{28} \text{ atoms/m}^3$$

Substituting in the expression for the drift velocity gives $v = \underline{0.12 \text{ mm/s}}$.

27.18^c As shown in Fig. 27-2, a metal rod of radius r_1 is concentric with a metal cylindrical shell of radius r_2 and length L. The space between rod and cylinder is tightly packed with a high-resistance material of resistivity ρ. A battery having a terminal voltage v_t is connected as shown. Neglecting resistances of rod and cylinder, derive expressions for **(a)** the total current I, **(b)** the current density J and the electric field E at any point P between rod and cylinder, and **(c)** the resistance R between rod and cylinder.

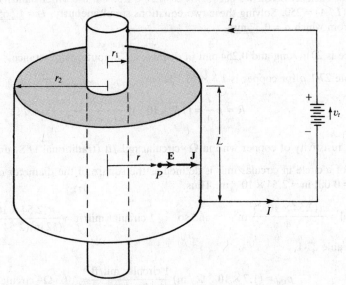

Fig. 27-2

I **(a)** Assuming radial flow of charge between rod and cylinder, we have at P

$$J = \frac{I}{2\pi r L} \quad \text{and} \quad E = \rho J = \frac{\rho I}{2\pi r L}$$

with both **J** and **E** in the direction of **r**. Then, by definition of the potential,

$$dv = -\mathbf{E} \cdot d\mathbf{s} = -E \, dr = -\frac{\rho I}{2\pi L} \frac{dr}{r}$$

and so, noting the polarity of v_t,

$$-v_t = \int_{r_1}^{r_2} dv = -\frac{\rho I}{2\pi L} \int_{r_1}^{r_2} \frac{dr}{r} = -\frac{\rho I}{2\pi L} \ln \frac{r_2}{r_1}$$

Solving for I,

$$I = \frac{2\pi L v_t}{\rho \ln (r_2/r_1)}$$

(b) From **a**,

$$J = \frac{I}{2\pi r L} = \frac{v_t}{\rho r \ln (r_2/r_1)} \quad \text{and} \quad E = \rho J = \frac{v_t}{r \ln (r_2/r_1)}$$

(c) From Ohm's law,

$$R = \frac{v_t}{I} = \frac{\rho \ln (r_2/r_1)}{2\pi L}$$

27.19 Compute the resistance of 180 m of silver wire having a cross section of 0.3 mm². (Assume that $t = 20 \,^\circ$C.)

I Resistance is given by $R = \rho(L/A)$. From Table 27-1 we have ρ for silver at $20\,^\circ$C $= 1.6 \times 10^{-8} \,\Omega \cdot$ m. Then $R = (1.6 \times 10^{-8} \,\Omega \cdot \text{m})(180 \text{ m})/(0.3 \times 10^{-6} \text{ m}^2) = \underline{9.6\ \Omega}$.

27.20 How long a piece of aluminum wire 1 mm in diameter is needed to give a resistance of 4 Ω? (Assume $t = 20 \,^\circ$C).

❚ $R = \rho(L/A)$. From Table 27-1 $\rho = 2.8 \times 10^{-8} \,\Omega \cdot m$. The cross-sectional area A is $\pi r^2 = 3.14 \,(0.5 \times 10^{-3} \,m)^2 = 7.85 \times 10^{-7} \,m^2$. Then $4 \,\Omega = [(2.8 \times 10^{-8} \,\Omega \cdot m)L]/(7.85 \times 10^{-7} \,m^2)$; and solving we get $L = \underline{112 \,m}$.

27.21 A 20-cm-long copper tube has an inner diameter of 0.85 cm and an outer diameter of 1.10 cm. Find its electric resistance when used lengthwise.

❚ $R = \rho(L/A)$. The cross-sectional area is $\pi[(1.10^2 - 0.85^2)]/(4 \times 10^4)] = 3.83 \times 10^{-5} \,m^2$; then with $L = 0.20 \,m$ and $\rho = 1.7 \times 10^{-8}$ from Table 27-1, $R = \underline{89 \,\mu\Omega}$.

27.22 A bar of copper having mass 1.5 kg is to be drawn into a wire having resistance 250 Ω at 20 °C. Determine the length L and diameter d of the wire. The density of copper is $8.9 \times 10^3 \,kg/m^3$.

❚ From mass = density × vol. we get $(8.9 \times 10^3)LA = 1.5$. On the other hand, Table 27-1 gives for $R_{20°C}$ $(1.72 \times 10^{-8})(L/A) = 250$. Solving these two equations simultaneously, $L = \underline{1.565 \,km}$ and $A = 0.1077 \,mm^2$. But $A = \pi d^2/4$, from which $d = \underline{0.37 \,mm}$.

27.23 A copper wire is 20 m long and 0.254 mm in diameter. Compute its resistance.

❚ From Table 27-1 ρ for copper is $1.7 \times 10^{-8} \,\Omega \cdot m$.

$$R = \rho \frac{l}{A} = (1.7 \times 10^{-8})\left(\frac{20}{\pi (0.000127)^2}\right) = 6.7 \,\Omega$$

27.24 Compute the resistivity of copper wire in $\Omega \cdot$ circular mils/ft (traditional U.S. units).

❚ The area of a circle in circular mils is defined as the square of the diameter of the circle expressed in mils, where 1 mil = 0.001 in = $2.54 \times 10^{-5} \,m$. Thus,

$$1 \text{ circular mil} = \frac{\pi (2.54 \times 10^{-5})^2}{4} \,m^2 \quad \text{and so} \quad 1 \text{ circular mil/ft} = \frac{\pi (2.54 \times 10^{-5})^2/4 \quad m^2}{(12 \text{ in})(2.54 \times 10^{-2} \,m/\text{in})} = 1.65 \times 10^{-9} \,m$$

Then, from Table 27-1,

$$\rho_{Cu} = (1.7 \times 10^{-8} \,\Omega \cdot m) \frac{1 \text{ circular mil/ft}}{1.65 \times 10^{-9} \,m} = 10.3 \,\Omega \cdot \text{circular mil/ft}$$

27.25 A coil of wire has a resistance of 25.00 Ω at 20 °C and a resistance of 25.17 Ω at 35 °C. What is its temperature coefficient of resistance?

❚ $R = R_0[1 + \alpha(T - T_0)]$, or $\alpha = \Delta R/(R_0 \,\Delta T)$, with $\Delta R = R - R_0 = 0.17 \,\Omega$ and $\Delta T = T - T_0 = 15 \,°C$. Then $\alpha = (0.17)/(25.00 \times 15) = \underline{4.5 \times 10^{-4} \,°C^{-1}}$.

27.26 A metal wire of diameter 2 mm and of length 300 m has a resistance of 1.6424 Ω at 20 °C, and 2.415 Ω at 150 °C. Find the values of α, R_0, ρ_0, where the zero subscript refers to 0 °C, and $\rho_{20°C}$. Identify the metal.

❚ $$R_{150°C} = 2.415 = R_0(1 + \alpha 150) \qquad R_{20°C} = 1.6424 = R_0(1 + \alpha 20)$$

Solving these relations simultaneously, $\alpha = \underline{3.9 \times 10^{-3} \,°C^{-1}}$ and $R_0 = \underline{1.5236 \,\Omega}$.
From $R_0 = \rho_0(L/A)$,

$$1.5236 = \frac{\rho_0(300)}{\pi(2 \times 10^{-3})^2/4} \quad \text{or} \quad \rho_0 = \underline{1.596 \times 10^{-8} \,\Omega \cdot m}$$

Then, $\qquad \rho_{20°C} = \rho_0(1 + \alpha 20) = (1.596 \times 10^{-8})[1 + (3.9 \times 10^{-3})(20)] = \underline{1.720 \times 10^{-8} \,\Omega \cdot m}$

Table 27-1 indicates that the metal is copper.

27.27 It is desired to make a 20.0-Ω coil of wire which has a zero thermal coefficient of resistance. To do this, a carbon resistor of resistance R_1 is placed in series with an iron resistor of resistance R_2. The proportions of iron and carbon are so chosen that $R_1 + R_2 = 20.00 \,\Omega$ for all temperatures near 20°C. How large are R_1 and R_2?

❚ We need $R_1(1 + \alpha_1 \,\Delta t) + R_2(1 + \alpha_2 \,\Delta t) = 20$. Because $R_1 + R_2 = 20$ when $\Delta t = 0$, we must have $R_1\alpha_1 = -R_2\alpha_2$ with $\alpha_1 = -0.5 \times 10^{-3}$ and $\alpha_2 = 5 \times 10^{-3}$. Solving the two equations $R_1 + R_2 = 20$ and $R_1 = 10R_2$ simultaneously leads to $R_1 = \underline{18.18}$ and $R_2 = \underline{1.82 \,\Omega}$.

27.28 A resistance thermometer measures temperature by the increase in resistance of a wire at high temperature. If the wire is platinum and has a resistance of 10 Ω at 20 °C and a resistance of 35 Ω in a hot furnace, what is the temperature of the furnace? (α for platinum is 0.0036 °C^{-1}.)

▮ We assume that α is constant over the needed temperature range. Then $\Delta R = \alpha R \Delta t$ leads to $(35 - 10) = 0.0036(10)\,\Delta t$. Solving we get $\Delta t = 25/0.036 = 694$ °C. Finally $694 + 20 = \underline{714\,°C}$ (furnace temperature).

27.29 A 75-W tungsten light bulb has a resistance of 190 Ω when lighted and 15 Ω when turned off. Estimate the temperature of the filament when the bulb is lighted.

▮ We can make only a very rough estimate since we must use α from Table 27-1 over far too wide a range. We have

$$R = R_{20}(1 + \alpha\,\Delta T) \qquad \text{or} \qquad \Delta T = \frac{R - R_{20}}{\alpha R_{20}}$$

Then

$$T - 20\,°C = \frac{(190 - 15)\,\Omega}{(4.5 \times 10^{-3}/°C)(15\,\Omega)} = 2590\,°C$$

from which $T \approx \underline{2600\,°C}$.

27.30 A 60-W bulb carries a current of 0.5 A when operating on 120 V. The temperature of its tungsten filament is then 1800 °C. Find the resistance at its operating temperature. Find its approximate resistance at 20 °C.

▮ As in Prob. 27.29, the temperature range is too high for much accuracy. Nonetheless we proceed. Use $V = IR$ to find $R = 240\,\Omega$. Then $R = R_{20}(1 + \alpha\,\Delta t)$ yields $240 = R_{20}[1 + (4.5 \times 10^{-3})(1780)]$, from which $R_{20} = \underline{26.6\,\Omega}$.

27.2 RESISTORS IN COMBINATION

27.31 Resistors R_1, R_2, and R_3 are (a) in series and (b) in parallel, as shown in Fig. 27-3(a) and (b). Derive the formula for the equivalent resistance R_{eq} of each network.

▮ (a) For the series network,

$$V_{ad} = V_{ab} + V_{bc} + V_{cd} = IR_1 + IR_2 + IR_3$$

since the current I is the same in each resistor. Dividing by I,

$$\frac{V_{ad}}{I} = R_1 + R_2 + R_3 \qquad \text{or} \qquad \underline{R_{eq} = R_1 + R_2 + R_3}$$

since V_{ad}/I is by definition the equivalent resistance R_{eq} of the network.
(b) The p.d. across each resistor is the same, whence

$$I_1 = \frac{V_{ab}}{R_1} \qquad I_2 = \frac{V_{ab}}{R_2} \qquad I_3 = \frac{V_{ab}}{R_3}$$

Since the line current I is the sum of the branch currents,

$$I = I_1 + I_2 + I_3 = \frac{V_{ab}}{R_1} + \frac{V_{ab}}{R_2} + \frac{V_{ab}}{R_3}$$

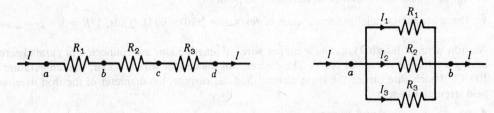

(a) Resistors in series   (b) Resistors in parallel

Fig. 27-3

Dividing by V_{ab},

$$\frac{I}{V_{ab}} = \frac{1}{R_1} + \frac{1}{R_2} + \frac{1}{R_3} \quad \text{or} \quad \frac{1}{R_{eq}} = \frac{1}{R_1} + \frac{1}{R_2} + \frac{1}{R_3}$$

since V_{ab}/I is by definition the equivalent resistance R_{eq} of the network.

27.32 Find the resistance equivalent to three resistances in parallel: $R_1 = 12\,\Omega$, $R_2 = 12\,\Omega$, and $R_3 = 6\,\Omega$.

▌ The equivalent resistance R is given by $1/R = 1/R_1 + 1/R_2 + 1/R_3 = \frac{1}{12} + \frac{1}{12} + \frac{1}{6} = \frac{1}{12} + \frac{1}{12} + \frac{2}{12} = \frac{4}{12}$. Thus $R = \underline{3\,\Omega}$.

27.33 What is the resistance between A and B in Fig. 27-4?

▌ For the two resistors in parallel,

$$\frac{1}{R} = \frac{1}{3} + \frac{1}{6} = \frac{2}{6} + \frac{1}{6} = \frac{3}{6} \quad \text{or} \quad R = 2\,\Omega$$

This is in series with $8\,\Omega$, so

$$R_{AB} = 2\,\Omega + 8\,\Omega = \underline{10\,\Omega}$$

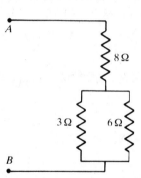

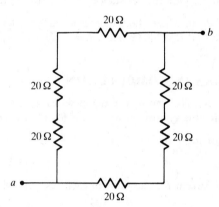

Fig. 27-4   **Fig. 27-5**

27.34 Three resistances of 12, 16, and $20\,\Omega$ are connected in parallel. What resistance must be connected in series with this combination to give a total resistance of $25\,\Omega$?

▌ The resistance R of the parallel combination is given by

$$\frac{1}{R} = \frac{1}{R_1} + \frac{1}{R_2} + \frac{1}{R_3} = \frac{1}{12} + \frac{1}{16} + \frac{1}{20} = \frac{20}{240} + \frac{15}{240} + \frac{12}{240} = \frac{47}{240} \quad \text{or} \quad R = 5.11\,\Omega$$

Then

$$R_x + R = 25 \quad \text{or} \quad R_x = 25 - 5.11 = \underline{19.89\,\Omega}$$

27.35 In Fig. 27-5, find the resistance from point a to point b.

▌ There are two parallel branches, each of resistance $3(20) = 60\,\Omega$. Thus, $1/R = \frac{1}{60} + \frac{1}{60} = \frac{2}{60}$, and $R = \underline{30\,\Omega}$.

27.36 An iron wire, of length 2 km, and a copper wire, of length 3 km, are connected in parallel across a source having a terminal voltage of 200 V. The diameter of the copper wire is 1 mm; the temperature of the wires is 100 °C. If each wire carries the same current, find the current, the diameter of the iron wire, and the electric field strength in each.

▌ We get the resistivities at 20 °C and the temperature coefficients from the Table 27-1. Hence, at 100 °C,

$$R_{Cu} = \frac{(1.7 \times 10^{-8})(3000)}{\pi(10^{-6})/4}[1 + (3.9 \times 10^{-3})(80)] = 85.24\,\Omega \quad \text{thus} \quad I_{Cu} = \frac{200}{85.24} = \underline{2.34\,A}$$

Since the current is assumed to be the same in each wire,

$$R_{Fe} = R_{Cu} = 85.24 = \frac{(10 \times 10^{-8})(2000)}{\pi d^2/4}[1 + (5.0 \times 10^{-3})(80)]$$

from which the diameter of the iron wire is $d = \underline{2.046 \text{ mm.}}$ The electric fields are

$$E_{Cu} = \frac{200 \text{ V}}{3000 \text{ m}} = \frac{1}{15} \text{ V/m} \qquad E_{Fe} = \frac{200 \text{ V}}{2000 \text{ m}} = \frac{1}{10} \text{ V/m}$$

These values also follow from $E = \rho I/A$.

27.37 A 50-cm-long metal rod consists of a copper sheath (inner diameter = 2 mm, outer diameter = 3 mm) with an iron core (see Fig. 27-6). What is the resistance of the rod? (*Hint*: Find the current that would flow through it when the potential difference is V.)

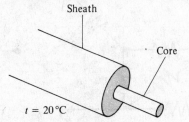

Fig. 27-6

▮ Potential V across the rod causes I to flow; by definition $I = V/R$ but $I = I_{Cu} + I_{Fe} = V/R_{Cu} + V/R_{Fe}$; $1/R = 1/R_{Cu} + 1/R_{Fe}$. Using Table 27-1, $R_{Cu} = \rho_{Cu}(L/A_{Cu}) = [(1.7 \times 10^{-8})(0.5)]/[\pi(1.5^2 - 1.0^2) \times 10^{-6}] = 0.00216 \,\Omega$. In like manner for Fe, $R = \rho(L/A) = 0.0159 \,\Omega$; then $1/R = 1/0.00216 + 1/0.0159$ yields $R = 1.91 \text{ m}\Omega$.

27.38 Find all the resistances that can be realized with a 6-, a 9-, and a 15-Ω resistor in various combinations. Not every combination need use all three resistors.

▮ Figure 27-7 shows all possible combinations and their equivalent resistances.

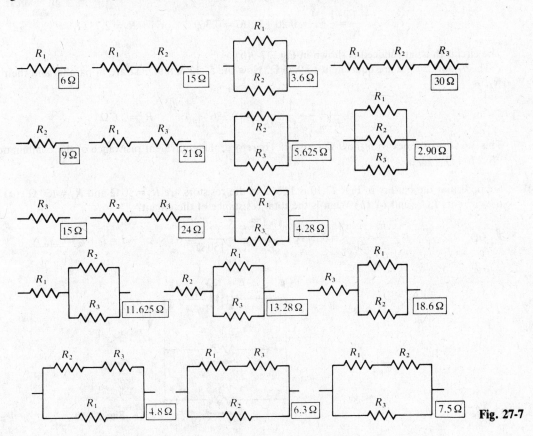

Fig. 27-7

27.39 Arrange an 8-, a 12-, and a 16-Ω resistor in a combination that has a total resistance of 8.89 Ω.

❚ It is clear that we cannot have any resistance in series with the remaining pair. This obsservation leads to the solution indicated in Fig. 27-8.

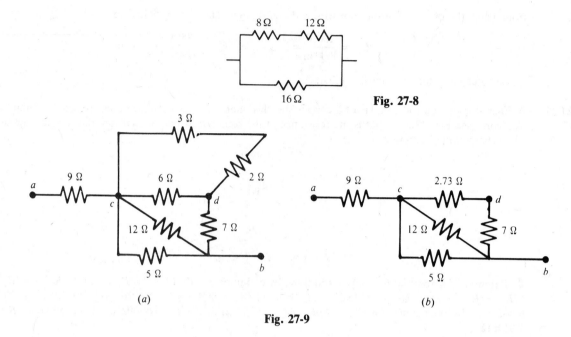

Fig. 27-8

Fig. 27-9

27.40 Find the equivalent resistance between points a and b for the combination shown in Fig. 27-9(a).

❚ The 3- and 2-Ω resistors are in series and are equivalent to a 5-Ω resistor. The equivalent 5 Ω is in parallel with the 6 Ω, and their equivalent, R_1, is

$$\frac{1}{R_1} = \frac{1}{5} + \frac{1}{6} = 0.20 + 0.167 = 0.369 \quad \text{or} \quad R_1 = 2.73 \, \Omega$$

The circuit thus far reduced is shown in Fig. 27-9(b).

The 7- and 2.73-Ω are equivalent to 9.73 Ω. Now the 5, 12 and 9.73 Ω are in parallel and their equivalent, R_2, is

$$\frac{1}{R_2} = \frac{1}{5} + \frac{1}{12} + \frac{1}{9.73} = 0.386 \quad \text{or} \quad R_2 = 2.6 \, \Omega$$

This 2.6 Ω is in series with the 9-Ω resistor. Therefore, the equivalent resistance of the combination is $9 + 2.6 = \underline{11.6 \, \Omega}$.

27.41 Suppose that the battery in Fig. 27-10 is 12 V and the resistors are $R_1 = 50 \, \Omega$ and $R_2 = 150 \, \Omega$. (a) What are the currents I, I_1, and I_2? (b) What is the total resistance of the circuit?

❚ (a) $I_1 = \dfrac{E}{R_1} = \dfrac{12 \, \text{V}}{50 \, \Omega} = 0.24 \, \text{A} \qquad I_2 = \dfrac{E}{R_2} = \dfrac{12 \, \text{V}}{150 \, \Omega} = 0.08 \, \text{A} \qquad I = I_1 + I_2 = \underline{0.32 \, \text{A}}$

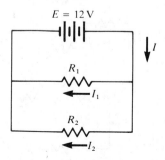

Fig. 27-10

(b)
$$\frac{1}{R} = \frac{1}{R_1} + \frac{1}{R_2} = \frac{1}{50\ \Omega} + \frac{1}{150\ \Omega} = \frac{4}{150\ \Omega} \qquad R = \underline{37.5\ \Omega}$$

As a check, $I = E/R = 12\ \text{V}/37.5\ \Omega = 0.32\ \text{A}$.

27.42 Suppose that the emf of the battery in Fig. 27-10 is 45 V and the resistor $R_1 = 300\ \Omega$. **(a)** What must the resistor R_2 be in order that the current I be 0.45 A? **(b)** What are the currents I_1 and I_2?

▌ (a) The total resistance must be

$$R = \frac{E}{I} = \frac{45\ \text{V}}{0.45\ \text{A}} = 100\ \Omega$$

But $\quad \dfrac{1}{R} = \dfrac{1}{R_1} + \dfrac{1}{R_2} \quad$ so $\quad \dfrac{1}{R_2} = \dfrac{1}{R} - \dfrac{1}{R_1} = \dfrac{1}{100\ \Omega} - \dfrac{1}{300\ \Omega} = \dfrac{2}{300\ \Omega} \quad$ or $\quad R_2 = \underline{150\ \Omega}$

(b)
$$I_1 = \frac{E}{R_1} = \frac{45\ \text{V}}{300\ \Omega} = \underline{0.15\ \text{A}} \qquad I_2 = \frac{E}{R_2} = \frac{45\ \text{V}}{150\ \Omega} = \underline{0.30\ \text{A}}$$

27.43 The three resistors in Fig. 27-11 are $R_1 = 25\ \Omega$, $R_2 = 50\ \Omega$, and $R_3 = 100\ \Omega$. **(a)** What is the total resistance of the circuit? **(b)** What are the currents I_1, I_2, and I_3 for a 12-V battery?

▌ (a) The sum of R_2 and R_3 in parallel is

$$\frac{1}{R'} = \frac{1}{R_2} + \frac{1}{R_3} = \frac{1}{50\ \Omega} + \frac{1}{100\ \Omega} = \frac{3}{100\ \Omega} \qquad \text{or} \qquad R' = 33.3\ \Omega$$

Since R' is in series with R_1, the total resistance of the circuit is $R = R' + R_1 = 33.3\ \Omega + 25\ \Omega = \underline{58.3\ \Omega}$.

(b)
$$I = \frac{E}{R} = \frac{12\ \text{V}}{58.3\ \Omega} = \underline{0.206\ \text{A}}$$

The potential V' across R_2 and R_3 is $V' = E - R_1 I = 12\ \text{V} - (25\ \Omega)(0.206\ \text{A}) = 6.85\ \text{V}$. Therefore,

$$I_2 = \frac{V'}{R_2} = \frac{6.85\ \text{V}}{50\ \Omega} = \underline{0.137\ \text{A}} \qquad I_3 = \frac{V'}{R_3} = \frac{6.85\ \text{V}}{100\ \Omega} = \underline{0.0685\ \text{A}}$$

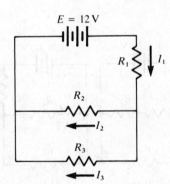

Fig. 27-11

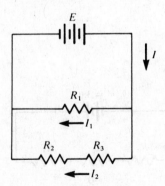

Fig. 27-12

27.44 The three resistors in Fig. 27-12 are $R_1 = 80\ \Omega$, $R_2 = 25\ \Omega$, and $R_3 = 15\ \Omega$. **(a)** What is the total resistance of the circuit? **(b)** What are the currents I and I_2, and the voltage across the battery, if $I_1 = 0.3$ A?

▌ (a) The sum of R_2 and R_3 in series is $R' = R_2 + R_3 = 25\ \Omega + 15\ \Omega = 40\ \Omega$. Since R' is in parallel with R_1, the total resistance of the circuit is

$$\frac{1}{R} = \frac{1}{R_1} + \frac{1}{R'} = \frac{1}{80\ \Omega} + \frac{1}{40\ \Omega} = \frac{3}{80\ \Omega} \qquad \text{or} \qquad R = \underline{26.7\ \Omega}$$

(b) $\quad E = R_1 I_1 = (80\ \Omega)(0.3\ A) = \underline{24\ \text{V}} \qquad I_2 = \dfrac{E}{R'} = \dfrac{24\ \text{V}}{40\ \Omega} = \underline{0.6\ \text{A}} \qquad I = I_1 + I_2 = \underline{0.9\ \text{A}}$

(*Check*: $I = E/R = 24\ V/26.7\ \Omega = \underline{0.9\ \text{A}}$)

27.45 Find I_1 and I_2 for the circuit of Fig. 27-13.

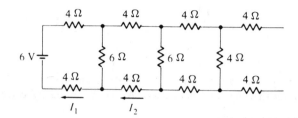

Fig. 27-13

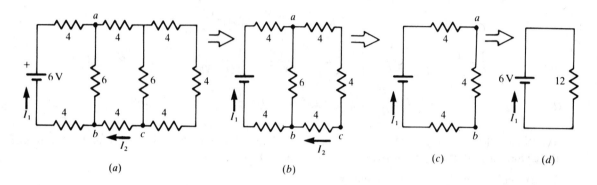

Fig. 27-14

▌ In Fig. 27-14 we show successive reduction of the circuit. From Fig. 27-14(d), $I_1 = \frac{6}{12} = \underline{0.5\,A}$. Fig. 27-14($c$), $V_{ab} = (\frac{1}{2})(4) = 2.0\,V$. Thus (Fig. 27-14($b$)) the drop across wire acb is 2 V, and $I_2 = \frac{2}{12} = \underline{0.167\,A}$.

27.46 Figure 27-15 shows the three resistors $R_1 = 5\,\Omega$, $R_2 = 15\,\Omega$, and $R_3 = 25\,\Omega$ in four different circuits. For each circuit find the currents I_1, I_2, and I_3 in each resistor, and the current I in the battery.

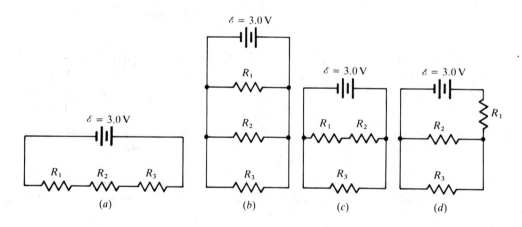

Fig. 27-15

▌ **(a)**
$$R = R_1 + R_2 + R_3 = \underline{45\,\Omega} \qquad I_1 = I_2 = I_3 = I = \frac{\varepsilon}{R} = \frac{3.0\,V}{45\,\Omega} = \underline{0.067\,A}$$

(b)
$$I_1 = \frac{\varepsilon}{R_1} = \frac{3.0\,V}{5.0\,\Omega} = \underline{0.6\,A} \qquad I_2 = \frac{\varepsilon}{R_2} = \frac{3.0\,V}{15.0\,\Omega} = \underline{0.2\,A}$$

$$I_3 = \frac{\varepsilon}{R_3} = \frac{3.0\,V}{25\,\Omega} = \underline{0.12\,A} \qquad I = I_1 + I_2 + I_3 = \underline{0.92\,A}$$

(c)
$$I_3 = \frac{\varepsilon}{R_3} = \frac{3.0\,V}{25\,\Omega} = \underline{0.12\,A} \qquad I_1 = I_2 = \frac{\varepsilon}{R_1 + R_2} = \frac{3.0\,V}{20\,\Omega} = \underline{0.15\,A} \qquad I = I_1 + I_3 = \underline{0.27\,A}$$

(d)
$$\frac{1}{R'} = \frac{1}{R_2} + \frac{1}{R_3} = \frac{1}{15} + \frac{1}{25} = \frac{8}{75} \qquad R' = 9.375\ \Omega$$

$$I_1 = I = \frac{\varepsilon}{R_1 + R'} = \frac{3.0\ \text{V}}{14.375\ \Omega} = \underline{0.209\ \text{A}}$$

$$I_2 = \frac{V_2}{R_2} = \frac{\varepsilon - I_1 R_1}{R_2} = \frac{1.96\ \text{V}}{15\ \Omega} = \underline{0.130\ \text{A}} \qquad I_3 = I - I_2 = \underline{0.079\ \text{A}}$$

27.47 It is known that the potential difference across the 6-Ω resistance in Fig. 27-16 is 48 V. Determine (a) the entering current I, (b) the potential difference across the 8-Ω resistance, (c) the potential difference across the 10-Ω resistance, (d) the potential difference from a to b. (*Hint:* The wire connecting c and d can be shrunk to zero length without altering the currents or potentials.)

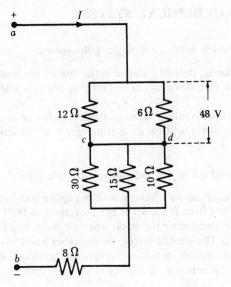

Fig. 27-16

▮ (a) Since the 6-Ω and 12-Ω resistors are in parallel, $I = I_6 + I_{12} = 48\ \text{V}/6\ \Omega + 48\ \text{V}/12\ \Omega = \underline{12\ \text{A}}$. (b) Since I passes through the 8-Ω resistor, we have $V_8 = (12\ \text{A})(8\ \Omega) = \underline{96\ \text{V}}$. (c) Let R_e be the equivalent resistance for the three parallel resistors of 10, 15, and 30 Ω. Then $1/R_e = \frac{1}{10} + \frac{1}{15} + \frac{1}{30} = \frac{6}{30}$, and $R_e = 5\ \Omega$. Since the voltages across each of the three resistors are the same, we have $V_e = V_{10} = IR_e = (12\ \text{A})(5\ \Omega) = \underline{60\ \text{V}}$. (d) $V_{ab} = V_6 + V_{10} + V_8 = 48\ \text{V} + 60\ \text{V} + 96\ \text{V} = \underline{204\ \text{V}}$.

27.48 For the circuit shown in Fig. 27-17 find R_{eq} together with I_1 and I_2. (*Hint:* Note that a, b, c, and d are all the same point from an electric standpoint. Redraw the diagram to show this.)

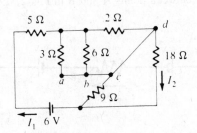

Fig. 27-17

▮ We show the equivalent diagram as well as reduced diagrams in Fig. 27-18. The equivalent resistance is 12 Ω. From Fig. 27-18(c), $I_1 = \frac{6}{12} = \underline{0.5\ \text{A}}$. The voltage drop across the 18 Ω is the same as that across the 6 Ω in Fig. 27-18(b), namely, $6(\frac{1}{2}) = 3\ \text{V}$. Then $I_2 = \frac{3}{18} = \underline{0.167\ \text{A}}$.

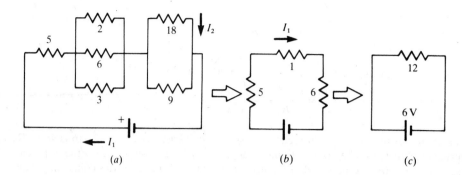

Fig. 27-18

27.3 EMF AND ELECTROCHEMICAL SYSTEMS

27.49 Define: *electromotive force, internal resistance, galvanometer.*

▌ The *electromotive force* (emf) of a source is the energy per unit charge converted from chemical or some other form of energy to electric energy. The emf $\mathcal{E}$ is one volt when each coulomb of charge receives one joule of energy.
   The term *internal resistance* refers to the resistance within an emf source.
   A *galvanometer* is an instrument for measuring very small electric currents; it underlies both voltmeters and ammeters.

27.50 How are electric potential difference and emf related in a circuit?

▌ The potential difference across a battery with negligible internal resistance is equal in magnitude to the emf. As a charge q passes from the low-voltage (negative) to the high-voltage (positive) terminal through the battery with emf $\mathcal{E}$, the nonelectric forces do work $q\mathcal{E}$ on it. If q is positive, this is positive work and the battery gives up energy. The electric forces, on the other hand, do an equal amount of negative work and the electrostatic system gains electric potential energy. As the charge q passes around the *external* circuit back to the negative terminal, this electrostatic energy is depleted and gets replenished by q going through the battery once more.

27.51 If a battery or other source of emf has internal resistance, then the terminal voltage does not equal the emf when current is flowing. What are the relations between emf and terminal voltage in such situations?

▌ The terminal voltage of a battery or generator when it delivers a current I is equal to the total electromotive force (emf or $\mathcal{E}$) minus the potential drop (or voltage drop) in its *internal resistance, r.*
(1) When delivering current (*on discharge*), terminal voltage = emf − (voltage drop in internal resistance) = $\mathcal{E} - Ir$.
(2) When receiving current (*on charge*), terminal voltage = emf + (voltage drop in internal resistance) = $\mathcal{E} + Ir$.
(3) When no current exists, terminal voltage = emf of battery or generator.

27.52 Find the potential difference between points A and B in Fig. 27-19 if R is 0.7 Ω. Which point is at the higher potential?

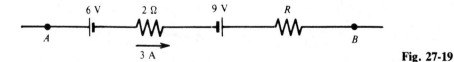

Fig. 27-19

▌ Clearly this is only part of a larger circuit that imposes the 3-A current on the system. We add the voltage difference across the four elements starting at A: $V_B - V_A = (-6 \text{ V}) - (3 \text{ A})(2 \text{ Ω}) + (9 \text{ V}) - (3 \text{ A})(0.7 \text{ Ω}) = -5.1 \text{ V}$. V_A is at the higher potential.
[Equivalently we could add the "voltage drops" from A to B: $V_A - V_B = 6 \text{ V} + (3 \text{ A})(2 \text{ Ω}) + (-9 \text{ V}) + (3 \text{ A})(0.7 \text{ Ω}) = 5.1 \text{ V}$, which is the same result.]

27.53 Repeat Prob. 27.52 if the current flows in the opposite direction and $R = 0.7\,\Omega$.

▎ We proceed with the first method of Prob. 27.52. $V_B - V_A = (-6\,\text{V}) + (3\,\text{A})(2\,\Omega) + (9\,\text{V}) + (3\,\text{A})(0.7\,\Omega) = \underline{11.1\,\text{V}}$. V_B is the higher potential.

27.54 The emf of a mercury-cadmium cell decreases $40\,\mu\text{V}$ for each 1-°C increase in temperature. If its emf is 1.0183 V at 20 °C, what would its emf be at 0 °C?

▎ Since the temperature is dropping,

$$\mathcal{E}_0 = \mathcal{E}_{20} + 20(40 \times 10^{-6}) = 1.0183 + 800 \times 10^{-6} = 1.0183 + 0.0008 = \underline{1.0191\,\text{V}}.$$

27.55 A battery usually has a small internal resistance of its own. This is indicated by the resistor r in Fig. 27-20. If the emf of the battery is 3.0 V, $r = 0.5\,\Omega$, and $R = 5\,\Omega$, what is the potential difference between the terminals a and b of the battery?

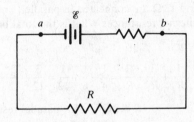

Fig. 27-20

▎ Let V be the potential difference $V_b - V_a$. Applying Ohm's law,

$$I = \frac{\mathcal{E}}{R + r} = \frac{3.0\,\text{V}}{5.5\,\Omega} = 0.545\,\text{A}. \qquad \text{Then} \qquad V = \mathcal{E} - rI = (3.0\,\text{V}) - (0.5\,\Omega)(0.545\,\text{A}) = \underline{2.73\,\text{V}}$$

27.56 The circuit in Fig. 27-20 has a current of 0.5 A when R is $10\,\Omega$ and a current of 0.27 A when R is $20\,\Omega$. Find (a) the internal resistance r and (b) the emf $\mathcal{E}$ of the battery.

▎ (a) For the two cases,

$$I_1 = \frac{\mathcal{E}}{R_1 + r} \quad \text{and} \quad I_2 = \frac{\mathcal{E}}{R_2 + r} \quad \text{so} \quad \frac{I_1}{I_2} = \frac{R_2 + r}{R_1 + r} \quad \text{or} \quad \frac{20\,\Omega + r}{10\,\Omega + r} = \frac{0.5\,\text{A}}{0.27\,\text{A}} = 1.85$$

Solving this last equation for r we get

$$20\,\Omega + r = 1.85(10\,\Omega + r) = 18.5\,\Omega + 1.85r \qquad r = \frac{1.5\,\Omega}{0.85} = \underline{1.76\,\Omega}$$

(b) $$\mathcal{E} = I(R + r) = (0.5\,\text{A} \times 11.76\,\Omega) = \underline{5.88\,\text{V}}$$

27.57 (a) A new 1.5-V dry cell furnishes 30 A when short-circuited. Find the internal resistance of the cell. (b) An old 1.5-V dry cell furnishes 10 A when short-circuited. Find its internal resistance.

▎ (a) The only resistance in the circuit is internal resistance, so:

$$E = IR \qquad 1.5 = 30R \qquad R = \underline{0.05\,\Omega}$$

(b) $$E = IR \qquad 1.5 = 10R \qquad R = \underline{0.15\,\Omega}$$

27.58 Compute the internal resistance of an electric generator which has an emf of 120 V and a terminal voltage of 110 V when supplying 20 A.

▎ We have $V = E - ir$, with $E = 120\,\text{V}$, $V = 110\,\text{V}$, $i = 20\,\text{A}$. Then $110 = 120 - 20r$, and $r = \underline{0.5\,\Omega}$.

27.59 A dry cell delivering 2 A has terminal voltage 1.41 V. What is the internal resistance of the cell if its open-circuit voltage is 1.59 V?

▎ The open-circuit voltage is simply the emf of the cell, so $V = E - ir$ with $V = 1.41\,\text{V}$, $i = 2\,\text{A}$, $E = 1.59\,\text{V}$. $1.41 = 1.59 - 2r$, and $r = \underline{0.09\,\Omega}$.

27.60 A storage battery has emf 25 V and internal resistance $0.20\,\Omega$. Compute its terminal voltage (a) when it is delivering 8 A and (b) when it is being charged with 8 A.

▮ (a) $V = E - Ir = 25\text{ V} - (8\text{ A})(0.20\ \Omega) = \underline{23.4\text{ V}}$. (b) When being charged, the terminal voltage supplies both the emf and potential drop in the internal resistance. Thus $V = E + Ir = 25\text{ V} + (8\text{ A})(0.20\ \Omega) = \underline{26.6\text{ V}}$.

27.61 A battery charger supplies a current of 10 A to charge a storage battery which has an open-circuit voltage of 5.6 V. If the voltmeter connected across the charger reads 6.8 V, what is the internal resistance of the battery at this time?

▮ Since the battery is charging, $V = E + Ir$, where $I = 10$ A, $E = 5.6$ V, and $V = 6.8$ V. Then $6.8 = 5.6 + 10r$, and $r = \underline{0.12\ \Omega}$.

27.62 An automobile battery has an emf of 6 V and an internal resistance of 0.01 Ω. When the starter draws 200 A from the battery, what is the terminal voltage of the battery?

▮ $V_t = \mathscr{E} - Ir = 6 - 200(0.01) = \underline{4\text{ V}}$

27.63 Three resistors of 4 Ω, 6 Ω, and 12 Ω are connected in parallel, and the combination is connected in series with a 1.5-V battery of 1-Ω internal resistance. What is the total battery current?

▮ For the three parallel resistors,

$$\frac{1}{R} = \frac{1}{R_1} + \frac{1}{R_2} + \frac{1}{R_3} = \frac{1}{4} + \frac{1}{6} + \frac{1}{12} = \frac{3}{12} + \frac{2}{12} + \frac{1}{12} = \frac{6}{12} \quad \text{and}$$

$$R = 2\ \Omega \quad \text{for parallel resistors}$$

Ohm's law becomes

$$\mathscr{E} = I(R + r) \quad \text{or} \quad 1.5 = I(2 + 1) \quad \text{and} \quad I = \underline{0.5\text{ A}}$$

27.64 Define the following: *anode, cathode, electrolysis, electrolyte, valence, chemical equivalent weight, faraday, electrochemical equivalent, polarization* in a cell, *fuel cell,* and *thermal emf*.

▮ The *anode* is the positive electrode in a cell.
The *cathode* is the negative electrode in a cell.
Electrolysis is the production of a chemical reaction by the flow of electric current through a solution.
An *electrolyte* is a substance that breaks up into positive and negative ions when dissolved in water.
Valence is the net charge in electronic units carried by an ion. The ion may be a charged atom or a charged group of atoms.
The *chemical equivalent weight* of an element is its atomic weight divided by its valence.
A *faraday* ($\mathscr{F}$) is the charge required to deposit one chemical equivalent weight of any element; thus, $1\ \mathscr{F} = N_A e = (6.03 \times 10^{26}\text{ kmol}^{-1})(1.6 \times 10^{-19}\text{ C/electron}) = 9.65 \times 10^7\text{ C/kmol}$ electrons.
The *electrochemical equivalent* of an element is its chemical equivalent weight divided by the faraday.
Polarization is the formation of gas bubbles on the electrodes of a cell, causing the emf to drop and the internal resistance to increase.
A *fuel cell* is one in which active materials are supplied and reaction products removed continuously. It can operate as long as these processes continue.
A *thermal electromotive force* is an emf resulting from the conversion of heat energy into electric energy when two dissimilar metals are connected in series and maintained at different temperatures.

27.65 Describe the process of electrolysis

▮ When an electrolyte is dissolved in water, it breaks up into positively and negatively charged ions, making the solution an electrical conductor. When a potential difference is applied to two electrodes in a solution, positive ions are collected at the cathode and negative ions at the anode, thus effecting a chemical transformation.

27.66 Write an equation embodying *Faraday's laws of electrolysis*.

▮ *Faraday's laws of electrolysis* state that (a) the mass of any substance liberated by a given charge is proportional to the charge passing through the electrolysis cell or to the product of the current and the time ($Q = It$); and that (b) the mass of the different elements liberated by a given charge is proportional to atomic weight divided by valence. These two laws may be combined to give a single equation: $m = QA/\mathscr{F}v$, where m = mass of element deposited, kg
$\qquad Q$ = charge transferred, C
$\qquad A$ = atomic weight of element, kg/kmol
$\qquad v$ = valence (a pure number)

27.67 What is a *battery*?

▮ In some electrolytic cells chemical energy can be usefully converted to electric energy. A battery consists of one or more such cells. If the reaction that converts chemical energy to electric energy is irreversible, the cell is said to be a *primary cell*; it cannot be recharged. On the other hand, if the chemical reaction is reversible, it is said to be a *secondary cell*; the cell can be discharged and charged many times.

27.68 What are common examples of a primary and a secondary cell?

▮ An example of a primary cell is the common dry cell (also known as a Leclanché dry cell) in which the positive terminal is carbon and the cathode is zinc. An example of a secondary cell is the widely used lead–acid storage cell, which can be recharged many times, since its chemical reaction is reversible. It delivers 2.2 V when fully charged:

$$Pb + PbO_2 + 2H_2SO_4 \rightleftarrows 2PbSO_4 + 2H_2O$$

27.69 Describe a *thermocouple*.

▮ A *thermocouple* is a simple series circuit in which there are two junctions of a pair of dissimilar materials. When one of the junctions is maintained at a temperature different from that at the other junction, an electromotive force results from the conversion of thermal energy to electric energy. This potential difference, called the *Seebeck emf*, amounts to a few millivolts. The Seebeck emf is itself the result of two other effects, the Thomson emf and the Peltier emf, of which the Peltier is much greater. Thermocouples are useful for measuring high temperatures (up to 1700 °C for platinum and platinum–10% rhodium thermocouples) at which ordinary thermometers would melt, via measurement of the associated emf.

27.70 The charge on a silver ion is e, and the molecular weight of each ion is 108 kg/kmol. When a current passes from one electrode to another through a silver nitrate solution, silver ions plate out as atoms on the negative electrode. If the current is 0.20 A, how much silver will plate out in 10 min? Assume that the current is carried through the solution by the silver ions.

▮ The total charge passing in 10 min is $Q = It = 0.20(600) = 120$ C; number of ions $= Q/e = 7.5 \times 10^{20}$. The mass per ion $= 108/N_A = 1.8 \times 10^{-25}$ kg, so mass $=$ (mass per ion)(number) $= (1.8 \times 10^{-25}$ kg$)(7.5 \times 10^{20}) = 1.35 \times 10^{-4}$ kg $= \underline{135 \text{ mg}}$.

27.71 A charge of 2×10^{-4} $\mathscr{F}$ is passed through an electrolytic cell containing ferric iron (Fe^{+++}). Assuming that the only cathode reaction is $Fe^{+++} + 3e^- \rightarrow Fe$, what mass of iron will be deposited? (Atomic weight of iron is 55.85.)

▮ By Prob. 27.66,

$$m = \frac{Q}{\mathscr{F}} \frac{A}{v} = (2 \times 10^{-4}) \frac{55.85}{3} = 3.72 \times 10^{-3} \text{ kg} = \underline{3.72 \text{ g}}$$

27.72 A current flowing through a divalent copper-plating bath and a silver-plating bath in series deposits 0.400 g of copper in 10 min. How much silver is deposited? Atomic weights are 108 and 63.54 for silver and copper, respectively.)

▮ Let m_1 be the mass of silver deposited; m_2, the mass of copper deposited; A_1, the gram atomic weight of silver; A_2, the gram atomic weight of copper; v_1, the valence of silver; and v_2, the valence of copper. From Prob. 27.66 we have

$$m_1 = \frac{Q}{9.65 \times 10^7} \frac{A_1}{v_1} \qquad m_2 = \frac{Q}{9.65 \times 10^7} \frac{A_2}{v_2}$$

Divide m_1 by m_2:

$$\frac{m_1}{m_2} = \frac{A_1 v_2}{A_2 v_1} \qquad \frac{m_1}{0.400} = \frac{108(2)}{63.54(1)} \qquad m_1 = \underline{1.36 \text{ g of silver deposited}}$$

27.73 An ammeter reads 0.90 A when connected in series with a silver-plating bath that deposits 2.60 g of silver in 39 min. By what percentage is the ammeter reading incorrect? (Atomic weight of silver = 108.)

▮ Using the results of Prob. 27.66 we have

$$m = \frac{Q}{9.65 \times 10^7} \frac{A}{v} \qquad Q = It \qquad m = \frac{It}{9.65 \times 10^7} \frac{A}{v}$$

$$2.60 \times 10^{-3} = \frac{I(39 \times 60)}{9.65 \times 10^7} \left(\frac{108}{1}\right) \qquad I = \frac{(2.60)(9.65 \times 10^4)}{39(60)(108)} = 0.993 \text{ A}$$

$$\frac{0.993 - 0.90}{0.993} = \underline{9.3\% \text{ error}}$$

27.74 (a) Compute the electrochemical equivalent of gold. Gold has atomic weight 197.0 and valence 3. (b) What constant current is required to deposit on the cathode 5 g of gold per hour?

▮ (a)   electrochemical equivalent $= \dfrac{\text{equivalent weight}}{\mathscr{F}} = \dfrac{(197.0/3) \text{ kg/kmol}}{9.65 \times 10^7 \text{ C/kmol}} = \underline{0.681 \text{ mg/C}}$

(b)   mass deposited = (electrochemical equivalent)(number of coulombs transferred)

$$5 \text{ g} = (6.81 \times 10^{-4} \text{ g/C})(I \times 3600 \text{ s}) \qquad I = \underline{2.04 \text{ A}}$$

27.75 The electrochemical equivalent Z of silver is 1.118 mg/C. About how much silver is deposited by a current of 10 A in 5 min?

▮ $Z = \dfrac{m}{Q}$ so $1.118 \times 10^{-6} = \dfrac{m}{It} = \dfrac{m}{10(5 \times 60)}$ $m = 3000(1.118 \times 10^{-6}) = 3354 \times 10^{-3} \text{ kg} = \underline{3.35 \text{ g}}$

27.76 A chromel-constantan thermocouple gives about 70 μV for each 1-°C difference in temperature between the two junctions. If 100 such thermocouples are made into a thermopile, what voltage is produced when the hot junctions are at 240 °C and the cold junctions are at 20 °C?

▮ The thermopile is the series connection of the thermocouples. Then, letting $\mathscr{E}_0 = 70 \, \mu$V, we have
$V = 100\mathscr{E}_0(T_2 - T_1) = 100(70 \times 10^{-6})(240 \, °\text{C} - 20 \, °\text{C}) = \underline{1.54 \text{ V}}$.

27.77 Describe the electric potential in living cells and the Nernst potential.

▮ The resting potential of a cell is caused by differences in the concentration of ions inside and outside the cell and by differences in the permeability of the cell wall to different ions. At equilibrium the potential difference V is given in terms of the concentrations c_1 and c_2 of permeable ions in the two compartments (see, e.g., Table 27-2) by the equation

$$V = V_1 - V_2 = \pm 2.3 \frac{kT}{e} \log \frac{c_1}{c_2}$$

This is the *Nernst equilibrium potential*. It is negative when the membrane is permeable to positive ions and positive when the membrane is permeable to negative ions. Here k is the Boltzmann constant and T is the absolute temperature. At body temperature 37 °C the quantity kT/e is

$$\frac{kT}{e} = \frac{(1.38 \times 10^{-23} \text{ J/K})(310 \text{ K})}{1.60 \times 10^{-19} \text{ C}} = 0.0267 \text{ V} = 26.7 \text{ mV}$$

so the Nernst potential is

$$V = V_1 - V_2 = \pm(61.4 \text{ mV}) \log \frac{c_1}{c_2}$$

27.78 If the cell wall were permeable to the negatively charged organic ions, A$^-$, in the cellular fluids (Table 27-2), what would the Nernst potential due to these ions be?

▮ $V_I - V_E = (61.4 \text{ mV}) \log \left[\dfrac{c_I}{c_E}\right] = (61.4 \text{ mV}) \log \dfrac{0.147}{0.044} = (61.4 \text{ mV}) \log 3.34 = \underline{+32.1 \text{ mV}}$

27.79 A 0.5 mol/L NaCl solution is separated from a 1.5 mol/L NaCl solution by a membrane permeable to Na$^+$ ions but impermeable to Cl$^-$ ions. What is the Nernst potential across the membrane when the temperature of

TABLE 27-2

	concentration, mol/L	
	extracellular	intracellular
ion	c_E	c_I
K^+	0.005	0.141
Na^+	0.142	0.010
	0.147	0.151
Cl^-	0.103	0.004
A^- *	0.044	0.147
	0.147	0.151

* Represents all other anions.

the solutions is 10 °C? ($k = 1.38 \times 10^{-23}$ J/K.)

$$\blacksquare \quad V_1 - V_2 = -2.3 \frac{kT}{e} \log \frac{c_1}{c_2} = -(2.3) \frac{(1.38 \times 10^{-23} \text{ J/K})(283 \text{ K})}{1.6 \times 10^{-19} \text{ C}} \log \frac{1.5}{0.5}$$

$$= -(0.056 \text{ V}) \log 3 = -(0.056 \text{ V})(0.477) = \underline{-26.7 \text{ mV}}$$

The minus sign indicates that the compartment with the higher concentration is at the lower potential.

27.4 ELECTRIC MEASUREMENT

27.80 In the circuit shown in Fig. 27-21, the ideal ammeter A registers 2 A. (**a**) Assuming XY to be a resistance, find its value. (**b**) Assuming XY to be a battery (with 2-Ω internal resistance) that is being charged, find its emf. (**c**) Under the conditions of part (**b**), what is the potential change from point Y to point X?

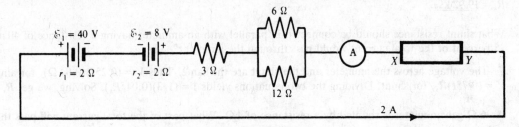

Fig. 27-21

$\blacksquare$ The ideal ammeter has zero resistance. (**a**) $\mathscr{E} = 40 - 8 = 32$ V. $R_T = r_1 + r_2 + 3\,\Omega + R_e + R_{XY}$, where $r_1 = 2\,\Omega$, $r_2 = 2\,\Omega$; $1/R_e = 1/12 + 1/6 = 1/4$, or $R_e = 4\,\Omega$. Thus $R_T = 11\,\Omega + R_{XY}$. From Ohm's law, $\mathscr{E} = IR_T$, or 32 V $= (2A)(11\,\Omega + R_{XY})$, and $R_{XY} = \underline{5\,\Omega}$. (**b**) Now $\mathscr{E} = 32$ V $+ \mathscr{E}_{XY}$ and $R_T = 11\,\Omega + 2\,\Omega = 13\,\Omega$. Then $(32 \text{ V} + \mathscr{E}_{XY}) = (2A)(13\,\Omega) = 26$ V, so $\mathscr{E}_{XY} = \underline{-6 \text{ V}}$. The minus sign indicates that X is the negative terminal, and indeed it is being charged. (**c**) $V_X - V_Y = -6\text{V} - (2\text{A})(2\,\Omega) = \underline{-10 \text{ V}}$

27.81 How can an ammeter and voltmeter be used jointly to measure resistance?

$\blacksquare$ The current is measured by inserting in series a (low-resistance) ammeter into the circuit. The potential difference is measured by connecting the terminals of a (high-resistance) voltmeter across the resistance being measured, i.e., in parallel. The resistance is computed by dividing the voltmeter reading by the ammeter reading, according to Ohm's law, $R = V/I$. (If an exact value of the resistance is required, the resistances of the voltmeter and ammeter must be considered part of the circuit.)

27.82 An ammeter is connected in series with an unknown resistance, and a voltmeter is connected across the terminals of the resistance. If the ammeter reads 1.2 A and the voltmeter reads 18 V, what is the value of the resistance? Assume ideal meters.

$\blacksquare$ The voltmeter reads the voltage across the resistance. The ammeter reads the current through the parallel combination of the voltmeter and the resistance. Since the voltmeter is ideal, its resistance is infinite and all the current goes through the resistor. Then $R = V/I = 18$ V/1.2 A $= \underline{15\,\Omega}$.

27.83 A certain galvanometer has a resistance of 400 Ω and deflects full scale for a current of 0.2 mA through it. How large a shunt resistance is required to change it to a 3 A ammeter?

▮ In Fig. 27-22 we label the galvanometer G and the shunt resistance R_s. At full scale deflection, the currents are as shown.

In the voltage drop from a to b across G is the same as across R_s. Therefore $(2.9998 \text{ A})R_s = (2 \times 10^{-4} \text{ A})(400 \text{ Ω})$, from which $R_s = \underline{0.027 \text{ Ω}}$.

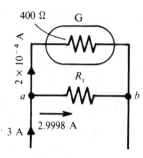

Fig. 27-22

27.84 A voltmeter which deflects full scale for a potential difference of 5.0 V across it is to be made by connecting a resistance R_x in series with a galvanometer. The 80 Ω galvanometer deflects full scale for a potential of 20 mV across it. Find R_x.

▮ When deflecting full scale, the current through the galvanometer is

$$I = \frac{V}{R} = \frac{20 \times 10^{-3} \text{ V}}{80 \text{ Ω}} = 2.5 \times 10^{-4} \text{ A}$$

When R_x is connected in series with the galvanometer, we wish I to be 2.5×10^{-4} A for a potential difference of 5 V across the combination. Hence $V = IR$ becomes $5 \text{ V} = (2.5 \times 10^{-4} \text{ A})(80 \text{ Ω} + R_x)$, from which $R_x = \underline{19.92 \text{ kΩ}}$.

27.85 What shunt resistance should be connected in parallel with an ammeter having a resistance of 40 mΩ so that 25 percent of the total current would pass through the ammeter?

▮ The voltage across the ammeter and the shunt are the same. Thus $V = (0.25I)(0.04 \text{ Ω})$, for ammeter and $V = (0.75I)R_s$, for shunt. Dividing the two equations yields $1 = (1/3)(0.04/R_s)$. Solving, we get $R_s = \underline{13.3 \text{ mΩ}}$.

27.86 A 36-Ω galvanometer is shunted by a resistance of 4 Ω. What part of the total current will pass through the instrument?

▮ Galvanometer and shunt are in parallel. Let I equal combined current. Then $V = aI(36 \text{ Ω})$, for galvanometer, with $0 \le a \le 1$; $V = (1 - a)I(4 \text{ Ω})$, for shunt. Thus $36a = 4(1 - a)$ and $a = 1/10$. 10 percent of the current passes through the galvanometer.

27.87 The circuit of Fig. 27-23(a) is used to measure an unknown resistance by use of a 20-kΩ voltmeter and a 50-mΩ ammeter. The meter readings are 9.00 V and 800 μA. (a) What is the unknown resistance? (b) What would the meters read if they had been connected as in Fig. 27-23(b)? Assume that the voltage across the resistor is still 9 V. (c) Which method is better?

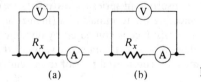

(a)   (b)   **Fig. 27-23**

▮ Here we check how good an estimate of R is given by the ratio of the voltmeter to ammeter readings. (a) The ratio $V/A = 9.00/(800 \times 10^{-6}) = 11.25 \text{ kΩ}$. The actual R_x can be found as follows. The current through the voltmeter $= 9.00/(20 \times 10^3) = 0.45 \text{ mA}$, so the current through R_x is $0.80 - 0.45 = 0.35 \text{ mA}$; thus $R_x = 9.00/(0.35 \text{ mA}) = 25.7 \text{ kΩ}$. Hence V/A is a very poor estimate. (b) The ammeter records only I through R or 0.35 mA, while the voltmeter records $9.00 \text{ V} + (0.35 \text{ mA})(0.050 \text{ Ω}) \approx 9.00 \text{ V}$; so $V/A = \underline{25.7 \text{ kΩ}}$. (c) The arrangement of part b is better.

27.88 A 1400-Ω resistor is connected in series with a 200-Ω resistor and a 12-V battery. (*a*) How much does the current in the circuit change when a 5000-Ω voltmeter is connected across (in parallel with) the 1400-Ω resistor? (*b*) Across the 200-Ω resistor? (*c*) What does the voltmeter read in *a*? (*d*) In *b*?

▌ Original $I = 12/(1400 + 200) = 7.50$ mA. (*a*) Placing 5000 Ω across the 1400 Ω changes the circuit resistance to 1093 + 200 Ω; $I = 9.28$ mA and $\Delta I = \underline{+1.78\text{ mA}}$. (*b*) Across the 200-Ω resistor the circuit resistance = 192.3 + 1400, giving 7.536 mA, so the current change is $\underline{+36\,\mu\text{A}}$. (*c*) The drop across 1400 Ω is $12 - 200(9.28 \times 10^{-3}) = \underline{10.15\text{ V}}$ (without voltmeter, potential drop = 10.5 V). (*d*) Voltmeter reads $12 - (1400)(7.536 \times 10^{-3}) = \underline{1.45\text{ V}}$ instead of 1.5 V without the voltmeter.

27.89 Describe a *Wheatstone bridge* and show how it can be used to find an unknown resistance *X*. What are the attractive features of measuring resistance this way?

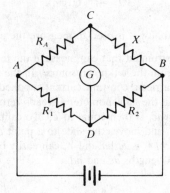

Fig. 27-24

▌ A *Wheatstone bridge* is the circuit of Fig. 27-24. The terms R_A, R_1, and R_2 represent accurately known variable resistors whose values are set so that the galvanometer *G* reads zero. Then, points *C* and *D* are at the same potential, and the current through the upper branch is I_C and through the lower branch, I_D. Using these facts we have $I_C R_A = I_D R_1$, and $I_C X = I_D R_2$. To find *X*, we divide the second equation by the first:

$$\frac{X}{R_A} = \frac{R_2}{R_1} \qquad X = \frac{R_2}{R_1} R_A$$

Note that since the galvanometer is being used as a "null" instrument, we have no errors associated with the resistance of the instrument. Similarly, the actual terminal voltage of the battery need not be known, so no errors from that measurement enter.

27.90 The Wheatstone bridge shown in Fig. 27-25 is being used to measure resistance *X*. At balance, the current through the galvanometer G is zero and resistances *L*, *M*, and *N* are 3, 2, and 10 Ω, respectively. Find the value of *X*.

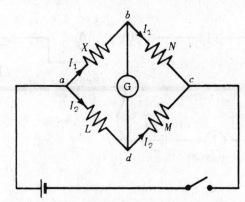

Fig. 27-25

▌ From Prob. 27.89, $X/10 = 3/2$, or $X = \underline{15\,\Omega}$.

27.91 The slidewire Wheatstone bridge shown in Fig. 27-26 is balanced when the uniform slide wire *AB* is divided as shown. Find the value of the resistance *X*.

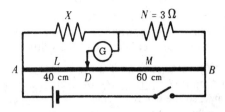

Fig. 27-26

▮ It is easy to see that this bridge is identical in form to that of Fig. 27-25, with the same symbols used for the four resistances. Then:

$$\frac{X}{3\Omega} = \frac{L}{M} = \frac{40\,\text{cm}}{60\,\text{cm}} \quad \text{or} \quad X = \underline{2\,\Omega}$$

27.92 Describe the *potentiometer* and how it is used to measure the emf of an unknown source.

▮ The potentiometer is an instrument that can measure the terminal potential difference with high accuracy without drawing any current from the unknown source. In the slide-wire version of Fig. 27-27, the wire ab is of uniform cross section and carries a constant current supplied by battery B. To measure the unknown emf $\mathcal{E}_x$, the contact c is moved until the galvanometer G reads zero. The potential difference between a and c, which is proportional to length $\overline{ac}$, is then equal to $\mathcal{E}_x$. To calibrate a potentiometer, one switches the contact from $\mathcal{E}_x$ to the standard cell $\mathcal{E}_s$ and moves the slide to a point d on wire ab to obtain another zero galvanometer reading. Then, $\mathcal{E}_x/\mathcal{E}_s = \overline{ac}/\overline{ad}$ and $\mathcal{E}_x$ can easily be computed from the known emf $\mathcal{E}_s$ of the standard cell and the measured lengths ac and ad.

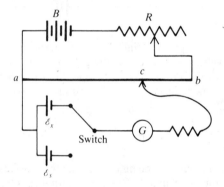

Fig. 27-27

27.93 The slide-wire potentiometer shown in Fig. 27-28 has a steady current flowing through the uniform-resistance wire. A balance is found at a slider position of 45 cm, as shown, when an unknown emf is measured. When a standard cell with emf 1.018 V is used, the slider position is 30 cm at balance. What is the unknown emf?

▮ Using the notation of Prob. 27.92, $\mathcal{E}_x/\mathcal{E}_s = \overline{ac}/\overline{ad}$, or $\mathcal{E}_x/1.018\,\text{V} = 45/30$, and $\mathcal{E}_x = \underline{1.527\,\text{V}}$.

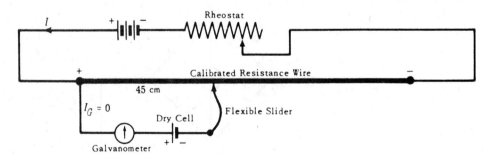

Fig. 27-28

27.5 ELECTRIC POWER

27.94 Find the power dissipated in a resistance R when a current I is flowing through it.

▮ In a resistor with current I passing through, the potential drop from the point of current entry to the point of exit is $V = IR$. For each charge q passing through the resistor, the electrostatic forces do work: Vq, and that much electrostatic energy is lost. Since for constant I the drift velocity of charges is constant, this energy appears only in the form of increased thermal energy in the resistor. The power dissipated is thus $P = Vq/t = VI$. Using Ohm's law we also have $P = I^2R = V^2/R$.

27.95 Find the power supplied to an electric system by nonelectric sources.

❚ For an ideal source the emf $\mathscr{E}$ equals the terminal voltage V, even when current is flowing. For example, in a battery, for each charge q that passes through the battery from negative to positive terminal, the chemical forces do work: $\mathscr{E}q$. This work appears in the form of the gained electrostatic energy Vq. If q passes in a time t while a steady current I is flowing, then $I = q/t$. Thus the power supplied by the chemical forces is $P = \mathscr{E}q/t = \mathscr{E}I$.

For a nonideal battery with interval resistance r and terminal voltage V, we have power supplied = electrostatic power gained plus power dissipated, or $\mathscr{E}I = VI + I^2r \Rightarrow V = \mathscr{E} - Ir$ as before.

27.96 A current of 0.50 A flows through a 200-Ω resistor. How much power is lost in the resistor?

❚ Power $= VI = I^2R = (0.50)^2(200) = 50$ W, and this lost power will appear as heat in the resistor. In this case 50 J of energy is lost each second, and so 50/4.184, or about 12 cal of heat, is generated each second.

27.97 A bulb rated 120 V/90 W is operated from a 120-V power source. Find the current flowing through it and its resistance.

❚ Since power $= VI$, we can find I as follows:

$$I = \frac{\text{power}}{V} = \frac{90\text{ W}}{120\text{ V}} = \frac{3}{4}\text{ A}$$

Since the potential drop across the bulb is 120 V and since the current through it is $\frac{3}{4}$ A, Ohm's law tells us that

$$R = \frac{V}{I} = \frac{120\text{ V}}{0.75\text{ A}} = \underline{160\ \Omega}$$

27.98 What is the power output of a 1.5-V ideal battery which is delivering a current of 0.2 A?

❚ The potential difference V is simply the emf of the battery, and so $P = I\mathscr{E} = (0.2\text{ A})(1.5\text{ V}) = \underline{0.3\text{ W}}$.

27.99 What is the resistance of a 1000-W 120-V toaster?

❚ Power expended in the resistance is $P = VI$, or $1000 = 120I$, and $I = 8.333$ A. From Ohm's law, $V = IR$, or $120 = 8.333R$, and $R = \underline{14.4\ \Omega}$. (Directly, $P = V^2/R$, or $1000 = 120^2/R$ and $R = \underline{14.4\ \Omega}$.)

27.100 Show that if two resistors are connected in parallel, the rates of heat production in each vary inversely as their resistances.

❚ Since V across each resistor is the same, $P_1 = V^2/R_1$; $P_2 = V^2/R_2$, demonstrating that $P \propto 1/R$.

27.101 Suppose that the emf of the ideal battery in Fig. 27-29 is 4.5 V and the resistors are $R_1 = 3\ \Omega$ and $R_2 = 6\ \Omega$. What are (a) the total resistance of the circuit, (b) the current in the circuit, and (c) the power dissipated in each resistor?

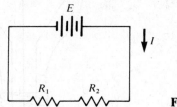

Fig. 27-29

❚ (a) $R = R_1 + R_2 = 3\ \Omega + 6\ \Omega = \underline{9\ \Omega}$. (b) $I = \dfrac{E}{R} = \dfrac{4.5\text{ V}}{9.0\ \Omega} = \underline{0.5\text{ A}}$. (c) $P_1 = R_1I^2 = (3\ \Omega)(0.5\text{ A})^2 = \underline{0.75\text{ W}}$.

$P_2 = R_2I^2 = (6\ \Omega)(0.5\text{ A})^2 = \underline{1.50\text{ W}}$. As a check $P = EI = (4.5\text{ V})(0.5\text{ A}) = 2.25\text{ W} = P_1 + P_2$.

27.102 A 120-V house circuit has the following light bulbs turned on: 40, 60, and 75 W. Find the equivalent resistance of these lights.

▮ House circuits are so constructed that each device is connected in parallel with the others. From $P = VI = V^2/R$, we have for the first bulb

$$R_1 = \frac{V^2}{P_1} = \frac{(120)^2}{40} = \underline{360\ \Omega}$$

Similarly, $R_2 = \underline{240\ \Omega}$ and $R_3 = \underline{192\ \Omega}$. Because they are in parallel,

$$\frac{1}{R_{eq}} = \frac{1}{360\ \Omega} + \frac{1}{240\ \Omega} + \frac{1}{192\ \Omega} \qquad \text{or} \qquad R_{eq} = \underline{82\ \Omega}$$

As a check, we note that the total power drawn from the line is $40 + 60 + 75 = 175$ W. Then, using $P = V^2/R$,

$$R_{eq} = \frac{V^2}{\text{total power}} = \frac{120^2}{175} = \underline{82\ \Omega}$$

27.103 What is the power dissipated in a 4-Ω light bulb connected to a 12-V battery? What is the power dissipated in a 2-Ω light bulb connected to the same battery? Which bulb is brighter?

▮ $$P_{4\Omega} = \frac{V^2}{R} = \frac{(12\ \text{V})^2}{4\ \Omega} = \underline{36\ \text{W}} \qquad P_{2\Omega} = \frac{(12\ \text{V})^2}{2\ \Omega} = \underline{72\ \text{W}}$$

Since the brightness of a bulb increases with the power, the 2-Ω bulb is brighter than the 4-Ω bulb.

27.104 An ideal battery with $\mathscr{E} = 12$ V is connected in series to two bulbs with resistances $R_1 = 2\ \Omega$ and $R_2 = 4\ \Omega$. What is the current in the circuit and the power dissipation in each bulb?

▮ The total resistance of the circuit is $R = R_1 + R_2 = 2\ \Omega + 4\ \Omega = 6\ \Omega$, and the current is

$$I = \frac{\mathscr{E}}{R} = \frac{12\ \text{V}}{6\ \Omega} = \underline{2\ \text{A}}$$

The potential across R_1 is $V_1 = R_1 I = (2\ \Omega)(2\ \text{A}) = 4$ V and the potential across R_2 is $V_2 = R_2 I = (4\ \Omega)(2\ \text{A}) = 8$ V. The sum of the potentials across the two resistors is equal to the emf of the battery. The power dissipation in the resistors is

$$P_1 = R_1 I^2 = (2\ \Omega)(2\ \text{A})^2 = \underline{8\ \text{W}} \qquad P_2 = R_2 I^2 = (4\ \Omega)(2\ \text{A})^2 = \underline{16\ \text{W}}$$

27.105 A three-way light bulb has two filaments which are connected to three wires, as shown in Fig. 27-30. By turning the socket switch, 120 V is put across either ab, bc, or ac. (a) If $R_1 = 144\ \Omega$ and $R_2 = 216\ \Omega$, what are the three possible power dissipations of the light bulb? (b) A different three-way light bulb can operate at 300, 100, and 75 W. What are the resistances of its two filaments?

$a\ b\ c$ **Fig. 27-30**

▌ **(a)** Case ab:

$$P_1 = \frac{V^2}{R_1} = \frac{(120 \text{ V})^2}{144 \text{ }\Omega} = \underline{100 \text{ W}}$$

Case bc:

$$P_2 = \frac{V^2}{R_2} = \frac{(120 \text{ V})^2}{216 \text{ }\Omega} = \underline{67 \text{ W}}$$

Case ac:

$$P_3 = \frac{V^2}{R_1 + R_2} = \frac{(120 \text{ v})^2}{360 \text{ }\Omega} = \underline{40 \text{ W}}$$

(b) The highest wattage occurs for the smallest resistance, so

Case ab:

$$R_1 = \frac{V^2}{P_1} = \frac{(120 \text{ V})^2}{300 \text{ W}} = \underline{48 \text{ }\Omega}$$

Case bc:

$$R_2 = \frac{V^2}{P_2} = \frac{(120 \text{ V})^2}{100 \text{ W}} = \underline{144 \text{ }\Omega}$$

Check:

Case ac:

$$R_3 = \frac{V^2}{P_3} = \frac{(120 \text{ V})^2}{75 \text{ W}} = 192 \text{ }\Omega = R_1 + R_2$$

27.106 Six Christmas tree lights are arranged in a parallel circuit, as shown in Fig. 27-31. Each bulb dissipates 10 W when operated at 120 V. **(a)** What is the resistance R of each bulb? **(b)** What is the resistance of the entire array of bulbs? **(c)** What is the total power consumption of the array? **(d)** What are the currents at points a, b, c, and d?

▌ **(a)** For each bulb we have

$$R = \frac{V^2}{P} = \frac{(120 \text{ V})^2}{10 \text{ W}} = \underline{1440 \text{ }\Omega}$$

(b)

$$\frac{1}{R_{\text{total}}} = \frac{1}{R} + \frac{1}{R} + \frac{1}{R} + \frac{1}{R} + \frac{1}{R} + \frac{1}{R} = \frac{6}{R} \qquad R_{\text{total}} = \frac{R}{6} = \underline{240 \text{ }\Omega}$$

(c)

$$P = \frac{V^2}{R_{\text{total}}} = \frac{(120 \text{ V})^2}{240 \text{ }\Omega} = \underline{60 \text{ W}}$$

(d) $\quad I_a = \dfrac{V}{R_{\text{total}}} = \dfrac{120 \text{ V}}{240 \text{ }\Omega} = \underline{0.5 \text{ A}} \qquad I_b = \dfrac{V}{R} = \dfrac{120 \text{ V}}{1440 \text{ }\Omega} = \underline{0.083 \text{ A}} \qquad I_c = I_a - 4I_b = \underline{0.167 \text{ A}} \qquad I_d = I_b = \underline{0.083 \text{ A}}$

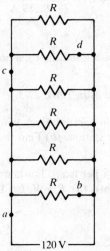

Fig. 27-31

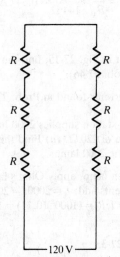

Fig. 27-32

27.107 Six bulbs are arranged in a series circuit, as shown in Fig. 27-32. (Old-fashioned Christmas tree lights were connected this way.) Each bulb has the same resistance R, and the entire array is designed to dissipate 60 W when operated at 120 V. **(a)** What is the resistance of each bulb? **(b)** What is the current in the circuit when it is operated at 120 V? **(c)** What would be the power dissipated in a single bulb operated at 120 V? **(d)** Compare this circuit with the circuit in Prob. 27.106 and discuss the disadvantages of arranging Christmas tree lights in a series.

❚ (a)
$$R_{total} = 6R = \frac{V^2}{P} = \frac{(120\ V)^2}{60\ W} = 240\ \Omega \qquad R = \underline{40\ \Omega}$$

(b)
$$I = \frac{V}{R_{total}} = \frac{120\ V}{240\ \Omega} = \underline{0.5\ A}$$

(c)
$$P = \frac{V^2}{R} = \frac{(120\ V)^2}{40\ \Omega} = \underline{360\ W}$$

(d) All the lights go out when any individual light in the series arrangement burns out.

27.108 Figure 27-33 shows a potential difference of 120 V placed across a circuit that has a lamp with resistance $R_1 = 144\ \Omega$ connected in series to a variable resistor R_2. The brightness of the lamp is controlled by changing the magnitude of R_2. Find the power dissipations in the lamp (a) when R_2 is zero and (b) when $R_2 = 144\ \Omega$. (c) What must R_2 be for the power dissipation in the lamp to be 50 W?

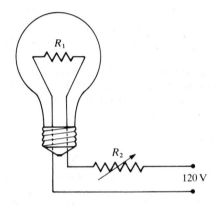

Fig. 27-33

❚ (a)
$$I = \frac{V}{R_1} = \frac{120\ V}{144\ \Omega} = 0.833\ A \qquad P_1 = R_1 I^2 = (144\ \Omega)(0.833\ A)^2 = \underline{100\ W}$$

(b)
$$I = \frac{V}{R_1 + R_2} = \frac{120\ V}{288\ \Omega} = 0.4165\ A \qquad P_1 = R_1 I^2 = (144\ \Omega)(0.4165\ A)^2 = \underline{25\ W}$$

(c)
$$I^2 = \frac{P_1}{R_1} = \frac{50\ W}{144\ \Omega} = 0.347\ A^2; \qquad I = 0.589\ A \qquad R = R_1 + R_2 = \frac{V}{I} = \frac{120\ V}{0.589\ A} = 203.7\ \Omega$$

$$R_2 = R - R_1 = \underline{59.7\ \Omega}$$

27.109 For each circuit in Fig. 27-15, find the power dissipation in each resistor and the total power output of the battery (see Prob. 27.46).

❚ With the currents found in Prob. 27.46 the powers are calculated as in Table 27-3.

27.110 A small power station supplies 2000 incandescent lamps connected in parallel. Each lamp requires 110 V and has a resistance of 220 Ω. (a) Find the total current supplied by the station. (b) Find the total power dissipated by the 2000 lamps.

❚ (a) For each lamp, apply Ohm's law. $I = V/R = 110/220 = 0.5\ A$ per lamp. For lamps in parallel, the individual currents add: $I_t = 2000I = 2000(0.5) = \underline{1000\ A}$ (total current). (b) $R_t = V/I_t = 110/1000 = 0.11\ \Omega$ for the lamps. $P_t = I_t^2 R_t = (1000^2)(0.11) = 1.1 \times 10^5\ W = \underline{110\ kW}$.

TABLE 27-3

	circuit a, W	circuit b, W	circuit c, W	circuit d, W
$P_1 = I_1 R_1^2$	0.022	1.80	0.112	0.218
$P_2 = I_2 R_2^2$	0.067	0.60	0.337	0.258
$P_3 = I_3 R_3^2$	0.112	0.36	0.360	0.152
$P_b = I\mathscr{E} = P_1 + P_2 + P_3$	0.201	2.76	0.809	0.628

27.111 A battery with an emf of 9.0 V maintains a current of 0.75 A in an external circuit for 30 min. What are **(a)** the power generated by the battery and **(b)** the total energy expended by the battery?

> **(a)** $P = \mathscr{E}I = (9.0 \text{ V})(0.75 \text{ A}) = \underline{6.75 \text{ W}}$ **(b)** $t = (30 \text{ min}) (60 \text{ s/min}) = 1800 \text{ s}$ Energy $E = Pt =$ $(6.75 \text{ W})(1800 \text{ s}) = \underline{12.1 \text{ kJ}}$

These results are true even if the battery has internal resistance, since we used $\mathscr{E}$ and not terminal voltage in calculating P. Of course, part of this energy is expended in the internal resistance.

27-112 An ideal battery dissipates 2.6 W when connected to a 125-Ω resistor. What are **(a)** the emf of the battery and **(b)** the current in the resistor?

> **(a)** From the relation $P = V^2/R$ we get $V^2 = PR = (2.6 \text{ W})(125 \text{ }\Omega) = 325 \text{ V}^2$. So, since emf $\mathscr{E} = V$, $\mathscr{E} = \sqrt{V^2} = \underline{18.0 \text{ V}}$.
>
> **(b)** $$I = \frac{\mathscr{E}}{R} = \frac{18.0 \text{ V}}{125 \text{ }\Omega} = \underline{0.144 \text{ A}}$$

27.113 A very old dry cell may have a rather large internal resistance. Suppose that a voltmeter reads the voltage of such a cell to be 1.40 V, while a potentiometer reads its voltage to be 1.55 V. What are **(a)** the emf of the cell, **(b)** its internal resistance, and **(c)** the maximum power a resistor can draw from the battery? Assume the voltmeter's resistance to be 280 Ω.

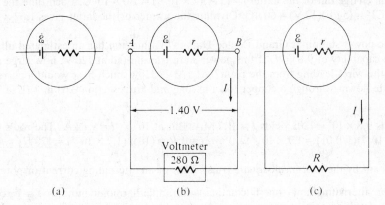

(a)   (b)   (c)   **Fig. 27-34**

> **(a)** The dry cell may be thought of as a pure voltage source $\mathscr{E}$ and a resistor r in series [Fig. 27-34(a)]. When no current is being drawn from it, the potential difference between its terminals is only that of the pure voltage source, or the emf of the battery. It is this quantity which the potentiometer measures, since this instrument draws no current from the cell being measured. Therefore the emf of the battery is $\underline{1.55 \text{ V}}$.
> **(b)** When the 280-Ω voltmeter is connected across the battery as shown in Fig. 27-34(b), current flows from the battery. Assuming the voltmeter to read correctly, we know that the voltage difference between the battery terminals, $V_B - V_A$, is just the voltmeter reading, 1.40 V. It will be equal to the IR drop through the voltmeter. Therefore $I(280) = 1.40$, from which $I = 0.0050$ A. If we write the series equation for Fig. 27-34(b), we find (using 1.55 for $\mathscr{E}$): $1.55 - Ir - I(280) = 0$. Knowing I to be 0.0050 A, we find the internal resistance r of the battery to be $\underline{30 \text{ }\Omega}$. **(c)** In the circuit of Fig. 27-34(c), the power drawn by (dissipated in) resistor R is
>
> $$P = I^2 R = \left(\frac{\mathscr{E}}{r + R}\right)^2 R = \frac{\mathscr{E}^2}{4r}\left[1 - \left(\frac{r - R}{r + R}\right)^2\right]$$
>
> It is seen that P attains its maximum value, $\mathscr{E}^2/4r$, for $R = r$. [Calculus, although unnecessary here, could also be used to maximize P.]

27.114 A 500-W heater is used to heat 250 mL of water from 20 to 100 °C. What is the minimum time in which this can be done?

> The electric energy generated is $Pt = 500t$ J, where t is time in seconds. To heat the water requires thermal energy: $mc \, \Delta T = (0.250 \text{ kg})(4.184 \text{ kJ/kg} \cdot \text{K})(80 \text{ K}) = 83.7 \text{ kJ}$. Thus $0.500t_{\min} = 83.7$, or $t_{\min} = \underline{167 \text{ s}}$.

27.115 It is desired to heat a cup of coffee (200 mL) by use of an immersion heater from 20 to 90 °C in 0.5 min. How much current would the heater draw from 120 V?

❚ The specific heat and mass of the coffee can be taken as that of water. The heat energy needed is $Q = cm \, \Delta T = (4.184)(0.200)(70) = 58.6 \text{ kJ}$; $P = \text{energy}/t = 58.6/30 = 1.95 \text{ kW}$. The electric power $P = VI$, so $I = 1950/120 = \underline{16.3 \text{ A}}$.

27.116 The resistance of a 1-m length of 14-gauge copper household wire is 8.1 mΩ. (*a*) How much energy loss occurs per second in 1 m of it when the current is 20 A? (*b*) The wire's cross-sectional area is 2.1 mm². What is the shortest time in which the current could heat the wire by 5 °C? (Specific gravity of copper = 8.92; specific heat = 0.093 kcal/kg · °C).

❚ (*a*) Use $P = I^2R = 20^2(8.1 \times 10^{-3}) = \underline{3.24 \text{ W}}$; (*b*) the mass of copper is $\rho V = (8.92 \times 10^3 \text{ kg/m}^3)[(1 \text{ m})(2.1 \times 10^{-6} \text{ m}^2)] = 0.0187 \text{ kg}$. Then set $Pt = Q = (0.093)(0.0187)(5.0)(4184 \text{ J/kcal})$ to find $t = \underline{11.2 \text{ s}}$.

27.117 Lead storage batteries have an *energy density* of about 80 kJ/kg; that is, 1 kg of such batteries can furnish about 80 kJ of energy. If a 12-V car battery has a mass of 15 kg and maintains a constant voltage of 12 V, for how many hours can the fully charged battery furnish a 5-A current before becoming discharged?

❚ The time t is found from $(12 \text{ V})(5 \text{ A})(t) = (8 \times 10^4 \text{ J/kg})(15 \text{ kg})$, so $t = (2 \times 10^4 \text{ s})/(3600 \text{ s/h}) = \underline{5.56 \text{ h}}$.

27.118 How many times could the fully charged 12-V battery mentioned in the previous problem recharge a 5-μF capacitor before becoming discharged?

❚ The total charge out of the battery $= (5 \text{ A})(2 \times 10^4 \text{ s}) = 1.0 \times 10^5 \text{ C}$. Each time the capacitor is charged it takes $Q = CV = (5 \text{ μF})(12 \text{ V}) = 60 \text{ μC}$. Dividing the former by the latter gives $\underline{1.67 \times 10^9}$.

27.119 The electric power station at Grand Coulee Dam in Washington has an estimated ultimate electric power production capability of 9.8 GW. If this power were transmitted at 120 V, how large a current would be carried by the wires leading from the plant? At 1 MV? How much heat would a current of 10 kA release each second while flowing through a copper rod 1 m long and cross section 0.01 m²? (Use Table 27-1 for needed data.)

❚ $P = VI$ is $9.8 \times 10^9 = 120I$ yields $I = \underline{81.7 \text{ MA}}$. But at 10^6 V, $I = \underline{9.8 \text{ kA}}$. The rod's resistance would be $R = (1.7 \times 10^{-8})(1/0.01) = 1.7 \times 10^{-6} \text{ Ω}$. Power $= I^2R = (10^4)^2(1.7 \times 10^{-6}) = \underline{170 \text{ J/s}} = \underline{41 \text{ cal/s}}$.

27.120 What is meant by the *root-mean-square* (rms) value of an alternating current or potential?

❚ When an alternating emf—one fluctuating with simple harmonic motion: $\mathscr{E} = \mathscr{E}_p \cos(\omega t + \delta)$—is the energy source in a circuit containing only resistors, the current fluctuates in the same way: $I = I_p \cos(\omega t + \delta)$. Here $\mathscr{E}_p$ is the peak emf, $\mathscr{E}$ is the instantaneous emf, ω is the angular frequency, and δ is the constant phase angle. I_p and I are peak and instantaneous current. $\mathscr{E}_{\text{rms}}$ is the square root of the time average of $\mathscr{E}^2$ over a complete cycle. It can be shown that the time average of $\cos^2(\omega t + \delta)$ over a complete period is equal to $\frac{1}{2}$. Then $(\mathscr{E}^2)_{\text{avg}} = \frac{1}{2}\mathscr{E}_p^2$ and $\mathscr{E}_{\text{rms}} = \mathscr{E}_p/\sqrt{2}$. Similarly $I_{\text{rms}} = I_p/\sqrt{2}$. For a circuit with total equivalent resistance R, we have $\mathscr{E} = IR$ or, canceling the cosine term on both sides, $\mathscr{E}_p = I_pR$. Then $\mathscr{E}_{\text{rms}} = I_{\text{rms}}R$. The average power dissipated is $(I^2R)_{\text{avg}} = (I^2)_{\text{avg}}R = \frac{1}{2}I_p^2R = I_{\text{rms}}^2R$. The average power supplied by the emf is $\bar{P} \equiv P_{\text{avg}} = (\mathscr{E}I)_{\text{avg}} = \mathscr{E}_pI_p(\cos^2(\omega t + \delta))_{\text{avg}} = \frac{1}{2}\mathscr{E}_pI_p = \mathscr{E}_{\text{rms}}I_{\text{rms}}$. Similar reasoning shows that for any individual resistance with alternating current I, $V = IR \Rightarrow V_{\text{rms}} = I_{\text{rms}}R$, where V is the alternating voltage across the resistance.

27.121 An alternating potential with a peak potential of 75 V is placed across a 15-Ω resistor. Find (a) the rms current and (b) the average power dissipation in the resistor.

❚ (*a*)
$$I_p = \frac{V_p}{R} = \frac{75 \text{ V}}{15 \text{ Ω}} = 5 \text{ A} \qquad I_{\text{rms}} = \frac{I_p}{\sqrt{2}} = \frac{5 \text{ A}}{1.41} = \underline{3.53 \text{ A}}$$

(*b*)
$$\bar{P} = RI_{\text{rms}}^2 = (15 \text{ Ω})(3.53 \text{ A})^2 = \underline{187.5 \text{ W}}$$

27.122 An alternating current with an rms current of 2.4 A passes through a 25-Ω resistor. Find (*a*) the peak potential across the resistor and (*b*) the average power dissipation in it.

❚ (*a*) $V_{\text{rms}} = RI_{\text{rms}} = (25 \text{ Ω})(2.4 \text{ A}) = \underline{60 \text{ V}}$ $\quad V_p = \sqrt{2} \, V_{\text{rms}} = (1.41)(60 \text{ V}) = \underline{84.6 \text{ V}}$ (*b*) $\bar{P} = RI_{\text{rms}}^2 = (25 \text{ Ω})(2.4 \text{ A})^2 = \underline{144 \text{ W}}$

27.123 During a power cutback, the rms potential drops to 108 V. (*a*) What is the peak voltage during the cutback? (*b*) If a toaster dissipates 850 W when operated at an rms voltage of 120 V, what does it dissipate during the cutback?

(a) $V_P = \sqrt{2}\, V_{\text{rms}} = (1.41)(108\text{ V}) = \underline{152\text{ V}}$

(b) $\bar{P} = \dfrac{V_{\text{rms}}^2}{R}$ so $R = \dfrac{V_{\text{rms}}^2}{\bar{P}} = \dfrac{(120\text{ V})^2}{850\text{ W}} = 16.9\ \Omega$

At 108 V the power is

$$\bar{P} = \frac{V_{\text{rms}}^2}{R} = \frac{(108\text{ V})^2}{16.9\ \Omega} = \underline{690\text{ W}}$$

27.6 MORE COMPLEX CIRCUITS, KIRCHHOFF'S CIRCUIT RULES, CIRCUITS WITH CAPACITANCE

27.124 State *Kirchhoff's laws*.

▌ *Kirchhoff's Point (or Junction) Rule*: The sum of all the currents coming into a point must equal the sum of all the currents leaving the point.

Kirchhoff's Loop (or Circuit) Rule: As one traces out a closed circuit, the algebraic sum of the potential changes encountered is zero. In this sum, a potential rise is positive and a potential drop is negative.

27.125 In Fig. 27-35, find I_1, I_2, and I_3 if switch k is open.

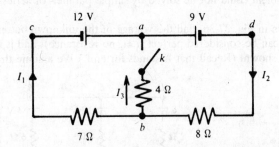

Fig. 27-35

▌ **(a)** When k is open, $I_3 = \underline{0}$, because no current can flow through the open switch. Apply the point rule to point a.

$$I_1 + I_3 = I_2 \qquad \text{or} \qquad I_2 = I_1 + 0 = I_1$$

Apply the loop rule to loop $acbda$. In volts,

$$-12 + 7I_1 + 8I_2 + 9 = 0 \tag{1}$$

To understand the signs used, remember that current always flows from high to low potential through a resistor.

Because $I_2 = I_1$, (1) becomes

$$15I_1 = 3 \qquad \text{or} \qquad I_1 = \underline{0.20\text{ A}}$$

Also, $I_2 = I_1 = \underline{0.20\text{ A}}$. Note that this is the same result that one would obtain by replacing the two batteries by a single 3-V battery.

27.126 In Fig. 27-35, find I_1, I_2, I_3 if switch k is closed.

▌ We refer to Prob. 27.125. With k closed, I_3 is no longer known to be zero. Applying the point rule to point a gives

$$I_1 + I_3 = I_2 \tag{2}$$

Applying the loop rule to loop $acba$ gives, in volts,

$$-12 + 7I_1 - 4I_3 = 0 \tag{3}$$

and to loop $adba$ gives

$$-9 - 8I_2 - 4I_3 = 0 \tag{4}$$

Applying the loop rule to the remaining loop, $acbda$, would yield a redundant equation, because, among the three loop equations, each voltage change would appear twice.

We must now solve (2), (3), and (4) for I_1, I_2, and I_3. From (4),

$$I_3 = -2I_2 - 2.25$$

Substituting this in (3) gives

$$-12 + 7I_1 + 9 + 8I_2 = 0 \qquad \text{or} \qquad 7I_1 + 8I_2 = 3$$

Also substituting for I_3 in (2) gives

$$I_1 - 2I_2 - 2.25 = I_2 \qquad \text{or} \qquad I_1 = 3I_2 + 2.25$$

Substituting this value in the previous equation,

$$21I_2 + 15.75 + 8I_2 = 3 \qquad \text{or} \qquad I_2 = \underline{-0.44 \text{ A}}$$

Using this in the equation for I_1 gives

$$I_1 = 3(-0.44) + 2.25 = -1.32 + 2.25 \underline{= 0.93 \text{ A}}$$

Note that the minus sign is a part of the value we have found for I_2. It must be carried along with its numerical value. Now we can use (2) to find

$$I_3 = I_2 - I_1 = (-0.44) - 0.93 = \underline{-1.37 \text{ A}}$$

Note that this problem could not be solved by simple parallel- or series-resistance techniques.

27.127 For the circuit shown in Fig. 27-36, find the voltage of the unknown battery $\mathscr{E}$. The ammeter, represented by the symbol —Ⓐ—, can be considered perfect (i.e., no resistance), and it reads the current in the wire to be $\frac{1}{2}$ A in the direction shown. (Recall that $\mathscr{E}$ stands for emf.) We assume that all emf's are ideal.

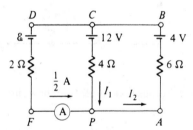

Fig. 27-36

▮ We first assign currents to all the wires. Since the current will be negative if we assign the wrong direction to it, we do not worry about its proper direction when we make this assignment. The currents have been labeled I_1 and I_2 on the diagram. Note that the current in parts of the wire P-A-B-C is I_2, while in C-D-F-P, it is $\frac{1}{2}$ A. Writing the point-rule equation for point P, we have $\frac{1}{2} + I_1 - I_2 = 0$. We now write the circuit equation for loop $CDFPC$: $-\mathscr{E} - (\frac{1}{2})(2) + 4I_1 + 12 = 0$. The ammeter is assumed perfect in writing this; i.e., it is assumed to have zero resistance. Similarly, for loop $CBAPC$, we have $-4 + 6I_2 + 4I_1 + 12 = 0$. It is possible to write other loop equations, but they will not be independent. Since we have covered all the circuit elements in our equations, another loop could contain no new information.

In any case, we now have three equations involving the three unknowns I_1, I_2, and $\mathscr{E}$. Solving simultaneously, we find

$$I_1 = -1.1 \text{ A} \qquad I_2 = -0.6 \text{ A} \qquad \mathscr{E} = \underline{6.6 \text{ V}}$$

27.128 Each of the cells shown in Fig. 27-37 has an emf of 1.50 V and a 0.075-Ω internal resistance. Find I_1, I_2, and I_3.

▮ Applying the point rule to point a,

$$I_1 = I_2 + I_3 \tag{1}$$

Applying the loop rule to loop $abcea$ gives, in volts,

$$-(0.075)I_2 + 1.5 - (0.075)I_2 + 1.5 - 3I_1 = 0$$

or
$$3I_1 + 0.15I_2 = 3 \tag{2}$$

Also, for loop $adcea$,

$$-(0.075)I_3 + 1.5 - (0.075)I_3 + 1.5 - 3I_1 = 0$$

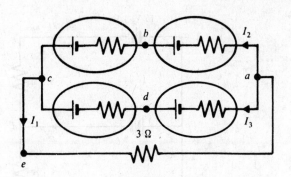

Fig. 27-37

or
$$3I_1 + 0.15I_3 = 3 \qquad (3)$$

Solve (2) for $3I_1$ and substitute in (3), yielding

$$3 - 0.15I_2 + 0.15I_3 = 3 \qquad \text{or} \qquad I_2 = I_3$$

as we might have guessed from the symmetry of the problem. Then (1) yields

$$I_1 = 2I_2$$

Substituting this in (2) gives

$$6I_2 + 0.15I_2 = 3 \qquad \text{or} \qquad I_2 = \underline{0.488 \text{ A}}$$

Then, $I_3 = I_2 = \underline{0.488 \text{ A}}$ and $I_1 = 2I_2 = \underline{0.976 \text{ A}}$.

27.129 For the circuit shown in Fig. 27-38, find the current in the 0.96-Ω resistor and the terminal voltages of the batteries.

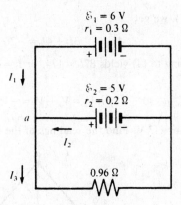

$\mathscr{E}_1 = 6 \text{ V}$
$r_1 = 0.3 \text{ }\Omega$

$\mathscr{E}_2 = 5 \text{ V}$
$r_2 = 0.2 \text{ }\Omega$

$0.96 \text{ }\Omega$

Fig. 27-38

❚ At point a in Fig. 27-38, we have

$$I_1 + I_2 - I_3 = 0 \qquad (1)$$

Choosing the two smaller loops, and moving counterclockwise, we have $6V - (0.3\,\Omega)I_1 - 5\text{ V} + (0.2\,\Omega)I_2 = 0$, or

$$3I_1 - 2I_2 = 10 \qquad (2)$$

and $5 \text{ V} - (0.2\,\Omega)I_2 - (0.96\,\Omega)I_3 = 0$, or

$$2I_2 + 9.6I_3 = 50 \qquad (3)$$

Substituting (1) into (3) to eliminate I_3 yields

$$9.6I_1 + 11.6I_2 = 50 \qquad (4)$$

Multiplying (2) by 5.8 yields

$$17.4I_1 - 11.6I_2 = 58 \qquad (5)$$

Adding (4) and (5) yields: $27I_1 = 108$, or $I_1 = \underline{4.0 \text{ A}}$. Substituting into (2) yields $I_2 = \underline{1.0 \text{ A}}$; and from (1), $I_3 = \underline{5.0 \text{ A}}$. $V_1 = \mathscr{E}_1 - I_1 r_1 = 6 \text{ V} - (4.0 \text{ A})(0.3 \,\Omega) = \underline{4.8 \text{ V}}$. $V_2 = \mathscr{E}_2 - I_2 r_2 = 5 \text{ V} - (1.0 \text{ A})(0.2 \,\Omega) = \underline{4.8 \text{ V}}$.

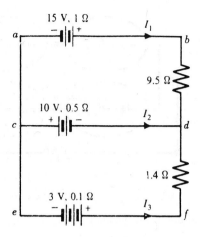

Fig. 27-39

27.130 Find I_1, I_2, I_3, and the potential difference from point b to point e in Fig. 27-39.

▮ For point d,

$$I_1 + I_2 + I_3 = 0 \qquad (1)$$

We move clockwise in the two small loops. For upper loop, $15\text{ V} - (1\ \Omega)I_1 - (9.5\ \Omega)I_1 + 10\text{ V} + (0.50\ \Omega)I_2 = 0$, or

$$21I_1 - I_2 = 50 \qquad (2)$$

For lower loop, $-10\text{ V} - (0.5\ \Omega)I_2 + (1.4\ \Omega)I_3 - 3\text{ V} + (0.1\ \Omega)I_3 = 0$, or

$$I_2 - 3I_3 = -26 \qquad (3)$$

Eliminating I_3 in (3) using (1), we get

$$3I_1 + 4I_2 = -26 \qquad (4)$$

Multiplying (2) by 4 and adding to (4) yields $87I_1 = 174$, or $I_1 = \underline{2\text{ A}}$. Substituting into (2) yields $I_2 = \underline{-8\text{ A}}$ and finally from (1), $I_3 = \underline{6\text{ A}}$.

$$V_e = V_a \qquad \text{so} \qquad V_e - V_b = V_a - V_b = -15\text{ V} + (1\ \Omega)(2\text{ A}) = \underline{-13\text{ V}}$$

27.131 In Fig. 27-40, $R = 10\ \Omega$ and $\mathscr{E} = 13$ V. Find the readings of the ideal ammeter and voltmeter.

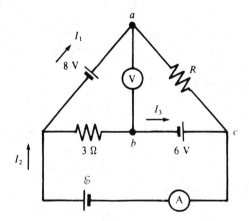

Fig. 27-40

▮ Since no internal resistances are specified for the emf's, we assume that they are zero. The voltmeter is ideal, so its resistance is essentially infinite and zero current flows through it. Similarly, voltage across ammeter = 0. For point c,

$$I_1 + I_3 - I_2 = 0 \qquad (1)$$

For the rectangle, $13\text{ V} - (3\ \Omega)I_3 + 6\text{ V} = 0$, or $3I_3 = 19$, and $I_3 = \underline{6.3\text{ A}}$. For the large triangle, $8\text{ V} - (10\ \Omega)I_1 - 6\text{ V} + (3\ \Omega)I_3 = 0$; or using the value of I_3 just obtained, $I_1 = \underline{2.1\text{ A}}$. Then, from Eq. (1), $I_2 = \underline{8.4\text{ A}} =$ ammeter reading. To obtain the voltmeter reading, we calculate $V_a - V_b = 8\text{ V} + (3\ \Omega)I_3 = \underline{27\text{ V}}$.

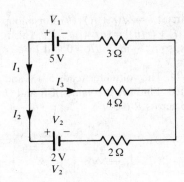

Fig. 27-41

27.132 Find the current in each branch of the circuit shown in Fig. 27-41.

▌ Use Kirchhoff's laws.

$$I_1 = I_3 + I_2 \tag{1}$$

Upper loop, $V_1 = 4I_3 + 3I_1$, or

$$5 = 4I_3 + 3I_1 \tag{2}$$

Large loop, $V_1 - V_2 = 2I_2 + 3I_1$, or $5 - 2 = 2I_2 + 3I_1 = 2(I_1 - I_3) + 3I_1$, where we have used (1). Thus $3 = 5I_1 - 2I_3$. Multiplying by 2 we get $6 = -4I_3 + 10I_1$. Writing (2) directly below and adding, yields

$$\begin{aligned} 6 &= -4I_3 + 10I_1 \\ 5 &= 4I_3 + 3I_1 \\ \hline 11 &= 13I_1 \end{aligned}$$

and $I_1 = \frac{11}{13}$ A. Then we solve for I_3 in a previous equation:

$$3 = \frac{55}{13} - 2I_3 \quad \text{or} \quad -\frac{16}{13} = -2I_3 \quad \text{and} \quad I_3 = \frac{8}{13} \text{ A}$$

Finally from (1), $I_2 = I_1 - I_3 = \frac{11}{13} - \frac{8}{13} = \frac{3}{13}$ A.

27.133 Find I_1, I_2, I_3, and I_4 for the circuit of Fig. 27-42.

▌ The loop equations are: $-6 + 6 - 10I_4 = 0$, so $I_4 = 0$. (This result tells us that I_1 and I_2 flow through their respective 3-Ω resistors.) $+6 - 3I_1 + 12I_3 = 0$; $-6 - 12I_3 - 3I_2 = 0$. The point equation is $I_2 = I_1 + I_3$. Solve simultaneously to find I_1, I_2, I_3, and I_4, $\underline{0.222}$, $\underline{-0.222}$, $\underline{-0.444}$, and $\underline{0}$ A.

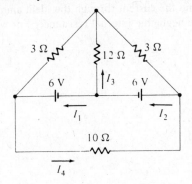

Fig. 27-42

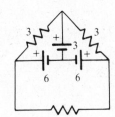

Fig. 27-43

27.134 The 12-Ω resistor in Fig. 27-42 is replaced by a 3-V battery with the positive terminal at the top (Fig. 27-43). Find I_1, I_2, I_3, and I_4.

▌ As in Prob. 27.133 $I_4 = 0$. Other loop equations are $+6 - 3I_1 - 3 = 0$, $-6 + 3 - 3I_2 = 0$, from which $I_1 = \underline{1.0 \text{ A}}$ and $I_2 = \underline{-1.0 \text{ A}}$. From $I_2 = I_1 + I_3$, find $I_3 = \underline{-2 \text{ A}}$. Note that $I_1 = -I_2$ is a consequence of symmetry in both problems.

27.135 Suppose that you have available a large number of identical batteries with emf $\mathscr{E}$ and internal resistance r. **(a)** What is the current produced through a resistor R by a single battery connected across R? **(b)** Repeat when n batteries are connected in parallel across R. **(c)** Repeat when n batteries are connected in series across R so as to aid each other.

▮ (a) For the simple series circuit $I = \mathscr{E}/(r + R)$. (b) For a loop with R and one emf write the loop rule as $\mathscr{E} - (I/n)r - IR = 0$, so $I = \mathscr{E}/[R + (r/n)] = (n\mathscr{E})(r + nR)$. This is the same as one emf with internal resistance r/n. (c) Elements are in series, so $n\mathscr{E} - nrI - IR = 0$ and $I = (n\mathscr{E})/(nr + R)$.

27.136 Refer to the circuit of Fig. 27-44. The voltmeter reads 5.0 V and the ammeter reads 2.0 A with the current flowing in the direction indicated. Find (a) the value of R and (b) the value of $\mathscr{E}$. (Note in writing the circuit equation that the voltage drop across R is 5 V.)

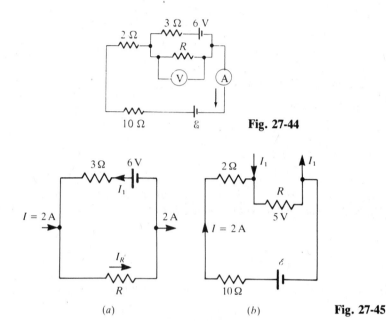

Fig. 27-44

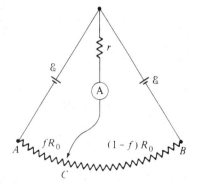

(a)   (b)   **Fig. 27-45**

▮ We isolate the two loops in Fig. 27-45. (a) The junction and loop rules for Fig. 27-45(a) yield $I_R = 2.0 + I_1$ and $+6 - 3I_1 - I_R R = 0$. The voltmeter reads $5.0 = I_R R$; we solve these two equations to find $I_1 = 0.33$ A and $I_R = 2.33$ A. Then $R = V/I_R = 5.0/2.33 = \underline{2.1\ \Omega}$. (b) Use the loop rule for Fig. 27-45(b): $\mathscr{E} - 2(10) - 2(2) - 5 = 0$; therefore $\mathscr{E} = \underline{29\ V}$.

27.137 In Fig. 27-46, the uniform-resistance wire between A and B has a total resistance R_0. The contact at C can apportion the wire into resistors fR_0 and $(1 - f)R_0$. (a) Find the current through the ideal ammeter for any $0 < f < 1$. Assume that the batteries are identical and have negligible internal resistance. For what value of f is the ammeter reading (b) maximum and (c) minimum?

Fig. 27-46

▮ (a) In Fig. 27-46, call I, the ammeter current directed down, I_1 the current up through the left-hand emf, and I_2 the current up through the right-hand emf. Loop and junction equations give $\mathscr{E} - Ir - I_1 fR_0 = 0$, $\mathscr{E} - Ir - I_2(1 - f)R_0 = 0$, and $I = I_1 + I_2$. Solving, $I = \mathscr{E}/(r + R_0 f - R_0 f^2)$. (b) The denominator is smallest when $f = 0$ or 1, making I a maximum. (c) Since $f - f^2 = \frac{1}{4} - (\frac{1}{2} - f)^2$, I is smallest for $f = \frac{1}{2}$.

27.138 Find the equivalent resistance of the circuit shown in Fig. 27-47(a), if each resistor has a resistance R.

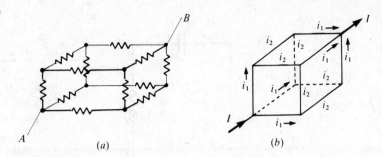

(a) (b) **Fig. 27-47**

▌ Symmetry about the diagonal from A to B, plus reversibility, gives the currents shown in Fig. 27-47(b). We have $i_1 = 2i_2$ from the point rule and also $I = 3i_1$. Then if an emf is connected across the system, $\mathscr{E} = IR_{eq}$. Tracing a path through the system from entrance to exit gives $\mathscr{E} = i_1 R + i_2 R + i_1 R = (5/2)i_1 R = (5/2)(I/3)R$. Equating this to IR_{eq} gives $R_{eq} = \underline{5R/6}$.

27.139 The currents are steady in the circuit of Fig. 27-48. Find I_1, I_2, I_3, I_4, I_5, and the charge on the capacitor.

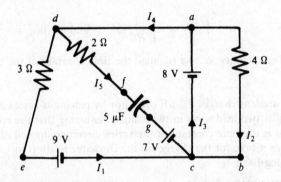

Fig. 27-48

▌ The capacitor passes no current when charged and so $I_5 = \underline{0}$. Consider loop $acba$. The loop rule gives $-8 + 4I_2 = 0$, or $I_2 = \underline{2\,A}$. Using loop $adeca$ gives $-3I_1 - 9 + 8 = 0$, or $I_1 = \underline{-0.33\,A}$.
    Apply the point rule at point c: $I_1 + I_5 + I_2 = I_3$, or $I_3 = \underline{1.67\,A}$; and at point a: $I_3 = I_4 + I_2$, or $I_4 = \underline{-0.33\,A}$. (We should have realized this at once, because $I_5 = 0$ and so $I_4 = I_1$.)
    To find the charge on the capacitor, we need the voltage across it, V_{fg}. Applying the loop rule to loop $dfgced$ gives $-2I_5 + V_{fg} - 7 + 9 + 3I_1 = 0$, or $0 + V_{fg} - 7 + 9 - 1.0 = 0$, from which $V_{fg} = -1$ V. The negative sign tells us that plate g is negative. The capacitor's charge is $Q = CV = (5\,\mu\text{F})(1\,\text{V}) = \underline{5\,\mu\text{C}}$.

27.140 Figure 27-49 shows a possible bridge for measuring capacitance. If no current flows through the galvanometer when the key K is closed, the bridge is balanced. Show that the balance condition is $C_2 = (R_1/R_2)C_1$. This bridge method has several practical difficulties which make it undesirable.

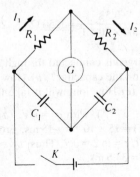

Fig. 27-49

▌ Let Q_1 and Q_2 be the magnitudes of charge on the capacitors. To be balanced, $I_1 = I_2$, $I_1 R_1 = Q_1/C_1$ and $I_2 R_2 = Q_2/C_2$. But also, since C_1 and C_2 are in series and no current flows through the galvanometer, $Q_1 = Q_2$. Therefore $I_1 R_1 C_1 = I_2 R_2 C_2$. Canceling I_1 with I_2 gives $C_2 = (R_1/R_2)C_1$.

27.141ᶜ In Fig. 27-50 a capacitor of capacitance C is charged with charge Q, in the open circuit shown. When the

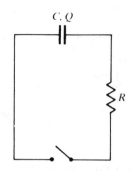

Fig. 27-50

switch is closed the capacitor will discharge through resistance R. Find an expression for $q(t)$, the charge on the capacitor at any time t.

▮ With the switch closed we must have for the loop equation $q/C + IR = 0$. Since $I = $ rate of discharge, $I = dq/dt$. Then, substituting in the loop equation, we get $dq/dt = -(1/RC)q$, or $dq/q = -(1/RC)\,dt$. Integrating,

$$\int_Q^q \frac{dq'}{q'} = -\int_0^t \frac{1}{RC}\,dt' \quad \text{and} \quad \ln\frac{q}{Q} = -\frac{t}{RC}$$

Then $q = Qe^{-t/RC}$. The quantity $RC \equiv \tau$ is called the time constant of the circuit; the larger τ, the longer the discharge time.

27.142 In the laboratory, a student charges a 2-μF capacitor by placing it across a 1.5-V battery. While disconnecting it, the student holds its two lead wires in two hands. Assuming that the resistance of the body between the hands is 60 kΩ, what is the time constant of the series circuit composed of the capacitor and the student's body? How long does it take for the charge on the capacitor to drop to $1/e$ of its orgiinal value? To $1/100$? (*Hint:* $\ln 100 \approx 2.30 \log 100$.)

▮ $\tau = RC = (6 \times 10^4)(2 \times 10^{-6}) = 0.12$ s. Now $Q = Q_0 \exp(-t/RC) = Q_0 \exp(-t/0.12)$. If $\exp(-t/0.12) = \exp(-1)$, then $t = 0.12$ s. If $\exp(-t/0.12) = 0.01$, then $\exp(t/0.12) = 100$, or $t/0.12 = \ln 100 \approx 2.30 \log 100 = 4.60$; so $t = \underline{0.55\ \text{s}}$.

27.143 In a certain electronic device, a 10-μF capacitor is charged to 2000 V. When the device is shut off, the capacitor is discharged for safety reasons by a so-called bleeder resistor of 1 $M\Omega$ placed across its terminals. How long does it take for the charge on the capacitor to decrease to 0.01 of its original value?

▮ We recall that $\ln x \approx 2.30 \log x$; $\tau = RC = 10$ s. We wish that $Q/Q_0 = \exp(-t/\tau) = 0.01$. From this, $t/10 = 2.30 \log 100$, which gives $t = \underline{46\ \text{s}}$.

27.144 A 400-μF capacitor is connected through a resistor to a battery. Find (**a**) the resistance R and (**b**) the emf of the battery $\mathcal{E}$ if the time constant of the circuit is 0.5 s and the maximum charge on the capacitor is 0.024 C.

▮ $$R = \frac{\tau}{C} = \frac{0.5\ \text{s}}{400 \times 10^{-6}\ \text{F}} = \underline{1250\ \Omega} \qquad \mathcal{E} = \frac{q}{C} = \frac{0.024\ \text{C}}{400 \times 10^{-6}\ \text{F}} = \underline{60\ \text{V}}$$

27.145 A 50-μF capacitor initially uncharged is connected through a 300-Ω resistor to a 12-V battery. (**a**) What is the magnitude of the final charge q_0 on the capacitor? (**b**) How long after the capacitor is connected to the battery will it be charged to $\frac{1}{2}q_0$? (**c**) How long will it take for the capacitor to be charged to $0.90q_0$?

▮ In charging we have $q = q_0(1 - e^{-t/RC})$, with $q_0 = CV$. (**a**) $q_0 = CV = (50 \times 10^{-6}\ \text{F})(12\ \text{V}) = \underline{600\ \mu\text{C}}$. (**b**) $\tau = RC = (300\ \Omega)(50 \times 10^{-6}\ \text{F}) = 15 \times 10^{-3}\ \text{s} = \underline{15\ \text{ms}}$. From the formula the charge reaches $\frac{1}{2}q_0$ in the time t such that $e^{-t/RC} = \frac{1}{2}$, or $t/(RC) = t/\tau = \ln 2 \approx 0.7$. Thus: $t_{0.5} \approx 0.7\tau = \underline{10.5\ \text{ms}}$. (**c**) Similarly, $e^{-t/\tau} = 0.1$ and $t/\tau = \ln 10 \approx 2.3$. Thus $t_{0.9} \approx 2.3\tau = \underline{34.5\ \text{ms}}$.

27.146 A 150-μF capacitor is connected through a 500-Ω resistor to a 40-V battery. (**a**) What is the final charge q_0 on a capacitor plate? (**b**) What is the time constant of the circuit? (**c**) How long does it take the charge on a capacitor plate to reach $0.8q_0$?

▮ (**a**) $q_0 = VC = (40\ \text{V})(150 \times 10^{-6}\ \text{F}) = \underline{6.0\ \text{mC}}$. (**b**) $\tau = RC = (500\ \Omega)(150 \times 10^{-6}\ \text{F}) = \underline{75\ \text{ms}}$. (**c**) Following the procedure in Prob. 27.145: $-t/\tau = \ln 0.2$ or $t/\tau = \ln 5 \approx 1.6$. Then $t \approx (1.6)(75\ \text{ms}) = \underline{120\ \text{ms}}$.

28.1 FORCE ON A MOVING CHARGE

28.1 What is a *magnetic field*?

▌ A region of space is said to be the site of a *magnetic field* if a test charge moving in the region experiences a force *by virtue of its motion with respect to an inertial frame*. This force may be described in terms of a field vector **B**, called the *magnetic induction* or the *magnetic flux density*, or simply, the *magnetic field*. If **v** is the velocity of the charge, then the force is given by

$$\mathbf{F} = q(\mathbf{v} \times \mathbf{B}) \tag{1}$$

relative to the given inertial frame. (For the cross product, review Prob. 1.89.) According to (1), the force has magnitude $F = qvB \sin \theta$, and its direction is as given by the familiar right-hand rule. The unit of **B** or B is the *weber per square meter* (Wb/m^2), renamed in the SI the *tesla* (T). From (1),

$$1\,\text{T} = 1\,\frac{\text{N} \cdot \text{s}}{\text{C} \cdot \text{m}} = 1\,\frac{\text{N}}{\text{A} \cdot \text{m}} = 1\,\frac{\text{V} \cdot \text{s}}{\text{m}^2} = 1\,\frac{\text{Wb}}{\text{m}^2}$$

When both a magnetic field **B** and an electric field **E** are present in a region, the force on a test charge is the vector sum of the electric and magnetic forces: $\mathbf{F}_{\text{tot}} = q[\mathbf{E} + (\mathbf{v} \times \mathbf{B})]$. This relation is known as the *Lorentz equation*.

28.2 Contrast the magnetic field with the electric field.

▌ (i) At each point of the magnetic field there is a certain direction of motion (namely, the direction of **B**) for which the magnetic force is zero. (ii) The magnetic force is perpendicular to the velocity of the moving charge and thus performs no work; hence there is no scalar potential function for the force. Moreover, the force is noncentral. Despite these apparent differences, the special theory of relativity shows that the magnetic and electric fields are two aspects of a single *electromagnetic field*.

28.3 The particle shown in Fig. 28-1 is positively charged. What is the direction of the force on it due to the magnetic field? Give its magnitude in terms of B, q, and v.

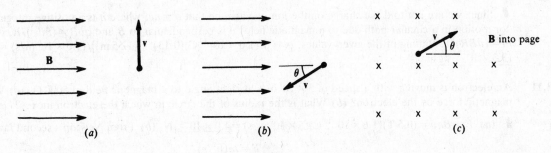

Fig. 28-1

▌ $F = qvB \sin \theta$; direction follows right-hand rule. (*a*) **v** is perpendicular to **B**, so $\theta = 90°$ and $F = qvB$. Direction (rotating **v** arrowhead toward **B**) is into paper. (*b*) Here $F = qvB \sin (\pi - \theta) = qvB \sin \theta$. Rotating **v** toward **B** through $(\pi - \theta)$, we have **F** out of paper. (*c*) Here **v** is perpendicular to **B**, so $F = qvB$. Rotating **v** into paper gives **F** in plane of paper 90° counterclockwise from **v**.

28.4 A He^{2+} ion travels at right angles to a magnetic field of 0.80 T with a velocity of 10^5 m/s. Find the magnitude of the magnetic force on the ion.

▌ $\qquad F = qvB \sin \theta \qquad \theta = 90° \qquad \sin \theta = 1 \qquad F = 2(1.60 \times 10^{-19})10^5(0.80) = \underline{2.56 \times 10^{-14}\,\text{N}}$.

28.5 A beam of protons (positively charged) is moving horizontally toward you. As it approaches, it passes through a magnetic field directed downward. This magnetic field deflects the beam to your _____.

▮ Right (direction of progress of right-hand screw rotating from **v** into **B**).

28.6 In Vermont the dip angle for the earth's magnetic field (the angle the field lines make with the horizontal) is 74°, and the component of the field parallel to the earth's surface is 0.16 G. If an electron is shot with a speed 10^6 m/s vertically upward there, how large a force acts on the electron and what is the direction of the force? How large an acceleration will this force cause? ($m_e = 9.1 \times 10^{-31}$ kg; $1 G = 10^{-4}$ T.)

▮ The force is caused by the northward component $B_N = 0.16$ G. Since the force on a negative charge moving up is equivalent to that on a positive charge moving down, the force is eastward with magnitude

$$F = evB_N = (1.6 \times 10^{-19}\,\text{C})(10^6\,\text{m/s})(0.16 \times 10^{-4}\,\text{T}) = \underline{2.56 \times 10^{-18}\,\text{N}}$$

Then
$$a = \frac{F}{m_e} = \frac{2.56 \times 10^{-18}}{9.1 \times 10^{-31}} = \underline{2.8 \times 10^{12}\,\text{m/s}^2}$$

28.7 Repeat Prob. 28.6 if the electron is shot horizontally northward.

▮ In this case, the downward component $B_N \tan 74°$ causes the force; hence,

$$F = (2.56 \times 10^{-18})(\tan 74°) = \underline{8.9 \times 10^{-18}\,\text{N}}$$
$$a = (2.8 \times 10^{12})(\tan 74°) = \underline{9.8 \times 10^{12}\,\text{m/s}^2}$$

28.8 An ion ($q = +2e$) enters a magnetic field with flux density 1.2 Wb/m² at velocity of 2.5×10^5 m/s perpendicular to the field. Determine the force on the ion.

▮ $F = 2(1.6 \times 10^{-19}\,\text{C})(2.5 \times 10^5\,\text{m/s})(1.2\,\text{Wb/m}^2) = \underline{9.6 \times 10^{-14}\,\text{N}}$. Direction perpendicular to **v** and **B**, by right-hand rule.

28.9 An electron is accelerated from rest through a potential difference of 3750 V. It enters a region where $B = 4$ mT perpendicular to its velocity. Calculate the radius of the path it will follow.

▮ Since the magnetic force is always perpendicular to **B** and to **v** it acts as a centripetal force. Furthermore, since there is no acceleration tangent to the path of motion after passing through the potential difference, both v and $F = evB$ are constant, which are the conditions for uniform circular motion. First we find v from $eV = \frac{1}{2}mv^2$, or $(1.6 \times 10^{-19}\,\text{C})(3750\,\text{V}) = \frac{1}{2}(9.1 \times 10^{-31}\,\text{kg})v^2$, or $v = 3.63 \times 10^7$ m/s. Then from Newton's second law, $qvB = (mv^2)/R$, where R is the radius of the circular motion. Then $R = (mv)/(qB) = [(9.1 \times 10^{-31})(3.63 \times 10^7)]/[(1.6 \times 10^{-19})(4 \times 10^{-3}\,\text{T})] = \underline{52\,\text{mm}}$.

28.10 What might be the mass of a positive ion that is moving at 10^7 m/s and is bent into a circular path of radius 1.55 m by a magnetic field of 0.134 Wb/m²? There are several possible answers.

▮ Since we are not told the charge on the ion, we assume that $q = ne$, where n is a positive integer. Then, for motion in a circular path due to a magnetic field, v is perpendicular to B and $qvB = (mv^2)/R$, or $m = (qBR)/v$. Putting in the given values, $m = [n(1.6 \times 10^{-19}\,\text{C})(0.134\,\text{T})(1.55\,\text{m})]/(1.0 \times 10^7\,\text{m/s}) = \underline{(3.3 \times 10^{-27}\,\text{kg})n}$.

28.11 An electron is moving with a speed of 5×10^7 m/s at right angle to a magnetic field of 5000 G. (*a*) What is the magnetic force on the electron? (*b*) What is the radius of the circle in which the electron moves?

▮ (*a*) $F = Bqv = (0.5\,\text{T})(1.6 \times 10^{-19}\,\text{C})(5 \times 10^7\,\text{m/s}) = \underline{4.0 \times 10^{-12}\,\text{N}}$. (*b*) From Newton's second law,

$$F = m(v^2/r)$$

so
$$r = \frac{mv^2}{F} = \frac{(9.11 \times 10^{-31}\,\text{kg})(5 \times 10^7\,\text{m/s})^2}{4.0 \times 10^{-12}\,\text{N}} = 5.7 \times 10^{-4}\,\text{m} = \underline{0.57\,\text{mm}}$$

28.12 (*a*) What is the force on a singly charged carbon ion moving with a speed of 3×10^5 m/s at right angles to a magnetic field of 7500 G? (*b*) What is the centripetal acceleration of the ion? (*c*) What is the radius of the circle in which the ion moves?

▮ (*a*) $q = e = 1.6 \times 10^{-19}$ C $\quad F = Bqv = (0.75\,\text{T})(1.6 \times 10^{-19}\,\text{C})(3 \times 10^5\,\text{m/s}) = \underline{3.6 \times 10^{-14}\,\text{N}}$

(*b*) $m_C = (12\,\text{u})(1.66 \times 10^{-27}\,\text{kg/u}) = 19.9 \times 10^{-27}\,\text{kg}$ $\quad a = \frac{F}{m_C} = \frac{3.6 \times 10^{-14}\,\text{N}}{19.9 \times 10^{-27}\,\text{kg}} = \underline{1.81 \times 10^{12}\,\text{m/s}^2}$

(c)
$$r = \frac{v^2}{a} = \frac{(3 \times 10^5 \text{ m/s})^2}{1.81 \times 10^{12} \text{ m/s}^2} = \underline{49.7 \text{ mm}}$$

28.13 If it is assumed that a circular path around the earth (radius = 6400 km) can be found upon which the earth's field is horizontal and constant at 0.50 G, how fast must a proton be shot in order to circle the earth? In what direction?

❚ The centripetal force $(mv^2)/r$ is furnished by the force evB_H. Then $v = (eB_H r)/m = [(1.6 \times 10^{-19}) \times (5 \times 10^{-5})(6.4 \times 10^6)]/(1.67 \times 10^{-27}) = \underline{3.1 \times 10^{10} \text{ m/s}}$. It must be shot westward, since the field points north.

28.14 Why is the solution to Prob. 28.13 faulty?

❚ The speed of the proton cannot exceed the speed of light. Evidently, the relativistic equation of motion should have been employed.

28.15 Alpha particles ($m = 6.68 \times 10^{-27}$ kg, $q = +2e$), accelerated through a potential difference V to 2-keV, enter a magnetic field $B = 0.2$ T perpendicular to their direction of motion. Calculate the radius of their path.

❚ The KE of a particle is conserved in the magnetic field:
$$\frac{1}{2} mv^2 = Vq \qquad \text{or} \qquad v = \sqrt{\frac{2Vq}{m}}$$

They follow a circular path in which
$$r = \frac{mv}{qB} = \frac{m}{qB}\sqrt{\frac{2Vq}{m}} = \frac{1}{B}\sqrt{\frac{2Vm}{q}} = \frac{1}{0.2 \text{ T}}\sqrt{\frac{2(1000 \text{ V})(6.68 \times 10^{-27} \text{ kg})}{3.2 \times 10^{-19} \text{ C}}} = \underline{32 \text{ mm}}$$

28.16 After being accelerated through a potential difference of 5000 V, a singly charged carbon ion moves in a circle of radius 21 cm in the magnetic field of a mass spectrometer. What is the magnitude of the field?

❚ Following the approach of Prob. 28.15,

$$m_C = (12 \text{ u})(1.66 \times 10^{-27} \text{ kg/u}) = 19.9 \times 10^{-27} \text{ kg} \qquad B^2 = \frac{2Vm_C}{r^2 q} = \frac{2(5000 \text{ V})(19.9 \times 10^{-27} \text{ kg})}{(0.21 \text{ m})^2(1.6 \times 10^{-19} \text{ C})}$$

$$= 2.82 \times 10^{-2} \text{ T}^2 \qquad B = \underline{0.168 \text{ T}}$$

28.17 A particle with charge q and mass m is shot with kinetic energy K into the region between two plates as shown in Fig. 28-2. If the magnetic field between the plates is B and as shown, how large must B be if the particle is to miss collision with the opposite plate?

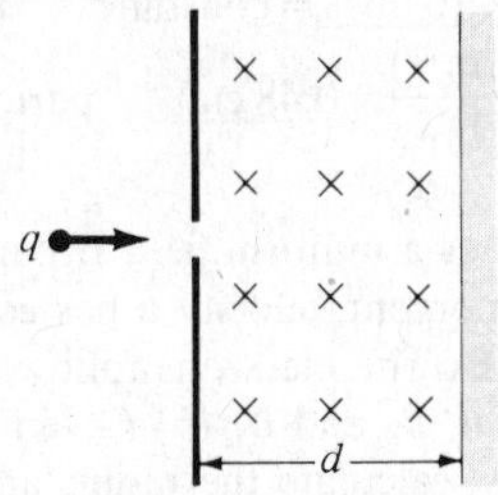

Fig. 28-2

❚ To just miss the opposite plate, the particle must move in a circular path with radius d so from $Bqd = mv$; and using $K = (mv^2)/2$, we have $B = (2mK)^{1/2}/(qd)$.

28.18 A cathode-ray beam (an electron beam; $m = 9.1 \times 10^{-31}$ kg, $q = -e$) is bent in a circle of radius 2 cm by a uniform field with $B = 4.5$ mT. What is the speed of the electrons?

❚ To describe a circle like this, the particle must be moving perpendicular to **B**. Then
$$v = \frac{rqB}{m} = \frac{(0.02 \text{ m})(1.6 \times 10^{-19} \text{ C})(4.5 \times 10^{-3} \text{ T})}{9.1 \times 10^{-31} \text{ kg}} = \underline{1.58 \times 10^7 \text{ m/s}}$$

28.19 In Fig. 28-3(a), a proton ($q = +e$, $m = 1.67 \times 10^{-27}$ kg) is shot with speed 8×10^6 m/s at an angle of 30° to an x-directed field $B = 0.15$ T. Describe the path followed by the proton.

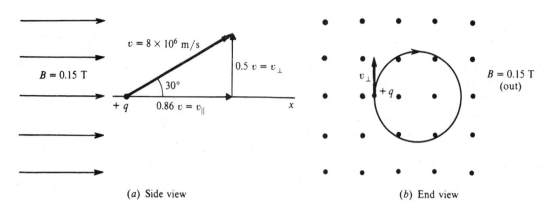

Fig. 28-3

▐ We resolve the particle velocity into components parallel to and perpendicular to the magnetic field. The magnetic force due to $v_\parallel$ is zero (sin $\theta = 0$); the magnetic force due to $v_\perp$ has no x component. Therefore, the x motion is uniform, at speed $v_\parallel = (0.86)(8 \times 10^6 \text{ m/s}) = \underline{6.88 \times 10^6 \text{ m/s}}$, while the transverse motion is circular, with radius

$$r = \frac{mv_\perp}{qB} = \frac{(1.67 \times 10^{-27} \text{ kg})(0.5 \times 8 \times 10^6 \text{ m/s})}{(1.6 \times 10^{-19} \text{ C})(0.15 \text{ T})} = \underline{0.28 \text{ m}}$$

The proton will spiral along the x axis; the radius of the spiral (or helix) will be 28 cm.

To find the *pitch* of the helix (the x distance traveled during one revolution), we note that the time taken to complete one circle is

$$\text{period} = \frac{2\pi r}{v_\perp} = \frac{2\pi(0.28 \text{ m})}{(0.5)(8 \times 10^6 \text{ m/s})} = \underline{4.4 \times 10^{-7} \text{ s}}$$

During that time, the proton will travel an x distance of

$$\text{pitch} = (v_\parallel)(\text{period}) = (6.88 \times 10^6 \text{ m/s})(4.4 \times 10^{-7} \text{ s}) = \underline{3.0 \text{ m}}$$

28.20 An electron is shot with speed 5×10^6 m/s out from the origin of coordinates. Its initial velocity makes an angle of 20° to the +x axis. Describe its motion if a magnetic field $B = 2.0$ mT exists in the +x direction.

▐ Proceed as in Prob. 28.19, except that the charge is negative. This causes the charge to spiral in the opposite rotational sense.

$$v_x = v \cos 20° = \underline{4.7 \times 10^6 \text{ m/s}} = \text{constant} \qquad v_\perp = v \sin 20° = \underline{1.7 \times 10^6 \text{ m/s}} \qquad \text{and}$$
$$r = \frac{mv_\perp}{qB} = \frac{(9.1 \times 10^{-31} \text{ kg})(1.7 \times 10^6 \text{ m/s})}{(1.6 \times 10^{-19} \text{ C})(0.0020 \text{ T})} = \underline{0.48 \text{ cm}} \qquad \text{period } T = \frac{2\pi r}{v_\perp} = \underline{1.78 \times 10^{-8} \text{ s}} \qquad \text{pitch} = v_x T = \underline{8.33 \text{ cm}}$$

28.21 A long solenoid, Fig. 28-4, establishes a uniform field **B** parallel to its axis. An electron gun mounted inside fires a stream of electrons whose emergent velocity **u** has components $u_1 = u \cos \beta$ and $u_2 = u \sin \beta$, respectively normal and parallel to **B**. The electrons follow a helical path as indicated. (a) Find expressions for the radius and pitch in terms of u, β, and B. (b) Given $B = 1$ mT, $\beta = 30°$, $u = 1.45 \times 10^7$ m/s, and $e/m = 1.76 \times 10^{11}$ C/kg for electrons, calculate the radius and the pitch of the helix.

▐ (a) $$R = \frac{mu \cos \beta}{eB} = \frac{u \cos \beta}{(e/m)(B)}, \qquad \text{pitch} = (u \sin \beta)T = \frac{(u \sin \beta)(2\pi R)}{u \cos \beta} = 2\pi R \tan \beta$$

(b) $$R = [(1.45 \times 10^7 \text{ m/s})(0.866)]/[(1.76 \times 10^{11} \text{ C/kg})(1.0 \times 10^{-3} \text{ T})] = \underline{71.3 \text{ mm}}$$
$$\text{pitch} = 2\pi R \tan \beta = (6.28)(0.0713 \text{ m})(0.577) = \underline{258 \text{ mm}}.$$

28.22 The magnetic field in a certain region of space is given by $\mathbf{B} = 0.080\mathbf{i}$ T. A proton is shot into the field with velocity $2 \times 10^5 \mathbf{i} + 3 \times 10^5 \mathbf{j}$ m/s. Give the radius and pitch for the helical path that the proton follows.

▐ Since $\mathbf{F} = e\mathbf{v} \times \mathbf{B}$, only the y component of **v** gives rise to a force. This component, $3 \times 10^5 \mathbf{j}$, causes the particle to move in a circle in the yz plane with a radius given by $(mv_y^2)/r = ev_y B$. Hence, $r = [(1.67 \times 10^{-27})(3 \times 10^5)]/[(1.6 \times 10^{-19})(0.08)] = 0.039$ m. The time taken to complete one circle is $(2\pi r)/v_y = 8.2 \times 10^{-7}$ s. During this time it will travel $(2 \times 10^5 \mathbf{i})t$; thus the pitch of the helix $= \underline{0.164 \text{ m}}$.

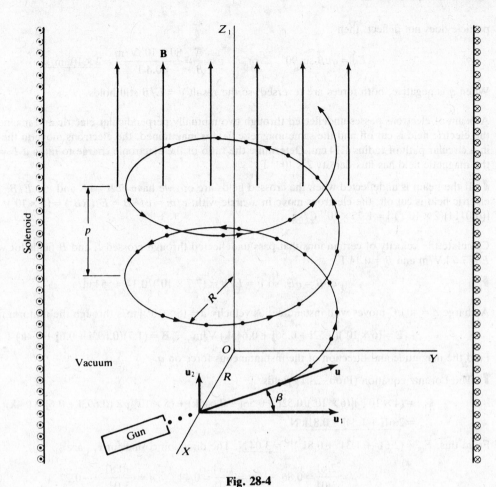

Fig. 28-4

28.23 As shown in Fig. 28-5, a beam of particles of charge q enters a region where an electric field is uniform and directed downward. Its value is 80 kV/m. Perpendicular to **E** and directed into the page is a magnetic field $B = 0.4$ T. If the speed of the particles is properly chosen, the particles will not be deflected by these crossed electric and magnetic fields. What speed is selected in this case? (This device is called a *velocity selector*.)

▎ The electric field causes a downward force Eq on the charge if it is positive. The right-hand rule tells that the magnetic force, $qvB \sin 90°$, is upward if q is positive. If these two forces are to balance so that the

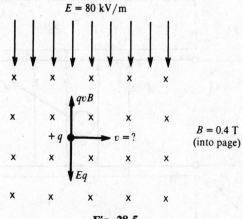

Fig. 28-5

particle does not deflect, then

$$Eq = qvB \sin 90° \quad \text{or} \quad v = \frac{E}{B} = \frac{80 \times 10^3 \text{ V/m}}{0.4 \text{ T}} = \underline{2 \times 10^5 \text{ m/s}}$$

When q is negative, both forces are reversed, so the result $v = E/B$ still holds.

28.24 A beam of electrons passes undeflected through two mutually perpendicular electric and magnetic fields. If the electric field is cut off and the same magnetic field is maintained, the electrons move in the magnetic field in a circular path of radius 1.14 cm. Determine the ratio of the electronic charge to mass if $E = 8$ kV/m and the magnetic field has flux density 2×10^{-3} T.

▮ If the beam is undeflected when the crossed fields are on, we have $evB = eE$ and $v = E/B$. When the electric field is cut off, the electrons move in a circle with: $e/m = v/RB = E/(RB^2) = (8 \times 10^3)/[(0.0114)(2 \times 10^{-3})^2] = \underline{1.75 \times 10^{11} \text{ C/kg}}$.

28.25 Calculate the velocity of certain ions that pass undeflected through crossed E and B fields for which $E = 7.7$ kV/m and $B = 0.14$ T.

▮ $\qquad qvB = qE$, so $v = E/B = (7.7 \times 10^3)/0.14 = \underline{55 \text{ km/s}}$

28.26 A charge $q = 40\ \mu$C moves with instantaneous velocity $\mathbf{u} = (5 \times 10^4)\mathbf{j}$ m/s through the uniform fields

$$\mathbf{E} = (6 \times 10^4)(0.52\mathbf{i} + 0.56\mathbf{j} + 0.645\mathbf{k}) \text{ V/m} \qquad \mathbf{B} = (1.7)(0.693\mathbf{i} + 0.6\mathbf{j} + 0.4\mathbf{k}) \text{ T}$$

Find the magnitide and direction of the instantaneous force on q.

▮ The Lorentz equation (Prob. 28.1) yields

$$\mathbf{F}_{\text{tot}} = (4 \times 10^{-5})[(6 \times 10^4)(0.52\mathbf{i} + 0.56\mathbf{j} + 0.645\mathbf{k}) + (5 \times 10^4)\mathbf{j} \times (0.693\mathbf{i} + 0.6\mathbf{j} + 0.4\mathbf{k})(1.7)]$$
$$= \underline{2.61\mathbf{i} + 1.34\mathbf{j} - 0.81\mathbf{k} \text{ N}}$$

From this, $F_{\text{tot}} = (2.61^2 + 1.34^2 + 0.81^2)^{1/2} = \underline{3.04 \text{ N}}$. The direction cosines of $\mathbf{F}_{\text{tot}}$ are

$$l = \frac{2.61}{3.04} = 0.86 \qquad m = \frac{1.34}{3.04} = 0.44 \qquad n = \frac{-0.81}{3.04} = -0.27$$

28.27 A capacitor is placed between the poles of a large magnet, as shown in Fig. 28-6. In the space between the plates of the capacitor there is a uniform electric field $\mathbf{E}$ and also a uniform magnetic field $\mathbf{B}$; the two fields are parallel and in the direction of Z. A small sphere, carrying charge $q = 3\ \mu$C, has instantaneous velocity

$$\mathbf{u} = (4 \times 10^4)(0.766\mathbf{j} + 0.643\mathbf{k}) \text{ m/s}$$

If $d = 40$ mm, $V = 2000$ V, and $B = 1.5$ T, what is the instantaneous force on the sphere?

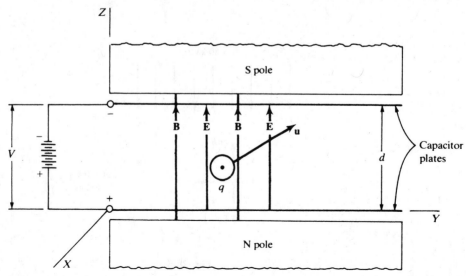

Fig. 28-6

❚ Here,

$$\mathbf{E} = \frac{V}{d}\mathbf{k} = (5 \times 10^4)\mathbf{k} \quad \text{V/m}$$

and the Lorentz equation gives

$$\mathbf{F}_{\text{tot}} = (3 \times 10^{-6})[(5 \times 10^4)\mathbf{k} + (4 \times 10^4)(0.766\mathbf{j} + 0.643\mathbf{k}) \times (1.5\mathbf{k})] = (3 \times 10^{-6})[(5 \times 10^4)\mathbf{k} + (4 \times 10^4)(0.766)(1.5)\mathbf{i}]$$

$$= \underline{0.1397\mathbf{i} + 0.15\mathbf{k}\,\text{N}}$$

28.28 The value of e/m_e can be obtained by using a specially designed vacuum tube illustrated in Fig. 28-7. It contains a heated filament F and an anode A which is maintained at a positive potential relative to the filament by a battery of known voltage V. Electrons evaporate from the heated filament and are accelerated to the anode, which has a small hole in the center for the electrons to pass through into a region of constant magnetic field B, which points into the paper. The electrons then move in a semicircle of diameter d, hitting the detector as shown. Prove that $e/m_e = 8V/(Bd)^2$.

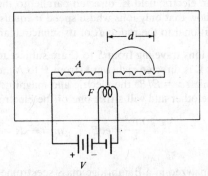

Fig. 28-7

❚ Refer to Fig. 28-7. Newton's second law implies that $(m_e v^2)/r = evB$, where $r = \frac{1}{2}d$ is the radius of the electron's orbit. Solving for the speed, we find that

$$v = \frac{eBr}{m_e} = \frac{eB}{2m_e}d \tag{1}$$

Since the electrons have been accelerated essentially from rest through a potential difference V, each has kinetic energy $\frac{1}{2}m_e v^2 = eV$, so that

$$v^2 = \frac{2eV}{m_e} \tag{2}$$

Combining Eqs. (1) and (2), we obtain

$$\left(\frac{eBd}{2m_e}\right)^2 = \frac{2eV}{m_e}$$

Solving for the charge-to-mass ratio, we have

$$\frac{e}{m_e} = \frac{8V}{B^2 d^2} \tag{3}$$

which was to be shown.

28.29 If in a mass spectrograph carbon ions move in a circle of radius $r_C = 9.0$ cm and oxygen ions move in a circle of radius $r_O = 10.4$ cm, what is the mass of an oxygen ion?

❚ We assume that V and B are the same for both ions, so that from Eq. (3) of Prob. 28.28, $m = (B^2 r^2 e)/(2V)$ and

$$\frac{m_O}{m_C} = \frac{r_O^2}{r_C^2} = \frac{(10.4 \text{ cm})^2}{(9.0 \text{ cm})^2} = 1.33$$

That is, the mass of an oxygen ion is 1.33 times the mass of a carbon ion. Since the mass of carbon is defined to be exactly 12 u, the mass of oxygen is $m_O = 1.33 m_C = (1.33)(12 \text{ u}) = \underline{16 \text{ u}}$.

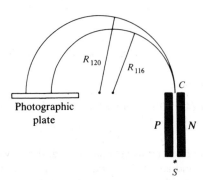

Fig. 28-8

28.30 Figure 28-8 represents the device designed by Bainbridge to accurately measure the masses of isotopes. S is the source of positively charged ions of the element being investigated. The ions all have the same charge e, but they have a range of speeds. Throughout the region a uniform magnetic field $\mathbf{B}$ is directed into the plane of the page. In addition, an electric field $\mathbf{E}$, directed parallel to the plane of the page, is set up between electrodes P and N. (**a**) Show that only ions whose speed v equals E/B will emerge at C. (**b**) Show that the mass m of an ion is proportional to the radius R of its semicircular trajectory.

▌ (**a**) Refer to Fig. 28-8. Ions traveling from S to C are subject to an electric force $\mathbf{F}_e = q\mathbf{E}$ and a magnetic force $\mathbf{F}_m = q(\mathbf{v} \times \mathbf{B})$. Here $\mathbf{E}$ is directed rightward (from P to N) and $\mathbf{v} \times \mathbf{B}$ points in the opposite direction and has magnitude vB. Unless $v = E/B$, the electric and magnitude forces do not cancel and the ions will be deflected to one side or the other and will strike one of the electrodes. (**b**) From

$$\frac{mv^2}{R} = evB, \qquad m = \frac{eB}{v}R = \frac{eB^2}{E}R$$

28.31 The element tin is being analyzed in a Bainbridge mass spectrometer such as that shown in Fig. 28-8. Among the isotopes present are those of masses 116, 117, 118, 119, and 120 u. The electric and magnetic fields are $E = 20\,\text{kV/m}$ and $B = 0.25\,\text{T}$. What is the spacing between the marks produced on the photographic plate by ions of tin-116 and ions of tin-120?

▌ From Prob. 28.30(*b*), the distance x from point C to the image of an isotope is given by $x = 2R = [2E/(eB^2)]\,m$. Hence,

$$\Delta x = \frac{2E}{eB^2}\,\Delta m = \frac{2(2.0 \times 10^4\,\text{V/m})}{(1.6 \times 10^{-19}\,\text{C})(0.25\,\text{T})^2}\,(4\,\text{u})(1.66 \times 10^{-27}\,\text{kg/u}) = 2.66 \times 10^{-2}\,\text{m} = \underline{26.6\,\text{mm}}$$

28.32 A particle with charge q and mass m orbits perpendicular to a uniform magnetic field $\mathbf{B}$. Show that its frequency of orbital motion is $(Bq)/(2\pi m)$ Hz. The fact that the frequency is independent of the particle's speed is important in particle accelerators called *cyclotrons*; this frequency is called the *cyclotron frequency*.

▌ The period is $(2\pi r)/v$, so $f = v/(2\pi r)$. Using $qvB = (mv^2)/r$, we have $f = (qB)/(2\pi m)$.

28.33 Describe a cyclotron and its operation.

▌ A cyclotron is a device for accelerating nuclear particles. The heart of the apparatus consists of a split metal pillbox. Figure 28-9 shows top and side views of the halves called Dees. A rapidly oscillating potential difference is applied between the Dees. This produces an oscillating electric field in the gap between the Dees, the region inside each Dee being essentially free of electric field. The Dees are enclosed in an evacuated container, and the entire unit is placed in a uniform magnetic field $\mathbf{B}$ whose direction is normal to the plane of the Dees. A charged particle of mass m and charge q in the gap between the Dees is accelerated by the electric field toward one of them. Inside the Dees, it moves with constant speed in a semicircle.

From Prob. 28.32 the period of uniform circular motion is $T = 1/f = (2\pi m)/(qB)$. For half a circle $t = T/2$ or $t = (\pi m)/(qB)$ and is independent of the speed. If the half-period of the oscillating electric field is equal to this time, the charged particle will again be accelerated when it next crosses the gap, because of the reversal of the direction of the electric field. Thus it will gain energy. This makes the next semicircle it traverses have larger radius, as shown in the figure. The energy gain can be repeated many times.

28.34 A cyclotron is accelerating deuterons which are nuclei of heavy hydrogen carrying a charge of $+e$ and having mass of 3.3×10^{-27} kg. (**a**) What is the required frequency of the oscillating electric field if $B = 1.5\,\text{T}$? (**b**) If

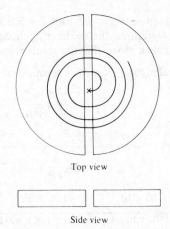

Top view

Side view **Fig. 28-9**

the deuterons are to acquire 15 MeV of kinetic energy and the difference of potential across the gap is 50 kV, how many times does the deuteron undergo acceleration?

▌ (a) The oscillation period of the electric field must equal the orbital period, so the required oscillation frequency is

$$v = \frac{1}{T} = \frac{qB}{2\pi m_0} = \frac{(1.60 \times 10^{-19})(1.5)}{2\pi(3.3 \times 10^{-27})} = 11.6 \text{ MHz}$$

(b) Each time the deuteron crosses the gap, it gains $50 \text{ keV} = 5 \times 10^4 \text{ eV}$. In order to gain a total of $15 \text{ MeV} = 15 \times 10^6 \text{ eV}$, the deuteron must undergo $(15 \times 10^6)/(5 \times 10^4) = \underline{300}$ gap crossings.

28.35 Refer to Prob. 28.34. What is the radius of the semicircle if the deuterons acquire 15 MeV of energy?

▌ Since $15 \text{ MeV} \ll m_0 c^2$, we may apply the newtonian formula obtained in Prob. 28.15:

$$r = \frac{1}{B} \sqrt{\frac{2Vm}{q}} = \frac{1}{1.5} \sqrt{\frac{2(15 \times 10^6)(3.3 \times 10^{-27})}{1.6 \times 10^{-19}}} = \underline{0.524 \text{ m}}$$

28.36 A cyclotron has been adjusted to accelerate deuterons. It is now to be adjusted to accelerate protons instead, which have almost exactly half the deuteron mass. (a) What change must be made if there is no change in the frequency of the oscillating potential difference applied between the Dees? (b) What change must be made if there is no change in the strength of the magnetic field applied normal to the Dees?

▌ (a) As given in Prob. 28.32, the cyclotron angular frequency $\omega_c = (qB)/m$, so we have

$$B = \frac{m\omega_c}{q} \tag{1}$$

Since the proton and the deuteron have the same charge ($q_p = q_d$) and $m_p = \frac{1}{2}m_d$, the magnetic field must be halved.
(b) Refer to Eq. (1). If B is to remain unchanged, the oscillation frequency of the field must be doubled.

28.37 How will each of the changes in Prob. 28.36 alter the maximum energy which the protons can acquire?

▌ Assuming that nonrelativistic mechanics is applicable throughout the motion, the maximum kinetic energy $K_{max} = \frac{1}{2}mv_{max}^2$, and $qv_{max}B = (mv_{max}^2)/R$, where R is the radius of the device. (The quantity R is an upper limit for the orbital radius of an accelerated particle.) Solving for K_{max}, we find that $K_{max} = \frac{1}{2}(q^2B^2R^2)/m$.
(a) The maximum kinetic energy is halved: $K_{max}(p) = [(\frac{1}{2})^2/(\frac{1}{2})]K_{max}(d) = \frac{1}{2}K_{max}(d)$.
(b) The maximum kinetic energy is doubled: $K_{max}(p) = [1/(\frac{1}{2})]K_{max}(d) = 2K_{max}(d)$.

28.38 In a cyclotron resonance experiment, the magnetic field is directed upward. The results indicate that the charged particles are circulating counterclockwise as viewed from above. What is the sign of the charge on the particles?

▌ Negative (force must be directed to center of circle).

28.39 "Thermal electrons", having kinetic energy $\frac{3}{2}kT$ with $T = 300$ K, move perpendicular to a constant magnetic field of magnitude $B = 0.10$ T. Determine the radius of their circular paths. When substituting numerical values, include their units and check to see that the result is indeed given in meters.

❚ Since the particles are nonrelativistic, Newton's second law implies that $(m_e v^2)/r = evB$, so that

$$r = \frac{m_e v}{eB} \qquad (1)$$

But $v = \sqrt{(3kT)/m_e}$, so Eq. (1) becomes

$$r = \frac{1}{eB}\sqrt{3m_e kT} \qquad (2)$$

With $m_e = 9.11 \times 10^{-31}$ kg, $e = 1.60 \times 10^{-19}$ C, $k = 1.38 \times 10^{-23}$ J/K, $T = 300$ K, and $B = 0.10$ T, Eq. (2) yields

$$r = \frac{\sqrt{3(9.11 \times 10^{-31}\text{ kg})(1.38 \times 10^{-23}\text{ J/K})(300\text{ K})}}{(1.60 \times 10^{-19}\text{ C})(0.1\text{ T})} = 6.65 \times 10^{-6}(\text{J} \cdot \text{kg})^{1/2}/\text{C} \cdot \text{T}$$

Since $1\text{ J} = 1\text{ kg} \cdot \text{m}^2/\text{s}^2$, we have $(\text{J} \cdot \text{kg})^{1/2} = \text{kg} \cdot \text{m/s}$. Furthermore, $1\text{ T} \equiv 1\text{ N} \cdot \text{s/C} \cdot \text{m}$, so that $1\text{ C} \cdot \text{T} = 1\text{ N} \cdot \text{s/m} = 1(\text{kg} \cdot \text{m/s}^2) \cdot \text{s/m} = 1\text{ kg/s}$. Therefore $1(\text{J} \cdot \text{kg})^{1/2}/\text{C} \cdot \text{T} = 1\text{ m}$, which verifies that $r = \underline{6.65\ \mu\text{m}}$.

28.40 Refer to Prob. 28.39. How long does it take the particles to complete one orbit?

❚ Using Eq. (1) in Prob. 28.39, we obtain

$$T = \frac{2\pi r}{v} = \frac{2\pi m_e}{eB} = \frac{2\pi(9.11 \times 10^{-31})}{(1.60 \times 10^{-19})(0.1)} = \underline{0.358\text{ ns}}$$

28.41 A device used to measure magnetic fields makes use of the Hall effect. When in a magnetic field of 200 G it gives a Hall voltage of 16 μV. If with the same current and orientation it gives a Hall voltage of 23 μV in an unknown field, what is the magnitude of the unknown field?

❚ Since the Hall voltage is directly proportional to B, the unknown field is $(200\text{ G})(\frac{23}{16}) = \underline{287.5\text{ G}}$.

28.42 In a certain Hall-type experiment a current of 0.25 A is sent through a metal strip having thickness 0.20 mm and width 5 mm. The Hall voltage is measured to be 0.15 mV when a magnetic field of 2000 G is used. **(a)** What is the number of charge carriers (assume $q = e$) per unit volume, and **(b)** what is the drift speed of these carriers?

❚ For the Hall effect $nq = (BJ)/E = (BJd)/V_H$, where n = number of charges per unit volume and J equals current density. With $J = 0.25/[(2 \times 10^{-4})(5 \times 10^{-3})] = 25 \times 10^4$ A/m^2, the calculation for n yields $\underline{1.04 \times 10^{25}/\text{m}^3}$. The drift velocity can be found from either $v = J/(ne)$; or $v = V_H/(Bd)$ to give $v = \underline{0.15\text{ m/s}}$.

28.2 FORCE ON AN ELECTRIC CURRENT

28.43 What is the force on a current in a magnetic field?

❚ Consider a wire of arbitrary shape, as ab in Fig. 28-10, carrying current I in a magnetic field **B** which may

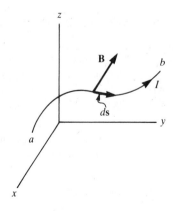

Fig. 28-10

vary from point to point. It follows directly from (1) of Prob. 28.1 that an element of length ds experiences a force $d\mathbf{F}$ given by

$$d\mathbf{F} = I(d\mathbf{s} \times \mathbf{B}) \tag{1}$$

Here the vector $d\mathbf{s}$ is of magnitude ds and is in the direction of motion of positive charge (conventional current). Actually, (1) gives the force on the charge carriers within the length ds. However, this force is converted, by collisions, into a force on the wire as a whole—a force which, moreover, is capable of doing work on the wire.

The net force on the wire ab is obtained by integrating (1) from a to b. In the special case of a straight wire of length L in a uniform field $\mathbf{B}$, we can see without integrating that

$$\mathbf{F} = I(\mathbf{L} \times \mathbf{B}) \tag{2}$$

or $F = ILB \sin \theta$; in component form,

$$F_x = I(B_z L_y - B_y L_z) \qquad F_y = I(B_x L_z - B_z L_x) \qquad F_z = I(B_y L_x - B_x L_y)$$

28.44 A wire bearing a current of 10 A lies perpendicular to a uniform magnetic field. A force of 0.2 N is found to exist on a section of the wire 80 cm long. Determine the magnetic induction $\mathbf{B}$.

▐ For a straight segment of wire of length l,

$$F = IlB \sin 90° \qquad \text{or} \qquad 0.2 = 10(0.80)B \qquad B = 0.025 \text{ T}$$

The direction of $\mathbf{B}$ will be normal to the plane of the force and the wire.

28.45 Calculate the force on a straight wire 11 cm long carrying a current of 12 A when the wire is in a 200-μT magnetic field perpendicular to the wire.

▐ $$F = IlB = 12(0.11)(0.0002) = \underline{260\ \mu\text{N}}$$

28.46 At the equator, the earth's magnetic field is nearly horizontal, directed from the southern to northern hemisphere. Its magnitude is about 0.50 G. Find the force (direction and magnitude) on a 20-m wire carrying a current of 30 A parallel to the earth (**a**) from east to west, (**b**) from north to south.

▐ Recall that $1 \text{ T} = 10^4 \text{ G}$.
(**a**) $F = ILB \sin \theta = 30(20)(5 \times 10^{-5})(1) = \underline{0.030 \text{ N}}$, down
(**b**) $F = 30(20)(5 \times 10^{-5})(0) = \text{zero}$

28.47 In Nebraska, the horizontal component of the earth's field is 0.20 G. If a vertical wire carries a current of 30 A upward there, what is the magnitude and direction of the force on 1 m of the wire?

▐ The vertical component of $\mathbf{B}$ is parallel to the current and does not contribute to the force; therefore, $F = ILB_H = (30 \text{ A})(1 \text{ m})(2 \times 10^{-5} \text{ T}) = 6 \times 10^{-4} \text{ N}$, west.

28.48 Find the force on each segment of the wire shown in Fig. 28-11, if $B = 0.15$ T. Assume that the current in the wire is 5 A.

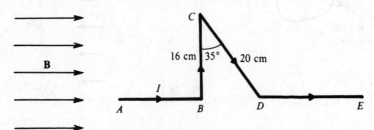

Fig. 28-11

▐ For each straight segment $\mathbf{F} = I\mathbf{L} \times \mathbf{B}$, where $\mathbf{L}$ is the directed line segment. In sections AB and DE, $\mathbf{L}$ and $\mathbf{B}$ are parallel, $\sin \theta = 0$, and $F = 0$. In section BC, $F = ILB = (5 \text{ A})(0.16 \text{ m})(0.15 \text{ T}) = \underline{0.12 \text{ N}}$, into page. In section CD, $F = (5 \text{ A})(0.20 \text{ m})(0.15 \text{ T}) \sin 65° = \underline{0.136 \text{ N}}$, out of page.

28.49ᶜ A wire in the form of a semicircle lies on the top of a smooth table. A downward-directed uniform magnetic field of magnitude B is confined to the region above the dashed line in Fig. 28-12. The ends of the semicircle

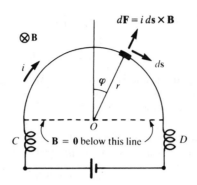

Fig. 28-12

are attached to springs C and D, whose other ends are fixed. The current i is introduced by attaching a battery to the ends of the springs. Show that the sum of the tension of the springs is $2Bir$, where r is the radius of the semicircle.

I There is a radially outward magnetic force $d\mathbf{F}$ on an element ds of the semicircle. The force has magnitude $dF = iB\, ds = iBr\, d\varphi$. When these forces are summed over the entire semicircle, the components parallel to the diameter of the semicircle cancel. The net magnetic force is along OX, of magnitude

$$F_m = \int dF \cos \varphi = 2iBr \int_0^{\pi/2} \cos \varphi\, d\varphi = 2iBr(\sin \pi/2 - \sin 0) = 2iBr$$

In order for the loop to be in equilibrium, the springs C and D must exert a force equal and opposite to $\mathbf{F}_m$, so the sum of the spring tensions is $2iBr$, as desired. In fact, rotational equilibrium requires either tension to equal iBr.

28.50ᶜ   Figure 28-13 shows an arbitrary (nonplanar) current loop arbitrarily oriented in a *uniform* magnetic field $\mathbf{B}$. Prove that the resultant force on the loop is zero.

I The force on the indicated current element $I\, d\mathbf{R}$ is $d\mathbf{F} = I\, d\mathbf{R} \times \mathbf{B}$. Integrate around the loop to obtain the net force, remembering that $\mathbf{B}$ is a constant vector:

$$\mathbf{F} = \oint I\, d\mathbf{R} \times \mathbf{B} = I\left(\oint d\mathbf{R}\right) \times \mathbf{B} = I(\mathbf{0} \times \mathbf{B}) = 0$$

where we have used the fact that $\mathbf{R}$ returns to its original value after one traversal of the loop.

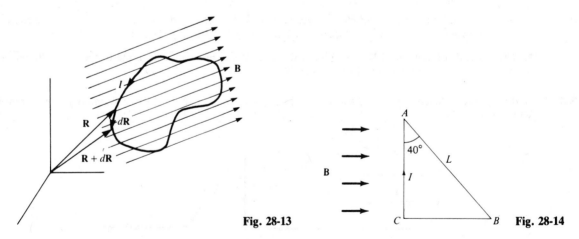

Fig. 28-13   Fig. 28-14

28.51   For the circuit shown in Fig. 28-14, find the magnitude and direction of the force on wire **(a)** AC, **(b)** BC, **(c)** AB, and **(d)** $ABCA$.

I **(a)** $F = I(L \cos 40°)B \sin 90° = 0.77ILB$ into the page. **(b)** $F = I(L \sin 40°)B \sin 0° = 0$. **(c)** $F = ILB \sin (90° - 40°) = 0.77ILB$ out of the page. **(d)** $\sum F = 0.77ILB$ in $+ 0 + 0.77ILB$ out $= 0$ (as must be, by Prob. 28.50).

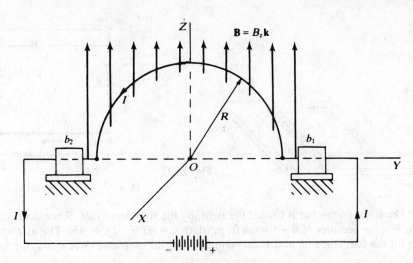

Fig. 28-15

28.52 Refer to Fig. 28-15. The semicircular wire of radius R, supported at b_1 and b_2 and carrying current I, is in a uniform magnetic field parallel to Z. The angle between the YZ plane and the plane of the semicircle is arbitrary. Find the total force on the wire.

▌ Rather than integrate along the wire, as in Prob. 28.51, we can exploit the result of Prob. 28.50. Hence we imagine current I also to flow along the Y axis from b_2 and b_1, completing the loop. The force on this straight segment would be $\mathbf{F}' = I(2R\mathbf{j} \times B_z\mathbf{k}) = 2IRB_z\mathbf{i}$. The actual force on the semicircle must be such as to cancel $\mathbf{F}'$; that is, $\mathbf{F} = -\mathbf{F}' = \underline{-2iRB_z\mathbf{i}}$.

28.53 As shown in Fig. 28-16(a), a long bar magnet has poles at its two ends. Near the poles, the magnetic field is radial. At the position of the circular current loop shown, the radial field is B. What is the magnitude and direction of the resultant force on the current loop?

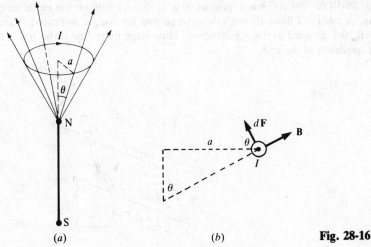

(a)   (b)   **Fig. 28-16**

▌ The situation, for a cross section of wire with current emerging from page, is as shown in Fig. 28-16(b). Force on a current segment $ds = BI\,ds$; by symmetry, forces in the plane of the current loop give zero when contributions around the loop are added. Vertically, integrating around the loop, $F = IB \sin \theta \times$ circumference $= \underline{2\pi aIB \sin \theta}$. This nonzero result reflects the fact that the field is nonconstant (in direction).

28.54 (a) As shown in Fig. 28-17, a metal bar of mass m can slide on two bare wires along a horizontal table. A current I is maintained in the wire and bar. The whole system is immersed in a vertical downward magnetic field B. If the frictional force acting on the bar is f and is small enough so that the bar will move, what is the acceleration of the bar? Will it move toward the right or toward the left? (b) Repeat if the magnetic field lines are parallel to the tabletop and make an angle of θ to the right of the bar.

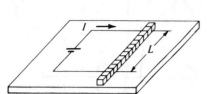

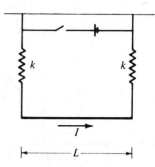

Fig. 28-17   ├────── L ──────┤   Fig. 28-18

▮ (a) The force on the bar is toward the right, by the right-hand rule. Since the friction force f opposes the motion, $F = ma$ becomes $ILB - f = ma$ from which $a = (ILB - f)/m$. (b) The magnetic force $ILB \sin \theta$ will tend to lift the bar. Since no horizontal component of the magnetic force exists, the bar will not slide.

28.55 As shown in Fig. 28-18, a bar of mass M is suspended by two wires. Assume that a uniform magnetic field **B** is directed into the page. Find the tension in each supporting wire when the current through the bar is I, as shown.

▮ The magnetic force is ILB and according to the RHR, it is directed upward. Equilibrium in the vertical direction yields $2T + ILB = Mg$, so that $T = (Mg - ILB)/2$.

28.56 As shown in Fig. 28-18, a bar of mass M is suspended by two springs. Assume that a magnetic field **B** is directed out of the page. Each spring has a spring constant k. Describe the bar's displacement when a current I is sent through it in the direction shown.

▮ From the RHR the magnetic force ILB is directed downward. This constant force shifts the equilibrium position downward by $(ILB)/(2k)$. (Two springs in parallel, each of constant k, are equivalent to one spring of constant $2k$.)

28.57 In Fig. 28-19 the bar AC has a mass of 50 g. It slides freely on the metal strips 40 cm apart at the edges of the incline. A current I flows through these strips and the bar, as indicated. There is a magnetic field $|B_y| = 0.20$ T directed in the $-y$ direction. How large must I be if the rod is to remain motionless? Neglect the slight overhang of the rod.

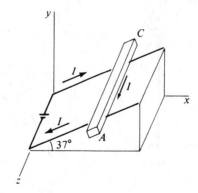

Fig. 28-19

▮ Magnetic force is along the positive x axis. If no motion is to occur along the incline, $\sum F = 0$:
$IL |B_y| \cos 37° = mg \sin 37°$ from which

$$I = \frac{mg \tan 37°}{L |B_y|} = \frac{(0.050)(9.8)(0.75)}{(0.40)(0.20)} = \underline{4.6 \text{ A}}$$

28.58 As measured in an inertial coordinate system, a uniform magnetic field of 0.3 T exists parallel to the Z axis. A straight wire, of length 250 mm and having direction cosines $l = 0.45$, $m = 0.56$, and $n = 0.6956$, carries a constant current of 50 A. Find the magnitude and direction of the net force on the wire.

▌ Write $\mathbf{B} = 0.3\mathbf{k}$ and $\mathbf{L} = (0.250)(0.45\mathbf{i} + 0.56\mathbf{j} + 0.6956\mathbf{k})$. Then, applying $\mathbf{F} = I\mathbf{L} \times \mathbf{B}$,

$$\mathbf{F} = 50[(0.250)(0.45\mathbf{i} + 0.56\mathbf{j} + 0.6956\mathbf{k}) \times (0.3\mathbf{k})] = 2.10\mathbf{i} - 1.688\mathbf{j} \quad \text{N}$$

Hence, $F = (2.1^2 + 1.688^2)^{1/2} = \underline{2.694 \text{ N}}$; direction cosines of $\mathbf{F}$ are

$$\alpha_1 = \frac{2.1}{2.694} = \underline{0.78} \qquad \alpha_2 = \frac{-1.688}{2.694} = \underline{-0.63} \qquad \alpha_3 = \underline{0}$$

28.59 A wire connects points a and b in space. It carries a current I from a to b and is in a uniform magnetic field $\mathbf{B}$. Prove that the force on the wire is $I\mathbf{L} \times \mathbf{B}$, where $\mathbf{L}$ is the vector displacement from a to b.

▌ This is another version of Prob. 28.50, and the proof goes the same way.

28.60 In Fig. 28-20, a three-sided frame if pivoted at AC and hangs vertically. Its sides are each of the same length and have a linear density of 0.10 kg/m. A current of 10.0 A is sent through the frame, which is in a uniform magnetic field of 10 mT directed upward. Through what angle will the frame be deflected?

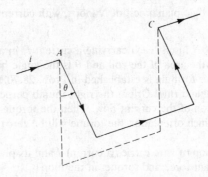

Fig. 28-20

▌ Referring to Fig. 28-20, we note that the magnetic forces on the slanting sides are parallel to AC and therefore do not produce any torque about the axis AC. The magnetic force on the horizontal side of the frame is

$$\mathbf{F}_{\text{mag}} = i\mathbf{L} \times \mathbf{B} \tag{1}$$

This force has magnitude $F_{\text{mag}} = iLB \sin 90° = iLB$ and points horizontally to the right in Fig. 28-20, perpendicular to $\mathbf{L}$ and to $\mathbf{B}$. The magnetic torque about AC has magnitude

$$T_m = iLB(L \cos \theta) = iL^2B \cos \theta \tag{2}$$

and tends to increase the angle θ. The gravitational torque about the axis AC is

$$T_g = -[(\lambda L)g(L \sin \theta) + 2(\lambda L)g(\tfrac{1}{2}L \sin \theta)]$$

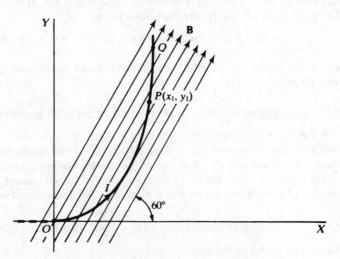

Fig. 28-21

where the minus sign indicates that gravity tends to turn the frame in the direction of decreasing θ. We have let λ denote the mass per unit length of the wire. In equilibrium, the torques evaluated in Eqs. (2) and (3) must cancel, which implies that $\tan \theta = (Bi)/(2\lambda g)$. Substituting the data, we obtain $\tan \theta = 0.0510$, or $\theta \approx 3°$.

28.61 A parabolic section of wire, Fig. 28-21, is located in the XY plane and carries current $I = 12$ A. A uniform field, $B = 0.4$ T, making an angle of 60° with X, exists throughout the plane. Compute the total force on the wire between the origin and the point $x_1 = 0.25$ m, $y_1 = 1.00$ m.

▮ In the notation of Prob. 28.59, we have $I = 12$ A, $\mathbf{L} = 0.25\mathbf{i} + 1.00\mathbf{j}$ m, and $\mathbf{B} = 0.2\mathbf{i} + 0.2\sqrt{3}\mathbf{j}$ T. Hence,

$$\mathbf{F} = I\mathbf{L} \times \mathbf{B} = 12 \begin{vmatrix} \mathbf{i} & \mathbf{j} & \mathbf{k} \\ 0.25 & 1.00 & 0 \\ 0.2 & 0.2\sqrt{3} & 0 \end{vmatrix} = -1.36\mathbf{k} \quad \text{N}$$

28.3 TORQUE AND MAGNETIC DIPOLE MOMENT

28.62 What torque is experienced by a planar coil of N loops, with current I in each loop, when it sits in a constant external magnetic field?

▮ The torque τ on a coil of N loops, each carrying a current I, in an external magnetic field B is $\tau = NIAB \sin \theta$, where A is the area of the coil and θ is the angle between the field lines and a perpendicular to the plane of the coil. (This formula is established in Prob. 28.70.) For the direction of rotation of the coil, we have the following right-hand rule: Orient the right thumb perpendicular to the plane of the coil, such that the fingers run in the direction of the current flow. Then the torque acts to rotate the thumb into alignment with the external field (in which orientation the torque will be zero).

28.63 As shown in Fig. 28-22, a loop of wire carries a current I and its plane is perpendicular to a uniform magnetic field $\mathbf{B}$. What are the resultant force and torque on the loop?

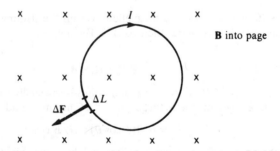

Fig. 28-22

▮ By Prob. 28.50, the resultant force on it is zero. From Fig. 28-22, the $\Delta\mathbf{F}$'s acting on the loop are trying to expand it, not rotate it. Therefore the torque on the loop is zero.

28.64 A coil of 20 turns has an area of 800 mm² and bears a current of 0.5 A. It is placed with its plane parallel to a magnetic field of intensity 0.3 T. Determine the torque on the coil.

▮ $\tau = 20IAB \sin 90° = 20(0.5)(800 \times 10^{-6})(0.3)(1) = 2.4 \times 10^{-3} \text{ N} \cdot \text{m}$

28.65 The 40-loop coil shown in Fig. 28-23 carries a current of 2 A in a magnetic field $B = 0.25$ T. Find the torque on it. How will it rotate?

▮ *Method 1* $\quad \tau = NIAB \sin \theta = (40)(2 \text{ A})(0.10 \text{ m} \times 0.12 \text{ m})(0.25 \text{ T})(\sin 90°) = \underline{0.24 \text{ N} \cdot \text{m}}$

(Remember that θ is the angle between the field lines and the perpendicular to the loop.) By the right-hand rule, the coil will turn about a vertical axis in such a way that side ad moves out of the page.

Method 2 Because sides dc and ab are in line with the field, the force on each of them is zero, while the force on each vertical wire is $f = ILB = (2 \text{ A})(0.12 \text{ m})(0.25 \text{ T}) = 0.060$ N out of the page on side ad and into the page on side bc. Thus we have two forces, of magnitude $40f$ and separation 10 cm, comprising a couple. The torque is

$$\tau = (\text{force})(\text{separation}) = (40 \times 0.060 \text{ N})(0.10 \text{ m}) = \underline{0.24 \text{ N} \cdot \text{m}}$$

and tends to rotate side ad out of the page.

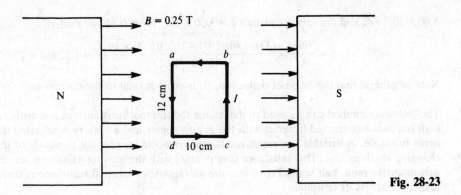

$B = 0.25$ T

N

12 cm

a   b

I

d 10 cm c

S

Fig. 28-23

28.66 The galvanometer of Fig. 28-24(a) is to have an internal resistance of 1.00 Ω. It is to be wound with No. 30 copper wire whose resistance is 338.6 Ω/km. The dimensions of the rectangular coil are 2.50 cm by 2.00 cm. (a) How many turns will the coil have? (b) The magnetic field in the galvanometer is to be 0.40 T. The spring constant k equals 5.00×10^{-6} N · m/rad. The galvanometer is to give full-scale deflection with a current of 1.00 mA. What is the full-scale deflection angle in radians?

▌ As can be seen from Fig. 28-24(b), for any position of the coil, its plane is parallel to the field B, near the coil wires that contribute to the torque.

(a) The length of a 1.00-Ω piece of No. 30 copper wire is 1 Ω/(338.6 Ω/km) = $(10^5/338.6)$ cm = 295.3 cm. Since the perimeter of the coil is 2(2.50 + 2.00) = 9.00 cm, the required number of turns N = 295.3/9.00 = 32.8 ≈ 33 turns.

(b) We equate the full-scale torques due to magnetic and spring forces: $N\mathscr{B}ia = k\theta$. Since the magnetic field $\mathscr{B} = 0.40$ T, the full-scale current $i = 1.00 \times 10^{-3}$ A, the coil area $a = (2.50 \times 10^{-2}$ m)$(2.00 \times 10^{-2}$ m) =

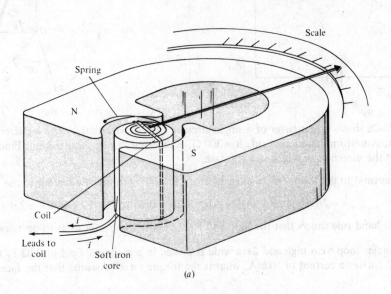

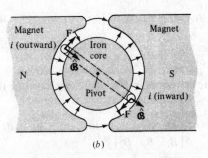

Fig. 28-24

$5.00 \times 10^{-4}\,\text{m}^2$, and the spring constant $k = 5.00 \times 10^{-6}\,\text{N} \cdot \text{m/rad}$, we find

$$\theta = \frac{N\mathcal{B}ia}{k} = \frac{(33)(0.40)(1.00 \times 10^{-3})(5.00 \times 10^{-4})}{(5.00 \times 10^{-6})} = 1.32\,\text{rad} = 75.6° \approx \underline{76°}$$

Note in general that the angle of deflection, θ, is proportional to the current i.

28.67 The following method can be used to determine the internal resistance of a sensitive galvanometer. A variable high resistance connected in series with the galvanometer and a battery is adjusted until the galvanometer reads full scale. A variable low resistance is now connected across the terminals of the galvanometer without changing anything else. This resistance is in parallel with the moving galvanometer coil. It is varied until the galvanometer reads half scale. Prove that the galvanometer internal resistance is than equal to the variable resistance across its terminals.

▮ By Prob. 28.66, galvanometer deflection is proportional to the current through the coil. The circuit is indicated schematically in Fig. 28-25. Since the low variable resistance r has a marked effect on the current through the galvanometer, r must be of the same order of magnitude as R_G, the unknown galvanometer coil resistance. Since the variable high resistance R is much larger than either R_G or r, the total resistance of the circuit is not significantly different from R and the current i flowing from the battery is essentially given by $i \approx V/R$, regardless of the value of r or R_G. Since the galvanometer and r are in parallel, we have $i_1 + i_2 = i$ and $i_1 R_G = i_2 r$. Hence when the low variable resistance r has been adjusted until $i_1 = i_2 = i/2$, we have $R_G = r$, as was to be shown.

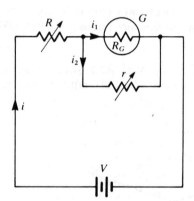

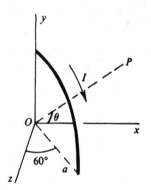

Fig. 28-25   **Fig. 28-26**

28.68 Figure 28-26 shows one quarter of a single circular loop of wire that carries a current of 14 A. Its radius is $a = 5\,\text{cm}$. A uniform magnetic field, $B = 300\,\text{G}$, is directed in the $+x$ direction. Find the torque on the entire loop and the direction in which it will rotate.

▮ The normal to the loop, OP, makes an angle $\theta = 60°$ with the $+x$ direction, the field direction. Hence,

$$\tau = NIAB \sin\theta = (1)(14\,\text{A})(\pi \times 25 \times 10^{-4}\,\text{m}^2)(0.03\,\text{T}) \sin 60° = \underline{2.9 \times 10^{-3}\,\text{N} \cdot \text{m}}$$

The right-hand rule shows that the loop will rotate about the y axis so as to decrease the angle labeled $60°$.

28.69 A rectangular loop 6 cm high and 2 cm wide is placed in a magnetic field of 0.02 T. If the loop contains 200 turns and carries a current of 50 mA, what is the torque on it? Assume that the face of the loop is parallel to the field.

▮ The torque is

$$\tau_m = nBIA = (200)(0.02\,\text{T})(50 \times 10^{-3}\,\text{A})(12 \times 10^{-4}\,\text{m}^2) = \underline{2.4 \times 10^{-4}\,\text{N} \cdot \text{m}}$$

28.70ᶜ Prove that the torque on an arbitrarily shaped planar coil of directed area $\mathbf{A}$, n turns and current I, in a uniform magnetic field $\mathbf{B}$, obeys $\boldsymbol{\tau} = nI\mathbf{A} \times \mathbf{B}$.

▮ Orienting the coordinate system as in Fig. 28-27, we have for the element of force on the coil

$$d\mathbf{F} = nI \begin{vmatrix} \mathbf{i} & \mathbf{j} & \mathbf{k} \\ dx & dy & 0 \\ B_x & B_y & B_z \end{vmatrix} = \mathbf{i}nIB_z\,dy - \mathbf{j}nIB_z\,dx + \mathbf{k}nI(B_y\,dx - B_x\,dy)$$

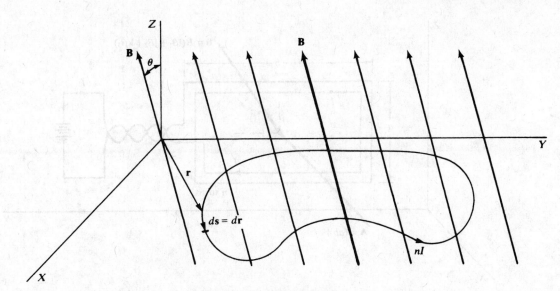

Fig. 28-27

Since the resultant force equals zero (Prob. 28.50), we may choose any convenient point—say, the origin—about which to calculate the torque. We have $d\tau = r \times dF$. Using the above expression for dF, this becomes

$$d\tau = \begin{vmatrix} i & j & k \\ x & y & 0 \\ nIB_z\,dy & -nIB_z\,dx & nI(B_y\,dx - B_x\,dy) \end{vmatrix}$$

$$= i(nIyB_y\,dx - nIyB_x\,dy) - j(nIxB_y\,dx - nIxB_x\,dy) + k(-nIxB_z\,dx - nIyB_z\,dy)$$

When this is integrated around the perimeter of the coil, all terms in $x\,dx = d(x^2/2)$ or $y\,dy = d(y^2/2)$ vanish, since $x^2/2$ and $y^2/2$ return to their starting values. We are left with

$$\tau = inIB_y \oint y\,dx + jnIB_x \oint x\,dy$$

But we know that the area of the coil is given by

$$A = \oint x\,dy = -\oint y\,dx$$

Thus, finally,

$$\tau = nIA(-iB_y + jB_x) = nIA(k \times B) = nI(A \times B) \tag{1}$$

where $A = Ak$ is the directed area of the planar coil. Clearly, (1) is independent of any particular coordinate system, and it implies that $\tau = nIAB\sin\theta$, where θ is the angle between A and B.

28.71 A rectangular coil of n turns, Fig. 28-28, is located in the YZ plane; its area is ab and it carries current I in each turn. In the region there is a uniform magnetic field B having direction cosines β_1, β_2, β_3. Given that $a = 100$ mm, $b = 150$ mm, $n = 20$, $I = 12$ A, $\beta_1 = 0.49$, $\beta_2 = 0.56$, $\beta_3 = 0.668$, and $B = 0.4$ T, find the magnitude and direction of the torque on the coil.

▮ We have $B = (0.4)(0.49i + 0.56j + 0.668k)$ T, and in the given position, the coil has directed area $A = (ab)i = 0.015i$ m². Then

$$\tau = (20)(12)[0.015i \times (0.4)(0.49i + 0.56j + 0.668k)] = -0.962j + 0.806k \quad N \cdot m$$

From this, $\tau = \underline{1.255\,N \cdot m}$, with direction cosines $\alpha_1 = \underline{0}$, $\alpha_2 = \underline{-0.766}$, $\alpha_3 = \underline{0.642}$.

28.72 Define the *magnetic dipole moment* of a current-carrying planar coil and express the potential energy of the coil associated with its orientation in a uniform magnetic field B.

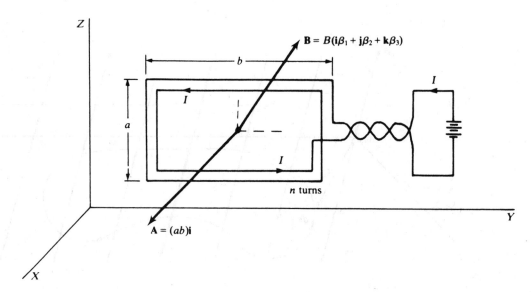

Fig. 28-28

▮ By definition $\boldsymbol{\tau} = \boldsymbol{\Pi} \times \mathbf{B}$, where $\boldsymbol{\Pi}$ is the magnetic moment of a coil or magnet in a uniform magnetic field, $\mathbf{B}$, consequently (Prob. 28.70), $\boldsymbol{\Pi} \equiv nI\mathbf{A}$. (Note that $\mathbf{M}$ and $\boldsymbol{\mu}$ are also commonly used symbols for magnetic moment.)

By rotating the coil from a position $\boldsymbol{\Pi} \perp \mathbf{B}$ to a position in which $\boldsymbol{\Pi}$ makes the angle θ with $\mathbf{B}$, and calculating the work done against the torque (the positive senses of τ and θ are opposite),

$$U_\theta = \int_{\pi/2}^{\theta} \tau\, d\theta = \int_{\pi/2}^{\theta} \Pi B \sin\theta\, d\theta = -\Pi B \cos\theta = -\boldsymbol{\Pi} \cdot \mathbf{B}$$

28.73 A planar coil of 12 turns carries 15 A. The coil is oriented with respect to the uniform magnetic field $\mathbf{B} = 0.2\mathbf{i} + 0.3\mathbf{j} - 0.4\mathbf{k}$ T such that its directed area is $\mathbf{A} = 0.04\mathbf{i} - 0.05\mathbf{j} + 0.07\mathbf{k}$ m^2. Find (**a**) the dipole moment of the coil, (**b**) the potential energy of the coil in the given orientation, and (**c**) the angle between the positive normal to the coil and the field.

▮ (**a**) $\qquad\qquad \boldsymbol{\Pi} = nI\mathbf{A} = (12)(15)(0.04\mathbf{i} - 0.05\mathbf{j} + 0.07\mathbf{k}) = \underline{7.2\mathbf{i} - 9.0\mathbf{j} + 12.6\mathbf{k}\quad \text{A} \cdot \text{m}^2}$

(**b**) $\qquad\qquad U_\theta = -\boldsymbol{\Pi} \cdot \mathbf{B} = -[(7.2)(0.2) + (-9.0)(0.3) + (12.6)(-0.4)] = \underline{+6.3\,\text{J}}$

(**c**) The desired angle θ is that between $\boldsymbol{\Pi}$ and $\mathbf{B}$. Since $U_\theta = -\Pi B \cos\theta$,

$$\cos\theta = -\frac{U_\theta}{\Pi B} = -\frac{6.3}{(7.2^2 + 9.0^2 + 12.6^2)^{1/2}(0.2^2 + 0.3^2 + 0.4^2)^{1/2}} = -0.6851$$

or $\theta = \underline{133.24°}$.

28.74 (**a**) Define the total magnetic flux Φ through an arbitrary surface S, and relate this to the flux in the electric-field case. (**b**) Calculate Φ for the coil of Prob. 28.73.

▮ (**a**) Just as electric flux ψ is associated with the electric field $\mathbf{E}$, so *magnetic flux* Φ is associated with the magnetic field $\mathbf{B}$. Thus, the magnetic flux through an elementary area dS is defined as $d\Phi = \mathbf{B} \cdot d\mathbf{S}$ Wb, and through surface S,

$$\Phi = \int_S \mathbf{B} \cdot d\mathbf{S}\quad \text{Wb}$$

Note that the SI unit of magnetic flux is the *weber*, where $1\,\text{Wb} = 1\,\text{T} \cdot \text{m}^2 = 1\,\text{V} \cdot \text{s} = 1\,\text{J/A}$. Not uncommonly, $\mathbf{B}$ (the "magnetic flux density") is given in Wb/m^2.

One can interpret Φ in terms of "lines of force" or "lines of flux," exactly as in the electric case. However, it can be shown that for magnetic fields Gauss' law becomes $\Phi_{\text{closed surface}} = 0$, which means that magnetic lines of force do not terminate on some sort of "magnetic charge" but close on themselves. (**b**) For a uniform magnetic field and a planar coil of n turns (with effective surface area nA),

$$\Phi = \int_S \mathbf{B} \cdot d\mathbf{S} = \mathbf{B} \cdot \int_S d\mathbf{S} = \mathbf{B} \cdot n\mathbf{A} = \mathbf{B} \cdot \frac{1}{I}\boldsymbol{\Pi} = -\frac{U_\theta}{I}$$

Substituting $U_\theta = 6.3\,J$ and $I = 15$ A, we get $\Phi = \underline{-0.42}$ Wb. The minus sign on Φ means flux in the direction *opposite to* that of **A**.

28.75 For the current loop shown in Fig. 28-14, find **(a)** the direction and magnitude of its **magnetic moment and (b)** the torque on the loop.

▌ **(a)** From the RHR magnetic moment **μ** is into the page with magnitude $IA = [I(L \sin 40°)(L \cos 40°)]/2 = 0.246IL^2$. **(b)** $\tau = \mu \times B = \mu B \sin 90° = 0.246IL^2B$.

28.76 The circular current loop of radius b shown in Fig. 28-29 is mounted rigidly on the axle, midway between the two supporting cords. In the absence of an external magnetic field the tensions in the cords are equal and are T_0 **(a)** What will be the tensions in the two cords when the vertical magnetic field **B** is present? **(b)** Repeat if the field is parallel to the axis.

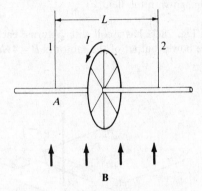

Fig. 28-29

▌ **(a)** $\tau = \mu B = \pi b^2 IB$, directed out of the page, so that the loop starts to rotate, decreasing the tension in cord 2 by an amount ΔT. But taking torques about A, obtain $L\,\Delta T = \mu B$, so $\Delta T = (\pi b^2 IB)/L$, and thus $T_1 = T_0 + \Delta T$ and $T_2 = T_0 - \Delta T$. **(b)** Since $\mu \times B = 0$, the tensions are the same as in the absence of a field.

28.77 In the Bohr model of the hydrogen atom the electron follows a circular path centered on the nucleus. Its speed is 2.2×10^6 m/s, and the radius of the orbit is 5.3×10^{-11} m. **(a)** Show that the effective current in the orbit is $ev/2\pi r$. **(b)** Show that $\mu = -(e/2m)\mathbf{L}$, where $L = mrv$ is the angular momentum of the electron in its orbit.

▌ **(a)** Since charge $-e$ passes a point once every revolution, $i = e/T$, where $T = (2\pi r)/v$, so that $i = (ev)/(2\pi r)$. **(b)** In magnitude, the dipole moment is

$$\mu = iA = \left(\frac{ev}{2\pi r}\right)(\pi r^2) = \frac{e}{2m}(mvr) = \frac{e}{2m}L$$

and because the electron is negatively charged, **μ** is antiparallel to **L**.

28.78 **(a)** A rigid circular loop of radius r and mass m lies in the xy plane on a flat table and has a current I flowing in it. At this particular place, the earth's magnetic field is $\mathbf{B} = B_x\mathbf{i} + B_y\mathbf{j}$. How large must I be before one edge of the loop will lift from the table? **(b)** Repeat if $\mathbf{B} = B_x\mathbf{i} + B_z\mathbf{k}$.

▌ **(a)** The torque on the loop must be equal to the gravitational torque exerted about an axis tangent to the loop. The gravitational torque $= mgr$ while magnetic torque $= |\mu \times B| = \mu B \sin 90° = \pi r^2 IB$. Equating gives $I = (mg)/[\pi r(B_x^2 + B_y^2)^{1/2}]$. **(b)** Only B_x causes a torque, so $I = (mg)/(\pi r B_x)$.

28.79 A circular loop of wire of radius r lies in the xy plane and carries a current I. Impinging on it is a magnetic field given by $\mathbf{B} = B_x\mathbf{i} + B_y\mathbf{j} + B_z\mathbf{k}$. Find the vector torque which acts on the coil due to the magnetic field. Two answers are possible; give both.

▌ Torque is $\mu \times B = I\pi r^2(\pm\mathbf{k}) \times \mathbf{B} = I\pi r^2[(\pm\mathbf{k}) \times (B_x\mathbf{i} + B_y\mathbf{j})] = \pm\pi r^2 I(B_x\mathbf{j} - B_y\mathbf{i})$. The $\pm$ arises since current can circulate in either direction.

28.80 Find the torque which acts on the rectangular current loop shown in Fig. 28-30 if a magnetic field $\mathbf{B} = B\mathbf{i}$ is impressed upon it. If allowed to move, will the loop turn so as to increase or decrease the angle ϕ?

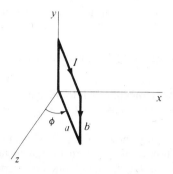

Fig. 28-30

▮ $\tau = \mu \times \mathbf{B} = [(IA \cos \phi)\mathbf{i} - (IA \sin \phi)\mathbf{k}] \times B\mathbf{i} = (-IAB \sin \phi)\mathbf{j}$, τ is in the $-y$-direction, thus decreasing ϕ, so the moment vector wants to align with the field.

28.81^c The electric motor shown in Fig. 28-31 has a coil with N turns, each of area a, and the current flowing through it is i. Show that the power output of the motor is $P = 4\nu NiaB$, where ν is the number of revolutions of the coil per second.

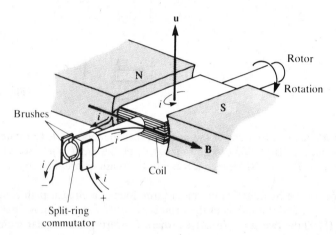

Fig. 28-31

▮ When μ makes angle θ with $\mathbf{B}$, the torque on the coil is $T(\theta) = NiaB \sin \theta$. The work done by this torque in each half-revolution is

$$\frac{1}{2}\Delta W = \int_0^\pi T(\theta) \, d\theta = NiaB \int_0^\pi \sin \theta \, d\theta = NiaB(-\cos \theta |_0^\pi) = 2NiaB$$

Hence the work output per revolution is $\Delta W = 4NiaB$. If there are ν revolutions per unit time, the power output $P = dW/dt = \nu \, \Delta W = \underline{4\nu NiaB}$.

28.82^c A uniformly charged disk whose total charge has magnitude $|q|$ and whose radius is r rotates with constant angular velocity of magnitude ω. Show that the magnetic dipole moment has the magnitude $\omega |q| r^2/4$.

▮ The surface charge density is $|q|/(\pi r^2)$. Hence the charge within a ring of radius R and width dR is

$$dq = \left(\frac{q}{\pi r^2}\right)(2\pi R \, dR) = \frac{2q}{r^2}(R \, dR)$$

The current carried by this ring is its charge divided by the rotation period:

$$di = \frac{dq}{(2\pi/\omega)} = \frac{q\omega}{\pi r^2}(R \, dR)$$

The magnetic moment contributed by this ring has magnitude $dM = a \, |di|$, where a is the area of the ring. Therefore

$$dM = \pi R^2 \, |di| = \frac{|q| \, \omega}{r^2}(R^3 \, dR)$$

Since the contributions $d\mathbf{M}$ from the various rings are all parallel, we have

$$M = \int dM = \int_{R=0}^{r} \frac{|q|\,\omega}{r^2}(R^3\,dR) = \frac{|q|\,\omega}{r^2}\left(\frac{R^4}{4}\Big|_{R=0}^{R=r}\right) = \frac{|q|\,\omega r^2}{4} \tag{1}$$

28.83 Refer to Prob. 28.82. (*a*) If the mass of the disk is m, what is the magnitude of its angular momentum? (*b*) What is the ratio of the magnitude of its magnetic dipole moment to the magnitude of its angular momentum? How does the sign of the charge govern the relation between the directions of these vector quantities?

▮ (*a*) *Assuming that the disk has a uniform mass distribution*, we find that its moment of inertia $I = \frac{1}{2}mr^2$. Hence its angular momentum has magnitude $L = \frac{1}{2}mr^2\omega$. (*b*) Using (1) from Prob. 28.82,

$$\frac{M}{L} = \left(\frac{|q|\,\omega r^2}{4}\right)\frac{1}{(\frac{1}{2}mr^2\omega)} = \frac{|q|}{2m}$$

If the charge $\mathbf{q}$ is positive, $\mathbf{M}$ and $\mathbf{L}$ are parallel. If the charge is negative, $\mathbf{M}$ and $\mathbf{L}$ are antiparallel. The relationship between $\mathbf{M}$ and $\mathbf{L}$ can be summarized by the equation

$$\mathbf{M} = \frac{q}{2m}\mathbf{L} \tag{1}$$

[Note: Compare Prob. 28.77. Equation (1) can be shown to be valid for any macroscopic object (uniformly charged or not!), provided only that the ratio of the local charge density to the local mass density is constant throughout the object.]

28.84ᶜ A spherical shell of radius a carries a uniform surface charge σ per unit area and spins with frequency f on an axis through its center. Show that the magnetic moment of the sphere is $\frac{8}{3}\pi^2 a^4 \sigma f$.

Assuming a uniform surface mass-density λ, we apply (1) of Prob. 28.83. The moment of inertia about the axis of rotation is (see Fig. 28-32)

$$I = \int_0^\pi (a\sin\theta)^2 \lambda(2\pi a\sin\theta)a\,d\theta = 2\pi\lambda a^4 \int_0^\pi \sin^3\theta\,d\theta = \frac{8}{3}\pi\lambda a^4$$

giving an angular momentum $L = I(2\pi f) = \frac{16}{3}\pi^2\lambda a^4 f$. Then

$$M = \frac{q}{2m}L = \frac{\sigma}{2\lambda}L = \frac{8}{3}\pi^2\sigma a^4 f$$

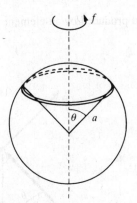

Fig. 28-32

28.85 An electron has an intrinsic (spin) magnetic dipole moment of magnitude 0.93×10^{-23} A·m². It has an intrinsic angular momentum of magnitude 0.53×10^{-34} kg·m²/s. What is the ratio of the magnitudes of its magnetic dipole moment to its angular momentum? Is this ratio equal to the value $e/2m_e$ predicted in Prob. 28.83 for a rotating macroscopic body with the same charge-to-mass ratio as an electron?

▮ The magnitude ratio of the magnetic moment to angular momentum is

$$\frac{M_e}{L_e} = \frac{0.93 \times 10^{-23}\text{ A·m}^2}{0.53 \times 10^{-34}\text{ kg·m}^2/\text{s}} = 1.8 \times 10^{11}\text{ C/kg} \qquad \text{while} \qquad \frac{e}{2m_e} = \frac{1.60 \times 10^{-19}\text{ C}}{2(9.11 \times 10^{-31}\text{ kg})} = 0.88 \times 10^{11}\text{ C/kg}$$

Hence M_e/L_e is approximately twice the expected value. This is an important result in the quantum physics of atomic systems.

28.86 Refer to Prob. 28.85. The intrinsic angular momentum could be attributed to a spinning motion of the electron. Assuming that the electron is a spinning sphere, calculate the speed of a point on the electron's equator from the value of its angular momentum magnitude and the value of its "classical radius," 2.8×10^{-15} m. Why isn't the model tenable?

▮ The moment of inertia of a homogeneous sphere about any axis through its center is $I = \frac{2}{5}mr^2$. Hence the angular momentum of a spinning sphere is $L = I\omega = \frac{2}{5}mr^2\omega$. Since the speed of a point on the sphere's equator is $v = r\omega$, the angular momentum can be written as $L = \frac{2}{5}mrv$. Solving this equation for v and using $L = 0.53 \times 10^{-34}$ kg·m^2/s, $m = 9.11 \times 10^{-31}$ kg, and $r = 2.8 \times 10^{-15}$ m, we find that

$$v = \frac{5L}{2mr} = \frac{5(0.53 \times 10^{-34})}{2(9.11 \times 10^{-31})(2.8 \times 10^{-15})} = 5.2 \times 10^{10} \text{ m/s}$$

which is nearly 200 times the speed of light. The model of an electron as a tiny homogeneous spinning sphere is not tenable because it requires (impossible) "superluminal" speeds.

28.4 SOURCES OF THE MAGNETIC FIELD; LAW OF BIOT AND SAVART

28.87 What is the magnetic field, **B**, that is created by a moving charge?

▮ Figure 28-33 shows a point charge q in motion with velocity **v** relative to inertial axes X, Y, Z. The instantaneous location of q with respect to the fixed observation point P is specified by the displacement vector **r** (or by $-$**r**). Then at P there exists a magnetic field, whose instantaneous value as determined in the inertial frame is

$$\mathbf{B} = \frac{\mu_0 q}{4\pi r^3}(\mathbf{v} \times \mathbf{r}) \quad (\text{T})$$

The constant μ_0, called the *permeability of empty space*, is, in the SI, not an experimental, but a defined quantity: $\mu_0 \equiv 4\pi \times 10^{-7}$ H/m. Here, the *henry* is the unit of inductance and 1 H/m = 1 T·m/A.
  In magnitude,

$$B = \frac{qv \sin \theta}{r^2} \times 10^{-7} \quad \text{T}$$

where θ is the angle between **v** and **r** (see Fig. 28-33). Observe that B is constant on circles centered on and perpendicular to the instantaneous line of motion of q, and that $B = 0$ along that line.

28.88 Derive the expression for the field produced by an element of current, from the expression for the field produced by a moving charge.

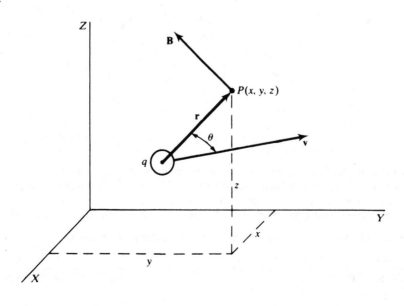

Fig. 28-33

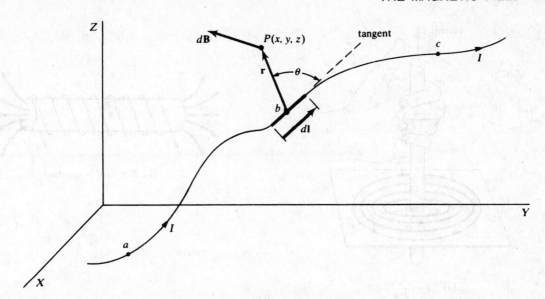

Fig. 28-34

▌ We consider an infinitesmal element of length dl of a wire carrying current I, as shown in Fig. 28-34. Let dq represent the total moving charges in dl at a given instant of time. If dt represents the time for all that charge to move out of the element dl, then $I = dq/dt$ and the drift velocity of the charge is: $\mathbf{v} = d\mathbf{l}/dt$. Thus, from $I\,dt = dq$ we have: $dq\,\mathbf{v} = I\,d\mathbf{l}$. Then from Prob. 28.87,

$$d\mathbf{B} = \frac{\mu_0\,dq}{4\pi r^3}(\mathbf{v} \times \mathbf{r}) = \frac{\mu_0 I}{4\pi r^3}(d\mathbf{l} \times \mathbf{r})$$

gives the field produced at an external point P, where $\mathbf{r}$ is the displacement from b, the site of the current element, to P. This last relationship is the *Biot–Savart law*.

28.89 Reexpress the law of Biot and Savart found in Prob. 28.88 without differentials and in terms of magnitude and direction.

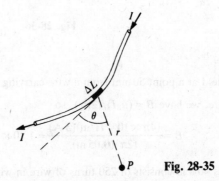

Fig. 28-35

▌ The current element of length ΔL shown in Fig. 28-35 contributes $\Delta \mathbf{B}$ to the field at P. The magnitude of $\Delta \mathbf{B}$ is

$$\Delta B = \frac{\mu_0 I\,\Delta L}{4\pi r^2}\sin\theta$$

where r and θ are defined in the figure. The direction of $\Delta \mathbf{B}$ is perpendicular to the plane determined by ΔL and r (the plane of the page). In the case shown, the right-hand rule tells us that $\Delta \mathbf{B}$ is out of the page.

When r is directed along ΔL, then $\theta = 0$ and thus $\Delta B = 0$, which means that the field due to a straight wire at a point on the line of the wire is zero.

28.90 Specify with the aid of diagrams the magnitude and direction of the magnetic field $\mathbf{B}$ at point P, which is (a) outside a long straight wire; (b) the center of a circular coil of N loops having radius a; (c) inside a long solenoid with n loops per meter; and (d) inside a toroid having N loops.

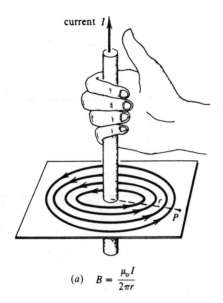

(a) $\quad B = \dfrac{\mu_o I}{2\pi r}$

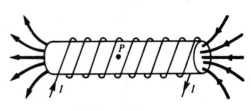

(c) $B = \mu_0 nI = $ constant

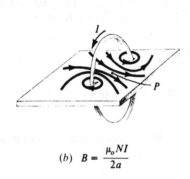

(b) $\quad B = \dfrac{\mu_o NI}{2a}$

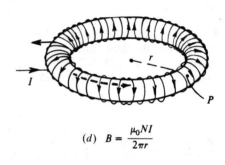

(d) $\quad B = \dfrac{\mu_0 NI}{2\pi r}$

Fig. 28-36

▮ See Fig. 28-36.

28.91 What is the magnetic field at a point 50 mm from a wire carrying a current of 3 A?

▮ Assuming a long wire, we have $B = (\mu_0 I)/(2\pi r)$, so

$$B = \frac{(4\pi \times 10^{-7}\,\text{H/m})(3\,\text{A})}{(2\pi)(0.05\,\text{m})} = 1.20 \times 10^{-5}\,\text{T} = \underline{0.12\,\text{G}}$$

28.92 A circular coil of radius 40 mm consists of 250 turns of wire in which the current is 20 mA. What is the magnetic field in the center of the coil?

▮ $$B = \frac{\mu_0 NI}{2a} = \frac{(4\pi \times 10^{-7}\,\text{H/m})(250)(20 \times 10^{-3}\,\text{A})}{2(0.040\,\text{m})} = 0.785 \times 10^{-4}\,\text{T} = \underline{0.785\,\text{G}}$$

28.93 Find the total magnetic field at point C produced by the two currents in Fig. 28-37(a). (*Hint*: Make a scale drawing to determine the directions of the component fields.)

▮ The magnitudes of the fields produced by wires 1 and 2 are

$$B_1 = (2 \times 10^{-7}\,\text{H/m})\frac{5\,\text{A}}{0.08\,\text{m}} = 12.5\,\mu\text{T} \qquad B_2 = (2 \times 10^{-7}\,\text{H/m})\frac{10\,\text{A}}{0.2\,\text{m}} = 10.0\,\mu\text{T}$$

In Fig. 28-37(b), circles of radii $r_1 = 2$ units and $r_2 = 5$ units are drawn. The vectors $\mathbf{B}_1$ and $\mathbf{B}_2$ are tangent to these circles at point C where they intersect. The sum $\mathbf{B}$ is found by vector addition using protractor and ruler or trigonometry: $B = \underline{9.0\,\mu\text{T}}$.

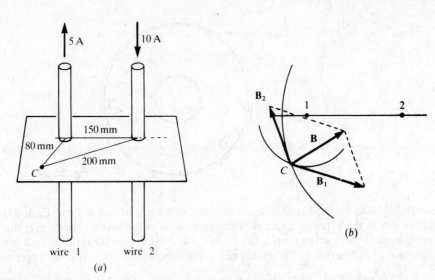

(a)

(b)

Fig. 28-37

28.94 Figure 28-38 shows two long parallel wires separated by a distance of 180 mm. There is a current of 8 A in wire 1 and a current of 12 A in wire 2. (a) Find the total magnetic field at the point A, which is on the line joining the wires and 30 mm from wire 1 and 150 mm from wire 2. (b) At what point on the line joining the wires is the magnetic field zero?

▌ (a) $B_1 = (2 \times 10^{-7} \text{ H/m}) \dfrac{8 \text{ A}}{0.030 \text{ m}} = 53.3 \ \mu\text{T}$ $B_2 = (2 \times 10^{-7} \text{ H/m}) \dfrac{12 \text{ A}}{0.150 \text{ m}} = 16.0 \ \mu\text{T}$

B_1 and B_2 are in opposite directions at A, so the magnitude of their sum is $B = B_1 - B_2 = \underline{37.3 \ \mu\text{T}}$.
(b) The total field is zero at the point between the wires where

$$\frac{I_1}{r_1} = \frac{I_2}{r_2} \quad \text{or} \quad \frac{r_2}{r_1} = \frac{I_2}{I_1} = \frac{12 \text{ A}}{8 \text{ A}} = 1.5$$

Since $r_1 + r_2 = 180$ mm, we obtain $r_1 = \underline{72 \text{ mm}}$.

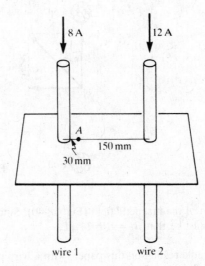

Fig. 28-38

28.95 A coil of radius 200 mm is to produce a field of 0.4 G in its center with a current of 0.25 A. How many turns must there be in the coil?

▌ From $B = (\mu_0 N I)/(2a)$ we have:

$$N = \frac{2aB}{\mu_0 I} = \frac{2(0.200 \text{ m})(0.4 \times 10^{-4} \text{ T})}{(4\pi \times 10^{-7} \text{ H/m})(0.25 \text{ A})} = \underline{51}$$

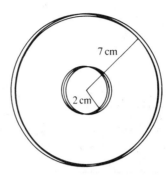

Fig. 28-39

28.96 Figure 28-39 shows a coil of radius 2 cm concentric with a coil of radius 7 cm. Each coil has 100 turns. With a current of 5 A in the larger coil, find the currents needed in the smaller coil to give the following values for the total magnetic field at the center: (a) 9.0 mT, (b) 2.0 mT, and (c) zero. In each case, determine whether the direction of the current in the smaller coil is the same as the current in the larger coil or opposite.

$$B_1 = \frac{\mu_0 N I_1}{2a_1} = \frac{(4\pi \times 10^{-7}\,\text{H/m})(100)(5\,\text{A})}{2(0.07\,\text{m})} = 4.49\,\text{mT} \qquad B_2 = \frac{\mu_0 N I_2}{2a_2} = \frac{(4\pi \times 10^{-7}\,\text{H/m})(100)I_2}{2(0.02\,\text{m})} = (3.14\,\text{mT/A})I_2$$

(a) Since $B > B_1$, $B_2 = B - B_1 = 4.51\,\text{mT}$, then $I_2 = (4.51\,\text{mT})/(3.14\,\text{mT/A}) = \underline{1.44\,\text{A}}$ and has the same direction as I_1. (b) Since $B < B_1$, $B_2 = B_1 - B = 2.49\,\text{mT}$, then $I_2 = (2.49\,\text{mT})/(3.14\,\text{mT/A}) = \underline{0.793\,\text{A}}$ and has the opposite direction as I_1. (c) $B_2 = 4.49\,\text{mT}$, so $I_2 = (4.49\,\text{mT})/(3.14\,\text{mT/A}) = \underline{1.43\,\text{A}}$ and has the opposite direction as I_1.

28.97 A circular coil having ten turns and a radius of 120 mm is placed with its plane parallel to the earth's magnetic field. A compass needle placed at the center of the coil points at an angle of 45° with the plane of the coil when a current of 0.45 A flows through the coil. Calculate the intensity of the earth's magnetic field.

First find the field intensity **B** produced at the center of the coil by the current.

$$B = \frac{2\pi NI}{10^7 a} = \frac{2\pi (10)(0.45)}{10^7 (0.12)} = 24\,\mu\text{T}$$

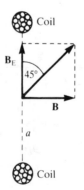

Fig. 28-40

Now draw a vector diagram of the magnetic field (Fig. 28-40). Since the compass needle points along the resultant field, $\mathbf{B}_E + \mathbf{B}$, it must be that $B_E = B = 24\,\mu\text{T}$.

28.98 Imagine a light compass needle resting on this paper, with a wire placed slightly above and parallel to the needle. If a current toward the north is established in the wire, how will the S pole of the needle become oriented?

In accordance with the right-hand rule, the magnetic field of the current below the wire is toward the west. The S pole of the compass points in a direction opposite to that of the magnetic field, that is, to the east.

28.99 A solenoid 0.5 m long has 2000 turns. The magnetic induction near the center of the solenoid is 0.08 T. What is the current in the solenoid?

▋ For a long solenoid, $B = \mu_0 nI$, or

$$I = \frac{B}{\mu_0 n} = \frac{0.08\ \text{T}}{(4\pi \times 10^{-7}\ \text{H/m})(4000\ \text{m}^{-1})} \approx \underline{16\ \text{A}}$$

28.100 A circular loop of 50 mm radius and 30 turns carries a current of 0.5 A. Find the magnetic induction at the center of the loop.

▋
$$B = \frac{2\pi IN}{10^7 a} = \frac{2\pi (0.5)(30)}{10^7 (0.050)} = \underline{190\ \mu\text{T}}.$$

28.101 A flat circular coil having 40 loops of wire on it has a diameter of 320 mm. What current must flow in its wires to produce a flux density of 300 μWb/m^2 at its center?

▋
$$B = \frac{\mu_0 NI}{2r} \quad \text{or} \quad 3 \times 10^{-4}\ \text{T} = \frac{(4\pi \times 10^{-7}\ \text{T} \cdot \text{m/A})(40)I}{0.320\ \text{m}}$$

which gives $I = \underline{1.9\ \text{A}}$.

28.102 Five very long, straight, insulated wires are closely bound together to form a small cable. Currents carried by the wires are $I_1 = 20$ A, $I_2 = -6$ A, $I_3 = 12$ A, $I_4 = -7$ A, $I_5 = 18$ A (negative currents are opposite in direction to the positive). Find B at a distance of 10 cm from the cable.

▋ By superposition the field is just the sum of the fields due to the individual currents. At $r = 10$ cm all the fields are either parallel or antiparallel as the currents are parallel or antiparallel. Then

$$B = \frac{\mu_0}{2\pi r}(I_1 + I_2 + I_3 + I_4 + I_5) = (2 \times 10^{-7}\ \text{H/m})37\ \text{A}/0.10\ \text{m} = 74\ \mu\text{T}$$

28.103 A horizontal wire is lined up in the north-south direction. A compass needle is placed above the wire. When the current is turned on, the North pole of the compass is deflected toward the west. In which direction are electrons in the wire moving?

▋ The westward deflection of the compass needle indicates that the field *above the wire* must be directed toward the west. By the right-hand rule, in order to produce this field, the current in the wire must flow from north to south. Since the motion of the (negative) conduction electrons is opposite to the current flow, the electrons are moving northward.

28.104 A long, straight, nonconducting string, painted with charge at a density of 40 μC/m, is pulled along its length at a speed of 300 m/s. What is the magnetic field at a normal distance of 5 mm from the moving string?

▋ The moving string behaves just like a current with $I = (40 \times 10^{-6}\ \text{C/m})(300\ \text{m/s}) = 1.2 \times 10^{-2}$ A. Then

$$B = \frac{\mu_0 I}{2\pi r} = \frac{(2 \times 10^{-7})(1.2 \times 10^{-2})}{0.005} = \underline{4.8 \times 10^{-7}\ \text{T}}$$

28.105 Two long and fixed parallel wires, A and B, are 10 cm apart in air and carry currents of 40 and 20 A, respectively, in opposite directions. Determine the resultant flux density (**a**) on a line midway between the wires and parallel to them, and (**b**) on a line 8 cm from wire A and 18 cm from wire B.

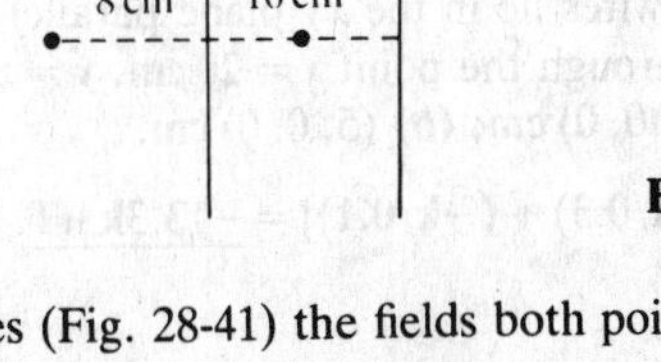

$$I_A = 40\ \text{A} \qquad I_B = 20\ \text{A}$$

$$8\ \text{cm} \qquad 10\ \text{cm}$$

Fig. 28-41

▋ (**a**) At the midpoint between the wires (Fig. 28-41) the fields both point into the page and hence reinforce:

$$B = B_A + B_B = \frac{(2 \times 10^{-7})(40 + 20)}{0.05} = \underline{2.4 \times 10^{-4}\ \text{T}}$$

(b) Here B_A points out of the page and B_B into the page, so:

$$B = (2 \times 10^{-7})\left(\frac{40}{0.08} - \frac{20}{0.18}\right) = \underline{7.8 \times 10^{-5}\,\text{T}}$$

out of the page.

28.106 An air-core solenoid with 2000 loops on it is 600 mm long and has a diameter of 20 mm. If a current of 5 A is sent through it, what will be the flux density within it?

❚ $$B = \mu_0 nI = (4\pi \times 10^{-7}\,\text{H/m})\left(\frac{2000}{0.6\,\text{m}}\right)(5\,\text{A}) = \underline{0.021\,\text{T}}$$

For a "long" (30 : 1) solenoid such as this, the field will be 21 mT everywhere inside, except right near the ends.

28.107 In the Bohr-model hydrogen atom the single electron orbits the nucleus in a circle of radius $a \approx 5.3 \times 10^{-11}$ m, making $f \approx 6.6 \times 10^{15}$ revolutions each second. Estimate the magnetic field at the nucleus.

❚ The revolving electron is equivalent to a circular loop of current I, where

$$I = ef = (1.6 \times 10^{-19})(6.6 \times 10^{15}) = 1.06 \times 10^{-3}\,\text{A} \qquad B_{\text{nucleus}} = \frac{(4\pi \times 10^{-7})(1.06 \times 10^{-3})}{2(5.3 \times 10^{-11})} \approx \underline{13\,\text{T}}.$$

28.108 A long wire carries a current of 20 A along the directed axis of a long solenoid. The field due to the solenoid is 4 mT. Find the resultant field at a point 3 mm from the solenoid axis.

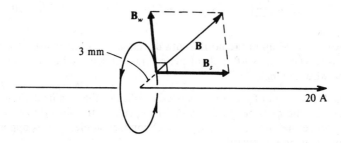

Fig. 28-42

❚ The situation is shown in Fig. 28-42. The field of the solenoid, B_s, is directed parallel to the wire. The field of the long straight wire, B_w, circles the wire and is perpendicular to B_s. We have $B_s = 4$ mT and

$$B_w = \frac{\mu_0 I}{2\pi r} = \frac{(4\pi \times 10^{-7}\,\text{T} \cdot \text{m/A})(20\,\text{A})}{2\pi(3 \times 10^{-3}\,\text{m})} = 1.33\,\text{mT}$$

Since B_s and B_w are perpendicular, their resultant, B, has magnitude

$$B = \sqrt{4^2 + 1.33^2} = \underline{4.2\,\text{mT}}$$

and makes angle $\tan^{-1}(1.33/4) = \underline{18.4°}$ with the directed axis.

28.109 Two long straight wires carry currents of 5 A out along the x axis and y axis, respectively. Find the direction and magnitude of **B** at the point (40, 20, 0) cm.

❚ We assume that x, y, z axes form a right-handed system, $\mathbf{B} = (\mu_0 I/2\pi)(\mathbf{k}/0.20 - \mathbf{k}/0.40) = \underline{2.5\mathbf{k}\,\mu\text{T}}$.

28.110 Two long, parallel, straight wires lie in the xy plane parallel to the y axis. One wire is coincident with the y axis, and the other passes through the point $x = 20$ cm, $y = z = 0$. Both wires carry currents of 5 A in the $+y$ direction. Find **B** at (a) (30, 0, 0) cm; (b) (5, 0, 0) cm.

❚ **(a)** $\mathbf{B} = [(\mu_0 I)/(2\pi)][(-\mathbf{k}/0.3) + (-\mathbf{k}/0.1)] = \underline{-13.3\mathbf{k}\,\mu\text{T}}$. **(b)** $\mathbf{B} = [(\mu_0 I)/(2\pi)][(-\mathbf{k}/0.05) + (+\mathbf{k}/0.15)] = \underline{-13.3\mathbf{k}\,\mu\text{T}}$.

28.111 For the two wires described in Prob. 28.110, find the components of **B** at the point (10, 0, 5) cm.

❚ In Fig. 28-43 the y axis goes into the page. The B due to each wire is $[(\mu_0 I)/(2\pi)]/(0.11) = 8.9\,\mu\text{T}$. Vertical components cancel and horizontal components add. Therefore $B_x = B = 2(8.9)\sin\theta = \underline{8.0\,\mu\text{T}}$.

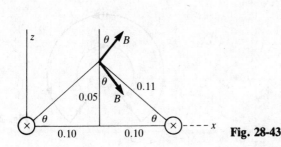

Fig. 28-43   Fig. 28-44

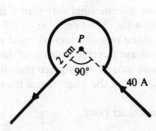

Fig. 28-45

28.112 For the two wires described in Prob. 28.110, find the components of **B** at the point $(0, 0, 40)$ cm.

❚ See Fig. 28-44. B due to the wire through the origin is $[(\mu_0 I)/(2\pi)]/(0.4) = 2.5\,\mu\text{T}$ in the x direction. B due to the other wire is $[(\mu_0 I)/(2\pi)]/\sqrt{0.20} = 2.24\,\mu\text{T}$ at an angle θ above the x axis as shown. Then $B_x = 2.5 + 2.24\cos\theta = 4.5\,\mu\text{T}$. $B_z = 2.24\sin\theta = 1.0\,\mu\text{T}$. $\mathbf{B} = \underline{(4.5\mathbf{i} + 1.0\mathbf{k})\,\mu\text{T}}$.

28.113 The wire shown in Fig. 28-45 carries a current of 40 A. Find the field at point P.

❚ Since P lies on the lines of the straight wires, they contribute no field at P. A circular loop of radius r gives a field of $B = \mu_0 I/2r$ at its center point. Here we have only three-fourths of a loop and so

$$B \text{ at point } P = \frac{3}{4}\frac{\mu_0 I}{2r} = \frac{3(4\pi\times 10^{-7}\,\text{H/m})(40\,\text{A})}{4(2)(0.02\,\text{m})} = \underline{0.94\,\text{mT}}$$

The field is out of the page.

29.114 Referring to Fig. 28-33, let $q = 35\,\mu\text{C}$, $r = 50$ mm, $v = 2\times 10^6$ m/s, and $\theta = 60°$. Find the magnitude of **B** at point P.

❚ $$B = \frac{qv\sin\theta}{r^2}\times 10^{-7} \qquad \text{so} \qquad B = \frac{(35\times 10^{-6})(2\times 10^6)(\sin 60°)}{10^7(50\times 10^{-3})^2} = \underline{2425\,\mu\text{T}}$$

28.115 Again referring to Fig. 28-33, let the displacement from q to $P(x, y, z)$ be $\mathbf{r} = 50\mathbf{i} + 80\mathbf{j} + 70\mathbf{k}$ mm. Let $q = 400\,\mu\text{C}$ and $\mathbf{v} = (3\mathbf{i} - 6\mathbf{j} + 9\mathbf{k})\times 10^6$ m/s. Find B_x, B_y, B_z at P and also find angle θ.

❚ The magnitude of $\mathbf{r}$ is $r = (50^2 + 80^2 + 70^2)^{1/2}(10^{-3}) = 0.1175$ m.

$$\mathbf{B} = \frac{\mu_0}{4\pi}\frac{q\mathbf{v}\times\mathbf{r}}{r^3} \qquad \text{so} \qquad \mathbf{B} = \frac{10^{-7}(400\times 10^{-6})}{0.1175^3}(10^6)(10^{-3})(3\mathbf{i} - 6\mathbf{j} + 9\mathbf{k})\times(50\mathbf{i} + 80\mathbf{j} + 70\mathbf{k})$$

$$= (-2.813\mathbf{i} + 0.592\mathbf{j} + 1.332\mathbf{k})(10^{-2})\,\text{T}$$

whence $B_x = \underline{-28130\,\mu\text{T}} \qquad B_y = \underline{5920\,\mu\text{T}} \qquad B_z = \underline{13320\,\mu\text{T}}$

The magnitude of **B** is then

$$B = [28130^2 + 5920^2 + 13320^2]^{1/2} = 31682\,\mu\text{T}$$

and that of **v** is $v = (3^2 + 6^2 + 9^2)^{1/2}(10^6) = 11.225\times 10^6$ m/s

Thus, $$\sin\theta = \frac{10^7 r^2 B}{qv} = \frac{10^7(0.1175^2)(31682\times 10^{-6})}{(400\times 10^{-6})(11.225\times 10^6)} = 0.974 \qquad \text{and} \qquad \theta = \underline{76.8°}$$

28.116 An equilateral triangle is formed from a piece of uniform-resistance wire. Current is fed into one corner and led out of the other. See Fig. 28-46. Show that the current flowing through the sides of the triangle produces no magnetic field at its center O (the intersection of the medians).

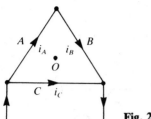

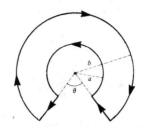

Fig. 28-46

Fig. 28-47

┃ We note that wires A and B are in series and that wire C is in parallel with them. Since each segment has the same resistance, the current $i_C = (2i)/3$, while $i_A = i_B = i/3$. The point O is equidistant from the sides A, B, and C. Hence wires A and B make contributions of equal magnitude to the field at O. According to the right-hand rule, these contributions are directed down into the page. Furthermore, wire C makes a contribution to the field at O whose magnitude is twice that of A or B. The right-hand rule implies that the contribution from C is directed up out of the page, hence it exactly cancels the other two contributions.

28.117 Refer to Fig. 28-47. Find B at the center point.

┃ The straight segments make no contribution to the field at the center; the curved segments give, by the Biot–Savart law,

$$\Delta B = \frac{\mu_0 I}{4\pi} \left[\frac{a(2\pi - \theta)}{a^2} - \frac{b(2\pi - \theta)}{b^2} \right] = \frac{\mu_0 I (b - a)(2\pi - \theta)}{4\pi ab}$$

28.118ᶜ A phonograph record of radius R, which carries a uniformly distributed charge Q, is rotating with constant angular speed ω. Show that the magnetic field at the center of the disk is given by $B = (\mu_0 \omega Q)/(2\pi R)$.

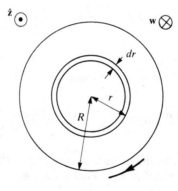

Fig. 28-48

┃ In Fig. 28-48, we show the record in clockwise rotation (as seen from above) which is standard for phonograph records. The ring of charge dq between radii r and $r + dr$ constitutes a current $di = (dq)/T$, where $T = (2\pi)/\omega$ is the rotation period of the disk. Because the disk is uniformly charged, we have

$$\frac{dq}{Q} = \frac{2\pi r \, dr}{\pi R^2} \qquad \text{so that} \qquad dq = \frac{2Qr \, dr}{R^2}$$

Using the result of Prob. 28.90(b), we find that the ring makes a magnetic field contribution

$$dB = \frac{\mu_0}{2} \frac{di}{r} = \frac{\mu_0}{2} \frac{\omega}{2\pi} \frac{2Qr \, dr}{rR^2} = \frac{\mu_0 \omega Q}{2\pi R^2} \, dr$$

The net field at the center is then

$$B = \frac{\mu_0 \omega Q}{2\pi R^2} \int_0^R dr = \frac{\mu_0 \omega Q}{2\pi R}$$

For $Q > 0$, the field has the direction $-\hat{\mathbf{z}}$.

28.5 MORE COMPLEX GEOMETRIES; AMPÈRE'S LAW

28.119 Calculate the magnetic field of a circular current loop at any point on its axis.

▌ Choosing a coordinate system as in Fig. 28-49, and noting that by symmetry the resultant field at P must be along Z, we have:

$$dB_z = (dB) \cos \omega = \left(\frac{\mu_0 I}{4\pi r^3} dl \, r \sin 90°\right) \cos \omega = \frac{\mu_0 I a^2}{4\pi r^3} d\phi = \frac{\mu_0 I a^2}{4\pi (a^2 + z^2)^{3/2}} d\phi \qquad (1)$$

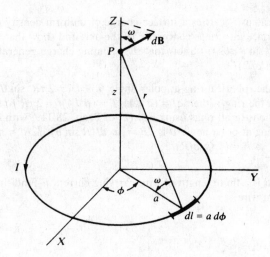

Fig. 28-49

Adding all the elements around the loop, we in effect just add all the $d\phi$'s to get

$$B_z = B = \frac{\mu_0 I a^2}{2(a^2 + z^2)^{3/2}} \qquad (2)$$

In particular, at the center of the loop ($z = 0$), $B_{\text{center}} = (\mu_0 I)/(2a)$, the well-known result.

28.120 Two circular coils shown in Fig. 28-50(a) have the same number of turns and carry current in the same sense. They have different diameters but subtend the same angle at P. (a) Which coil makes the larger contribution to the magnetic field at P? (b) If the smaller coil is midway between P and the larger one, what is the ratio of the larger contribution at P to the smaller?

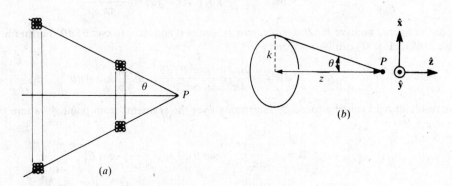

Fig. 28-50

▌ (a) Referring to Fig. 28-50(b) and using Eq. (2) of Prob. 28.120, we find that the magnetic field at point P, due to a coil of N circular loops at the position shown, is given by

$$\mathbf{B} = \frac{N\mu_0 i}{2} \frac{k^2}{(z^2 + k^2)^{3/2}} \hat{\mathbf{z}}$$

But $k/(z^2 + k^2)^{1/2} = \sin \theta$, so

$$\mathbf{B} = \hat{\mathbf{z}} \frac{N\mu_0 i}{2k} \sin^3 \theta$$

which shows that <u>the coil with the smaller radius makes the larger contribution</u>.
(b) If $k_2 = 2k_1$, then we find

$$\frac{B_1}{B_2} = \frac{k_2}{k_1} = \underline{2}$$

28.121ᶜ A spherical shell of radius R carries a surface charge of uniform density σ and spins on its axis with frequency f. **(a)** Section the surface into rings concentric to the axis and show that the current carried by each ring of width $R\, d\theta$ is $2\pi\sigma fR^2 \sin \theta\, d\theta$. **(b)** Show that the spinning charge generates the following field at the center of the sphere: $B = (4\pi\mu_0\sigma fR)/3$.

▌ **(a)** The area of the spherical ring at polar angle θ is $dA = 2\pi R^2 \sin \theta\, d\theta$, and the total charge on it is $\sigma\, dA$. The current in the ring is then $dI = (\sigma\, dA)/T = (\sigma\, dA)f = 2\pi R^2 f\sigma \sin \theta\, d\theta$. **(b)** To find B at the center we add the contributions of all rings using Eq. (2) of Prob. 28.119, with $a = R \sin \theta$ and $(a^2 + z^2)^{1/2} = R$. The contribution of the ring at polar angle θ is $dB = [\mu_0\, dI (R \sin \theta)^2]/2R^3 = \pi\mu_0\sigma fR \sin^3 \theta\, d\theta$. Integrating $\sin^3 \theta\, d\theta$ from 0 to π yields $\frac{4}{3}$, so $B = \underline{(4\pi\mu_0\sigma fR)/3}$.

28.122ᶜ **(a)** In Fig. 28-51, AB is a finite length of wire carrying current i. Find the field at P. **(b)** Deduce the results for an infinite straight wire.

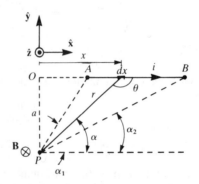

Fig. 28-51

▌ **(a)** The Biot–Savart law states that the contribution $d\mathbf{B}$ to the field at P (due to a wire segment dx centered at x) is given by

$$d\mathbf{B} = \frac{\mu_0}{4\pi} \frac{i\, dx}{r^3} \hat{\mathbf{x}} \times (-\hat{\mathbf{x}}x - \hat{\mathbf{y}}a) = -\frac{\hat{\mathbf{z}}\mu_0 i}{4\pi} \frac{a\, dx}{r^3} \tag{1}$$

From the figure, we have $\tan \theta = -a/x$, so $x = -a \cot \theta$ and $dx = (a \csc^2 \theta)\, d\theta$. Furthermore, $r = a/\sin \theta = a \csc \theta$. Thus Eq. (1) can be rewritten as

$$d\mathbf{B} = -\frac{\hat{\mathbf{z}}\mu_0 i}{4\pi} \frac{(a^2 \csc^2 \theta)\, d\theta}{a^3 \csc^3 \theta} = -\frac{\hat{\mathbf{z}}\mu_0 i}{4\pi a} \sin \theta\, d\theta \tag{2}$$

The resultant field must be found by integrating over the segments from point A (where $\theta = \pi - \alpha_1$) to point B (where $\theta = \pi - \alpha_2$):

$$\mathbf{B} = -\frac{\hat{\mathbf{z}}\mu_0 i}{4\pi a} \int_{\pi - \alpha_1}^{\pi - \alpha_2} \sin \theta\, d\theta = -\frac{\hat{\mathbf{z}}\mu_0 i}{4\pi a} [-\cos \theta] \Big|_{\pi - \alpha_1}^{\pi - \alpha_2}$$

$$= -\frac{\hat{\mathbf{z}}\mu_0 i}{4\pi a} [-\cos (\pi - \alpha_2) + \cos (\pi - \alpha_1)] = -\frac{\hat{\mathbf{z}}\mu_0 i}{4\pi a} (\cos \alpha_2 - \cos \alpha_1) \tag{3}$$

since $\cos (\pi - \alpha) = -\cos \alpha$. Equation (3) is the desired result. **(b)** In the case of an infinite straight wire, we have $\alpha_1 = \pi$ and $\alpha_2 = 0$, so that $\cos \alpha_1 = -1$ and $\cos \alpha_2 = 1$. Then $\mathbf{B} = -\hat{\mathbf{z}}[(\mu_0 i)/(2\pi a)]$, as expected.

28.123 Use the result of Prob. 28.122 to show that the magnetic field at the center of a square frame of side l illustrated in Fig. 28-52(a), and carrying current i, is $[(\mu_0 i)/(\pi l)]2\sqrt{2}$.

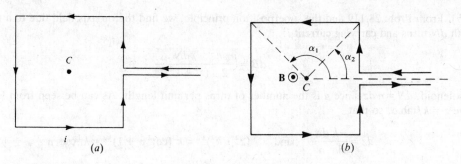

(a)   (b)   **Fig. 28-52**

▮ Figure 28-52(b) shows a square loop of side l, with angles labeled as in Fig. 28-51, so that the contribution to the magnetic field at C from the upper side can be evaluated. According to the result of Prob. 28.122, the magnetic field due to the upper side has magnitude

$$B_1 = \frac{\mu_0 i}{4\pi a}(\cos \alpha_2 - \cos \alpha_1) \tag{1}$$

With $a = l/2$, $\alpha_1 = 135°$, and $\alpha_2 = 45°$, Eq. (1) yields

$$B_1 = \frac{\mu_0 i}{4\pi(l/2)}\left[\left(\frac{\sqrt{2}}{2}\right) - \left(-\frac{\sqrt{2}}{2}\right)\right] = \frac{\mu_0 i \sqrt{2}}{2\pi l} \tag{2}$$

By applying the right-hand rule and considering the symmetry of the arrangement, we conclude that the net magnetic field $\mathbf{B}_4$ at C due to the square loop is *directed up out of the page* and has a magnitude

$$B_4 = 4B_1 = \frac{2\sqrt{2}\mu_0 i}{\pi l} \tag{3}$$

28.124   A length L of wire carrying current i is to be bent into a circle or a square, each of one turn. In which case is B at the center of the figure greater? What is the ratio of B_g (greater) to B_s (smaller)?

▮ A square of perimeter L has side $l = L/4$, so Eq. (3) of Prob. 28.123 yields

$$B_4 = \frac{2\sqrt{2}\mu_0 i}{\pi(L/4)} = \frac{8\sqrt{2}\mu_0 i}{\pi L} = \frac{3.60\mu_0 i}{L}$$

A circle of circumference L has radius $r = L/(2\pi)$. Then,

$$B_c = \frac{\mu_0 i}{2r} = \frac{\mu_0 i}{2[L/(2\pi)]} = \frac{\pi\mu_0 i}{L} = \frac{3.14\mu_0 i}{L}$$

Therefore the field is stronger at the center of the square than it is at the center of the circle (of equal *perimeter*). The ratio is

$$\frac{B_g}{B_s} = \frac{B_4}{B_c} = \frac{8\sqrt{2}\mu_0 i}{\pi L}\frac{L}{\pi\mu_0 i} = \frac{8\sqrt{2}}{\pi^2} = \underline{1.15}$$

28.125^c   Show that $\mathbf{B}$ at point P on the axis of a solenoid of finite length equals $(\mu_0/2)ni(\cos \alpha_2 - \cos \alpha_1)$, the angles being defined in Fig. 28-53(a) and n being the number of turns per unit length.

▮ We denote the radius of the solenoid by k and we use the coordinates z and α as indicated in Fig.

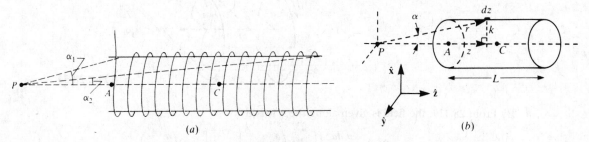

(a)   (b)

Fig. 28-53

28-53(b). From Prob. 28.119 and the superposition principle, we find the on-axis field due to a (flat) circular coil with dN turns and carrying current i:

$$dB = \frac{\mu_0 i k^2}{2} \frac{dN}{(z^2 + k^2)^{3/2}} \tag{1}$$

For a solenoid, $dN = n\,dz$ since n is the number of turns per unit length. As can be seen from Fig. 28-53(b), we have $z = k/\tan \alpha$, so that

$$dz = \frac{-k\,d\alpha}{\sin^2 \alpha} \qquad \text{and} \qquad (z^2 + k^2)^{3/2} = k^3 (\cot^2 \alpha + 1)^{3/2} = k^3 \csc^3 \alpha$$

Since all contributions to the net field at P are along the same line, we may integrate Eq. (1) to find

$$B = \int_{Z_1}^{Z_2} \frac{\mu_0 i k^2}{2} \frac{n\,dz}{(z^2 + k^2)^{3/2}} = \frac{\mu_0 n i k^2}{2} \int_{\alpha_1}^{\alpha_2} \frac{-k\,d\alpha}{\sin^2 \alpha (k^3 \csc^3 \alpha)}$$
$$= \frac{\mu_0 n i}{2} \int_{\alpha_1}^{\alpha_2} (-\sin \alpha)\,d\alpha = \frac{\mu_0 n i}{2} (\cos \alpha \,|_{\alpha_1}^{\alpha_2}) = \frac{\mu_0 n i}{2} (\cos \alpha_2 - \cos \alpha_1) \tag{2}$$

28.126 Refer to Prob. 28.125. What is the value of B at point A on the axis of the solenoid just outside the end?

❚ We let L denote the length of the solenoid. If P coincides with A, the angles are $\alpha_1 = \pi/2$ and $\alpha_2 = \tan^{-1}(k/L) = \cos^{-1}(L/\sqrt{L^2 + k^2})$. Therefore the field at A is

$$B_A = \frac{\mu_0 n i}{2} \left(\frac{L}{\sqrt{L^2 + k^2}} - 0 \right) = \frac{\mu_0 n i L}{2\sqrt{L^2 + k^2}}.$$

28.127 In Prob. 28.125 what is the value of B at C on the axis in the center (see Fig. 28-53)?

❚ If P coincides with C, then

$$\cos \alpha_1 = \frac{-L/2}{\sqrt{(L/2)^2 + k^2}} \qquad \text{and} \qquad \cos \alpha_2 = \frac{L/2}{\sqrt{(L/2)^2 + k^2}}$$

Therefore the field at C is

$$B_C = \frac{\mu_0 n i}{2} \left[\frac{L/2}{\sqrt{(L/2)^2 + k^2}} - \frac{-L/2}{\sqrt{(L/2)^2 + k^2}} \right] = \frac{\mu_0 n i L}{2\sqrt{(L/2)^2 + k^2}} = \frac{\mu_0 n i L}{\sqrt{L^2 + 4k^2}}$$

28.128 Show that the result in Prob. 28.125 for B inside a long solenoid agrees with the usual formula.

❚ As $L \to \infty$, then for any point inside the solenoid, the angles α_1 and α_2 approach π and 0, respectively. Hence Eq. (2) implies

$$B \to \frac{\mu_0 n i}{2} [\cos 0 - \cos (\pi)] = \mu_0 n i$$

28.129ᶜ *Helmholtz coils* (Fig. 28-54) are sometimes used to obtain a nearly uniform magnetic field in situations where a solenoid would be impractical. Verify that on the axis, midway between the coils, the first three derivatives of B with respect to x vanish.

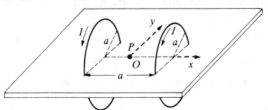

Fig. 28-54

❚ By Prob. 28.119, the field is given along the x axis by

$$B(x) = \frac{\mu_0 I a^2}{2} \left\{ \left[a^2 + \left(\frac{a}{2} + x \right)^2 \right]^{-3/2} + \left[a^2 + \left(\frac{a}{2} - x \right)^2 \right]^{-3/2} \right\}$$

As this is an even function of x, $B'(0) = B'''(0) = 0$. Moreover,

$$B''(x) = 6\,\mu_0 Ia^2 x\left\{\left[a^2+\left(\frac{a}{2}+x\right)^2\right]^{-7/2}(a+x)-\left[a^2+\left(\frac{a}{2}-x\right)^2\right]^{-7/2}(a-x)\right\}$$

which shows that $B''(0) = 0$. Consequently, near O, we have $B(x) \approx B_0 + B_4 x^4$, an almost-constant function.

28.130 A multilayer coil is to be wound on a fixed form with the wires in contact. See the cross section in Fig. 28-55. The insulation is thin, and the space provided is to be filled. How will the strength of the magnetic field produced depend on the diameter d of the wire chosen if the power consumed by the coil is fixed?

▌ Since all geometrical factors are fixed, the field is proportional to Ni, where N is the total number of turns and i is the current. The number of turns in each layer is proportional to $1/d$, and the number of layers is also proportional to $1/d$. Therefore $N \propto 1/d^2$.

In this case, $i^2 R$ is a constant, where R is the electric resistance. Hence i is proportional to $R^{-1/2}$. But R is proportional to l/d^2 for any given conducting material. Since $l \propto N \propto 1/d^2$, $l/d^2 \propto 1/d^4$, so $i \propto R^{-1/2} \propto d^2$. Therefore $B \propto Ni \propto (1/d^2)(d^2) \propto d^0$. That is, at constant power, B is independent of d.

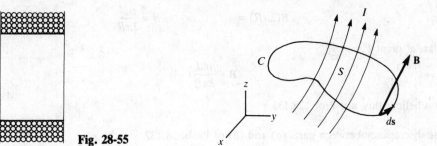

Fig. 28-55　　　Fig. 28-56

28.131ᶜ State *Ampère's circuital law*. How is it related to the Biot–Savart law?

▌ Figure 28-56 shows an open surface S bounded by a closed curve C. Total current I, either concentrated or distributed, crosses S; in the latter case, $I = \int_S \mathbf{J} \cdot d\mathbf{S}$, where $\mathbf{J}$ is the current density vector. This current generates a magnetic field which, according to Ampère's law, satisfies $\oint \mathbf{B} \cdot d\mathbf{s} = \mu_0 I$, where the line integral is around C.

Since both Ampère's law and the Biot–Savart law specify the connection between electric currents and magnetic fields, they must be equivalent (as can be shown mathematically by use of vector calculus).

28.132ᶜ A straight wire, along Z in Fig. 28-57, carries a current I. Find the field at P due to section ab only (length, $l_1 + l_2$) using (*a*) the Biot–Savart law, and (*b*) Ampère's circuital law.

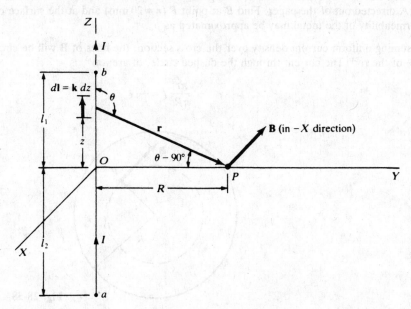

Fig. 28-57

▮ (a) It is evident that $d\mathbf{B}$ at P due to any current element $I\,d\mathbf{l}$ has the direction $-\mathbf{i}$. Hence

$$d\mathbf{B} = \frac{\mu_0 I \sin \theta}{4\pi r^2}\, dz\, (-\mathbf{i})$$

But

$$r^2 = R^2 + z^2 \qquad \text{and} \qquad \sin \theta = \cos (\theta - 90°) = \frac{R}{r}$$

Thus,

$$d\mathbf{B} = \frac{\mu_0 I R}{4\pi (R^2 + z^2)^{3/2}}\, dz\, (-\mathbf{i})$$

and

$$\mathbf{B} = \frac{\mu_0 I R}{4\pi}(-\mathbf{i}) \int_{-l_2}^{+l_1} \frac{dz}{(R^2 + z^2)^{3/2}} = \frac{\mu_0 I}{4\pi R}\left[\frac{l_1}{(R^2 + l_1^2)^{1/2}} + \frac{l_2}{(R^2 + l_2^2)^{1/2}}\right](-\mathbf{i})$$

From the cylindrical symmetry of the problem it is evident that at all points of the circle $x^2 + y^2 = R^2$, $\mathbf{B}$ is tangential and of constant magnitude.

(b) Application of Ampère's law to the circle $x^2 + y^2 = R^2$, together with the symmetry considerations of (a), gives

$$B(2\pi R) = \mu_0 I \qquad \text{or} \qquad B = \frac{\mu_0 I}{2\pi R}$$

In particular, at point P,

$$\mathbf{B} = \frac{\mu_0 I}{2\pi R}(-\mathbf{i})$$

(Before you believe this, see Prob. 28.133.)

28.133 Explain the discrepancy between parts (a) and (b) of Prob. 28.132.

▮ Electric current cannot really appear abruptly at point a and disappear abruptly at point b; there must be a return path from b to a. If the integration in (a) were extended over the entire closed path, a different result for $\mathbf{B}$ would be obtained. Likewise, the problem no longer being cylindrically symmetric, a different result would also be found in (b); and these two new results would agree. For example, suppose that the path closes at infinity. Letting $l_1 \to \infty$ and $l_2 \to \infty$ in (a) and holding R fixed (noting that the return path at ∞ can be ignored),

$$\frac{l_1}{(R^2 + l_1^2)^{1/2}} \to 1 \qquad \frac{l_2}{(R^2 + l_2^2)^{1/2}} \to 1 \qquad \text{and so} \qquad \mathbf{B} \to \frac{\mu_0 I}{2\pi R}(-\mathbf{i})$$

This result agrees with that of (b), which, for an *infinite* straight wire, is correct as it stands.

28.134 Figure 28-58 shows the cross section of a long metal rod of radius $R = 30$ mm. The rod carries a current $I = 5$ kA directed out of the paper. Find B at point P ($r = 20$ mm) and at the surface of the rod, assuming that the permeability of the metal may be approximated as μ_0.

▮ Assuming uniform current density over the cross section, the lines of $\mathbf{B}$ will be circles concentric with the surface of the rod. The current through the dashed circle, of area πr^2, is

$$\frac{\pi r^2}{\pi R^2}I = \frac{r^2}{R^2}I$$

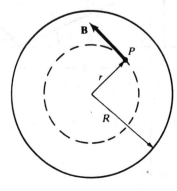

Fig. 28-58

and Ampère's circuital law gives, by symmetry,

$$B(2\pi r) = \mu_0 \frac{r^2}{R^2} I \quad \text{or} \quad B = \left(\frac{\mu_0 I}{2\pi R^2}\right) r$$

Substituting numerical values,

$$B_P = \underline{22.2 \text{ mT}} \qquad B_{\text{surface}} = \frac{3}{2} B_P = \underline{33.3 \text{ mT}}.$$

28.135 By applying Ampère's circuital law to an appropriate contour, show that **B** is uniform within a very long solenoid *of arbitrary cross section*.

▌ In Fig. 28-59 the dashed breaks allow us to show the end faces of the solenoid. At any point in the interior the field must be parallel to the axis of the solenoid since perpendicular components will be canceled by contributions from the solenoid symmetrically to the left and right of the point. Furthermore, the parallel component must be the same anywhere along a given parallel line since ideally the solenoid is infinitely long. Then we choose a rectangular path, as shown, inside the solenoid, one side, of length L, being along the axis. Since no current is enclosed by the path, and using our earlier conclusions, Ampère's law yields

$$B_b L - B_a L = 0 \quad \text{or} \quad B_b = B_a$$

Since this is true for any choice of rectangle, we have our result.

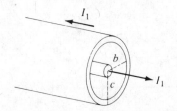

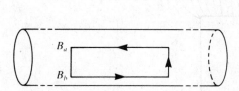

Fig. 28-59

Fig. 28-60

28.136 In the coaxial cable of Fig. 28-60, a straight wire of radius a carries a current I_1 along the axis of a metal tube with inner radius b and outer radius c. The tube carries a current I_1 in a direction opposite to that in the wire. Find B for $a < r < b$. Repeat for $r > c$.

▌ Use Ampère's circuital law for $a < r < b$, $\mu_0 I_1 = 2\pi r B$, so $B = (\mu_0 I_1)/(2\pi r) = [(2 \times 10^{-7}) I_1]/r$. For $r > c$, one has $\mu_0 (I_1 - I_1) = 2\pi r B$ from which $B = \text{const}\, (I_1 - I_1) = 0$.

28.137 Repeat Prob. 28.136 if the current in the outer tube is I_2. Repeat if I_1 and I_2 are in the same direction.

▌ As before, $B = [(2 \times 10^{-7}) I_1]/r$ for $r < b$ but now for $r > c$, $B = [(2 \times 10^{-7})(I_1 \mp I_2)]/r$.

28.138ᶜ Refer to Fig. 28-61. Suppose that the current density in the wire varies with r according to $J = kr^2$, where k is a constant. Find the value of B for **(a)** $r > a$ and **(b)** $r < a$.

▌ Choose a circular path centered on the conductor and apply Ampère's law. **(a)** To find the current passing through the area enclosed by the path integrate $J\, dA = (kr^2)(2\pi r\, dr)$ from 0 to a to find $I = (\pi ka^4)/2$. The field is then $B = (\mu_0 I)/2\pi r = (\mu_0 ka^4)/4r$. **(b)** Integrate $J\, dA$ from 0 to r; $I = (\pi kr^4)/2$, then $B = (\mu_0 kr^3)/4$.

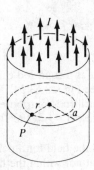

Fig. 28-61

28.139 A long straight metal tube has inner radius a and outer radius b. It carries lengthwise a current I spread uniformly over its cross section for $a \leqslant r \leqslant b$. Find the magnetic field in each of the following regions: **(a)** $r < a$; **(b)** $a \leqslant r \leqslant b$; **(c)** $r > b$.

▋ Apply the circuital law to circular paths in each region. **(a)** $B2\pi r = \mu_0(0)$, so $B = 0$; no current is enclosed by the path. **(b)** $B2\pi r = [\mu_0 I(r^2 - a^2)]/(b^2 - a^2)$; only a fraction is enclosed. **(c)** $B2\pi r = \mu_0 I$ giving $B = (\mu_0 I)/(2\pi r)$; all the current is enclosed.

28.140 Show that fringing of the magnetic field must occur on *both* sides in Fig. 28-62.

▋ We apply Ampère's law to the rectangular path $ABCD$. Since the path encloses no current, we find that $\Sigma \equiv \oint \mathbf{B} \cdot d\mathbf{l} = 0$. By hypothesis, $\mathbf{B}$ is perpendicular to sides BC and DA, and $B = 0$ along CD. Therefore $\Sigma = \int_{AB} \mathbf{B} \cdot d\mathbf{l}$. But $\mathbf{B}$ is antiparallel to side AB and is nonzero there. This implies that $\Sigma \neq 0$, in violation of Ampère's law. Thus our hypothesis that $\mathbf{B}$ drops sharply to zero (outside the region between the pole pieces) must be false.

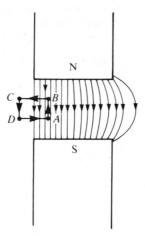

Fig. 28-62

28.141 As shown in Fig. 28-63(a), a long, straight metal rod has a very long hole of radius a drilled parallel to the rod axis. If the rod carries a current I, show that B has the value **(a)** $(\mu_0 I a^2)/[(2\pi c)(b^2 - a^2)]$ on the axis of the rod and **(b)** $(\mu_0 I c)/[(2\pi)(b^2 - a^2)]$ on the axis of the hole. (*Hint*: Superpose the fields from two cylindrical conductors to obtain the required field.)

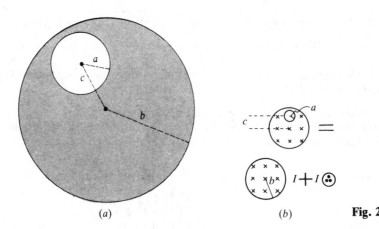

Fig. 28-63

▋ In the rod $J = I/[\pi(b^2 - a^2)]$. Treat as a solid rod carrying current $J\pi b^2$ one way and a second rod with current $J\pi a^2$ moving the other way [see Fig. 28-63(b)]. The actual field is the sum of these two current-carrying rods. **(a)** Only the small rod contributes to the field, so $B = [\mu_0(J\pi a^2)]/(2\pi c)$. **(b)** On the hole axis only the larger rod contributes; $B = [\mu_0(J\pi c^2)]/(2\pi c) = (\mu_0 I c)/[2\pi(b^2 - a^2)]$.

28.142 For the situation of Prob. 28.141, find the field for a point P on the line of centers at a radius **(a)** $R > b$ from the center of the rod and **(b)** $R < b$ but not within the hole.

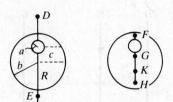

Fig. 28-64

▌ *(a)* At points D and E (Fig. 28-64), fields oppose each other. Using ideas of Prob. 28.141, the actual field $B = [\mu_0(J\pi b^2)]/(2\pi R) - [\mu_0(J\pi a^2)]/[2\pi(R \pm c)]$. *(b)* At points F and H, $B = [\mu_0(J\pi R^2)]/(2\pi R) - [\mu_0(J\pi a^2)]/[2\pi(R \pm c)]$. At point G, $B = (\mu_0 JR)/2 + (\mu_0 Ja^2)/[2(c - R)]$ since the fields of the two current-carrying rods add; at K the fields subtract so $B = [(\mu_0 J)/2][R - a^2/(c + R)]$ with $J = I/[\pi(b^2 - a^2)]$.

28.143 Show that the force per unit length f between two very long parallel wires is $f = -(\mu_0 I_1 I_2)/(2\pi r)$, where I_1 and I_2 are the currents in the wire and r is the distance between the wires. (The minus sign indicates that the wires attract each other when their currents are in the same direction, and that they repel each other when their currents are in opposite directions.)

▌ Refer to Fig. 28-65. The field at a distance r from a long wire carrying a current I_1 is $B = (\mu_0 I_1)/(2\pi r)$. This field is at right angles to a second wire that is parallel to the first. Therefore, the force on a length l of a second wire at r carrying the current I_2 is $F_m = BI_2 l$ and the force per unit length is

$$f = \frac{F_m}{l} = BI_2 = \frac{\mu_0 I_1 I_2}{2\pi r}$$

Using the right-hand rules for fields and forces, it is seen that the force on wire 2 is repulsive when I_1 and I_2 have opposite directions and attractive when I_1 and I_2 have the same direction. This can be expressed in the usual sign convention by

$$f = -\frac{\mu_0 I_1 I_2}{2\pi r} \tag{1}$$

since f will be negative (attractive) when I_1 and I_2 have the same sign, and will be positive (repulsive) when I_1 and I_2 have opposite signs.

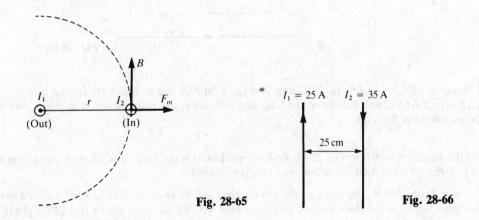

Fig. 28-65   **Fig. 28-66**

28.144 What is the force per unit length between the wires in Fig. 28-66?

▌ Since the currents have opposite directions, we have, using the results of Prob. 28.143,

$$f = -\frac{\mu_0 I_1 I_2}{2\pi r} = -\frac{\mu_0(25\text{ A})(-35\text{ A})}{2\pi(0.25\text{ m})} = +7.0 \times 10^{-4}\text{ N/m}.$$

This force is repulsive.

28.145 Consider the three long, straight, parallel wires shown in Fig. 28-67. Find the force experienced by a 25 cm length of wire C.

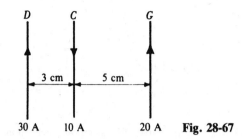

Fig. 28-67

┃ The fields due to wires D and G at wire C are

$$B_D = \frac{\mu_0 I}{2\pi r} = \frac{(4\pi \times 10^{-7}\,\text{T} \cdot \text{m/A})(30\,\text{A})}{2\pi(0.03\,\text{m})} = 2 \times 10^{-4}\,\text{T}$$

into the page, and

$$B_G = \frac{(4\pi \times 10^{-7}\,\text{T} \cdot \text{m/A})(20\,\text{A})}{2\pi(0.05\,\text{m})} = 0.8 \times 10^{-4}\,\text{T}$$

out of the page. Therefore the field at the position of wire C is

$$B = 2 \times 10^{-4} - 0.8 \times 10^{-4} = 1.2 \times 10^{-4}\,\text{T}$$

into the page. The force on a 25 cm length of C is

$$F = ILB \sin \theta = (10\,\text{A})(0.25\,\text{m})(1.2 \times 10^{-4}\,\text{T})(\sin 90°) = \underline{3 \times 10^{-4}\,\text{N}}$$

Using the right-hand rule at wire C tells us that the force on wire C is toward the right.

28.146 For the situation shown in Fig. 28-68, find the force experienced by side MN of the rectangular loop. Also, find the torque on the loop.

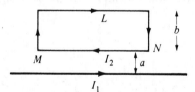

Fig. 28-68

┃ Field at MN due to I_1 is $(\mu_0 I_1)/(2\pi a)$. The force on MN due to I_1 is away from I_1 and is $I_2 LB = (\mu_0 I_1 I_2 L)/(2\pi a)$. Forces on the loop due to I_1 act in the plane of the loop giving zero torque. These forces compress the loop.

28.147 For the situation shown in Fig. 28-68, find the net force on the long, straight wire due to the current in the loop. (*Hint*: Assume that the action–reaction law is valid.)

┃ From Prob. 28.146, the force on the top element will be $(\mu_0 I_1 I_2 L)/[2\pi(a + b)]$ directed toward I_1. From Newton's third law the force exerted on the straight wire by the loop will be $[(\mu_0 I_1 I_2 L)/(2\pi)][1/a - 1/(a + b)]$ and is repulsive.

28.148 Explain why the permeability of empty space turns out to be exactly $4\pi \times 10^{-7}$ (in SI units).

┃ Equation (1) of Prob. 28.143 is used to define the ampere, as follows: When r is set at 1 m and f is measured to be 2×10^{-7} N/m (a convenient value), for $I_1 = I_2 = I$, then $I = 1$ A, exactly. This forces the value $\mu_0 = 4\pi \times 10^{-7}$.

28.149 The two infinite plates shown in cross section in Fig. 28-69(a) carry j amperes of current out of the page per unit width of plate. Find the magnetic field at points P and Q.

┃ Consider only one plate as shown in Fig. 28-69(b). The field on the two sides is directed as indicated.

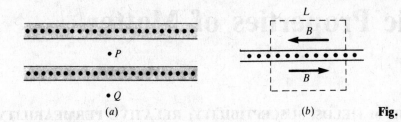

Fig. 28-69

Apply the circuital law to the dotted path. Then $2BL = \mu_0 jL$, which gives

$$B = \frac{\mu_0 j}{2}. \tag{1}$$

Now superpose the fields of two such plates. At P they cancel and $B = 0$; at Q they add, so $B = \mu_0 j$.

28.150 In Prob. 28.149, find the force per unit area on the lower plate because of the current in the upper plate.

▌ The upward force on a square of side L of the lower plate is $F = BIL$, with $I = jL$. Using Eq. (1) of Prob. 28.149, $F = [(\mu_0 j)/2](jL)L = (\mu_0 j^2 L^2)/2$. The force per unit area $= F/L^2 = (\mu_0 j^2)/2$.

28.151 A flat circular coil with 10 loops of wire on it has a diameter of 20 mm and carries a current of 0.5 A. It is mounted inside a long solenoid that has 200 loops on its 250 mm length. The current in the solenoid is 2.4 A. Compute the torque required to hold the coil with its axis perpendicular to that of the solenoid.

▌ Let the subscripts s and c refer to the solenoid and coil, respectively. Then $\tau = N_c I_c A_c B_s \sin 90°$. But $B_s = \mu_0 n I_s = \mu_0 (N_s/L_s) I_s$, which gives

$$\tau = \mu_0 N_c N_s I_c I_s (\pi r_c^2)/L_s = (4\pi \times 10^{-7} \text{ H/m})(10)(200)(0.5 \text{ A})(2.4 \text{ A})\pi(0.01 \text{ m})^2/(0.25 \text{ m})$$
$$= \underline{3.8 \times 10^{-6} \text{ N} \cdot \text{m}}$$

CHAPTER 29
Magnetic Properties of Matter

29.1 THE H AND M FIELDS; SUSCEPTIBILITY; RELATIVE PERMEABILITY

29.1 Define: *pole strength*; *magnetic declination*; *inclination*; *relative permeability*; *Curie temperature*; *hysteresis*.

▎ The *pole strength* of a bar magnet is the ratio of the magnetic moment to the distance between the two poles (approximately but a little less than the length of the magnet).

The *magnetic declination* at any point on the earth's surface is the angle between the north-pointing axis of a horizontal compass needle and true north.

The *inclination* (or *angle of dip*) is the angle between the axis of a free compass needle and the horizontal.

The *relative permeability* of a material is the ratio of the magnetic induction **B** in a toroid filled with that material to the corresponding value of **B** in the same toroid when the region is evacuated (or filled with air).

The *Curie temperature* is the temperature above which a ferromagnetic material completely loses its ferromagnetic properties.

Hysteresis is a phenomenon of magnetic materials that do not completely lose their magnetization when a magnetizing field **H** is applied and then removed. In that event, **B** and **H** are not proportional, and the relationship between them is a *hysteresis curve*.

29.2 Relate the magnetic induction **B** to the *magnetization* **M** and the *magnetic intensity* **H**.

▎ Magnetic phenomena in empty space can be described in terms of a single vector quantity, the magnetic induction **B**. In a material medium, **B** is still the vector field that accounts for the force on a moving charge. Now, however, we introduce two new vectors, **H** and **M**, such that $B = \mu_0 H + \mu_0 M$.

H, called the *magnetic field strength* or the *magnetic intensity*, satisfies the Biot–Savart law or Ampère's law in the form

$$d\mathbf{H} = \frac{I}{4\pi r^3}(d\mathbf{l} \times \mathbf{r}) \qquad \text{or} \qquad \oint \mathbf{H} \cdot d\mathbf{s} = I$$

where I denotes the true current in the medium, i.e., the macroscopic flow of charge through the medium. (The **B** field satisfies its Biot–Savart law for *any* source of current, including the magnetization current caused by the alignment of elemental atomic magnetic dipole moments.) In vacuum, **H** is simply B/μ_0. Thus **H** (or $\mu_0 H$) inside a medium is the part of **B** that is independent of the microscopic properties of the medium. The SI unit of **H** is A/m. The **H** field goes by various names (e.g., magnetizing field) just as the **B** field does.

M, called the *magnetization* or the *magnetic polarization* or the *magnetic dipole moment per unit volume*, is just what the third name implies: the sum of the magnetic dipole moments due to electronic spin, nuclear spin, and orbital motions of electrons—per unit volume of the material. In empty space, $\mathbf{M} = 0$. Thus **M** (or $\mu_0 M$) is the part of **B** that depends on the microscopic properties of the medium. Values of **M** can be given in A/m or J/T · m³. We note the following equivalences: 1 A/m = 1 N/Wb = 1 Wb/H · m = 1 J/T · m³.

29.3 Describe the magnetic susceptibility and permeability of materials in terms of **B**, **H**, and **M**.

▎ For an isotropic medium in which **B**, **H**, and **M** are all in the same direction, we have

$$B = \mu_0(H + M) = \mu_0(H + \chi_m H) = \mu H \qquad \text{where} \qquad \mu = \mu_0(1 + \chi_m)$$

The dimensionless quantity χ_m is called the *magnetic susceptibility* of the medium; and μ, which like μ_0 carries the units H/m, is the *permeability* of the material. The ratio $\mu_{\text{rel}} \equiv \mu/\mu_0$, a pure number, is called the *relative permeability* and is often given the symbol K_m.

Materials are classified according to their susceptibilities, as follows:

Diamagnetic materials, for which χ_m is a very small *negative* constant.

Paramagnetic materials, for which χ_m is small and positive and is inversely proportional to the absolute temperature.

Ferromagnetic materials, for which χ_m is positive and may be much greater than 1. Moreover, χ_m depends in a complicated manner on H, so that M is not proportional to H in these materials.

29.4 Find the magnetizing field H and the magnetic flux density B at (a) a point of 105 mm from a long straight wire bearing a current of 15 A and (b) the center of a 2000-turn solenoid which is 0.24 m long and bears a current of 1.6 A.

▌ (a) $H = \dfrac{B}{\mu_0} = \dfrac{2I/10^7 r}{4\pi/10^7} = \dfrac{I}{2\pi r} = \dfrac{15}{(2\pi)(0.105)} = \underline{22.7 \text{ A/m}}.$ $B = \dfrac{(2)(15)}{10^7 \times 0.105} = \underline{28.57 \ \mu\text{T}}$

(b) $H = nI = \dfrac{2000}{0.24} 1.6 = \underline{13\,333 \text{ A/m}}$ $B = \dfrac{4\pi}{10^7}\left(\dfrac{2000}{0.24}\right) 1.6 = \underline{0.0168 \text{ T}}$

29.5ᶜ For the iron with the magnetization curve of Fig. 29-1 ($B_0 \equiv \mu_0 H$), what is the maximum differential relative permeability? At what value of B_0 does the maximum occur?

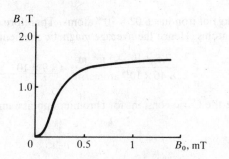

Fig. 29-1

▌ Differential permeability μ' is defined as $\mu' = dB/dH$, and differential relative permeability is thus

$$K_m' = \dfrac{\mu'}{\mu_0} = \dfrac{1}{\mu_0}\dfrac{dB}{dH} = \dfrac{dB}{dB_0}$$

We are looking for the maximum slope in Fig. 29-1. This occurs at the inflection point, $B_0 \approx 0.13$ mT, and measurements on the graph give the value to be $K_{m,(\text{max})}' \approx 3000$.

29.6 An electromagnet has a solenoidal winding 225 mm long with a total of 900 turns. What is the magnetizing field H near the center of the winding and far from any poles if the current is 0.8 A? What is the magnetic induction B at this point if the iron has a relative permeability of 350?

▌ $H = nI = \dfrac{900}{0.225} 0.8 = \underline{3200 \text{ A/m}}$ $B = \mu_0 K_m H = (4\pi \times 10^{-7})(350)(3200) = \underline{1.41 \text{ T}}$

29.7 An air-core solenoid wound with 20 turns per centimeter carries a current of 0.18 A. Find H and B at the center of the solenoid.

▌ $H = nI = (2000 \text{ m}^{-1})(0.18 \text{ A}) = 360 \text{ A/m}$ $B = \mu_0 H = (4\pi \times 10^{-7} \text{ H/m})(360 \text{ A/m}) = \underline{0.45 \text{ mT}}$

29.8 Repeat Prob. 29.7 after an iron core of absolute permeability 6×10^{-3} H/m is inserted in the solenoid.

▌ $H = 360 \text{ A/m}$ (unchanged) $B = \mu H = (6 \times 10^{-3} \text{ H/m})(360 \text{ A/m}) = \underline{2.16 \text{ T}}$

29.9 A toroid wound on a plastic form ($K_m = 1$) has a mean radius of 10 cm and 200 turns of wire. How large a current must flow in it if B is to have a value of 0.01 T? What is the value of H?

▌ For a toroid,

$$B = \dfrac{4\pi nI}{10^7} \quad \text{In our case} \quad n = \dfrac{200}{2\pi(0.1)} = 318.5 \text{ turns/m}$$

and thus

$$0.01 = \dfrac{4\pi}{10^7}(318.5)I \quad \text{and} \quad I = \underline{25 \text{ A}} \quad H = nI = 318.5(25) = \underline{7960 \text{ A/m}}$$

29.10 A toroid of mean circumference 0.5 m has 500 turns, each bearing a current of 0.15 A. (a) Find H and B if the toroid has an air core. (b) Find B and the magnetization M if the core is filled with iron of relative permeability 5000.

▌ For a toroid $H = nI$, and we use $B = (4\pi/10^7)(K_m H) = \mu H$. Thus

(a) $\qquad H = \dfrac{500 \text{ turns}}{0.5 \text{ m}} 0.15 \text{ A} = \underline{150 \text{ A/m}}$, and $B = (4\pi \times 10^{-7} \text{ H/m})(150 \text{ A/m}) = \underline{0.188 \text{ mT}}$

(b) $\qquad\qquad\qquad\qquad B = 5000(0.188 \text{ mT}) = \underline{0.94 \text{ T}}$

Using $B/\mu_0 = H + M$,

$$\frac{0.94}{4\pi \times 10^{-7}} = 150 + M \qquad M = 7.5 \times 10^5 \text{ A/m}$$

29.11 For the situation in Prob. 29.10, find the average magnetic moment per iron atom if the density of iron is 7850 kg/m³.

▌ One kilomole (55.85 kg) of iron has 6.02×10^{26} atoms. Therefore in 1 m³ there are $(7850)(6.02 \times 10^{26})/55.85 = 8.46 \times 10^{28}$ atoms. Hence the average magnetic moment per iron atom is

$$\frac{7.5 \times 10^5 \text{ A} \cdot \text{m}^2/\text{m}^3}{8.46 \times 10^{28} \text{ atoms/m}^3} = \underline{8.9 \times 10^{-24} \text{ A} \cdot \text{m}^2}$$

29.12 Using Fig. 29-2, compute the Curie constant for chromium potassium alum.

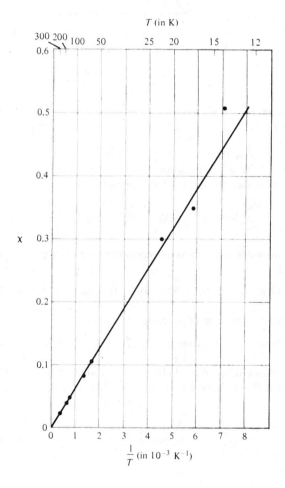

Fig. 29-2

▌ According to Curie's law, the susceptibility $\chi = C/T$, so that $C = \chi T$. Using the line fitted to the data in Fig. 29-2, we see that $\chi = 0.50$ for $1/T = 8.0 \times 10^{-3} \text{ K}^{-1}$, or for $T = 125$ K. According to these numerical values, the Curie constant for chromium potassium alum is $C = (0.50)(125) = \underline{63 \text{ K}}$.

29.13 Using Fig. 29-3, compute the Curie constant and the Curie temperature for gadolinium.

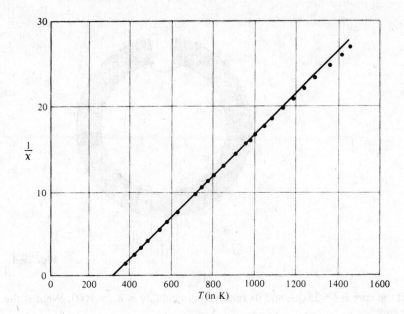

Fig. 29-3

According to the Curie–Weiss law, $\chi = C/(T - T_c)$, so that

$$\frac{1}{\chi} = \frac{T - T_c}{C} \tag{1}$$

Hence T_c is the T- intercept of the best-fit straight line in Fig. 29-3. As can be seen, this yields $\underline{T_c = 310\,K}$. Furthermore, we find that $1/\chi = 26.8$ for $T = 1400\,K$ and that $1/\chi = 2.0$ for $T = 400\,K$. Since Eq. (1) implies that

$$\frac{\Delta(1/\chi)}{\Delta(T)} = \frac{1}{C} \tag{2}$$

we find

$$C = \frac{\Delta(T)}{\Delta(1/\chi)} = \frac{(1400 - 400)\,K}{(26.8 - 2.0)} \approx \underline{40\,K}$$

29.14 A solenoid that is 0.6 m long is wound with 1800 turns of copper wire. An iron rod having a relative permeability of 500 is placed along the axis of the solenoid. What is the magnetic intensity B in the rod when a current of 0.9 A flows through the wire? What are the magnetizing field H and the magnetic moment per unit volume M of the iron? Find the average contribution per iron atom to the magnetization.

$$H = nI = \frac{1800}{0.6}\,0.9 = \underline{2700\,A/m}$$

$$B = K_m \mu_0 H = 500(4\pi \times 10^{-7})(2700) = \underline{1.69\,T}$$

$$M = (K_m - 1)H = (499)(2700) = \underline{1.35\,MA/m}$$

From Prob. 29.11,

$$\frac{\text{magnetic moment}}{\text{atom}} = \frac{1.35 \times 10^6\,A \cdot m^2/m^3}{8.48 \times 10^{28}\,\text{atoms/m}^3} = \underline{1.59 \times 10^{-23}\,A \cdot m^2}$$

29.15 An iron ring has a cross section of 80 mm² and an average diameter of 0.25 m. It is wound with 900 turns of copper wire. If the iron in the core has a relative permeability of 250, what is the magnetic intensity B in the iron when a current of 4 A exists in the windings?

$$n = \frac{900}{0.25\pi} = \frac{3600}{\pi}\,\text{turns/m} \qquad B = \frac{4\pi}{10^7}K_m nI = \frac{4\pi}{10^7}(250)\,\frac{3600}{\pi}(4) = \underline{1.44\,T}$$

29.16 The number of turns for an iron-core toroid is $N = 20$, and the current it carries is $i = 10$ A. The average

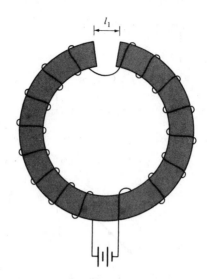

Fig. 29-4

length of the core is $l = 25$ cm, and its relative permeability is $K_m = 1000$. What is the internal magnetic field B in the core?

$$B = K_m\mu_0\frac{Ni}{l} = \frac{(1000)(4\pi \times 10^{-7}\text{ H/m})(20)(10\text{ A})}{(0.25\text{ m})} = \underline{1.01\text{ T}} \tag{1}$$

29.17 Refer to Prob. 29.16. Suppose that a piece of iron of length $l_1 = 1.0$ cm is sawed out to make the magnetic field in the iron accessible, as in Fig. 29-4. Find the value of B in the iron core and show it is weakened considerably.

By Ampère's law, with $l_2 = l - l_1$

$$H_1 l_1 + H_2 l_2 = Ni$$

Since the lines of flux of B are continuous and the gap is small, $B_1 = B_2 = B$. Thus

$$\frac{B}{\mu_0}l_1 + \frac{B}{K_m\mu_0}l_2 = Ni$$

$$B = \frac{\mu_0 Ni}{l_1 + \dfrac{l_2}{K_m}} = \frac{(4\pi \times 10^{-7})(20)(10)}{0.01 + (0.24/1000)} = \underline{2.45 \times 10^{-2}\text{ T}}$$

which is less than 3 percent of the original value given in Eq. (1) of Prob. 29.16.

29.18 A solenoid is 40 cm long, has cross-sectional area 8 cm^2 and is wound with 300 turns of wire that carry a current of 1.2 A. The relative permeability of its iron core is 600. Compute (*a*) B for an interior point and (*b*) the flux through the solenoid.

(*a*) For a long solenoid,

$$B_v = \frac{\mu_0 NI}{L} = \frac{(4\pi \times 10^{-7}\text{ T}\cdot\text{m/A})(300)(1.2\text{ A})}{0.40\text{ m}} = 1.13 \times 10^{-3}\text{ T}$$

and so

$$B = k_m B_v = (600)(1.13 \times 10^{-3}\text{ T}) = \underline{0.68\text{ T}}$$

(*b*) Because the field lines are perpendicular to the cross-sectional area inside the solenoid,

$$\Phi = B_\perp A = BA = (0.68\text{ T})(8 \times 10^{-4}\text{ m}^2) = \underline{5.4 \times 10^{-4}\text{ Wb}}$$

29.19 The flux through a certain toroid changes from 0.65 to 0.91 mWb when the air core is replaced by another material. What are the relative permeability and the permeability of the material?

The air core is essentially the same as a vacuum core. Since $k_m = B/B_v$ and $\Phi = B_\perp A$,

$$k_m = \frac{\Phi}{\Phi_v} = \frac{0.91\text{ mWb}}{0.65\text{ mWb}} = \underline{1.40}$$

This is the relative permeability. The magnetic permeability is

$$\mu = k_m \mu_0 = (1.40)(4\pi \times 10^{-7}\ \text{H/m}) = \underline{5.6\pi \times 10^{-7}\ \text{H/m}}$$

29.20 Derive the magnetic circuit equation and show how it is analogous to Ohm's law.

▮ Consider a toroidal iron-core solenoid of such dimensions that **B** and **H** may be taken as uniform over a cross section. Ampère's law for **H** gives $Hl = NI$, where l is the mean circumferential length of the core and N is the total number of turns in the winding. The flux in the core is given by $\Phi = BA = \mu HA$, where A is the cross-sectional area. Thus the relation between winding current and core flux is

$$NI = \Phi \frac{l}{\mu A} \qquad \text{or} \qquad \text{mmf} = \Phi \mathcal{R} \tag{1}$$

where we have titled NI the *magnetomotive force* (mmf) and defined $l/(\mu A)$ to be the *reluctance* $\mathcal{R}$ of the core. It is customary to specify mmf in ampere-turns, which are, of course, the same as A; reluctance is measured in reciprocal henries (H^{-1}).

As (1) has the form of Ohm's law, with

$$\text{mmf} \leftrightarrow \text{emf} \qquad \text{flux} \leftrightarrow \text{current} \qquad \text{reluctance} \leftrightarrow \text{resistance}$$

we can view the core as the analog of a resistive electric circuit. However, for ferromagnetic substances, the analogy breaks down in one important respect: whereas $R = l/(\sigma A)$ is independent of emf and is known beforehand, $\mathcal{R} = l/(\mu A)$ depends, through μ, on the mmf. Consequently, (1) becomes a nonlinear equation in H, which must be solved by iteration or graphically.

29.21 A toroid with 1500 turns is wound on an iron ring 360 mm² in cross-sectional area, of 0.75-m mean circumference and of 1500 relative permeability. If the windings carry 0.24 A, find (*a*) the magnetizing field H, (*b*) the mmf, (*c*) the magnetic induction B, (*d*) the magnetic flux, and (*e*) the reluctance of the circuit.

▮ (*a*)
$$H = nI = \frac{1500}{0.75}\,0.24 = \underline{480\ \text{A/m}}.$$

(*b*)
$$\text{mmf} = Hl = (480\ \text{A/m})(0.75\ \text{m}) = \underline{360\ \text{A}}$$

(*c*)
$$B = \frac{4\pi\,KnI}{10^7} = \frac{4\pi}{10^7}(1500)(480) = \underline{0.905\ \text{T}}$$

(*d*)
$$\Phi = BA = (0.905\ \text{Wb/m}^2)\left(\frac{360\ \text{m}^2}{10^6}\right) = \underline{3.26 \times 10^{-4}\ \text{Wb}}.$$

(*e*)
$$\text{Reluctance} = \frac{\text{mmf}}{\Phi} = \frac{360}{3.26 \times 10^{-4}} = 1.1 \times 10^6\ \text{H}^{-1} = \underline{1.1\ \mu\text{H}^{-1}}$$

29.22 Data regarding the magnetic circuit shown in Fig. 29-5 are cross-sectional areas $A_1 = 1200\ \text{mm}^2$, $A_2 = 800\ \text{mm}^2$; lengths $l_1 = 210\ \text{mm}$, $l_2 = 430\ \text{mm}$, and length of air gap $l_a = 2\ \text{mm}$; relative permeability of lower leg of iron

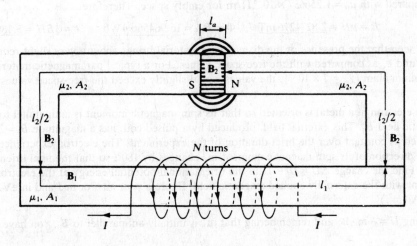

Fig. 29-5

$(\mu_{rel})_1 = 200$, for remaining iron $(\mu_{rel})_2 = 300$; total turns $N = 1000$; $I = 2.5$ A. Compute (a) the mmf and (b) total reluctance of the circuit.

▮ (a)   mmf $= NI = \underline{2500\ \text{ampere-turns}}$

(b) Reluctances in series or parallel are combined like electric resistances in series or parallel. Thus,

$$\mathcal{R}_1 = \frac{l_1}{\mu_1 A_1} = \frac{210 \times 10^{-3}}{(200 \times 4\pi \times 10^{-7})(1200 \times 10^{-6})} = 0.6963\ \mu\text{H}^{-1} \qquad \mathcal{R}_2 = \frac{l_2}{\mu_2 A_2} = 1.426\ \mu\text{H}^{-1}$$

$$\mathcal{R}_a = \frac{l_a}{\mu_0 A_2} = 1.989\ \mu\text{H}^{-1} \qquad \mathcal{R}_{tot} = \mathcal{R}_1 + \mathcal{R}_2 + \mathcal{R}_a = \underline{4.11\ \mu\text{H}^{-1}}$$

where we have assumed the air gap to have effective area A_2 (no fringing). Observe that the reluctance of the air gap is almost equal to that of the entire iron path.

29.23 Referring to Prob. 29.22 find (a) total flux threading the circuit and (b) flux density in each part.

▮ (a) $\Phi = \dfrac{\text{mmf}}{\mathcal{R}_{tot}} = \dfrac{2500}{4.11} = \underline{608\ \mu\text{Wb}}$

(b) $B_1 = \dfrac{\Phi}{A_1} = \dfrac{608 \times 10^{-6}}{1200 \times 10^{-6}} = \underline{0.507\ \text{T}} \qquad B_2 = B_a = \dfrac{\Phi}{A_2} = \dfrac{608 \times 10^{-6}}{800 \times 10^{-6}} = \underline{0.76\ \text{T}}$

29.24 What is the expression for the magnetic energy density in a material of permeability μ?

▮ The general expression for the energy per unit volume in a magnetic field is $\epsilon = \frac{1}{2}\mathbf{B} \cdot \mathbf{H}$ J/m^3. For a medium in which $\mathbf{B}$ and $\mathbf{H}$ are parallel, this becomes

$$\epsilon = \tfrac{1}{2}BH = \frac{1}{2\mu} B^2 = \frac{\mu}{2} H^2 \qquad \text{J/m}^3$$

29.25 A toroidal coil has $N = 1200$ turns; average length of core, $l = 80$ cm; cross-sectional area, $A = 60$ cm^2; current, $I = 1.5$ A. Compute B, H, total flux Φ, and energy density ϵ. Assume an empty core.

▮ Since the cross-section diameter $\ll l$, $\mathbf{B}$ is approximately uniform over any cross section, of magnitude

$$B = \mu_0 nI = (4\pi \times 10^{-7})\frac{1200}{0.80} 1.5 = \underline{2.8274334\ \text{mT}}$$

Then, $H = \dfrac{B}{\mu_0} = nI = \underline{2250\ \text{A/m}} \qquad \Phi = BA = \underline{16.964598\ \mu\text{Wb}} \qquad \epsilon = \tfrac{1}{2}BH = \underline{3.1808626\ \text{J/m}^3}$

29.26 Repeat Prob. 29.25 for a bismuth core $(\chi_m = -2 \times 10^{-6})$.

▮ H, which depends only on I, is the same as in the empty core; thus, $H = 2250$ A/m. The permeability of Bi is

$$\mu = \mu_0(1 + \chi_m) = (4\pi \times 10^{-7})(1 - 2 \times 10^{-6}) = 1.2566345 \times 10^{-6}\ \text{H/m}$$

as compared with $\mu_0 = 1.256637 \times 10^{-6}$ H/m for empty space. Therefore,

$$B = \mu H = \underline{2.8274276\ \text{mT}} \qquad \Phi = BA = \underline{16.964566\ \mu\text{Wb}} \qquad \epsilon = \tfrac{1}{2}BH = \underline{3.180856\ \text{J/m}^3}$$

It is seen that the presence of the diamagnetic material brings about a very slight reduction in the values of B, Φ, and ϵ, as compared with the free-space values. For a typical paramagnetic material, e.g., iron ammonium alum $(\chi_m = 7 \times 10^{-4})$, the values would slightly exceed the free-space values.

29.27 A free electron in a metal is oriented so that its spin magnetic moment is antiparallel to an externally applied magnetic field B_0. This external field, produced by a pulsed coil, has a magnitude $B_0 = 35.0$ T which can be regarded as constant over the brief duration of the experiment. The electron experiences a "spin flip." That is, the direction of its spin magnetic moment $\mathbf{m}_B$ changes by 180°, so that the final orientation of $\mathbf{m}_B$ is parallel to B_0. Find the change $\Delta U = U_f - U_i$ in the orientational potential energy of the electron-spin magnetic moment with the externally applied magnetic field. Express your answer in J and in eV. ($m_B = 9.27 \times 10^{-24}$ A $\cdot$ m^2.)

▮ Using $U = -\mathbf{m} \cdot \mathbf{B}$, and remembering that $\mathbf{m}_B$ is initially antiparallel to B_0, you have for the initial energy U_i

$$U_i = -\mathbf{m}_B \cdot \mathbf{B}_0 = m_B B_0 = (9.27 \times 10^{-24} \text{ A} \cdot \text{m}^2)(35.0 \text{ T}) = 3.24 \times 10^{-22} \text{ J}$$

And since the final orientation of $\mathbf{m}_B$ is parallel to B_0, you have for the final energy $U_f = -3.24 \times 10^{-22}$ J. The change in orientational energy is thus

$$\Delta U = U_f - U_i = \underline{-6.48 \times 10^{-22} \text{ J}} = \underline{-4.06 \times 10^{-3} \text{ eV}}$$

29.28 Refer to Prob. 29.27. Suppose that the metal of which the free electron is a part is at room temperature ($T = 300$ K). Is it likely that saturation will take place, that is, that the electron-spin magnetic moments of the many free electrons in the metal will all be substantially aligned parallel to the externally applied field B_0? Is it likely that saturation will take place if the metal is cooled to a temperature $T = 0.10$ K?

❙ At $T = 300$ K,

$$\frac{|\Delta U|}{kT} = \frac{6.5 \times 10^{-22} \text{ J}}{4.2 \times 10^{-21} \text{ J}} = 0.15$$

that is, there is much more energy associated with the electron's random thermal motion than with its tendency to line up with the field. Hence, it is fair to guess that the electron-spin magnetic moments of the free electrons in the metal are not substantially all aligned parallel to the externally applied magnetic field B_0.

On the other hand, at $T = 0.10$ K,

$$\frac{|\Delta U|}{kT} = \frac{6.5 \times 10^{-22} \text{ J}}{1.4 \times 10^{-24} \text{ J}} = 460$$

This value is very much greater than 1. So you can safely guess that the tendency toward alignment of the spin magnetic moments measured by $|\Delta U|$ dominates the tendency toward random alignment measured by kT, and saturation therefore takes place.

29.29 (*a*) Show that magnetic dipole moments can be given in J/T. (*b*) In the Bohr hydrogen atom, the orbital angular momentum of the electron is quantized in units of $h/2\pi$, where $h = 6.626 \times 10^{-34}$ J · s is *Planck's constant*. Calculate, in J/T, the smallest allowed magnitude of the atomic dipole moment. (This quantity is known as the *Bohr magneton*.)

❙ (*a*) $U_\theta = -\mathbf{\Pi} \cdot \mathbf{B}$ gives the unit J/T for $\mathbf{\Pi}$, since U_θ is in J and $\mathbf{B}$ is in T. On the atomic or subatomic scale, dipole moments show up through their contributions to the atomic energy. (*b*) The dipole moment, Π, is directly proportional to the angular momentum, mvr, of the electron in its circular orbit; in fact,

$$\Pi = IA = \frac{e}{2\pi r/v} \pi r^2 = \frac{e}{2m} mvr$$

Thus, the Bohr magneton is given by

$$\Pi_{\min} = \frac{e}{2m} \frac{h}{2\pi} = \frac{1.602 \times 10^{-19} \text{ C}}{2(9.109 \times 10^{-31} \text{ kg})} \frac{6.626 \times 10^{-34} \text{ J} \cdot \text{s}}{2\pi} = 9.27 \times 10^{-24} \text{ C} \cdot \text{J} \cdot \text{s/kg} = \underline{9.27 \times 10^{-24} \text{ J/T}}$$

29.30 A toroidal coil is wound on a ring made of a good grade of iron for which, for $NI = 12\,000$ ampere-turns and a mean radius $R = 150$ mm, $B = 1.8$ T. Compute H, M (assumed constant), χ_m, and the equivalent ampere-turns contributed by the magnetic dipoles of the iron.

❙

$$H = \frac{NI}{2\pi R} = \frac{12\,000}{2\pi(0.150)} = \underline{12.732 \text{ kA/m}}$$

From $B = \mu H$,

$$\mu = \frac{1.8}{12.732 \times 10^3} = 1.4137 \times 10^{-4} \text{ H/m}$$

Then, from $\mu = \mu_0(1 + \chi_m)$, $\chi_m = 111.5$ and

$$M = \chi_m H = (111.5)(12.732 \times 10^3) = \underline{1420 \text{ kA/m}}.$$

Noting that $B = \mu_0(H + M) = \mu_0(1 + \chi_m)H$, while without the core, $B_0 = \mu_0 H$, we have that the equivalent number of ampere-turns contributed by the magnetic dipoles of the iron is

$$\chi_m(NI) = (111.5)(12\,000) = \underline{1.338 \times 10^6}$$

which is more than 100 times the mmf of the coil current. This effective current is called the *magnetization current*, or *Amperean surface current*. Thus, to establish $B = 1.8$ T in an air core would require $12\,000 + 1.338 \times 10^6 = \underline{1.35 \times 10^6}$ ampere-turns.

29.31 The number of atoms per cubic meter of iron is about 8.5×10^{28} (Prob. 29.11). Assuming that each has a magnetic dipole moment of one Bohr magneton, and that for $NI = 3000$ ampere-turns on the ring of Prob. 29.30 one-quarter of the dipoles are aligned with **H** compute M, χ_m, μ, B.

▌ Taking the value of the Bohr magneton from Prob. 29.29 we have

$$M = \frac{1}{4}(8.5 \times 10^{28})(9.27 \times 10^{-24}) = 1.97 \times 10^5 \text{ J/T} \cdot \text{m}^3 = \underline{197 \text{ kA/m}} \qquad H = \frac{3000}{2\pi(0.150)} = \underline{3.1831 \text{ kA/m}}$$

$$\chi_m = \frac{M}{H} = \underline{61.9} \qquad \mu = \mu_0(1 + \chi_m) = 7.9 \times 10^{-5} \underline{\text{H/m}} \qquad B = \mu H = \underline{0.25 \text{ T}}.$$

29.32^c A long straight wire of circular cross section is made of nonmagnetic material (that is, $K_m = 1$ to a good approximation). It is of radius a. The wire carries a current i which is uniformly distributed over its cross section. Compute the energy per unit length stored in the magnetic field contained within the wire.

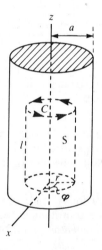

Fig. 29-6

▌ A section of the wire is shown in Fig. 29-6. We use a standard right-handed coordinate system (r, φ, z). Since the current is entirely axial, the Biot–Savart law implies that $B_z = 0$ everywhere. Symmetry considerations require that the r and φ components of **B** depend only upon r: $\mathbf{B} = B_r(r)\hat{\mathbf{r}} + B_\varphi(r)\hat{\boldsymbol{\varphi}}$. Now we apply Gauss' law for magnetic fields to the coaxial cylindrical surface S, which has length l and radius r, as shown in the figure. We obtain

$$0 = \int_S \mathbf{B} \cdot d\mathbf{a} = B_r(r)2\pi rl$$

Hence $B_r(r) = 0$, so **B** is purely azimuthal. Applying Ampere's law to the coaxial circular path C, we obtain

$$\oint_C \mathbf{B} \cdot d\mathbf{l} = 2\pi r B_\varphi(r) = \mu_0 i(r) \tag{1}$$

Here $i(r)$ is the current passing through the surface enclosed by C. Since the current density is uniform, we have

$$\frac{i(r)}{i} = \frac{\pi r^2}{\pi a^2} \qquad \text{or} \qquad i(r) = \frac{r^2 i}{a^2} \tag{2}$$

Equations (1) and (2) imply that $B_\varphi(r) = (\mu_0 ir)/(2\pi a^2)$. Now we must integrate the magnetic energy density over the volume of the wire. The magnetic energy dU_m contained in a cylindrical shell of length l, inner radius r, and outer radius $r + dr$ is

$$dU_m = \varepsilon_m \, dV = \frac{B_\varphi^2(r)}{2\mu_0} 2\pi r l \, dr = \frac{\mu_0 i^2 l}{4\pi a^4} r^3 \, dr$$

Hence the magnetic energy per unit length contained in the wire is

$$\frac{U_m}{l} = \frac{1}{l} \int dU_m = \frac{\mu_0 i^2}{4\pi a^4} \int_0^a r^3 \, dr = \frac{\mu_0 i^2}{4\pi a^4} \left(\frac{r^4}{4} \Big|_0^a \right) = \frac{\mu_0 i^2}{16\pi}$$

29.33 The external magnetic field of a spherical object of radius R, surrounded by vacuum and carrying an (idealized) central point magnetic dipole, contains magnetic energy $U = (\pi/2)[(B_p^2 R^3)/(2\mu_0)]$. The quantity B_p is the maximum magnetic field strength at the surface of the object—that is, the value of B at the object's magnetic poles.

   (*a*) The earth's external magnetic field is not that of a pure centered point dipole, nor is the earth surrounded by a perfect vacuum. Nevertheless, a reasonable estimate of the energy content U in the external field is $U \approx (B_p^2 R^3)/(2\mu_0)$. Evaluate this, using $B_p = 6.0 \times 10^{-5}$ T and $R = 6.4 \times 10^6$ m.

   (*b*) Compare the energy U to the total annual usage of electric energy in the United States. It was 1.7×10^{12} kW · h in 1972.

   ▌ (*a*) With $B_p = 6.0 \times 10^{-5}$ T, $R = 6.4 \times 10^6$ m, and $\mu_0 = 4\pi \times 10^{-7}$ N/A^2, we have

$$U \approx \frac{B_p^2 R^3}{2\mu_0} = \frac{(6.0 \times 10^{-5})^2 (6.4 \times 10^6)^3}{2(4\pi \times 10^{-7})} = \underline{3.8 \times 10^{17} \text{ J}}$$

   (*b*) The electric energy usage in the Unites States for 1972 was

$$E = 1.7 \times 10^{12} \text{ kW} \cdot \text{h} = (1.7 \times 10^{12})(10^3 \text{ J/s})(3600 \text{ s}) = \underline{6.1 \times 10^{18} \text{ J}}$$

29.2 MAGNETS; POLE STRENGTH

29.34 Describe how the torque on a bar magnet placed in a magnetic field can be expressed in terms of pole strength.

   ▌ When a bar magnet is placed with its axis perpendicular to a uniform magnetic field of intensity **B**, it experiences a torque $\tau = Bm = Bpl$, where m is the magnetic moment, l is the length of the magnet, and p is the pole strength of the magnet. Fundamentally, this torque arises from the interaction between the magnetic induction and the infinitesimal current loops associated with the orbits and spin of certain of the magnet's electrons. An alternative approach, frequently convenient, is to endow the magnet with a pair of equal and oppositely signed *poles* at its two ends, and to attribute the torque to the equal and opposite forces exerted by the field B on these poles. When a pole of strength p is placed in a magnetic field of intensity B, the pole experiences a force $F = Bp$. In general, $\tau = mB \sin \theta$, where θ = angle between **m** and **B**.

29.35 An iron bar magnet of length 10 cm and cross section 1.0 cm^2 has a magnetization of 10^2 A/m. Calculate the magnet's magnetic pole strength.

   ▌ The magnetic moment of a bar magnet of length $2d$ has magnitude $m = 2 |p| d$, where $|p|$ is the pole strength. In terms of the magnetization **M**, the magnetic moment of the bar magnet is $\mathbf{m} = \mathbf{M}(2ad)$, where a is the cross-sectional area. Hence $|p| = m/2d = aM$. With $a = 1.0$ cm^2 = 1.00×10^{-4} m^2 and $M = 1.00 \times 10^2$ A/m, we obtain $|p| = (1.00 \times 10^{-4})(1.00 \times 10^2) = \underline{1.00 \times 10^{-2} \text{ A} \cdot \text{m}}$.

29.36 The field of a bar magnet can be considered to be caused by a surface current flowing on the surface of the magnet. If a bar magnet is to act like a solenoid whose interior field is 0.3 T, how large a surface current must flow on each centimeter length of the bar?

   ▌ In Fig. 29-7 take segment length $l = 1.0$ cm. Since B is axial and ~0 outside the solenoid, the circuital law yields Bl equal to $\mu_0 I_m$, so $I_m = 0.3(10^{-2})/(4\pi \times 10^{-7}) = \underline{2390 \text{ A/m}} = 23.9$ A/cm.

29.37 What is the relationship between the magnetization (or Amperean) surface current I_m in a long bar magnet and the pole strength?

   ▌ From Prob. 29.35, $p = aM$. Applying the circuital law for **B** to the magnet (Fig. 29.7), and noting that $\mathbf{B} = \mu_0(\mathbf{H} + \mathbf{M})$ and that **H** cannot contribute since there are no true currents, we get $Ml = I_m$: Then $p = aI_m/l$.

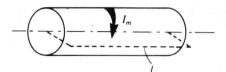

Fig. 29-7

29.38 If a long bar magnet of length l and pole strength p is cut in half, what is the pole strength of the magnets thus formed?

▎ From Prob. 29.37 $p = i_a a$, where i_a is the surface current per unit length. Thus p does not depend on the length of the magnet but only on the alignment of the microscopic dipole whorls. Thus p for each half is the same as for the original magnet.

29.39 The needle of a small magnetic pocket compass has length $l = 3$ cm. It has rectangular cross section, and its width w and thickness t are both small compared with its length. When the needle is disturbed slightly, it oscillates about its equilibrium north-south orientation with a frequency $v = 2$ Hz. If the horizontal component of the earth's magnetic field is $\mathscr{B}_h = 5 \times 10^{-5}$ T, what is the value of the Amperean surface current i_a which circulates around the axis of the needle? Take the density of the needle to be $\rho = 8 \times 10^3$ kg/m³.

▎ Consider the compass needle as a torsion pendulum about a vertical axis in which the restoring torque is $T = mB_h \sin\theta \approx mB_h\theta$, where θ is the (assumed small) angle between the magnetic field lines of B_h and the dipole moment **m** of the needle. Since m and B_h are constants, the torque T obeys the rotational form of Hooke's law, with torsion constant $k = mB_h$.

The oscillation frequency of a torsion pendulum is given by $v = (1/2\pi)(k/I)^{1/2}$, where k is the torsion constant and I is the moment of inertia about the vertical axis through the center of the needle. The moment of inertia of a thin bar about a perpendicular axis through its center is given by $I = \frac{1}{12}Ml^2$, where M is the mass of the bar and l is its length. The moment of inertia is thus $I = \frac{1}{12}\rho wtl^3$.

Substitute the values of the torsion constant k and the moment of inertia I into the expression for the frequency v and solve for m:

$$m = \frac{\pi^2 v^2 \rho wtl^3}{3B_h}$$

The Amperean surface current is, from Prob. 29.37,

$$Ml = \frac{m}{wt} = \frac{\pi^2 v^2 \rho l^3}{3B_h} = \frac{\pi^2 (2\text{ Hz})^2 (8 \times 10^3\text{ kg/m}^3)(3 \times 10^{-2}\text{ m})^3}{(3)(5 \times 10^{-5}\text{ T})} = \underline{57\text{ kA}}.$$

—a large current.

29.40 What is Coulomb's law for magnetic poles, and what is the magnetic field due to a pole?

▎ If two magnetic poles of strengths p_1 and p_2 (amperes × meters) interact, the force between them (in newtons) is given by Coulomb's law:

$$F = \frac{\mu_0}{4\pi} \frac{p_1 p_2}{r^2} = \frac{p_1 p_2}{10^7 r^2}$$

where r is the distance (in meters) between the poles. The magnetic induction B produced at a distance r from a pole of strength p is thus given by

$$B = \frac{p}{10^7 r^2} \quad \text{T}$$

If the induction arises from contributions of several poles, the resultant **B** is found by adding vectorially the contributions of the individual poles. Note that Coulomb's law for poles gives the **B** field only outside the magnet. Inside the magnet the **B** field is accurately given only by the Amperean surface current approach.

29.41 A toroid has an iron core. Measurement shows that the internal magnetic field in the iron core is $B_{\text{int}} = 0.60$ T when the current in the winding of 1000 turns per meter is 1.0 A. (*a*) What is B_0, the magnetic field applied externally to the core (the field present if there were no iron core)? (b) What is B_d, the field due to magnetization ("demagnetizing field")? (c) What is i_s/l, the equivalent Amperean surface current per unit length?

▌ (a) According to Prob. 28.90, the externally applied field along the axis of the toroid is $B_0 = \mu_0 ni$. With $n = 1000 \text{ m}^{-1}$ and $i = 1.0 \text{ A}$, we find $B_0 = (4\pi \times 10^{-7})(1000)(1.0) = \underline{1.26 \text{ mT}}$.
(b) The "demagnetizing field" $\mathbf{B}_d = \mathbf{B}_{\text{int}} - \mathbf{B}_0$. Since $\mathbf{B}_0$ and $\mathbf{B}_{\text{int}}$ are parallel, the magnitude B_d is given by $B_d = B_{\text{int}} - B_0 = 0.60 \text{ T} - 1.26 \times 10^{-3} \text{ T} = \underline{0.599 \text{ T}}$.
(c) From the analog of $B_0 = \mu_0 ni$ for the Amperean current—i.e., $B_d = (\mu_0 i_s)/l$—we obtain

$$\frac{i_s}{l} = \frac{B_d}{\mu_0} = \frac{0.599}{(4\pi)(10^{-7})} = \underline{476 \text{ kA/m}}.$$

29.42 Refer to Prob. 29.41. (a) If there were no iron core, what current would be necessary in the winding for B to equal 0.60 T? (b) What is K_m, the relative permeability of iron, under the given conditions?

▌ (a) In order to have $B_0' = \mu_0 ni' = 0.60 \text{ T}$, we would need

$$i' = \frac{B_0'}{\mu_0 n} = \frac{0.60}{(4\pi)(10^{-7})(10^3)} = \underline{477 \text{ A}}$$

(b) The relative permeability $K_m = B_{\text{int}}/B_0$, so with $B_{\text{int}} = 0.60 \text{ T}$ and $B_0 = 1.257 \times 10^{-3} \text{ T}$, we obtain $K_m = \underline{477}$. Note that $K_m = i'/i$.

29.43 A bar magnet has a magnetic moment of $5 \text{ A} \cdot \text{m}^2$ and a length of 0.1 m. It is placed in a uniform magnetic field of intensity 0.4 T, with its axis making an angle of 60° with the field. Find the torque on the magnet. Find the torque on the magnet.

▌ $$\tau = Bm \sin \theta = 0.4(5) \sin 60° = 2(0.866) = \underline{1.732 \text{ N} \cdot \text{m}}$$

29.44 Find the pole strength of the bar magnet in Prob. 29.43.

▌ $$m = pl \quad \text{or} \quad 5 = p(0.1) \quad \text{and} \quad p = \underline{50 \text{ A} \cdot \text{m}}$$

29.45 Two very long identical bar magnets are placed in line with their N poles 60 mm apart. If the repulsive force is 0.2 N, what is the pole strength of each magnet?

▌ Let each N pole be of strength p.

$$F = \frac{p^2}{10^7 r^2} \quad \text{so} \quad 0.2 = \frac{p^2}{10^7(0.06)^2} \quad \text{and} \quad p^2 = (2 \times 10^6)(0.06)^2 \quad \text{Finally} \quad \underline{p = 85 \text{ A} \cdot \text{m}}$$

29.46 The poles of a bar magnet have a strength of $20 \text{ A} \cdot \text{m}$ each and are 25 cm apart. Find the magnitude and direction of the magnetic induction **B** at a point 25 cm from both poles.

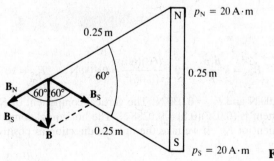

Fig. 29-8

▌ The situation is depicted in Fig. 29-8.

$$B_N = \frac{p_N}{10^7 r^2} = \frac{20}{10^7(0.25)^2} = 3.2 \times 10^{-5} \text{ T} \qquad B_S = \frac{p_S}{10^7 r^2} = \frac{20}{10^7(0.25)^2} = 3.2 \times 10^{-5} \text{ T}$$

Vectors $\mathbf{B}_N$ and $\mathbf{B}_S$ and $\mathbf{B}$ form the three sides of an equilateral triangle and are all equal in magnitude. The

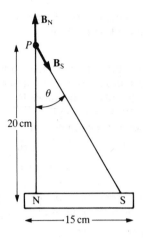

Fig. 29-9

resultant magnetic induction is $\mathbf{B} = \underline{3.2 \times 10^{-5}\,\text{T}}$; its direction is parallel to the axis of the magnet and away from the N pole.

29.47 A magnet 15 cm long with poles of strength 250 A · m lies on a table. Find the magnitude of the magnetic intensity B at a point P 20 cm directly above the north pole of the magnet (Fig. 29-9).

▮ The magnetic fields are given by

$$B_N = \frac{p_N}{10^7 r^2} = \frac{250}{10^7 (0.2)^2} = 625\,\mu\text{T} \qquad B_S = \frac{p_S}{10^7 r^2} = \frac{250}{10^7 (0.25)^2} = 400\,\mu\text{T}$$

The vertical component of B_S is $B_S \cos\theta = 320\,\mu\text{T}$, and the horizontal component is $B_S \sin\theta = 240\,\mu\text{T}$. The vertical component of the resultant intensity is $625 - 320 = 305\,\mu\text{T}$, and the horizontal component is $240\,\mu\text{T}$. The resultant B is $\sqrt{305^2 + 240^2} = \underline{390\,\mu\text{T}}$.

29.48 Two small identical magnets are 8 cm long and have pole strengths of 60 A · m. When they are restrained so they cannot rotate, it is possible to make one "float" 6 cm above the other (Fig. 29-10). Find the force which the lower magnet exerts on the upper one (which is the weight of the upper magnet, of course).

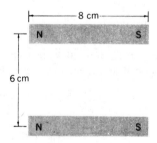

Fig. 29-10

▮
$$F_{NN} = \kappa \frac{p_1 p_2}{r^2} = 10^{-7} \frac{(60)(60)}{(0.06)^2} = 0.100\,\text{N} \qquad F_{NS} = 10^{-7} \frac{(60)(60)}{(0.1)^2} = 0.036\,\text{N}$$

Similarly $F_{SS} = 0.100\,\text{N}$ and $F_{SN} = 0.036\,\text{N}$. The vertical component of F_{NS} is $(0.036)(0.6) = 0.0216\,\text{N}$, and the horizontal component is $(0.036)(0.8) = 0.0288\,\text{N}$. The horizontal component of F_{SN} exactly balances the horizontal component of F_{NS}. If we take the upward direction as positive, the vertical force on the upper magnet is given by

$$F = 0.100 + 0.100 - 0.0216 - 0.0216 = \underline{0.1568\,\text{N}}$$

29.49 A magnet has a pole strength of 50 A · m. It is placed at right angles to the earth's magnetic field of 20 μT and a torque of 60 μN · m acts on it. Find the magnetic moment of the magnet and the distance between the poles. What would the torque be if the magnet were rotated so its axis made an angle of 53.1° (sin 53.1° = 0.8) with the field?

▮ Let m be the magnetic moment. Then

$$\tau = mB \sin(\mathbf{m}, \mathbf{B}) \qquad \text{and} \qquad 60\,\mu\text{N} \cdot \text{m} = m(20\,\mu\text{T})1$$

Thus $m = \underline{3\,\text{A} \cdot \text{m}^2}$. From $m = pl$; $l = (\frac{3}{50})\text{m} = \underline{60\,\text{mm}}$. At 53.1°, $\tau = (3)(20 \times 10^{-6}) \sin 53.1° = \underline{48\,\mu\text{N} \cdot \text{m}}$.

29.50 What torque is necessary to hold the axis of a magnet at an angle of 50° to the magnetic meridian where the horizontal component of the earth's magnetic field is $20\,\mu\text{T}$? The length of the magnet is 125 mm, and its pole strength is $40\,\text{A} \cdot \text{m}$.

▮ The magnetic moment $m = pl$, so $\tau = mB \sin \theta = (40 \times 0.125)(20 \times 10^{-6})(\sin 50°) = \underline{76.6\,\mu\text{N} \cdot \text{m}}$.

29.51 A bar magnet has poles of strength $12\,\text{A} \cdot \text{m}$ and a magnetic moment of $m = 0.96\,\text{A} \cdot \text{m}^2$. This magnet is placed perpendicular to the earth's magnetic field, which has a horizontal component of $25\,\mu\text{T}$. (**a**) Find the torque acting to rotate the magnet about a vertical axis. (**b**) Find the magnitude of the resultant horizontal magnetic induction at point P on the south-north axis 0.2 m from the north pole of the magnet (Fig. 29-11).

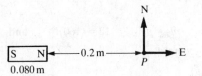

Fig. 29-11

▮ (**a**) $\tau = mB \sin \theta = (0.96)(25 \times 10^{-6})(1) = \underline{24\,\mu\text{N} \cdot \text{m}}$
(**b**) $m = pl \qquad 0.96 = 12l \qquad l = \underline{80\,\text{mm}}$.

$$B_{\text{earth}} = 25\,\mu\text{T} \qquad \text{(north)}$$

$$B_{\text{mag}} = 10^{-7}\frac{p_N}{r_N^2} - 10^{-7}\frac{p_S}{r_S^2} = 10^{-7}\frac{12}{0.2^2} - 10^{-7}\frac{12}{0.28^2} = 10^{-7}(300 - 153.06) = 14.69\,\mu\text{T} \qquad \text{(east)}$$

resultant $B = \sqrt{25^2 + 14.69^2} = \underline{29.0\,\mu\text{T}}$

29.52 Two magnetic poles have strengths of 50 and $90\,\text{A} \cdot \text{m}$, respectively. At what distance in air will the force of attraction between them be 1 mN?

▮ The poles must have opposite polarities. Solving $F = (10^{-7})[(p_1 p_2)/r^2]$, we find $r = \underline{0.671\,\text{m}}$.

29.53 Find the magnetic induction at a point 120 mm from the center of a magnet on the perpendicular bisector of the line joining the poles if the magnet is 100 mm long and its pole strength is $40\,\text{A} \cdot \text{m}$.

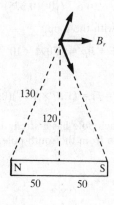

Fig. 29-12

▮ See Fig. 29-12.

$$120^2 + 50^2 = 16\,900 \qquad \sqrt{16\,900} = 130\,\text{mm}$$

Vertical components of $\mathbf{B}_N + \mathbf{B}_S$ cancel; horizontal components add.

$$B_N = B_S = \frac{(10^{-7})(40)}{0.13^2} \qquad \text{so} \qquad B_r = \frac{4 \times 10^{-6}}{1.69 \times 10^{-2}}(2)\frac{5}{13} = \underline{182\,\mu\text{T}}$$

parallel to magnet in direction of N to S.

29.54 A long bar of cobalt steel with a mass of 40 g when placed horizontally over a similar bar, both being equally magnetized, remains suspended at a distance of 9 mm above it (e.g., Fig. 29-10). What is the pole strength at each end of each bar? *Hint*: Assume that the bars are so long that only the nearest poles interact significantly.

❚ From equilibrium,

$$F = (0.040)(9.8) = 2\frac{p^2 \times 10^{-7}}{(9 \times 10^{-3})^2} \qquad p^2 = \frac{(0.392)(81 \times 10^{-6})}{2 \times 10^{-7}} = 158.8 \qquad p = \underline{12.6\,\text{A}\cdot\text{m}}$$

29.55 The axes of two magnets are collinear. One has poles of strength 80 A · m separated by 125 mm, and the second has a magnetic moment of 12 A · m^2 with poles of strength 160 A · m. Find the attractive force between the magnets if the north pole of one is 45 mm from the south pole of the second.

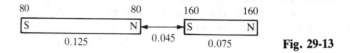

Fig. 29-13

❚ $\qquad m_2 = p_2 l_2 \qquad 12 = (160)l_2 \qquad \text{and} \qquad l = 0.075\,\text{m}$

so the situation is as in Fig. 29-13.

$$F_{\text{SS}} = 10^{-7}\frac{(80)(160)}{0.17^2} = 44.3\,\text{mN} \quad (\text{repulsive}) \qquad F_{\text{NN}} = 10^{-7}\frac{(80)(160)}{0.12^2} = 88.9\,\text{mN} \quad (\text{repulsive})$$

$$F_{\text{S}_1\text{N}_2} = 10^{-7}\frac{(80)(160)}{0.245^2} = 21.32\,\text{mN} \quad (\text{attractive}) \qquad F_{\text{N}_1\text{S}_2} = 10^{-7}\frac{(80)(160)}{0.045^2} = 632\,\text{mN} \quad (\text{attractive})$$

resultant $F = 632 + 21.32 - 44.3 - 88.9 = \underline{520\,\text{mN}}$ (attractive).

29.56 A small test magnet has poles of 75 A · m strength separated by 25 mm. Find the force on each pole when this magnet is placed at the center of a circular loop with a radius $a = 375$ mm in which a current of 12.5 A is flowing. (Assume that the magnetic intensity is uniform over the volume occupied by the magnet.) What is the resultant force on a pole? Find the torque on the magnet if its axis is perpendicular to the magnetic field.

❚ We assume that the strength of the magnet is not affected by the external field. From Prob. 28.90 the field due to the loop at its center is

$$B = \frac{2\pi\,I}{10^7 a} = \frac{(2\pi)(12.5)}{(10^7)(0.375)} = 2.094 \times 10^{-5}\,\text{T}$$

Since the magnet is small compared with the loop,

$$F = Bp = (2.094 \times 10^{-5})(75) = 1.57\,\text{mN}$$

in magnitude for each pole.

$$\tau = Fl = (1.57 \times 10^{-3})(0.025) = \underline{39\,\mu\text{N}\cdot\text{m}}$$

29.57 A bar magnet is 128 mm long, and each pole has a strength of 64 A · m. Find the magnitude of the intensity of the magnetic field at a point 96 mm from the south pole, measured at right angles to the axis of the magnet.

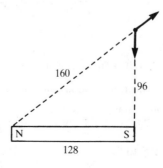

Fig. 29-14

▌ See Fig. 29-14.

$$B_S = \frac{(10^{-7})(64)}{0.096^2} = 694 \,\mu T \qquad B_N = \frac{(10^{-7})(64)}{0.16^2} = 250 \,\mu T \qquad (B_N)_x = 0.8 B_N = 200 \,\mu T \qquad (B_N)_y = 0.6 B_N = 150 \,\mu T$$

$$B_x = 200 \,\mu T \qquad B_y = -694.4 + 150 = -544.4 \,\mu T \qquad B = \sqrt{200^2 + 544.4^2} = \underline{580 \,\mu T}$$

29.58 A north pole of 60 A · m strength is placed 150 mm from a south pole of 90 A · m strength. How far from the north pole, on a line drawn through the two poles, will the resultant field due to these poles be zero?

```
        90              60
   ┌─────────┐     ┌─────────┐              P
   │    S    │ 150 mm │    N    │           •←•→•
   └─────────┘     └─────────┘         B_S      B_N
                                   ←────── d ──────→        Fig. 29-15
```

▌ The situation is as shown in Fig. 29-15. We ignore the effect of any other poles.

$$\frac{60}{d^2} = \frac{90}{(d + 0.15)^2} \qquad 2(d^2 + 0.30d + 0.0225) = 3d^2 \qquad d^2 - 0.60d - 0.0450 = 0$$

$$d = \frac{0.6 \pm \sqrt{0.36 + 0.18}}{2} = \frac{0.6 \pm 0.7348}{2} \quad (\text{need} + \text{sign}) \qquad = \frac{1.3348}{2} = \underline{0.6674 \text{ m}}$$

29.59 A bar magnet 75 mm long with a pole strength of 20 A · m is horizontal, at right angles to the earth's magnetic field, with its north pole pointing west. Find the intensity of the magnetic field in a horizontal plane at a point 225 mm west of the north pole. Take the horizontal component of the earth's magnetic field as 20 μT.

```
                    N
                    ↑
                 20 μT = B_E
   W ←───────────0.225───────────→ ┌───┐
                                    │N S│
                                    └───┘
                                    0.075        Fig. 29-16
```

▌ The situation is as shown in Fig. 29-16.

$$B_N = 10^{-7}\frac{20}{0.225^2} = 39.51 \,\mu T \quad (\text{west}) \qquad B_S = 10^{-7}\frac{20}{0.3^2} = 22.22 \,\mu T \quad (\text{east})$$

$$B_{\text{mag}} = 39.51 - 22.22 = 17.29 \,\mu T \quad (\text{west}) \qquad \text{Resultant } B = \sqrt{20^2 + 17.29^2} = 26.44 \,\mu T$$

at an angle $\theta = 49.2°$ N of W.

29.60 A magnet with poles of 40 A · m strength separated by a distance of 25 mm is placed in the uniform field inside a solenoid with its axis at right angles to the lines of force. If the solenoid is 0.4 m long and has 600 turns bearing a current of 3.5 A, what torque does the field exert on the magnet?

▌ As in Prob. 29.56 we ignore the effect of the external field on the strength of the magnet. The dipole moment of the magnet is

$$m = pl = (40 \text{ A} \cdot \text{m})(0.025 \text{ m}) = 1 \text{ A} \cdot \text{m}^2$$

For the solenoid we have

$$B = \frac{4\pi nI}{10^7} = \frac{4\pi}{10^7}\frac{600}{0.4}3.5 = 6.60 \text{ mT} \qquad \text{Then} \qquad \tau = mB \sin \theta = (1 \text{ A} \cdot \text{m}^2)(6.60 \text{ mT})(1) = \underline{6.60 \text{ mN} \cdot \text{m}}$$

29.61 The poles of a magnet are 95 mm apart, and each has a strength of 19 A · m. What is the magnitude of the magnetic intensity at a point 76 mm from one pole and 57 mm from the other? (Hint: 57 : 76 : 95 = 3 : 4 : 5.)

▌ $B_1 = 10^{-7}\frac{19}{0.076^2} = 328.9 \,\mu T \qquad B_2 = 10^{-7}\frac{19}{0.057^2} = 584.8 \,\mu T \qquad \text{resultant } B = \sqrt{328.9^2 + 584.8^2} = \underline{671 \,\mu T}$

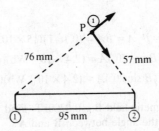

Fig. 29-17

CHAPTER 30
Induced emf; Generators and Motors

30.1 CHANGE IN MAGNETIC FLUX, FARADAY'S LAW, LENZ'S LAW

30.1 State *Lenz's law*. What is its relation to conservation of energy?

▌ Lenz's law states that an induced current, through its magnetic field, always opposes the flux change inducing the emf that causes the current. In other words, a flux change through a loop is opposed by the magnetic flux induced in the loop, or the change in a current is opposed by the resulting induced emf. If Lenz's law, a corollary of the conservation of energy law, were not true, the induced magnetic field would add to the original field, and the reinforced field would induce a still larger current in the loop—in clear violation of the conservation of energy law.

30.2 Show that the dimensions of $\Delta\Phi_m/\Delta t$ and of V are the same.

▌ The SI unit for B is the tesla, and $1\,T = 1\,N\cdot s/C\cdot m = 1\,V\cdot s/m^2$ (since $1\,N/C = 1\,V/m$). Hence $\Phi_m =$ induction × area is measured in $V\cdot s$, and so $\Delta\Phi_m/\Delta t$ is measured in V.
According to Faraday's law, emf will thus be measured in V. This should not obscure the fact that no potential function can be associated with emf, which performs nonzero work around a closed path.

30.3 In Fig. 30-1 there is a $+x$-directed magnetic field of 0.2 T. Find the magnetic flux through each face of the box.

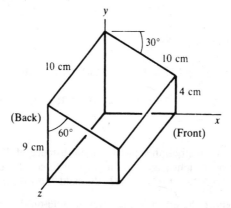

Fig. 30-1

▌ Let $\hat{\mathbf{n}}$ represent the outward unit normal vector to a given face of the box and A the area of that face. Then the outward flux through the face is $\Phi = \mathbf{B}\cdot\hat{\mathbf{n}}A = BA\cos\theta$, where $\theta =$ angle $(\mathbf{B}, \hat{\mathbf{n}})$.
Clearly $\Phi = 0$ through the two side faces ($\hat{\mathbf{n}}$ in $\pm z$ direction) and the bottom face ($\hat{\mathbf{n}}$ in $-y$ direction). Through the front and back faces $\hat{\mathbf{n}}$ is along $+x$ and $-x$, respectively, so

$$\Phi_{\text{front}} = (0.2\,T)(40\times10^{-4}\,m^2)(1) = \underline{0.8\,mWb} \qquad \Phi_{\text{back}} = (0.2\,T)(90\times10^{-4}\,m^2)(-1) = \underline{-1.8\,mWb}.$$

(The minus sign indicates flux is inward through surface.) For the top surface $\theta = 60°$ and $\Phi_{\text{top}} = (0.2\,T)(100\times10^{-4}\,m^2)(1/2) = \underline{1.0\,mWb}$.

30.4 The quarter-circle loop shown in Fig. 30-2 has an area of 15 cm². A magnetic field, with $B = 0.16$ T, exists in the $+x$ direction. Find the flux through the loop in each orientation shown.

▌ We know that $\Phi = B_\perp A$.

(a) $$\Phi = B_\perp A = BA = (0.16\,T)(15\times10^{-4}\,m^2) = \underline{240\,\mu Wb}$$

(b) $$\Phi = (B\cos20°)A = (2.4\times10^{-4}\,Wb)(\cos20°) = \underline{226\,\mu\,Wb}$$

(c) $$\Phi = (B\sin20°)A = (2.4\times10^{-4}\,Wb)(\sin20°) = \underline{82\,\mu\,Wb}$$

30.5 A loop of wire is placed in a magnetic field $\mathbf{B} = 0.0200\mathbf{i}$ T. Find the flux through the loop if its area vector is $\mathbf{A} = 30\mathbf{i} + 16\mathbf{j} + 23\mathbf{k}$ cm². What is the angle between $\mathbf{B}$ and $\mathbf{A}$?

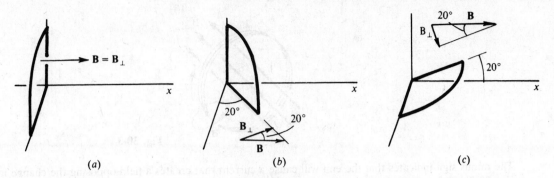

(a)   (b)   (c)

Fig. 30-2

▮ $\Phi = \mathbf{B} \cdot \mathbf{A} = (0.0200\mathbf{i}\ \text{T})(30\mathbf{i} + 16\mathbf{j} + 23\mathbf{k}\ \ 10^{-4}\ \text{m}^2) = \underline{60\ \mu\text{Wb}}$. In terms of magnitudes, $\Phi = BA \cos\theta$, with $B = 0.0200$ T and $A = (30^2 + 16^2 + 23^2)^{1/2} = 41.0$ cm^2. Then 6.0×10^{-5} Wb $= (0.02\ \text{T})(41 \times 10^{-4}\ \text{m}^2) \cos\theta$, and $\cos\theta = 0.73$, or $\theta = \underline{43°}$.

30.6   The magnetic field in a certain region is given by $\mathbf{B} = 40\mathbf{i} - 18\mathbf{k}$ G. How much flux passes through a 5.0-cm^2-area loop in this region if the loop lies flat on the xy plane?

▮ In this case $\mathbf{A} = 5.0 \times 10^{-4}\mathbf{k}$ m^2 and $\mathbf{B} = 40\mathbf{i} - 18\mathbf{k}\ \ 10^{-4}$ T. Then $\Phi = \mathbf{B} \cdot \mathbf{A} = \underline{-900\ \text{nWb}}$.

30.7   A circular coil of wire (N loops, radius a) lies on a table. On top of it and bisecting it lies a long straight wire carrying a current $i = i_0 \sin \omega t$. Find the induced emf in the coil. Repeat if the straight wire is perpendicular to the tabletop.

▮ In both cases the answer is zero, since the net flux through the coil at any instant is zero. In the first case, by symmetry, as many flux lines enter as leave the plane of the coil. In the second case, the field lines are in the plane of the coil and hence don't pass through it.

30.8   A long straight wire carrying a current $i = i_0 \sin \omega t$ lies on the axis of a solenoid (n turns per meter and radius a). Find the emf induced in the solenoid by the current in the straight wire.

▮ $\mathbf{B}$ due to the long wire is perpendicular to the cross-sectional-area vector, $\Phi = \mathbf{B} \cdot \mathbf{A} = 0$, so the induced emf $= 0$.

30.9   A flux of 900 μWb is produced in the iron core of a solenoid. When the core is removed, a flux (in air) of 0.5 μWb is produced in the same solenoid by the same current. What is the relative permeability of the iron?

▮ $\Phi = BA$ inside the solenoid. Thus

$$K_m = \frac{B_I}{B_A} = \frac{\Phi_I}{\Phi_A} = \frac{900}{0.5} = \underline{1800}$$

30.10   A solenoid 600 mm long has 5000 turns on it and is wound on an iron rod of 7.5 mm radius. Find the flux through the solenoid when the current in it is 3 A. The relative permeability of the iron is 300.

▮ For a solenoid, $B = K_m\mu_0 nI$, where n = number of windings per unit length. Hence,

$$\Phi = BA = K_m\mu_0 nIA = (300)(4\pi \times 10^{-7})\frac{5000}{0.600}(3)(56.2\pi \times 10^{-6}) = \underline{1.66\ \text{m Wb}}$$

30.11   The perpendicular component of the external magnetic field through a 10-turn coil of radius 50 mm increases from 0 to 18 T in 3 s, as shown in Fig. 30-3. If the resistance of the coil is 2 Ω, what is the magnitude of the induced current? What is the direction of the current?

▮ The initial flux ϕ_1 is zero, and the final flux ϕ_2 is

$$\phi_2 = NB_n A = (10)(18\ \text{T})(25\pi \times 10^{-4}\ \text{m}^2) = 1.41\ \text{Wb}$$

and so $\Delta\phi = \phi_2 - \phi_1 = 1.41$ Wb. The induced emf is

$$\mathscr{E} = -\frac{\Delta\phi}{\Delta t} = -\frac{1.41\ \text{Wb}}{3\ \text{s}} = -0.47\ \text{V}$$

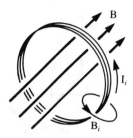

Fig. 30-3

The minus sign indicates that the emf will cause a current that creates a field opposing the change in **B** (see Fig. 30-3). The magnitude of the induced current is

$$I_i = \frac{\mathscr{E}}{R} = \frac{0.47\,\text{V}}{2\,\Omega} = \underline{0.235\,\text{A}}$$

30.12 A 50-loop circular coil has a radius of 30 mm. It is oriented so that the field lines of a magnetic field are parallel to a normal to the area of the coil. Suppose that the magnetic field is varied so that B increases from 0.10 to 0.35 T in a time of 2 ms. Find the average induced emf in the coil.

❚ $\Delta\Phi = B_{\text{final}}A - B_{\text{initial}}A = (0.25\,\text{T})(\pi r^2) = (0.25\,\text{T})\pi(0.030\,\text{m})^2 = 7.1 \times 10^{-4}\,\text{Wb}$ for each loop of the coil

$$|\mathscr{E}| = N\left|\frac{\Delta\Phi}{\Delta t}\right| = (50)\frac{7.1 \times 10^{-4}\,\text{Wb}}{2 \times 10^{-3}\,\text{s}} = \underline{17.7\,\text{V}}$$

30.13 In Fig. 30-4 the rectangular loop of wire is being pulled to the right, away from the long straight wire through which a steady current i flows upward. Does the current induced in the loop flow in the clockwise sense or in the counterclockwise sense?

❚ The magnetic field inside the loop in Fig. 30-4 is directed into the paper. As the loop is pulled away from the wire to regions of weaker field, the magnetic flux through the loop decreases in magnitude. As a consequence of Lenz's law, the magnetic field produced by the induced current must counteract the decrease in flux. Therefore it must be directed into the plane of the figure (within the loop). Hence the induced current must be *clockwise*.

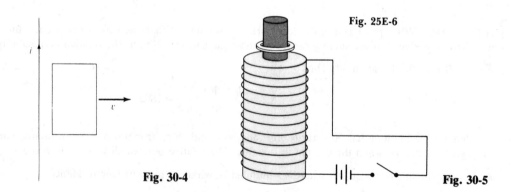

Fig. 30-4   **Fig. 25E-6**   **Fig. 30-5**

30.14 An aluminum ring is placed around the projecting core of a powerful electromagnet. See Fig. 30-5. When the circuit is closed the ring jumps up to a surprising height. Explain.

❚ The closing of the circuit is accompanied by a rapid increase in the magnitude of the magnetic flux through the ring, which induces a current in it. According to Lenz's law, the magnetic field due to the induced current must be directed opposite to the rapidly strengthening field due to the coil. This implies that the current in the ring circulates in the sense *opposite* to the current in the windings of the coil. But two opposite dipoles are like two antiparallel bar magnets: they repel each other, sending the ring upward.

30.15 A pendulum consists of a pivoted rod at the lower end of which there is a metal ring. The pendulum swings in the plane of the ring. The ring is raised and released. At the bottom of its swing, the ring enters a magnetic field normal to the plane of the ring, in the gap of a strong horseshoe magnet. On entering this region, it soon comes to a stop. Why?

❚ A current is induced in the ring as it enters the magnetic field. By Lenz's law, the magnetic force on the induced current opposes the motion which induced the current. It is this force which halts the translational motion of the ring. The kinetic energy of the macroscopic motion is converted to thermal energy by Joule heating in the ring. Note that (other things being equal) the more highly conducting the ring material is, the more rapidly the motion is stopped and the energy is dissipated.

30.16 For the situation of Prob. 30.15, if a small piece of the ring is cut out and the experiment repeated, the pendulum keeps swinging for some time. Explain.

❚ If the ring is cut, no current can flow in it. Hence there is no magnetic force to halt the translation and no Joule heating to dissipate the ring's kinetic energy. The pendulum will swing freely till ordinary friction dissipates the energy.

30.17 The magnet in Fig. 30-6 induces an emf in the coils as the magnet moves toward the right or the left. Find the directions of the induced currents through the resistors when the magnet is moving (*a*) toward the right and (*b*) toward the left.

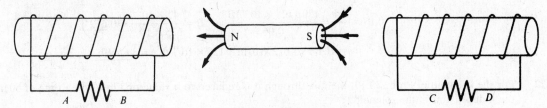

Fig. 30-6

❚ (*a*) Consider first the coil on the left. As the magnet moves to the right, the flux through the coil, which is directed generally to the left, decreases. To compensate for this, the induced current in the coil will flow so as to produce a flux toward the left through itself. Apply the right-hand rule to the loop on the left end. For it to produce flux inside the coil directed toward the left, the current must flow through the resistor from *B* to *A*.

Now consider the coil on the right. As the magnet moves toward the right, the flux inside the coil, also generally to the left, increases. The induced current in the coil will produce a flux toward the right to cancel this increased flux. Applying the right-hand rule to the loop on the right end, we find that the loop generates flux to the right inside itself if the current flows from *C* to *D* through the resistor.

(*b*) In this case the flux change caused by the magnet's motion is opposite to what it was in (*a*). Using the same type of reasoning, we find that the induced currents flow through the resistors from *A* to *B* and from *D* to *C*.

30.18 The magnet in Fig. 30-7 rotates as shown on a pivot through its center. At the instant shown, what are the directions of the induced currents?

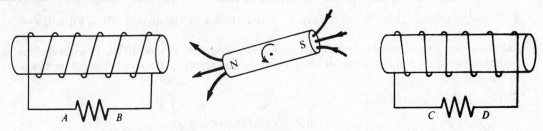

Fig. 30-7

❚ The situation is similar to that in Prob. 30.17, except that the coil on the right is oppositely wound. Here, both coils experience a decrease in flux to the left as the magnet rotates from the parallel position shown to the perpendicular position. The induced currents in both coils thus tend to create a field toward the left so the currents flow <u>from *B* to *A*</u> and <u>from *C* to *D*</u>.

30.19 A square loop of wire 75 mm on a side lies with its plane perpendicular to a uniform magnetic field of 0.8 T. (*a*) Find the magnetic flux through the loop. (*b*) If the coil is rotated through 90° in 0.015 s in such a way that there is no flux through the loop at the end, find the average emf induced during the rotation.

▌ (a) $\Phi = BA \cos (\mathbf{B}, \mathbf{A}) = (0.8 \text{ Wb/m}^2)[(75 \times 10^{-3})^2 \text{ m}^2] = \underline{4.5 \text{ mWb}}$

(b) $|e_{avg}| = \dfrac{\Delta \Phi}{\Delta t} = \dfrac{4.5 \times 10^{-3} \text{ Wb}}{1.5 \times 10^{-2} \text{ s}} = \underline{0.3 \text{ V}}$

30.20   The magnetic flux through a spark coil of 1000 turns changes from 0.5 Wb to zero in 0.01 s. Determine the emf induced in the coil.

▌ Applying Faraday's law, with Φ the flux through a single loop and $N\Phi$ the flux linking the entire coil, $|\mathscr{E}| = N |\Delta\Phi/\Delta t| = [(1000)(0.5)]/(0.01) = \underline{50 \text{ kV}}$.

30.21   A small "search" coil with an area of 125 mm^2 has 50 turns of very fine wire. This coil is placed between the pole pieces of a small magnet and then suddenly jerked out. If the average induced emf is 0.07 V when the coil is pulled to a field-free region in 60 ms, what is the magnetic intensity between the poles? What was the original flux through each turn?

▌ Here we let Φ = the total flux linking the coil and Φ_0 = the flux through a single loop; $\Phi = N\Phi_0$.

$$0.07 = \frac{\Delta\Phi}{\Delta t} = \frac{(50)(125 \times 10^{-6})B}{0.06} \qquad B = \frac{42 \times 10^{-4}}{625 \times 10^{-5}} = \underline{0.672 \text{ T}}$$

$$\Phi_0 = BA = (0.672 \text{ Wb/m}^2)(125 \times 10^{-6} \text{ m}^2) = \underline{84 \ \mu\text{Wb}}$$

30.22   For the solenoid of Prob. 30.10, the flux through it is reduced to a value of 1 mWb in a time of 50 ms. Find the induced emf in the solenoid.

▌ From Prob. 30.10 the flux through the solenoid was originally 1.66 mWb. Recalling that there are 5000 turns we have $N\,\Delta\Phi = (5 \times 10^3)(1.66 \times 10^{-3} - 1.00 \times 10^{-3}) = 3.3 \text{ Wb}$. Then $|\mathscr{E}| = N |\Delta\Phi/\Delta t| =$ 3.3 Wb/0.050 s = $\underline{66 \text{ V}}$.

30.23   Calculate the induced emf in a 150-cm^2 circular coil having 100 turns when the field strength B passing through the coil changes from 0.0 to 0.001 T in 0.1 s at a constant rate.

▌ Use Faraday's law $\mathscr{E} = -[(N\,\Delta\Phi)/\Delta t]$.

$$\mathscr{E} = -\frac{(150 \times 10^{-4})(100)(0.001 - 0.0)}{0.1} = \underline{-0.015 \text{ V}}$$

Note that in this example the induced emf is constant within the time interval considered.

30.24   A flat coil with radius 8 mm has 50 loops of wire on it. It is placed in a magnetic field, $B = 0.30$ T, so that the maximum flux goes through it. Later, it is rotated in 0.02 s to a position such that no flux goes through it. Find the average emf induced between the terminals of the coil.

▌ $\mathscr{E} = -[(N\,\Delta\Phi)/\Delta t]$, with $\Phi = BA$. Then $\mathscr{E} = -[(50)(64\pi \times 10^{-6} \text{ m}^2)(0.00 \text{ T} - 0.30 \text{ T})]/0.02 \text{ s} = \underline{0.15 \text{ V}}$.

30.25   Calculate the average induced emf in a coil having an area of 150 cm^2 and 100 turns when the field strength perpendicular to the coil is constant at 0.001 T and the coil is rotated through 90° about an axis perpendicular to the field in 0.1 s.

▌ From $\mathscr{E}_{avg} = -[(N\,\Delta\Phi)/\Delta t]$,

$$\mathscr{E}_{avg} = -\frac{(150 \times 10^{-4})(100)(0.000 - 0.001)}{0.1} = +\underline{0.015 \text{ V}}$$

30.26   A coil of 50 loops is pulled in 0.02 s from between the poles of a magnet, where its area intercepts a flux of 310 μWb, to a place where the intercepted flux is 100 μWb. Determine the average emf induced in the coil.

▌ $|\mathscr{E}| = N \left| \dfrac{\Delta\Phi}{\Delta t} \right| = 50 \dfrac{210 \times 10^{-6} \text{ Wb}}{0.02 \text{ s}} = \underline{0.525 \text{ V}}$

30.27   A coil of 275 turns with an area of 0.024 m^2 is placed with its plane perpendicular to the earth's field and is rotated in 0.025 s through a quarter turn, so that its plane is parallel to the earth's field. What is the average emf induced if the earth's field has an intensity of 80 μT? What was the original flux through each turn?

❚ From Faradays law, the magnitude of average induced emf is

$$\mathscr{E}_{avg} = \frac{(275)(0.024)(8 \times 10^{-5})}{0.025} = \underline{0.0211 \text{ V}} \qquad \Phi = BA = (8 \times 10^{-5} \text{ Wb/m}^2)(0.024 \text{ m}^2) = \underline{1.92 \ \mu\text{Wb}}.$$

30.28 The secondary of an induction coil has 12 000 turns. If the flux linking the coil changes from 740 to 40 μWb in 180 μs, how great is the induced emf?

❚ The magnitude of the induced emf is

$$\mathscr{E}_{avg} = \frac{(12\,000)(700 \times 10^{-6}) \text{ Wb}}{1.8 \times 10^{-4} \text{ s}} = \underline{46.7 \text{ kV}}$$

30.29 The magnetic induction B in the core of a spark coil changes from 1.4 to 0.1 T in 0.21 ms. If the cross-sectional area of the core is 450 mm², what is the original flux through the core. What is the average emf induced in the secondary coil if it has 6000 turns?

❚ $$\Phi_0 = BA = (1.4 \text{ Wb/m}^2)(450 \times 10^{-6} \text{ m}^2) = \underline{630 \ \mu\text{Wb}}$$

Magnitude of emf,

$$\mathscr{E}_{avg} = n\frac{\Delta\Phi_t}{t} = 6000 \times \frac{(1.4 - 0.1)(450 \times 10^{-6})}{2.1 \times 10^{-4}} = \underline{16.7 \text{ kV}}$$

30.2 MOTIONAL EMF; INDUCED CURRENTS AND FORCES

30.30 A conductor of length 0.19 m is passing perpendicular to a magnetic field of intensity 0.003 Wb/m² at a velocity of 11.5 m/s, as shown in Fig. 30-8. Calculate the induced emf.

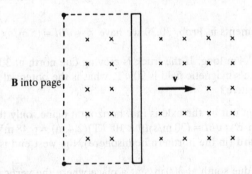

B into page

Fig. 30-8

❚ A wire of length l moving at a velocity v sweeps out an area lv per second. The flux cut per second is equal to BA/t, or Blv; thus, $\mathscr{E} = (0.003 \text{ Wb/m}^2)(0.19 \text{ m})(11.5 \text{ m/s}) = 6.6 \text{ mV}$. For the case shown, Lenz's law implies the emf is counterclockwise through the imaginary circuit, hence it points from bottom to top in the conductor.

Another method   Every charge q in the wire is moving with velocity **v** and experiences a force $F = qvB$, which by the right-hand rule points upward in the conductor for positive q. The work per unit charge (voltage), if these charges were free to move a distance l along the conductor, is $V = (Fl)/q = Blv$. (Note that this does not contradict the statement that the magnetic field does no work on a moving charge, because the charges are constrained to stay in the wire and the constraint forces actually perform this work.)

30.31 For the situation in Prob. 30.30 there is no real complete circuit. What prevents charges in the wire from moving through the wire?

❚ This is similar to the case of a battery on open circuit. The charges move until there is a sufficient pileup of opposite charge near the two ends to create an *electrostatic* force that just cancels the emf force. This balancing force sets up an electrostatic voltage difference (terminal voltage) that exactly balances the emf. In the case of Prob. 30.30 the positive voltage terminal is on top.

30.32 A straight wire of length L moves with constant velocity **U** (no rotation) through a uniform magnetic field **B**, Fig. 30-9. Find the voltage difference between the ends of the wire.

❚ In Fig. 30-9, let **L** be the vector from a to b and let **W** be the vector from d to a, so that $d\mathbf{W}/dt = \mathbf{U}$. Then

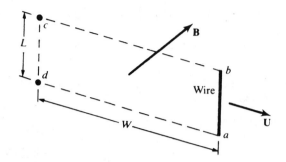

Fig. 30-9

the directed area of the parallelogram may be written as $\mathbf{A} = \mathbf{W} \times \mathbf{L}$, and the instantaneous flux through the parallelogram is $\Phi = \mathbf{B} \cdot \mathbf{A} = \mathbf{B} \cdot (\mathbf{W} \times \mathbf{L})$.

Faraday's law now gives (remember that $\mathbf{B}$ and $\mathbf{L}$ are constants)

$$V_i = -\frac{d\Phi}{dt} = -\mathbf{B} \cdot \left(\frac{d\mathbf{W}}{dt} \times \mathbf{L}\right) = -\mathbf{B} \cdot (\mathbf{U} \times \mathbf{L}) = \mathbf{B} \cdot (\mathbf{L} \times \mathbf{U}).$$

The sign of V_i gives the sense of circulation of charge in the imaginary circuit relative to the direction of $\mathbf{A}$.

30.33 An emf of 3.5 V is obtained by moving a wire 1.1 m long at a rate of 7 m/s perpendicular to the wire and to a uniform magnetic field. What is the intensity of the field?

▮ Using the result of Prob 30.30, or the more general result of Prob. 30.32, $\mathscr{E} = vBl$, so 3.5 V = (7 m/s)(B)(1.1 m), and $B = \underline{0.455\ \text{T}}$.

30.34 A horizontal wire 0.8 m long is falling at a speed of 5 m/s perpendicular to a uniform magnetic field of 1.1 T, which is directed from east to west. Calculate the magnitude of the induced emf. Is the north or south end of the wire positive?

▮ Following the arguments in Prob. 30.30 we have $\mathscr{E} = vBl = (5\ \text{m/s})(1.1\ \text{T})(0.8\ \text{m}) = \underline{4.4\ \text{V}}$. North end is +.

30.35 An axle of a truck is 2.4 m long. If the truck is moving due north at 30 m/s at a place where the vertical component of the earth's magnetic field is 90 μT, what is the potential difference between the two ends of the axle. Which end is positive?

▮ Since the area swept out by the axle is in a horizontal plane, only the vertical component of the earth's field contributes. Then $\mathscr{E} = vBl = (30\ \text{m/s})(9 \times 10^{-5}\ \text{T})(2.4\ \text{m}) = \underline{6.48\ \text{mV}}$. Since the vertical component of the earth's field is downward (in the northern hemisphere), the west end is +.

30.36 A jet aircraft is flying due south at 300 m/s at a place where the vertical component of the earth's magnetic field is 80 μT. Find the potential difference between wing tips if they are 25 m apart. Which tip has the higher potential?

▮ As in Prob. 30.35, $\mathscr{E} = vBl = (300\ \text{m/s})(8 \times 10^{-5}\ \text{T})(25\ \text{m}) = \underline{0.6\ \text{V}}$. This time, east end is + (in northern hemisphere).

30.37^c In Fig. 30-10(a), a wire perpendicular to a long straight wire is moving parallel to the latter with a speed $v = 10$ m/s in the direction of the current flowing in the latter. The current is 10 A. What is the magnitude of the potential difference between the ends of the moving wire? What is the sign of the potential difference?

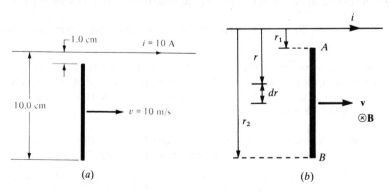

(a) (b) **Fig. 30-10**

❚ The notation is indicated in Fig. 30-10(*b*). According to the right-hand rule, the magnetic field due to the current-carrying wire is directed down into the plane of the figure all along the wire segment *AB*. The magnitude of the field is $B = (\mu_0 i)/(2\pi r)$. Since **v** is perpendicular to **B**, we have $|\mathbf{v} \times \mathbf{B}| = Bv$. In order for the Lorentz force to be zero (as required for the equilibrium of a mobile charge carrier in the wire segment), there must be an electric field $\mathbf{E}_{\text{opp}} = -\mathbf{v} \times \mathbf{B}$. Referring to the figure we see that E_{opp} is directed from *A* toward *B* along the wire, so that the electric potential is higher at *A* than at *B*: $V_A > V_B$. Specifically, we have

$$V_A - V_B = \int_{r_1}^{r_2} |\mathbf{E}_{\text{opp}}(r)| \, dr = \int_{r_1}^{r_2} Bv \, dr = \frac{\mu_0 i v}{2\pi} \int_{r_1}^{r_2} \frac{dr}{r} = \frac{\mu_0 i v}{2\pi} \ln\left(\frac{r_2}{r_1}\right)$$

Using $i = 10$ A, $v = 10$ m/s, and $r_2/r_1 = (10.0 \text{ cm})/(1.0 \text{ cm}) = 10$, we find that

$$V_A - V_B = (2 \times 10^{-7})(10)(10)(\ln 10) = \underline{46.1 \, \mu\text{V}}$$

30.38 The rod shown in Fig. 30-11 rotates about point *C* as pivot with the constant frequency 5 rev/s. Find the potential difference between its two ends, 80 cm apart, because of the magnetic field, $B = 0.3$ T.

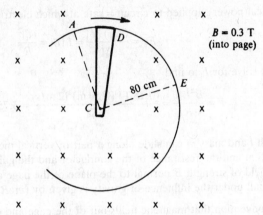

Fig. 30-11

❚ Consider a fictitious loop *CADC*. As time goes on, its area and the flux through it will increase. The induced emf in this loop will equal the potential difference we seek.

$$|\mathscr{E}| = N \left|\frac{\Delta\Phi}{\Delta t}\right| = (1)\frac{B \, \Delta A}{\Delta t}$$

It takes one-fifth second for the area to change from zero to that of a full circle, πr^2. Therefore,

$$|\mathscr{E}| = B\frac{\Delta A}{\Delta t} = B\frac{\pi r^2}{0.20 \text{ s}} = 0.3 \text{ T} \frac{\pi(0.8 \text{ m})^2}{0.20 \text{ s}} = \underline{3.0 \text{ V}}$$

30.39 As shown in Fig. 30-12, a metal rod makes contact with a partial circuit and completes the circuit. The circuit area is perpendicular to a magnetic field with $B = 0.15$ T. If the resistance of the total circuit is 3 Ω, how large a force is needed to move the rod as indicated with a constant speed of 2 m/s?

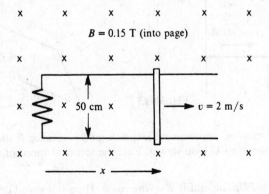

Fig. 30-12

▌ The induced emf in the rod causes a current to flow counterclockwise in the circuit. Because of this current in the rod, it experiences a force to the left due to the magnetic field. In order to pull the rod to the right with constant speed, this force must be balanced by the puller.

Method 1   The induced emf in the rod is

$$|\mathscr{E}| = BLv = (0.15\ \text{T})(0.50\ \text{m})(2\ \text{m/s}) = 0.15\ \text{V}$$

Then
$$I = \frac{|\mathscr{E}|}{R} = \frac{0.15\ \text{V}}{3\ \Omega} = 0.050\ \text{A}\qquad\text{from which}$$

$$F = ILB \sin 90° = (0.050\ \text{A})(0.50\ \text{m})(0.15\ \text{T})(1) = \underline{3.75 \times 10^{-3}\ \text{N}}$$

Method 2   The emf induced in the loop is

$$|\mathscr{E}| = N\left|\frac{\Delta\Phi}{\Delta t}\right| = (1)\frac{B\ \Delta A}{\Delta t} = \frac{B(L\ \Delta x)}{\Delta t} = BLv$$

as before. Now proceed as follows:

mechanical power supplied to circuit = rate at which electric work is done on charges

$$Fv = \frac{(\Delta q)\,|\mathscr{E}|}{\Delta t} = I\,|\mathscr{E}| = \frac{|\mathscr{E}|^2}{R}$$

Substitute for $|\mathscr{E}|$ and solve for F to find

$$F = \frac{B^2 L^2 v}{R} = \frac{(0.15\ \text{T})^2 (0.50\ \text{m})^2 (2\ \text{m/s})}{3\ \Omega} = \underline{3.75 \times 10^{-3}\ \text{N}}$$

30.40   A conductor of length l and mass m can slide along a pair of vertical metal guides connected by a resistor R, as in Fig. 30-13. Friction and the resistance of the conductor and the guides are negligible. There is a uniform horizontal magnetic field of strength B normal to the plane of the page and directed outward. Show that the final steady speed of fall under the influence of gravity is given by $(mgR)/(B^2 l^2)$.

▌ We use the sign convention that magnetic fields out of the page and counterclockwise currents are positive. As the bar falls, the flux through the circuit increases. Since $\mathscr{E} = -(d\Phi_m)/dt$, the induced emf is negative (i.e., clockwise) and has magnitude $|\mathscr{E}| = Blv$, where v is the speed of the descending bar. Hence the current in the circuit is $i = \mathscr{E}/R = -(Blv)/R$. The negative sign indicates clockwise current, so the current flows to the *left* in the bar. The resulting magnetic force on the bar has magnitude

$$F_m = ilB = \frac{B^2 l^2 v}{R}\tag{1}$$

and is directed upward. When the bar has reached its terminal speed, the net force on it is zero, so the upward magnetic force must equal the weight of the bar:

$$F_m = mg\tag{2}$$

Solving Eqs. (1) and (2) for the terminal speed, we find that $v_t = (mgR)/(B^2 l^2)$, as was to be shown.

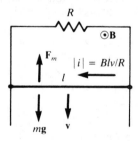

Fig. 30-13

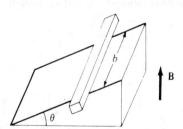

Fig. 30-14

30.41   In Fig. 30-14 a bar of mass m, approximate length b, and resistance R slides down the incline while making contact with the resistanceless U loop shown. Find the terminal speed of the bar as it slides in the vertical magnetic field B.

▌ Induced current $I = \mathscr{E}/R$ and emf is $\mathscr{E} = Bbv \cos\theta$. Here v is speed down the incline. Magnetic force $F = IbB$ is, from Lenz's law, to the right. At equilibrium, the component of F up the incline,

$IbB \cos \theta = (b^2 B^2 v \cos^2 \theta)/R$ equals the component of weight down the incline, $mg \sin \theta$. Solving, $v = (mgR \sin \theta)/(bB \cos \theta)^2$.

30.42 The bar of approximate length b shown in Fig. 30-14 slides from rest down the incline along the contacts as shown. Assume negligible friction and infinite resistance in the circuit. If **B** is vertical, find an equation for the induced emf between the ends of the sliding rod as a function of time. Is the emf clockwise or counterclockwise when viewed from above?

❚ First find the flux through the loop at any instant. Call x the distance along incline from the bar to the bottom, $\Phi = \mathbf{B} \cdot \mathbf{A} = Bbx \cos \theta$. Then emf $= (Bb \cos \theta)(-dx/dt)$. To see how x varies, use the relations: acceleration of rod down incline $= a = -g \sin \theta$ and $dx/dt = v = at = -gt \sin \theta$. Thus, emf $= +Bbgt \sin \theta \cos \theta$; that is, in the positive sense of **A**, or counterclockwise as viewed from above.

30.43 In Fig. 30-15(a) there is a magnetic field in the $+x$ direction with $B = 0.20$ T. The loop has an area of 5 cm² and rotates about line CD as axis. Point A rotates toward positive x values from the position shown. If line AE rotates through 50° from its indicated position in a time of 0.2 s, (a) what is the change in flux through the coil, (b) what is the average induced emf in it, and (c) does the induced current flow from A to C or C to A in the upper part of the coil?

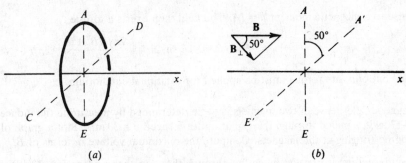

(a)   (b)

Fig. 30-15

❚ (a)   initial flux $= B_\perp A = BA = (0.20 \text{ T})(5 \times 10^{-4} \text{ m}^2) = 100 \, \mu\text{Wb}$

final flux $= (B \cos 50°)A = (1 \times 10^{-4} \text{ Wb})(\cos 50°) = 64 \, \mu\text{Wb}$

$\Delta\Phi = 0.64 \times 10^{-4} \text{ Wb} - 1 \times 10^{-4} \text{ Wb} = -36 \, \mu\text{Wb}$

(b) $|\mathcal{E}| = N \left| \dfrac{\Delta\Phi}{\Delta t} \right| = (1) \dfrac{0.36 \times 10^{-4} \text{ Wb}}{0.2 \text{ s}} = 180 \, \mu\text{V}$

(c) The flux through the loop from left to right decreased. The induced current will tend to set up flux from left to right through the loop. By the right-hand rule, the current flows from A to C. Alternatively, a torque must be set up that tends to rotate the loop back into its original position. The appropriate right-hand rule again gives a current flow from A to C.

30.44ᶜ A circular coil with N loops and area A rotates with frequency f about a diameter as axis. Impressed upon it is a constant magnetic field B directed perpendicular to the axis of rotation. Find the flux through the coil at time t, and obtain from it the emf generated in the coil, assuming **B** and **A** aligned at $t = 0$.

❚ $\Phi = \mathbf{B} \cdot \mathbf{A} = BA \cos \theta = \underline{BA \cos 2\pi ft}$. Then, $\mathcal{E} = -[(N \, d\phi)/dt] = \underline{2\pi f NBA \sin 2\pi ft}$.

30.45 A circular loop of radius r is fixed to a rotation axis along the z direction, as shown in Fig. 30-16, so that the plane of the loop is always perpendicular to the xy plane. The loop is rotating about the constant angular velocity ω_L directed along the z axis; at $t = 0$, the loop lies in the yz plane. Given a uniform and constant externally applied magnetic field $\mathbf{B} = \mathcal{B}(\cos \alpha \hat{\mathbf{x}} + \sin \alpha \hat{\mathbf{z}})$, evaluate the magnetic flux $\Phi_m(t)$ through the loop and the emf $\mathcal{E}(t)$ induced in the loop.

❚ Since the z component of the field does not contribute to the flux, we are back to Prob. 30.44, with $N = 1$, $A = \pi r^2$, $f = \omega_L/2\pi$, and $B = \mathcal{B} \cos \alpha$. Thus, $\Phi(t) = \pi r^2 \mathcal{B} \cos \alpha \cos \omega_L t$ and $\mathcal{E}(t) = \pi r^2 \omega_L \mathcal{B} \cos \alpha \sin \omega_L t$.

30.46 Assume that the coil of Prob. 30.44 has (large) resistance R. What torque must be applied to keep the coil in steady rotation?

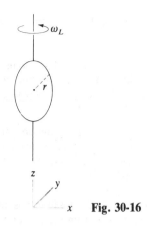

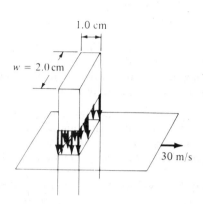

Fig. 30-16

Fig. 30-17

▮ The induced current in the coil is

$$I = \frac{\mathscr{E}}{R} = \frac{2\pi fNBA}{R} \sin 2\pi ft$$

giving rise to a magnetic moment $\mu = IA$. The field then exerts a torque

$$\tau_B = \mu B \sin \theta = IAB \sin 2\pi ft = \frac{2\pi fNB^2A^2}{R} \sin^2 2\pi ft$$

An equal but opposite torque must be applied for rotational equilibrium.

30.47 The magnetic field between two magnets is to be determined by measuring the induced voltage in the loop of Fig. 30-17 as it is pulled through the gap at uniform speed $v = 30$ m/s. Plot a graph of voltage vs. time, assuming no fringing at the magnets. Compute the maximum voltage in terms of B.

▮ The induced emf is nonzero when flux through the loop changes (see Fig. 30-18). When the front end of the loop passes through the flux, $d\Phi/dt = B(dA/dt) = Bwv = 0.60B$ for the time it takes for the end to traverse the field, which is $\Delta t = 0.01/30 = 3.3 \times 10^{-4}$ s. When the back end passes through the magnet, the flux change is equal and opposite, and the second emf pulse is as shown.

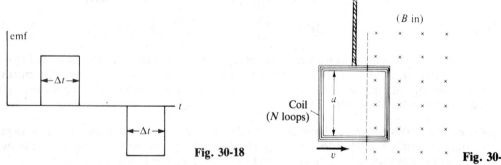

Fig. 30-18

Fig. 30-19

30.48 A coil of wire with N loops swings with speed v into a uniform magnetic field B, as shown in Fig. 30-19. If the ends of the coil are connected together and the resistance of the coil is R, find the force exerted on the coil by the field when the coil is in the position shown. What does the force do to the coil?

▮ The induced emf at the instant shown is $-N(d\Phi/dt) = NavB$ (counterclockwise) and causes a current $(NavB)/R$. The right-hand side of the loop feels a force $NBIa = (N^2a^2vB^2)/R$ stopping the motion. (Compare Prob. 30.15.)

30.49 Magnetic transducers are often used to monitor small vibrations. For example, the end of a vibrating bar is attached to a coil which in turn vibrates in and out of a uniform magnetic field B, as shown in Fig. 30-20. Show that the speed of the end of the bar, dx/dt, is related to the emf induced in the coil by $\mathscr{E} = NBb\,(dx/dt)$.

▮ We assume the coil to be rigid. The flux through the coil is $-Bbx$, so that $\mathscr{E} = -N(d\Phi/dt) = NBb(dx/dt)$.

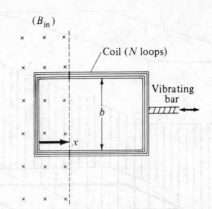

(B_{in})

Coil (N loops)

Vibrating bar

b

x

Fig. 30-20

30.50 The resistance of the coil in Fig. 30-20 is R, and its two ends are connected together. (*a*) What is the induced current in the coil as it moves to the right with speed v? (*b*) How large a force is required to keep it moving with constant speed? (*c*) Find the power output of the force. (*d*) Compare (*c*) to the i^2R-loss in the coil.

▎ (*a*) Use the results of Prob. 30.49: $I = \mathscr{E}/R = (NBbv)/R$. (*b*) Since a current element Ib is in a perpendicular magnetic field, the external force $NIbB = (N^2B^2b^2v)/R$. (*c*) The power $= Fv = IbBv = (N^2B^2b^2v^2)/R$. (*d*) This is precisely I^2R.

30.3 TIME-VARYING MAGNETIC AND INDUCED ELECTRIC FIELDS

30.51ᶜ A magnetic field given by $B = B_x = 0.2 \cos \omega t$ T is impressed upon a flat loop of area A and resistance R. If the fixed angle between the area vector and the x axis is α, what are both the induced emf and the current in the loop.

▎ $\Phi = \mathbf{B} \cdot \mathbf{A} = B_x A \cos \alpha$; $\mathscr{E} = -d\Phi/dt = 0.2\,A\omega \sin \omega t \cos \alpha$ and $I = \mathscr{E}/R$. Note that $\mathscr{E}$ is the same as that in a loop spinning with radian frequency ω in a constant field (Prob. 30.45).

30.52ᶜ The magnetic field perpendicular to a 3.0-cm²-area 40-loop coil is changing in the following way: $B = 250 - 0.60t$ mT, with t in seconds. What is the induced emf in the coil?

▎ $\Phi = \mathbf{B} \cdot \mathbf{A} = (750 - 1.80t) \times 10^{-7}$ Wb, then $\mathscr{E} = -N(d\Phi/dt) = 40(1.80 \times 10^{-7}) = \underline{7.2\ \mu V}$.

30.53ᶜ A magnetic field B is perpendicular to the 7.0-cm² area of a 50-loop coil. The ends of the essentially resistanceless coil are connected across a 30-Ω resistor. Find the current through the resistor if $B = 75e^{-200t}$ G, where t is measured in seconds.

▎ We recall that 10^4 G $= 1$ Wb/m². The flux is $7.0[75 \exp(-200t)] \times 10^{-8}$ Wb, so $\mathscr{E} = -50(525)(-200) \times \exp(-200t) \times 10^{-8} = 0.0525 \exp(-200t)$ V. Dividing this emf by $R = 30\ \Omega$ gives $I = \underline{1.75 \exp(-200t)}$ mA.

30.54ᶜ A rectangular loop of N turns, Fig. 30-21, moves to the right with velocity $\mathbf{u}$ between the poles of an electromagnet, which sets up a uniform time-dependent field $B = B_0 \sin \omega t$ between the pole faces. Calculate the emf induced in the loop.

▎ The total flux is $N\Phi = N(B_0 \sin \omega t)lx$, so that the instantaneous emf is

$$v_i = -\frac{d}{dt}(N\Phi) = -NB_0 l(\omega x \cos \omega t + \dot{x} \sin \omega t) = NB_0 l(u \sin \omega t - \omega x \cos \omega t) \tag{1}$$

If the loop moves with *constant* velocity, then, in the above expression, $x = x_0 + ut$.

30.55 In Prob. 30.54, let $B_0 = 0.1$ T, $\omega = 400$ rad/s, $l = 150$ mm, $N = 10$, $u = 5$ m/s $=$ constant, $x_0 = 300$ mm. Evaluate v_i at times corresponding to (*a*) $\omega t = 0$, (*b*) $\omega t = \pi/4$, (*c*) $\omega t = \pi/2$, (*d*) $\omega t = 3\pi/4$.

▎ We substitute the given values into Eq. (1) of Prob. 30.54 to get $v_i = 0.150\,[5 \sin \omega t - 400(0.300 + 5t) \times \cos \omega t]$ V. Then for the cases given we have (*a*) -18 V (clockwise); (*b*) -12.6 V; (*c*) $+0.75$ V; and (*d*) $+13.5$ V.

30.56ᶜ A planar loop consisting of $N = 10$ turns of flexible wire is placed inside a long solenoid having n' turns per meter and carrying current $I = I_0 \sin \omega t$. The area A enclosed by the loop can be changed by pulling the sides

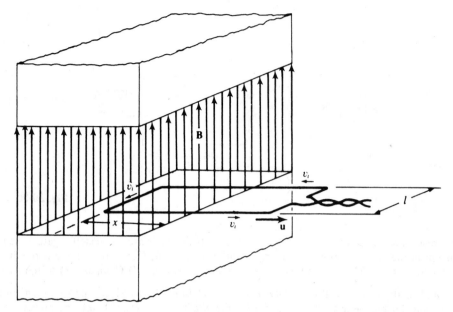

Fig. 30-21

apart; however, the plane of the loop always remains normal to the axis of the solenoid. For $n' = 3000 \text{ m}^{-1}$, $I = 12 \sin 150t$ A, and assuming that

$$A = 0.15 \text{ m}^2 \qquad \frac{dA}{dt} = 2 \text{ m}^2/\text{s}$$

at $t = 5$ ms, compute the emf induced in the loop at this instant.

▐ For a long solenoid $B = \mu_0 n' I = \mu_0 n' I_0 \sin \omega t$ within the solenoid, so that the flux linkage with the loop is

$$N\Phi = NBA = NA\mu_0 n' I_0 \sin \omega t \qquad \text{and} \qquad v_i = -\frac{d}{dt}(N\Phi) = -\mu_0 I_0 n' N\left(A\omega \cos \omega t + \frac{dA}{dt} \sin \omega t\right). \qquad (1)$$

Substituting the data,

$$v_i = -(4\pi \times 10^{-7})(12)(3000)(10)[(0.15)(150) \cos (150)(0.005) + 2 \sin (150)(0.005)] = \underline{-8.064 \text{ V}}$$

in accordance with Lenz's law.

30.57 Suppose that in Prob. 30.56 the deformable loop is replaced by a rigid circular loop, of radius r, with its center on the axis of the solenoid. All other conditions remain the same. Find **(a)** the induced emf v_i in the loop, and **(b)** the induced electric field E_i at any point on the loop.

▐ **(a)** Setting

$$A = \pi r^2 \qquad \frac{dA}{dt} = 0$$

in (1) of Prob. 30.56, we obtain $v_i = -\mu_0 I_0 n' N \pi r^2 \omega \cos \omega t$. **(b)** Since the loop is stationary, $\mathbf{u} = 0$ at each point of it, and the induced emf is due only to the time-varying field. The induced electric field is then $v_i = E_i(2\pi r)$ since, by symmetry, $\mathbf{E}_i$ is constant in magnitude and tangential around the circle. Thus

$$E_i = \frac{v_i}{2\pi r} = \frac{-\mu_0 I_0 n' N r \omega}{2} \cos \omega t \qquad (1)$$

Note that the induced electric field is *not* conservative like the electrostatic field. That is why the work done by E_i on a unit charge going around the loop is not zero.

30.58 Make a unit-dimensional check of (1) of Prob. 30.57.

▐ Recall that $\omega = [\text{s}^{-1}]$, the radian being a dimensionless unit. Thus, $\cos \omega t$, along with N, is a pure number. Furthermore,

$$\mu_0 = \left[\frac{\text{N} \cdot \text{s}^2}{\text{C}^2}\right] \qquad \text{Hence} \qquad E_i = \left[\frac{\text{N} \cdot \text{s}^2}{\text{C}^2}\right]\left[\frac{\text{C}}{\text{s}}\right][\text{m}^{-1}][\text{m}][\text{s}^{-1}] = \left[\frac{\text{N}}{\text{C}}\right] = \left[\frac{\text{V}}{\text{m}}\right]$$

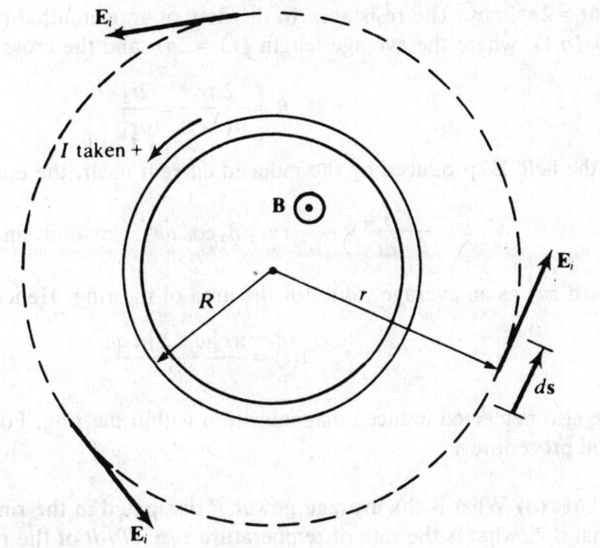

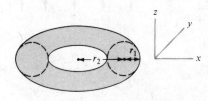

Fig. 30-22

30.59ᶜ In Fig. 30-22, the circular ring indicates a cross section of a long air-core solenoid, of radius $R = 0.10$ m. It has $n' = 2000$ turns per meter and carries an alternating current $I = I_0 \sin \omega t = 25 \sin 300t$ A. The dashed line indicates a concentric circular wire of radius $r = 0.12$ m. Compute **(a)** the induced emf v_i in the wire and **(b)** the induced electric field $\mathbf{E}_i$ at any point in the wire. **(c)** If the circular wire were removed, would $\mathbf{E}_i$ remain unchanged at all points on the dashed circle?

▮ **(a)** Outside the solenoid $B = 0$, so that the flux linking the wire is the flux through the cross section of the solenoid: $\Phi_s = BA = (\mu_0 n' I)(\pi R^2) = \pi R^2 \mu_0 n' I_0 \sin \omega t$. Then

$$v_i = -\frac{d\Phi_s}{dt} = -\pi R^2 \mu_0 n' I_0 \omega \cos \omega t = \underline{-0.5922 \cos 300t} \quad \text{V}$$

This result might have been obtained at once by setting $N = 1$ and $r = R$ in Prob. 30.57(a).
(b) As in Prob. 30.57(b),

$$E_i = \frac{v_i}{2\pi r} = -0.7854 \cos 300t \quad \text{V/m}$$

Here the minus sign implies that when I is positive and increasing, $\mathbf{E}_i$ is in the opposite direction to that indicated in Fig. 30-22. **(c)** Yes.

30.60ᶜ A long solenoid that has 800 loops per meter carries a current $i = 3 \sin (400t)$ A. Find the electric field inside the solenoid at a distance of 2 mm from the solenoid axis. Consider only the field tangential to a circle having its center on the axis of the solenoid.

▮ $B = \mu_0 n' i$ for a long solenoid with $n' = 800$ m^{-1} for our case. The flux through the area $A = \pi(2 \times 10^{-3})^2$ is $\Phi = BA = A\mu_0 (800)(3 \sin 400t) = 0.03\mu_0 \sin 400t$. Emf induced $= -d\Phi/dt = -12\mu_0 \cos 400t$. E is tangent to and constant around the path so $2\pi(2 \times 10^{-3})E = -12\mu_0 \cos 400t$; thus $E = \underline{-960\mu_0 \cos 400t}$ V/m.

30.61ᶜ A fixed metal ring of density ρ and conductivity σ in the shape of a toroid is shown in Fig. 30-23. The ring lies in the xy plane and is immersed in a spatially uniform but sinusoidally varying magnetic field $\mathbf{B}(t) = \mathcal{B}_0 \cos (\omega t)\hat{\mathbf{z}}$. **(a)** Find the mass M and electric resistance R of the ring. **(b)** Find the induced current $i(t)$ in the ring, taking the positive sense to be counterclockwise as viewed from above.

▮ **(a)** Referring to Fig. 30-23, we determine that the volume of the toroid is $(\pi r_1^2)(2\pi r_2) = 2\pi^2 r_1^2 r_2$. Hence the

Fig. 30-23

mass is given by $M = 2\pi^2 \rho r_1^2 r_2$. The resistance to the flow of an azimuthal (circular about z axis) electric current is $R = \langle l \rangle /(\sigma A)$, where the average length $\langle l \rangle = 2\pi r_2$ and the cross-sectional area $A = \pi r_1^2$:

$$R = \frac{2\pi r_2}{\sigma(\pi r_1^2)} = \frac{2r_2}{\sigma r_1^2}$$

(b) If we ignore the field $\mathbf{B}_c$ produced by the induced current itself, the emf around the ring is

$$-\frac{d\Phi_m}{dt} = -\frac{d}{dt}(\pi r_2^2 \mathcal{B}_0 \cos \omega t) = \pi r_2^2 \omega \mathcal{B}_0 \sin \omega t$$

where we have used πr_2^2 as an average value for the area of the ring. Hence the induced current is

$$i(t) = \frac{\pi r_2^2 \omega \mathcal{B}_0 \sin \omega t}{R}$$

(NOTE: We have also neglected induced magnetization within the ring. For a nonmagnetic material such as brass this is a valid procedure.)

30.62ᶜ Refer to Prob. 30.61. **(a)** What is the average power P dissipated in the ring? **(b)** If the specific heat capacity of the ring material is c, what is the rate of temperature rise dT/dt of the ring, assuming that no heat is lost?

▌ **(a)** The instantaneous power $P(t)$ dissipated as Joule heat is

$$i^2(t)R = \frac{\pi^2 r_2^4 \omega^2 \mathcal{B}_0^2 \sin^2 \omega t}{R}$$

Since $\langle \sin^2 \omega t \rangle = \frac{1}{2}$, the time-averaged power is

$$P = \frac{\pi^2 r_2^4 \omega^2 \mathcal{B}_0^2}{2R} = \frac{\pi^2 r_1^2 r_2^3 \omega^2 \sigma \mathcal{B}_0^2}{4}$$

(b) The deposition of thermal energy dH in the wire is accompanied by a temperature rise $dT = dH/cM$, so

$$\frac{dT}{dt} = \frac{1}{cM}\frac{dH}{dt} = \frac{P}{cM} = \frac{r_2^2 \omega^2 \sigma \mathcal{B}_0^2}{8c\rho}$$

30.63 Make a unit-dimensional check of the result of Prob. 30.62(b).

▌ Because $1\,\text{T} = 1\,\text{J/A} \cdot \text{m}^2$, we may write

$$\frac{r_2^2 \omega^2 \sigma \mathcal{B}_0^2}{c\rho} = \frac{[\text{m}^2][\text{s}^{-2}][\Omega^{-1} \cdot \text{m}^{-1}][\text{J}^2 \cdot \text{A}^{-2} \cdot \text{m}^{-4}]}{[\text{J} \cdot \text{kg}^{-1} \cdot \text{K}^{-1}][\text{kg} \cdot \text{m}^{-3}]} = \left[\frac{\text{K}}{\text{s}}\right]\left[\frac{\text{J}}{\text{A}^2 \cdot \Omega \cdot \text{s}}\right] \quad (1)$$

But $1\,\Omega = 1\,\text{V/A} = 1\,(\text{J/A} \cdot \text{s})/\text{A} = 1\,\text{J/A}^2 \cdot \text{s}$; hence the second factor on the right of (1) equals 1, a pure number, and we are left with the correct units, K/s.

30.64ᶜ A metal disk (radius $= b$, thickness $= w$) is placed in a solenoid with its axis coincident with the axis of the solenoid. The solenoid produces a magnetic field $B = B_0 \sin 2\pi f t$, as shown in Fig. 30-24. **(a)** Find the induced emf in the dotted ring. **(b)** Find the lengthwise resistance of the ring if its width is dr and the resisitivity of the material is ρ. **(c)** An eddy current flows around the ring. What is its value? **(d)** What is the total power loss due to eddy currents in the disk? Ignore all fringing effects and the self-induced currents in the disk.

▌ **(a)** From Faraday's law the emf $\mathcal{E} = -[d(BA)]/dt$ with $A = \pi r^2$ and $B = B_0 \sin 2\pi f t$, so $\mathcal{E} = -2\pi^2 r^2 f B_0 \cos 2\pi f t$. **(b)** Use $R = (\rho L)/A = (2\pi r \rho)/(w\,dr)$. **(c)** $I = \mathcal{E}/R = -\{[\pi f B_0 w(\cos 2\pi f t)r]/\rho\}\,dr$.

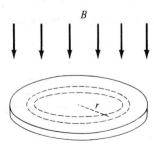

B

Fig. 30-24

(d) For one ring $dP = I^2R = \{[2\pi^3B_0^2f^2w(\cos 2\pi ft)^2]/\rho\}r^3\,dr$; integrate from 0 to b to obtain $P = [(\pi^3wb^4f^2B_0^2)/2\rho](\cos 2\pi ft)^2$ or time averaging, $\bar{P} = (\pi^3wb^4f^2B_0^2)/4\rho$. [This is analogous to the formula found in Prob. 30.62(a): $P \propto (\omega B_0)^2$.]

30.65 A cylindrical vacuum tube is placed inside a solenoid so that the solenoid fits snugly around it. **Show that the force on an electron within the tube has two distinct parts resulting from the increasing flux in the solenoid. What are they?**

❚ One force is due to the charge motion in a magnetic field, $e\mathbf{v} \times \mathbf{B}$. The other is due to the electric field induced by the changing flux. Taking a circle of radius r concentric with the axis, one has, in magnitude, emf $= 2\pi rE = \pi r^2\,(dB/dt)$. Therefore, this force is $eE = [er(dB/dt)]/2$, with direction tangent to the concentric circle about the axis that passes through the location of the electron.

30.66 A circular ring of diameter 20 cm has a resistance of 0.01 Ω. How much charge will flow through the ring if it is turned from a position perpendicular to a uniform magnetic field of 2.0 T to a position parallel to the field?

❚ Since $iR = V = -\Delta\Phi_m/\Delta t$ and the current $i = \Delta q/\Delta t$, we have $R\,\Delta q = -\Delta\Phi_m$, or

$$|\Delta q| = \frac{|\Delta\Phi_m|}{R} = \frac{|\Phi_{mf} - \Phi_{mi}|}{R}$$

Since the final flux is zero, and the initial flux has magnitude

$$|\Phi_{mi}| = \frac{\pi d^2 B}{4} \qquad \text{we have} \qquad |\Delta q| = \frac{\pi d^2 B}{4R} = \frac{\pi(0.20)^2(2.0)}{(4)(0.01)} = \underline{6.28\,\text{C}}$$

30.67 A 5-Ω coil, of 100 turns and diameter 6 cm, is placed between the poles of a magnet so that the flux is maximum through its area. When the coil is suddenly removed from the field of the magnet, a charge of 10^{-4} C flows through a 595-Ω galvanometer connected to the coil. Compute B between the poles of the magnet.

❚ As the coil is removed, the flux changes from BA, where A is the coil area, to zero. Therefore,

$$|\mathscr{E}| = N\left|\frac{\Delta\Phi}{\Delta t}\right| = N\frac{BA}{\Delta t}$$

We are told that $\Delta q = 10^{-4}$ C. But, by Ohm's law,

$$|\mathscr{E}| = IR = \frac{\Delta q}{\Delta t}R$$

where $R = 600$ Ω, the total resistance. If we now equate these two expressions for $|\mathscr{E}|$ and solve for B, we find that

$$B = \frac{R\,\Delta q}{NA} = \frac{(600\,\Omega)(10^{-4}\,\text{C})}{(100)(\pi)(9 \times 10^{-4}\,\text{m}^2)} = \underline{0.212\,\text{T}}$$

30.68 How much charge will flow through a 200-Ω galvanometer connected to a 400-Ω circular coil of 1000 turns wound on a wooden stick 20 mm in diameter, if a magnetic field $B = 0.0113$ T parallel to the axis of the stick is decreased suddenly to zero?

❚ R for the circuit of coil plus galvanometer $= 200\,\Omega + 400\,\Omega = 600\,\Omega$.

$$\mathscr{E} = IR = \frac{\Delta q}{\Delta t}R = -N\frac{\Delta\Phi}{\Delta t} \qquad \text{so}$$

$$\Delta q = -\frac{N\Delta\Phi}{R} = -\frac{NA\,\Delta B}{R} = -\frac{1000(\pi \times 10^{-4}\,\text{m}^2)(0.00\,\text{T} - 0.0113\,\text{T})}{600\,\Omega} = \underline{+5.9\,\mu\text{C}}$$

(The sign merely indicates that the charge passes through the galvanometer in the same sense as the induced current.)

30.69 Modern estimates of the physical conditions within sunspots include these typical values: number of free particles per unit volume $n \approx 10^{22}$ m^{-3}; gas temperature $T \approx 4000$ K; magnetic field strength $\mathscr{B} \approx 0.1$ T.
(a) Use the ideal-gas law to estimate the typical gas pressure within a sunspot. Compare it to atmospheric pressure. (b) Assuming that the gas is ideal and monatomic, find the typical thermal energy density ρ_t in a sunspot. (c) Find the typical magnetic energy density ρ_m in a sunspot.

❚   **(a)** $p = nkT \approx (10^{22})(1.38 \times 10^{-23})(4 \times 10^3) \approx \underline{550 \text{ Pa}}$ $\dfrac{0.55 \text{ kPa}}{101 \text{ kPa/atm}} = \underline{5.4 \times 10^{-3} \text{ atm}}$

(b) The thermal energy density in an ideal monatomic gas is given by

$$\rho_t = \tfrac{3}{2} nkT = \tfrac{3}{2} p \approx \tfrac{3}{2} (550 \text{ J/m}^3) = \underline{830 \text{ J/m}^3}$$

(c) The magnetic energy density is

$$\rho_m = \frac{\mathcal{B}^2}{2\mu_0} \approx \frac{(0.1 \text{ T})^2}{2(4\pi \times 10^{-7} \text{ H/m}^2)} = \underline{4000 \text{ J/m}^3}$$

30.70   Refer to Prob 30.69. **(a)** Which predominate in sunspots, pressure forces or magnetic forces. **(b)** Away from sunspots, typical conditions in the lower solar atmosphere are given by $n \approx 10^{22} \text{ m}^{-3}$, $T \approx 6000$ K, and $\mathcal{B} \approx 10^{-4}$ T. Which forces are dominant here?

❚   **(a)** Since $\rho_m \approx 5\rho_t$, <u>magnetic forces should dominate</u>. **(b)** For the "quiet" solar atmosphere, we find that

$$\rho_t' = \frac{3}{2} nkT' \approx 1.2 \times 10^3 \text{ J/m}^3 \qquad \rho_m' = \frac{(\mathcal{B}')^2}{2\mu_0} = 4 \times 10^{-3} \text{ J/m}^3$$

Hence the <u>pressure forces should dominate</u>.

30.71ᶜ   A highly conductive ring of radius R is perpendicular to and concentric with the axis of a long solenoid, as shown in Fig. 30-25. The ring has a narrow gap of width δ in its circumference. The solenoid has cross-sectional area a and a uniform internal field of magnitude $\mathcal{B}$.

     (a) What is the induced emf around the loop if $\mathcal{B}$ is constant?

     (b) Beginning at $t = 0$, the solenoid current is steadily increased, so that the field magnitude is $\mathcal{B}(t) = \mathcal{B}_0 + \beta t$, where $\beta > 0$. What is the induced emf $\mathcal{E}(t)$ around the ring? Assuming that no charge can flow across the gap, which face of the gap (F_1 or F_2) will accumulate an excess of positive charge?

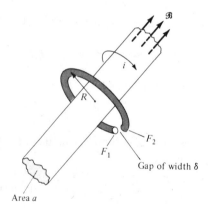

Area a **Fig. 30-25**

❚   **(a)** If $\mathcal{B}$ is constant, then $(d\Phi_m)/dt = 0$, so the induced emf $\underline{\mathcal{E} = 0}$. **(b)** The induced emf is given by

$$\mathcal{E}(t) = -\frac{d\Phi_m}{dt} = -\frac{d}{dt}(\mathcal{B}a) = \underline{-a\beta < 0}$$

The emf being constant and negative, the induced electric field points around the ring from F_2 toward F_1 (that is, leftward on the near side of the ring in Fig. 30-25). Hence face F_1 will develop an excess of positive charge.

30.72   Refer to Prob. 30.71. The accumulation of charge on the gap faces will cease when the total electric field within the ring material is zero. When this happens, what will be the electric field in the gap? Does your expression depend upon R?

❚   When the net electric field is zero within the ring, the electrostatic field within the ring just balances the induced electric field (which is constant around the ring, including the gap). The electrostatic field, however, must do zero work in bringing a charge around the *entire* loop, so it must be huge and point from F_1 to F_2 through the gap. If we ignore the small contribution from the induced field in the gap, the electric field must have magnitude

$$E = \frac{|\mathcal{E}|}{\delta} = \frac{a\beta}{\delta} \tag{1}$$

This expression is <u>independent of R</u> (as long as the radius of the ring exceeds the radius $\sqrt{a/\pi}$ of the solenoid).

30.73 Refer to Prob. 30.72. (*a*) A spark will jump the gap if the electric field magnitude exceeds the breakdown field E_b necessary to ionize air. Find the minimum gap width δ_{min} that can tolerate the magnitude found in (1) of Prob. 30.72. (*b*) Obtain a numerical value of δ_{min} for the case $a = 0.10 \text{ m}^2$, $\beta = 1.0 \times 10^3$ T/s, and $E_b = 3.0 \times 10^6$ V/m.

▌ (*a*) $\delta_{min} = a\beta/E_b$; (*b*) $\delta_{min} = (0.10)(1.0 \times 10^3)/(3.0 \times 10^6) = \underline{33.3 \ \mu\text{m}}$

30.4 ELECTRIC GENERATORS AND MOTORS

30.74 Figure 30-26 illustrates *Faraday's disk*, the first generator. A copper disk of radius R rotates about O with clockwise angular speed ω. The lowest part of the disk dips into a trough of mercury at A. A voltmeter V makes contact with the mercury at D and sliding contact with the metal axle. The disk is in a uniform magnetic field of magnitude $\mathcal{B}$, directed into the plane of the page. In what direction does current flow through the voltmeter?

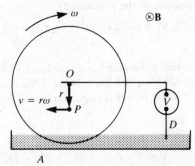

Fig. 30-26

▌ A positively charged particle at P, moving with the disk, experiences a magnetic force directed downward toward the mercury. Thus the induced current flows *counterclockwise* in the circuit. This means that the current in the voltmeter flows upward, <u>from D to O.</u> (The potential at D is higher than that at O.)

30.75ᶜ For the Faraday's disk of Prob. 30.74, (*a*) show that the emf is given by $|V| = \frac{1}{2}\mathcal{B}\omega R^2$; and (*b*) if $\mathcal{B} = 0.50$ T, $\omega = 1200$ rev/min, and $R = 10$ cm, evaluate the emf.

▌ (*a*) Under open-circuit conditions, the field magnitude is $E(r) = \mathcal{B}v = \mathcal{B}r\omega$, where r is the distance from O. Then

$$|V| = V_D - V_O = \int_0^R \mathcal{B}r\omega \, dr = \mathcal{B}\omega \left(\frac{1}{2}r^2 \Big|_0^R \right) = \frac{1}{2}\mathcal{B}\omega R^2$$

as was to be shown. (*b*) With $\mathcal{B} = 0.50$ T, $R = 0.10$ m, and $\omega = 1200$ rev/min $= 40\pi$ rad/s, we find $|V| = \frac{1}{2}(0.50)(40\pi)(0.10)^2 = \underline{0.314 \text{ V}}$.

30.76 A hand-operated generator is easy to turn when it is not connected to any electric device. However, it becomes quite difficult to turn when it is connected, particularly if the device has a low resistance. Explain.

▌ When the generator is not connected, there is no current in its coils and the effort required to turn it is only that required to overcome mechanical friction in its bearings. However, when the generator delivers current, there are magnetic forces on the wires in the coil due to the presence of the induced current in the magnetic field. By Lenz's law, the direction of the induced current is such that the magnetic torque on the coils opposes the rotation of the coils. The torque is proportional to the current and is therefore large if the load has low resistance.

30.77 What is the relationship between the emf delivered by the armature of a generator and the terminal voltage of the armature?

▌ When delivering current I_a, the armature of a generator has a resistance r_a and a power loss such that the potential difference of the actual output delivered by the generator is

$$V_t = \mathcal{E} - I_a r_a \tag{1}$$

[Note that in general emf's, voltages and currents involved with generators will be time dependent. In the case of harmonically varying emf's the relationships between emf's, voltages and currents in this chapter can be thought of as involving their root-mean-square (rms) values.]

30.78 Describe *series-wound* and *shunt-wound* generators.

▎ For a given electric generator the magnetic field in which the conductor moves may be supplied by a permanent magnet, by an electromagnet from a separate power source, or by a *self-excited* electromagnet for which the magnetizing current is itself derived from the generated emf. Self-excited generators have field coils that can be connected either in series or in parallel with the armature, giving rise to *series-wound* and *shunt-wound* generators, respectively.

(A compound-wound generator has both a shunt-wound and a series-wound field coil. To control output, one can add varying amounts of resistance to the field-coil circuit. Changes in the field current affect the magnetic field strength and, hence, the output of the generator. A generator can have stationary field coils and a rotating armature, or vice versa.)

30.79 A generator develops an emf of 120 V and has a terminal potential difference of 115 V when the armature current is 25 A. What is the resistance of the armature?

▎ We use Eq. (1) of Prob. 30.77, $V_T = \mathscr{E} - I_a r_a$, so $115 = 120 - (25.0)(r_a)$, and $r_a = \underline{0.2\,\Omega}$.

30.80 The armature of a separately excited generator has a resistance of $0.16\,\Omega$. When run at its rated speed, it yields 132 V on open circuit and 126 V on full load. What is the current at full load? How much power is delivered to the external circuit? What power is needed to drive the generator if its overall efficiency is 85 percent?

▎ The emf is the open-circuit voltage:

$$\mathscr{E} = 132\,\text{V} \qquad V_t = \mathscr{E} - Ir_A \qquad \text{so} \qquad 126 = 132 - 0.16I \qquad I = \underline{37.5\,\text{A}}$$

The power delivered to the external circuit is the product of the terminal voltage and current, so

$$P_e = V_t I = (126)(37.5) = \underline{4725\,\text{W}}$$

$$P_e = 0.85 P_{\text{supp}} \qquad P_{\text{supp}} = \frac{4725}{0.85} = \underline{5560\,\text{W}} = \underline{7.45\,\text{hp}} \qquad (1\,\text{hp} = 745.7\,\text{W})$$

30.81 The brush potential of a separately excited generator when it is delivering 5 A is 125 V. When the generator delivers 15 A, the potential difference across the brushes falls to 122 V. What are the induced emf and the resistance of the armature?

▎ We use Eq. (1) of Prob. 30.77, $V_t = \mathscr{E} - Ir_A$, for each of the two cases, getting

$$125 = \mathscr{E} - 5r_A \qquad \text{and} \qquad 122 = \mathscr{E} - 15r_A$$

Subtracting to eliminate $\mathscr{E}$ yields

$$10r_A = 3 \qquad r_A = \underline{0.3\,\Omega} \qquad \text{Then} \qquad 125 = \mathscr{E} - (5)(0.3) \qquad \text{and} \qquad \mathscr{E} = \underline{126.5\,\text{V}}$$

30.82 The armature of a generator has a resistance of $0.2\,\Omega$. When the current through the armature is 5 A, the terminal potential is 224 V. What will the terminal potential be when the current is 40 A, assuming that the field strength and the speed remain unchanged? What is the emf induced?

▎ First we solve for $\mathscr{E}$:

$$V_t = \mathscr{E} - Ir_A \qquad 224 = \mathscr{E} - [(5) \times (0.2)] \qquad \text{and} \qquad \mathscr{E} = \underline{225\,\text{V}}$$

Next we determine the new terminal voltage:

$$V_t = 225 - [(40) \times (0.2)] = 225 - 8 = \underline{217\,\text{V}}$$

30.83 A shunt generator has armature resistance $0.06\,\Omega$ and shunt-field resistance $100\,\Omega$. What power is developed in the armature when it delivers 40 kW at 250 V to an external circuit? See Fig. 30-27.

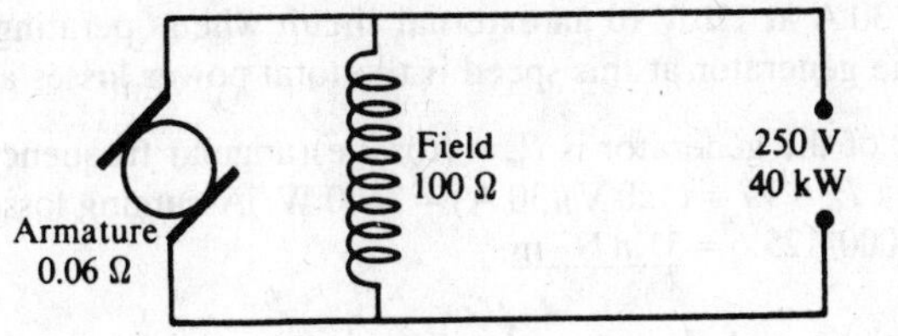

Fig. 30-27

$$\text{Current in external circuit} = I_x = \frac{I_x V}{V} = \frac{40\,000\ \text{W}}{250\ \text{V}} = 160\ \text{A} \qquad \text{field current} = I_f = \frac{V_f}{R_f} = \frac{250\ \text{V}}{100\ \Omega} = 2.5\ \text{A}$$

$$\text{armature current} = I_a = I_x + I_f = 162.5\ \text{A}$$

$$\text{total induced emf} = |\mathscr{E}| = (250\ \text{V} + I_a R_a \text{ drop in armature}) = 250\ \text{V} + (162.5\ \text{A})(0.06\ \Omega) = 260\ \text{V}$$

$$\text{power developed by armature} = I_a |\mathscr{E}| = (162.5\ \text{A})(260\ \text{V}) = \underline{42.2\ \text{kW}}$$

Another method

$$\text{power loss in armature} = I_a^2 R_a = (162.5\ \text{A})^2(0.06\ \Omega) = 1.6\ \text{kW}$$
$$\text{power loss in field} = I_f^2 R_f = (2.5\ \text{A})^2(100\ \Omega) = 0.6\ \text{kW}$$

$$\text{power developed} = (\text{power delivered}) + (\text{power loss in armature}) + (\text{power loss in field})$$
$$= 40\ \text{kW} + 1.6\ \text{kW} + 0.6\ \text{kW} = \underline{42.2\ \text{kW}}.$$

30.84 A shunt-wound generator has an armature resistance of 0.5 Ω, and the field coils have a resistance of 200 Ω. Its terminal potential difference is 120 V when a current of 8 A is delivered to an external circuit. Find (a) the current in the field coils, (b) the armature current, and (c) the induced emf.

▌ Let I_f = field current, I_a = armature current.

(a) $$I_f = \frac{V_t}{R_f} = \frac{120}{200} = \underline{0.6\ \text{A}}$$

(b) $$I_a = I_{\text{load}} + I_f = \underline{8.6\ \text{A}}$$

(c) $$V_t = \mathscr{E} - I_a r_a \quad \text{or} \quad 120 = \mathscr{E} - [(8.6)(0.5)] \quad \text{and} \quad \mathscr{E} = \underline{124.3\ \text{V}}$$

30.85 The armature of a shunt-wound generator has a resistance of 0.12 Ω. The terminal potential difference of the generator is 118 V when the armature current is 60 A. Find the emf of the generator and the power dissipated in heat in the armature.

▌ $$V_t = \mathscr{E} - Ir \quad \text{so} \quad 118 = \mathscr{E} - [(60)(0.12)] \quad \mathscr{E} = 118 + 7.2 = 125.2\ \text{V}$$
$$P_{\text{heat}} = I^2 R = (60^2)(0.12) = \underline{432\ \text{W}}$$

30.86 A 75-kW, 230-V shunt generator has a generated emf of 243.5 V. If the field current is 12.5 A at rated output, what is the armature resistance?

▌ $\mathscr{E} = 243.5$ V, while $V_t = 230$ V. $V_t = \mathscr{E} - I_a R_a$, so we need I_a to find R_a. Referring to the general diagram, Fig. 30-27, we have $I_a = I_f + I_x$, where the field current $I_f = 12.5$ A and the external load current I_x obeys $I_x V_t = 75 \times 10^3$ W. Then $I_x = 75\,000/230 = 326.1$ A, and $I_a = 338.6$ A. Finally $230 = 243.5 - 338.6 R_a$, and $R_a = \underline{0.040\ \Omega}$.

30.87 A shunt-wound generator delivers 48 A at a brush potential of 120 V. The field coils have a resistance of 60 Ω, and the armature has a resistance of 0.14 Ω. If the stray-power loss is 500 W, what is the efficiency of the generator?

▌ The terminal voltage is $V_t = 120$ V. We calculate the delivered power output, and the power losses in the field coils, $P_{\text{FC}} = I_F^2 R_F$, and the armature, $P_a = I_a^2 R_a$. The efficiency is power output/power input where the denominator includes the stray losses. Thus

$$P_{\text{output}} = V_t I_a = (120)(48) = 5760\ \text{W} \qquad P_{\text{FC}} = (120/60)^2(60) = 240\ \text{W} \qquad P_{\text{armature}} = (48)^2(0.14) = 322.56\ \text{W}$$

$$P_{\text{stray}} = 500\ \text{W} \qquad \text{Eff} = \frac{5760}{5760 + 240 + 322.56 + 500} = \underline{0.84}$$

30.88 A dynamo delivers 30 A at 120 V to an external circuit when operating at 1200 rev/min. What torque is required to drive the generator at this speed if the total power losses are 400 W?

▌ The input power of the generator is $P_{in} = $ (torque)(angular frequency) $= (125.6 \text{ rad/s})\tau$. But we are given the power output via $P_0 = VI = (120 \text{ V})(30 \text{ A}) = 3600 \text{ W}$. Assuming losses of 400 W, we have $P_{in} = P_0 + 400 = 4000 \text{ W}$. Then $\tau = 4000/125.6 = \underline{31.8 \text{ N} \cdot \text{m}}$.

30.89 A two-pole generator has a drum armature with a cylindrical core 20 cm long and 10 cm in diameter, and 200 armature conductors arranged in two parallel paths. The flux density in the air gap is 0.5 Wb/m². If the armature is driven at 30 rev/s, what is the magnitude of the average emf generated?

▌ The flux in the air gap is $\Phi = BA = (0.5 \text{ Wb/m}^2)(0.2 \text{ m} \times 0.1 \text{ m}) = 0.01 \text{ Wb}$. Each conductor cuts 0.01 Wb twice per revolution. Then in 1 s, or 30 rev, each conductor cuts $2(30)(0.01) = 0.6 \text{ Wb}$. As there are 200 conductors in two parallel paths, the number of conductors in series in each path is $200/2 = 100$. (The emf's in series are additive.)

$$|\mathcal{E}| = \text{flux lines cut per s by 100 conductors in series} = (100)(0.60 \text{ Wb/s}) = \underline{60 \text{ V}}$$

30.90 A six-pole generator with fixed-field excitation develops an emf of 100 V when operating at 1500 rev/min. At what speed must it rotate to develop 120 V?

▌ The flux cut per second, and hence the emf, varies directly as the speed.

$$\text{speed} = \frac{120}{100}(1500 \text{ rev/min}) = \underline{1800 \text{ rev/min}}$$

30.91 Determine the separate effects on the induced emf of a generator if (a) the flux per pole is doubled, and (b) the speed of the armature is doubled.

▌ Basically, $|\mathcal{E}| \propto \Delta\Phi/\Delta t$. If the flux per pole is doubled, all else being equal, $\Delta\Phi$ doubles, so $|\mathcal{E}|$ doubles. If the speed of the armature doubles, Δt halves, and all else being equal, $|\mathcal{E}|$ doubles.

30.92 Figure 30-28 represents a primitive motor. A metal wire slides on a horseshoe-shaped metal loop of width 0.25 m. These have negligible resistance, but there is a 1.0-Ω resistor in the circuit as well as a 6.0-V battery. There is a uniform magnetic field directed into the plane of the page of magnitude 0.50 T. The slide wire is pushed to the right by the magnetic force. A force of 0.25 N to the left is required to keep it moving with constant speed to the right. (a) What is the current i in the circuit? (b) What is the voltage drop across the 1.0-Ω resistor? (c) What is the back emf generated by the moving wire?

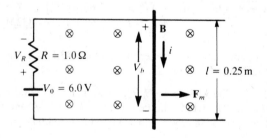

Fig. 30-28

▌ (a) Since the slide wire is perpendicular to the magnetic field, the magnetic force on the wire has magnitude $F_m = i\mathcal{B}l$. The magnetic force is balanced by a leftward force of 0.25 N, so we have $i = F_m/(\mathcal{B}l) = 0.25/[(0.50)(0.25)] = \underline{2.0 \text{ A}}$.
(b) With a current $i = 2.0 \text{ A}$, the voltage drop V_R across the resistor is $V_R = iR = (2.0)(1.0) = \underline{2.0 \text{ V}}$.
(c) The back emf $V_b = V_0 - V_R = 6.0 - 2.0 = \underline{4.0 \text{ V}}$.

30.93 Refer to Prob. 30.92. (a) With what constant speed is the wire moving? (b) What mechanical power does the motor produce? (c) Show that the electric power converted to mechanical power is equal to the power in part (b). What is the efficiency of the motor?

▌ (a) The back emf is equal to $\mathcal{B}$ times the rate of increase of the enclosed area: $V_b = \mathcal{B}lv$. Thus the speed is given by $v = V_b/(\mathcal{B}l) = 4.0/[(0.5)(0.25)] = \underline{32 \text{ m/s}}$. (b) The mechanical power P is the product of the rightward force F_m and the speed v: $P = F_m v = (0.25)(32) = \underline{8.0\,\text{W}}$. (c) The battery is expending electric energy at the rate $V_0 i = (6.0)(2.0) = 12.0\,\text{W}$. Of this total, the resistor dissipates $V_R i = i^2 R = (2.0^2)(1.0) = 4.0\,\text{W}$. The remainder, $12.0 - 4.0 = 8.0\,\text{W}$, is converted from electric to macroscopic mechanical form. Hence the conversion efficiency is $8.0\,\text{W}/12.0\,\text{W} = \underline{67\%}$.

30.94 In a motor a conductor 125 mm long carries a current of 3 A. If it lies perpendicular to a uniform magnetic field of 1.2 T, what is the force on the conductor? What torque is produced if the conductor is 50 mm from the axis of the motor and parallel to it and the field points from the axis to the wire?

   ▌ $F = IlB = (3)(0.125)(1.2) = \underline{0.45\,\text{N}}$ $\text{torque} = rF = (0.05)(0.45) = \underline{22.5 \times 10^{-3}\,\text{N}\cdot\text{m}}$.

30.95 Find the torque on a rectangular loop of wire in a motor bearing a current of 9 A when its plane is parallel to the magnetic lines of a field with an intensity of 1.25 T. The dimensions of the loop are 0.25 m parallel to the flux and 12 cm at right angles to it.

   ▌ The magnetic moment is given by

$$M = IA = (9)(0.25)(0.12) = 0.27\,\text{A}\cdot\text{m}^2 \quad \text{then} \quad \text{torque} = MB \sin 90° = (0.27)(1.25)(1) = \underline{0.3375\,\text{N}\cdot\text{m}}.$$

30.96 A rectangular loop of wire in a motor, 8 by 15 cm, bears a current of 6 A. The plane of the coil is parallel to a uniform magnetic field of 1.2 T, with its 15-cm length perpendicular to the lines of flux and the other side parallel to them. Find the force on each side of the loop. Using this force, find the torque on the loop. Find the magnetic moment of the loop and use its value to check the value of the torque.

   ▌ Labeling the forces by the length of the side involved, we have

$$F_{15} = IlB = (6)(0.15)(1.2) = \underline{1.08\,\text{N}}$$

(in opposite directions for the two sides). $F_8 = 0$. The torque about an axis through the center of the loop perpendicular to the 8 cm sides is

$$\text{torque} = (2)(0.04)(1.08) = \underline{0.0864\,\text{N}\cdot\text{m}}$$

The magnetic moment is

$$M = IA = (6)(0.15)(0.08) = 0.072\,\text{A}\cdot\text{m}^2 \quad \text{torque} = MB \sin(\mathbf{M}, \mathbf{B}) = (0.072)(1.2)(1) = \underline{0.0864\,\text{N}\cdot\text{m}}.$$

30.97 The flux density between the poles of a certain motor is 1.1 T. A conductor carrying 8 A lies in this magnetic field so that it is perpendicular to the field. Find the force on it if 0.1 m of the conductor is in the field. If the wire moves 15 mm perpendicular to the field in 0.002 s, and if the field is uniform over this distance, what is the work done on the wire? Show that the same work is predicted by finding the induced back emf and multiplying it by the charge transferred in the 0.002 s.

   ▌ For the force,

$$F = IlB = (8)(0.1)(1.1) = \underline{0.88\,\text{N}} \quad \text{then} \quad \text{work} = Fs \cos\theta = (0.88)(0.015)(1) = \underline{0.0132\,\text{J}}$$

The back emf is

$$\mathcal{E} = vBl = \left(\frac{0.015}{0.002}\right)(1.1)(0.1) = 0.825\,\text{V}$$

and the total charge passing in 0.002 s is

$$Q = 8\,\text{A} \times 0.002\,\text{s} = 0.016\,\text{C} \quad \text{then} \quad \text{work} = \mathcal{E}Q = (0.825)(0.016) = \underline{0.0132\,\text{J}}.$$

30.98 A motor running at full load on a 118-V line develops a back emf of 109 V and draws a current of 8 A through the armature. What is the mechanical power output of the motor, disregarding frictional losses? What is the armature resistance?

   ▌ Let $\mathcal{E}_b$ be the back emf developed in the motor.

$$P_{\text{out}} = \mathcal{E}_b I_A = (109)(8) = \underline{872\,\text{W}}$$

For the resistance we have $V_t = \mathcal{E}_b + IR_A$. (Compare Prob. 30.77; a motor is just a generator run in reverse.)

$$8 = \frac{118 - 109}{R_A} \quad \text{and} \quad R_A = \frac{9}{8} = \underline{1.125\,\Omega}$$

30.99 A motor has back emf 110 V and armature current 90 A when running at 1500 rev/min. Determine the power and the torque developed within the armature.

▌

$$\text{power} = (\text{armature current})(\text{back emf}) = (90 \text{ A})(110 \text{ V}) = \underline{9900 \text{ W}}$$

$$\text{torque} = \frac{\text{power}}{\text{angular speed}} = \frac{9900 \text{ W}}{(2\pi \times 25) \text{ rad/s}} = \underline{63.0 \text{ N} \cdot \text{m}}$$

30.100 A motor generates a 210-V back emf when the applied voltage is 220 V and the armature current is 5.7 A. Find (**a**) the total mechanical power output and (**b**) the power dissipated in the motor.

▌ The back emf is less than the applied voltage because of dissipative loss.
(**a**) $P_{\text{mech}} = \mathcal{E}_b I_a = (210)(5.7) = \underline{1197 \text{ W}}$.
(**b**) The dissipative power loss is $V_t I_a - P_{\text{mech}}$, so $P_{\text{dis}} = (220 - 210)(5.7) = \underline{57 \text{ W}}$.

30.101 A motor armature develops a torque of 100 N · m when it draws 40 A from the line. Determine the torque developed if the armature current is increased to 70 A and the magnetic field strength is reduced to 80% of its initial value.

▌ The torque developed by the armature of a given motor is proportional to the armature current and to the field strength so torque = $(100 \text{ N} \cdot \text{m})(70/40)(0.80) = \underline{140 \text{ N} \cdot \text{m}}$.

30.102 The shunt motor shown in Fig. 30-29 has armature resistance 0.05 Ω and is connected to 120-V mains. (**a**) What is the armature current at the starting instant, i.e., before the armature develops any back emf? (Ignore the rheostat.) (**b**) What starting rheostat resistance R, in series with the armature, will limit the starting current to 60 A; (**c**) With no starting resistance, what back emf is generated when the armature current is 20 A? (**d**) If this machine were running as a generator, what would be the total induced emf developed by the armature when the armature is delivering 20 A at 120 V to the shunt field and external circuit?

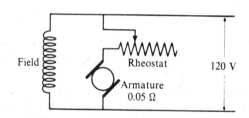

Fig. 30-29

▌ (**a**)
$$\text{armature current} = \frac{\text{impressed voltage}}{\text{armature resistance}} = \frac{120 \text{ V}}{0.05 \text{ Ω}} = \underline{2400 \text{ A}}$$

(**b**)
$$\text{armature current} = \frac{\text{impressed voltage}}{0.05 \text{ Ω} + R} \quad \text{or} \quad 60 \text{ A} = \frac{120 \text{ V}}{0.05 \text{ Ω} + R}$$

from which $R = \underline{1.95 \text{ Ω}}$.
(**c**) back emf = (impressed voltage) − (voltage drop in armature resistance) = 120 V − (20 A)(0.05 Ω) = $\underline{119 \text{ V}}$.
(**d**) induced emf = (terminal voltage) + (voltage drop in armature resistance) = 120 V + (20 A)(0.05 Ω) = $\underline{121 \text{ V}}$.

See Prob. 30.98 with regard to parts (**c**) and (**d**).

30.103 The stationary coil producing the magnetic field and the rotating coil of a motor are connected in parallel. The current in the rotating coil is 1.4 A, and its resistance R is 5.0 Ω. The voltage applied to the motor is 220 V. (**a**) What is the back emf? (**b**) What is the mechanical power output?

▌ (**a**) The back emf is $V_b = V − iR$, where i is the rotor current and R is the (ohmic) resistance of the coil. Thus, $V_b = 220 − (1.4)(5) = \underline{213 \text{ V}}$. (**b**) The mechanical power output is the rate at which electric energy is expended against the back voltage: $P_{\text{out}} = V_b i = (213)(1.4) = \underline{298 \text{ W}}$. (Note that the currents and voltages in a motor will be time-dependent. The values given in this chapter can be considered root-mean-square (rms) in the case of harmonically varying emfs.)

30.104 Refer to Prob. 30.103. (**a**) If an added mechanical load slows the motor by 5 percent, what is now the back emf? (**b**) What is the current in the rotor now? (**c**) What is the new mechanical power output? (**d**) How do the results of parts (**a**), (**b**), and (**c**), show that an electric automobile does not require a gear shift?

❚ (a) The back emf is due to the rotating coil in the motor acting as a generator. V_b is proportional to the angular speed. Therefore a 5 percent reduction in the angular speed implies a 5 percent reduction in V_b. That is, $V_b' = 0.95V_b = (0.95)(213) = \underline{202\ V}$. (b) Solving for the current, we find that

$$i' = \frac{V - V_b'}{R} = \frac{220 - 202}{5} = \underline{3.6\ A}$$

(c) The new mechanical power output is

$$P_{out}' = V_b'i' = (202)(3.6) = \underline{727\ W} \qquad \text{[Compare with Prob. 30.103(b).]}$$

(d) When an electric automobile slows down, say, for a hill, the back emf decreases. This results in an increase in the rotor current, which provides increases in both the torque and the useful mechanical power output <u>without</u> the need for a change in gear ratios in the drive train.

30.105 The shunt motor shown in Fig. 30-30 has armature resistance $0.25\ \Omega$ and field resistance $150\ \Omega$. It is connected across 120-V mains and is generating a back emf of 115 V. Compute (a) the armature current I_a, the field current I_f, and the total current I_t taken by the motor; (b) the total power taken by the motor; (c) the power lost in heat in the armature and field circuits; and (d) the electric efficiency of this machine (when only heat losses in the armature and field are considered).

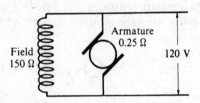

Field
150 Ω

Armature
0.25 Ω

120 V

Fig. 30-30

❚ (a)
$$I_a = \frac{(\text{impressed voltage}) - (\text{back emf})}{\text{armature resistance}} = \frac{(120 - 115)\ V}{0.25\ \Omega} = \underline{20\ A}$$

$$I_f = \frac{\text{impressed voltage}}{\text{field resistance}} = \frac{120\ V}{150\ \Omega} = \underline{0.80\ A} \qquad I_t = I_a + I_f = \underline{20.80\ A}$$

(b) $\qquad\qquad\qquad\qquad$ power input $= (120\ V)(20.80\ A) = \underline{2496\ W}$

(c) $\qquad I_a^2 R_a$ loss in armature $= (20\ A)^2(0.25\ \Omega) = \underline{100\ W} \qquad I_f^2 R_f$ loss in field $= (0.80\ A)^2(150\ \Omega) = \underline{96\ W}$

(d) $\qquad\qquad\qquad$ power output $=$ (power input) $-$ (power losses) $= 2496 - (100 + 96) = 2300\ W$

or, $\qquad\qquad\qquad$ power output $=$ (armature current)(back emf) $= (20\ A)(115\ V) = 2300\ W$

$$\text{efficiency} = \frac{\text{power output}}{\text{power input}} = \frac{2300\ W}{2496\ W} = 0.921 = \underline{92.1\%}$$

30.106 A shunt-wound motor operates from a 120-V circuit. The field coils have a resistance of $240\ \Omega$, while the armature resistance is $0.5\ \Omega$. (a) If the motor draws 4.5 A from the line, what is the armature current? (b) What is the power output of the motor? (Assume no other losses.)

❚ (a)
$$I_{\text{field}} = \frac{V_T}{R_{\text{field}}} = \frac{120}{240} = 0.5\ A$$

$$\text{then } I_{\text{arm}} = I_{\text{tot}} - I_{\text{field}} = \underline{4.0\ A}$$

(b)
$$P_{out} = P_{in} - I_f^2 R_f - I_a^2 R_a$$
$$= (120\ V)(4.5\ A) - (0.5\ A)^2(240\ \Omega) - (4.0\ A)^2(0.5\ \Omega)$$
$$= \underline{472\ W}$$

30.107 A shunt motor takes a total current of 9 A from 120-V mains. The resistance of the armature is $1.5\ \Omega$, and that of the field coils is $240\ \Omega$. Find the current in the field coils and in the armature, the back emf induced, and the mechanical power output of the motor.

❚
$$I_F = \frac{120}{240} = 0.5\ A \qquad I_A = 9 - 0.5 = 8.5\ A$$

Then from $V_t = \mathcal{E}_B + I_A R_A$ we have

$$\mathcal{E}_B = (120\text{ V}) - (8.5\text{ A})(1.5\ \Omega) = \underline{107.25\text{ V}} \qquad P_{\text{out}} = \mathcal{E}_B I_A = (107.25)(8.5) = \underline{911.6\text{ W}}$$

30.108 The armature of a shunt-wound motor can withstand up to 8 A before it overheats and is damaged. If the armature resistance is 0.5 Ω, what minimum back emf must be generated to avoid motor damage when the motor is connected to a 120-V line?

$$\mathcal{E} = V_t - I_a r_a = 120 - (8.0)(0.5) = \underline{116\text{ V}}$$

30.109 A shunt-wound motor draws 2.9 A from a 120-V line. The field coils have a resistance of 300 Ω, and the armature resistance is 0.8 Ω. Find the armature current and the back emf. What is the mechanical power output of the motor? How much power is dissipated in I^2R heating in the motor?

$$I_F = \frac{120}{300} = \underline{0.4\text{ A}} \qquad I_A = 2.9 - 0.4 = \underline{2.5\text{ A}} \qquad V_t = \mathcal{E}_b + I_A r_A \qquad 120 = \mathcal{E}_b + (2.5)(0.8) \qquad \mathcal{E}_b = \underline{118\text{ V}}$$

$$P_{\text{out}} = \mathcal{E}_b I_A = (118)(2.5) = \underline{295\text{ W}} \qquad P_A = (2.5)^2(0.8) = 5\text{ W} \qquad P_F = (0.4)^2(300) = 48\text{ W}$$

$$\text{power dissipated in } I^2R \text{ heating} = 5\text{ W} + 48\text{ W} = \underline{53\text{ W}}$$

30.110 In the circuit of Fig. 30-31 a battery of 64-V emf and 1-Ω internal resistance is providing electric energy to operate a motor which has an armature resistance of 0.5 Ω. If $R_3 = 12\ \Omega$, $R_4 = 6\ \Omega$, $R_1 = 2.5\ \Omega$, and $R_2 = 3\ \Omega$, what is the back emf of the motor when the armature draws 2 A? What is the mechanical power output of the motor?

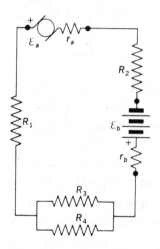

Fig. 30-31

$\sum \mathcal{E} = IR_e$, where $\sum \mathcal{E} = 64\text{ V} - \mathcal{E}_a$ and the circuit equivalent resistance $R_e = r_a + r_b + R_1 + R_2 + R_{34}$, with

$$\frac{1}{R_{34}} = \frac{1}{12} + \frac{1}{6} = \frac{1}{4} \qquad \text{and} \qquad R_{34} = 4\ \Omega$$

Then

$$2 = \frac{64 - \mathcal{E}_a}{0.5 + 1 + 2.5 + 3 + 4} \qquad 22 = 64 - \mathcal{E} \qquad \text{and} \qquad \mathcal{E}_a = \underline{42\text{ V}} \qquad P_{\text{out}} = \mathcal{E}_a I = (42)(2) = \underline{84\text{ W}}$$

30.111 In the circuit of Fig. 30-31, $\mathcal{E}_a = 80\text{ V}$, $r_a = 1.5\ \Omega$, $R_2 = 3\ \Omega$, $\mathcal{E}_b = 32\text{ V}$, $r_b = 0.5\ \Omega$, $R_3 = 15\ \Omega$, $R_4 = 10\ \Omega$, and the current through the generator is 3 A. Find R_1 and the terminal potential differences of the generator and battery.

Note that here $\mathcal{E}_a$ is a generator emf. Then

$$3 = \frac{80 - 32}{1.5 + 0.5 + R_1 + 3 + R_{34}} \qquad \frac{1}{R_{34}} = \frac{1}{15} + \frac{1}{10} = \frac{2+3}{30} = \frac{1}{6} \qquad R_{34} = 6\ \Omega \qquad 3 = \frac{48}{11 + R_1} \qquad R_1 = \underline{5\ \Omega}$$

$$V_{t,\text{gen}} = \mathcal{E}_{\text{gen}} - I_A R_A = 80 - (3)(1.5) = \underline{75.5\text{ V}}$$

$$V_{t,\text{batt}} = \mathcal{E}_b + I_b r_b = 32 + (3)(0.5) = \underline{33.5\text{ V}}$$

30.112 In the circuit of Fig. 30-31 $\mathcal{E}_a = 100$ V, $r_a = 0.8\ \Omega$, $R_1 = 6\ \Omega$, $R_2 = 5\ \Omega$, $\mathcal{E}_b = 36$ V, $r_b = 0.2\ \Omega$, $R_3 = 20\ \Omega$, and $R_4 = 5\ \Omega$. Find the current in the main circuit and the terminal potential differences of the generator and battery.

$$I = \frac{\sum \mathcal{E}}{\sum R} = \frac{100 - 36}{0.8 + 6 + 5 + 0.2 + 4} = \frac{64}{16} = \underline{4\ A}$$

$$\frac{1}{R_{34}} = \frac{1}{20} + \frac{1}{5} = \frac{1}{4} \qquad V_{t,\,\text{gen}} = \mathcal{E}_a - Ir_a = 100 - (4)(0.8) = \underline{96.8\ V} \qquad V_{t,\,\text{batt}} = \mathcal{E}_b + Ir_b = 36 + (4)(0.2) = \underline{36.8\ V}$$

30.113 In the circuit of Fig. 30-31, $\mathcal{E}_a$ is the back emf of a motor. It has a value of 30 V. The armature resistance and the resistance of the battery are both 1 Ω, and all other resistances are 8 Ω. Find the emf of the battery if the motor draws 2 A. What are the terminal potential and the mechanical power output of the motor?

$$\frac{1}{R_{34}} = \frac{1}{8} + \frac{1}{8} = \frac{1}{4} \qquad R_{34} = 4\ \Omega$$

Then from $I = \sum \mathcal{E} / \sum R$,

$$2 = \frac{\mathcal{E}_{\text{batt}} - 30}{1 + 1 + 8 + 8 + 4} \qquad \text{and} \qquad \mathcal{E}_{\text{batt}} = 30 + (2)(22) = \underline{74\ V}$$

The terminal voltage for the motor is

$$V_{\text{tm}} = \mathcal{E}_a + Ir_a = 30 + (2)(1) = \underline{32\ V}$$

The mechanical output of the motor is

$$P_{\text{out}} = \mathcal{E}_a I = (30)(2) = \underline{60\ W}$$

30.114 A shunt-wound motor with an armature resistance of 0.3 Ω operates on a 120-V circuit. (*a*) Find the starting resistance which must be temporarily connected in series with the armature to limit the starting current to 10 A. (*b*) Find the current drawn and the mechanical power output when the back emf generated is 118 V.

(*a*) Initially there is zero back emf so

$$10 = \frac{120}{0.3 + R_S} \qquad \text{and} \qquad R_S = \underline{11.7\ \Omega}$$

(*b*) $$I = \frac{120 - 118}{0.3} = \underline{6.67\ A} \qquad P_{\text{out}} = \mathcal{E}_a I = (118)(6.67) = \underline{787\ W}$$

30.115 A shunt-wound motor has an armature with a resistance of 0.2 Ω and operates from a 220-V line. What starting resistance is required if the armature current must be limited to 11 A? When the motor is operating normally with the starting resistor out, the armature draws 4 A. Find the back emf and the mechanical power output.

At start $\mathcal{E}_a = 0$ and

$$11 = \frac{220}{0.2 + R_S} \qquad R_S = \underline{19.8\ \Omega}$$

With the resistor out and $I = 4$ A,

$$4 = \frac{220 - \mathcal{E}_a}{0.2} \qquad \mathcal{E}_a = \underline{219.2\ V} \qquad P_{\text{out}} = \mathcal{E}_a I = (219.2)(4) = \underline{876.8\ W}$$

30.116 The armature of a motor connected to a 220-V line draws 25 A at the instant the connection is made, when the motor is at rest. There is a starting resistance of 8.5 Ω in the circuit. What is the resistance of the armature? What is the counter emf when the current drawn is 15 A (with the starting resistance shorted out)?

At start,

$$25 = \frac{220}{8.5 + r_A} \qquad \text{and} \qquad r_A = \underline{0.3\ \Omega}$$

When current is 15 A, we have

$$15 = \frac{220 - \mathcal{E}_a}{0.3} \qquad \mathcal{E}_a = \underline{215.5\ V}$$

CHAPTER 31
Inductance

31.1 SELF-INDUCTANCE

31.1 Describe what is meant by a *self-induced emf* and define *self-inductance* in terms of it.

▎ A coil can induce an emf in itself. If the current in a coil changes, the flux through the coil due to the current also changes. As a result, the changing current in a coil induces an emf in that same coil.

Because an induced emf $\mathscr{E}$ is proportional to $\Delta\Phi/\Delta t$ and because $\Delta\Phi$ is proportional to Δi, where i is the current that causes the flux, $\mathscr{E} = -(\text{constant})(\Delta i/\Delta t)$. Here i is the current through the same coil in which $\mathscr{E}$ is induced. The negative sign indicates that the self-induced emf, $\mathscr{E}$, is a back emf and opposes the change in current.

The proportionality constant depends upon the geometry of the coil. We represent it by L and call it the *self-inductance* of the coil. Then

$$\mathscr{E} = -L\frac{\Delta i}{\Delta t} \tag{1}$$

For $\mathscr{E}$ in V, i in A, and t in s, L is in *henries*, where $1\,\text{H} = 1\,\text{V}\cdot\text{s/A} = 1\,\text{J/A}^2 = 1\,\text{Wb/A} = 1\,\text{T}\cdot\text{m}^2/\text{A}$.

31.2 Define self-inductance directly in terms of flux linkage and obtain Eq. (1) of Prob. 31.1.

▎ In Fig. 31-1, current I establishes a magnetic field in and around coil C as indicated. Each line of flux threads (or "links") some or all of the turns. The total linkage is $N\Phi$, where N is the total number of turns and Φ is the *average* flux linking a turn.

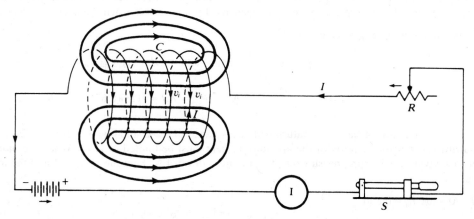

Fig. 31-1

With C in empty space and no magnetic material nearby, the flux established, and thus the flux linkage, is directly proportional to I:

$$N\Phi = LI \tag{1}$$

The proportionality factor L is referred to as the *self-inductance* (or simply *inductance*). If, over a time Δt, I changes (as by sliding the contact on R), the corresponding change in Φ is given by (1) as

$$N\,\Delta\Phi = L\,\Delta I \qquad \text{or} \qquad -\mathscr{E} = N\frac{\Delta\Phi}{\Delta t} = L\frac{\Delta I}{\Delta t}$$

31.3ᶜ Prove that the energy stored in a coil due to its self-inductance is $U = \frac{1}{2}LI^2$.

▎ Suppose that upon closing switch S in Fig. 31-1 the current in the circuit is built up from an initial value zero to a final value I. Then the work done (by the battery) in driving charge through C, against the induced emf, is

$$U = \int_0^q (-\mathscr{E})\,dq = \int_0^t \left(L\frac{dI}{dt}\right)(I\,dt) = L\int_0^I I\,dI = \frac{1}{2}LI^2$$

This amount of energy is actually stored in the final magnetic field around the inductor, i.e., principally in the space enclosed by the turns of C.

31.4 A long air core solenoid has a cross-sectional area A and N loops of wire on its length, d. (**a**) Find its self-inductance. (**b**) What is its inductance if the core material has permeability μ?

▮ (**a**) We can write

$$|\mathscr{E}| = N \left| \frac{\Delta\Phi}{\Delta t} \right| \quad \text{and} \quad |\mathscr{E}| = L \left| \frac{\Delta i}{\Delta t} \right|.$$

Upon equating these two expressions for $|\mathscr{E}|$,

$$L = N \left| \frac{\Delta\Phi}{\Delta i} \right|$$

If the current changes from zero to I, then the flux changes from zero to Φ. Therefore, $\Delta i = I$ and $\Delta\Phi = \Phi$ in this case. The self-inductance, assumed constant for all cases, is then

$$L = N \frac{\Phi}{I} = N \frac{BA}{I}$$

But, for an air core solenoid, $B = \mu_0 nI = \mu_0(N/d)I$. Substitution gives $\underline{L = (\mu_0 N^2 A)/d}$.
(**b**) If the material of the core has permeability μ instead of μ_0, then B, and therefore L, will be increased by the factor μ/μ_0. In that case, $L = (\mu N^2 A)/d$. An iron core solenoid has a much higher self-inductance than an air core solenoid has.

31.5 A solenoid 30 cm long is made by winding 2000 loops of wire on an iron rod whose cross section is 1.5 cm². If the relative permeability of the iron is 600, what is the self-inductance of the solenoid? What average emf is induced in the solenoid as the current in it is decreased from 0.6 to 0.1 A in a time of 0.03 s?

▮ From Prob. 31.4, with $k_m = \mu/\mu_0$,

$$L = \frac{k_m \mu_0 N^2 A}{d} = \frac{(600)(4\pi \times 10^{-7}\,\text{H/m})(2000)^2(1.5 \times 10^{-4}\,\text{m}^2)}{0.30\,\text{m}} = \underline{1.51\,\text{H}}$$

Then

$$|\mathscr{E}| = L \left| \frac{\Delta i}{\Delta t} \right| = 1.51\,\text{H}\frac{0.5\,\text{A}}{0.03\,\text{s}} = \underline{25\,\text{V}}$$

31.6 A hollow solenoid has a self-inductance of 3.5 mH when it is air-filled. But when an iron rod is slipped into it so as to fill its interior, the self-inductance rises to 1.3 H. What is k_m for the iron being used?

▮ L is proportional to the flux, which is proportional to B. Therefore, $k_m = L/L_0 = 1.3/(3.5 \times 10^{-3}) = \underline{371}$.

31.7 An emf of 8 V is induced in a coil when the current in it changes at the rate of 32 A/s. Compute the inductance of the coil.

▮ $\qquad |\mathscr{E}| = L \left| \frac{\Delta i}{\Delta t} \right| \qquad$ so we have $\qquad 8\,\text{V} = L(32\,\text{A/s}) \qquad$ and $\qquad L = \underline{0.25\,\text{H}}$

31.8 A steady current of 2.5 A creates a flux of 140 μWb in a coil of 500 turns. What is the inductance of the coil?

▮ From Prob. 31.2, $N\Phi = LI$. Then $500(1.4 \times 10^{-4}\,\text{Wb}) = L(2.5\,\text{A})$, and $L = \underline{28\,\text{mH}}$.

31.9 The iron core of a solenoid has a length of 400 mm and a cross section of 500 mm² and is wound with 1000 turns of wire per meter of length. Compute the inductance of the solenoid, assuming the relative permeability of the iron to be constant at 500.

▮ From Prob. 31.4, $L = k_m[(\mu_0 N^2 A)/d]$. For our case $k_m = 500$, $N = (1000)(0.400) = 400$, $A = 5 \times 10^{-4}\,\text{m}^2$ and $d = 0.40\,\text{m}$. Then $L = (500)\{[(4\pi \times 10^{-7})(400)^2(5 \times 10^{-4})]/0.400\} = \underline{126\,\text{mH}}$.

31.10 A coil of resistance 15 Ω and inductance 0.6 H is connected to a steady 120-V power source. At what rate will the current in the coil rise (**a**) at the instant the coil is connected to the power source, and (**b**) at the instant the current reaches 80 percent of its maximum value?

▮ The effective driving voltage in the circuit is the 120-V power supply minus the induced back emf,

$L(\Delta i/\Delta t)$. This equals the p.d. in the resistance of the coil:

$$120\,\text{V} - L\frac{\Delta i}{\Delta t} = iR$$

(a) At the first instant, i is essentially zero. Then

$$\frac{\Delta i}{\Delta t} = \frac{120\,\text{V}}{L} = \frac{120\,\text{V}}{0.6\,\text{H}} = \underline{200\,\text{A/s}}$$

(b) The current reaches a maximum value of $(120\,\text{V})/R$ when the current finally stops changing (i.e., when $\Delta i/\Delta t = 0$). We are interested in the case when

$$i = 0.8\frac{120\,\text{V}}{R}$$

Then, substitution in the loop equation gives

$$120\,\text{V} - L\frac{\Delta i}{\Delta t} = 0.8\frac{120\,\text{V}}{R}R$$

from which
$$\frac{\Delta i}{\Delta t} = \frac{(0.2)(120\,\text{V})}{L} = \frac{(0.2)(120\,\text{V})}{0.6\,\text{H}} = \underline{40\,\text{A/s}}$$

31.11 A coil of inductance 0.2 H and 1.0-Ω resistance is connected to a 90-V source. (a) At what rate will the current in the coil grow at the instant the coil is connected to the source? (b) What is the current when it is growing at a rate of 100 A/s?

▮ The loop equation for the circuit is $(90\,\text{V}) - (0.2\,\text{H})(\Delta i/\Delta t) - (1.0\,\Omega)\,i = 0$.

(a) At the instant of connection $i = 0$, so

$$\frac{\Delta i}{\Delta t} = \frac{90}{0.2} = \underline{450\,\text{A/s}}$$

(b) Here $\Delta i/\Delta t = 100$ A/s. Then $i = \{90\,\text{V} - [(0.2)(100)]\,\text{V}\}/1.0\,\Omega = \underline{70\,\text{A}}$.

31.12 (a) In a certain coil the flux is changing at the rate $N\dot\Phi = 30$ Wb/s, while I increases at the rate of 60 A/s; compute L and the induced emf, v_i. (b) If instead $\dot I = -100$ A/s, what is v_i? (Here a dot over a variable means the rate of change of that variable with time.)

▮ (a) The emf is given by $v_i = -L\dot I = -N\dot\Phi$. For our case we have $30 = L\,(60)$, or $L = \underline{0.5\,\text{H}}$, and $v_i = -N\dot\Phi = \underline{-30\,\text{V}}$, opposite in direction to I.

(b) $$v_i = -L\dot I = -(0.5)(-100) = 50\,\text{V} \qquad \text{in the direction of } I$$

31.13 The air-core toroidal coil shown in Fig. 31-2 has a circular cross section and is uniformly wound with $N = 500$

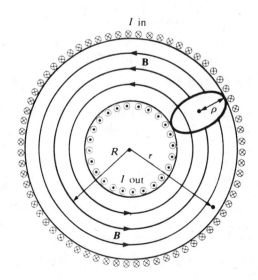

Fig. 31-2

turns of insulated wire. Its average radius is $R = 100$ mm, and the cross-sectional radius is $p = 20$ mm. Compute approximately the self-inductance of the coil.

▌ The flux lines are concentric circles. Applying Ampère's circuital law around a circle of radius r, we obtain $B(2\pi r) = \mu_0 NI$, or $B = (\mu_0 NI)/(2\pi r)$. Thus, B varies as $1/r$ over the cross section of the coil. However, if R is large with respect to ρ (as it is for the values given), we may write $1/r \approx 1/R = $ constant over the cross section, so that $B \approx \mu_0 n'I = $ constant, $[n' \equiv N/(2\pi R)]$. To this approximation, the total flux linkage is

$$N\Phi = NBA = (2\pi R n')(\mu_0 n'I)(\pi\rho^2) = 2\pi^2 \mu_0 n'^2 R\rho^2 I$$

For the numerical data, $L = (N\Phi)/I = \underline{0.6283 \text{ mH}}$.

31.14   Show that the result for the inductance of a toroid is the same as that for a solenoid of the same cross section, A, and a length, d, equal to the average circumference of the toroid.

▌ Taking the result of Prob. 31.13 for the toroid $L = 2\pi^2 \mu_0 n'^2 R\rho^2$, and noting $n' = N/(2\pi R)$, $A = \pi\rho^2$, and $d = 2\pi R$, we have $L = \mu_0 n'NA = (\mu_0 N^2 A)/d$, which is the proper result for the solenoid (Prob. 31.4).

31.15ᶜ   The coaxial cable shown in Fig. 31-3 has an inner wire of radius a and an outer metal sheath with inner radius b. A current i flows down the inner wire and back through the sheath. What is the self-inductance of a unit length of cable?

▌ Between inner wire and sheath only the inner wire contributes to B, and $B = (\mu_0 i)/(2\pi r)$. We consider a rectangular area between the surface of the inner wire and the outer sheath, and parallel to the cable axis. We break this area into long parallel strips of width dr and length s. Flux through area $s\,dr$ is $Bs\,dr$ and total flux between $r = a$ and $r = b$ per unit length is

$$\Phi' = \frac{1}{s}\int_a^b \frac{\mu_0 is}{2\pi r}\,dr = \frac{\mu_0 i}{2\pi}\ln(b/a)$$

Then $L' = \Phi'/i = (\mu_0/2\pi)\ln(b/a)$.

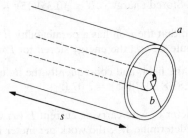

Fig. 31-3

Fig. 31-4

31.16   Two long parallel wires have radii equal to a and are a distance d between centers (Fig. 31-4). Their far ends are connected together so the wires form two sides of the same loop. For a section far from the ends, show that the self-inductance per meter is $(\mu_0/\pi)\ln[(d-a)/a]$ if the flux within the wires is neglected.

▌ Because each wire in Fig. 31-4 sets up the same flux as the center wire in Fig. 31-3, we need only double the result of Prob. 31.15 and replace b by $d - a$: $L' = (\mu_0/\pi)\ln[(d-a)/a]$.

31.17ᶜ   It is desired to send an alternating current $5\sin 400t$ A through a solenoid that has a self-inductance of 3 mH. If the resistance of the solenoid is assumed to be negligible, what voltage must be impressed across the solenoid terminals to provide this current?

▌ Since the induced emf opposes the impressed voltage, $V = L(di/dt) = (3 \times 10^{-3})(5)(400)\cos 400t = \underline{6\cos 400t \text{ V}}$.

31.18   A steady current of 2 A in a coil of 400 turns causes a flux of 10^{-4} Wb to link (pass through) the loops of the coil. Compute (a) the average back emf induced in the coil if the current is stopped in 0.08 s, (b) the inductance of the coil, and (c) the energy stored in the coil.

▌   (a) $$|\mathscr{E}| = N\left|\frac{\Delta\Phi}{\Delta t}\right| = 400\frac{(10^{-4} - 0)\text{ Wb}}{0.08\text{ s}} = \underline{0.5\text{ V}}$$

(b)
$$|\mathscr{E}| = L \left| \frac{\Delta i}{\Delta t} \right| \quad \text{or} \quad L = \left| \frac{\mathscr{E} \, \Delta t}{\Delta i} \right| = \frac{(0.5 \text{ V})(0.08 \text{ s})}{(2 - 0) \text{ A}} = \underline{0.02 \text{ H}}$$

(c)
$$\text{energy} = \tfrac{1}{2}LI^2 = \tfrac{1}{2}(0.02 \text{ H})(2 \text{ A})^2 = \underline{0.04 \text{ J}}$$

31.19   A coil of 0.48 H carries a current of 5 A. Compute the energy stored in it.

❚ energy stored $= \tfrac{1}{2}LI^2 = \tfrac{1}{2}(0.48 \text{ H})(5 \text{ A})^2 = \underline{6.0 \text{ J}}$

31.20   What back emf is induced in a coil of self-inductance 0.008 H when the current in the coil is changing at the rate of 110 A/s? What energy is stored in the inductor when the current is 6 A?

❚ For the emf we have $|e| = |-L(\Delta I/\Delta t)| = (0.008 \text{ H})(110 \text{ A/s}) = 0.88 \text{ V}$. For the energy stored, energy $= \tfrac{1}{2}LI^2 = \tfrac{1}{2}(0.008)(6^2) = \underline{0.144 \text{ J}}$.

31.21   The current in a circuit changes from 24 A to zero in 3 ms. If the average induced emf is 260 V, what is the coefficient of self-inductance of the circuit? How much energy was initially stored in the magnetic field of the inductor?

❚ We use $e = -L(\Delta I/\Delta t)$, or $260 = -L(-24/0.003)$ to get $L = 260/8000 = \underline{32.5 \text{ mH}}$. The energy is then obtained from energy $= \tfrac{1}{2}LI^2 = [(32.5 \times 10^{-3})(24)^2]/2 = \underline{9.36 \text{ J}}$.

31.22   An electromagnet has stored 648 J of magnetic energy when a current of 9 A exists in its coils. What average emf is induced if the current is reduced to zero in 0.45 s?

❚ $648 \text{ J} = \tfrac{1}{2}LI^2 = \tfrac{1}{2}L(9 \text{ A})^2$, and $L = 16 \text{ H}$. Now we use: $|\mathscr{E}| = L \,|\Delta i/\Delta t| = (16 \text{ H})(9 \text{ A}/0.45 \text{ s}) = \underline{320 \text{ V}}$.

31.23   **(a)** In a certain coil a flux linkage of 3.5 "weber-turns" is established by a current of 10 A. Determine the inductance of the coil. **(b)** In **(a)**, I is made to decrease at the rate of 100 A/s. What is the voltage of self-induction and what is its direction? **(c)** For a current of 15 A, how much energy is stored in the coil?

❚ **(a)** It is given that $N\Phi = 3.5 \text{ Wb}$. Hence, using $L = (N\Phi)/I$, $L = 3.5 \text{ Wb}/10 \text{ A} = \underline{0.35 \text{ H}}$. **(b)** The emf is $v_i = -(0.35)(-100) = \underline{+35 \text{ V, in the direction of the current}}$. **(c)** Stored energy is $U = \tfrac{1}{2}(0.35)(15)^2 = \underline{39.375 \text{ J}}$.

31.24   Suppose that the toroid in Prob. 31.13 has an iron core and that the iron has a permeability $\mu = 100\mu_0$ (assumed constant, which is never exactly the case). Compute L, and the energy stored for $I = 15 \text{ A}$.

❚ Increasing the permeability by a factor of 100 increases the field, and consequently the inductance, by that same factor. Thus, $L = 62.83 \text{ mH}$ and energy $= \tfrac{1}{2}LI^2 = \tfrac{1}{2}(62.83 \times 10^{-3})(15)^2 = \underline{7.07 \text{ J}}$.

31.25   A very long air-core solenoid is wound with n' turns per meter of wire carrying current I. **(a)** Calculate the self-inductance per meter, L', of the solenoid, and from it determine W', the work per meter required to establish current I. **(b)** The energy stored per unit volume when a magnetic field exists is given by

$$U' = \frac{B^2}{2\mu_0} \tag{1}$$

Use this to show that the magnetic energy stored per meter in the core of the solenoid is exactly $\tfrac{1}{2}L'I^2$.

❚ **(a)** The flux per turn is $\Phi = BA = \mu_0 n' I A$, where A is the cross-sectional area, and there are n' turns in a 1-m length. Then

$$L' = \frac{n'\Phi}{I} = \underline{\mu_0 n'^2 A} \text{ H/m} \quad \text{and} \quad W' = \frac{1}{2}L'I^2 = \underline{\frac{1}{2}\mu_0 n'^2 A I^2} \text{ J/m}$$

(b) Applying (1) to 1 m of the core, we have, since B is constant, that the energy per unit length U' is

$$U' = \frac{1}{2\mu_0} B^2 V' = \frac{1}{2\mu_0} (\mu_0^2 n'^2 I^2)(A \times 1) = \frac{1}{2}\mu_0 n'^2 A I^2 = W'$$

31.26   Show that for a long narrow toroid of average circumference $2\pi r$, cross-sectional area A, number of turns N, and permeability of core $K_m\mu_0$, the energy stored per unit volume in the magnetic field of the toroid is $\tfrac{1}{2}BH$.

❚ From Probs. 31.13 and 31.14 applied to this case, with $K_m\mu_0$ replacing μ_0, we have $L = (N^2\mu_0 K_m A)/(2\pi r)$.

Then energy $= \frac{1}{2}LI^2 = \frac{1}{2}[(N^2\mu_0 K_m A)/(2\pi r)]I^2$, and with $n = N/(2\pi r)$, energy $= \frac{1}{2}(\mu_0 K_m nI)(nI)(2\pi rA)$. From Ampère's law for **B** and **H** (see Chap. 29), energy $= \frac{1}{2}BH$[Volume]. Thus: energy/volume $= \frac{1}{2}BH$.

31.27 A toroid of 0.5-m circumference and 480-mm^2 cross-sectional area has 2500 turns bearing a current of 0.6 A. It is wound on an iron ring with relative permeability of 350. Find the magnetizing field H, the magnetic induction B, the flux Φ, the coefficient of self-inductance L, and the energy stored in the magnetic field.

❚ From Ampère's law (see Chap. 29),

$$H = nI = \left(\frac{2500}{0.5}\right)(0.6) = \underline{3000 \text{ A/m}} \quad \text{and} \quad B = \frac{4\pi}{10^7}K_m nI = \left(\frac{4\pi}{10^7}\right)(350)\left(\frac{2500}{0.5}\right)(0.6) = \underline{1.32 \text{ T}}$$

Then $\Phi = BA = (1.32 \text{ Wb/m}^2)(480 \times 10^{-6} \text{ m}^2) = 633 \text{ } \mu\text{Wb}$. $LI = N\Phi$ yields $0.6L = (2500)(6.33 \times 10^{-4})$, and $L = \underline{2.64 \text{ H}}$. Energy $= \frac{1}{2}LI^2 = \frac{1}{2}(2.64)(0.6^2) = \underline{0.475 \text{ J}}$. Alternatively, energy $= (\frac{1}{2}BH)V = \frac{1}{2}(1.32)(3000)(0.5)(4.80 \times 10^{-4}) = \underline{0.475 \text{ J}}$.

31.28 Show that the energy stored in a linear magnetic circuit may be written as $U = \frac{1}{2}\Phi\mathcal{R}$.

❚ From Probs. 31.2 and 29.20 we have

$$U = \frac{1}{2}LI^2 = \frac{1}{2}N\Phi I = \frac{1}{2}\Phi(\text{mmf}) = \frac{1}{2}\Phi^2\mathcal{R}$$

Alternatively, assuming a constant energy density ϵ,

$$U = \epsilon V = \left(\frac{1}{2\mu}B^2\right)(lA) = \frac{1}{2}(BA)^2\left(\frac{l}{\mu A}\right) = \frac{1}{2}\Phi^2\mathcal{R}$$

31.2 MUTUAL INDUCTANCE; THE IDEAL TRANSFORMER

31.29 Define the *mutual inductance* between the two circuits in terms of flux linkage.

❚ In Fig. 31-5, *part* of the flux established by I_1 in coil C_1 threads *some* of the turns of coil C_2. The total linkage of flux from C_1 with C_2 is $N_2\Phi_{12}$, where Φ_{12} is the *average number* of lines from C_1 linking each of the N_2 turns. Since the flux established in C_1 is directly proportional to I_1,

$$N_2\Phi_{12} = M_{12}I_1 \quad \text{or} \quad M_{12} = \frac{N_2\Phi_{12}}{I_1} \tag{1}$$

where the proportionality factor, M_{12}, is constant, assuming that the coils are fixed relative to each other.

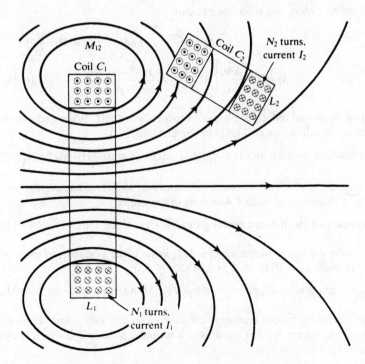

Fig. 31-5

Similarly, interchanging the roles of C_1 and C_2,

$$N_1\Phi_{21} = M_{21}I_2 \qquad \text{or} \qquad M_{21} = \frac{N_1\Phi_{21}}{I_2} \tag{2}$$

It can be shown that $M_{12} = M_{21} = M$. Thus, the two coils have a single *mutual inductance, M*; like self-inductance, mutual inductance is measured in henries.

31.30 Describe the emf induced in one coil due to a change in current in another coil.

❚ When the flux from one coil threads through another coil, an emf can be induced in either one by the other. The coil that contains the power source is called the *primary coil*. The other coil, in which an emf is induced by the changing current in the primary, is called the *secondary coil*. The induced secondary emf, $\mathcal{E}_s$, is proportional to the time rate of change of the flux linkage. Then, from Prob. 31.29, and disregarding signs,

$$\mathcal{E}_s = N_s \frac{\Delta\Phi_{ps}}{\Delta t} = M\frac{\Delta i_p}{\Delta t} \tag{1}$$

where M is the mutual inductance of the two-coil system. In calculus terms, $\mathcal{E}_s = M(di_p/dt)$.

31.31 Discuss the sign of the mutually induced emf.

❚ When two circuits are involved, the sign of mutually induced emf will depend on the particular choices for the positive sense of current in the two circuits. Even in a single circuit with two coils, the sign of the mutually induced emf will depend on the relative sense of the windings on the two coils. One way of dealing with this is to *define* $\mathcal{E}_s = -M(\Delta i_p/\Delta t)$, and let M take on positive and negative values. Another way is to let M refer to the *magnitude* of the mutual inductance and then let $\mathcal{E}_s = \pm M(\Delta i_p/\Delta t)$, with the sign chosen by inspection for the particular geometry and choice of circuit senses. This is the approach adopted here.

31.32 When the current in a certain coil is changing at a rate of 3 A/s, it is found that an emf of 7 mV is induced in a nearby coil. What is the mutual inductance of the combination?

❚ $$\mathcal{E}_s = M\frac{\Delta i_p}{\Delta t} \qquad \text{or} \qquad M = \mathcal{E}_s \frac{\Delta t}{\Delta i_p} = (7 \times 10^{-3}\,\text{V})\frac{1\,\text{s}}{3\,\text{A}} = \underline{2.33\,\text{mH}}$$

31.33 Two coils are wound on the same iron rod so that the flux generated by one passes through the other also. The primary coil has N_p loops on it and, when a current of 2 A flows through it, the flux in it is 2.5×10^{-4} Wb. Determine the mutual inductance of the two coils if the secondary coil has N_s loops on it.

❚ Assume the secondary coil is an open circuit.

$$|\mathcal{E}_s| = N_s\left|\frac{\Delta\Phi_s}{\Delta t}\right| \qquad \text{and} \qquad |\mathcal{E}_s| = M\left|\frac{\Delta i_p}{\Delta t}\right|$$

give $$M = N_s\left|\frac{\Delta\Phi_s}{\Delta i_p}\right| = N_s\frac{(2.5 \times 10^{-4} - 0)\,\text{Wb}}{(2-0)\,\text{A}} = \underline{(1.25 \times 10^{-4}N_s)\,\text{H}}$$

31.34 The coefficient of mutual inductance between two coils is 8 mH. What emf is induced in the second coil if the current is changing at the rate of 4 kA/s in the first coil?

❚ The magnitude of induced emf is $e_2 = M(\Delta i_1/\Delta t) = (8 \times 10^{-3}\,\text{H})(4000\,\text{A/s}) = \underline{32\,\text{V}}$.

31.35 Two circuits have a coefficient of mutual inductance of 16 mH. What average emf is introduced in the secondary by a change from 40 to 4 A in 6 ms in the primary?

❚ The magnitude of the induced emf is given by $|e_2| = M\,|\Delta i_1/\Delta t| = (16 \times 10^{-3}\,\text{H})(36\,\text{A}/0.006\,\text{s}) = \underline{96\,\text{V}}$.

31.36 Two circuits have a mutual inductance of 0.3 H. What is the average emf induced in one coil when the current in the other changes from 10 to 40 A in 0.01 s?

❚ The magnitude of induced emf is given by $|\mathcal{E}| = M\,|\Delta I/\Delta t| = (0.3)(40.0 - 10.0)/(0.01) = \underline{900\,\text{V}}$.

31.37 A 2000-loop solenoid is wound uniformly on a long iron rod with length d and cross section A. The relative permeability of the iron is k_m. On top of this is wound a 50-loop coil which is used as a secondary. Find the mutual inductance of the system.

▌ The flux through the solenoid due to the primary is $\Phi = BA = (k_m\mu_0 nI_p)A = k_m\mu_0 I_p A(2000/d)$. This same flux goes through the secondary. We have

$$|\mathcal{E}_s| = N_s \left|\frac{\Delta\Phi}{\Delta t}\right| \quad \text{and} \quad |\mathcal{E}_s| = M\left|\frac{\Delta i_p}{\Delta t}\right|$$

from which

$$M = N_s\left|\frac{\Delta\Phi}{\Delta i_p}\right| = N_s\frac{\Phi - 0}{I_p - 0} = (50)\frac{k_m\mu_0 I_p A(2000/d)}{I_p} = \underline{\frac{10^5 k_m\mu_0 A}{d}}$$

31.38 Describe an ideal transformer in terms of induction.

▌ An ideal transformer is a mutual inductor with all its flux lines common to both the *primary* and the *secondary,* as·coils 1 and 2 are called. For a transformer with no losses, the ratio of the emf induced in the secondary to the emf induced in the primary is equal to the number of turns of the secondary N_2 divided by the number of turns of the primary N_1.

31.39 The mutual inductance between the primary and secondary of a transformer is 0.3 H. Compute the induced emf in the secondary when the primary current changes at the rate of 4 A/s.

▌

$$|\mathcal{E}_s| = M\left|\frac{\Delta i_p}{\Delta t}\right| = (0.3\text{ H})(4\text{ A/s}) = \underline{1.2\text{ V}}$$

31.40 The primary and secondary of an ideal transformer have 2000 and 100 turns, respectively. If the maximum primary voltage is 120 V, what is the maximum value of the emf induced in the secondary?

▌ From Prob. 31.38, $V_2 = (V_1)(N_2/N_1) = (120)(100/2000) = \underline{6\text{ V}}$. This is an example of a "step-down" transformer.

31.41 When the current in the primary of a small transformer is changing at the rate of 600 A/s, the induced emf in the secondary is 8 V. What is the coefficient of mutual inductance?

▌

$$e_2 = M\frac{\Delta i_1}{\Delta t} \quad \text{or} \quad 8\text{ V} = M(600\text{ A/s}) \quad \text{and} \quad M = \frac{8\text{ V}}{600\text{ A/s}} = \underline{13.3\text{ mH}}$$

31.42 The secondary of an ideal transformer has 275 times as many turns as the primary. It is used in a 110-V circuit. What is the voltage across the secondary? If the current in the secondary is 50 mA, what is the primary current?

▌ Since the flux linkage is the same, the ratio of emfs = ratio of turns. Thus

$$\frac{V_S}{V_P} = \frac{N_S}{N_P} = 275 \quad V_S = (275)(110\text{ V}) = \underline{30.25\text{ kV}}$$

This is a "step-up" transformer. When the secondary circuit is closed and current flows, the situation is generally complex. If the load in the secondary circuit is a pure resistance, then the current produced by the secondary emf is in phase with it, and the power output is $V_S I_S$. If there are no losses in the primary coil then we have $V_P I_P \approx V_S I_S$ and $I_P/I_S = V_S/V_P = N_S/N_P$. For our case,

$$\frac{I_P}{I_S} = \frac{N_S}{N_P} = 275 \quad I_P = (275)\left(\frac{50}{1000}\right) = \underline{13.75\text{ A}}$$

31.43 An ideal transformer has 550 turns on the primary and 30 turns on the secondary. What is the maximum output potential difference if the maximum input voltage is 3.3 kV? If the transformer is assumed to have an efficiency of 100 percent, what maximum primary current is required if a maximum current of 11 A is drawn from the secondary?

▌ We have

$$\frac{V_S}{V_P} = \frac{N_S}{N_P} \quad \text{or} \quad \frac{V_S}{3300} = \frac{30}{550} \quad V_S = \underline{180\text{ V}}$$

For 100 percent efficiency we can assume that

$$\frac{I_P}{I_S} = \frac{N_S}{N_P} \quad I_P = \frac{30}{550}11 = \underline{0.6\text{ A}}$$

31.44 Two neighboring coils, A and B, have 300 and 600 turns, respectively. A current of 1.5 A in A causes 1.2×10^{-4} Wb to pass through A and 0.9×10^{-4} Wb through B. Determine (a) the self-inductance of A, (b) the mutual inductance of A and B, and (c) the average induced emf in B when the current in A is interrupted in 0.2 s.

▌ (a) For the self-inductance we have $L_A I_A = N_A \Phi_A$, where Φ_A is the self-induced flux. Thus $(1.5\ \text{A})L_A = 300(1.2 \times 10^{-4}\ \text{Wb})$ and $L_A = \underline{24\ \text{mH}}$. (b) For the mutual inductance $M I_A = N_B \Phi_B$, where Φ_B is the flux through B due to I_A. Thus $(1.5\ \text{A})M = 600(0.9 \times 10^{-4}\ \text{Wb})$ and $M = \underline{36\ \text{mH}}$. (c) $\mathscr{E}_b = M(\Delta I_A / \Delta t) = (36 \times 10^{-3})(1.5/0.2) = 0.27\ \text{V}$.

31.45 (a) With a current of 10 A in C_1, Fig. 31-5, there is a flux linkage of 0.5 "weber-turns" with C_2. What is the mutual inductance? (b) With a current of 10 A in C_2, what is the linkage with C_1? (c) If I_1 is increasing at the rate of 200 A/s, what electromotive force v_2 is induced in C_2?

▌ (a) From Eq. (1) of Prob. 31.29 with $N_2 \Phi_{12} = 0.5$ Wb, $M = M_{12} = 0.5/10 = 0.05\ \text{H} = \underline{50\ \text{mH}}$.
(b) From Eq. (2) of Prob. 31.29, $N_1 \Phi_{21} = (0.05)(10) = \underline{0.5\ \text{Wb}}$. That is, for $I_1 = I_2$, the linkages are equal.
(c) Induced emf:

$$v_2 = M \frac{dI_1}{dt} = (0.05)(200) = \underline{10\ \text{V}}$$

31.46 As is seen from Fig. 31-6, I_1 and I_2 can be independently increased, decreased, or held constant by varying resistances R_1 and R_2. Let $L_1 = 50$ mH, $L_2 = 40$ mH, $M = 15$ mH. (a) I_1 is made to increase at the rate of 120 A/s; I_2 is held constant. Compute the voltages v_1 and v_2 induced in each coil. (b) I_1 is made to decrease at the rate of 120 A/s; I_2 is held constant. Compute v_1 and v_2. (c) I_1 is increased at the rate of 120 A/s and I_2 is decreased at the rate of 200 A/s. Compute v_1 and v_2.

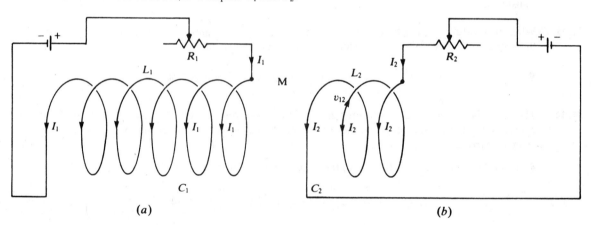

Fig. 31-6

▌ Each coil has both self-induced and mutually induced emf. With the current senses as chosen for each circuit we have (see Prob. 31.31)

$$v_1 = -L_1 \frac{dI_1}{dt} - M \frac{dI_2}{dt} \qquad v_2 = -L_2 \frac{dI_2}{dt} - M \frac{dI_1}{dt}$$

The sign in front of M is negative since an increase in I_1 creates an emf in circuit 2 opposing the sense of I_2 (and vice versa). Then

(a) $\qquad v_1 = -L_1 \frac{dI_1}{dt} + 0 = -(0.050)(120) = \underline{-6\ \text{V}} \qquad v_2 = 0 - M \frac{dI_1}{dt} = -(0.015)(120) = \underline{-1.8\ \text{V}}$

(b) $\qquad v_1 = -L_1 \frac{dI_1}{dt} + 0 = -(0.050)(-120) = \underline{+6\ \text{V}} \qquad v_2 = 0 - M \frac{dI_1}{dt} = -(0.015)(-120) = \underline{+1.8\ \text{V}}$

(c) $\qquad v_1 = -L_1 \frac{dI_1}{dt} - M \frac{dI_2}{dt} = -(0.050)(120) - (0.015)(-200) = \underline{-3\ \text{V}}$

$$v_2 = -L_2 \frac{dI_2}{dt} - M \frac{dI_1}{dt} = -(0.040)(-200) - (0.015)(120) = \underline{+6.2\ \text{V}}$$

31.47 A toroid with an iron core has an average circumference of 0.4 m and a cross-sectional area of 320 mm². When the current in the windings is 0.6 A, the magnetic induction B is 0.8 T. The relative permeability of the coil is 250. A four-turn fluxmeter connected to a galvanometer is used to determine B in the coil. (**a**) Calculate the flux through the solenoid. (**b**) Find the number of turns on the toroidal winding. (**c**) When the key is opened, the current in the winding and the flux drop to zero. Find the coefficient of mutual inductance between the fluxmeter and the toroidal winding.

▮ (**a**) We are given $B = 0.8$ T $= 0.8$ Wb/m². Then $\Phi = BA = (0.8 \text{ Wb/m}^2)(320 \times 10^{-6} \text{ m}^2) = \underline{256\ \mu\text{Wb}}$. (**b**) We are given $B = (4\pi/10^7)nK_mI$ (see Probs. 31.13 and 31.14), and $0.8 = (4\pi/10^7)(N/0.4)(250)(0.6)$, which yields $N = \underline{1698 \text{ turns}}$. (**c**) We are given $\mathscr{E} = M(\Delta i_t/\Delta t) = N'(\Delta \Phi'/\Delta t)$ where N', Φ', refer to turns on, and flux through, the fluxmeter. Equivalently: $Mi = N'\Phi'$, so $0.6M = 4(256 \times 10^{-6})$ and $M = \underline{1.71 \text{ mH}}$.

31.48ᶜ Figure 31-7 shows an arrangement for measuring the mutual inductance M_{12} of a pair of coils, 1 and 2. Coil 1 is connected to a battery, an ammeter A, and a switch. Coil 2 is connected to a ballistic galvanometer G which measures the total charge flowing through it during a current pulse. The switch is closed, the galvanometer reading is observed, and the steady reading of the ammeter is noted. Show that $M_{12} = (q_2R_2)/i_1$, where q_2 is the charge read by the galvanometer, R_2 is the total resistance of the galvanometer and coil 2, and i_1 is the ammeter reading.

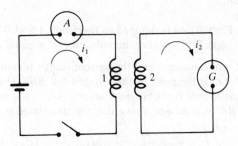

Fig. 31-7

▮ The emf induced in coil 2 drives a current i_2 through the galvanometer, so we have $V_2 = i_2R_2$. Since the current i_2 is the time derivative of the charge that passes through the galvanometer, we have $i_2 = dq_2/dt$. Furthermore, with the positive senses chosen in Fig. 31-7, the emf induced in coil 2 is given by

$$V_2 = -M_{12}\frac{di_1}{dt}$$

Therefore we have

$$\frac{dq_2}{dt}R_2 = -M_{12}\frac{di_1}{dt}$$

Integrating this equation from $t = 0$ until a steady current is flowing through the ammeter, we obtain $q_2R_2 = -M_{12}i_1$. Solving for the mutual inductance, we find that $\underline{M_{12} = -(q_2R_2)/i_1}$, as was to be shown. The minus sign is present because an increasing clockwise current in the ammeter circuit induces an emf in coil 2 that drives a counterclockwise current around the galvanometer circuit. (This assumes that the coils are side by side and are wound the same way, as shown in the figure.) No matter what the relative positions and winding directions are, the (magnitude of the) mutual inductance equals $(qR_2)/i$, where $q \equiv |q_2|$ and $i \equiv |i_1|$.

31.49 In Fig. 31-8, coil AA' (solid line) and coil BB' (dashed line) are wound on a long plastic tube in the same sense. Ends A and B are joined together and a source of current i is connected to A' and B'. Show that the induced emf is equal to $-(L_A + L_B - 2M_{AB})(di/dt)$ where L is self-inductance, and M is mutual inductance.

▮ The self-inductance of each coil is independent of the method of connection. If terminals A and B are connected, the sense of the current flow around coil AA' is opposite to that around coil BB'. The emf

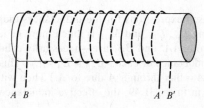

$A\ B$ $A'\ B'$ **Fig. 31-8**

562 □ CHAPTER 31

induced across coil BB' must be

$$V_B = -L_B \frac{di_B}{dt} + M_{AB} \frac{di_A}{dt}$$

where M_{AB} is the (magnitude of the) mutual inductance between the coils. Since the coils are connected in series $i_A = i_B = i$, so that

$$V_B = -(L_B - M_{AB}) \frac{di}{dt}$$

Similarly, we find that the emf induced across coil AA' is

$$V_A = -(L_A - M_{AB}) \frac{di}{dt}$$

The emf across the entire series coil assembly is the sum of V_A and V_B:

$$V = V_A + V_B = -(L_A + L_B - 2M_{AB}) \frac{di}{dt}$$

which is the desired result.

31.50 Suppose the situation in Prob. 31.49 is changed so that ends A and B' are joined together and the current source is connected to A' and B. Show that the induced emf is equal to $-(L_A + L_B + 2M_{AB})(di/dt)$.

■ If terminals A and B' are connected and the current source is connected to A' and B, the sense of the current flow around coil AA' is the same as that around coil BB'. Therefore, for a given sign of the current, the direction of the magnetic field contribution from coil AA' is the same as that from coil BB'. This means that the emf across coil BB' is increased above that for an isolated coil BB'. Specifically,

$$V'_B = -L_B \frac{di_B}{dt} - M_{AB} \frac{di_A}{dt} = -(L_B + M_{AB}) \frac{di}{dt}$$

Similarly, the emf induced across coil AA' is

$$V'_A = -(L_A + M_{AB}) \frac{di}{dt}$$

The emf across the entire series assembly is the sum of V'_A and V'_B:

$$V' = V'_A + V'_B = -(L_A + L_B + 2M_{AB}) \frac{di}{dt}$$

which is the desired result.

31.51 If the two equal length coils of Fig. 31-8 have the same number N of turns and $L_A = 0.010$ H, what is the effective inductance **(a)** in Prob. 31.49, and **(b)** in Prob. 31.50?

■ Since the lengths, radii, and numbers of turns of the two coils are equal, they must have equal inductances:

$$L_B = \frac{\mu_0 N_B^2 \pi r_B^2}{l_B} = \frac{\mu_0 N_A^2 \pi r_A^2}{l_A} = L_A \qquad N_A = N_B = N$$

The number of turns per unit length in coil AA' is $n_A = N_A/l_A = N/l_B = n_B$. Since the flux linking both coils is the same, the mutual inductance is

$$M_{AB} = \pi \mu_0 n_A n_B l_A r_A^2 = \frac{\mu_0 N^2 \pi r_A^2}{l_A} = L_A = L_B$$

(Note: Although we have used expressions for the inductances that are applicable only to long tightly wound solenoids, the *equality* among L_A, L_B, and M_{AB} is much more general. Any two identical inductors which are physically superposed—as coils AA' and BB' are—necessarily exhibit a mutual inductance equal to the common value of the self-inductance of each: $M_{AB} = L_A = L_B$. This follows from the definitions and the fact that flux through B due to A = flux through A due to A.) Then with respect to our two cases: **(a)** For the series connection described in Prob. 31.49, the effective inductance is $L_a = L_A + L_B - 2M_{AB} = L_A + L_A -$

$2L_A = \underline{0}$. (b) For the series connection described in Prob. 31.50, the effective inductance is $L_b = L_A + L_B + 2M_{AB} = L_A + L_A + 2L_A = 4L_A$. With $L_A = 0.01$ H, we have $L_b = \underline{0.04\ H}$.

31.52ᶜ In Fig. 31-9, evaluate the mutual inductance between the rectangular loop and the long straight wire.

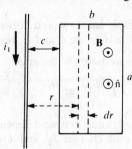

Fig. 31-9

❚ We suppose that the straight wire in Fig. 31-9 is carrying a downward current i_1. Then the magnetic field due to the straight wire has magnitude $B_1 = (\mu_0 i_1)/(2\pi r)$ at distance r from the wire. According to the right-hand rule, $\mathbf{B}_1$ points directly up from the page within the rectangular loop. The magnetic flux Φ_{12} that links the loop is given by

$$\Phi_{12} = \int \mathbf{B}_1 \cdot d\mathbf{A}_2 = \int B_1\, dA_2$$

Since $dA_2 = a\, dr$, and r ranges from $r_{min} = c$ to $r_{max} = c + b$, we find that

$$\Phi_{12} = \int_c^{c+b} \left(\frac{\mu_0 i_1}{2\pi r}\right) \cdot a\, dr = \frac{\mu_0 i_1 a}{2\pi} \int_c^{c+b} \frac{dr}{r} = \frac{\mu_0 i_1 a}{2\pi} \ln\left(\frac{c+b}{c}\right) = \frac{\mu_0 i_1 a}{2\pi} \ln\left(1 + \frac{b}{c}\right)$$

Therefore, the mutual inductance

$$M = M_{12} = \frac{\Phi_{12}}{i_1} = \underline{\frac{\mu_0 a}{2\pi} \ln\left(1 + \frac{b}{c}\right)}$$

31.53ᶜ A coil of N loops is wound tightly on the center section of a long solenoid with n loops per meter. The cross-sectional areas of both are essentially the same, A. Find the induced emf in the coil due to a current $i = i_0 \cos \omega t$ in the solenoid. What is the mutual inductance of the combination? (Assume that all coils and sense of current are in the same direction.)

❚ B inside the solenoid is $\mu_0 ni$. Then the flux through the coil is $\mu_0 niA$ and the induced emf in it is $-\mu_0 nAN(di/dt) = \mu_0 nNA 2\pi f i_0 \sin 2\pi ft$. But from the definition of mutual inductance, emf $= -M(di/dt)$. Comparison shows $M = \underline{\mu_0 nAN}$.

31.54 A small flat coil (N loops and area A_c) is placed inside a long solenoid (n loops per meter and area A_s) with its area vector making an acute angle θ to the solenoid axis. Find the induced emf in the coil due to a current $i = i_0 \cos \omega t$ in the solenoid. What is the mutual inductance of the combination? By making θ variable, it is possible to construct a variable mutual inductance this way. (Assume that both coils are wound in the same sense and the positive sense of current is chosen in the same way.)

❚ B in the solenoid is $\mu_0 ni$. The flux through the coil is $\mu_0 niA_c \cos \theta$. From this, the induced emf $= -N\mu_0 nA_c \cos \theta(di/dt) = \mu_0 nNA_c 2\pi f i_0 \cos \theta \sin 2\pi ft$. Since emf $= -M(di/dt)$, comparison gives $M = \underline{\mu_0 nA_c N \cos \theta}$.

31.55ᶜ For the situation shown in Fig. 31-10, if the current in the long straight wire is i, what is the net flux through the rectangular loop? What is the mutual inductance of the combination? What are the answers to these two questions if $b = c$?

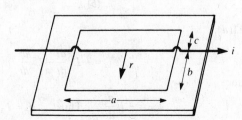

Fig. 31-10

❚ This is similar to Prob. 31.52 except the wire now straddles the loop. Note that the flux on one side is equal and opposite to that on the other for $0 \le r \le c$. Only the flux through the portion $c < r < b$ is not canceled. Then $\Phi = [(\mu_0 ia)/(2\pi)] \int_c^b dr/r = [(\mu_0 ia)/(2\pi)] \ln (b/c)$. Thus $M = [(\mu_0 a)/(2\pi)] \ln (b/c)$. If $b = c$, both answers are zero.

31.56ᶜ Find an expression for the total magnetic energy stored in two coils with inductances L_1 and L_2 and mutual inductance M, when the currents in the coils are I_1 and I_2, respectively.

❚ While the currents are building up we have for the emfs:

$$\mathcal{E}_1 = -L_1 \frac{di_1}{dt} \pm M \frac{di_2}{dt} \qquad \mathcal{E}_2 = -L_2 \frac{di_2}{dt} \pm M \frac{di_1}{dt} \qquad (1)$$

where the $\pm$ appear consistently in both equations, and depend on the coil geometry and circuit sense. Then the external work done in time dt to push charges dq_1 and dq_2 through each circuit, respectively, is

$$dW = -\mathcal{E}_1 \, dq_1 - \mathcal{E}_2 \, dq_2 = L_1 \frac{di_1}{dt} dq_1 \mp M \frac{di_2}{dt} dq_1 + L_2 \frac{di_2}{dt} dq_2 \mp M \frac{di_1}{dt} dq_2$$

Noting that $i_1 = dq_1/dt$ and $i_2 = dq_2/dt$, we have: $dW = L_1 i_1 \, di_1 + L_2 i_2 \, di_2 \mp (M i_1 \, di_2 + M i_2 \, di_1) = L_1 i_1 \, di_1 + L_2 i_2 \, di_2 \mp M \, d(i_1 i_2)$. Integrating from 0 to final current we have

$$U = \int dW = L_1 \int_0^{I_1} i_1 \, di_1 + L_2 \int_0^{I_2} i_2 \, di_2 \mp M \int_0^{I_1 I_2} d(i_1 i_2) = \tfrac{1}{2} L_1 I_1^2 + \tfrac{1}{2} L_2 I_2^2 \mp M I_1 I_2 \qquad (2)$$

31.57 Show that the mutual inductance of two coils must be smaller than the geometric mean of their self-inductances.

❚ With $x \equiv I_2/I_1$, (2) of Prob. 31.56 may be written $U = \tfrac{1}{2} I_1^2 (L_2 x^2 \mp 2Mx + L_1)$. Because the energy U is intrinsically positive, the quadratic polynomial must have a negative discriminant: $4M^2 - 4L_1 L_2 < 0$ or $M < \sqrt{L_1 L_2}$.

32.58 Suppose that coils C_1 and C_2, Fig. 31-5, have self-inductances $L_1 = 200$ mH and $L_2 = 120$ mH, and mutual inductance $M = 50$ mH. For $I_1 = 20$ A and $I_2 = 15$ A, compute the total energy of inductance.

❚ Using the geometry in Fig. 31-5 and observing the sense of current in the two coils and the positive linkage of coil C_2 by the flux due to C_1, we have the $-M$ sign in Eq. (1) of Prob. 31.56. Then from Eq. (2) of Prob. 31.56, $U = \tfrac{1}{2} L_1 I_1^2 + \tfrac{1}{2} L_2 I_2^2 + M I_1 I_2$, we have $U = \tfrac{1}{2}[(0.200)(20)^2 + (0.120)(15)^2 + 2(0.050)(20)(15)] = \underline{68.5 \text{ J}}$.

31.59 Find the equivalent inductance for two inductances when placed in (**a**) series and (**b**) parallel. Assume that the inductances are well separated so that mutual inductance can be ignored.

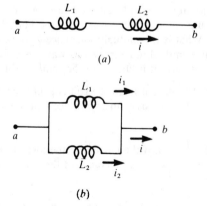

(a)

(b)   **Fig. 31-11**

❚ (**a**) From Fig. 31-11(*a*),

$$\mathcal{E}_{ab} = -L_1 \frac{di}{dt} - L_2 \frac{di}{dt}$$

For equivalent inductance

$$\mathcal{E}_{ab} = -L \frac{di}{dt} \qquad \text{Thus} \qquad L = L_1 + L_2$$

(b) From Fig. 31-11(b),

$$\mathcal{E}_{ab} = -L_1 \frac{di_1}{dt} = -L_2 \frac{di_2}{dt} \qquad i = i_1 + i_2 \qquad \text{and} \qquad \frac{di}{dt} = \frac{di_1}{dt} + \frac{di_2}{dt}$$

For equivalent inductance:

$$\mathcal{E}_{ab} = -L \frac{di}{dt} \qquad \text{Then} \qquad \frac{\mathcal{E}_{ab}}{L} = \frac{\mathcal{E}_{ab}}{L_1} + \frac{\mathcal{E}_{ab}}{L_2} \qquad \text{or} \qquad \frac{1}{L} = \frac{1}{L_1} + \frac{1}{L_2}$$

Thus inductors in series and parallel add like resistances, as long as mutual inductance can be ignored.

31.60 What is the inductance between terminals A and B in the network shown in Fig. 31-12(a), assuming the three inductors are well separated?

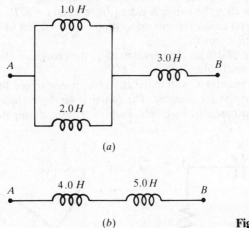

(a)

(b) **Fig. 31-12**

▌ We index the inductors shown in Fig. 31-12(a) from upper left to right, so that $L_1 = 1.0$ H, $L_2 = 2.0$ H, and $L_3 = 3.0$ H. From Prob. 31.59, the equivalent inductance of two well-separated inductors in parallel is given by

$$\frac{1}{L_p} = \frac{1}{L_1} + \frac{1}{L_2} \qquad \text{or} \qquad L_p = \frac{L_1 L_2}{L_1 + L_2} = \frac{(1.0)(2.0)}{(1.0 + 2.0)} = \frac{2}{3} \text{H}$$

Since the parallel assembly of inductors 1 and 2 is in series with inductor 3, the equivalent inductance is

$$L_s = L_p + L_3 = \tfrac{2}{3} \text{H} + 3 \text{H} = 3\tfrac{2}{3} \text{H} = \underline{3.67 \text{ H}}$$

31.61 If the separation between the two inductors in Fig. 31-12(b) is decreased to the point at which it becomes small, and the inductors are wound in the same sense, would you expect the inductance between A and B to be larger or smaller than it is when they are well separated? Why?

▌ If the inductors of Fig. 31-12(b) are solenoids wound in the same sense, the inductance between A and B is somewhat larger when the coils are placed close together (and share a common axis) than it is when they are far apart. This is because some of the magnetic field due to each coil will thread the other coil in the same direction as that other coil's own field. On the other hand, if the inductors are solenoids wound in opposite senses, the inductance between A and B will be somewhat smaller when the coils are placed close together (and share a common axis) than it is when they are far apart. This is because some of the magnetic field due to each coil will thread the other coil in the direction opposite to that of the other coil's own field.

CHAPTER 32
Electric Circuits

32.1 R-C, R-L, L-C, AND R-L-C CIRCUITS; TIME RESPONSE

32.1 Describe the charging process for an R-C circuit [see Fig 32-1(a)], and the time constant of the circuit. Repeat for the discharging process when the battery is removed.

▮ Consider the circuit shown in Fig. 32-1(a). The capacitor is initially uncharged. If the switch is now closed, the current i in the circuit and the charge q on the capacitor vary as shown in Fig. 32-1(b). Writing the loop rule for this circuit gives, calling the p.d. across the capacitor v_c, $-iR - v_c + \mathcal{E} = 0$, or $i = (\mathcal{E} - v_c)/R$.

At the first instant after the switch is closed, $v_c = 0$ and $i = \mathcal{E}/R$. As time goes on, v_c increases and i decreases. The time taken for the current to drop to $e^{-1} \approx 0.368$ of its initial value is RC, which is called the *time constant* of the R-C circuit.

Also shown in Fig. 32-1(b) is the variation of q, the charge on the capacitor, with time. At $t = RC$, q has attained 0.632 of its final value.

When a charged capacitor C with initial charge q_0 is discharged through a resistor R, its discharge current follows the same curve as for charging. The charge q on the capacitor follows a curve similar to that for the discharge current. At time RC, $i = 0.368i_0$ and $q = 0.368q_0$ during discharge.

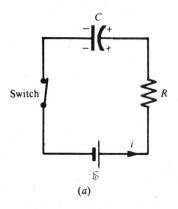

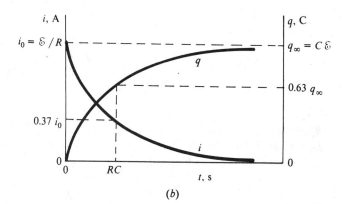

(a) (b)

Fig. 32-1

32.2 A certain series circuit consists of a 12-V battery, a switch, a 1-MΩ resistor, and a 2-μF capacitor, initially uncharged. If the switch is now closed, find (a) initial current in the circuit, (b) time taken for the current to drop to 0.37 of its initial value, (c) charge on the capacitor then, (d) final charge on the capacitor.

▮ (a) The loop rule applied to the circuit of Fig. 32-1(a) at any instant gives $12\text{ V} - iR - v_c = 0$, where v_c is the p.d. across the capacitor. At the first instant, q is essentially zero and so $v_c = 0$. Then $12\text{ V} - iR - 0 = 0$, or $i = 12\text{ V}/10^6\,\Omega = \underline{12\,\mu A}$. (b) From Prob. 32.1, the current drops to 0.37 of its initial value when $t = RC = (10^6\,\Omega)(2 \times 10^{-6}\,\text{F}) = \underline{2\,\text{s}}$. (c) At $t = 2$ s the charge on the capacitor has increased to 0.63 of its final value. [See (d) below.] (d) The final value for the charge occurs when $i = 0$ and $v_c = 12$ V. Therefore, $q_{\text{final}} = Cv_c = (2 \times 10^{-6}\,\text{F})(12\,\text{V}) = \underline{24\,\mu C}$.

32.3 A 5-μF capacitor is charged to a potential of 20 kV. After being disconnected from the power source, it is connected across a 7-MΩ resistor to discharge. What is the initial discharge current and how long will it take for the capacitor voltage to decrease to 37 percent of the 20 kV?

▮ The loop equation for the discharging capacitor is $v_c - iR = 0$, where v_c is the p.d. across the capacitor. At the first instant, $v_c = 20$ kV, so

$$i = \frac{v_c}{R} = \frac{20 \times 10^3\,\text{V}}{7 \times 10^6\,\Omega} = \underline{2.86\,\text{mA}}$$

The potential across the capacitor, as well as the charge on it, will decrease to 0.37 of its original value in one time constant. The required time is $RC = (7 \times 10^6\,\Omega)(5 \times 10^{-6}\,\text{F}) = \underline{35\,\text{s}}$.

32.4 A series circuit consisting of an uncharged 2-μF capacitor and a 10-MΩ resistor is connected across a 100-V power source. What are the current in the circuit and the charge on the capacitor (**a**) after one time constant, (**b**) when the capacitor has acquired 90 percent of its final charge?

❚ As in Prob. 32.1, $\mathscr{E} - v_c - iR = 0$. At $t = 0$, $v_c = 0$ and $i_{max} = \mathscr{E}/R = (100\,\text{V})/(10^7\,\Omega) = 10\,\mu\text{A}$. Also the maximum charge occurs when $i \approx 0$ and $\mathscr{E} = v_{c,max} = q_{max}/C$, and $v_{c,max} = 100\,\text{V}$; $q_{max} = (2 \times 10^{-6}\,\text{F})(100\,\text{V}) = 200\,\mu\text{C}$. (**a**) After one time constant $i = 0.368i_{max} = \underline{3.68\,\mu\text{A}}$; also $q = 0.632q_{max} = \underline{126\,\mu\text{C}}$. (**b**) At 90 percent of final charge $q = 0.9q_{max} = \underline{180\,\mu\text{C}}$. Then $v_c = q/C = \underline{90\,\text{V}}$, and $i = (\mathscr{E} - v_c)/R = (100 - 90)/(10^7\,\Omega) = \underline{1\,\mu\text{A}}$.

32.5 A charged capacitor is connected across a 10-kΩ resistor and allowed to discharge. The potential difference across the capacitor drops to 0.37 of its original value after a time of 7 s. What is the capacitance of the capacitor?

❚ From Prob. 32.1, q decays to $0.37q_0$ in about one time constant. Since $v_c = q/C$, it drops in proportion to q. Then, from data given, $RC = 7\,\text{s}$, or $C = (7\,\text{s})/(10^4\,\Omega) = \underline{700\,\mu\text{F}}$.

32.6ᶜ Give a full analysis of the simple R-C circuit (Fig. 32-2, with L absent). Assume that v is constant, and $q = 0$ at $t = 0$ (the instant the switch is closed).

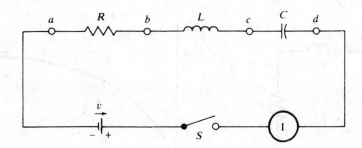

Fig. 32-2

❚ For the first-order differential equation,

$$R\dot{q} + \frac{1}{C}q = v \tag{1}$$

the transient function is $q_1(t) = e^{-t/RC}$, as may be verified by substitution in $R\dot{q}_1 + (1/C)q_1 = 0$. Hence, $q(t) = c_1 e^{-t/RC} + Cv$. The initial condition gives $0 = c_1 + Cv$, or $c_1 = -Cv$. Thus, finally,

$$q(t) = Cv(1 - e^{-t/RC}) \tag{2}$$

The remaining results follow from (2):

current: $\qquad\qquad I = \dot{q} = (v/R)e^{-t/RC}$

voltage across resistor: $\qquad v_{ab} = RI = ve^{-t/RC}$

voltage across capacitor: $\qquad v_{cd} = q/C = v(1 - e^{-t/RC}) \tag{3}$

power from battery: $\qquad\quad P = Iv = (v^2/R)e^{-t/RC}$

energy in capacitor: $\qquad\quad U_C = \int v_{cd}\,dq = \frac{1}{C}\int q\,dq = \frac{q^2}{2C} = \frac{1}{2}Cv^2(1 - e^{-t/RC})^2$

At $t \to \infty$, the various quantities approach their steady-state values at a rate which depends on the size of RC; the smaller RC, the faster the approach. We have:

$$q \to Cv \qquad I \to 0 \qquad v_{ab} \to 0 \qquad v_{cd} \to v \qquad P \to 0 \qquad U_C \to \tfrac{1}{2}Cv^2$$

32.7 The parameter RC in (2) of Prob. 32.6 is the time constant of the R-C circuit. (**a**) The time constant must have the dimensions of time, to make the exponent in (2) a pure number. Verify that this is so by means of a unit-dimensional analysis. (**b**) What should be the size of the time constant if the current in the circuit is to decrease by 50 percent in the first millisecond?

▮ (a)

$$RC = [\Omega][F] = \left[\frac{V}{A}\right]\left[\frac{C}{V}\right] = \left[\frac{C}{A}\right] = \left[\frac{A \cdot s}{A}\right] = [s]$$

(b) From $I = I_0 e^{-t/RC}$, we obtain, with time measured in ms, $\frac{1}{2} = e^{-1/RC}$, or $RC = 1/\ln 2 = \underline{1.44 \text{ ms}}$. This could be attained with, say, $R = 144\,\Omega$, $C = 10\,\mu\text{F}$.

32.8^c Consider an R-C circuit with no applied emf. **(a)** Find the charge on the capacitor at any time, if the initial charge was q_0. **(b)** Show that the rate of loss of energy in the capacitor equals the rate of heat loss in the resistor.

▮ (a) As found in Prob. 27.141, $q = q_0 e^{-t/RC}$. **(b)** Differentiating the result in **(a)**, $|i| = |q_0/RC| \, e^{-t/RC}$ and the power dissipated in the resistor is $P_R = Ri^2 = [q_0^2/(RC^2)]e^{-2t/RC}$. At any instant the energy stored in the capacitor is $U_c = \frac{1}{2}Cv^2 = \frac{1}{2}q^2/C = \frac{1}{2}(q_0^2/C)e^{-2t/RC}$. Differentiating with respect to time, $dU_c/dt = -[q_0^2/(RC^2)]e^{-2t/RC}$, where the minus sign indicates that energy is being lost as time increases. Then $P_R + dU_c/dt = 0$ or $|dU_c/dt| = P_R$ as required for conservation of energy.

32.9 Describe the current buildup process for an R-L circuit [see Fig. 32-3(a)], and the associated time constant.

▮ When the switch in the circuit is first closed, the current in the circuit rises as shown in Fig. 32-3(b). The current does not jump to its final value because the changing flux through the coil induces a back emf in the coil, which opposes the rising current. After L/R seconds, the current has risen to 0.632 of its final value i_∞. This time, $t = L/R$, is called the *time constant* of the R-L circuit. After a long time, the current is changing so slowly that the back emf in the inductor, $L(\Delta i/\Delta t)$, is negligible. Then $i = i_\infty = \mathscr{E}/R$.

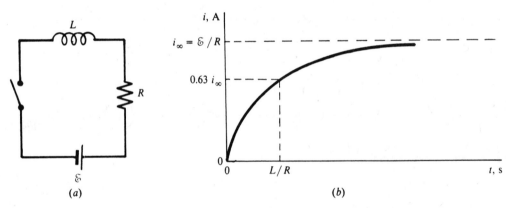

Fig. 32-3

32.10 A coil has an inductance of 1.5 H and a resistance of 0.6 Ω. If the coil is suddenly connected across a 12-V battery, find the time required for the current to rise to 0.63 of its final value. What will be the final current through the coil?

▮ The time required is the time constant of the circuit.

$$\text{time constant} = \frac{L}{R} = \frac{1.5\,\text{H}}{0.6\,\Omega} = \underline{2.5\,\text{s}}.$$

At long times, the current will be steady and so no back emf will exist in the coil. Under such conditions,

$$I = \frac{\mathscr{E}}{R} = \frac{12\,\text{V}}{0.6\,\Omega} = \underline{20\,\text{A}}$$

32.11 A constant potential difference of 60 V is suddenly applied to a coil which has a resistance of 30 Ω and a self-inductance of 8 mH. At what rate does the current begin to rise? What is the current at the instant the rate of change of current is 500 A/s? What is the final current?

▮ The circuit is as shown in Fig. 32-4 and obeys the loop equation $\mathscr{E} - L(\Delta i/\Delta t) - Ri = 0$. At $t = 0$, $i = 0$,

$$60 - 0.008\frac{\Delta i}{\Delta t} = 0 \qquad \frac{\Delta i}{\Delta t} = \frac{60}{0.008} = \underline{7500\,\text{A/s}}$$

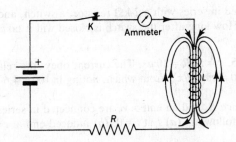

Fig. 32-4

When $\Delta i/\Delta t = 500$ A/s our equation yields

$$60 - (0.008)(500) - 30i = 0 \qquad 30i = 60 - 4 \qquad i = \underline{1.867 \text{ A}}$$

For the final current we note that $\Delta i/\Delta t = 0$, and

$$\mathscr{E} - L(0) - 30I_F = 0 \qquad I_F = \tfrac{60}{30} = \underline{2 \text{ A}}$$

32.12 When a long iron core solenoid is connected across a 6-V battery, the current rises to 0.63 of its maximum value after a time of 0.75 s. The experiment is then repeated with the iron core removed. Now the time required to reach 0.63 of the maximum is 2.5 ms. Calculate (a) the relative permeability of the iron and (b) L for the air core solenoid if the maximum current is 0.5 A.

▌ (a) With the iron core in the time constant is $L_1/R = 0.75$ s. With the iron core removed the inductance has changed while the resistance stays the same. The new time constant is: $L_2/R = 0.0025$ s. Dividing the second result by the first we have $L_1/L_2 = 300$. Since the number of turns and dimensions of the solenoid are the same before and after, we have $L_1/L_2 = \Phi_1/\Phi_2 = B_1/B_2 = K_m = \underline{300}$, as required. (b) The maximum current obeys $\mathscr{E} = iR$, or 6 V = (0.5 A)R, yielding $R = 12 \, \Omega$. Then from $L_2/R = 0.0025$ s, we have $L_2 = \underline{0.03 \text{ H}}$.

32.13 The circuit of Fig. 32-5 consists of a 50-V battery, a 20-Ω resistor, a low-resistance 25-mH inductor, and a key. Find the rate at which current begins to rise when the key is closed, the current at the instant the rate of change of current is 400 A/s, and the final steady current.

▌ From the loop equation, $\mathscr{E} - L(\Delta i/\Delta t) - Ri = 0$. Initially $i = 0$, so

$$50 - 0.025 \frac{\Delta i}{\Delta t} - R(0) = 0 \qquad \text{at} \qquad t = 0 \qquad \text{and} \qquad \left.\frac{\Delta i}{\Delta t}\right|_{t=0} = \frac{50}{0.025} = \underline{2000 \text{ A/s}}$$

When $\Delta i/\Delta t = 400$ A/s,

$$50 - (0.025)(400) - 20i = 0 \qquad 20i = 50 - 10 \qquad i = \underline{2 \text{ A}}$$

For steady current,

$$50 - L(0) - 20I_F = 0 \qquad \text{and} \qquad I_F = \tfrac{50}{20} = \underline{2.5 \text{ A}}$$

32.14 The series circuit of Fig. 32-5 consists of a 30-V battery, a 6-Ω resistor, an 8-mH inductor, and a switch. Apply Kirchhoff's second law to the circuit to relate the current i and its rate of change $\Delta i/\Delta t$ to the emf, resistance, and inductance of the circuit at any time after the switch is closed. At what rate does the current begin to rise when the switch is closed? Find the back emf induced in the inductor when the current is 3 A and the energy stored in the inductor at that instant.

▌ As in Prob. 32.13, $\mathscr{E} - L(\Delta i/\Delta t) - Ri = 0$. At $t = 0$, $i = 0$, so

$$30 - L\left.\frac{\Delta i}{\Delta t}\right|_{t=0} - R(0) = 0 \qquad \left.\frac{\Delta i}{\Delta t}\right|_{t=0} = \frac{30}{0.008} = \underline{3750 \text{ A/s}}$$

When $i = 3$ A,

$$30 - 0.008 \frac{\Delta i}{\Delta t} - (3)(6) = 0 \qquad L \frac{\Delta i}{\Delta t} = 30 - 18 = \underline{12 \text{ V}}$$

For the energy, energy $= \frac{1}{2}Li^2 = \frac{1}{2}(8/1000)(3^2) = \underline{0.036 \text{ J}}$.

32.15ᶜ Give a full analysis of the simple R-L circuit (Fig. 32-2, with C absent). Assume that v is constant, and $I = 0$ at $t = 0$ (the instant the switch is closed).

▌ $$L\ddot{q} + R\dot{q} = v \qquad \text{or} \qquad L\dot{I} + RI = v$$

Inasmuch as this equation is formally identical to (1) of Prob. 32.6, and the initial conditions correspond, the solution is yielded by (2) of Prob. 32.6 as

$$I(t) = \frac{v}{R}(1 - e^{-Rt/L}) \tag{1}$$

From this, and the further condition that $q = 0$ at $t = 0$,

charge: $$q = \int_0^t I\,dt = \frac{v}{R}t - \frac{vL}{R^2}(1 - e^{-Rt/L})$$

voltage across resistor: $$v_{ab} = RI = v(1 - e^{-Rt/L})$$

voltage across inductor: $$v_{bc} = v - v_{ab} = ve^{-Rt/L}$$

power from battery: $$P = Iv = \frac{v^2}{R}(1 - e^{-Rt/L})$$

energy in inductor: $$U_L = \frac{1}{2}LI^2 = \frac{Lv^2}{2R^2}(1 - e^{-Rt/L})^2$$

The time constant of the R-L circuit is L/R, as discussed in Prob. 32.9.

32.16 A 20-mH coil is connected in series with a 2-kΩ resistor, a switch, and a 12-V battery. What is the time constant of this circuit? How long after the switch is closed will it take for the current to reach 99 percent of its final value?

▌ Following Prob. 32.15, $\tau = L/R = \underline{10\ \mu s}$. The current obeys the relation $i = i_0[1 - \exp(-t/\tau)]$ and we want $-(i - i_0)/i_0 = 0.01$, so $100 = \exp(t/\tau)$, from which, noting $\ln 100 = 4.6$ we have $t = 4.6\tau = \underline{46\ \mu s}$.

32.17 An inductor L, a resistance R, and an emf $= V_0$ are connected in series with a switch. The switch is pushed closed at $t = 0$. Find the following: (a) I at $t = 0$; (b) induced emf at $t = 0$ and (c) at $t = \frac{1}{2}(L/R)$; (d) I at $\frac{1}{2}(L/R)$.

▌ (a) At $t = 0$ the current has not had a chance to build up, so $i = 0$; or recall that $i = I_0\{1 - \exp[(-Rt)/L]\}$ with $I_0 = V_0/R$, so at $t = 0$, $i = 0$. (b) $|\mathscr{E}| = V_0$ since $iR = 0$ at $t = 0$ so the applied voltage appears across the inductance; or, since $|\mathscr{E}| = +L(di/dt) = V_0 \exp[(-Rt)/L]$. Thus at $t = 0$, $|\mathscr{E}| = V_0$. (c) At $t = L/2R$, we have $|\mathscr{E}| = V_0 e^{-1/2} = 0.607V_0$. (d) $i = (V_0/R)(1 - e^{-1/2}) = \underline{(0.393V_0)/R}$.

32.18ᶜ In a series circuit consisting of an inductance coil L, a resistor R, a switch, and a battery V, find the following at a time equal to the time constant after the switch is closed: induced emf in the inductor, power output of battery, power loss in the resistor, rate at which energy is being stored in the inductor.

▌ We let $\tau = L/R = $ time constant. Maximum current in the circuit is V/R so $i = (V/R)[1 - \exp(-t/\tau)]$. When $t = \tau$, $i = (0.63V)/R$. Induced emf $= -L(di/dt) = -[(LV)/(R\tau)]\exp(-t/\tau) = -V\exp(-t/\tau)$. At $t = \tau$ this becomes $-0.37V$. Power output of battery is Vi and when $t = \tau$, this is $(0.63V^2)/R$. At that time, power loss in the resistor $= i^2R$ or $(0.40V^2)/R$. The rest of the power output of the battery, $(0.23V^2)/R$, is stored in the inductor. The latter result can also be obtained from $d[(Li^2)/2]/dt$.

32.19 Show that the current in the circuit through the battery of Fig. 32-6 rises instantly to its final value of V/R when the switch S is closed, provided that the time constants of the two branches are equal. The internal resistance of the battery and of the connecting wires is negligible.

▌ Let $i_L(t)$ denote the current (positive rightward) in the RL branch and $i_C(t)$ denote the current (positive rightward) in the RC branch. Then the current (positive upward) through the battery is $i(t) = i_L(t) + i_C(t)$.

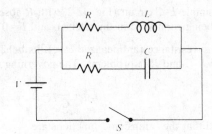

Fig. 32-6

Since the battery is across the two branches in parallel the current through the *RL* branch is unaffected by the presence of the *RC* branch, so we may directly apply Eq. (1) of Prob. 32.15:

$$i_L(t) = \frac{V}{R}[1 - e^{-(R/L)t}] \tag{1}$$

Similarly, the current in the *RC* branch is given by Eq. (3) of Prob. 32.6:

$$i_C(t) = \frac{V}{R} e^{-t/RC} \tag{2}$$

Hence the current through the battery is

$$i(t) = i_L(t) + i_C(t) = \frac{V}{R}[1 - e^{-(R/L)t} + e^{-t/RC}] \tag{3}$$

The time constant of the *RL* branch is $\tau_L \equiv L/R$; the time constant of the *RC* branch is $\tau_C \equiv RC$. If these have the common value $\tau_L = \tau_C = \tau$, then Eq. (3) becomes

$$i(t) = \frac{V}{R}[1 - e^{-t/\tau} + e^{-t/\tau}] = \frac{V}{R}$$

for all $t > 0$. This establishes the desired result.

32.20 A ballistic galvanometer can be used to measure a pulse of charge sent through it. Suppose that a coil with N loops is connected across the terminals of such a galvanometer. The total resistance of coil and galvanometer is R. If the coil is now suddenly thrust into the region between the poles of a U-type magnet, the flux through the coil will suddenly change from zero to ϕ_{max}. Show that the total charge pulse Q which flows through the galvanometer is given by $Q = N\phi_{max}/R$. By measuring Q, one can determine ϕ and, therefore, B for the magnet.

▌ Over the time period Δt of the process, the average current in the galvanometer is

$$i_{avg} = \frac{\mathscr{E}_{avg}}{R} = \frac{1}{R}\frac{N\phi_{max}}{\Delta t} \quad \text{and so} \quad Q = i_{avg}\,\Delta t = \frac{N\phi_{max}}{R}.$$

32.21 A ballistic galvanometer with a resistance of 240 Ω gives a full-scale deflection for 500 μC of electricity. A coil of 320 turns and 160-Ω resistance is to be constructed to study fields up to 1.4 T by observing deflections produced when the coil is suddenly removed from the field. What is the maximum area allowable for the coil if full-scale deflection is not to be exceeded?

▌ Substitute in the result of Prob. 32.20,

$$5 \times 10^{-4}\,C = \frac{(320)[(1.4\,\text{T})A]}{(240 + 160)\Omega}$$

Solving, $A = \underline{446\,\text{mm}^2}$.

32.22 A coil of 100 turns with a radius of 6 mm and a resistance of 40 Ω is placed between the poles of an electromagnet and suddenly removed. A charge of 32 μC is sent through a ballistic galvanometer connected to the coil. The resistance of the galvanometer is 160 Ω. What is the intensity of the magnetic field?

▌ From Prob. 32.20,

$$Q = \frac{N\Phi}{R} = \frac{NBA}{R}. \quad 32 \times 10^{-6} = \frac{100B(36\pi \times 10^{-6})}{40 + 160} \quad \text{and} \quad B = \frac{(200)(32 \times 10^{-6})}{(100)(36\pi \times 10^{-6})} = \underline{0.566\,\text{T}}$$

32.23^c Give a full analysis of the simple L-C circuit (Fig. 32-2, with R absent). Assume that v is constant, $q = q_0$ at $t = 0$ (C is charged before being connected in the circuit), and $I = 0$ at $t = 0$.

▌ When the circuit has zero resistance (an idealized case), its behavior is essentially different from the exponential decay found in R-C and R-L problems. The governing differential equation, is now

$$L\ddot{q} + \frac{1}{C}q = v \tag{1}$$

Setting $v = 0$ in (1), we find that the "transient" functions are

$$q_1(t) = \sin \omega_0 t \qquad q_2(t) = \cos \omega_0 t$$

where $\omega_0 \equiv (LC)^{-1/2}$ is called the *resonant frequency*. These functions, and their derivatives, do not die out as time increases; rather they represent undamped oscillations (SHM). The solution of (1) is then of the form

$$q(t) = c_1 \sin \omega_0 t + c_2 \cos \omega_0 t + Cv$$

The initial conditions require

$$q_0 = c_2 + Cv \qquad \text{and} \qquad I_0 = 0 = \omega_0 c_1$$

whence $c_1 = 0$, $c_2 = q_0 - Cv$, and

$$q(t) = (q_0 - Cv)\cos \omega_0 t + Cv \tag{2}$$

From (2),

current: $\qquad\qquad\qquad\qquad I = \dot{q} = -\omega_0(q_0 - Cv)\sin \omega_0 t$

voltage across capacitor: $\qquad v_{cd} = \dfrac{q}{C} = \left(\dfrac{q_0}{C} - v\right)\cos \omega_0 t + v$

voltage across inductor: $\qquad v_{bc} = v - v_{cd} = -\left(\dfrac{q_0}{C} - v\right)\cos \omega_0 t$

power from battery: $\qquad\quad P = Iv = -\omega_0 v(q_0 - Cv)\sin \omega_0 t$

energy in capacitor: $\qquad\quad U_C = \dfrac{q^2}{2C} = \dfrac{1}{2C}[(q_0 - Cv)\cos \omega_0 t + Cv]^2$

Note that I is an alternating current, just as if it were the (steady-state) current produced by an ac generator running at angular velocity ω_0.

32.24 An undriven L-C circuit has parameters $L = 1.5$ H, $C = 4\,\mu$F. The initial charge on the condensor is 4 mC. Find (**a**) the frequency of oscillation, and (**b**) the current as a function of time.

▌ In this case $v = 0$ in Prob. 32.23. Also the system starts ($t = 0$) with the capacitor charged with $q_0 = 4 \times 10^{-3}$ C. (**a**) $\omega_0 = 1/\sqrt{LC} = 1/[(1.5)(4 \times 10^{-6})]^{1/2} = 408$ s^{-1}. For the frequency, $f_0 = \omega_0/2\pi = \underline{65.0\text{ Hz}}$.

(**b**) $\qquad\qquad q = q_0 \cos \omega_0 t \qquad \text{and} \qquad I = -\omega_0 q_0 \sin \omega_0 t \qquad \text{so} \qquad I = \underline{-1.63 \sin 408t \qquad \text{A}}$

32.25 It is desired to set up an undriven L-C circuit in which the capacitor is originally charged to a difference of potential of 100.0 V. The maximum current is to be 10.0 A, and the oscillation frequency is to be 1000 Hz. What are the required values of L and C?

▌ In an L-C circuit, the maximum current is given by

$$i_{max} = \omega_0 q_0 = \omega_0 CV_0 \tag{1}$$

Here ω_0 is the resonant frequency and q_0 is the maximum charge on the capacitor. We have used $q_0 = CV_0$, where V_0 is the original potential difference across the capacitor. Since the oscillation frequency is to be 1000 Hz, we must have $\omega_0 = 2000\pi$ rad/s. With $i_{max} = 10.0$ A and $V_0 = 100.0$ V, we solve Eq. (1) for the required capacitance:

$$C = \frac{i_{max}}{\omega_0 V_0} = \frac{10.0}{(2000\pi)(100.0)} = 1.59 \times 10^{-5}\text{ F} = \underline{15.9\,\mu\text{F}}$$

The resonant frequency of an L-C circuit is $\omega_0 = 1/\sqrt{LC}$, so $L = (\omega_0^2 C)^{-1}$. Using Eq. (1) we can write this as

$$L = \frac{1}{\omega_0^2 C} = \frac{1}{\omega_0^2}\left(\frac{1}{i_{max}/(\omega_0 V_0)}\right) = \frac{V_0}{\omega_0 i_{max}} \tag{2}$$

With the given numerical values, we obtain

$$L = \frac{100.0}{(2000\pi)(10.0)} = 1.59 \times 10^{-3}\,\text{H} = \underline{1.59\,\text{mH}}$$

32.26ᶜ Show that energy is conserved in the circuit of Prob. 32.23.

❚ Multiply (1) of Prob. 32.23 by $\dot{q} = I$:

$$LI\frac{dI}{dt} + \frac{1}{C}q\frac{dq}{dt} = vI \qquad \text{or} \qquad \frac{d}{dt}\left(\frac{1}{2}LI^2\right) + \frac{d}{dt}\left(\frac{q^2}{2C}\right) = P_{\text{in}}$$

i.e., the input power is the sum of the powers delivered to the inductor and in the capacitor.

32.27ᶜ Consider the circuit shown in Fig. 32-7. Prior to $t = 0$, the switch is in position A and the capacitor is uncharged. At $t = 0$, the switch is instantaneously moved to position B. **(a)** Determine the current in the L-C circuit for $t > 0$. **(b)** Find the charge q on the lower capacitor plate for $t \geq 0$.

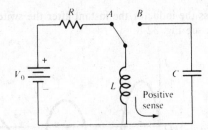

Fig. 32-7

❚ **(a)** At $t = 0$, there is a steady current downward through the inductor: $i(0) = V_0/R$. For $t \geq 0$, Kirchhoff's voltage law for the L-C circuit is

$$L\frac{di(t)}{dt} + \frac{q(t)}{C} = 0 \tag{1}$$

Here $q(t)$ is the charge on the lower plate of the capacitor. Since $i(t) = dq/dt$, Eq. (1) implies that

$$\frac{d^2 i(t)}{dt^2} = -\frac{i(t)}{LC} = -\omega_0^2 i(t) \tag{2}$$

where $\omega_0 \equiv 1/\sqrt{LC}$. The general solution of Eq. (2) is a sinusoid:

$$i(t) = i_{\max}\cos(\omega_0 t + \varphi)$$

Using the initial conditions on the charge and current, we find that $i_{\max} = V_0/R$ and $\varphi = 0$, so the current is

$$i(t) = \frac{V_0}{R}\cos(\omega_0 t) \tag{3}$$

(b) We are told that the charge on the lower capacitor plate is zero initially: $q(0) = 0$. The charge at time $t \geq 0$ is given by

$$q(t) = q(0) + \int_0^t \frac{dq(t')}{dt'}\,dt' = 0 + \int_0^t i(t')\,dt' = \frac{V_0}{R}\int_0^t \cos(\omega_0 t')\,dt'$$

$$= \frac{V_0}{\omega_0 R}[\sin(\omega_0 t')|_0^t] = \frac{V_0}{\omega_0 R}\sin(\omega_0 t) \tag{4}$$

32.28ᶜ Briefly analyze the R-L-C circuit of Fig. 32-2 (v constant). Assume the same initial conditions as in Prob. 32.23, and suppose the circuit parameters to be such that $R^2 < 4L/C$.

❚ Depending on the relative values of R, L, and C, the transient response of the R-L-C circuit can range between the extremes of exponential decay and undamped oscillation. The relationship assumed above will be seen to lead to damped oscillations (the so-called *underdamped case*).

The circuit equation

$$L\ddot{q} + R\dot{q} + \frac{1}{C}q = v \tag{1}$$

will have transient solutions (solutions corresponding to $v = 0$) of the form

$$q_1(t) = e^{-\alpha t}\sin \omega t \qquad q_2(t) = e^{-\alpha t}\cos \omega t$$

Substituting in (1) (with $v = 0$), we evaluate the positive real constants α and ω as

$$\alpha = \frac{R}{2L} \qquad \omega = \left(\frac{1}{LC} - \frac{R^2}{4L^2}\right)^{1/2}$$

[Clearly the condition for underdamped motion requires that ω be real, or $1/(LC) > R^2/(4L^2)$, or $R^2 < (4L)/C$.] Then the solution of the full equation is $q(t) = c_1 q_1(t) + c_2 q_2(t) + Cv$, which, after c_1 and c_2 are evaluated from the initial conditions, becomes

$$q = \frac{q_0 - Cv}{\omega}e^{-\alpha t}(\alpha \sin \omega t + \omega \cos \omega t) + Cv \tag{2}$$

From (2), the current is found as

$$I = \dot{q} = -\frac{q_0 - Cv}{\omega}(\alpha^2 + \omega^2)e^{-\alpha t}\sin \omega t = \frac{-v_L(0)}{\omega L}e^{-Rt/2L}\sin \omega t \tag{3}$$

where $v_L(0)$ is the voltage across the inductor the instant after the switch is closed. The current is an exponentially damped sinusoid; see Fig. 32-8.

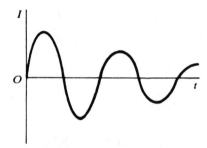

Fig. 32-8

32.29 (a) In Prob. 32.15, verify that $L/R = $ [s]. (b) In Prob. 32.23, verify that $\omega = $ [s^{-1}].

❚ (a)
$$[L/R] = [\mathrm{H}/\Omega] = \left[\frac{(\mathrm{V}\cdot\mathrm{s}/\mathrm{A})}{(\mathrm{V}/\mathrm{A})}\right] = [\mathrm{s}].$$

(b)
$$[\omega] = [(LC)^{-1/2}] = [(\mathrm{H}\cdot\mathrm{F})^{-1/2}] = [\{(\mathrm{V}\cdot\mathrm{s}/\mathrm{A})(\mathrm{C}/\mathrm{V})\}^{-1/2}] = [(\mathrm{s}^2)^{-1/2}] = [\mathrm{s}^{-1}].$$

32.30 Show that systems (a) and (b) of Fig. 32-9 are equivalent.

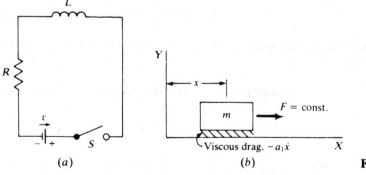

(a)   (b)   **Fig. 32-9**

❚ The two differential equations (a) $L\ddot{q} + R\dot{q} = v$ and (b) $m\ddot{x} + a_1\dot{x} = F$ have the same mathematical form. Hence, the solution to (b) that satisfies the initial conditions $x = \dot{x} = 0$ at $t = 0$ may be found by substitution of x, F, a_1, m for q, v, R, L, respectively, in the expression for the charge obtained in Prob. 32.15.

$$x = \frac{F}{a_1}t - \frac{Fm}{a_1^2}(1 - e^{-a_1 t/m})$$

Likewise, the kinetic energy of the block at time t is

$$E_k = \frac{1}{2} m\dot{x}^2 = \frac{mF^2}{2a_1^2} \left(1 - e^{-a_1 t/m}\right)^2$$

32.31ᶜ Show that in an undriven L-R-C circuit the stored energy E decays at a rate given by $dE/dt = -i^2 R$.

▌ Set $v = 0$ in (1) of Prob. 32.28 and multiply through by $\dot{q} = i$:

$$Li\frac{di}{dt} + Ri^2 + \frac{1}{C}q\frac{dq}{dt} = 0 \qquad \text{or} \qquad \frac{dE}{dt} + Ri^2 = 0$$

where $E \equiv \frac{1}{2}Li^2 + q^2/(2C) = $ total stored energy.

32.32 When connected in series, L_1, C_1 have the same resonant frequency as L_2, C_2, also connected in series. Prove that if all these circuit elements are connected in series, the new circuit will have the same resonant frequency as either of the circuits first mentioned.

▌ The resonant frequency of the first circuit is given by

$$\omega_{r1} = \frac{1}{\sqrt{L_1 C_1}} \tag{1}$$

According to the exercise statement, this is the same as the resonant frequency of the second circuit:

$$\frac{1}{\sqrt{L_2 C_2}} = \omega_{r2} = \omega_{r1} \tag{2}$$

If the circuit elements are connected in series, the resultant circuit has inductance $L' = L_1 + L_2$, and capacitance $C' = (C_1 C_2)/(C_1 + C_2)$. The resonant frequency of the new circuit is given by

$$(\omega_r')^2 = \frac{1}{L'C'} = \frac{C_1 + C_2}{C_1 C_2 (L_1 + L_2)} \tag{3}$$

Since $\omega_{r2} = \omega_{r1}$, we have $L_2 C_2 = L_1 C_1$, so that $C_1 C_2 (L_1 + L_2) = C_2^2 L_2 + C_1 C_2 L_2 = L_2 C_2 (C_1 + C_2)$. Therefore, Eq. (3) becomes

$$(\omega_r')^2 = \frac{C_1 + C_2}{L_2 C_2 (C_1 + C_2)} = \frac{1}{L_2 C_2} = \omega_{r2}^2$$

That is, the new resonant frequency ω_r' equals the common resonant frequency ω_{r2} of circuits 1 and 2.

32.33 Find criteria for the underdamped motion of the mechanical systems shown in Fig. 32-10.

▌ In system (a), the sphere, of mass m, experiences the downward force of gravity, mg; the upward spring force, $-k(x - l_0)$, where l_0 is the natural length of the spring; and a viscous drag, $-a_1\dot{x}$ $(a_1 > 0)$. Its equation

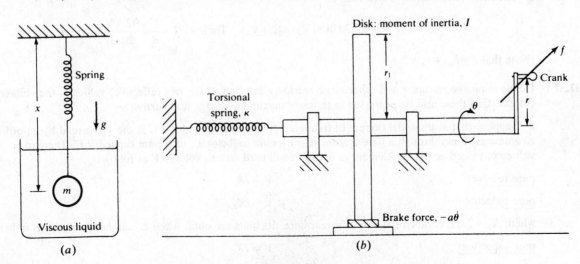

Fig. 32-10

of motion is therefore

$$m\ddot{x} = mg - k(x - l_0) - a_1\dot{x} \quad \text{or} \quad (a) \quad m\ddot{x} + a_1\dot{x} + kx = mg + kl_0$$

In system (b), the disk is acted on by three torques: $\tau_1 = fr$, where force f is applied to the crank handle as shown; $\tau_2 = -r_1(a\dot{\theta}) \equiv -b\dot{\theta}$, due to the viscous brake force; and $\tau_3 = -\kappa\theta$, the restoring torque in the spring. Thus its equation of motion is

$$I\ddot{\theta} = fr - b\dot{\theta} - \kappa\theta \quad \text{or} \quad (b) \quad I\ddot{\theta} + b\dot{\theta} + \kappa\theta = fr$$

It is seen that both (a) and (b) are analogs of the R-L-C circuit, whose differential equation is

$$L\ddot{q} + R\dot{q} + \frac{1}{C}q = v$$

Hence, by comparison with Prob. 32.28, the criteria for underdamped motion are (a) $a_1^2 < 4mk$ and (b) $b^2 < 4I\kappa$.

32.2 AC CIRCUITS IN THE STEADY STATE

32-34 A voltmeter reads 80 V when connected across the terminals of a sinusoidal power source with $f = 1000$ Hz. Write the equation for the instantaneous voltage provided by the source.

▮ A sinusoidal voltage can be expressed in the form $v = v_0 \sin \omega t = v_0 \sin 2\pi ft$ [or $v = v_0 \cos \omega t$ or more generally $v = v_0 \cos(\omega t + \phi)$ with ϕ constant], where $\omega = 2\pi f = 2\pi(1000 \text{ Hz}) = 6280 \text{ s}^{-1}$, and v_0 is the maximum value that the voltage can take. The voltmeter reads *not* the maximum but the *effective* or *root-mean-square* (rms) voltage: $v_{rms} = V = v_0/\sqrt{2}$. This is obtained by taking the time average of v^2 over a complete period of oscillation, $\langle v^2 \rangle_T = v_0^2/2$, and taking the square root. (A similar definition holds for the rms current.) Then

$$v_0 = \sqrt{2}\,V = \sqrt{2}(80 \text{ V}) = \underline{113 \text{ V}} \quad \text{and} \quad v = \underline{(113 \text{ V}) \sin 2000\pi t}.$$

32.35 A sinusoidal 60-cycle ac voltage is read to be 120 V by an ordinary voltmeter. (**a**) What is the maximum value the voltage takes on during a cycle? (**b**) What is the equation for the voltage?

▮ (**a**)
$$V = \frac{v_0}{\sqrt{2}} \quad \text{or} \quad v_0 = \sqrt{2}\,V = \sqrt{2}(120 \text{ V}) = \underline{170 \text{ V}}$$

(**b**)
$$v = v_0 \sin 2\pi ft = \underline{(170 \text{ V}) \sin 120\pi t}$$

where t is in seconds.

32.36 A voltage $v = (60 \text{ V}) \sin 120\pi t$ is applied across a 20-Ω resistor. What will an ac ammeter in series with the resistor read?

▮ The rms voltage across the resistor is

$$V = \frac{v_0}{\sqrt{2}} = 0.707v_0 = (0.707)(60 \text{ V}) = \underline{42.4 \text{ V}} \quad \text{Then} \quad I = \frac{V}{R} = \frac{42.4 \text{ V}}{20 \ \Omega} = \underline{2.12 \text{ A}}$$

(Note that $I \equiv i_{rms} = i_0/\sqrt{2}$.)

32.37 Define *inductive reactance* and *capacitative reactance* in terms of the rms (effective) values of the voltage and current. Can these also be expressed in terms of maximum voltage and current?

▮ Suppose that a sinusoidal current of frequency f with effective value I (I is the value read by an ordinary ac ammeter) flows through a pure resistor R, or a pure inductor L, or a pure capacitor C. Then an ac voltmeter placed across the element in question will read an rms voltage V as follows:

pure resistor: $V = IR$

pure inductor: $V = IX_L$

where $X_L = 2\pi fL$ is called the *inductive reactance*. Its units are ohms when L is in henries and f is in hertz.

pure capacitor: $V = IX_C$

where $X_C = 1/2\pi fC$ is called the *capacitive reactance*. Its units are ohms when C is in farads.

Since $I = i_0/\sqrt{2}$ and $V = v_0/\sqrt{2}$ for any sinusoidal current or voltage, we clearly also have

$$v_0 = i_0 R \qquad v_0 = i_0 X_L \qquad v_0 = i_0 X_C$$

for the three cases discussed above.

Note: Unless otherwise stated, the remaining problems in this chapter deal with rms values of sinusoidal voltages and currents.

32.38   Describe the phase relations between voltage across and current through (a) a pure resistance, (b) a pure inductance, and (c) a pure capacitance.

▮   (a) When an ac voltage is applied to a pure resistance, the voltage across the resistance and the current through it attain their maximum values at the same instant and their zero values at the same instant; the voltage and current are said to be *in phase*. (b) When an ac voltage is applied to a pure inductance, the voltage across the inductance reaches its maximum value one-fourth cycle ahead of the current, i.e., when the current is zero. The back emf of the inductance causes the current through the inductance to lag behind the voltage by one-fourth cycle (or 90°), and the two are 90° *out of phase*. (c) When an ac voltage is applied to a pure capacitor, the voltage across it lags 90° behind the current flowing through it. Current must flow before the voltage across (and charge on) the capacitor can build up.

32.39   Define the *impedance* of a series *R-L-C* circuit in terms of rms voltage and current; and define the *phase angle* between voltage and current.

▮   The impedance (Z) of a series circuit containing resistance, inductance, and capacitance is given by

$$Z = \sqrt{R^2 + (X_L - X_C)^2} \qquad (1)$$

with Z in ohms. If a voltage V is applied to such a series circuit, then an Ohm's law relates V to the current I through it:

$$V = IZ \qquad (\text{or} \quad v_0 = i_0 Z).$$

The phase angle ϕ between V and I is given by

$$\tan \phi = \frac{X_L - X_C}{R} \qquad \text{or} \qquad \cos \phi = \frac{R}{Z}$$

[See Prob. 32.85 for a more sophisticated development.]

32.40   Describe the vector representation for the rms voltages across the elements of an *R-L-C* circuit.

▮   Vector representations in a series *R-L-C* circuit are possible because (1) of Prob. 32.39 can be interpreted as the Pythagorean theorem for a right triangle. As shown in Fig. 32-11(a), Z is the hypotenuse of the right triangle, while R and $(X_L - X_C)$ are its two legs. The angle labeled ϕ is the phase angle between the current and the voltage.

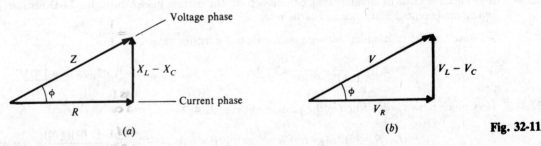

Fig. 32-11

A similar relation applies to the voltages across the elements in the series circuit. As shown in Fig. 32-11(b), it is

$$V^2 = V_R^2 + (V_L - V_C)^2$$

A more sophisticated analysis shows that these vectors, as well as others for current, when set rotating with the circuit frequency allow a complete time-dependent analysis of the circuit (see Prob. 32.86).

32.41   An ac current in a 10-Ω resistance produces heat at the rate of 360 W. Determine the effective values of the current and voltage.

▮ The instantaneous power loss in the resistor is $P_{inst} = i^2R = (i_0^2 \sin^2 2\pi ft)R$. The time average power is $P = (i_0^2/2)R = I^2R$, where $I = i_{rms}$. Then $360\text{ W} = I^2(10\,\Omega)$, and $I = \underline{6\text{ A}}$. Similarly $V = IR = (6\text{ A})(10\,\Omega) = \underline{60\text{ V}}$. As a check, since current across a resistor and voltage are in phase, $P = VI = (60\text{ V})(6\text{ A}) = 360\text{ W}$. *Note*: In terms of maximum values, $P = (i_0^2 R)/2 = (v_0 i_0)/2$.

32.42 What is the general expression for the average power in an R-L-C circuit in terms of the applied rms voltage across the circuit and the rms current through the circuit.

▮ Suppose that an ac voltage V is impressed across an impedance of any type. It gives rise to a current I through the impedance, and the phase angle between V and I is ϕ. The power loss in the impedance is given by power $= VI \cos \phi$. The quantity $\cos \phi$ is called the *power factor*. It is unity for a pure resistor; but it is zero for a pure inductor or capacitor (no power loss occurs in a pure inductor or capacitor). In terms of maximum values, power $= (v_0 i_0 \cos \phi)/2$.

32.43 Calculate the inductive reactance of a coil of 0.2-H inductance to a frequency of 60 Hz.

▮
$$X_L = 2\pi fL = (6.28)(60)(0.2) = \underline{75.4\,\Omega}$$

32.44 A 40-Ω resistor is connected across a 15-V variable frequency electronic oscillator. Find the current through the resistor when the frequency is (a) 100 Hz and (b) 100 kHz.

▮ For a pure resistance $V = IR$, and there is no frequency dependence. Then the answer to both parts is, given the rms voltage of 15 V, $I = (15\text{ V})/(40\,\Omega) = \underline{0.375\text{ A}}$.

32.45 Solve Prob. 32.44 if the 40-Ω resistor is replaced by a 2-mH inductor.

▮
$$V = IX_L \qquad X_L = \omega L = 2\pi fL$$

(a) $f = 100$ Hz, so $X_L = 6.28(100\text{ Hz})(2 \times 10^{-3}\text{ H}) = 1.256\,\Omega$. Then $I = (15\text{ V})/(1.256\,\Omega) = \underline{11.9\text{ A}}$. (b) X_L goes up by a factor of 1000, so I drops by the same factor. $I = \underline{11.9\text{ mA}}$.

32.46 Solve Prob. 32.44 if the 40 Ω-resistor is replaced by a 0.3-μF capacitor.

▮
$$V = IX_C \qquad X_C = \frac{1}{\omega C} = \frac{1}{2\pi fC}$$

(a) $X_C = 1.0/[6.28(100\text{ Hz})(0.3 \times 10^{-6}\text{ F})] = 5308\,\Omega$. Then $I = (15\text{ V})/(5308\,\Omega) = \underline{2.83\text{ mA}}$. (b) For frequency increasing 1000-fold, X_C decreases 1000-fold, and hence current increases 1000-fold. Thus $I = \underline{2.83\text{ A}}$.

32.47 What is the power factor of a circuit that draws 5 A at 160 V and whose power consumption is 600 W?

▮ Given the rms values of voltage and current and average power consumption, we have from Prob. 32.42 $\cos \phi = $ power factor $= $ power/$IV = 600/[(5)(160)] = \underline{0.75}$.

32.48 The effective value of an alternating current is 5 A. The current passes through a 24-Ω resistor. Find the maximum potential difference across the resistor.

▮ Recalling the relationship between maximum and effective values,
$$i_0 \equiv I_{max} = \frac{I_{eff}}{0.707} = \frac{5.0}{0.707} = \underline{7.07\text{ A}} \qquad v_0 \equiv V_{max} = I_{max}R = (7.07)(24.0) = \underline{170\text{ V}}$$

32.49 How much power is dissipated in the resistor of Prob. 32.48?

▮
$$P = (I_{eff})^2 R = (5)^2(24) = \underline{600\text{ W}} \qquad \text{or} \qquad P = I_{eff}V_{eff} = \frac{I_{max}V_{max}}{2} = \frac{(7.07)(170)}{2} = \underline{600\text{ W}}$$

32.50 A coil has resistance 20 Ω and inductance 0.35 H. Compute its reactance and its impedance to an alternating current of 25 cycles/s.

▮ For reactance: $\qquad R = \underline{20\,\Omega} \qquad X_L = 2\pi fL = 6.28(25\text{ Hz})(0.35\text{ H}) = \underline{55\,\Omega}$

For impedance: $\qquad Z = (R^2 + X_L^2)^{1/2} = (20^2 + 55^2)^{1/2} = \underline{58.5\,\Omega}$

32.51 A current of 30 mA is taken by a 4-μF capacitor connected across an alternating current line having a frequency of 500 Hz. Compute the reactance of the capacitor and the voltage across the capacitor.

▮ We assume that $R = L = 0$.

$$X_C = [2\pi fC]^{-1} = [6.28(500)(4 \times 10^{-6})]^{-1} = \underline{80\ \Omega} \qquad Z = X_C \qquad \text{and} \qquad V = IZ = (30 \times 10^{-3}\ \text{A})(80\ \Omega) = \underline{2.4\ \text{V}}$$

32.52 A 120-V ac voltage source is connected across a 2-μF capacitor. Find the current to the capacitor if the frequency of the source is (a) 60 Hz and (b) 60 kHz. (c) What is the power loss in the capacitor?

▮ (a)
$$X_C = \frac{1}{2\pi fC} = \frac{1}{2\pi(60\ \text{s}^{-1})(2 \times 10^{-6}\ \text{F})} = 1330\ \Omega \qquad I = \frac{V}{X_C} = \frac{120\ \text{V}}{1330\ \Omega} = \underline{0.090\ \text{A}}.$$

(b) Now $X_C = 1.33\ \Omega$, so $I = \underline{90\ \text{A}}$. Note that the impeding effect of a capacitor varies inversely with the frequency.

(c)
$$\text{average power} = VI \cos \phi = VI \cos 90° = \underline{0}$$

32.53 A 120-V ac voltage source is connected across a pure 0.70-H inductor. Find the current through it if the frequency of the source is (a) 60 Hz and (b) 60 kHz. (c) What is the power loss in it?

▮ (a)
$$X_L = 2\pi fL = 2\pi(60\ \text{s}^{-1})(0.7\ \text{H}) = 264\ \Omega \qquad I = \frac{V}{X_L} = \frac{120\ \text{V}}{264\ \Omega} = \underline{0.455\ \text{A}}.$$

(b) Now $X_L = 264 \times 10^3\ \Omega$, so $I = \underline{0.455 \times 10^{-3}\ \text{A}}$. Note that the impeding effect of an inductor varies directly with the frequency.

(c)
$$\text{average power} = VI \cos \phi = VI \cos 90° = \underline{0}$$

32.54 A coil has an inductance of 0.10 H and a resistance of 12 Ω. It is connected to a 110-V, 60-Hz line. Determine (a) the reactance of the coil, (b) the impedance of the coil, (c) the current through the coil, (d) the phase angle between current and supply voltage, (e) the power factor of the circuit, and (f) the reading of a wattmeter connected in the circuit.

▮ (a)
$$R = 12\ \Omega \qquad X_L = 2\pi fL = 6.28(60)(0.10) = \underline{37.7\ \Omega}$$

(b)
$$Z = [R^2 + X_L^2]^{1/2} = \underline{39.6\ \Omega}$$

(c)
$$I = \frac{V}{Z} = \frac{110\ \text{V}}{39.6\ \Omega} = \underline{2.78\ \text{A}}$$

(d)
$$\cos \phi = \frac{R}{Z} = \frac{12}{39.6} = 0.303 \qquad \phi = 72.4° = \text{angle by which voltage leads current}$$

(See Probs. 32.38 and 32.39.)

(e)
$$\cos \phi = 0.303 = \text{power factor}$$

(f)
$$P = \text{average power} = VI \cos \phi = (110\ \text{V})(2.78\ \text{A})(0.303) = \underline{92.6\ \text{W}}$$

32.55 A 10-μF capacitor is in series with a 40-Ω resistance and the combination is connected to a 110-V, 60-Hz line. Calculate (a) the capacitive reactance, (b) the impedance of the circuit, (c) the current in the circuit, (d) the phase angle between current and supply voltage, and (e) the power factor for the circuit.

▮ (a)
$$X_C = (2\pi fC)^{-1} = [6.28(60)(10 \times 10^{-6})]^{-1} = \underline{265\ \Omega}$$

(b)
$$Z = [R^2 + X_C^2]^{1/2} = [40^2 + 265^2]^{1/2} = \underline{268\ \Omega}$$

(c)
$$I = \frac{V}{Z} = \frac{110\ \text{V}}{268\ \Omega} = \underline{0.410\ \text{A}}.$$

(d)
$$\cos \phi = \frac{R}{Z} = \frac{40}{268} = 0.149 \qquad \phi = 81.4°$$

which is the angle by which voltage lags current, since we are dealing with a totally capacitative reactance (see Prob. 32.38). More formally $\cos \phi = 0.149 \Rightarrow \phi = \pm 81.4°$. The correct choice can be determined from $\tan \phi = (X_L - X_C)/R = -(X_C)/R$, and $\phi = -81.4°$. The voltage "leads" by $-81.4°$ or lags by $81.4°$.

(e)
$$\text{power factor} = \cos \phi = 0.149$$

32.56 In an ac series circuit the inductive reactance is 20 Ω, the capacitive reactance is 60 Ω, the resistance is 30 Ω, and the effective current is 2 A. What is the impedance of this circuit?

▮
$$Z = \sqrt{R^2 + (X_L - X_C)^2} = \sqrt{30^2 + (20 - 60)^2} = \underline{50\ \Omega}$$

32.57 Find the power factor in Prob. 32.56.

▎ From Prob. 32.39,

$$\tan \theta = \frac{X_L - X_C}{R} = \frac{20 - 60}{30} \qquad \theta = -53° \qquad \text{(voltage lags current)}$$

$$\text{power factor} = \cos(-53°) = \cos 53° = \underline{0.6}$$

32.58 How much power in watts is dissipated in the resistance of Prob. 32.56?

▎ The average power can be calculated two ways:
(1) Average power is dissipated only in the resistor. (The inductor and capacitor store and release energy only over the course of a cycle with no losses.) Then $P = I^2 R = (2^2)(30) = \underline{120 \text{ W}}$.
(2) $P = VI \cos \theta$, and using Prob. 32.57 as well as 32.56, $P = (ZI)I \cos \theta = (50 \, \Omega)(2 \text{ A})^2(0.6) = \underline{120 \text{ W}}$.

32.59 What is the maximum voltage across the inductive reactance of Prob. 32.56?

▎ Recalling $V_{max} = \sqrt{2}V_{rms}$ and $I_{max} = \sqrt{2}I_{rms}$ we have $V_{max} = I_{max}X_L = (I_{rms}X_L)/0.707 = [(2.0)(20.0)]/(0.707) = \underline{56.6 \text{ V}}$.

32.60 An ac series circuit consists of a 75-V source of emf $\mathscr{E}$, a total resistance of $14\,\Omega$, and an inductive reactance of $48\,\Omega$ at a frequency of 60 Hz. Find (a) the impedance of the circuit and (b) the current.

▎ (a) $\qquad\qquad Z^2 = R^2 + (X_L - X_C)^2 = (14)^2 + (48 - 0)^2 \, \Omega^2 \qquad Z = 50\,\Omega$

(b) $$I = \frac{\mathscr{E}}{Z} = \frac{75 \text{ V}}{50\,\Omega} = 1.5 \text{ A}$$

32.61 Referring to the circuit of Prob. 32.60, find (a) the angle by which the current lags behind the emf and (b) the inductance of the circuit.

▎ (a) The angle between the source voltage and the current is given by

$$\tan \theta = \frac{X_L - X_C}{R} = \frac{48 - 0}{14} = 3.43 \qquad \theta = \underline{73.7°}$$

the angle by which current lags behind emf.

(b) $\qquad\qquad X_L = 2\pi v L \qquad 48 = 2\pi(60)L \qquad L = \underline{0.127 \text{ H}}$

32.62 A series circuit has a resistance of $60\,\Omega$ and an impedance of $135\,\Omega$. Calculate the power consumed in the circuit when the total potential difference is 120 V.

▎ The impedance is given by

$$Z^2 = R^2 + (X_L - X_C)^2 \quad \text{so} \quad Z = \sqrt{60^2 + (X_L - X_C)^2} = 135 \quad \text{Then} \quad |X_L - X_C| = \sqrt{135^2 - 60^2} = \underline{120.9\,\Omega}$$

$$\text{Next} \quad |\tan\theta| = \frac{|X_L - X_C|}{R} = \frac{120.9}{60} \quad \text{and} \quad |\theta| = 63.6° \qquad P = \frac{V^2}{Z}\cos\theta$$

and recalling $\cos\theta = \cos(-\theta)$ we have $P = (120^2/135)\cos(63.6°) = \underline{47.4 \text{ W}}$. Whether current or voltage leads depends on the sign of $(X_L - X_C)$, which cannot be determined from the data.

32.63 A coil having inductance 0.14 H and resistance $12\,\Omega$ is connected across a 110-V, 25-Hz line. Compute (a) the current in the coil, (b) the phase angle between the current and supply voltage, (c) the power factor, and (d) the power loss in the coil.

▎ (a) $\quad X_L = 2\pi fL = 2\pi(25)(0.14) = 22.0\,\Omega \qquad Z = \sqrt{R^2 + (X_L - X_C)^2} = \sqrt{12^2 + (22 - 0)^2} = 25.1\,\Omega$

$$I = \frac{V}{Z} = \frac{110 \text{ V}}{25.1\,\Omega} = \underline{4.4 \text{ A}}$$

(b) $$\tan\phi = \frac{X_L - X_C}{R} = \frac{22 - 0}{12} = 1.83 \quad \text{or} \quad \phi = \underline{61.3°}$$

The voltage leads the current by 61.3°.

(c) $$\text{power factor} = \cos\phi = \cos 61.3° = \underline{0.48}$$

(d) $\quad$ power loss $= VI \cos \phi = (110 \text{ V})(4.4 \text{ A})(0.48) = \underline{230 \text{ W}}$

Or, since power loss occurs only because of the resistance of the coil,

$$\text{power loss} = I^2 R = (4.4 \text{ A})^2 (12 \text{ }\Omega) = \underline{230 \text{ W}}$$

32.64 A capacitor is in series with a resistance of 30 Ω and is connected to a 220-V ac line. The reactance of the capacitor is 40 Ω. Determine **(a)** the current in the circuit, **(b)** the phase angle between the current and the supply voltage, and **(c)** the power loss in the circuit.

▌ (a) $\quad Z = \sqrt{R^2 + (X_L - X_C)^2} = \sqrt{30^2 + (0 - 40)^2} = 50 \text{ }\Omega \qquad I = \dfrac{V}{Z} = \dfrac{220 \text{ V}}{50 \text{ }\Omega} = \underline{4.4 \text{ A}}$

(b) $\quad \tan \phi = \dfrac{X_L - X_C}{R} = \dfrac{0 - 40}{30} = -1.33 \quad$ or $\quad \phi = \underline{-53°}$

The negative sign tells us that the voltage *lags* the current by 53°. The angle ϕ in Fig. 32-11 would lie below the horizontal axis.

(c) $\quad$ *Method* 1 $\qquad$ power $= VI \cos \phi = (220)(4.4) \cos(-53°) = (220)(4.4) \cos 53° = \underline{580 \text{ W}}$

Method 2 $\quad$ the power loss occurs only in the resistor, not in the pure capacitor.

$$\text{power} = I^2 R = (4.4 \text{ A})^2 (30 \text{ }\Omega) = \underline{580 \text{ W}}$$

32.65 A series circuit consisting of a 100-Ω noninductive resistor, a coil of 0.10-H inductance and negligible resistance, and a 20-μF capacitor is connected across a 110-V, 60-Hz power source. Find **(a)** the current and **(b)** the power loss.

▌ (a) For the entire circuit, $Z = \sqrt{R^2 + (X_L - X_C)^2}$, with

$$R = 100 \text{ }\Omega \qquad X_L = 2\pi f L = 2\pi(60 \text{ s}^{-1})(0.10 \text{ H}) = 37.7 \text{ }\Omega \qquad X_C = \dfrac{1}{2\pi f C} = \dfrac{1}{2\pi(60 \text{ s}^{-1})(20 \times 10^{-6} \text{ F})} = 132.7 \text{ }\Omega$$

from which $\qquad Z = \sqrt{(100)^2 + (38 - 133)^2} = 138 \text{ }\Omega \qquad$ and $\qquad I = \dfrac{V}{Z} = \dfrac{110 \text{ V}}{138 \text{ }\Omega} = \underline{0.8 \text{ A}}$.

(b) All the power loss occurs in the resistor.

$$\text{power} = I^2 R = (0.8 \text{ A})^2 (100 \text{ }\Omega) = \underline{64 \text{ W}}.$$

32.66 Referring to the circuit of Prob. 32.65, find **(a)** the phase angle between the current and the source voltage and **(b)** voltmeter readings across the three elements.

▌ (a) $\quad \tan \phi = \dfrac{X_L - X_C}{R} = \dfrac{-95 \text{ }\Omega}{100 \text{ }\Omega} = -0.95 \quad$ or $\quad \phi = \underline{-43.5°}$

The voltage lags the current.

(b) $\qquad V_R = IR = (0.79 \text{ A})(100 \text{ }\Omega) = \underline{79 \text{ V}} \qquad V_C = IX_C = (0.79 \text{ A})(132.7 \text{ }\Omega) = \underline{105 \text{ V}}$

$$V_L = IX_L = (0.79 \text{ A})(37.7 \text{ }\Omega) = \underline{30 \text{ V}}$$

Note that $V_C + V_L + V_R$ does not equal the source voltage. From Fig. 32-11(*b*), the correct relationship is $V = \sqrt{V_R^2 + (V_L - V_C)^2} = \sqrt{79^2 + (-75)^2} = 109 \text{ V}$, which checks within the limits of rounding-off errors.

32.67 A 5-Ω resistance is in a series circuit with a 0.2-H pure inductance and a 4×10^{-8} F pure capacitance. The combination is placed across a 30-V, 1780-Hz power supply. Find **(a)** the current in the circuit and **(b)** the phase angle between source voltage and current.

▌ (a) $\quad X_L = 2\pi f L = 2\pi(1780 \text{ s}^{-1})(0.2 \text{ H}) = 2240 \text{ }\Omega \qquad X_C = \dfrac{1}{2\pi f C} = \dfrac{1}{2\pi(1780 \text{ s}^{-1})(4 \times 10^{-8} \text{ F})} = 2240 \text{ }\Omega$

$$Z = \sqrt{R^2 + (X_L - X_C)^2} = R = 5 \text{ }\Omega \qquad I = \dfrac{V}{Z} = \dfrac{30 \text{ V}}{5 \text{ }\Omega} = \underline{6 \text{ A}}.$$

(b) $\qquad \tan \phi = \dfrac{X_L - X_C}{R} = 0 \quad$ or $\quad \underline{\phi = 0°}$

32.68 For the circuit in Prob. 32.67, find **(a)** the power loss in the circuit and **(b)** the voltmeter reading across each element of the circuit.

▌ **(a)**   power = $VI \cos \phi = (30 \text{ V})(6 \text{ A})(1) = \underline{180 \text{ W}}$ or   power = $I^2 R = (6 \text{ A})^2 (5 \text{ }\Omega) = \underline{180 \text{ W}}$

(b) $V_R = IR = (6 \text{ A})(5 \text{ }\Omega) = \underline{30 \text{ V}}$ $V_C = IX_C = (6 \text{ A})(2240 \text{ }\Omega) = \underline{13.44 \text{ kV}}$

$V_L = IX_L = (6 \text{ A})(2240 \text{ }\Omega) = \underline{13.44 \text{ kV}}$

This circuit is in resonance because $X_C = X_L$. Note how very large the voltages across the inductor and capacitor can become, even though the source voltage is low.

32.69   As shown in Fig. 32-12, a series circuit connected across a 200-V, 60-cycle line consists of a capacitor of capacitive reactance 30 Ω, a noninductive resistor of 44 Ω, and a coil of inductive reactance 90 Ω and resistance 36 Ω. Determine **(a)** the current in the circuit and **(b)** the potential difference across each unit.

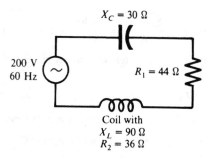

$X_C = 30 \text{ }\Omega$

200 V
60 Hz

$R_1 = 44 \text{ }\Omega$

Coil with
$X_L = 90 \text{ }\Omega$
$R_2 = 36 \text{ }\Omega$

Fig. 32-12

▌ **(a)** $Z = \sqrt{(R_1 + R_2)^2 + (X_L - X_C)^2} = \sqrt{(44 + 36)^2 + (90 - 30)^2} = 100 \text{ }\Omega$ $I = \dfrac{V}{Z} = \dfrac{200 \text{ V}}{100 \text{ }\Omega} = \underline{2 \text{ A}}$

(b)   p.d. across capacitor = $IX_C = (2 \text{ A})(30 \text{ }\Omega) = \underline{60 \text{ V}}$ p.d. across resistor = $IR_1 = (2 \text{ A})(44 \text{ }\Omega) = \underline{88 \text{ V}}$

impedance of coil = $\sqrt{R_2^2 + X_L^2} = \sqrt{(36)^2 + (90)^2} = 97 \text{ }\Omega$ p.d. across coil = $(2 \text{ A})(97 \text{ }\Omega) = \underline{194 \text{ V}}$

32.70   For the circuit of Prob. 32.69, find the **(a)** power factor of the circuit, **(b)** power absorbed by the circuit, and **(c)** power dissipated in the coil.

▌ **(a)**   power factor = $\cos \phi = \dfrac{R}{Z} = \dfrac{80}{100} = \underline{0.80}$.

(b)   power used = $VI \cos \phi = (200 \text{ V})(2 \text{ A})(0.80) = \underline{320 \text{ W}}$ or $= I^2 R = (2 \text{ A})^2 (80 \text{ }\Omega) = \underline{320 \text{ W}}$

(c)   power dissipated in coil = $I^2 R_2 = (2 \text{ A})^2 (36 \text{ }\Omega) = \underline{144 \text{ W}}$

32.71   A circuit having a resistance, an inductance, and a capacitance in series is connected to a 110-V ac line. For it, $R = 9.0 \text{ }\Omega$, $X_L = 28 \text{ }\Omega$, and $X_C = 16 \text{ }\Omega$. Compute **(a)** the impedance of the circuit, **(b)** the current, **(c)** the phase angle between the current and the supply voltage, and **(d)** the power factor of the circuit.

▌ **(a)** $Z = [R^2 + (X_L - X_C)^2]^{1/2} = \underline{15 \text{ }\Omega}$

(b) $I = \dfrac{V}{Z} = \dfrac{110 \text{ V}}{15 \text{ }\Omega} = \underline{7.33 \text{ A}}$

(c) $\tan \phi = \dfrac{X_L - X_C}{R} = \dfrac{12}{9} = 1.33$ and $\phi = 53.1°$

by which voltage leads.

(d) $\cos \phi = \text{power factor} = \underline{0.60}$ or $\cos \phi = \dfrac{R}{Z} = \dfrac{9}{15} = 0.60$

32.72   A circuit has a resistance 11 Ω, a coil of inductive reactance 120 Ω, and a capacitor with 120-Ω reactance, all connected in series with a 110-V, 60-Hz power source. What is the potential difference across each circuit element?

▌ First we find $I = \mathscr{E}/Z$, with $\mathscr{E} = 110 \text{ V}$ and $Z = [11 \text{ }\Omega^2 + (120 \text{ }\Omega - 120 \text{ }\Omega)^2]^{1/2} = 11 \text{ }\Omega$, yielding $I = 10 \text{ A}$. Then

$V_R = IR = (10 \text{ A})(11 \text{ }\Omega) = \underline{110 \text{ V}}$ $V_L = IX_L = (10 \text{ A})(120 \text{ }\Omega) = \underline{1200 \text{ V}}$ $V_C = IX_C = (10 \text{ A})(120 \text{ }\Omega) = \underline{1200 \text{ V}}$

(Check: $\mathscr{E}^2 = V_R^2 + (V_L - V_C)^2 = (110 \text{ V})^2$, and $\mathscr{E} = 110 \text{ V}$.)

32.73   A 120-V, 60-Hz power source is connected across an 800-Ω noninductive resistance and an unknown capacitance in series. The voltage drop across the resistor is 102 V. (*a*) What is the voltage drop across the capacitor? (*b*) What is the reactance of the capacitor?

▍ We first obtain the current using $V_R = IR$, or 102 V $= I(800 \Omega)$, and $I = 0.128$ A. Next we obtain V_C using $\mathscr{E}^2 = V_R^2 + V_C^2$ or $(120 \text{ V})^2 = (102 \text{ V})^2 + V_C^2$, and (*a*) $V_C = \underline{63 \text{ V}}$. Then $V_C = IX_C$ and 63 V $= (0.128 \text{ A})X_C$, so (*b*) $X_C = \underline{494 \Omega}$.

32.74 A coil of negligible resistance is connected in series with a 90-Ω resistor across a 120-V, 60-Hz line. A voltmeter reads 36 V across the resistance. Find the voltage across the coil and the inductance of the coil.

▍ $V_R = IR$, or 36 V $= I(90 \Omega)$, and $I = 0.40$ A. We could proceed to find V_L using the technique of Prob. 32.73. Instead we first find X_L by finding Z: $\mathscr{E} = IZ$, or 120 V $= (0.40 \text{ A})Z$, and $Z = 300 \Omega$. Then $Z^2 = R^2 + X_L^2$, or $300^2 = 90^2 + X_L^2$, and $X_L = 286 \Omega$. Then we first find L from $X_L = 2\pi f L$, or $(286 \Omega) = 6.28(60 \text{ Hz})L$, and $L = \underline{0.76 \text{ H}}$. Next we find $V_L = IX_L = (0.40 \text{ A})(286 \Omega) = \underline{114 \text{ V}}$.

32.75 A step-up transformer is used on a 120-V line to furnish 1800 V. The primary has 100 turns. How many turns are on the secondary?

▍ From Prob. 31.38:

$$\frac{V_1}{V_2} = \frac{N_1}{N_2} \qquad \text{or} \qquad \frac{120 \text{ V}}{1800 \text{ V}} = \frac{100 \text{ turns}}{N_2}$$

from which $N_2 = \underline{1500 \text{ turns}}$.

32.76 A transformer used on a 120-V line delivers 2 A at 900 V. What current is drawn from the line? Assume 100 percent efficiency.

▍ From Prob. 31.42:

$$\text{power in primary} = \text{power in secondary} \qquad \text{(for resistive load and 100 percent efficiency)}$$
$$I_1(120 \text{ V}) = (2 \text{ A})(900 \text{ V}) \qquad I_1 = \underline{15 \text{ A}}$$

32.77 A step-down transformer operates on a 2.5-kV line and supplies a load with 80 A. The ratio of the primary winding to the secondary winding is 20:1. Assuming 100 percent efficiency, determine the secondary voltage, V_2, the primary current, I_1, and the power output, P_2.

▍ $V_2 = (\frac{1}{20})V_1 = \underline{125 \text{ V}} \qquad I_1 = (\frac{1}{20})I_2 = \underline{4 \text{ A}} \qquad P_2 = V_2I_2 = \underline{10 \text{ kW}}$

The last expression is correct only if it is assumed that the load is resistive, so that the power factor is unity.

32.78 A step-down transformer is used on a 2200-V line to deliver 110 V. How many turns are on the primary winding if the secondary has 25 turns?

▍ The voltage must drop by a factor of 20, so the primary-to-secondary-turn ratio is 20 and $N_p = \underline{500}$.

32.79 A step-down transformer is used on a 1650-V line to deliver 45 A at 110 V. What current is drawn from the line? Assume 100 percent efficiency.

▍ We assume, as in Probs. 32.76 and 32.77, a resistive load. Then $V_1I_1 = V_2I_2$, or $1650I_1 = (110)(45)$, and $I_1 = \underline{3.0 \text{ A}}$.

32.80 A step-up transformer operates on a 110-V line and supplies a load with 2 A. The ratio of the primary and secondary windings is 1:25. Determine the secondary voltage, the primary current, and the power output. Assume a resistive load and 100 percent efficiency.

▍ Let 1 and 2 refer to primary and secondary coils, respectively. $V_1/V_2 = N_1/N_2$ yields $V_2 = 25(110 \text{ V}) = \underline{2750 \text{ V}}$. $V_1I_1 = V_2I_2$ yields $I_1 = (2750 \text{ V})(2 \text{ A})/(110 \text{ V}) = \underline{50 \text{ A}}$. Also $P = V_1I_1 = (110 \text{ V})(50 \text{ A}) = \underline{5500 \text{ W}}$.

32.81 Calculate the resonant frequency of a circuit of negligible resistance containing an inductance of 40 mH and a capacitance of 600 pF.

▍ In an R-L-C circuit the resonant frequency is the one for which the impedance is minimized. Since $Z = [R^2 + (X_L - X_C)^2]^{1/2}$, minimum Z occurs for $X_L = X_C$, or $2\pi f_0 L = 1/(2\pi f_0 C)$, and $4\pi^2 f_0^2 = 1/(LC)$. From this we deduce that $f_0 = 1/(2\pi\sqrt{LC})$, which is also the natural frequency of a pure L-C circuit. (See Prob.

32.23 and note $\omega_0 = 2\pi f_0$.) For our case,

$$f_0 = \frac{1}{2\pi\sqrt{LC}} = \frac{1}{2\pi\sqrt{(40 \times 10^{-3}\,\text{H})(600 \times 10^{-12}\,\text{F})}} = \underline{32.5\,\text{kHz}}$$

32.82 An experimenter has a coil of inductance 3 mH and wishes to construct a circuit whose resonant frequency is 1 MHz. What should be the value of the capacitor used?

▌ From Prob. 32.81, $f_0 = [2\pi\sqrt{LC}]^{-1}$, and $4\pi^2 LC = 1/f_0^2$. Then $4(3.14)^2(3 \times 10^{-3}\,\text{H})C = 10^{-12}\,\text{s}^{-2}$, and $C = \underline{8.45\,\text{pF}}$.

32.83 A transmitter operates at 1 MHz. The oscillating circuit has a capacitance of 200 pF. What is the inductance and the capacitive reactance of the resonant circuit?

▌ For the inductance, $L = 1/(4\pi^2 f^2 C) = 1/[(39.4)(10^{12})(2.0 \times 10^{-10})] = \underline{130\,\mu\text{H}}$. For the capacitive reactance, $X_C = 1/(2\pi f C) = 1/[(6.28)(10^6)(2.0 \times 10^{-10})] = \underline{800\,\Omega}$.

32.84 The impedance of a series R-C-L circuit is 8 Ω when $v = 60$ Hz at resonance and 10 Ω at 80 Hz. Calculate the values of L and C.

▌ At resonance, $v = 1/(2\pi\sqrt{LC})$ $\qquad (2\pi)^2(60)^2 = 1/(LC)$.

$$L = \frac{1}{(2\pi)^2(60)^2 C} \tag{1}$$

Also since $X_L = X_C$, $Z = R = 8\,\Omega$. At $v = 80$ Hz, we then have

$$Z^2 = 8^2 + (X_L - X_C)^2 = 10^2 \quad \text{and} \quad X_L - X_C = 2\pi(80)L - \frac{1}{(2\pi)(80)C} = 6\,\Omega$$

Using (1)

$$X_L - X_C = \frac{2\pi(80)}{(2\pi)^2(60)^2 C} - \frac{1}{2\pi(80)C} = 6\,\Omega \quad \text{and} \quad C = \frac{80}{2\pi(60)^2(6)} - \frac{1}{(2\pi)(80)(6)} = \underline{0.00027\,\text{F}}$$

Substituting into (1) we get $L = \underline{0.0261\,\text{H}}$. Note that we knew $X_L > X_C$ at 80 Hz since X_L increases with frequency while X_C decreases.

32.3 TIME BEHAVIOR OF AC CIRCUITS

32.85ᶜ For the circuit of Fig. 32-13 find an expression for the steady-state current through the circuit. In so doing find an expresssion for the circuit impedance, Z, defined by $Z \equiv \mathcal{E}_{max}/i_{max} = \mathcal{E}_{rms}/i_{rms}$.

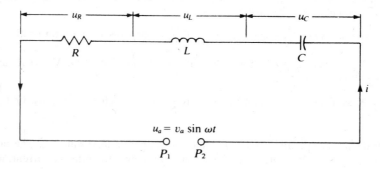

Fig. 32-13

▌ The differential equation of the circuit is (1) of Prob. 32.28, with $v = v_a \sin \omega t$ $\ (\mathcal{E}_{max} = v_a)$:

$$L\ddot{q} + R\dot{q} + \frac{1}{C}q = v_a \sin \omega t \tag{1}$$

The transient solution of (1) is that found in Prob. 32.28. We are concerned here only with the steady-state solution, which we assume to be of the form

$$q = -\frac{v_a}{\omega Z}\cos(\omega t - \phi) \quad \Rightarrow \quad i = \frac{v_a}{Z}\sin(\omega t - \phi) \tag{2}$$

Substitution of (2) into (1) determines the unknown constants Z and ϕ (as well as verifies the form of solution). Then

$$Z = \left[R^2 + \left(\omega L - \frac{1}{\omega C} \right)^2 \right]^{1/2} \equiv [R^2 + (X_L - X_C)^2]^{1/2}$$

$$\tan \phi = \left(\omega L - \frac{1}{\omega C} \right) \Big/ R \equiv (X_L - X_C)/R \tag{3}$$

where we have defined the *impedance* Z (Ω) and the *phase angle* ϕ (rad) in terms of the resistance R (Ω), the *inductive reactance* $X_L \equiv \omega L$ (Ω), and the *capacitative reactance* $X_C \equiv (\omega C)^{-1}$ (Ω).

32.86 Referring to Prob. 32.85, find the instantaneous voltages across the elements, and show these results in a *phasor diagram*.

▌ Using the steady-state current (2), with $i_{\max} = I \equiv v_a/Z$, the three voltages may be written as

$$u_R = v_R \sin(\omega t - \phi) \qquad u_L = v_L \cos(\omega t - \phi) \qquad u_C = -v_C \cos(\omega t - \phi) \tag{1}$$

where $v_R \equiv IR$ (V), $v_L \equiv IX_L$ (V), and $v_C \equiv IX_C$ (V).

Relations (1) are conveniently pictured in a *phasor diagram*, Fig. 32-14. The various waves are represented by vectors from the origin, $\mathbf{I}$, $\mathbf{v}_R$, $\mathbf{v}_L$, $\mathbf{v}_C$, $\mathbf{v}_a$, whose lengths are, respectively, I, v_R, v_L, v_C, v_a. Vectors $\mathbf{v}_R$, $\mathbf{v}_L$, $\mathbf{v}_C$, $\mathbf{v}_a$, respectively, make the fixed angles 0°, 90°, 270°, ϕ with the vector $\mathbf{I}$. If the five vectors are regarded as a rigid unit that rotates counterclockwise about O with constant angular velocity ω, it is seen that the Y component of each rotating vector, or *phasor*, gives the instantaneous value of the corresponding wave. Note that $\mathbf{I}$ and $\mathbf{v}_R$ are "in phase"; $\mathbf{v}_L$ "leads" $\mathbf{I}$ by 90°, $\mathbf{v}_C$ "lags" $\mathbf{I}$ by 90°; and $\mathbf{v}_a$ "leads" or "lags" $\mathbf{I}$ according as ϕ is positive or negative. Note also that the circuit equation (1) is reflected in the phasor equality

$$\mathbf{v}_a = \mathbf{v}_R + \mathbf{v}_L + \mathbf{v}_C \tag{2}$$

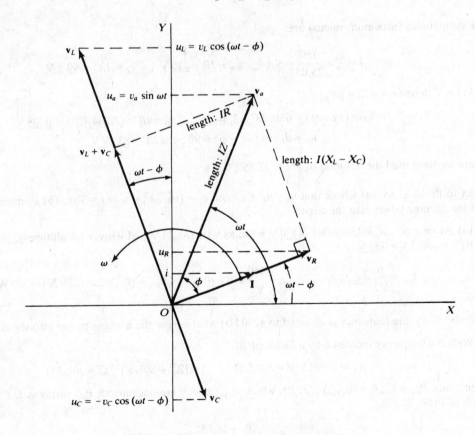

Fig. 32-14

32.87ᶜ Express the steady-state, average power input to the circuit of Fig. 32-13.

▌ During one period of the applied voltage—say, from $t = 0$ to $t = 2\pi/\omega \equiv T$—conservation of energy requires that the energy into the circuit, $P_{\text{avg}}T$, be stored in the inductor and capacitor or dissipated in the resistor. Thus

$$P_{\text{avg}} T = \frac{1}{2} Li^2 \Big|_0^T + \frac{q^2}{2C}\Big|_0^T + \int_0^T i^2 R\, dt$$

But, in the steady state, i and q are periodic with period T, and so the first two terms on the right-hand side vanish, giving

$$P_{\text{avg}} = R\left(\frac{1}{T}\int_0^T i^2\, dt\right) \equiv R(I_{\text{rms}})^2$$

Since i is a sinusoid, we can easily show that

$$I_{\text{rms}} = \frac{1}{\sqrt{2}} I = \frac{1}{\sqrt{2}} \frac{v_a}{Z} = \frac{v_{a,\text{rms}}}{Z}$$

(Here I = maximum current, in the notation of Prob. 32.85.) So $P_{\text{avg}} = (RI_{\text{rms}} v_{a,\text{rms}})/Z = I_{\text{rms}} v_{a,\text{rms}} \cos \phi$, where ϕ is the phase angle.

32.88 A sinusoidal voltage of frequency $f = 60$ Hz and peak value 150 V is applied to a series $R\text{-}L$ circuit, where $R = 20\ \Omega$ and $L = 40$ mH. (**a**) Compute the period T, ω, X_L, Z, ϕ. (**b**) Compute the amplitudes I, v_R, v_L and find the instantaneous values i, u_R, u_L, at $t = T/6$.

■ (**a**) $\quad T = \frac{1}{f} = \frac{1}{60}$ s $\quad \omega = 2\pi f = \underline{377\ \text{rad/s}} \quad X_L = \omega L = (377)(0.040) = \underline{15.08\ \Omega}$

$$Z = (R^2 + X_L^2)^{1/2} = \underline{25.05\ \Omega} \qquad \phi = \arctan\frac{X_L}{R} = \arctan 0.754 = \underline{37°}$$

(**b**) Amplitudes (maximum values) are

$$I = \frac{v_a}{Z} = \frac{150}{25.05} = \underline{6\ \text{A}} \qquad v_R = IR = \underline{120\ \text{V}} \qquad v_L = IX_L = \underline{90.5\ \text{V}}$$

At $t = T/6$ or $\omega t = \pi/3 = 60°$,

$$i = I \sin(\omega t - \phi) = 6 \sin 23° = \underline{2.344\ \text{A}} \qquad u_R = iR = (2.344)(20) = \underline{46.88\ \text{V}}$$
$$u_L = v_L \cos(\omega t - \phi) = 90.5 \cos 23° = \underline{83.29\ \text{V}}$$

where we have used the notation of Prob. 32.85.

32.89 Refer to Prob. 32.88. (**a**) Check that $u_a = u_R + u_L$ and $v_a = (v_R^2 + v_L^2)^{1/2}$, at $t = T/6$. (**b**) Compute I_{rms}, $v_{a,\text{rms}}$, and the average power into the circuit.

■ (**a**) At $\omega t = 60°$, $u_a = 150 \sin 60° = \underline{130\ \text{V}} = (46.88\ \text{V}) + (83.29\ \text{V})$. Moreover (at all times), $[120^2 + 90.5^2]^{1/2} = 150.3\ \text{V} \approx \underline{150\ \text{V}}$.

(**b**) $\quad I_{\text{rms}} = \frac{6}{\sqrt{2}} = \underline{4.24\ \text{A}} \qquad v_{a,\text{rms}} = \frac{150}{\sqrt{2}} = \underline{106.07\ \text{V}} \qquad P_{\text{avg}} = (I_{\text{rms}})^2 R = (18)(20) = \underline{360\ \text{W}}$

32.90 In Prob. 32.88, the frequency is changed to 1200 Hz; what is now the average power into the circuit?

■ With the frequency increased by a factor of 20,

$$X_L = 20 \times 15.08 = 301.6\ \Omega \qquad Z = [20^2 + 301.6^2]^{1/2}\ \Omega = 302.3\ \Omega$$

Then, since $P_{\text{avg}} = I_{\text{rms}}^2 R = (v_{a,\text{rms}})^2/(Z^2 R)$, where $v_{a,\text{rms}}$ and R are unchanged, P_{avg} varies as $1/Z^2$ so, from the result of Prob. 32.89,

$$P_{\text{avg}} = \frac{20^2 + 15.08^2}{20^2 + 301.6^2} 360 = \underline{2.47\ \text{W}}$$

32.91 In the circuit of Fig. 32-13, let $R = 20\ \Omega$, $L = 0.16$ H, and $C = 30\ \mu\text{F}$, and let the applied voltage be $u = 250 \sin 400t$ (V). (**a**) Compute X_L, X_C, Z, ϕ, I, v_R, v_L, v_C. (**b**) Verify that $v_a = [v_R^2 + (v_L - v_C)^2]^{1/2}$.

(a) $X_L = (400)(0.16) = \underline{64\ \Omega}$ $I = \dfrac{250}{27.817} = \underline{8.987\ A} = $ maximum current

$X_C = \dfrac{1}{(400)(30 \times 10^{-6})} = \underline{83.333\ \Omega}$ $v_R = (8.987)(20) = \underline{179.75\ V}$

$Z = [(20)^2 + (64 - 83.333)^2]^{1/2} = \underline{27.817\ \Omega}$ $v_L = (8.987)(64) = \underline{575.17\ V}$

$\phi = \arctan \dfrac{64 - 83.333}{20} = \underline{-0.7685\ rad}$ $v_C = (8.987)(83.333) = \underline{748.91\ V}$

(b) $[179.75^2 + (575.17 - 748.91)^2]^{1/2} = \underline{250.0}$

32.92   Refer to Prob. 32.91. **(a)** Write expressions for i, u_R, u_L, u_C and show that $u_a = u_R + u_L + u_C$. **(b)** Compute P_{avg}.

(a) $i = 8.987 \sin (400t + 0.7685)$ A $u_L = 575.17 \cos (400t + 0.7685)$ V

$u_R = 179.75 \sin (400t + 0.7685)$ V $u_C = -748.91 \cos (400t + 0.7685)$ V

Using the trigonometric identity

$$A \sin \theta + B \cos \theta = (A^2 + B^2)^{1/2} \sin (\theta + \delta) \qquad \text{where} \qquad \tan \delta = \dfrac{B}{A}$$

and the last result of Prob. 32.91,

$$u_R + u_L + u_C = 179.75 \sin (400t + 0.7685) + (575.17 - 748.91) \cos (400t + 0.7685)$$
$$= 250.0 \sin (400t + 0.7685 + \delta) = u_a$$

where the last equality follows from

$$\delta = \arctan \dfrac{575.17 - 748.91}{179.75} = \arctan \dfrac{64 - 83.333}{20} = -0.7685$$

(b) $P_{avg} = (I_{rms})^2 R = \left(\dfrac{8.987}{\sqrt{2}}\right)^2 20 = \underline{807.66\ W}$

32.93   For a certain angular frequency, ω_0, of the applied voltage, the R-L-C circuit will have unity power factor (a condition known as *voltage resonance*). Because

$$P_{avg} = I_{rms} v_{a,rms} \cos \phi = \dfrac{v_{a,rms}^2}{R} \cos^2 \phi$$

the circuit consumes maximum power, for a given v_a, at voltage resonance. Rework Prob. 32.91(a) at voltage resonance.

We have $\cos \phi = 1$ only if

$$X_L = X_C \qquad \text{or} \qquad \omega L = \dfrac{1}{\omega C} \qquad \text{or} \qquad \omega = \omega_0 \equiv \sqrt{\dfrac{1}{LC}}$$

Then $\omega_0 = \sqrt{\dfrac{1}{(0.16)(30 \times 10^{-6})}} = 456.4\ rad/s$

and $X_L = (456.4)(0.16) = \underline{73.03\ \Omega} = X_C$ $I = \dfrac{250}{20} = \underline{12.5\ A}$

$Z = R = \underline{20\ \Omega}$ $v_R = v_a = \underline{250\ V}$

$\phi = \underline{0}$ $v_L = (12.5)(73.03) = \underline{913\ V} = v_C$

Note that the maximum voltage between terminals of the inductor (or the capacitor) is $913/250 = 3.65$ times the maximum applied voltage.

32.94^c   For the circuit of Fig. 32-15, find the instantaneous current through each element.

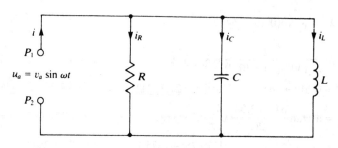

Fig. 32-15

▮ The three current equations are

$$Ri_R = u_a \qquad L\frac{di_L}{dt} = u_a \qquad \frac{1}{C}i_C = \frac{du_a}{dt} \tag{1}$$

The steady-state solutions of (1) are

$$i_R = \frac{v_a}{R}\sin\omega t \equiv I_R\sin\omega t \qquad i_L = -\frac{v_a}{\omega L}\cos\omega t \equiv -\frac{v_a}{X_L}\cos\omega t \equiv -I_L\cos\omega t$$

$$i_C = v_a\omega C\cos\omega t \equiv \frac{v_a}{X_C}\cos\omega t \equiv I_C\cos\omega t \tag{2}$$

where the reactances X_L and X_C are as defined in earlier problems.

32.95 For the circuit of Fig. 32-15 find the total instantaneous current through the source, and find expressions for the phase angle of this current and the impedance of the circuit.

▮ For the total current, we have

$$i = i_R + i_L + i_C = v_a\left[\frac{1}{R}\sin\omega t + \left(\frac{1}{X_C} - \frac{1}{X_L}\right)\cos\omega t\right] \equiv I\sin(\omega t + \phi) \tag{1}$$

where $I \equiv v_a/Z$, and the impedance and phase angle are then given by (see identity in Prob. 32.92)

$$\frac{1}{Z} \equiv \left[\left(\frac{1}{R}\right)^2 + \left(\frac{1}{X_C} - \frac{1}{X_L}\right)^2\right]^{1/2}$$

$$\tan\phi \equiv \left(\frac{1}{X_C} - \frac{1}{X_L}\right)\bigg/\frac{1}{R} \tag{2}$$

32.96 Referring to the circuit of Fig. 32-15 and the results of Probs. 32.94 and 32.95, draw a phasor diagram to illustrate the relationship of the instantaneous currents to the instantaneous emf.

▮ In Fig. 32-16, the five phasors are $\mathbf{v}_a$, $\mathbf{I}_R$, $\mathbf{I}_L$, $\mathbf{I}_C$, $\mathbf{I}$, where $\mathbf{I}_R$ is in phase with $\mathbf{v}_a$, $\mathbf{I}_C$ leads $\mathbf{v}_a$ by 90°, $\mathbf{I}_L$ lags $\mathbf{v}_a$ by 90°, and $\mathbf{I}$ leads or lags $\mathbf{v}_a$ by $|\phi|$. As before, Y components give instantaneous values. The phasor equality $\mathbf{I} = \mathbf{I}_R + \mathbf{I}_L + \mathbf{I}_C$ reflects conservation of charge in the circuit.

32.97 Referring to Prob. 32.96 find an expression for the average power delivered by the source.

▮ The formulas for average power from the alternator, given that i makes a phase angle ϕ with v_a is $P_{avg} = I_{rms}v_{a,rms}\cos\phi = (I_{R,rms})^2 R$, as for the series circuit, but where now $\cos\phi = Z/R$.

32.98 Suppose that in Fig. 32-15 the applied voltage is $v_a = 150\sin 400t$ (V), $R = 20\,\Omega$, $L = 0.06\,$H, $C = 18\,\mu$F. Give a complete analysis of the circuit.

▮
$$X_L = \omega L = (400)(0.060) = \underline{24\,\Omega} \qquad X_C = \frac{1}{\omega C} = \frac{1}{(400)(18\times 10^{-6})} = \underline{139\,\Omega}$$

$$\frac{1}{Z} = \left[\frac{1}{20^2} + \left(\frac{1}{139} - \frac{1}{24}\right)^2\right]^{1/2} \qquad \text{or} \qquad Z = \underline{16.5\,\Omega}$$

$$I_R = \frac{150}{20} = \underline{7.5\,\text{A}} \qquad I_L = \frac{150}{24} = \underline{6.25\,\text{A}} \qquad I_C = \frac{150}{139} = \underline{1.08\,\text{A}} \qquad I = \frac{150}{16.5} = \underline{9.09\,\text{A}}$$

Alternatively, $I = [I_R^2 + (I_C - I_L)^2]^{1/2}$.

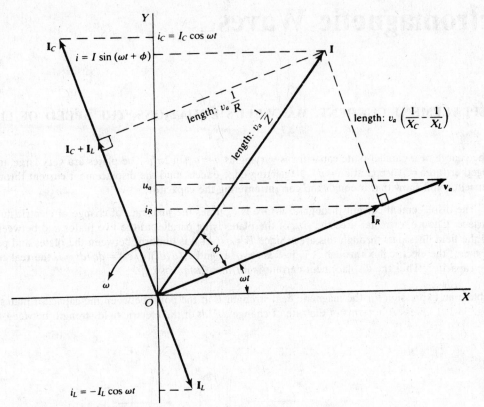

Fig. 32-16

Instantaneous values of current are:

$$i_R = \underline{7.5 \sin 400t} \ \text{A} \qquad i_L = \underline{-6.25 \cos 400t} \ \text{A} \qquad i_C = \underline{1.08 \cos 400t} \ \text{A}$$

The phase angle is given by

$$\tan \phi = R\left(\frac{1}{X_C} - \frac{1}{X_L}\right) = -0.689 \quad \text{or} \quad \phi = \underline{-0.604 \ \text{rad}} \qquad i = \underline{9.09 \sin (400t - 0.604)} \ \text{A}$$

The average power taken by the circuit is $P_{\text{avg}} = (I_{R,\text{rms}})^2 R = (7.5/\sqrt{2})^2 (20) = \underline{562.5 \ \text{W}}$.

CHAPTER 33
Electromagnetic Waves

33.1 DISPLACEMENT CURRENT, MAXWELL'S EQUATIONS, THE SPEED OF LIGHT

33.1ᶜ The charge of a parallel-plate capacitor is varying as $q = q_0 \sin 2\pi f t$. The plates are very large and close together (area $= A$, separation $= d$). Neglecting edge effects, find the displacement current through the capacitor and show that it equals the current entering the capacitor.

❚ The displacement current through a surface is ϵ_0 times the time rate of change of electric flux through the surface. Choose a circular area the size of the plates lying parallel to the two plates and between them so that all the field lines pass through this area. Since $E = q/(A\epsilon_0)$ is uniform between the plates and perpendicular to them, the electric flux through A is just $EA = q/\epsilon_0$ and $\epsilon_0(d/dt)(q/\epsilon_0) = dq/dt = i$, the real current entering the capacitor. Thus the displacement current is $dq/dt = 2\pi f q_0 \cos 2\pi f t$.

33.2ᶜ Obtain an expression for the magnetic field strength B at the point between the capacitor plates indicated in Fig. 33-1. Express B in terms of the rate of change dE/dt of the electric field strength between the plates.

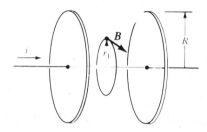

Fig. 33-1

❚ Consider a circle of radius r_1 inside the plates in a plane parallel to them. Applying Ampère's law, as modified by Maxwell to include displacement current, to this closed curve and to the plane enclosed by the circle, you have

$$\int_{\substack{\text{closed} \\ \text{curve}}} \mathbf{B} \cdot d\mathbf{l} = \mu_0 \left(\int_{\substack{\text{enclosed} \\ \text{surface}}} \mathbf{J} \cdot d\mathbf{a} + \epsilon_0 \int_{\substack{\text{enclosed} \\ \text{surface}}} \frac{\partial \mathbf{E}}{\partial t} \cdot d\mathbf{a} \right) \tag{1}$$

Between the plates, $\mathbf{J} = 0$ and $\mathbf{E}$ is spatially uniform; therefore (1) becomes

$$B 2\pi r_1 = \mu_0 \epsilon_0 \frac{dE}{dt} \pi r_1^2$$

or
$$B = \frac{\mu_0 \epsilon_0 r_1}{2} \frac{dE}{dt} \qquad \text{for} \qquad r_1 \le R \tag{2}$$

33.3 Refer to Prob. 33.2. Evaluate B for $r_1 = R = 0.100$ m and $dE/dt = 1.00 \times 10^{10}$ V/m · s.

❚ With $\mu_0 = 4\pi \times 10^{-7}$ T · m/A and $\epsilon_0 = 8.85 \times 10^{-12}$ C²/N · m², Eq. (2) of Prob. 32.2 gives

$$B = \frac{(4\pi \times 10^{-7} \text{ T} \cdot \text{m/A})(8.85 \times 10^{-12} \text{ C}^2/\text{N} \cdot \text{m}^2)(1.00 \times 10^{-1} \text{ m})(1.00 \times 10^{10} \text{ V/m} \cdot \text{s})}{2} = \underline{5.56 \times 10^{-9} \text{ T}}$$

33.4ᶜ Write the complete set of *Maxwell's equations in integral form*, and briefly explain the significance of each one. Assume that a charge-density distribution and current-density distribution exist in the region of interest, and that $\mu = \mu_0$, $\epsilon = \epsilon_0$ for the medium involved.

❚ The set can be written

$$\int_{\substack{\text{closed} \\ \text{surface}}} \mathbf{E} \cdot d\mathbf{a} = \frac{1}{\epsilon_0} \int_{\substack{\text{enclosed} \\ \text{volume}}} \rho \, dv \tag{1}$$

$$\int_{\substack{\text{closed}\\\text{surface}}} \mathbf{B} \cdot d\mathbf{a} = 0 \tag{2}$$

$$\int_{\substack{\text{closed}\\\text{curve}}} \mathbf{E} \cdot d\mathbf{l} = -\int_{\substack{\text{enclosed}\\\text{surface}}} \frac{\partial \mathbf{B}}{\partial t} \cdot d\mathbf{a} \tag{3}$$

$$\int_{\substack{\text{closed}\\\text{curve}}} \mathbf{B} \cdot d\mathbf{l} = \mu_0\left(\int_{\substack{\text{enclosed}\\\text{surface}}} \mathbf{J} \cdot d\mathbf{a} + \epsilon_0\int_{\substack{\text{enclosed}\\\text{surface}}} \frac{\partial \mathbf{E}}{\partial t} \cdot d\mathbf{a}\right) \tag{4}$$

Equation (1) is just Gauss' law for the electric field $\mathbf{E}$; it is tantamount to Coulomb's law.

Equation (2) is Gauss' law for the magnetic field $\mathbf{B}$; its basis is the experimental observation that there are no magnetic monopoles, and hence no net field lines can emanate from a closed surface.

Equation (3) is just the integral form of Faraday's law of induction.

Equation (4) is Ampère's circuital law, with conduction and displacement currents on an equal footing.

The complete experimental basis for electromagnetism is summarized in these four equations, together with the equation defining $\mathbf{E}$ and $\mathbf{B}$ in terms of the force $\mathbf{F}$ acting on a test charge q moving at velocity $\mathbf{v}$. This is the Lorentz force, given by

$$\mathbf{F} = q\mathbf{E} + q\mathbf{v} \times \mathbf{B} \tag{5}$$

33.5^c For free space, write down *Maxwell's equations in differential form.*

❙ For a region of space free of charges and currents, (1) through (4) of Prob. 33.4 have the respective differential forms:

$$\frac{\partial E_x}{\partial x} + \frac{\partial E_y}{\partial y} + \frac{\partial E_z}{\partial z} = 0 \tag{1}$$

$$\frac{\partial B_x}{\partial x} + \frac{\partial B_y}{\partial y} + \frac{\partial B_z}{\partial z} = 0 \tag{2}$$

$$\frac{\partial E_z}{\partial y} - \frac{\partial E_y}{\partial z} = -\frac{\partial B_x}{\partial t} \tag{3a}$$

$$\frac{\partial E_x}{\partial z} - \frac{\partial E_z}{\partial x} = -\frac{\partial B_y}{\partial t} \tag{3b}$$

$$\frac{\partial E_y}{\partial x} - \frac{\partial E_x}{\partial y} = -\frac{\partial B_z}{\partial t} \tag{3c}$$

$$\frac{\partial B_z}{\partial y} - \frac{\partial B_y}{\partial z} = \epsilon_0\mu_0\frac{\partial E_x}{\partial t} \tag{4a}$$

$$\frac{\partial B_x}{\partial z} - \frac{\partial B_z}{\partial x} = \epsilon_0\mu_0\frac{\partial E_y}{\partial t} \tag{4b}$$

$$\frac{\partial B_y}{\partial x} - \frac{\partial B_x}{\partial y} = \epsilon_0\mu_0\frac{\partial E_z}{\partial t} \tag{4c}$$

33.6^c How do Maxwell's equations imply that electromagnetic waves exist in free space and that light is such a wave?

❙ By suitably differentiating and combining the eight differential equations of Prob. 33.5 (see Prob. 33.23), one can obtain a single second-order equation:

$$\frac{\partial^2\psi}{\partial x^2} + \frac{\partial^2\psi}{\partial y^2} + \frac{\partial^2\psi}{\partial z^2} = \epsilon_0\mu_0\frac{\partial^2\psi}{\partial t^2}$$

where ψ represents any component of $\mathbf{E}$ or of $\mathbf{B}$. But this is the well-known differential equation of three-dimensional wave motion (see Prob. 33.21), with speed of propagation

$$v = \frac{1}{\sqrt{\epsilon_0\mu_0}} = \frac{1}{\sqrt{(8.854 \times 10^{-12}\,\text{C}^2/\text{N}\cdot\text{m}^2)(4\pi \times 10^{-7}\,\text{H/m})}} = 2.998 \times 10^8\,\text{m/s} = c \quad \text{(experimental speed)}$$

33.7 Check the units given v in Prob. 33.6.

❙ Since $1\,\text{H} = 1\,\text{J/A}^2 = 1\,\text{N}\cdot\text{m}\cdot\text{s}^2/\text{C}^2$,

$$[v] = \frac{1}{\sqrt{(\text{C}^2/\text{N}\cdot\text{m}^2)(\text{N}\cdot\text{s}^2/\text{C}^2)}} = \frac{1}{\sqrt{\text{s}^2/\text{m}^2}} = \text{m/s}$$

33.8 A laser beam is sent to the moon and reflected back to earth by a mirror placed on the moon by an astronaut. If the moon is 240 000 mi from earth, how long does it take the light to make the round trip?

▌ Let S be the round-trip distance. Then $S = ct$, and $2(240\,000) = (186\,000 \text{ mi/s})t$, and $t = \underline{2.58 \text{ s}}$.

33.9 An astronomical unit of distance is the light-year, which is the distance light travels in a year. Compute this distance in kilometers.

▌ 1 light year $= (2.998 \times 10^5 \text{ km/s})(1 \text{ yr})(365.25 \text{ d/yr})(24 \text{ h/d})(3600 \text{ s/h}) = \underline{9.47 \times 10^{12} \text{ km}}$

33.10 What minimum frequency of rotation is necessary for an eight-sided mirror used in measuring the speed of light with Michelson's arrangement (Fig. 33-2) if the distance from the fixed mirror to the rotating one is 35.4 km?

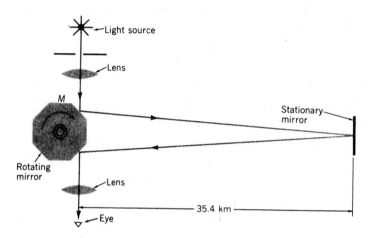

Fig. 33-2

▌ The mirror must rotate $\frac{1}{8}$ turn in the time it takes light to travel to the stationary mirror and back. Letting $f =$ frequency of rotation, we thus have

$$t = \frac{70.8 \text{ km}}{3 \times 10^5 \text{ km/s}} = \frac{1}{8f} \quad \text{so} \quad f = 529.7 \text{ rev/s}$$

33.11 In measuring the velocity of light by Michelson's method, a 12-sided mirror is rotated at 500 rev/s. If the velocity of light is 300 000 km/s, what is the distance from the rotating mirror to the fixed mirror?

▌ Proceeding as in Prob. 33.10, the round-trip time is

$$\text{time} = \frac{1}{12}\left(\frac{1}{500}\right) = \frac{1}{6000} \text{ s} \quad S = ct = (300\,000)\left(\frac{1}{6000}\right) = 50 \text{ km} \quad \frac{S}{2} = \underline{25 \text{ km}}$$

33.12 At a certain location on earth the sun appears to cross the meridian at 12 noon. At what angular position was the sun when it emitted the observed light?

▌ The sun is approximately 92×10^6 mi from earth. Since $c = 186\,000$ mi/s, light takes about 495 s ($8\frac{1}{4}$ min) to reach earth. The sun's angular position was somewhat east of the meridian. To get the exact longitude, measured east from the meridian, we note that the earth turns once in 24 h $= 86\,400$ s. Then if θ is the angle in question: $\theta/360 = 495/86\,400$, or $\theta = \underline{2.06°}$ east of the meridian.

33.13 Fizeau devised a method of measuring the speed of light in which a beam of light passed between two teeth of a toothed wheel, was reflected back by a mirror, and arrived back at the wheel just in time to pass through the next slot between the teeth. Fizeau's wheel had 720 teeth, and the mirror was 8.6 km from the toothed wheel. He observed no light returning through the wheel when it was rotating at 12.6 rev/s and maximum brightness of returning light at 25.2 rev/s. Find Fizeau's value for the speed of light.

▌ We equate the round-trip time for light to $\frac{1}{720}$ of the period of rotation, so

$$t = \frac{2 \times 8.6 \text{ km}}{c} = \frac{1}{25.2 \times 720 \text{ s}^{-1}} \quad \text{and} \quad c = \underline{3.12 \times 10^5 \text{ km/s}}$$

33.2 MATHEMATICAL DESCRIPTION OF WAVES IN ONE AND THREE DIMENSIONS

33.14 Show that $f(x - vt)$ is a progressive wave moving in the positive x direction with an unchanging profile.

❚ Set up a coordinate system s' moving to the right with the disturbance at a speed v, as in Fig. 33-3. At $t = 0$, the two systems s and s' overlapped and so $x' = x - vt$. In s' the wave function is independent of time. Therefore, in s' the profile is unchanging and in s it moves to the right undistorted with speed v.

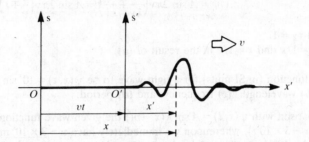

Fig. 33-3

33.15c Show that $\psi(x, t) = f(x \mp vt)$ is a solution of the one-dimensional differential wave equation.

❚ Here f is a function of x', where $x' \equiv x \mp vt$ is in turn a function of x and t. Thus, using the chain rule,

$$\frac{\partial \psi}{\partial x} = \frac{\partial f}{\partial x'} \frac{\partial x'}{\partial x} = \frac{\partial f}{\partial x'} \qquad \text{and} \qquad \frac{\partial \psi}{\partial t} = \frac{\partial f}{\partial x'} \frac{\partial x'}{\partial t} = \mp v \frac{\partial f}{\partial x'}$$

and so

$$\frac{\partial^2 \psi}{\partial x^2} = \frac{\partial^2 f}{\partial x'^2} \qquad \text{while} \qquad \frac{\partial^2 \psi}{\partial t^2} = \frac{\partial}{\partial t}\left(\mp v \frac{\partial f}{\partial x'}\right) = \mp v \frac{\partial}{\partial x'}\left(\frac{\partial f}{\partial t}\right)$$

Hence

$$\frac{\partial^2 \psi}{\partial t^2} = \mp v \frac{\partial}{\partial x'}\left(\mp v \frac{\partial f}{\partial x'}\right) = v^2 \frac{\partial^2 f}{\partial x'^2} = v^2 \frac{\partial^2 \psi}{\partial x^2} \qquad \text{or} \qquad \frac{\partial^2 \psi}{\partial x^2} = \frac{1}{v^2} \frac{\partial^2 \psi}{\partial t^2}$$

33.16c If $\psi_1(x, t)$ and $\psi_2(x, t)$ are both solutions of the differential wave equation, show that $\psi_1(x, t) + \psi_2(x, t)$ is also a solution.

❚ Since both ψ_1 and ψ_2 are solutions,

$$\frac{\partial^2 \psi_1}{\partial x^2} = \frac{1}{v^2} \frac{\partial^2 \psi_1}{\partial t^2} \qquad \text{and} \qquad \frac{\partial^2 \psi_2}{\partial x^2} = \frac{1}{v^2} \frac{\partial^2 \psi_2}{\partial t^2}$$

Adding these together yields

$$\frac{\partial^2 \psi_1}{\partial x^2} + \frac{\partial^2 \psi_2}{\partial x^2} = \frac{1}{v^2}\left(\frac{\partial^2 \psi_1}{\partial t^2} + \frac{\partial^2 \psi_2}{\partial t^2}\right) \qquad \text{or} \qquad \frac{\partial^2}{\partial x^2}(\psi_1 + \psi_2) = \frac{1}{v^2}\frac{\partial^2}{\partial t^2}(\psi_1 + \psi_2)$$

The above result is the *superposition principle* for the one-dimensional wave equation.

33.17 Show that for a harmonic wave, $\psi(x, t) = A \sin(kx \mp \omega t)$, the repetitive nature in space, $\psi(x, t) = \psi(x \pm \lambda, t)$, requires that $k = 2\pi/\lambda$.

❚ We know that the sine function repeats itself when the argument increases or decreases by 2π. Thus

$$A \sin k(x - vt) = A \sin k[(x \pm \lambda) - vt] = A \sin[k(x - vt) \pm 2\pi]$$

The second equality gives $|k\lambda| = 2\pi$, or since both k and λ are positive, $k = 2\pi/\lambda$.

33.18 Verify that the harmonic wave function $\psi(x, t) = A \sin(kx - \omega t)$ is a solution of the one-dimensional differential wave equation with $v = \omega/k$.

❚ Since $\psi(x, t) = A \sin k(x - vt) = f(x - vt)$, the result follows at once from Prob. 33.15.

33.19 Prove that a progressive harmonic wave can be described alternatively by

$$(\boldsymbol{a}) \;\; \psi = A \sin 2\pi\left(\frac{x}{\lambda} \mp \frac{t}{\tau}\right) \qquad (\boldsymbol{b}) \;\; \psi = A \sin 2\pi v\left(\frac{x}{v} \mp t\right) \qquad (\boldsymbol{c}) \;\; \psi = A \sin 2\pi(\kappa x \mp vt)$$

where $\kappa \equiv 1/\lambda$, and λ, τ are wavelength and period, respectively.

▌ (a) Starting with $\psi = A \sin k(x \mp vt)$, use $k = 2\pi/\lambda$ to obtain

$$\psi = A \sin 2\pi\left(\frac{x}{\lambda} \mp \frac{vt}{\lambda}\right)$$

But $v/\lambda = \nu = 1/\tau$.
(b) From the result of (a):

$$\psi = A \sin 2\pi\nu\left(\frac{x}{\lambda\nu} \mp \frac{t}{\tau\nu}\right) = A \sin 2\pi\nu\left(\frac{x}{v} \mp t\right)$$

since $\lambda\nu = v$ and $\tau\nu = 1$.
(c) Substitute $\lambda = 1/\kappa$ and $\tau = 1/\nu$ in the result of (a).

33.20 Given the wave function (in SI units) for a light wave to be $\psi(x, t) = 10^3 \sin \pi(3 \times 10^6 x - 9 \times 10^{14} t)$, determine (a) the speed, (b) wavelength, (c) frequency, and (d) period.

▌ (a) By comparison with $\psi(x, t) = A \sin k(x - vt)$ the given wave function can be written as $\psi(x, t) = 10^3 \sin 3 \times 10^6 \pi(x - 3 \times 10^8 t)$, whereupon we immediately have $v = \underline{3 \times 10^8 \text{ m/s}}$.
(b) By (a), $k = 3 \times 10^6 \pi \text{ m}^{-1}$. Therefore, $\lambda = 2\pi/k = \underline{666 \text{ nm}}$.

(c)
$$\nu = \frac{v}{\lambda} = \frac{3 \times 10^8 \text{ m/s}}{\frac{2}{3} \times 10^{-6} \text{ m}} = \underline{4.5 \times 10^{14} \text{ Hz}}$$

(d)
$$\tau = 1/\nu = \underline{2.2 \times 10^{-15} \text{ s}}$$

33.21 How may the one-dimensional wave equation be generalized to three dimensions, and what would the generalization of harmonic waves be?

▌ We can generalize the differential wave equation to three dimensions by noting that the space variables should appear symmetrically. That is, the equation should not change if we interchange the space variables, as long as the coordinate system remains right-handed. In any event

$$\frac{\partial^2 \psi}{\partial x^2} + \frac{\partial^2 \psi}{\partial y^2} + \frac{\partial^2 \psi}{\partial z^2} = \frac{1}{v^2}\frac{\partial^2 \psi}{\partial t^2}$$

is the appropriate three-dimensional form in cartesian coordinates.

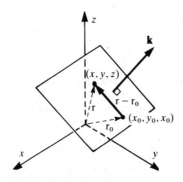

Fig. 33-4

We now write the equation for a plane passing through an arbitrary point (x_0, y_0, z_0) and perpendicular to a given direction delineated by the *propagation vector* **k**, as in Fig. 33-4. The vector $\mathbf{r} - \mathbf{r}_0$ will sweep out the desired plane provided that

$$(\mathbf{r} - \mathbf{r}_0) \cdot \mathbf{k} = 0 \qquad \text{or} \qquad \mathbf{k} \cdot \mathbf{r} = \text{constant}$$

This is the equation of a plane and so $\psi(\mathbf{r}) = A \sin (\mathbf{k} \cdot \mathbf{r})$ is a function defined on a family of planes all perpendicular to **k**. Over each of these $\mathbf{k} \cdot \mathbf{r} = \text{constant}$, and so $\psi(\mathbf{r})$ is a constant. As we move from plane to plane, $\psi(\mathbf{r})$ varies sinusoidally. As before, to convert this into a progressive *harmonic plane wave*, we simply rewrite it as $\psi(\mathbf{r}, t) = A \sin (\mathbf{k} \cdot \mathbf{r} \mp \omega t)$. The minus sign corresponds to motion in the positive **k** direction, the plus sign to motion in the negative **k** direction. In general an arbitrary phase constant can be added to the argument of the sine function, as in one-dimension.

33.22 In three dimensions one can also have spherical waves, which diverge symmetrically from a point source (or converge toward a point). Describe spherical harmonic waves.

❚ The form of the *harmonic spherical wave* is most easily arrived at by solving the differential wave equation in spherical coordinates. That procedure leads to

$$\psi(r, t) = \frac{\mathcal{A}}{r} \sin k(r \mp vt)$$

where the constant $\mathcal{A}$ is known as the *source strength*. Observe that the amplitude $\mathcal{A}/r$ varies inversely with distance from the origin. This is a requirement of energy conservation. Again, the minus and plus signs in the phase respectively correspond to waves diverging from and converging toward the origin. The expression at any instant represents a cluster of concentric spheres, over each of which r is constant and therefore $\psi(r, t)$ is constant.

(Instead of a harmonic wave we could equally well have considered a spherical *pulse*. For example, imagine a point source which rather than oscillating harmonically just turns on, builds up, and then shuts off. The disturbance, although short-lived, would move out in all directions as a spherical pulse of some sort.)

33.23^c Starting with the differential form of Maxwell's equations in free space, show that E_x indeed obeys the three-dimensional wave equation (see Prob. 33.5) with speed $v = 1/\sqrt{\epsilon_0\mu_0} \equiv C$.

❚ Referring to the four sets of equations in Prob. 33.5, we take the time derivative of (4a), yielding

$$\frac{\partial^2 B_z}{\partial y\, \partial t} - \frac{\partial^2 B_y}{\partial z\, \partial t} = \epsilon_0\mu_0 \frac{\partial^2 E_x}{\partial t^2}$$

We then use Eqs. (3b) and (3c) to eliminate the time derivatives of B_z and B_y, yielding

$$\frac{\partial^2 E_x}{\partial y^2} - \frac{\partial^2 E_y}{\partial y\, \partial x} + \frac{\partial^2 E_x}{\partial z^2} - \frac{\partial^2 E_z}{\partial z\, \partial x} = \epsilon_0\mu_0 \frac{\partial^2 E_x}{\partial t^2}$$

From Eq. (1), taking the derivative with respect to x, we get

$$\frac{\partial^2 E_x}{\partial x^2} = -\frac{\partial^2 E_y}{\partial y\, \partial x} - \frac{\partial^2 E_z}{\partial z\, \partial x}$$

Comparing the terms involving E_y and E_z in the last two equations we get, finally,

$$\frac{\partial^2 E_x}{\partial x^2} + \frac{\partial^2 E_x}{\partial y^2} + \frac{\partial^2 E_x}{\partial z^2} = \epsilon_0\mu_0 \frac{\partial^2 E_x}{\partial t^2}$$

This is the result in Prob. 33.5, which when compared with the wave equation in Prob. 33.6 yields $v = 1/\sqrt{\epsilon_0\mu_0}$.

33.24 By Prob. 33.21, $\mathbf{k} \cdot \mathbf{r} = $ constant is the equation of a plane normal to $\mathbf{k}$ and passing through some point (x_0, y_0, z_0). Determine the form of the constant and write out the harmonic wave function in cartesian coordinates.

❚ The equation of the plane is $(\mathbf{r} - \mathbf{r}_0) \cdot \mathbf{k} = 0$. In cartesian coordinates $\mathbf{r} = [x, y, z]$, $\mathbf{r}_0 = [x_0, y_0, z_0]$, and $\mathbf{k} = [k_x, k_y, k_z]$. Hence

$$(x - x_0)k_x + (y - y_0)k_y + (z - z_0)k_z = 0 \quad \text{or} \quad xk_x + yk_y + zk_z = x_0k_x + y_0k_y + z_0k_z$$

The left side of this last equation is $\mathbf{k} \cdot \mathbf{r}$, while the right side is the constant in question. The harmonic wave function $\psi(\mathbf{r}, t) = A \sin(\mathbf{k} \cdot \mathbf{r} \mp \omega t)$ becomes

$$\psi(x, y, z, t) = A \sin(xk_x + yk_y + zk_z \mp \omega t)$$

33.25^c Write the planar harmonic wave function in cartesian coordinates in terms of direction cosines α, β, γ, where $k_x = \alpha k$, $k_y = \beta k$, $k_z = \gamma k$, and $\alpha^2 + \beta^2 + \gamma^2 = 1$. Then show that the function is a solution of the three-dimensional differential wave equation.

❚ Beginning with $\psi(\mathbf{r}, t) = A \sin(k_x x + k_y y + k_z z - \omega t)$, we replace k_x, k_y and k_z by the corresponding direction cosine terms. As shown in Fig. 33-5

$$k_x = k \cos\theta_1 = k\alpha \qquad k_y = k \cos\theta_2 = k\beta \qquad k_z = k \cos\theta_3 = k\gamma$$

where $\quad k_x^2 + k_y^2 + k_z^2 = k^2(\alpha^2 + \beta^2 + \gamma^2) = k^2 \quad$ Hence $\quad \psi(\mathbf{r}, t) = A \sin[k(\alpha x + \beta y + \gamma z) - \omega t]$

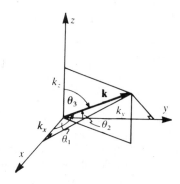

Fig. 33-5

Now to check that this is a solution of the wave equation, take the appropriate derivatives:

$$\frac{\partial^2\psi}{\partial x^2} = -\alpha^2 k^2 \psi \qquad \frac{\partial^2\psi}{\partial y^2} = -\beta^2 k^2 \psi \qquad \frac{\partial^2\psi}{\partial z^2} = -\gamma^2 k^2 \psi \qquad \text{and} \qquad \frac{\partial^2\psi}{\partial t^2} = -\omega^2 \psi$$

Adding the first three equations and making use of $\alpha^2 + \beta^2 + \gamma^2 = 1$ and $k = 2\pi/\lambda = \omega/v$, we obtain

$$\frac{\partial^2\psi}{\partial x^2} + \frac{\partial^2\psi}{\partial y^2} + \frac{\partial^2\psi}{\partial z^2} = -k^2 \psi = \frac{1}{v^2}\frac{\partial^2\psi}{\partial t^2}$$

33.26   (*a*) Find a solution of the three-dimensional wave equation that exhibits the direction of the *disturbance*, as well as the direction of propagation. (*b*) If the disturbance resides in a plane, known as the *plane of vibration*, the wave is said to be *plane polarized* or *linearly polarized*. Write an expression for a linearly polarized harmonic plane wave.

❚   (*a*) The direction of the displacement in a transverse wave can be fixed by making the amplitude a vector: $\psi(\mathbf{r}, t) = \mathbf{A}\sin(\mathbf{k}\cdot\mathbf{r} - \omega t)$, where now $\psi(x, t)$ is spoken of as the *wave vector*. The vectors $\mathbf{A}$ and $\mathbf{k}$ determine the plane of vibration at any instant in time. (*b*) For a linearly polarized harmonic plane wave, $\mathbf{A}$ is constant and $\mathbf{k}\cdot\mathbf{A} = 0$. Figure 33-6 shows several planar wave fronts normal to $\mathbf{k}$. It depicts a harmonic plane wave, so that $\psi(\mathbf{r}, t)$ varies sinusoidally from one plane to the next. Furthermore it's linearly polarized, so that at all points on any planar wave front the amplitude vector is identical and the corresponding planes of vibration are all parallel. If the amplitude vector is a function of time which varies sufficiently rapidly and randomly, the wave is said to be *unpolarized*.

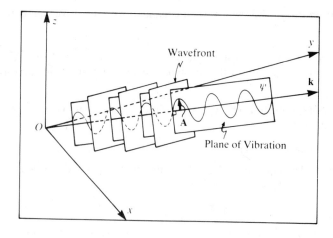

Fig. 33-6

33.27   Because of the spatial repetition in a harmonic plane wave, we can expect that $\psi(\mathbf{r}, 0) = \psi(\mathbf{r} + \lambda\mathbf{k}/k, 0)$. In other words, the profile at one location in space is identical to the profile a distance λ farther along in the direction of the unit propagation vector, $\mathbf{k}/k$. Use this to show that $|\mathbf{k}| = k = 2\pi/\lambda$.

❚   For any fixed time t, $\psi(\mathbf{r}, t) = A\sin(\mathbf{k}\cdot\mathbf{r} - \omega t)$ and $\psi[(\mathbf{r} + \lambda\mathbf{k}/k), t] = A\sin(\mathbf{k}\cdot[\mathbf{r} + \lambda\mathbf{k}/k] - \omega t) = A\sin(\mathbf{k}\cdot\mathbf{r} + \lambda k - \omega t)$. Equating the two yields, for the minimum repeat distance λ, $\lambda k = 2\pi$, or $k = 2\pi/\lambda$, as for one dimension.

33.28 What is meant by a *wave front*?

▎ A surface over which the phase of a wave is constant is called a *wave front*. Thus, for a plane wave, a wave front is given by $\mathbf{k} \cdot \mathbf{r} - \omega t = \alpha$. This is the equation of a plane translating with speed $\omega/k = v$ in the direction of $\mathbf{k}$; at time $t = 0$, the plane is at a distance $|\alpha|/k$ from the origin. Similarly (see Prob. 33.22), a (expanding) spherical wave has wave fronts $r - vt = \beta$, which are spherical shells expanding with speed v.

33.29 (*a*) Draw a sketch of a planar wave front propagating in the positive x direction. (*b*) Write an expression for a harmonic plane wave of this sort moving along the x axis.

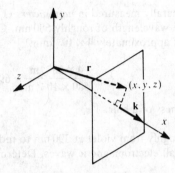

Fig. 33-7

▎ (*a*) See Fig. 33-7. (*b*) The wave function in general is $\psi(\mathbf{r}, t) = A \sin (\mathbf{k} \cdot \mathbf{r} - \omega t)$, or in cartesian coordinates (see Prob. 33.24), $\psi(\mathbf{r}, t) = A \sin (k_x x + k_y y + k_z z - \omega t)$. But here $\mathbf{k}$ is along x, so that $k_x = k$, $k_y = k_z = 0$, and $\psi(\mathbf{r}, t) = A \sin (kx - \omega t)$.

33.30 To photograph a distant source of spherical light waves you would set the camera lens to focus at ∞, i.e., to receive incoming plane waves. Why?

▎ See Fig. 33-8. Over the relatively tiny area of a remote detector these wave fronts appear planar.

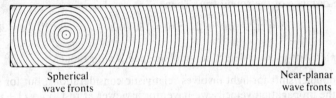

Spherical
wave fronts

Near-planar
wave fronts   **Fig. 33-8**

33.31 A certain radio station transmits at a frequency of 900 kHz. What is the wavelength of its waves? How many wave crests pass a point each second if the point is 50 km from the station?

▎ The wavelength $\lambda = c/f = (3.00 \times 10^8)/(9.00 \times 10^5) = \underline{333\ m}$. The wave crests past a point correspond to the number of wavelengths per second, the frequency $\underline{9.00 \times 10^5\ Hz}$ (regardless of distance).

33.32 When you are tuned to a radio station which is 161 km (100 mi) away, how long does it take for an electromagnetic signal to travel from the station to you? If the station operates at 10^6 Hz, how many wave crests does it send out each second? How far does a wave crest travel in a second through free space? What is λ for this wave?

▎ Travel time $t = s/c = (1.61 \times 10^5)/(3 \times 10^8) = \underline{540\ \mu s}$. It sends out one crest per cycle or $\underline{10^6\ \text{crests per second}}$ and moves a distance $\underline{3 \times 10^8\ m}$ in 1 s, the distance between crests is $\lambda = (3 \times 10^8)/10^6 = \underline{300\ m}$.

33.33 A standard radio broadcasting station has an assigned frequency between 535 and 1605 kHz. The VHF (*very-high-frequency*) television stations (channels 2 to 13) have frequencies between 54 and 216 MHz, while the UHF (*u = ultra*) stations (channels 14 to 83) have frequencies between 470 and 890 MHz. What is the wavelength corresponding to each of the frequencies mentioned?

❙ In all cases the waves move with the speed of light, so

$$\lambda = \frac{v}{\nu} = \frac{3 \times 10^8 \text{ m/s}}{\nu} \quad \text{yields}$$

ν, MHz	λ, m	ν	λ
0.535	561	216	1.39
1.605	187	470	0.638
54	5.56	890	0.337

33.34 The wavelength of light is generally measured in *nanometers* (1 nm = 10^{-9} m). For example, yellow, which is just about midspectrum, has a wavelength of roughly 580 nm. Compare this with the thickness of a human hair (from the head), which is approximately 4×10^{-2} mm.

$$\frac{4 \times 10^{-5} \text{ m}}{580 \times 10^{-9} \text{ m}} = 69$$

i.e., 69 wavelengths per thickness of a hair.

33.35 Light ranges in wavelength roughly from violet at 390 nm to red at 780 nm. Its speed in vacuum is about 3×10^8 m/s, as is the case for all electromagnetic waves. Determine the corresponding frequency range.

❙ Since $v = \nu\lambda$,

$$\nu_{\text{vio}} = \frac{3 \times 10^8 \text{ m/s}}{390 \times 10^{-9} \text{ m}} = 7.7 \times 10^{14} \text{ Hz} \quad \text{and} \quad \nu_{\text{red}} = \frac{3 \times 10^8 \text{ m/s}}{780 \times 10^{-9} \text{ m}} = 3.8 \times 10^{14} \text{ Hz}$$

The frequency range is then from 380 to 770 THz (1 terahertz = 10^{12} Hz = 1 THz).

33.36 The speed of light is 186 000 mi/s. A radar system sends out pulses of extremely short radio waves, which move with the speed of light. How many microseconds after a pulse is transmitted will an echo from an airplane 18.6 mi away from the radar station be received?

❙ Let S be the distance of the round trip. Then

$$S = ct \quad \text{or} \quad 2(18.6) = 186\,000t \quad t = 2 \times 10^{-4} \text{ s} = \underline{200\ \mu\text{s}}.$$

33.37 How is the doppler shift used to detect automobile speeds with radar?

❙ The correct doppler shift for light involves relativistic considerations, but for source velocities small compared with the propagation velocity we have, for a viewer at rest, $v_L \approx (1 + v/c)v_s$. Here v is the velocity of the source (positive toward viewer) and v_L is the frequency perceived by the viewer. If light is reflected from a moving mirror (automobile), the reflected frequency is that of a source moving with the velocity of the image, or twice the velocity of the mirror.

   When the doppler effect is used to measure the speed of an automobile, the reflected wave is beat against the transmitted frequency to give the small frequency difference.

$$F = v_r - v \approx \frac{2vv}{c} \quad \text{or} \quad v \approx \frac{Fc}{2v}$$

which gives the automobile's speed in terms of the observed beat frequency, the speed of light, and the known transmitted frequency.

33.38 A radar set for determining the speed of approaching cars sends out waves of frequency 1000 MHz. If the beat frequency observed is 150 Hz, what is the speed of the car?

❙ From Prob. 33.37,

$$v = \frac{(150 \text{ Hz})(186\,000 \text{ mi/s})(3600 \text{ s/h})}{2(10^9 \text{ Hz})} = 50.2 \text{ mi/h}$$

33.39 Briefly discuss waves *in two dimensions*.

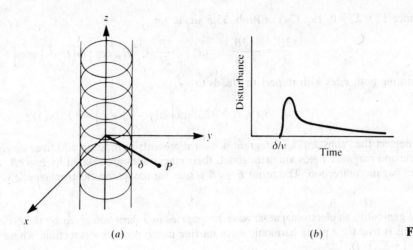

(a)   *(b)*   **Fig. 33-9**

▌ The two-dimensional wave equation has the expected form

$$\frac{\partial^2 \psi}{\partial x^2} + \frac{\partial^2 \psi}{\partial y^2} = \frac{1}{v^2}\frac{\partial^2 \psi}{\partial t^2} \tag{1}$$

but physically significant solutions have different properties from those of odd-dimensional waves. To see this, we note that in one dimension we have solutions of the form $f(x \pm vt)$ representing a disturbance traveling undistorted along the x axis, emanating from a point source at, say, the origin. In three dimensions, the corresponding undistorted waves due to a point source at the origin are the spherical waves $(1/r)f(r - vt)$. These diminish in amplitude but maintain their shape. The analogue for a point source exists in two dimensions. The emitted waves always distort, typically picking up a long trailing tail. See Fig. 33-9.

33.3 THE COMPONENT FIELDS OF AN ELECTROMAGNETIC WAVE; INDUCED EMF

33.40^c A plane electromagnetic wave is one where the electric and magnetic fields are constant on planes perpendicular to the direction of propagation. Show that such a wave must have its electric field *transverse* to the propagation direction.

▌ If the wave propagates in the z direction, the electric field must be independent of x and y; that is, $\mathbf{E} = \mathbf{E}(z, t)$. Then (1) of Prob. 33.5 leads to $\partial E_z/\partial z = 0$, since $\mathbf{E}$ is not a function of either x or y. This means that $E_z = $ constant and is therefore of no interest. We are concerned only with the electromagnetic wave, which must vary along z. Thus the wave can possess only x and y components, and so $\mathbf{E}$ is transverse.

33.41^c Suppose that we have a linearly polarized electromagnetic plane wave (Prob 33.26) whose electric field is of the form $\mathbf{E} = E_x(z, t)\mathbf{i}$. Show that $\mathbf{B} = B_y(z, t)\mathbf{j}$.

▌ The electric field is polarized along the x axis, as indicated by the presence of only an $\mathbf{i}$ term. Moreover, the wave propagates in the z direction. Since $E_y = E_z = 0$ and $E_x = E_x(z, t)$, Eqs. (3) of Prob. 33.5 become

$$0 = -\frac{\partial B_x}{\partial t} \qquad \frac{\partial E_x}{\partial z} = -\frac{\partial B_y}{\partial t} \qquad 0 = -\frac{\partial B_z}{\partial t}$$

This means that B_x and B_z are constant in time and of no concern to us. Hence B_y is the only time-varying term, and so $\mathbf{B} = B_y(z, t)\mathbf{j}$ is the wave's magnetic component. Note that $\mathbf{E}$ and $\mathbf{B}$ are perpendicular to each other and to the propagation direction as well.

33.42^c Given a harmonic plane electromagnetic wave whose $\mathbf{E}$ field has the form

$$E_z(y, t) = E_{0z} \sin\left[\omega\left(t - \frac{y}{c}\right) + \epsilon\right] \qquad \text{(where } \epsilon \text{ is a constant phase angle)}$$

determine the corresponding $\mathbf{B}$ field and make a sketch of the wave.

▮ Since $E_x = E_y = 0$, Eq. (3a) of Prob. 33.5 yields

$$\frac{\partial E_z}{\partial y} = -\frac{\partial B_x}{\partial t} \quad \text{or} \quad \frac{\partial B_x}{\partial t} = \frac{\omega}{c} E_{0z} \cos\left[\omega\left(t - \frac{y}{c}\right) + \epsilon\right]$$

Integrating both sides with respect to t leads to

$$B_x(y, t) = \frac{1}{c} E_{0z} \sin\left[\omega\left(t - \frac{y}{c}\right) + \epsilon\right] = \frac{1}{c} E_z(y, t)$$

(We neglect the "constant" of integration, as it represents at most an additive, *steady* magnetic field.) The electric and magnetic fields are orthogonal, their magnitudes are related by $E = cB$, and both are normal to the propagation direction. The result $E = cB$ is true for any plane electromagnetic wave propagating through space.

33.43ᶜ Quite generally, an electromagnetic wave propagates in a direction given by the cross product $\mathbf{E} \times \mathbf{B}$. Prove that this is true for a plane harmonic wave moving in the positive x direction, whose $\mathbf{E}$ field is $E(x, t) = E_z(x, t)$.

▮ Maxwell's equations (3) of Prob. 33.5 become

$$0 = \frac{\partial B_x}{\partial t} \qquad -\frac{\partial E_z}{\partial x} = -\frac{\partial B_y}{\partial t} \qquad 0 = \frac{\partial B_z}{\partial t}$$

These imply, as in Prob. 33.42, that $B_x = B_z = 0$, $B_y = -E_z/c$; in vector form, $\mathbf{E} = E_z\mathbf{k}$, $\mathbf{B} = -(E_z/c)\mathbf{j}$, and $\mathbf{E} \times \mathbf{B} = -(E_z^2/c)\mathbf{k} \times \mathbf{j}$. Since $-\mathbf{k} \times \mathbf{j} = \mathbf{i}$, all is in agreement.

33.44 Write the equation for the traveling electric field wave from a 1.20-MHz radio station. Assume that you are far enough from the station that the wave is a plane wave traveling out along the $+x$ axis. The rms field in the wave is 3.0 mV/m. Repeat if the wave is traveling in the $-x$ direction.

▮ Use the standard form $E = E_0 \sin \omega[t - (x/c)]$. We have $\omega = 2\pi(1.20 \times 10^6) = 7.54 \times 10^6$ rad/s, $c = 3 \times 10^8$ m/s, and $E_0 = \sqrt{2}(3.0) = 4.24$ mV/m. Hence,

$$E = 4.24 \sin\left[(7.54 \times 10^6)\left(t - \frac{x}{3 \times 10^8}\right)\right] \quad \text{mV/m}$$

Replace x by $-x$ to reverse the direction of propagation.

33.45 Imagine an electromagnetic plane wave in vacuum whose $\mathbf{E}$ field (in SI units) is given by

$$E_x = 10^2 \sin \pi(3 \times 10^6 z - 9 \times 10^{14} t) \qquad E_y = 0 \qquad E_z = 0$$

Determine the speed, frequency, wavelength, period, initial phase, and $\mathbf{E}$-field amplitude and polarization.

▮ The wave function has the basic form $E_x(z, t) = E_{0x} \sin k(z - vt)$. Consequently, it can be reformulated as $E_x = 10^2 \sin [3 \times 10^6 \pi(z - 3 \times 10^8 t)]$, whereupon we see that $k = \underline{3 \times 10^6 \pi \text{ m}^{-1}}$ and $v = \underline{3 \times 10^8 \text{ m/s}}$. Since $k = 2\pi/\lambda = 3 \times 10^6 \pi$, $\lambda = \underline{666 \text{ nm}}$. Furthermore,

$$\nu = \frac{v}{\lambda} = \frac{3 \times 10^8}{\frac{2}{3} \times 10^{-6}} = \underline{4.5 \times 10^{14} \text{ Hz}}$$

The period τ is $\tau = 1/\nu = \underline{2.2 \times 10^{-15} \text{ s}}$, while the initial phase is evidently <u>zero</u>. The field amplitude is just $E_{0x} = \underline{10^2 \text{ V/m}}$. The wave is linearly polarized in the x direction and propagates along the z axis. This wave corresponds to red light.

33.46 Write an expression for the magnetic field associated with the wave of Prob. 33.45.

▮ The wave propagates in the z direction while the $\mathbf{E}$ field oscillates along x. In other words, the $\mathbf{E}$ field resides in the xz plane. Accordingly, since $\mathbf{B}$ is normal to both $\mathbf{E}$ and the propagation direction, it must reside in the yz plane. Thus, $B_x = 0$, $B_z = 0$, and $\mathbf{B} = B_y(z, t)\mathbf{j}$. From Prob. 33.42, $E = cB$, the application of which leads to

$$B_y(z, t) = \underline{0.33 \times 10^{-6} \sin \pi(3 \times 10^6 z - 9 \times 10^{14} t) \quad \text{T}}$$

33.47 At a certain place on the earth the electric field portion of a plane electromagnetic wave is found to vary in the following way, expressed in SI units: $E = 2 \times 10^{-4} \cos (5 \times 10^6 t)$. If the place in question is at $x = 0$ and the wave is moving in the $-x$ direction through air, what is the equation of the wave in terms of x and t? What is the equation of the accompanying magnetic field wave?

▌ Here we use the cosine form for the wave, with initial phase $= 0$. The standard form is $E = 2 \times 10^{-4} \cos [2\pi ft + (2\pi fx)/v]$, where the plus sign is for a wave traveling in the $-x$ direction. But $2\pi f = 5 \times 10^6$ and $v = 3 \times 10^8$, so $E = 2.0 \times 10^{-4} \cos (5 \times 10^6 t + 0.017x)$ V/m. Since $B = E/c$, $B = 6.7 \times 10^{-13} \cos (5 \times 10^6 t + 0.017x)$ T. The directions of E and B are mutually perpendicular and in the yz plane, with $\mathbf{E}$, $\mathbf{B}$, and $-\hat{\mathbf{x}}$ forming a right-handed triplet.

33.48 The electric field portion of a particular electromagnetic wave is given by the equation (all variables in SI units) $E = 10^{-4} \sin (6 \times 10^5 t - 0.01x)$. Find the frequency of the wave and its speed. What is the maximum value of the electric field in the wave? What is the equation for the accompanying magnetic field wave?

▌ The general wave form is $E = E_0 \sin [2\pi ft - (2\pi x)/\lambda]$. Comparison shows $2\pi f = 6 \times 10^5$, giving $f = $ 95.5 kHz. Also, $(2\pi x)/\lambda = 0.01x$, so $\lambda = 628$ m. But $\lambda = v/f$, so $v = 6.0 \times 10^7$ m/s. The amplitude is 10^{-4} V/m, the maximum value of E. For the magnetic equation $B = E/v = 16.7 \times 10^{-13} \sin (6 \times 10^5 t - 0.01x)$ T. $\mathbf{E}$, $\mathbf{B}$, and the positive x axis form a right-handed triplet of directions. Note that $v < c$ implies a wave traveling in a medium.

33.49 A plane electromagnetic harmonic wave of frequency 600×10^{12} Hz (green light), propagating in the positive x direction in vacuum, has an electric field amplitude of 42.42 V/m. The wave is linearly polarized such that the plane of vibration of the electric field is at $45°$ to the xz plane. Write expressions for $\mathbf{E}$ and $\mathbf{B}$.

▌ The amplitude of the $\mathbf{E}$ field, E_0, is equal to $(E_{0y}^2 + E_{0z}^2)^{1/2}$, where, because of the polarization at $45°$, $E_{0y} = E_{0z}$. Thus $E_0 = 42.42 = \sqrt{2} E_{0y}$ and $E_{0y} = E_{0z} = 30$ V/m. Writing the phase in the form $\omega(t - x/c)$, the electric field becomes

$$E_x = 0 \qquad E_y = E_z = 30 \sin \left[2\pi 600 \times 10^{12} \left(t - \frac{x}{3 \times 10^8} \right) \right]$$

where, of course, $\omega = 2\pi v$. Inasmuch as $E = cB$,

$$B_x = 0, \qquad B_z = -B_y = 10^{-7} \sin \left[2\pi 600 \times 10^{12} \left(t - \frac{x}{3 \times 10^8} \right) \right]$$

Consequently,

$$\mathbf{E} = E_y \mathbf{j} + E_z \mathbf{k} = E_y (\mathbf{j} + \mathbf{k}) \qquad \mathbf{B} = B_y \mathbf{j} + B_z \mathbf{k} = B_z (-\mathbf{j} + \mathbf{k})$$

Then $\mathbf{E} \cdot \mathbf{B} = E_y B_z (-1 + 1) = 0$, so $\mathbf{E}$ is perpendicular to $\mathbf{B}$, and $\mathbf{E} \times \mathbf{B} = E_y B_z (\mathbf{i} + \mathbf{i}) = (E_y^2/c)2\mathbf{i}$ points along the positive x axis, as required.

33.50 The magnetic field of an electromagnetic wave obeys the following relation in a certain region: $B = 10^{-12} \sin (5 \times 10^6 t)$ where all quantities are in SI units. How large an emf would the field induce in a 300-turn coil of 20-cm^2 area oriented perpendicular to the field?

▌ The flux through the coil $\phi = \mathbf{B} \cdot \mathbf{A} = 20 \times 10^{-16} \sin (5 \times 10^6 t)$. But $\mathscr{E} = -N \, d\phi/dt = -300(2 \times 10^{-15})(5 \times 10^6) \cos (5 \times 10^6 t) = -3.0 \times 10^{-6} \cos (5 \times 10^6 t)$ V.

33.51 Television frequencies are of the order of 100 MHz, while radio frequencies are of order 1 MHz. Using these as typical frequencies, find the ratio of the emf generated in a loop antenna by a television wave to that generated by a radio wave if both have equal electric field intensities.

▌ The emf, $\mathscr{E}$, is proportional to dB/dt, and since $B = B_0 \sin (2\pi ft + \phi)$, dB/dt has maximum value proportional to f. Since the two waves have the same intensities, B_0 is the same for both, and the ratio of the maximum emf's $\mathscr{E}_{10}/\mathscr{E}_{20} = f_1/f_2 = 100$. Since $\mathscr{E}_0 = \sqrt{2} \mathscr{E}_{rms}$ for either wave, we have $\mathscr{E}_{1,rms}/\mathscr{E}_{2,rms} = 100$.

33.52 At the limit of detectability, the B field of a radio wave is given in SI units by the following equation: $B = 3 \times 10^{-14} \sin (6 \times 10^6 t)$. Find the emf the field induces in a coil antenna with 200 loops and 30-mm^2 cross-sectional area wound on a core with permeability of 100.

▌ The flux through the coil $\Phi = k_m BA = 100B(3.0 \times 10^{-5}) = 9.0 \times 10^{-17} \sin (6 \times 10^6 t)$. Then $\mathscr{E} = -200(d\Phi/dt) = -1.08 \times 10^{-7} \cos (6 \times 10^6 t)$ V.

33.53 At a particular place, the electric field due to a distant radio station is (in SI units) $E = 10^{-4} \cos (1 \times 10^7 t)$. What is the frequency of the wave? How large a maximum potential difference would be induced by it between the ends of a 3-m-long wire oriented in the direction of the field? How large a maximum emf would the accompanying magnetic field wave induce in a single loop with area of 35 cm²?

▮ At any instant the plane-wave field is constant over the wire length, s. Since the wave is of the form $\cos 2\pi ft$, we have $2\pi f = 1 \times 10^7$, giving $f = \underline{1.60 \times 10^6 \, \text{Hz}}$. Since $V = -Es$, we have $V_{max} = 10^{-4}(3) = \underline{3 \times 10^{-4} \, \text{V}}$. We can find B_0 from $E_0/c = 10^{-4}/c = 3.33 \times 10^{-13}$; then $\mathscr{E}_{max} = (d\Phi/dt)_{max} = A 2\pi f B_0 = 2\pi(35 \times 10^{-4})(1.60 \times 10^6)(3.33 \times 10^{-13}) = \underline{11.7 \, \text{nV}}$.

33.4 ENERGY AND MOMENTUM FLUXES

33.54 Discuss the energy associated with light waves, and define the *Poynting vector*.

▮ Electromagnetic energy flows in the direction in which the wave advances (see Prob. 33.43), i.e., in the direction of $\mathbf{E} \times \mathbf{B}$. Accordingly, the energy per unit area per unit time flowing perpendicularly into a surface in free space is given by the *Poynting vector* $\mathbf{S}$, where $\mathbf{S} = c^2 \epsilon_0 \mathbf{E} \times \mathbf{B} = (\mathbf{E} \times \mathbf{B})/\mu_0$. The SI units of $\mathbf{S}$ are W/m².

33.55 Define *radiant flux density* and *irradiance*.

▮ At optical frequencies $\mathbf{E}$, $\mathbf{B}$, and $\mathbf{S}$ all oscillate at exceedingly rapid rates and it remains impractical to measure an instantaneous value of $\mathbf{S}$ directly. Instead, one determines its average value $\langle S \rangle$ over a convenient time interval. This, in turn, is known as the *radiant flux density*. When energy is incident on a surface, the flux density is called *irradiance*, symbolized by $I \equiv \langle S \rangle$.

33.56 A laser emits a 2-mm-diameter beam of highly collimated light at a power level, or *radiant flux*, of 100 mW. Neglecting any divergence of the beam, compute its irradiance.

▮ The cross-sectional area of the beam is $\pi(10^{-3})^2$ and so

$$I = \frac{100 \times 10^{-3}}{\pi(10^{-3})^2} = \underline{31.8 \, \text{kW/m}^2}.$$

33.57 A harmonic electromagnetic wave in free space is described by $\mathbf{E} = \mathbf{E}_0 \cos(kx - \omega t)$. Show that $I = (c\epsilon_0/2)E_0^2$.

▮ The $\mathbf{B}$ field has the form $\mathbf{B} = \mathbf{B}_0 \cos(kx - \omega t)$ and therefore

$$\mathbf{S} = c^2 \epsilon_0 \mathbf{E} \times \mathbf{B} = c^2 \epsilon_0 \mathbf{E}_0 \times \mathbf{B}_0 \cos^2(kx - \omega t) \qquad \text{hence} \qquad \langle S \rangle = c^2 \epsilon_0 |\mathbf{E}_0 \times \mathbf{B}_0| \langle \cos^2(kx - \omega t) \rangle$$

Calculating the average over a time interval of length T', we find that

$$\langle \cos^2(kx - \omega t) \rangle = \frac{1}{T'} \int_t^{t+T'} \cos^2(kx - \omega t') \, dt'$$

For T' large this average is the same as the average over a single period T, for which we have the well-known result $\langle \cos^2(kx - \omega t) \rangle = \frac{1}{2}$. Consequently, since $E_0 = cB_0$, and $\mathbf{E} \perp \mathbf{B}$,

$$\langle S' \rangle = I = \frac{c\epsilon_0}{2} E_0^2 = \frac{1}{2c\mu_0} E_0^2 \qquad (1)$$

33.58 A plane electromagnetic wave moving through free space has an $\mathbf{E}$ field (also referred to as the *optical field*) given by $E_x = 0$, $E_y = 0$ and

$$E_z = 100 \sin\left[8\pi \times 10^{14}\left(t - \frac{x}{3 \times 10^8}\right)\right]$$

Calculate the corresponding flux density.

▮ From Prob. 33.57, $I = (c\epsilon_0/2)E_0^2$. Then, since $\epsilon_0 = 8.8542 \times 10^{-12} \, \text{C}^2 \cdot \text{N}^{-1} \cdot \text{m}^{-2}$ and $E_0 = 100 \, \text{V/m}$,

$$I = \frac{(3 \times 10^8)(8.85 \times 10^{-12})100^2}{2} = \underline{13.3 \, \text{W/m}^2}$$

33.59 A plane harmonic electromagnetic wave is propagating in space along the y axis. If the $\mathbf{E}$ field is linearly polarized in the yz plane and if $\lambda_0 = 500 \, \text{nm}$, what is an expression for the corresponding $\mathbf{B}$ field when the irradiance is 53.2 W/m²?

▮ We can determine E_0 from the irradiance:

$$I = \frac{c\epsilon_0 E_0^2}{2} \qquad 53.2 = \frac{(3 \times 10^8)(8.85 \times 10^{-12})E_0^2}{2} \qquad E_0 = 200 \text{ V/m}$$

Then, from $B_0 = E_0/c = 66.7 \times 10^{-8} \text{ T}$, it follows that

$$B_x = 66.7 \times 10^{-8} \sin \frac{2\pi}{500 \times 10^{-9}} (y - 3 \times 10^8 t) \qquad B_y = B_z = 0$$

33.60 A 60-W monochromatic point source radiating equally in all directions in vacuum is being monitored at a distance of 2.0 m. Determine the amplitude of the **E** field at the detector.

▮ If A is the area of a sphere of radius r surrounding the source and I is the irradiance at that distance, then the power radiated by the source is given by $IA = I(4\pi r^2)$, or equivalently, $\langle S \rangle (4\pi r^2)$. Thus

$$60 = \left(\frac{c\epsilon_0}{2} E_0^2\right) 4\pi r^2 = \frac{4\pi r^2}{2\mu_0 c} E_0^2 \qquad \text{Hence} \qquad E_0 = \left(\frac{30\mu_0 c}{\pi r^2}\right)^{1/2} = \left[\frac{30(4\pi \times 10^{-7})(3 \times 10^8)}{\pi(2)^2}\right]^{1/2} = \underline{30 \text{ V/m}}$$

33.61 Measurements of the temperature rise of an absorbing plate oriented normal to the sun's rays show that the energy flux delivered by sunlight to the surface of the earth on a clear day is about 1.0 kW/m^2. Determine the amplitude of the electric and magnetic fields in the sunlight if the sun were a monochromatic source.

▮ From Eq. (1) of Prob. 33.57,

$$E_0 = \sqrt{2\mu_0 c \langle S \rangle} = \sqrt{2(4\pi \times 10^{-7} \text{ H/m})(3.0 \times 10^8 \text{ m/s})(1.0 \times 10^3 \text{ W/m}^2)} = \underline{900 \text{ V/m}}$$

Then

$$B_0 = \frac{E_0}{c} = \frac{900 \text{ V/m}}{3.0 \times 10^8 \text{ m/s}} = \underline{3.0 \ \mu\text{T}}$$

33.62 A certain laser beam has a cross-sectional area of 2.0 mm^2 and a power of 0.80 mW. Find the intensity of the beam as well as E_0 and B_0 in it. Assume that the beam consists of a single sinusoidal plane wave.

▮ $I = P/A = (8.0 \times 10^{-4})/(2.0 \times 10^{-6}) = 400 \text{ W/m}^2$. Proceed as in Probs. 33.59 to 33.61 to find $E_0 = [2I/(c\epsilon_0)]^{1/2} = \underline{550 \text{ V/m}}$ and $B_0 = E_0/c = \underline{1.83 \ \mu\text{T}}$.

33.63 A certain 0.90-MHz radio transmitter sends out 50 W of radiation. Make the (poor) assumption that the radiation is uniform on a sphere with the transmitter at its center. (*a*) What is the intensity of the wave at a distance of 15 km? (*b*) What are the amplitudes of the E and B waves at this distance?

▮ (*a*) The area of the sphere is $A = 4\pi R^2 = 4\pi(15 \times 10^3)^2 = 2.83 \times 10^9 \text{ m}^2$, so $I = P/A = 50/A = 1.77 \times 10^{-8} \text{ W/m}^2$. (*b*) From the intensity relation of a plane wave, $I = (c\epsilon_0 E_0^2)/2$, we find $E_0 = [2I/(c\epsilon_0)]^{1/2} = 3.65 \times 10^{-3} \text{ V/m}$. The amplitude of B is $B_0 = E_0/c = \underline{1.22 \times 10^{-11} \text{ T}}$.

33.64 Show that the average energy density in an electromagnetic plane wave is $\langle \rho_e \rangle = \langle S \rangle / c$, where **S** is the Poynting vector. What is the momentum density in the wave?

▮ The irradiance, $I = \langle S \rangle$, is the average radiant energy flowing per second through a unit cross-sectional area perpendicular to the energy flow. If we imagine such a cross section of area A, the energy flowing through in a time Δt will be the energy in a volume $= (c \Delta t)A$, since the energy flows at speed c. The average energy density, $\langle \rho_e \rangle$, is the same everywhere in the wave region, so total energy through A in time Δt is $\langle \rho_e \rangle c \Delta t A$. Then the energy/unit area/unit time is obtained by dividing by $A \Delta t$, yielding $\langle S \rangle = c \langle \rho_e \rangle$.

The same result can be obtained by recalling that the average energy density in a harmonic electric field is $\epsilon_0 E_0^2/4$ and in a harmonic magnetic field is $B_0^2/4\mu_0$. Then using $B_0 = E_0/c$, we get $\langle \rho_e \rangle = (\epsilon_0 E_0^2)/4 + E_0^2/(4c^2\mu_0) = (\epsilon_0 E_0^2)/2$. Finally we recall (Prob. 35.37) that $\langle S \rangle = (c\epsilon_0 E_0^2)/2$, and the result follows.

With somewhat more complicated reasoning it can be shown that radiation carries momentum, and can therefore exert pressure. The momentum density $\langle \rho_m \rangle = \langle \rho_e \rangle / c = \langle S \rangle / c^2$. [Quantum theory gives, for each photon, energy $\mathscr{E} = h\nu = hc/\lambda = pc$; and $\langle \rho_e \rangle = c \langle \rho_m \rangle$ follows at once.]

33.65 Use the value $\langle S \rangle = 1.0 \text{ kW/m}^2$ for the average energy flux in sunlight at the earth's surface to evaluate $\langle \rho_e \rangle$ and $\langle \rho_m \rangle$ for sunlight.

▮ Employing the results of Prob. 33.64 we have

$$\langle \rho_e \rangle = \frac{\langle S \rangle}{c} = \frac{1.0 \times 10^3 \text{ W/m}^2}{3.0 \times 10^8 \text{ m/s}} = \underline{3.3 \times 10^{-6} \text{ J/m}^3} \qquad \langle \rho_m \rangle = \frac{\langle S \rangle}{c^2} = \frac{1.0 \times 10^3 \text{ W/m}^2}{(3.0 \times 10^8 \text{ m/s})^2} = \underline{1.1 \times 10^{-14} \text{ kg} \cdot \text{m/s} \cdot \text{m}^3}.$$

33.66ᶜ A plane-parallel vacuum capacitor is being charged. The plates are circular and of radius R. At a certain instant the magnitude of the electric field between the plates is $\mathscr{E}$. **(a)** Express in terms of $\mathscr{E}$ the magnitude and direction of the electric field on the three parts of a closed surface having the form of a circular cylinder in the region between the plates of radius slightly less than their radius and of length slightly less than their separation. (Ignore "edge effects." That is, assume that the electric field terminates abruptly just outside the capacitor and is uniform everywhere inside it.) **(b)** Evaluate the magnitude and direction of the magnetic field on the three parts of the cylinder in terms of $\mathscr{E}$.

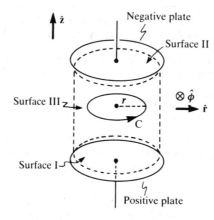

Fig. 33-10

❚ The capacitor and the three-part cylindrical surface are shown in Fig. 33-10. We use a cylindrical coordinate system; the local unit vectors are $\mathbf{r}$, $\boldsymbol{\varphi}$, and $\mathbf{z}$. Surface I is the base of the cylinder near the positive plate, surface II is the base of the cylinder near the negative plate, and surface III is the wall of the cylinder. **(a)** Ignoring edge effects, the electric field everywhere between the plates has magnitude $\mathscr{E}(t)$ and is parallel to the symmetry axis, pointing from the positive plate to the negative plate:

$$\mathscr{E}(r, t) = \mathbf{z}\mathscr{E}(t) \tag{1}$$

On surface I, $\mathscr{E}$ is perpendicular to the surface and points into the cylindrical volume. On surface II, $\mathscr{E}$ is perpendicular to the surface and points outward from the cylindrical volume. On surface III, $\mathscr{E}$ is parallel to the surface. **(b)** The source of the magnetic field is the displacement current. Since $d\mathscr{E}/dt$ is purely axial, the field has no axial component. In fact, since $d\mathscr{E}/dt = \mathbf{z}(d\mathscr{E}/dt)$ is a function only of t, $\mathscr{B}$ is purely azimuthal, with a magnitude that depends only upon t and r: $\mathscr{B} = \boldsymbol{\varphi}\mathscr{B}_\varphi(r, t)$. Applying Prob. 33.2 to the concentric circular path C shown in the figure, we have

$$[\mathscr{B}_\varphi(r, t)]2\pi r = \mu_0\epsilon_0\pi r^2\frac{d\mathscr{E}}{dt} \quad \text{or} \quad \mathscr{B}(r, t) = \hat{\boldsymbol{\varphi}}\frac{\mu_0\epsilon_0 r}{2}\frac{d\mathscr{E}}{dt} \tag{2}$$

For surface III, take $r \approx R$ in (2).

33.67 Refer to Prob. 33.66. Evaluate the Poynting vector on the three parts of the cylinder.

❚ Using Eqs. (1) and (2) and $\hat{\mathbf{z}} \times \hat{\boldsymbol{\varphi}} = -\hat{\mathbf{r}}$, we find that

$$\mathbf{S} = \frac{1}{\mu_0}(\mathscr{E} \times \mathscr{B}) = -\hat{\mathbf{r}}\frac{\epsilon_0 r}{2}\mathscr{E}(t)\frac{d\mathscr{E}}{dt} \tag{1}$$

with $r \approx R$ for surface III.

33.68ᶜ Refer to Probs. 33.66 and 33.67 and **(a)** calculate the rate at which energy is flowing through the closed surface formed by the three parts of the cylinder and **(b)** show that the value obtained in **(a)** equals the rate of change of the energy content of the electric field in the volume enclosed by the surface.

❚ **(a)** The rate of energy loss in a region is the outward flux of $\mathbf{S}$ over the bounding surface. Equivalently, the rate of energy increase, P, in a region is the inward flux of $\mathbf{S}$ over the bounding surface:

$$P = \int_{\substack{\text{closed}\\\text{surface}}} \mathbf{S} \cdot (-\hat{\mathbf{n}})\, da \tag{1}$$

where $\hat{\mathbf{n}}$ is the unit outward normal. In the present case, the surface integral includes three terms. Since $\mathbf{S}$ is parallel to the surface (and hence is perpendicular to $-\hat{\mathbf{n}}$) on surfaces I and II, only surface III contributes.

Since $\hat{\mathbf{n}} = \hat{\mathbf{r}}$ on surface III, we have from (1) of Prob. 33.67

$$P = \int_{\text{surface III}} \left\{ -\hat{\mathbf{r}} \left[\frac{\epsilon_0 R}{2} \mathscr{E}(t) \frac{d\mathscr{E}(t)}{dt} \right] \right\} \cdot (-\hat{\mathbf{r}}) \, da = \frac{\epsilon_0 R}{2} \mathscr{E}(t) \frac{d\mathscr{E}(t)}{dt} \int_{\text{surface III}} da = \epsilon_0 \pi R^2 h \mathscr{E}(t) \frac{d\mathscr{E}}{dt} \quad (2)$$

where h is the separation between the plates. (b) The value obtained for P in (2) can be written as

$$P = (\pi R^2 h) \frac{d}{dt} \left[\frac{1}{2} \epsilon_0 \mathscr{E}^2(t) \right] \quad (3)$$

But $\pi R^2 h$ is the volume between the plates, while the energy density ρ_e contained in the electric field is $\rho_e = \frac{1}{2} \epsilon_0 \mathscr{E}^2$. Hence P equals the rate of increase of the electric energy contained in the capacitor.

(This problem demonstrates that even in nonradiative situations the integral of the Poynting vector over a closed surface gives the rate of accumulation of electromagnetic energy inside. Only in the case of true radiation, however, does the Poynting vector represent the physical flow of electromagnetic energy across a boundary.)

33.69   In the classical picture, what is the ultimate source of radiant energy?

▮   According to Maxwell's equations, the radiation **E** and **B** fields are caused by time-dependent current and charge densities, that is, by *accelerated charged particles*. For a nonrelativistic particle ($v \ll c$) it can be shown that the total radiant flux emitted at any instant is given by $P = (q^2 a^2)/(6\pi\epsilon_0 c^3)$, where q is the charge on the particle and a is its instantaneous acceleration. In SI units P is in watts. Once such radiation is emitted, it travels through free space in accordance with the free-space wave equation. If a varies harmonically in time (SHM), the electromagnetic wave is harmonic of the same frequency. In general the radiation emitted by an accelerating charge is not isotropic.

33.70   A proton of kinetic energy $K = 50.0$ MeV is traveling in a circular orbit of radius $R = 1.00$ m through the uniform magnetic field of a cyclotron. Evaluate the energy it loses to radiation in one trip around the orbit.

▮   The rest energy of a proton is 938 MeV, so a nonrelativistic treatment is valid. In the circular orbit, the proton undergoes an acceleration

$$a = \frac{v^2}{R} \quad \text{or} \quad a = \frac{2mv^2}{2mR} = \frac{2K}{mR}$$

where m is its mass. So the electromagnetic energy radiated per second is (Prob. 33.69)

$$P = \frac{e^2 a^2}{6\pi\epsilon_0 c^3} = \frac{2e^2 K^2}{3\pi\epsilon_0 c^3 m^2 R^2}$$

where e is the charge of the proton. The energy radiated per orbit traversal is P multiplied by the orbital period $2\pi R/v = 2\pi R\sqrt{m/2K}$. This energy can come only from the proton's kinetic energy, so the loss of kinetic energy by radiation is

$$\Delta K = -\frac{4e^2 K^{3/2}}{3\sqrt{2}\epsilon_0 c^3 m^{3/2} R} = -\frac{4(1.60 \times 10^{-19} \text{ C})^2 [(50.0)(1.60 \times 10^{-13} \text{ J})]^{3/2}}{3\sqrt{2}(8.85 \times 10^{-12} \text{ C}^2/\text{N} \cdot \text{m}^2)(3.00 \times 10^8 \text{ m/s})^3 (1.67 \times 10^{-27} \text{ kg})^{3/2}(1.00 \text{ m})}$$

The energy lost to radiation per orbit, ΔK, is completely negligible compared to the energy of the proton, $K = 50.0$ MeV. So the effect can be ignored in designing the cyclotron.

33.71   Estimate the power radiated by an atom emitting light of wavelength $\lambda = 500$ nm. Assume that a single electron is responsible for the radiation and that it acts like a harmonic oscillator, with its oscillation amplitude equal to the typical atomic dimension 0.1 nm.

▮   The frequency of the light emitted,

$$v = \frac{c}{\lambda} = \frac{3.0 \times 10^8 \text{ m/s}}{5.0 \times 10^{-7} \text{ m}} = 6.0 \times 10^{14} \text{ Hz}$$

is also the oscillation frequency of the electron. Using it and the fact that $a_{max} = \omega^2 A = 4\pi^2 v^2 A$, while $\langle a^2 \rangle = a_{max}^2/2$, we get, from Prob. 33.69,

$$P = \frac{4\pi^3 (6.0 \times 10^{14} \text{ Hz})^4 (1.6 \times 10^{-19} \text{ C})^2 (1.0 \times 10^{-10} \text{ m})^2}{3(8.85 \times 10^{-12} \text{ C}^2/\text{N} \cdot \text{m}^2)(3.0 \times 10^8 \text{ m/s})^3} = \underline{5.7 \times 10^{-12} \text{ W}}$$

33.72 Find the power emitted by an electron falling freely in the earth's gravitational field.

▮ We apply the radiant flux equation of Prob. 33.69 with $q = -e = -1.60 \times 10^{-19}$ C and $a = g = 9.80$ m/s^2:

$$P = \frac{(1.60 \times 10^{-19})^2(9.80)^2}{6\pi(8.85 \times 10^{-12})(3 \times 10^8)^3} = \underline{5.46 \times 10^{-52} \text{ W}}$$

33.73 For an electron starting from rest and dropping 1.0 km in the earth's gravitational field, find the ratio of radiated energy to the increase in kinetic energy. (See Prob. 33.72.)

▮ For a particle falling from rest, we have $s = \frac{1}{2}gt^2$, so that the fall time $t = \sqrt{2s/g} = \sqrt{(2 \times 10^3)/9.80} = 14.3$ s. Hence the radiated energy is $E_{rad} = Pt = (5.46 \times 10^{-52})(14.3) = 7.80 \times 10^{-51}$ J. With $m = m_e = 9.11 \times 10^{-31}$ kg, the increase in kinetic energy is given by

$$\Delta K = mgs = (9.11 \times 10^{-31})(9.80)(1.0 \times 10^3) = 8.93 \times 10^{-27} \text{ J} \quad \text{so that} \quad \frac{E_{rad}}{\Delta K} = \frac{7.80 \times 10^{-51} \text{ J}}{8.93 \times 10^{-27} \text{ J}} = \underline{8.74 \times 10^{-25}}$$

33.74 For the electron in Prob. 33.73, what source supplies the energy being radiated?

▮ The only available source of energy is gravity. Some force must obviously be acting, opposed to gravity, so that some gravitational energy is expended against that force. Thus, the actual acceleration is somewhat less than g. Nonetheless, the ratio found in Prob. 33.73 indicates that little error is made in ignoring this force for this case. The force is called the *radiation reaction* or *radiation damping force*.

CHAPTER 34
Light and Optical Phenomena

34.1 REFLECTION AND REFRACTION

34.1 Define the *index of refraction* of a material medium.

▌ Maxwell's equations predict a propagation velocity of $c = 1/\sqrt{\epsilon_0 \mu_0}$ for electromagnetic waves in vacuum. In contrast, a wave moving through a material medium travels at a speed $v = 1/\sqrt{\epsilon \mu}$. Here ϵ and μ are the permittivity and permeability of the medium. The *absolute index of refraction n* is then defined by

$$n \equiv \frac{c}{v} = \sqrt{\frac{\epsilon \mu}{\epsilon_0 \mu_0}}$$

Generally the magnetic properties of the media have little effect on v, since in materials of concern to us, $\mu \approx \mu_0$.

An incoming electromagnetic wave applies an electric field to the medium, which as a result becomes electrically polarized. That, in turn, contributes to ϵ, which then determines n. All this is dependent on the driving frequency of the incident wave. Figure 34-1 illustrates the frequency dependence of n for various substances of interest.

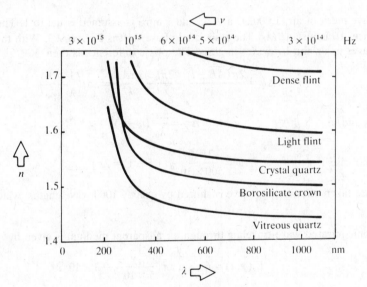

Fig. 34-1

34.2 Light having a free-space wavelength of $\lambda_0 = 500$ nm passes from vacuum into diamond ($n_d = 2.4$). Under ordinary circumstances the frequency is unaltered as light traverses different substances. Assuming this to be the case, compute the wave's speed and wavelength in the diamond.

▌ Since $n = c/v$, $v = c/n = 3 \times 10^8/2.4 = \underline{1.25 \times 10^8 \text{ m/s}}$. As for the wavelength, $n = c/v = (\lambda_0 v_0)/(\lambda v) = \lambda_0/\lambda$, inasmuch as $v_0 = v$. Hence, $\lambda = 500/2.4 = \underline{208 \text{ nm}}$.

34.3 (a) What is the speed of light in crystalline quartz? (b) The speed of light in zircon is 1.52×10^8 m/s. What is the index of refraction of zircon? Take the index of refraction of quartz to be 1.553.

▌ (a)
$$v = \frac{c}{n} = \frac{3.0 \times 10^8 \text{ m/s}}{1.553} = \underline{1.93 \times 10^8 \text{ m/s}}$$

(b)
$$n = \frac{c}{v} = \frac{3.0 \times 10^8 \text{ m/s}}{1.52 \times 10^8 \text{ m/s}} = \underline{1.97}$$

34.4 (a) When light enters a medium of index of refraction n, its frequency does not change, but its wavelength and speed do. Show that the wavelength λ' in the medium is $\lambda' = \lambda/n$, where λ is the wavelength of the light in vacuum. **(b)** What is the wavelength in water of a blue light whose wavelength in air is 420 nm? ($n = 1.33$ for water.)

(a)
$$\lambda' = \frac{v}{f} = \frac{c}{nf} = \frac{\lambda}{n}$$

(b)
$$\lambda' = \frac{\lambda}{n} = \frac{420 \text{ nm}}{1.33} = \underline{316 \text{ nm}}$$

34.5 Light of a certain color has 1800 waves to the millimeter in air. What is its frequency? Find the number of waves per millimeter when the light is traveling in water and in plate glass. What is the wavelength in each medium? ($n = 1.333$ and 1.52 for water and plate glass, respectively.)

$$\lambda = \frac{1 \text{ mm}}{1800} = 5.556 \times 10^{-7} \text{ m} = 556 \text{ nm} \qquad v = \frac{3 \times 10^8}{5.556 \times 10^{-7}} = \underline{5.4 \times 10^{14} \text{ Hz}}$$

Number of waves per millimeter in water $= 1800 \times 1.333 = \underline{2399 \text{ mm}^{-1}}$

Number of waves per millimeter in glass $1800 \times 1.52 = \underline{2736 \text{ mm}^{-1}}$

$$\lambda_{\text{water}} = \frac{555.6 \text{ nm}}{1.333} = 417 \text{ nm} \qquad \lambda_{\text{glass}} = \frac{555.6 \text{ nm}}{1.52} = \underline{365 \text{ nm}}$$

34.6 Suppose that a light wave propagates from a point A to another point B, and we introduce into its path a glass plate ($n_g = 1.5$) of thickness $l = 1$ mm. By how much will that alter the phase of the wave at B if $\lambda_0 = 500$ nm?

The refractive index of air (1.000293 at 0 °C and 1 atm) is assumed equal to 1. The number of waves in air over the distance $\overline{AB}$ is just $\overline{AB}/\lambda_0$. The associated phase shift is $2\pi(\overline{AB}/\lambda_0)$. With the glass inserted there are $(\overline{AB} - l)/\lambda_0$ waves in air and l/λ waves in glass. The phase difference is then

$$\Delta\varphi = \frac{2\pi(\overline{AB} - l)}{\lambda_0} + \frac{2\pi l}{\lambda} - \frac{2\pi\overline{AB}}{\lambda_0} = 2\pi l\left(\frac{1}{\lambda} - \frac{1}{\lambda_0}\right)$$

But $1/\lambda = n/\lambda_0$ and so
$$\Delta\varphi = \frac{2\pi l}{\lambda_0}(n - 1)$$

In this particular case
$$\Delta\varphi = \frac{2\pi 10^{-3}}{500 \times 10^{-9}}(1.5 - 1) = \underline{2\pi 10^3 \text{ rad}}$$

If the data are taken as exact, the wave is shifted by exactly 1000 wavelengths, which is the same thing as no shift at all.

34.7 A plane harmonic infrared wave traveling through a transparent medium is given by

$$E_x(y, t) = E_{0x} \sin 2\pi\left(\frac{y}{5 \times 10^{-7}} - 3 \times 10^{14}t\right)$$

in SI as usual. Determine the refractive index of the medium at that frequency, and the vacuum wavelength of the disturbance.

The phase is familiar in the form $k(y - vt)$; accordingly, we rewrite the above as $E_{0x} \sin \phi$, where

$$\phi = \frac{2\pi}{5 \times 10^{-7}}(y - 15 \times 10^7 t)$$

Clearly, $\lambda = 5 \times 10^{-7}$ m and $v = 1.5 \times 10^8$ m/s. Thus $n = c/v = \underline{2}$ and $\lambda_0 = n\lambda = \underline{1000 \text{ nm}}$.

34.8 Light from a sodium lamp ($\lambda_0 = 589$ nm) passes through a tank of glycerin (refractive index 1.47) 20 m long in a time t_1. If it takes a time t_2 to traverse the same tank when filled with carbon disulfide (index 1.63), determine the difference $t_2 - t_1$.

Since $v = c/n$,
$$t_1 = \frac{20}{c/n} = \frac{20(1.47)}{c} \qquad \text{and} \qquad t_2 = \frac{20(1.63)}{c}$$

Accordingly,
$$t_2 - t_1 = \frac{20}{c}(1.63 - 1.47) = \underline{1.07 \times 10^{-8} \text{ s}}$$

34.9 A beam of 436-nm light is incident perpendicular to a glass plate that is 2.0 cm thick and has $n = 1.66$.
(a) How long does it take a point on the beam to pass through the plate? (b) How many wavelengths thick is the plate in terms of the beam within it?

▌ (a) v within the plate $= c/1.66 = 1.807 \times 10^8$ m/s, so $t = d/v = 0.020/v = \underline{0.111\,\text{ns}}$. (b) The wavelength $\lambda' = (436\,\text{nm})/1.66$; the number of complete waves $= d/\lambda' = \underline{76\,147}$.

34.10 At what angle must a ray of light be incident on acetone to be refracted into the liquid at 25°? ($n = 1.358$.)

▌ We use Snell's law: $n_1 \sin \theta_1 = n_2 \sin \theta_2$ with angles measured from normal. Then

$$\sin \theta_1 = \frac{n_2 \sin \theta_2}{n_1} = \frac{(1.358)(\sin 25°)}{1.00} = 0.574 \quad \text{or} \quad \underline{\theta_1 = 35°}$$

34.11 What is the speed of light in water? Find the angle of refraction of light incident on a water surface at an angle of 48° to the normal. ($n = 1.333$.)

▌
$$v = \frac{c}{n} = \frac{3 \times 10^8\,\text{m/s}}{1.333} = \underline{2.25 \times 10^8\,\text{m/s}}$$

From Snell's law,

$$1 \sin 48° = 1.333 \sin \theta_w \quad \text{and} \quad \sin \theta_w = 0.5575; \quad \underline{\theta_w = 33.9°}$$

34.12 The index of refraction of n-propyl alcohol is 1.39. Find the speed of light in that medium and the angle of refraction if light comes from air with an angle of incidence of 55°.

▌
$$v = \frac{c}{n} = \frac{3 \times 10^8}{1.39} = \underline{2.16 \times 10^8\,\text{m/s}} \qquad \frac{\sin 55°}{\sin \theta_{al}} = 1.39 \qquad \sin \theta_{al} = 0.58932 \qquad \underline{\theta_{al} = 36.1°}$$

34.13 Light passes from air into a liquid and is deviated 19° when the angle of incidence is 52°. What is the index of refraction of the liquid?

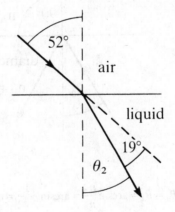

Fig. 34-2

▌ The situation is as shown in Fig. 34-2.

$$\theta_2 = 52° - 19° = 33° \qquad n = \frac{\sin 52°}{\sin 33°} = \frac{0.7880}{0.5446} = \underline{1.45}$$

34.14 At what angle of incidence should a beam of light strike the surface of a still pond if the angle between the reflected ray and the refracted ray is to be 90°? ($n = 1.33$ for water.)

▌ For angles of incidence and refraction θ, ϕ we must have $\theta + \phi = 90°$, or $\phi = 90° - \theta$. From Snell's law, $\sin \theta = n \sin (90° - \theta) = n \cos \theta$, so $\tan \theta = n$, or $\theta = \tan^{-1} n = \tan^{-1} 1.33 = \underline{53°}$.

34.15 Find the sine of the critical angle of incidence for an air–water interface.

▌ For light passing from water to air the angle of refraction is greater than the angle of incidence, so for angles of incidence greater than $i_c < 90°$, the refraction angle will be greater than 90°, implying no refraction.

The angle i_c is the critical angle and corresponds to a refraction angle of exactly 90°. Then

$$1 \sin 90° = 1.333 \sin i_c \qquad \sin i_c = \underline{0.750}$$

34.16 A ray of light passes from crown glass to water. What is the critical angle of incidence? If the angle of incidence in the glass is 55°, what is the angle of refraction? ($n = 1.333$ and 1.52 for water and crown glass, respectively.)

▮ For critical angle we assume light travels from glass to water. Then from the definition of critical angle,

$$1.52 \sin \theta_c = 1.333 \sin 90° \qquad \sin \theta_c = 0.877 \qquad \theta_c = \underline{61.3°}$$

For the second part of the problem, we have an angle of incidence $< \theta_c$, so we have refraction in the water with

$$1.52 \sin 55° = 1.333 \sin \theta_w \qquad \theta_w = \underline{69.1°}$$

34.17 Assume that the index of refraction of a glass sphere is $n_1 = 1.76$. For rays originating within the sphere, find the critical angle if the sphere is immersed in (a) air ($n_2 = 1$) and (b) water ($n_2 = 1.33$).

▮ For the critical angle, $\sin \theta_c = n_2 / n_1$, and so

(a)
$$\theta_c = \arcsin \frac{1}{1.76} = \underline{34.6°}$$

(b)
$$\theta_c = \arcsin \frac{1.33}{1.76} = \underline{49.1°}$$

34.18 To some extent, the brilliance of diamonds is attributable to total internal reflection. Calculate the critical angle for a diamond–air surface.

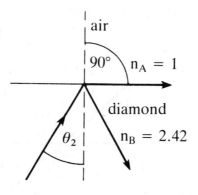

Fig. 34-3

▮ From Fig. 34-3,

$$\theta_c = \theta_2 = \arcsin \frac{n_A}{n_B} = \arcsin \frac{1}{2.42} = \arcsin 0.4132 = \underline{24°}$$

34.19 If the speed of light in ice is 2.3×10^8 m/s, what is its index of refraction? What is the critical angle of incidence for light going from ice to air?

▮ From the data we can determine the index of refraction of ice, $n = c/v = (3 \times 10^8)/(2.3 \times 10^8) = 1.304$. Then for the critical angle,

$$1 \sin 90° = 1.304 \sin i_c \qquad \text{and} \qquad \sin i_c = 0.7667 \qquad i_c = \underline{50.1°}$$

34.20 The index of refraction of a flint glass is 1.64. Find the speed of light in the glass, the angle of refraction in glass if light is incident from water at an angle of 44°, and the critical angle of incidence for a glass–water interface.

▮ $$v = \frac{c}{n} = \frac{3 \times 10^8}{1.64} = \underline{1.83 \times 10^8 \text{ m/s}} \qquad \text{Then} \qquad \frac{\sin 44°}{\sin \theta_{gl}} = \frac{v_w}{v_{gl}} \qquad \text{or} \qquad 1.333 \sin 44° = 1.64 \sin \theta_{gl} \qquad \text{and}$$

$$\sin \theta_{gl} = 0.56462 \qquad \theta_{gl} = \underline{34.4°}.$$

For the critical angle,

$$1.333 \sin 90° = 1.64 \sin i_c \qquad \text{and} \qquad \sin i_c = 0.8128 \qquad i_c = \underline{54.4°}$$

34.21 In what direction does the fish in Fig. 34-4 see the setting sun? The index of refraction of water is $n_2 = \tfrac{4}{3}$ and of air is $n_1 \approx 1$.

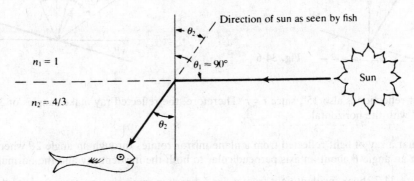

Fig. 34-4

▌ The rays of the setting sun come in nearly tangent to the surface of the water. By Snell's law, with $\theta_1 = 90°$,

$$1 = \tfrac{4}{3} \sin \theta_2 \qquad \text{or} \qquad \theta_2 = \arcsin \tfrac{3}{4} = \underline{48.6°}$$

Note that θ_2 is the critical angle for the reversed ray (from the fish to the sun). The fish perceives the sun at $90° - \theta_2 = \underline{41.4°\ \text{above the horizontal}}$.

34.22 When a fish looks up at the surface of a perfectly smooth lake, the surface appears dark except inside a circular area directly above it. Calculate the angle ϕ that this illuminated region subtends.

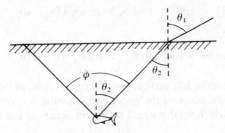

Fig. 34-5

▌ As can be seen from Fig. 34-5 the maximum value of θ_2 for observing light from the air corresponds to $\theta_1 = 90°$. Thus θ_2 is the critical angle and is given by

$$\sin \theta_2 = \frac{n_1 \sin 90°}{n_2} = \frac{(1.00)(1.0)}{1.333} = 0.75 \qquad \text{or} \qquad \theta_2 = 48.6°$$

The figure shows that $\phi = 2\theta_2$, so $\phi = (2)(48.5°) = \underline{97.2°}$.

34.23 A cylindrical tin can open at the top has a height of 0.4 m. In the center of its bottom surface is a tiny black dot and the can is completely filled with water ($n = 1.33$). Calculate the radius of the smallest opaque circular disk that would prevent the dot from being seen, if the disk were floated centrally on the surface of the water.

▌ This problem is the reverse of Prob. 34.22. Light coming up out of the water from the (immediate neighborhood of the) black dot is refracted into air only if the angle of incidence from water is less than the critical angle. This has already been determined in Prob. 34.22 as 48.5°. Thus, if the height of water is 0.4 m, then we must have for the radius of our disk $r = 0.4 \tan 48.6° = \underline{0.45\ \text{m}}$.

34.24 A plane mirror lies face up, making an angle of 15° with the horizontal. A ray of light shines down vertically on the mirror. What is the angle of incidence? What will the angle between the reflected ray and the horizontal be?

▌ See Fig. 34-6. The angle of incidence is the angle made with the normal to the mirror. Thus $i = 15°$. The

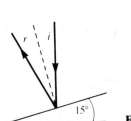

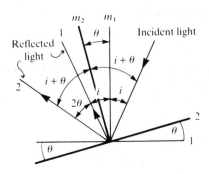

Fig. 34-6   **Fig. 34-7**

angle of reflection is also 15°, since $i = r$. Therefore, the reflected ray makes angles of 30° with the vertical and 60° with the horizontal.

34.25   Show that a ray of light reflected from a plane mirror rotates through an angle 2θ when the mirror is rotated through an angle θ about an axis perpendicular to both the incident ray and the normal to the surface.

▮ Figure 34-7 shows incident light from a fixed direction impinging on a mirror initially in position 1 and then rotated to position 2 with corresponding normals m_1 and m_2. Reflected ray 1 makes an angle of $2i$ with the incident light while reflected ray 2 makes an angle of $2(i + \theta) = 2i + 2\theta$. Thus the two reflected rays make an angle of 2θ with each other.

34.26   Light is passing from air into a liquid and is deviated 19° when the angle of incidence is 52°. Under what conditions will total reflection occur at this interface?

▮ When light travels from the liquid to the air at angles of incidence greater than the critical angle, we have total internal reflection at the interface. To find the critical angle, we need n_1, the index of refraction in the liquid. Now for the liquid

$$\theta_1 = 52° - 19° = 33° \quad \text{and} \quad 1 \sin 52° = n_1 \sin 33° \quad \text{so} \quad 0.7880 = 0.54464 n_1 \quad n_1 = \underline{1.447}$$

Then for the critical angle,

$$1 \sin 90° = 1.447 \sin \theta_{cl} \quad \sin \theta_{cl} = 0.69116 \quad \text{and} \quad \theta_{cl} = \underline{43.7°}$$

34.27   A ray of light is incident on the left vertical face of a glass cube of refractive index n_2, as shown in Fig. 34-8. The plane of incidence is the plane of the page, and the cube is surrounded by liquid (n_1). What is the largest angle of incidence θ_1 for which total internal reflection occurs at the top surface?

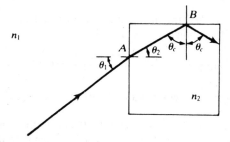

Fig. 34-8

▮ Applying Snell's law at A,

$$n_1 \sin \theta_1 = n_2 \sin \theta_2 \tag{1}$$

But $\theta_2 = 90° - \theta_c$, and so

$$\cos \theta_2 = \sin \theta_c = \frac{n_1}{n_2} \tag{2}$$

Elimination of θ_2 between (1) and (2) gives

$$\sin \theta_1 = \sqrt{\left(\frac{n_2}{n_1}\right)^2 - 1} \tag{3}$$

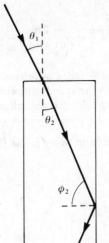

Fig. 34-9

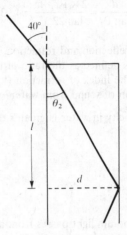

Fig. 34-10

34.28 Figure 34-9 shows a ray of light incident at an angle θ_1 on one end of an optical fiber. Its angle of refraction is θ_2, and it strikes the side of the fiber at an angle ϕ_2. If the index of refraction of the fiber is 1.30, what is the largest angle of incidence θ_1 that a ray can have and still be totally reflected from the side of the fiber?

  ▮ We can apply formula (3) of Prob. 34.27, with $n_1 = 1$, to obtain $\sin \theta_1 = \sqrt{1.30^2 - 1} = 0.831$, or $\theta_1 = \underline{56.2°}$.

34.29 The optical fiber in Fig. 34-10 is 2 m long and has a diameter of 20 μm. If a ray of light is incident on one end of the fiber at an angle $\theta_1 = 40°$, how many reflections does it make before emerging from the other end? (The index of refraction of the fiber is 1.30.)

  ▮ See Fig. 34-10. From Prob. 34.28 the condition for total internal reflection is fulfilled. Then

$$\sin \theta_2 = \frac{n_1}{n_2} \sin \theta_1 = \frac{\sin 40°}{1.3} = 0.495 \qquad \text{and} \qquad \theta_2 = 29.7°$$

The length along fiber to the first reflection is

$$l = \frac{d}{\tan \theta_2} = \frac{2 \times 10^{-5}\,\text{m}}{0.570} = 3.51 \times 10^{-5}\,\text{m}$$

Therefore noting that, by the law of reflection, l also represents the length along the fiber between successive reflections, the number of reflections is

$$\frac{L}{l} = \frac{2\,\text{m}}{3.51 \times 10^{-5}\,\text{m}} = \underline{57\,000}$$

34.30 In practice, optical fibers have a coating of glass ($n_3 = 1.512$) to protect the optical surface of the fiber. If the fiber itself has an index of refraction $n_2 = 1.700$, what is the critical angle for total reflection of a ray inside the fiber?

  ▮ The critical angle is determined by the interface between fiber and coating. Then

$$\sin \theta_c = \frac{n_3}{n_1} = \frac{1.512}{1.700} = 0.889 \qquad \theta_c = \underline{62.7°}$$

34.31 Repeat Probs. 34.28 and 34.29 for the coated fiber described in Prob. 34.30.

  ▮ We proceed first as in Prob. 34.28 to find the maximum angle of incidence, θ_1. $\phi_2 > \theta_c = 62.7°$ from Prob. 34.30. Then $\theta_2 < 90° - 62.7° = 27.3°$. Therefore $\sin \theta_1 < n_2 \sin 27.3° = (1.700)(0.459) = 0.780$, and $\theta_1 < 51.3°$. We next proceed to find the number of reflections, assuming that the dimensions of the fiber and the angle of incidence are as in Prob. 34.29.

$$\sin \theta_2 = \frac{n_1}{n_2} \sin \theta_1 = \frac{\sin 40°}{1.700} = 0.378 \qquad \theta_2 = 22.1°$$

The length along fiber between reflections is then

$$l = \frac{d}{\tan \theta_2} = \frac{2 \times 10^{-5} \text{ m}}{\tan 22.1°} = 4.93 \times 10^{-5} \text{ m} \qquad \text{number of reflections} = \frac{L}{l} = \frac{2 \text{ m}}{4.93 \times 10^{-5} \text{ m}} = \underline{40\ 600}$$

34.32 The laws of reflection and refraction are the same for sound as for light. The index of refraction of a medium (for sound) is defined as the ratio of the speed of sound in air (343 m/s) to the speed of sound in the medium. (*a*) What is the index of refraction (for sound) of water ($v = 1498$ m/s)? (*b*) What is the critical angle θ_c for total reflection of sound from water?

▌ Since velocity in water is greater than in air the sound index of refraction in water is less than 1.

(*a*)
$$n_{\text{water}} = \frac{v_{\text{air}}}{v_{\text{water}}} = \frac{343 \text{ m/s}}{1498 \text{ m/s}} = \underline{0.229}$$

(*b*)
$$\sin \theta_c = \frac{n_{\text{water}}}{n_{\text{air}}} = 0.229 \qquad \theta_c = \underline{13.2°}$$

34.33 A beam of sodium light passes from air into water and then into flint glass, all with parallel surfaces. If the angle of incidence in the air is 45°, what is the angle of refraction in the glass? ($n = 1.333$ and 1.63 for water and flint glass, respectively.)

▌ Refer to Fig. 34-11. Use Snell's law for the air–water surface.

$$n_w = \frac{\sin \theta_1}{\sin \theta_2} \qquad 1.333 = \frac{\sin 45°}{\sin \theta_2} = \frac{0.7071}{\sin \theta_2} \qquad \sin \theta_2 = \frac{0.7071}{1.333} = 0.5305 \qquad \text{and} \qquad \theta_2 = 32°$$

Now use Snell's law for the water–glass surface:

$$n_{BA} = \frac{n_g}{n_w} = \frac{\sin \theta_2}{\sin \theta_3} \qquad \frac{1.63}{1.333} = \frac{0.5305}{\sin \theta_3} \qquad \sin \theta_3 = \frac{0.5305(1.333)}{1.63} = 0.4338 \qquad \text{and} \qquad \theta_3 = \underline{26°}$$

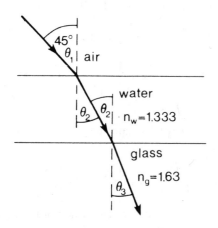

Fig. 34-11

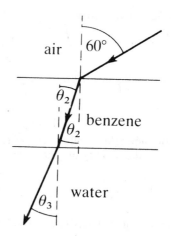

Fig. 34-12

34.34 A layer of benzene (index of refraction = 1.50) floats on water. If the angle of incidence of the light entering the benzene from air is 60°, what is the angle the light makes with the vertical in the benzene and in the water?

▌ Here the light rays bend inward in the benzene and outward in the water, as shown in Fig. 34-12. Use Snell's law first at the air–benzene surface and then at the benzene–water surface.

$$n_b = \frac{\sin \theta_1}{\sin \theta_2} \qquad 1.50 = \frac{\sin 60°}{\sin \theta_2} \qquad \sin \theta_2 = \frac{0.8660}{1.50} = 0.5773 \qquad \text{and} \qquad \theta_2 = 35°$$

Next

$$\frac{n_w}{n_b} = \frac{\sin \theta_2}{\sin \theta_3} \qquad \frac{1.333}{1.50} = \frac{0.5773}{\sin \theta_3} \qquad \sin \theta_3 = \frac{0.5773(1.50)}{1.333} = 0.6496 \qquad \theta_3 = \underline{41°}$$

34.35 A glass dish with a plane parallel bottom and refractive index 1.51 is half filled with water. Then carbon disulfide is poured on top of the water. Finally, a flat cover of the same type of glass is placed on top of the dish. A beam of light making an angle of 50° with the vertical is incident on the horizontal cover. Find the angles which the beam makes with the vertical as it passes through glass, carbon disulfide, water, glass, and air.

▮ We apply Snell's law successively to each surface.

$$1 \sin 50° = 1.51 \sin \theta_g \qquad 1.51 \sin \theta_g = 1.63 \sin \theta_{CS_2} \qquad \text{and} \qquad 1.63 \sin \theta_{CS_2} = 1.333 \sin \theta_w$$

Then

$$\sin \theta_g = 0.5073 \qquad \sin \theta_{CS_2} = 0.4700 \qquad \sin \theta_w = 0.5747 \qquad \text{and} \qquad \theta_g = 30.5°; \qquad \theta_{CS_2} = 28.0° \qquad \theta_w = 35.1°$$

Furthermore,

$$1.333 \sin \theta_w = 1.51 \sin \theta_g' \qquad 1.51 \sin \theta_g' = \sin \theta_A' \qquad \text{Then} \qquad \sin \theta_g' = 0.5073 \qquad \sin \theta_A' = 0.766$$

and $\theta_g' = 30.5 = \theta_g$; $\theta_A' = 50° = \theta_A$, original angle of incidence (see Prob. 34.36).

34.36 Imagine a stratified system consisting of planar layers of transparent materials of different thicknesses. Show that the propagation direction of the emerging beam is determined by only the incident direction and the refractive indices of the initial and final layers (n_1 and n_f).

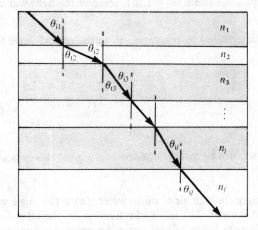

Fig. 34-13

▮ Referring to Fig. 34-13, we obtain from Snell's law:

$$n_1 \sin \theta_{i1} = n_2 \sin \theta_{t2}$$
$$n_2 \sin \theta_{i2} = n_3 \sin \theta_{t3}$$
$$\cdots \cdots \cdots \cdots \cdots$$
$$n_l \sin \theta_{il} = n_f \sin \theta_{tf}$$

Because $\theta_{t2} = \theta_{i2}$, $\theta_{t3} = \theta_{i3}$, etc., these equations lead to

$$n_1 \sin \theta_{i1} = n_2 \sin \theta_{i2} = \cdots = n_l \sin \theta_{il} = n_f \sin \theta_{tf} \qquad \text{whence} \qquad n_1 \sin \theta_{i1} = n_f \sin \theta_{tf}$$

Note that if $n_1 = n_f$, as for a stack of plates immersed in air, $\theta_{i1} = \theta_{tf}$ and the incoming and outgoing rays are parallel.

34.37 The bottom of a glass vessel is a thick plane plate that has an index of refraction of 1.50. The vessel is filled with water. What is the largest angle of incidence in the water for which a ray can pass through the glass bottom and into the air below? How does this compare with the critical angle of incidence for a water–air interface which is 48.6°? What is the angle of refraction in the glass?

▮ We work our way up from the air at the bottom. For largest angle of incidence $\theta_a = 90°$ for light from glass. Then, for angle in glass,

$$1 \sin 90° = 1.50 \sin \theta_{gl} \qquad \sin \theta_{gl} = 0.66667 \qquad \theta_{gl} = \underline{41.8°}$$

Then, for angle in water,

$$1.50 \sin \theta_{gl} = 1.333 \sin \theta_w \qquad \sin \theta_w = 0.7500 \qquad \theta_w = 48.6°$$

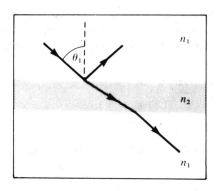

Fig. 34-14

This is the same as for a direct air–water interface, because the angle relation between first and third layers doesn't depend on the second layer, as shown in Prob. 34.36. One exception is if the light doesn't pass through the second layer because of total internal reflection at the first interface.

34.38 Figure 34-14 shows two plates of glass ($n_1 = 1.50$) separated by a liquid film (n_2). Show that if the liquid is water ($n_2 = 1.33$), a ray of light incident on the upper glass–liquid surface at an angle $\theta_1 = 64°$ will be totally reflected; but if the liquid is alcohol ($n_2 = 1.36$), some of the light will be refracted through to the lower glass plate.

▮ Total reflection is determined by comparing θ_1 to the critical angle for the interface, θ_c, which is given by $\sin \theta_c = n_2/n_1$.

For glass–water: $\qquad\qquad\qquad\qquad \dfrac{1.33}{1.50} = 0.887 \qquad \theta_c = 62.5°$

For glass–alcohol: $\qquad\qquad\qquad\quad \dfrac{1.36}{1.50} = 0.907 \qquad \theta_c = 65°.$

Thus light incident at 64° will be totally reflected at the glass–water boundary but not at the glass–alcohol boundary.

34.39 A beam of light is passing through air to oil to water. (*a*) If the angle of incidence in the air is 40°, find the angle of refraction in the water. The index of refraction for the oil is 1.45. (*b*) If possible, find the angle of incidence in air such that the beam will not enter the water. (*c*) Suppose that the direction of the beam is reversed. If possible, find the angle of incidence in water so that the beam will not enter the air.

▮ At the air–oil interface, $\sin 40° = 1.45 \sin \theta_{oil}$; at the oil–water interface, $1.45 \sin \theta_{oil} = 1.33 \sin \theta_w$, so $\theta_w = \sin^{-1}[(\sin 40°)/1.33] = \underline{28.9°}$. (*b*) This is not possible, since $\theta_a > \theta_w$ and θ_a will reach 90° before θ_w does. (*c*) In this case, $\sin 90° = 1.45 \sin \theta_{oil} = 1.33 \sin \theta_w$ so $(\theta_w)_{critical} = \sin^{-1}(1/1.33) = \underline{48.6°}$, as for an air–water interface.

34.40 A narrow beam of light strikes a glass plate ($n = 1.60$) at an angle of 53° to the normal. If the plate is 20 mm thick, what will be the lateral displacement of the beam after it emerges from the plate?

▮ The situation is described by Fig. 34-15, with CE the lateral displacement; $\overline{AB} = 20$ mm.

$$\sin r = \frac{\sin 53°}{1.60} \qquad r = 30° \qquad \overline{BD} = 20 \tan 53° = 26.5 \qquad \overline{BC} = 20 \tan 30° = 11.5$$

$$\overline{CD} = 26.5 - 11.5 = 15.0 \qquad \overline{CE} = \overline{CD} \cos 53° = \underline{9.0 \text{ mm}}$$

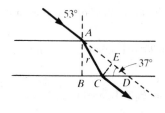

Fig. 34-15

34.41 Two identical beakers, one filled with water ($n = 1.361$) and the other filled with mineral oil ($n = 1.47$), are viewed from directly above. Which beaker appears to contain the greater depth of liquid, and what is the ratio of the apparent depths?

❙ Figure 34-16 indicates the bottom of a beaker by point P. For an observer looking in the $-Y$ direction, the image of P, labeled P', will be determined by a vertical ray 1 and the nearly vertical ray 3 (or 2). Since θ_1 and θ_2 are small, Snell's law gives

$$n_1\theta_1 \approx n_2\theta_2 \qquad \text{or} \qquad \frac{\theta_1}{\theta_2} \approx \frac{n_2}{n_1}$$

But
$$\overline{AD} = y \tan\theta_1 = y' \tan\theta_2 \qquad \text{or} \qquad \frac{\tan\theta_1}{\tan\theta_2} \approx \frac{\theta_1}{\theta_2} = \frac{y'}{y}$$

Consequently, for given y and n_2, y' is inversely proportional to n_1. The water-filled beaker appears deeper, by a factor of $1.47/1.361 = \underline{1.08}$.

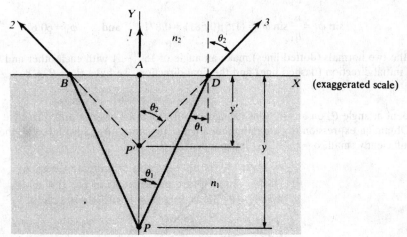

(exaggerated scale)

Fig. 34-16

34.42 A person looks into a swimming pool at the 5-ft-deep level. How deep does it look to the person?

❙ Following Prob. 34.41, we find that $n = $ actual depth/apparent depth, so $1.333 = (5\,\text{ft})/y$, and $y = \underline{3.75\,\text{ft}}$ (apparent depth).

34.43 Light enters a glass prism having a refracting angle of 60°. If the angle of incidence is 30° (incident ray is parallel to base) and the index of refraction of the glass is 1.50, what is the angle the ray leaving the prism makes with the normal?

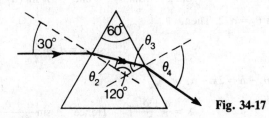

Fig. 34-17

❙ Analyze the refractions in sequence, from the left in Fig. 34-17.

$$n = \frac{\sin\theta_1}{\sin\theta_2} \qquad 1.50 = \frac{\sin 30°}{\sin\theta_2} \qquad \sin\theta_2 = \frac{0.5}{1.50} = 0.3333 \qquad \text{and} \qquad \theta_2 = 19.5°$$

As can be seen from the figure, $\theta_2 + \theta_3 + 120° = 180°$, or $\theta_2 + \theta_3 = 60°$, the prism angle. Then $19.5° + \theta_3 = 60°$, and $\theta_3 = 40.5°$. Next

$$n = \frac{\sin\theta_4}{\sin\theta_3} \qquad 1.50 = \frac{\sin\theta_4}{\sin 40.5°} \qquad 0.6494(1.50) = \sin\theta_4 \qquad \text{and} \qquad \theta_4 = \underline{77°}$$

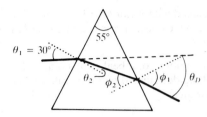

Fig. 34-18

34.44 A ray of light is incident at 30° on a prism with apex angle $A = 55°$ and index of refraction 1.50. Calculate **(a)** the angles the ray makes with the sides of the prism and **(b)** the angle of deviation.

❚ The angles indicated in Fig. 34-18 are obtained from Snell's law and the fact that $\theta_2 + \phi_2 + (180° - A) = 180°$, or $\theta_2 + \phi_2 = A$ (see Fig. 34-17).

(a) $\qquad \sin \theta_2 = \dfrac{n_1}{n_2} \sin \theta_1 = \dfrac{\sin 30°}{1.50} = 0.333 \qquad \theta_2 = \underline{19.5°} \qquad \phi_2 = A - \theta_2 = 55° - 19.5° = \underline{35.5°}$

(b) $\qquad\qquad \sin \phi_1 = \dfrac{n_2}{n_1} \sin \phi_2 = (1.5)(0.581) = 0.871 \qquad \text{and} \qquad \phi_1 = \underline{60.6°}$

Noting that the two normals (dotted lines) make an angle of $180° - A$ with each other and that θ_D is the angle between the initial direction (dashed line) and the final direction, we have $\theta_D = \theta_1 + \phi_1 - A = 30° + 60.6° - 55° = \underline{35.5°}$.

34.45 Light is incident at angle θ_1 on an isosceles triangular prism of apex angle α and refractive index $n > 1$ (Fig. 34-19). **(a)** Obtain an expression for the minimum angular deviation, $\delta_{\min}$, produced by the prism. **(b)** Show that if α is sufficiently small, $\delta \approx (n - 1)\alpha$, independent of θ_1.

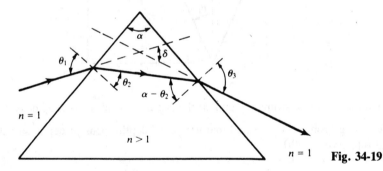

Fig. 34-19

❚ **(a)** Because the prism is symmetrical and because light paths are reversible, it is easy to see that the deviation δ will be a minimum when $\theta_1 = \theta_3 = \kappa$. Under this condition, Snell's law gives at the two interfaces

$$\sin \kappa = n \sin \theta_2 \qquad \text{and} \qquad n \sin (\alpha - \theta_2) = \sin \kappa$$

which imply that $\theta_2 = \alpha/2$. Then,

$$\sin \kappa = n \sin \frac{\alpha}{2}$$

But $\delta_{\min} = \theta_1 + \theta_3 - \alpha = 2\kappa - \alpha$, or

$$\kappa = \frac{\alpha + \delta_{\min}}{2} \qquad \text{Hence} \qquad \sin \frac{\alpha + \delta_{\min}}{2} = n \sin \frac{\alpha}{2}$$

This equation implicitly determines $\delta_{\min}$. **(b)** When α is very small, so is $\delta_{\min}$; and the result of **(a)** becomes

$$\frac{\alpha + \delta_{\min}}{2} \approx n \frac{\alpha}{2} \qquad \text{or} \qquad \delta_{\min} \approx (n - 1)\alpha$$

Furthermore, for α small enough, all deviations δ are close to $\delta_{\min}$, so that $\delta \approx (n - 1)\alpha$.

34.46 A prism having an index of refraction of 1.60 is set for minimum deviation of the incident light. If the refracting angle of the prism is 45°, find the angle of minimum deviation, D (Fig. 34-20).

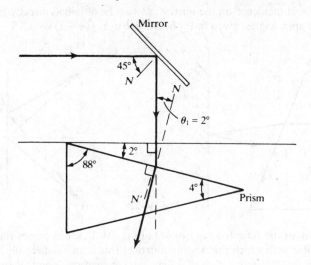

Fig. 34-20

▌ Using the results of Prob. 34.45(a), we have

$$n = \frac{\sin[(A+D)/2]}{\sin(A/2)} \qquad 1.60 = \frac{\sin[(45°+D)/2]}{\sin(45°/2)} \qquad 1.60(\sin 22.5°) = \sin\frac{45°+D}{2} \quad \text{or} \quad 0.6122 = \sin\frac{45°+D}{2}$$

Taking the inverse sine, $37.8° = (45° + D)/2$ and $D = \underline{30.6°}$.

34.47 A ray of light is passed through a prism having a refracting angle of 50°. Rotation of the prism causes the ray to be deviated various amounts, the least of which is 30°. Determine the index of refraction of the glass of which the prism is made.

▌ Letting $D = 30°$ and $A = 50°$, and using the results of Prob 34.45(a), we have

$$n = \frac{\sin[(A+D)/2]}{\sin A/2} = \frac{\sin 40°}{\sin 25°} = \underline{1.52}$$

34.48 A ray of light strikes a plane mirror at an angle of incidence 45°, as shown in Fig. 34-21. After reflection, the ray passes through a prism of refractive index 1.50, whose apex angle is 4°. Through what angle must the mirror be rotated if the total deviation of the ray is to be 90°?

▌ By Prob. 34.45(b), the prism will, to the first order, cause a fixed deviation, $\delta = (1.50 - 1)(4°) = 2°$, whatever the orientation of the mirror. In the given orientation, the mirror by itself produces the desired 90° deviation; hence it must be rotated so as to cancel the additional 2°. Thus, it must be rotated through an angle of $\frac{1}{2}(2°) = \underline{1°}$, counterclockwise in Fig. 34-21.

34.49 A ray of light makes an angle of incidence of 40° with a glass prism of index of refraction 1.52 and refracting angle $A = 56°$. Through what angle is the ray deviated by the prism? Does this represent minimum deviation?

▌ Following the procedure of Probs. 34.43 and 34.44, we have (letting primed angles in air and glass refer to the second surface)

$$1 \sin 40° = 1.52 \sin \theta_g \qquad \sin \theta_g = 0.42289 \qquad \theta_g = 25.017° \qquad \theta_g' = 56 - 25.017 = 30.983°$$

This is clearly *not* going to lead to minimum deviation for which, by symmetry, $\theta_g' = \theta_g$. Continuing,

$$1.52 \sin 30.983° = 1 \sin \theta_a' \qquad \sin \theta_a' = 0.78247 \qquad \theta_a' = 51.487° \qquad \text{deviation} = 40° + 51.487° - 56° = \underline{35.5°}$$

34.50 Given a 60° prism of flint glass, find the angle of minimum deviation. At what angle of incidence must the light strike the prism? Find the deviation if the angle of incidence is 5° less. (For flint glass $n = 1.63$.)

▮ Using the notation of Prob. 34.47, we have

$$1.63 = \frac{\sin\left[(A + D)/2\right]}{\sin A/2} = \frac{\sin\left[(A + D)/2\right]}{\sin 30°} \quad \text{or} \quad \sin\frac{A + D}{2} = 0.815 \quad \frac{A + D}{2} = 54.587° \quad \text{and}$$

$$D = 109.2° - 60° = \underline{49.2°}$$

Using the formula relating the initial angle of incidence, the final angle of refraction, and A and D (see Prob. 34.44), and the symmetry condition for minimum deviation, we have

$$2\theta_a = A + D \qquad \theta_a = 54.587° \qquad \text{New } \theta_a = 49.587° \qquad 1 \sin 49.587 = 1.63 \sin \theta_g \qquad \theta_g = 27.847$$

$$\theta'_g = 60° - 27.847° = 32.153° \qquad 163 \sin 32.153° = 1 \sin \theta'_a \qquad \theta'_a = 60.164°$$

$$\text{deviation} = 60.164° + 49.587° - 60° = \underline{49.75°}$$

34.51 A prism having an apex angle 4° and refractive index 1.50 is located in front of a vertical plane mirror as shown in Fig. 34-22. A horizontal ray of light is incident on the prism. (*a*) What is the angle of incidence at the mirror? (*b*) Through what total angle is the ray deviated?

▮ The first refraction takes place at the second prism interface where the angle of incidence in the glass is $\theta_g = 4°$. Then the angle of refraction in air is given by

$$n \sin \theta_g = \sin \theta_a \qquad \text{or} \qquad 1.5 \sin 4° = \sin \theta_a \qquad \text{and} \qquad \theta_a = 6.0°$$

In the figure this is 6° below the normal to the right side of the prism, which is itself 4° above the horizontal. Thus the refracted beam makes an angle of $\underline{2°}$ with the horizontal, which is the angle of incidence at the mirror. The reflected beam is 2° below the horizontal, to the left, so the angle of deviation from the original beam is: $\underline{-178°}$.

Note: The angle of incidence on the mirror can also be obtained directly from the prism angle of deviation formula for small apex angles, given in Prob. 34.45(*b*): $\delta \approx (n - 1)\alpha = (1.5 - 1)(4°) = 2°$, as found above.

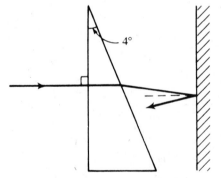

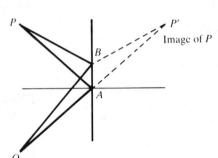

Fig. 34-22 **Fig. 34-23**

34.52 Show that the path of the reflected ray from O of Fig. 34-23 which passes through point P after reflection is the shortest possible path which reaches the mirror. (This is an example illustrating *Fermat's principle of least time*, which states that the actual path of a light ray between two points is such that it takes less time for the light to traverse this path than it would to traverse any other path which varies slightly from the actual path. See Prob. 34.53.)

▮ Let OBP be any other path from O to mirror to P. $PAO = P'AO$ and $PBO = P'BO$. $P'BO > P'AO$. The shortest distance from P' to O is a straight line. Since the speed of light c is the same for both paths, OPA takes the shortest time.

34.53^c Define *optical path length* and state *Fermat's principle*.

▮ Suppose that a ray in going from S to P traverses distances $s_1, s_2, s_3, \ldots, s_m$ in media of indices $n_1, n_2, n_3, \ldots, n_m$, respectively. The total time of flight is then

$$t = \sum_{i=1}^{m} \frac{s_i}{v_i} = \frac{1}{c} \sum_{i=1}^{m} n_i s_i$$

This last summation is referred to as the *optical path length*, or O.P.L.

Fermat's principle can be stated as follows: *A ray of light in traversing a route from any one point to another follows the path, or paths, which takes the least time to negotiate.* (Although often correct as it stands, this statement is not the whole truth and will need some modification.)

In terms of O.P.L. the above statement can then be reworded as: *A ray traverses a route which corresponds to the shortest optical path length.*

To give the modern and most general statement of Fermat's principle, we recall the notion of a *stationary value*. The function $f(x)$ is said to have a stationary value at $x = x_0$ if its derivative, df/dx, vanishes at $x = x_0$. A stationary value could correspond to a maximum, a minimum, or a point of inflection with a horizontal tangent. In any case, $f(x)$ varies slowly in the vicinity of a stationary value $f(x_0)$, so that $f(x) \approx f(x_0)$ for $x \approx x_0$. Generalizing the concept of stationary value we can express Fermat's principle as follows. *A ray of light in going from any one point to another follows, regardless of the media involved, a route which corresponds to a stationary value of the optical path length.* This applies equally well for inhomogeneous media, for which we have

$$\text{O.P.L.} = \int_S^P n(s)\, ds$$

and we examine all possible pathways between points S and P to find the "stationary" one(s).

34.54^c Use Fermat's principle to arrive at the law of reflection.

❚ As shown in Fig. 34-24, a ray leaves S, strikes the interface at an unspecified point B, and reflects off to P. Assuming the medium to be homogeneous and of index n, we have

$$\text{O.P.L.} = n\overline{SB} + n\overline{BP} = n(h^2 + x^2)^{1/2} + n[b^2 + (a-x)^2]^{1/2}$$

Here the O.P.L. is a function of the variable x, and light will take only the route for which

$$\frac{d(\text{O.P.L.})}{dx} = 0 \quad \text{i.e.,} \quad nx(h^2 + x^2)^{-1/2} - n(a-x)[b^2 + (a-x)^2]^{-1/2} = 0$$

But this is equivalent to $\quad n \sin \theta_i - n \sin \theta_r = 0 \quad$ and so $\quad \theta_i = \theta_r$

Thus, if a ray goes from S to P via reflection at B, Fermat's principle demands that B be located such that the angle of incidence equals the angle of reflection.

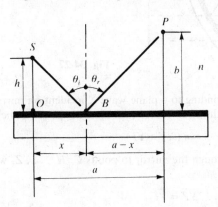

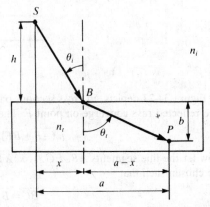

Fig. 34-24 **Fig. 34-25**

34.55^c Apply Fermat's principle to the case of refraction in order to derive Snell's law.

❚ In Fig. 34-25 a ray goes from S to P via refraction at point B on the interface. The scheme is to locate B such that the derivative of the O.P.L. is zero. Thus we write

$$\text{O.P.L.} = n_i \overline{SB} + n_t \overline{BP} = n_i(h^2 + x^2)^{1/2} + n_t[b^2 + (a-x)^2]^{1/2}$$

The variable is x and so

$$\frac{d(\text{O.P.L.})}{dx} = 0 = n_i x (h^2 + x^2)^{-1/2} - n_t(a-x)[b^2 + (a-x)^2]^{-1/2}$$

This has the form $0 = n_i \sin \theta_i - n_t \sin \theta_t$, which is obviously equivalent to Snell's law. The value of x corresponding to a stationary O.P.L. is the one for which Snell's law applies. Other locations of B mean different values of x, none of them corresponding to a stationary value of the O.P.L.

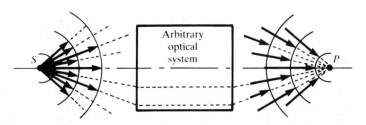

Fig. 34-26

34.56   A spherical wave diverging from a point S is to enter some arbitrary optical system from which it is to emerge as a wave converging to a point P (as in Fig. 34-26). What does Fermat's principle tell us about the optical path lengths for the various rays going from S to P?

▌ Rays from S to P will presumably traverse a great many different paths through the system. Suppose that one such path corresponds to the minimum O.P.L. between S and P. Fermat's principle implies that light would traverse that minimum O.P.L. route and no other. But other routes must obviously be taken because rays leave S in many directions. It follows that a minimum (or maximum) O.P.L. cannot be *uniquely* attained. In other words, *all rays from S through the system to P must traverse identical optical path lengths*. This is true for all sorts of focusing systems (such as lenses and mirrors).

34.57   A collimated beam incident parallel to the symmetry axis of a certain concave mirror is reflected into a converging beam. Use Fermat's principle to show that the mirror is paraboloidal.

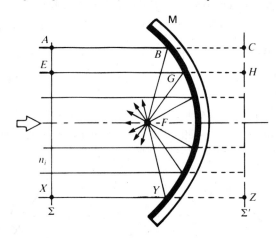

Fig. 34-27

▌ Figure 34-27 depicts, in cross section, parallel rays corresponding to a plane wave Σ incident on mirror M. The reflected rays converge on point F. The optical path lengths of all routes to F must be the same; hence

$$n_i(\overline{AB} + \overline{BF}) = n_i(\overline{EG} + \overline{GF}) = \cdots = n_i(\overline{XY} + \overline{YF})$$

Now let the line segments $\overline{AB}, \overline{EG}, \ldots, \overline{XY}$ be prolonged through the mirror to points $C, H, \ldots, Z$, which are chosen such that

$$\overline{BC} = \overline{BF}, \qquad \overline{GH} = \overline{GF}, \ldots, \overline{YZ} = \overline{YF}$$

The two sets of equalities above imply that $\overline{AB} + \overline{BC} = \overline{EG} + \overline{GH} = \cdots = \overline{XY} + \overline{YZ}$, which tells us that the distance between Σ and the line Σ' through $C, H, \ldots, Z$ is constant. We have thus constructed a *straight* line Σ' such that the points of M are equidistant from it and from the point F. By definition, then, M is a parabola (with *focus* F and *directrix* Σ'). Note the implication that a spherical mirror can only approximately focus parallel light to a point.

34.58   Describe the equation relating object to image distance when light is refracted at a spherical interface between two media.

▌ When light from a point object O is refracted at a spherical surface (Fig. 34-28), the equation relating the object distance p and the image distance q is

$$\frac{n_A}{p} + \frac{n_B}{q} = \frac{n_B - n_A}{R} \tag{1}$$

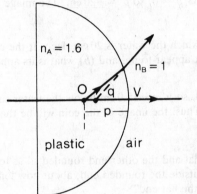

Fig. 34-28

where R is the radius of curvature of the refracting surface. In the figure, R is positive. If the curvature of the refracting surface were reversed, R would be negative. Radius of curvature R is positive for a surface convex to the incident light and negative for a surface concave to the incident light.

Formula (1) holds for α very small (*paraxial rays*).

34.59 A plastic hemisphere has a radius of curvature of 8 cm and an index of refraction of 1.6. On the axis halfway between the plane surface and the spherical one (4 cm from each) is a small flaw. How far from the surface does the flaw appear to be when veiwed along the axis of the spherical surface?

Fig. 34-29   **Fig. 34-30**

▌ For the curved side $R = -8$ cm. Using (1) of Prob. 34.58 we obtain

$$\frac{1.6}{4} + \frac{1}{q} = \frac{1 - 1.6}{-8} \qquad \text{from which} \qquad q = \underline{-3.25 \text{ cm}}$$

The minus sign means that the image is on the same side of the interface as the incident light (*virtual image*).

34.60 A plastic hemisphere (radius $= R$, $n = 1.42$) is sealed to the end of a long cylinder of the same material, as shown in Fig. 34-30. A narrow beam of parallel light is sent through the system from left to right. How far from its point of entry is the beam brought to focus?

▌ Let p and p' be object and image distances, respectively. Here, $p = \infty$, so the spherical refraction equation is $1.42/p' = (1.42 - 1.00)/R$ and $p' = \underline{3.38R}$.

34.61 A small fish is 7 cm from the far side of a spherical fish bowl 28 cm in diameter. Neglecting the effect of the glass walls of the bowl, where does an observer see the image of the fish?

▌ See Fig. 34-31.

$$\frac{n_A}{p} + \frac{n_B}{q} = \frac{n_B - n_A}{R}$$

with $R = -14$ cm. Then

$$\frac{1.333}{21} + \frac{1}{q} = \frac{1 - 1.333}{-14}$$

giving $q = \underline{-25.2 \text{ cm}}$ (from the near side of the bowl).

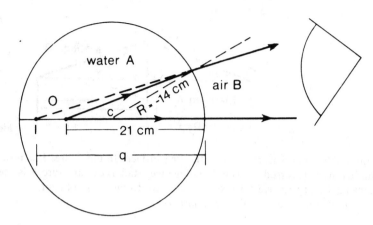

Fig. 34-31

34.62 A fish 40 cm beneath the surface of a pond sees a bug flying directly overhead. If the bug is actually 15 cm above the water, how far above the surface does the bug appear to be as seen by the fish?

▮ Letting $q = p'$ in (1) of Prob. 34.58 and $R = \infty$, we get $\frac{1}{15} + 1.33/p' = 0$, so $p' = \underline{-20\,cm}$. The image is above the water surface.

34.63 A 2.5-cm-diameter coin rests flat on the bottom of a fishbowl in which the water is 20 cm deep. If the coin is viewed directly from above, (a) how far below the surface does it appear to be, and (b) what is its apparent diameter?

▮ We proceed as in Prob. 34.62. (a) This time $1.33/p + 1/p' = 0$, so $p' = \underline{-15.0\,cm}$ below the water surface. (b) For a plane refracting surface, the lateral magnification is 1. Thus, the image of the coin will be the same size as the coin itself.

34.64 A plastic rod with index of refraction 1.50 has one end polished flat and the other end rounded so as to have a radius of curvature of 20 cm. If a light source is placed 50 cm outside the rounded end, about how long must the rod be if the image is to appear at the same distance beyond the flat end?

▮ We have two interfaces through which light travels, so we use the spherical refraction equation twice: $1/50 + 1.5/p' = 0.5/20$, giving $p' = 300$ cm. Then $1.5/(L - 300) + 1/50 = -0.5/\infty$ and $L = \underline{225\,cm}$. Note that for the flat surface $p = L - 300$ is negative, indicating that the "object" for the flat surface is on the side opposite to where the light is coming from (*virtual object*)

34.65 Repeat Prob. 34.64 if both ends of the rod are rounded so as to have 20-cm radii of curvature.

▮ As in Prob. 34.64, $p' = 300$ cm. Then $1.5/(L - 300) + 1/50 = -0.5/\pm 20$ depending upon whether the end curves inward (+), or outward (−). The answers are $L = \underline{267}$ and $\underline{600\,cm}$.

34.66 A glass sphere with 10-cm radius has a 5-cm-radius spherical hole at its center. A narrow beam of parallel light is directed radially into the sphere. Where, if anywhere, will the sphere produce an image? The index of refraction of the glass is 1.50.

▮ Successive application of the spherical refraction formula to the four surfaces in Fig. 34-32 yields the final image $\underline{15\,cm}$ from the left edge, on the side from which the light comes (virtual image). The image formed by

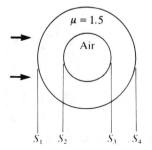

S_1 S_2 S_3 S_4 **Fig. 34-32**

the surface S_1 will be given by $1/\infty + 1.5/p' = (1.5 - 1)/10$, so $p'_1 = +30$. This image acts as an object for S_2 (object distance -25). Continuing in this manner will yield the final result.

34.67 A fish is near the center of a spherical water-filled fishbowl. Where would a child's nose, which is at a distance of the bowl's radius from the surface of the bowl, appear to be to the fish? Call the radius of the bowl R. Where would the child see the fish to be?

    ❚ Using the spherical refraction equation with $p = R$ gives $1/R + 1.33/p' = 0.33/R$; thus $p' = -2R$. The image in air is $\underline{2R\text{ from the surface}}$. Since the fish is at the center, rays from the fish strike the bowl perpendicularly, so they are not deviated. The fish appears at the center of the bowl at its actual location.

34.68 The plano-concave lens shown in Fig. 34-33 has $\mu = 1.56$ and $R = 80$ cm. It rests on a sheet of paper that has a 3.0-mm-diameter circle on it. When viewed directly from above, how far above its actual position does the circle appear to be?

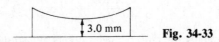

    Fig. 34-33

    ❚ The spherical refraction equation for the glass–air interface is $1.56/0.300 + 1/p' = (1.00 - 1.56)/80$, giving $p' = -0.192$ cm. The circle appears $0.300 - 0.192 = \underline{0.108\text{ cm}}$ above its original position.

34.2 DISPERSION AND COLOR

34.69 Define the following phenomena related to light emission: *continuous spectrum, bright-line spectrum, band spectrum, Fraunhofer lines,* and *fluorescence.*

    ❚ A *continuous spectrum* is one that contains all the wavelengths of visible light, such as that emitted by an incandescent source.

    A *bright-line spectrum* consists of a number of bright lines at wavelengths characteristic of the elements emitting them when excited in the gaseous state.

    Band spectra are characteristic emissions produced by excited molecules. The bands are actually groups of lines very close together.

    Fraunhofer lines are dark lines in the sun's spectrum produced by absorption at specific wavelengths by gases in the atmospheres of the sun and of the earth.

    Fluorescence is the process by which bodies absorb shorter wavelengths of light and subsequently emit light of longer wavelengths.

34.70 It is suspected that a gas mixture containing helium, neon, and argon has been contaminated with carbon monoxide. Suggest a method using spectra to determine whether the CO impurity is present.

    ❚ Excite emission spectra from the gas mixture with an electric discharge. Analyze the spectra with a grating spectrograph. Carbon monoxide, if present, can easily be detected because of the band spectra it emits. Helium, neon, and argon all emit bright-line spectra.

34.71 A beam of white light is incident on a block of flint glass at an angle of 55°. What is the angular separation in the glass of two rays of light, one of wavelength 486 nm and the other of wavelength 656 nm, if their respective indices of refraction are 1.670 and 1.650?

    ❚ We apply Snell's law to each wavelength:

$$1\sin 55° = 1.670 \sin \theta_{486} \qquad \sin \theta_{486} = 0.49051 \qquad \theta_{486} = 29.374° \qquad 1\sin 55° = 1.650 \sin \theta_{656}$$

$$\sin \theta_{656} = 0.49646 \qquad \theta_{656} = 29.766° \qquad \Delta\theta = 29.766° - 29.374° = \underline{0.392°}.$$

34.72 What angular dispersion between the C (red) and F (blue) spectral lines is produced by a flint-glass prism whose refracting angle is 12°? Assume that the prism is being used under conditions of minimum deviation. For flint glass, $n_C = 1.644$ and $n_F = 1.664$.

    ❚ From Prob. 34.45 the index of refraction n is related to the prism refracting angle A and minimum

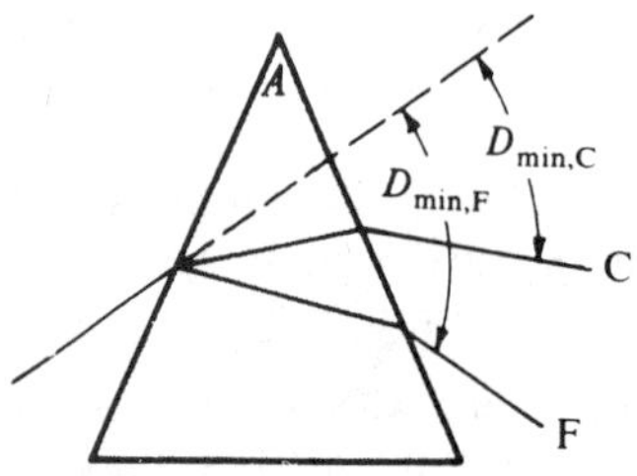

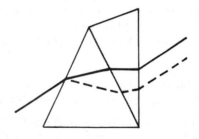

Fig. 34-34

deviation angle D_{min} (see Fig. 34-34) by

$$n = \frac{\sin \frac{1}{2}(A + D_{min})}{\sin \frac{1}{2}A}$$

and, for small A and hence small D_{min} (such that $\sin \theta \approx \theta$), the above equation becomes

$$n = \frac{A + D_{min}}{A} \qquad \text{or} \qquad D_{min} = A(n - 1).$$

In this particular case, only the ratio of angles occurs and so the angles can be measured in either degrees or radians, so long as we use the same angular measure throughout. This small angle approximation is valid in our case with $A = 12° = 0.21$ rad, and we have

dispersion between C and F lines = (deviation of F line) -- (deviation of C line)

$$= D_{min,\,F} - D_{min,\,C} = A(n_F - 1) - A(n_C - 1)$$

$$= A(n_F - n_C) = (12°)(1.664 - 1.644) = \underline{0.24°}$$

34.73 By joining prisms of two different glasses (Fig. 34-35), it is possible to deviate two wavelengths by the same amount. We then say that the prism is *achromatic* for these two wavelengths. The same principle is used to make achromatic lenses. It is required to fit a 10° crown-glass prism with a flint-glass prism so as to achromatize the wavelength interval between the C and F spectra lines. (**a**) What must be the angle of the flint-glass prism? (**b**) What is the deviation produced by the prism combination, as computed for the D line? (The D line is located in the yellow region of the spectrum and is taken as the mean ray between the C and F lines.) Indices of the crown glass for the C, D, F lines are $n_C = 1.514$, $n_D = 1.517$, $n_F = 1.523$; indices of the flint glass for C, D, F lines are $n'_C = 1.644$, $n'_D = 1.650$, $n'_F = 1.664$.

Fig. 34-35

❚ (**a**) As in Prob. 34.72

dispersion by crown-glass prism $= A(n_F - n_C)$ dispersion by flint-glass prism $= A'(n'_F - n'_C)$

Achromatism obtains when the dispersion produced by the crown-glass prism equals (and so annuls) that produced by the inverted flint-glass prism. Then

$$A(n_F - n_C) = A'(n'_F - n'_C) \qquad (10°)(1.523 - 1.514) = A'(1.664 - 1.644) \qquad A' = \underline{4.5°}$$

(**b**)   resultant deviation of D line by prism combination

= (deviation of D line by crown-glass prism) − (deviation of D line by flint-glass prism)

$$= A(n_D - 1) - A'(n'_D - 1) = (10°)(1.517 - 1) - (4.5°)(1.650 - 1) = \underline{2.24°}$$

34.74 It is desired to fit a 12° crown-glass prism with a flint-glass prism so that the combination will produce dispersion, but without deviation of the mean D line. The indices of the crown and flint glass for the D line are, respectively, 1.520 and 1.650. What should be the angle of the flint-glass prism?

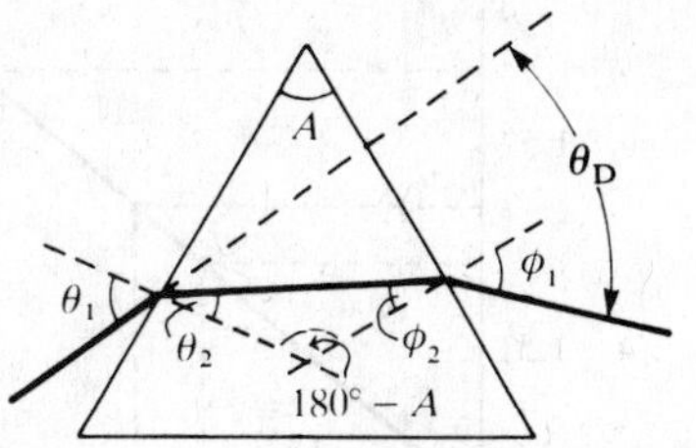

Fig. 34-36

I The situation is similar to that of Prob. 34.73 (see Fig. 34-35), but now the required condition is (using the small-angle formula of Prob. 34.72)

deviation of D line by crown-glass prism = deviation of D line by flint-glass prism

$$A(n_D - 1) = A'(n'_D - 1) \qquad (12°)(1.520 - 1) = A'(1.650 - 1) \qquad A' = 9.6°$$

34.75 Compute the *dispersive power* of a light flint glass in the region between the C and F spectral lines. The D line is taken as the mean ray between lines C and F. Indices of the glass for C, D, F lines: $n_C = 1.571$, $n_D = 1.575$, $n_F = 1.585$.

I By definition, the dispersive power of the flint glass between the Fraunhofer lines C and F

$$= \frac{\text{dispersion by small-angle prism between the C and F lines}}{\text{deviation of the mean D line}}$$

$$= \frac{A(n_F - n_C)}{A(n_D - 1)} = \frac{n_F - n_C}{n_D - 1} = \frac{1.585 - 1.571}{1.575 - 1.000} = \underline{0.0243}$$

34.76 The index of refraction of heavy flint glass is 1.68 at 434 nm and 1.65 at 671 nm. Calculate the difference in the angle of deviation of blue (434 nm) and red (671 nm) light incident at 65° on one side of a heavy-flint-glass prism with apex angle 60°.

I Here we cannot use the small-angle approximation, or the minimum deviation formula. Instead we go back to basics and follow the procedure of Prob. 34.43. (See Fig. 34-36.) For the case with $n_2 = 1.65$, Snell's law applied to the first surface gives

$$\sin \theta_2 = \frac{n_1 \sin \theta_1}{n_2} = \frac{(1.00)(\sin 65°)}{1.65} = 0.549 \qquad \text{or} \qquad \theta_2 = 33.3°$$

From geometry $\phi_2 = A - \theta_2 = 60° - 33.3° = 26.7°$. Snell's law applied to the second surface thus gives

$$\sin \phi_1 = \frac{n_2 \sin \phi_2}{n_1} = \frac{(1.65)(\sin 26.7°)}{1.00} = 0.741 \qquad \text{or} \qquad \phi_1 = 47.8°$$

Figure 34-36 shows a ray traced through the prism. The angle of deviation is

$$\theta_D = \theta_1 + \phi_1 - A = 65° + 47.8° - 60° = 52.8°$$

We now repeat the above steps with $n_2 = 1.68$ instead of 1.65. The results are

$$\theta_2 = 32.6° \qquad \phi_2 = 27.4° \qquad \phi_1 = 50.7° \qquad \theta_D = 55.7°$$

Consequently, the difference in θ_D for blue and red light is $\Delta \theta_D = 55.7° - 52.8° = \underline{2.9°}$.

34.77ᶜ The constants A and B in the Cauchy equation for a certain glass are 1.5020 and 5.2×10^3 nm². What is the index of refraction of the glass at 500 nm? What is the dispersion at this wavelength?

I The Cauchy equation gives the refractive index as $\mu = A + B/\lambda^2 = 1.5020 + 5200/250\,000 = 1.5228$; the dispersion is $d\mu/d\lambda = -2B/\lambda^3 = \underline{8.3 \times 10^{-5} \text{ nm}^{-1}}$.

34.78 Suppose that the index of refraction of glass, μ, has been measured as a function of λ and found to be

λ, nm	656.3	589.0	486.2
μ	1.514	1.517	1.524

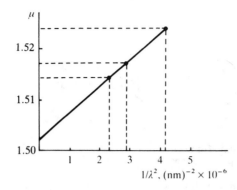

Fig. 34-37

If these data obey the Cauchy relation, how should they be plotted to obtain a straight line? Make such a plot and determine A and B. What should μ be at 500 nm?

❚ Plot μ versus $1/\lambda^2$, since $\mu = A + B/\lambda^2$. A straight line of slope B and intercept A results. From the plot, Fig. 34-37, $A = \underline{1.5020}$ and $B = \underline{5200\ nm^2}$; so $\mu = 1.5020 + [5200/(2.50 \times 10^5)] = \underline{1.5228}$.

34.79 A lens is made from the glass for which data are given in Prob. 34.77. If the lens images the sun at a distance of 12.000 cm when red light (656 nm) from the sun is used, where will an image of the sun appear when 486-nm light is used? This defect of a lens is called *chromatic aberration*.

❚ Using the Cauchy relation of Prob. 34.77 one obtains $\mu_R = 1.5141$ and $\mu_B = 1.5240$. The lensmaker's equation (see Prob. 35.67) yields expressions for $1/f_B$ and $1/f_R$; $(1/f_R)/(1/f_B)$ yields $f_B = [(\mu_R - 1)f_R]/(\mu_B - 1) = [0.5141(12)]/0.5240 = \underline{11.77\ cm}$, where we have used the fact that light from a distant object converges to the focal point of the lens.

34.80 (*a*) What is the wavelength of the spectral color that is the complement of 485 nm? (*b*) Is there a spectral color that is the complement of 520 nm?

❚ A straight line produced from 485 nm through C on an ICI chromaticity diagram (Fig. 34-38) intersects the other side of the spectral curve at 588 nm. Thus 588 nm is the spectral complement of 485 nm.
The complement of 520 nm is red-purple, which is not a spectral color.

34.81 (*a*) What is the color P with ICI chromaticity coordinates $(x, y) = (0.2, 0.5)$? (*b*) What are the coordinates of the color which when mixed in equal proportions with P gives standard white (C)?

❚ Use Fig. 34-39, a simplification of Fig. 34-38. (*a*) P is in the <u>green</u> region of the diagram. (*b*) To find the complement of P, draw a line from P to C, and extend it an equal distance beyond C to Q. The coordinates of Q are seen to be $\underline{(0.42, 0.14)}$.

34.82 What proportions of the spectral colors with wavelengths 480, 510, and 610 nm are required to produce standard white (C)?

❚ We refer to Fig. 34-38, which is roughly reproduced as in Fig. 34-40. Draw one line between the points 480 and 510 nm and another between 610 nm and C. Extend the second line until it intersects the first at P. Then P is the color composed of 510 and 480 nm, which when mixed with 610 nm yields C. The fractions of 510 and 480 nm in P are

$$f_{510} = \frac{d_1}{d_1 + d_2} = 0.28 \qquad f_{480} = \frac{d_2}{d_1 + d_2} = 0.72$$

The fractions of P and 610 nm in C are

$$f_P = \frac{d_2'}{d_1' + d_2'} = 0.59 \qquad f_{610} = \frac{d_1'}{d_1' + d_2'} = 0.41$$

Since P is 28 percent 510 nm, C is composed of $(0.28)(0.59) = 0.165$ part 510 nm and $(0.72)(0.59) = 0.425$ part 480 nm. Then 480, 510, and 610 nm are in the ratio $\underline{0.425 : 0.165 : 0.41}$.

34.83 What proportions of the spectral colors with wavelengths 485, 520, and 600 nm are required to produce the green color whose chromaticity coordinates are $(0.25, 0.40)$?

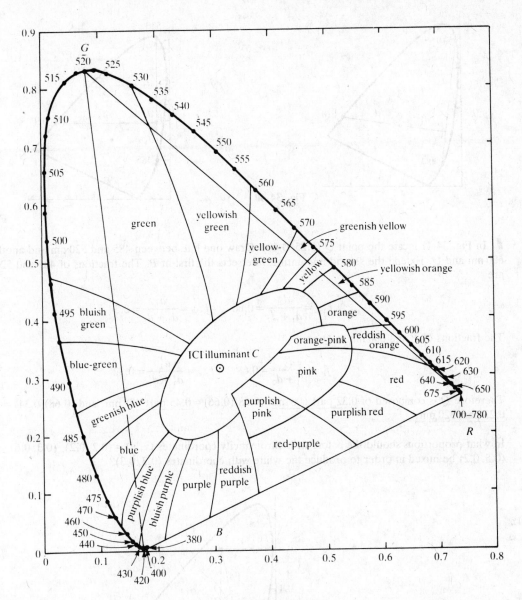

Fig. 34-38

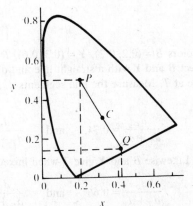

Fig. 34-39

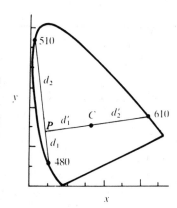

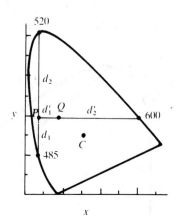

Fig. 34-40

Fig. 34-41

❚ In Fig. 34-41 locate the point Q (0.25, 0.40). Draw one line between 485 and 520 nm and another between 600 nm and Q. Extend the second line until it intersects the first at P. The fractions of 485 and 520 nm in P are

$$f_{485} = \frac{d_2}{d_1 + d_2} = 0.66 \qquad f_{520} = \frac{d_1}{d_1 + d_2} = 0.34$$

The fractions of P and 600 nm in Q are

$$f_P = \frac{d_2'}{d_1' + d_2'} = 0.68 \qquad f_{600} = \frac{d_1'}{d_1' + d_2'} = 0.32$$

Therefore Q is composed of 0.32 part 600 nm, (0.68)(0.66) = 0.45 part 485 nm, and (0.68)(0.34) = 0.23 part 520 nm.

34.84 In what proportions should the colors with chromaticity coordinates $(x, y) = (0.2, 0.2)$, $(0.3, 0.6)$, and $(0.5, 0.2)$ be mixed in order to produce the white with coordinates $(0.3, 0.3)$?

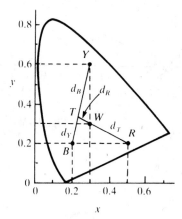

Fig. 34-42

❚ In Fig. 34-42 locate the colors $B = (0.2, 0.2)$, $Y = (0.3, 0.6)$, $R = (0.5, 0.2)$, and $W = (0.3, 0.3)$ on the chromaticity diagram. Connect B and Y with a straight line and draw a second line from R to W and extend it until it intersects the first line at T. Measure the line segments d_R, d_T, d_B, and d_Y. T and R give W, when mixed in the proportions

$$\frac{d_T}{d_T + d_R} = 0.74 \qquad \text{and} \qquad \frac{d_R}{d_T + d_R} = 0.26$$

That is, $W = 0.74T + 0.26R$. Likewise B and Y give T when mixed in the proportions

$$\frac{d_B}{d_B + d_Y} = 0.66 \qquad \text{and} \qquad \frac{d_Y}{d_B + d_Y} = 0.34$$

That is, $T = 0.66B + 0.34Y$. Substituting this last expression for T into the above expression for W, we get

$$W = (0.74)(0.66B + 0.34Y) + 0.26R = 0.49B + 0.25Y + 0.26R$$

That is, B, Y, and R, when mixed in the proportions $\underline{0.49:0.25:0.26}$, give W.

34.3 PHOTOMETRY AND ILLUMINATION

34.85   Light from a projection lantern provides an illumination of 12 000 lm/m² on a wall perpendicular to the beam and at a distance of 5 m from the source. What intensity must an isotropic source have to give this same illumination at a distance of 5 m?

▮   For an isotropic source and normal incidence, $E = I/r^2$. Therefore, $I = r^2 E = (5 \text{ m})^2(12\,000 \text{ lm/m}^2) = \underline{3 \times 10^5 \text{ cd}}$.

34.86   What is the luminous intensity of a 200-W tungsten lamp whose efficiency is 18 lm/W?

▮   The total luminous flux is $F = (200 \text{ W})(18 \text{ lm/W}) = 3600 \text{ lm}$. Assuming the flux to fill a solid angle of 4π steradians, the luminous intensity is $I = (3600 \text{ lm})/(4\pi \text{ sr}) = \underline{286 \text{ cd}}$.

34.87   A 40-W, 110-V lamp is rated 11.0 lm/W. At what distance from the lamp is the maximum illumination 5 lm/m²?

▮   Assuming the lamp to be an isotropic light source, we have $E_{\max} = F/(4\pi r^2) = I/r^2$. From the data, $F = (40 \text{ W})(11.0 \text{ lm/W}) = \underline{440 \text{ lm}}$, and $I = F/(4\pi) = 35.0 \text{ cd}$. Also $E_{\max}$ is given as 5 lm/m². Then $5.0 = 35.0/r^2$, and $r = \underline{2.65 \text{ m}}$.

34.88   A small light source having a uniform luminous intensity of 10 cd is situated at the center of a spherical surface of radius 1 m. What is the illumination on the spherical surface?

▮   A 10-cd source emits $4\pi \times 10 = 40\pi$ lm.

$$E = \frac{F}{A} = \frac{F}{4\pi R^2} = \frac{40\pi}{4\pi(1^2)} = \underline{10 \text{ lm/m}^2}$$

34.89   How many lumens pass through an area of 0.3 m² on a sphere of radius 4 m, if an 800-cd isotropic source is at its center? What is E for the area?

▮   The area or the sphere is perpendicular to the flow of light. The illumination is then $E = I/r^2 = (800 \text{ cd})/(4.0 \text{ m})^2 = \underline{50 \text{ lm/m}^2}$ and the flux through the area is $\Delta F = EA = (50)(0.3) = \underline{15 \text{ lm}}$.

34.90   An isotropic point source has an intensity of 200 cd. (**a**) How much luminous flux is emitted by the source? (**b**) How much flux strikes a 2-cm² area on a tabletop 80 cm directly below the source? (**c**) What is the illuminance at that point on the tabletop?

▮   (**a**) For an isotropic source $F = 4\pi I = (4\pi \text{ sr})(200 \text{ cd}) = \underline{2512 \text{ lm}}$. (**b**) For the small area $\Delta A = 2 \text{ cm}^2$ at a perpendicular distance of $r = 80 \text{ cm}$ from the source, the solid angle subtended at the source is $\Delta \Omega = \Delta A/r^2 = 2/80^2 = 3.125 \times 10^{-4} \text{ sr}$. Then the total flux hitting the area is $\Delta F = I \, \Delta \Omega = (200 \text{ cd})(3.125 \times 10^{-4}) = \underline{0.0625 \text{ lm}}$. (**c**) $E = E_{\max} = I/r^2 = (200 \text{ cd})/(0.80 \text{ m})^2 = \underline{313 \text{ lm/m}^2}$. Alternatively, from part (**b**), $E = \Delta F/\Delta A = (0.0625 \text{ lm})/(2.0 \times 10^{-4} \text{ m}^2) = \underline{313 \text{ lm/m}^2}$.

34.91   Determine the illumination on a surface 7-ft distant from a 125-cd source if (**a**) the surface is normal to the light rays, and (**b**) the surface makes an angle of 15° with the rays.

▮   (**a**) $E = E_{\max} = I/r^2 = (125 \text{ cd})(7 \text{ ft})^2 = \underline{2.55 \text{ lm/ft}^2}$. (**b**) $E = E_{\max} \cos \theta$, where θ is the angle between the normal to the surface and the incoming rays. For our case $\theta = 85°$ and $E = (2.55 \text{ lm/ft}^2) \cos 85° = \underline{0.222 \text{ lm/ft}^2}$.

34.92   Compute the illumination at the edge of a circular table of radius 1 m, if a 200-cd source is suspended 3 m above its center.

▮ $E = (I/r^2) \cos \theta$, where (Fig. 34-43) $r = (3^2 + 1^2)^{1/2} = 3.16 \text{ m}$, and $\cos \theta = (3 \text{ m})/(3.16 \text{ m}) = 0.949$. Then, $E = (200 \text{ cd}/10.0 \text{ m}^2)(0.949)$, or $E = \underline{19.0 \text{ lm/m}^2}$.

34.93   A frosted light bulb, which may be taken as an isotropic source, gives an illuminance of 8 lm/m² at a distance of 5 m. Compute (**a**) the total luminous flux from the bulb and (**b**) its luminous intensity.

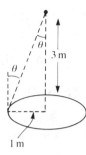

Fig. 34-43

❚ Given no other information, we can assume that the illuminance given is the maximum illuminance, i.e., for perpendicular incidence. Then (*a*) $F = E(4\pi r^2) = (8 \text{ lm/m}^2)(4\pi)(25 \text{ m}^2) = \underline{2513 \text{ lm}}$. (*b*) $I = F/(4\pi) = \underline{200 \text{ cd}}$, or $I = Er^2 = \underline{200 \text{ cd}}$.

34.94 Compute the illumination *E* of a small surface at a distance of 120 cm from an isotropic point source of luminous intensity 72 cd, if (*a*) the surface is normal to the luminous flux and (*b*) the normal to the surface makes an angle of 30° with the light rays.

❚ (*a*)
$$E = \frac{I}{r^2} = \frac{72 \text{ lm/sr}}{(1.2 \text{ m})^2} = \underline{50 \text{ lm/m}^2}$$

As in the case of other angle measures, the steradian is not a unit in the usual sense and does not carry through in equations.

(*b*)
$$E = E_{\text{max}} \cos \theta = \frac{I}{r^2} \cos \theta = (50 \text{ lm/m}^2)(\cos 30°) = \underline{43 \text{ lm/m}^2}$$

34.95 A light meter reads the illumination received from the sun as 10^5 lm/m^2. If the distance from the earth to the sun is 1.5×10^{11} m, what is the luminous intensity of the sun?

❚ We assume that the light meter is held with its face to the sun for maximum illumination. Then
$$I = Er^2 = (10^5 \text{ lm/m}^2)(1.5 \times 10^{11} \text{ m})^2 = 2.25 \times 10^{27} \text{ lm/sr} = \underline{2.25 \times 10^{27} \text{ cd}}.$$

34.96 A street lamp 6 m above a sidewalk dissipates 100 W and has a luminous efficiency of 50 lm/W. Assuming that the source is isotropic, find the illuminance at two locations : (*a*) on the sidewalk directly beneath the lamp and (*b*) at a point on the sidewalk 8 m from the base of the lamp.

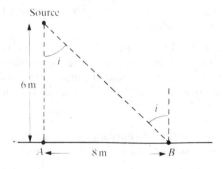

Fig. 34-44

❚ The situation is sketched in Fig. 34-34. (*a*) The flux *F* from the lamp is $(100 \text{ W})(50 \text{ lm/W}) = 5000 \text{ lm}$, and the luminous intensity *I* is $F/(4\pi) = [5000/(4\pi)]\text{cd} = 398 \text{ cd}$. Directly below the lamp the illuminance *E* is
$$E = \frac{I}{R^2} = \frac{398}{36} = \underline{11.1 \text{ lm/m}^2}$$

(*b*) The point on the sidewalk 10 m from the lamp is 8 m from the point under the lamp; for it,
$$\cos i = 0.6 \quad \text{and} \quad E = \frac{I \cos i}{R^2} = \frac{(398)(0.6)}{100} = \underline{2.4 \text{ lm/m}^2}$$

34.97 An air-cooled searchlight bulb has an efficiency of 25 lm/W. The searchlight is powered by a generator of 80 percent efficiency to which power is supplied by a belt passing around a pulley of 16 cm diameter. If the difference in tension of the two portions of the belt is 98 N, what is the angular speed of the generator required to give 40 lm/m² of illumination of the ground? The reflector system of the searchlight concentrates all the light into a circle on the ground 50 m in diameter.

▌ The total flux F of the searchlight is $EA = (40 \text{ lm/m}^2)(\pi)(25 \text{ m})^2 = 78\,500 \text{ lm}$. The wattage output of the searchlight is then $(78\,500 \text{ lm})/(25 \text{ lm/W}) = 3140 \text{ W}$. Since the generator is 80 percent efficient, the power input at the generator belt, P, is given by $(3140 \text{ W}) = 0.80P$ and $P = 3925 \text{ W}$. But $P = \Gamma\omega$, where Γ is the net torque on the belt and $\omega = 2\pi f$ is the angular velocity.

$$\Gamma = (T_2 - T_1)R = (98 \text{ N})(0.08 \text{ m}) = 7.84 \text{ N} \cdot \text{m} \quad \text{and hence} \quad \omega = \frac{P}{\Gamma} = \frac{3925 \text{ W}}{7.84 \text{ N} \cdot \text{m}} = \underline{501 \text{ rad/s}}$$

$$\text{or} \quad f = \omega/2\pi = \underline{79.8 \text{ rev/s}}$$

34.98 How much should a 60-W lamp be lowered to double the illumination on an object which is 60 cm directly under it?

▌ Directly facing a source, a surface's illumination is $E = I/r^2$. Then for two different distances, $E_1/E_2 = r_2^2/r_1^2$. For $E_2 = 2E_1$ and $r_1 = 60$ cm, we get $r_2^2 = \frac{1}{2}(60 \text{ cm})^2$ and $r_2 = (60 \text{ cm})/\sqrt{2} = 42.4$ cm. Then the lamp should be lowered $r_1 - r_2 = \underline{17.6 \text{ cm}}$.

34.99 At what distance from a screen will a 27-cd lamp provide the same illumination as a 75-cd lamp 15 ft from the screen?

▌ $E_1 = E_2 \Rightarrow I_1/r_1^2 = I_2/r_2^2$, so $27/r_1^2 = 75/15^2 = 0.333$; $r_1^2 = 27/0.333 = 81 \text{ ft}^2$; $r_1 = \underline{9 \text{ ft}}$.

34.100 An unknown lamp placed 90 cm from a photometer screen gives the same illumination as a standard lamp of 32 cd placed 60 cm from the screen. Compute the luminous intensity I_1 of the lamp under test.

▌ $$\frac{I_1}{I_2} = \frac{r_1^2}{r_2^2} \quad \text{or} \quad I_1 = 32 \text{ cd} \frac{(90 \text{ cm})^2}{(60 \text{ cm})^2} = \underline{72 \text{ cd}}$$

34.101 Two lamps of 5 and 20 cd are 150 cm apart. At what point between them will the same illumination be produced by each?

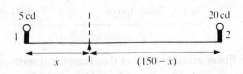

Fig. 34-45

▌ We have $E_1 = E_2$ at the location of interest, shown by the vertical dashed line in Fig. 34-45. Then $I_1/x^2 = I_2/(150 - x)^2$, or $5/x^2 = 20/(150 - x)^2$. Then $(150 - x)^2 = 4x^2$, or $x^2 + 100x - 7500 = (x + 150)(x - 50) = 0$. Thus $x = \underline{50 \text{ cm}}$ is the solution between the two lamps and is the distance from the 5-cd source. The other solution, $x = -150$, corresponds to a distance 150 cm to the left of the 5-cd source, at which location the illuminance of the two sources is again the same.

34.102 A long, straight, fluorescent lamp gives an illumination of E_1 at a radial distance of r_1. Find E_2 at r_2. Assume the length of the bulb to be much greater than r_1 and r_2, so that the effects of the ends of the bulb can be ignored.

▌ Consider two short cylindrical shells, of length L and radii r_1 and r_2, coaxial with the lamp. The flux through each is the same. Assume negligible flux through the ends. Then

$$A_1E_1 = A_2E_2 \quad \text{or} \quad 2\pi r_1 LE_1 = 2\pi r_2 LE_2$$

from which
$$E_2 = E_1 \frac{r_1}{r_2}$$

Note that the illumination decreases as $1/r$, unlike the $1/r^2$ relation that applied to a point source.

CHAPTER 35
Mirrors, Lenses, and Optical Instruments

35.1 MIRRORS

35.1 Give the equation governing thin lenses and thin spherical mirrors.

▮ The basic equation can be written as

$$\frac{1}{p} + \frac{1}{q} = \frac{1}{f} \tag{1}$$

where p = distance of object, q = distance of image, and f = focal length, and distances are measured to the lens or mirror along the symmetry axis. This formula is valid only for thin lenses and mirrors whose dimensions are small compared with the radii of curvature involved, and for light rays coming in at shallow angles to the symmetry axis (paraxial rays).

The sign conventions for p, q, f are given in Table 35-1. Light rays diverge from a point on a *real object* but converge to a point on a *virtual object*. Similarly, light rays converge toward a point on a *real image* but diverge from a point on a *virtual image*. Other symbols commonly used instead of (p, q) are: (p, p'), (s, s'), (s_0, s_i), (d_o, d_i), etc.

TABLE 35-1

quantity	sign +	sign −
f	Converging lens, concave mirror	Diverging lens, convex mirror
p	Real object	Virtual object
q	Real image	Virtual image

35.2 What is the formula for linear magnification of thin lenses and mirrors.

▮ If h_o and h_i refer to the heights (algebraic) of object and image, respectively, then using the notation of Prob. 35.1,

$$\frac{h_i}{h_o} = \frac{-q}{p} \tag{1}$$

h_o and h_i are positive when object and image, respectively, are erect and they are negative when object and image are inverted. Other symbols commonly used for (h_0, h_i) are (h, h'), (S_o, S_i), (l_o, l_i), etc.

35.3 Describe the image that a plane mirror forms of a real object (an illuminated material body).

▮ In Fig. 35-1, $P_1 P_2$ represents a real object at distance d_o from the reflecting surface of mirror M.

Draw any number of "real rays," as a, b, c, from P_1 to the mirror. Corresponding reflected rays (constructed by applying the law of reflection) are a', b', c'. Projected backward as dashed lines, the *virtual rays* intersect at P_1', which locates one point on the virutal image $P_1' P_2'$. When rays from any point on the real object are treated in the same way, a corresponding point is located on $P_1' P_2'$. Hence, the entire image can be located.

The image is erect. As can be shown from simple geometry, $\overline{P_1 P_2}$ (length of object, l_0) is equal to $\overline{P_1' P_2'}$ (length of image, l_i). Likewise d_o and d_i are equal in magnitude. An eye placed as shown would see an arrow $P_1' P_2'$ *behind* the mirror.

We check the result using the formulas of Probs. 35.1 and 35.2. For a plane mirror, $f = \infty$. Thus, (1) of Prob. 35.1 gives $d_i = -d_o < 0$ and (1) of Prob. 35.2 gives $l_i = l_o > 0$.

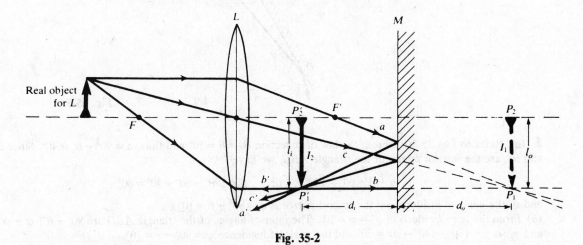

Fig. 35-1

35.4 Describe the image formed by a plane mirror of a virtual object.

❚ A virtual object is an intercepted real image (produced by some auxiliary optical system). Thus, in Fig. 35-2, the auxiliary converging lens L forms the real image I_1 which serves as a virtual object for plane mirror M. The ray construction shows the image to be real (the reflected rays a', b', c' converge toward it), erect (same orientation as I_1), and of the same size and distance from the mirror as I_1.

We can check the result using the formulas of Probs. 35.1 and 35.2. Assuming a focal length of +15 cm for the lens, a real object at a distance of 45 cm from the lens, and a separation of 18 cm between lens and mirror, we find

$$\frac{1}{45} + \frac{1}{x} = \frac{1}{15} \qquad \text{or} \qquad x = 22.5 \text{ cm}$$

i.e., I_1 is $22.5 - 18 = 4.5$ cm behing M. Then, using $d_o = -4.5$ cm (virtual object), call $l_o > 0$, and $f = \infty$, we obtain

$$d_i = \underline{+4.5 \text{ cm}} \text{ (real, same distance)} \qquad l_i = l_o \text{ (erect, same size)}$$

We could say that the net effect of M has been to move I_1 9.0 cm closer to the lens.

Fig. 35-2

35.5 Two mirrors are arranged at right angles to each other, as shown in Fig. 35-3. A ray of light incident on the horizontal mirror at an angle θ is reflected toward the vertical mirror. Show that after being reflected from the vertical mirror, the ray emerges parallel to the incoming ray.

❚ The incident and the second reflected ray make alternate angles 2θ with the first reflected ray; therefore, they are parallel.

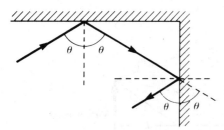

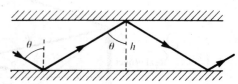

Fig. 35-3   **Fig. 35-4**

35.6   Two mirrors, each 1.6 m long, are facing each other. The distance between the mirrors is 20 cm. A light ray is incident on one end of one of the mirrors at an angle of incidence of 30°. How many times is the ray reflected before it reaches the other end?

❚ As is shown by Fig. 35-4, the ray travels the horizontal distance $d = h \tan \theta = (20 \text{ cm})(\tan 30°) = 11.54 \text{ cm}$ between each reflection. Since

$$\frac{L}{d} = \frac{160 \text{ cm}}{11.54 \text{ cm}} = 13.86$$

there are <u>14</u> reflections, counting the first one.

35.7   In Fig. 35-5(a) a ray of light is incident at 50° on the middle of one of a pair of mirrors arranged at 60° to each other. (a) Calculate the angle at which the ray is incident on the second mirror. (b) Calculate the angle at which the ray is incident on the first mirror after being reflected from the second mirror.

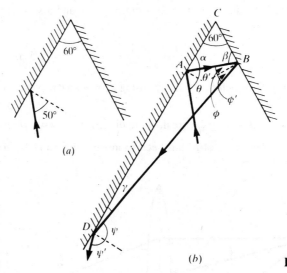

Fig. 35-5

❚ (a) Refer to Fig. 35-5(b). From the law of reflection $\theta' = \theta = 50°$ and thus, $\alpha = 90° - \theta' = 40°$. Since α, β, and 60° are the interior angles of the triangle ABC, we have

$$\alpha + \beta + 60° = 180° \quad \text{or} \quad \beta = 180° - 60° - 40° = 80°$$

and so the angle of incidence on the second mirror is $\phi = 90° - \beta = \underline{10°}$.
(b) From the law of reflection $\phi' = \phi = 10°$. The interior angles of the triangle ABD are $90° + \theta'$, $\phi + \phi'$, and γ, so $\gamma = 180° - 140° - 20° = 20°$ and the angle of incidence $\psi = 90° - \gamma = \underline{70°}$.

35.8   When you stand with your nose 20 cm in front of a plane mirror, for what distance must you focus your eyes in order to see your nose in the mirror? If your right eye is blue and your left eye is green, your image behind the plane mirror will have a green right eye and blue left eye. Explain.

❚ The image is 20 cm beind the mirror, so the eye should focus at <u>40 cm</u>. In the plane mirror each part of the object has its image behind the mirror along a perpendicular to the mirror. Hence the left eye becomes the right eye in the image, and vice-versa. In general, left and right are interchanged.

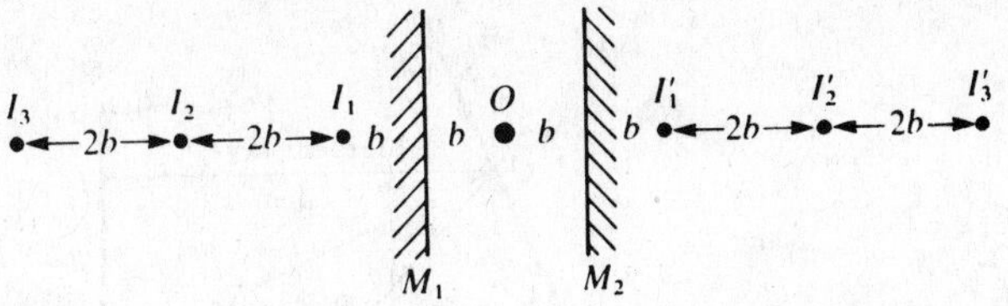

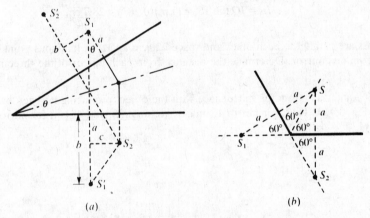

Fig. 35-6

35.9 If an object is placed between two parallel mirrors, an infinite number of images results. Suppose that the mirrors are a distance $2b$ apart and the object is put at the midpoint between the mirrors. Find the distances of the images from the object.

▌ The images are shown in Fig. 35-6. I_1 and I_1' are the direct images of O. I_1 acts as an object for M_2 giving image I_2', and so on. The images are at $2nb$ from the object with n an integer.

35.10 Two plane mirrors intersect at an angle θ. An object is placed on the bisector of the angle between them. Find the location of the closest four images if **(a)** $\theta = 30°$ and **(b)** $\theta = 120°$. Show the images on a diagram.

▌ **(a)** The situation for a vertex angle of 30° is shown in Fig. 35-7(a). S_1 is the image of source S in the top mirror and is distance a behind the mirror. S_1 is a source for the bottom mirror, giving image S_1' distance b below the mirror. The situation is repeated in the other mirror to give S_2 and S_2'. Relevant distances b and c are found in terms of a from the geometry of the problem to be $b = a(2\cos\theta + 1)$ and $c = (b - a)\tan\theta$.
(b) For a vertex angle of 120°, there is one image located distance a behind each mirror. No other images appear because the initial images lie in the planes of the respective mirrors [see Fig. 35-7(b)].

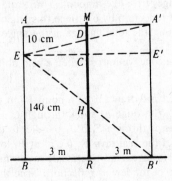

Fig. 35-7

35.11 A boy is 1.50 m tall and can just see his image in a vertical plane mirror 3 m away. His eyes are 1.40 m from the floor level. Determine the vertical dimension and elevation of the mirror.

▌ In Fig. 35-8, let AB represent the boy. His eyes are at E. Then $A'B'$ is the image of AB in mirror MR, and DH represents the shortest mirror necessary for the eye to view the image $A'B'$.
    Triangles DEC and $DA'M$ are congruent and so $\overline{CD} = \overline{DM} = 5$ cm. Triangles HRB' and HCE are congruent and so $\overline{RH} = \overline{HC} = 70$ cm. The dimension of the mirror is $\overline{HC} + \overline{CD} = \underline{75\ cm}$ and its elevation is $\overline{RH} = \underline{70\ cm}$.

Fig. 35-8

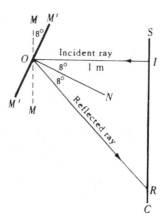

Fig. 35-9

35.12 A shown in Fig. 35-9, a light ray *IO* is incident on a small plane mirror which is attached to a galvanometer coil. The mirror reflects this ray upon a straight scale *SC* which is 1 m distant and parallel to the undeflected mirror *MM*. When the current through the instrument has a certain value, the mirror turns through an angle of 8° and assumes the position *M'M'*. Across what distance on the scale will the spot of light move?

▌ When the mirror turns through 8°, the normal to it also turns through 8° and the incident ray makes an angle of 8° with the normal *NO* to the deflected mirror *M'M'*. Because the incident ray *IO* and the reflected ray *OR* make equal angles with the normal, angle *IOR* is twice the angle through which the mirror has turned, or 16°.

$$\overline{IR} = \overline{IO} \tan 16° = (1 \text{ m})(0.287) = \underline{28.7 \text{ cm}}$$

35.13 Two plane mirrors are parallel to each other and spaced 20 cm apart. A luminous point is placed between them and 5 cm from one mirror. Determine the distance from each mirror of the three nearest images in each.

▌ Each image is equidistant from the mirror face with the object producing it. Then by construction the images are as in Fig. 35-10. Thus, for mirror 1, images are at <u>5, 35, and 45 cm</u>; for mirror 2 at <u>15, 25, and 55 cm</u>.

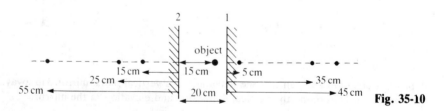

Fig. 35-10

35.14 A ray of light makes an angle of 25° with the normal to a plane mirror. If the mirror is turned through 6°, making the angle of incidence 31°, through what angle is the reflected ray rotated?

▌ The incoming ray is fixed. By rotating the mirror in the plane of incidence the angle of incidence changes, and the angle of reflection changes by a like amount. Thus, rotating the mirror by 6° to increase the angle of incidence increases the angle of reflection by 6°. Since this angle is measured off the normal to the mirror which itself has rotated through 6°, the total change in the angle of the reflected ray is <u>12°</u>.

35.15 Describe the image formed by a concave mirror when a real object is situated outside the center of curvature, *C*.

▌ In Fig. 35-11, real rays *a, b, c* are drawn from a point P_1 on the real object P_1P_2. For convenience *a* is drawn parallel to the optic axis, *b* through *F*, and *c* through *C*. From the geometry of the mirror and the law of reflection, reflected ray *a'* passes through *F*, *b'* is parallel to the optic axis, and *c'* returns through *C* along the path of *c*. The intersection of the real rays *a', b', c'* at P_1' locates one point on the real image $P_1'P_2'$. Rays *a, b, c* may seem very special. However, any (paraxial) ray from P_1, after reflection, passes through P_1', as may be shown by an application of the law of reflection. Note that the image is inverted.

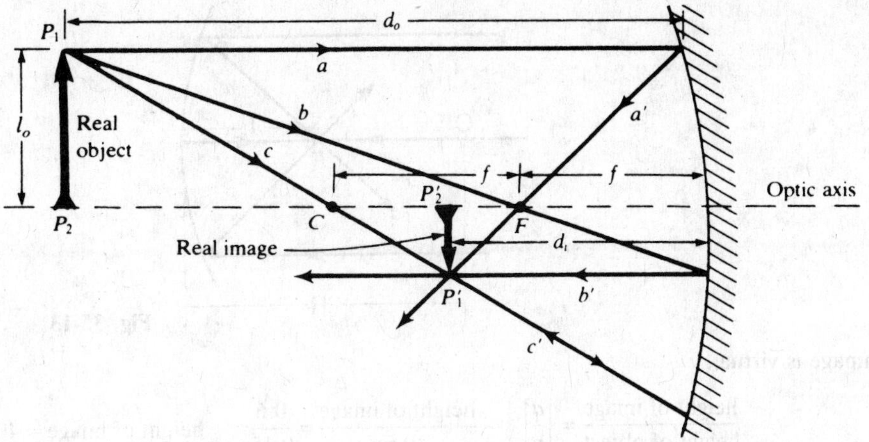

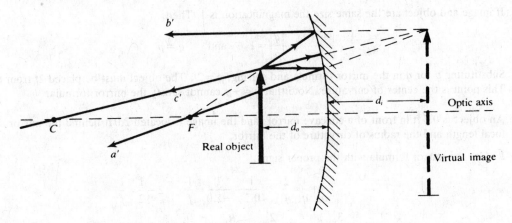

Fig. 35-11

Let $f = +20$ cm, $d_o = +45$ cm, $l_o = +5$ cm. Then

$$\frac{1}{45} + \frac{1}{d_i} = \frac{1}{20} \quad \text{or} \quad d_i = +36 \text{ cm} \quad \text{(a real image)}$$

$$\frac{45}{36} = -\frac{5}{l_i} \quad \text{or} \quad l_i = -4 \text{ cm} \quad \text{(an inverted, minified image)}$$

35.16 Describe the image formed by a concave mirror when a real object is situated inside the focal point, F.

❚ See Fig. 35-12. Constructing rays as in Prob. 35.15, we find the image to be erect, virtual (behind the mirror), and magnified.

Letting $f = +20$ cm, $d_o = +15$ cm, $l_o = +5$ cm, we have from the lens equations $d_i = -60$ cm (virtual) and $l_i = +20$ cm (erect, magnified).

Fig. 35-12

35.17 A concave mirror has a radius of curvature of 0.80 m. Where does this mirror bring sunlight to a focus?

❚ Sunlight consists of rays parallel to the principal axis. After reflection, the rays must all pass through the principal focus (Fig. 35-13): $f = R/2 = 0.80/2 = 0.40$ m (focal length).

35.18 An object is placed 0.15 m from a concave mirror of focal length 0.20 m. (a) Where is the image produced? (b) If the object is 10 cm high, how high is the image?

❚ (a)

$$\frac{1}{p} + \frac{1}{q} = \frac{1}{f} \qquad \frac{1}{0.15} + \frac{1}{q} = \frac{1}{0.20} \qquad q = -0.6 \text{ m}$$

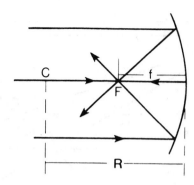

Fig. 35-13

The image is virtual.

(b) $\quad \dfrac{\text{height of image}}{\text{height of object}} = \left|\dfrac{q}{p}\right| \qquad \dfrac{\text{height of image}}{10 \text{ cm}} = \dfrac{0.6}{0.15} \qquad \text{height of image} = \underline{40 \text{ cm}}$

35.19 An object 10 cm high is 50 cm from a concave mirror of 20 cm focal length. Find the image distance, height, and direction.

▌ $\qquad \dfrac{1}{p} + \dfrac{1}{q} = \dfrac{1}{f} \quad \text{or} \quad \dfrac{1}{50} + \dfrac{1}{q} = \dfrac{1}{20} \quad \text{and} \quad q = \dfrac{100}{3} = \underline{33.3 \text{ cm}}$

$\qquad \dfrac{h_i}{h_o} = -\dfrac{q}{p} \approx \dfrac{-33.3}{50} = -0.666 \quad \text{and} \quad h_i = -6.66 \text{ cm}$

Thus image is real, <u>6.66 cm high and inverted</u>.

35.20 From the mirror formula, determine where the object must be placed if the image is to be the same size as the object. (The focal length of the concave mirror is f.)

▌ $\qquad\qquad\qquad \dfrac{1}{p} + \dfrac{1}{q} = \dfrac{1}{f}$

If image and object are the same size the magnification is 1. Then,

$\qquad\qquad\qquad \left|\dfrac{q}{p}\right| = 1 \quad \text{and} \quad q = p$

Substituting p for q in the mirror formula and solving, $p = 2f$. The object must be placed $2f$ from the mirror. This point is the center of curvature. Note that $q = -p$ cannot satisfy the mirror formula.

35.21 An object is 0.5 ft in front of a concave mirror, and the image is located 2.0 ft behind the mirror. Find the focal length and the radius of curvature of the mirror.

▌ Use the mirror formula with the proper signs.

$\qquad \dfrac{1}{p} + \dfrac{1}{q} = \dfrac{1}{f} \qquad \dfrac{1}{0.5} + \dfrac{1}{-2.0} = \dfrac{1}{f} \qquad 2 - \dfrac{1}{2} = \dfrac{1}{f}$

$\qquad \dfrac{3}{2} = \dfrac{1}{f} \qquad f = \dfrac{2}{3} \text{ ft} \qquad \dfrac{R}{2} = f = \dfrac{2}{3} \text{ ft} \qquad R = \underline{1.33 \text{ ft}}$

35.22 The diameter of the sun subtends an angle of approximately 32 min (32′) at any point on the earth. Determine the position and diameter of the solar image formed by a concave spherical mirror of radius 400 cm. Refer to Fig. 35-14.

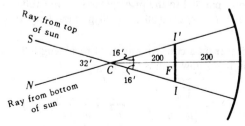

Fig. 35-14

❚ Since the sun is very distant, p is very large and $1/p$ is practically zero.

$$\frac{1}{p}+\frac{1}{q}=\frac{2}{r} \quad \text{or} \quad 0+\frac{1}{q}=\frac{2}{400}$$

Solving, $q = 200$ cm. The image is at the principal focus F, 200 cm from the mirror.

The diameter of the sun and its image II' subtend equal angles at the center of curvature C of the mirror.

$$\overline{I'I} = 2(\overline{I'F}) = 2(\overline{CF}\tan 16') = 2(200\text{ cm})(0.00465) = \underline{1.86\text{ cm}}$$

35.23 How far should an object be from a concave spherical mirror of radius 36 cm to form a real image one-ninth its size?

❚ The focal length is $R/2 = 18$ cm. Then

$$\frac{1}{p}+\frac{1}{q}=\frac{1}{18} \quad \text{and} \quad \frac{1}{9}=\left|\frac{q}{p}\right| \quad \text{with } q \text{ and } p \text{ positive (real object and image)}$$

Then $p = 9q$, so

$$\frac{1}{p}+\frac{9}{p}=\frac{1}{18} \quad \text{or} \quad \frac{10}{p}=\frac{1}{18} \quad \text{and} \quad p = \underline{180\text{ cm}}$$

35.24 It is desired to cast the image of a lamp, magnified 5 times, upon a wall 12 ft distant from the lamp. What kind of spherical mirror is required and what is its position?

❚ To cast the image on a wall it must be real so q is positive, as is p, the object distance. Thus the image is inverted and $q/p = 5$. Since $q > p$ we have, from the information given, $q - p = 12$ ft, or $5p - p = 12$ ft, and $p = 3$ ft. The mirror is thus $\underline{3\text{ ft}}$ from the lamp and $\underline{15\text{ ft}}$ from the wall. Furthermore $(1/p) + (1/q) = (1/f)$, or $\frac{1}{3}+\frac{1}{15} = 1/f$, and $f = \underline{2.5\text{ ft}}$. Thus the mirror is $\underline{\text{concave}}$ of radius $\underline{5\text{ ft}}$.

35.25 An object 12 mm high is placed 0.5 m from a concave mirror with a radius of curvature of 0.2 m. Find the focal length and the location, height, and orientation of the image.

❚ $$f=\frac{r}{2}=\frac{0.2\text{ m}}{2}=\underline{0.1\text{ m}} \quad \text{Then} \quad \frac{1}{p}+\frac{1}{q}=\frac{1}{f} \quad \frac{1}{0.5}+\frac{1}{q}=\frac{1}{0.1} \quad \frac{1}{q}=10-2=8 \quad q=\underline{0.125\text{ m}}$$

from the mirror. For height we have

$$\frac{-q}{p}=\frac{h_i}{h_o} \quad \frac{-0.125}{0.5}=\frac{h_i}{0.012} \quad \text{and} \quad h_i=-0.003\text{ m}=\underline{-3\text{ mm}}$$

The image is real and inverted.

35.26 As the position of an object reflected in a concave shell mirror of 0.25 m focal length is varied, the position of the image varies. Plot the image distance as a function of the object distance, letting the latter change from 0 to $+\infty$. Where is the image real? Where virtual?

❚ See Fig. 35-15. The image is real for q positive, or $p > f$, and virtual for q negative, or $p < f$.

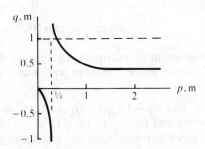

Fig. 35-15

35.27 An object 8 mm high is located 125 mm in front of a concave mirror which has a focal length of 200 mm. Find the position, size, and character of the image.

❚ Since $p < f$, the image is virtual. Quantitatively,

$$\frac{1}{125} + \frac{1}{q} = \frac{1}{200} \qquad \frac{1}{q} = \frac{5}{1000} - \frac{8}{1000} \quad \text{and}$$

$$q = \frac{-1000}{3} = \underline{-333\,\text{mm}} \qquad \text{Then} \qquad \frac{-q}{p} = \frac{h}{8} = \frac{333}{125} \qquad h = \underline{21.3\,\text{mm}}$$

The image is errect and virtual.

35.28 What magnification will be obtained by using a concave mirror with a focal length 18 in, if the mirror is held 12 in from the face?

❚ Applying the mirror equation we have $\frac{1}{12} + 1/q = \frac{1}{18}$, or $q = -36$ in. The image is virutal and upright with magnification $= |36/12| = \underline{3}$.

35.29 A man is shaving with his chin 0.4 m from a concave magnifying mirror. If the linear magnification is 2.5, what is the radius of curvature of the mirror?

❚ For magnification of the face we are looking for a virtual upright image. $|q/p| = 2.5$ and thus $q = -2.5p = \underline{-1.0\,\text{m}}$. Next we find f using

$$\frac{1}{0.4} + \frac{1}{-1.0} = \frac{1}{f} \qquad 2.5 - 1 = \frac{1}{f} \quad \text{and}$$

$$f = \frac{1}{1.5} \qquad \text{Finally} \qquad R = 2f = 2\left(\frac{1}{1.5}\right) = \underline{1.33\,\text{m}}$$

35.30 A doctor looks through a small hole at the vertex of a concave mirror to examine a sore throat. If the radius of curvature of the mirror is 462 mm and the light source is 1 m from the mirror, how far from the throat should the mirror be if the light source is to be imaged on the inflamed area?

❚ The focal length is

$$f = \frac{r}{2} = \frac{462}{2} = 231\,\text{mm} \qquad \text{or} \qquad f = 0.231\,\text{m}$$

Then, noting $p = 1$ m, we have

$$\frac{1}{1} + \frac{1}{q} = \frac{1}{0.231} = 4.329 \qquad \text{and} \qquad \frac{1}{q} = 3.329 \qquad \text{or} \qquad q = \underline{300.4\,\text{mm}}$$

35.31 To give a mangified image of a cavity a dentist holds a small mirror with a focal length of 12 mm a distance of 9 mm from a tooth. What is the linear magnification obtained?

❚ We have

$$\frac{1}{9} + \frac{1}{q} = \frac{1}{12} \qquad \frac{1}{q} = \frac{3}{36} - \frac{4}{36} = -\frac{1}{36} \qquad \text{and} \qquad q = \underline{-36\,\text{mm}}$$

The image is upright with

$$\text{magnification} = |36/9| = \underline{4}.$$

35.32 An object is placed in front of a concave mirror having a radius of curvature of 0.3 m. If you want to produce first a real image and then a virtual image 3 times as high as the object, find the object distance required in each case.

❚ For the real image $q = 3p$, and $1/p + 1/(3p) = 1/f = 1/0.15$, or $p = \underline{0.200\,\text{m}}$
For the virtual image $q = -3p$ and $1/p + 1/(-3p) = 1/f = 1/0.15$, or $p = \underline{0.100\,\text{m}}$.
In a facial magnifier, one wants a virtual image not only because it is upright, but because a magnified real image is behind one's face and one can't see it.

35.33 An object is 375 mm from a concave mirror of 250 mm focal length. Find the image distance. If the object is moved 5 mm farther from the mirror, how far does the image move?

$\blacksquare$ We find the image distance for each case. $1/375 + 1/q = 1/250$; $q = 750$ mm for $p = 375$ mm. Next, for p 5 mm farther, $1/380 + 1/q = 1/250$; $q = 730.77$ mm for $p = 380$ mm. Thus $\Delta q = -19.23$ mm, so the image moves in <u>19.23 mm</u>.

35.34 A short match of length L lies along the axis of a concave spherical mirror of focal length f with its most distant point a distance p from the vertex. Calculate the length L' of the image and find the *longitudinal magnification* L'/L. Show that if $L \ll p - f$, the longitudinal magnification is equal to the square of the lateral linear magnification.

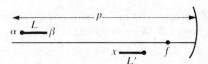

Fig. 35-16

$\blacksquare$ We proceed the same way as in Prob. 35.33. The two object points are at p and $(p - L)$ as shown. For end α: $1/p + 1/q_1 = 1/f$; $q_1 = (fp)/(p - f)$. For end β: $1/(p - L) + 1/q_2 = 1/f$; $1/q_2 = (p - L - f)/[(p - L)f]$, and $q_2 = [(p - L)f]/(p - L - f)$. Thus, using a common denominator, $L' = q_2 - q_1 = Lf^2/[(p - f)^2 - L(p - f)]$. For $L \ll p - f$ we get $L' \approx (Lf^2)/(p - f)^2$. But, from above, $f/(p - f) = q_1/p$, so $L' \approx (q_1^2/p^2)L$, which is the desired result, with lateral linear magnification for the object taken at position p.

35.35 A convex mirror has a radius of curvature of 90 cm. (*a*) Where is the image of an object 70 cm from the mirror formed? (*b*) What is the magnification?

$\blacksquare$ (*a*) $f = R/2 = (-90)/2 = -45$ cm. Focal length f must be negative for a convex mirror.

$$\frac{1}{p} + \frac{1}{q} = \frac{1}{f} \qquad \frac{1}{70} + \frac{1}{q} = -\frac{1}{45} \qquad q = -27.4 \text{ cm}$$

The image is <u>27.4 cm behind the mirror</u>, and it is virtual.
(*b*) $-q/p = 27.4/70 = \underline{0.39}$ (magnification), and the image is upright.

35.36 The magnitude of the focal length of a convex mirror is 12 cm. If an object 6 cm high is placed 24 cm from the mirror, what will the image distance (in cm) be?

$\blacksquare$ We use the mirror equation $1/p + 1/q = 1/f$, which yields, recalling that f is negative,

$$\frac{1}{24} + \frac{1}{q} = \frac{1}{-12} \qquad \text{or} \qquad \frac{1}{q} = -\frac{1}{12} - \frac{1}{24} = -\frac{3}{24} \qquad \text{and} \qquad q = \underline{-8 \text{ cm}}$$

35.37 What is the focal length of a convex spherical mirror which produces an image $\frac{1}{6}$ the size of an object located 12 cm from the mirror?

$\blacksquare$ We have $|q/p| = \frac{1}{6}$. Since for a real object the image in a convex mirror is always virtual, we have $q = -p/6 = -\frac{12}{6} = -2$ cm. Then

$$\frac{1}{12} + \frac{1}{-2} = \frac{1}{f} \qquad \text{or} \qquad \frac{1}{12} - \frac{6}{12} = \frac{-5}{12} = \frac{1}{f} \qquad \text{and} \qquad f = \underline{-2.4 \text{ cm}}$$

35.38 An object is placed 3 ft from a convex mirror which has a focal length of 1.2 ft. Find the image distance and the ratio of the height of the image to the height of the object.

$\blacksquare$ We apply the mirror equation to get

$$\frac{1}{3} + \frac{1}{q} = \frac{1}{-1.2} \qquad \frac{1}{q} = -\frac{5}{6} - \frac{2}{6} = -\frac{7}{6} \qquad \text{and} \qquad q = \underline{-0.857 \text{ m}} \qquad \frac{h_i}{h_o} = \left| \frac{6/7}{3} \right| = \underline{0.286}$$

35.39 A man's eye is 175 mm from the center of a spherical reflecting Christmas tree ornament which is 100 mm in diameter. Find the image position and the linear magnification.

▮ Recalling that object distance is measured to the reflecting surface and that $f = R/2$, we have

$$p = 175 - 50 = 125 \text{ mm} \quad \text{and} \quad f = -25 \text{ mm} \quad \text{Then} \quad \frac{1}{125} + \frac{1}{q} = \frac{1}{-25}$$

$$\frac{1}{q} = -\frac{5}{125} - \frac{1}{125} \quad \text{and} \quad q = -\frac{125}{6} = \underline{-20.8 \text{ mm}} \quad \text{linear magnification} = \left|\frac{q}{p}\right| = \frac{20.8}{125} = \underline{0.167}$$

Note that the condition on the dimensions of the mirror and the requirement of paraxial rays (Prob. 35.1) are not satisfied, so the image will be quite distorted.

35.40 An object 28 mm high is 0.48 m from a convex mirror with a radius of curvature of 0.32 m. Locate the image. Is it real or virtual? Is it erect or inverted? What is its size?

▮ For a convex mirror we have

$$f = \frac{r}{2} = \frac{-0.32}{2} = -0.16 \text{ m} \quad \text{Then} \quad \frac{1}{0.48} + \frac{1}{q} = \frac{1}{-0.16} \quad \text{or} \quad \frac{1}{q} = \frac{-3}{0.48} - \frac{1}{0.48} = -\frac{4}{0.48} \quad \text{and} \quad q = \underline{-0.12 \text{ m}}$$

Magnification $= -q/p$, so $h/28 = 120/480$; $h = \underline{7 \text{ mm}}$. The image is virtual and erect.

35.41 A furtune-teller uses a polished sphere of 8 in radius. If her eye is 10 in from the sphere, where is the image of the eye? Where would her eye have to be for the image to be at the back surface of the sphere?

▮ The sphere acts as a convex mirror with $f = R/2 = -\frac{8}{2} = -4$ in. Then, for the first part, with $p = 10$ in,

$$\frac{1}{10} + \frac{1}{q} = \frac{1}{-4} \quad \frac{1}{q} = -\frac{5}{20} - \frac{2}{20} \quad \text{and} \quad q = -\frac{20}{7} = \underline{-2.857 \text{ in}}$$

For the second part we require $q = -16$ in. Then

$$\frac{1}{p} + \frac{1}{-16} = \frac{1}{-4} \quad \text{or} \quad \frac{1}{p} = \frac{1}{16} - \frac{4}{16} = \frac{-3}{16} \quad \text{and} \quad p = -5.33 \text{ in}$$

This cannot happen for a real object, so there is no location for her eye that will create the required image. This could also be seen by noting that for $p > 0$, q must be negative and $|q| < |f|$. Thus no real object has an image farther than $|f|$ behind the mirror.

35.42 Prove the the image size is always less than the object size for a real object in front of a convex mirror.

▮
$$f = -|f| \quad \text{so} \quad \frac{1}{p} + \frac{1}{q} = -\frac{1}{|f|}$$

Since $p > 0$ we have $q < 0$, and

$$\frac{1}{p} - \frac{1}{|q|} = \frac{-1}{|f|} \quad \text{Thus} \quad \frac{1}{p} < \frac{1}{|q|} \quad \text{or} \quad p > |q|.$$

Finally, magnification $= |q/p| < 1$.

35.43 An object is placed 42 cm in front of a concave mirror of focal length 21 cm. Light from the concave mirror is reflected onto a small plane mirror 21 cm in front of the concave mirror. Where is the final image?

▮ First, use the mirror equation for a concave mirror.

$$\frac{1}{p} + \frac{1}{q} = \frac{1}{f} \quad \frac{1}{42} + \frac{1}{q} = \frac{1}{21} \quad q = 42 \text{ cm}$$

The image distance is 42 cm to the left of the concave mirror. Now use the mirror formula for the plane mirror, with $f = \infty$ and $p = 21 \text{ cm} - 42 \text{ cm} = -21 \text{ cm}$, since the image produced by the concave mirror alone is the object for the plane mirror (see Fig. 35-17).

$$\frac{1}{p} + \frac{1}{q} = \frac{1}{f} \quad \frac{1}{-21} + \frac{1}{q} = \frac{1}{\infty} \quad q = \underline{21 \text{ cm}}$$

The position of the final image I is 21 cm in front of the plane mirror.

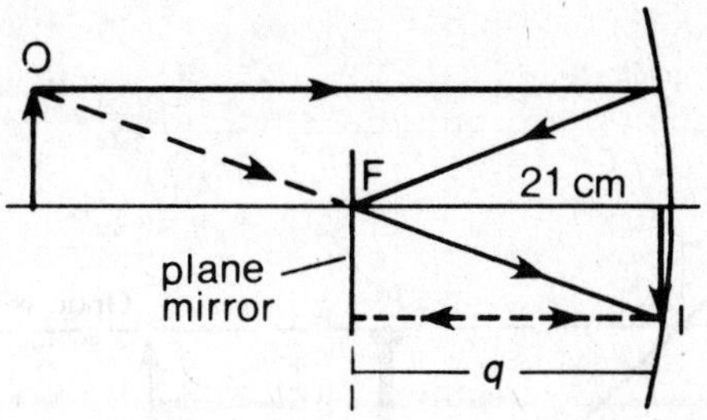

Fig. 35-17

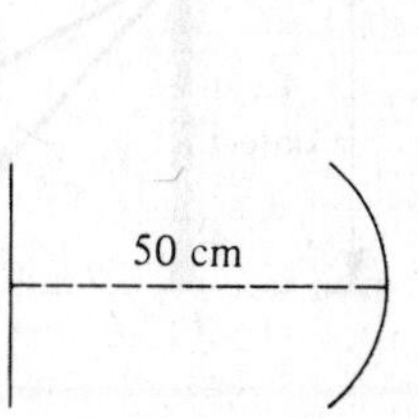

Fig. 35-18

35.44 In Fig. 35-18 a plane mirror is 50 cm from a concave mirror that has $f = 30$ cm. Find several of the real images that the system forms for a tiny light bulb placed 10 cm to the right of the plane mirror.

▌ Let PM = plane mirror and CM = concave mirror. Real images in front of PM are marked with parentheses and all distances are in centimeters. Start with image in CM: $p = 40$ and from $1/p + 1/p' = \frac{1}{30}$, $p' = 120$, which becomes a negative object $p = -70$ for PM; using $1/p + 1/p' = 0$ for PM gives $p' = 70$. This image is a negative object for CM with $p = -20$, so $-\frac{1}{20} + 1/p' = \frac{1}{30}$ gives $p' = +12$, which is (38) in front of PM. This serves as an object $p = 38$ for PM, yielding $p' = -38$. This is a new source $p = 88$ for CM, giving $p' = 45.5$, which is (4.5). The pattern continues: the 4.5 is p for the PM giving $p' = -4.5$ and is $p = 54.5$ for the CM. Then $\frac{1}{54.5} + 1/p' = \frac{1}{30}$ gives 66.7, which is a $p = -16.7$ for the PM, in turn, giving $p' = (16.7)$, and so on. Now take the first image of the source in PM: $p = 10$, so we get $p' = -10$, which is a $p = 60$ for the CM, giving $p' = 60$. This in turn is $p = -10$ for PM, giving $p' = (10)$. Since this is a $p = 40$ for CM, the cycle above repeats. In summary, real images have been found at 38, 4.5, 16.7, 10, ...

35.45 Two identical concave mirrors (A and B) face each other on the same axis. Their centers are coincident. A tiny bulb is placed at the focal point of mirror A. Find several of the real images that the system forms of the bulb. Give your answer in terms of distances from A.

▌ The mirrors are $4f$ apart and for A, $p = f$, which gives an image at ∞. This image at ∞ serves as a source for B, leading to an image at f which is ($3f$) away from A. Again as in Prob. 35.44, we continue the pattern: $p = 3f$ for A, giving $1/3f + 1/p' = 1/f$, or $p' = (1.5f)$. This image is $p = 2.5f$ for B, yielding $p' = 1.67f$, or ($2.33f$) from A. Since $p = 2.33$ for A, $p' = (1.75f)$, and so on. Now start with B, the first image, since $p = 3f$ is at $p' = 1.5f$, or ($2.5f$) from A. This in turn gives ($p = 2.5f$ for A) an image at $p' = (1.67f)$. The numbers in parentheses above are the locations of some of the real images.

35.2 THIN LENSES

35.46 Define the following properties of lenses: *converging, diverging, principal focus, focal length, power, diopter,* and *F number*.

▌ A *converging* lens is one that focuses parallel rays to a point.

A *diverging* lens is one that spreads parallel rays.

The *principal focus* is the point on each side of a lens through which all rays parallel to the principal axis pass (or appear to have passed) depending on which side the light enters.

The term *focal length* refers to the distance from the lens to either principal focus.

The *power of a lens* is the reciprocal of the focal length in meters.

A *diopter* is the unit for the power of a lens, equal to the power of a lens of focal length one meter.

The *F number* of a (camera) lens is the ratio of the focal length to the diameter of the lens opening.

35.47 Describe the basic flaws that make the image of a simple lens imperfect.

▌ *Distortion* is a lens defect that makes the image of perpendicular lines appear to be curved—outward like a pincushion or inward like a barrel. Variations of the magnification with distance from the principal axis produce this effect.

Spherical aberration is a lens defect that occurs because sections of the lens at different distances from the axis have different focal lengths.

Chromatic aberration is a lens defect that occurs because the index of refraction is greater for violet light than for red light, making the focal length different for different colors.

Astigmatism is a lens defect that occurs when rays of light from a point object are far from the principal axis and do not focus to a point image but to two focal lines, at slightly different distances from the lens.

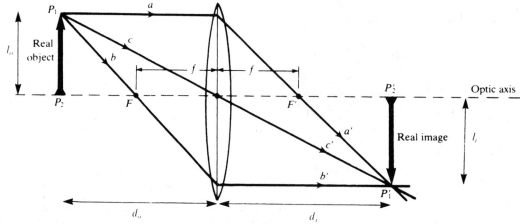

Fig. 35-19

35.48 Describe the image formed by a thin converging (positive) lens when a real object is situated outside the front focal point, F.

❚ In Fig. 35-19, real rays a, b, c from P_1 on real object P_1P_2 are, for convenience, drawn parallel to the optic axis, through F, and through the center of the lens. Refracted ray a' passes through the back focal point F', b' is parallel to the optic axis, and c' is the continuation of c. These rays, of course, follow the law of refraction. The intersection of real rays a', b', c' locates point P_1' on the real image $P_1'P_2'$. Other points on $P_1'P_2'$ can be located in the same way. The image is inverted.

Letting $f = +20$ cm, $d_o = +30$ cm, $l_o = +6$ cm, we find from the equations in Probs. 35.1 and 35.2 that $d_i = +60$ cm (real) and $l_i = -12$ cm (inverted). For the assumed values, the image is magnified.

35.49 Repeat Prob. 35.48 for a real object inside F.

❚ Figure 35-20 gives the ray construction; the image is virtual, erect, and magnified. Letting $f = +15$ cm, $d_o = +10$ cm, $l_o = 2$ cm in the equations of Probs. 35.1 and 35.2, we find that $d_i = -30$ cm (virtual) and $l_i = +6$ cm (erect, magnified).

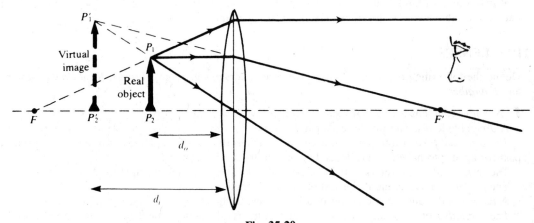

Fig. 35-20

35.50 Describe the image of a real object formed by a thin diverging (negative) lens.

❚ Figure 35-21 gives the ray construction, which is independent of the relationship between d_o and f. The image is virtual, erect, and minified.

Checking with $f = -30$ cm, $d_o > 0$, $l_o = 10$ cm,

$$\frac{1}{d_o} + \frac{1}{d_i} = -\frac{1}{30} \quad \text{or} \quad d_i = -\frac{30d_o}{30 + d_o} \text{ (virtual)} \quad \text{and} \quad \frac{30 + d_o}{30} = -\frac{10}{l_i}$$

$$\text{or} \quad l_i = \frac{10}{1 + (d_o/30)} \text{ (erect, minified)}$$

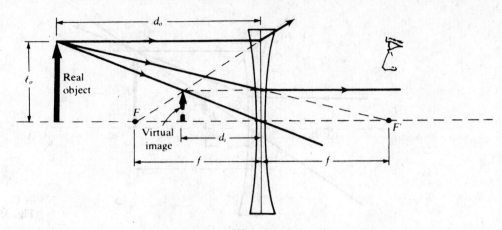

Fig. 35-21

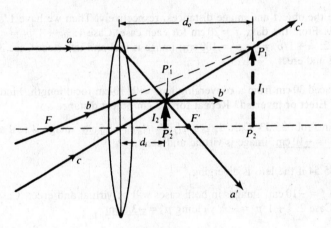

Fig. 35-22

35.51 Describe the image formed by a positive lens of a virtual object.

▌ In Fig. 35-22, the positive lens, L_2, intercepts the real image I_1 produced by a lens L_1 (not shown); I_1 is the virtual object for L_2. Applying the thin lens equation, with $d_o < 0$ (virtual object) and $f > 0$, we obtain

$$-\frac{1}{|d_o|} + \frac{1}{d_i} = \frac{1}{f} \quad \text{or} \quad d_i = \left(\frac{|d_o|}{|d_o| + f}\right)f$$

which shows that the image is real and situated inside the back focal point of the lens. Moreover,

$$\frac{l_i}{l_o} = \frac{-d_i}{d_o} \quad \text{and} \quad l_i = \left(\frac{f}{|d_o| + f}\right)l_o$$

and so the image is erect and minified. These conclusions are borne out by Fig. 35-22.

35.52 Describe the image formed by a negative lens when a virtual object is situated inside the back focal point.

▌ In Fig. 35-23, the negative lens, L_2, intercepts the real image I_1 produced by a lens L_1 (not shown). I_1 is the virtual object for L_2. Applying $1/d_o + 1/d_i = 1/f$, with $f < d_o < 0$, we obtain

$$d_i = \frac{d_o f}{d_o - f} > 0 \quad \text{(real)}$$

and applying $l_i/l_o = -d_i/d_o$

$$l_i = -\frac{l_o f}{d_o - f} > 0 \quad \text{(erect)}$$

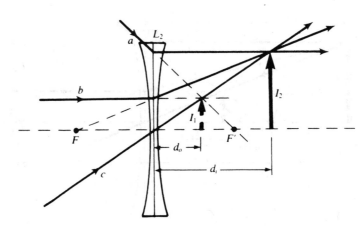

Fig. 35-23

35.53 A converging lens images the sun at a distance of 20 cm from the lens. What is the focal length of the lens? If an object is placed 100 cm from the lens, where will its image be formed? Is it real? Upright? Repeat for object distances of 25 and 10 cm.

▌ Let p, p' be the object and image distances, respectively. Then we have $1/p + 1/p' = 1/f$, and $h_i/h_o = (-p')/p$. From the data, $f = 20$ cm for each case. Case 1: $\frac{1}{100} + 1/p' = \frac{1}{20}$; $p' = \underline{25\text{ cm}}$; image is real and inverted. Case 2: $\frac{1}{25} + 1/p' = \frac{1}{20}$; $p' = \underline{100\text{ cm}}$; image is real and inverted. Case 3: $\frac{1}{10} + 1/p' = \frac{1}{20}$; $p' = \underline{-20\text{ cm}}$; image is virtual and erect.

35.54 An object is placed 30 cm from a converging lens with 10-cm focal length. Find the position of the image. Is it real or virtual? Erect or inverted? Repeat for a 5-cm object distance.

▌ Proceed as in Prob. 35.53. Case 1: $\frac{1}{30} + 1/p' = \frac{1}{10}$; $p' = \underline{15\text{ cm}}$; image is real and inverted. Case 2: $\frac{1}{5} + 1/p' = \frac{1}{10}$; $p' = \underline{-10\text{ cm}}$; image is virtual and erect.

35.55 Repeat Prob. 35.54 if the lens is diverging.

▌ In this case, $f = -10$ cm. Images in both cases will be virtual and erect. Case 1: $\frac{1}{30} + 1/p' = -\frac{1}{10}$, so that $p' = \underline{-7.5\text{ cm}}$. Case 2: $\frac{1}{5} + 1/p' = -\frac{1}{10}$ yielding $p' = \underline{-3.3\text{ cm}}$.

35.56 Using the lens equation, show that if a convex lens produces the same size image as the object, both object distance and image distance are twice the focal length.

▌ In this case $|p/q|$ = height of object/height of image = 1. We must have p and q positive for the lens equation to have a solution, so

$$p = q \qquad \frac{1}{p} + \frac{1}{q} = \frac{1}{f} \quad \text{or} \quad \frac{2}{p} = \frac{1}{f} \quad \text{and} \quad p = 2f \quad \text{Thus} \quad q = p = \underline{2f}$$

In this case the locations of the object and the image, each at $2f$, are called the <u>conjugate foci</u>.

35.57 Where must an object be placed in the case of a converging lens of focal length f if the image is to be virtual and 3 times as large as the object?

▌ We are given $3 = h_i/h_o = -q/p$, with p positive (real object) and q negative (virtual image). Then $q = -3p$.

$$\frac{1}{p} + \frac{1}{q} = \frac{1}{f} \quad \text{becomes} \quad \frac{1}{p} - \frac{1}{3p} = \frac{1}{f} \quad \text{or} \quad \frac{2}{3p} = \frac{1}{f} \quad \text{and} \quad p = \frac{2f}{3}$$

35.58 A diverging lens forms an image one-third the size of an object that is 24 cm from the lens. Determine the focal length of the lens.

▌ Since the image is $\frac{1}{3}$ the size of the object and the image is always virtual for a diverging lens with real object,

$$\frac{q}{p} = -\frac{1}{3} \quad \text{and} \quad \frac{1}{p} + \frac{1}{q} = \frac{1}{f} \quad \text{yields} \quad \frac{1}{24} + \frac{1}{-8} = \frac{1}{f}$$

$$\frac{1}{24} - \frac{3}{24} = \frac{1}{f} \qquad -\frac{2}{24} = \frac{1}{f} \qquad f = \underline{-12\text{ cm}}$$

35.59 A double-convex lens is used to project a slide. The slide is 2 in high and 10 in from the lens. The image is 90 in from the lens. What is the focal length of the lens?

▌ A projected image is real, so,

$$\frac{1}{p}+\frac{1}{q}=\frac{1}{f} \quad \text{becomes} \quad \frac{1}{10}+\frac{1}{90}=\frac{1}{f} \quad \text{or} \quad \frac{10}{90}=\frac{1}{f} \quad \text{and} \quad f=\underline{9 \text{ in}}$$

35.60 An object is 6 cm high and 30 cm from a double-convex lens. Its image is 90 cm from the lens and on the same side as the object. What is the focal length of the lens?

▌ The image must be virtual and q is negative. Thus

$$\frac{1}{p}+\frac{1}{q}=\frac{1}{f} \quad \frac{1}{30}+\frac{1}{-90}=\frac{1}{f} \quad \text{or} \quad \frac{3}{90}-\frac{1}{90}=\frac{1}{f}$$

$$\frac{2}{90}=\frac{1}{f} \quad \text{and} \quad f=\underline{45 \text{ cm}}$$

35.61 A virtual image that is half as far from the lens as the object is to be formed by a diverging lens for which $f=-30$ cm. (a) Where should the object be placed? (b) What will be the magnification?

▌ Let p, p' be the object and image distances; S, S' the corresponding heights. (a) For a real object, the image is virtual, so if $p'=-p/2$, then $1/p-2/p=-\frac{1}{30}$ yields $p=\underline{30 \text{ cm}}$. (b) $S'/S=-p'/p=\frac{30}{60}=0.50$. The image is thus erect and $\frac{1}{2}$ the object height.

35.62 If a diverging lens is to be used to form an image which is half the size of the object, where must the object be placed?

▌ $q=-p/2$, since image is virtual and object is real. The thin lens formula gives $1/p-2/p=-1/|f|$; $\underline{p=|f|}$.

35.63 Compute the position and focal length of the converging lens which will project the image of a lamp, magnified 4 diameters, upon a screen 10 m from the lamp.

▌ From $p+q=10$ and $q=4p$, we find $p=\underline{2}$ and $q=\underline{8}$. Then

$$\frac{1}{f}=\frac{1}{p}+\frac{1}{q}=\frac{1}{2}+\frac{1}{8}=\frac{5}{8} \quad \text{or} \quad f=\frac{8}{5}=\underline{+1.6 \text{ m}}$$

35.64 A lens of focal length f projects upon a screen the image of a luminous object magnified M times. Show that the lens distance from the screen is $f(M+1)$.

▌ The image is real, since it can be shown on a screen, and so $q>0$. We then have

$$M=\left|\frac{q}{p}\right|=q\left(\frac{1}{p}\right)=q\left(\frac{1}{f}-\frac{1}{q}\right)=\frac{q}{f}-1 \quad \text{or} \quad q=\underline{f(M+1)}$$

35.65 An alternative form of the lens and mirror equations is $s_i s_o = f^2$, where s_o is the distance of the object from the focal point and s_i is the distance of the image from the focal point. (a) Derive this relation. (b) Show s_o and s_i in a sketch for a lens and for a mirror.

▌ (a) Let p, p' be the object and image distances. In the thin lens equation let $p=s_o+f$ and $p'=s_i+f$; then $1/f=1/(s_o+f)+1/(s_i+f)$. Use a common denominator: $1/f=[(s_i+f)+(s_o+f)]/[(s_o+f)(s_i+f)]$, and cross-multiply to get $f(s_i+s_o+2f)=(s_o+f)(s_i+f)$. This yields $s_i f+s_o f+2f^2=s_o s_i+s_i f+s_o f+f^2$, and we cancel common terms to find that $s_i s_o=f^2$. (b) See Fig. 35-24. Note that s_i and s_o always have the same sign. If the object is within the focal length for either case shown, then s_o is negative and $|s_o|<f$. Thus s_i is negative

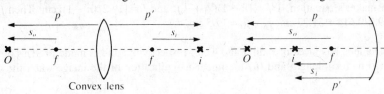

Convex lens

Concave mirror

Fig. 35-24

and $|s_i| > f$, so that the image is virtual. Similar arguments can be made for diverging lenses and convex mirrors.

35.66 In certain cases when an object and screen are separated by a distance D, two positions x of the converging lens relative to the object will give an image on the screen. Show that these two values are

$$x = \frac{D}{2}\left(1 \pm \sqrt{1 - \frac{4f}{D}}\right)$$

Under what conditions will no image be found?

❚ We have $p + p' = D$ and $x = p$. Then $1/p + 1/p' = 1/f$ is $1/x + 1/(D - x) = 1/f$. Forming a common denominator we have $D/[x(D - x)] = 1/f$, or $x(D - x) = fD$. Then $x^2 - Dx + fD = 0$, which leads to the stated result. For $f > D/4$, the radical is imaginary and no image is formed.

35.3 LENSMAKER'S EQUATION; COMPOSITE LENS SYSTEMS

35.67 What is the lensmaker's equation?

❚ The focal length of a lens depends on the index of refraction of the lens material and on the radius of curvature of each spherical surface. The *lensmaker's equation* enables one to compute the focal length f, if the index of refraction n and the radii of curvature R_1 and R_2 are known: $1/f = (n - 1)[(1/R_1) - (1/R_2)]$. If the lens is immersed in a substance of index of refraction n' the formula is generalized to $1/f = (n/n' - 1)[(1/R_1) - (1/R_2)]$. In all cases we determine the signs of R_1 and R_2 by assuming that light enters through surface 1 and using the sign conventions of Prob. 34.58. Thus for a double convex lens $R_1 > 0$, $R_2 < 0$, etc.

35.68 An amateur lens grinder wants to grind a converging lens of crown glass ($n = 1.52$) with the same curvature on both sides and a focal length of 25 cm. What radius of curvature must he grind on each face?

❚ Use the lensmaker's equation.

$$\frac{1}{f} = (n - 1)\left(\frac{1}{R_1} - \frac{1}{R_2}\right) \qquad \text{with} \qquad R_1 = -R_2 \qquad \text{or}$$

$$\frac{1}{25} = (1.52 - 1)\left(\frac{1}{R_1} - \frac{1}{-R_1}\right) = (0.52)\left(\frac{2}{R_1}\right) \qquad \text{and} \qquad R_1 = 25(0.52)(2) = \underline{26 \text{ cm}} \text{ radius of curvature for each side}$$

35.69 A lens has a convex surface of radius 20 cm and a concave surface of radius 40 cm, and is made of glass of refractive index 1.54. Compute the focal length of the lens, and state whether it is a converging or a diverging lens.

❚ For definiteness assume that side 1 is convex and side 2 is concave.

$$\frac{1}{f} = (n - 1)\left(\frac{1}{R_1} - \frac{1}{R_2}\right) \qquad \text{with} \qquad R_1 = 20 \text{ cm} \qquad R_2 = 40 \text{ cm}$$

Then $1/f = (0.54)(\frac{1}{20} - \frac{1}{40}) = \frac{0.54}{40} = 0.0135$, and $f = \underline{74.1 \text{ cm}}$. Since $f > 0$ the lens is converging.
   Note: Had we reversed the lens, we would have $R_1 = -40$ cm and $R_2 = -20$ cm, leading to the same result.

35.70 A parallel beam of white light strikes a double convex lens having faces of radii $+32$ and $+48$ cm. The refractive indices of the glass for the A (red region) and H (violet region) spectral lines are, respectively, 1.578 and 1.614. Determine the distance between the focal points of the red and violet radiations.

❚ From the lensmaker's equation, $1/f = (n - 1)(\frac{1}{32} + \frac{1}{48})$, and $f = [19.2/(n - 1)]$ cm. Then $f_A = 19.2/0.578 = 33.22$ cm; $f_H = 19.2/0.614 = 31.27$ cm; $f_A - f_H = \underline{1.95 \text{ cm}}$.

35.71 A convexo-concave lens has faces of radii 3 and 4 cm, respectively, and is made of glass of refractive index 1.6. Determine (**a**) its focal length and (**b**) the linear magnification of the image when the object is 28 cm from the lens.

❚ First we find f: $1/f = (n - 1)[(1/R_1) - (1/R_2)] = (1.6 - 1)(\frac{1}{3} - \frac{1}{4}) = 0.6(\frac{1}{12}) = 0.05$, and $f = \underline{20 \text{ cm}}$. Next we solve for the image distance:

$$\frac{1}{p}+\frac{1}{q}=\frac{1}{f} \quad \text{or} \quad \frac{1}{28}+\frac{1}{q}=\frac{1}{20} \quad \frac{1}{q}=0.0143 \quad \text{and} \quad q=70 \text{ cm}$$

The linear magnification is $|q/p|=\frac{70}{28}=\underline{2.5}$.

35.72 A symmetric lens with a focal length of 5 cm is made of glass with an index of refraction of 1.50. What is the radius of curvature of each surface of the lens?

$\blacksquare$ $R_1=-R_2=r$, so $1/f=(n-1)(2/r)$, and $r=2(n-1)f=2(1.50-1)(5 \text{ cm})=\underline{5 \text{ cm}}$.

35.73 A plano-concave lens has a spherical surface of radius 12 cm, and its focal length is −22.2 cm. Compute the refractive index of the lens material.

$\blacksquare$ $1/f=(n-1)[(1/R_1)-(1/R_2)]$ with $R_1=-12$ cm, $R_2=\infty$, $f=-22.2$ cm. Thus $(n-1)=R_1/f=12/22.2=0.54$, and $n=\underline{1.54}$.

35.74 A plano-convex lens has a focal length of 30 cm and an index of refraction of 1.50. Find the radius of the convex surface.

$\blacksquare$ Use the lensmaker's equation with $R_1=\infty$.

$$\frac{1}{f}=(n-1)\left(\frac{1}{R_1}-\frac{1}{R_2}\right) \quad \frac{1}{30}=(1.50-1)\left(\frac{1}{\infty}-\frac{1}{R_2}\right) \quad \frac{1}{30}=\frac{0.50}{-R_2} \quad R_2=\underline{-15 \text{ cm}}$$

The negative sign means that the surface is convex to the right.

35.75 Two symmetric double convex lenses, A and B, have the same focal length, but the radii of curvature differ so that $R_A=0.9R_B$. If $n_A=1.63$, find n_B.

$\blacksquare$ Since the focal lengths are the same, $(n_A-1)(2/R_A)=(n_B-1)(2/R_B)$, or $(n_B-1)=(R_B/R_A)(n_A-1)=0.63/0.9=0.70$, and $n_B=\underline{1.70}$.

35.76 It is desired to make a lens with two concave faces. The lens material has $n=1.54$, and the radii of the two faces are to be the same. What must these radii be to give a 30-cm focal length?

$\blacksquare$ We use the lensmaker's equation with $f=-30$ cm, $R_1=-R$, $R_2=R$, and index of refraction $n=1.54$. Then $-\frac{1}{30}=(1.54-1)[(-1/R)-(1/R)]$; $R=\underline{32.4 \text{ cm}}$.

35.77 A double convex lens has faces of radii 18 and 20 cm. When an object is 24 cm from the lens, a real image is formed 32 cm from the lens. Determine (a) the focal length of the lens and (b) the refractive index of the lens material.

$\blacksquare$ (a)
$$\frac{1}{f}=\frac{1}{p}+\frac{1}{q}=\frac{1}{24}+\frac{1}{32}=\frac{7}{96} \quad \text{or} \quad f=\frac{96}{7}=\underline{+13.7 \text{ cm}}$$

(b)
$$\frac{1}{f}=(n-1)\left(\frac{1}{r_1}-\frac{1}{r_2}\right) \quad \text{or} \quad \frac{1}{13.7}=(n-1)\left(\frac{1}{18}+\frac{1}{20}\right) \quad \text{or} \quad n=\underline{1.69}$$

where we have used the fact that r_2 is negative.

35.78 A double convex glass lens has faces of radii 8 cm each. Compute its focal length in air and when immersed in water. Refractive index of the glass, 1.50; of water, 1.33.

$\blacksquare$ We must use the generalized lensmaker's equation (Prob. 35.67) with $n=1.50$, and $n'=1.0$ for air and 1.33 for water. Then for air, $1/f=[(n/n')-1][(1/R_1)-(1/R_2)]=[(1.50/1.00)-1](\frac{2}{8})=0.50/4=0.125$, and $f=\underline{8.0 \text{ cm}}$. For water, $1/f=[(1.50/1.33)-1](\frac{1}{4})=0.128/4=0.0320$, and $f=\underline{31.3 \text{ cm}}$.

35.79 A glass lens ($n=1.50$) has a focal length of +10 cm in air. Compute its focal length in water ($n=1.33$).

$\blacksquare$ Using, for air and water,

$$\frac{1}{f}=\left(\frac{n_1}{n_2}-1\right)\left(\frac{1}{r_1}-\frac{1}{r_2}\right) \quad \text{with} \quad n_1=1.50 \quad \text{and} \quad n_2=1.0 \quad \text{or} \quad 1.33$$

For air:
$$\frac{1}{10} = (1.50 - 1)\left(\frac{1}{r_1} - \frac{1}{r_2}\right)$$

For water:
$$\frac{1}{f_W} = \left(\frac{1.50}{1.33} - 1\right)\left(\frac{1}{r_1} - \frac{1}{r_2}\right)$$

Divide one equation by the other to obtain $f_W = 5/0.128 = \underline{39\ cm}$.

35.80 A double convex lens has faces with 20 cm radii for each. The index of refraction of the glass is 1.50. Compute the focal length of this lens (**a**) in air and (**b**) when immersed in carbon disulfide ($n = 1.63$).

▌ Using $\dfrac{1}{f} = \left(\dfrac{n_1}{n_2} - 1\right)\left(\dfrac{1}{r_1} - \dfrac{1}{r_2}\right)$ with $n_1 = 1.50$ and $n_2 = 1.0$ or 1.63

(**a**)
$$\frac{1}{f} = (1.50 - 1)\left(\frac{1}{20} + \frac{1}{20}\right) \quad \text{or} \quad f = +20\ \text{cm}$$

(**b**)
$$\frac{1}{f} = \left(\frac{1.50}{1.63} - 1\right)\left(\frac{1}{20} + \frac{1}{20}\right) \quad \text{or} \quad f = -125\ \text{cm}$$

Here, the focal length is negative and so the lens is diverging, even though it is double convex! This is because the surrounding medium has a greater index of refraction, reversing the convergence or divergence of rays at the interface.

35.81 A certain lens is shaped as shown in Fig. 35-25 and has radii of 80 and 120 cm. Its material has $n = 1.64$. What will be its focal length when it is immersed in oil for which $n = 1.45$?

▌ We have $1/f = [(n_1/n_2) - 1][(1/R_1) - (1/R_2)]$, where $R_1 = -120$ cm, $R_2 = -80$ cm, $n_2 = 1.45$, and $n_1 = 1.64$. Then $1/f = [(1.64/1.45) - 1][(-1/120) - (-1)/80]$, leading to $f = 18.3$ m. The choice of radii used was made so that the lens is a positive lens thicker at the center.

Fig. 35-25

35.82 An object 6 cm high is placed 40 cm from a thin converging lens of 8-cm focal length. A second converging lens of 12 cm focal length is placed 20 cm from the first lens as shown in Fig. 35-26. Find the position, size, and character of the final image.

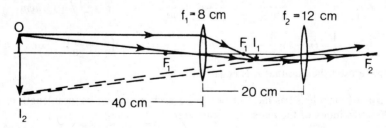

Fig. 35-26

▌ Use the thin-lens equation first for the lens with $f_1 = 8$ cm, and then, after determining p_2, for the second lens with $f_2 = 12$ cm.

$$\frac{1}{p_1} + \frac{1}{q_1} = \frac{1}{f_1} \qquad \frac{1}{40} + \frac{1}{q_1} = \frac{1}{8} \qquad \frac{1}{q_1} = \frac{5}{40} - \frac{1}{40} = \frac{4}{40}$$

$$q_1 = 10\ \text{cm} \qquad 20 - q_1 = 10\ \text{cm} = p_2 \qquad \frac{1}{p_2} + \frac{1}{q_2} = \frac{1}{f_2}$$

$$\frac{1}{10} + \frac{1}{q_2} = \frac{1}{12} \qquad \frac{1}{q_2} = \frac{5}{60} - \frac{6}{60} = -\frac{1}{60} \qquad q_2 = -60\ \text{cm}$$

The final image is 60 cm to the left of lens 2; it is inverted and virtual. Its size is gotten by using the magnification formula in succession.

$$6 \text{ cm} \left| \frac{q_1}{p_1} \right| \left| \frac{q_2}{p_2} \right| = 6 \left(\frac{10}{40} \right) \left(\frac{60}{10} \right) = \underline{9 \text{ cm}} \text{ (height of final image)}$$

35.83 An object is placed 60 cm in front of a diverging lens of focal length -15 cm. A distance of 10 cm behind this lens is a converging lens of focal length 20 cm. Where is the final image, and what is the overall magnification?

▌ We apply the lens formula twice.

$$\frac{1}{p_1} + \frac{1}{q_1} = \frac{1}{f_1} \qquad \frac{1}{60} + \frac{1}{q_1} = \frac{1}{-15} \qquad \frac{1}{q_1} = -\frac{4}{60} - \frac{1}{60} = -\frac{5}{60} \qquad q_1 = \underline{-12 \text{ cm}}$$

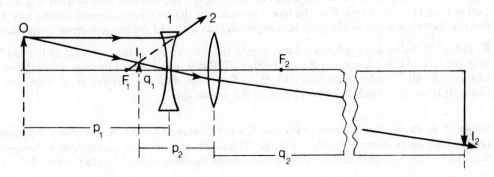

Fig. 35-27

Use I_1 as the object for lens 2, and $p_2 = 12 + 10 = 22$ cm (see Fig. 35-27).

$$\frac{1}{p_2} + \frac{1}{q_2} = \frac{1}{f_2} \qquad \frac{1}{22} + \frac{1}{q_2} = \frac{1}{20} \qquad \frac{1}{q_2} = \frac{1}{20} - \frac{1}{22} = \frac{11}{220} - \frac{10}{220} = \frac{1}{220} \qquad q_2 = \underline{220 \text{ cm}} \text{ to the right of lens 2}$$

Magnification, lens $1 = -q_1/p_1 = \frac{12}{60} = \frac{1}{5}$. Magnification, lens $2 = -q_2/p_2 = -220/22 = -10$. Overall magnification $= (q_1/p_1)(q_2/p_2) = (-1/5)(10) = \underline{-2}$. The final image is inverted and twice as large.

35.84 A convex lens of focal length 12 cm is placed in contact with a plane mirror. If an object is placed 20 cm from the lens, where is the final image formed?

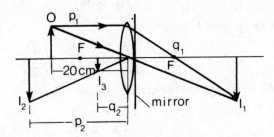

Fig. 35-28

▌ See Fig. 35-28. Because of the mirror, light goes through the lens twice. First we have

$$\frac{1}{p_1} + \frac{1}{q_1} = \frac{1}{f} \qquad \frac{1}{20} + \frac{1}{q_1} = \frac{1}{12} \qquad \frac{1}{q_1} = \frac{5}{60} - \frac{3}{60} = \frac{2}{60} \qquad \text{and} \qquad q_1 = \underline{30 \text{ cm}}.$$

Image I_1 formed by the lens would be 30 cm behind the mirror location if the mirror were removed. After reflection in the mirror, the image of I_1 is at I_2, 30 cm in front of the lens. Since the light now goes to the left, I_2 acts as a virtual object for the lens and p_2 is negative. Use the lens equation again.

$$\frac{1}{p_2} + \frac{1}{q_2} = \frac{1}{f} \qquad \frac{1}{-30} + \frac{1}{q_2} = \frac{1}{12} \qquad \frac{1}{q_2} = \frac{1}{12} + \frac{1}{30} = \frac{5}{60} + \frac{2}{60} = \frac{7}{60} \qquad q_2 = \underline{8.6 \text{ cm}} \text{ to the left of the lens}$$

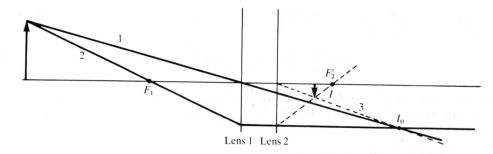

Fig. 35-29

35.85 An object is placed 12 cm in front of a lens of focal length 5 cm. Another lens of focal length 3 cm is placed 2 cm in back of the first lens. Find the image produced by this two-lens system by tracing rays. (*Hint:* First find the image produced by the front lens alone, and use it to find the image formed by the second lens.)

▮ In Fig. 35-29, the scale is 5 mm = 1 cm. Rays 1 and 2 locate the image I_0 formed by the first lens alone. With the second lens present, ray 2 goes through F'_2. Ray 3, which passes through the center of the second lens, is undeviated, and so must pass through I_0. The intersection I locates the image formed by the two lenses. It is located 11 mm ⇒ 2.2 cm in back of the second lens.

35.86 Figure 35-30 shows a combination of two thin lenses (a *compound lens*). A real object is presented to L_1. (**a**) Find the image distance d_i in terms of the object distance d_o and the parameters of the system. (**b**) Show that as $s \rightarrow 0$ the system becomes equivalent to a single thin lens of focal length f, where $1/f = 1/f_1 + 1/f_2$.

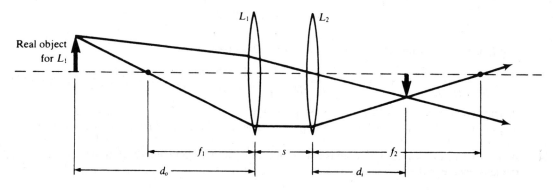

Fig. 35-30

▮ (**a**) First locate the image from L_1 alone:

$$\frac{1}{d_o} + \frac{1}{x} = \frac{1}{f_1} \quad \text{or} \quad x = \frac{f_1 d_o}{d_o - f_1}$$

This image serves as the (real or virtual) object for L_2 alone:

$$\frac{1}{s - x} + \frac{1}{d_i} = \frac{1}{f_2}$$

whence

$$d_i = \frac{f_2(s - x)}{s - x - f_2} = f_2 \frac{s - [f_1 d_o/(d_o - f_1)]}{s - [f_1 d_o/(d_o - f_1)] - f_2} \tag{1}$$

Depending on the relative sizes of d_o and the parameters, d_i may be positive (real image) or negative (virtual image).

Observe that no assumptions have been made about the signs of f_1 and f_2; thus (1) holds for any pair of thin lenses. (**b**) As $s \rightarrow 0$, (1) becomes

$$d_i = \frac{f_2 f_1 d_o/(d_o - f_1)}{[f_1 d_o/(d_o - f_1)] + f_2} \quad \text{or} \quad \frac{1}{d_i} = \frac{1}{f_2} + \frac{1}{f_1} - \frac{1}{d_0} \quad \text{or} \quad \frac{1}{d_o} + \frac{1}{d_i} = \frac{1}{f}$$

which last is the gaussian formula for a thin lens of focal length $f = (f_1^{-1} + f_2^{-1})^{-1}$.

35.87   Two lenses, each of focal length +6 in, are in contact. Find the focal length of the combination.

▮ From Prob. 35.86(b) we have the effective focal length given by $1/f = 1/f_1 + 1/f_2 = \frac{1}{6} + \frac{1}{6} = \frac{2}{6} = \frac{1}{3}$, and $f = \underline{3\text{ in}}$.

35.88   What is the "power" of a lens and why is it a useful concept?

▮ The reciprocal of the focal length f of a lens (in meters) is known as the *power* of the lens (in *diopters*).

$$P = (1/f)\,\text{m}^{-1}$$

Problem 35.86(b) shows that for thin lenses in contact, powers are additive.

35.89   Two lenses, of power 5 diopters and 7 diopters, are placed in contact. Find the focal length of the combination.

▮ $P = P_1 + P_2 = 5 + 7 = 12$ diopters. $f = 1/P = \frac{1}{12} = 0.083\text{ m} = \underline{8.3\text{ cm}}$.

35.90   What is the power of two lenses of focal length 20 cm and −50 cm in contact?

▮
$$\frac{1}{f} = \frac{1}{f_1} + \frac{1}{f_2} = \frac{1}{20} + \frac{1}{-50} = \frac{5}{100} - \frac{2}{100} = \frac{3}{100}$$

$$f = 33.3\text{ cm} = 0.333\text{ m} \qquad P = \frac{1}{f} = \frac{1}{0.333} = \underline{3\text{ diopters}}.$$

35.91   Two thin lenses, of focal lengths −12 cm and −30 cm, are in contact. Compute the focal length and power of the combination.

▮ $1/f = 1/f_1 + 1/f_2 = -\frac{1}{12} - \frac{1}{30} = -0.1167\text{ cm}^{-1}$. Then $f = \underline{-8.57\text{ cm}}$; $P = \underline{-11.67\text{ diopters}}$.

35.4 OPTICAL INSTRUMENTS: PROJECTORS, CAMERAS, THE EYE

35.92   A slide projector has a projecting lens of focal length 20 cm. If the slide is 25 cm from this lens, what is the distance to the screen for a clear image?

▮ From the lens equation,

$$\frac{1}{p} + \frac{1}{q} = \frac{1}{f} \qquad \frac{1}{25} + \frac{1}{q} = \frac{1}{20}$$

$$\frac{1}{q} = \frac{5}{100} - \frac{4}{100} = \frac{1}{100} \qquad q = 100\text{ cm} = \underline{1\text{ m}}$$

35.93   A slide projector has a lens whose focal length is 15 cm. If a 5 cm by 5 cm slide is to appear 1 m by 1 m when projected, how far from the lens should the screen be placed?

▮ The linear magnification is to be $m = 100/5 = 20 = |q/p|$. Since object and image are real, p and q are positive, so $q = 20p$. Then $1/p + 1/q = 1/f$, becomes $20/q + 1/q = \frac{1}{15}$; $21/q = \frac{1}{15}$, and $q = \underline{315\text{ cm}}$.

35.94   A slide projector forms a 0.75-m-wide image of a 35-mm-wide slide on a screen 6 m from the lens. (a) What is the focal length of the lens? (b) How far should the screen be from the lens to obtain a 1.0-m-wide image of the slide?

▮ For a typical slide projector the object distance, $s \ll s'$, the image distance. Then $1/s' \ll 1/s$ and $1/s \approx 1/f$, or $s \approx f$. (a) The magnification is given by $s'/s = m = (0.75 \times 10^3\text{ mm})/35\text{ mm} = 21.4$, or since $s = s'/m$, we can approximate $f = s'/m = 600\text{ cm}/21.4 = \underline{28\text{ cm}}$. (b) $m = 1000\text{ mm}/35\text{ mm} = 28.6$ in the new situation, so now the image distance is $s' = mf = (28.6)(28\text{ cm}) = 800\text{ cm} = \underline{8.0\text{ m}}$.

35.95   A pinhole camera with film 75 mm high is to be used to take a picture of a tree 10 m high. The film is 150 mm from the pinhole. How far should the camera be from the tree to include the full height of the tree?

▮ The pinhole camera has a simple opening with no lens. The image is formed because light from any point on the object can reach only one point on the film since only a single ray enters the pinhole from that point (see Fig. 35-31). By similar triangles, $\frac{75}{150} = 10/d$, $d = \underline{20\text{ m}}$.

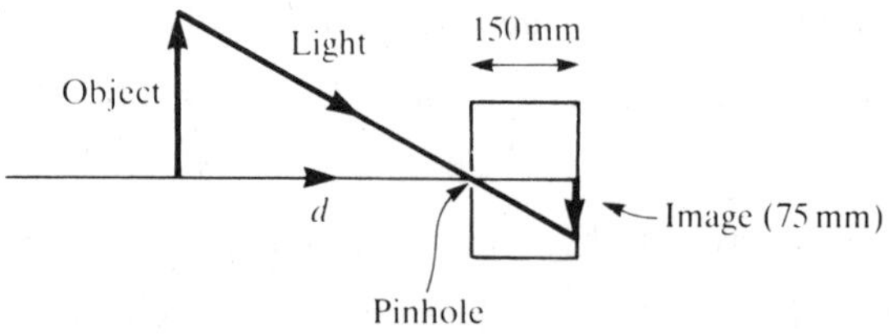

Fig. 35-31

35.96 Define the *F number* of a lens.

▮
$$F \text{ number} \equiv \frac{\text{focal length}}{\text{diameter of aperture}}$$

Thus, for a camera lens of variable aperture, the smaller the *F* number, the more light admitted in a given time and the 'faster'' the lens.

35.97 What is the ratio of the brighness of the image formed by a camera lens when the lens is set at *F*-1.8 to that formed when the lens is set at *F*-11?

▮ For a lens of a given focal length f, the diameter of the aperture is related to the *F* number by $d = f/F$ and so the area *A* of the aperture is

$$A = \frac{\pi d^2}{4} = \frac{\pi f^2}{4F^2}$$

The brightness is proportional to A, so the ratio of the brightness at two *F* values is

$$\frac{A}{A'} = \frac{F'^2}{F^2} = \frac{11^2}{1.8^2} = \underline{37.3}$$

35.98 A camera is fitted with a bellows in order to be able to vary the lens-to-film distance from 7 to 12 cm. (*a*) With a lens of focal length 50 mm, what are the nearest and farthest distances an object can be from the lens and still be focused on the film? (*b*) What are the magnifications of the images of an object at the nearest and farthest distances?

▮ Let s and s' refer to object and image distances. (*a*) The nearest distance s occurs when s' is largest, so

$$\frac{1}{s} = \frac{1}{f} - \frac{1}{s'} = \frac{1}{5.0 \text{ cm}} - \frac{1}{12 \text{ cm}} = \frac{7}{60 \text{ cm}} \qquad \text{or} \qquad s = \underline{8.57 \text{ cm}}$$

The farthest distance s occurs when s' is smallest, so

$$\frac{1}{s} = \frac{1}{f} - \frac{1}{s'} = \frac{1}{5.0 \text{ cm}} - \frac{1}{7 \text{ cm}} = \frac{2}{35 \text{ cm}} \qquad \text{or} \qquad s = \underline{17.5 \text{ cm}}$$

(*b*) At the nearest distance:
$$m = \frac{s'}{s} = \frac{12 \text{ cm}}{8.57 \text{ cm}} = \underline{1.4}$$

At the farthest distance:
$$m = \frac{s'}{s} = \frac{7 \text{ cm}}{17.5 \text{ cm}} = \underline{0.4}$$

35.99 A camera with a telephoto lens of focal length 125 mm takes a picture of a 1.8-m-tall woman standing 5.0 m away. (*a*) What must be the distance s' between the film and the lens to get a properly focused picture? (*b*) What is the magnification of the image? (*c*) What is the size of the woman's image on the film?

▮ (*a*)
$$\frac{1}{s'} = \frac{1}{f} - \frac{1}{s} = \frac{1}{125 \text{ mm}} - \frac{1}{5000 \text{ mm}} = \frac{39}{5000 \text{ mm}} \qquad \text{so} \qquad s' = \underline{128 \text{ mm}}$$

(*b*)
$$m = \frac{s'}{s} \approx \frac{f}{s} = \frac{125 \text{ mm}}{5000 \text{ mm}} = \underline{0.025}$$

(*c*)
$$h' = mh = (0.025)(1800 \text{ mm}) = \underline{45 \text{ mm}}$$

35.100 Why is the *F* number useful?

▮ As in Prob. 35.99, when the object distance, s, is large compared with the image distance, s', we have

$s' \approx f$, the focal length. Then the height of the image, h', obeys $h' = (s'/s)h \approx f(h/s)$, where h is the object height. Thus for a given object at a given distance the image height is proportional to f. The amount of light entering the lens is proportional to the area of the lens (see Prob. 35.97) and hence to the square of the diameter D. The light energy per unit image area (intensity) is thus proportional to D^2/s'^2 or to D^2/f^2. Since the F number is defined as f/D, the light intensity on the film varies inversely as the square of the F number. Thus small F numbers mean greater light intensity on the film.

35.101 What is the focal length of a positive symmetric lens with an index of refraction of 1.62 and a radius of curvature of 20 cm?

▌ From $1/f = (n - 1)(2/r)$ we have

$$f = \frac{r}{2(n - 1)} = \frac{20 \text{ cm}}{2(1.62 - 1)} = \underline{16.1 \text{ cm}}$$

35.102 Show that the minimum possible F number of a symmetric lens is $\frac{1}{4}(n - 1)$.

▌ We use the expression for focal length, f, from Prob. 35.101. The maximum diameter of a lens with a radius of curvature r is $d = 2r$. (This is the case where the lens is a whole sphere—a very bad shape in practice.) Therefore the smallest F number (largest aperture) is

$$F = \frac{f}{d} = \frac{r/[2(n - 1)]}{2r} = \frac{1}{4(n - 1)}$$

35.103 A lens of focal length 500 mm is mounted in front of a 50-mm camera lens. (a) What is the focal length of the combination, assuming zero distance between the lenses? (b) If the lens-to-film distance can be varied from 50.0 to 52.2 mm, what are the closest and farthest object distances at which the camera can focus with this attachment?

▌ (a)
$$\frac{1}{f} = \frac{1}{f_1} + \frac{1}{f_2} = \frac{1}{500} + \frac{1}{50} \qquad f = \underline{45.45 \text{ mm}}$$

(b) Closest distance:
$$\frac{1}{s} = \frac{1}{f} - \frac{1}{s'} = \frac{1}{45.45} - \frac{1}{52.2} \qquad s = 351 \text{ mm} = \underline{0.351 \text{ m}}$$

Farthest distance:
$$\frac{1}{s} = \frac{1}{f} - \frac{1}{s'} = \frac{1}{45.45} - \frac{1}{50.0} \qquad s = 499 \text{ mm} = \underline{0.499 \text{ m}}$$

35.104 An achromatic lens is composed of a symmetric positive lens made of flint glass ($n = 1.65$) and a symmetric negative lens made of crown glass ($n = 1.52$). If both components have a radius of curvature of 10 cm, what is the focal length of the combination?

▌ Following Prob. 35.101 we have

$$f_1 = \frac{r_1}{2(n_1 - 1)} = \frac{10 \text{ cm}}{2(0.65)} = 7.69 \text{ cm} \qquad f_2 = \frac{r_2}{2(n_2 - 1)} = \frac{-10 \text{ cm}}{2(0.52)} = -9.62 \text{ cm}$$

$$\frac{1}{f} = \frac{1}{f_1} + \frac{1}{f_2} = \frac{1}{7.69 \text{ cm}} - \frac{1}{9.62 \text{ cm}} = 0.0261 \text{ cm}^{-1} \qquad f = \underline{38.3 \text{ cm}}$$

35.105 An achromatic lens having a focal length of +20 cm is to be constructed by combining a crown-glass lens with a flint-glass lens. What must be the focal lengths (f_1 and f_2) of the component lenses if the dispersive powers (δ_1 and δ_2) of the crown and flint glass used are 0.0158 and 0.0324, respectively? (*Hint:* Achromatism obtains when $(\delta_1/f_1) + (\delta_2/f_2) = 0$.)

▌
$$\frac{1}{f} = \frac{1}{f_1} + \frac{1}{f_2} \qquad \text{or} \qquad \frac{1}{20} = \frac{1}{f_1} + \frac{1}{f_2}$$

Also, the condition for achromatism is $\delta_1/f_1 + \delta_2/f_2 = 0$, or $0.0158/f_1 + 0.0324/f_2 = 0$. Then, $f_2 = -2.04f_1$, and substituting into the first equation,

$$\frac{1}{20} = \frac{1}{f_1} - \frac{1}{2.04f_1} = \frac{1.04}{2.04f_1} \qquad \text{and} \qquad f_1 = \underline{10.2 \text{ cm}} \text{ (crown)} \qquad f_2 = -2.04(10.2) = \underline{-20.8 \text{ cm}} \text{ (flint)}$$

35.106 Describe briefly the following components of the eye: *cornea, iris, retina, rods and cones, optic nerve*, and *rhodospin*.

▌ The *cornea* is the curved-front transparent portion of the eyeball.

The *iris* is the colored portion of the eye that controls the size of the pupil.

The *retina* is the light-sensitive surface at the back of the eyeball. It has photosensitive cells called *rods* and *cones*.

The *optic nerve* carries signals from the retina to the brain. The location on the retina where the optic nerve leaves is the *blind spot*.

Rhodopsin is the photosensitive pigment in the rods.

35.107 Define: *hyperopia, myopia, far point, near point*.

▌ *Hyperopia*, or *farsightedness*, is an eye defect caused, for example, by too short an eyeball. The eye can see distant objects clearly. A converging lens corrects the defect.

Myopia, or *nearsightedness*, is an eye defect caused, for example, by too long an eyeball. The eye can see close objects clearly. A diverging lens corrects the defect.

The *far point* is the greatest distance an eye can see distinctly. For a normal eye, this distance is infinity.

The *near point* is the shortest distance an eye can see distinctly. For a normal eye, the distance is 25 cm.

35.108 A farsighted individual who can read clearly without glasses at a distance of no less than 75 cm puts on glasses of power 2.5 diopters. What is his new near point?

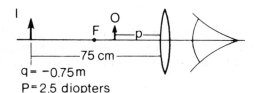

Fig. 35-32

▌ The lens has the effect of taking an object at the new near point and making it appear to the eye as if it were at the 75-cm distance, so the eye can see it clearly. This image is on the same side of the lens and is hence virtual (and upright)—see Fig. 35-32. The focal length is $f = 1/P$, or $f = 1/2.5 = 0.4$ m. Then, from $1/p + 1/q = 1/f$, we get $1/p + 1/(-0.75) = 1/0.4$; $1/p = 2.5 + 1.33 = 3.83$, and $p = \underline{0.26\ m}$ (near point).

35.109 A farsighted man cannot see objects clearly unless they are at least 36 in from his eyes. What is the focal length in inches of the lens that will just extend his range of clear vision down to 9 in?

▌ We proceed as in Prob. 35.108, with image virtual at 36 in.

$$\frac{1}{p} + \frac{1}{q} = \frac{1}{f} \quad \text{or} \quad \frac{1}{9} + \frac{1}{-36} = \frac{1}{f} \quad \frac{3}{36} = \frac{1}{f} \quad \text{and} \quad f = \underline{12\ in}$$

35.110 A nearsighted man cannot see objects clearly unless they are within 2 m of his eyes. What is the approximate focal length of a lens that will just enable him to see very distant objects?

▌ Here the corrective lens must take very distant objects and bring them to the 2-m point. Again the image is virtual. The object distance is taken at infinity. Then $1/p + 1/q = 1/f$ becomes $1/\infty + 1/(-2) = 1/f$, and $f = -2\ m = \underline{-200\ cm}$, a diverging lens.

35.111 A man who wears glasses of power 3 diopters must hold a newspaper at least 25 cm away to see the print clearly. How far away would the newspaper have to be if he took off the glasses and still wanted clear vision?

▌ The virtual image formed by the lens is at the naked eye's true near point, as in Fig. 35-32. We thus have $1/p + 1/q = 1/f$, with $p = 25$ cm and $f = 1/P = \frac{1}{3}$ m = 33.3 cm. Then $\frac{1}{25} + 1/q = 1/33.3$, $1/q = 0.03 - 0.04 = -0.01$, and $q = -100$ cm. Without glasses, he must hold the paper $\underline{1\ m}$ away.

35.112 Define *angular magnification* and the *magnifying power* of a lens, and show that the latter is given by $M = 1 + (25\ cm)/f$, with f in cm.

▌ When an object is observed with the naked eye, the maximum angle it subtends at the eye, while still being distinct, is the angle when placed at the near point, or about 25 cm from the eye. When looking at the object through an optical instrument, it can be made to subtend a much larger angle at the eye. The ratio of this angle to that with the object at the near point, in front of the naked eye, is called the *angular magnification*. When an object is observed through a simple converging lens and placed within the focal length of the lens, an upright virtual image is created. This image can be made to appear anywhere from the lens to infinity by adjusting the location of the object, but it can be examined by the eye only between the near point and infinity. The largest angle the image subtends occurs when it appears at the near point. This gives the maximum angular magnification, or *magnifying power,* of the lens. To calculate it we have, for object and image distances s and s', $1/s + 1/s' = 1/f$, with $s' = -25$ cm; so $1/s = \frac{1}{25} + 1/f$. The angle α' subtended at the lens (and at the eye just behind it) obeys $|h'/s'| = \tan \alpha'$ or, equivalently, $h/s = \tan \alpha'$, where h and h' are object and image heights. For small angles (in radians) $\tan \alpha' \approx \alpha'$ and $\alpha' = h/s$. Then, since for the naked eye at the near point, $\alpha = h/(25\text{ cm})$, we get $\alpha'/\alpha = 25/s$. Using the lens formula result, we get $\alpha'/\alpha = 1 + 25/f$, which is the magnifying power.

35.113 (a) What is the magnifying power of a lens with a power of 25 diopters? (b) How far must the object be from this lens to obtain this magnification?

▌ We use the results of Prob. 35.112. (a) The focal length of the lens is

$$f = \frac{1}{25 \text{ diopters}} = 0.04 \text{ m} = 4 \text{ cm}$$

so the magnifying power of the lens is

$$M = 1 + \frac{25 \text{ cm}}{f} = 1 + \frac{25 \text{ cm}}{4 \text{ cm}} = \underline{7.25}.$$

(b) The virtual-image distance $|s'|$ is 25 cm in front of the lens, so the object distance s is

$$\frac{1}{s} = \frac{1}{|s'|} + \frac{1}{f} = \frac{1}{25 \text{ cm}} + \frac{1}{4 \text{ cm}} = \frac{29}{100 \text{ cm}} \qquad \text{or} \qquad s = \underline{3.45 \text{ cm}}$$

35.114 (a) What is the focal length of the reading glasses required by a person whose near point is 90 cm? (b) How far would this person have to hold a book to be able to read it while wearing reading glasses with a power of 2.0 diopters?

▌ (a) When the object distance s is 25 cm, a virtual image is to be formed at $|s'| = 90$ cm, so the focal length f of the lens must be

$$\frac{1}{f} = \frac{1}{s} + \frac{1}{s'} = \frac{1}{25 \text{ cm}} - \frac{1}{90 \text{ cm}} = \frac{13}{450 \text{ cm}} \qquad \text{or} \qquad f = \underline{34.6 \text{ cm}}$$

(b) When the focal length is

$$f = \frac{1}{2 \text{ diopters}} = 0.50 \text{ m} = 50 \text{ cm}$$

the virtual image will still be at $s' = 90$ cm if the object distance is

$$\frac{1}{s} = \frac{1}{|s'|} + \frac{1}{f} = \frac{1}{90 \text{ cm}} + \frac{1}{50 \text{ cm}} = \frac{14}{450 \text{ cm}} \qquad \text{or} \qquad s = \underline{32.1 \text{ cm}}$$

35.115 (a) What is the power of the eyeglasses worn by a person whose far point is 5 m? (b) Locate the virtual image of an object 2 m in front of the eyeglasses.

▌ (a) When the object distance is infinite, a virutal image is to be formed at the far point, so $s = \infty$ and $s' = -5$ m, and the power of the lens is

$$\frac{1}{f} = \frac{1}{s} + \frac{1}{s'} = 0 - \frac{1}{5 \text{ m}} = \underline{-0.2 \text{ diopter}}$$

(b) $$\frac{1}{s'} = \frac{-1}{s} + \frac{1}{f} = \frac{-1}{2 \text{ m}} + \frac{1}{-5 \text{ m}} = \frac{-7}{10 \text{ m}} \qquad \text{or} \qquad s' = \underline{-1.43 \text{ m}}$$

35.116 (a) What is the power (in diopters) of a lens system consisting of a 0.5-diopter lens mounted in front of a 2.0-diopter lens? Assume zero distance between the lenses. (b) What is the focal length of this lens combination?

▮ (a)
$$\frac{1}{f}=\frac{1}{f_1}+\frac{1}{f_2}=0.5 \text{ diopter}+2.0 \text{ diopters}=\underline{2.5 \text{ diopters}}$$

(b)
$$f=\frac{1}{2.5 \text{ diopters}}=0.4 \text{ m}=\underline{40 \text{ cm}}$$

35.117 At age 40 a woman requires eyeglasses with lenses of 2 diopters power in order to read a book at 25 cm. At 45 she finds that while wearing these glasses she must hold a book 40 cm from her eyes. What power lenses does she require at 45 to read a book at 25 cm?

▮ With the $f=\frac{1}{2}$-m glasses, and the book held at 40 cm, the image is at
$$\frac{1}{s'}=\frac{-1}{s}+\frac{1}{f}=\frac{-1}{40}+\frac{1}{50}=\frac{-10}{2000} \qquad s'=-200 \text{ cm}$$

To form an image at this same distance when $s=25$ cm, the lens must have the focal length
$$\frac{1}{f}=\frac{1}{s}+\frac{1}{s'}=\frac{1}{25}-\frac{1}{200}=\frac{7}{200} \qquad f=28.57 \text{ cm} \qquad \frac{1}{f}=\frac{1}{0.2857 \text{ m}}=\underline{3.5 \text{ diopters}}$$

35.118 By graphical construction locate the image of an object 10 cm in front of the principal plane of the relaxed eye in Fig. 35-33.

▮ Recall that the eye cannot be treated by the simple thin-lens formula because the index of refraction of the vitreous humor is not equal to 1. There are thus two distinct focal lengths, and we must use the principal plane and nodal point for construction of images. Figure 35-33 is drawn full-size. Ray 1, which is parallel to the axis, is refracted at the principal plane and passes through the back focal point. Ray 2, which passes through the front focal point, is refracted parallel to the axis at the principal plane. Ray 3 passes undeviated through the nodal point. The image point, located by the intersection of these rays, is about 0.5 cm in back of the retina (F').

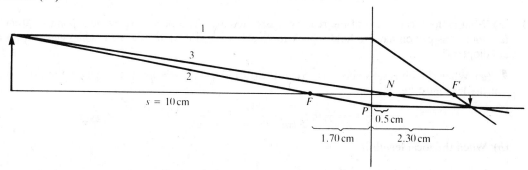

Fig. 35-33

35.119 What is the size of the retinal image of a 20-m-tall building 50 m away?

▮ Since the ray through the nodal point is undeviated, and the nodal point is about 1.70 cm from the retina in the normal eye (see Prob. 35.118), we have for the linear magnification
$$m=\frac{1.70 \text{ cm}}{s}=\frac{1.70 \text{ cm}}{5000 \text{ cm}}=3.4\times10^{-4}$$

and the image height is $h'=mh=(3.4\times10^{-4})(20 \text{ m})=6.8\times10^{-3} \text{ m}=\underline{6.8 \text{ mm}}$.

35.120 How far away is a 25-m-tall tree if its image on the retina is 1.0 cm?

▮ We proceed with the same basic approach as in Prob. 35.119.
$$m=\frac{h'}{h}=\frac{1.0 \text{ cm}}{2500 \text{ cm}}=4\times10^{-4} \qquad s=\frac{1.70 \text{ cm}}{m}=\frac{1.70 \text{ cm}}{4\times10^{-4} \text{ cm}}=4250 \text{ cm}=\underline{42.5 \text{ m}}$$

35.121 Show that the relation between the object and image distance for the eye is given by $1/\bar{s}+1/\bar{s}'=1$, with $\bar{s}=s/f$ and $\bar{s}'=s'/f'$. Here f and f' are the front and back focal lengths, respectively, and s and s' are the object and image distances measured from the principal plane.

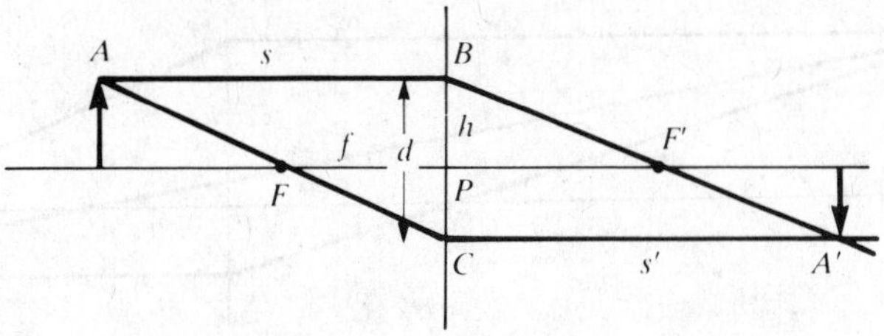

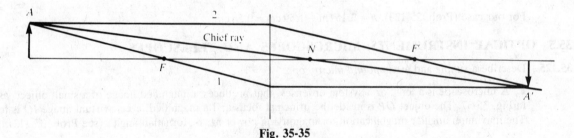

Fig. 35-34

I The situation is as shown in Fig. 35-34, using the two focal points F, F', and the principal plane, BC. From the similar triangles ABC and FPC we have

$$\frac{s}{d} = \frac{f}{d-h} \quad \text{or} \quad \frac{f}{s} = \frac{d-h}{d} = 1 - \frac{h}{d}$$

From the similar triangles $A'BC$ and $F'PB$ we have

$$\frac{s'}{d} = \frac{f'}{h} \quad \text{or} \quad \frac{f'}{s'} = \frac{h}{d}$$

Adding these equations we get

$$\frac{f}{s} + \frac{f'}{s'} = 1 \quad \text{or} \quad \frac{1}{\bar{s}} + \frac{1}{\bar{s}'} = 1 \qquad (1)$$

where $\bar{s} = s/f$ and $\bar{s}' = s'/f'$.

35.122 Consider an eye in which the front focal point is 3 cm from the principal plane and the back focal point is 4 cm from the principal plane. Locate the position of the nodal point graphically. (*Hint*: The chief ray must intersect the image point determined by the rays passing through the focal points.)

I In constructing Fig. 35-35, first locate the image point A' of an object using rays 1 and 2 which pass through F and F'. The chief ray passes undeviated from A to A' and intersects the axis at the nodal point N, which is found to be 1 cm in back of the principal plane.

Fig. 35-35

35.123 The ratio f'/f of the back and front focal lengths of the eye is always 1.35, the index of refraction of vitreous humor. Locate the positions of the focal points when the eye is focused on an object 25 cm in front of the principal plane. Assume that the principal plane is 2.3 cm in front of the retina. (See Probs. 35.121 and 35.122.)

I To be in focus the image distance must be $s' = 2.3$ cm so that the object is focused on the retina. Thus we have

$$s = 25 \text{ cm} \qquad s' = 2.3 \text{ cm} \qquad f' = 1.35f \qquad \text{so}$$

$$\bar{s} = \frac{s}{f} = \frac{25 \text{ cm}}{f} \qquad \bar{s}' = \frac{s'}{f'} = \frac{2.3 \text{ cm}}{1.35f} = \frac{1.70 \text{ cm}}{f}$$

Then from Eq. (1) of Prob. 35.121 we get

$$1 = \frac{1}{\bar{s}} + \frac{1}{\bar{s}'} = \frac{f}{25 \text{ cm}} + \frac{f}{1.70 \text{ cm}} = f\frac{26.70 \text{ cm}}{42.5 \text{ cm}^2} \qquad f = \frac{42.5 \text{ cm}^2}{26.70 \text{ cm}} = \underline{1.59 \text{ cm}}$$

$$f' = 1.35f = (1.35)(1.59 \text{ cm}) = \underline{2.15 \text{ cm}}$$

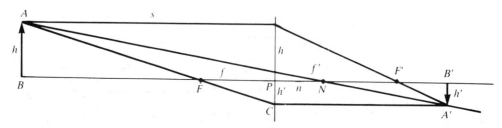

Fig. 35-36

35.124 Refer to Prob. 35.123 and find the position of the nodal point.

▮ Let n be the distance from the principal plane to the nodal point. Examine Fig. 35-36, which in construction follows Fig. 35-35. From the similar traingles ABN and $A'B'N$ we have

$$\frac{h'}{s'-n} = \frac{h}{s+n} \quad \text{or} \quad \frac{h'}{h} = \frac{s'-n}{s+n}$$

From the similar triangles ABF and CPF we have

$$\frac{h}{s-f} = \frac{h'}{f} \quad \text{or} \quad \frac{h'}{h} = \frac{f}{s-f}$$

Combining the two expressions for h'/h we get

$$\frac{s'-n}{s+n} = \frac{f}{s-f} \quad (s'-n)(s-f) = f(s+n) \quad s's - ns - s'f + nf = fs + nf$$

$$n = -f + s' - \frac{fs'}{s} = -f + s'\left(1 - \frac{1}{s}\right)$$

But from Eq. (1) of Prob. 35.121 we have

$$1 - \frac{1}{s} = \frac{1}{s'} = \frac{f'}{s'} \quad \text{so} \quad n = -f + s'\left(\frac{f'}{s'}\right) = f' - f$$

For our case (Prob. 35.123), $n = 2.15 \text{ cm} - 1.59 \text{ cm} = \underline{0.56 \text{ cm}}$.

35.5 OPTICAL INSTRUMENTS: MICROSCOPES AND TELESCOPES

35.125 Describe a *simple* and a *compound microscope*.

▮ A microscope is a lens, or a system of lenses, that produces a magnified image of a small object, as shown in Fig. 35-37. The object OP is inside the principal focus and a magnified, erect, virtual image IQ is formed. The maximum angular magnification, or magnifying power M, is, for small angles (see Prob. 35.112),

$$M = \frac{\beta}{\alpha} = \frac{IQ}{OP} = \frac{QE}{PE}$$

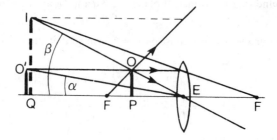

Fig. 35-37

A compound microscope has a very short focal-length objective lens that forms a real, enlarged, inverted image of a very small object. An eyepiece serves as a simple microscope with which to view this image. Since the image formed by the object is inverted, the final image is also inverted.

35.126 What is the angular magnification of a simple lens of focal length 5 cm used as a magnifier producing an image 25 cm from the eye?

❚ From the lens equation, noting that the image is virtual, we have $1/p + 1/q = 1/f$, or $1/p + 1/(-25) = \frac{1}{5}$; $1/p = \frac{5}{25} + \frac{1}{25} = \frac{6}{25}$, and object distance, $p = \frac{25}{6} = 4.17$ cm. From Prob. 35.125 we have angular magnification $M = |q/p| = 25/4.17 = \underline{6}$. Note that this is the same as the linear magnification with image at the near point. We could also use the formula of Prob. 35.112—$M = 1 + 25/f = 1 + 25/5 = \underline{6}$.

35.127 A microscope has an objective of focal length 0.3 cm and an ocular of focal length 2.0 cm. (a) Where must the image formed by the objective be for the ocular to produce a virtual image 25 cm in front of the ocular? (b) If the lenses are 20 cm apart, what is the distance of the objective from the object on the slide? (c) What is the total magnification of the microscope? (d) What distance would the object have to be from a single lens that gave the same magnification? What would its focal length have to be?

❚ Let 1 and 2 refer to the objective and ocular lenses, respectively. s and s' refer to object and image distances for a given lens. (a) The image for the ocular is $s_2' = -0.25$ cm. Then

$$\frac{1}{s_2} = \frac{1}{f_2} - \frac{1}{s_2'} = \frac{1}{2} + \frac{1}{25} = \frac{27}{50} \qquad s_2 = 1.85 \text{ cm in front of the ocular}$$

(b)
$$s_1' = 20 - s_2 = 18.15 \text{ cm} \qquad \text{so} \qquad \frac{1}{s_1} = \frac{1}{f_1} - \frac{1}{s_1'} = \frac{1}{0.3} - \frac{1}{18.15} = \frac{17.85}{0.3 \times 18.15} \qquad \text{and}$$

$$s_1 = \underline{0.305} \text{ cm in front of the objective}$$

(c) The total angular magnification, M, is the height of the final image, h_2', which is at the near point (25 cm from the eye) divided by the object height h_1. Then $M = h_2'/h_1 = (h_2'/h_2)(h_2/h_1)$, where h_2 is the image height for the objective as well as the object height for the ocular. Thus for this case the angular magnification is just the overall linear magnification, or

$$M = m_1 m_2 = \left|\frac{s_1'}{s_1}\right| \left|\frac{s_2'}{s_2}\right| = \frac{18.15}{0.305}\left(\frac{25}{1.85}\right) = \underline{804}$$

(d) From $m = |s'/s|$ we have

$$s = \frac{25 \text{ cm}}{m} = \frac{25 \text{ cm}}{804} = 0.031 \text{ cm} \qquad \text{Then} \qquad \frac{1}{s} + \frac{1}{s'} = \frac{1}{f} \Rightarrow f \approx s = \underline{0.031 \text{ cm}}$$

35.128 A lens of focal length 0.5 cm is placed 12 cm in front of a lens of focal length 4 cm. An object is placed 0.53 cm in front of the 0.5-cm lens. (a) Locate the real image formed by the 0.5-cm lens. (b) Locate the virtual image formed by the 4-cm lens of the real image. (c) What is the overall magnification achieved by this arrangement of lenses?

❚ (a) The image distance s_1' of the real image formed by the objective ($f_1 = 0.5$ cm) is

$$\frac{1}{s_1'} = \frac{1}{f_1} - \frac{1}{s_1} = \frac{1}{0.5 \text{ cm}} - \frac{1}{0.53 \text{ cm}} = \frac{0.03}{(0.5)(0.53 \text{ cm})} \qquad \text{or} \qquad s_1' = \underline{8.83 \text{ cm}}$$

This is the distance $s_2 = 12 \text{ cm} - s_1' = 3.17$ cm in front of the ocular ($f_2 = 4$ cm).
(b) The image distance s_2' of the virtual image formed by the ocular is

$$\frac{1}{s_2'} = \frac{-1}{s_2} + \frac{1}{f_2} = \frac{-1}{3.17 \text{ cm}} + \frac{1}{4 \text{ cm}} = \frac{-0.83}{(3.17)(4) \text{ cm}} \qquad \text{or} \qquad s_2' = \underline{-15.28 \text{ cm}}.$$

This is 15.28 cm in front of the ocular, or the distance $s_2' - 12 \text{ cm} = 3.28$ cm in front of the objective.
(c) The magnification of the real image formed by the objective is

$$m_1 = \frac{s_1'}{s_1} = \frac{8.83 \text{ cm}}{0.53 \text{ cm}} = 16.7$$

The magnification of the virutal image formed by the ocular is

$$m_2 = \left|\frac{s_2'}{s_2}\right| = \frac{15.28 \text{ cm}}{3.17 \text{ cm}} = 4.82$$

The overall linear magnification is $M = m_1 m_2 = (16.7)(4.82) = \underline{80.5}$. This represents the "angular magnification" for someone who can examine the object at a distance of 15.28 cm in front of the eyepiece with the naked eye, not a likely prospect, since the near point for the normal eye is 25 cm. Thus angular magnification for a microscope refers to the case where the image appears at or beyond the near point. The angular and linear magnifications are the same when the final image is at the near point.

35.129 A dissecting microscope is designed to have a large distance s_1 between the object and the objective. Suppose that the focal length of the objective is 5.0 cm, the focal length of the ocular is 4.0 cm, and the distance between these lenses is 17.0 cm. Find (a) s_1, (b) the total magnification.

▮ We use the notation of Prob. 35.127. The image for the ocular is virtual and at $s_2' = -25$ cm, so the eye can examine it at the near point.

(a)
$$\frac{1}{s_2} = \frac{-1}{s_2'} + \frac{1}{f_2} = \frac{1}{25} + \frac{1}{4} = \frac{29}{100} \qquad s_2 = 3.45 \text{ cm} \qquad \text{then} \qquad s_1' = 17 - 3.45 = 13.55 \qquad \text{and}$$

$$\frac{1}{s_1} = \frac{1}{f_1} - \frac{1}{s_1'} = \frac{1}{5} - \frac{1}{13.55} = \frac{8.55}{(5)(13.55)} \qquad s_1 = \underline{7.92 \text{ cm}}$$

(b)
$$M = m_1 m_2 = \frac{s_1'}{s_1} \times \frac{s_2'}{s_2} = \frac{13.55}{7.92}\left(\frac{25}{3.45}\right) = \underline{12.4}$$

is the linear magnification. This is also the angular magnification since the calculation is for the final image at the near point. Note that s_2' can be moved much further out by increasing s_2 only slightly. This has a small effect on the angular magnification but is more comfortable for viewing.

35.130 Describe the basic lens structure of the telescope and opera glass.

▮ An astronomical telescope consists of an objective lens that forms a real, inverted, reduced image of a distant object at (or very near to) its principal focus. This image is then viewed through a simple microscope. In a terrestrial telescope, the image must be inverted once more if the distant scene is to be viewed erect. To accomplish this, a third lens may be introduced between the objective and the ocular; an alternative approach, used in the prism binocular, involves a pair of total reflecting prisms that invert the image. The angular magnification of the astronomical telescope is

$$\text{angular magnification} = \left|\frac{f_o}{f_e}\right| = \left|\frac{\text{focal length of objective}}{\text{focal length of eyepiece}}\right|$$

A galilean telescope, or opera glass, consists of a converging objective and a diverging eyepiece. It forms an erect, virtual image of a distant object. The angular magnification for the galilean telescope is given by the same equation as for the astronomical telescope above. The advantage of the galilean telescope is that it is shorter than the astronomical telescope.

35.131 An opera glass has an objective lens of focal length 15 cm. If the eyepiece is a diverging lens 10 cm away from the objective when a distant object is viewed, what is the focal length of the eyepiece?

▮ Refer to Fig. 35-38; assume that the object and the final image are at infinity (for viewing by the relaxed eye). Then for lens 1,

$$p_1 = \infty \qquad \frac{1}{p_1} + \frac{1}{q_1} = \frac{1}{f_1} \qquad \frac{1}{\infty} + \frac{1}{q_1} = \frac{1}{15} \qquad q_1 = 15 \text{ cm}$$

For lens 2, $p_2 = 10 - 15 = -5$ cm, and q_2 equals infinity:

$$\frac{1}{p_2} + \frac{1}{q_2} = \frac{1}{f_2} \qquad \frac{1}{-5} + \frac{1}{\infty} = \frac{1}{f_2} \qquad \text{and} \qquad f_2 = \underline{-5 \text{ cm}}$$

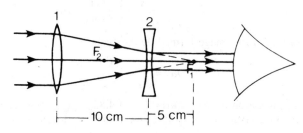

Fig. 35-38

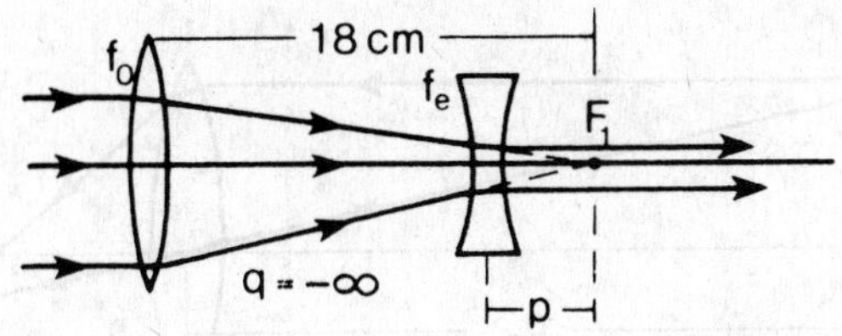

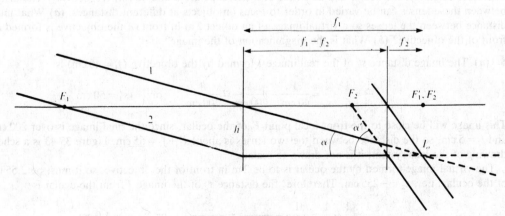

Fig. 35-39

35.132 An opera glass is constructed of an objective lens of 18 cm focal length and a diverging ocular lens with a focal length of −6 cm. (*a*) How far apart must these lenses be to form a final image of a distant object at infinity? (*b*) What is the angular magnification under these conditions?

▌ The situation is depicted in Fig. 35-39. (*a*) The objective forms an image of a distant object at its principal focus 18 cm from the lens. This image serves as the object for the ocular. If the final image is to be at infinity,

$$\frac{1}{p}+\frac{1}{q}=\frac{1}{f_e} \qquad \frac{1}{p}+\frac{1}{-\infty}=-\frac{1}{6} \qquad p=\underline{-6\text{ cm}}$$

The minus sign means that the ocular is inserted before the objective len's image is formed. Therefore, the distance between the lenses must be $(18-6)$ cm = 12 cm. (*b*) The angular magnification under these conditions is given by $M=|f_o/f_e|=\frac{18}{6}=\underline{3}$.

35.133 An astronomical telescope has an objective lens of focal length 150 cm and an eyepiece of focal length 10 cm. For viewing distant objects, what must the spacing between the objective and the eyepiece be?

▌ Since the object can be considered to be at infinity, the first image (real and inverted) is at the focal point of the objective. This real image becomes the object for the ocular which creates a virtual image at infinity. Thus the real image is just within the focal length of the ocular. The distance between the lenses is thus just $f_o+f_e=150$ cm + 10 cm = $\underline{160\text{ cm}}$.

35.134 What is the angular magnification produced by an astronomical telescope with an objective of focal length 150 cm and an eyepiece of focal length 10 cm? What must the focal length of the eyepiece be if the magnification is tripled?

▌ Angular magnification = $|f_o/f_e|$ = 150/10 = $\underline{15}$. To triple this magnification, f_e must be $\frac{1}{3}$ its original value, or $f_e=\underline{3.33\text{ cm}}$.

35.135 A galilean telescope consists of an objective of focal length f_1 and an ocular of focal length $-f_2$ (negative lens) separated by a distance $d=f_1-f_2$. Trace parallel rays through such a lens system when $f_1=10$ cm and $f_2=-2$ cm. Show that the magnification is $|f_1/f_2|$ and that the virtual image is erect.

▌ Figure 35-40 is to scale: 5 mm = 1 cm. Rays 1 and 2 locate the image I_o that would be formed by the objective alone. From I_o rays are traced through the second lens. Using the notation of Fig. 35-40, we define

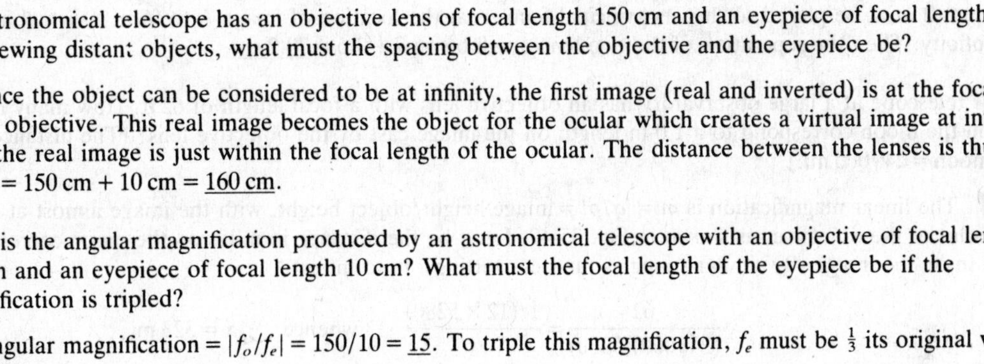

Fig. 35-40

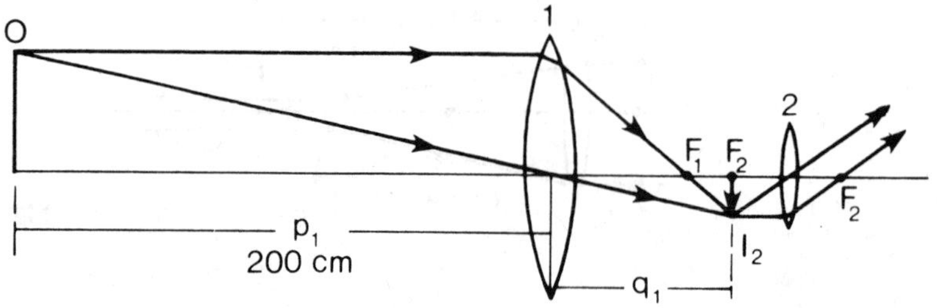

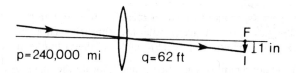

Fig. 35-41

the angular magnification as

$$\frac{\alpha'}{\alpha} \approx \frac{\tan \alpha'}{\tan \alpha} = \left|\frac{h/f_2}{h/f_1}\right| = \left|\frac{f_1}{f_2}\right| \qquad \text{for small angles}$$

and the image is erect since the original rays, which enter from above the axis, emerge below the axis.

35.136 An astronomical telescope has an objective of 50-cm focal length. The eyepiece has a focal length of 3.5 cm. How far must these lenses be separated when viewing an object 200 cm away from the objective?

▌ See Fig. 35-41. Here we cannot assume that the object is at infinity. Instead we have

$$\frac{1}{p_1} + \frac{1}{q_1} = \frac{1}{f_1} \qquad \frac{1}{200} + \frac{1}{q_1} = \frac{1}{50} \qquad \frac{1}{q_1} = \frac{4}{200} - \frac{1}{200} = \frac{3}{200} \qquad \text{and} \qquad q_1 = \underline{66.7 \text{ cm}}$$

Lens 2 must be placed so that its principal focus is at the location of image I_1 to form a virtual image at infinity. Then the separation of the two lenses will be $66.7 + 3.5 = \underline{70.2 \text{ cm}}$.

35.137 A telescope at a large observatory has an objective lens with a focal length of 62 ft. How many miles observed on the moon correspond to a 1.0-in length on the image cast by the objective lens? (The distance to the moon = 240 000 mi.)

▌ The linear magnification is $m = |q/p| =$ image height/object height, with the image almost at the focal point of the objective, as shown in Fig. 35-42. Let x be the distance in miles on the moon corresponding to 1 in on the image. By similar triangles (and converting 1 in to miles),

$$\frac{62}{(240\,000)(5280)} = \frac{1/(12 \times 5280)}{x} \qquad \text{whence} \qquad x = \underline{323 \text{ mi}}$$

p=240,000 mi   q=62 ft

F
1 in
I

Fig. 35-42

35.138 The focal lengths of the ocular and objective of a telescope are 5 and 40 cm, respectively. The distance between these lenses can be varied in order to focus on objects at different distances. (**a**) What must be the distance between the lenses so a virtual image of an object 2 m in front of the objective is formed also 2 m in front of the objective? (**b**) What is the magnification of the image?

▌ (**a**) The image distance s_1' of the real image I formed by the objective ($f_1 = 40$ cm) is

$$\frac{1}{s_1'} = \frac{1}{f_1} - \frac{1}{s_1} = \frac{1}{40 \text{ cm}} - \frac{1}{200 \text{ cm}} = \frac{4}{200 \text{ cm}} \qquad \text{or} \qquad s_1' = 50 \text{ cm}$$

This image will be close to the front focal point F_2 of the ocular, since the final image is over 200 cm away, and $f_2 = 5$ cm, so the distance between the two lenses is about $s_1' + f_2 = \underline{55 \text{ cm}}$. Figure 35-43 is a schematic drawing of the lens system; it is not drawn to scale.

The virtual image formed by the ocular is to be 2 m in front of the objective, so it must be 2.55 m in front of the ocular; i.e., $s_2' = -255$ cm. Therefore, the distance s_2 of the image I from the ocular is

$$\frac{1}{s_2} = \frac{1}{f_2} - \frac{1}{s_2'} = \frac{1}{5 \text{ cm}} + \frac{1}{255 \text{ cm}} = \frac{52}{255 \text{ cm}} \qquad \text{or} \qquad s_2 = 4.9 \text{ cm}$$

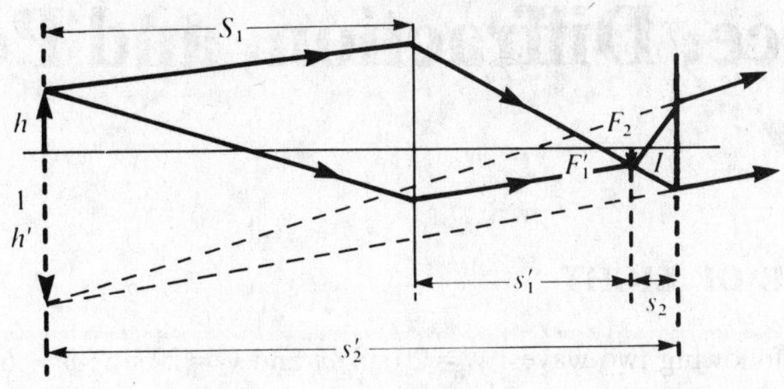

Fig. 35-43

Thus the distance between the lenses is more accurately given by $d = s_1' + s_2 = \underline{54.9\ cm}$.
(b) The magnification of the real image formed by the objective is

$$m_1 = \left|\frac{s_1'}{s_1}\right| = \frac{50\ cm}{200\ cm} = 0.25$$

The magnification of the virtual image formed by the ocular is

$$m_2 = \left|\frac{s_2'}{s_2}\right| = \frac{255\ cm}{4.9\ cm} = 52.0$$

The overall linear magnification is $M = m_1 m_2 = (0.25)(52.0) = \underline{13.0}$. The angular magnification for a telescope is not defined in terms of the near point, as for a microscope. For astronomical observations it is defined with image and object located at infinity as in Probs. 35.130–35.135. For terrestrial observations it is most sensibly defined as α'/α, where α' is the angle the image subtends at the eye and α is the angle the object, at its actual location when viewed, subtends at the naked eye. For our case $\alpha' = 13h/255$ and $\alpha = h/255$, where h is the object height. Then $\alpha'/\alpha = 13$. (We have assumed $\tan \alpha' \approx \alpha'$.) This is the same as the linear magnification in this case because image and object distances were chosen to be the same.

CHAPTER 36
Interference, Diffraction, and Polarization

36.1 INTERFERENCE OF LIGHT

36.1   Superimpose the following two waves: $y_1 = 20 \sin \omega t$ and $y_2 = 20 \sin (\omega t + 60°)$.

▌ $y = 20 \sin \omega t + 20 \sin (\omega t + 60°)$. We can think of y_1 and y_2 as projections on the y axis of vectors of length 20 rotating with angular velocity ω and making angles ωt and $\omega t + 60°$, respectively, with the x axis. The projection of the vector sum of these two vectors on the y axis is then our result $y = y_1 + y_2$. As shown in Fig. 36-1, $y = A \sin (\omega t + \phi)$, where from the law of cosines, $A^2 = 20^2 + 20^2 + 2(20)(20) \cos 60° = 3(20^2)$; $A = 34.6$; and from the isosceles triangle, $2\phi = 60°$; $\phi = 30°$. Thus, $y = \underline{34.6 \sin (\omega t + 30°)}$. Since $34.6 > 20$, this is an example of *partially constructive* interference.

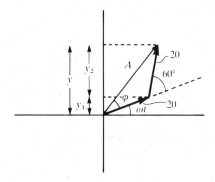

Fig. 36-1

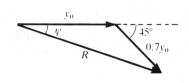

Fig. 36-2

36.2   Superimpose the following two waves: $y_1 = 30 \sin \omega t$ and $y_2 = 30 \cos \omega t$.

▌ We follow the general procedure of Prob. 36.1, writing $\cos \omega t = \sin (\omega t + \pi/2)$. Thus y_2 "leads" y_1 by 90° and in the notation of Prob. 36.1, $A^2 = 30^2 + 30^2 = 2(30^2)$; $A = 42.4$. For the right isosceles triangle, $\phi = 45°$, and $y = y_1 + y_2 = \underline{42.4 \sin (\omega t + 45°)}$. Again we have partially constructive interference.

36.3   Graph the *phasors* (rotating vectors; see Prob. 36.1) of the following two waves and find their resultant by graphical means: $y_1 = y_0 \sin \omega t$ and $y_2 = 0.7y_0 \sin [\omega t - (\pi/4)]$.

▌ From Fig. 36-2, $y = y_1 + y_2 = R \sin (\omega t - \psi)$. Graphically, R is $1.5y_0$ and $\psi = 18°$, so $y = \underline{1.5y_0 \sin (\omega t - 18°)}$.

36.4   (*a*) For the two waves given in Prob. 36.3, use trigonometry to find the amplitudes of the rectangular components R_x and R_y of their resultant. (*b*) Determine R in the form $R = A \sin (\omega t + \theta)$.

▌ From Fig. 36-2 we take x and y components at the instant shown. Then $y_{2x} \equiv a = (0.70y_0)/2^{1/2}$ so (*a*) $R_x = y_0(1 + 0.495) = 1.495y_0$ and $R_y = -0.495y_0$. (*b*) Then $R = (R_x^2 + R_y^2)^{1/2} = 1.57y_0$ and $\psi = \tan^{-1} a/(a + y_0) = 18°$ so $y = \underline{1.57y_0 \sin (\omega t - 18°)}$.

36.5   When are two light sources of a common frequency said to be *coherent*? What is the relation between the state of coherence and the phenomenon of interference?

▌ If there is a fixed difference between the phase of the wave emitted by source 1, at source 1, and the wave from source 2, at source 2, then the two sources are coherent. Interference (constructive or destructive) can occur only between overlapping waves from coherent sources.

36.6   Describe the interference pattern due to Young's double-slit arrangement.

▌ When coherent light of wavelength λ, coming from two different sources with fixed phase difference $\Delta\phi'$, meet at a point, the phase difference at that point is $\phi = (2\pi/\lambda)(\Delta r) + \Delta\phi'$, where Δr is the difference in path length traveled by the two waves from the sources to the point. For phase differences which are multiples of 2π we have maximum intensity. Thus, in the Young's double-slit arrangement, Fig. 36-3(*a*),

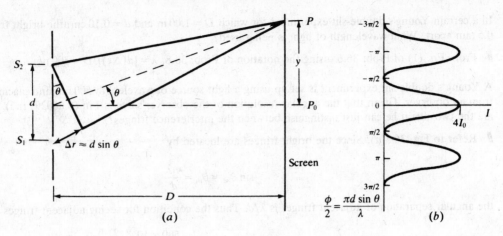

Fig. 36-3

where the two slits act as in-phase sources ($\Delta\phi' = 0$) of equal intensities and we assume $d \ll D$, intensity maxima are located by (for any integer m)

$$\Delta r = m\lambda \quad \text{or} \quad d \sin\theta = m\lambda \quad \text{or} \quad y = \frac{mD\lambda}{d} \qquad (1)$$

for θ small. The value of the maximum intensity is $4I_0$ since the amplitude is doubled. In Fig. 36-3(a), the minima are located by $\phi =$ odd multiples of π, so

$$\Delta r = (m + \tfrac{1}{2})\lambda \quad \text{or} \quad d \sin\theta = (m + \tfrac{1}{2})\lambda \quad \text{or} \quad y = \frac{(m + \tfrac{1}{2})D\lambda}{d} \qquad (2)$$

That is, they are midway between the maxima; the value of the minimum intensity is 0. [In general for any point on the screen it can be shown that $I = 4I_0 \cos^2(\phi/2)$; see Fig. 36-3b.]

36.7 The interference pattern of two identical slits separated by a distance $d = 0.25$ mm is observed on a screen at a distance of 1 m from the plane of the slits. The slits are illuminated by monochromatic light of wavelength 589.3 nm (sodium D) traveling perpendicular to the plane of the slits. Bright bands are observed on each side of the central maximum. Calculate the separation between adjacent bright bands.

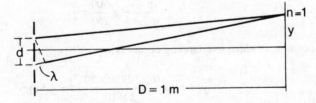

Fig. 36-4

▌ The situation is shown in Fig. 36-4. For the first bright band $n = 1$. The path difference is one wavelength.

$$n\lambda = \frac{dy}{D} \qquad 5893 \times 10^{-10} = \frac{0.25 \times 10^{-3}y}{1} \qquad y = \frac{5893 \times 10^{-10}}{0.25 \times 10^{-3}} = 2.36 \times 10^{-3}\ \text{m} = \underline{2.36\ \text{mm}}$$

separation of adjacent bright fringes.

36.8 Light from a sodium vapor lamp (589 nm) forms an interference pattern on a screen 0.8 m from a pair of slits. The bright fringes in the pattern are 0.35 cm apart. What is the slit separation?

▌ We follow Eq. (1) of Prob. 36.6, with $y = x_n$ the position of the nth maximum on the screen, measured from the central maximum. The distance between adjacent fringes is

$$\Delta x = x_{n+1} - x_n = \frac{D\lambda}{d}(n+1) - \frac{D\lambda}{d}n = \frac{D\lambda}{d} \qquad \text{so} \qquad d = \frac{D\lambda}{\Delta x} = \frac{(0.8\ \text{m})(589 \times 10^{-9}\ \text{m})}{0.35 \times 10^{-2}\ \text{m}} = 1.35 \times 10^{-4}\ \text{m} = \underline{0.135\ \text{mm}}$$

36.9 In a certain Young's double-slit experiment for which $D = 1.00\,\text{m}$ and $d = 0.10\,\text{cm}$, the bright fringes are 0.5 mm apart. What wavelength of light is being used?

▮ From Eq. (1) of Prob. 36.6, using the notation of Prob. 36.8, $\lambda = [d(\Delta x)]/D = \underline{500\,\text{nm}}$.

36.10 A Young's double-slit experiment is set up using a light source of wavelength 500 nm and placing slits 2 m from an observer. Given that the angular resolution of the observer's eye is 1 in (0.000291 rad), how far apart are the two slits if he can just distinguish between the interference fringes?

▮ Refer to Fig. 36-3(a). Since the bright fringes are located by

$$\sin \theta_m \approx \theta_m = \frac{m\lambda}{d}$$

the angular separation of adjacent fringes is λ/d. Thus the condition for seeing adjacent fringes as distinct is

$$\frac{\lambda}{d_{max}} = 0.000291 \qquad \text{or} \qquad d_{max} = \frac{500 \times 10^{-9}}{0.000291} = \underline{1.72\,\text{mm}}$$

36.11 Laser light (630 nm) incident on a pair of slits produces an interference pattern in which the bright fringes are separated by 8.3 mm. A second light produces an interference pattern in which the bright fringes are separated by 7.6 mm. What is the wavelength of this second light?

▮ From $\lambda = (d/D)\Delta x$, where d and D are fixed,

$$\lambda' = \frac{\Delta x'}{\Delta x}\lambda = \frac{7.6}{8.3}630 = \underline{577\,\text{nm}}$$

36.12 In a Young's double-slit experiment, the slits are 2 mm apart and are illuminated with a mixture of two wavelengths, $\lambda = 750\,\text{nm}$ and $\lambda' = 900\,\text{nm}$. At what minimum distance from the common central bright fringe on a screen 2 m from the slits will a bright fringe from one interference pattern coincide with a bright fringe from the other?

▮ Refer to Fig. 36-3(a). The mth bright fringe of the λ pattern and the m'th bright fringe of the λ' pattern are located at

$$y_m = \frac{mD\lambda}{d} \qquad \text{and} \qquad y'_{m'} = \frac{m'D\lambda'}{d}$$

Equating these distances gives

$$\frac{m}{m'} = \frac{\lambda'}{\lambda} = \frac{900}{750} = \frac{6}{5}$$

Hence, the first position at which overlapping occurs is

$$y_6 = y'_5 = \frac{(6)(2)(750 \times 10^{-9})}{2 \times 10^{-3}} = \underline{4.5\,\text{mm}}$$

36.13 In Young's interference experiment two slits are illuminated with orange light of wavelength 6000 Å. The interference pattern is observed on a screen very far from the slits. If the central bright fringe is numbered zero, what must be the path difference for the light from the two slits at the fourth bright fringe?

▮ Path difference $= 4\lambda = 4(6000\,\text{Å}) = \underline{24\,000\,\text{Å}}$ or $\underline{2.4\,\mu\text{m}}$

36.14 Two wavelengths of light λ_1 and λ_2 are sent through a Young's double-slit apparatus simultaneously. What must be true concerning λ_1 and λ_2 if the third-order λ_1 bright fringe is to coincide with the fourth-order λ_2 fringe?

▮ Use the notation of Prob. 36.8. The distance along the screen for the third order fringe is $x_1 = [(3\lambda_1)D]/d$; for λ_2 it is $x_2 = [(4\lambda_2)D]/d$. Since they fall at the same location, $x_1 = x_2$ and $\lambda_2 = \underline{\frac{3}{4}\lambda_1}$.

36.15 Two wavelengths λ_1 and λ_2 are used in the double-slit experiment. If one is 430 nm, what value must the other have for the fourth-order bright fringe of one to fall on the sixth-order bright fringe of the other?

❚ Apply the condition for constructive interference twice: $(4\lambda_1)/d = x/D$ and $(6\lambda_2)/d = x/d$, so that $4\lambda_1 = 6\lambda_2$. If $\lambda_2 = 430$ nm, then $\lambda_1 = \underline{645\ nm}$; if $\lambda_1 = 430$ nm, then $\lambda_2 = \underline{287\ nm}$.

36.16 Two identical radiators have a separation $d = \lambda/8$, where λ is the wavelength of the waves emitted by either source. The phase difference $\Delta\phi'$ of the sources is $\pi/4$. Find the intensity distribution in the radiation field as a function of the angle θ which specifies the direction from the radiators to the distant observation point P [see Fig. 36-3(a)].

❚ Since $\Delta r = (\lambda/8)\sin\theta$, the phase difference at P is

$$\phi = \frac{2\pi}{\lambda}\Delta r + \Delta\phi' = \frac{\pi}{4}(\sin\theta + 1) \quad \text{and} \quad I(\theta) = 4I_0\cos^2\frac{\phi}{2} = 4I_0\cos^2\left[\frac{\pi}{8}(\sin\theta + 1)\right]$$

36.17 A flake of glass of index of refraction 1.5 is placed over one of the openings of a double-slit apparatus. There is a displacement of the interference pattern through seven successive maxima toward the side where the flake was placed. If the wavelength of the diffracted light is $\lambda = 600$ nm, how thick is the flake?

❚ With λ denoting the wavelength in air, the wavelength within the glass is $\lambda' = \lambda/n$, where n is the refractive index of the glass. Letting t denote the thickness of the flake, the number of wavelengths contained in it (for "turning angles" $\theta \ll 1$) is $t/\lambda' = (nt)/\lambda$. In the thickness of air displaced by the flake, there would only have been t/λ wavelengths. Therefore the presence of the flake adds $[(nt)/\lambda] - (t/\lambda) = [(n-1)t]/\lambda$ wavelengths to the effective path of the light which passes through the flake. This causes the entire interference pattern to shift toward the side with the flake by $\Sigma = [(n-1)t]/\lambda$ fringes (i.e., one fringe per wavelength). Hence,

$$t = \frac{\Sigma\lambda}{n-1} = \frac{7(600\ \text{nm})}{0.5} = \underline{8400\ \text{nm}}$$

36.18 The two loudspeakers shown in Fig. 36-5 give off identical 200-Hz waves. As speaker A is moved back along the line indicated, an observer at P hears a strong tone when the speaker is at the positions marked S but a weak tone when the speaker is at positions W. How far apart are the W positions? P is farther away than shown.

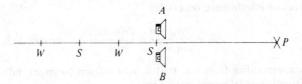

Fig. 36-5

❚ The sound emitted by the two speakers is in phase. Destructive interference occurs when $L_A - L_B = \lambda/2 + n\lambda$ with n an integer, and in effect the waves have a $\lambda/2$ path difference. Successive quiet points are λ apart; $\lambda = v/f = 330/200 = \underline{1.65\ m}$.

36.19 In order to measure the speed of sound, the two loudspeakers shown in Fig. 36-5 are driven at 400 Hz by the same oscillator. As speaker B is moved away from the observer at P, the sound heard at P is strongest when the speaker is at the positions marked S. These positions are found to be 82 cm apart. What is the speed of sound one would calculate from these data? Distance $PA \gg AB$.

❚ Constructive interference occurs when $L_B = L_A + n\lambda$, where n is an integer, and the waves are in phase. Loud positions are λ apart; $\lambda = v/f$, and $v = 0.82(400) = \underline{328\ m/s}$.

36.20 Consider a thin slab of thickness t, as shown in Fig. 36-6. A beam of light, a, in medium 1 at almost-normal incidence to the interface with the slab, medium 2, is split into a reflected ray, b, and refracted ray, d. The refracted ray is (partially) reflected on the bottom interface and finds its way back into medium 1 as a ray parallel to b. Find the interference conditions for rays b and d.

❚ We compute the difference in phase, ϕ, between the two rays at P. Ray d has traveled an extra distance $2t$ (assuming nearly normal incidence), in a medium where the wavelength is $\lambda_2 = (n_1/n_2)\lambda_1$. Moreover, one of the rays (b, if $n_2 > n_1$) has suffered a 180° phase change in reflection. Consequently,

$$\phi = \frac{2\pi}{\lambda_2}(2t) - \pi = 2\pi\left(\frac{2n_2 t}{n_1\lambda_1} - \frac{1}{2}\right)$$

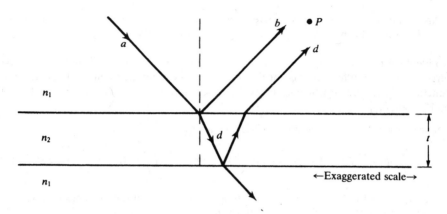

Fig. 36-6

The condition for maximum constructive interference is then

$$\phi = 2\pi m \qquad \text{or} \qquad \frac{2n_2 t}{n_1 \lambda_1} = m + \tfrac{1}{2} \qquad (m = 0, 1, 2, \ldots)$$

whereas the condition for maximum destructive interference is

$$\phi = 2\pi(m + \tfrac{1}{2}) \qquad \text{or} \qquad \frac{2n_2 t}{n_1 \lambda_1} = m + 1 \qquad (m = 0, 1, 2, \ldots)$$

36.21 By an anodizing process, a transparent film of aluminum oxide, of thickness $t = 250$ nm and index of refraction $n_2 = 1.80$, is deposited on a sheet of polished aluminum. What is the color of utensils made from this sheet when observed in white light? Assume normal incidence of the light.

▌ We must find which colors in the visible region, having vacuum wavelengths from 400 nm (violet) to 700 nm (red), will interference constructively and which destructively. From Prob. 36.20, with $n_1 = 1$ (air), maximum constructive interference occurs for

$$\lambda_1 = \frac{2(n_2/n_1)t}{m + \tfrac{1}{2}} = \frac{(900 \text{ nm})}{m + \tfrac{1}{2}} \qquad (m = 0, 1, 2, \ldots)$$

Only the value corresponding to $m = 1$, that is, $\lambda_1 = 600$ nm (orange), falls in the visible range.
Maximum destructive interference occurs for

$$\lambda_1 = \frac{2(n_2/n_1)t}{m + 1} = \frac{(900 \text{ nm})}{m + 1} \qquad (m = 0, 1, 2, \ldots)$$

and of these values, only $\lambda_1 = 450$ nm (violet) is in the visible range.
We infer that the red-orange-yellow end of the spectrum will be strongly reflected, while the violet-blue end will be greatly diminished in intensity as compared with the illuminating white light.

36.22 A radar antenna is located atop a high cliff on the edge of a lake. This antenna operates on a wavelength of 400 m. Venus rises above the horizon and is tracked by the antenna. The first minimum in the signal reflected off the surface of Venus is recorded when Venus is 35° above the horizon. Find the height of the cliff.

▌ The radar antenna receives signals directly from Venus and by reflection from the surface of the lake (Fig. 36-7). Venus may be considered infinitely far away, so that the disturbances at B and D have the same phases at every instant. The path difference between the reflected and direct rays is

$$\overline{BE} - \overline{DE} = \frac{y}{\sin 35°}(1 - \sin 20°)$$

and there is a phase shift of π due to the reflection at B. Thus the phase difference at E is

$$\phi = \frac{2\pi y}{\lambda \sin 35°}(1 - \sin 20°) + \pi$$

and for an intensity minimum this must equal $\pi, 3\pi, 5\pi, \ldots$. Ruling out $\phi = \pi$, which would imply $y = 0$, we

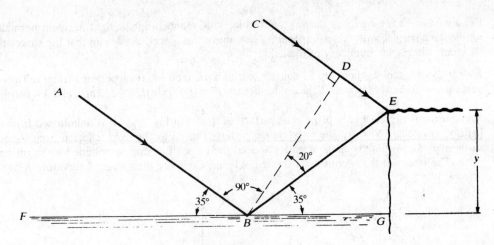

Fig. 36-7

have $\phi = 3\pi$, whence

$$y = \frac{\lambda \sin 35°}{1 - \sin 20°} = \frac{(400)(0.5736)}{1 - 0.3420} = \underline{349 \text{ m}}$$

36.23 A soap film has an index of refraction of 1.333. What is the smallest thickness of this film that will give an interference maximum when light of wavelength $\lambda = 500$ nm is incident normally upon it?

❚ Inside the soap film the wavelength of light λ_f is shorter than it is in air by a factor of the index of refraction of the soap film, since the velocity of light is less inside the film.

$$\lambda_f = \frac{\lambda_a}{n} = \frac{500}{1.333} = 375 \text{ nm}$$

The film thickness is $\frac{1}{4}\lambda_f$ for constructive interference, including the phase change on reflection at the top surface: $\frac{1}{4}\lambda_f = \frac{375}{4} = \underline{94 \text{ nm}}$.

36.24 As shown in Fig. 36-8, two flat glass plates touch at one edge and are separated at the other edge by a spacer. Using vertical viewing and light with $\lambda = 589$ nm, five dark fringes (D) are obtained from edge to edge. What is the thickness of the spacer?

❚ The pattern is caused by interference between a beam reflected from the upper surface of the air wedge and a beam reflected from the lower surface of the wedge. The two reflections are of different nature in that reflection at the upper surface takes place at the boundary of a medium (air) of lower refractive index, while reflection at the lower surface occurs at the boundary of a medium (glass) of higher refractive index. In such cases, the act of reflection by itself involves a phase displacement of 180° between the two reflected beams. This explains the presence of a dark fringe at the left-hand edge.

As we move from a dark fringe to the next dark fringe, the beam that traverses the wedge must be held back by a path-length difference of λ. Because the beam travels twice through the wedge (down and back up), the wedge thickness changes by only $\frac{1}{2}\lambda$ as we move from fringe to fringe.

$$\text{spacer thickness} = 4(\tfrac{1}{2}\lambda) = 2(589 \text{ nm}) = \underline{1178 \text{ nm}}$$

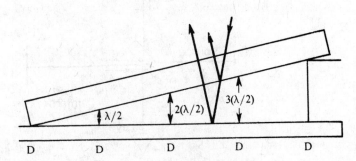

Fig. 36-8

36.25 For a wedge-shaped air space such as that of Fig. 36-8, monochromatic light, incident normally, produces 28 visible dark fringes, with the last appearing just above the spacer. Assuming that the spacer thickness is 9000 nm, what is the frequency of the light?

▌ For 28 dark fringes there are 27 equally spaced horizontal intervals between fringes. Thus the change in vertical wedge height between adjacent fringes is $\lambda/2 = (9000 \text{ nm})/27 = 333 \text{ nm}$, and $\lambda = \underline{666 \text{ nm}}$.

36.26 A convex lens is placed on a plane glass surface as shown in Fig. 36-9. It is illuminated from above with red light of wavelength 6700 Å. The interference pattern, caused by the light reflected from the convex and plane surfaces of the air film, consists of a dark spot at the point of contact surrounded by bright and dark *Newton's rings*. The radius of the twentieth dark ring is 11 mm. Compute the radius of curvature R of the lens.

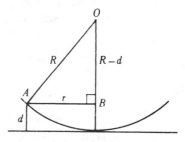

Fig. 36-9

▌ As in Prob. 36.24, the gap thickness changes by $\frac{1}{2}\lambda$ as we move from one fringe to the next fringe of like type. Hence, its thickness d at the twentieth dark ring is $d = 20(\frac{1}{2} \times 6.7 \times 10^{-7} \text{ m}) = 6.7 \times 10^{-6} \text{ m}$.

Let us now refer to Fig. 36-9. Let O be the center of curvature of the lens and r be the radius of the twentieth dark ring. Then, in right triangle ABO,

$$R^2 = r^2 + (R - d)^2 = r^2 + R^2 - 2Rd + d^2$$

which gives $2Rd = r^2 + d^2$. Because d^2 is negligibly small compared with r^2,

$$R = \frac{r^2}{2d} = \frac{(11 \times 10^{-3} \text{ m})^2}{2(6.7 \times 10^{-6} \text{ m})} = \underline{9.03 \text{ m}}$$

36.27 Newton's rings are formed by light of 400-nm wavelength. (a) Between the third and sixth bright fringe what is the change in thickness of the air film? (b) If the radius of curvature of the curved surface is 5.0 m, what is the radius of the third bright fringe?

▌ (a) Between the third and the sixth bright fringe there must be a path of difference of three wavelengths. For Newton's rings, the ray crosses the air film twice to produce interference. The change in thickness is then one-half the path difference: $\frac{1}{2}(3\lambda) = \frac{3}{2}(400) = \underline{600 \text{ nm}}$. (b) From Prob. 36.26 and Fig. 36-9 we have $r^2 \approx 2Rd$, and noting that the central fringe is dark, we have $d_3 = (2.5\lambda)/2$ for the third bright fringe. Then $r_3 = (2.5\lambda R)^{1/2} = \underline{2.2 \text{ mm}}$.

36.28 A Michelson interferometer (Fig. 36-10) is adjusted until good fringes are obtained with monochromatic light. When the movable mirror is shifted 0.015 mm, a shift of 50 fringes is observed. What is the wavelength of the light used? Include a diagram.

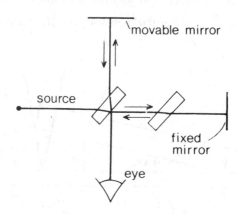

Fig. 36-10

❚ Each time the movable mirror of the Michelson interferometer is moved a half wavelength, the path length of the rays reflected from that mirror is changed by one wavelength, and a shift in the pattern of one fringe is observed. Therefore,

$$\frac{1}{2}\lambda = \frac{0.015 \times 10^{-3}\,\text{m}}{50} \qquad \lambda = 6.00 \times 10^{-7}\,\text{m} = \underline{600\,\text{nm}}$$

36.29 When one leg of a Michelson interferometer is lengthened slightly, 150 dark fringes sweep through the field of view. If the light used has $\lambda = 480$ nm, how far was the mirror in that leg moved?

❚ Darkness is observed when the light beams from the two legs are 180° out of phase. As the length of one leg is increased by $\frac{1}{2}\lambda$, the path length (down and back) increases by λ and the field of view changes from dark to bright to dark. When 150 fringes pass, the leg is lengthened by an amount

$$(150)(\tfrac{1}{2}\lambda) = (150)(240\,\text{nm}) = 36\,000\,\text{nm} = \underline{0.036\,\text{mm}}$$

36.30 Generalize the result of Prob. 36.6 to the case of N identical equally spaced slits (a *diffraction grating*).

❚ It is easy to see that maxima will occur at the same locations, $d \sin \theta = m\lambda$, since light from all the slits will be in phase. The intensity drops dramatically away from these peaks, however, because if the wave from one slit is even a little out of phase with that from an adjacent slit, successive slits are farther and farther out of phase, leading to substantial destructive interference over the N slits. (See Prob. 36.40.) The pattern of Fig. 36-3(b) now changes to that of Fig. 36-11, with small ripples between the main maxima. {The intensity at all points can be shown to obey $I = I_0[\sin^2(N\phi/2)]/[\sin^2(\phi/2)]$, where I_0 is the intensity through a single slit.}

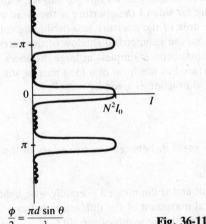

$$\frac{\phi}{2} = \frac{\pi d \sin \theta}{\lambda} \qquad \textbf{Fig. 36-11}$$

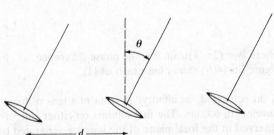

Fig. 36-12

36.31 If a number of radio telescopes are placed in a straight line, they can be used as a unit to obtain high resolution. In effect, they act as the reverse of a diffraction grating and respond strongly to only a very limited angular range. Referring to Fig. 36-12, show that they select for observation an angle and wavelength related via $n\lambda = d \sin \theta$. (The individual telescopes are mounted on a track so that d may be altered.)

❚ We want maximum constructive interference between the rays reaching each radio telescope. The rays from the distant source are parallel and in phase over a plane perpendicular to the rays. The path length difference for adjacent beams is $d \sin \theta$. For greatest reinforcement, this must be equal to $n\lambda$. Hence maximum response is obtained at $n\lambda = d \sin \theta$.

36.32 For the radio-telescope array described in Prob. 36.31, show that the resolution angle (measured from the maximum to the first minimum) is given by

$$\Delta \theta = \frac{\lambda}{Nd \cos \theta}$$

where N is the number of telescopes in the array.

❚ At a major maximum, all the beams are in phase. The first minimum occurs when the path change across the N telescopes is an additional λ so that the path change between adjacent telescopes is λ/N and the phase

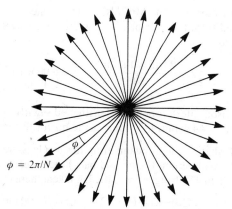

Fig. 36-13

difference is $\phi = (2\pi)/N$, as shown in the N-phasor diagram of Fig. 36-13. The amplitudes clearly add up to zero by symmetry, so the minimum condition is $n\lambda + \lambda/N = d \sin(\theta + \Delta\theta)$. Expand the sine, recalling that $\Delta\theta$ is small, and subtract $n\lambda = d \sin \theta$ to find $\Delta\theta = \lambda/(Nd) \cos \theta$.

36.2 DIFFRACTION AND THE DIFFRACTION GRATING

36.33 Describe *diffraction*, and express the formula for the *Fraunhofer diffraction pattern* of a thin slit.

▮ For plane light waves (i.e., the wavefronts are planes) incident normally on an opaque surface with an aperture, *Huygens' principle* states that the radiation field on the far side of the aperture is the same as would be produced by identical sources distributed uniformly over the area of the aperture and oscillating coherently in phase. These secondary sources would illuminate regions within the geometrical shadow of the barrier; hence the name *diffraction* ("breaking" or "bending"). The phenomenon is simplest at large distances behind the barrier, where it is called *Fraunhofer diffraction*. If the aperture has the form of a long narrow slit, of width w [Fig. 36-14(a)], the intensity distribution in Fraunhofer diffraction is given by

$$I = I_0 \frac{\sin^2 (\phi/2)}{(\phi/2)^2} \qquad (1)$$

where $\phi \equiv (2\pi/\lambda) w \sin \theta$ is the phase difference, at observation angle θ, between the two edges of the slit. Figure 36-14(b) shows the graph of (1).

36.34 A slit is located "at infinity" in front of a lens of focal length 1 m and is illuminated normally with light of wavelength 600 nm. The first minima on either side of the central maximum of the diffraction pattern observed in the focal plane of the lens are separated by 4 mm. What is the width of the slit?

▮ A fundamental property of any focusing system, e.g., a lens, is that all rays traverse the same optical path length through it (Prob. 34.56). Thus the lens does not alter the diffraction pattern of the slit. In fact, since rays through the center of the lens are undeviated, the net effect of the lens is to put the slit a distance $\gtrsim f$ in front of the viewing screen, producing a pattern as if the slit were at infinity.

From Eq. (1), Prob. 36.33, or Fig. 36-14(b), the angle of the first minimum on the positive side of the central maximum is given by

$$\frac{\pi w \sin \theta}{\lambda} = \pi \qquad \text{or} \qquad \sin \theta = \frac{\lambda}{w}$$

where w is the slit width. But

$$\sin \theta \approx \tan \theta = \frac{2 \text{ mm}}{1 \text{ m}} = 0.002 \qquad \text{Therefore,} \qquad w = \frac{\lambda}{0.002} = \frac{600 \times 10^{-9}}{0.002} = \underline{0.3 \text{ mm}}$$

36.35 Describe the overall effect of a *diffraction grating*.

▮ A diffraction grating consists of a large number, N, of parallel slits, ruled lines, or grooves. For normal incidence its Fraunhofer pattern can be shown to be the N-slit interference pattern, Fig. 36-11, modulated by the single-slit diffraction pattern, Fig. 36-14. The actual intensity pattern observed is in effect the product of the N-slit intensity pattern with the single-slit intensity pattern. Thus, there are strong, equally spaced

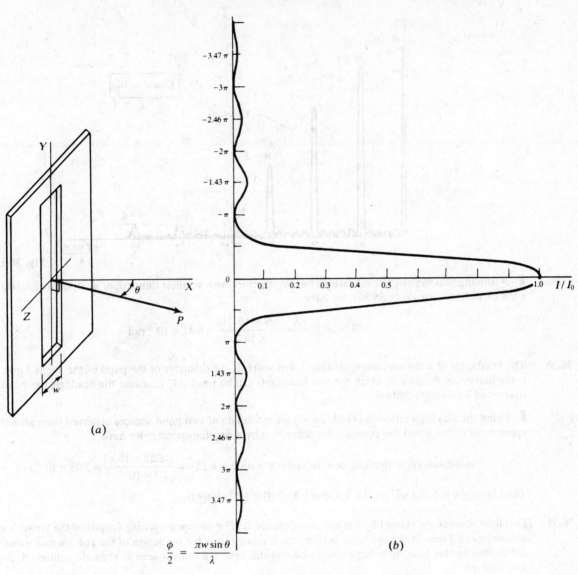

$$\frac{\phi}{2} = \frac{\pi w \sin \theta}{\lambda}$$

(a)

(b)

Fig. 36-14

interference peaks located by

$$d \sin \theta = m\lambda \qquad (1)$$

(the *grating equation*), but the heights of these peaks conform to the diffraction envelope, as shown in Fig. 36-15. The value of m (0, 1, 2, ...; we need only consider $\theta \geq 0$) corresponding to a given peak is called its *order*. Certain orders may be absent from the grating pattern, depending on the relation between the slit width w and the slit spacing d.

36.36 A parallel beam of blue light (420 nm) is incident on a small aperture. After passing through the aperture, the beam is no longer parallel but diverges at 1° to the incident direction. What is the diameter of the aperture?

▎ From Fig. 36-14(b) we see that the central maximum has an angular width corresponding to $[(\pi w)(\sin \theta)]/\lambda = \pm \pi$. Substituting $\theta = \pm 1° = (\pi/180)$ rad we have

$$w = \frac{\lambda}{\sin \theta} = \frac{420 \times 10^{-9} \text{ m}}{\sin (\pi/180)} \approx \frac{420 \times 10^{-9} \text{ m}}{\pi/180} = 24 \ \mu\text{m}$$

36.37 Light enters the eye through the *pupil*, a transparent aperture about 7 mm in diameter. What is the diffraction angle θ that results when a parallel beam of yellow light (589 nm) passes through the pupil?

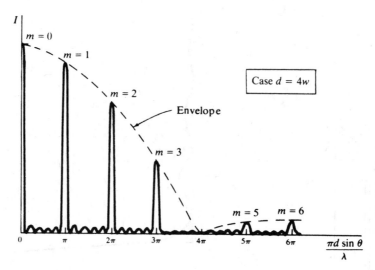

Fig. 36-15

▮ Assuming that the angular diffraction for any aperture with smallest dimension, d, is approximately that of a slit of width d (see Prob. 36.36), we have

$$\theta \approx \sin \theta = \frac{\lambda}{d} = \frac{589 \times 10^{-9} \, \text{m}}{7 \times 10^{-3} \, \text{m}} = 8.41 \times 10^{-5} \, \text{rad}$$

36.38 The headlights of a distant automobile are 1.4 m apart. If the diameter of the pupil of the eye is 3 mm, what is the maximum distance at which the two headlights can be resolved? Consider the headlights as point sources of wavelength 500 nm.

▮ Using the Rayleigh criterion (Prob. 36.46) for resolution of two point sources observed through a circular aperture of radius a and the equation for diffraction by a circular opening, we have

$$\text{minimum angle that can be resolved} = \theta \approx \sin \theta = 1.22 \frac{\lambda}{a} = \frac{1.22(5 \times 10^{-7})}{3 \times 10^{-3}} = 2.03 \times 10^{-4} \, \text{rad}$$

Then setting $\theta = 1.4/d$ we get $d = 1.4/\theta = 1.4/(2.03 \times 10^{-4}) = \underline{6900 \, \text{m}}$.

36.39 Two light sources are viewed by the eye at a distance L. The entrance opening (pupil) of the viewer's eye has a diameter of 3 mm. If the eye were perfect, the limiting factor for resolution of the two sources would be diffraction. In that limit, how large could s be and still have the sources seen as separate entities, if $L = 2500$ m?

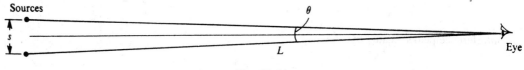

Fig. 36-16

▮ Proceed as in Prob. 36.38. In the limiting case, $\theta = \theta_c$, where $\sin \theta_c = (1.22)(\lambda/D)$. But, from Fig. 36-16, $\sin \theta_c$ is nearly equal to s/L, because s is so much smaller than L. Substitution of this value gives

$$L = 2500 \, \text{m} = \frac{sD}{1.22\lambda} \approx \frac{(s)(3 \times 10^{-3} \, \text{m})}{(1.22)(5 \times 10^{-7} \, \text{m})} \quad \text{or} \quad s = \underline{0.51 \, \text{m}}$$

We have taken $\lambda = 500$ nm, about the middle of the visible range.

36.40^c When monochromatic light from a long, very narrow source—say, a slit in a screen—falls normally on a diffraction grating, each principal maximum in the Fraunhofer pattern is a bright image of the source. Express the angular width of such a *spectral line* in terms of the incident wavelength and the width of the grating.

▮ Disregarding the secondary maxima and minima (see Fig. 36-15) we can assume that a principal maximum fills the angle $\Delta\theta$ between two consecutive minima. From the intensity equation in Prob. 36.30 we infer that the central maximum extends from $(N\phi)/2 = -\pi$ to $(N\phi)/2 = +\pi$; that is, $\Delta\phi = (4\pi)/N$. This same phase difference characterizes the other principal maxima. But

$$\phi = \frac{2\pi d}{\lambda} \sin \theta \quad \text{and so} \quad \Delta\phi = \frac{2\pi d}{\lambda} \Delta(\sin \theta) = \frac{2\pi d}{\lambda} \cos \theta \, \Delta\theta = \frac{4\pi}{N}$$

or

$$\Delta\theta = \frac{2\lambda}{Nd \cos \theta} \qquad (1)$$

Equation (1) applies to the mth-order peak when $\cos \theta$ is evaluated from the grating equation. Note that this is the same result deduced for a related situation in Prob. 36.32, by means of phasors. (The factor-of-2 discrepancy is due to the fact that $\Delta\theta$ in Prob. 36.32 is the half-angular width.)

36.41 A diffraction grating is ruled with 6000 lines per centimeter. The first order of a spectral line is observed to be diffracted at an angle of 30°. What is the wavelength of this radiation?

❙ The nth-order line obeys $n\lambda = d \sin \theta$. For our case $n = 1$ and $d = \frac{1}{6000}$ cm, so $\lambda = \frac{1}{6000} \sin 30° = 833 \times 10^{-7}$ cm = <u>833 nm</u>.

36.42 Figure 36-14(b) indicates that the first secondary peak in the single-slit Fraunhofer diffraction pattern is 0.047 times as high as the central peak. Verify this value.

❙ From Prob. 36.33 the first two intensity minima (zeros) are located at $\phi = 2\pi$ and $\phi = 4\pi$. Supposing that the ϕ value for the first secondary peak is midway between these two values, that is, $\phi = 3\pi$, we have

$$I = I_0 \frac{\sin^2 (3\pi/2)}{(3\pi/2)^2} = 0.045 I_0$$

The error resulting from our assumption is only about 4 percent.

36.43 A grating having 15 000 lines per inch produces spectra of a mercury arc. The green line of the mercury spectrum has a wavelength of 5461 Å. What is the angular separation between the first-order green line and the second-order green line?

❙ Use the grating formula and solve for θ for both $n = 1$ and $n = 2$, with distances in meters:

$$n\lambda = d \sin \theta \qquad 5461 \times 10^{-10} = \frac{1}{(39.37)(15\,000)} \sin \theta_1 \quad \text{or} \quad 0.3225 = \sin \theta_1 \quad \text{and} \quad \theta_1 = 18.8° \text{ first order}$$

Next, for $n = 2$,

$$2(5461 \times 10^{-10}) = \frac{1}{(39.37)(15\,000)} \sin \theta_2 \quad \text{or} \quad 0.6449 = \sin \theta_2 \quad \text{and} \quad \theta_2 = 40.2° \text{ second order}$$

Finally $\theta_2 - \theta_1 = 40.2° - 18.8° = \underline{21.4°}$ separation.

36.44 A diffraction grating having 7000 lines per centimeter is illuminated normally with red light from a helium–neon gas laser. If its second order spectral line is at 62.4°, what is the wavelength of the red laser light?

❙ Use the grating equation with $n = 2$,

$$n\lambda = d \sin \theta \qquad 2\lambda = \frac{1}{7000} \sin 62.4° = \frac{0.8862}{7000} \quad \text{and} \quad \lambda = 6330 \times 10^{-8} \text{ cm} = \underline{633 \text{ nm}}$$

36.45 A light source emits a mixture of wavelengths from 450 to 600 nm. When a diffraction grating is illuminated normally by this source, it is noted that two adjacent spectra barely overlap at an angle of 30°. How many lines per meter are ruled on the grating?

❙ According to the grating equation, the long-wavelength limit of the mth-order spectrum and the short-wavelength limit of the $(m + 1)$th-order spectrum will just coincide if

$$d \sin \theta = m(600 \times 10^{-9}) = (m + 1)(450 \times 10^{-9})$$

whence $m = 3$. Then, since $\theta = 30°$,

$$\frac{1}{d} = \frac{\sin 30°}{3(600 \times 10^{-9})} = \underline{277\,778 \text{ lines per meter}}$$

36.46ᶜ Reconsider Prob. 36.40 when the source emits a continuous mixture of wavelengths. The *resolving power*, at wavelength λ, of the grating is defined as $R \equiv \lambda/\delta$, where δ is the smallest wavelength difference for which the spectral lines $\lambda - \frac{1}{2}\delta$ and $\lambda + \frac{1}{2}\delta$ are resolvable. Find R_m, the resolving power in the mth order.

▌ According to *Rayleigh's criterion*, two peaks are just resolvable when their angular separation is half the angular width of either peak. This minimal separation is given by (1) of Prob. 36.40 as

$$(\Delta\theta)_{min} = \frac{\lambda}{Nd\cos\theta} \qquad \text{or} \qquad d\cos\theta(\Delta\theta)_{min} = \frac{\lambda}{N}$$

Differentiation of the grating equation, $d\sin\theta = m\lambda$, gives

$$d\cos\theta\,\Delta\theta = m\,\Delta\lambda \qquad \text{whence} \qquad d\cos\theta(\Delta\theta)_{min} = m\delta$$

Comparing the two expressions for $d\cos\theta(\Delta\theta)_{min}$ gives $R_m = mN$.
 Note that the resolving power is the same for all wavelengths.

36.47 A transmission grating is used with light incident normal to its plane. The width of each slit is one-third the spacing between slits. By considering single-slit diffraction, show that the third-order ($j = 3$) multislit diffraction maxima are missing from the diffraction pattern of the grating.

▌ As discussed in Prob. 36.35, the intensity profile in the overall diffraction pattern due to N slits of finite width is the *product* of the "ideal" N-slit intensity profile and the single-slit profile. According to Eq. (1) of Prob. 36.35, the jth maximum of the ideal N-slit profile occurs at an angle θ_j such that

$$\sin\theta_j = \frac{j\lambda}{d} \tag{1}$$

where d is the slit spacing. However, if each slit has width $w = d/3$, then Fig. 36-14b [or (1) of Prob. 36.33] implies that the single slit profile has zeros at angles given by

$$\sin\theta_m = \frac{m\lambda}{w} = \frac{3m\lambda}{d} \tag{2}$$

for $m = 0, \pm1, \pm2, \pm3$. Equations (1) and (2) show that the *overall* profile must have a zero at every third maximum of the ideal N-slit pattern. In particular, the third-order maxima ($j = \pm3$) are removed by the first minima ($m = \pm1$) of the single-slit pattern.

36.48 A beam of light of wavelength λ falls on a diffraction grating of line spacing D at angle of incidence ϕ measured from the normal to the plane of the grating. Show that the maxima in the diffraction pattern occur at angles θ which are determined by the equation $j\lambda = D(\sin\theta - \sin\phi)$, where $j = 0, \pm1, \pm2, \pm3, \ldots$.

▌ The situation is indicated in Fig. 36-17. The initial beam direction is assumed to be perpendicular to the

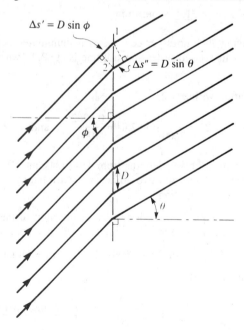

Fig. 36-17

lines that form the grating; the incident beam, the normal line to the grating, and the diffracted beam all lie in the plane of the figure. Furthermore, we assume that the source-to-grating and grating-to-screen distances are both much larger than the grating, so that the angles φ and θ indicated in the figure have values that do not vary from slit to slit across the grating. As the figure shows, the source-to-grating distance is greater for slit 1 than for the adjacent slit 2 by $\Delta s' = D \sin \varphi$. The grating-to-screen distance is greater for slit 2 than for slit 1 by $\Delta s'' = D \sin \theta$. Hence the overall source-to-screen distance is greater for slit 2 than for slit 1 by an amount

$$l_2 - l_1 = \Delta s'' - \Delta s' = D(\sin \theta - \sin \varphi) \tag{1}$$

The central maximum of the diffraction pattern occurs for $l_2 = l_1$. Counting upward in the figure from the central maximum ($m = 0$), the mth maximum occurs for $l_2 = l_1 + m\lambda$. Counting downward, the mth maximum occurs for $l_2 = l_1 - m\lambda$. Using (1), these results can be summarized by stating that maxima occur at angles θ such that $j\lambda = D(\sin \theta - \sin \varphi)$, where $j = 0, \pm1, \pm2, \pm3, \ldots$, as desired.

36.49 A parallel beam of x-rays is diffracted by a rock salt crystal. The first-order strong reflection is obtained when the glancing angle (the angle between the crystal face and the beam) is 6°50′. The distance between reflection planes in the crystal is 2.81 Å. What is the wavelength of the x-rays?

❚ This is an application of *Bragg's reflection law*, $m\lambda = 2d \sin \phi$, where ϕ is the glancing angle, not the angle of incidence. For $m = 1$,

$$\lambda = \frac{2d \sin \phi_1}{1} = \frac{(2)(2.81 \text{ Å})(0.119)}{1} = \underline{0.67 \text{ Å}}$$

36.3 POLARIZATION OF LIGHT

36.50 *Brewster's law* says that light reflected from a smooth dielectric surface will be completely linearly polarized normal to the plane of incidence, if the angle of incidence has a certain value, θ_p. What is the physical basis for this law?

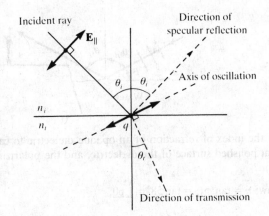

Fig. 36-18

❚ Consider the incident wave as a superposition of $\mathbf{E}_\parallel$ (in the plane of incidence) and $\mathbf{E}_\perp$ (normal to that plane). Figure 36-18 isolates $\mathbf{E}_\parallel$. In the classical model, the force $\mathbf{E}_\parallel$ sets charges in the surface of the dielectric into oscillation, thereby generating the transmitted field. There will also be a reflected field, propagated at the required angle θ_i, *unless* the angle of incidence is such as to make the axis of oscillation, which is perpendicular to the transmitted ray, coincide with the direction of reflection. This follows from the fact that a classical linear oscillator radiates zero intensity in the direction of its acceleration. Thus, at the critical angle of incidence θ_p (the *Brewster angle*), only $\mathbf{E}_\perp$ is reflected.

36.51 With reference to Prob. 36.50, obtain a formula for the Brewster angle.

❚ From Fig. 36-18, the condition for zero reflection of $\mathbf{E}_\parallel$ may be expressed as $\theta_p + 90° + \theta_t = 180°$, or $\theta_t = 90° - \theta_p$, or

$$\sin \theta_t = \cos \theta_p \tag{1}$$

Together with Snell's law,

$$n_i \sin \theta_p = n_t \sin \theta_t \qquad (2)$$

(1) implies the desired relation:

$$\tan \theta_p = \frac{n_t}{n_i} \qquad (3)$$

36.52 Linearly polarized light is incident at Brewster's angle on the surface of a medium. What can be said about the refracted and reflected beams if the incident beam is polarized (*a*) parallel to the plane of incidence and (*b*) perpendicular to the plane of incidence?

▌ (*a*) At Brewster's angle the parallel component is completely refracted; thus no light is reflected at all. (*b*) Some of the incident light is reflected and some is refracted. Both the reflected and the refracted beams will be polarized perpendicular to the plane of incidence.

36.53 Light is reflected from a lead glass plate with index of refraction = 1.96. At what angle of incidence must light strike the plate so that the reflected light is completely plane polarized?

▌ Use Brewster's law. $n = \tan \phi_{1p} = 1.96$ and $\phi_{1p} = \underline{63°}$. At this angle the reflected light is plane-polarized perpendicular to the plane of incidence.

36.54 At what angle β above the horizon is the sun when a person observing its rays reflected in water ($n_2 = 1.33$) finds them linearly polarized along the horizontal? See Fig. 36-19.

▌ In order for the reflected rays to be linearly polarized, the angle of incidence must be Brewster's angle.

$$\tan \theta_{1p} = \frac{n_2}{n_1} = \frac{1.33}{1.00} \qquad \text{or} \qquad \theta_{1p} \approx 53°$$

and $\beta = 90° - \theta_{1p} \approx \underline{37°}$.

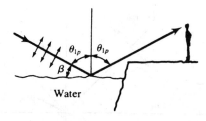

Fig. 36-19

36.55 It is desired to know the index of refraction of an opaque dielectric to calculate the dielectric constant. Light is reflected from a flat polished surface of the dielectric, and the polarizing angle is found to be 58°. What is the index of refraction?

▌ Use Brewster's law: $n = \tan \phi_{1p} = \tan 58° = \underline{1.60}$.

36.56 Two coherent *x*-directed plane-polarized beams of equal amplitude are combined. One has its **E** vector in the *y* direction, while that for the other is in the *z* direction. If the two beams are in phase, what is the state of polarization of the resultant beam? What is the amplitude of the resultant beam if the original beams had amplitude E_0?

▌ The two perpendicular waves of equal amplitude are drawn in Fig. 36-20 at a given position *x* and time *t*. The two waves combine to give a vector of amplitude $1.414E_0$; note that since the two amplitudes oscillate in phase, $\theta = 45°$ is constant, so the electric field is plane-polarized.

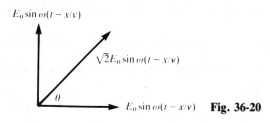

Fig. 36-20

36.57 Find the direction of polarization when at some point the electric field is given by

$$E_x = 20 \cos \omega t \quad \text{(V/m)} \qquad E_y = 40 \cos (\omega t + \pi) \quad \text{(V/m)}$$

▌ The situation is shown in Fig. 36-21 at an instant when $E_x > 0$. E_y is 180° out of phase with E_x, which means it points along $-y$ when E_x points along $+x$; their magnitudes are always in the ratio 1:2. Their resultant $\mathbf{E}$ thus has magnitude: $\sqrt{40^2 + 20^2} = 20\sqrt{5}$ and is linearly polarized at $\theta = \tan^{-1}(\frac{40}{20}) = 63.4°$ below the x axis.

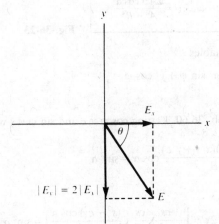

Fig. 36-21

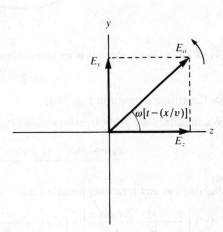

Fig. 36-22

36.58 *Circularly polarized light* is obtained by combining two coherent, equal-amplitude, plane-polarized beams whose planes of polarization are perpendicular to each other and which are 90° out of phase. Show that when the two beams $E_y = E_0 \sin \{\omega[t - (x/v)]\}$ and $E_z = E_0 \cos \{\omega[t - (x/v)]\}$ are combined, their resultant at a particular point in space is an electric field vector whose magnitude is constant but whose direction rotates on a circle perpendicular to the x axis. *Elliptically polarized light* can be obtained by adding beams of unequal amplitudes (see Prob. 36.60).

▌ Suppose that we consider a vector of length E_0 rotating in the zy plane with angular velocity ω such that the angle it makes with the z axis at any time is $\omega[t - (x/v)]$, as shown in Fig. 36-22. Then its components are $E_z = E_0 \cos \{\omega[t - (x/v)]\}$ and $E_y = E_0 \sin \{\omega[t - (x/v)]\}$. Clearly, the rotating constant-magnitude vector is the vector sum of the two linearly polarized waves given in the problem, at all times.

36.59 Find the transmitted intensity when circularly polarized light of intensity I' is incident on a perfect polarizer.

▌ Consider the circularly polarized light of Prob. 36.58 with y axis along the polarizer direction. Then, since the circularly polarized light has constant amplitude E_0, $I' \sim |E_0|^2$. The transmitted magnitude is $E_y = E_0 \sin \{\omega[t - (x/v)]\}$, and $I \sim \langle E_y^2 \rangle = \frac{1}{2} |E_0|^2$, where $\langle \rangle$ means "time average." Thus $I = \frac{1}{2} I'$.

36.60 Prove that a particle undergoing two simple harmonic vibrations of the same frequency, at right angles and out of phase, traces an elliptical path.

▌ Suppose the vibrations take place along the X and Y axes, with a phase difference α; that is,

$$x = A \sin \omega t \qquad y = B \sin (\omega t - \alpha)$$

Then

$$\frac{y}{B} = \sin \omega t \cos \alpha + \cos \omega t \sin \alpha = \frac{x}{A} \cos \alpha + \sqrt{1 - \frac{x^2}{A^2}} \sin \alpha \qquad \text{and so} \qquad \left(\frac{y}{B} - \frac{x}{A} \cos \alpha\right)^2 = \left(1 - \frac{x^2}{A^2}\right) \sin^2 \alpha$$

which gives upon expansion

$$\frac{x^2}{A^2} + \frac{y^2}{B^2} - \frac{2xy}{AB} \cos \alpha = \sin^2 \alpha \tag{1}$$

This is the equation of an ellipse whose major and minor axes are inclined to the X and Y axes.

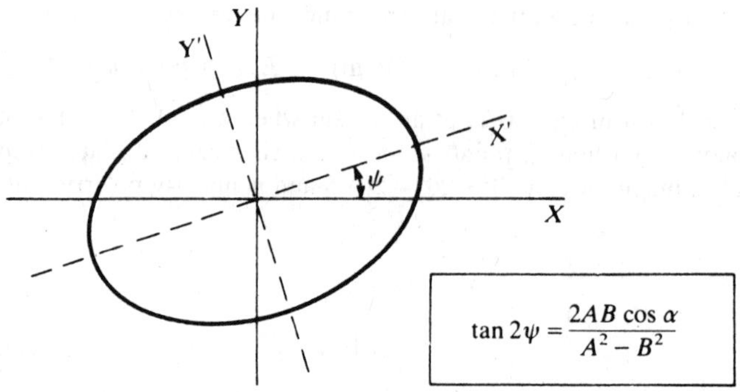

$$\tan 2\psi = \frac{2AB \cos \alpha}{A^2 - B^2}$$

Fig. 36-23

36.61 In (1) of Prob. 36.60, show that an appropriate change of variables

$$x = x' \cos \psi - y' \sin \psi \qquad y = x' \sin \psi + y' \cos \psi$$

yields the ellipse graphed in Fig. 36-23.

▌ We substitute the given transformation into Eq. (1) of Prob. 36.60. Letting $\cos \psi \equiv c$ and $\sin \psi \equiv s$ we get

$$\frac{(x'c - y's)^2}{A^2} + \frac{(x's + y'c)^2}{B^2} - \frac{2(x'c - y's)(x's + y'c)}{AB} \cos \alpha = \sin^2 \alpha$$

Multiplying out and arranging terms we have

$$(x')^2 \left[\frac{c^2}{A^2} + \frac{s^2}{B^2} - \frac{2cs \cos \alpha}{AB} \right] + (y')^2 \left[\frac{s^2}{A^2} + \frac{c^2}{B^2} + \frac{2sc \cos \alpha}{AB} \right] + 2x'y' \left[-\frac{cs}{A^2} + \frac{cs}{B^2} + \frac{(s^2 - c^2) \cos \alpha}{AB} \right] = \sin^2 \alpha$$

This is the equation of an ellipse with symmetry axes along X', Y' if the cross-term (last bracket) can be shown to be zero. We require

$$cs \left(\frac{1}{A^2} - \frac{1}{B^2} \right) = \frac{(s^2 - c^2) \cos \alpha}{AB} \Rightarrow \frac{cs}{c^2 - s^2} = \frac{AB \cos \alpha}{A^2 - B^2}$$

But

$$\cos \psi \sin \psi = \frac{\sin 2\psi}{2} \qquad \text{and} \qquad \cos^2 \psi - \sin^2 \psi = \cos 2\psi \qquad \text{so} \qquad \tan 2\psi = \frac{2AB \cos \alpha}{A^2 - B^2} \qquad \text{as required.}$$

36.62 Derive Malus' law, $I = I_0 \cos^2 \theta$.

▌ The incident plane-polarized wave, $\mathbf{E} = \mathbf{E}_0 \sin \omega t$, making angle θ with the transmission axis, is equivalent to two plane-polarized waves,

$$\mathbf{E}_{\parallel} = (\mathbf{E}_0 \cos \theta) \sin \omega t \qquad \mathbf{E}_{\perp} = (\mathbf{E}_0 \sin \theta) \sin \omega t$$

respectively, along and perpendicular to the transmission axis. Since intensity is proportional to the square of the amplitude, and since only $\mathbf{E}_{\parallel}$ is transmitted,

$$\frac{I}{I_0} = \frac{(\mathbf{E}_0 \cos \theta) \cdot (\mathbf{E}_0 \cos \theta)}{\mathbf{E}_0 \cdot \mathbf{E}_0} = \cos^2 \theta$$

36.63 If a beam of polarized light has one-tenth of its initial intensity after passing through an analyzer, what is the angle between the axis of the analyzer and the initial amplitude of the beam?

▌ We apply Malus' law.

$$I = 0.1 I_0 = I_0 \cos^2 \theta \qquad \text{so} \qquad \cos^2 \theta = 0.1 \qquad \text{Then} \qquad \cos \theta = \sqrt{0.1} = 0.316 \qquad \text{and} \qquad \theta = \underline{71.6°}$$

36.64 The amplitude of a beam of polarized light makes an angle of 65° with the axis of a Polaroid sheet. What fraction of the beam is transmitted through the sheet?

▌ The ratio of the transmitted intensity I' to the incident intensity I is, by Malus' law,

$$\frac{I'}{I} = \cos^2 \theta = \cos^2 65° = \underline{0.179}$$

36.65 Unpolarized light of intensity I' is incident upon a stack of two filters whose transmission axes make an angle θ. Express the intensity I of the emerging beam.

▮ The unpolarized light behaves like two equal intensity polarized waves polarized at right angles to each other. We can assume one is along the incident transmission axis, and hence the transmitted light has intensity $\frac{1}{2}I'$. This light now passes through the second filter and by Malus' law $I = (\frac{1}{2}I')\cos^2\theta$.

36.66 (a) Ordinary light incident on one Polaroid sheet falls on a second Polaroid whose plane of vibration makes an angle of 30° with that of the first Polaroid. If the Polaroids are assumed to be ideal, what is the fraction of the original light transmitted through both Polaroids? (b) If the second Polaroid is rotated until the transmitted intensity is 10 percent of the incident intensity, what is the new angle?

▮ (a) From Malus' law for the second Polaroid, $I_1/I_2 = \cos^2\phi = \cos^2 30° = 0.866^2 = 0.75$. Three-fourths of the light striking the second Polaroid is transmitted. The first Polaroid removes half the unpolarized light. Therefore, $\frac{1}{2} \times \frac{3}{4} = \frac{3}{8}$ of the original light transmitted. (b) Now the final intensity is 20 percent of the intensity of the light entering the second Polaroid, so $I_1/I_2 = 0.2 = \cos^2\phi$, and $\phi = 63.4°$.

36.67 Two Polaroid sheets are aligned with their axes parallel. One of the sheets is then rotated through an angle of 60°. What is the ratio of the light intensity transmitted in the first instance to that in the second instance?

▮ Let I_0 = intensity of light through first Polaroid. All this intensity passes through the second Polaroid when their axes are parallel. When they make an angle with each other, we use Malus' law: $I/I_0 = \cos^2\phi = \cos^2 60° = 0.5^2 = 0.25$. Then $I_0/I = \underline{4}$.

36.68 Polarized light of initial intensity I_0 passes through two analyzers—the first with its axis at 45° to the amplitude of the initial beam and the second with its axis at 90° to the initial amplitude (Fig. 36-24). What is the intensity of the light that emerges from this system and what is the direction of its amplitude?

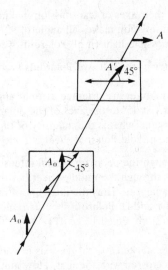

Fig. 36-24

▮ Since the angle between the axis of the first analyzer and the initial amplitude $\mathbf{A}_0$ is 45°, the intensity after passing through the first analyzer is $I' = I_0 \cos^2 45° = 0.5I_0$. The transmitted amplitude $\mathbf{A}'$ is oriented at 45° with respect to the axis of the second analyzer, so the final intensity is $I = I' \cos^2 45° = 0.5I' = 0.25I_0$ and the final amplitude $\mathbf{A}$ is oriented at 90° with respect to the initial amplitude $\mathbf{A}_0$.

Note that if only the second analyzer were in place, *no* light would get through, since $\mathbf{A}_0$ is perpendicular to its transmission axis. This problem thus demonstrates the vector nature of the light disturbance.

36.69 The axes of a polarizer and an analyzer are oriented at 30° to each other. (a) If unpolarized light of intensity I_0 is incident on them, what is the intensity of the transmitted light? (b) Polarized light of intensity I_0 is incident on this polarizer-analyzer system. If the amplitude of the light makes an angle of 30° with the axis of the polarizer, what is the intensity of the transmitted light?

▮ (a) Half the light passes through the polarizer, so the intensity of the polarized light incident on the analyzer is $I = \frac{1}{2}I_0$, and the intensity I' of the light passing through the analyzer is $I' = I \cos^2\theta = \frac{1}{2}I_0 \cos^2 30° = \underline{0.375\,I_0}$. (b) After passing through the polarizer the intensity is $I = I_0 \cos^2\theta = I_0 \cos^2 30° = 0.75I_0$. This light is now polarized at 30° to the axis of the analyzer, so the intensity of the light passing through the analyzer is $I' = I \cos^2 \theta = 0.75I_0 \cos^2 30° = \underline{0.563I_0}$.

36.70 Polarized light of intensity I_0 is incident on a pair of Polaroid sheets. Let θ_1 and θ_2 be the angles between the incident amplitude and the axes of the first and second sheets, respectively. Show that the intensity of the transmitted light is $I = I_0 \cos^2 \theta_1 \cos^2 (\theta_1 - \theta_2)$.

▌ The intensity of the light after passing through the first polarizer is $I = I_0 \cos^2 \theta_1$. This light is polarized in the direction of the axis of the first sheet, and so its axis makes the angle $\theta_2 - \theta_1$ with the axis of the second sheet (see Fig. 36-25). Consequently, the intensity of the light after passing through the second polarizer is

$$I' = I \cos^2 (\theta_2 - \theta_1) = I_0 \cos^2 \theta_1 \cos^2 (\theta_2 - \theta_1)$$

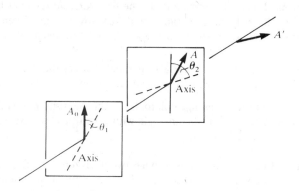

Fig. 36-25

36.71 Two Polaroids are aligned with their axes of transmission making an angle of 45°. They are followed by a third Polaroid whose axis of transmission makes an angle of 90° with the first Polaroid. If all three are ideal, what fraction of the maximum possible light (if all Polaroids were at the same angle) passes through all three?

▌ The reduction factor for either of the last two Polaroids is $\cos^2 45° = 1/2$; so the overall factor is $(\frac{1}{2})^2 = \frac{1}{4}$.

36.72 The axes of a polarizer and an analyzer are oriented at right angles to each other. A third Polaroid sheet is placed between them with its axis at 45° to the axes of the polarizer and analyzer. (*a*) If unpolarized light of intensity I_0 is incident on this system, what is the intensity of the transmitted light? (*b*) What is the intensity of the transmitted light when the middle Polaroid sheet is removed?

▌ (*a*) The light that passes through the polarizer has an intensity of $\frac{1}{2}I_0$ and is polarized at 45° to the middle sheet. Thus the light that passes through the middle sheet has the intensity $I = \frac{1}{2}I_0 \cos^2 45° = 0.25I_0$ and is polarized 45° to the axis of the analyzer. Thus the intensity of the light passing through the analyzer is $I' = I \cos^2 45° = \underline{0.125I_0}$. (*b*) If the middle Polaroid sheet is removed, we have a crossed polarizer-analyzer and no light gets through. (Compare Prob. 36.68.)

36.73 Four perfect polarizing plates are stacked so that the axis of each is turned 30° clockwise with respect to the preceding plate; the last plate is crossed with the first. How much of the intensity of an incident unpolarized beam of light is transmitted by the stack?

▌ The first plate transmits one half of the incident intensity. Each succeeding plate makes a vector resolution at angle 30°, transmitting the fraction $\cos 30°$ of the amplitude, or $\cos^2 30° = \frac{3}{4}$ of the intensity, leaving the preceding plate. The fraction of the initial intensity transmitted by the stack is then $\frac{1}{2}(\frac{3}{4})^3 = \underline{0.211}$.

36.74 A quarter-wave plate is made from a material whose indices of refraction for light of free-space wavelength $\lambda_0 = 589$ nm are $n_\perp = 1.732$ and $n_\parallel = 1.456$. What is the minimum necessary thickness of the plate for this wavelength?

▌ The optical path length (Prob. 34.53) of the ordinary wave in a plate of thickness l is $n_\perp l$ and the optical path length of the extraordinary ray is $n_\parallel l$. Since the two rays must emerge from the plate with a 90° phase difference, the optical paths must differ by $(k + \frac{1}{4})\lambda_0$, $k = 0, 1, 2, \ldots$. The minimum thickness thus satisfies

$$\frac{\lambda_0}{4} = l(n_\perp - n_\parallel) \quad \text{or} \quad l = \frac{\lambda_0}{4(n_\perp - n_\parallel)} = \frac{589 \text{ nm}}{4(1.732 - 1.456)} = 534 \text{ nm}$$

36.75 Unpolarized monochromatic light is incident on a polarizer and analyzer with axes at right angles to each other. A quarter-wave plate is placed between them so that the polarizer axis is at 45° to the two axes of the plate. (**a**) Describe the nature of the light after it passes the quarter-wave plate. (**b**) What fraction of the incident intensity passes through the analyzer?

▌ (**a**) The light incident on the quarter-wave plate is linearly polarized at 45° to each of the two axes of the plate. We can resolve this into two waves in phase each along a plate axis. If E_1 is the amplitude incident on the plate, each component will have an amplitude $E_1/\sqrt{2}$ (see Prob. 36.56). After traveling through the plate the two waves will be out of phase by one-quarter of a wavelength, or by 90°. Then they can be recombined to give, as a net result, a circularly polarized wave of amplitude $E_1/\sqrt{2}$ (see Prob. 36.58). (**b**) If I_0 is the incident intensity, then $I_1 = \frac{1}{2}I_0$ is the intensity passing through the polarizer. The intensity passing through *each* axis of the quarter-wave plate is $\frac{1}{2}I_1$, since intensity goes as the square of the amplitude. The total intensity beyond the plate is thus I_1. (This can be seen directly in terms of the circularly polarized wave, whose constant magnitude $E_1/\sqrt{2}$ is undiminished in time-averaging.) Finally, following Prob. 36.59, the intensity passing through the analyzer is $\frac{1}{2}I_1 = \frac{1}{4}I_0$.

CHAPTER 37
Special Relativity

37.1 LORENTZ TRANSFORMATION, LENGTH CONTRACTION, TIME DILATION, AND VELOCITY TRANSFORMATION

37.1 Consider two inertial frames, $\mathscr{L}$ and $\mathscr{L}'$, with axes parallel and origins O, O' coinciding at $t = t' = 0$, and with $\mathscr{L}'$ moving with uniform velocity v along the x axis of $\mathscr{L}$. Letting $\gamma \equiv 1/\sqrt{1 - (v^2/c^2)}$, the Lorentz transformation $\mathscr{L} \to \mathscr{L}'$ is

$$x' = \gamma(x - vt) \tag{1}$$
$$y' = y \tag{2}$$
$$z' = z \tag{3}$$
$$t' = \gamma\left(t - \frac{vx}{c^2}\right) \tag{4}$$

From the principle of equivalence of inertial frames, infer the inverse Lorentz transformation $\mathscr{L}' \to \mathscr{L}$.

▌ Because the *form* of the transformation equations must be the same for all inertial observers, the observer in $\mathscr{L}'$ must be able to obtain the equations giving (x, y, z, t) in terms of (x', y', z', t') simply by replacing v with $v' = -v$ in Eqs. (1) to (4) and simultaneously interchanging primed and unprimed quantities. Therefore we obtain

$$x = \gamma'(x' - v't) = \gamma(x' + vt') \tag{5}$$
$$y = y' \tag{6}$$
$$z = z' \tag{7}$$
$$t = \gamma'\left(t' - \frac{v'x'}{c^2}\right) = \gamma\left(t' + \frac{vx'}{c^2}\right) \tag{8}$$

where we have used the fact that

$$\gamma' \equiv \frac{1}{\sqrt{1 - [(v')^2/c^2]}} = \frac{1}{\sqrt{1 - [(-v)^2/c^2]}} = \frac{1}{\sqrt{1 - (v^2/c^2)}} = \gamma$$

37.2 Obtain the result of Prob. 37.1 by algebraic manipulation of the direct Lorentz transformation.

▌ We need to solve Eqs. (1) and (4) of Prob. 37.1 for x and t in terms of x' and t'. From Eq. (1), we have

$$x = \frac{x'}{\gamma} + vt \tag{9}$$

and from Eq. (4), we have

$$t = \frac{t'}{\gamma} + \frac{vx}{c^2} \tag{10}$$

Substituting Eq. (10) into Eq. (9), we find that

$$x = \frac{x'}{\gamma} + \frac{vt'}{\gamma} + \frac{v^2x}{c^2}$$

Solving this for x, we find that

$$x = \frac{1}{1 - (v^2/c^2)}\left(\frac{x' + vt'}{\gamma}\right) = \gamma(x' + vt') \tag{11}$$

Substituting Eq. (11) into Eq. (10), we obtain

$$t = \frac{t'}{\gamma} + \frac{\gamma vx'}{c^2} + \frac{\gamma v^2 t'}{c^2} = \gamma\left[t'\left(\frac{1}{\gamma^2} + \frac{v^2}{c^2}\right) + \frac{vx'}{c^2}\right] = \gamma\left(t' + \frac{vx'}{c^2}\right) \tag{12}$$

Equations (11) and (12) agree with Eqs. (5) and (8) of Prob. 37.1; the inversion of the y' and z' equations is trivial.

37.3 For the situation of Prob. 37.1, suppose that at the instant origin O' coincides with O (at $t = t' = 0$), a flashbulb is exploded at this common origin. According to observers in $\mathscr{L}$, a spherical wavefront expands outward from O at speed c. Show that, even though $\mathscr{L}'$ is moving relative to $\mathscr{L}$ with velocity v, observers in $\mathscr{L}'$ note an exactly similar wavefront expanding outward from O'.

▌ The equation of the wavefront in $\mathscr{L}$ is

$$x^2 + y^2 + z^2 = c^2 t^2 \tag{1}$$

The Lorentz transformation of (1) is (Eqs. (5) to (8), Prob. 37.1)

$$\left[\frac{x' + vt'}{\sqrt{1 - (v^2/c^2)}}\right]^2 + (y')^2 + (z')^2 = c^2 \left[\frac{t' + (vx'/c^2)}{\sqrt{1 - (v^2/c^2)}}\right]^2$$

which reduces to

$$(x')^2 + (y')^2 + (z')^2 = c^2(t')^2 \tag{2}$$

Equation (2) represents a spherical wavefront expanding outward from O' at speed c.

Actually, the reasoning goes the other way round. The $\mathscr{L}$ and $\mathscr{L}'$ observers must see exactly the same kind of wave, by the postulates of special relativity. This (almost) forces the relations between the coordinates in the two frames to have a particular form—that of the Lorentz transformations. See Prob. 37.5.

37.4 When inertial reference frames $\mathscr{L}$ and $\mathscr{L}'$ coincide, let a flash of light be produced at the common origin. Each observer is justified in considering himself at the center of an expanding sphere of light. Experiment has revealed that each obtains the same value c for the speed of light. The galilean transformation, $x' = x - vt$, does not give this result. Therefore try a modification, $x' = \gamma(x - vt)$, where γ is to be determined. The principle of equivalence requires that this equation hold for the inverse transformation, $x = \gamma(x' - v't') = \gamma(x' + vt')$. In this equation, we use the assertion that $v' = -v$. But for generality, the possibility has been allowed that t' may be different from t.

If x and x' are the intersections of the sphere with the axis at times t and t', respectively: **(a)** To what is x'/t' equal? **(b)** To what is x/t equal? **(c)** Use the results of parts **(a)** and **(b)** to eliminate x and x' in the transformation equations and thus to determine γ.

▌ We note that the problem statement includes the tacit assumption that $\gamma' = \gamma$ when $v' = -v$.
(a) At time t', the light sphere observed by $\mathscr{L}'$ intersects the positive x' axis at a coordinate $x' = ct'$. Therefore $x'/t' = c$. **(b)** At time t, the light sphere observed by $\mathscr{L}$ intersects the positive x axis at a coordinate $x = ct$. Therefore $x/t = c$. **(c)** Since the equations obtained in parts **(a)** and **(b)** refer to the same point event, we use $x' = ct'$ and $x = ct$ in our equations to get

$$ct' = \gamma(ct - vt) \tag{1}$$

and

$$ct = \gamma(ct' + vt') \tag{2}$$

Multiplying Eqs. (1) and (2) we have

$$c^2 tt' = \gamma^2(c^2 - v^2)tt'$$

Solving this for γ, we find

$$\gamma = \frac{1}{\sqrt{1 - (v^2/c^2)}} \tag{3}$$

37.5 Continuing with the results of Prob. 37.4, use the relation for γ, express t' in terms of t and x.

▌ When Eq. (3) is used in the transformation equations (Prob. 37.4), we obtain

$$x' = \frac{1}{\sqrt{1 - (v^2/c^2)}}(x - vt) \tag{4}$$

and

$$x = \frac{1}{\sqrt{1 - (v^2/c^2)}}(x' + vt') \tag{5}$$

These equations relate arbitrary event-coordinate observations of observers O and O'. Using Eq. (4) to eliminate x' from Eq. (5), we find that

$$x = \frac{(x-vt)}{1-(v^2/c^2)} + \frac{vt'}{\sqrt{1-(v^2/c^2)}}$$

Solving this for t', we find (after some algebra) that

$$t' = \frac{1}{\sqrt{1-(v^2/c^2)}}\left(t - \frac{vx}{c^2}\right)$$

which is the standard Lorentz transformation equation for t'.

37.6 As measured by $\mathscr{L}$ a flashbulb goes off at $x = 100$ km, $y = 10$ km, $z = 1$ km at $t = 0.5$ ms. What are the coordinates x', y', z', and t' of this event as determined by a second observer, $\mathscr{L}'$, moving relative to $\mathscr{L}$ at $-0.8c$ along the common xx' axis?

▌ From the Lorentz transformations,

$$x' = \frac{x-vt}{\sqrt{1-(v^2/c^2)}} = \frac{100\text{ km} - (-0.8\times3\times10^5\text{ km/s})(5\times10^{-4}\text{ s})}{\sqrt{1-0.8^2}} = \underline{367\text{ km}}$$

$$t' = \frac{t-(v/c^2)x}{\sqrt{1-(v^2/c^2)}} = \frac{5\times10^{-4}\text{ s} - [(-0.8)(100\text{ km})]/(3\times10^5\text{ km/s})}{\sqrt{1-(0.8)^2}} = \underline{1.28\text{ ms}}.$$

$$y' = y = \underline{10\text{ km}} \qquad z' = z = \underline{1\text{ km}}$$

37.7 At time $t' = 0.4$ ms, as measured in $\mathscr{L}'$, a particle is at the point $x' = 10$ m, $y' = 4$ m, $z' = 6$ m. (Note that this constitutes an event.) Compute the corresponding values of x, y, z, t, as measured in $\mathscr{L}$, for **(a)** $v = +500$ m/s, **(b)** $v = -500$ m/s, and **(c)** $v = 2\times10^8$ m/s.

▌ From the inverse Lorentz transformation, Prob. 37.1,

(a) $$x = \frac{10+(500)(4\times10^{-4})}{\sqrt{1-(500^2/c^2)}} \approx 10+(500)(4\times10^{-4}) = \underline{10.2\text{ m}} \qquad y = \underline{4\text{ m}} \qquad z = \underline{6\text{ m}}$$

$$t = \frac{(4\times10^{-4})+[(500)(10)]/c^2}{\sqrt{1-(500^2/c^2)}} \approx \underline{4\times10^{-4}\text{ s}}$$

(b) $$x = \frac{10-(500)(4\times10^{-4})}{\sqrt{1-(500^2/c^2)}} \approx 10-(500)(4\times10^{-4}) = \underline{9.8\text{ m}} \qquad y = \underline{4\text{ m}} \qquad z = \underline{6\text{ m}}$$

$$t = \frac{(4\times10^{-4})-[(500)(10)]/c^2}{\sqrt{1-(500^2/c^2)}} \approx \underline{4\times10^{-4}\text{ s}}$$

(c) $$x = \frac{10+(2\times10^8)(4\times10^{-4})}{\sqrt{1-(2/2.997925)^2}} = \frac{8.001\times10^4}{0.744943} = \underline{107.4\text{ km}} \qquad y = \underline{4\text{ m}} \qquad z = \underline{6\text{ m}}$$

$$t = \frac{(4\times10^{-4})+[(2\times10^8)(10)]c^2}{\sqrt{1-(v^2/c^2)}} = \underline{5.37\times10^{-4}\text{ s}}$$

37.8 At $t = 1$ ms (in $\mathscr{L}$), an explosion occurs at $x = 5$ km. What is the time of the event for the $\mathscr{L}'$ observer, if for him it occurs at $x' = 35.354$ km?

▌ From the Lorentz transformation for x',

$$35.354\times10^3 = \frac{(5\times10^3)-v(10^{-3})}{\sqrt{1-(v^2/c^2)}} \qquad\text{or}\qquad v = -3\times10^7\text{ m/s}$$

Then, from the Lorentz transformation for t',

$$t' = \frac{10^{-3}+[(3\times10^7)(5\times10^3)]/c^2}{\sqrt{1-(v^2/c^2)}} = \underline{1.0067\text{ ms}}.$$

37.9 A spaceship of (rest) length 100 m takes $4\,\mu$s to pass an observer on earth. What is its velocity relative to the earth?

▌ The observer measures the length of the spaceship to be $l = 100\sqrt{1-(v^2/c^2)}$ (m), where v is the velocity of the ship relative to him. He must then find the flyby time to be l/v. Thus,

$$100\sqrt{\frac{1}{v^2}-\frac{1}{c^2}}=4\times10^{-6}$$

from which $v=0.083c$.

Alternatively, from the viewpoint of the spaceship, it takes a time

$$\Delta t=\frac{4\times10^{-6}}{\sqrt{1-(v^2/c^2)}}\quad(s)$$

for the earth observer to pass by. Equating this to $100/v$ (s) yields the same value for v.

37.10 A spaceship passes the earth at $t=t'=0$ with relative velocity v. At time t_1 on earth clocks, a super spaceship leaves the earth with relative velocity $V>v$ to catch up with the first one. This will happen at earth time t_2, where $vt_2=V(t_2-t_1)$ or $t_2=Vt_1/(V-v)$.
(a) Is t_1-0 a proper or dilated time interval? (b) What does the clock on the "slow" spaceship read when the earth clock read t_1? (c) How far from the spaceship is the earth then? (d) Is t_2-0 a proper or dilated time interval? (e) What does the clock on the slow spaceship register when the ship is overtaken? (f) In the frame of reference of the slow spaceship, how much time has elapsed since the pursuit started? (g) In the same reference frame, how large a distance was covered in the pursuit?

▌ (a) The interval t_1-0 is a proper time because it is determined by events at the same location in the frame of reference fixed to the earth. (b) The spaceship's clock reads an improper time interval since the two events occur at different places in the spaceship's rest frame. Using the Lorentz time transformation, the spaceship's clock reads $t_1'=\gamma[t_1-(v0/c^2)]=\underline{\gamma t_1}$. (c) The velocity of the earth relative to the spaceship is $-v$. Therefore, as reckoned by the spaceship, the earth has receded to a distance given by $|-v|\,t_1'=\underline{\gamma vt_1}$. (d) The interval t_2-0 is improper because the events occur at different places in the earth's frame of reference. (e) The following events occur at the same place in the spaceship's frame of reference. At time $t'=0$, the earth passes the spaceship. At a later time $t=t_2'$, the super spaceship passes the spaceship. Using the differenced Lorentz transformation $\Delta t=\gamma\{\Delta t'+[(v\,\Delta x')/c^2]\}$ with $\Delta x'=0$, we have $\Delta t=\gamma\,\Delta t'$. But $\Delta t'=t_2'-0=t_2'$ while $\Delta t=t_2-0=t_2$. Therefore $t_2'=t_2/\gamma$. The interval $t_2'-0$ is a proper time interval. (f) Combining the results of parts (b) and (e), the "pursuit" required an elapsed time $T_p'\equiv t_2'-t_1'=\underline{(t_2/\gamma)-\gamma t_1}$, as reckoned in the spaceship's reference frame. (g) In the spaceship's reference frame, the super spaceship had to travel the distance $\underline{\gamma vt_1}$ found in part (c).

37.11 Divide the result of part (g) by the result of part (f) in Prob. 37.10 to obtain the velocity of the fast ship relative to the slow one. Compare this result with the one that could be obtained by using the formulas for the Lorentz velocity transformation.

▌ Combining the results of parts (f) and (g), we find the velocity V' of the fast ship, as reckoned by the slow one:

$$V'=\frac{\gamma vt_1}{(t_2/\gamma)-\gamma t_1}\tag{1}$$

As shown in the statement of Prob. 37.10, the time t_2 is given by

$$t_2=\frac{Vt_1}{V-v}\tag{2}$$

Combining Eqs. (1) and (2), we obtain

$$V'=\frac{\gamma vt_1}{Vt_1/[\gamma(V-v)]-\gamma t_1}=\frac{(V-v)v}{(V/\gamma^2)-(V-v)}$$

Since $1/\gamma^2=1-(v^2/c^2)$, this reduces to

$$V'=\frac{(V-v)v}{(-Vv^2/c^2)+v}=\frac{V-v}{1-(Vv/c^2)}$$

With the proper attention to differences in notation, this is found to agree with the Lorentz velocity transformation (see Prob. 37.26).

37.12 A clock is fastened to the $\mathcal{L}'$ frame at some point (x',y',z'). At a certain moment it reads exactly 200 s. Later the same clock reads 300 s (two events at the same location in $\mathcal{L}'$). Find the time interval between these two events, as measured in the "stationary" frame $\mathscr{L}$, if (a) $v = 2 \times 10^5$ m/s, and (b) $v = 2 \times 10^8$ m/s.

▌ (a)
$$\Delta t = \frac{100}{\sqrt{1 - (v^2/c^2)}} = \underline{100.000\ 022\ 3\ \text{s}}$$

(b)
$$\Delta t = \underline{134.2385\ \text{s}}$$

In (a) the time intervals are practically equal, but in (b) the result indicates that the $\mathscr{L}'$ clock runs quite slowly as compared with the one in $\mathscr{L}$.

37.13 A beam of radioactive particles is measured as it shoots through the laboratory. It is found that, on the average, each particle "lives" for a time of 20 ns; after that time, the particle changes to a new form. When at rest in the laboratory, the same particles "live" 7.5 ns on the average. How fast are the particles in the beam moving?

▌ Some sort of timing mechanism within the particle determines how long it "lives." This internal clock, which gives the proper lifetime, must obey the time dilation relation. We have

$$t_p = t_0\sqrt{1 - (v/c)^2} \quad \text{or} \quad 0.75 \times 10^{-8} = (2 \times 10^{-8})\sqrt{1 - (v/c)^2}$$

Note that t_p is the time that the moving clock ticks out during the 2×10^{-8} s ticked out by the laboratory clock. Squaring each side of the equation and solving for v gives $v = \underline{0.927c = 2.78 \times 10^8\ \text{m/s}}$.

37.14 A certain strain of bacteria doubles in number each 20 days. Two of these bacteria are placed on a spaceship and sent away from the earth for 1000 earth days. During this time, the speed of the ship was $0.9950c$. How many bacteria would be aboard when the ship lands on the earth?

▌ The time for doubling, t_0, as seen on earth is given by $t_0 = t_p/\sqrt{1 - (v^2/c^2)}$, where $t_p = 20$ days is the proper doubling time. Then $t_0 = 20/\sqrt{1 - 0.9950^2} = 200$ days. Thus in 1000 earth days there are 5 doublings. Starting from 2 bacteria there are 2^6, or $\underline{64}$ bacteria.

37.15 As seen by inertial observer $\mathscr{L}'$ a certain event 1 takes place at $x_1' = -L'/2$ at time $t_1' = L'/2c$. Another event 2 takes place at $x_2' = L'/2$ at time $t_2' = L'/2c$, so that for $\mathscr{L}'$ the two events are simultaneous. Show that for another inertial observer $\mathscr{L}''$ moving along the x' axis at velocity V with respect to $\mathscr{L}'$ the events are not simultaneous and $\Delta t'' = -(\gamma L'V)/c^2$, where $\gamma = 1/\sqrt{1 - (V^2/c^2)}$.

▌ For convenience we assume that $t'' = t' = 0$ when the origins of the primed and unprimed frames coincide. Because $\mathscr{L}''$ is moving to the right with respect to $\mathscr{L}'$, we find the transformation equations

$$x'' = \gamma(x' - Vt') \tag{1}$$

and

$$t'' = \gamma\left(t' - \frac{Vx'}{c^2}\right) \tag{2}$$

where $\gamma \equiv [1 - (V^2/c^2)]^{-1/2}$. We are informed that in the primed frame, event 1 occurs at $x_1' = -L'/2$ and $t_1' = L'/2c$, while event 2 occurs at $x_2' = L'/2$ and $t_2' = t_1' = L'/2c$. Using Eq. (2) we find that in the double primed frame, event 1 occurs at time t_1'' given by

$$t_1'' = \gamma\left[\frac{L'}{2c} - \frac{V(-L'/2)}{c^2}\right] = \frac{\gamma L'}{2c}\left(1 + \frac{V}{c}\right)$$

Event 2 occurs at time t_2'' given by

$$t_2'' = \gamma\left[\frac{L'}{2c} - \frac{V(L'/2)}{c^2}\right] = \frac{\gamma L'}{2c}\left(1 - \frac{V}{c}\right)$$

Therefore $t_2'' - t_1'' = [(\gamma L')/2c][(-2V)/c] = \underline{-(\gamma L'V)/c^2}$. Event 2 occurs before event 1 in the unprimed frame.

37.16 Refer to Prob. 37.15. (a) Determine the spatial separation of the two events, $\Delta x''$, for observer $\mathscr{L}''$. (b) Does $\Delta x''$ represent the length of an object? Explain.

▌ (a) Equation (1) of Prob. 37.15 implies that

$$\Delta x'' \equiv x_2'' - x_1'' = \gamma(\Delta x' - V\,\Delta t')$$

where $\Delta x' \equiv x_2' - x_1'$ and $\Delta t' \equiv t_2' - t_1'$. For the given events, $\Delta x' = L'$ and $\Delta t' = 0$. Therefore $\Delta x'' = \underline{\gamma L'}$.
(b) The spatial separation $\Delta x''$ found in part (a) does <u>not</u> represent the length of an object because the positions x_1'' and x_2'' were not measured simultaneously $(t_2'' \neq t_1'')$.

37.17 A man drives a nail at point $P_1(4000 \text{ m}, -100 \text{ m}, 200 \text{ m})$ of $\mathscr{L}$; the clock at P_1 records $t = 15\ \mu\text{s}$. At the same moment—as read on another, synchronized clock located at $P_2(3000 \text{ m}, 500 \text{ m}, 150 \text{ m})$—his brother drives a nail at P_2. (Two events occur simultaneously in $\mathscr{L}$.) Compute the time of each event as determined in $\mathscr{L}'$, if $v = 2 \times 10^8$ m/s.

▮ Applying Eq. (4) of Prob. 37.1,

$$t_1' = \frac{(15 \times 10^{-6}) - [(2 \times 10^8)(4000)]/c^2}{\sqrt{1 - (v^2/c^2)}} = \underline{8.187\ \mu\text{s}} \qquad t_2' = \frac{(15 \times 10^{-6}) - [(2 \times 10^8)(3000)]/c^2}{\sqrt{1 - (v^2/c^2)}} = \underline{11.174\ \mu\text{s}}.$$

The events do not occur simultaneously in $\mathscr{L}'$. Hence, from the point of view of $\mathscr{L}$, the $\mathscr{L}'$ clocks are not synchronized, even though the $\mathscr{L}'$ clocks have been synchronized in the $\mathscr{L}'$ frame.

37.18 An astronomer observed that a group of protons from the sun (part of the solar wind) passed the earth at time t_1. Later, she discovers that Jupiter has emitted a large burst of radio noise at time $t_2 = t_1 + \Delta t$. A second astronomer O' riding in a rocket traveling from earth to Jupiter at speed $|V|$, observes the same two events. Assume that the earth is directly between the sun and Jupiter, 6.3×10^8 km from Jupiter. Let $|V| = 0.50c$ and $\Delta t = 900$ s. Calculate the time interval $\Delta t'$ measured by observer O' in the rocket. Could the protons from the sun have triggered the radio burst from Jupiter?

▮ Let event 1 be the arrival of the protons at the earth (at time t_1 in the earth's frame). Let event 2 be the emission of a radio burst by Jupiter: this event occurs at Jupiter at time $t_2 = t_1 + \Delta t$. The spatial separation of the events is $\Delta x \equiv x_2 - x_1$. The transformation of time intervals from the earth frame O to the rocket frame O' is given by

$$\Delta t' = \gamma\left(\Delta t - \frac{|V|\,\Delta x}{c^2}\right) \qquad (1)$$

From the data, $\gamma = 2/\sqrt{3}$ and

$$\Delta t' = \frac{2}{\sqrt{3}}\left[900 - \frac{0.5(6.3 \times 10^{11})}{3 \times 10^8}\right] = \underline{-173\ \text{s}}$$

This means that in the rocket frame, the radio burst was emitted at Jupiter *before* the outbound protons passed the earth. Therefore the protons could not have caused the radio burst in the rocket's frame (or in any other frame).

37.19 Refer to Prob. 37.18. Is there a second rocket reference frame in which the two events were simultaneous? If so, what is its speed with respect to the earth? If not, why not?

▮ The two events are simultaneous in a frame traveling with velocity V^* if

$$\Delta t^* = \gamma^*\left(\Delta t - \frac{V^*\,\Delta x}{c^2}\right) = 0 \qquad (1)$$

With the given numerical values for Δx and Δt, we find that

$$\frac{V^*}{c} = \frac{c\,\Delta t}{\Delta x} = \frac{(3 \times 10^8 \text{ m/s})(900 \text{ s})}{6.3 \times 10^{11} \text{ m}} = \frac{3}{7}$$

which *is* realizable. Events 1 and 2 are simultaneous in a frame of reference traveling from the earth toward Jupiter at speed $\underline{(3c)/7}$.

37.20 Refer to Prob. 37.18. (*a*) Assume that a radio noise burst *is* triggered by a burst of protons. What limit can be placed on Δt? (*b*) Suppose that the two events were separated by $\Delta t = 60$ min. What was the speed of an observer who measured a proper time interval between these events? Calculate the time interval measured by the observer.

▮ (*a*) For any burst that the protons could trigger, the time interval between events 1 and 2 must be nonnegative in every frame of reference. Referring to Eq. (1) of Prob. 37.18, we see that this requires that $\Delta t - [(V\,\Delta x)/c^2]$ be a positive quantity for all $V < c$. This requires Δt to equal or exceed $\Delta x/c$. Any burst that the protons trigger must be emitted at least $(6.3 \times 10^{11} \text{ m})/(3 \times 10^8 \text{ m/s}) = 2.1 \times 10^3$ s after the protons pass the earth. That is, $\underline{\Delta t \geq \Delta t_{\min} = 2100 \text{ s} = 35.0 \text{ min}}$. (*b*) Proper time intervals are intervals between events that occur at a given location. Suppose that two events are separated spatially by $\Delta x \equiv x_2 - x_1$ and occur a time interval $\Delta t \equiv t_2 - t_1$ apart in one inertial frame (the unprimed frame). Then the inertial observer O' who measures a proper time interval between them is the one who travels with constant velocity V from location x_1 at time t_1 to location x_2 at time t_2. With the given numerical values, we have $V = \Delta x/\Delta t = (6.3 \times 10^{11}\,\text{m})/(3600\,\text{s}) = 1.75 \times 10^8\,\text{m/s} = \underline{0.583c}$. Then $\sqrt{1 - (V^2/c^2)} = 0.812$ and the proper time $\Delta t' = \Delta t\sqrt{1 - (V^2/c^2)} = (3600\,\text{s})(0.812) = 2.92 \times 10^3\,\text{s} = \underline{48.7\,\text{min}}$. Any inertial observer O'' who is moving with respect to observer O' will measure a time interval $\Delta t''$ that exceeds $\Delta t'$.

37.21 A person in a spaceship holds a meterstick as the ship shoots past the earth with a speed v parallel to the earth's surface. What does the person in the ship notice as the stick is rotated from parallel to perpendicular to the ship's motion?

▮ The stick behaves normally; it does not appear to change its length, because it has no translational motion relative to the observer in the spaceship. However, an observer on earth would measure the stick to be $(1\,\text{m})\sqrt{1 - (v/c)^2}$ long when parallel to the ship's motion, and 1 m long when perpendicular to the ship's motion.

37.22 Consider the two reference frames $\mathscr{L}$ and $\mathscr{L}'$ of Prob. 37.1. A wooden slab is at rest in frame $\mathscr{L}'$ and is centered on the origin of $\mathscr{L}'$. Observer $\mathscr{L}'$ finds that it is square, with sides exactly 1 m long. Describe the shape and orientation of the slab, as measured in frame $\mathscr{L}$ at $t = 0$, when frames $\mathscr{L}$ and $\mathscr{L}'$ coincide and the edges of the slab in $\mathscr{L}'$ are parallel to the x' and y' axes.

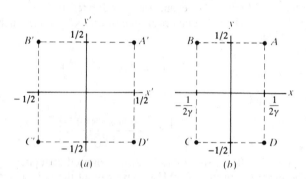

Fig. 37-1

▮ The Lorentz transformations include the equations $x' = \gamma(x - vt)$ and $y' = y$. Therefore if a point P is at rest in the primed frame, so that its coordinates are (x_P', y_P') for all times t', then the point's unprimed coordinates at time $t = 0$ will be $(x_P, y_P) = [(x_P'/\gamma), y_P']$. Hence, we have Fig. 37-1(a) for the geometry in $\mathscr{L}'$ (for all t') and Fig. 37-1(b) for the geometry in $\mathscr{L}$ at the instant $t = 0$.

37.23 Repeat Prob. 37.22 if the edges of the slab in $\mathscr{L}'$ are inclined at 45° to the x' and y' axes.

▮ See Fig. 37-2.

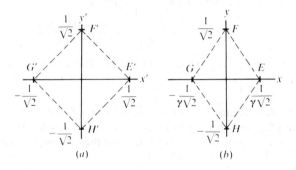

Fig. 37-2

37.24 The insignia painted on the side of a spaceship is a circle with a line across it at 45° to the vertical. As the ship shoots past another ship in space, with a relative speed of $0.95c$, the second ship observes the insignia. What angle does the observed line make to the vertical?

▮ Let L be the length of the line as seen in the first spaceship. The horizontal and vertical extents of the line are each $L/\sqrt{2}$. In the second ship the horizontal extent is shortened to $(L/\sqrt{2})\sqrt{1 - (v^2/c^2)}$ while the vertical extent is unchanged. Then

$$\tan \theta = \frac{\text{horizontal}}{\text{vertical}} = \sqrt{1 - \frac{V^2}{c^2}} \quad \text{or} \quad \tan \theta = \sqrt{1 - 0.95^2} = 0.31 \quad \text{and} \quad \theta = \underline{17.3^\circ}$$

37.25 A circular hoop of radius a' is at rest in the $x'y'$ plane of system O' moving at constant velocity V with respect to the inertial frame O.
(*a*) Show that the measurements made in frame O will indicate that the hoop is elliptical in shape, with its semimajor axis parallel to the y axis and of length $a = a'$, and its semiminor axis of length $b = a'\sqrt{1 - (V^2/c^2)}$.
(*b*) The *eccentricity* e of an ellipse is defined by $e \equiv \sqrt{1 - (b/a)^2}$. Derive a simple expression for the eccentricity of the ellipse in part (*a*).

❚ (*a*) For convenience, we suppose that the hoop is centered on the origin of the primed system. The equation that describes the circle is therefore

$$(x')^2 + (y')^2 = (a')^2 \tag{1}$$

If the circle is observed in the unprimed frame at time t, the Lorentz transformations imply that $x' = \gamma(x - Vt)$ and $y' = y$. Therefore Eq. (1) becomes

$$\gamma^2(x - Vt)^2 + y^2 = (a')^2$$

or

$$\frac{(x - Vt)^2}{(a'/\gamma)^2} + \frac{y^2}{(a')^2} = 1 \tag{2}$$

This is a standard form for the equation of an ellipse centered at $x_c = Vt$ and $y_c = 0$, with semimajor axis $a = a'$ parallel to the y axis and with semiminor axis $b = a'/\gamma = a'\sqrt{1 - (V^2/c^2)}$ along the x axis. (*b*) Applying the definition of the eccentricity and the result of part (*a*), we have

$$e = \sqrt{1 - \left(\frac{a'}{a'\gamma}\right)^2} = \frac{|V|}{c}$$

37.26^c Derive the transformation of longitudinal velocities from $\mathscr{L}$ to $\mathscr{L}'$ in Prob. 37.1.

❚ Taking the differentials of the Lorentz coordinate transformations (1) and (4), one finds that

$$dx' = \frac{dx - v\,dt}{\sqrt{1 - (v^2/c^2)}} \qquad dt' = \frac{dt - (v/c^2)\,dx}{\sqrt{1 - (v^2/c^2)}}$$

Dividing dx' by dt' gives

$$u_x' = \frac{dx'}{dt'} = \frac{dx - v\,dt}{dt - (v/c^2)\,dx} = \frac{(dx/dt) - v}{1 - [(v/c^2)(dx/dt)]} = \frac{u_x - v}{1 - (v/c^2)u_x}$$

37.27^c Repeat Prob. 37.26 for the y and z components of velocity.

❚ For the y direction:

$$dy' = dy \qquad dt' = \frac{dt - (v/c^2)\,dx}{\sqrt{1 - (v^2/c^2)}} \qquad \text{Then} \qquad u_y' = \frac{dy'}{dt'} = \frac{dy\sqrt{1 - (v^2/c^2)}}{dt - (v/c^2)\,dx}$$

Dividing top and bottom by dt and recalling that $dx/dt = u_x$, we have

$$u_y' = \frac{u_y\sqrt{1 - (v^2/c^2)}}{1 - [(vu_x)/c^2]} \qquad \text{In exactly similar fashion} \qquad u_z' = \frac{u_z\sqrt{1 - (v^2/c^2)}}{1 - [(vu_x)/c^2]},$$

37.28 Rocket A travels to the right and rocket B travels to the left, with velocities $0.8c$ and $0.6c$, respectively, relative to the earth. What is the velocity of rocket A measured from rocket B?

❚ Let observers O, O' be associated with the earth and rocket B, respectively. Then, letting u_x be the velocity of rocket A relative to the earth,

$$u_x' = \frac{u_x - v}{1 - (v/c^2)u_x} = \frac{0.8c - (-0.6c)}{1 - [(-0.6c)(0.8c)]/c^2} = \underline{0.946c}$$

The problem can also be solved with other associations. For example, let observers O and O' be associated with rocket A and rocket B, respectively. Let u_x, u_x' represent velocity of earth. Then

$$u_x' = \frac{u_x - v}{1 - (v/c^2)u_x} \quad \text{or} \quad 0.6c = \frac{-0.8c - v}{1 - (v/c^2)(-0.8c)}$$

Solving, $v = -0.946c$, which agrees with the above answer. (The minus sign appears because v is the velocity of O' with respect to O, which, with the present association, is the velocity of rocket B with respect to rocket A.)

37.29 At what speeds will the galilean and Lorentz expressions for u_x' differ by 2 percent?

⬛ The galilean transformation is $u_{xG}' = u_x - v$ and the Lorentz transformation is

$$u_{xR}' = \frac{u_x - v}{1 - (v/c^2)u_x} = \frac{u_{xG}'}{1 - (v/c^2)u_x}$$

Rearranging,

$$\frac{u_{xR}' - u_{xG}'}{u_{xR}'} = \frac{vu_x}{c^2}$$

Thus, if the product vu_x exceeds $0.02c^2$, the error in using the galilean transformation instead of the Lorentz transformation will exceed 2 percent.

37.30 As a rocket ship sweeps past the earth with speed v, it sends out a pulse of light ahead of it. How fast does the light pulse move according to people on the earth?

⬛ *Method 1* c (by the second postulate of special relativity).
Method 2 According to the velocity addition formula, the observed speed will be (since $u = c$ in this case)

$$\frac{v + u}{1 + (uv/c^2)} = \frac{v + c}{1 + (v/c)} = \frac{(v + c)c}{c + v} = c$$

37.31 We consider the frames of reference in Prob. 37.1 with $v = 2 \times 10^8$ m/s. An electron has velocity $\mathbf{u}'$ relative to $\mathcal{L}'$, the components of which are

$$u_x' = 6 \times 10^7 \text{ m/s} \quad u_y' = 4 \times 10^7 \text{ m/s} \quad u_z' = 3 \times 10^7 \text{ m/s}$$

Find the velocity components as measured in $\mathcal{L}$, as well as the magnitude of $\mathbf{u}$.

⬛ From Probs. 37.26 and 37.27, with $v \to -v$ for the inverse transformation, we have

$$u_x = \frac{(6 \times 10^7) + (2 \times 10^8)}{1 + [(6 \times 10^7)(2 \times 10^8)]/c^2} = \underline{2.29 \times 10^8 \text{ m/s}} \quad u_y = \frac{(4 \times 10^7)\sqrt{1 - (v^2/c^2)}}{1 + [(6 \times 10^7)(2 \times 10^8)]/c^2} = \underline{2.63 \times 10^7 \text{ m/s}}$$

Likewise, $u_z = 1.97 \times 10^7$ m/s. Then,

$$u = (u_x^2 + u_y^2 + u_z^2)^{1/2} = \underline{2.32 \times 10^8 \text{ m/s}}$$

37.32 The speed of light in still water is c/n, where $n \approx 1.33$ (the index of refraction). If the water is moving with speed $|V| \ll c$ in the same direction as the light is traveling, show that the speed observed in the laboratory is equal to $c/n + |V|(1 - 1/n^2)$. This result was obtained by the French physicist Fizeau about the middle of the nineteenth century and can be explained only relativistically.

⬛ We let the primed frame be the reference frame fixed to the water. Then the Lorentz velocity transformation relating the lab velocity v_x to the velocity v_x' in the primed frame is

$$v_x = \frac{v_x' + V}{1 + [(v_x'V)/c^2]} \tag{1}$$

(Here we have made the usual choice that the primed frame is moving in the direction of increasing x.) For a flash of light traveling in the same direction as the water flow, we have

$$v_x' = \frac{c}{n} \tag{2}$$

(We must realize that the "speed of light in still water" is to be read as "the speed of light with respect to the water.") According to Eqs. (1) and (2) the velocity in the laboratory is given by

$$v_x = \frac{(c/n) + |V|}{1 + [(c|V|)/(nc^2)]}$$

Since $|V| \ll c$, this can be expanded and approximated:

$$v_x = \left(\frac{c}{n} + |V|\right)\left[1 - \frac{|V|}{nc} + \left(\frac{|V|}{nc}\right)^2 - \cdots\right] = \frac{c}{n} + |V| - \frac{|V|}{n^2} - \frac{|V|^2}{nc} + \frac{|V|^2}{n^3c} + \cdots \approx \frac{c}{n} + |V|\left(1 - \frac{1}{n^2}\right)$$

37.33 A flash of light has components of velocity v_x, v_y, and v_z in the unprimed coordinate system. That is, $v_x^2 + v_y^2 + v_z^2 = c^2$. Using the Lorentz velocity transformation, calculate the speed of the light in the reference frame O' which is moving with speed V relative to O in the direction of the x and x' axes. The y and y' axes are parallel.

▮ Using the Lorentz velocity transformation given in Probs. 37.26 and 37.27 we find the square of the speed in frame O':

$$(v')^2 = (v'_x)^2 + (v'_y)^2 + (v'_z)^2 = \left\{\frac{v_x - V}{1 - [(Vv_x)/c^2]}\right\}^2 + \left\{\frac{\sqrt{1 - (V^2/c^2)}v_y}{1 - [(Vv_x)/c^2]}\right\}^2 + \left\{\frac{\sqrt{1 - (V^2/c^2)}v_z}{1 - [(Vv_x)/c^2]}\right\}^2$$

$$= \frac{1}{\{1 - [(Vv_x/c^2)]\}^2}\left[(v_x - V)^2 + \left(1 - \frac{V^2}{c^2}\right)(v_y^2 + v_z^2)\right] = \left(1 - \frac{Vv_x}{c^2}\right)^{-2}\left[v_x^2 - 2v_xV + V^2 + v_y^2 + v_z^2 - V^2\frac{v_y^2 + v_z^2}{c^2}\right]$$

Since $v_x^2 + v_y^2 + v_z^2 = c^2$, we have

$$(v')^2 = \left(1 - \frac{Vv_x}{c^2}\right)^{-2}\left[c^2 - 2v_xV + V^2\left(1 - \frac{v_y^2 + v_z^2}{c^2}\right)\right]$$

But $v = c$ implies that $1 - [(v_y^2 + v_z^2)/c^2] = v_x^2/c^2$. Therefore we obtain

$$(v')^2 = \left(1 - \frac{Vv_x}{c^2}\right)^{-2}\left(c^2 - 2v_xV + \frac{V^2v_x^2}{c^2}\right) = \left(1 - \frac{Vv_x}{c^2}\right)^{-2}c^2\left[1 - \frac{2Vv_x}{c^2} + \left(\frac{Vv_x}{c^2}\right)^2\right]$$

$$= \left(1 - \frac{Vv_x}{c^2}\right)^{-2}c^2\left(1 - \frac{Vv_x}{c^2}\right)^2 = c^2$$

That is, $\underline{v' = c}$: the speed of light is the same in all inertial frames.

37.34 Consider three inertial reference frames O, O', and O''. Let O' move with velocity V with respect to O, and let O'' move with velocity V' with respect to O'. Both velocities are in the same direction. Write the transformation equations relating x, y, z, t with x', y', z', t' and also those relating x', y', z', t' with x'', y'', z'', t''. Combine these equations to obtain the relations between x, y, z, t and x'', y'', z'', t''.

▮ Adopting the conventional choice of coordinate axes, the x, x', and x'' axes are parallel to the relative motions, frame O'' moves with velocity $V'\hat{x}'$ with respect to frame O' and frame O' moves with velocity $V\hat{x}$ with respect to frame O. The equations for the Lorentz transformation from O to O' are

$$x' = \gamma(x - Vt) \tag{1}$$

$$y' = y \tag{2}$$

$$z' = z \tag{3}$$

$$t' = \gamma\left(t - \frac{Vx}{c^2}\right) \tag{4}$$

where $\gamma \equiv 1/\sqrt{1 - (V^2/c^2)}$. The equations for the Lorentz transformation from O' to O'' are

$$x'' = \gamma'(x' - V't') \tag{5}$$

$$y'' = y' \tag{6}$$

$$z'' = z' \tag{7}$$

$$t'' = \gamma'\left(t' - \frac{V'x'}{c^2}\right) \tag{8}$$

where $\gamma' \equiv 1/\sqrt{1 - [(V')^2/c^2]}$. Using Eqs. (1) and (4) to eliminate x' and t' from Eqs. (5) and (8), we find that

$$x'' = \gamma'\left[\gamma(x - Vt) - V'\gamma\left(t - \frac{Vx}{c^2}\right)\right] \qquad t'' = \gamma'\left[\gamma\left(t - \frac{Vx}{c^2}\right) - \frac{V'\gamma(x - Vt)}{c^2}\right]$$

or

$$x'' = \gamma\gamma'\left(1 + \frac{VV'}{c^2}\right)x - t\gamma\gamma'(V + V') \tag{9}$$

$$t'' = \gamma\gamma'\left(1 + \frac{VV'}{c^2}\right)t - \frac{x}{c^2}\gamma\gamma'(V + V') \tag{10}$$

Equations (9) and (10), plus the easily obtained equations $y'' = y$ and $z'' = z$, constitute the desired equations giving (x'', y'', z'', t'') in terms of (x, y, z, t).

37.35 Refer to Prob.37.34. Show that (9) and (10) are equivalent to a direct Lorentz transformation from O to O'' in which the relative velocity V'' of O'' with respect to O is given by

$$V'' = \frac{V + V'}{1 + [(VV')/c^2]}$$

▮ Referring to Eqs. (9) and (10), we see that if we define

$$\gamma'' \equiv \gamma\gamma'\left(1 + \frac{VV'}{c^2}\right) \tag{1}$$

and

$$V'' = \frac{V + V'}{1 + [(VV')/c^2]} \tag{2}$$

the transformation from O to O'' has the standard form

$$x'' = \gamma''(x - V''t) \tag{3}$$
$$y'' = y \tag{4}$$
$$z'' = z \tag{5}$$
$$t'' = \gamma''\left(t - \frac{V''x}{c^2}\right) \tag{6}$$

All that remains is to verify that the definitions given in (1) and (2) above are consistent. That is, we must show that $\gamma'' = 1/\sqrt{1 - [(V'')^2/c^2]}$. We make this confirmation by showing that $1/(\gamma'')^2$ equals $1 - [(V'')^2/c^2]$. From (1),

$$\frac{1}{(\gamma'')^2} = \frac{1}{\gamma^2}\frac{1}{\gamma'^2}\frac{1}{[1 + (VV'/c^2)]^2} = \frac{[1 - V^2/c^2][1 - (V')^2/c^2]}{[1 + (VV'/c^2)]^2} = \frac{1 - [V^2 + (V')^2]/c^2 + V^2(V')^2/c^4}{[1 + (VV'/c^2)]^2}$$
$$= \frac{1 + (2VV')/c^2 + V^2(V')^2/c^4 - [V^2 + 2VV' + (V')^2]/c^2}{[1 + (VV'/c^2)]^2} = 1 - \frac{1}{c^2}\left[\frac{V + V'}{1 + (VV'/c^2)}\right]^2 = 1 - \frac{(V'')^2}{c^2}$$

where the last step uses (2).

37.36 Refer to Probs. 37.34 and 37.35. (*a*) Show that (2) of Prob. 37.35 is in agreement with the Lorentz velocity transformation of Prob. 37.26. (*b*) M, M', and M'' are metersticks lying along the parallel x, x', and x'' axes. They are at rest in O, O', and O'', respectively. Construct a table that shows the lengths observers in each frame would assign to each meterstick.

▮ (*a*) The velocity of O'' with respect to O' is $v_x' = V'$, and the velocity of O' with respect to O is V. If we apply the Lorentz velocity transformation, with proper attention to the notation, we conclude that the velocity v_x of O'' with respect to O should be given by

$$v_x = \frac{v_x' + V}{1 + [(v_x'V)/c^2]} = \frac{V' + V}{1 + [(V'V)/c^2]}$$

This is identical with the quantity V'' which characterizes the transformation from O to O'', as found in Prob. 37.35. (*b*) See Table 37-1.

TABLE 37-1

Metersticks

Observers	M	M'	M''
O	1	$\frac{1}{\gamma}$	$\frac{1}{\gamma''}$
O'	$\frac{1}{\gamma}$	1	$\frac{1}{\gamma'}$
O''	$\frac{1}{\gamma''}$	$\frac{1}{\gamma'}$	1

37.37^c An accelerating object is observed from two inertial frames, O and O'. The x and x' axes are collinear with the path of the object and O' has a velocity V relative to O which lies along the x axis. Show that the accelerations as measured in the two frames are related by

$$a' = a\left(1 - \frac{V^2}{c^2}\right)^{3/2} \bigg/ \left(1 - \frac{vV}{c^2}\right)^3$$

where v is the instantaneous velocity of the object as observed by O.

▌ In order to relate $a' \equiv dv'/dt'$ to $a \equiv dv/dt$, we must use the Lorentz velocity transformation

$$v' = \frac{v - V}{1 - [(vV)/c^2]} \tag{1}$$

We omit the subscripts x on the velocities v and v' (and on the accelerations a and a') because the probem posed is essentially one-dimensional. According to the chain rule

$$a' \equiv \frac{dv'}{dt'} = \frac{dv'}{dv}\frac{dv}{dt}\frac{dt}{dt'} = \frac{dv'/dv}{dt'/dt}\, a \tag{2}$$

Using Eq. (1), we find that

$$\frac{dv'}{dv} = \frac{(1)(1 - vV/c^2) - (-V/c^2)(v - V)}{\{1 - [(vV)/c^2]\}^2} = \frac{1 - V^2/c^2}{\{1 - [(vV)/c^2]\}^2} \tag{3}$$

To evaluate dt'/dt we must begin with the Lorentz transformation equation for the time:

$$t' = \gamma\left(t - \frac{Vx}{c^2}\right) \tag{4}$$

We find that

$$\frac{dt'}{dt} = \gamma\left[1 - \left(\frac{V}{c^2}\right)\left(\frac{dx}{dt}\right)\right] = \gamma\left(1 - \frac{Vv}{c^2}\right) \tag{5}$$

Using Eqs. (3) and (5) in Eq. (2), we obtain the desired transformation:

$$a' = \frac{1 - (V^2/c^2)}{\{1 - [(vV)/c^2]\}^2}\frac{a}{\gamma\{1 - [(Vv)/c^2]\}} = \frac{[1 - (V^2/c^2)]^{3/2}}{\{1 - [(vV)/c^2]\}^3}\, a.$$

37.38 A spaceship traveling toward the earth with speed $|v_s|$ has a long rod sticking out at right angles to its direction of travel. When a light at the spaceship end of the rod flashes, the light pulse which travels along the rod is reflected back to the spaceship by a mirror at the end of the rod. The returning light pulse activates a very fast acting mechanism which makes the light flash again. Let the frequency of flashes, as determined from the spaceship, be v.
(a) If the time dilation were the only effect to take into account, what would be the period of the flash on earth clocks? Light from the flash also travels toward the earth. This period of flashing will be less than the period found in part (a) since the spaceship moves toward the earth between flashes and the second flash has less far to travel than the first one. (b) How far toward the earth will an earth observer say the light has traveled in the time between flashes? (c) How far will this observer say the ship has traveled between flashes? (d) What is the distance between successive flashes for an earth observer? (e) What is the frequency of flashes as measured by an earth observer? (The result expresses the *relativistic Doppler effect* for an approaching source.)

▌ (a) The period of the flash would be increased by a factor $\gamma \equiv 1/\sqrt{1 - (|v_s|^2/c^2)}$, so the period would be γ/v. (b) Since the light travels at speed c, the earthbound observer will reckon that a light flash travels a distance $c(\gamma/v) = \underline{\gamma(c/v)}$ in each interval between flashes. (c) Since the spaceship is traveling at speed $|v_s|$, the earthbound observer will reckon that the ship travels a distance $|v_s|\,(\gamma/v) = \underline{\gamma(|v_s|/v)}$ between successive flashes. (d) Using the results of parts (b) and (c), as the ship travels toward the earth, the earthbound observer finds that the successive flashes are separated by a distance $D = \gamma(c/v) - \gamma(|v_s|/v) = \underline{(\gamma/v)(c - |v_s|)}$. (e) The time interval between the arrival of successive flashes, as reckoned on the earth, is simply the distance D divided by c. Hence the received frequency is

$$v' = \frac{c}{D} = v\frac{c}{\gamma(c - |v_s|)} = v\sqrt{\frac{1 + (|v_s|/c)}{1 - (|v_s|/c)}} \tag{1}$$

37.39 Find the analogue of (1), Prob. 37.38, for a receding source.

▌ A repetition of the argument of Prob.37.38 shows that the desired formula is only (1) with the signs reversed under the radical. Thus, the single formula

$$v' = v \sqrt{\frac{1 + (v_s/c)}{1 - (v_s/c)}} \tag{1}$$

where the source *velocity* v_s is positive for approach and negative for recession, covers either case.

37.40 A certain transition in potassium produces light of frequency 8.0×10^{14} Hz. When this transition occurs in a distant galaxy, the light reaching the earth has the measured frequency 5.0×10^{14} Hz (a "red shift"). Determine the radial motion of the galaxy with respect to the earth.

▌ Substitute $v' = 5.0$ and $v = 8.0$ in (1) of Prob. 37.39 and solve for v_s/c, obtaining $v_s/c = -0.438$. The galaxy is receding at speed $0.438c$.

37.41 A spaceship coasting in interstellar space encounters an alien space probe which has a radio transmitter. As the probe approaches, the frequency initially received by the ship is 130 MHz. As the probe recedes into the distance, the frequency eventually drops to 60 MHz. What is the intrinsic frequency v_0 of the probe's transmitter? What is the relative speed of the two ships?

▌ The spaceship receives signals at frequency

$$v_a = v_0 \sqrt{\frac{1 + (v/c)}{1 - (v/c)}} \tag{1}$$

as the probe approaches, and at frequency

$$v_r = v_0 \sqrt{\frac{1 - (v/c)}{1 + (v/c)}} \tag{2}$$

as it recedes. Here, v is the constant relative speed. Multiplying Eq. (2) by Eq. (1), we find that $v_a v_r = v_0^2$. Therefore $v_0 = \sqrt{v_a v_r} = \sqrt{(60)(130)} = 88.3\,\text{MHz}$. Substituting into either (1) or (2), we solve and obtain $v = 7c/19$.

37.42 Refer to Prob. 37.41; find the received frequency of the signals emitted by the probe at the instant of its closest approach to the ship.

▌ For these particular signals, the wavefronts are not "pushed together" (as they are when the ship-to-probe distance is decreasing), nor are they "pulled apart" (as they are when the ship-to-probe is increasing). The only factor which operates to change the frequency is time dilation. (Cf. Prob. 37.38(a).) Therefore the frequency v^* at which those signals are received is given by $v^* = v_0/\gamma = v_0 \sqrt{1 - (v^2/c^2)}$. With $v_0 = 88.3$ MHz and $v/c = 7c/19$, we find that $v^* = 82.1\,\text{MHz}$.

37.43 Figure 37-3 shows a Michelson interferometer. Let the distance to each mirror from the inclined lightly

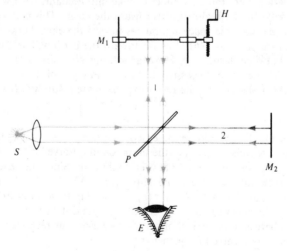

Representation
of fringe pattern

Fig. 37-3

silvered mirror P be L. Assume that there is a stationary medium, the ether, through which the apparatus moves with velocity V in the direction of mirror M_2. Assume also that newtonian physics is valid, so that the galilean transformations apply. (a) If the speed of light is c, what is the speed of light relative to M_2 as it approaches M_2 from P? (b) How long does it take the light to travel from P to M_2? (c) What is the speed of light relative to P on the return trip? (d) How long does it take the light to travel from M_2 to P? What is the total time for the round trip from P to M_2 and back to P?

▮ (a) Since we are assuming that the speed of light with respect to the ether is c and that the galilean transformations apply, when the light is traveling from P to M_2, its speed is $c - V$ relative to the apparatus.
(b) The time t_A required for the light to travel from P to M_2 is given by $t_A = L/(c - V)$.
(c) When the light is traveling from M_2 back to P, its speed is $c + V$ relative to the apparatus.
(d) The time t_B required for the return trip from M_2 to P is given by $t_B = L/(c + V)$. The time t_2 required for the entire trip (P to M_2 to P) is given by

$$t_2 = t_A + t_B = \frac{L}{c - V} + \frac{L}{c + V} = \frac{2Lc}{c^2 - V^2}$$

37.44 Refer to Prob. 37.43. (a) Show that the speed of light relative to M_1 is $\sqrt{c^2 - V^2}$ and that this is also the speed relative to P on the return trip. (b) Calculate the total time for the round trip to the mirror M_1 and back to P.

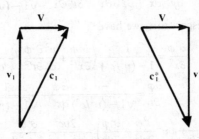

Fig. 37-4

▮ (a) The situation is shown in the (galilean) velocity addition diagrams, Fig. 37-4. The light reaching M_1 from P must have a velocity vector $\mathbf{v}_1$, with respect to the apparatus. Because the apparatus is moving through the ether with velocity $\mathbf{V}$, the velocity $\mathbf{v}_1 = \mathbf{c}_1 - \mathbf{V}$, where $\mathbf{c}_1$ is the velocity of the light with respect to the ether. Because $|\mathbf{c}_1| = c$ and $\mathbf{v}_1$ is perpendicular to $\mathbf{V}$, the magnitude of $\mathbf{v}_1$ is given by $|\mathbf{v}_1| = \sqrt{c^2 - V^2}$. Similarly, $|\mathbf{v}_1^*| = \sqrt{c^2 - V^2}$.
(b) Since the total distance traveled is $2L$, the round-trip time is given by $t_1 = 2L/\sqrt{c^2 - V^2}$.

37.45 Refer to Probs. 37.43 and 37.44. (a) If $|V|/c \ll 1$, show that the difference in time between the round trip to M_2 and the round trip to M_1 is equal to LV^2/c^3. (b). What distance does light travel in this time difference? (c) If the wavelength of the light is λ, to what fraction of a wavelength does this distance correspond? (d) Calculate this fraction if $L = 10$ m, $\lambda = 600$ nm and $|V| = 3 \times 10^4$ m/s. Take c equal to 3×10^8 m/s. (e). If the apparatus is rotated through 90°, how much of a fringe shift is expected? (Michelson and Morley could have detected a shift of 0.01 fringe.)

▮ (a) The difference in round-trip times is given by

$$t_2 - t_1 = \frac{2Lc}{c^2 - V^2} - \frac{2L}{\sqrt{c^2 - V^2}} = \frac{2L}{c}\left[\left(1 - \frac{V^2}{c^2}\right)^{-1} - \left(1 - \frac{V^2}{c^2}\right)^{-1/2}\right] = \frac{2L}{c}\left[\left(1 + \frac{V^2}{c^2} + \cdots\right) - \left(1 + \frac{1}{2}\frac{V^2}{c^2} + \cdots\right)\right] \approx \frac{LV^2}{c^3}$$

as claimed. (b) The time difference $t_2 - t_1$ corresponds to a path difference $d = c(t_2 - t_1) = (LV^2)/c^2$.
(c) The path difference $c(t_2 - t_1)$ is equal to $(LV^2)/(\lambda c^2)$ wavelengths. (d) With $L = 10$ m, $\lambda = 6 \times 10^{-7}$ m, and $V = 3 \times 10^4$ m/s, we find that

$$\frac{d}{\lambda} = \frac{10}{6 \times 10^{-7}}\frac{(3 \times 10^4)^2}{(3 \times 10^8)^2} = \frac{1}{6}$$

(e) When the apparatus is rotated through 90°, the sign of the path difference is reversed, so the fringe shift should be $(2)(\frac{1}{6}) = \frac{1}{3}$ fringe.

37.46^c Show that the electromagnetic wave equation,

$$\frac{\partial^2 \phi}{\partial x^2} + \frac{\partial^2 \phi}{\partial y^2} + \frac{\partial^2 \phi}{\partial z^2} - \frac{1}{c^2}\frac{\partial^2 \phi}{\partial t^2} = 0$$

is invariant under a Lorentz transformation.

▮ The equation will be invariant if it retains the same form when expressed in terms of the new variables x', y', z', t'. To express the wave equation in terms of the primed variables we first find from the Lorentz transformations that

$$\frac{\partial x'}{\partial x} = \frac{1}{\sqrt{1 - (v^2/c^2)}} \qquad \frac{\partial x'}{\partial t} = -\frac{v}{\sqrt{1 - (v^2/c^2)}}$$

$$\frac{\partial t'}{\partial x} = -\frac{v/c^2}{\sqrt{1 - (v^2/c^2)}} \qquad \frac{\partial t'}{\partial t} = \frac{1}{\sqrt{1 - (v^2/c^2)}}$$

$$\frac{\partial y'}{\partial y} = \frac{\partial z'}{\partial z} = 1 \qquad \frac{\partial x'}{\partial y} = \frac{\partial x'}{\partial z} = \frac{\partial y'}{\partial x} = \cdots = 0$$

From the chain rule, and using the above results, we have

$$\frac{\partial \phi}{\partial x} = \frac{\partial \phi}{\partial x'}\frac{\partial x'}{\partial x} + \frac{\partial \phi}{\partial y'}\frac{\partial y'}{\partial x} + \frac{\partial \phi}{\partial z'}\frac{\partial z'}{\partial x} + \frac{\partial \phi}{\partial t'}\frac{\partial t'}{\partial x} = \frac{1}{\sqrt{1 - (v^2/c^2)}}\frac{\partial \phi}{\partial x'} + \frac{-v/c^2}{\sqrt{1 - (v^2/c^2)}}\frac{\partial \phi}{\partial t'}$$

Differentiating again with respect to x, we have

$$\frac{\partial^2 \phi}{\partial x^2} = \frac{1}{1 - (v^2/c^2)}\left(\frac{\partial^2 \phi}{\partial x'^2} + \frac{v^2}{c^4}\frac{\partial^2 \phi}{\partial t'^2}\right) - \frac{2v}{c^2 - v^2}\frac{\partial^2 \phi}{\partial x' \partial t'}$$

Similarly we have

$$\frac{\partial \phi}{\partial t} = \frac{-v}{\sqrt{1 - (v^2/c^2)}}\frac{\partial \phi}{\partial x'} + \frac{1}{\sqrt{1 - (v^2/c^2)}}\frac{\partial \phi}{\partial t'}$$

$$\frac{\partial^2 \phi}{\partial t^2} = \frac{1}{1 - (v^2/c^2)}\left(v^2\frac{\partial^2 \phi}{\partial x'^2} + \frac{\partial^2 \phi}{\partial t'^2}\right) - \frac{2vc^2}{c^2 - v^2}\frac{\partial^2 \phi}{\partial x' \partial t'} \qquad \frac{\partial^2 \phi}{\partial y^2} = \frac{\partial^2 \phi}{\partial y'^2} \qquad \frac{\partial^2 \phi}{\partial z^2} = \frac{\partial^2 \phi}{\partial z'^2}$$

Substituting these in the wave equation, we obtain

$$\frac{\partial^2 \phi}{\partial x^2} + \frac{\partial^2 \phi}{\partial y^2} + \frac{\partial^2 \phi}{\partial z^2} - \frac{1}{c^2}\frac{\partial^2 \phi}{\partial t^2} = \frac{\partial^2 \phi}{\partial x'^2} + \frac{\partial^2 \phi}{\partial y'^2} + \frac{\partial^2 \phi}{\partial z'^2} - \frac{1}{c^2}\frac{\partial^2 \phi}{\partial t'^2}$$

so that the equation is invariant under Lorentz transformations. Recall that the wave equation is not invariant under galilean transformations.

37.2 MASS-ENERGY RELATION; RELATIVISTIC DYNAMICS

37.47 A particle is traveling at a speed v such that $v/c = 0.9900$. Find m/m_0 for the particle.

▮ The relativistic mass is given by $m = m_0/\sqrt{1 - (v^2/c^2)}$. Then $m/m_0 = 1/\sqrt{1 - 0.99^2} = \underline{7.1}$.

37.48 A 2000-kg car is moving at 15 m/s. How much larger than its rest mass is its mass at this speed? (*Hint:* For x very small, $1/\sqrt{1-x} \approx 1 + \frac{1}{2}x$.)

▮ $m = m_0[1 - (v^2/c^2)]^{-1/2} \approx m_0[1 + \frac{1}{2}(v^2/c^2)]$, or $(m - m_0) = \frac{1}{2}m_0(v^2/c^2) = \frac{1}{2}(2000)[15/(3.0 \times 10^8)]^2 = \underline{2.5 \times 10^{-12}\text{ kg}}$.

37.49 Compute the *rest energy* of an electron, i.e., the energy equivalent of its rest mass, 9.11×10^{-31} kg.

▮ Rest energy $= m_0 c^2 = (9.11 \times 10^{-31}\text{ kg})(3 \times 10^8\text{ m/s})^2 = \underline{8.2 \times 10^{-14}\text{ J}}$. Recalling 1 MeV $= 1.6 \times 10^{-13}$ J, we have rest energy $= \underline{0.512\text{ MeV}}$.

37.50 The rest masses of the proton, neutron, and deuteron are

$$m_p = 1.67261 \times 10^{-27}\text{ kg} \qquad m_n = 1.67492 = 10^{-27}\text{ kg} \qquad m_d = 3.34357 \times 10^{-27}\text{ kg}$$

The deuteron (the nucleus of heavy hydrogen) consists of a proton and a neutron. How much energy should be liberated in the formation of a deuteron from a free proton and a free neutron, initially at rest "at infinity"?

▮ The rest mass lost in the formation of a deutron is

$$(m_p + m_n) - m_d = (1.67261 + 1.67492 - 3.34357) \times 10^{-27} = 3.96 \times 10^{-30} \text{ kg}$$

which is equivalent to a rest-energy loss

$$\Delta E_0 = 3.96 \times 10^{-30} c^2 = 3.56 \times 10^{-13} \text{ J} = 2.22 \text{ MeV}$$

Hence, by energy conservation, the surroundings must gain exactly 2.22 MeV of energy. This same amount of energy, the *binding energy* of the deuteron, would have to be supplied to the deuteron to tear it apart into an infinitely separated proton and neutron.

37.51 Determine the mass and speed of an electron having kinetic energy of 100 keV (1.6×10^{-14} J).

▮ The kinetic energy is $K = mc^2 - m_0 c^2$, or 1.6×10^{-14} J $= (m - m_0)(3 \times 10^8 \text{ m/s})^2$, and $m - m_0 = 1.78 \times 10^{-31}$ kg. In Prob. 37.49 we are given $m_0 = 9.11 \times 10^{-31}$ kg; so $m = (9.11 + 1.78) \times 10^{-31}$ kg $= \underline{1.089 \times 10^{-30} \text{ kg}}$. To get velocity we note that $m = m_0 / \sqrt{1 - (v^2/c^2)}$, or $1 - (v^2/c^2) = (m_0/m)^2 = 0.700$. Then $v/c = 0.548$, and $v = \underline{1.64 \times 10^8 \text{ m/s}}$.

37.52 Find the speed that a proton must be given if its mass is to be twice its rest mass of 1.67×10^{-27} kg. What energy must be given the proton to achieve this speed?

▮ Use the mass-increase formula.

$$m = m_0 \Big/ \sqrt{1 - \frac{v^2}{c^2}} \qquad 2m_0 = m_0 \Big/ \sqrt{1 - \frac{v^2}{c^2}} \qquad 4\Big(1 - \frac{v^2}{c^2}\Big) = 1; \qquad \frac{v^2}{c^2} = \frac{3}{4} \qquad v = \underline{0.866c}$$

$$\Delta W = (m - m_0)c^2 = (2m_0 - m_0)c^2 = (1.67 \times 10^{-27} \text{ kg})(3 \times 10^8 \text{ m/s})^2 = 1.50 \times 10^{-10} \text{ J} = \underline{938 \text{ MeV}}$$

37.53 How much energy must be given to an electron to accelerate it to $0.95c$?

▮ Use the mass-increase formula.

$$m = \frac{m_0}{\sqrt{1 - (v^2/c^2)}} = \frac{9.1 \times 10^{-31} \text{ kg}}{\sqrt{1 - 0.95^2}} = 29.1 \times 10^{-31} \text{ kg}$$

$$\Delta W = (m - m_0)c^2 = [(29.1 - 9.1)(10^{-31})](9 \times 10^{16}) = 1.8 \times 10^{-13} \text{ J} = \underline{1.125 \text{ MeV}}$$

37.54 A 2-kg object is lifted from the floor to a tabletop 30 cm above the floor. By how much did the mass of the object increase because of its increased PE?

▮ We use $\Delta E = (\Delta m)c^2$, with $\Delta E = mgh$. Therefore

$$\Delta m = \frac{\Delta E}{c^2} = \frac{mgh}{c^2} = \frac{(2 \text{ kg})(9.8 \text{ m/s}^2)(0.3 \text{ m})}{(3 \times 10^8 \text{ m/s})^2} = \underline{6.5 \times 10^{-17} \text{ kg}}$$

37.55 The rest mass of a proton is $m_0 = 1.672614 \times 10^{-27}$ kg. Its laboratory speed after having fallen through a high difference in potential, ΔV, is $u = 2 \times 10^8$ m/s. (**a**) Evaluate ΔV. (**b**) Find the total energy of the proton.

▮ (**a**) The kinetic energy of the proton is

$$K = (m - m_0)c^2 = \Big[\frac{1}{\sqrt{1 - (u^2/c^2)}} - 1\Big] m_0 c^2 = e\,\Delta V$$

from which

$$\Delta V = \frac{[(1/0.744943) - 1](1.672614 \times 10^{-27})(2.997925 \times 10^8)^2}{1.602 \times 10^{-19}} = \underline{321 \text{ MV}}$$

(**b**) The rest energy of the proton is $m_0 c^2 = 938.3$ MeV, and so the total energy is

$$E = 938.3 + 321 = \underline{1259.3 \text{ MeV}}$$

37.56 Show that KE $= (m - m_0)c^2$ reduces to KE $= \frac{1}{2} m_0 v^2$ when v is very much smaller than c.

▮
$$\text{KE} = (m - m_0)c^2 = \Big(\frac{m_0}{\sqrt{1 - (v/c)^2}} - m_0\Big)c^2 = m_0 c^2 [(1 - v^2/c^2)^{-1/2} - 1]$$

Let $b = -v^2/c^2$ and expand $(1 + b)^{-1/2}$ by the binomial theorem. Then

$$(1 + b)^{-1/2} = 1 + (-\tfrac{1}{2})b + \frac{(-\tfrac{1}{2})(-\tfrac{3}{2})}{2!}b^2 + \cdots = 1 + \frac{1}{2}\frac{v^2}{c^2} + \frac{3}{8}\frac{v^4}{c^4} + \cdots$$

and
$$KE = m_0c^2\left[\left(1 + \frac{1}{2}\frac{v^2}{c^2} + \frac{3}{8}\frac{v^4}{c^4} + \cdots\right) - 1\right] = \frac{1}{2}m_0v^2 + \frac{3}{8}m_0v^2\frac{v^2}{c^2} + \cdots$$

If v is very much smaller than c, the terms after $\frac{1}{2}m_0v^2$ are negligibly small.

37.57 Find the increase in mass of 100 kg of copper ($C = 0.389$ kJ/kg · K) if its temperature is increased 100 °C.

▮ $$\Delta E = mC\,\Delta T = (100)(0.389)(100) = 3890 \text{ kJ} \qquad \Delta m = \frac{\Delta E}{c^2} = \underline{4.33 \times 10^{-11} \text{ kg}}$$

37.58 If 1 g of matter could be converted entirely into energy, what would be the value of the energy so produced, at $0.01 per kW · h?

▮ We make use of $\Delta E = (\Delta m)c^2$.

$$\text{energy gained} = (\text{mass lost})c^2 = (10^{-3} \text{ kg})(3 \times 10^8 \text{ m/s})^2 = 9 \times 10^{13} \text{ J}$$

$$\text{value of energy} = (9 \times 10^{13} \text{ J})\left(\frac{1 \text{ kW · h}}{3.6 \times 10^6 \text{ J}}\right)\left(\frac{\$0.01}{\text{kW · h}}\right) = \underline{\$250\,000}$$

37.59 A rectangular tank is fixed in $\mathscr{L}'$, with its edges parallel to the coordinate axes. If the tank is filled with liquid of density ρ', as measured in $\mathscr{L}'$, what is the density as measured in $\mathscr{L}$?

▮ The volume, V, of the tank as seen in $\mathscr{L}$ is smaller as a consequence of Lorentz contraction along the x axis. Since the y and z dimensions are not affected, we have $V = V'\sqrt{1 - (v^2/c^2)}$. The total mass of the liquid as seen in $\mathscr{L}$ is increased as a consequence of the kinetic energy it has moving at speed v and $M = M'/\sqrt{1 - (v^2/c^2)}$. Then $\rho = M/V = (M'/V')\{1/[1 - (v^2/c^2)]\} = \rho'/[1 - (v^2/c^2)]$

37.60 Refer to Prob. 37.59. Show that the $\mathscr{L}'$ observer and the $\mathscr{L}$ observer agree as to the rate at which mass is being lost from the tank by evaporation.

▮ If $\Delta M'$ is the mass that evaporates in time $\Delta t'$ as seen in $\mathscr{L}'$ we have for the corresponding quantities in $\mathscr{L}$ $\Delta M = \Delta M'/\sqrt{1 - (v^2/c^2)}$; $\Delta t = \Delta t'/\sqrt{1 - (v^2/c^2)}$, where $\Delta t'$ is the proper time interval. Then $(\Delta M/\Delta t) = (\Delta M'/\Delta t')$ as required.

37.61 Our galaxy is about 10^5 light years across, and the most energetic particles traversing it have energies of about 10^{20} eV. Find (**a**) the relativistic energy of the particle, assuming it is a proton; (**b**) the mass-increase factor; and (**c**) the time for the proton to cross the galaxy, as seen in the frame of the galaxy and in the frame of the proton.

▮ (**a**) The energy in joules is $(10^{20} \text{ eV})(1.6 \times 10^{-19} \text{ J/eV}) = \underline{16 \text{ J}}$.
(**b**) To a very good degree of approximation, the relativistic mass is given by

$$KE = 16 \text{ J} = (m - m_0)c^2 \approx mc^2 \qquad m = \frac{16}{9 \times 10^{16}} = \underline{1.78 \times 10^{-16} \text{ kg}}$$

or approximately 10^{11} times the rest mass.

(**c**) $$m = \gamma m_0 \qquad \gamma = \frac{m}{m_0} = 1.06 \times 10^{11}$$

The proton sees the galaxy contracted in the direction in which it is traveling by

$$L = \gamma^{-1}L^* = (0.94 \times 10^{-11})(10^5) = 0.94 \times 10^{-6} \text{ light years}$$

Approximating the velocity of the proton as c, $t = L/c = 0.94 \times 10^{-6}$ years, which is about $\underline{30 \text{ s}}$! In the earth's frame $t^* = \gamma t = 10^5$ years, as expected.

37.62 Conservation of linear momentum of an isolated system holds in special relativity; indeed, it is this principle that leads to the nonconstancy of mass. (**a**) Derive the relation between momentum and total energy. (**b**) Calculate the momentum of a 1 MeV electron.

▮ (**a**) We start with the relativistic mass formula $m = m_0/\sqrt{1 - (u^2/c^2)}$. We square this and then multiply both sides by $c^4[1 - (u^2/c^2)]$, obtaining

$$m^2c^4 - m^2u^2c^2 = m_0^2c^4 \qquad \text{or} \qquad (mc^2)^2 = (m_0c^2)^2 + (muc)^2$$

But $mc^2 = E$, the total energy; $m_0c^2 = E_0$, the rest energy; $mu = p$, the magnitude of the momentum vector.

Thus the desired relation is

$$E^2 = E_0^2 + (pc)^2 \tag{1}$$

Because of the form of (1) momenta are often specified in MeV/c, on the atomic level, where energies are given in eV or MeV. (See Prob. 37.63.) **(b)** The total energy of the electron is its rest energy, 0.511 MeV, plus its kinetic energy, 1 MeV. Then (1) gives $1.511^2 = 0.511^2 + (pc)^2$, from which $pc = 1.42$ MeV, or $p = \underline{1.42\ \text{MeV}/c}$.

37.63 Compute the conversion factor between the ordinary units of momentum and MeV/c.

▮ $1\ \text{MeV}/c = (1.6 \times 10^{-13}\ \text{J})/(3 \times 10^8\ \text{m/s}) = \underline{5.33 \times 10^{-22}\ \text{kg} \cdot \text{m/s}}$.

37.64 Show that rest mass is not conserved in a symmetrical, perfectly inelastic collision.

▮ Let two identical bodies, each with rest mass m_0, approach each other at equal speeds u, collide, and stick together. By momentum conservation, the conglomerate body must be at rest, with energy $M_0 c^2$. The initial energy of the system was $2[m_0 c^2 / \sqrt{1 - (u^2/c^2)}]$. Hence,

$$\frac{2m_0 c^2}{\sqrt{1 - (u^2/c^2)}} = M_0 c^2 \qquad \text{or} \qquad M_0 = \frac{2m_0}{\sqrt{1 - (u^2/c^2)}}$$

It is seen that the final rest mass, M_0, exceeds the initial rest mass, $2m_0$.

37.65^c In newtonian mechanics the relation $dE/dt = \mathbf{F} \cdot \mathbf{v}$ is valid, where E is the total energy of a particle that is moving with velocity $\mathbf{v}$ and is acted on by a net force $\mathbf{F}$. Show that this relation is also valid in relativistic mechanics. (Assume that Newton's second law is valid under special relativity.)

▮ As shown in Prob. 37.62, the particle's instantaneous momentum $\mathbf{p}(t)$ and instantaneous energy $E(t)$ are related by

$$E^2 = c^2 \mathbf{p} \cdot \mathbf{p} + (m_0 c^2)^2 \tag{1}$$

Differentiating Eq. (1) with respect to time, we find that

$$2E \frac{dE}{dt} = 2c^2 \mathbf{p} \cdot \frac{d\mathbf{p}}{dt}$$

or

$$\frac{dE}{dt} = \frac{c^2 \mathbf{p}}{E} \cdot \frac{d\mathbf{p}}{dt} \tag{2}$$

But $\mathbf{p} = m\mathbf{v} = \gamma m_0 \mathbf{v}$ and $E = mc^2 = \gamma m_0 c^2$, so $(c^2 \mathbf{p})/E = \mathbf{v}$. Using this in Eq. (2), we obtain

$$\frac{dE}{dt} = \frac{d\mathbf{p}}{dt} \cdot \mathbf{v} = \mathbf{F} \cdot \mathbf{v} \tag{3}$$

since $\mathbf{F} = d\mathbf{p}/dt$.

37.66^c Show that the components of the velocity of a particle of energy E and momentum $\mathbf{p}$ are given by

$$v_x = \frac{\partial E}{\partial p_x} \qquad v_y = \frac{\partial E}{\partial p_y} \qquad v_z = \frac{\partial E}{\partial p_z}$$

These relations apply in both the relativistic and newtonian domains.

▮ The energy is given in terms of the momentum by

$$E = \sqrt{c^2 p^2 + (m_0 c^2)^2} = \sqrt{c^2(p_x^2 + p_y^2 + p_z^2) + (m_0 c^2)^2}$$

Differentiating with respect to p_x, we find that

$$\frac{\partial E}{\partial p_x} = \frac{1}{2} \frac{2c^2 p_x}{\sqrt{c^2(p_x^2 + p_y^2 + p_z^2) + (m_0 c^2)^2}} = \frac{c^2 p_x}{E} = \frac{c^2 \gamma m_0 v_x}{\gamma m_0 c^2} = v_x$$

The proofs for the y and z components follow the same procedures.

Since the newtonian result is just the low-velocity limit of the relativistic result, the above relations hold in the newtonian domain. They may also be easily derived from $E = p^2/(2m_0) = p_x^2/(2m_0) + p_y^2/(2m_0) + p_z^2/(2m_0)$. Then $\partial E/\partial p_x = p_x/m_0 = v_x$, etc.

37.67 A beam of protons passing a given point moves at 5.00×10^7 m/s and carries a charge per unit length of 2.00×10^{-12} C/m, as measured by an observer in the laboratory frame of reference. An isolated proton moves at the same velocity, parallel to and at a distance of 10 mm from the beam. Compute the electric and magnetic forces on the proton as measured by an observer in the laboratory.

▮ In the laboratory frame, the electric field **E** points away from the beam. With a charge per unit length $\lambda = 2.00 \times 10^{-12}$ C/m and a beam-to-proton distance $r = 0.0100$ m, the field magnitude is (Prob. 25.43)

$$E = \frac{\lambda}{2\pi\varepsilon_0 r} = \frac{2\lambda}{4\pi\varepsilon_0 r} = \frac{(2)(8.99 \times 10^9)(2.00 \times 10^{-12})}{1.00 \times 10^{-2}} = 3.596 \text{ N/C}$$

Hence the <u>electric</u> force on the proton is $F_e = eE = (1.60 \times 10^{-19})(3.596) = \underline{5.75 \times 10^{-19} \text{ N, away from the beam.}}$ The magnetic field lines circle the beam. Since the current carried by the beam is $i = \lambda v_p = (2.00 \times 10^{-12})(5.00 \times 10^7) = 100 \ \mu\text{A}$, the field strength at the location of the proton is (Prob. 28.90)

$$B = \frac{\mu_0 i}{2\pi r} = \frac{(2 \times 10^{-7})(1.00 \times 10^{-4})}{1.00 \times 10^{-2}} = 2.00 \text{ nT}$$

Since the proton's velocity $\mathbf{v}_p$ is perpendicular to **B**, the <u>magnetic</u> force $\mathbf{F}_m = e\mathbf{v}_p \times \mathbf{B}$ has magnitude

$$F_m = ev_p B = (1.60 \times 10^{-19})(5.00 \times 10^7)(2.00 \times 10^{-9}) = \underline{1.60 \times 10^{-20} \text{ N}}$$

This force is directed <u>toward the beam.</u>

37.68 Refering to Prob. 37.67; compute the electric and magnetic forces on the proton as measured by an observer moving with the proton.

▮ In the proton's rest frame, the beam's charge per unit length is $\lambda' = \lambda\sqrt{1 - (v_p^2/c^2)}$, since the beam particles are also at rest in this frame:

$$\lambda' = (2.00 \times 10^{-12})\sqrt{1 - \left(\frac{5.00 \times 10^7}{3.00 \times 10^8}\right)^2} = 1.972 \times 10^{-12} \text{ C/m}$$

The beam-to-proton distance $r' = r$, since transverse distances are unaffected by the Lorentz transformation. Hence the electric field due to the beam at the position of the proton is

$$E' = \frac{\lambda'}{2\pi\varepsilon_0 r'} = \frac{2\lambda'}{4\pi\varepsilon_0 r} = \frac{2(8.99 \times 10^9)(1.972 \times 10^{-12})}{1.00 \times 10^{-2}} = 3.546 \text{ N/C}$$

Hence the <u>electric</u> force on the proton is $F'_e = eE' = (1.60 \times 10^{-19})(3.546) = \underline{5.67 \times 10^{-19} \text{ N, away from the beam.}}$ The <u>magnetic</u> force $\mathbf{F}'_m$ is evidently <u>zero</u> in the proton's rest frame.

37.69 Refer to Probs. 37.67 and 37.68. Compute the electric and magnetic forces on the proton as measured by an observer moving at a velocity of magnitude 1.00×10^8 m/s in the direction of the beam.

▮ To find the (common) velocity v''_p of the protons in a frame of reference moving at speed $V = 1.00 \times 10^8$ m/s in the direction of the beam, we use the Lorentz velocity transformation:

$$v''_p = \frac{v_p - V}{1 - [(v_p V)/c^2]} = \frac{(5.00 \times 10^7) - (1.00 \times 10^8)}{1 - [(5.00 \times 10^7)(1.00 \times 10^8)]/(3.00 \times 10^8)^2} = -5.294 \times 10^7 \text{ m/s}$$

Due to the Lorentz contraction, the beam's charge per unit length λ'' is larger than its rest-frame value λ' by a factor $\gamma'' = 1/\sqrt{1 - [(v''_p)^2/c^2]}$:

$$\lambda'' = \frac{\lambda'}{\sqrt{1 - [(v''_p)^2/c^2]}} = \frac{1.972 \times 10^{-12}}{\sqrt{1 - [(5.294 \times 10^7)/(3.00 \times 10^8)]^2}} = 2.003 \times 10^{-12} \text{ C/m}$$

Hence the electric field is directed away from the beam; and at a distance $r'' = r$, it has magnitude

$$E'' = \frac{2\lambda''}{4\pi\varepsilon_0 r''} = \frac{2(8.99 \times 10^9)(2.003 \times 10^{-12})}{1.00 \times 10^{-2}} = \underline{3.602 \text{ N/C}}$$

Thus the <u>electric</u> force on the proton is $F''_e = eE'' = (1.60 \times 10^{-19})(3.602) = \underline{5.76 \times 10^{-19} \text{ N, away from the beam.}}$ The magnetic field lines are circles directed opposite to those found in part (*a*) and the field magnitude $B'' = (\mu_0 i'')/(2\pi r'')$. The current magnitude is

$$i'' = \lambda'' |v''_p| = (2.003 \times 10^{-12})(5.294 \times 10^7) = 106 \ \mu\text{A}$$

so the magnetic field magnitude at distance r'' is

$$B'' = \frac{\mu_0 i''}{2\pi r''} = \frac{\mu_0 i''}{2\pi r} = \frac{(2\times10^{-7})(1.06\times10^{-4})}{1.00\times10^{-2}} = \underline{2.120\,\text{nT}}$$

Finally, $F_m = ev_p'' B'' = (1.6\times10^{-19})(5.294\times10^{7})(2.12\times10^{-9}) = 1.79\times10^{-20}\,\text{N}$ toward the beam.

37.70 A large cyclotron is designed to accelerate deuterons to 450 MeV of kinetic energy. This means that their speed will become a substantial fraction of c. Hence their mass will become substantially larger than the rest mass. If the magnetic field is everywhere of the same value, this requires that the frequency of the oscillating potential difference applied between the Dees be decreased during the acceleration of a group of deuterons. What is the ratio of the final to the initial frequency? The rest mass of a deuteron is 3.3×10^{-27} kg.

▌ For a particle circling at relativistic speed in a magnetic field, the relativistic force, $\mathbf{F} = d\mathbf{p}/dt$, is still correctly given by the Lorentz formula:

$$\mathbf{F} \equiv \frac{d\mathbf{p}}{dt} = q\mathbf{v}\times\mathbf{B} \tag{1}$$

The momentum is given by $\mathbf{p} = \gamma m_0 \mathbf{v}$, where $\gamma \equiv [1-(v^2/c^2)]^{-1/2}$. If the particle is circling with angular velocity $\boldsymbol{\omega}_c$, we have, since the magnitude of $\mathbf{p}$ is constant,

$$\frac{d\mathbf{p}}{dt} = \boldsymbol{\omega}_c \times \mathbf{p} = \gamma m_0 (\boldsymbol{\omega}_c \times \mathbf{v}) \tag{2}$$

Equations (1) and (2) imply that, in magnitude,

$$\omega_c = \frac{qB}{\gamma m_0} = \frac{qB}{m} \tag{3}$$

where $m \equiv \gamma m_0$ is the relativistic mass. Since the deuterons are initially nonrelativistic, Eq. (3) implies that

$$\frac{\omega_{cf}}{\omega_{ci}} = \frac{\gamma_i}{\gamma_f} = \frac{1}{\gamma_f} \tag{4}$$

In terms of the final kinetic energy K_f, we have

$$\gamma_f = \frac{K_f + m_0 c^2}{m_0 c^2} \tag{5}$$

The rest energy of the deuteron is

$$m_0 c^2 = \frac{(3.3\times10^{-27}\,\text{kg})(3.00\times10^{8}\,\text{m/s})^2}{1.60\times10^{-13}\,\text{J/MeV}} = 1.86\times10^{3}\,\text{MeV}$$

Then with $K_f = 450$ MeV, Eqs. (4) and (5) yield

$$\frac{\omega_{cf}}{\omega_{ci}} = \frac{1}{\gamma_f} = \frac{1860}{450+1860} = \underline{0.81}$$

CHAPTER 38
Particles of Light and Waves of Matter

38.1 PHOTONS AND THE PHOTOELECTRIC EFFECT

38.1 Describe briefly the photon concept, including expressions for the energy and momentum of a photon.

❚ The photon picture portrays the emission and absorption of electromagnetic energy ($\mathscr{E}$) in the form of quanta having the value $\mathscr{E} = h\nu$. In other words, a photon has an energy $\mathscr{E}$ which is proportional to its frequency ν. The constant of proportionality $h = 6.6256 \times 10^{-34}$ J · s is known as *Planck's constant*. It can be shown (Prob. 38.11) that electromagnetic energy and momentum (p) are related by $p = \mathscr{E}/c$. Inasmuch as $\mathscr{E} = h\nu$, we then have $p = h/\lambda$.

38.2 Find the energy of the photons in a beam whose wavelength is 526 nm.

❚ $\mathscr{E} = h\nu = (hc)/\lambda = [(6.63 \times 10^{-34}$ J · s$)(3 \times 10^8$ m/s$)]/(526 \times 10^{-9}$ m$) = 3.78 \times 10^{-17}$ J. Since 1 eV $= 1.6 \times 10^{-19}$ J, $\mathscr{E} = \underline{2.36\,\text{eV}}$.
Calculations like this are facilitated by using suitable mixed units for hc:

$$hc = (6.63 \times 10^{-34}\,\text{J} \cdot \text{s})(3 \times 10^8\,\text{m/s})(1.6 \times 10^{-19}\,\text{J/eV}) = 1240\,\text{eV} \cdot \text{nm} = 1.240\,\text{MeV} \cdot \text{pm} = \cdots \quad (1)$$

Thus, $\mathscr{E} = (1240\,\text{eV} \cdot \text{nm})/(526\,\text{nm}) = 2.36\,\text{eV}$.

38.3 Determine the vacuum wavelength corresponding to a γ-ray energy of 10^{19} eV.

❚ From (1) of Prob. 38.2,

$$\lambda = \frac{hc}{\mathscr{E}} = \frac{1240\,\text{eV} \cdot \text{nm}}{10^{19}\,\text{eV}} = \underline{1.240 \times 10^{-25}\,\text{m}}$$

38.4 Compute the frequency, vacuum wavelength, and energy in joules of a photon having an energy of 2 eV.

❚ $\lambda = \dfrac{hc}{\mathscr{E}} = \dfrac{1240\,\text{eV} \cdot \text{nm}}{2\,\text{eV}} = \underline{620\,\text{nm}}$ $\nu = \dfrac{c}{\lambda} = \dfrac{3 \times 10^8\,\text{m/s}}{620 \times 10^{-9}\,\text{m}} = \underline{480\,\text{THz}}$ $2\,\text{eV} = \underline{3.2 \times 10^{-19}\,\text{J}}$

38.5 What is the photon energy in joules corresponding to a 60-Hz wave emitted from a power line? How does this compare with the energy range for light?

❚ Using the expression $\mathscr{E} = h\nu$, we have $\mathscr{E} = (6.6 \times 10^{-34})(60) = \underline{39.6 \times 10^{-33}\,\text{J}}$ as the energy of a 60-Hz photon. Light extends from 3.8×10^{14} Hz to 7.7×10^{14} Hz, or, from 25.1×10^{-20} J to 50.8×10^{-20} J. Thus, 60 Hz is less energetic than is light by a factor of about 10^{13}.

38.6 Show that for x-rays, the wavelength λ is given by $\lambda = 12.40/E$, where E is the energy in keV and λ is in Å.

❚ Since 1 nm $= 10$ Å, $hc = 1240\,\text{eV} \cdot \text{nm} = 12.40\,\text{keV} \cdot \text{Å}$.

38.7 In order to break a chemical bond in the molecules of human skin, causing sunburn, a photon energy of about 3.5 eV is required. To what wavelength does this correspond?

❚ $\lambda = \dfrac{1240\,\text{eV} \cdot \text{nm}}{3.5\,\text{eV}} = 354\,\text{nm}$ (ultraviolet)

38.8 Imagine a source emitting 100 W of green light at a wavelength of 500 nm. How many photons per second are emerging from the source?

❚ The power multiplied by the given time interval is the emitted energy, that is, (100 W)(1 s) = 100 J. Denoting the photon flux by N, we have

$$N = \frac{100}{h\nu} = \frac{100\lambda}{hc} = \frac{100(500 \times 10^{-9})}{(6.6 \times 10^{-34})(3 \times 10^8)} = \underline{25 \times 10^{19}\,\text{photons/s}}.$$

38.9 A desk lamp illuminates a desk top with violet light of wavelength 412 nm. The amplitude of this electromagnetic wave is 63.2 V/m. Find the number N of photons striking the desk per second per unit area, assuming that the illumination is normal.

▌ From $\mathscr{E} = (hc)/\lambda$ and the energy density formula, $(\epsilon_0 E_0^2)/2$, for a wave with electric field amplitude E_0 (Prob. 33.64)

$$N = \frac{\lambda \epsilon_0 E_0^2}{2h} = \frac{(412 \times 10^{-9})(8.85 \times 10^{-12})(63.2^2)}{2(6.63 \times 10^{-34})} = 1.10 \times 10^{19} \text{ photons/s} \cdot \text{m}^2$$

38.10 A sensor is exposed for 0.1 s to a 200-W lamp 10 m away. The sensor has an opening that is 20 mm in diameter. How many photons enter the sensor if the wavelength of the light is 600 nm? Assume that all the energy of the lamp is given off as light.

▌ The energy of a photon of the light is

$$E = \frac{hc}{\lambda} = \frac{(6.63 \times 10^{-34})(3 \times 10^8)}{600 \times 10^{-9}} = 3.3 \times 10^{-19} \text{ J}$$

The lamp uses 200 W of power. The number of photons emitted per second is therefore

$$n = \frac{200}{3.3 \times 10^{-19}} = 6.1 \times 10^{20} \text{ photons/s}$$

Since the radiation is spherically symmetrical, the number of photons entering the sensor per second is n multiplied by the ratio of the aperture area to the area of a sphere of radius 10 m:

$$(6.1 \times 10^{20}) \frac{\pi(0.010^2)}{4\pi(10^2)} = 1.53 \times 10^{14} \text{ photons/s}$$

and the number of photons that enter the sensor in 0.1 s is $(0.1)(1.53 \times 10^{14}) = \underline{1.53 \times 10^{13}}$.

38.11 What is the momentum-energy relation for photons?

▌ From Prob. 37.62, $\mathscr{E}^2 = (pc)^2 + (m_0 c^2)^2$. But a photon has zero rest mass, so $\mathscr{E} = pc$.

38.12 What is the momentum of a single photon of red light ($v = 400 \times 10^{12}$ Hz) moving through free space?

▌ The momentum is given by $p = h/\lambda = (hv)/c$. Hence

$$p = \frac{(6.6 \times 10^{-34})(400 \times 10^{12})}{3 \times 10^8} = \underline{8.8 \times 10^{-28} \text{ kg} \cdot \text{m/s}}.$$

38.13 What wavelength must electromagnetic radiation have if a photon in the beam is to have the same momentum as an electron moving with a speed 2×10^5 m/s?

▌ The requirement is that $(mv)_{\text{electron}} = (h/\lambda)_{\text{photon}}$. From this,

$$\lambda = \frac{h}{mv} = \frac{6.63 \times 10^{-34} \text{ J} \cdot \text{s}}{(9.1 \times 10^{-31} \text{ kg})(2 \times 10^5 \text{ m/s})} = \underline{3.64 \text{ nm}}$$

This wavelength is in the x-ray region.

38.14 A stream of photons impinging normally on a completely absorbing screen in vacuum exerts a pressure $\mathscr{P}$. Show that $\mathscr{P} = I/c$, where I is the irradiance.

▌ The time rate of change of momentum equals the force, that is, $\Delta p/\Delta t = F$. The force per unit area, F/A, is the pressure, and so

$$\mathscr{P} = \frac{1}{A} \frac{\Delta p}{\Delta t}$$

But $\mathscr{E} = cp$ for photons, which means that $\Delta \mathscr{E} = c \Delta p$ and

$$\mathscr{P} = \frac{1}{Ac} \frac{\Delta \mathscr{E}}{\Delta t}$$

Since irradiance by definition is energy per unit area per unit time, we obtain $\underline{\mathscr{P} = I/c}$.

38.15 A collimated beam of light of flux density 30 kW/m² is incident normally on a 100-mm² completely absorbing screen. Using the results of Prob. 38.14, determine both the pressure exerted on and the momentum transferred to the screen during a 1000-s interval.

▮ The pressure is simply

$$\mathcal{P} = \frac{I}{c} = \frac{3 \times 10^4}{3 \times 10^8} = \underline{10^{-4} \text{ Pa}} = \frac{F}{A} = \frac{(\Delta p / \Delta t)}{A} \quad \text{and} \quad \Delta p = \mathcal{P} A \, \Delta t = (10^{-4})(10^{-4})(10^3) = \underline{10^{-5} \text{ kg} \cdot \text{m/s}}$$

(One could actually build an interplanetary sailboat using solar pressure.)

38.16 What is the largest momentum we can expect for a microwave photon?

▮ Microwave frequencies go up to 3×10^{11} Hz. Therefore, since $p = h/\lambda = (h\nu)/c$,

$$p = \frac{(6.6 \times 10^{-34})(3 \times 10^{11})}{3 \times 10^8} = \underline{6.6 \times 10^{-31} \text{ kg} \cdot \text{m} \cdot \text{s}^{-1}}$$

38.17 How many red photons ($\lambda = 663$ nm) must strike a totally *reflecting* screen per second, at normal incidence, if the exerted force is to be 0.225 lb?

▮ We know that $F = \Delta p / \Delta t$ and, in this case, Δp is *twice* the incident momentum. Thus, if N is the number of incoming photons per second, $F = N[(2h)/\lambda]$. Since 0.225 lb = 1 N, we have $F = 1$, and

$$N = \frac{\lambda}{2h} = \frac{663 \times 10^{-9}}{2(6.63 \times 10^{-34})} = \underline{5 \times 10^{26} \text{ photons/s}}$$

38.18 What potential difference must be applied to stop the fastest photoelectrons emitted by a nickel surface under the action of ultraviolet light of wavelength 2000 Å? The work function of nickel is 5.00 eV.

▮
$$\text{energy of photon} = \frac{hc}{\lambda} = \frac{1240 \text{ eV} \cdot \text{nm}}{200 \text{ nm}} = 6.20 \text{ eV}$$

Then, from the photoelectric equation, the energy of the fastest emitted electron is 6.20 eV − 5.00 eV = 1.20 eV. Hence a retarding potential of $\underline{1.20 \text{ V}}$ is required.

38.19 The work function of sodium metal is 2.3 eV. What is the longest-wavelength light that can cause photoelectron emission from sodium?

▮ At threshold, the photon energy just equals the energy required to tear the electron loose from the metal, namely, the work function W_{min}.

$$W_{min} = \frac{hc}{\lambda} \quad \text{or} \quad \lambda = \frac{1240 \text{ eV} \cdot \text{nm}}{2.3 \text{ eV}} = \underline{540 \text{ nm}}$$

38.20 What is the work function of sodium metal if the photoelectric threshold wavelength is 680 nm?

▮
$$\text{work function} = W_{min} = h\nu_{min} = \frac{hc}{\lambda_{max}} = \frac{1240 \text{ eV} \cdot \text{nm}}{680 \text{ nm}} = \underline{1.83 \text{ eV}}$$

38.21 Will photoelectrons be emitted by a copper surface, of work function 4.4 eV, when illuminated by visible light?

▮ As in Prob. 38.19,

$$\text{threshold } \lambda = \frac{hc}{W_{min}} = \frac{1240 \text{ eV} \cdot \text{nm}}{4.4 \text{ eV}} = 282 \text{ nm}$$

Hence visible light (400 to 700 nm) cannot eject photoelectrons from copper.

38.22 Light of wavelength 600 nm falls on a metal having photoelectric work function 2 eV. Find (*a*) the energy of a photon, (*b*) the kinetic energy of the most energetic photoelectron, and (*c*) the stopping potential.

▮ (*a*)
$$E = \frac{hc}{\lambda} = \frac{1240 \text{ eV} \cdot \text{nm}}{600 \text{ nm}} = 2.07 \text{ eV}$$

(b) $$K_{max} = E - W_{min} = 2.07 - 2 = 0.07 \text{ eV}$$

(c) $$eV_s = K_{max} = 0.07 \text{ eV} \quad \text{or} \quad V_s = 0.07 \text{ V}$$

38.23 It takes 4.2 eV to remove one of the least tightly bound electrons from a metal surface. When ultraviolet photons of a single frequency strike a metal, electrons with kinetic energies from zero to 2.6 eV are ejected. What is the energy of the incident photons?

▌ $$\text{WF} = 4.2 \text{ eV} \quad \text{KE}_{max} = 2.6 \text{ eV} \quad \text{and} \quad E_{photon} = \text{WF} + \text{KE}_{max} = 4.2 + 2.6 = \underline{6.8 \text{ eV}}$$

38.24 A photon of energy 4.0 eV imparts all its energy to an electron that leaves a metal surface with 1.1 eV of kinetic energy. What is the work function W_{min} of the metal?

▌ $$W_{min} = h\nu - K_{max} = 4.0 - 1.1 = \underline{2.9 \text{ eV}}$$

38.25 Determine the maximum KE of photoelectrons ejected from a potassium surface by ultraviolet light of wavelength 2000 Å. What retarding potential difference is required to stop the emission of electrons? The photoelectric threshold wavelength for potassium is 4400 Å.

▌ $$\text{Work function} = W_{min} = \frac{hc}{\lambda_{max}} = \frac{12400 \text{ eV} \cdot \text{Å}}{4400 \text{ Å}} = 2.82 \text{ eV}$$

$$\text{then} \quad \frac{hc}{\lambda} - \frac{hc}{\lambda_{max}} = \text{KE}_{max} = \frac{12\,400}{2000} - 2.82 = 6.20 - 2.82 = \underline{3.38 \text{ eV}}$$

The retarding potential must be just large enough to stop electrons with KE_{max}, so $V = \underline{3.38 \text{ V}}$.

38.26 A surface has light of wavelength $\lambda_1 = 550$ nm incident on it, causing the ejection of photoelectrons for which the stopping potential is $V_{s1} = 0.19$ V. Suppose that radiation of wavelength $\lambda_2 = 190$ nm were incident on the surface. Calculate **(a)** the stopping potential V_{s2}, **(b)** the work function of the surface, and **(c)** the threshold frequency for the surface.

▌ **(a)** Since $eV = \text{KE}_{max} = h\nu - W_{min}$, with W_{min} a constant of the surface, we have $e(V_{s2} - V_{s1}) = h(\nu_2 - \nu_1)$ or

$$V_{s2} = V_{s1} + \frac{h}{e}(\nu_2 - \nu_1) = V_{s1} + \frac{hc}{e}\left(\frac{1}{\lambda_2} - \frac{1}{\lambda_1}\right) = 0.19 + 1240\left(\frac{1}{190} - \frac{1}{550}\right) = \underline{4.47 \text{ V}}$$

(b) $$W_{min} = \frac{hc}{\lambda_1} - eV_{s1} = \frac{1240}{550} - 0.19 = \underline{2.07 \text{ eV}}$$

(c) $$h\nu_c = W_{min} \quad \text{or} \quad \nu_c = \frac{W_{min}}{h} = \frac{(2.07)(1.602 \times 10^{-19})}{6.63 \times 10^{-34}} = \underline{498 \text{ THz}}$$

38.27 In the photoionization of atomic hydrogen, what will be the maximum kinetic energy of the ejected electron when a 60-nm photon is absorbed by the atom? The ionization energy of H is 13.6 eV.

▌ The ionization energy is minimum energy to remove the electron from the atom. Then

$$K_{max} = \frac{hc}{\lambda} - 13.6 \text{ eV} = \frac{1240 \text{ eV} \cdot \text{nm}}{60 \text{ nm}} - 13.6 \text{ eV} = \underline{7.1 \text{ eV}}$$

38.2 COMPTON SCATTERING; X-RAYS; PAIR PRODUCTION AND ANNIHILATION

38.28 Suppose that a 3.64 nm photon going in the $+x$ direction collides head-on with a 2×10^5 m/s electron moving in the $-x$ direction. If the collision is perfectly elastic, what are the conditions after collision.

▌ From the law of conservation of momentum,

$$\text{momentum before} = \text{momentum after} \quad \frac{h}{\lambda_0} - m\nu_0 = \frac{h}{\lambda} - m\nu$$

But, from Prob. 38.13, $h/\lambda_0 = m\nu_0$ in this case. Hence, $h/\lambda = m\nu$. Also, for a perfectly elastic collision,

$$\text{KE before} = \text{KE after} \quad \frac{hc}{\lambda_0} + \frac{1}{2}m\nu_0^2 = \frac{hc}{\lambda} + \frac{1}{2}m\nu^2$$

Using the facts that $h/\lambda_0 = mv_0$ and $h/\lambda = mv$, we find that

$$v_0(c + \tfrac{1}{2}v_0) = v(c + \tfrac{1}{2}v)$$

Therefore $v = v_0$ and the electron rebounds with its original speed. Because $h/\lambda = mv = mv_0$, the photon also rebounds and with its original wavelength.

38.29 The *Compton equation* can be written as $\lambda' - \lambda = [h/(m_0c)](1 - \cos \phi)$. The factor $h/(m_0c)$, called the *Compton wavelength*, is the wavelength shift that occurs when $\phi = 90°$. Evaluate it for scattering from (**a**) electrons and (**b**) protons. In the case of electrons, what percentage change is this if the incident radiation has a wavelength (**c**) 500 nm (visible light) and (**d**) 0.050 nm (0.5-Å x-rays)?

▎ Inserting the rest masses of the electron (9.1×10^{-31} kg) and proton (1.67×10^{-27} kg) into h/mc yields Compton wavelengths of 0.024 Å and 1.32×10^{-15} m, respectively. For the electron $(0.024/5000)(100) = 4.8 \times 10^{-4}$ percent, while for the 0.5-Å x-ray, $(0.024/0.500)(100) = 4.8$ percent.

38.30 A photon ($\lambda = 0.400$ nm) strikes an electron at rest and rebounds at an angle of 150° to its original direction. Find the speed and wavelength of the photon after the collision.

▎ The speed of a photon is always the speed of light in vacuum, c. To obtain the wavelength after collision, we use the equation for the Compton effect:

$$\lambda' = \lambda + \frac{h}{mc}(1 - \cos \phi) = 4 \times 10^{-10}\text{ m} + \frac{6.63 \times 10^{-34}\text{ J} \cdot \text{s}}{(9.1 \times 10^{-31}\text{ kg})(3 \times 10^8\text{ ms})}(1 - \cos 150°)$$

$$= 4 \times 10^{-10}\text{ m} + (2.43 \times 10^{-12}\text{ m})(1 + 0.866) = 0.4045\text{ nm}$$

38.31 Suppose that a beam of 0.2-MeV photons is scattered by the electrons in a carbon target. (**a**) What is the wavelength associated with these photons? (**b**) What is the wavelength of those photons scattered through an angle of 90°? (**c**) What is the energy of the scattered photons that emerge at an angle of 60° relative to the incident direction?

▎ (**a**)
$$\lambda = \frac{hc}{E} = \frac{1240\text{ MeV} \cdot \text{fm}}{0.2\text{ MeV}} = 6200\text{ fm}$$

(**b**)
$$\lambda' = \lambda + \frac{h}{m_ec}(1 - \cos \theta) = 6200 + (2430)(1 - 0) = 8630\text{ fm}$$

(**c**)
$$\frac{hc}{\lambda'} = \frac{hc}{\lambda + [h/(m_ec)](1 - \cos \theta)} = \frac{1240}{6200 + (2430)(1 - \tfrac{1}{2})} = 0.168\text{ MeV}$$

38.32 Verify Compton's equation (Prob. 38.29) when the photon is back-scattered ($\phi = 180°$).

▎ Assuming the electron to be at rest initially, we have from conservation of momentum

$$\frac{h}{\lambda} = P - \frac{h}{\lambda'} \quad \text{or} \quad P = h\left(\frac{1}{\lambda} + \frac{1}{\lambda'}\right)$$

and from conservation of relativistic energy

$$\frac{hc}{\lambda} + E_0 = \frac{hc}{\lambda'} + E \quad \text{or} \quad E = E_0 + hc\left(\frac{1}{\lambda} - \frac{1}{\lambda'}\right)$$

Substituting these expressions for the electron's final momentum and final energy into the momentum-energy relation for the electron, we obtain

$$\left[E_0 + hc\left(\frac{1}{\lambda} - \frac{1}{\lambda'}\right)\right]^2 = E_0^2 + \left[hc\left(\frac{1}{\lambda} + \frac{1}{\lambda'}\right)\right]^2 \qquad 2E_0hc\left(\frac{1}{\lambda} - \frac{1}{\lambda'}\right) - 2h^2c^2\frac{1}{\lambda\lambda'} = 2h^2c^2\frac{1}{\lambda\lambda'} \qquad E_0(\lambda' - \lambda) = 2hc$$

whence $\lambda' - \lambda = (2hc)/E_0 = (2h)/(m_ec)$, which is Compton's equation with $\cos \phi = -1$.

38.33 A photon strikes a free electron at rest and is scattered straight backward. If the electron's speed after collision is αc, where $\alpha \ll 1$, show that the electron's kinetic energy is a fraction α of the photon's initial energy.

▎ Since $v = \alpha c$ and $\alpha \ll 1$, the electron is nonrelativistic. Conservation of momentum yields $h/\lambda + h/\lambda' = mv$; multiplying by c, $(hc)/\lambda + (hc)/\lambda' = mvc = (mv^2)/\alpha$. The energy equation is $(hc)/\lambda - (hc)/\lambda' = (mv^2)/2$.

Adding we find that $(hc)/\lambda = E_\gamma = (mv^2)/4 + (mv^2)/(2\alpha)$. Since $\alpha \ll 1$, $(mv^2)/4$ is neglected and $(mv^2)/2 = \alpha E_\gamma$.

38.34 A photon with $\lambda = 0.5$ nm strikes a free electron head-on and is scattered straight backward. If the electron is initially at rest, what is its speed after the collision?

$\blacksquare$ Momentum conservation gives $h/\lambda = m_0 v_x - h/\lambda'$, assuming that the scattered electron is nonrelativistic. λ is given as 0.5 nm and λ' for 180° scattering is 0.5 nm $+ (2h)/(m_0 c) = 0.5048$ nm, so $v_x = (h/m_0)(1/\lambda + 1/\lambda') = [(6.63 \times 10^{-34})/(9.1 \times 10^{-31})](3.98 \times 10^9) = \underline{2.9 \times 10^6 \text{ m/s}}$.

38.35 A photon with $\lambda = 0.5$ nm is moving along the x axis when it strikes a free electron (initially at rest) and is scattered so as to move along the y axis. What are the x and y components of the electron's velocity after collision?

$\blacksquare$ In the x direction, conservation of momentum yields $h/\lambda = m_0 v_x$ and in y direction $h/\lambda' + m_0 v_y = 0$. From the Compton scattering relation for $\phi = 90°$, $\lambda' = \lambda + 0.024$ Å $= 5.0 + 0.024 = 5.024$ Å. Solving for v_x and v_y from above, with m_0 the electron rest mass, gives $v_x = +1.46 \times 10^6$ m/s and $v_y = -1.46 \times 10^6$ m/s; the electron is nonrelativistic.

38.36 Derive the Compton equation, using the relativistic expressions for the energy and momentum of the electron.

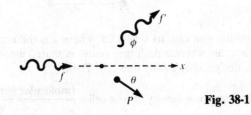

Fig. 38-1

$\blacksquare$ For notation, see Fig. 38-1.

$$\text{conservation of } x \text{ momentum:} \qquad \frac{hf}{c} = \frac{hf'}{c}\cos\phi + P\cos\theta \qquad (1)$$

$$\text{conservation of } y \text{ momentum:} \qquad 0 = \frac{hf'}{c}\sin\phi - P\sin\theta \qquad (2)$$

$$\text{energy conservation:} \qquad hf + m_0 c^2 = hf' + (P^2 c^2 + m_0^2 c^4)^{1/2} \qquad (3)$$

(*i*) Solve (1) for $P\cos\theta$ and (2) for $P\sin\theta$. Square and add; $h^2 f^2 - 2h^2 ff' \cos\phi + h^2 f'^2 = c^2 P^2$.
(*ii*) Square (3) to obtain: $h^2 f^2 + h^2 f'^2 - 2h^2 ff' + 2hm_0 c^2 (f - f') = c^2 P^2$.
(*iii*) Subtract: $(f - f')/ff' = [h/(m_0 c^2)](1 - \cos\phi)$.
(*iv*) But $f = c/\lambda$ and so $(f - f')/(ff') = (\lambda' - \lambda)/c$.
(*v*) Therefore, $\lambda' - \lambda = [h/(m_0 c)](1 - \cos\phi)$.

38.37 In what amounts to the inverse of the photoelectric effect, x-ray photons are produced when a tungsten target is bombarded by accelerated electrons. If an x-ray machine has an accelerating potential of 60 kV, what is the shortest wavelength present in its radiation?

$\blacksquare$ Since the work function of tungsten is very much smaller than the accelerating potential, we may suppose that the entire kinetic energy of an electron is lost to create a single photon of maximum energy:

$$eV = \frac{hc}{\lambda_{\min}} \qquad \lambda_{\min} = \frac{hc}{eV} = \frac{1240 \text{ keV} \cdot \text{pm}}{60 \text{ keV}} = \underline{20.7 \text{ pm}}$$

38.38 An electron is shot down an x-ray tube. Its energy just before it strikes the target of the tube is 30 keV. If it loses all this energy in a single collision with a very massive atom, what is the wavelength of the single x-ray photon that is emitted?

$$\blacksquare \qquad E = \frac{hc}{\lambda} = \frac{1.24 \text{ keV} \cdot \text{nm}}{\lambda} \qquad \lambda = \frac{1.24}{30} = 0.0413 \text{ nm} = \underline{0.413 \text{ Å}}$$

38.39 Determine the cutoff wavelength (in Å) of x-rays produced by 50-keV electrons in a Coolidge tube.

▌ The minimum wavelength is the cutoff wavelength; so

$$\lambda_{\text{cutoff}} = \frac{hc}{eV} = \frac{12.4 \text{ keV} \cdot \text{Å}}{50 \text{ keV}} = 0.248 \text{ Å}$$

38.40 Estimate the value of Planck's constant, if the cutoff wavelength is measured to be 1.18 Å for 10-keV electrons striking a target in a Coolidge tube.

▌

$$h = \frac{eV\lambda}{c} = \frac{(1.6 \times 10^{-19})(10\,000)(1.18 \times 10^{-10})}{3 \times 10^8} = 6.3 \times 10^{-34} \text{ J} \cdot \text{s}$$

38.41 Determine the grating spacing (for x-ray diffraction) of a substance that crystallizes in a face-centered cubic lattice.

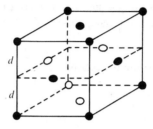

Fig. 38-2

▌ Figure 38-2 depicts the unit cell; its side is $2d$, where d is the required spacing. Now, each corner molecule is shared among eight cells; whereas each face center is shared among only two. Consequently, there are $\frac{1}{8}(8) + \frac{1}{2}(6) = 4$ molecules peculiar to any one cell. Then,

$$\rho \equiv \text{density of substance} = \text{density of unit cell} = \frac{(\text{molecules per cell})(\text{mass per molecule})}{\text{volume of cell}} = \frac{4(M/N_A)}{(2d)^3}$$

where $M \equiv$ molecular weight, $N_A \equiv$ Avogadro's number. Solve for d to obtain $d = [M/(2N_A\rho)]^{1/3}$.

38.42 Calculate the grating spacing d of KCl, given that its molecular weight is 74.6 kg/kmol and its density is 1980 kg/m³.

▌ The formula of Prob. 38.41 applies.

$$d = \sqrt[3]{\frac{74.6}{(2)(6.02 \times 10^{26})(1980)}} = 3.14 \times 10^{-10} \text{ m} = \underline{3.14 \text{ Å}}$$

38.43 Gypsum, which has a grating spacing of 7.600 Å at 18 °C, is sometimes used in working with long-wavelength x-rays. Find the wavelength of the $K\alpha$ characteristic line of sodium if first-order Bragg reflection occurs at a glancing angle of 51°35′.

▌ For Bragg reflection,

$$n\lambda = 2d \sin\theta \qquad (1)(\lambda) = 2(7.600)(\sin 51°35') = (15.200)(0.7835) = \underline{11.9 \text{ Å}}.$$

38.44 The grating spacing of rock salt is about 2.8 Å. If second-order Bragg "reflection" occurs for a given line at 30°, what is the wavelength of the "reflected" radiation?

▌ We use the Bragg reflection formula with $n = 2$.

$$n\lambda = 2d \sin\theta \qquad \text{or} \qquad 2\lambda = (2)(2.8)(\sin 30°) \qquad \text{and} \qquad \lambda = \underline{1.4 \text{ Å}}$$

38.45 X radiation of wavelength 1.5 Å is incident on an NaCl crystal with d-spacing 2.8 Å. What is the highest order that the crystal can diffract?

▌ $\sin\theta = (n\lambda)/(2d)$. The quantity on the right side of this equation cannot exceed 1; therefore, for $n = 3$, $\sin\theta = [(3)(1.5)]/5.6 < 1$. For $n = 4$, $\sin\theta = [(4)(1.5)]/5.6 > 1$. The maximum order possible is $\underline{3}$.

38.46 X radiation of wavelength 1.2 Å strikes a crystal of d-spacing 2.2 Å. Where does the diffraction angle of the second order from the 100-plane occur?

❚ For Bragg reflection the scattering angle is given by $\sin \theta = (n\lambda)/(2d) = [(2)(1.2)]/4.4$, and $\theta = \underline{33°}$.

38.47 The attenuation coefficient of aluminum for soft X-rays is 1.73/cm. Compute the fraction of these X-rays transmitted by an aluminum sheet 1.156 cm thick.

❚ The exponential law of attenuation is

$$I = I_0 e^{-\mu z} \qquad \frac{I}{I_0} = e^{-1.73(1.156)} = e^{-2} = \frac{1}{e^2} = \frac{1}{2.718^2} = 0.135 \qquad \text{or} \qquad \underline{13.5\% \text{ transmitted}}$$

38.48 One-half the intensity of a homogeneous x-ray beam is removed by an aluminum filter 5 mm thick. What is the fraction of this homogeneous beam that would be removed by 15 mm of aluminum?

❚ $\dfrac{I_1}{I_0} = e^{-\mu z}$ $\quad \dfrac{1}{2} = e^{-5\mu}$ $\quad \dfrac{I_2}{I_0} = e^{-15\mu} = (e^{-5\mu})^3 = \left(\dfrac{1}{2}\right)^3$ $\quad = \dfrac{1}{8}$ (fraction remaining) $\quad 1 - \dfrac{1}{8} = \dfrac{7}{8}$ (fraction removed)

38.49 Show that when a positron and an electron (both essentially at rest) annihilate, creating two photons, either photon has the Compton wavelength (Prob. 38.29).

❚ Total energy before annihilation is $2m_e c^2$; and after, $2[(hc)/\lambda]$ (momentum conservation requires that the photon energies be equal). Then, by conservation of energy,

$$2m_e c^2 = 2\frac{hc}{\lambda} \qquad \text{or} \qquad \lambda = \frac{h}{m_e c} = 2430 \text{ fm}$$

38.50 Show that the threshold wavelength for the production of a positron-electron pair is half the Compton wavelength.

❚ The incident photon must have at least enough energy to create a pair having zero kinetic energy (we ignore any kinetic energy of the recoil nucleus). Thus,

$$\frac{hc}{\lambda} \geq 2m_e c^2 \qquad \text{or} \qquad \lambda \leq \frac{h}{2m_e c} = \underline{1215 \text{ fm}}$$

38.51 After pair annihilation, two 1-MeV photons move off in opposite directions. Find the kinetic energy of the electron and positron.

❚ The two photons have the same energy, and hence the same momentum, which we are told are in opposite directions. Thus by momentum conservation the electron and positron must have had equal and opposite momentum. Hence their kinetic energies are equal. By energy conservation: $2 \text{ MeV} = 2(m_0 c^2) + 2 \text{ KE} = 2(0.51) \text{ MeV} + 2 \text{ KE}$ and KE of each $= \underline{0.49 \text{ MeV}}$.

38.3 DE BROGLIE WAVES AND THE UNCERTAINTY PRINCIPLE

38.52 A proton is accelerated through a potential difference of 1000 V. What is its de Broglie wavelength?

❚ We have $\lambda = h/(mv)$. But $m \approx m_0$ (nonrelativistic), so v may be found from $(mv^2)/2 = 1000e$. Substitution gives $\underline{\lambda = 9.1 \times 10^{-13} \text{ m}}$.

38.53 Find the de Broglie wavelength of a thermal neutron of mass 1.67×10^{-27} kg traveling at a speed of 2200 m/s.

❚ Use the de Broglie wave equation.

$$\lambda = \frac{h}{mv} = \frac{6.63 \times 10^{-34}}{1.67 \times 10^{-27}(2200)} = \underline{0.18 \text{ nm}}$$

38.54 What is the de Broglie wavelength for a particle moving with speed 2×10^6 m/s if the particle is (a) an electron, (b) a proton, and (c) a 0.2-kg ball?

❚ We make use of the definition of the de Broglie wavelength:

$$\lambda = \frac{h}{mv} = \frac{6.63 \times 10^{-34} \text{ J} \cdot \text{s}}{m(2 \times 10^6 \text{ m/s})} = \frac{3.3 \times 10^{-40} \text{ m} \cdot \text{kg}}{m}$$

Substituting the required values for m, one finds that the wavelength is 3.6×10^{-10} m for the electron, 2×10^{-13} m for the proton, and 1.65×10^{-39} m for the 0.2-kg ball.

38.55 An electron falls from rest through a potential difference of 100 V. What is its de Broglie wavelength?

▌ Its speed will still be far below c, so relativistic effects can be ignored. The KE gained, $\frac{1}{2}mv^2$, equals the electric PE lost, Vq. Therefore, after solving for v,

$$v = \sqrt{\frac{2Vq}{m}} = \sqrt{\frac{2(100 \text{ V})(1.6 \times 10^{-19} \text{ C})}{9.1 \times 10^{-31} \text{ kg}}} = 5.9 \times 10^6 \text{ m/s}$$

$$\text{Then} \qquad \lambda = \frac{h}{mv} = \frac{6.63 \times 10^{-34} \text{ J} \cdot \text{s}}{(9.1 \times 10^{-31} \text{ kg})(5.9 \times 10^6 \text{ m/s})} = \underline{0.123 \text{ nm}}$$

38.56 If the de Broglie wavelength of an electron is 1 Å, what are its velocity and its kinetic energy?

▌ $$\lambda = \frac{h}{mv} \qquad 10^{-10} = \frac{6.63 \times 10^{-34}}{9.11 \times 10^{-31} v} \qquad v = \frac{6.63 \times 10^{-34} \times 10^{10}}{9.11 \times 10^{-31}} = \underline{7.28 \times 10^6 \text{ m/s}}$$

$$\text{KE} = \frac{1}{2}mv^2 = \frac{1}{2}(9.11 \times 10^{-31})(7.28 \times 10^6)^2 = 2.41 \times 10^{-17} \text{ J} = \underline{151 \text{ eV}}$$

38.57 It is proposed to send a beam of electrons through a diffraction grating with the distance between slits being d. The electrons have a speed of 400 m/s. How large must d be if a strong beam of electrons is to emerge at an angle of 25° to the straight-through beam?

▌ The first-order maximum is at $\lambda = d \sin \theta = d \sin 25°$. From the de Broglie's relation, $h/(mv) = \lambda = (6.63 \times 10^{-34})/(9.1 \times 10^{-31} \times 400)$, or $\lambda = 1.82 \ \mu\text{m}$. Then $d = \lambda/(\sin 25°) = \underline{4.3 \ \mu\text{m}}$. If the maximum were nth order, then $n\lambda = d \sin \theta$ and $d = n(4.3 \ \mu\text{m})$.

38.58 Rock salt forms a cubic lattice with the sides of each cube having a length $a = 5.63$ Å. An electron beam is incident on a rock salt crystal as shown in Fig. 38-3(a). (a) What is the spacing between the lattice planes indicated by the dashed lines? (b) Through what smallest potential should the electrons be accelerated if they are to be reflected strongly off the planes indicated by the dashed lines?

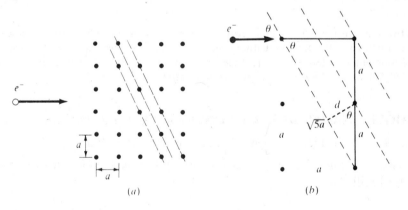

Fig. 38-3

▌ (a) From the geometry of Fig. 38-3(b),

$$\frac{d}{a} = \cos \theta = \frac{1}{\sqrt{5}} \qquad \text{so} \qquad d = \frac{a}{\sqrt{5}} = 2.518 \text{ Å}$$

(b) From the Bragg relation $n\lambda = 2d \sin \theta$, the longest wavelength is $\lambda = 2(2.518 \times 10^{-10})(2/5^{1/2}) = 4.5$ Å. This corresponds to an energy $E = h^2/(2m\lambda^2) = 1.2 \times 10^{-18}$ J $= 7.44$ eV, so the potential difference is $\underline{7.44 \text{ V}}$.

38.59 The average kinetic energy of a free electron in a metal is $(3kT)/2$ at high temperatures. (a) At what temperature would the electron's average de Broglie wavelength be 0.5 nm? (b) Repeat for a helium atom that has an energy $(3kT)/2$ and a mass $4m_p$.

▌ In both cases $(mv^2)/2 = p^2/(2m) = (3kT)/2$. Solving for T and setting $p = h/\lambda$ yields $h^2/(3k\lambda^2 m)$. After

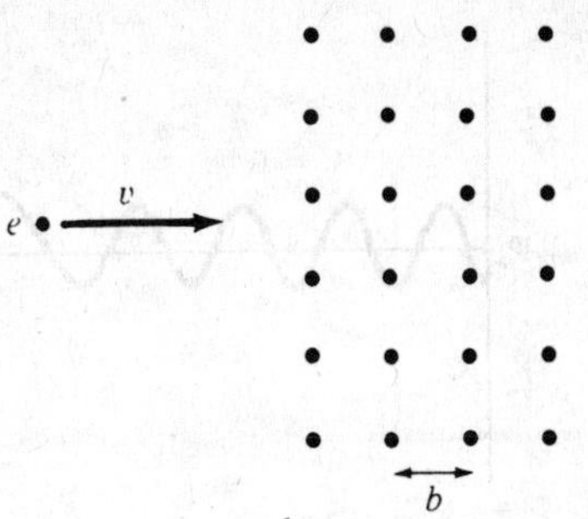

Fig. 38-4

substitution, $T = 4.25 \times 10^{-26}/m$. Using the electron mass and the helium mass $(4 \times 1.67 \times 10^{-27}$ kg), we have $T_e = \underline{46\,600\,K}$ and $T_{He} = \underline{6.3\,K}$.

38.60 Consider a beam of electrons shot toward a crystal, as shown in Fig. 38-4. The crystal spacing is b, as indicated. For what de Broglie wavelengths will the electron beam be strongly reflected straight back upon itself? For what electron kinetic energies? It is found by experiment that electrons having these energies are unable to move through such a crystal in the direction shown. Evaluate the energies in eV for $b = 2 \times 10^{-10}$ m.

❚ Strong reflection will occur if $2b = n\lambda$, because the beams reflected from successive planes will then reinforce. Therefore, $\lambda = (2b)/n$. For nonrelativistic electrons $E_n = (p_n^2)/(2m) = h^2/(2m\lambda_n^2) = (n^2 h^2)/(8b^2 m)$, which gives $E_n = \underline{9.4n^2\,eV}$.

38.61 The nuclei of atoms have radii of order 10^{-15} m. Consider a hypothetical situation of a proton confined to a narrow tube with length 2×10^{-15} m. What will be the de Broglie wavelengths which will resonate in the tube? To what momentum does the longest wavelength correspond? If relativistic effects are assumed negligible, to what energy (in eV) does this correspond?

❚ This is a simple standing-wave problem. For symmetric boundary conditions, the allowed λ's are $\lambda_n = (2L)/n = [(4 \times 10^{-15})/n]$ m; since $\lambda_n = h/p_n$, $p_n = (nh)/2L = n(1.66 \times 10^{-19})$. Nonrelativistic kinetic energy $= E_n = p_n^2/(2m) = n^2(8.2 \times 10^{-12}\,J)$ or $\underline{n^2(51\,Mev)}$. $n = 1$ corresponds to longest wavelength.

38.62 At what energy will the nonrelativistic calculation of the de Broglie wavelength of an electron be in error by 5 percent?

❚ For the nonrelativistic case, the de Broglie wavelength is

$$\lambda_{nr} = \frac{hc}{pc} = \frac{hc}{\sqrt{2m_0 c^2 K}}$$

For the relativistic case,

$$(K + m_0 c^2)^2 = (pc)^2 + (m_0 c^2)^2 \quad \text{or} \quad pc = \left[2m_0 c^2 K\left(1 + \frac{K}{2m_0 c^2}\right)\right]^{1/2}$$

and the de Broglie wavelength is

$$\lambda_r = \frac{hc}{pc} = \frac{hc}{\{2m_0 c^2 K[1 + K/(2m_0 c^2)]\}^{1/2}}$$

For our case, $\lambda_{nr} - \lambda_r = 0.05\lambda_r$; $\lambda_{nr}/\lambda_r = 1.05$.

$$\frac{\lambda_{nr}}{\lambda_r} = \sqrt{1 + \frac{K}{2m_0 c^2}} \qquad 1.05 = \sqrt{1 + \frac{K}{2(0.511\,MeV)}}$$

Solving, $K = \underline{0.105\,MeV}$.

38.63 Show that the de Broglie wavelength of a particle is approximately the same as that of a photon with the same energy, when the energy of the particle is much greater than its rest energy.

❚ $\qquad E^2 = p^2 c^2 + E_0^2 \quad \text{or} \quad p = \frac{E}{c}\sqrt{1 - \left(\frac{E_0}{E}\right)^2} \approx \frac{E}{c} \quad \text{if} \quad E \gg E_0. \quad \text{So} \quad \lambda = \frac{h}{p} \approx \frac{hc}{E}$

For a photon, $E = h\nu = (hc)/\lambda_\gamma$, so $\lambda_\gamma = (hc)/E \approx \lambda$.

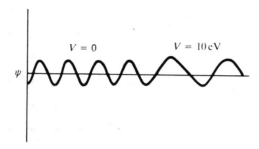

$V = 0$ $V = 10\,eV$

ψ

Fig. 38-5

38.64 An electron is confined to a tube of length L. The electron's potential energy in one half of the tube is zero, while the potential energy in the other half is 10 eV. If the electron has a total energy $E = 15$ eV, what will be the ratio of the de Broglie wavelength of the electron in the 10-eV region of the tube to that in the other half of the tube? Sketch the wave function ψ for the electron along the tube.

▌ The wave function is sketched in Fig. 38-5. $\lambda = h/p = h/(2mK)^{1/2}$; with K the kinetic energy of the electron, $\lambda_1/\lambda_2 = (K_2/K_1)^{1/2} = (15/5)^{1/2} = \underline{1.73}$.

38.65 A particle of mass m is confined to a one-dimensional line of length L. From arguments based on the wave interpretation of matter, show that the energy of the particle can have only discrete values and determine these values.

▌ If the particle is confined to a line segment, say from $x = 0$ to $x = L$, the probability of finding the particle outside this region must be zero. Therefore, the wave function ψ must be zero for $x \leq 0$ or $x \geq L$, since the square of ψ gives the probability for finding the particle at a certain location. Inside the limited region the wavelength of ψ must be such that ψ vanishes at the boundaries $x = 0$ and $x = L$, so that it can vary continuously to the outside region. Hence only those wavelengths will be possible for which an integral number of half wavelengths fit between $x = 0$ and $x = L$, that is, $L = (n\lambda)/2$, where n is an integer, called the *quantum number*, with values $n = 1, 2, 3, \ldots$. From the de Broglie relationship $\lambda = h/p$ we then find that the particle's momentum can have only discrete values given by

$$p = \frac{h}{\lambda} = \frac{nh}{2L}$$

Since the particle is not acted upon by any forces inside the region, its potential energy will be a constant which we set equal to zero. Therefore the energy of the body is entirely kinetic and will have the discrete values obtained from

$$E = K = \frac{1}{2}mv^2 = \frac{p^2}{2m} = \frac{[(nh)/(2L)]^2}{2m} \quad \text{that is,} \quad E_n = n^2\frac{h^2}{8mL^2} \quad n = 1, 2, 3, \ldots$$

This very simple problem illustrates one of the basic features of the probability interpretation of matter; namely, that the energy of a bound system can take on only discrete values, with zero energy not being a possible value.

38.66 Assume that the uncertainty in the position of a particle is equal to its de Broglie wavelength. Show that the uncertainty in its velocity is equal to or greater than $1/(4\pi)$ times its velocity.

▌ Use the Heisenberg uncertainty principle with $\Delta x = \lambda = h/(mv_x)$.

$$\Delta x\, \Delta(mv_x) \geq \frac{h}{4\pi} \qquad \frac{h}{mv_x}\Delta(mv_x) \geq \frac{h}{4\pi}$$

Since m is constant,

$$\frac{h}{mv_x}(m\,\Delta v_x) \geq \frac{h}{4\pi} \qquad \frac{h}{v_x}\Delta v_x \geq \frac{h}{4\pi} \qquad \Delta v_x \geq \frac{1}{4\pi}v_x$$

38.67 If the uncertainty in the time during which an electron remains in an excited state is 10^{-7} s, what is the least uncertainty (in J) in the energy of the excited state?

▌ Let W be the energy of the excited state.

$$\Delta W\,\Delta t \geq \frac{h}{4\pi} \qquad (\Delta W)(10^{-7}) \geq \frac{6.63 \times 10^{-34}}{4\pi} \qquad \Delta W \geq \frac{6.63 \times 10^{-27}}{4\pi} \geq \underline{0.528 \times 10^{-27}\,J}$$

38.68 A golf ball has a mass of 50 g and travels horizontally with a velocity of 80 m/s. What limit of uncertainty is placed on a measurement of its position by the Heisenberg uncertainty principle, if the uncertainty in measuring the velocity is 0.01 m/s? Assume that the uncertainty in the mass is negligible.

▌ We must have

$$\Delta x \, \Delta(mv_x) \geq \frac{h}{4\pi}$$

Since m is constant,

$$(\Delta x)(m \, \Delta v_x) \geq \frac{h}{4\pi} \quad (\Delta x)(0.050)(0.01) \geq \frac{6.63 \times 10^{-34}}{4\pi} \quad \Delta x \geq \underline{1.06 \times 10^{-31} \text{ m}}$$

For large objects the uncertainty principle imposes no practical limit on position measurements.

38.69ᶜ The radius of a typical atomic nucleus is about 5×10^{-15} m. Assuming the position uncertainty of a proton in the nucleus to be 5×10^{-15} m, what will be the smallest uncertainty in the proton's momentum? In its energy in eV?

▌ The condition $\Delta x \, \Delta p = h/(4\pi)$ yields $\Delta p = h/[4\pi(5 \times 10^{-15})] = 1.05 \times 10^{-20}$ kg · m/s. Before writing down an energy relation note that if $\Delta p = m_0 v$, we find $v = \Delta p/(1.67 \times 10^{-27}) = 6.3 \times 10^6$ m/s, so the proton is nonrelativistic. Then $E = p^2/(2m)$ and $dE = (2p \, dp)/2m$; p cannot be smaller than Δp, ΔE must be no less than $\Delta p^2/m = 6.6 \times 10^{-14}$ J, or $\underline{412 \text{ keV}}$.

38.70 If Planck's constant were $h' = 660$ J · s, what would be the de Broglie wavelength of a 100-kg football player running at 5 m/s? Determine the least uncertainty of his location according to an opposing player, if the uncertainty in his momentum equaled his momentum.

▌ We have $\lambda = h'/(mv) = 660/[100(5)] = \underline{1.32 \text{ m}}$. For uncertainty, $\Delta p \, \Delta x = h'/4\pi$, so $\Delta x = 660/[4\pi(500)] = \underline{10 \text{ cm}}$.

38.71 From the relation $\Delta p \, \Delta x \geq h/(4\pi)$, show that for a particle moving in a circle, $\Delta L \, \Delta \theta \geq h/(4\pi)$. The quantity ΔL is the uncertainty in the angular momentum and $\Delta \theta$ is the uncertainty in the angle.

▌ Since the particle moves in a circle, the uncertainty principle will apply to directions tangent to the circle. Thus,

$$\Delta p_s \, \Delta s \geq \frac{h}{4\pi}$$

where s is measured along the circumference of the circle. The angular momentum is related to the linear momentum by

$$L = mvR = p_s R$$

therefore $\Delta p_s = \Delta L/R$. The angular displacement is related to the arc length by $\theta = s/R$; therefore $\Delta s = R \, \Delta \theta$. Hence

$$\Delta p_s \, \Delta s = (\Delta L/R)(R \, \Delta \theta) = \Delta L \, \Delta \theta \geq h/(4\pi)$$

CHAPTER 39
Modern Physics: Atoms, Nuclei, Solid-State Electronics

39.1 ATOMS AND MOLECULES

39.1 Sodium atoms emit a spectral line with a wavelength in the yellow, 589.6 nm. What is the difference in energy between the two energy levels involved in the emission of this spectral line?

▮ $$\Delta E = hf = \frac{hc}{\lambda} = \frac{1240\,\text{eV} \cdot \text{nm}}{589.6\,\text{nm}} = \underline{2.1\,\text{eV}}$$

39.2 A certain molecule has an energy-level diagram for its vibrational energy in which two levels are 0.0141 eV apart. Find the wavelength of the emitted line for the molecule as it falls from one of these levels to the other.

▮ $$\Delta E = \frac{hc}{\lambda} \quad \text{so} \quad \lambda = \frac{1.240\,\text{eV} \cdot \mu\text{m}}{0.0141\,\text{eV}} = \underline{88\,\mu\text{m}}$$

39.3 x-rays are emitted as the electrons deep within atoms having many electrons fall to lower energy states. What is the difference in energy between two levels if a transition between them gives rise to 0.5 Å x-rays?

▮ $$\Delta E = \frac{hc}{\lambda} = \frac{12.4\,\text{keV} \cdot \text{Å}}{0.5\,\text{Å}} = \underline{24.8\,\text{keV}}$$

39.4 Given that the value of the Rydberg constant R (for hydrogen) is about $1.097 \times 10^7\,\text{m}^{-1}$, find the wavelength of the first line of the Balmer series.

▮ The first line of the Balmer series arises from an electron transition from the $n = 3$ level to the $n = 2$ level of the hydrogen atom.

$$\frac{1}{\lambda} = R\left(\frac{1}{n_B^2} - \frac{1}{n_A^2}\right) = (1.097 \times 10^7)\left(\frac{1}{2^2} - \frac{1}{3^2}\right) = 1.097(0.1389) \times 10^7 \qquad \lambda = \underline{656.3\,\text{nm}}$$

39.5 Given that the ionization energy of the hydrogen atom is 13.6 eV, calculate the wavelength of the first line of the Lyman series.

▮ The energy of a state of the hydrogen atom with quantum number n is

$$E_n = -\frac{13.6}{n^2} \quad \text{eV}$$

The first line of the Lyman series arises from a transition from the $n = 2$ state to the $n = 1$ state. The energy of this photon is

$$E = E_2 - E_1 = -\frac{13.6}{2^2} + \frac{13.6}{1^2} \qquad E = E_2 - E_1 = -\frac{13.6}{4} + 13.6 = 10.2\,\text{eV}$$

Now use Planck's equation to find λ:

$$\lambda = \frac{hc}{E} = \frac{1240\,\text{eV} \cdot \text{nm}}{10.2\,\text{eV}} = \underline{122\,\text{nm}}$$

39.6 If the ionization energy for a hydrogen atom is 13.6 eV, what is the energy of the level with quantum number $n = 3$?

▮ The energy of a quantum state for H is

$$E_n = -\frac{13.6}{n^2} \quad \text{eV} \qquad \text{For} \qquad n = 3, \qquad E_3 = -\frac{13.6}{3^2} = \underline{-1.51\,\text{eV}}$$

39.7 The series limit wavelength of the Balmer series is emitted as the hydrogen atom falls from the $n = \infty$ state to the $n = 2$ state. What is the wavelength of this line?

▌ $E_n = \dfrac{-13.6\,\text{eV}}{n^2}$ $E_2 = \dfrac{-13.6\,\text{eV}}{2^2} = -3.4\,\text{eV}$ $E_\infty = 0$ Then $\Delta E = E_\infty - E_2 = 0 - (-3.4\,\text{eV}) = 3.4\,\text{eV}$

We find the corresponding wavelength in the usual way from $\Delta E = (hc)/\lambda$. The result is <u>365 nm</u>.

39.8 When a hydrogen atom is bombarded, the atom is excited to its higher energy states. As it falls back to the lower energy levels, light is emitted. What are the three longest-wavelength spectral lines emitted by the hydrogen atom as it falls to the $n = 1$ state from higher energy states?

▌ We are interested in the following transitions

$$n = 2 \to n = 1 \qquad \Delta E_{2,1} = -3.4 - (-13.6) = 10.2\,\text{eV}$$
$$n = 3 \to n = 1 \qquad \Delta E_{3,1} = -1.5 - (-13.6) = 12.1\,\text{eV}$$
$$n = 4 \to n = 1 \qquad \Delta E_{4,1} = -0.85 - (-13.6) = 12.75\,\text{eV}$$

To find the corresponding wavelengths we can use $\Delta E = hf = (hc)/\lambda$. For example, for the $n = 2$ to $n = 1$ transition,

$$\lambda = \frac{hc}{\Delta E_{2,1}} = \frac{1240\,\text{eV} \cdot \text{nm}}{10.2\,\text{eV}} = \underline{122\,\text{nm}} \qquad (\text{cf. Prob. 39.5})$$

The other lines are found in the same way to be <u>102 nm</u> and <u>97 nm</u>. These are the first three lines of the Lyman series.

39.9 What is the radiation of greatest wavelength that will ionize unexcited hydrogen atoms?

▌ The incident photons must have enough energy to raise the atom from the $n = 1$ level to the $n = \infty$ level when absorbed by the atom. Because $E_\infty - E_1 = 13.6\,\text{eV}$, we can use $E_\infty - E_1 = (hc)/\lambda$ to find the wavelength as 91 nm. Wavelengths shorter than this would not only remove the electron from the atom but would give the removed electron KE.

39.10 (*a*) Compute the first three energy levels of doubly ionized lithium. What is its ionization potential? (*b*) Find the wavelengths of the first three lines of the K series ($n_l = 1$; $n_h = 2, 3, 4$).

▌ (*a*) With $Z = 3$ the Bohr energy formula becomes

$$E_n = -\frac{122.40\,\text{eV}}{n^2} \tag{1}$$

Hence $E_1 = -122.40\,\text{eV}$ (ionization potential is 122.40 V) and

$$E_2 = -30.60\,\text{eV} \qquad E_3 = -13.60\,\text{eV} \qquad E_4 = -7.65\,\text{eV}$$

(*b*) According to (1), the wavelengths for doubly ionized lithium ($Z = 3$) will be $\frac{1}{9}$ the corresponding wavelengths for hydrogen (Prob. 39.8): $\lambda_1 = 13.5\,\text{nm}$, $\lambda_2 = 11.3\,\text{nm}$, $\lambda_3 = 10.8\,\text{nm}$.

39.11 The energy levels for singly ionized helium atoms (atoms from which one of the two electrons has been removed) are given by $E_n = (-54.4/n^2)\,\text{eV}$. Construct the energy-level diagram for this system.

▌ See Fig. 39-1.

39.12 Suppose that means were available for stripping 28 electrons from $_{29}$Cu in a vapor of this metal. (*a*) Compute the first three energy levels for the remaining electron. (*b*) Find the wavelengths of the spectral lines of the series for which $n_l = 1$; $n_h = 2, 3, 4$. What is the ionization potential for the last electron?

▌ (*a*) Here $Z = 29$ so the allowed energies will be $29^2 = 841$ times the corresponding energies for $_1$H:

$$E_1 = 841(-13.60) = \underline{-11.438\,\text{keV}} \qquad E_2 = \frac{E_1}{4} = \underline{-2.859\,\text{keV}} \qquad E_3 = \frac{E_1}{9} = \underline{-1.271\,\text{keV}}.$$

(*b*) Using $\Delta E = E_n - E_1 = (hc)/\lambda = (12.40\,\text{keV} \cdot \text{Å})/\lambda$,

$$\lambda_1 = \underline{1.44\,\text{Å}} \qquad \lambda_2 = \underline{1.22\,\text{Å}} \qquad \lambda_3 = \underline{1.15\,\text{Å}} \qquad \lambda_\infty = \underline{1.08\,\text{Å}}$$

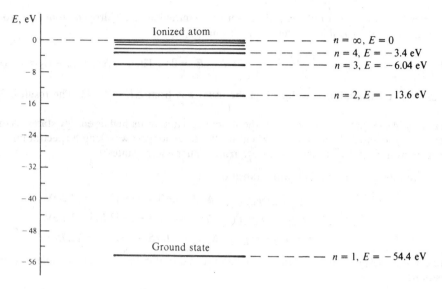

Fig. 39-1

These wavelengths are in the x-ray region. The ionization potential is the voltage corresponding to λ_∞.

$$\frac{hc}{\lambda_\infty} = \frac{12.40 \text{ keV} \cdot \text{Å}}{1.08 \text{ Å}} = \underline{11.5 \text{ keV}}$$

that is, an ionization potential of $\underline{11.5 \text{ kV}}$.

39.13 An electron revolves in a circle around a nucleus with positive charge Ze. How is the electron's velocity related to the radius of its orbit?

❚ Equating the Coulomb force to the (electron mass) (centripetal acceleration),

$$\frac{k(e)(Ze)}{r^2} = \frac{mv^2}{r} \quad \text{or} \quad v^2 = \frac{kZe^2}{mr} \quad \text{where} \quad k = \frac{1}{4\pi\epsilon_0}$$

In the above we have assumed that the nucleus is stationary, i.e., its mass is effectively infinite.

39.14 How is the total energy of the electron in Prob. 39.13 related to the radius of its orbit?

❚ The electric potential energy of the electron is

$$U = qV = (-e)V = -e\frac{k(Ze)}{r} = -\frac{kZe^2}{r}$$

The kinetic energy of the electron is found by using the result of Prob. 39.13:

$$K = \frac{1}{2}mv^2 = \frac{1}{2}m\frac{kZe^2}{mr} = \frac{kZe^2}{2r}$$

The total energy is then $\quad E = K + U = \frac{kZe^2}{2r} - \frac{kZe^2}{r} = -\frac{kZe^2}{2r} = \frac{1}{2}U$

39.15 Show how the Bohr postulate that orbital angular momentum $L = mvr$ be an integral multiple of $h/(2\pi)$ leads to quantization of orbital radius, velocity, and energy.

❚ $$mv_n r_n = \frac{nh}{2\pi} \quad \text{or} \quad m^2 v_n^2 r_n^2 = \frac{n^2 h^2}{4\pi^2} \qquad (1, 2)$$

From Prob. 39.13 $v_n^2 = (kZe^2)/(mr_n)$ and eliminating v_n^2 in (1), $mkZe^2 r_n = (n^2 h^2)/(4\pi^2)$. Then

$$r_n = \frac{n^2 h^2}{4\pi^2 mkZe^2} \qquad (3)$$

and $$v_n = \left(\frac{4\pi^2 k^2 Z^2 e^4}{n^2 h^2}\right)^{1/2} = \frac{2\pi kZe^2}{nh} \qquad (4)$$

From Prob 39.14,
$$E_n = -\frac{kZe^2}{2r_n} = -\frac{2\pi^2 mk^2 Z^2 e^4}{n^2 h^2} = -\frac{mZ^2 e^4}{8\epsilon_0^2 n^2 h^2} \tag{5}$$

In terms of $n = 1$ values for hydrogen ($Z = 1$), subscripted below with a zero

$$r_n = \frac{n^2 r_0}{Z} \qquad r_0 = \frac{h^2}{4\pi^2 mke^2} \tag{6}$$

$$v_n = v_0 \frac{Z}{n} \qquad v_0 = \frac{2\pi ke^2}{h} = \frac{e^2}{2h\epsilon_0} \tag{7}$$

$$E_n = -\frac{Z^2}{n^2} E_0 \qquad E_0 = \frac{-me^4}{8\epsilon_0^2 h^2} = -hcR_\infty \tag{8}$$

where R_∞ is the Rydberg constant for an infinitely massive nucleus.

39.16   Calculate the radius r of the third electron orbit of the hydrogen atom.

▮ From Eq. (3) of Prob. 39.15, using $k = 1/4\pi\epsilon_0$,

$$r_n = \frac{\epsilon_0 n^2 h^2}{\pi me^2} = \frac{(8.85 \times 10^{-12})(3^2)(6.63 \times 10^{-34})^2}{\pi(9.11 \times 10^{-31})(1.60 \times 10^{-19})^2} = \underline{0.478 \text{ nm}}$$

39.17   Find the radius of the second Bohr orbit of the singly ionized helium atom and calculate the velocity of an electron in this orbit.

▮ By (3) of Prob. 39.15, the allowed radii of the Bohr orbits in He$^+$ are

$$r_n = \frac{\epsilon_0 n^2 h^2}{\pi me^2 Z}$$

For the second orbit, $n = 2$ and for He, $Z = 2$.

$$r_2 = \frac{(8.85 \times 10^{-12})(2^2)(6.63 \times 10^{-34})^2}{\pi(9.11 \times 10^{-31})(1.60 \times 10^{-19})^2(2)} = \underline{0.106 \text{ nm}}$$

To find the velocity v, use the Bohr quantum condition $mvr = (nh)/(2\pi)$:

$$mvr_2 = \frac{2h}{2\pi} \qquad v = \frac{h}{\pi m r_2} = \frac{6.63 \times 10^{-34}}{\pi(9.11 \times 10^{-31})(1.06 \times 10^{-10})} = \underline{2.19 \times 10^6 \text{ m/s}}.$$

39.18   The combination of fundamental quantities $e^2/(2\epsilon_0 hc) \equiv \alpha$ is known as the *fine structure constant*. Show that α is a pure number, of magnitude $\frac{1}{137}$.

▮ Unit-dimensionally,

$$\alpha = \left[\frac{C^2}{(F/m)(J \cdot s)(m/s)}\right] = \left[\frac{C^2}{F \cdot J}\right] = \left[\frac{C/F}{J/C}\right] = \left[\frac{V}{V}\right] = [1]$$

Numerically, $$\frac{1}{\alpha} = \frac{2(8.85432 \times 10^{-12})(6.6262 \times 10^{-34})(2.997925 \times 10^8)}{(1.6021 \times 10^{-19})^2} = 137.05$$

39.19   Express the electron velocity in the ground state of H in terms of the fine-structure constant.

▮ From Prob. 39.18 and (7) of Prob. 39.15, $v_0 = \alpha c$.

39.20   For hydrogen, show that when $n \gg 1$ the frequency of the emitted photon in a transition from n to $n-1$ equals the rotational frequency in state n.

▮ The rotational frequency in state n is (see Prob. 39.15)

$$\frac{\omega}{2\pi} = \frac{v_n}{2\pi r_n} = \frac{(2\pi ke^2)/(nh)}{(2\pi n^2 h^2)/(4\pi^2 kme^2)} = \frac{4\pi^2 k^2 me^4}{n^3 h^3}$$

The frequency of the emitted photon is

$$v = c\frac{1}{\lambda} = cR_\infty\left[\frac{1}{(n-1)^2} - \frac{1}{n^2}\right] = cR_\infty \frac{2n-1}{n^2(n-1)^2}$$

For $n \gg 1$,

$$\nu \approx cR_\infty \frac{2n}{n^2 n^2} = c \frac{2\pi^2 k^2 e^4 m}{h^3 c} \frac{2}{n^3} = \frac{4\pi^2 k^2 m e^4}{n^3 h^3}$$

which is the same as the rotational frequency given above. [According to classical theory, radiation emitted from a rotating charge will have a frequency equal to the rotational frequency.]

39.21 A gas of monatomic hydrogen is bombarded with a stream of electrons that have been accelerated from rest through a potential difference of 12.75 V. Which spectral lines should be emitted?

▌ If an atom absorbs 12.75 eV of energy, it will be lifted from the ground level ($n = 1$) to the fourth level ($n = 4$). Hence, in falling back to the ground level there is the possibility of exciting the first three Lyman, the first two Balmer, and the first Paschen lines.

39.22 A continuous band of radiation having all wavelengths from about 1000 to 10 000 Å is passed through a gas of monatomic hydrogen. The radiation is then examined with a spectrograph. What results are found?

▌ The wavelengths absorbed by the hydrogen atoms in the range 1000 to 10 000 Å are 1215 and 1025.7 Å in the Lyman series, the entire Balmer series, and all but the first four lines in the Paschen series. Since the gas will be primarily in the ground state, only the Lyman lines will be significantly depleted in the spectrograph.

39.23 Using the reduced mass of $^1_1$H, compute the radius of (a) the smallest Bohr orbit and (b) the fifth Bohr orbit.

▌ When we take account of the finite mass of the nucleus, M, the equations of Prob. 39.15 must be modified by replacing everywhere the electron mass m by the *reduced mass*

$$m' = \frac{mM}{m + M} = \frac{m}{1 + (m/M)}$$

Then letting r_1^0 be the ideal and r_{1H} the true hydrogen first radius we have (see Prob. 39.15):

(a) $$r_{1H} = \left(1 + \frac{m}{M}\right) r_1^0 = \left(1 + \frac{1}{1836}\right)(0.5293) = \underline{0.5296 \text{ Å}}$$

(b) $$r_5 = 25 r_{1H} = \underline{13.24 \text{ Å}}$$

39.24 Compute the electron's speed and angular momentum in the two orbits of Prob. 39.23.

▌ (a) Total angular momentum, L_n, depends only on the quantum number, n. Thus

$$L_1 = 1\left(\frac{h}{2\pi}\right) = 1.055 \times 10^{-34} \text{ J} \cdot \text{s}$$

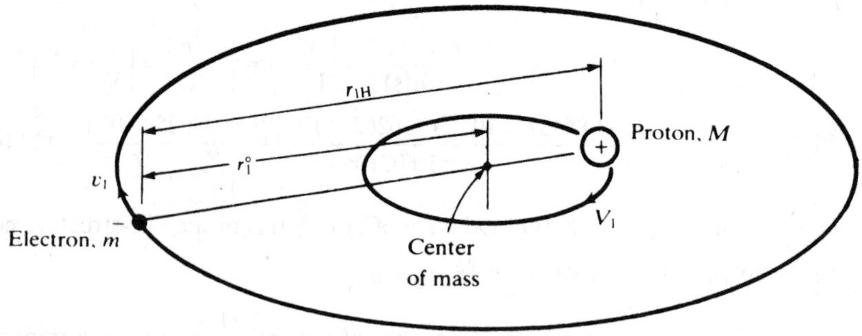

Fig. 39-2

But, from Fig. 39-2, and the definition of the center of mass,

$$L_1 = m v_1 r_1^0 + M V_1 (r_{1H} - r_1^0) = m v_1 r_{1H} = m v_1 \left(1 + \frac{m}{M}\right) r_1^0$$

Hence, $$v_1 = \frac{L_1}{m[1 + (m/M)] r_1^0} = \frac{h/2\pi}{m[1 + (m/M)][(\epsilon_0 h^2)/(\pi m e^2)]} = \frac{v_1^0}{1 + (m/M)} = \frac{c/137.0}{1 + (1/1836)} = \underline{2189 \text{ km/s}}$$

(b) $$L_5 = 5L_1 = 5.277 \times 10^{-34} \text{ J} \cdot \text{s} \qquad v_5 = \frac{v_1}{5} = \underline{437.8 \text{ km/s}}$$

39.25 Find the values of the first four energy levels of ${}_1^1\text{H}$ including the reduced-mass correction.

$$E_1 = \frac{m'}{m}E_1^0 = \frac{E_1^0}{1+(m/M)} = -\frac{13.60}{1+(1/1836)} = -13.598 \text{ eV} \qquad E_3 = \frac{E_1}{9} = -1.5109 \text{ eV}$$

$$E_2 = \frac{E_1}{4} = -3.3995 \text{ eV} \qquad E_4 = \frac{E_1}{16} = -0.8499 \text{ eV}$$

39.26 Determine the mass ratio of deuterium and hydrogen if, respectively, their H_α lines have wavelengths of 6561.01 and 6562.80 Å.

▮ In terms of the reduced mass of the atom, the Rydberg formula is

$$\frac{1}{\lambda} = \frac{R_\infty Z^2}{1+(m/M)}\left(\frac{1}{n_l^2} - \frac{1}{n_u^2}\right)$$

Thus, $\lambda_H/\lambda_D = [1+(m/M_H)]/[1+(m/M_D)] \approx 1 + (m/M_H) - m/M_D$ or $0.0002728 = m/M_H - m/M_D = [(M_D-M_H)/M_D](m/M_H)$. Noting that $m/M_H = 1/1836$, we get

$$0.5009 = 1 - \frac{M_H}{M_D} \quad \text{and} \quad \frac{M_H}{M_D} = \underline{0.499}.$$

39.27 An electron and its antiparticle, the positron, can form a bound system, called *positronium*. What should be the ionization potential of positronium?

▮ The positron has the same mass as the electron and the charge of the proton, so the Bohr theory predicts the same energy levels as hydrogen, modified by the reduced mass. In this case $m' = m^2/2m = m/2$, and since the energy levels are proportional to the mass, the energy levels of positronium are half the value as for hydrogen. Thus the ionization potential is $\frac{1}{2}(13.6 \text{ V}) = \underline{6.8 \text{ V}}$.

39.28 Estimate the wavelength of x-rays a tungsten atom should emit in a transition from $n=2$ to $n=1$. The value found from experiment turns out to be 0.021 nm. ($Z=74$).

▮ Use hydrogenlike energies $E_n = (-13.6Z^2)/n^2$ to estimate the electron energy in the $n=1$ and 2 levels. Use $Z = 74$ for tungsten. Since $E_2 - E_1 = \frac{3}{4}(7.5 \times 10^4) = 5.59 \times 10^4$ eV, and since $(hc)/\lambda = E_2 - E_1$, we find $\lambda = \underline{0.022 \text{ nm}}$.

39.29 Estimate the energy needed to eject from a lead atom the electron with $n=1$. What wavelength x-rays would be required to do this? ($Z=82$).

▮ For $Z=82$ use $E_n = (-13.6Z^2)/n^2$ to estimate the binding energy of a $n=1$ electron; it is 91 keV. This corresponds to an x-ray of $\lambda = 1240/(91 \times 10^3) = \underline{0.0136 \text{ nm}}$.

39.30 *Moseley's law* states that the square root of the $K\alpha$ x-ray frequency should be proportional to the atomic number Z of the element from which the x-rays come. Justify this result in terms of the hydrogenlike energy levels.

▮ Start with $(hc)/\lambda = E_2 - E_1 = (\text{const})Z^2(\frac{1}{1} - \frac{1}{4})$, but $c/\lambda = f$, so $f^{1/2}$ depends linearly on Z.

39.31 X-rays of wavelength 1.37 nm incident on an atom cause photoemission of its electrons. If the emitted electrons have an energy of 83 eV, what is the energy of the level from which the electrons were ejected?

▮ Using $E = (hc)/\lambda$ for the x-ray, we find that the x-ray energy $= 1240/1.37 = 905$ eV. The binding energy for the electron in question is $905 - 83 = 822$ eV; the electron energy level is $\underline{-822 \text{ eV}}$.

39.32 Suppose that the electron were to have no spin so there would be no spin quantum number. List Z for the first three univalent elements if the Pauli exclusion principle still held.

▮ In the $n=1$ level only 1 electron can exist ($n=1$, $l=0$, $m_l=0$). For the $n=2$ level we can have four electrons ($n=2$, $l=0$, $m_l=0$; $n=2$, $l=1$, $m_l=0$, ±1). The sixth electron would be in the $n=3$ level. The first three univalent elements would be 1, 2. and 6. The $n=3$ level could have nine electrons ($n=3$, $l=0$, $m_l=0$; $l=1$, $m_l=0$, ±1; $l=2$, $m_l=0$, ±1, ±2) so that $Z=15$ would be the fourth univalent element in this hypothetical case.

39.33 The two nitrogen atoms in a nitrogen molecule act much like two equal masses connected by a spring. The molecule's natural vibrational frequency is 7×10^{13} Hz. What is the difference in energy between its allowed states of vibration? In order to have this much energy, through how large a potential difference would an electron have to fall?

▮ In the quantized harmonic oscillator the energies are $(n + \frac{1}{2})hf_0$, with gaps $hf_0 = (6.63 \times 10^{-34})(7 \times 10^{13}) = 4.64 \times 10^{-20}$ J $= \underline{0.29\text{ eV}}$. Hence, $V = \underline{0.29\text{ V}}$.

39.34 The natural frequency of vibration for the CO molecule as the springlike bond between its two atoms stretches and compresses is 6.5×10^{13} Hz. With what energies (in eV) can this molecule vibrate?

▮ Energies $= (n + \frac{1}{2})hf_0 = (n + \frac{1}{2})(6.63 \times 10^{-34})(6.5 \times 10^{13}) = (n + \frac{1}{2})(4.31 \times 10^{-20})$ J $= \underline{0.27(n + \frac{1}{2})\ \text{eV}}$.

39.35 The spring constant for the stretching vibration of the H_2 molecule is 570 N/m. What are the allowed vibrational energies for this molecule? Thermal energy at room temperature is about $\frac{1}{40}$ eV. Find the ratio of the thermal energy to the gap energy between vibrational levels.

▮ We use the frequency formula for a harmonic oscillator of force constant k; and since both masses are in motion, we employ the reduced mass $\mu = m/2$: $2\pi f_c = (k/\mu)^{1/2} = (2k/m)^{1/2}$. Then $f_c = 1.32 \times 10^{14}$ Hz, and $E_n = (n + \frac{1}{2})\ hf_c = (n + \frac{1}{2})(8.7 \times 10^{-20})$J $= \underline{(n + \frac{1}{2})(0.55)\text{ eV}}$. The required ratio is $0.025/0.55 = \underline{0.046}$.

39.36 The distance between nuclei in the N_2 molecule is 1.12 Å. Determine the moment of inertia of this molecule about its center. What are the rotational energy levels for this molecule? At about what level (that is, j value) will the rotational energy equal thermal energy, about $\frac{1}{40}$ eV?

▮ We have $I = \sum mr^2 = 2[14(1.67 \times 10^{-27})](0.56 \times 10^{-10})^2 = 1.47 \times 10^{-46}$ kg $\cdot$ m^2. The quantized rotational energy formula is $E_{rot} = [h^2/(8\pi^2 I)]j(j + 1) = 3.8 \times 10^{-23}j(j + 1)$ J $= 2.4 \times 10^{-4}j(j + 1)\ $ eV. We want 2.4×10^{-4} $j(j + 1) = 1/40$, so $\underline{j = 10}$.

39.37 Compare the spacings of the vibrational and rotational energy levels in diatomic molecules.

▮ From Probs. 39.35 and 39.36, $\Delta E_{vib} \approx 1$ eV, whereas from Prob. 39.36, $\Delta E_{rot} \approx 10^{-3}$ eV (for low j values).

39.2 NUCLEI AND RADIOACTIVITY

39.38 What is the number of neutrons in the nucleus of $^{23}_{11}$Na?

▮ $A - Z = 23 - 11 = \underline{12}$.

39.39 How many protons, neutrons, and electrons are there in (a) ^{3}He, (b) ^{12}C, and (c) ^{206}Pb?

▮ (a) The atomic number of He is 2; therefore the nucleus must contain 2 protons. Since the mass number of this isotope is 3, the sum of the protons and neutrons in the nucleus must equal 3; therefore there is 1 neutron. The number of electrons in the atom is the same as the atomic number, 2.
(b) The atomic number of carbon is 6; hence the nucleus must contain 6 protons. The number of neutrons in the nucleus is equal to $12 - 6 = 6$. The number of electrons is the same as the atomic number, 6.
(c) The atomic number of lead is 82; hence there are 82 protons in the nucleus and 82 electrons in the atom. The number of neutrons is $206 - 82 = 124$.

39.40 Carbon has two stable isotopes. Natural carbon is 98.89% carbon 12 and 1.11% carbon 13. Calculate the average atomic weight of carbon. (The atomic weight of ^{12}C is exactly 12.)

▮ $$\text{average atomic weight} = \frac{0.9889(12) + 0.0111(13)}{1.000} = \underline{12.011}$$

39.41 The radius of a carbon nucleus is about 3×10^{-15} m and its mass is 12 u. Find the average density of the nuclear material. How many more times dense than water is this?

▮ $\rho = \dfrac{m}{V} = \dfrac{m}{(4\pi r^3)/3} = \dfrac{(12\text{ u})(1.66 \times 10^{-27}\text{ kg/u})}{4\pi(3 \times 10^{-15}\text{ m})^3/3} = \underline{1.8 \times 10^{17}\text{ kg/m}^3}$ $\dfrac{\rho}{\rho_{water}} = \dfrac{1.8 \times 10^{17}}{1000} = \underline{1.8 \times 10^{14}}$

39.42 How large is the largest nucleus?

▮ On the *liquid-drop model*, the radius, in fm, and the mass number are related through $r = 1.2\sqrt[3]{A}$. For the largest nuclei, A is something over 240; thus, $\sqrt[3]{240} = 6.2$ and $r \approx 7.4$ fm $\approx \underline{10^{-14}\text{ m}}$.

39.43 An α particle (He^{2+}) comes to within 80 fm of a gold nucleus (^{79}Au). Assuming that the gold nucleus and the α particle are point charges, find the maximum repulsive force.

▮ The charge on He^{2+} is $2e$ and on the ^{79}Au nucleus it is $79e$. Use Coulomb's law.

$$F = k\frac{Q_1Q_2}{r^2} = (9 \times 10^9)\frac{(2)(79)(1.6 \times 10^{-19})^2}{(8 \times 10^{-14})^2} = \underline{5.7\,N}$$

39.44 Show that if two ions of the same charge and velocity but of different mass pass through a uniform transverse magnetic field, the radii of the paths are proportional to the masses. (This is the principle of the *mass spectrograph*.) Find an expression for Δm if Δr is a change in radius.

▮ $mv^2/r = qvB$, so $r \propto m$; $\Delta m = (qB/v)\,\Delta r$.

39.45 Find the frequency of rotation of a proton in a cyclotron whose magnetic field is 1 T. The mass of the proton is 1.67×10^{-27} kg.

▮ $v = \omega/(2\pi) = v/(2\pi r)$. Then, from $r = (mv)/qB$ we get

$$v = \frac{Bq}{2\pi m} = \frac{1(1.60 \times 10^{-19})}{2\pi(1.67 \times 10^{-27})} = \underline{15.3\,MHz}$$

39.46 A proton synchrotron has a magnetic field of 1.1 T. When the speed of the accelerated protons is $0.800c$, what is the radius of the circular path they follow?

▮ At $0.800c$ the relativistic mass of the proton must be used.

$$m = \frac{m_0}{\sqrt{1-(v^2/c^2)}} = \frac{1.67 \times 10^{-27}}{\sqrt{1-[(0.800c)^2/c^2]}} = \frac{1.67 \times 10^{-27}}{\sqrt{1-0.64}} = \frac{1.67 \times 10^{-27}}{0.6} = 2.78 \times 10^{-27}\,kg$$

$$\frac{mv^2}{r} = qvB \quad so \quad r = \frac{mv}{qB} = \frac{2.78 \times 10^{-27}(0.8 \times 3 \times 10^8)}{1.60 \times 10^{-19}(1.1)} = \underline{3.79\,m}$$

39.47 Show that 1 u (atomic mass unit) is equivalent to 931.5 MeV of energy. ($1\,u = 1.6605 \times 10^{-27}$ kg.)

▮ $E = mc^2 = (1.6605 \times 10^{-27})(2.998 \times 10^8)^2 = 14.924 \times 10^{-11}\,J = \dfrac{14.924 \times 10^{-11}\,J}{1.6022 \times 10^{-13}\,J/MeV} = \underline{931.5\,MeV}$

39.48 What is the binding energy of ^{12}C?

▮ One atom of ^{12}C consists of 6 protons, 6 electrons, and 6 neutrons. The mass of the uncombined protons and electrons is the same as that of six ^1H atoms (if we ignore the very small binding energy of each proton-electron pair). The component particles may thus be considered as six ^1H atoms and six neutrons. A mass balance may be computed as follows.

$$
\begin{array}{lll}
\text{mass of six } ^1H \text{ atoms} = 6 \times 1.0078 & = 6.0468\,u \\
\text{mass of 6 neutrons} = 6 \times 1.0087 & = 6.0522\,u \\
\text{total mass of component particles} & = 12.0990\,u \\
\text{mass of } ^{12}C & = 12.0000\,u \\
\text{loss in mass on forming } ^{12}C & = 0.0990\,u \\
\text{binding energy} = (931 \times 0.0990)\,MeV & = \underline{92\,MeV}
\end{array}
$$

39.49 Calculate (*a*) the binding energy of $^{16}_8O$ and (*b*) the binding energy per nucleon. The mass of neutral $^{16}_8O$ is $M_O = 15.994915$ u.

▮ (*a*) $BE = (ZM_H + Nm_n) - M_O = 8(1.007825) + 8(1.008665) - 15.994915 = +0.137005\,u$ or $\underline{127.62\,MeV}$

(*b*) $\dfrac{127.62}{16} = \underline{7.98\,MeV/nucleon}$

39.50 The binding energy per nucleon for ^{238}U is about 7.5 MeV, whereas it is about 8.5 MeV for nuclei of half that mass. If a ^{238}U nucleus were to split into two equal-size nuclei, about how much energy would be released in the process?

▮ There are 238 nucleons involved. Each nucleon will release about $8.5 - 7.5 = 1.0$ MeV of energy when the nucleus undergoes fission. The total energy liberated is therefore about <u>238 MeV</u>.

39.51 In a particular fission reaction, a $^{235}_{92}U$ nucleus captures a slow neutron; the fission products are three neutrons, a $^{142}_{57}La$ nucleus, and a fission product $_ZX$. Determine Z.

▮ Charge is conserved; so $Z = 92 - 57 = \underline{35}$.

39.52 When an atom of ^{235}U undergoes fission in a reactor, about 200 MeV of energy is liberated. Suppose that a reactor using uranium-235 has an output of 700 MW and is 20 percent efficient. (**a**) How many uranium atoms does it consume in one day? (**b**) What mass of uranium does it consume each day?

▮ (**a**) Each fission yields 200 MeV $= (200 \times 10^6)(1.6 \times 10^{-19})$ J of energy. Only 20 percent of this is utilized efficiently and so

$$\text{energy generated per fission} = (200 \times 10^6)(1.6 \times 10^{-19})(0.20) = 6.4 \times 10^{-12} \text{ J}$$

Because the reactor's output is 700×10^6 J/s, the number of fissions required per second is

$$\text{fissions/s} = \frac{7 \times 10^8 \text{ J/s}}{6.4 \times 10^{-12} \text{ J}} = 1.1 \times 10^{20} \text{ s}^{-1} \quad \text{and} \quad \text{fissions/day} = (86\,400 \text{ s/d})(1.1 \times 10^{20} \text{ s}^{-1}) = \underline{9.5 \times 10^{24} \text{ d}^{-1}}$$

(**b**) There are 6.02×10^{26} atoms in 235 kg of uranium-235. Therefore the mass of uranium-235 consumed in one day is

$$\text{mass} = \left(\frac{9.5 \times 10^{24}}{6.02 \times 10^{26}}\right)(235 \text{ kg}) = \underline{3.7 \text{ kg}}$$

39.53 Uranium-238 ($^{238}_{92}U$) is radioactive and decays by emitting the following particles in succession before reaching a stable form: α, β, β, α, α, α, α, α, β, β, α, β, β, and α (β stands for "beta particle," e^-). What is the final stable nucleus?

▮ The original nucleus emitted 8 alpha particles and 6 beta particles. When an alpha particle is emitted, Z decreases by 2, since the alpha particle carries away a charge of $+2e$. A beta particle carries away a charge of $-1e$ and so the charge on the nucleus must increase to $(Z + 1)e$. We then have for the final nucleus

$$\text{final } Z = 92 + 6 - (2)(8) = 82 \qquad \text{final } A = 238 - (6)(0) - (8)(4) = \underline{206}$$

The final stable nucleus is $^{206}_{82}Pb$.

39.54 The following fusion reaction takes place in the sun and furnishes much of its energy:

$$4^1_1H \rightarrow {}^4_2He + 2_{+1}^0e + \text{energy}$$

where $_{+1}^0e$ is a positron, a positive electron. How much energy is released as 1 kg of hydrogen is consumed? The masses of ^1H, ^4He, and $_{+1}^0e$ are, respectively, 1.007 825, 4.002 604, and 0.000 549 u, where atomic electrons are included in the first two values.

▮ The mass of the reactants, 4 protons, is 4 times the atomic mass of hydrogen (^1H), less the mass of 4 electrons.

$$\text{reactant mass} = (4)(1.007\,825 \text{ u}) - 4m_e = 4.031\,300 \text{ u} - 4m_e$$

where m_e is the mass of the electron (or positron). The reaction products have a combined mass

$$\text{product mass} = (\text{mass of } {}^4_2He \text{ nucleus}) + 2m_e = (4.002\,604 \text{ u} - 2m_e) + 2m_e = 4.002\,604 \text{ u}$$

The mass loss is therefore

$$(\text{reactant mass}) - (\text{product mass}) = (4.0313 \text{ u} - 4m_e) - 4.0026 \text{ u}$$

Substituting $m_e = 0.000\,549$ u, the mass loss is found to be 0.0265 u.

But 1 kg of ^1H contains 6.02×10^{26} atoms. For each four atoms that undergo fusion, 0.0265 u is lost. The mass lost when 1 kg undergoes fusion is therefore

$$\text{mass loss/kg} = (0.0265 \text{ u})(6 \times 10^{26}/4) = \underline{4.0 \times 10^{24} \text{ u}} = (4.0 \times 10^{24} \text{ u})(1.66 \times 10^{-27} \text{ kg/u}) = \underline{0.0066 \text{ kg}}$$

Then, from the Einstein relation,

$$\Delta E = (\Delta m)c^2 = (0.0066 \text{ kg})(3 \times 10^8 \text{ m/s})^2 = \underline{6.0 \times 10^{14} \text{ J}}$$

39.55 Find the Q value ($Q = \Delta m_0 = \Delta KE$) of the nuclear reaction $_2^4He + _7^{14}N \rightarrow _8^{17}O + _1^1H$. The rest masses of the neutral atoms $_2^4He$, $_7^{14}N$, and $_8^{17}O$ are 4.002603, 14.003074, and 16.999133 u, respectively.

▌ After nine electronic masses are added to either side of the reaction, the given atomic masses may be used, along with the mass of atomic $_1^1H$, 1.007825 u.

$$Q = (4.002603 + 14.003074) - (16.999133 + 1.007825) = -0.001281\,u = -1.19\,MeV$$

This reaction is *endothermic* ($Q < 0$); 1.19 MeV of kinetic energy is *lost* in the reaction. Clearly, then, the reaction cannot take place unless the bullet (the α particle, $_2^4He$) has at least 1.19 MeV of kinetic energy (as measured, so it turns out, in the center-of-mass frame).

39.56 The following *proton-proton cycle* of nuclear reactions has been suggested as a possible source of stellar energy:

$$_1^1H + _1^1H \rightarrow _1^2H + _1^0e + \nu \qquad _1^2H + _1^1H \rightarrow _2^3He + \gamma \qquad _2^3He + _2^3He \rightarrow _2^4He + 2_1^1H$$

The net result of these reactions is that four protons ($_1^1H$) are combined to produce an α particle ($_2^4He$), two positrons ($_1^0e$), two γ-ray photons, and two neutrinos (ν). Find the Q value of the cycle. The rest mass of atomic helium is 4.002603 u.

▌ The overall reaction is $4_1^1H \rightarrow _2^4He + 2_1^0e + 2\gamma + 2\nu$, in which we may use the masses of neutral atoms, provided that we add two more electrons to the right side. Then, recalling that γ and ν have zero rest mass, we have

$$Q = 4M_H - (M_{He} + 4m_e) = 4.031300 - (4.002603 + 0.002196) = 0.0265\,u = \underline{24.69\,MeV}$$

39.57 Neutrons are frequently detected by allowing them to be captured by boron-10 nuclei in a counter (similar to a Geiger counter) filled with $B^{10}F_3$ gas. The reaction is

$$_0n^1 + _5B^{10} \rightarrow _3Li^7 + _2He^4$$

Compute the energy released in this reaction, if the relevant rest masses are neutron, 1.00866 u; boron-10, 10.01294 u; lithium-7, 7.01600 u; helium-4, 4.00260 u

▌ We have

$$_0n^1 + _5B^{10} \rightarrow _3Li^7 + _2He^4 \qquad 1.00866 + 10.01294 \rightarrow 7.01600 + 4.00260 \qquad 11.02160 \rightarrow 11.01860$$

$$11.02160 - 11.01860 = 0.00300\,u \text{ converted to energy} \qquad 0.00300 \times 931 = \underline{2.79\,MeV}$$

39.58 Find the Q of the reaction $_7^{14}N(d, \alpha)_6^{12}C$.

▌ $_7^{14}N$ has mass of 14.00307; $_1^2H$ has mass of 2.01410

total is 16.01717 u

$_6^{12}C$ has mass of 12.000000; $_2^4He$ has mass of 4.002603

total is 16.002603 u

net loss in mass $= (16.01717 - 16.002603) = 0.01457\,u \qquad Q = (0.01457\,u)(931.48\,MeV/u) = \underline{13.6\,MeV}$

39.59 Calculate the energy released in the (α, n) reaction for $_4^9Be$. The mass of $_4^9Be$ is 9.012183 u.

▌ The equation for the nuclear reaction is

$$_4^9Be + _2^4He \rightarrow _6^{12}C + _0^1n + Q$$

The masses are, from standard sources,

$_4^9Be$:	9.012183	$_6^{12}C$:	12.000000
$_2^4He$:	4.002603	$_0^1n$:	1.008665
	13.014786 u		13.008665 u

net loss in mass $= 13.014786 - 13.008665 = 0.006121\,u \qquad Q = (0.006121)(931) = \underline{5.70\,MeV}$

39.60 Calculate the Q value of the reaction

$$^2_1H + ^3_1H \rightarrow ^1_0n + ^4_2He$$

if the rest masses of the neutral atoms ^2_1H, ^3_1H, and ^4_2He are 2.014102, 3.016049, and 4.002603 u, respectively.

▌ Add two electronic masses to either side of the reaction so that atomic masses may be used. The neutron rest mass is 1.008665 u, and so

Q = (rest mass of reactants) − (rest mass of products) = $(2.014102 + 3.016049) - (1.008665 + 4.002603)$ =
0.018883 u = <u>17.6 MeV</u>

This reaction is *exothermic* ($Q > 0$); kinetic energy, in the amount 17.6 MeV, is released by the reaction.

39.61 Complete and balance the equations for the following nuclear reactions by replacing the question mark with the correct symbol:
(*a*) $^{222}_{86}Rn \rightarrow ? + ^4_2He$ (*b*) $^2_1H + ^2_1H \rightarrow ? + ^1_1H$
(*c*) $^{239}_{93}Np \rightarrow ? + ^0_{-1}\beta$ (electron)   (*d*) $^{22}_{11}Na \rightarrow ? + ^0_{+1}\beta$ (positron)

▌ (*a*) $^{222}_{86}Rn \rightarrow ^{218}_{84}Po + ^4_2He$ (*b*) $^2_1H + ^2_1H \rightarrow ^3_1H + ^1_1H$
(*c*) $^{239}_{93}Np \rightarrow ^{239}_{94}Pu + ^0_{-1}\beta + \bar{\nu}$ (antineutrino)   (*d*) $^{22}_{11}Na \rightarrow ^{22}_{10}Ne + ^0_{+1}\beta + \nu$ (neutrino)

39.62 Complete the following nuclear equations:

(*a*) $^{14}_7N + ^4_2He \rightarrow ^{17}_8O + ?$

(*b*) $^9_4Be + ^4_2He \rightarrow ^{12}_6C + ?$

(*c*) $^9_4Be(p, \alpha)?$

▌ (*a*) The sum of the subscripts on the left is $7 + 2 = 9$. The subscript of the first product on the right is 8. Hence the second product on the right must have a subscript (net charge) of 1. The sum of the superscripts on the left is $14 + 4 = 18$. The superscript of the first product is 17. Hence the second product on the right must have a superscript (mass number) of 1. The particle with a nuclear charge 1 and a mass number 1 is the proton, ^1_1H.
(*b*) The nuclear charge of the second product particle (its subscript) is $(4 + 2) - 6 = 0$. The mass number of the particle (its subscript) is $(9 + 4) - 12 = 1$. Hence the particle must be the neutron, ^1_0n.
(*c*) The reactants, ^9_4Be and ^1_1H, have a combined nuclear charge of 5 and a mass number of 10. In addition to the α particle, a product will be formed of charge $5 - 2 = 3$, and mass $10 - 4 = 6$. This is ^6_3Li.

39.63 Complete the following nuclear equations

(*a*) $^{30}_{15}P \rightarrow ^{30}_{14}Si + ?$

(*b*) $^3_1H \rightarrow ^3_2He + ?$

(*c*) $^{43}_{20}Ca(\alpha, ?)^{46}_{21}Sc$

▌ (*a*) The nuclear charge of the second particle is $15 - 14 = +1$. The mass is $30 - 30 = 0$. Hence the particle must be a positron, $^0_{+1}e$.
(*b*) The nuclear charge of the second particle is $1 - 2 = -1$. Its mass number is $3 - 3 = 0$. Hence the particle must be a beta particle (an electron), $^0_{-1}e$.
(*c*) The reactants, $^{43}_{20}Ca$ and ^4_2He, have a combined nuclear charge of 22 and mass number 47. The ejected product will have a charge $22 - 21 = 1$, and mass $47 - 46 = 1$. This is a proton and should be represented in the parentheses by p.

39.64 Define the following quantities related to radiation: gamma; alpha, and beta-rays; X-rays; becquerel, roentgen; rad; rem; and isotopes.

▌ *Gamma rays* are a form of electromagnetic radiation, usually of very short wavelength, emitted by excited nuclei.
Alpha particles are helium nuclei.
Beta particles are electrons emitted by radioactive nuclei.
X-rays are high-energy electromagnetic photons, usually produced by the rapid deceleration of high-energy electrons or by electron jumps to vacancies in the inner electron shells of an atom.
The *becquerel* (Bq) is a unit of radioactivity, equal to one nuclear disintegration per second.

The *roentgen* is a unit of radiation exposure equal to that quantity of X- or gamma radiation that, in dry air, produces a total ionization of 0.333 nC per 1.293×10^{-6} kg of air.

The *rad* is a unit equal to the absorbed dose of any radiation that releases 100 erg (10^{-5} J) of energy per gram of absorbing material.

The *rem* is the *r*oentgen *e*quivalent unit for *m*an. The rem dose is found by multiplying the dose in rads by the relative biological effectiveness (RBE).

Isotopes are atoms of the same Z but of different A.

39.65^c Derive the statistical law of radioactive decay.

> ▌ Starting with a large number, N_0, of a species of unstable nucleus, let $N(t)$ be the expected number of nuclei still undecayed after a time t. [That is, $N(t)$ would be the average number of nuclei observed if the experiment were repeated many times.] Consider what might happen in the infinitesimal time interval $(t, t + \Delta t)$. Since any one of the $N(t)$ nuclei has probability $\lambda \Delta t$ of decaying, the *expected* number that do decay is $N(t) \times (\lambda \Delta t)$. [Compare: "If 1000 fair dice are thrown, the expected number showing 5 is $1000(\frac{1}{6})$."] Consequently, one has

$$N(t + \Delta t) = N(t) - N(t) \times (\lambda \Delta t) \quad \text{or} \quad \frac{N(t + \Delta t) - N(t)}{\Delta t} = -\lambda N(t)$$

which becomes, in the limit as $\Delta t \to 0$,

$$\frac{dN}{dt} = -\lambda N$$

Integrating,

$$\int_{N_0}^{N} \frac{dN}{N} = -\lambda \int_{0}^{t} dt \qquad \ln \frac{N}{N_0} = -\lambda t \qquad N = N_0 e^{-\lambda t}$$

When N_0 is very large, as it almost always is, the expected number, N, of undecayed nuclei may be identified with the number actually observed.

39.66^c Show that the average lifetime of an unstable nucleus is $T_{avg} = 1/\lambda$.

> ▌ From Prob. 39.65, the fraction of nuclei that have lifetime t (i.e., that decay between times t and $t + dt$) is given by

$$\frac{N(t) \times (\lambda \, dt)}{N_0} = e^{-\lambda t} \lambda \, dt \qquad \text{Hence} \qquad T_{avg} = \int_{0}^{\infty} t e^{-\lambda t} \lambda \, dt = \frac{1}{\lambda}$$

39.67 The half-life of radium is 1620 years. How many radium atoms decay in 1 s in a 1-g sample of radium? The atomic weight of radium is 226 kg/kmol.

> ▌ A 1-g sample is (0.001/226) kmol and so it contains

$$N = \left(\frac{0.001}{226} \text{ kmol}\right)(6.02 \times 10^{26} \text{ atoms/kmol}) = 2.66 \times 10^{21} \text{ atoms}$$

The decay constant is

$$\lambda = \frac{0.693}{T_{1/2}} = \frac{0.693}{(1620 \text{ y})(3.16 \times 10^7 \text{ s/y})} = 1.35 \times 10^{-11} \text{ s}^{-1}$$

Then

$$\Delta N = \lambda N \, \Delta t = (1.35 \times 10^{-11} \text{ s}^{-1})(2.66 \times 10^{21})(1 \text{ s}) = \underline{3.6 \times 10^{10}}$$

is the number of disintegrations per second in 1 g of radium.

The above result led to the definition of the *curie* (Ci) as a unit of activity:

$$1 \text{ Ci} = 3.7 \times 10^{10} \text{ disintegrations/s}$$

39.68 Cobalt-60 (^{60}Co) is often used as a radiation source in medicine. It has a half-life of 5.25 years. How long, after a new sample is delivered, will the activity have decreased (**a**) to about $\frac{1}{8}$ its original value and (**b**) to about $\frac{1}{3}$ its original value?

> ▌ The activity is proportional to the number of undecayed atoms ($\Delta N / \Delta t = \lambda N$). (**a**) In each half-life, half of the remaining sample decays. Because $\frac{1}{2} \times \frac{1}{2} \times \frac{1}{2} = \frac{1}{8}$, three half-lives, or $\underline{15.75 \text{ years}}$, are required for the sample to decay to one-eighth its original strength. (**b**) Using the fact that the original material present

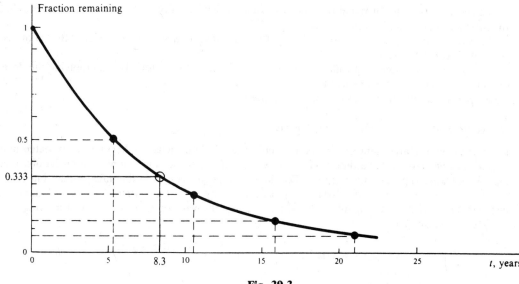

Fig. 39-3

decreases by one-half during each 5.25 years, we can plot the graph shown in Fig. 39-3. From it, the sample decays to 0.33 its original value after a time of about 8.3 years.

39.69 Plutonium decays by the following reaction with a half-life of 24 000 years

$$^{239}_{94}\text{Pu} \rightarrow ^{235}_{92}\text{U} + ^{4}_{2}\text{He}$$

If plutonium is stored for 73 200 years, what fraction of it remains?

⫶ $\dfrac{73\,200}{24\,000} \approx 3$ half-lives $\quad (\frac{1}{2})^3 = \frac{1}{8}$ (fraction of Pu left)

39.70 The half-life of ^{215}At is 100 μs. If a sample initially contains 6 mg of the element, what is its activity (a) initially and (b) after 200 μs?

⫶ (a) The number of radioactive atoms initially present is

$$N_0 = \frac{(6 \times 10^{-3}\,\text{g})}{215\,\text{g/mol}}(6.03 \times 10^{23}\,\text{atoms/mol}) = 1.68 \times 10^{19}\,\text{atoms}$$

and the decay constant of ^{215}At is

$$\lambda = \frac{\ln 2}{T_{1/2}} = \frac{0.693}{100 \times 10^{-6}} = 6930\,\text{s}^{-1}$$

Hence the initial activity is $\quad A_0 = \lambda N_0 = (6930)(1.68 \times 10^{19}) = 1.16 \times 10^{23}\,\text{Bq}$

(b) At $t = 2T_{1/2}$, $\quad A = A_0 e^{-\lambda t} = (1.16 \times 10^{23})(\frac{1}{2})^2 = 2.9 \times 10^{22}\,\text{Bq}$

39.71ᶜ If a radioactive nuclide is produced at the constant rate of n per second (say, by bombarding a target with neutrons), find (a) the (expected) number of nuclei in existence t seconds after the number is N_0 and (b) the maximum (expected) number of these radioactive nuclei.

⫶ (a) The population N is simultaneously increasing at rate n and decreasing, by decay, at rate λN. Thus, the net rate of increase is $dN/dt = n - \lambda N$. Separating the variables and integrating, we have

$$\int_{N_0}^{N} \frac{dN}{n - \lambda N} = \int_{0}^{t} dt \qquad \frac{1}{\lambda} \ln \frac{n - \lambda N_0}{n - \lambda N} = t \qquad N = \frac{n}{\lambda} + \left(N_0 - \frac{n}{\lambda}\right)e^{-\lambda t}$$

(b) From the result of (a), it is seen that $N(t)$ starts at N_0 and asymptotically decreases or increases to n/λ, according as $N_0 > n/\lambda$ or $n/\lambda > N_0$. The maximum number of nuclei is thus the larger of N_0 and n/λ.

It is supposed here that n/λ, like N_0, is very large; otherwise the law of radioactive decay, and therefore the result of (a), ceases to be valid at some moment.

39.72ᶜ A substance X disintegrates into a radioactive substance Y. Find N_y, given that the initial amounts of X and Y are N_{x_0} and N_{y_0}, respectively.

▌ X decays at the rate $dN_x/dt = -\lambda_x N_x = -\lambda_x N_{x_0} e^{-\lambda_x t}$. For each X nucleus that disintegrates, one of Y is formed. So Y is formed at the rate $\lambda_x N_x$ nuclei per second. But at the same time Y is being formed, it is disintegrating at the rate $\lambda_y N_y$ nuclei per second. The net increase in Y nuclei per second is therefore

$$\frac{dN_y}{dt} = \lambda_x N_x - \lambda_y N_y = \lambda_x N_{x_0} e^{-\lambda_x t} - \lambda_y N_y$$

To integrate this, transpose $\lambda_y N_y$ and multiply both sides by $e^{\lambda_y t}\, dt$:

$$e^{\lambda_y t}\, dN_y + \lambda_y N_y e^{\lambda_y t}\, dt = \lambda_x N_{x_0} e^{(\lambda_y - \lambda_x)t}\, dt \qquad N_y e^{\lambda_y t} = \frac{\lambda_x N_{x_0}}{\lambda_y - \lambda_x} e^{(\lambda_y - \lambda_x)t} + C$$

When $t = 0$, $N_y = N_{y_0}$, which gives

$$C = N_{y_0} - \frac{\lambda_x N_{x_0}}{\lambda_y - \lambda_x} \qquad \text{Thus} \qquad N_y = \left(N_{y_0} - \frac{\lambda_x N_{x_0}}{\lambda_y - \lambda_x}\right)e^{-\lambda_y t} + \left(\frac{\lambda_x N_{x_0}}{\lambda_y - \lambda_x}\right)e^{-\lambda_x t}$$

39.73 A beam of gamma rays has a cross-sectional area of 2 cm² and carries 7×10^8 photons through the cross section each second. Each photon has an energy of 1.25 MeV. The beam passes through a 0.75-cm thickness of flesh ($\rho = 0.95$ g/cm³) and loses 5 percent of its intensity in the process. What is the average dose (in rad) applied to the flesh each second?

▌ Dose is measured as the energy absorbed per kilogram of flesh in this case. We have

$$\text{number of } \gamma\text{-rays absorbed/s} = (7 \times 10^8 \text{ s}^{-1})(0.05) = 3.5 \times 10^7 \text{ s}^{-1}$$
$$\text{energy absorbed/s} = (3.5 \times 10^7 \text{ s}^{-1})(1.25 \text{ MeV}) = 4.4 \times 10^7 \text{ MeV/s}$$

We need the mass of the flesh in which this energy was absorbed.

$$\text{mass} = \rho V = (0.95 \text{ g/cm}^3)[(2 \text{ cm}^2)(0.75 \text{ cm})] = 1.43 \text{ g}$$

We then have
$$\text{dose/s} = \frac{\text{energy/s}}{\text{mass}} = \frac{(4.4 \times 10^7 \text{ MeV/s})(1.6 \times 10^{-13} \text{ J/MeV})}{1.43 \times 10^{-3} \text{ kg}} \times \frac{100 \text{ rad}}{1 \text{ J/kg}} = \underline{0.49 \text{ rad/s}}.$$

39.74 An alpha particle beam passes through flesh and deposits 0.2 J of energy in each kilogram of flesh. The RBE for these particles is 12 rem/rad. Find the dose in rad and in rem.

▌ $$\text{dose in rad} = \left(\frac{\text{absorbed energy}}{\text{mass}}\right) \times \left(\frac{100 \text{ rad}}{\text{J/kg}}\right) = (0.2 \text{ J/kg})\left(\frac{100 \text{ rad}}{\text{J/kg}}\right) = \underline{20 \text{ rad}}$$
$$\text{dose in rem} = (\text{RBE}) \times (\text{dose in rad}) = (12)(20) = \underline{240 \text{ rem}}$$

39.75 Describe briefly the following particle-physics terms: π-meson, muon, neutrino; antiparticle, hadron, and lepton.

▌ A *π-meson* (pion) is a particle that can be positive, negative, or neutral. A π^+- or a π^--meson has a mass of 273 electron masses. A neutral π-meson has a mass of 264 electron masses.

A *muon* (μ) is a particle having a mass of 207 electron masses. It is formed in the decay of a π^+- or a π^--meson and can be either positive or negative. It behaves just like a heavy electron.

A *neutrino* is a neutral particle of almost zero rest mass that is emitted in beta and in π^+- or π^--meson decays. Six kinds of neutrinos are known, three of which are antineutrinos.

An *antiparticle* is a particle of antimatter corresponding to a given particle in every respect except that charge and certain other discrete properties change sign. The positron, for example, is the antiparticle to the electron. In collisions between a particle and its antimatter counterpart, both are annihilated.

A *hadron* is a particle that exhibits the strong nuclear force (e.g., protons, neutrons, mesons).

A *lepton* is a particle that does not exhibit the strong nuclear force (e.g., electrons, muons, neutrinos).

39.3 SOLID-STATE ELECTRONICS

39.76 Define briefly: *Semiconductor*; *p-type*; *n-type*; and *phonon*.

▌ A *semiconductor* is a material in which some electrons are available for conduction at room temperatures.

As the temperature of a semiconductor increases, its conductivity increases. The number of electrons available for conduction can be increased enormously if *acceptor* or *donor impurities* are added to the material.

A *p-type* material contains added acceptor impurities that do not provide enough electrons for complete covalent bonding with neighboring host atoms. The holes left by missing electrons act as positive charges that can "hop" from atom to atom.

An *n-type* material contains added donor impurities that carry one more electron than the neighboring atoms require for complete bonding. At room temperature, the extra electron is often in the conduction band.

A *phonon* is a quantized vibrational (sound) wave in a material that carries energy $h\nu$, where h is Planck's constant and ν is the frequency.

39.77 Calculate the number density of free carriers in silver, assuming that each atom contributes one carrier. The density of silver is 10.5×10^3 kg/m³ and the atomic weight is 107.8.

$$n = (6.02 \times 10^{26} \text{ carriers/kmol})\left(\frac{1 \text{ kmol}}{107.8 \text{ kg}}\right)(10.5 \times 10^3 \text{ kg/m}^3) = 5.85 \times 10^{28} \text{ carriers/m}^3$$

39.78 With reference to Prob. 39.77, calculate the average drift velocity of electrons when a current of 1.5 A is passing through a silver wire of 1.0 mm² cross-sectional area.

By the usual argument, $\Delta q = (v_d \, \Delta t)Ane$, or

$$v_d = \frac{I}{Ane} = \frac{1.5 \text{ A}}{(1.0 \times 10^{-6} \text{ m}^2)(5.85 \times 10^{28} \text{ m}^{-3})(1.6 \times 10^{-19} \text{ C})} = \underline{0.16 \text{ mm/s}}$$

39.79 The conduction electrons in a metal such as silver may be heated as though composing a gas confined in the metallic lattice; write λ and $\bar{v}$ for the electron's mean free path and mean thermal velocity, respectively. Obtain a formula for the electric conductivity of the metal.

Consider a rectilinear volume of the metal, of length l (in the x direction) and cross-sectional area A. If there is an applied longitudinal field E, a free electron in the box experiences a constant acceleration $a_x = (eE)/m_e$ in the periods between successive collisions with the metal lattice. Now, the average time between collisions is $\bar{t} = \lambda/\bar{v}$. Hence, the electron acquires a (mean) drift velocity

$$v_x \equiv v_d = a_x\bar{t} = \frac{eE\lambda}{m_e\bar{v}}$$

But $v_d = I/(Ane)$ (Prob. 39.78); so we obtain Ohm's law, $J \equiv I/A = \sigma E$, with

$$\sigma = \frac{ne^2\lambda}{m_e\bar{v}} \qquad (1)$$

39.80 Refer to Probs. 39.77 and 39.79. Estimate the conductivity of silver at 0 °C. Take the mean free path in the electron gas to be 25 nm.

Suppose that the electron gas is ideal and forget the distinction between mean and rms speeds. Then, $\frac{1}{2}m_e\bar{v}^2 = \frac{3}{2}kT$, or $m_e\bar{v} = \sqrt{3m_e kT}$, so that (1) of Prob. 39.79 becomes

$$\sigma = \frac{ne^2\lambda}{\sqrt{3m_e kT}} \approx \frac{(6 \times 10^{28})(1.6 \times 10^{-19})^2(25 \times 10^{-9})}{\sqrt{3(10^{-30})(1.38 \times 10^{-23})(273)}} \approx \frac{4 \times 10^{-17}}{10^{-25}} = 4 \times 10^8 \text{ S/m}$$

Considering the crudeness of our guess for λ, our result checks quite well with the measured value, 0.7×10^8 S/m.

39.81 Find the magnitude of the Hall voltage induced across a silver wire of square cross section 1 mm on a side, when it carries a current of 1.5 A and a transverse magnetic field B of strength 0.1 T is applied.

Let the wire run in the x direction and the magnetic field in the z direction. In the Hall effect, the Lorentz force, which is due to B_z and the induced electric field E_y, vanishes,

$$eE_y = ev_x B_z \qquad (1)$$

allowing the conduction electrons to move down the wire at drift speed v_x. From Prob. 39.78,

$$v_x = \frac{I}{Ane} = \frac{I}{l^2ne} \qquad (2)$$

and the Hall voltage is given by

$$E_y = \frac{V_H}{l} \tag{3}$$

Substitution of (2) and (3) in (1) yields

$$V_H = \frac{IB_z}{nel} = \frac{(1.5\text{ A})(0.1\text{ T})}{(5.85 \times 10^{28}\text{ m}^{-3})(1.6 \times 10^{-19}\text{ C})(1 \times 10^{-3}\text{ m})} = 1.6 \times 10^{-8}\text{ V}$$

39.82 Check the units in the result of Prob. 39.81.

┃ Use the facts that $1\text{ T} = 1\text{ Wb/m}^2 = 1\text{ V}\cdot\text{s/m}^2$ and $1\text{ C} = 1\text{ A}\cdot\text{s}$.

39.83 Define the following electronics terms: *rectifier*; *thermionic diode*; *semiconductor diode*; *triode*; *amplification factor*; *voltage gain*; and *transistor*.

┃ A *rectifier* is a device that provides a direct current output from an alternating current input.
A *thermionic diode* is a vacuum tube that contains an electron-emitting cathode and a plate.
A *semiconductor diode* is a *pn* junction that allows current to pass essentially in one direction.
A *triode* is a vacuum tube that contains a cathode, a grid, and a plate.
The *amplification factor* μ of a triode is the ratio of the increase in plate potential to the corresponding decrease in grid potential:

$$\mu = \frac{\Delta V_p}{-\Delta V_g} \quad (I_p \text{ held constant})$$

The *voltage gain* of a triode serving as an amplifier is the ratio of the variation in potential difference across the load resistor to the variation in potential difference of the grid:

$$\text{voltage gain} = \frac{V_{\text{out}}}{V_{\text{in}}} = \frac{R_L \, \Delta i_p}{\Delta V_g}$$

where R_L is the resistance in the load resistor and Δi_p and Δv_g are the variations in plate current and grid potential, respectively.
A *transistor* is a sandwich made of three semiconductors: the *emitter*, the *base*, and the *collector*. The base is a very thin piece, opposite in type to the emitter and the collector.

39.84 The load resistor R_L in the plate circuit of a triode is 12 kΩ. When the grid potential V_g oscillates between −2.96 and −2.88 V, the plate current I_p varies from 0.57 to 0.81 mA. What is the voltage gain for this triode amplifier?

┃ $\qquad \Delta V_L = R_L \, \Delta i_p = (12\,000)[(0.81 - 0.57) \times 10^{-3}] = 2.88\text{ V} \qquad \Delta V_g = -2.96 + 2.88 = -0.08\text{ V}$

$$\text{voltage gain} = \frac{\Delta V_L}{-\Delta V_g} = \frac{2.88}{0.08} = \underline{36}$$

39.85 In the transistor circuit of Fig. 39-4 the base current is 35 μA. What is the value of the resistor R_b?

┃ $\qquad V = I_b R_b \qquad 9 = 35 \times 10^{-6} R_b \qquad R_b = \underline{257\text{ k}\Omega}.$

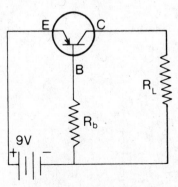

Fig. 39-4

39.86 In a triode, the plate potential increases by 24 V to hold constant plate current when the grid potential changes from -1.8 to -3 V. Find the amplification factor of the triode.

▮
$$\mu = \frac{\Delta V_p}{-\Delta V_g} = \frac{24}{-[-3-(-1.8)]} = \frac{24}{1.2} = \underline{20}.$$

39.87 In an *npn* transistor circuit, the collector current is 15 mA. If 95 percent of the electrons emitted reach the collector, what is the base current?

▮ First find the emitter current I:

$I_c = 0.95I = 15\,\text{mA}$ and $I = 15.79\,\text{mA}$ Then, $I_b = I - I_c = 15.79 - 15 = \underline{0.79\,\text{mA}}$ base current

SCHAUM'S INTERACTIVE OUTLINE SERIES

Schaum's Outlines and Mathcad™ Combined. . .
The Ultimate Solution.

NOW AVAILABLE! Electronic, interactive versions of engineering titles from the Schaum's Outline Series:

- *Electric Circuits*
- *Electromagnetics*
- *Feedback and Control Systems*
- *Thermodynamics For Engineers*
- *Fluid Mechanics and Hydraulics*

McGraw-Hill has joined with MathSoft, Inc., makers of Mathcad, the world's leading technical calculation software, to offer you interactive versions of popular engineering titles from the Schaum's Outline Series. Designed for students, educators, and technical professionals, the *Interactive Outlines* provide comprehensive on-screen access to theory and approximately 100 representative solved problems. Hyperlinked cross-references and an electronic search feature make it easy to find related topics. In each electronic outline, you will find all related text, diagrams and equations for a particular solved problem together on your computer screen. Every number, formula and graph is interactive, allowing you to easily experiment with the problem parameters, or adapt a problem to solve related problems. The *Interactive Outline* does all the calculating, graphing and unit analysis for you.

These "live" *Interactive Outlines* are designed to help you learn the subject matter and gain a more complete, more intuitive understanding of the concepts underlying the problems. They make your problem solving easier, with power to quickly do a wide range of technical calculations. All the formulas needed to solve the problem appear in real math notation, and use Mathcad's wide range of built in functions, units, and graphing features. This interactive format should make learning the subject matter easier, more effective and even fun.

For more information about *Schaum's Interactive Outlines* listed above and other titles in the series, please contact:

Schaum Division
McGraw-Hill, Inc.
1221 Avenue of the Americas
New York, New York 10020
Phone: 1-800-338-3987

To place an order, please mail the coupon below to the above address or call the 800 number.

--✂---

Schaum's Interactive Outline Series
using Mathcad®

(Software requires 80386/80486 PC or compatibles, with Windows 3.1 or higher, 4 MB of RAM, 4 MB of hard disk space, and 3 1/2" disk drive.)

AUTHOR/TITLE	Interactive Software Only ($29.95 ea)		Software and Printed Outline ($38.95 ea)	
	ISBN	Quantity Ordered	ISBN	Quantity Ordered
MathSoft, Inc./DiStefano: Feedback & Control Systems	07-842708-8	_____	07-842709-6	_____
MathSoft, Inc./Edminister: Electric Circuits	07-842710-x	_____	07-842711-8	_____
MathSoft, Inc./Edminister: Electromagnetics	07-842712-6	_____	07-842713-4	_____
MathSoft, Inc./Giles: Fluid Mechanics & Hydraulics	07-842714-2	_____	07-842715-0	_____
MathSoft, Inc./Potter: Thermodynamics For Engineers	07-842716-9	_____	07-842717-7	_____

NAME_____ ADDRESS_____

CITY _____ STATE_____ ZIP_____

ENCLOSED IS ❑ A CHECK ❑ MASTERCARD ❑ VISA ❑ AMEX (✓ ONE)

ACCOUNT #_____ EXP. DATE _____

SIGNATURE_____

MAKE CHECKS PAYABLE TO McGRAW-HILL, INC. PLEASE INCLUDE LOCAL SALES TAX AND $1.25 SHIPPING/HANDLING

SCHAUM'S SOLVED PROBLEMS SERIES

- ■ Learn the best strategies for solving tough problems in step-by-step detail
- ■ Prepare effectively for exams and save time in doing homework problems
- ■ Use the indexes to quickly locate the types of problems you need the most help solving
- ■ Save these books for reference in other courses and even for your professional library

To order, please check the appropriate box(es) and complete the following coupon.

❑ **3000 SOLVED PROBLEMS IN BIOLOGY**
ORDER CODE 005022-8/**$16.95 406 pp.**

❑ **3000 SOLVED PROBLEMS IN CALCULUS**
ORDER CODE 041523-4/**$19.95 442 pp.**

❑ **3000 SOLVED PROBLEMS IN CHEMISTRY**
ORDER CODE 023684-4/**$20.95 624 pp.**

❑ **2500 SOLVED PROBLEMS IN COLLEGE ALGEBRA & TRIGONOMETRY**
ORDER CODE 055373-4/**$14.95 608 pp.**

❑ **2500 SOLVED PROBLEMS IN DIFFERENTIAL EQUATIONS**
ORDER CODE 007979-x/**$19.95 448 pp.**

❑ **2000 SOLVED PROBLEMS IN DISCRETE MATHEMATICS**
ORDER CODE 038031-7/**$16.95 412 pp.**

❑ **3000 SOLVED PROBLEMS IN ELECTRIC CIRCUITS**
ORDER CODE 045936-3/**$21.95 746 pp.**

❑ **2000 SOLVED PROBLEMS IN ELECTROMAGNETICS**
ORDER CODE 045902-9/**$18.95 480 pp.**

❑ **2000 SOLVED PROBLEMS IN ELECTRONICS**
ORDER CODE 010284-8/**$19.95 640 pp.**

❑ **2500 SOLVED PROBLEMS IN FLUID MECHANICS & HYDRAULICS**
ORDER CODE 019784-9/**$21.95 800 pp.**

❑ **1000 SOLVED PROBLEMS IN HEAT TRANSFER**
ORDER CODE 050204-8/**$19.95 750 pp.**

❑ **3000 SOLVED PROBLEMS IN LINEAR ALGEBRA**
ORDER CODE 038023-6/**$19.95 750 pp.**

❑ **2000 SOLVED PROBLEMS IN Mechanical Engineering THERMODYNAMICS**
ORDER CODE 037863-0/**$19.95 406 pp.**

❑ **2000 SOLVED PROBLEMS IN NUMERICAL ANALYSIS**
ORDER CODE 055233-9/**$20.95 704 pp.**

❑ **3000 SOLVED PROBLEMS IN ORGANIC CHEMISTRY**
ORDER CODE 056424-8/**$22.95 688 pp.**

❑ **2000 SOLVED PROBLEMS IN PHYSICAL CHEMISTRY**
ORDER CODE 041716-4/**$21.95 448 pp.**

❑ **3000 SOLVED PROBLEMS IN PHYSICS**
ORDER CODE 025734-5/**$20.95 752 pp.**

❑ **3000 SOLVED PROBLEMS IN PRECALCULUS**
ORDER CODE 055365-3/**$16.95 385 pp.**

❑ **800 SOLVED PROBLEMS IN VECTOR MECHANICS FOR ENGINEERS**
Vol I: STATICS
ORDER CODE 056582-1/**$20.95 800 pp.**

❑ **700 SOLVED PROBLEMS IN VECTOR MECHANICS FOR ENGINEERS**
Vol II: DYNAMICS
ORDER CODE 056687-9/**$20.95 672 pp.**